Managing Water: Coping with Scarcity and Abundance

PROCEEDINGS OF THEME A

Water for A Changing Global Community
The 27th Congress of the
International Association for
Hydraulic Research

Hosted by the
Water Resources Engineering Division of the
American Society of Civil Engineers

Series Editors: Forrest M. Holly Jr. and Adnan Alsaffar
Theme Editor: Marshall English
Theme Co-Editor: Andras Szollosi-Nagy

San Francisco, California
August 10-15, 1997

Published by the
ASCE *American Society of Civil Engineers*
345 East 47th Street
New York, New York 10017-2398

Abstract:

The proceedings, *Managing Water: Coping with Scarcity and Abundance*, consists of papers presented at the 27th Congress of the International Association of Hydraulic Research (IAHR) entitled "Water for a Changing Global Community." The topics covered include flood and drought estimation and control; regional and multinational river basin management; improved management of water distribution systems; analysis and management of excess water; and water scarcity and the future of irrigated agriculture.

Library of Congress Cataloging-in-Publication Data

International Association for Hydraulic Research. Congress (27th : 1997 : San Francisco, Calif.)
Managing water : coping with scarcity and abundance : proceedings of theme A : the 27th Congress of the International Association for Hydraulic Research : San Francisco, California, August 10-15, 1997 / volume editor, Marshall English, volume co-editor, Andras Szollosi-Nagy.
p.   cm. -- (Water for a changing global community)
"Hosted by the Water Resources Engineering Division of the American Society of Civil Engineers."
Includes bibliographical references and indexes.
ISBN 0-7844-0271-X
1. Flood control--Congresses. 2. Hydrology--Congresses. 3. Water-supply--Congresses. I. English, Marshall J. II. Szöllösi-Nagy, András. III. Title. IV. Series: International Association for Hydraulic Research. Congress (27th : 1997 : San Francisco, Calif.).
Water for a changing global community.
TC530.I626 1997   97-17973
627--dc21   CIP

Any statements expressed in these materials are those of the individual authors and do not necessarily represent the views of ASCE, which takes no responsibility for any statement made herein. No reference made in this publication to any specific method, product, process or service constitutes or implies an endorsement, recommendation, or warranty thereof by ASCE. The materials are for general information only and do not represent a standard of ASCE, nor are they intended as a reference in purchase specifications, contracts, regulations, statutes, or any other legal document.
ASCE makes no representation or warranty of any kind, whether express or implied, concerning the accuracy, completeness, suitability, or utility of any information, apparatus, product, or process discussed in this publication, and assumes no liability therefore. This information should not be used without first securing competent advice with respect to its suitability for any general or specific application. Anyone utilizing this information assumes all liability arising from such use, including but not limited to infringement of any patent or patents.

Photocopies. Authorization to photocopy material for internal or personal use under circumstances not falling within the fair use provisions of the Copyright Act is granted by ASCE to libraries and other users registered with the Copyright Clearance Center (CCC) Transactional Reporting Service, provided that the base fee of $4.00 per article plus $.25 per page is paid directly to CCC, 222 Rosewood Drive, Danvers, MA 01923. The identification for ASCE Books is 0-7844-0271-X/97/$4.00 + $.25 per page. Requests for special permission or bulk copying should be addressed to Permissions & Copyright Dept., ASCE.

Copyright © 1997 by the American Society of Civil Engineers,
All Rights Reserved.
Library of Congress Catalog Card No: 97-17973
ISBN 0-7844-0271-X
Manufactured in the United States of America.

GENERAL PREFACE

The XXVIIth Congress of the International Association of Hydraulic Research (IAHR) was held at the Hyatt Regency Embarcadero Hotel in San Francisco, California August 10-15, 1997. IAHR is a worldwide independent organization of engineers and scientists interested in or working in fields related to hydraulics and its practical application. The Congress was hosted by ASCE Water Resources Engineering Division. The overall theme of the Congress, "Water for a Changing Global Community", was developed in four sub-themes encompassing broad aspects of hydraulics and water- resources research and engineering applications:

Theme A : Managing Water: Coping with Scarcity and Abundance.

Theme B : Environmental and Coastal Hydraulics: Protecting the Aquatic Habitat.

Theme C : Groundwater : An Endangered Resource.

Theme D : Energy and Water : Sustainable Development.

These Themes were complemented by the following Specialty Seminars: Multidirectional Waves and Their Interaction with Structures; Modeling of Turbulent flows; Wind Energy and Windflow Around Structures; Continuing Education and Training; Management of Hydraulic Research; and Hydroinformatics for Control, Management and Risk Assessment. The Congress also included the John F. Kennedy Student Paper Competition.

Over 600 papers were presented in 130 sessions during the five days of the Congress. The papers of each Theme and its keynote lecture are grouped in individual volumes of these Proceedings, with Theme B occupying two volumes. The sixth volume is devoted to the papers of the John F. Kennedy Student Paper Competition, as well as brief summaries of the six Specialty Seminars which occupied 18 sessions.

Papers solicited by topic convenors and those received from the general call for papers underwent an external review process organized by the Theme Editors using several designated reviewers for each paper. Revisions were requested for many papers, and final selections were based on technical quality, relevance to the themes of the Congress, and available sessions for presentation. All papers included in this Proceedings were printed directly from camera-ready manuscripts prepared by the authors, who retain responsibility for the presentation and technical content.

In the interest of international cooperation and broad dissemination of technical developments, IAHR has granted permission for all the papers included in these Proceedings to be eligible for discussion in the ASCE Journals of the Water Resources Engineering Division: *Journal of Hydraulic Engineering, Journal of Hydrologic Engineering and Journal of Irrigation and Drainage Engineering.* The papers are also eligible to be nominated for ASCE Awards. More detailed and substantive manuscripts arising from the work reported in Proceedings papers may be submitted to the above ASCE Journals, as well as to IAHR's *Journal of Hydraulic Research*, for possible publication..

The General Editors would like to acknowledge the financial sponsors of the Congress, including at the time of this writing: U.S. Bureau of Reclamation, Iowa Institute of Hydraulic Research, Bechtel Corporation, WEST Consultants Inc., Parsons Brinckerhoff Quade & Douglas Inc., Golder Associates, Ayres Associates, and Harza Corporation. Grateful appreciation is also extended to The Department of Bioresource Engineering of Oregon State University, The Center for Computational and Hydroscience and Engineering of The University of Mississippi, Bechtel Environmental Inc., St. Anthony Falls Hydraulic Laboratory of the University of Minnesota, The College of Engineering of San Francisco State University, Bechtel Corporation, and particularly the Iowa Institute of Hydraulic Research and Dept. of Civil and Environmental Engineering of the University of Iowa for extensive logistic, travel, and office support of Congress planning and Proceedings preparation. Members of the Local Organizing Committee, including the Theme Editors Marshall English, Sam S.Y. Wang, Angelos Findikakis, and John Gulliver, as well as Phil Burgi, Richard Denton, George Hecker, Young Kim, Tatsuaki Nakato, Clifford Pugh, Patrick Ryan and Hsieh Wen Shen are all deeply thanked for their unselfish efforts and camaraderie during the three-year planning effort. Mary Anne Vigander is thanked for her help in organizing the Accompanying Persons Program. Hollie Boyle and Shiela Menaker of ASCE provided invaluable support for the Congress and Proceedings, as did the staff of the IAHR Secretariat in Delft. Twila Meder of the Iowa Institute of Hydraulic Research cannot be thanked enough for her unselfish devotion to the organization of this Congress. Finally, Gerhard Jirka must be acknowledged for his leadership in successfully attracting the XXVIIth Congress to the United States and initiating Congress planning.

Forrest M. Holly Jr., Chair of the Local Organizing Committee
Adnan Alsaffar, ASCE/WED Program Chair
Iowa City, Iowa, April 1997

Theme A Editor's Preface

Theme A of the XXVIIth IAHR Congress is "Managing Water: Coping with Scarcity and Abundance". The ageless problems of scarcity and abundance are still vital issues, as California's recent history of crippling droughts and devastating floods should remind us. Theme A comprises 155 technical papers. Nearly half of the papers are organized in eleven sessions encompassing (i) flood and drought estimation and control, (ii) regional and multinational river basin management, and (iii) improved management of water distribution systems. Another twelve sessions are associated with analysis and management of excess water, including hydrology, hydrodynamics and morphology of catchments, rivers and coastal waters, stream bed scour and design of stormwater infiltration structures. Two sessions deal with the ways in which water scarcity will affect the future of irrigated agriculture. Additional sessions in support of this theme deal with decision making to account for risk and uncertainty, decision support algorithms, and reliability based design and analysis. Taken as a whole, these various topics are characterized by broad scope and technical excellence.

Space does not permit us to individually recognize the several hundred people who contributed in one way or another to the Theme A program. We must first acknowledge the efforts of the topic convenors who, in the course of the past three years, proposed and organized sessions dealing with 15 specific topics which collectively became the Theme A program. We also wish to express our appreciation to the authors whose technical papers constitute the Proceedings. And we especially want to thank the dedicated individuals who, under the pressure of tight deadlines, reviewed and critiqued the technical papers to ensure a quality program. We salute you all with a heartfelt and humble "thank you!" We also wish to acknowledge with a "well done' the efforts of Dr. Forrest Holly, Dr. Adnan Alsaffar and Ms. Twila Meder for overseeing, coordinating and sometimes untangling the problems associated with the technical program of the XXVIIth IAHR Congress.

Marshall English, Theme A Editor
Andras Szollosi-Nagy, Co-Editor
Corvallis, Oregon, April 1997

CONTENTS

THEME A - MANAGING WATER: COPING WITH SCARCITY AND ABUNDANCE

Theme Keynote Address: HAROLD D. FREDERICKSEN, USA "Are Current Water Management Policies Equal to the Task? 1

Session A.2
APPLICATIONS OF NONLINEAR DYNAMICS, CHAOS, AND FRACTALS IN COASTAL, AND ESTUARINE MORPHODYNAMICS
Sponsor: IAHR Section I.1 Fluid Mechanics
Convenor: Giovanni Seminara, Università di Genova, Italy

River Morphology: A Manifestation of Nonlinear Dynamics
H.J. DE VRIEND, Delft University of Technology, Delft, The Netherlands 10

Edge Wave Excitation by Random Sea Waves
G. VITTORI, University of Genoa, Genoa, Italy, P. BLONDEAUX, University of L'Aquila, L'Aquila, Italy 16

Venice Gates in Waves
P. SAMMARCO, H.H. TRAN, O. GOTTLIEB, and C.C. MEI, Massachusetts Institute of Technology, Cambridge, MA 22

Cyclic Bar Behaviour in a Nonlinear Model of a Tidal Inlet
H.M. SCHUTTELAARS and H.E. DE SWART, Utrecht University, Utrecht, The Netherlands 28

Dynamics of Self-Organization and the Fluvial Landscape: A Nonreductionist Perspective
ANDREA RINALDO, Università de Padova, Padova, Italy, and IGNACIO RODRIGUEZ-ITURBE, Texas A & M University, College Station,TX 34

Transport and Separation of Particles by Chaotic Advection
AHMET C. OMURTAG, VICTOR G. STICKEL JR., SHAWN M. GOMEZ, EDWARD F. LEONARD, and RENE CHEVRAY, Columbia University, New York, NY 40

Non-Linear Methods for Analysis of Long-Term Beach and Nearshore Morphology
HOWARD N. SOUTHGATE, HR Wallingford Ltd., Oxfordshire, UK 46

Internally Generated (AUTOCYCLIC) Fluctuations in Fluvial Systems
CHRIS PAOLA, University of Minnesota, Minneapolis, MN 52

Session A.3
STORMWATER MANAGEMENT
  Sponsors: IAHR Section II.2 IAHR/IAWQ Joint Committee on Urban Storm Drainage
  Convenor: Govert Geldof, Tauw Civiel en Bouw Bv, The Netherlands

Flood Water Harvest
  TIMOTHY L. BUCHANAN, RICHARD H. FRENCH, and STEVE A. MIZELL,
  Water Resources Center, Las Vegas, NV 58

Rainwater Collection and Utilization as a Potential Resource for Urban Areas
  SAAD A. ALGHARIANI, Alfateh University, Tripoli, Libya 64

Theoretical Analysis on Air-Entrained Flow in Vertical Drop Shafts of the Channel in Urban Drainage System
  KUNIHIRO OGIHARA and TADASHI KUDOU, Toyo University, Saitama, Japan 69

Optimum Design of Self-Regulating Spring Steel Throttles for Sewer Overflow Tanks
  DETLEF AIGNER and Hans-B. Horlacher, Technische Universität Dresden, Dresden, Germany 75

Infiltration of Urban Stormwater into Soils as an Integrated Part of the Urban Drainage System
  ALINA NOWAKOWSKA-BLASZCZYK, Warsaw University of Technology, Warsaw, Poland, and PAWEL BLASZCZYK, Institute of Environmental Protection, Warsaw, Poland 81

Infiltration as an Urban Source Control for Metal Elements and Solids
  JOHN J. SANSALONE, University of Cincinnati, Cincinnati, OH 86

The Infiltration of Rainwater Through Reservoir Structures
  GEORGES RAIMBAULT and MICHEL LEGRET, Laboratoire Central des Ponts et Chaussées, Bouguenais, France 92

Design of Stormwater Infiltration Trenches Based on Uncertain Information
  PETER STEEN MIKKELSEN, Technical University of Denmark, Lyngby, Denmark 98

Evaporation at the Surface of Permeable Pavement and its Impacts on the Urban Thermal Environment
  VU THANH CA and TAKASHI ASAEDA, Saitama University, Saitama, Japan ..104

Session A.6
SCOUR AT BRIDGES
  Sponsor: ASCE Task Committee on Bridge Management for Scour Safety
  Convenor: Jorge Pagan-Ortiz, U.S. Federal Highway Administration

Bridge Scour Evaluation Program in the United States
JOHNNY L. MORRIS, FHWA, Atlanta, GA, and JORGE E. PAGAN-ORTIZ, FHWA, Washington, D. C. 110

Findings of the I-5 Bridge Failure
EVERETT V. RICHARDSON, Ayres Associates, Fort Collins, CO, J. STERLING JONES, FHWA, McLean, VA, and JAMES C. BLODGETT, USGS, Sacramento, CA 117

The Need for Research on Scour at Bridge Crossings
A.C. PAROLA, D.J. HAGERTY; University of Louisville, Louisville, KY; D.S. MUELLER; USGS, Louisville, KY, B.W. MELVILLE, University of Auckland, Auckland, New Zealand, G. PARKER, University of Minnesota, Minneapolis, MN, AND J.S. USHER, University of Louisville, Louisville, KY 124

Streamlining the Permit Approval Process for the Installation of Scour Countermeasures at Bridges
STANLEY R. DAVIS, Maryland State Highway Administration, Baltimore, MD, JORGE E. PAGAN-ORTIZ, FHWA, Washington, D.C., and LINDA KELBAUGH, Brightwater, Inc., Ellicott City, MD 130

Evaluation of Field and Laboratory Research on Scour at Bridge Piers in the United States
DAVID S. MUELLER, USGS, Louisville, KY, and J. STERLING JONES, Federal Highway Administration, McLean, VA 135

Research Needs in Geomorphology Pertaining to Bridge Scour
RICHARD L. VOIGT, Jr., University of Minnesota, Minneapolis, MN, CARLOS M.TORO-ESCOBAR, University of Minnesota, Minneapolis, MN, and GARY PARKER, University of Minnesota, Minneapolis, MN 141

Quantitative Techniques for Stream Stability Analysis
P.F. LAGASSE, S.A. SCHUMM, and L.W. ZEVENBERGEN, Ayres Associates, Fort Collins, CO 147

Geomorphic Factors for Tidal Waterway Hydraulic Modeling
L.W. ZEVENBERGEN, J.D. SCHALL, and J.H. HUNT, Ayres Associates, Fort Collins, CO 154

Pier Scour in Resistant Material: Current Research on Erosive Power
S.P. SMITH, Colorado Department of Transportation, Denver, CO, G.W. ANNANDALE, Golder Associates Inc., Lakewood, CO, P. A. JOHNSON, Pennsylvania State University, University Park, PA, J. S. Jones, FHWA, McLean, VA, and E.R. UMBRELL, GKY & Associates, Inc., Springfield, VA 160

Pier Riprap Protection
B.W. MELVILLE, C. S. LAUCHLAN, and S.E. COLEMAN, The University of Auckland, Auckland, New Zealand 166

Riprap Particle Stability by Moment Analysis
DAVID C. FROEHLICH, Parsons Brinckerhoff, Raleigh, NC 172

Design Riprap to Protect Scour Around Circular Piers
TAE-HOON YOON, Hanyang University, Seoul, Korea, and SUNG BUM YOON, Hanyang University, Ansan, Korea 178

Failure Behavior of Riprap Layer Around Bridge Pier
FOO-HOAT LIM and YEE-MENG CHIEW, Nanyang Technological University, Nanyang, Singapore 184

Influence of Lateral Momentum Transfer on Bridge Abutment Scour
S. KOUCHAKZADEH and R.D. TOWNSEND, University of Ottawa, Ontario, Canada 190

Local Scaling of Bridge Abutment Scour in Compound Channels
T.W. STURM and A. CHRISOCHOIDES, Georgia Institute of Technology, Atlanta, GA 196

Vorticity Distribution of Horseshoe Vortex on Scoured Bed
TOMONAO KOBAYASHI, TOSHIFUMI AIBARA, and HIROKAZU HARADA, Tokyo Rika University, Chiba, Japan 202

Flow Patterns Around Bridge Piers and Offshore Structures
KAMIL H.M. ALI, OTHMAN A. KARIM, and BRIAN A. O'CONNOR, University of Liverpool, Liverpool, United Kingdom 208

Building Scour in Floodplains
ALEXANDER KOHLI and WILLI H. HAGER, VAW, ETH-Zentrum, Zurich, Switzerland 214

Physical and Computational Modeling of Bridge Scour at Oregon Inlet, North Carolina
CONOR SHEA and MICHAEL PORTS, Parsons Brinckerhoff Quade & Douglas, Inc., Baltimore, MD 220

Three-Dimensional Hydraulic Analysis for Calculation of Scour at Bridge Piers with Fender Systems
EDWARD J. KENT, Earth Tech, Concord, NH, and JOHN E. RICHARDSON, Flow Science, Los Alamos, NM 226

Prediction of Flow Generated by Turbulent Offset Jets
KAMIL H.M. ALI, OTHMAN A. KARIM, and BRIAN A. O'CONNOR, University of

Liverpool, Liverpool, United Kingdom 232

Experiments on Flow Upstream of a Cylinder
W.H. GRAF and B. YULISTIYANTO, Ecole, Polytechnique Fédérale, Lausanne,
Switzerland 238

Numerical Analysis of Three-Dimensional Turbulent Flows Around Submerged Groins
J. PENG, Y. KAWAHARA, and N. TAMAI, University of Tokyo, Tokyo, Japan . . . 244

Session A.8
FUTURE OF IRRIGATED AGRICULTURE
Sponsor: ASCE Committee on Irrigation and Drainage
Convenor: Marshall English, Oregon State University, USA

The Future of Irrigation
JAN VAN SCHILFGAARDE and THOMAS J. TROUT, USDA-Agricultural
Research Service, Fresno, CA 250

Partial Irrigation: A Fundamental Change
MARSHALL ENGLISH, Oregon State University, Corvallis, OR 256

Areal Evapotranspiration Estimation Using Several Alternative Formulations of the Crae Hypothesis
LAKSHMAN NANDAGIRI, Karnataka Regional Engineering College,
Karnataka, India 262

A Radial Model of Surface Flow: Application to Level-Basin Irrigation
P. GARCIA-NAVARRO, Universidad de Zaragoza, Zaragoza, Spain, and
E. PLAYAN, Estación Exp. Aula Dei (DGA-CSIC), Zaragoza, Spain 268

Knowledge-Based Advisory System for Predictive Control of Irrigation Management Strategies
MUHAMMAD ABID BODLA, International Waterlogging and Salinity
Research Institute, Lahore, Pakistan 274

Limitations and Consequences of Recycling Drainage Water for Irrigation Over the Long Term
S.R. GRATTAN, University of California, Davis, CA 279

Management of Drainage Water, Salt, and Selenium as Resources
V. CERVINKA, California Department of Food & Agriculture, Sacramento, CA,
J.DIENER, Red Rock Ranch, Five Points, CA, C. FINCH, Westside Resource
Conservation District, Five Points, CA, M. MARTIN, USDA-Natural Resources
Conservation Department, Fresno, CA, F. MENEZES, California Department of
Water Resources, Fresno, CA, R. MUNOZ, University of California, Davis, CA,

and D. PETERS 285

Reuse System of Irrigation Return Flow and Water Quality Management in Low Lying Creeks-Paddy Area
 MASAHARU KURODA, and TETSURO FUKUDA, Kyushu University, Fukuoka, Japan 290

Wastewater Reuse for Irrigation in Arid Regions
 N. ISMAIL, Arab Academy for Science & Technology, Alexandria, Egypt 296

Session A.9
OPTIMAL CANAL SYSTEM OPERATION
 Sponsor: ASCE Task Committee on Canal Automation
 Convenor: Bert Clemmens, U.S. Water Conservation Lab

Implementation of Canal Automation in Central Arizona
 A.J. CLEMMENS, E. BAUTISTA, and R.J. STRAND, U. S. Water Conservation Laboratory, Phoenix, AZ 302

Comparison of Constant Level and Constant Volume Control Method for Open Channel Flow
 FUBO LIU and JAN FEYEN, Katholieke Universiteit Leuven, Leuven, Belgium .303

Optimal Control of Sudden Water Release from a Reservoir
 BRETT F. SANDERS and NIKOLAOS D. KATOPODES, University of Michigan, Ann Arbor, MI 314

Flap Gate for Hydraulic Head Control
 FELIX RAEMY and WILLI H. HAGER, VAW, ETH-Zentrum, Zurich, Switzerland 320

Local Supervisory and Control Algorithms for a Laboratory Slide Gate
 VICTOR M. RUIZ C., RAFAEL HERRERA, and EDMUNDO PEDROZA, Instituto Mexicano de Tecnología del Agua, Morelos, Mexico 326

Accuracy of Flow Measurements in the Imperial Irrigation District
 BRIAN WAHLIN, JOHN REPLOGLE, and ALBERT CLEMMENS, U.S. Water Conservation Laboratory, Phoenix, AZ 332

Calibration of Open Channel Flow Computer Simulations
 JOHN B. PARRISH, University of Iowa, Iowa City, IA, and RAM DHAN KHALSA, USBR, Grand Junction 338

Session A.12
Multinational River Basin Management
 Convenor: Vahid Alavian, Rankin International Inc., USA

International River Basins: Forging a Consensus
BARBARA A. MILLER, VAHID ALAVIAN, RANKIN International, Inc., Knoxville, TN, GEOFFERY MATTHEWS, World Bank, Washington, D. C., and LAURA L. COLE, Tennessee Valley Authority, Knoxville, TN 917

Management of the Euphrates and Tigris River Basins: Need for a Rational Approach
MEHMETCIK BAYAZIT, Istanbul Technical University, Istanbul, Turkey 343

Negotiating Middle East Water Management: Lessons from Other International River Basins
DEBORAH SHMUELI and NURIT KLIOT, University of Haifa, Haifa, Israel ..349

The July 1996 Flood on a Trans DMZ River in Korea
HYOSEOP WOO, Korea Institute of Construction Technology, Seoul, Korea 357

International River Basin Management-Imperative for Sustainable Water Development
JELISAVETA MUSKATIROVIC, Institute for Development of Water Resources, Belgrade, Yugoslavia 363

River Basin Management: Battleground for Economic Growth
CHRISTOPHER D. UNGATE, TVA, Chattanooga, TN 369

Slovenian Water Management Strategy-Towards EU Water Policy
FRANCI STEINMAN, and PRIMOZ BANOVEC, University of Ljubljana, Ljubljana, Slovenia 375

The Okavango River Basin
DANIEL P. MILLER, Stanley Consultants, Muscatine, IA 381

Managing Extremes: The IJC Experience
LISA BOURGET, International Joint Commission, Washington, D. C., and MURRAY CLAMEN, International Joint Commission, Ottawa, Canada 387

Comprehensive Assessment of the Freshwater Resources of the World
PAUL H. KIRSHEN, Stockholm Environment Institute, Boston, MA and Tufts University, Medford, MA, and KENNETH M. STRZEPEK, University of Colorado, Boulder, CO 393

Session A.14
DECISION SUPPORT ENVIRONMENTS
Sponsor: IAHR Section I.2 Hydroinformatics

Convenor: M.B. Abbott, International Institute for Hydraulic and Environmental Engineering, The Netherlands

Intranetted Management of Water Resources
M.B. ABBOTT, and S. SHIPTON, Infrastructure Hydraulics Environment, Delft, The Netherlands *399*

Aquarius: A General Model for Efficient Water Allocation in River Basins
GUSTAVO E. DIAZ, and THOMAS C. BROWN, Colorado State University, CO *.405*

Implementation of Object-Oriented Programming for Dynamic Simulation of Open Channel Flow
JOHN A. HINCKLEY, Jr., SHAW L. YU, University of Virginia, Charlottesville, VA, and PERRY J. LAPOTIN, U.S. Army Cold Regions Research and Engineering Laboratory, Hanover, NH .... *.411*

The Caspian Sea Storm Surge Coastal Protection
M. GALANT, A. ZVEGINTSEV, and P. LYSSENKO, VOLGA Ltd., Moscow, Russia .... *.417*

Session A.15
RELIABILITY-BASED DESIGN AND ANALYSIS IN WATER RESOURCES
Sponsors: IAHR Section I.4 Probabilistic Methods, ASCE Committee on Probabilistic Approaches to Hydraulics
Convenors: Steve Melching, U.S. Geological Survey; Yeou-Koung Tung, Hong Kong University of Science and Technology

A New Model for Reliability-Based Optimal Design of Water Distribution Networks
CHENGCHAO XU, and IAN C. GOULTER, Central Queensland University, Rockhampton, Australia .... *.423*

Reliability Assessment of Water Distribution Networks Using the First Order Reliability Method
CHENGCHAO XU, and IAN C. GOULTER, Central Queensland University, Rockhampton, Australia .... *.429*

Parameter Dimension Estimation for Water Distribution Networks
CHENGCHAO XU, and KEVIN S. TICKLE, Central Queensland University, Rockhampton, Australia .... *.435*

Water Quality Modeling and Risk Analysis of the Keelung River, Taiwan
JAN-TAI KUO, National Taiwan University, Taipei, Taiwan, ALBERT Y. KUO, The College of William and Mary, Gloucester Point, VA, and WEN-SHIANG CHUNG, National Taiwan University, Taipei, Taiwan .... *.441*

Reliability Analysis for Water Quality Management in the Han River
KUN-YEUN HAN, and SANG-HO KIM, Kyungpook National University, Daegu, Korea 447

Uncertainty in the Design of Stream Channel Restoration
PEGGY A. JOHNSON, Pennsylvania State University, University Park, PA, and MASSIMO RINALDI, University of Florence, Florence, Italy 453

Risk-Based Design of Roadway Crossing Structures Considering Intangible Factors
YEOU-KOUNG TUNG, Hong Kong University of Science & Technology, Kowloon, Hong Kong, A. MAINARD WACKER, Wyoming Highway Department, Cheyenne, WY, and VICTOR HASFURTHER, University of Wyoming, Laramie, WY 458

Case Study: Lesson and Probabilistic Determination on Eroded Earth Spillway Rehabilitation
S. SAMUEL LIN, JOSEPH S. HAUGH, Virginia Dam Safety Program, Richmond, VA, and L. LYNN CLEMENTS, Rapidan Service Authority, Ruckersville, VA 464

Reservoir Reliability Design Under Interannual Climatic and Hydrologic Variability
OSCAR MESA, GERMÁN POVEDA, LUIS CARVAJAL, and JOSÉ SALAZAR, Universidad Nacional de Colombia, Medellin, Columbia 470

Statistics of Maximum Flows
EDUARDO VARAS, Universidad Catolica de Chile, Santiago, Chile 476

Session A.16
RISK AND UNCERTAINTY IN ANALYSIS OF WATER-RESOURCES SYSTEMS
Sponsors: IAHR Section I.4 Probabilistic Methods and ASCE Committee on Probabilistic Approaches to Hydraulics
Convenors: Jose Salas, Colorado State University, USA, and Steven Buchberger, University of Cincinnati, USA ....

Risk Estimation of Monthly Rainfall in Semiarid Regions
BONIFACIO FERNÁNDEZ, Pontificia Universidad Católica de Chile, Santiago, Chile 482

Poisson Pulse Queuing Model for Residential Water Demands
STEVEN G. BUCHBERGER, and TRENT G. SCHADE, University of Cincinnati, Cincinnati, OH 488

An Evaluation of Data Needs to Support Flood Frequency Estimation at Regulated Sites
S. ROCKY DURRANS, SASA TOMIC, and STEPHEN J. NIX, The University of Alabama, Tuscaloosa, AL 494

Protection of the Environment Against Floods-A Statistical Problem?
REINHARD POHL, Dresden University of Technology, Dresden, Germany 500

On the Uncertainty of the Risk of Failure of Hydraulic Structures
JOSE D. SALAS, Colorado State University, Fort Collins, CO, and JUN H. HEO, Yonsei University, Seoul, South Korea 506

Long-Term Memory in Hydrologic Series?
A. R. RAO, and D. BHATTACHARYA, Purdue University, West Lafayette, IN ...512

Precipitation Variability and Curve Numbers
JOSEPH A. VAN MULLEM, Natural Resources Conservation Service, Bozeman, MT 518

Errors and Variability of Reservoir Yield Estimation as a Function of the Coefficient Variation of Annual Inflows
JOSE NILSON BEZERRA CAMPOS, Universidade Federal do Ceará, Fortaleza, Brazil, FRANCISCO DE ASSIS DE SOUZA FILHO, Universidade de Fortaleza, Fortaleza, Brazil, and Jose Carlos de Araújo, Universidade Federal de Ouro Preto, Ouro Preto, Brazil 524

Multivariable Marima Modeling of Water Resources Time Series: Applications to River Nile Ten-Day Discharges
AHMED H. EL-SAYED, Zagzig, University, Egypt, and MOHAMED ERRIH, University of Sciences and Technology, Oran, Algeria 530

Uncertainty Analysis of a Water Resources System in a Competition and Privatization Environment
RICARDO SMITH, ISAAC DYNER, SANTIAGO MONTOYA, and CARLOS FRANCO, National University of Colombia, Medellin, Colombia 536

Water Resource System Operation Under Hydrologic Uncertainty: The Mae Klong River Basin, Thailand
TAWATCHAI TINGSANCHALI, and VEERAKCUDDY RAJASEKARAM, Asian, Institute of Technology, Pathumthani, Thailand 542

Stochastic Multiobjective Optimization of Multireservoir Systems
Y. C. WANG, Newjec Engineering Consulting Corporation, Tokyo, Japan 548

Coping With Uncertainty in the Economical Optimization of a Dike Design
P.H A.J.M. VAN GELDER, J.K. VRIJLING, and K.A. H. SLIJKHUIS, Delft University of Technology, Delft, The Netherlands 554

Sensitivity and Uncertainty Analysis of a Regional Simulation Model for the Natural System in South Florida
A. M. WASANTHA LAL, JAYANTHA OBEYSEKERA, and RANDY VAN ZEE, South

Florida Water Management District, West Palm Beach, FL 560

A Simple Extension to the 'abc' Watershed Model
RICHARD M. VOGEL, and ANTIGONI ZAFIRAKOU-KOULOURIS, Thaddeus
Green, Tufts University, Medford, MA 566

The Effect of Climate Change on Danube Flow Regime
B. GAUZER, and ÖSTAROSOLSZKY, VITUKI, Budapest, Hungary 572

Reliability of Operation of the Volga Water-Resource System Under Global Climate Change Conditions
A.L. VELIKANOV, Russian Academy of Sciences, Moscow, Russia 578

Network Flow Model Applied to Geum River Basin
G.B. YEON, Chung Cheong College, Cheong-Ju, Korea, S.B. SHIM, Chung Buk University, Cheong-Ju, Korea, and K. H. YOON, KICT, Seoul, Korea 583

Session A.17
Flood and Drought Estimation, Control, Management, and Mitigation
Convenor: Hsieh Wen Shen, University of California, USA

Reservoir Management During Drought
DAVID STEPHENSON, University of the Witwatersrand, Johannesburg, South Africa 589

Stochastic Modeling of Hydrological Droughts
HSIEH WEN SHEN, and JENQ TZONG SHIAU, University of California, Berkeley, CA 595

Comparison of National Weather Service Operational Mean Areal Precipitation Estimates Derived from NEXRAD Radar vs. Rain Gage Networks
BRYCE FINNERTY, and DENNIS JOHNSON, National Weather Service, Silver Spring, MD 601

Managing Water Scarcity Through Man-Made Rivers
SAAD A. ALGHARIANI, Alfateh University, Tripoli, Libya 607

Studies on Stochastic Distribution Laws of Dryness/Wetness in Time and Space in Yunnan
WENBIN XU, YOUQUE ZHANG, and SAIZHEN ZHANG, Yunnan Polytechnic University, Kunming, China . . . .*

Influence of Over-Basin Diversion Reservoir on Water Management
Youn-Jan Lin, Ming Hsin Institute of Technology, Taiwan, ROC, Chang-Shian Chen,

---
* Manuscript not available at time of printing

Feng Chia University, Taiwan, ROC, Edward S. C. Young, Constants Corp., Ltd., Taiwan, ROC 615

Flood Risk Analysis in the Upper Adriatic Sea Due to Sea Level Rise and Land Subsidence
G. GAMBOLATI, P. TEATINI, L. TOMASI, University of Padova, Padova, Italy, M. GONELLA, MED Ingegneria S.r.l., Ferrara, Italy, and C.S. YU, C. Decouttere, Catholic University of Leuven, Heverlee, Belgium 621

Simulation of Floods in Ping River Using IIS Distributed Hydrological Model (IISDHM)
R. JHA, S. HERATH, and K. MUSIAKE, University of Tokyo, Tokyo, Japan 627

Regional Model of Glaciers Runoff and its Application for Hydrological Forecasts
V.G. KONOVALOV, Central Asian Research Hydrometeorological Institute, Tashkent, Uzbekistan 633

Coupling 1-D and 2-D Models for Floods Management
A. PAQUIER, and B. SIGRIST, Cemagref, Lyon, France 639

Forecasting Drought Risk for a Water Supply Storage System Using Bootstrap Position Analysis
GARY TASKER, USGS, Reston, VA, and PAUL DUNNE, USGS, West Trenton, NJ 645

Flood Attenuation Effects of Natural Flood Basins in the Sacramento Valley, California
J.C. VICK, and P.B. WILLIAMS, Philip Williams & Associates, Ltd., San Francisco, CA 651

Hydrodynamic Approaches to Design Balancing Ponds
M.J.J. PIROTTON, University of Liege, Liege, Belgium 657

The Threat of Flooding and the Problem of Protection of Territories in the Delta of the Ural River in View of the Rising of the Caspian Sea Level
V.F. POLONSKY, L.P. OSTROUMOVA, State Oceanography Institute, Moscow, Russia, Y.G. VIKULOV, and R.R. MULIKOV, Department of Ecology and Bioresources, Atyrau, Kazakhstan 663

New Dimension to Spillway Gate Operation for Reducing the Flood Intensity
SUNIL Y. KUTE and M.J. DEODHAR, K.K. Wagh College of Engineering, Nashik, India 669

Session A.18
BENEFITS OF PROVIDING FLEXIBLE WATER DELIVERY
Sponsor: ASCE Task Committee on Benefits of Providing Flexible Water Delivery

Convenor: Paul Cross, Lake Chelan Reclamation District, USA

A Flexible Irrigation Water Supply: Why and How
PAUL R. CROSS, Lake Chelan Reclamation District, Manson, WA 673

Irrigation District Modernization for the Western U. S.
C.M. BURT, S.W. STYLES, California Polytechnic State University, San Luis Obispo, CA, M. FIDELL, California Polytechnic State University, San Luis Obispo, CA, and E. REIFSNIDER, USBR, Sacramento, CA 677

Experience in Operating a Large Limited Rate Arranged System - Westlands Irrigation District
STEPHEN H. OTTEMOELLER, Westlands Water District, Fresno, CA 683

Design and Congestion Considerations for Flexible Irrigation Supply Systems, Part 1: Concepts
JOHN L. MERRIAM, and STUART STYLES, California Polytechnic State University, San Luis Obispo, CA 689

Design and Congestion Considerations for Flexible Irrigation Supply Systems, Part 2: Application
JOHN L. MERRIAM, and STUART STYLES, California Polytechnic State University, San Luis Obispo, CA 695

Use of Reservoirs and Large Capacity Distribution Systems to Simplify Flexible Operations, Part 1: System Capacity
JOHN L. MERRIAM, and STUART STYLES, California Polytechnic State University, San Luis Obispo, CA 700

Use of Reservoirs and Large Capacity Distribution Systems to Simplify Flexible Operations, Part 2: Reservoir Capacity and Case Studies
JOHN L. MERRIAM, and STUART STYLES, California Polytechnic State University, San Luis Obispo, CA 706

Pima-Maricopa Flexible Irrigation Project
SHANE LINDSTROM, Gila River Indian Community, Sacaton, AZ 711

Alleviation of Surface and Subsurface Drainage Problems by Flexible Delivery Schedules
STUART STYLES, California Polytechnic State University, San Luis Obispo, CA 717

Session A.19
HYDRODYNAMICS OF CHANNELS AND WATERWAYS
Convenor: Jeffrey B. Bradley, WEST Consultants, USA

Entrance Flow and the Achievement of Uniform Fully-Developed Open Channel Flow
HECTOR R. BRAVO, and JOHN W. MEINECKE, *University of Wisconsin, Milwaukee, WI* 723

Positive Front of a Dambreak Wave
GUIDO LAUBER, and WILLI H. HAGER, *VAW, ETH-Zentrum, Zurich, Switzerland* 729

Analysis of Flow Field in the Meandering Channel
KOUICHI OZAWA, *Ritsumeikan University, Kyoto, Japan*, and NOBUYUKI TAMAI, *Tokyo University, Tokyo, Japan* 734

Unsteady Flow Characteristics in a Compound Channel
B. L. JAYARATNE, N. TAMAI, Y. KAWAHARA, *University of Tokyo, Tokyo, Japan*, and K. KAN, *Shibaura Institute of Technology, Japan* 740

The Mixing Mechanism in Turbulent Two-Stage Meandering Channel Flows
P. RAMESHWARAN, and B.B. WILLETTS, *University of Aberdeen, Aberdeen, Scotland* 746

Estimation of Flow Resistance in Ice Covered Channels
ZENGNAN DONG, ZEYU MAO, Yongtian Wang, *Tsinghua University, Beijing, China*, GAOFENG MU, and JING WANG, *Xinjiang Institute of Water Resources and Hydroelectric Research, China* 752

Turbulent Structure of Open and Ice-Covered Flow in a Channel
D.S. KUZNETSOV, *Moscow State University, Moscow, Russia*, and E.I. DEBOL'SKAYA, *Water Problems Institute of Russian Academy of Sciences, Moscow, Russia* 758

Fluid Mixing and Boundary Shear Stress in Compound Meandering Channel
T. ISHIGAKI, Y. MUTO, N. TAKEO, and H. IMAMOTO, *Kyoto University, Kyoto, Japan* 763

A Simple Model for Gravel-Bed Roughness
V.I. NIKORA, D.G. GORING, and B.J.F. BIGGS, *National Institute of Water and Atmospheric Research, Christchurch, New Zealand* 769

Analytical Model for Hydraulic Roughness of Submerged Vegetation
D. KLOPSTRA, H. J. BARNEVELD, J.M. VAN NOORTWIJK, *HKV Consultants, Lelystad, The Netherlands*, and E. H. VAN VELZEN, *Ministry of Transport, Public Works, and Water Management, Arnhem, The Netherlands* 775

1-D or 2-D Model for River Hydraulic Studies?
R. WALTON, J.B. BRADLEY, and T.R. GRINDELAND, *WEST Consultants, Inc., Seattle, WA* 781

Steady Flow Over an Obstacle with Contraction and Sill
ADRIAN W.K. LAW, Nanyang Technological University, Singapore 787

Lateral Velocity Variations in a Compound Channel - A Practical Approach
*H.S. TU, S. TAKAKI, and H. TAKAMATU, Pacific Consultants Co. Ltd., Tokyo,
Japan* 793

Comparison of Water Surface Profiles from Physical and Numerical Models in Mixed Regime Flow
*MICHAEL E. MULVIHILL, U. S. Army Corps of Engineers, Los Angeles, CA,
Loyola Marymount University, Los Angeles, CA, and SCOTT E. STONESTREET,
U. S. Army Corps of Engineers, Los Angeles, CA* 799

Spur Dike Effects on the River Nile Morphology After High Aswan Dam
*M.M. SOLIMAN, K.M. ATTIA, KOTB, A.M. TALAAT, and A.F. AHMED, Ain Shams
University, Cairo, Egypt* 805

Two-Dimensional Floodwave Analysis Resulting From Breached Levee
*KUN-YEUN HAN, JAE-HONG PARK, Kyungpook National University, Daegu,
Korea, and JONG-TAE LEE, Kyonggi University, Seoul, Korea* 811

Two-Dimensional Large Eddy Simulation for Shallow Recirculating Flow
*M. NASSIRI, S. BABARUTSI, and V.H. CHU, McGill University, Quebec,
Canada* 817

Numerical Analysis of Horizontal Vortices in Compound Open Channel Flows by the Two-layered Flow Model
*I. KIMURA, Wakayama National College of Technology Wakayama, Japan,
T. HOSODA, Y. MURAMOTO, Kyoto University, Kyoto, Japan, and R. YASUNAGA,
Tokyo Electric Power Co. Ltd., Tokyo, Japan* 823

2-D Models for Flows in the River with Submerged Groins
*SHIROU AYA, Osaka Institute of Technology, Osaka, Japan, ICHIRO FUJITA,
Gifu University, Gifu, Japan, and NOBUYUKI MIYAWAKI, CTI Engineering Co.
Ltd., Fukuoka, Japan* 827

Three Dimensional Modeling of Flow and Transport Mechanisms in Meandering Two-Stage Channel Flows
*J. RUSSELL MANSON, Bucknell University, Lewisburg, PA, and GARETH
PENDER, Glasgow University, Glasgow, United Kingdom* 835

Session A.20
CATCHMENT HYDRAULICS AND HYDROLOGY

Modeling for Study of the Hydrologic Processes in Watershed

ADILSON PINHEIRO, PHILIPPE MAISON, and BERNARD CAUSSADE,
Institut de Mecanique des Fluides de Toulouse, Toulouse, France 841

Watershed-Change Induced Uncertainty on Runoff Frequency for Water Resources Management
STEFANO PAGLIARA, University di Pisa, Italy, Ben Chie Yen, University of Illinois, Urbana, IL 847

A Physically Based Model for Large River Basins
W.T. SLOAN, J. EWEN, C.G. KILSBY, C.S. FALLOWS, and P.E. O'CONNELL, University of Newcastle Upon Tyne, United Kingdom 853

Water and Energy Balances and Their Spatial Distribution in a Catchment of Complex Land Use
Y. JIA, and N. TAMAI, University of Tokyo, Tokyo, Japan 859

Continuous Distributed-Parameter Hydrologic Modeling with CASC2D
FRED L. OGDEN, SHARIKA U.S. SENARATH, University of Connecticut, Storrs, CT 864

Session A.21
HYDRAULICS AND WATER QUALITY OF PIPE SYSTEMS

Optimization of Piped Systems
B. B. SHARP, Burnell Research Laboratory, Victoria, Australia 870

Mathematical Modelling of Water Quality in Distribution Systems
DOMENICO PIANESE, FRANCESCO PIROZZI, and LUCIO TAGLIALATELA, University of Naples, Naples, Italy 875

Influence of Distinct Processes on the Quality of Water in Distribution Systems
DOMENICO PIANESE, FRANCESCO PIROZZI, and LUCIO TAGLIALATELA, University of Naples, Naples, Italy 881

Reliability of Algorithms for Water Quality Analysis in Hydraulic Networks
SAMI ELMAALOUF, The Levantine Engineers Society, Los Angeles, CA, and YOUNG C. KIM, California State University, Los Angeles, CA 887

Session A.22
REGIONAL WATER RESOURCE MANAGEMENT

Bohol-Cebu Water Supply Project Central Visayas, The Philippines
DAVID W. PRASIFKA, Brown & Root, Inc., Houston, TX ....*

---
* Manuscript not available at time of printing

Enhancement of Irrigation Systems in Developing Countries. A "Holistic" Approach
SAMI ELMAALOUF, The Levantine Engineers Society, Los Angeles, CA, and YOUNG C. KIM, California State University, Los Angeles, CA 893

A Study on Water Management of an Irrigation Scheme in Sri Lanka
SHAHANE DE COSTA, Open University of Sri Lanka, Dehiwela, Sri Lanka ...899

The Use of Aquifers in Saudi Arabia to Reclaim and Store Wastewaters
ACHI M. ISHAQ, and AMIR ALI KHAN, King Fahd University of Petroleum & Minerals, Dhahran, Saudi Arabia 905

The Salt Water Intrusion in the Maryout Aquifer
RAWYA M. KANSOH, Alexandria University, Alexandria, Egypt 911

Subject Index 923

Author Index 931

ARE CURRENT WATER MANAGEMENT POLICIES EQUAL TO THE TASK?

HARALD D. FREDERIKSEN
Consultant Eugene, Oregon USA

ABSTRACT

The developing countries face enormous difficulties managing their water resources in a manner that will fulfill the aspirations of their people. Populations and economies continue to grow in countries that have reached present limits of their water supply. Flood losses, pollution of surface and groundwater and environmental damages mount. Countries are ill-prepared for the next drought. The support for their developing economies strain the already inadequate national budgets. Time is short for implementing actions that will avoid more serious problems introduced by the world's addition of one billion inhabitants in the next ten years and two billion in twenty years. A country's water management policies, rather than technology, will largely determine its degree of success. Many countries have adopted wise policies regarding several matters. Yet, when matched against the conditions in such areas as drought management, reallocation of water, financing of water related services and water supply augmentation, the entire body of policies displays deficiencies or is only partially enforced. Extended international debate of some new concepts are delaying adoption of viable actions. It is judged that water management policies should be framed in a manner that better reflects the countries' conditions; their social, economic, environmental and security goals; and the countries' time, money and political constraints.

INTRODUCTION

It was requested that these opening remarks for the Theme A sessions focus on the world's water situation and the adequacy of water management policies. The title of the Session; Managing Water; Coping with Scarcity and Abundance, describes much of what constitutes water management in many countries today. The consequences are reported daily. Indeed, it seems appropriate to commence this session with the question -- why are so many countries having to cope, rather than having adequate water management in place?

Mankind has been engaged in water management for thousands of years. Today's water needs and the institutional and technical means for meeting them have been evolving within society since humans first organized. It is not a suddenly encountered frontier. The technical aspects, such as hydrology, groundwater geology, waste treatment, computer simulation and evaluation of environmental impacts are adequately understood for basic management.

Water management institutions have evolved at a similar pace. Local irrigators and villagers adopted the same organizational principles, whether in Nepal, Morocco or North America. Wherever water is well managed, one finds an enforceable supply right that underpins the activities that depend on the supply. Examples of sound institutions, from legislation to the structure of service entities, are available.

If one judges that there is ample knowledge of the physical, technical and institutional aspects for sound management, why are some countries only coping with water problems? There is no question, disrupted economies and money constraints have hobbled many countries. Ironically, advances in disease control have led to population growths that burden most. These contribute to the shortcomings. But a close examination suggests that a country's current water management situation, good or bad, is to a large extent a result of the total body of policies adopted and implemented. And since policy formulation and implementation are effected through the political process, a fully informed public is essential. Political leaders and government officials need the best counsel. The research and water management community has immense influence and responsibilities in respect to both.

These remarks will deal with a limited number of water policy areas. A background on the nature of the world's water problems precedes comments on the water quantities that serve as the base for today's policies. With that in mind, I would like to pose questions in six policy areas, though several others are of equal importance.

SOME CURRENT PROBLEMS

A country is fortunate if abundance merely means that quantities exceed demand. But for countries in the monsoon regions, it may mean floods with resulting losses followed by extreme shortages during the low flow season. The social and economic costs of both can be immense and the protective measures compete for the same budget. Examples of the world's water situation provide a snapshot of the physical problems and constraints to policy formulation.

The world's population will increase by one billion over the next ten years and two billion additional shortly thereafter. Several of the growing countries already face serious water shortages -- many are plagued by endemic flooding. The Nile is fully committed, with Egypt contesting upstream uses that would further reduce its supplies. The lower 600 kilometers of the Yellow River ran dry in 1995.

India provides examples of the range of problems developing countries face. The Ganges discharge into Bangladesh often falls below the provisions of the 1977 India / Bangladesh water sharing agreement, recently replaced. Old irrigation diversions ring the basin. Increased urban use and massive development of the groundwater underlying the plain has greatly reduced dry season inflow to the lower reaches of the river. After years of international debate, proposed external financing was withdrawn for additional storage on Nepal's tributaries. Recently, areas in Bangladesh served since the mid-nineteenth century were severely shorted and its ecologically rich wetlands in the west suffer from saltwater intrusion. Conflicts occur as migrating Bangladeshis attempt to enter India and Myanmar. Today, almost 400 million people, with a high population growth rate, depend on this one hydrologic system! And supply augmentation policy remains in debate.

Most major rivers of India essentially have no discharge to the ocean during the dry seasons -- basin efficiency is near 100 percent during the non-monsoon season. An exception is the Narmada River. Here India is wrestling with international opposition to dams. Its 1967 national policy to alleviate severe urban water shortages and offer rural employment to several million new inhabitants in the western dry regions has been frustrated. When completed, the project will have taken more than fifty years and cost additional billions of dollars in foregone benefits to effect change in the basin.

Thailand confronts similar regional shortages and floods -- and debate. In the Mideast, besides the Nile, disputes remain on the Tigris-Euphrates, the Jordan Valley tributaries and aquifers that extend across international boundaries. Tensions arise even under long existing agreements. Last year Mexico pleaded for a portion of the US farmers' emergency storage to save Mexico's Rio Grande agriculture. The US farmers had withheld this water from their own normal deliveries the year before. That the condition arose because the two countries applied different drought management policies for carry-over storage of their allocated quantities, did little to calm the rhetoric.

The African droughts, many of which are endemic water shortages, are other examples reported in the press. However, few realize that the 1987-88 drought in central and eastern US was the nation's costliest natural disaster up to that time. Mr. Stanley Shangnon, Principle Scientist and Emeritus of the Illinois State Water Survey reported at a drought conference that economic costs totaled US$ 40 billion. Some cities and villages were short. The nation lost 31 percent of its total annual grain production, with Midwest maize and barley harvest at 52 and 45 percent of normal. Fortunately, the US held substantial grain reserves. The 1987-88 experience prompted government bodies throughout the country to immediately prepare drought plans and enact the implementing legislation.

Water quality problems are far more serious than reflected in current policy. Conditions where sewerage and solid wastes render water unusable may be found in every urban area in the developing world and many of the developed. Not only is water made unusable, so is critically needed ground water storage adjacent to cities.

In each of the cited problem areas, one finds deficiencies in the basic water policies

FRAMEWORK FOR POLICY FORMULATION

National goals and near-term objectives, should frame a country's water policies. Typically, goals encompass social well being, national and regional economic development, environmental quality and national security. Poverty alleviation receives prominent mention. Internal political stability is inherent in all. Indeed, nations attempt to devise most resources management policies to meet their broader goals, but implementation, if not the policies, exhibit weaknesses.

Recognizing the role water plays in meeting a country's goals, the sub-area of water policy should not be considered as an area on to itself. For example, the much talked about classification of water as an economic good may be fine, but it seems to cloud reasoning, and in the minds of some over-rides a nation's broader goals. Maybe it would be better to let a country tailor its water policies to best meet all its goals -- economic and non-economic -- and worry less about definitions of water.

An example of the many policy interdependencies is the linkage of land and water management. Land use determines water use. Land use determines the potential for pollution of ground water and surface waters. Local land use, jointly with water use, determines the employment opportunities and environmental quality of the area. Is it reasonable to set water allocation polices independent of land resource policies, rural-urban migration policies, regional development policies or even the conjunctive management of water?

One means for evaluating the broader impacts of a contemplated policy is to analyze the implications of the "with and without" results. Is it better to adopt a policy to develop the Mekong Basin hydro or better to expand fossil and nuclear generation? More broadly, should a congested country build a reservoir to augment water supplies, and hence, economic opportunities in rural regions to dampen the influx to urban centers? Or is it better to retain the site for its present inhabitants and attempt to deal with the urban influx in the cities? As a part of this exercise, it may be useful to contemplate where the decade's one billion additional people can locate and what they might do.

Besides national goals, other factors should weigh heavily. As illustrated, few countries have the luxury of abundant water. Even fewer have low stable populations. Essentially none have ample national budgets. All have water problems requiring urgent attention. And these problems are but the first wave of even more difficult situations in the following decades.

Policy actions should be applied to produce real change within the next ten to fifteen years. And as was mentioned, some governments are acting, but delays continue with formulating a sound package. Some countries appear hampered by the policies of outside entities or await the promises of new concepts, even if ill suited to their situation. Countries would benefit more from analysis tailored to their situation.

POLICIES ON HYDROLOGIC CRITERIA

The hydrologic data base and criteria used in water management are usually not considered widely reviewed policy issues. But is that wise as ever larger populations and national economies are affected by each natural event? It may be informative to examine data on one region's past climate.

Dr. Scott Stine, California State University at Hayward identified extended periods of drought in California based on carbon dating tree stumps rooted well below today's water levels in some California lakes and marshes. The trees matured during periods of reduced precipitation, one of which lasted at least 200 years ending in 1112 and a second that lasted at least 140 years ending in 1350. The Laboratory of Tree Ring Research at the University of Arizona analyzed trees on Mt. Whitney and Dr. Wigand of the Desert Research Institute in Reno analyzed vegetative materials from Eastern Oregon that confirmed droughts of the same period and duration. These widespread occurrences coincided with Europe's Medieval Warm Epoch.

The Tree Ring Laboratory data from the Sacramento basin do not show any such extended droughts from 1600 to the present. California commenced measuring stream flows in 1850 and since then the longest period of flow substantially below normal was 10 years beginning in the late 1920s, the basis used for assessing the reliability of current supplies. Periods of precipitation slightly above average may have existed when some of the local water commitments were made, while periods of lower flow governed others.

Greater extremes have occurred in the last two decades. California recorded its most severe one-year drought in 1977 when runoff was only 20 percent of the long term average. Since the mid-1970s, California has recorded its worst one year (1977), two year (1976-77), three year (1990-92) and six year (1987-92) droughts of the past 150 years. California's precipitation from 1987 through 1992 was 75 percent of normal, however, runoff was less than 50 percent. Ironically, the devastating floods of January 1997 gave way to one of the driest February/March periods. Far longer runoff records are found in China, Europe and other countries that provide information on floods and drought flow in those regions.

It would seem sensible for countries to re-evaluate the hydrologic record and the characteristics of events upon which they base current development policies and water services -- supply, drainage and flood control. In particular, it would seem prudent to reevaluate the reliability levels of urban and rural water services.

DROUGHT MANAGEMENT POLICIES

Drought management policies should have a much higher profile. Large populations and national economies are at risk. And drought policies also link intimately with broader water management policies, particularly water allocation. Industrial and residential users require a supply of high reliability if human and economic losses are to be avoided. Only a portion of the average flow can meet the criteria and the remaining should be allocated to uses that can tolerate supplies of lower reliability.

Water supply management in California during recent droughts offers a simple example. The state's major urban centers draw from systems that also serve agriculture. During the drought periods, agricultural deliveries were greatly reduced and cities used the remaining supply to augment their local emergency actions. Santa Barbara was an exception. Little agricultural water was available in their watershed and they were not connected to the California Aqueduct. The result; more severe curtailment of supply during the emergency and the immediate construction of a stand-by desalter as drought insurance until they could connect to the state's system.

The city of Madras, with its many millions is another example. The domestic unit water consumption is low, limiting flexibility, and there is no significant low priority use within the primary supply watershed. A recent link to a distant irrigation reservoir functions as a marginal interim source, but that source is small and may suffer the same drought. The city's population and economy remain at serious risk until a fully reliable supply is functioning. Several expanding cities in the developing countries routinely ration water under normal conditions. Many deliver only 10 to 15 liters per day (lpd) per inhabitant, precluding substantial cutbacks without severe hardship.

Governments can sustain tolerable conditions during limited droughts only if their policies support the unswerving maintenance of supply reliability. Policy should call for the preparation and frequent updating of national, basin and local drought plans. And the public should participate so they understand the situation and accept the provisions. This important policy area is rarely discussed.

FINANCIAL POLICIES

Common sense would say that policy should require firm financing mechanisms in advance of launching any undertaking. But in fact, many aren't adequately funded and those essential to nation; its water related services, suffer most. Potential sources of funds for services are the government's general budget and/or customer charges. However, government budgets are inadequate and will become even less reliable.

An increasing number of countries have policies to recover costs of all services from beneficiaries. Unfortunately, few are applied. Worldwide deliberations focus most on the basis for water supply charges -- marginal pricing, market pricing, opportunity cost pricing, willingness to pay pricing and cost recovery. Irrigation subsidies attract attention, but rarely do subsidies to beneficiaries of flood control and drainage. Few mention the subsidized waste disposal given urban residents and industries by allowing them to dump untreated wastes into the nations' water supplies.

Today's funding constraints leave little choice. Industrial plants and expensive housing protected from inundation should pay for flood control just as readily as the rice farmer across the road should loose his irrigation subsidies. Policies of cost recovery are widely applied in urban water supply throughout the world. Levying taxes upon beneficiaries to pay for drainage and flood protection was fundamental to European schemes beginning in the 11th and 12th century.

A policy of financial self-sufficiency has other advantages. It removes the entity's reliance on the government's fluctuating general budget allowing more efficient scheduling of maintenance and expansion. In January of this year, the recently elected party granted Punjab farmers free electricity and irrigation water. If the legislation holds, one can imagine the future of those facilities and services.

Private sector management concessions may improve services and reduce costs which should increase customers' satisfaction. Investments in treatment plants can help with near term financing. But in all cases, of course, governments and customers will have to pay.

Countries should realistically examine their total funding needs, evaluate potential sources and then adopt effective policies. Recovery of costs from the beneficiaries of all services appears to be the only viable option, whether services are rendered by government, investor-owned or customer-owned entities. Revenue or general obligation bonds underpinned by property taxes, are proven financing mechanisms used by self-sufficient government and customer-owned entities. But it requires the supporting policies and legislation.

WATER ALLOCATION AND REALLOCATION POLICIES

It is no longer a question for many countries of whether to reallocate supplies. Demand exceeds supply and irrigation water is already being expropriated. "Droughts" are striking the same basins with greater frequency. The potential quantities available for reallocation will be discussed later.

Though the problem is widespread, reallocation by market mechanisms dominates the dialogue. The concept appears desirable, but can it be functioning in time and would results be consistent with country objectives? Can significant quantities of water be reallocated through markets without third party impacts? What are the politics of pitting small farmers and villagers against golf courses and urban centers? People would do well to study the past year's contentious incidents in Thailand and Ecuador.

There are institutional and physical considerations that mitigate against its application. Most countries lack effective water rights data and an administrative system sufficient to underpin water trade. They lack plumbing, the means of measurement and independent brokers and monitors. How long would it take to put everything in place and move substantial quantities through a market?

It would seem better that current efforts in developing countries focus on the allocation policies that meet the country's objectives; physical conditions and their political, financial and time constraints. The questions are not easy to answer. What are the criteria for making a reallocation? How should they be implemented? What should be the compensation to the user losing water? Can cities ensure additional long-term reliable supplies in advance of expansion? Should someone do so on their behalf? Should irrigated lands that recharge urban groundwater aquifers have priority, ahead of other lands, to retain supply? Drought management requirements will preclude consideration on some lands while others become the prime choice.

It would seem to be essential to initiate public discussion. Society should fully understand their country's predicament, the constraints, the options and the consequences. It would appear that only then could political leaders act and administrators implement substantial reallocations in a sensible manner.

WATER SUPPLY AUGMENTATION POLICIES

Many countries are reaching the present limits of the traditional means for increasing water supply. Efforts continue on other ways, such as improving the quality of heavily polluted waters and desalination. Plant breeding to increase salt tolerance and crop selection to maximize product per unit of water continue. But demand management offers little opportunity where residential deliveries are only 10 to 15 lpd per inhabitant and farmers are already stretching supply on their lands -- demand management is already in full effect. Re-allocation may meet pressing needs, but of course will yield no additional water. Let us focus on means to secure additional water in the amounts, in the time frame and in the location that may help solve problems described earlier.

How much water can be saved from irrigation improvements as opposed to curtailment? The vast majority of irrigation water is used on grain and forage crops. Drip irrigation has little practical application. And though 40 to 60 percent of surface irrigation diversions may flow to groundwater or surface drainage, most of this serves downstream agricultural and urban users and is not lost. US Soil Conservation Service measurements show that the US sector-wide losses is only 13 percent. More reliable deliveries consistent with crop maturity can increase usable product per unit of water. As with rain, irrigation shortages at certain stages of maturity can impair crop yields and waste much of the quantity of water consumed during the initial growth. This is a problem in many developing countries, but it is likely that any gains through improved deliveries will not exceed the increasing food demands.

Reasonably suitable effluent from inland cities likewise are reused. As noted earlier, basin utilization by means of recycling is essentially 100 percent in a majority of the large rivers in the developing countries with the highest demands. In California the Sacramento basin efficiency is essentially 100 percent, confirmed by the need to release reservoir water from the head of the basin to maintain adequate outflow to the Bay. Only recovery of return flows discharging directly to the sea or to saline sinks offers significant opportunities.

As mentioned, the policy to re-allocate water from current uses receives by far the greatest support of people outside the developing countries. So how much water can be productively reallocated? There are limits in addition to constraints explained under drought management. Much of the world's rice is raised in monsoon regions and the portion grown during the rainy season primarily utilizes water in surplus. The quantity of irrigation water that can be reallocated to meet the critical periods of urban shortages is the irrigation carried out during the low flow periods. Quantities and duration vary greatly depending on location and the amount drawn out of storage whose releases can be rescheduled. Irrigators in arid regions are only assured full supply in four out of five years in most developing countries. In practice it is less. The result; a smaller quantity is available during the years of greatest need by urban users. And as evident by the basin reuse data, only the consumptive fraction will net out as 'freed' water.

I would like to again refer to California's situation where firm numbers on allocations and flows are available. The average runoff is 87 bcm (71 maf) augmented by 7 bcm from outside rivers. Of this quantity 36 percent is lost to the sea, 28 percent is set aside by legislation for environmental purposes, 28 percent is used by agriculture, 7 percent is used by urban and 1 percent by misc. I am not questioning present allocations. But the public in many countries, as in California, believes that 80 percent of supply is used by agriculture and that all has a reliability suitable to urban supply. There may be substantial quantities, but the only quantity that counts is the quantity physically accessible to a priority user at the time of need. Statements in several popular books and articles, I fear, mislead the public and distort discussion and water policies.

Waste treatment can free considerable water precisely where and when it is needed most; during the critical low flow season when dilution is inadequate and urban demand for highly reliable supply is greatest. Surabaya uses 25 cumecs of basic supply to flush sewerage and solid wastes to the sea. This is equivalent to the consumptive use of perhaps 30,000 hectares of rice or 40,000 hectares (100,000 acres) of upland irrigation . The impacts of urban waste are immense. Countries should give more attention to waste management policies.

In many countries it is still feasible to construct additional reservoirs to augment supply. India has huge flood flows and agriculture is far from consuming 93 percent of all its renewable water, as stated in a 1995 report to the Consultative Group on International Agricultural Research (CGIAR). Irrigation use is well below 30 percent. indeed, dams are under construction in developing countries, most financed by the countries' own means to avoid the delays and constraints of international participation.

Is it wise policy to oppose dams if the environmental objectives are met? A 1996 Wall Street Journal reported that the retiring commissioner of the US Bureau of Reclamation informed a group of international officials that the US government no longer promotes dam building in other countries. He urged them to "avoid repeating our mistakes" and consider alternatives to large dams. However, New York, Atlanta, Denver, San Francisco, Los Angeles, Seattle and other US cities haven't demolished their reservoirs. Indeed, several seek more water from reservoirs built for other purposes, as would be the case in most of the suggested reallocations everywhere in the world. Large reservoirs will remain an important component of water management systems in both the developing and the developed countries.

It would seem responsible for the water management community to reevaluate its water augmentation policies and advise the world's political leaders to consider all options and select those that, indeed, can meet their needs.

RESETTLEMENT POLICIES
Resettlement issues play a major role in the water debate. There is no question, resettlement should be done properly. But one must recognize that resettlement concerns should extend beyond reservoir lands and levees. People will be relocated whether we build reservoirs, cut supplies to cities or reallocate water from agriculture. At the same time, nations will seek opportunities to settle their portion of the two billion arriving in teh next twenty years.

Present policies are incomplete. Some call for a land for land policy when resettling people off their lands. Should farmers losing water be given additional land equivalent to the resulting loss in their farm's productivity? Will rural villagers receive compensation when the region's irrigated agriculture is curtailed? The policy is to compensate villagers within proposed reservoirs. Resettlement, for whatever reason is not a pleasant matter. But it won't become any easier for countries and agencies that haven't resolved these policy questions.

CLOSING REMARKS

The world faces increasingly difficult water management problems. It would seem that researchers, managers and political leaders would do well to re-consider their present focus in several areas. Policies should be better tailored to the problems confronting each country and the time available to solve them. Prolonged deliberations should not be allowed to postpone the politically difficult decisions.

RIVER MORPHOLOGY: A MANIFESTATION OF NONLINEAR DYNAMICS

H.J. DE VRIEND
Delft University of Technology, Delft, The Netherlands

ABSTRACT
The dynamic character of river morphology is considered at various scale levels, from microscale turbulence and sediment motion, via bedforms, cross-sectional profile, longitudinal profile and channel alignment, through to the channel pattern at the drainage basin level. Linear and nonlinear dynamic behaviour is considered at each scale level, and dynamic interactions across the levels are discussed.

INTRODUCTION
The morphology of rivers involves a wide variety of space and time scales. Individual sand grains are of millimetre-size and move at the time scale of turbulence, either intermittently along the bed, or continuously in suspension. Bed ripples and dunes are of metre-size and develop at a time scale of hours, whereas the cross-sectional shape of a channel involves spatial scales of tens to hundreds of metres and may change at a time scale of days (flood) to years (low water). The longitudinal bed profile in a channel at a scale of many kilometres may take years to centuries to adjust to a disturbance, and the channel alignment changes at time scales of decades, centuries, or even more. The pattern of tributaries at a river basin scale varies at geological time scales. At each of these scales, the morphological behaviour looks rather organized and predictable, with comparable patters for most river systems.
On the other hand, river morphology is the result of a dynamic interaction of water and sediment motion with the alluvial topography. The hydrodynamic and sediment transport processes are basically nonlinear, so rivers must be nonlinear dynamic systems.
This may explain the various scale-levels in their morphological behaviour. Yet, various important aspects of this behaviour are described surprisingly well with linear or linearized models. It is therefore worth exploring when the system's nonlinearity is important and how it manifests itself.
This paper, which is meant as a basis for discussion, treats the various scale levels one by one, and subsequently goes into the interactions across these levels.

BEDFORMS
The term "bedforms" is used herein exclusively for smaller-scale rhythmic features, such as ripples and dunes.
Engelund (1970) showed, via a linear stability analysis, that the onset of ripple formation can be explained from the interaction of the near-bed water and sediment motion and the bed topography. Apparently, ripple formation is primarily a boundary-layer phenomenon, in which the free water surface does not play a prominent role. Kennedy (1963) had shown some years before, via a similar analysis based on a model of ideal fluid motion, that the interaction with the free water surface is essential to dune formation. Apparently, dune formation is a matter of the entire water column, including the free surface.

Both analyses are based on models for small perturbations to a uniform basic state (uniform flow over a plane bed). Hence they can only refer to the *onset* of the bedform formation, and they can only describe the wave length, the growth rate and migration speed of each harmonic mode of the bed perturbation. Once the bedforms have reached a finite amplitude, the basic assumptions of the analysis are no longer valid. As the amplitude increases, nonlinear effects become important, such as the deformation of the originally harmonic perturbations into oblique triangles, due to the height-dependence of the propagation speed. The same phenomenon causes smaller ripples to overtake larger ones and merge with them at the slip face. Thus the propagation speed of finite-amplitude ripples may deviate from that of infinitesimal harmonic perturbations.

However, not only the linear approximation fails for finite-amplitude ripples, but also the concept of the model equations becomes invalid. Phenomena start playing a role which are not included in these equations (flow separation, slip-face formation, etc.). Hence it must be doubted whether a nonlinear analysis of these equations will lead to useful information on the amplitude of bedforms.

Attempts to describe finite-amplitude bedform evolution with process-oriented simulation models (i.e. fully nonlinear numerical models which describe the water and sediment motion in detail) have not been very successful, probably because it is not only the mean water motion which plays a part, but also its turbulent variation. Another class of models relates bedform dimensions to the dimensions of coherent structures in the turbulent flow (e.g. Yalin, 1992).

Fredsoe (1982) uses an essentially different, simple and straightforward model for the equilibrium state of finite-amplitude bedforms. It expresses their height and propagation speed in terms of the variation of the sediment transport rate with the flow velocity and the water depth.

The aforementioned analyses presume long-crested bedforms, which does not always comply with reality. Ripple fields are often short-crested, with a variety of curvilinear features. Pattern formation in ripple and dune fields is an insufficiently explored field, which deserves further attention. Even if, at first sight, the pattern looks regular and long-crested, it usually includes dislocations, which indicates nonlinear behaviour.

CROSS-SECTIONAL PROFILE (STRAIGHT CHANNELS)

Rivers, like many other mobile-bed systems (estuaries, turbidity currents over an alluvial fan; cf. Parker, 1996), tend to form channels, rather than spreading evenly over a plane slope. The mechanism of this channel formation is still poorly understood. A vast body of morphological modelling expertise is based on the assumption that the channel width is a given quantity. The "width-equation" has long been considered as the principal missing link in river morphology.

Yet, the stable cross-sectional shape of straight alluvial channels with erodible banks has been addressed by many authors. An extensive class of models is based on extremal hypotheses (minimum energy loss, minimum shear stress, maximum transport, etc.; see Lamberti, 1988, for a review), often without a thorough foundation in the basic hydrodynamic and sediment transport processes. Contrastingly, process-based simulation models of cross-sectional shape formation are rare (e.g. Parker, 1978a, 1978b; Lamberti and Schippa, 1996). Although they give encouraging agreement with laboratory data, their relevance to natural rivers is not obvious, if it were only because these are seldomly straight and/or stable. Meandering rivers have a single channel, but it is not straight and keeps on migrating, and the definition of "the" channel in the case of a braided river is not unambiguous.

Nevertheless, these models may be assumed to give an indication of the timescale of response of the cross-sectional shape. Lamberti and Schippa (1996) give an expression of this time scale which for shallow channels boils down to inverse proportionality to the total transport rate (like all morphological time scales) and proportionality to the third power of the width and the square of the aspect ratio. So the cross-sectional dimensions determine this time scale. For a river of 100 m wide, with an aspect ratio of 30 and a transport rate per unit width of 10^{-4} m^2/s, the estimated time scale amounts 80 days, which is not very different from the time scale of cross-sectional shape adjustment in bends (see below).

Beyond a certain critical aspect ratio, stable straight channels only exist in an averaged sense, because the bed tends to be perturbed by migrating alternate bars. This phenomenon can be explained to a considerable extent with a linear stability analysis, which gives estimates of the wave length, the propagation speed and the initial growth rate. Colombini et al. (1986) and Schielen et al. (1992) apply nonlinear analysis techniques to alternate bars, in order to predict their amplitude. The latter authors, who perform a somewhat more extensive analysis than the former, find that at short time scales (a few wave periods of the bars) the pattern looks much like regular waves, though of a triangular shape. At larger time scales (a few tens of wave periods) a slow amplitude modulation appears, and at very large time scales (hundreds of wave periods) the picture becomes quasi-periodic and looks rather irregular. So this manifestation of the system's nonlinear character appears only at larger time scales.

If the channel becomes even shallower, multiple bar systems tend to evolve. The system seems to become increasingly nonlinear and increasingly irregular (braided rivers).

CROSS-SECTIONAL PROFILE (BENDS)

The cross-sectional deformation in bends is a much more extensively investigated phenomenon than that in straight channels. The first publications on this issue date back to the second half of the last century, but the first authors to abandon the assumption of downstream uniformity was Van Bendegom (1947), followed by Engelund (1974). Struiksma et al. (1985) performed a linear perturbation analysis of a river bend model, which made clear that the equilibrium transverse bed slope may exhibit a damped oscillatory response to downstream variations in the channel curvature. This phenomenon, which is controlled by the ratio of the length scale of inertial flow adjustment and a length scale referring to the cross-stream sediment redistribution capacity, is confirmed by laboratory experiments and field data. Struiksma and Crosato (1989) showed theoretically and experimentally that channel curvature is not a necessary condition for this phenomenon to occur: it can just as well be triggered by an obstacle in a straight channel.

Tubino and Seminara (1990) analyse the weakly nonlinear interaction of alternate bars ("free bars") and the pointbar/pool configuration due to curvature ("forced bars") and conclude that under certain conditions the free bars are suppressed by the forced ones (also see Seminara and Tubino, 1989). Schielen (1995) gives a nonlinear solution of the equilibrium bed topography in weakly meandering channels.

Various applications of fully nonlinear numerical models of river bend morphology confirm the findings from these analyses, and show that a linear analysis often yields a good first estimate. This gives confidence in the validity the weakly nonlinear analyses, but further prototype validation is needed.

A special complication arises in rivers with floodplains in which the flood flow does not follow the main channel. This gives rise to a very complicated flow pattern, right at the

moment of maximum morphological activity. This complication needs further analysis and incorporation in the state-of-the-art numerical models.
If the flood flow follows the main channel, the equilibrium bed profile is stage-dependent. This may mean that the equilibrium state is hardly ever reached, because the stage keeps on changing. Whether or not this is the case depends on the time scale of the profile evolution compared with the time scale of the discharge variation. Numerical model applications with real-life variations of the discharge have revealed that the cross-sectional profile gets close to the equilibrium state during flood conditions, when the time scale is relatively short (high transport rate). During and after the waning of the flood, the profile adjustment to the low flow conditions is considerably slower (low transport rate).

LARGE-SCALE LONGITUDINAL PROFILE

The classical model "per unit width" of the longitudinal profile evolution of river channels is essentially nonlinear (e.g. Jansen, 1979). It rests upon backwater curves and sediment transport formulae, which are both nonlinear. Yet, this nonlinearity does not give rise to non-unique solutions or chaotic behaviour, but at most to shock and expansion waves.
De Vries (1975) gives a method to estimate the morphological time scale of rivers, based on the so-called (linear) diffusion approximation of a 1-D morphological model. Since length and time scales are coupled in a diffusion process, the time scale can only be defined in combination with a length dimension. If this is chosen in the order of 100 km, the time scale is in the order of centuries. This explains why the bed of the river Rhine is still responding to the training works of the last century.
The time scale of this large-scale profile evolution is often comparable with that of changes in the alignment of the river (e.g. meandering). Since bends affect the backwater curve, and meander cut-offs even more so, this means that in natural rivers the longitudinal profile evolution can hardly be considered in isolation.

ALIGNMENT: MEANDERING

Various authors have identified alternate bar formation as the basic cause of meandering. Alternate bars in straight channels, however, are migrating and usually move too fast to cause substantial bank erosion. Although their migration speed may be influenced by the channel alignment, they also tend to be suppressed by curvature effects (Seminara and Tubino, 1989).
A more likely cause of the onset of meander formation is the stable spatial oscillation pattern which can be triggered by a disturbance of the channel profile (cf. Struiksma and Crosato, 1989). This causes initial meandering at the wave length of the oscillation, such that the curvature effect enhances the bars. This constitutes a positive feedback mechanism, which can be considered as resonance. Once they exist, however, meanders tend to increase in amplitude and evolved length, thus moving away from the resonance conditions. This is readily illustrated by the occurrence of consecutive point bars in a single meander bend: apparently the wave length of the stable oscillation has become smaller than the meander length.
Crosato (1987) describes a 1-D meander model based on a linear theory of stable oscillations, combined with a bank erosion model (erosion proportional to the excess velocity near the outer bank). The tests which were carried out with this model, with a profile constriction at the upstream end, showed growing meanders, which disappeared again when the upstream constriction was removed. The growth rate of the meanders in this model is reduced once the meander length starts to deviate from the wave length of

the stable oscillations, i.e. the system moves away from the resonance conditions. Another mechanism of meander amplitude stabilization is not included in this model, viz. the formation of a recirculation cell and a bar near the concave bank if the channel curvature becomes too strong (cf. Parker, 1996).
This meander model is not entirely satisfactory at various points, e.g. the triggering (observations in nature suggest that meandering is something inherent to the system, which needs no external forcing) and the shape of the meanders (insufficiently irregular). Probably, nonlinear effects need to be taken into account to improve the model. Mosselman (1992) describes a 2-D fully nonlinear evolution model for a relatively short river section. The assumption of constant width, whence the inner bank always follows the outer bank, turns out to be a drawback of this model. However, if this assumption is relaxed to the extent that the migration of either bank is determined by different processes, such that they may behave more or less independently, it may well be that the channel width is a result of the bank migration history, rather than of stable channel formation. Looking at the underlying sediment redistribution mechanisms, viz. curvature-induced secondary flow vs. lateral dispersion, the former may well dominate the latter in finite-amplitude meanders.
Another class of meander models can be called behaviour-oriented. They are based on mathematical equations which describe the observed phenomena without going explicitly into the water and sediment motion (e.g. Howard and Knutson, 1984; Liverpool and Edwards, 1995). They are essentially nonlinear and yield quite realistic-looking meander patterns. Apparently, nonlinearity is essential to natural meander patterns.

CHANNEL PATTERN AT BASIN SCALE

The channel pattern at the scale of the drainage basin exhibits a dendric structure which looks like a fractal (e.g. Rigon et al., 1996). There is competition between channels for drainage area, such that successful channels get bigger and hence more successful, until they reach some ideal size. The length L of a channel in such a system appears to be an increasing function of the drainage area A which discharges through this channel (L proportional to $A^{0.6}$).
The fractal structure of the channel system suggests chaotic behaviour and self-organization (Rigon et al., 1996), with a seemingly regular pattern and repetition of similar features over a wide range of scales. Whether this is really the case remains a point of discussion.

SCALE INTERACTIONS

The morphological behaviour of rivers seems to involve a cascade of more or less well-separated aggregation levels (levels of space and time scales). At each level, the behaviour may become extremely complex and difficult, if not impossible, to predict in a deterministic sense. The manifestations of this complex behaviour, however, are usually rather regular when considered from a higher aggregation level. For instance, the turbulent water motion is chaotic, but its effects on the mean water motion are described rather well with deterministic models. Bedform patterns can be extremely complicated, but their overal effects on the water motion and the sediment transport can be modelled with relatively simple roughness predictors. The channel pattern in meandering or braided rivers is very complex and variable, but the channel belt usually looks quite regular and stable.
So the behaviour at each scale level is not only determined by the higer scale levels, which determine the extrinsic conditions, but also by the lower scale levels, which determine a number of model parameters. Thus the various scales are mutually interac-

ting, with positive and negative feedbacks. The resonant interaction of initial meander growth with the oscillatory behaviour of the (quasi-)equilibrium bed topography is an example of positive feedback, the stabilization of the meander shape under the influence of concave-bank recirculation is one of negative feedback.

CONCLUSION

The morphological evolution of rivers takes place at a number of more or less separated aggregation levels. At most of these levels, important aspects of the system behaviour can be described by linear approximations, but only in part of the parameter domain. In other parts, the system behaviour is nonlinear and sometimes unpredictable in a deterministic sense.

REFERENCES

Colombini, M, Seminara, G. and Tubino, M., 1987. *J. Fluid Mech.*, 181: 213-232.
Crosato, A., 1987. *Euromech 215 Conference*, Genova, p. 158-161.
De Vries, M., 1975. *Proc. IAHR-Congress*, Sao Paolo.
Engelund, F., 1970. *J. Fluid Mech.*, 42: 225-244.
Engelund, F., 1974. *J. Hydr. Div., ASCE*, 100(HY11), p 1631.
Fredsoe, J., 1982. *J. Hydr. Div., ASCE*, 108(HY8): 932-947.
Howard, A.D. and T.R. Knutson, 1984. *Water Resources Res.*, 20(11): 1659-1667.
Jansen, P.Ph. (ed.), 1979. *Principles of River Engineering*. Pitman, London, 509 pp.
Kennedy, J.F., 1963. *J. Fluid Mech.*, 16: 521-544.
Lamberti, A., 1988. *Proc. Int. Conf. on River Regime*, Wallingford, U.K., p. 121-134.
Lamberti, A. and Schippa, L., 1996. *Excerpta*, 10.
Liverpool, T.B. and Edwards, S.F., 1995. *Phys. Rev. Letters*, 75(16): 3016-3019.
Mosselman, E., 1992. *Mathematical modelling of morphological processes in rivers with erodible cohesive banks*. Doct. thesis, Delft Univ. of Techn., 134 pp.
Parker, G., 1978a. *J. Fluid Mech.*, 89: 109-125.
Parker, G., 1978b. *J. Fluid Mech.*, 89: 127-146.
Parker, G., 1996. In: P.J. Ashwort et al. (Editors): *Coherent flow Structures in open channels*. Wiley & Sons, New York, p. 423-458.
Rigon, R., Rodriguez-Iturbe, I., Maritan, A., Giacommetti, A., Tarboton, D.G. and Rinaldo, A., 1996. *Water Resources Res.*, 32(11): 3367-3374.
Schielen, R., Doelman, A. and De Swart, H.E., 1992. *J. Fluid Mech.*, 252: 325-356.
Schielen, R., 1995. *Nonlinear stability analysis and pattern formation in morphological models*. University of Utrecht, Ph.D. Thessis, 150 pp.
Seminara, G., 1995. In: A. Doelman and A. van Harten (Editors): *Nonlinear dynamics and pattern formation in the natural environment*. Pitman Res. Notes Math. no. 335, p. 269-294.
Seminara, G. and Tubino, M., 1989. In: S. Ikeda and G. Parker (Editors): *River Meandering*, Water Resources Mon. 12, AGU, Washington DC, p. 267-320.
Struiksma, N. and Crosato, A., 1989. In: S. Ikeda and G. Parker (Editors): *River Meandering*, Water Resources Mon. 12, AGU, Washington DC, p. 153-180.
Struiksma, N., Olesen, K.W., Flokstra, C. and De Vriend, H.J., 1985. *J. Hydr. Res.*, 23(1): 57-79.
Tubino, M. and Seminara, G., 1990. *J. Fluid Mech.*, 214: 131-159.
Van Bendegom, L., 1947. *De Ingenieur*, 59(4). English translation: Nat. Res. Council Canada (1963), *Techn. Transl.* 1054.
Yalin, M.S., 1992. *River Mechanics*. Pergamon Press, 220 pp.

Edge Wave Excitation by Random Sea Waves

G. VITTORI
Hydraulic Institute, University of Genoa
Via Montallegro, 1 - 16145 Genoa - Italy
P. BLONDEAUX
D.I.S.A.T., University of L'Aquila
67040 Monteluco di Roio - L'Aquila - Italy

Abstract

The excitation of edge waves by a random wave approaching a straight beach of constant slope is investigated. The sea wave is assumed to be characterized by a narrow band amplitude spectrum centered around some value ω_0^*. It is found that edge waves of frequency $\omega_0^*/2$ may be excited. Comparing the obtained results with those characteristic of the monochromatic wave case it is found that the unstable regions in the parameter space widen but the equilibrium amplitude of the edge waves is usually smaller and decreases as the width of the spectrum increases.

1 INTRODUCTION

The excitation of edge waves of frequency $\omega_0^*/2$ by a monochromatic incident wave of frequency ω_0^* is a well known phenomenon first discussed by Guza & Davis (1974). They considered a long wave normally approaching a straight beach of small constant slope β and by using the shallow water approximation they showed that the incident wave can transfer energy to small disturbances in the form of subharmonic edge waves the amplitude of which grows exponentially. Later Minzoni & Whitham (1977) analysed the same instability problem using the full three-dimensional water wave theory and showed that the non-uniformity of the shallow water approximation far from the shore (Whitham (1976) and Minzoni (1976)) is mild and does not affect the main results for small beach angles. The linear analyses by Guza & Davis (1974) and Minzoni & Whitham (1977) predict an infinite growth of the edge wave amplitude, but when nonlinear effects become important, an equilibrium can be attained as described by Guza & Bowen (1976) and Minzoni & Whitham (1977).

Even though a monochromatic wave train is commonly used as a first approximation of the sea surface, actual records of the sea level show the spectral nature of sea waves. We then address the problem of edge wave excitation by a random wave characterized by a narrow-band amplitude spectrum centered around ω_0^*. In this case the sea surface can be described by a monochromatic wave, the amplitude of which is slowly modulated by a random function of time and space. It will be shown that the effect of the latter is to wide the unstable regions in the parameter space. However the average value attained by the amplitude of the subharmonic edge wave for large time is usually smaller than that characterizing the monochromatic wave case.

2 THE PROBLEM AND THE BASIC SOLUTION

Let us consider a straight beach of constant slope β in a rectangular cove of width L^*. Let us introduce a Cartesian coordinate system with the x^*-axis lying on the still water surface being directed offshore and the y^*-axis coincident with the coastline (see figure 1). The vertical walls bounding the cove are located at $y^* = 0$ and $y^* = L^*$ respectively. Our aim is to investigate the excitation of edge waves when a random sea wave approaches the beach along the x^*-direction and is fully reflected at the coastline. All the wave components are assumed to be characterized by wavelengths much larger than the local depth h^* and by small amplitudes. Hence the shallow water equations are used to describe the local wave dynamics and the solution is expanded using the wave steepness as a small perturbation parameter. In order to simplify the analysis let us assume that the random incoming wave is characterized by a narrow band amplitude spectrum that takes significant values only when the angular frequency ω^* is close to some value ω_o^* and then rapidly vanishes as ω^* moves away from ω_o^*.

Before writing down the equations, we introduce a time scale $(\omega_o^*)^{-1}$, length scales $g^*\beta/\omega_o^{*2}$ in the horizontal directions and $g^*\beta^2/\omega_o^{*2}$ in the vertical one and a velocity scale $g^*\beta^2/\omega_o^*$. The nondimensional shallow water equations can be combined with the irrotational assumption to provide a single equation for the velocity potential ϕ (Mei (1989))

$$-\frac{\partial^2 \phi}{\partial t^2} + x\left[\frac{\partial^2 \phi}{\partial x^2} + \frac{\partial^2 \phi}{\partial y^2}\right] + \frac{\partial \phi}{\partial x} = \left(\frac{\partial \phi}{\partial x}\right)^2 \frac{\partial^2 \phi}{\partial x^2} + \left(\frac{\partial \phi}{\partial y}\right)^2 \frac{\partial^2 \phi}{\partial y^2} + 2\frac{\partial \phi}{\partial x}\frac{\partial \phi}{\partial y}\frac{\partial^2 \phi}{\partial x \partial y}$$
$$+ \frac{\partial}{\partial t}\left[\left(\frac{\partial \phi}{\partial x}\right)^2 + \left(\frac{\partial \phi}{\partial y}\right)^2\right] + \left[\frac{\partial^2 \phi}{\partial x^2} + \frac{\partial^2 \phi}{\partial y^2}\right]\left[\frac{\partial \phi}{\partial t} + \frac{1}{2}\left(\frac{\partial \phi}{\partial x}\right)^2 + \frac{1}{2}\left(\frac{\partial \phi}{\partial y}\right)^2\right] \quad (1)$$

Moreover the vertical walls force $\partial \phi/\partial y$ to vanish at $y = 0$ and $y = L$ and the solution of (1) should be matched with that of the deep water region for x tending to infinity.

As already pointed out in the introduction, the aim of the present work is to investigate the stability of the narrow band random wave approaching the coast with respect to longshore perturbations in the form of edge waves. Let us assume that the width of the amplitude spectrum is of $O(a)$, the small parameter a being defined by

$$a = a^* \omega_o^{*2}/g\beta^2 \quad (2)$$

where a^* is the value at the shoreline of the amplitude of a fully reflected monochromatic wave characterized by the same time average specific energy as that of the random wave field. Since the amplitude spectrum vanishes when $|\omega^* - \omega_o^*|/\omega_o^*$ is larger than $O(a)$, let us introduce the nondimensional angular frequency Ω defined by

$$\Omega = \frac{\omega - 1}{a} \quad (3)$$

If the velocity potential ϕ_i of the incident wave reflected at the coast, is expanded in terms of the small parameter a, it is easy to verify that at the leading order of approximation

$$\phi_i = a\phi_o + O(a^2) = a\left[\frac{i}{2}\int_o^\infty E(\Omega)J_o(2(1+a\Omega)\sqrt{x})e^{i\Omega\tau}d\Omega\right]e^{it} + c.c. + O(a^2) \quad (4)$$

where $E(\Omega)$ is the dimensionless amplitude spectrum of the wave field in the nearshore region.

Because the interaction between the incoming wave and the edge waves is significant only in a region close to the coast, the width of which is of order $g^*\beta/\omega_o^{*2}$, the velocity potential (4) can be approssimated by

$$\phi_i \cong a\phi_o \cong a\mathcal{A}(\tau)\left[\frac{i}{2}J_o(2\sqrt{x})\right]e^{it} + c.c. \qquad (5)$$

where the modulating function $\mathcal{A}(\tau)$ is provided by

$$\mathcal{A}(\tau) = \int_o^\infty E(\Omega)e^{i\Omega\tau}d\Omega \qquad (6)$$

3 SUBHARMONIC RESONANCE OF EDGE WAVES

The linearized version of equation (1) admits also of eigensolutions (edge waves) which exponentially decay moving offshore

$$\phi_e = \epsilon B\left[ie^{-kx}L_n(2kx)\cos(ky)\right]e^{i\sigma t} + c.c. \qquad (7)$$

subject to the dispersion relation (eigenrelation)

$$\sigma^2 = k(2n+1) \qquad (8)$$

In (7) ϵ is an arbitrary small parameter, B is an unspecified constant and L_n is the Laguerre polynomial of order n. Even though different modes can be considered, for simplicity we shall consider only the lowest mode ($n = 0$) which is also the most unstable as shown in Guza & Davis (1974).

A clear account of the mechanism of subharmonic resonance, in which an edge wave of frequency $\omega_o^*/2$ is resonated by a normally incoming wave of frequency ω_o^*, is given in Mei (1989). If both ϕ_i and ϕ_e are initially present, the quadratic terms involving the pair (ϕ_i, ϕ_e) will give rise to a simple harmonic forcing term proportional to $G(x)\cos(ky)e^{i(\sigma\pm 1)t}$ where $G(x)$ is a certain function of x. When $\sigma \pm 1$ is equal to $\pm\sigma$ ($\sigma = \pm 1/2$), unless $G(x)$ is in some sense orthogonal to the homogeneous solution e^{-kx}, the edge waves are resonated to large amplitudes. As resonance develops, the edge waves are no longer small in comparison with the standing wave. At equilibrium the cubic terms should be as important as the quadratic terms, which implies that $0(a\epsilon) = 0(\epsilon^3)$ or $\epsilon = 0(a^{1/2})$. To account for this, eventually we allow ϕ_e to be $0(a^{1/2})$ and we introduce the slow time $\tau = at$

$$\phi_e = a^{1/2}\phi_1 + a\phi_2 + a^{3/2}\phi_3 + ... \qquad (9)$$

If (9) is substituted into (1), at $O(a^{1/2})$ the function ϕ_1 is provided by (7) where now the amplitude B depends on the slow time scale τ. For sake of generality let us suppose that the resonance condition is not exactly fulfilled and let us consider an angular frequency σ slightly different from $1/2$

$$\sigma = 1/2(1 + a\mu) \qquad (10)$$

where μ is an $O(1)$ detuning parameter. At order a, after some tedious but straightforward algebra, it can be shown that

$$\phi_2 = B^2 2i\sigma k\pi\phi_{20}(x)e^{i2\sigma t} + c.c. \qquad (11)$$

where
$$\phi_{20}(x) = J_o(4\sqrt{kx})[E_2(kx) - E_2(\infty) - iE_1(\infty)] - Y_o(4\sqrt{kx})E_1(kx) \quad (12)$$

$$E_1(kx) = \int_o^{kx} J_o(1\sqrt{\xi})e^{-2\xi}d\xi \; ; \qquad E_2(kx) = \int_o^{kx} Y_o(1\sqrt{\xi})e^{-2\xi}d\xi \quad (13)$$

As discussed previously at $0(a^{3/2})$ a solvability condition is required (Fredholm alternative) which provides a differential equation for $B(\tau)$ in the form

$$\frac{dB}{d\tau} + a_1\mathcal{A}(\tau)e^{-i\mu\tau}\overline{B} + a_2\overline{B}B^2 = 0 \quad (14)$$

where
$$a_1 \cong ik^2 0.27 \qquad a_2 = ik^3\{\sigma[2 - 8\pi(0.057 + i0.0092)] - 0.75k/\sigma\} \quad (15)$$

It is interesting to point out that the main difference between the monochromatic and random wave cases is due to the presence of the modulating function $\mathcal{A}(\tau)$ which depends on the form of the amplitude spectrum of the incoming wave.

4 DISCUSSION OF THE RESULTS

In the monochromatic wave case (Mei, 1989) $\mathcal{A} \equiv 1$ and equation (14) can be reduced to an equation with constant coefficients by introducing the variable

$$C(\tau) = B(\tau)e^{i(\mu/2)\tau} \quad (16)$$

$$\frac{dC}{d\tau} + a_1\overline{C} - \frac{i\mu}{2}C + a_2\overline{C}C^2 = 0 \quad (17)$$

This is an equation stating that the change of the edge wave amplitude is due to its interaction with the incoming wave (the second term) corrected by the third term which takes into account the not perfect tuning between the edge wave and the basic wave. Moreover $C(\tau)$ is affected by the radiation due to nonlinear effects which causes a damping of the amplitude (fourth term). The equilibrium amplitude can be determined by solving the cubic algebraic equation obtained by (17) setting $dC/d\tau = 0$. The analysis of the stability of the fixed points of (17) is straightforward and the results are shown in figure 2 where the bifurcation pattern is plotted for a fixed value of a. It appears that a subcritical bifurcation may occur in region II, while the bifurcation is supercritical in region III.

These results are drammatically changed when a random incoming wave is considered i.e. when the modulating function $\mathcal{A}(\tau)$ is present in (14). Indeed, because of the presence of $\mathcal{A}(\tau)$ no fixed point of (17) can be found. The closed form solution of (14) cannot be found and the edge wave amplitude has been obtained by the numerical integration of (14) for given $\mathcal{A}(\tau)$ and initial values $C(0)$ after introducing (16).

When we want to look for functions with a strong central tendency and vanishing small ordinates at large distances, we almost inevitably think of the Gaussian distribution. Hence let us consider an amplitude spectrum $E(\Omega)$ in the form

$$\mid E(\Omega) \mid = \frac{E_o}{\sqrt{\pi}S}e^{-(\frac{\Omega}{\sqrt{2}S})^2} \quad (18)$$

where $\mid E \mid$ denotes the modulus of E and $S = S^*/a\omega_o^*$ is a measure of the width of the spectrum. The value of the constant E_o appearing in (18) has been fixed in such a way that the time average of the specific energy of the basic wave at the shoreline is equal to one. In obtaining the numerical solution of (14), the function $\mathcal{A}(\tau)$ has been evaluated using the finite counterpart of the integral appearing in (6):

$$\mathcal{A}(\tau) = \sum_{l=-\mathcal{L}-1}^{\mathcal{L}} E_l e^{i\Omega_l \tau} = \sum_{l=-\mathcal{L}-1}^{\mathcal{L}} \mid E_l \mid e^{i[\Omega_l \tau + \theta_l]} \quad (19)$$

where the values of Ω_l are equal to $(2l+1)\Delta\Omega$ ($\mathcal{L}$ and $\Delta\Omega$ being parameters of the numerical code), the moduli of E_l are evaluated according to (18), while the phases θ_l are random numbers uniformly distribuited in the range $(0, 2\pi)$. It is well known (Lyon (1970), Yang (1973)) that (6) is approximated remarkably well by (19) even with small values of $\mathcal{L}$. Presently values of $\mathcal{L}$ as large as 600 are considered. Since the phases of the different components appearing in (19) are random numbers, different behaviours of the modulating function $\mathcal{A}$ are found depending on the choice of the set of values of θ_l. Typical time developments of $|C(\tau)|$ at resonance are plotted in figure 3 for particular sets of values of θ_l: starting from small values, random oscillations are superimposed to a net growth which takes place during an initially transient period. Then nonlinear effects become important and a regime configuration is attained by the edge waves even though their amplitude is still characterized by the presence of random time fluctuations superimposed to an average value. To compare present findings with those of the monochromatic wave case, the time development of $|C|$ for $\mathcal{A} \equiv 1$ is also plotted. Rather than looking at the solution of (14) for a particular modulating function $\mathcal{A}(\tau)$, it is more interesting to discuss the results in terms of statistical averages on a large number of realizations characterized by different sets of value of θ_l. Figure 3 clearly shows that a constant value is attained for large values of τ by the ensemble average amplitude of the edge waves. Moreover comparing the asymptotic value with that characteristic of the monochromatic case, it clearly appears that a random wave with the same time average specific energy as a monochromatic wave causes smaller edge waves.

The results of figure 3 are obtained setting the detuning parameter equal to zero, i.e. when perfect resonance takes place. The bifurcation diagram of figure 2 summarizes the main differences between the edge wave behaviour produced by a monochromatic wave and that caused by a narrow band irregular wave when μ is different from zero. The unstable region widens when random waves are considered, but the maximum value of the equilibrium amplitude significantly decreases. This decrease becomes more marked as the width of the wave spectrum increases and it is a reasonable guess that the equilibrium amplitude tends to vanish for large band spectra, i.e. edge waves are not triggered by random waves as tests performed in physical models seem to suggest.

Acknowledgements

This paper is partly based on work in the PACE-project, in the framework of the EU-sponsored Marine Science and Technology Programme (MAST-III), under contract n. MAS3-CT95-0002.

References

[1] Guza R.T., Bowen A.J., 1976. Finite amplitude edge waves. J. Mar. Res., 34, 269-293.

[2] Guza R.T., Davis R.E., 1974. Excitation of edge waves by waves incident on a beach. J. Geophys. Res., 79, 1285-1291.
[3] Lyon R.H., 1970. Statistics of combined sine waves. J. Acoust. Soc. Amer., 44, 145-149.
[4] Mei C.C., 1989. The applied dynamics of ocean surface waves. World Scientific.
[5] Minzoni A.A., 1976. Nonlinear edge waves and shallow-water theory. J. Fluid Mech., 74, 369-374.
[6] Minzoni A.A., Whitham G.B., 1977. On the excitation of edge waves on beaches. J. Fluid Mech., 79, 273-287.
[7] Whitham G.B., 1976. Nonlinear effects in edge waves. J. Fluid Mech., 74, 353-368.
[8] Yang J.N., 1973. On the normality of simulated random process. J. Sound Vib., 26, 417-428.

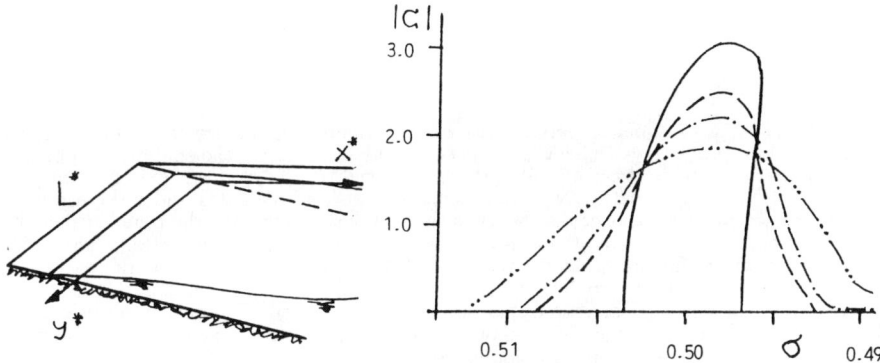

Figure 1 - Sketch of the problem.

Figure 2 - Bifurcation diagram ($a = 0.2$). Monochromatic wave (—). Random wave: $S = 0.0125$ (- - -), $S = 0.025$ (-·-·-), $S = 0.05$ (-··-··-).

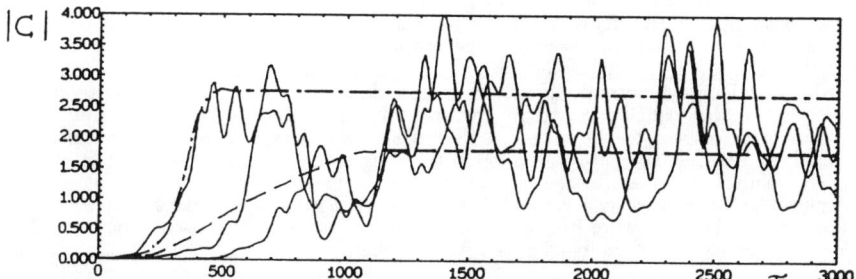

Figure 3 - Time development of $|C(\tau)|$ ($\sigma = 1/2, a = 0.2$). Three different realizations characterized by different values of θ_ℓ (—). Ensemble average value (- - -). Monochromatic case (-·-·-).

VENICE GATES IN WAVES

P. SAMMARCO, H. H. TRAN, O. GOTTLIEB & C. C. MEI
Department of Civil & Environmental Engineering,
Massachusetts Institute of Technology
Cambridge, MA, USA

Abstract

In this paper we summarize recent work on the nonlinear dynamics of the mobile gates proposed for protecting Venice City from storm tides. Assuming the equilibrium position of the gates to be vertical and all gates are hinged on the common axis spanning an inlet, we outline the theory for subharmonic resonance of the gates under the attack of incident waves. The basis of the theory is that the system coupling the mobile gates and the surrounding water, admits natural modes with certain eigenfrequencies. These mode have the special features that their energy is trapped to the vicinity of the gates and can only be excited by a nonlinear mechanism. The derivation of the nonlinear equation governing the envelope of gate oxcillations is then sketched. Physical implications are deduced. In particular the gate response to monocromatic and narrow banded seas are discussed. Experimental confirmation is also presented.

INTRODUCTION

Land subsidence due to industrial use of groundwater, and sea-level rise due to global warming have plagued the city of Venice with frequent floods in recent decades. A design of storm barriers has therefore been in consideration since 1988 to span the three inlets of the Venice Lagoon. Each barrier consists of a battery of mobile gates hinged along a bottom axis To allow normal navigation in calm days, the gates are ballasted to rest horizontally on the seabed. In stormy weather they are raised by buoyancy to a 40° inclination with respect to the seabed, so as to act as a dam. To reduce the cost of supporting structures, the gates are allowed to swing to and fro in unison in response to the normally incident sea waves. Laboratory experiments in Netherlands and Italy with sinusoidal incident waves have revealed however that neighboring gates oscillate out of phase, at one half the wave frequency, in a variety of ways. These oscillations would reduce the effectiveness of the barrier as a flood barrier.

Subharmonic resonance is known to occur in edge waves which are trapped modes on a sloping beach. Mei et al (1994, *Proc. Roy Soc. Lond*, A 444, 463-79), showed that articulated gates indeed support trapped modes when coupling of gate motion and the surrounding water is accounted for. As in the case of edge waves, it is necessary to account for nonlinearity in order that all stages of excitation be understood.

We have studied a model geometrically similar to the actual design, except that the gates are assumed to be vertically upright in the neutral state of static equilibrium. In addition to the linearized theory of eigenmodes, a nonlinear study has also been carried out. The equation governing the envelope of the gate oscillation is of Stuart-Landau type. For a simple harmonic incident wave the equilibrium response and implications on the design will be discussed. Moreover, a sinusoidally modulated incident waves is chosen as the simplest model of narrow-banded sea spectrum to examine various bifurcations corresponding to modulational resonances. Extensive experiments are performed to check the theoretical predictions of time invariant and modulational bifurcations and chaos.

OUTLINE OF THE THEORY

The theory is first developed on the basis of irrotational flow. The gates are assumed to be rectangular boxes of thickness $2a$, width $b/2$ and height slightly larger than the water depth h. The wave amplitude is assumed to be small so that the entire problem is weakly nonlinear.

A linearized theory gives the eigenmodes where wave oscillations are trapped near the gates. The eigenvalue condition for the natural frequency ω_0 through the parameter $G = g/\omega_0^2 b$ is :

$$-I + GC = I_a(G), \qquad (1)$$

where $I_a(G)$ is the added hydrodynamic inertia. The solution of (1) has been confirmed experimentally (Mei et al, 1994) for a model gate in the laboratory. For simplicity we consider hollow gates made of a light and homogeneous block of constant density ρ_g. S and I is then varied by varying gate density ρ_g and gate thickness a, for fixed water depth h. It is found that G increases for decreasing a or increasing I. Thus the eigenperiods can be increased by decreasing the displacement volume or increasing the inertia. On the other hand, for fixed S or I, an increase in a, meaning increasing displacement, shortens the natural period. Thus, heavier gates oscillate slower. For fixed inertia of the gate, an increase in the displaced volume induces an increase in the eigen-frequency. The effect of increasing gate width is to decrease the eigen-frequency. Finally, for fixed gate characteristics, deeper water is accompanied by higher natural frequencies; this effect is more pronounced for heavier gates. To avoid unwanted resonances, a solution is to decrease the natural frequency below the range of the local incident wave frequencies. This can be accomplished either by decreasing the displaced volume or by increasing the gate inertia and/or gate width $b/2$. Another solution is to render the gates so buoyant that their natural frequency is far higher than the above the incident sea spectrum. However an articulated gate array has other natural modes at lower frequencies, which may still be excited by the incident wave.

To construct the nonlinear theory, an important starting point is that at or near resonance the incident wave amplitude is an order smaller than the trapped wave; and that the time scale of resonance growth is much longer than the period of the trapped wave. Specifically let the gate rotation be $\theta = O(\epsilon) \ll 1$, then the dimensionless amplitude of the incident waves is $A/b = O(\epsilon)$, and the time scale of resonant growth is $O(\omega\epsilon^2)$. We therefore introduce the slow time $t_2 = \epsilon^2 t$ and multiple scale expansions for the nondimensional fluid velocity potential, free surface elevation and gate rotation. Moreover, at each order the unknowns are expanded in terms of harmonics, which are integral multiples of the eigenfrequency of the trapped mode. At the first order the linearized eigenvalue problem is recovered. At the second order, one must solve a three-dimensional radiation problem forced by quadratic nonlinearities from the free surface and from the gates, as well as a two-dimensional problem of diffraction-radiation for all gates oscillating in unison. At the third order we invoke the solvability of the inhomogeneous problem for the first harmonic; the homogeneous solution being the trapped wave potential. The evolution equation which describes the evolution of the amplitude of the trapped wave θ for a given incident wave amplitude A is then found.

Dissipation is inevitable in the boundary layers on the gates surface, at the bottom surface and at the vertical walls of a laboratory flume. To account for this effect, a term linear in the angular velocity of the gate is added. At higher amplitudes, flow separation occurs near the corners of the laboratory model as well the prototype. An extra term proportional to the square of the gate displacement is also added to the evolution equation. While the order of magnitude of the friction coefficient can be estimated crudely, the accurate value must be decided empirically.

Thus the final evolution equation for the gate rotation θ' is

$$-\frac{i}{\omega} \theta_t = \frac{\Delta\omega}{\omega} \theta + \theta^2 \theta^* (c_N + i\, c_R) + \frac{A}{b} \theta^* c_F + (1+i)\, c_L \theta + i c_Q |\theta| \theta \qquad (2)$$

which is of Stuart-Landau form. The coefficients c_N, c_R and c_F represent respectively the effects of nonlinearity, damping due to radiation of second harmonic waves, and coupling between incident waves and gate oscillations. They depend on the nondimensional thickness (a) and first and second moments of gate inertia (S, I) of the gate, and on the water depth h. Finally c_L and c_Q are the coefficients of linear and quadratic damping, which must in general depend on the scales of the laboratory model or of the prototype. In our study we only check the theory against laboratory models either in a flume with two gates or or in a wide basin with 13 gates. To reduce the degree of empiricism, we determine the damping coefficients from a number of free oscillation tests, whereby the neighboring gates are first given equal and opposite displacements and then simultaneously released, without incident waves. The envelope equation without incident waves can be solved exactly and the damping coefficients can be best-fitted to the experimental data.

In the flume experiment, two hollow gates of plexiglass walls, padded with styrofoam for added byoyancy, are hinged along a common axis across the flume at a central station. The top of the box is connected through two alluminum rods to a potentiometer for measuring gate rotation The distance d of the center of mass from the center of rotation is measured by a strain gage. The inertia I is evaluated from the period of the small amplitude oscillations in air of the gate hanging upside down. For this setup the nondimensional eigen-frequency is calculated to be $G = 1.305$, which correspond to a frequency of $\omega_0/2\pi = 0.713$ Hz. Next the following coefficients are calculated theoretically $c_N = 2.49$, $c_R = 1.47$, $c_F = 0.90$. The best-fit values of c_L and c_Q are $c_L = 0.009$, $c_Q = 0.291$. With these best-fit coefficients, the unforced $\theta'(t)$ predicted by (1) then agrees with each measured time series of free oscillations nearly perfectly. We also performed similar experiments for 13 gates in a wide basin, and obtained the damping coefficients.

UNIFORM INCIDENT WAVES

The Stuart-Landau equation is a second order dynamical system for two real unknowns (real and imaginary parts of θ). The fixed (equilibrium) points in the phase plane correspond to the states of equilibrium, which depend on the detuning $\Delta\omega/\omega$. Linearization about the equilibrium state allows one to predict the threshold of instability and the frequency band width. We have shown that, once resonated, heavier gates oscillates slower and with larger amplitude. For a fixed inertia, an increase of gate displacement induces a decrease in the resonant response. Similar effect can be achieved by an increase of water depth. As the gate inertia increases, the response increases, but the bandwidth of unstable frequencies decreases. Thus heavier gates is increasingly more difficult to excite, though the resonance response may be larger.

Detailed comparison with laboratory experiments have confirmed theoretical predictions In particular the jump phenomenon associated with the hysteretic amplitude-frequency relation, is confirmed for the first time, as shown in Figure 1.

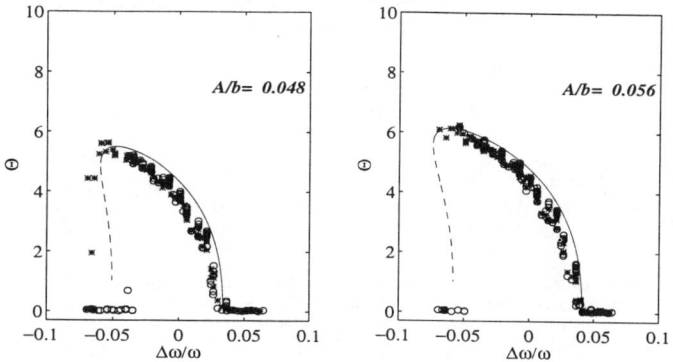

Figure 1: Comparision of theory vs. experiments for steady incident waves. Dots: increasing frequencies; circles : decreasing frequencies.

PERIODICALLY MODULATED INCIDENT WAVES

Since the frequency band of seawaves in nature is usually finite, we have also examined the effects of an idealized narrow band consisting of a central (carrier) frequency, slightly detuned from resonance, $2\omega = 2(\omega_0 + \Delta\omega)$ and two sidebands $2\omega + \Omega$ and $2\omega - \Omega$, with $\Omega = \epsilon^2 \Omega_2 \ll 1$. The envelope of the incident waves is therefore modulated sinusoidally in time with the period $2\pi/\Omega$. Since the time rate of change of the wave amplitude A is of much higher order, the evolution equation must be

$$-\frac{i}{\omega_0}\theta_t = \frac{\Delta\omega}{\omega_0}\theta + (c_N + ic_R)\theta^2\theta^* + c_F\left(\frac{\overline{A}}{b} + \frac{\tilde{A}}{b}\cos\Omega t\right)\theta^* + (1+i)c_L\theta + ic_Q|\theta|\theta. \qquad (3)$$

where $\overline{A}/b$ denotes the nomalized amplitude of the constant part and $\tilde{A}$ the amplitude of the modulating part. Due to the time dependence of a coefficient, the above dynamical system is non-autonomous; complicated bifurcations such as period-doubling and chaos are expected as the properties of the incident waves are varied.

Because the damping coefficents all have small or moderate numerical values, we first perform analytical approximations for small wave modulation $a \equiv \tilde{A}/\overline{A}$. Local bifurcations are analyzed by multiple-scale approximations to investigate modulational resonances. Prelude to global chaos is analyzed by the Melnikov method for Smale's horse-shoe tangles of either homoclinic or heteroclinic orbits in the phase plane. The approximate criteria for local and global bifurcations give preliminary guidance to experiments.

Extensive numerical investigation of the above parameter plane corroborates with the analytical predictions, and further delineates the regions of period-quadrupling and chaos.

We summarize the general picture for a fixed modulational frequency in the incident sea, hence fixed sideband width from the carrier wave. Firstly, in the absence of incident waves, there can be natural oscillations of the envelope. For a wavetrain with small periodic modulation a, the gate envelope is synchronous with the wave modulation. Increase of a induces bifurcations through a sequence of period-doublings until modulational chaos is attained. For higher a above a band of chaos, the gate envelope returns to the state of subharmonic resonance, which is distinguished from the period-doubling at lower a by a phase orbit symmetric about the origin. In the spectrum of the gate rotation, this state corresponds to a downshift of the central frequency from half the incident carrier wave of a quantity equal to half the modulational frequency. The response is in this case much larger than the non-shifted non-resonant response occurring for the same modulational amplitude but at larger modulational frequency.

Numerical findings indicate that for sufficiently high frequency of modulation, no temporal resonances of any kind is possible. This means that large side bands do not alter the resonance phenomenon induced by the carrier wave frequency. Let σ and W be suitably normalized versions of Ω and $\Delta\omega/\omega$, and λ be the normalized natural frequency of the envelope $\lambda = 2(1-W)^{1/2}$. The route to chaos via period-doubling bifurcations occurs only for sufficiently narrow wave sidebands, when σ is around twice the natural frequency of the envelope λ. Since $\lambda \to 0$ as $W \to 1$, no bifurcation occurs if the detuning W from the carrier wave is close to the margin of instability. On the other hand, as $W \to -1$ (negative detuning $\Delta\omega$), the amplitude a threshold for homoclinic or heteroclinic tangles, and period doubling also rises. Therefore the carrier wave must not be detuned too much from twice the eigen-frequency, in order for the gates to be resonated. The sidebands must also be close to the carrier frequency, in order for the gate envelope to be resonated and respond chaotically.

These numerical/theoretical findings are corroborated by laboratory experiments, as illustrated in Figure 2 where X, Y are the real, imaginary parts of θ. (The difference in orientatios is due to the choice of initial point in the time series and is immaterial.)

In the present study, attention has been limited to a single mode. Since adjacent gate modes can be close according to the linearized theory, nonlinear interactions between modes is an aspect worth investigating both theoretically and experimentally. In the proposed design, the Vencie Gates are inclined at 40° from the horizon . It is likely that gate evolution equation will remain of Stuart-Landau type. The coefficients of the Stuart-Landau equation must then be obtained by numerically solving a number of radiation-diffraction problems.

ACKNOWLEDEMENT

Funding were provided by grants from US Office of Naval Research (Accelerated Research

Initiative on Nonlinear Ocean Waves directed by Dr.Thomas Swean. Nos. N00014-92-J-1754, and N00014-95-1-8040 (AASERT)), US National Science Foundation (No.CTS-9115689), and Consorzio Venezia Nuova for laboratory equipments.

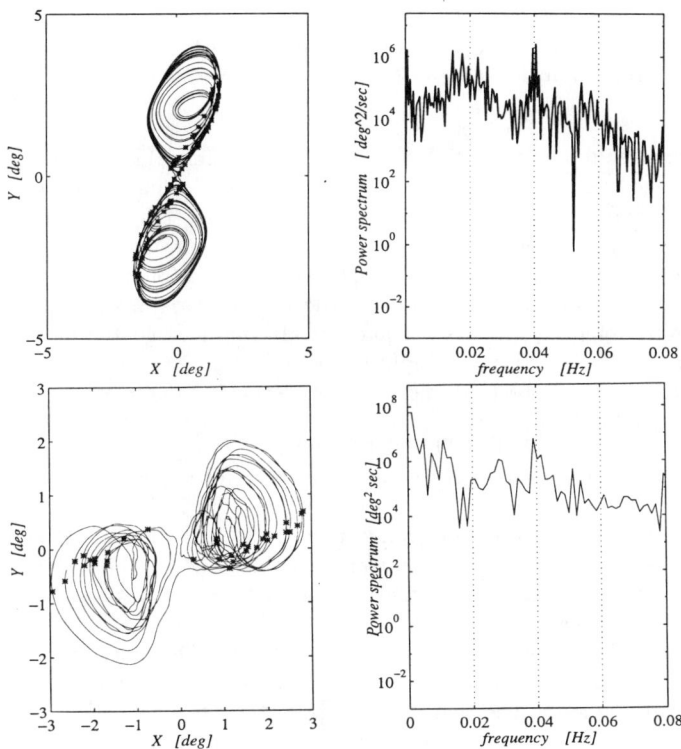

Figure 2: Comparision of theory vs. experiments. Periodically modulated incident waves. Left: phase trajectories and Poincare mas; right: power spectra. Top : predicted; bottom: measured.

Cyclic bar behaviour in a nonlinear model of a tidal inlet

H.M. SCHUTTELAARS   H.E. DE SWART
Institute for Marine & Atmospheric Research, Utrecht Univ.,
Princetonplein 5, 3584 CC Utrecht, The Netherlands

ABSTRACT: Results are presented of a model which simulates the behaviour of depth-averaged tidal currents, transport of suspended sediment and bottom changes in an embayment. The basic state of this model, a spatially uniform tide over a bottom with constant slopes, turns out to be unstable with respect to bottom perturbations if the width-to-depth ratio and bottom friction are sufficiently large. The long-term dynamics of the bed forms is investigated with a set of coupled nonlinear amplitude equations. It appears that the model can have multiple equilibria and finite-amplitude periodic solutions. The latter have characteristics similar to observed migrating bed forms in tidal inlets.

1 INTRODUCTION

Semi-enclosed tidal basins are present in many parts of the world. They are connected with the adjacent ocean by relatively narrow tidal inlets. In these areas water motions (driven by tides, wind, waves, etc.) are often strong enough to erode sediment from the bottom and transport this material. The subsequent spatial gradients in the sediment fluxes cause morphological changes such as the formation and migration of sand banks.

An example is the Frisian Inlet and adjacent basin. This inlet was formed some 6000 years ago after the last ice age and at present it is approximately 10 km wide and 20 km long. The channels are approximately 10 m deep and have a width of about 1 km. Water motions in this area are primarily driven by tides (the tidal range is ±2.5 m near the entrance), the most important constituent is the semi-diurnal lunar tide. The sediment in the channels of the inlet consist for 85% of noncohesive material with an average grainsize of 1.6×10^{-6} m. In the region between the barrier islands cyclic behaviour of channels and sand banks is observed, i.e., a repetitive formation and migration of channels in the eastward (transverse) direction (Oost, 1995). The typical migration speeds are 200 m year^{-1}, the time scales are of the order of 20-40 years. In the present literature three types of models are used to simulate morpholoigcal phenomena in tidal inlets. Semi-empirical models are based on observed statistical relationships between state variables of the system, for a review see De Vriend (1996). Process-oriented models on the other hand attempt to describe all the constituent processes based on physical principles, see Wang et al. (1995) and De Vriend & Ribberink (1996). However, the results of such models are difficult to interpret in terms of physical mechanisms. As an alternative idealized models have been developed to analyse the behaviour specific processes in isolation. The motivation for such studies is that their results may help to understand the output of complex models and field data. Examples of idealized morphodynamic models for tidal embayments are described in De Jong& Heemink (1995), Schuttelaars & De Swart (1996a,b) and Seminara & Tubino (1996). In this paper an idealized nonlinear morphodynamic model for a tidal inlet is analysed. It will be demonstrated that for specific parameter values the model is able to simulate cyclic bar behaviour, which resembles the observed periodic formation and migration of sand banks in the Frisian Inlet. In section 2 the model assumptions are briefly discussed. A basic state solution of this model, which represents a spatially uniform tide over a constantly sloping bottom, and its linear stability properties are briefly discussed in section 3. They serve as a starting point for the nonlinear analysis discussed in section 4. We end in section 5 with some conclusions.

2 THE MODEL

Here the model discussed in Schuttelaars & De Swart (1996a,b) is used. An idealized tidal embayment is considered, i.e., a half-open rectangular channel. All side-walls are fixed, only the bottom is erodeable. Water motions are driven by prescribed water level elevations (of tidal origin) at the entrance $x = 0$. The channel is assumed to be short and narrow: its width B is much smaller than its length L, which in turn is much smaller than the tidal wave-length.

The water motions are governed by the depth-averaged shallow water equations. The transport of sediment is modelled by a concentration equation which includes sediment erosion and deposition terms. Finally the bottom evolution can be computed from mass conservation of the sediment. It is assumed that the morphological time scale is large compared with the tidal period. This implies that the tidal motions can be computed for a fixed bed topography whereas bathymetric changes are determined by the net, tidally averaged sediment fluxes. In the present model advective contributions in the equations of motion are neglected, hence the sediment transport is fully due to diffusive processes. This is not realistic for most natural tidal embayments, but the analysis is considered to be a first step in gaining more understanding about the basic physical mechanisms controlling the morphodynamics of such systems.

3 BASIC STATE AND LINEAR STABILITY ANALYSIS

This model allows for a solution which describes a standing tidal wave over a bottom with a constant slope in the longitudinal (x) direction of the embayment. The amplitude of both the free surface elevations and the velocity field is constant in the domain (pumping mode). The model predicts a net sediment import by the embayment if its length is suddenly reduced. This is consistent with field observations carried out in the Frisian Inlet. It is demonstrated in Schuttelaars & De Swart (1996b) that this basic state can be unstable with respect to bottom perturbations which have a structure in the transverse direction. A typical result

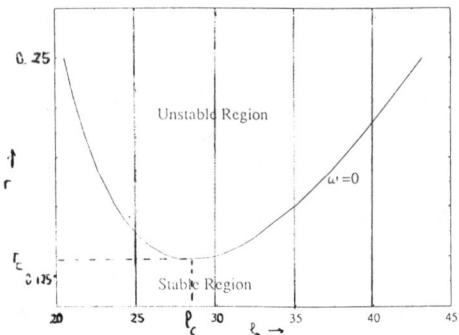

Figure 1: neutral curve in the $r - l_n$ space, separating exponentially growing and decaying bottom perturbations. Here r is the bottom friction coefficient and l_n a transverse wavenumber. For further information see the text.

of this linear stability analysis is shown in figure 1. It shows the neutral curve $\omega = 0$, where ω is the exponential growth rate of bottom perturbations, as a function of the bottom friction parameter r and the transverse wavenumber l_n. Here r is the ratio of the tidal period and frictional time scale and $l_n = n\pi L/B$, with n an integer. In case of a fixed length-to-width ratio L/B the wavenumber can only attain discrete values, as indicated by the vertical lines. The selected parameter values are representative for the Frisian Inlet system. From figure 3 it can be seen that there is a critical friction parameter r_c below which the basic state is stable: all perturbations exponentially decay in time. The same statement applies if, in case of r, the width-to-depth ratio is considered. In that sense the model shows similar behaviour as models for alternating bars in rivers and estuaries (Colombini et al., 1987; Schielen et al., 1993; Seminara & Tubino, 1996).

4 NONLINEAR ANALYSIS

A basic limitation of the linear stability analysis is that it only yields information on the initial behaviour of infinitesimally small bottom perturbations. In order to study the long-term finite-amplitude behaviour of the bottom a nonlinear study must be carried out. In Schuttelaars (1996) results of a weakly nonlinear analysis are presented, i.e., in figure 3 l_2 is chosen to be the critical wavenumber l_c and the friction parameter is close to its critical value: $(r - r_c)/r_c = \epsilon^2 \ll 1$. Thus there is just one mode unstable which has a small growth rate. The real-valued amplitude A (scaled with the water depth) of this mode turns out to be governed by the Landau equation

$$\frac{dA}{d\tau} = A(1 - \beta A^2), \qquad (1)$$

where β is a known coefficient which has a value between 0 and 1 and $\tau = \epsilon^2 t$ a slow time scale. Here we consider a situation that more modes are unstable and that the friction parameter is large (strongly nonlinear case). This is done by expanding, for a fixed value of the bottom friction parameter r, the bottom in the eigenmodes which follow from the linear stability analysis:

$$h = \sum_{n=0}^{N} A_n(t) \, h_n(x) \, \cos(l_n y). \qquad (2)$$

Here N is the truncation number and $h_n(x)$ are known functions. Projection of the equations of motion on these eigenmodes yields a set of coupled nonlinear differential equations which describe the time evolution of the amplitudes $A_n(t)$. They can be analysed by using techniques originating from the theory of dynamical systems. We consider the case that in figure 1 the modes with wavenumbers l_2 and l_3 have positive growth rates, all other modes have $\omega < 0$. This situation is more or less representative for the Frisian Inlet system. The value of the truncation number is $N = 15$, it has been verified that adding more modes in the expansion for the bottom does not change the results. In figure 2a a bifurcation diagram of the amplitude A_2 of the steady solutions of the model is shown. As can be seen for $r < r_c = 0.125$ the basic state solution is stable. At $r = r_c$ a pitchfork bifurcation occurs and two new branches of equilibria develop. Thus the model allows for multiple equilibria: for fixed parameter values the final state depends on the initial conditions. A contour plot of the nonlinear equilibrium bottom profile (for $r = 0.136$) is shown in figure 2b. It is characterized by large values of the $A_2 - A_5$ amplitudes. The pattern resembles a deep channel separated by two sandy shoals near the side-walls. Moreover a shallow secondary channel can be observed near the lower side-wall at the end of the mebyament. The second stable equilibrium for this case is obtained by reversing the sign of the bottom values in figure 2b. It should be remarked that there are also other branches of (unstable) equilibria which have zero amplitudes for the odd eigenmodes. They have not been investigated in detail yet. If the bottom friction parameter reaches the value $r_H = 0.137$ the nonlinear equilibria become unstable due to a Hopf bifurcation, resulting in new branches of periodic solutions. Thus for $r > r_H$ the bar behaviour is unsteady. In figure 3 time series of the amplitudes A_2 and A_3 are shown for $r = 0.138$, for which a stable periodic solution occurs. From inspection of the contour plots at several phases of the period (not shown here) it appears that this solution resembles cyclic bar behaviour: during each cycle a new channel is formed in the middle of the embayment, migrates to the side-wall and merges with another shoal. The dimensional period of this process is of the order of 20 years. These characteristics are similar to those of observed bars in the Frisian Inlet.

5 CONCLUSIONS

The model discussed in this paper describes the long-term nonlinear behaviour of bed forms in an idealized tidal embayment. It appears that bottom perturbations

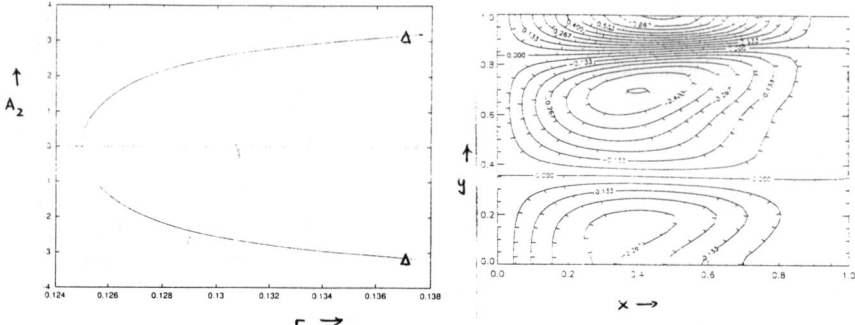

Figure 2: a. Amplitude A_2 of the steady solutions of the nonlinear model as a function of the friction parameter. At $r = r_c = 0.125$ and $r = r_H = 0.137$ a pitchork bifurcation and a Hopf bifurcation occur, respectively. b. Contour plot of the nonlinear equilibrium bottom profile for $r = 0.136$.

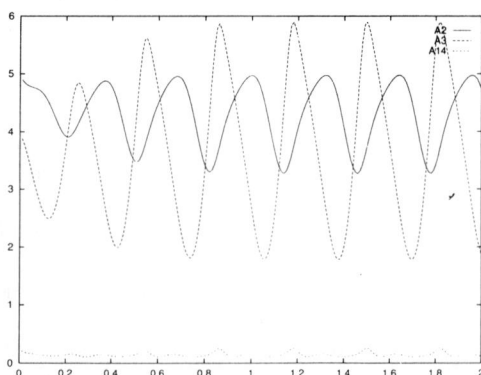

Figure 3: Time series of the amplitudes A_2, A_3 and A_{14} for $r = 0.138$. For this friction parameter stable periodic solutions exist. Note that the high mode has a small amplitude.

can spontaneously develop due to a morphologic instability mechanism which involves the interaction between the tidal currents and the erodeable bottom. The linear stability analysis yields the spatial structure of the eigenmodes with the largest growth rates. Projection of the equations of motion onto these modes yields a set of nonlinear amplitude equations. The results indicate that finite amplitude bed forms are obtained with a complicated spatial pattern. For fixed values of the model parameters multiple equilibria can exist. Far from equilibrium (large friction case) periodic solutions are obtained which correspond to cyclic bar behaviour. This type of behaviour is also observed in tidal inlets, such as the Frisian Inlet (Oost, 1995). So far the model simulations have not revaled the presence of channel separation or aperiodic behaviour of bed forms. However, the results presented here only give a first indication of the possible dynamics of the model. Presently sensitivity studies are carried out to investigate the effect of changing the width-to-depth ratio and length-to-width ratio of the embayment. This is expected to yield more information about the dynamics of the model.

ACKNOWLEDGEMENTS : This research was supported by NWO-grant nr. NLS 61-261 and has partly been carried out within the PACE-project, in the framework of the EU-sponsored Marine Science and Technology Programme (MAST-III), under contract no. MAS3-CT95-0002.

6 REFERENCES

Colombini, M., G. Seminara and M. Tubino, 1987. Finite-amplitude alternate bars. *J. Fluid Mech.* **181**, 213-232.
De Jong, K. & A.W. Heemink 1995. A long-term morphodynamic model for estuaries and tidal rivers'. In: Central board of irrigation and power, *Management of Sediment, Philosophy, Aims, and Techniques*, New Delhi, India: 687-698.
De Vriend, H.J. 1996. Mathematical modelling of meso-tidal barrier island coasts. Part I: empirical and semi-empirical models. In: Liu, P.L.-F. (ed.), *Advances in coastal and ocean engineering*, World Scientific: 115-149.
De Vriend, H.J. & J.S. Ribberink 1995. Ibid, part II: process-based simulation models. In: Liu, P.L.-F. (ed.), *Advances in coastal and ocean engineering*, World Scientific: 151-197.
Oost, A. 1995. *Dynamics and sedimentary development of the Dutch Wadden Sea with emphasis on the Frisian Inlet*. PhD thesis, Geologica Ultraiectina, Fac. of Earth Science, Univ. Utrecht, contr. 126.
Schuttelaars, H.M. 1996. Nonlinear equilibrium bottom profiles in a tidal embayment. *Proc. PECS'96 Conference, The Hague*, sumb..
Schuttelaars, H.M. & H.E. de Swart 1996a. An idealized long-term morphodynamic model of a tidal inlet. *Eur. J. Mech. B/ Fluids* 15: 55-80.
Schuttelaars, H.M., & H.E. de Swart 1996b. Formation of channels and shoals in a short tidal embayment. Univ. Utrecht Rep., to appear.
Seminara, G. & M. Tubino 1996. Bed formation in tidal channels: analogy with fluvial bars. *Proc. PECS'96 Conference, The Hague*, subm.
Wang, Z.B., T. Louters & H.J.de Vriend 1995. Morphodynamic modelling for a tidal inlet in the Wadden Sea. *Mar. Geol.* 126: 289-300.

Dynamics of self-organization and the fluvial landscape: a nonreductionist perspective

ANDREA RINALDO[†,§] and IGNACIO RODRIGUEZ-ITURBE[§]

[†] Istituto di Idraulica "G. Poleni", Universitá di Padova, via Loredan 20, I-35131 Padova (Italy)
[§] Department of Civil Engineering, Texas A & M University, College Station, Texas, U.S.A.

Self-organized criticality (SOC) refers to the tendency of large dissipative systems with many degrees of freedom to build up a state poised at criticality that is characterized by a wide range of length and time scales. SOC is now a common name for a general theory of the dynamics of fractal growth, whose main features are recalled in this lecture, especially with reference to applications within the context of earth sciences. In this lecture we suggest that principles of critical self-organization are at work in the development of the fluvial landscape. We also show that optimal channel networks are spatial models of self-organized criticality. This reinforces earlier suggestions [Rodriguez-Iturbe et al., 1992; Rodriguez-Iturbe and Rinaldo, 1997] that natural fractal structures like river networks may indeed arise as a joint consequence of optimality and randomness. Specifically, we suggest that natural fractal structures in the fluvial landscape are dynamically accessible optimal states, corresponding to locally optimal niches of a complex fitness landscape where evolution can settle in a stable manner. Such relative stability is achieved with respect to perturbations and is nonetheless reminiscent if the dynamic history, including an imprinting of its initial conditions and long-lived signatures of boundary conditions, here surrogating geologic constraints.

Two related problems have interested scientists for a long time. One is the fundamental dynamic reason behind Mandelbrot's observation that many structures in nature - such as river networks or coastlines - are fractal, i.e. looking 'alike' on many length scales. The other is the origin of the widespread phenomenon called $1/f$ noise, originally referring to the particular property of a time signal, be it the light curve of a quasar or the record of river flows, which has components of all durations, i.e. without a characteristic time scale. The name $1/f$ refers to the power law decay with exponent -1 of the power spectrum $S(f)$ of certain self-affine records and is conventionally extended to all signals whose spectrum decays algebraically, i.e. $S(f) \propto f^{-\alpha}$. Power-law decay of spectral features is also viewed as a fingerprint of spatially scale-free behaviour, commonly defined as critical. In this framework criticality of a system postulates the capability of communicating information throughout its entire structure, connections being distributed on all scales.

The causes and the possible relation for the abundance found in nature of fractal forms and $1/f$ signals have puzzled scientists for years.

Per Bak and collaborators have addressed the link of the above problems, suggesting that the abundance in nature of spatial and temporal scale-free behaviours may reflect a universal tendency of large, driven dynamical systems with many degrees of freedom to evolve into a stable critical state, far from equilibrium, characterized by the absence of characteristic spatial or temporal time scales. The key idea and its successive applications [Bak et al., 1987, 1996], which we will review in the development of the lecture, address such universal tendency and bear important implications on our understanding of complex natural processes. The common dynamic denominator underlying fractal growth is now central to our interests in landform evolution.

The resistance to Bak's idea of universality was (and still is in some circles) noteworthy. Science, and geomorphology in particular, is largely committed to the reductionist approach. The reductionist tenet is that if one is capable to dissect and understand the processes to their smallest pieces then the capability to explain the general picture, including complexity, is granted. However, the reductionist approach, affected as it is by the need of specifying so many detailed processes operating in nature and the tuning of many parameters, though suited to describe individual forms, is an unlikely candidate to explain the ubiquity of scaling forms and the recursive characters of processes operating in very different conditions. Are scale-free, recursive characters of the evolution of complex systems tied to the detailed specification of the dynamics? Or, on the contrary, do they appear out of some intrinsic property of the evolution itself? We believe, following Bak, that the invisible hand guiding evolution of large interactive systems should be found in some general properties of the dynamics rather than in some unlikely fine-tuning of its elementary ingredients.

One crucial feature of the organization of fractal structures in large dynamical systems is the power-law structure of the probability distributions characterizing their geometrical properties. This behaviour, characterized by events and forms of all sizes, is consistent with the fact that many complex systems in nature evolve in an intermittent, burst-like way rather than in a smooth, gradual manner. The distribution of earthquake magnitudes obeys Gutemberg-Richter law which is a power-law of energy release. Fluctuations in economics also follow power-law distributions with long tails describing intermittent large events, as first elucidated by Mandelbrot's famous example of the variation of cotton prices. Punctuations dominate biological evolution where many species become extinct and new species appear interrupting periods of stasis.

Lévy distributions (characterized by algebraic decay of tails, i.e. of the probability of large events) describe mathematically the probabilistic structure of such events. They differ fundamentally from Gaussian distributions - which have exponential decay of tails and therefore vanishing probabilities of large fluctuations - although both are limiting distributions when many independent random variables are added together. In essence, if the distribution of individual events decays sufficiently rapidly, say with non-diverging second moment, the limiting distribution is Gaussian. Thus the largest fluctuations appear because many individual events happen to concert their action in the same direction. If, instead, the individual events have a diverging second moment - or even diverging average size - the limiting distribution could be Lévy because its large fluctuations are formed by individual events rather than by the sum of many events.

When studying large, catastrophic events in a large system with interacting agents one can try to identify an individual event as the particular source. Rather than the recognition of the achievement of a critical state, a 'Gaussian' observer may discard the event as atypical - as noted by Mandelbrot [1983] - when studying the statistics of fluctuations because the remaining events trivially follow Gaussian statistics. A rather common reaction to catastrophic concerted actions is to find specific reasons for large events. Economists tend to look for specific mechanisms for large stock fluctuations, geophysicists look for specific configurations of fault zones leading to catastrophic earthquakes, biologists look for external sources, such as meteors hitting the earth, in order to explain large extinction events, physicists view the large scale structure of the universe as the consequence of some particular dynamics. In essence, as Bak put it, one reluctantly views large events as statistical phenomena.

Bak noted that there is another explanation, unrelated to specific events and embedded in the mechanisms of self-organization into critical states. In such states each large event has a specific source, a particular addition of a grain of sand landing on a specific spot of a sand-pile triggering a large avalanche, the burning of a given tree igniting a large forest fire, the rupture of a fault segment yielding the big earthquake, or the slowing of a particular car starting a giant traffic jam. Nevertheless even if each of the above particular initiating events were prevented, large events would eventually start for some other reason at some other place of the evolving system. In critical systems no local attempt to control large fluctuations can be successful unless for directing events to some other part of the system.

What are the signatures and the origins of the process of self-organization? Bak suggested that SOC systems have one key feature in common: the dynamics is governed by sites with extremal values of the 'signal', be it the slope of a sandpile or the age of the oldest tree in a burning forest, rather than by some average property of the field. In these systems nothing happens before some threshold is reached. When the least stable part of the system reaches its threshold, a burst of activity is triggered in the system yielding minor or major consequences depending on its state. Complexity arises through the unpredictable consequences of the bursts of activity suggesting that the dynamics of Nature may often be driven by atypical, extremal features. This is suggestive, among other things, of Kauffman's [1993] example of biological evolution as driven by exceptional mutations leading to species with a superior ability to proliferate or to Bak's example of the introduction of program trading causing the crash of stock prices in October 1987. In both cases a new fact leads to breakthroughs propagating throughout an entire concerted system because it generates chain reactions of global size. Another feature of self-organizing processes is that, in order to have a chance to appear ubiquitously, they must be robust with respect to initial conditions or to the presence of quenched disorder and should not depend on parameter tuning.

This lecture investigates whether the dynamics of the fluvial landscape, by conforming to the above general features, may be viewed as a particular case of self-organized criticality. Indeed we conclude, following Rinaldo et al. [1993, 1995, 1996], that this is the case. Although other processes interact with fluvial erosion in the making of a river basin [Dietrich et al., 1993; Montgomery and Dietrich, 1992; Howard, 1994], we suggest that the imprinting of fluvial processes dominates aggregation which results (within cutoffs) in scale-invariant landforms. The description of coordinated scaling exponents [Maritan et al., 1996; Rigon et al., 1996] acts then as a discriminant geomorphological test.

References

Bak, P., C. Tang and K. Wiesenfeld, Self-organized criticality: An explanation of 1/f noise, *Phys. Rev. Lett.*, 59, 381-385, 1987.

Bak, P., *How Nature works*, Copernicus-Springer, New York, 1996

Dietrich, W.E., C.J. Wilson, D.R. Montgomery and J. McKean, Analysis of erosion thresholds, channel networks and landscape morphology using a digital terrain model, *J. Geology*, 3, 161-180, 1993.

Howard, A.D., A detachment- limited model of drainage basin evolution, *Water Resour. Res.*, 30(7), 2261-2285, 1994.

Kauffman, S., *The Origins of Order*, Oxford Univ. Press, New York, 1993.

Mandelbrot, B.B., *The Fractal Geometry of Nature*, Freeman, New York, 1983.

Maritan, A., A. Rinaldo, A. Giacometti, R. Rigon and I. Rodriguez-Iturbe, Scaling in river networks, *Phys. Rev. E*, 53, 1510-1522, 1996.

Montgomery, D.R., and W.E. Dietrich, Channel initiation and the problem of landscape scale, *Science*, 255, 826-830, 1992.

Rigon, R. , I. Rodriguez-Iturbe, A. Giacometti, A. Maritan, D. Tarboton and A. Rinaldo, On Hack's law, 32, 3367-3374, 1996.

Rinaldo, A. , I. Rodriguez Iturbe, R. Rigon, E. Ijjasz Vasquez and R.L. Bras, Self-organized fractal river networks, *Phys. Rev. Lett.*, 70, 822-826, 1993.

Rinaldo, A., W.E. Dietrich, R. Rigon, G.K. Vogel and I. Rodriguez-Iturbe, Geomorphological signatures on varying climate, *Nature*, 374, 632-636, 1995.

Rinaldo, A., A. Maritan, A. Flammini, F. Colaiori, R. Rigon, I. Rodriguez-Iturbe and J.R. Banavar, Thermodynamics of fractal networks, *Phys. Rev. Lett.*, 76, 3364-3368, 1996.

Rodriguez-Iturbe, I., A. Rinaldo, R. Rigon, R.L. Bras and E. Ijjasz-Vasquez, Energy dissipation, runoff production and the three dimensional structure of channel networks, *Water Resour. Res.*, 28(4), 1095-1103, 1992.

Rodriguez-Iturbe, I. , and A. Rinaldo, *Fractal River Basins: Chance and Self-Organization*, Cambridge Univ. Press, in press, 1997.

TRANSPORT AND SEPARATION OF PARTICLES BY CHAOTIC ADVECTION

AHMET C. OMURTAG[†], VICTOR G. STICKEL, JR.[†], SHAWN M. GOMEZ[‡],
EDWARD F. LEONARD[‡], AND RENE CHEVRAY[†]
[†] Department of Mechanical Engineering
[‡] Department of Chemical Engineering
Columbia University
New York, New York 10027
USA

ABSTRACT

In this paper we examine the effects of fluid inertia and particle inertia on both transport and separation in the chaotic mixing of a flow consisting of a continuous phase and a disperse phase. The flow considered here occurs in the gap between eccentric cylinders. Mixing efficiency is found to decrease somewhat with increasing Reynolds number. By controlling flow parameters, we also find that finite-sized particles in the disperse phase may be "captured" by attractors of the system.

I. INTRODUCTION

Chaotic Advection can be described as a process in which trajectories in a flow, which is not necessarily turbulent, become chaotic, i.e. initially nearby trajectories in the flow diverge exponentially in time. When discrete particles are introduced into a chaotically advecting medium, they are often distributed rapidly over the accessible space. Under certain conditions, it is also possible for the particles entrained in this flow to become segregated to a specific region, causing local inhomogeneity or changes in boundary geometry. These processes are exemplified in many natural phenomena. One possible example is the action of tidal waves and tidal currents in particle and sediment transport. Here, the incoming and outgoing tides generate periodic streamline patterns which could induce chaotic Lagrangian trajectories producing effects similar to the bounded flow examined here. Particle transport in such conditions has implications in areas ranging from our understanding of industrial mixing processes or the dispersion of pollutants in natural media, to the design of bridges and other structures. Chaotic advection in fluids has been studied starting with Aref's[1] initial work relating two-dimensional incompressible flows with Hamiltonian dynamical systems. Following this landmark work, experimental visualization of laminar chaotic mixing was demonstrated by Chaiken[2], et al. through the use of eccentric rotating cylinders. Since then, investigations concerning specific applications of chaotic advection have been performed mainly for enhancement of fluid mixing and heat transfer. Researchers, such as Jones[3] and Dutta[4], have demonstrated that chaotic mixing and diffusion mutually enhance each other. It has also been shown that chaotic advection can provide efficient mixing of "delicate fluids," such as biofluids, which require low shear rates (Omurtag[5], et al.). Shear dispersion and chaotic advection of passive scalars have been considered for enhanced isotope separation[6]. More recently, it has been proposed that chaotic advection of finite-sized particles be used as a means for particle segregation or "classification" (Omurtag[7], et al.). By controlling the flow parameters for a two-component mixture, consisting of a disperse phase and a continuous phase, the disperse phase may be separated and confined to a smaller domain within the flow.

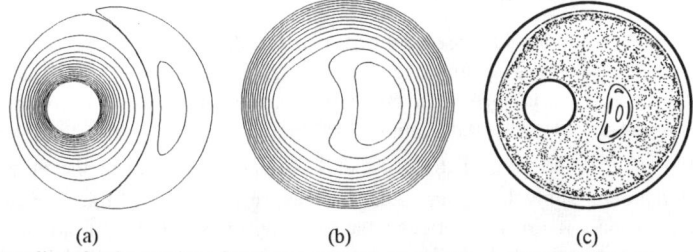

(a) (b) (c)
Fig.1 Streamlines for the rotation of (a) inner and (b) outer cylinders. (c) Poincaré section.

Specifically, by choosing a certain combination of Reynolds number, Re (for the continuous phase), and Stokes number, St (for the disperse phase), the particles in the disperse phase may be "captured" by attractors generated as a result of the chaotic flow. In this paper, the authors investigate the effects of fluid inertia and particle inertia on both transport and separation in the chaotic mixing of a two component substance.

The eccentric annular system provides a convenient setting to investigate these phenomena. Figures 1a-b show the streamlines that occur when the inner and the outer cylinder rotates, respectively. When the cylinders are rotated alternately, it is possible for the trajectories in the domain to become chaotic. In this case, a *Poincaré section* (Fig. 1c) can be constructed by initializing particles at various key locations inside the annular region and recording their subsequent positions at the end of each cycle. A cycle refers to a rotation of the inner cylinder followed by a rotation of the outer cylinder, through a specific angle of rotation. For infinitesimal particles stirred in a chaotic flow, a Poincaré section typically contains a mixture of chaotic and regular regions marked, respectively, by randomly distributed points and by closed curves. Particles initialized in one region cannot enter another since in order to do this they would have to cross the outermost closed curve ('KAM curve') surrounding a regular region; and all points on a curve are mapped back onto the curve itself. These have the implication that the fluid in a poorly mixed island is permanently isolated from the rest. The Reynolds number of the flow and the particles' inertia and density have important effects on these phenomena some of which are described here.

II. NUMERICAL MODEL

The viscous Navier-Stokes are solved for an eccentric annular region using a streamfunction-vorticity based finite difference technique. No-slip boundary conditions on the inner and outer cylinder surfaces are applied. Here, we have performed the computations transforming the governing equations and the boundary conditions from the Cartesian coordinate system to a bipolar system. The domain is discretized as a 31x61 non-uniformly spaced grid. The solution of the vorticity transport equation is obtained by using a finite difference based ADI method. For obtaining particle trajectories, a dimensionless equation of motion is used (Maxey and Riley[8]):

$$\dot{\mathbf{u}} = \frac{1}{St}\{\mathbf{u}_f[\mathbf{x}(t),t] - \mathbf{u}\} + \frac{1}{q+\frac{1}{2}}(\frac{D\mathbf{u}_f}{Dt} + \frac{1}{2}\dot{\mathbf{u}}_f)$$

(1)

where the Stokes number has been defined as $St = \tau_p\ (U_f/L)$ and the particle response time is $\tau_p = (q+1/2)d^2/(18\nu)$. In Eq. (1) the particle position and velocity are represented by $\mathbf{x}(t)$ and $\mathbf{u}(t)$, respectively. The interpolated Eulerian fluid velocity field is given by $\mathbf{u}_f[\mathbf{x}(t),t]$, d is the diameter of the particle, ν is the fluid kinematic viscosity, $q = \rho_p/\rho_f$ is the ratio of the density of the particle to that of the fluid, and U_f and L are characteristic velocity and length scales, respectively, of the fluid flow. The distinction is made in Eq. (1) between the time derivative, D/Dt, following a fluid element ('material derivative'), and that following a particle, which is denoted by the over-dot. The first term on the right hand side of Eq. (1) represents the Stokes drag. The second term represents the pressure gradient force on the particle and the added mass effect. Equation (1) leads to a four-dimensional, non-autonomous, dissipative dynamical system, $dy/dt = f[y(t),t]$, where y is the state vector containing two coordinates and the velocities in the corresponding directions. The rate of change of volume in phase space for this dynamical system is $\nabla \cdot \mathbf{f} = -2/St$. For small values of the inertia parameter, St, Eq.(1) displays *stiffness* requiring special treatment in the numerical integration. A semi-implicit discretization technique due to Press[9], et al. combined with Richardson extrapolation has been used here.

III. RESULTS

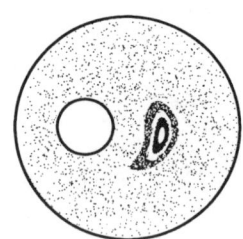

Fig. 2 Poincaré section for heavy finite-sized particles. 2000 cycles and 6 particles have been used.

SHORT-TIME EFFECTS: Poincaré sections for the eccentric annular system have previously been constructed by Chaiken[2] et al., Liu[11] et al., and others. Our model verifies that particle inertia allows particles to pass through the KAM curves confining fluid elements. Fig.2 shows a Poincaré section generated by heavy particles ($q=2$) with inertia ($St=0.001$) moving away from the central island. In this case, when a particle initially in the island reaches its boundary, it starts to move chaotically while slowly drifting toward the outer cylinder. Light particles with the same inertia (not shown) enter from the chaotic region into the island and those on the KAM curves spiral toward the center of the island at which an attracting fixed point of the Poincare map is located. In Fig. 3 the *concentration variance*, s^2, is shown as a function of cycles for various Reynolds numbers. Here, s^2 is defined as

$$s^2 = \sum_{i=1}^{N}(c_i-\bar{c})^2/\bar{c}^2$$ where, $c_i = n_i/A$, is the number density per unit area for the *i*th

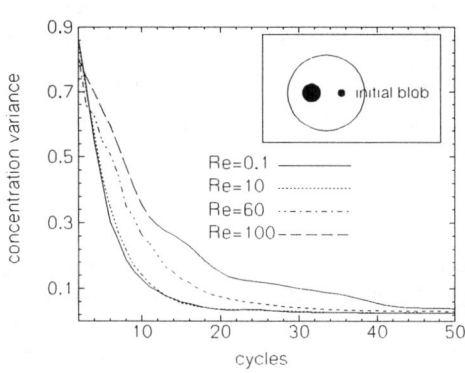

Figure 3 Concentration Variance versus Cycle Number for Various *Re*

domain and $\bar{c}$ is the mean number density per unit area. The concentration variance was calculated using an initial circular blob of 1500 infinitesimal particles located as shown in the inset. In all instances s^2 decreases monotonically as the cycle number increases until the mixture reaches a minimum s^2 value, s^2_{min}. Mixing is rapid at first with the majority of desegregation being achieved within the first 20 cycles. Looking at the effect of *Re* on the flow, it is seen to be quite similar for the regime $0.1 \leq Re \leq 10$ with the mixing curves tracing nearly identical paths. For this regime complete mixing is accomplished within 20 cycles. As *Re* increases, however, mixing per cycle decreases leading to longer times to achieve s^2_{min}. Thus, we see a stronger dependence of s^2 on *Re* for *Re*>10. In fact, for *Re*=100, s^2_{min} is not reached until greater than 40 cycles.

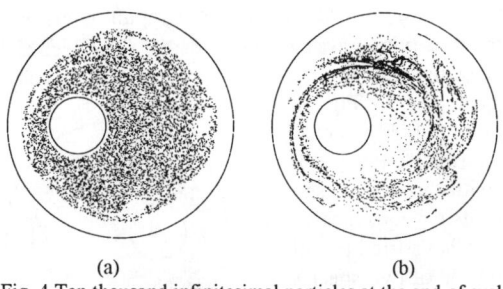

(a) (b)
Fig. 4 Ten thousand infinitesimal particles at the end of cycle 50 with (a) Stokes flow, (b) Re=60.

This behavior is due to the fact that at lower *Re*, boundary information is propagated more rapidly throughout the flow domain relative to the characteristic time of the flow. In other words, for low *Re* values, the flow has had a chance to completely distribute the effects of boundary motion to successive fluid layers before the next cycle begins, thereby maximizing the mixing effect. As *Re* increases, the effects of boundary motion may become passed to adjacent fluid layers on a time scale equivalent to that of the characteristic flow time. In this case, the mixing effect for one complete cycle may not be fully achieved before the next cycle begins. Thus, we get a degradation in overall mixing per cycle due to inertial effects.

LIMIT SETS: As a dissipative dynamical system evolves in time, the trajectory in phase space will in general approach a fixed point (sink), a singly periodic orbit (limit cycle), a two-period (quasi-periodic) orbit, or a fractal object (strange attractor)[12]. The particle paths themselves are attracted to the projection of one of these limit sets on the x-y plane. It was found that the location and type of limit set and the extent of its basin of attraction depend upon the unsteady fluid velocity field and the inertia and density of the particle. For the unsteady chaotic flows in the eccentric annulus the simplest kind of limit set was found to be limit cycles giving rise to singly periodic orbits. Fig. 5 shows the trajectory of particles while the phase point moves on a period-5 solution. Motion on this simple attractor is regular while the Lagrangian fluid flow of the system (i.e. $St=0$) is chaotic. This verifies that particle inertia can introduce a qualitatively different behavior by eliminating chaos from the system. Fig.6 shows a Poincaré section produced by an orbit moving on a strange attractor. The trajectory (not shown) of a particle moving on the attractor appears irregular but the Poincaré section reveals some patterns and fractal structure. Realization of fractals in physical space was reported earlier in numerical studies of unbounded and cavity time-periodic flows as well as in experiments. The power spectra of the coordinates or of the velocity components of a particle moving on the attractor in Fig. 6 were found to exhibit broadband features typical of chaotic behavior. In addition, the full set of Lyapunov exponents of this non-autonomous dynamical system was computed by simultaneously advecting, for very long times, five identical non-interacting particles initially nearby in 4-dimensional phase space. In all cases, the sum of the Lyapunov exponents equals the rate of phase volume contraction, $-2/St$. The Kaplan-Yorke fractal dimension of the chaotic attractor (associated with Fig.6b) is $d_f = 1.576 \pm 0.001$, based on the Lyapunov exponents (0.057,-0.099) corresponding to the two physical coordinates. The remaining Lyapunov exponents are always negative showing the strong contraction in velocity space.

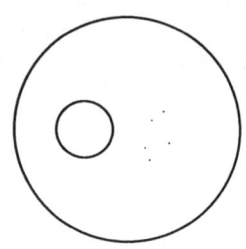

Fig. 5 Period five solution in the eccentric annular domain

The chaotic attractor presented in Fig. 6 is produced from the simple attractor in Fig. 5 by a change of 0.0001 in the density ratio at the "critical" value $q=1.2699$. Numerical experience with this system shows that the transition to chaotic limit sets is very sensitive to the value of the density ratio, q. We have tried to utilize this finding in simulating the separation of two species of particles with slightly different densities. A blob of particles with $St=1.0$ was initialized for this purpose within the island belonging to the fluid flow. When the system is stirred, particles having a density ratio less than the critical value, $q=1.2699$, are attracted to a limit cycle similar to that in Fig. 5. Heavier particles, on the other hand, drift toward and converge upon the strange attractor shown in Fig. 6. For the nth particle, the square

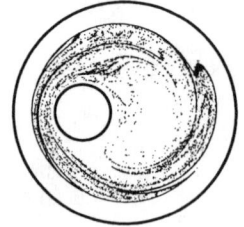

Fig. 6 Strange attractor in the eccentric annular domain

separation is defined as $\sigma_n^2 = |(x_n)_f - (x_n)_i|^2$ where the subscripts i and f refer to the initial and current states, respectively. The evolution of the mean square separation for the blob, σ^2, which is an average of σ_n^2 over all particles, is plotted in Fig. 7. The behavior of σ^2 indicates that particles heavier than the critical value eventually leave the initial neighborhood of the blob. It is found that the critical value of q depends on the Stokes number and hence can be adjusted by varying the stirring speed. Results indicate that it may be possible to separate particles with as small a density difference as desired by this or a similar procedure in a bounded time-periodic flow.

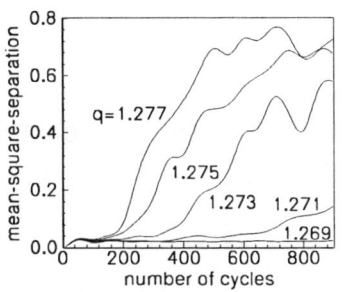

Fig. 7 Mean-square separation of a blob of 500 finite-sized particles

Acknowledgment. The support of the NSF/Whitaker Foundation under Grant No. CUO1627201 is gratefully acknowledged.

BIBLIOGRAPHY

[1] Aref, H. (1984) Stirring by Chaotic Advection. *J. Fluid Mech.* **143**, 1.
[2] Chaiken, J., Chevray, R., Tabor, M. & Tan, Q.M. (1986) Experimental study of Lagrangian turbulence in a Stokes flow. *Proc. R. Soc. Lond.* A **408**, 165.
[3] Jones, S.W. (1991) The enhancement of mixing by chaotic advection. *Phys. Fluids A* **3**, 1081
[4] Dutta, P. & Chevray, R. (1991) Effect of diffusion on chaotic advection in a Stokes flow. *Phys. Fluids* A **3**, 1440.
[5] Omurtag, A., Stickel, V. and Chevray, R.(1996) Chaotic Advection in a Bioengineering System. *Proc. Eleventh ASCE Engineering Mechanics Conference.*
[6] Aref, H. & Jones, S. W. (1989) Enhanced separation of diffusing particles by chaotic advection. *Phys. Fluids* A **1**, 470..
[7] Omurtag, A., Dutta, P. and Chevray, R. (1996) Attractors of finite-sized particles: an application to enhanced separation. To appear in *Physics of Fluids*.
[8] Maxey, M.R. & Riley, J.J. (1983) Equation of motion for a small rigid sphere in a non-uniform flow. *Phys. Fluids* **26**, 883.
[9] Press, W., Teukolsky, S.A., Vetterling, W.T. and Flannery, B.P. (1992) *Numerical Recipes*, Cambridge University Press, Cambridge, MA.
[11] Liu, M., Peskin, R.L. and Muzzio, F.J. (1994) Fractal structure of a dissipative particle-fluid system in a time-dependent chaotic flow. *Physical Review* E **50**, 4245
[12] Lichtenberg, A.J. & Lieberman, M.A. (1992) *Regular and Chaotic Dynamics*, Springer-Verlag: New York.

Non-Linear Methods for Analysis of Long-Term Beach and Nearshore Morphology

HOWARD N SOUTHGATE
HR Wallingford Ltd, Wallingford, Oxfordshire, OX10 8BA, UK

INTRODUCTION

The prediction of coastal morphology on time scales of up to decades is an important consideration in the design of sea defences and other coastal engineering works. In recent years, computer models of various types have been developed for this purpose. Mostly, these have focussed on the representation of small-scale wave, current and sediment transport processes which are then combined to determine morphological development in a time-stepping manner. This process-based approach turns out to be limited, by computing time and result accuracy, to predictions over total timespans of up to days, weeks and possibly months. Models for longer term morphodynamic predictions involve simplifications of many of the physical processes and operate with larger space and/or time discretisations. Although these models can be valuable for predictions of long-term and large-scale trends in beach behaviour, the shorter term variabilities about these trends are either not predicted at all, or suffer from oversimplified representations of processes and morphological feedback mechanisms.

This paper focusses on the problem of predicting these short-to-medium term (days, weeks, months) fluctuations about long-term (years, decades) trends in beach morphology. This is an important issue, for both scientific understanding and coastline management, since these fluctuations are often considerably larger than the trends. The problem can be tackled in two ways, through a 'bottom-up' approach involving development of models of the underlying physical processes (which have the difficulties mentioned above), and via a 'top-down' approach involving the analysis of field data to provide some understanding of the causes of the observed morphodynamic behaviour.

This paper is concerned with the latter approach, and in particular with methods of non-linear analysis of time series and spatial patterns that have been developed over the past 15 years or so. Before considering these analysis methods, their requirements for data and the available data sources will first be reviewed.

DATA REQUIREMENTS AND SOURCES

Two obvious requirements for an analysis of beach level fluctuations are that the data set should cover a 'long-term' period (ideally several decades), but must also be sampled at 'short-term' time intervals (days or a few weeks). As will be discussed in the next section, many methods of non-linear analysis require large amounts of data (perhaps upwards of several thousands in a single time series). Important information about the dynamics of beach morphology can be deduced from analyses of time series of beach and seabed levels at single locations, but greater understanding will be obtained from a spatial coverage of data (for example, on the response of beach profiles, or the propagation of bedforms). Again, although much can be gleaned from purely bathymetric measurements, simultaneous measurements of hydrodynamic forcing conditions (waves, winds, currents and water levels), and intermediate quantities such as sediment concentrations, will significantly add to our understanding.

It will come as no surprise that field data sets satisfying all, or only some, of the above requirements are very rare. The collection of even modest amounts of data is an expensive and time-consuming task, involving large amounts of manpower and, in the more comprehensive field exercises, deployment of many instruments often in hostile conditions with major planning and logistical problems. A survey of present coastal morphodynamic data sets was undertaken by Wallace (1993) who considered about 150 data sets world-wide, describing 24 in more detail. It was concluded that most data sets were so lacking in spatio-temporal coverage and resolution, and supporting hydrodynamic measurements, that they provided insufficient data for comparison with morphodynamic model predictions (it should be said, however, that providing data for morphodynamic model predictions was not the main motivation for most of these data collection exercises).

Similar conclusions can be reached for the present purpose of analysis of beach and nearshore seabed level fluctuations. Even the basic requirement (bathymetric measurements over a long period, taken at short time intervals), is met by very few data sets. Probably the most comprehensive data set has been compiled at Duck, on the Atlantic facing coast of North Carolina, USA (Lee et al, 1995). Beach and nearshore levels have been measured at about 50 locations on 4 profile lines out to a depth of 8m at roughly 2 week intervals since 1981. Wave conditions have been monitored during this time, and there have been a number of intensive periods of measurements of hydrodynamic and sediment parameters each lasting several weeks. Another potentially useful data set is from the central Lincolnshire coast of the UK where beach level measurements were made at about 10 locations on 18 profiles in the intertidal zone at monthly intervals between 1959 and 1990 (Southgate and Beltran, 1995).

The use of remote sensing techniques holds out the possibility of much larger data sets at less cost than conventional ground-based methods (although at present with less accuracy). Land-based radar and photogrammetric techniques have been in use for several years, and instruments for aircraft and satellite remote sensing of coastal bathymetry and other hydrodynamic and sediment parameters have recently been developed and deployed. Potentially far greater spatial coverage and resolution from remote sensing is possible than from ground-based methods; pixel sizes on a remotely sensed satellite or aircraft image can be as low as a few metres, and a series of images can cover tens or hundreds of kilometres of coastline. Regular monitoring to provide higher time resolution and coverage of data should be possible with far less manpower and cost. These techniques are just starting to be available for supplying coastal data, and it can be expected that these sources of data will be increasingly used as instruments are developed with better spatial resolution and accuracy.

DATA ANALYSIS TECHNIQUES

INTRODUCTION

System responses fall into two descriptive categories, forced and self-organised. In a forced response there is a direct relationship between forcing and response, and the complexity of the response is the result of a corresponding complexity in the forcing conditions. This category occurs for linear responses, and non-linear responses which move to an equilibrium state. Self-organisation is characterised by complex responses which are not directly related to the forcing conditions, and can occur even for very simple forcing conditions. These responses occur for non-linear dissipative systems maintained away from equilibrium, usually involving important feedback mechanisms, and are often associated with periodic (unrelated to any forcing periodicity) and chaotic behaviour.

Our knowledge of coastal morphodynamics indicates a strong non-linear relationship between hydrodynamic forcing and bed level response, and that there is feedback from bed levels as they develop (in time) to the hydrodynamic forcing parameters. Both these facts suggest that self-organisation could play an important role in spatial and temporal morphodynamic patterns. The wide variety and scales of bedforms, obvious to the naked eye, provide some initial circumstantial evidence for this. However, we would also expect to see significant forced responses, particularly during and immediately after major human interference or strong wave events (beaches erode during storms, followed by more gradual accretion), and possibly also as background forced noise correlated to the random sequencing of wave events. Furthermore, we might expect to see periodic forced responses resulting from periodic forcing (e.g. tidal, spring-neap and annual cycles), and long-term systematic forced responses resulting from factors such as sea level rise.

Although non-linear mechanisms for self-organised responses have been known for about 30 years, and applied in a number of environmental sciences, the explanations for coastal morphodynamic behaviour have until very recently been based on forced responses. This is apparent in the methods used for analysis of coastal bathymetric data (and other types of coastal process data). Linear methods have traditionally been used (e.g. spectral analysis to determine periodic responses) or empirical curve fitting to deduce overall trends. The variability in the data (often very large) is usually dismissed as being the result of random variations in forcing conditions, measurement error, or the influence of complex, unidentified processes. However, the possibility of self-organisation indicates that a substantial amount of this data variability can be a genuine dynamical response resulting from a relatively simple representation of the forcing conditions. The possibility of identifying self-organising mechanisms, and determining their importance relative to forced mechanisms, has been made possible (to some extent) by the great activity in the development of non-linear analysis of time series and spatial patterns during the past 15 years.

LINEAR ANALYSIS
Even if the underlying physical mechanisms are suspected to be non-linear, important information can be obtained from appropriate types of linear analysis. For example, significant data reduction can be made by transforming the data with Empirical Orthogonal Eigenfunction methods and related techniques. Furthermore, weak (i.e. data-rich and theory-poor) linear models can make comparably good predictions from short times series of data to those from equivalently weak non-linear models. As an illustration of this, Penland (1989) notes that meteorological time series data, which typically have less than 1000 data values, are 'highly multivariate and frustratingly short' (sentiments that can be shared about coastal morphodynamic time series data), and concludes that 'we do not expect more complicated models to significantly improve the predictions made by a linear Markov process unless there is independent knowledge of the physical processes involved'.

NON-LINEAR ANALYSIS: LOW-DIMENSIONAL SYSTEMS
Non-linear analysis of time series took a major step forward in the early 1980s with the concept of 'time-delay embedding', first put forward by Packard et al (1980) and Takens (1981). The idea is that the character of the overall dynamics of multivariate systems is reflected in the response of just one variable, and, by using this concept, the time series of this one variable is sufficient to determine the properties of the trajectory (such as Lyapunov exponents, K-S entropies and attractor dimensions) for the whole system. The most commonly used technique exploiting the time-delay embedding concept is by Grassberger and Procaccia (1983) for calculating the correlation dimension (a type of fractal dimension) of the attractor.

A problem with the Grassberger-Procaccia method is that it requires a large amount of data, of the order of 10^D values (D is the correlation dimension) in a single time series. The method becomes impractical for short time series (including presently available coastal bathymetric data), and this has prompted research into techniques which require fewer data values. Several investigators (e.g. Sugihara and May (1990), Rubin (1992), Sugihara (1994); further references given in these papers) have used methods involving several trial models, usually 'weak' and including a range of different generic features (e.g. linear or non-linear, random or chaotic etc.). These models generate their own data sequences with characteristic prediction properties; the models whose prediction properties most closely resemble those of the actual data then indicate the generic type of dynamic behaviour. Rubin (1992) has extended this approach to analyse 2-dimensional spatial patterns, and applied the technique to digitised photographs of wind-generated sand ripples. The method involves splitting the data into two halves, and using one half as a fitting set, and the other half to determine the predictive properties. Rubin selected six trial models involving a variety of linear and non-linear, deterministic and random, chaotic and periodic, model components. He concluded that the best predictions were derived from a non-linear model generating chaotic data, and by a linear model involving sine waves with correlated phase noise. Of these two, the non-linear model was favoured on the basis of the separate consideration that branchings in the ripple patterns (clearly seen in the data) could not be simulated by a linear model. This illustrates two general conclusions, that deducing system behaviour from non-linear data analysis requires: 1) consideration of trial (weak) models and 2) some knowledge of the underlying physical processes. Hence 'bottom-up' approaches can play an important part in 'top-down' data analysis.

NON-LINEAR ANALYSIS: HIGH-DIMENSIONAL SYSTEMS

For systems which are high-dimensional in relation to the length of available time series data, we are forced into using statistical models. It has been found empirically that many natural systems, known to be high dimensional with strongly interacting elements, show fractal (self-similar or self-affine) distributions of the dynamic variables. One well-known example is the Gutenberg-Richter law for earthquake magnitudes. Fractal analysis methods for spatial patterns have been widely used in the earth sciences, and their application to time series data is described in, for example, Hastings and Sugihara (1993). These methods have been applied by Southgate and Beltran (1995 and 1996) to the Lincolnshire and Duck beach level data, with stronger evidence for fractal behaviour at Lincolnshire.

The physics underlying the (apparently ubiquitous) presence of fractal distributions in nature is not yet firmly established. Although trajectories on low-dimensional chaotic attractors can show fractal statistics, there is some evidence for a generically different type of response in high-dimensional systems. One intriguing property of temporal fractal distributions is that they imply long-range correlations; the strength of correlations does not diminish with time. Bak, Tang and Weisenfeld (1987) have put forward the idea of 'self-organised criticality' as a physical explanation of this behaviour. They draw the analogy with the critical states which occur during phase transitions of matter (solid to liquid etc.). In these states, long-range correlations occur, but such states do not persist; they require fine-tuning of the external conditions. Bak, Tang and Weisenfeld suggested that complex, high-dimensional systems exist close to an analogous critical state, but are maintained there by self-organisational processes, and the critical state is therefore robust with respect to the external conditions (within certain limits).

CONCLUSIONS

Understanding of coastal morphodynamics has traditionally been based on the forced responses to hydrodynamic conditions and the implicit assumption of an 'equilibrium morphology' for a given set of forcing conditions. However, recent studies of non-linear systems indicate the possibility of much more dynamic behaviour, with systems being maintained away from equilibrium and showing self-organised responses. Coastal morphodynamics is a good candidate to show this type of behaviour because of the strong underlying non-linear processes and feedback mechanisms. In this paper a brief overview has been given of the sources of long-term morphodynamic data and recent methods of non-linear analysis of time series and spatial patterns. The relative importance of forced and self-organised mechanisms will fundamentally affect how we make predictions of future morphology.

ACKNOWLEDGEMENTS

This paper is based on work in the PACE project, in the framework of the EU-sponsored Marine Science and Technology Programme (MAST-3), under contract no. MAS3-CT95-0002. The work was co-funded by the UK Ministry of Agriculture, Fisheries and Food (Flood and Coastal Defence Division) as part of the CAMELOT Commission (FD1001).

REFERENCES

- Bak P, Tang C and Weisenfeld K (1987), Phys. Rev. Lett., **59**, pp381- 384.
- Grassberger P and Procaccia I (1983), Phys. Rev. Lett., **50**, pp346-349.
- Hastings H M and Sugihara G (1993), "Fractals. A user's guide for the natural sciences", Oxford University Press.
- Lee G H, Nicholls R J, Birkemeier W A and Leatherman S P (1995), J. Coastal Res., **11**, No. 4, pp1157-1166.
- Packard N H, Crutchfield J P, Farmer J D, and Shaw R S (1980), Phys. Rev. Lett., **45**, pp712-716.
- Penland C (1989), Monthly Weather Review, **117**, pp2165-2185.
- Rubin D M (1992), Chaos, **2**, No. 4, pp525-535.
- Southgate H N and Beltran L M (1995), Proc. Coastal Dynamics '95, Gdansk, ASCE, pp1006-1017.
- Southgate H N and Beltran L M (1996), Proc. Physics of Estuarine and Coastal Seas '96, The Hague (to appear).
- Sugihara G (1994), Phil. Trans. R. Soc. Lond. A, **348**, pp477-495.
- Sugihara G and May R M (1990), Nature, **344**, pp734-741.
- Takens F (1981), "Detecting strange attractors in turbulence", In *Lecture Notes in Mathematics*, No. 898, Springer-Verlag: New York, pp366-381
- Wallace H M (1993), "Coastal sand transport and morphodynamics: a review of field data", Report SR355, HR Wallingford.

INTERNALLY GENERATED ("AUTOCYCLIC") FLUCTUATIONS IN FLUVIAL SYSTEMS

CHRIS PAOLA
Dept. Geology & Geophysics, Univ. Minnesota, 310 Pillsbury Dr. SE, Minneapolis MN 55455-0219 USA, cpaola@maroon.tc.umn.edu

ABSTRACT

An "autocycle" is a cycle or fluctuation that is produced by internal processes within a transport system, in the absence of external forcing. Here we investigate the characteristics of autocyclic flucutations produced in two highly simplified but representative sediment-transport systems: (1) a two-dimensional cellular model of a braided stream; and (2) a directional variant of the original "sand-pile" system developed for the study of self-organized criticality (SOC), but with the addition of sediment input at the upstream end. Both models produce fluctuations in local sediment flux and rate of deposition entirely through their internal dynamics. The fluctuations are typically broadband and do not have any preferred cycle period. When external cycles are imposed occur on time scales that are much longer than those of the longest autocycles the autocycles appear as superimposed noise. However, when allocyclic forcing is imposed within the bandwidth of the internally generated fluctuations, the external signal can be lost. This is not a "masking" effect (i.e. a linear superposition of the autocylic and allocyclic signals); rather, the nonlinearity of the processes that generate the fluctuations causes the external signal to be destroyed as it propagates through the system.

INTRODUCTION

Autocyclicity is familiar in physical stratigraphy, where it usually refers to vertical cycles in grain size produced by channel shifting, but it is actually a quite general idea that deserves to be better known in other fields. In this paper we investigate two theoretical transport systems that produce autocycles in sediment flux. Both systems are nonlinear and display at least some of the characteristics of self-organized criticality (SOC). In both cases, fluctuations in sediment flux arise through purely internal mechanisms. A constant external flux is applied, and the system produces broad-band fluctuations in sediment flux through its own internal dynamics. Hence the autocycles are not analogous to the well known phenomenon of generation of harmonics in nonlinear systems, which requires a fluctuating input from which the harmonics are generated. For those interested in how input signals are recorded in sedimentary strata, the most important effect of autocyclic fluctuations is that they can destructively interfere with, as opposed to merely masking, input signals within their band of characteristic frequencies.

AUTOCYCLIC FLUCTUATIONS IN SEDIMENT FLUX

CASE 1. BRAIDED RIVERS. Autocyclic fluctuations in sediment flux in experimental and field braided rivers have been described by a number of workers (Ashmore 1991; Goff and Ashmore 1994; Hoey 1992; Hoey and Sutherland 1991). Similar fluctuations were also observed in the cellular computational model of braiding described by Murray and Paola (1994). The autocycles show a typical brown-noise frequency spectrum (Fig. 1) and are associated with short-term sediment storage in bar complexes and release as the complexes are dissected by shifts in flow pattern.

Murray and Paola (in press) describe an experiment in which short-term changes in sediment flux were imposed at the upstream end of their computational braided-river model. The short-term variations were not detected at a test location near the downstream end of the model. Murray and Paola attribute this to the degradation of the input signal as each pulse of additional sediment became smeared out among successive bar complexes as it worked its way downstream.

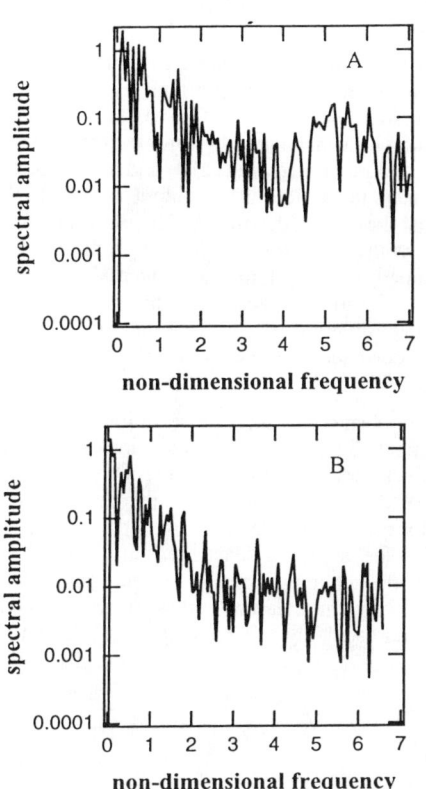

Case 2. SELF-ORGANIZED AVALANCHES. The braided-river model is a somewhat cumbersome tool for study of signal degradation because it takes a good deal of computer time to examine the behavior of long-period fluctuations. It is also of fundamental interest to know whether this behavior is peculiar to braided streams or whether it is characteristic of some general class of transport systems. The length scales on which the flow network in braided streams reconfigures itself has been shown to have a power-law distribution, an indicator of self-organized criticality (Sapozhnikov and Foufoula-Georgiou in press).

Figure 1. Power spectra of fluctuations in total load observed in (A) a laboratory experiment (Ashmore 1985) and (B) the computational braided river model proposed by Murray and Paola (1994). The frequency is nondimensionalized using a characteristic time $T = B^2H/Q_s$ where B is the mean summed channel width, H is the standard deviation of the surface topography, and Qs is the mean volumetric sediment flux.

Since this shifting and reconfiguration of the flow are also ultimately responsible for autocyclic sediment-flux variations, it seems plausible that other SOC systems may be autocyclic as well. We investigated the behavior of a variant of the original 'sandpile' model for SOC (Bak et al. 1987; Kadanoff et al. 1989). The model adopted here is essentially the same but allows for flow only downhill and laterally. To look for autocyclicity, one must treat the system differently than in the original SOC studies. Instead of running the sandpile to its critical state and then probing it by dropping particles onto its surface at random, we apply a constant particle flux at one end of the system (defined as the upstream end) and then study the flux of particles out the other end (the downstream end). The results of such an experiment are shown in Figure 2A. Evidently, this system also generates strong autocyclic variations in transport even in the presence of constant external sediment supply. These variations show a broad-band brown-noise spectrum similar to that found in braided rivers (Fig. 3A).

It is of particular interest to know how these autocycles interact with externally imposed fluctuations. Figure 2B shows the result of imposing a high-frequency cycle on the sediment feed rate to the sandpile system. The period, 20 time steps, is higher than the dominant frequencies of the autocyclic variability. The output time series is not obviously affected by the cycling of the sediment feed (Fig. 2B), and the frequency spectrum indicates that the cyclic input signal is not visible in the output. Note that this is not a 'masking' effect; the input signal has been destroyed. The mechanism is similar to that observed in the experiments with high-frequency forcing in the braided-river model: the input cycles become entrained in the system's own internal variability, in this case the formation of avalanches of various sizes, and the material soon becomes smeared or diffused among many different transport events. In a braided stream this would correspond to cyclically added sediment being sliced and resliced as bars are formed and then incised by shifts in flow path.

From an engineering point of view, the implication of these observations is that, in braided rivers, well defined cycles of sediment flux may be absorbed into the natural variability of the river and produce no effects downstream, even over relatively short distances. From a stratigraphic point of view, the implication is that there are categories of high-frequency signals that might be imposed on sedimentary systems that would not leave any sedimentary record at all—the signal is destroyed and cannot be recovered. An important goal for continued research in this area is the delineation of the full range of systems that show this 'destructive interference', between input signals and autocyclic fluctuations, and a better understanding of the magnitudes and time scales of autocyclic transport fluctuations.

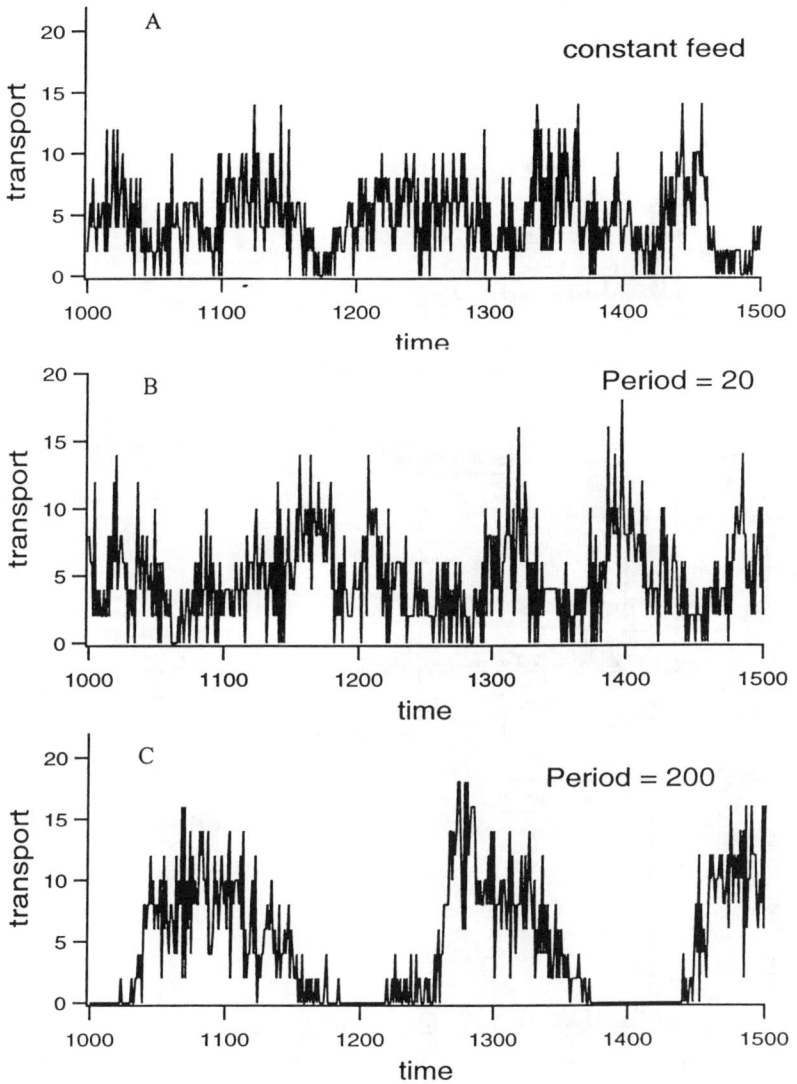

Figure 2. Time series of sediment transport generated by the sandpile model with imposed sediment feed. In the top panel, the feed rate is constant. In the middel panel, the feed rate is turned on and off with a period of 20 time steps. In the bottom panel, the feed period is 200 time steps. The mean feed rate is the same in all three panels.

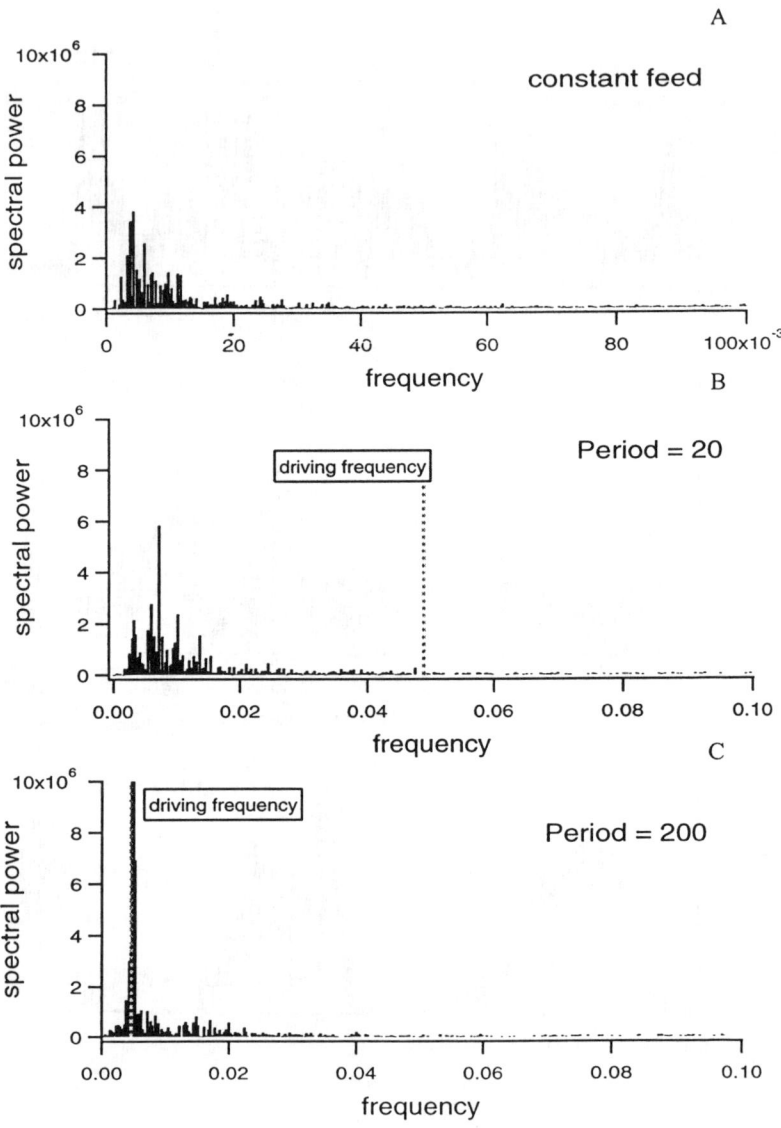

Figure 3. Frequency spectra corresponding to the three time series shown in Figure 2. Note that for the high-frequency case there is no signal at the driving frequency.

BIBLIOGRAPHY

Ashmore, P. (1991). "Channel morphology and bed load pulses in braided, gravel-bed streams." *Geograf. Annal.*, 73A, 37-52.

Ashmore, P. E. (1985). "Process and form in gravel braided streams: laboratory modelling and field observations," Ph. D., University of Alberta.

Bak, P., Tang, C., and Wiesenfeld, K. (1987). "Self-organized criticality: an explanation for $1/f$ noise." *Phys. Rev. Lett.*, 59(4), 381-384.

Goff, J. R., and Ashmore, P. (1994). "Gravel transport and morphological change in braided Sunwapta River, Alberta, Canada." *Earth Surf. Proc. Landforms*, 19, 195-212.

Hoey, T. (1992). "Temporal variations in bedload transport rates and sediment storage in gravel-bed rivers." *Prog. Phys. Geogr.*, 16, 319-338.

Hoey, T. B., and Sutherland, A. J. (1991). "Channel morphology and bedload pulses in braided rivers: a laboratory study." *Earth Surf. Proc. Landforms*, 16, 447-462.

Kadanoff, L. P., Nagel, S. R., Wu, L., and Zhou, S.-M. (1989). "Scaling and universality in avalanches." *Phys. Rev. A*, 39(12), 6524-6537.

Murray, A. B., and Paola, C. (1994). "A cellular model of braided rivers." *Nature*, 371, 54-57.

Murray, A. B., and Paola, C. (in press). "Properties of a cellular braided stream model." *Earth Surf. Proc. Landf.*

Sapozhnikov, V. B., and Foufoula-Georgiou, E. (in press). "Experimental evidence of dynamic scaling and indications of self-organized criticality in braided rivers." *Water Res. Res.*

FLOOD WATER HARVEST

TIMOTHY L. BUCHANAN
RICHARD H. FRENCH
STEVE A. MIZELL
Water Resources Center
Las Vegas, Nevada, USA

ABSTRACT

The potential of using the flood control detention basins in the Las Vegas Valley to harvest flood water for groundwater recharge and supplementing the potable water supply is being investigated. The initial step in this study was to develop a creditable method of estimating the runoff from the watersheds tributary to the basins.

INTRODUCTION

Flooding in the Las Vegas Valley, Nevada is episodic with long periods of precipitation insufficient to cause runoff separated by short periods with sufficient precipitation to cause significant flows. It is hypothesized that flood flows to the detention basins in or on the perimeter of the Valley (Figure 1) could be used to recharge the groundwater system and supplement the potable water supply of Valley. In June 1996, the Las Vegas Valley Water District initiated a two year project to investigate the technical feasibility of harvesting flood waters from the detention basins for groundwater recharge. The initial focus of the research has been directed to identifying basins which could be used from the viewpoint of in-flow water quality and developing a model to estimate the quantity of water potentially available for harvest.

WATER QUALITY CONSIDERATIONS

While the Clark County Regional Flood Control District monitors the quality of flood waters at several downstream discharge points (Figure 1), there are limited water quality data related to flood flows within the Valley, in general, or into the detention basins, in specific. Therefore, in eliminating detention basins from further consideration because of water quality concerns a qualitative approach was used. That is, from the viewpoint of water quality, the detention basins in the Las Vegas Valley can be categorized as either internal or external to development, Figure 1. Note, in Figure 1, some of the small basins on the east side of the Valley are not shown. The internal basins either receive runoff primarily from developed areas or areas that will likely be developed in the near future. In contrast, the external basins

FLOOD WATER HARVEST 59

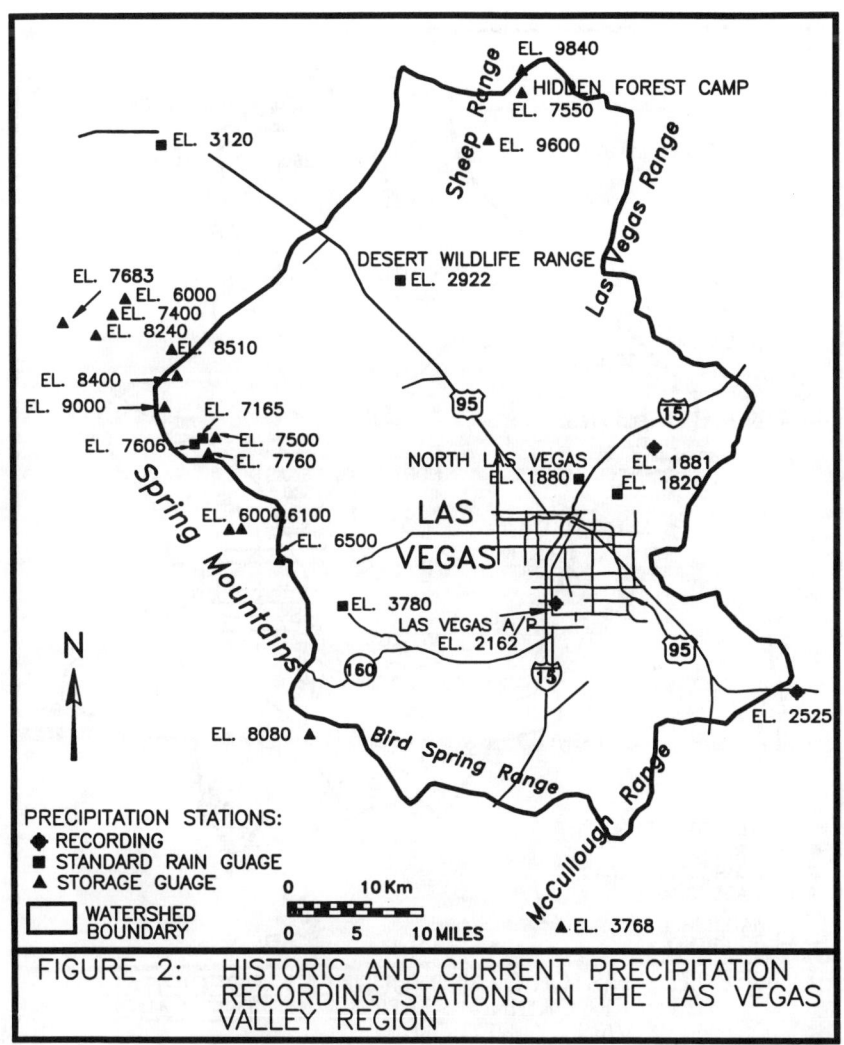

FIGURE 2: HISTORIC AND CURRENT PRECIPITATION RECORDING STATIONS IN THE LAS VEGAS VALLEY REGION

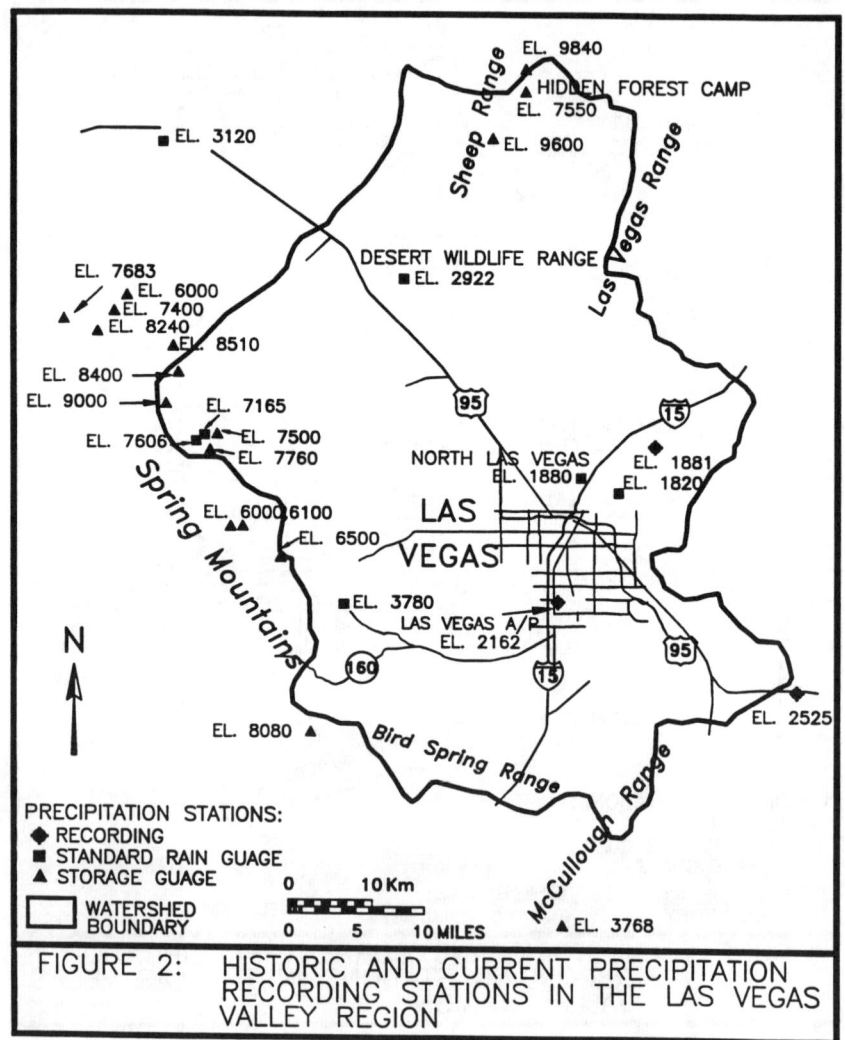

FIGURE 2: HISTORIC AND CURRENT PRECIPITATION RECORDING STATIONS IN THE LAS VEGAS VALLEY REGION

FLOOD WATER HARVEST 61

primarily receive runoff from undeveloped areas that, because of the ownership and land use, will remain undeveloped for the foreseeable future. For example, although some developed areas are tributary to the Red Rock Detention Basin, most of the tributary drainage basin is within the boundaries of the Red Rock Recreation Area and not available for development. In Figure 1, the internal and external basins are identified. All internal basins were eliminated from further consideration because of concerns regarding the quality of the inflowing water.

QUANTITY CONSIDERATIONS

A number of approaches to estimating the potential volume of water that may inflow to the detention basins were considered. Among the criteria used to evaluate these approaches were the following. First, the approach adapted must take into account the limited rainfall and runoff data available. Typical of the recently settled and developed arid southwestern United States, rainfall data are plentiful in comparison to runoff data; for example, there are 58 years of daily precipitation data at the Las Vegas McCarran Airport, and there are virtually no runoff data. Second, the approach used should focus attention on the results and their analysis rather than the approach itself and the assumptions on which the approach is based. Third, the approach should be conservative. That is, a critical factor in evaluating the viability of the hypothesis is economics - the cost of the water harvested relative to the cost of obtaining water from other sources. Therefore, an approach that over estimates the volume of water available for harvest is not appropriate. Fourth, the amount of runoff from a precipitation event is a primary variable and the runoff hydrograph is a secondary variable.

Given the foregoing criteria and the data available, it was decided to use the 58-year record of precipitation at the McCarran Airport gaging station as a base to estimate precipitation in the watersheds above the detention basins. The airport record contains not only extreme convective storm events; but also, frontal events having durations ranging from hours to multiple days. Other approaches, such as stochastic simulation, were considered and discarded because of the assumptions required for implementation and the possibility that the results obtained could not be demonstrated to be conservative.

With the airport precipitation data as a base, a method to translate this record to the ungaged watersheds above the basins was needed. In Figure 2, the locations of 26 additional precipitation records in and near the Las Vegas Valley are shown. With regard to these records, the following observations are pertinent. First, typical of the mountainous western United States, the long term precipitation records (such as the airport) are associated with low altitude (valley) stations and short, erratic records with the high altitude stations near the perimeter of the Valley watershed. Second, depths of precipitation at higher elevations are generally greater than those at low elevations (French, 1983; Osborn, 1984), and there is a strong correlation between the average annual and seasonal depth of precipitation and elevation. Third, Southern Nevada is an area where orographic effects and the moisture sources are important and result in areas of excess and deficit precipitation (French, 1983).

For each of the gaging locations shown in Figure 2, a regression relation was developed relating the annual depth of precipitation at the remote location to the annual depth of precipitation at McCarran airport using the coincident periods of record. In addition, regression relations relating annual precipitation to elevation were developed. With the foregoing relations defined, the procedure for estimating precipitation in an ungaged watershed above a detention basin is as follows. First, annual precipitation at two remote gaging locations in the vicinity of the basin are estimated using the regression relations between these locations and McCarran Airport. The elevations of the remote stations should bracket the elevation of the basin in question as tightly as possible. Second, the annual precipitation at the basin is estimated by factoring the precipitation at the two remote locations by precipitation depth -elevation gradient. Third, the first two steps of this procedure generally result in two estimates of precipitation in the watershed of interest and a single value is estimated using a standard inverse distance weighting scheme,

The validity of the methodology was tested by using it to generate a 48 year record of annual precipitation at the Desert Wildlife Range (DWR) station . The remote gaged stations used in this experiment were North Las Vegas and Hidden Forest Camp, Figure 2. The results of this comparison were as follows: 1) the actual annual precipitation at DWR was 4.20 (107 mm) in with a standard deviation of 2.27 in (57.6 mm) while the estimated annual precipitation was 4.19 in (106 mm) with a standard deviation of 1.96 in (49.8 mm); and 2) the standard error of estimate was 1.33 in (33.8 mm).

It is then assumed that the difference in annual precipitation between the various locations in the Valley are primarily due to differences in the amount of precipitation per event. That is, therefore a synthetic record of events in the watershed of interest can be developed by multiplying the depths of precipitation measured at the airport by the ratio of the annual precipitation in the watershed of interest to the annual precipitation at the airport.

The approach provides a point precipitation estimate that is then reduced by a depth-area-reduction factor based on the area of the watershed. Consideration also must be given to precipitation events that occur during the winter months (October through April) at the higher elevations. That is, during the winter months precipitation occurring at the higher elevations will be in the form of snow which will likely recharge the groundwater system rather than cause runoff to reach the detention basins. Therefore, it is assumed that precipitation occurring during the winter months at elevations greater than 7,500 ft will be in the form of snow and will not flow to the detention basins (Thomas *et al* 1994). This is a conservative assumption in that snowmelt likely contributes a small amount of runoff to the detention basins on the west side of the Valley.

Once the runoff to the basins is estimated, an operational model of the basin is used to estimate the volume of water that could be infiltrated while not altering the performance of the basin whose primary function is flood control.

CONCLUSION

The use of detention basins to harvest flood waters for groundwater recharge has the potential to supplement the potable water supply in the Las Vegas Valley. However, the potential volume of water available for harvest is not yet known. While the current study will provide an estimate of the volume there will be legal issues (water rights, for example) and economic issues that remain to be addressed.

ACKNOWLEDGMENT

The research on which this paper is based was sponsored by the Las Vegas Valley (Nevada) Water District. The opinions and interpretations discussed in the paper are not necessarily those of the sponsor.

REFERENCES

French, R.H., 1983. Precipitation in Southern Nevada. ASCE, *Journal of Hydraulic Engineering*, Vol. 109, No. 7, pp. 1023-1036.

Thomas, B., Thomas, E., Hjalmarson, H.W., and Waltemeyer, S.D., 1994. Methods of estimating magnitude and frequency of floods in the southwestern United States. Open File Report 93-419, U.S. Geological Survey, Tucson, Arizona.

Osborn, H.B., 1984. Estimating precipitation in mountainous regions. ASCE, *Journal of Hydraulic Engineering*, Vol. 110, No. 12, pp. 1859-1863.

RAINWATER COLLECTION AND UTILIZATION AS A POTENTIAL RESOURCE FOR URBAN AREAS

SAAD A. ALGHARIANI
Alfateh University, Tripoli, Libya

ABSTRACT

The traditional water harvesting methods in North Africa and the Middle East have been used since ancient times for domestic purposes. The potential prospects of this clean and simple technology for satisfying part of the domestic water demands of urban areas in the dry regions is assessed and evaluated. The growing urban community of Tripoli, Libya, was selected for this purpose. The results obtained from this study indicate that considerable gains in water conservation can be achieved at reasonable cost. In addition to providing clean water for domestic uses in water scarcity urban centers in the dry regions the technology may be useful for the more humid areas in reducing storm drainage facilities and associated costs. It is concluded that rainwater collection and utilization offers a viable option to the other less competitive alternatives of water supply developments. The practice should be introduced and expanded wherever hydroclimatic conditions allow its potential success.

INTRODUCTION

Rainwater harvesting and utilization for domestic purposes has been a long cultural tradition in North Africa, the Middle East and several other parts of the world [1]. This practice has been based on the collection and direction of rainfall run-off from roofs of buildings and paved or compacted ground to be stored in storage tanks of different designs and structures for later uses. As a result of population growth and urban expansion the municipalities of the mega-cities in the developing countries of Africa, Asia, and Latin America are struggling hard to provide their citizens with drinking water of acceptable quality at a reasonable cost. In many cases in the dry areas the available water resources are either insufficient to meet the escalating urban water demands or of low quality to make them acceptable for human uses without expensive treatment. Rainfall collection and utilization may offer a viable option to augment available water supplies in areas of reasonable annual precipitation, especially where these areas lack comprehensive centrally operated networks of water distribution systems. When considered from an urban water resources management perspective, rainwater collection is useful both in the urban centers of the dry areas to provide drinking water and in the mega-cities of the humid areas to reduce storm runoff and provide some degree of protection against potential floods.

This paper presents and discusses the results of a field study on the assessment of the potential technical and economic feasibilities of introducing and expanding rainwater collection systems for domestic uses in urban areas. The city of Tripoli, Libya has been selected as a pilot project case study from which indicative criteria are derived and generalized to other urban areas of similar nature in Libya and possibly other parts of the world.

RAINWATER COLLECTION POTENTIAL

Libya is located in the heart of the North African Sahara desert. Seasonal winter rainfall is limited to a narrow strip along the Mediterranean coast and ranges from a high of more than 400 mm/yr in some few locations to less than 100 mm/yr. in other areas. Most of the urban communities including Tripoli, the largest city in the country, have been experiencing drinking water shortages for more than twenty years. Local groundwater aquifers are polluted by saline seawater intrusions, surface water resources are lacking and seawater desalting is expensive to install and maintain. Rainwater collection, however, is not uncommon among the local people. Many householders practice water harvesting technologies in several areas of the country, especially where connections to water distribution systems are not possible. Most of the people in towns and rural villages collect and store rainwater run-off in cisterns to satisfy up to 70% of their domestic water demands. It is believed that a much greater potential for rainwater collection and utilization from building roofs is possible, especially if this practice is considered in the design of new housing and service structures. It is difficult, however, to remodel existing structures to fit the new practice. In addition, rainwater run-off from paved streets, highways, parking lots and other impervious areas in commercial districts and industrial complexes can be collected and beneficially used with minimum treatment. Collected rainwater from all sources is assumed to be used for secondary purposes augmenting the basic supply from present water distribution systems.

ESTIMATED RAINWATER COLLECTION FOR THE TRIPOLI AREA

Rainfall measurements for four locations representing the total Tripoli urban area during the last 50 years were collected and analyzed. Total annual rainfall averaged 286 mm/yr, distributed over six months of the rainy season. Four sites of concrete pavements were selected to measure rainwater runoff from single rain storms. Run-off coefficients ranged from 30-95% depending on the time of rainfall season and storm intensity and duration. The weighted time and storm average of the runoff coefficient corresponding to the average annual rainfall for three years has been estimated at 80%. This value is similar to other values used in the literature for some areas in the Middle East [2].

A detailed survey revealed that out of 79.5 square kilometers total Tripoli urban area, 47.7 square kilometers are roofed and potentially usable for rainwater collection. The annual rainfall runoff collection yield is calculated from the equation.

$$Y = C.A.R$$

in which Y is the collected run-off volume in cubic meters per year ; C is the runoff coefficient; A is the collection area in square meters and R is the average annual rainfall in millimeters.

RESULTS AND DISCUSSION

Table 1 summarises the results of the calculations for the year 1995. These results indicate that it is possible to collect and store runoff volumes close to 11 million cubic meters of rainwater per year.(MCM/Y). This amount is readily available for domestic uses with minimal treatment. The corresponding population of the Tripoli urban area for 1995 was estimated at 1.25 million people, averaging 7 persons per household. Thus the average per household share of the potentially available rainwater is around 61 cubic meters per year, or 24 liters per caput per day, which is more than sufficient to meet the per caput daily requirements of 7 liters for drinking and cooking purposes. The collected high quality rainwater can be mixed with the available lower quality waters for washing and other uses. The potential collection of storm runoff from roads, parking lots and other paved areas are estimated at 4.9 MCM/Y. With some degree of treatment this volume can be used to meet the irrigation requirements of parks and gardens or may be usable for some industries.

Table 1. Calculated rainwater collection yields from roofed public and residential buildings and other paved roads, parking lots and impervious areas.

Urban land use	Total area sq. km	Roofed area sq. km.	Runoff yield MCM/Y	m^3 /household per year
Residential areas	64.13	37.30	8.53	47.6
Public buildings	15.36	10.39	2.38	13.3
Roads, parking lots, other paved areas	31.80	-	4.9*	-
Total	111.29	47.69	15.81	60.9

*Unsuitable for direct domestic uses without substantial treatment

In addition to direct usage, the potential rainwater collection will significantly reduce storm runoff volumes. From table 1 the potential reductions in storm drainage from the roofed area of 47.7 km² amount to 80% of its expected volume from the roofed buildings and 55% from the paved open areas. The collection systems also provide a buffering storage effect reducing peak flow storm drainage. Thus storm drainage networks can be reduced to less than their present size with appreciable saving in construction and maintenance costs. It has been roughly estimated [3] that the savings in storm drainage costs can cover more than 50% of the construction costs of a comprehensive rainwater collectin system in the Tripoli urban area.

TECHNICAL AND ECONOMIC FEASIBILITIES

The collection and storage facilities can be constructed of both renforced concrete fiber glass and/or steel, erected either above or below groundsurface. Renforced concrete storage tanks of 60-80 m^3 capacity has been more acceptable with the local householders. Their cost of construction including fittings, pumping and maintainance for an operational life expectancy of 50 years gives an average unit cost of water close to 0.25 US Dollars per cubic meter. This is highly competitive with sea water desalting which costs no less than 3-5 US Dollars per cubic meter under the prevailing working conditions in Libya. It is even cheaper and of much better quality than the only other available alternative of the Libyan Man-made River water estimated to cost up to 0.83 US Dollars per cubic meter [4].

EXPANDING THE CONCEPT

The water shortage problems of the Tripoli urban area is not unique in Libya. Almost all the Libyan cities and urban communities falling within the 150 - 400 rainfall isohyets experience similar water problems. These urban areas share a more or less common climatic, architectural, cultural and socioeconomic features. Assuming a rainwater collection potential similar to that of the Tripoli area their total population of more than 3 million inhabitants can make use of up to 40 MCM/Y of excellent quality rainwater to meet their domestic needs.

The concept and practice can and should be introduced and expanded in the water scarcity areas of the arid zone as these areas are expected to cope with increasing water shortages in the future under conditions of escalating costs of water development projects and expanding population growth and urbanization. The countries of the humid zone may also benefit from this practice in reducing storm drainage and floods as well as in obtaining clean pollution-free drinking water.

SUMMARY AND CONCLUSIONS

The urban areas of the arid zone countries which are deprived of perennial rivers and largely dependent on meager nonrenewable groundwater supplies continue to experience an ever increasing water shortage problems and water quality deterioration. Seawater desalting facilities are expensive to install and maintain. The results of this investigation indicates that rainwater runoff collection and storage from building roofs and paved ground in the Tripoli urban area can provide considerable amounts of good quality water for domestic uses at a reasonable cost. Collection and storage facilities should be included in the design and construction of new buildings in the urban areas under similar conditions. Old buildings can be retrofitted by these facilities at a comparatively low cost. Technical guidance and appropriate water conservation regulations should be provided in this respect. Rainwater collection practices are highly promising and should be extended and expanded to cover all urban communities in Libya and other countries of similar hydroclimatic conditions in the Middle East and North Africa. Urban areas in the more humid countries may benefit from this practice through reducing storm runoff and associated costs of storm drainage facilities .

ACKNOWLEDGEMENTS

The basic information and data used for writing this article were gathered by the author and two of his senior students under a special study program. The assistance provided by the Tripoli area municipality and its personnel is gratefully acknowledged. The opinions and views presented here are the sole responsibility of the author and the cited references.

REFERENCES

[1] Frasier, G.W. Water Harvesting/Runoff Farming Systems for Agricultural Production. Proceedings of the FAO Expert Consultation on Water Harvesting for Improved Agricultural Production, Cairo, Egypt, November 1993, pp. 57-71.

[2] Preul, H.C. Rainfall-runoff Water Harvesting Prospects for Greater Amman and Jordan. Water International, Vol.19,No.2. June 1994, pp.82-85.

[3] Alghariani, S.A. Comparative Costs of Rainwater Harvesting Facilities and Storm Drainage Networks. Unpublished Report, Department of Soil and Water, College of Agriculture, Alfateh University, Tripoli, Libya, July 1995, 35 pages.

[4] Algharian, S.A. Man-Made Rivers: A New Approach to Water Resources Development in the Dry Areas. Proceedings of the 1st International Conference on (Rivertech'96: New/Emerging Concepts for Rivers), Chicago, Illinois, USA, September 22-26,1996.

Theoretical Analysis of air-entrained flow in vertical drop shafts of the channel in urban drainage system

KUNIHIRO OGIHARA

Dr. & Prof. of Civil Engineering, Toyo University
2100 Kujirai, Kawagoe, Saitama JAPAN 350

TADASHI KUDOU

Ms. of Civil Engineering, Toyo University
2100 Kujirai, Kawagoe, Saitama JAPAN 350

In the vertical drop shaft of urban drainage system, water flows with spiral motion and air is entrained. This phenomenon is very interesting in the similitude of motions between air and water such as air is pulled into the upper part of drop shaft. There is no scale similarity between water and air, because the relation of water and air is the same condition as surface tension, density and viscosity in model and prototype. Theoretical analysis and a few consideration on the model test are shown in this paper.

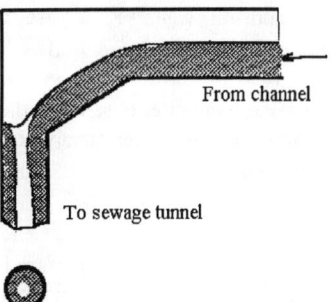

Fig.1 General View of Channel

THEORETICAL ANALYSIS

BASIC EQUATIONS
The water flows into the vertical drop shaft by spiral motion and air is entrained by the friction between water and air boundary. So coordinate system, velocity and friction forces are taken as in Fig.2 and internal friction in the water is neglected compared with the boundary friction between air and pipe wall. And the velocity is uniform in water and the velocity of the rotated direction is not so large compared with the vertical

$$w\frac{\partial v}{\partial z} = -\frac{1}{\rho}\frac{1}{r}\frac{\partial p}{\partial \theta} - \frac{1}{\rho}\frac{\tau_{w\theta} + \tau_{a\theta}}{t}, \quad w\frac{\partial w}{\partial z} = g - \frac{1}{\rho}\frac{\partial p}{\partial z} - \frac{1}{\rho}\frac{\tau_{wz} + \tau_{az}}{t} \quad (1)$$

direction. Under these assumptions the equation of motion and continuity equation are written as Equation (1) and (2).

$$\frac{\partial(wt)}{\partial z} = 0, \quad wt = q \qquad (2)$$

The friction forces between air and wall are given as follows.

$$\tau_{w\theta} = f_w \frac{\rho v^2}{2}, \tau_{a\theta} = f_a \frac{\rho v^2}{2}$$

$$\tau_{wz} = f_w \frac{\rho w^2}{2}, \tau_{az} = f_a \frac{\rho w^2}{2} \qquad (3)$$

Here $\tau_{w\theta}, \tau_{wz}$ are friction forces at wall for circulated and vertical direction respectively. And $\tau_{a\theta}, \tau_{az}$ are friction forces at boundary between air and water for circular and vertical directions respectively. ρ_a, ρ are densities of air and water respectively. f_a, f_w are: coefficient of friction on boundary between air and water and on wall. w_a, w are vertical velocity of air and water. And v is velocity of circular water flow.

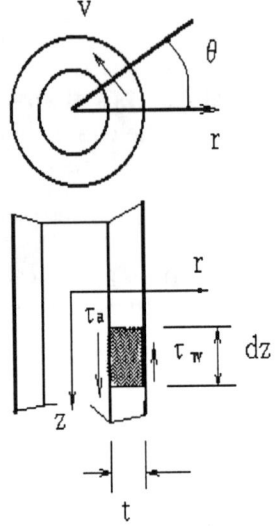

Fig.2 Coordinate System

The pressure in water is same as the pressure in air and so it becomes constant in the horizontal section. After steady condition is assumed, the equations for pressure are written as follows.

$$\frac{\partial p}{\partial \theta} = 0, \quad \frac{\partial p}{\partial z} = -\rho g I \qquad (4)$$

Here term I is gradient of pressure head and g is acceleration of gravity. The effect of gravity is not taken into account in air flow. So the equations of motion (1) are written as equation (5).

$$w\frac{\partial v}{\partial z} = -\frac{1}{\rho}\frac{\tau_{w\theta} + \tau_{a\theta}}{t}, \quad w\frac{\partial w}{\partial z} = g(1+I) - \frac{1}{\rho}\frac{\tau_{wz} + \tau_{az}}{t} \qquad (5)$$

Now next assumption has been introduced that the motion of air is uniform and depends on both the gradient of pressure in vertical direction and the friction between the water boundary. So the basic equations of friction force for air flow and the equation of equivalence of friction force and pressure are given as follows.

$$\tau_{az} = f_a \frac{\rho w_a^2}{2}, \quad \frac{\partial p}{\partial z}(r-t) = -2\tau_{az} \qquad (6)$$

As the pressure gradient of the second equation is related to the second equation of Eq.(4), then the second equation in (6) becomes equation (7).

$$\rho g I (r - t) = 2\tau_{az} \qquad (7)$$

SOLUTION OF BASIC EQUATIONS

The solution for the first equation of (5) is derived by taking account the equations of the continuity and friction forces, for the initial velocity v_0 at $z=0$. This gives the velocity change of rotation in the vertical shaft.

$$v = \frac{v_0}{1+(f_w + f_a)\dfrac{v_0 z}{2q}} \tag{8}$$

Next to solve the second equation in Eq.(5), two parameters a and b are introduced and the new function shown as equation (9) is derived by the integration in this step.

$$a = g(1+I), \quad b = \frac{f_w + f_a}{2q}, \quad a' = \sqrt[3]{\frac{a}{b}} \tag{9}$$

$$Y\left(\frac{w}{a'}\right) = \left[\frac{2}{\sqrt{3}}\tan^{-1}\frac{2}{\sqrt{3}}\left(\frac{w}{a'}+\frac{1}{2}\right) + \ln\frac{\left(1+\dfrac{w}{a'}+\left(\dfrac{w}{a'}\right)^2\right)}{\left(\dfrac{w}{a'}-1\right)^2}\right] \tag{10}$$

The solution for boundary conditions $z=0, w=w_0$ is derived as follows.

$$z = \frac{1}{6a'b}\left[Y\left(\frac{w}{a'}\right) - Y\left(\frac{w_0}{a'}\right)\right], \quad a'b = \sqrt[3]{\frac{g(1+I)(f_w+f_a)^2}{4q^2}} \tag{11}$$

And from the limiting condition that the vertical velocity direction is constant, the final constant velocity w_e is given as equation (12). That limit velocity is given as the velocity at z is infinity, namely the function $Y(w/a')$ is infinity in Eq.(11) and this is derived $w/a'=1$ in equation (10).

$$w_e = \sqrt[3]{\frac{2g(1+I)q}{f_w + f_a}} \tag{12}$$

Another limiting condition can be determined by the variable of vertical velocity becomes zero in the second equation (5).

$$w_e = \sqrt{\frac{2g(1+I)t}{f_w + f_a}} \tag{13}$$

The solution for air motion is from the first equation of (6) and equation (7), is derived as equation (14).

$$w_a = \sqrt{\frac{\rho g I (r-t)}{\rho_a f_a}} \tag{14}$$

RELATION OF DISCHARGE

The total water discharge Q_w is calculated from equation (12) by the relations $q = w_e t$ and $Q_w = 2\pi(r-t)q$.

$$Q_w = 2\pi(r-t)q = 2\pi(r-t)t\sqrt{\frac{2g(1+I)t}{f_w+f_a}} \quad (15)$$

And total air discharge Qa is given from equation (14).

$$Q_a = \pi(r-t)^2 w_a = \pi(r-t)^2 \sqrt{\frac{\rho g I(r-t)}{\rho_a f_a}} \quad (16)$$

The ratio of air and water discharge is derived by equations (15) and (16).

$$\frac{Q_a}{Q_w} = \frac{(r-t)}{2t}\sqrt{\frac{1}{2}\frac{\rho}{\rho_a}\frac{I}{1+I}\frac{r-t}{t}\frac{(f_w+f_a)}{f_a}} \quad (17)$$

This relation can be derived as another expression by the equation of water discharge q in Eq. (15).

$$\frac{Q_a}{Q_w} = \pi(r-t)^2 \frac{w_a}{2\pi(r-t)q} = \frac{(r-t)}{2q}\sqrt{\frac{\rho g I(r-t)}{\rho_a f_a}} \quad (18)$$

LIMIT CONDITION OF RATIO OF DISCHARGE

When the water thickness is so small to the diameter of the pipe, the term (r-t) is written as (D/2). So two equations (17) and (18) are transformed as follows.

$$\frac{Q_a}{Q_w} = \frac{D}{8t}\sqrt{\frac{\rho}{\rho_a}\frac{I}{1+I}\frac{D}{t}\frac{(f_w+f_a)}{f_a}} \quad (19)$$

$$\frac{Q_a}{Q_w} = \frac{D}{4q}\sqrt{\frac{\rho g I D}{2\rho_a f_a}} \quad (20)$$

Now introduce the non dimensional parameter $\overline{Q_w}$ for water discharge by Froude scale as the first equation of Eq. (21), and the equation (20) is rewritten as equation (22).

$$\overline{Q_w} = \frac{Q_w}{\sqrt{D}^5\sqrt{g}}, \quad A = \frac{\pi}{4\sqrt{2}}\sqrt{\frac{\rho I}{\rho_a f_a}} \quad (21)$$

$$\frac{Q_a}{Q_w} = \frac{A}{\overline{Q_w}} \quad (22)$$

To compare this equation and equation (19), it can be understood that the difference is derived by the difference of the used parameters such as water discharge in equation (22) or the coefficient of friction in equation (19).

MODEL TEST

Basic model test has been done to take the data such as air discharge, water discharge, and pressure difference. The diameters of the used drop shaft pipe are three types as 80,100 and 120mm and the length of them are also three types as 100,200 and 300 cm. And the main purpose of the model test is to take the similarity of model on the air transport by water flow between in different model scale.

BASIC TENDENCY OF AIR DISCHARGE

The relations of air discharge and water discharge are shown in Fig. 3 in the case of diameter is 80 mm and pipe lengths are 100,200 and 300 cm. The measurements have been done three times in same case and mean value is shown in this graph. Those are almost same tendency in this graph, such as air discharge is increased by the increased water discharge and it becomes constant in high water discharge. And the effect of pipe length is to make increase the air discharge, but there is not small difference in the cases of 200 cm and 300 cm length.

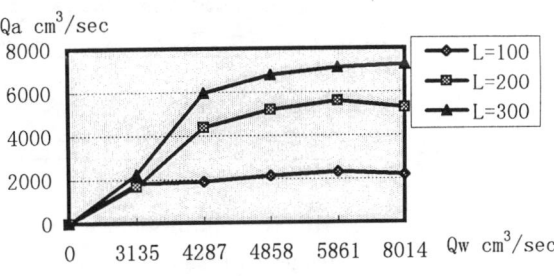

Fig. 3 Air discharge and water discharge

VELOCITY CHANGE OF WATER IN PIPE

One example of the velocity distribution of water in pipe are shown in the case of 80 mm pipe diameter is shown in Fig. 4. The numbers of horizontal axis are the measurement point number which is made 20 cm distance from inlet. Three curves shows the cases of different discharge of water as 4287,5861 and 8014 cm³/sec. The velocity tends to constant value toward the pipe end namely downward of pipe.

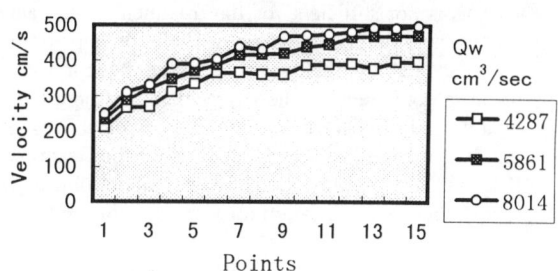

Fig.4 Velocity change toward down stream

CONSIDERATIO ON THE RESULTS OF MODEL TEST

First the parameter $(I+1)/(f_a+f_w)$ can be calculated from water discharge in unit width and final vertical velocity by Eq. (12). The result is shown

Q_w cm³/sec	4287	5861	8014	10958
d=80 mm	188.6	227.1	197.6	
d=100 mm	150.9	259.3	219.8	188.4

Tab.1 Parameter $(I+1)/(f_a+f_w)$

in Tab.1 and the value of parameter are almost same value in diameter 80 and 100 mm. But the values of parameter have peak value at discharge 5861 cm³/sec.
The second analysis is on the parameter I/f_a which is calculated from Eq. (22) by the ratio of air and water discharge and non-dimensional Froude discharge parameter.

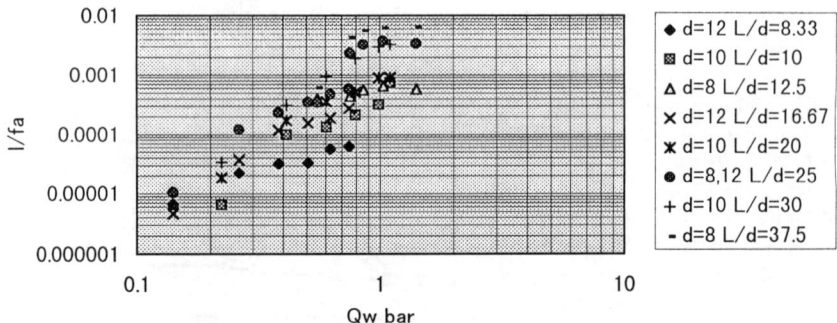

Fig.5 Parameter I/f_a and non-dimensional Froude discharge

The result is shown in Fig. 5 and the following tendency is derived. The parameter value increases with the non-dimensional Froude discharge number becomes larger and also the value L/d (ratio of pipe length and diameter) becomes larger. But it has very large difference as 100 times in the difference of cases L/d. It is not clear in this stage which parameter f_a or I is effect on this.

CONCLUDING REMARK

This analysis is not sufficient for the treatment of the air motion in drop shaft. This point makes the difference of value of I/fa in the pipe difference and the equation (22) might be changed by taking account the equation of motion to air flow. The experiment is continuing in the cases of longer pipe lines such as from 5m to 10m drops and it can be expected to bring the new data. This research has been made since 1994 and is continuing to March 1998 under the Grant-in-Aid for Scientific Research from The ministry of Education, Science and Culture (Monbushou) of Japan. The authors make much appreciation for these supporting body.

REFERENCES

(1) Anderson, A. G.,and Dahlin, W. Q. (1975). "Drop shafts for the tunnel and reservoir Plan." Project No. 154, St. Anthony Falls Hydraulic Laboratory, University of Minnesota, Mimeapolis, Minn.

(2) Ervine, D. A.,aid Kolknan, P. A. (1980). "Air entrainment and transport in closed conduit hydraulic structures." Report No. 5330, Delft Hydraulics Laboratory, Delft, The Netherlands.

(3) Jain, S. C. (1987). "Free-surface swirling flows in a vertical drop shaft." J. Hydr. Eng., ASCE, I 13(1O), 1277-1289.

(4) Jain, S. C., and Kenedy, J. F. (1983). "Vortex-flow drop structures for the Milwaukee Metropolitan Sewerage District inline storage system." IIHR Report No. 264, Iowa Institute of Hydraulic Research; The University of Iowa, Iowa City, Iowa.

(5) Laushey, L. M., and Mavis, F. T. (1953). "Air entrained by water flowing down vertical shafts." Proc. of 5th Congress, IAHR, Minneapolis. Minn. 483-487.

(6) Subhash C. Jain, "Air transport in vortex flow drop", Jr. of Hydraulic Engineering, ASCE, vol.114, no.12, December, 1988

(7) K. Ogihara and Tadashi Kudou, "Similitude on air-entrained flow in vertical drop shafts in the channel in urban drainage system", APD-IAHR Congress, 1996,9

OPTIMUM DESIGN OF SELF-REGULATING SPRING STEEL THROTTLES FOR SEWER OVERFLOW TANKS

DETLEF AIGNER and HANS-B. HORLACHER
member of IAHR
Mailing address: Technische Universität Dresden
D-01062 Dresden, Germany
Tel.: 049 351 4634397

ABSTRACT
In this paper a general equation was depeloped in order to calculate velocity coefficients for sharp-edged gates in a wide range for water level and angle variation. One important application of this equation was to design a fexible spring-steel throttles which can be used in sewer retention tanks. These devices offer ecological and economical advantages. The outflow is nearly constant and independent of variable water levels.

INTRODUCTION
Today, there are about 10 000 combined sewer overflow tanks in the Federal Republic of Germany. In order to achieve a country wide treatment of rainwater, the number of sewer overflow tanks should be approximately doubled. About half of all sewer overflow tanks are so-called clarifier tanks, which have a clarifier outlet. When it rains, solids of the waste water should temporarily settle in the clarifier tanks. After the rain, the sludge at the bottom of the tanks will be washed away to the waste water treatment plant by flushing boxes or other cleaning facilities.
In order that good settling effects are guaranteed in the clarifier tanks, the dis-charge and flow velocity are restricted, in a similar way to sewage tanks on waste water treatment plants.
The average flow velocity must not exceed 5 cm/s. Such tanks achieve a settlement of solids up to 80%, which is a significant degree of effectiveness. These overflow tanks can be equipped with a newly developed self-regulating outlet slit. This device keeps the overflow nearly constant by variable waterlevels in the tank. Due to this, the tanks can be used more effectively and decisive advantage are achieved from an ecological view point because there is less untreaded waste water released in the river. For an optimum design of these outlet slits discharge characteristics have to be investigated.
For sharp-edged gates (figure 1) or for outlet openings (s. figure 2) the discharge per unit width can be calculated by taking into consideration the jet contraction ψ, which is an essential component of the velocity coefficient μ.

$$q = \mu \cdot a \cdot \sqrt{2 \cdot g \cdot h_o} \qquad (1)$$

For many cases energy losses (due to the accelerating flow) and inflow velocity (at $a/h_0 \cong 0$) are negligible, so that velocity coefficients can be directly regarded as coefficients of contraction ($\mu = \psi$).

The scientific investigation of self-regulating outflow structures (figure 2), the so-called spring-steel throttle (Aigner, Horlacher 1995; Brombach, Horlacher 1994), led to the problem of determining velocity coefficients de-

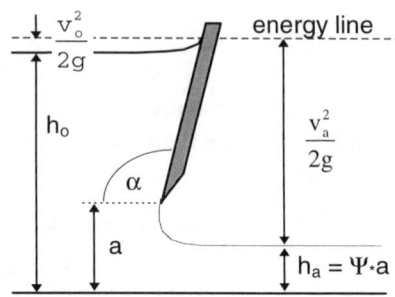

Figure 1: System figure of the gate

pendenting on the angle between the tip of the steel sheet and the bottom of this outlet construction. Since the tip of the steel sheet changes its angle with increasing water levels, a simple formula for this flow situation has to be developed.

The coefficients of contraction, which were theoretically and empirically determined for the discharge of gates, mostly end at a vertical position of the gate. For spring steel sheet throttles however, angles in the range of $\alpha = 0°...180°$ were of interest. Therefore, both theoretical and experimental investigations were made for determining velocity and contraction coefficients.

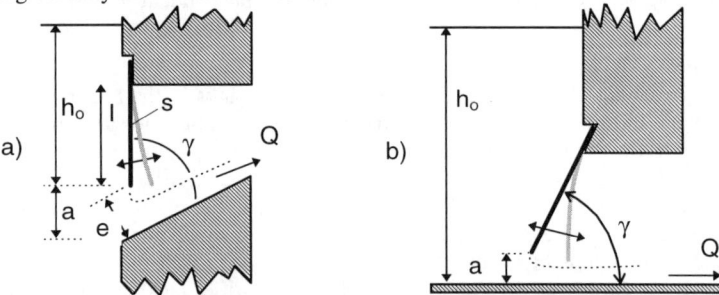

Figure 2: Principal design of self regulating outlet throttles, installation in the tank wall a) or at the bottom b)

In the following report first findings concering velocity coefficients were examined and supplemented with own measurements. Then, on the basis of these results a practical formula for velocity coefficients was set up so that flexible spring steel throttles can be designed.

THEORETICAL CONSIDERATIONS

The application of kinematic relationships to determine contraction coefficients for gates according to the theory of streamlines after Werner (1963) offers the possiblity of calculating coefficients dependent on geometrical quantities and physical simplifications (neglecting friction and gravity). Investigations made by Franke (1956) showed that the influence of gravity on the jet is low, especially at openings of $a < h_0/3$. The comparison of measurements carried out by Gentilini (1941) with theoretical results shows larger deviations only from $h_0 < 3a$.

Werner (1963) derived the boundary condition of a free jet under a gate as a problem of plane potential flow (mixed boundary value problem). On the basis of these derivations an algorithm could be developed to determine theoretically coefficients of contraction with respect to different flow situations. Then velocity coefficients of gates and spring steel throttles for a range of angles of $0° < \alpha < 180°$ and $h_0/a \geq 1$ can be calculated.

Following assumptions were made:
- Continuity equation:
$$v_0/v_a = h_a / h_0 \tag{2}$$
- Series expansion and cyclotomy with m and n as positive integers, where:
$$\frac{180}{\alpha} = \frac{m}{n} \tag{3}$$
- Relative water level:
$$h = h_a/h_0 \tag{4}$$

The complex flow vector $Z = z/h_a$, referred to the quantity h_a, results as:
$$Z_{n,h} = X_{n,h} - i \cdot Y_{n,h} \tag{5}$$

The integration constants can be derived from the boundary conditions for the X-value becoming $IK_x = 0$ and the Y-value becoming $IK_y = 1 + Y_{0,h}$ for $\alpha = 90°$ and $\alpha = 0°$, respectively.

$$X_{n,h} = \frac{1}{\pi} \sum_{k=0}^{m-1} X(n,h,k) \quad \text{and} \quad Y_{n,h} = 1 + \frac{1}{\pi} \sum_{k=0}^{m-1} Y(0,h,k) - \frac{1}{\pi} \sum_{k=0}^{m-1} Y(n,h,k)$$

$$X(n,h,k) = \cos(\frac{2\pi k n}{m}) \cdot \left[h \cdot \ln|N1_{k,n}| + \frac{1}{h} \cdot \ln|N2_{k,n}| - 2 \cdot \ln|N3_k| \right] +$$
$$+ \sin(\frac{2\pi k n}{m}) \cdot \left[h \cdot \arg(N1_{k,n}) + \frac{1}{h} \cdot \arg(N2_{k,n}) - 2 \cdot \arg(N3_k) \right] \tag{6}$$

$$Y(n,h,k) = \cos(\frac{2\pi k n}{m}) \cdot \left[h \cdot \arg(N1_{k,n}) + \frac{1}{h} \cdot \arg(N2_{k,n}) - 2 \cdot \arg(N3_k) \right] +$$
$$+ \sin(\frac{2\pi k n}{m}) \cdot \left[h \cdot \ln|N1_{k,n}| + \frac{1}{h} \cdot \ln|N2_{k,n}| - 2 \cdot \ln|N3_k| \right]$$

$$N1_{k,n} = \cos(\frac{\pi}{m}) - h^{-\frac{1}{n}} \cdot \cos(\frac{2\pi k}{m}) + i \cdot \left[-\sin(\frac{\pi}{m}) - h^{-\frac{1}{n}} \cdot \sin(\frac{2\pi k}{m}) \right]$$

$$N2_{k,n} = \cos(\frac{\pi}{m}) - h^{\frac{1}{n}} \cdot \cos(\frac{2\pi k}{m}) + i \cdot \left[-\sin(\frac{\pi}{m}) - h^{\frac{1}{n}} \cdot \sin(\frac{2\pi k}{m}) \right]$$

$$N3_k = \cos(\frac{\pi}{m}) - \cos(\frac{2\pi k}{m}) + i \cdot \left[-\sin(\frac{\pi}{m}) - \sin(\frac{2\pi k}{m}) \right]$$

The diagrams in figures 3 show coefficients of contraction $\psi = 1/Y_{n,h}$, determined by equation (6). For $a/h_0 = 0$ the discharge coefficients μ are identical to the coefficients of contraction ψ_0 and dependent only on the angle α. For this case, errors due to neglecting of gravitational acceleration or of energy losses are minimal. These values can be used for practical applications (Pulina/Voigt 1990). The practical use of equation (6) is difficult, an approximate solution can to be derived, for $a/h_0 \cong 0$.

$$\psi_0 = 1{,}3 - 0{,}8 \cdot \sqrt{1 - \left(\frac{\alpha - 205}{220}\right)^2} \quad \text{for } 0 \leq \alpha \leq 180°, \quad \frac{a}{h_0} \approx 0 \tag{7}$$

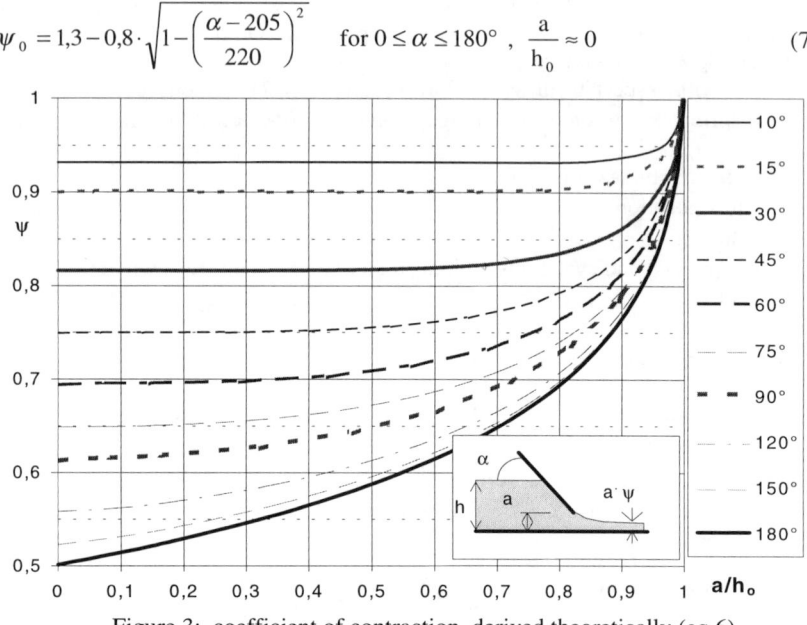

Figure 3: coefficient of contraction, derived theoretically (eq.6)

EVALUATION OF EXPERIMENTAL RESULTS AND EMPIRICAL EQUATIONS

Existing literature were evaluated and compared with own measurements within the framework of these investigations concerning velocity coefficients. The experiments were carried out in the Hubert Engels Laboratory of the Technical University of Dresden, in a 30-cm-wide glass flume especially for the sharp-edged gates for anges of 90° to 180° and slit widths between 25 to 100 mm. The water level was measured by an ultrasonic probe, and the flow by an inductive discharge meter (IDM). The measurements were very time-consuming, since they could only be performed under steady flow conditions. Above all it was attempted to document the range of smaller relative water levels h_0/a. These values are represented in Figure 4. The velocity coefficients were determined from equation(1).

The determination of the coefficients of contraction ψ were difficult and did not yield to any satisfactory results. In general, the measured coefficients of contraction were only a little lower than the theoretical ones and could be set up as relatively constant values over the whole water-level range. At smaller water levels vortices occurred temporarily, and at water levels close to the slit width an undulatory discharge occurred at upstream side.

The most comprehensive measurements known were carried out by Gentilini (1941) and were quoted in many textbooks (e. g. Miller 1994, Cozzo 1978). The measurements of Gentilini were performed with opening heights of 30 to 90 mm, which are within same range of our measurements. There for the values of Gentilini were used as basis of an empirical equation.

The influence of the angle on coefficients of contraction ψ_0 for $a/h_0 = 0$ was very well considered in equation (7). It therefore holds also for the velocity coefficient μ_0^*. If a/h_0 becomes greater then 0, both coefficients move in to different directions: ψ increasing and tending towards 1, and μ decreasing. Due to this influence an approximatie equation of Voigt (1971) was used to adapt the experimental results. Best agreements were obtained by equation (8), with ψ_0 according to equation (7).

$$\psi = \dfrac{1}{1 + \left(\dfrac{1}{\psi_0} - 1\right) \cdot \sqrt{1 - \left(\dfrac{a}{h_0}\right)^{\frac{210}{\alpha}}}} \qquad (8)$$

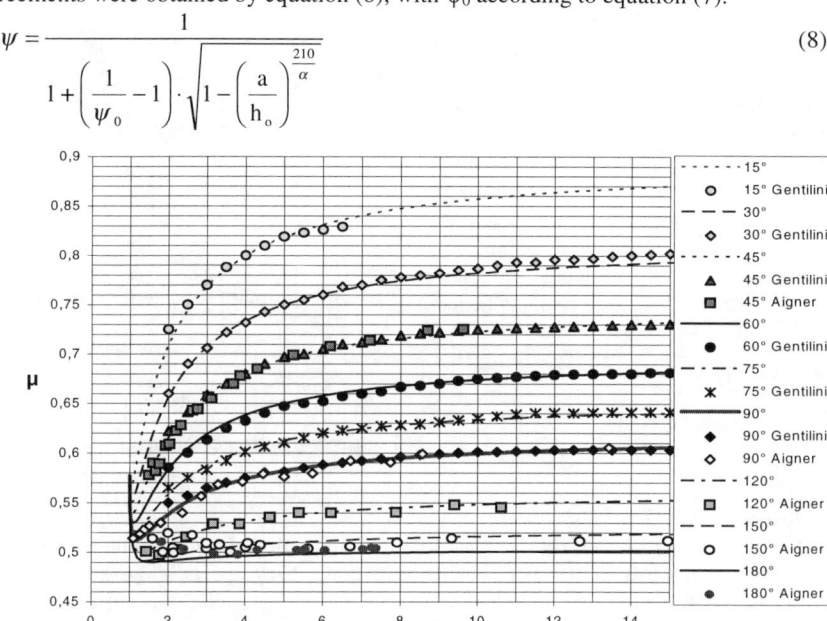

Figure 4: Equation (11) compared with experimental results for $\alpha = 15°$ bis $180°$

The velocity coefficients μ can be derived from an energy balance (s. figure 1), taking into account a head loss $h_v = h_{v1} + h_{v2}$.

$$\dfrac{v_0^2}{2g} + h_0 - h_{v1} = \dfrac{v_a^2}{2g} + \psi \cdot a + h_{v2} \qquad (9)$$

With the continuity $v_a \cdot \psi \cdot a = v_0 \cdot h_0$, equation (1) and relation (9) can be rearranged and results in:

$$\mu = \dfrac{\psi}{\sqrt{1 + \dfrac{\psi}{\dfrac{h_0}{a} - 0{,}5}}} \qquad (10)$$

This equation (10) is represented in figure 4 and compared with experimental results. Very good agreement with measurements was achieved.

APPLICATION

The main objectiv of the scientific investigation of a self-regulating outlet structure for rain-water retention tanks, the spring-steel throttle (figure 2), was to develop an equation to determine discharge coefficients as a function of the angle α and of the relative water level a/h_0.

The spring steel sheet with characteristic quantities of sheet thickness s, length l and angle γ has to be chosen and arranged in such a way that by bending with rising water levels an opening function was produced which approximately fulfils the following equation (11), so that the discharge remains nearly constant (s. figure 5).

$$a(\gamma, \frac{h_0}{a}, s, l) = \frac{q}{\mu\left(\gamma, \frac{h_0}{a}\right) \cdot \sqrt{2g \cdot h_0}} \quad (11)$$

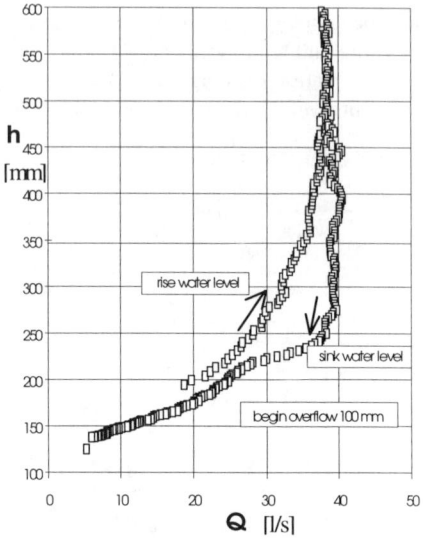

Figure 5: discharge characteristics of a self-regulating outlet slit (s.fig.2a)

REFERENCES

AIGNER,D.; HORLACHER, H.-B. : Self-regulating outlet slit in combined sewer overflow tanks. VI-th Conference, Hydroengineering. May 1996, Szldarska Poreba, Poland. BROMBACH, H.; HORLACHER, H.-B.: Selbregulierender Auslaufschlitz für Regenüberlaufbecken. Wasserwirtschaft, 86,(1994), pp.128-132 Cozzo, G.: Una formula per il calcoto del coefficiente défflusso delle luci sotto paratoie. L'ENERGIA ELETTRIUCA Nr. 11-12 1978 FRANKE,P.: Theoretische Betrachtung zur Strahlkontraktion beim Ausfluß unter Schützen. Die Bautechnik. 33.Jg, S.73-77, Heft 3-März 1956 Gentilini, B.: Effluso dalle luci soggiacenti alle paratoie piane inclinate e a settore, L'Energia Elettrica 1941,H.6, S.361 MILLER, D.S.: Discharge charakteristics. Hydraulic Structures Design Manuals. IAHR 8 1994 PULINA, B.; VOIGT, A.: Untersuchungen beim Umbau und Neubau von Wehranlagen an Bundeswasserstraßen. Mitt. der BfW, Karlsruhe 1990, Nr. 67 VOIGT, H.: Abflußberechnung gleichzeitig über- und unterströmter Stauelemente. Diss. TU Dresden, 1971 WERNER, W.: Ableitung einer kinematischen Beziehung zur Berechnung des Durchflusses unter Planschützen nach der Theorie freier Stromlinien. Wissenschaftliche Zeitschrift der TU Dresden. 12(1963) Heft 6 , S. 1693-1699

Infiltration of urban stormwater into soils as an Integral Part of the Urban Drainage System

ALINA NOWAKOWSKA-BLASZCZYK[1] and PAWEL BLASZCZYK[2]
[1]Department of Environmental Engineering
Warsaw University of Technology
[2]Institute of Environmental Protection
Warsaw, Poland

ABSTRACT

The procedures for designing urban stormwater infiltration in Poland are presented and were developed from studies of rainfall, runoff and infiltrated water quality, and infiltration characteristics of various soils found in the Warsaw area. The assessment of runoff quality in comparison to the regulations for discharge of sewage effluent indicated that runoff from most urban surfaces was of an acceptable quality and could be infiltrated directly into the soils. This was particularly true for roof runoff, and even some street runoff could be infiltrated after pretreatment by removal of suspended solids. The most severely contaminated street and parking lot runoff should be discharged into sewers and treated at the wastewater treatment plant.

INTRODUCTION

Before adopting design and construction practices for infiltration of urban stormwater into soils, the research on quantity and quality of rainfall, stormwater runoff and its infiltration should be performed in every country, to account for such specific conditions as the climate, housing standards and environmental laws and regulations. In Poland, until 1991, the law did not allow infiltration of stormwater runoff into the soils, because such runoff was regarded as wastewater. It was necessary to establish that the quality of stormwater runoff from various types of urban surfaces, such as roofs, pavements and parking lots, varies from poor to very good, and the less polluted stormwater may be infiltrated into the soils without adverse impacts on urban soils and ground waters.

STUDIES

During the period from 1986 to 1996, studies of stormwater infiltration into soils were undertaken in Poland in several steps (Nowakowska-Blaszczyk et al. 1992; Nowakowska-Blaszczyk et al. 1993), including the following:
- Investigations of the existing stormwater infiltration systems in Poland
- Literature reviews

- Laboratory and field investigations of infiltration
- Meteorological observations
- Research on quality and quantity of precipitation, storm runoff and filtrate.

Meteorological observations and sample collection were performed in the centre of the City of Warsaw, in the University area. The quality of urban stormwater runoff was studied for the following urban surfaces: (a) tar-paper roofs, (b) sheet-metal roofs, (c) clay-tile roofs, (d) busy downtown streets with parking along curbs, and (f) municipal parking lots.

Infiltration of stormwater runoff was studied for the following soils (Nowakowska-Blaszczyk et al. 1992; Nowakowska-Blaszczyk et al. 1993): (a) sand with good natural permeability, native to the Warsaw area, (b) thick sandy clay, with limited permeability, native to the Warsaw area, and (c) grass turf planted on topsoil at the University site.

All investigations were carried out systematically. In studies of infiltration, runoff from roofs covered with tar paper was used, because such roofs are typical for Poland. Complete chemical analyses were carried out for such constituents as metals and specific soluble organic compounds (Blaszczyk et al. 1992). To explain the degree and process of pollution of stormwater along its route to the receiving waters, the characterization of stormwater quality was based on a number of water quality indicators and compared with the quality requirements for the Class II rivers, and sewage effluents discharged to surface waters or disposed on soils. Those indicators were the following: suspended solids (SS), nitrates as N, phosphates as P, biochemical oxygen demand (BOD), chemical oxygen demand (COD), and lead, as shown in Fig. 1 (Nowakowska-Blaszczyk and Zakrzewski, 1996).

RESULTS

The comparison of stormwater pollution to the requirements for discharge of sewage effluents and stormwater to receiving waters indicates that the quality of roof runoff, for all the roofing materials studied, met such requirements. This finding allows to discharge roof runoff directly onto the soils or lawns, if there are sufficient permeable areas. The runoff from busy downtown streets should be pretreated to remove suspended solids prior to infiltration, but other constituent concentrations were below the required limits and did not require special treatment.

It is preferable to discharge runoff from very polluted streets and municipal car parks into the sewer system and convey it to the sewage treatment plant. Infiltration of runoff from tar-paper roofs into a natural sandy soil or grass on topsoil resulted in transport of some pollutants through the soils (BOD and COD; Pb only in the sandy soil). The effluent from a grassed area showed a decrease in the concentration of lead, because of some lead uptake by grass.

APPLICATION

In Warsaw, the infiltration systems are used mainly in the areas without storm sewer systems. The systems for infiltration of runoff from streets comprise underground trenches filled with gravel and perforated pipes serving for distribution of stormwater runoff. There are trenches under the road gutter (Fig. 2) and under the middle of the road (Fig. 3). Stormwater runoff from the roofs and footpaths is directed to the surface trenches filled with sand and stones, located in the middle of lawn strips between the rows of houses. In many houses, the discharge of roof runoff is reduced by installation of small detention reservoirs (large flower pots) below the gutter pipe outlet.

CONCLUSION

Comparisons of stormwater quality in drainage conveyance elements and the quality of stormwater runoff from roofs and even downtown streets indicate that surface runoff, especially from roofs, is considerably less polluted. The comparison of stormwater runoff quality with the requirements for discharge of wastewater effluents indicated that roof runoff could be locally infiltrated without any pretreatment. Runoff from downtown streets, with car parking on sides, should be pretreated (by removal of suspended solids) before infiltration into the ground. The runoff from very polluted parking lots and roads should be discharged directly into the conventional sewer system and conveyed to the wastewater treatment plant.

REFERENCES

NOWAKOWSKA-BLASZCZYK, A., ZAKRZEWSKI, J., CHUDZIK, B. and SITEK, M. (1993). Investigations (1987-1990 and 1992-1993) on Pollution of: Storm Water, Rainfall, Runoff and Infiltration to the Soil. Warsaw University of Technology (in Polish). R & D Central Programmes No. 11.10 and No.13.1. Ministry of Environmental Protection and Ministry of Housing, Warsaw, Poland.

NOWAKOWSKA-BLASZCZYK, A., ZAKRZEWSKI, J. CHUDZIK, B. and SITEK, M. (1992). Principles and Standards for Designing of Infiltration of Storm Water into Soil (in Polish). A manual prepared for the Ministry of Environmental Protection. Warsaw.

BLASZCZYK, P., NOWAKOWSKA-BLASZCZYK, A. and ZAKRZEWSKI, J.(1992). Infiltration of Storm Water into Soil. Investigations and Principles of Design. Proc. International Conference on Innovative Technology NOVATECH 92, GRAIE, Lyon, France, pp.447-454.

NOWAKOWSKA-BLASZCZYK, A. and ZAKRZEWSKI, J. (1996). The sources and phases of increase of pollution in runoff waters in route to receiving waters. In: Sieker, F. and H.-R. Verworn (Eds.), Proc. International Conference on Urban Storm Drainage, Hannover, Germany, Sept.9-13, 1996, pp.49-53.

COPING WITH SCARCITY AND ABUNDANCE

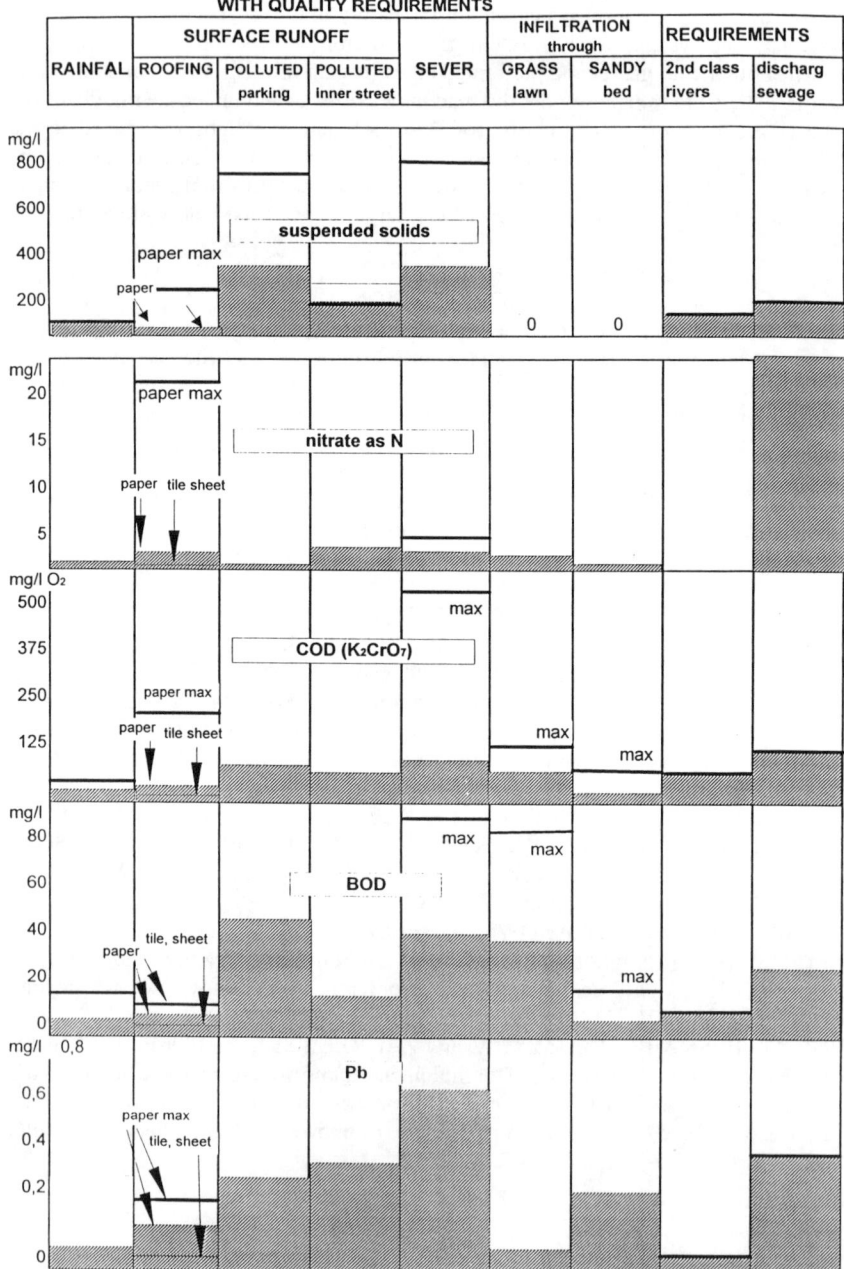

FIG. 1 THE POLLUTION OF RAINFALL, RUNOFF AND FILTRATE COMPARED WITH QUALITY REQUIREMENTS

INFILTRATION OF STORM-WATER INTO SOILS 85

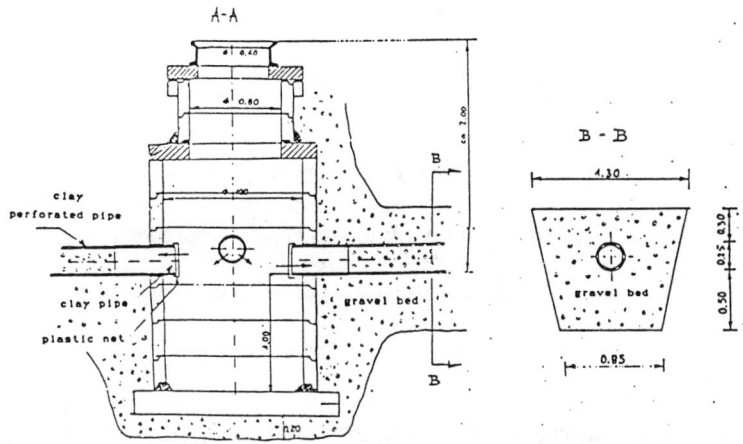

Fig. 2. Underground filtration system ("Florida)

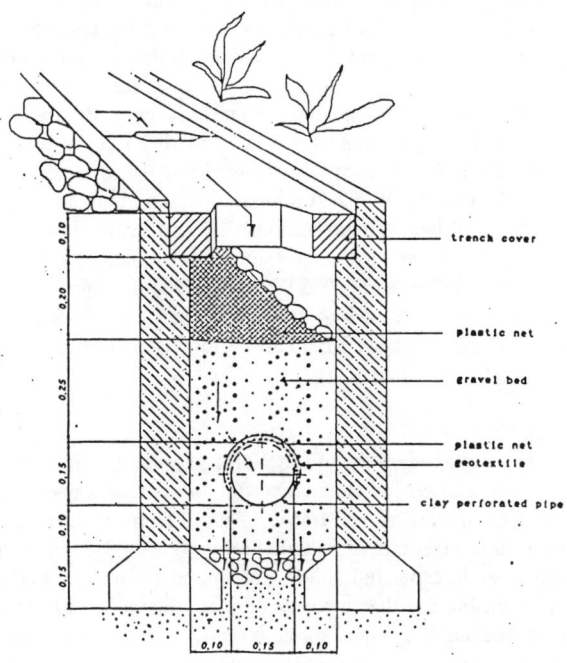

Fig.3. Trench under the gutter ("Wilga VII")

Infiltration as an urban source control for metal elements and solids

JOHN J. SANSALONE
University of Cincinnati
Department of Civil & Environmental Engineering
PO Box 210071
Cincinnati, OH 45221-0071 USA

ABSTRACT
Urban surface waters often contain high loads of metal elements, particulates and dissolved solids. Metal elements are a significant concern because they do not degrade in the environment and can exert severe acute and/or chronic stress on receiving waters and sediments. Numerous passive treatment strategies have been proposed to control anthropogenic constituents in urban surface waters. One novel and promising design combines the best attributes from porous pavement and infiltration trenches while functioning as a replacement for pavement underdrains. This design, known as a partial exfiltration trench (PET), intercepts and infiltrates lateral sheet runoff from pavement surfaces before it becomes concentrated gutter flow. The PET consists of an iron oxide coated sand (IOCS) media capped by a cementitious porous surface. Bench scale PET column experiments have been carried out to compare the metal element capacity of the IOCS against control media of silica sand. Metal elements investigated include Zn, Cd, Pb and Cu run separately and in mixed combination at pH levels of 6.5 and 8.0. When run in combination, Zn broke through first, closely followed by Cd; Cu and Pb were strongly retained in the columns. The IOCS capacity was greater than silica sand at a pH of 6.5 and was significantly improved when the infiltrating runoff pH was increased to 8.0 using the cementitious porous surface. Bench scale results indicate that the PET can perform satisfactorily for at least a ten year design life in a humid climate.

INTRODUCTION
By the year 2000 approximately 50% of the world population will live in urban areas. With this growth is an increase in vehicular traffic, urban roadways and pavement and pollutants generated from vehicles and traffic. Runoff from these impervious surfaces contains significant concentrations of soluble and particulate-bound metal elements in addition to dissolved and suspended solids. Nonpoint sources of metal element and solids are now responsible for the majority of metal element and solids loadings to receiving waters in the USA (Wanielista & Yousef, 1993). The highly impervious nature of these areas coupled with the practice of efficient surface drainage results in a corresponding flushing of pollutants from these areas during runoff events.

INFILTRATION AS URBAN SOURCE CONTROL

A variety of best management practices (BMPs) have already been used as passive control strategies for runoff from urban areas and pavement in the USA, Europe and Japan. In view of the nonpoint pollutant loads that originate in urban areas and highway corridors and the emerging NPDES stormwater criteria, it is likely that use of control strategies to control nonpoint pollution will grow in the future. The objective of these control strategies, for metal elements which are not degraded, is to immobilize and contain a diffuse, relatively mobile water pollutant before it exerts a toxicity impact on aquatic, terrestrial, animal or human life. Passive control strategies can be nominally classified as infiltration methods, detention/retention methods, wetlands or vegetated swales and buffers. Infiltration methods have included specific strategies including infiltration basins, infiltration trenches and porous pavement.

One novel and promising infiltration design, a PET, combines the best attributes from porous pavement and infiltration trenches while functioning as a replacement for pavement underdrains. This design intercepts and infiltrates lateral sheet runoff from pavement surfaces before it becomes concentrated gutter flow. The primary differences between the PET and current infiltration trench and underdrain designs is the installation of porous pavement directly above the trench to promote infiltration, and the iron-oxide coating of the granular backfill to enhance sorptive capacity. A PET also functions to intercept subgrade interflow which is significant for many roadways and pavement. Depending on the surrounding soil hydraulic conductivity the PET is capable of exfiltrating a significant amount of infiltrated runoff back to the surrounding soil. A schematic of a PET is presented in Figure 1. This paper examines the potential suitability of the PET for retaining metal elements, using bench scale column simulations.

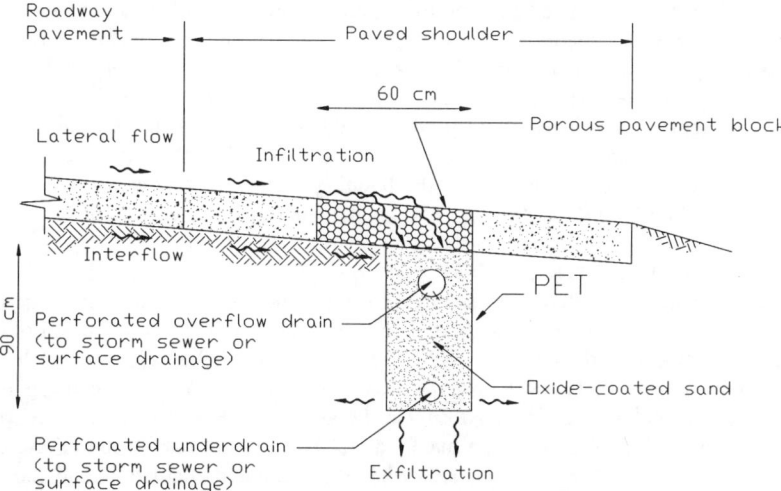

Figure 1. Schematic of a partial exfiltration trench (PET).

BACKGROUND
The geochemical transport and eventual fate of a majority of reactive substances, including infiltrated metal elements, is controlled by the reaction of these substances with solid surfaces. The solid-liquid interface of these solids, which include natural iron oxides and IOCS are a reservoir for reactive substances, which are characterized by large surface area to solid volume ratios (Stumm, 1992). In the presence of water these oxide surfaces are generally covered with surface hydroxyl groups, protons and coordinated water molecules. The formation of surface complexes of metal ions by hydrated iron oxides involves the coordination of the metal ions with surface hydroxyl oxygen atoms. A number of bonding reactions have been postulated to explain these results including monodentate bonding, bidentate bonding and hydrolysis-sorption (Benjamin & Leckie, 1981; Stumm, 1992).

The net surface charge of the hydrous iron oxide is strongly pH dependent and therefore the sorption of a metal ion to the hydrous iron oxide surface is strongly pH dependent. These mineral surfaces are amphoteric. The surface point of zero charge (PZC) in its most simple definition is the pH at which the net surface charge is zero. Therefore for iron oxides, which have a PZC between 7 and 8, a solution pH of 8 or greater would indicate the predominance of surface bonding sites would be negatively charged and available as metal ion bonding sites. On the other hand, a solution pH of less than 7 would indicate the predominance of surface bonding sites would be positively charged and not available as metal ion bonding sites.

Under conditions where a number of metal elements are present in solution the competitive order of sorption has been explained using covalent theory, electrostatics or the tendency of a metal element to undergo hydrolysis followed by sorption. According to covalent theory, for divalent metal ions, the order of bonding preference would be: Cu > Ni > Co > Pb > Cd > Zn > Mg > Sr. However, on the basis of electrostatics, the bonding preference would be for the metals with the greatest charge-to-radius ratio. This would produce a different order of preference for this group of divalent metals: Ni > Mg > Cu > Co > Zn > Cd > Sr > Pb (McBride, 1994). Based on the tendency to hydrolyze, the bonding preference of metal ions to iron oxides would be : Pb > Cu > Zn > Co > Ni > Cd > Mn (Dixon & Weed, 1989).

EXPERIMENTAL METHODOLOGY
The performance of a PET was simulated using bench scale column experiments. Acrylic columns were run under saturated conditions at two pH levels. The pH of 6.5 is representative of urban runoff pH. The pH of 8.0 is the level to which the cementitious porous surface will raise the urban runoff pH of 6.5. The metal elements utilized in each column run included Zn, Cd, Pb and Cu. These metal elements were chosen because Zn, Cd and Cu are mainly dissolved in urban rainfall-runoff and can exceed surface water discharge standards while Pb is mainly particulate-bound and can exceed surface water discharge standards for snowmelt runoff (Sansalone & Buchberger, 1997). A summary of the experimental configurations is given in Table 1.

Table 1
Summary of experimental column configurations

Column	I.D. (mm)	Packed length (mm)	Media	d_{50} (mm)	η^3	Flow rate [4] (mL/min)	EBCT [5] (min)	Influent pH
Control	38.1	612	silica sand [6]	0.5	0.35	50	4.4	6.5 & 8.0
IOCS[1]	38.1	612	IOCS [6]	0.52	0.37	50	4.5	6.5 & 8.0
BSPET[2]	38.1	612	IOCS & PP	0.5	0.39	50	4.6	6.5 & 8.0

[1]: Acid-washed Ottawa 2030 sand coated to dryness with 1.6 M $Fe(NO_3)_3 \cdot 9H_2O$ @ 110 °C
[2]: 522 cm of IOCS with 80 cm cementitious porous pavement aggregate layer for pH elevation
[3]: packed media porosity (specific gravity of silica sand = 2.63 and IOCS = 2.74)
[4]: scaled from lateral pavement sheet flow data at urban experimental site in Cincinnati, OH
[5]: Empty Bed Contact Time
[6]: Specific surface area (SSA) of silica sand : 0.01 m^2/g and IOCS :15.0 m^2/g

The influent metal element concentration and aqueous matrix were varied to evaluate the influence of concentration and partitioning. Experimental runs were conducted using a DI water matrix and Zn, Cd, Pb and Cu all at a constant concentration to assess the competitive order of sorption. Experimental runs were also conducted using actual stormwater to assess the breakthrough capacity for dissolved and particulate-bound metal elements. An experimental matrix is presented in Table 2.

Table 2
Experimental matrix for competitive sorption column runs

Runs	Influent characteristics						Column media		
	Zn [µg/L]	Cd [µg/L]	Pb [µg/L]	Cu [µg/L]	pH	Solute matrix	Silica sand	IOCS	BSPET
zpcc65-5	5000	5000	5000	5000	6.5	DI water			
zpcc80-5	5000	5000	5000	5000	8.0	DI water			----
zpcc65-S	10,000	1000	1000	1000	6.5	runoff			

RESULTS

The breakthrough curves for run zpcc65-5 (pH=6.5) are presented on the left side and for zpcc80-5 (pH=8.0) on the right side of Figure 2. The breakthrough capacity for silica sand is negligible for all metal elements. This can be attributed to the very low SSA of silica sand. The negligible capacity holds for all metal elements at both pH levels. The IOCS capacity is greater than silica sand at a pH of 6.5 although breakthrough is still relatively rapid for all metals except Pb. An influent pH of 6.5 is below the PZC of 7 to 8 for iron oxides indicates the predominance of surface sites are positively charged. However once the pH is raised to 8.0 (lower plot on right) the capacity of the IOCS is significantly improved. For IOCS the order of breakthrough is

Cd > Zn > Cu > Pb which indicates the bonding preference may be based on the tendency of a metal element to hydrolyze and then sorb to the iron oxide. The BSPET run in which the cementitious porous aggregate was used to elevate the pH from 6.5 to 8.0 had a capacity for Cd and Zn which was similar to run zpcc80-5 at pH 8.0 and a much greater capacity for Cu and Pb. This additional capacity may in fact be due to sorption and precipitation of Cu and Pb onto the cementitious material. A summary of the metal element mass sorbed per dry mass of media is presented in Table 3 for the BSPET experimental runs using actual stormwater. Since runoff contained solids, breakthrough capacity for the particulate-bound fraction is also tabulated.

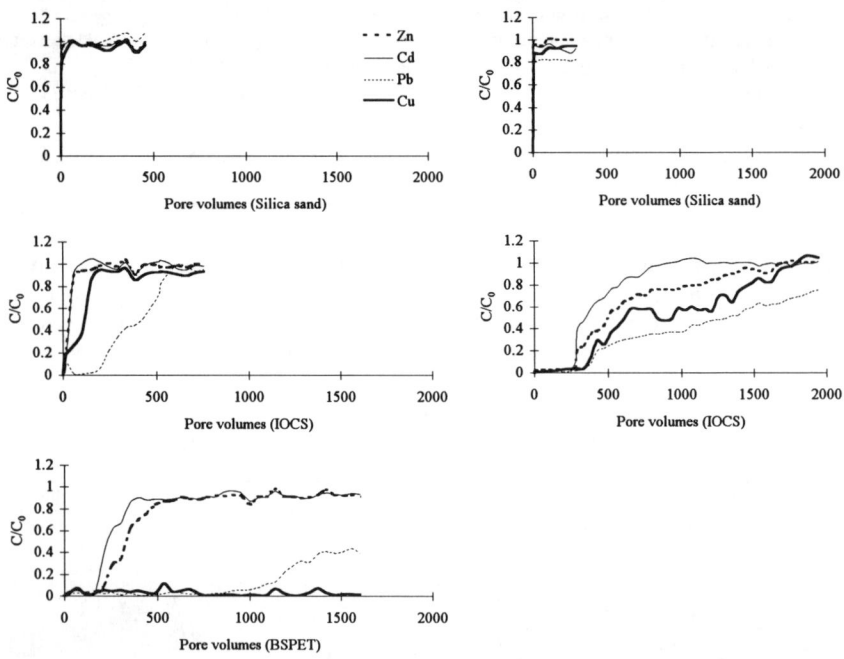

Figure 2. Column breakthrough curves at pH 6.5 (on left) and at pH 8.0 (on right).

Table 3
Metal element mass sorbed per dry mass of media at $C/C_0 = 0.90$ breakthrough

	mg sorbed / mass in kg of IOCS at 90% breakthrough				
Run	Zn	Cd	Cu	Pb	fractionation
Zpcc65-S	11836	522	>>2045	>>2785	dissolved
Zpcc65-S	633	33	2965	1202	particulate-bound

The results presented in Table 3 can be used to assess the design life of a prototype PET loaded by lateral pavement sheet flow. Application of a design life methodology and annual metal element mass loadings taken from two years of data collection at an urban experimental station in Cincinnati, Ohio indicate the design capacity of the IOCS of a prototype PET loaded by urban lateral sheet flow is at least 15 years.

CONCLUSIONS
The critical design elements of a PET are sufficient hydrologic loading capacity, sufficient breakthrough capacity for targeted metal elements and capability to filter finer infiltrated solids which would be otherwise discharged through the PET underdrain Column simulations were carried out to assess the capacity of IOCS to immobilize Zn, Cd, Cu and Pb. Column simulations using stormwater were considered most representative for evaluating the capacity of a prototype PET. A methodology to evaluate the potential capacity of a prototype PET from application of column simulation results. Hydrologic parameters and loadings as well as metal element fraction loadings, all derived from the experimental site were used to estimate annual loadings to a prototype PET. Results from application of this methodology to a PET design at the experimental site suggest that the design life of a PET in terms of metal element breakthrough is controlled by breakthrough of particulate-bound metal elements. Results indicate that the factored design life of a PET is approximately 15 to 20 years dictated by breakthrough of particulate-bound Zn. This result underscores the importance of solids control through the use of a geosynthetic liner around the underdrain would augment particle retention and therefore extend the life of the PET.

A 17 year PET design life is comparable to the service life of the typical asphalt pavement on urban highways. One alternative to rejuvenate PET capacity would be to replace the IOCS. This could be accomplished using the same IOCS on-site. Using a mobile process the IOCS would be excavated, washed at a low pH and replaced back in the PET as clean IOCS. The PET design would allows in-situ backwashing only under conditions where soil hydraulic conductivity was very low, as in clayey soils.

REFERENCES
Benjamin, M. and Leckie J. (1981) Multiple-Site Adsorption of Cd, Cu, Zn, and Pb on
    Amorphous Iron Oxyhydroxide, Journal of Colloid and Interface Science, Vol 79,
    No 1, Jan pp. 209-221.
Dixon J. and Weed S. (1989) Minerals in the Soil Environments, Soil Science Society of
    America, Madison, WI, pp. 1264.
McBride, M. (1994) Environmental Chemistry of Soils Oxford Univ. Press, New York .
Sansalone, J.J. and Buchberger, S.G., (1997). Partitioning and first flush of metals in
    urban roadway storm water. Journal of Environmental Engineering, American
    Society of Civil Engineers, Vol. 123, No. 2, February.
Stumm, W. (1992) Chemistry of the Solid-Water Interface, John Wiley , New York.
Wanielista, M. and Yousef, Y. (1993). "Stormwater Management.", John Wiley &
    Sons, Inc., New York, 1-579.

THE INFILTRATION OF RAINWATER THROUGH RESERVOIR STRUCTURES

Georges RAIMBAULT, Michel LEGRET
Laboratoire Central des Ponts et Chaussées - Bouguenais - France

ABSTRACT

The reservoir structure technique has been developed in France in order to reduce flooding risks and to protect receiving media. Before infiltration or drainage, stormwater is temporarily stored in the structure of streets, parking areas, sport grounds by using open-textured sud-base materials to hold water. In order to fulfil their hydraulic and mechanical functions these structures have to be specially designed and must be adapted to local constraints. For instance, on sloping surfaces, the sub-base "reservoir" must be partitioned, so that all the water will not be collected in the lowest part. In order to reduce clogging problems and the effects of heavy traffic, new building procedures are proposed. At last if the reservoir structures reduce the runoff flow rates, they improve the quality of runoff water.

INTRODUCTION : THE CONCEPT OF RESERVOIR STRUCTURE

Traditional rainwater drainage techniques first appeared inadequate in France for both hydrological reasons (the limited capacity of drainage systems) and economic reasons (the high cost of new main drains in old urban centres) [1]. More recently, changes in regulations to protect receiving media from rainwater pollution have led to the development of techniques which attempt to solve the hydrological, economic and environmental problems. A whole range of techniques which are referred to either as "alternative" because they provide alternatives to systems with main drains or "compensatory" because they compensate for the negative effects of urbanization, has been developed. The solution which is adopted depends on the type of project involved. Reservoir structures are a technique of this type which uses porous materials which have both a mechanical and a hydrological function. The first function (which is associated with the word "structure") enables the construction to withstand various types of loads (ranging from pedestrians to heavy freight vehicles). The second ("reservoir") enables this method to temporarily retain rainwater thereby reducing or eliminating runoff and improving the quality of discharged water. The concept of reservoir structure mainly relates to porous pavements which first underwent trials in the 1970s [9], but other applications exist, for example in school playgrounds or sports fields, etc.

POROUS MATERIALS AND MECHANICAL APPROACH

In order for reservoir structures to perform their temporary retention function, they must contain porous materials, at least in their lower layers. When rainwater is expected to infiltrate directly into the structure the surface materials must be permeable. The studies and tests which have been carried out in the last fifteen years have led to use of the following new porous materials:
- porous bitumen-bound granular materials: these are open textured materials with a lower bitumen content than porous asphalts (effective porosity: 10-20%)
- crushed untreated granular materials: these are materials with a minimum aggregate particle size of 10 mm and a maximum particle size of at least three times the minimum particle size (effective porosity: 30-40%).
- porous blocks which are mainly used in pedestrian zones (effective porosity: 10-20%).
- cellular plastic materials with a porosity of more than 95%.

Generally, water is stored in a layer of crushed materials which contains no fines or in cellular plastic materials at the base of the structure. The role of the other materials is merely to allow water to seep through from the surface. Mechanically, the porous materials are generally weaker than materials without voids. The complex modulus of porous bituminous mixes is half that of a traditional material, but both materials give similar results in fatigue tests. Crushed unbound materials must be manufactured with hard aggregates and are considered to have mechanical properties which are average for untreated materials. Standard cross sections for porous pavements have been developed which take account of expected traffic and the characteristics of the subgrade.

The choice of surface material depends on the pavement damaging power of the traffic. If the structure is expected to withstand manoeuvring heavy freight vehicles, porous materials whose aggregates might tend to be stripped by shearing stresses caused by the tyres must be avoided. In this case, an impermeable surfacing can be used and rainwater can be transferred to the underlying porous materials by means of gully holes and a spatial distribution system.

DESIGN AND HYDROLOGICAL EFFECTIVENESS
DESIGN
The design of reservoir structures must be considered at a very early stage in the preparation of a project. A proper approach to design entails breaking down the barriers between areas of competence and introducing team work. Such structures must be considered in the master plan because they must be downstream of the impermeable structures whose water they are intended to store. In addition, this storage is much more easily achieved in the case of roads which are parallel to contour lines than in the case of sloping roads. Design is linked to topography [6]. If the sites are sloping, it is frequently necessary to break down the structure into several reservoirs (Figure 1).

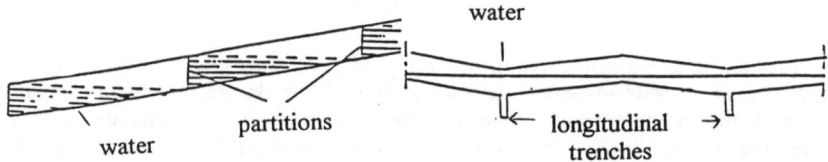

Figure 1 : The partitioning of a reservoir structure

Figure 2 : Cross section of a reservoir structure on a flat site.

In order to avoid water being retained for excessively long periods, in the case of flat sites, it may be necessary to create slopes at the base of the structures and discharge water through trenches which are perpendicular to them (Figure 2).
The available volume for each individual reservoir must be computed by means of design charts which provide the specific depths of water which must be stored on the basis of the specific discharge rate (here "specific" = per unit of surface area supplying the reservoir). In more complex cases where more precise rainfall data are available hydrological operation can be fully simulated.

When the structure is designed in order to receive runoff from upstream catchment areas or from the roofs of nearby buildings, the water must be spatially distributed between porous materials [3]. The maximum flows passing through the drainage systems can be very large. There is a model which can be used to compute the flow which passes through a drain on the basis of the head of water at its entrance and its diameter.

Water is discharged either by infiltration into the soil below the reservoir structure, by a drainage system, or by a combination of both principles. In the first case infiltration is frequently assumed to be equal to the permeability of the soil divided by a safety coefficient which takes account of the risks of compaction during construction works. In the second case, the flow discharged through the drainage system is generally controlled. This system will occasionally use the same pipes as the spatial distribution network mentioned in the previous paragraph. In this case, the emptying of the reservoirs will assist self-cleaning. These drainage networks must be accessible through manholes for maintenance purposes. The third case is the most common. All the water infiltrates away during periods of normal rainfall, but in periods of exceptional rainfall most of the rainwater is discharged through infiltration and the remainder is removed by the drains.

Because of the risks of porous surfacings clogging up, safety gullies may be incorporated which link directly with the base of the porous structure.

EFFECTIVENESS

In order to show the hydrological effectiveness of reservoir structures we shall briefly describe below the main results from an experimental pavement near Nantes (France) which consists of a 6 cm layer of porous asphalt, two 10 cm layers of bitumen-bound granular material, a 35 cm layer of 10/80 crushed aggregate and a geotextile separating membrane between the subgrade and the crushed material [7].

Water can infiltrate into the subgrade and be discharged through a drain when the porous structure is filled, but rain does not necessarily lead to flow through the drain. The ability of water to infiltrate into the subgrade varies a great deal depending on its degree of saturation. The average maximum value recorded over a 3 year period, during rainfall, was 5.8 mm/h, which is equivalent to a permeability of $19.8 \cdot 10^{-7}$ m/s. The lowest average infiltration recorded during rainfall which had led to discharge through the drain was 0.5 mm/h ($1.7 \cdot 10^{-7}$ m/s), in other words more than ten times less than the above maximum value. A minimum of 87% of all rainwater seeped into the subgrade. These findings illustrate that even when a reservoir structure is constructed on a subgrade which is relatively impermeable (i.e. with a permeability of between 1 and 5.10^{-8} m/s) a major proportion of the rainwater can still seep away.

The maximum flows recorded at the drain outlet show that in the absence of any flow control device maximum flow always remains less than 4/l/s/ha of supply surface.

RESERVOIR STRUCTURES AND POLLUTION

WATER QUALITY.
The quality of the water collected by the drain on the experimental site described above was monitored in order to compare the pollutant concentrations in the water which had percolated through a reservoir structure with those in the runoff water from a nearby housing estate (Table II et [5]).

	pH	COD mg.l^{-1}	SM mg.l^{-1}	SVM (%)	Pb µg/l^{-1}	Zn µg.l^{-1}	Cu µg.l^{-1}	Cd µg.l^{-1}	Hc mg.l^{-1}
Reservoir structure No. of samples Average	7.5	32 <22.1	38 11.9	38 32.1	22 5.4	39 46.3	39 14.6	39 0.49	39 <0.02
Reference catchment area									
No. of samples		28	31	31	30	32	32	32	32	3
Average	7,5	<23.4	33.2	37.3	26.3	165	11.3	1.48	<0.02

Table II. Comparison between pollution parameters for two sites

This comparison reveals the following points:
. The amount of suspended matter is low in both cases, but considerably lower at Rue de la Classerie (11,9 mg/l) that at the outlet of the reference catchment area (33,2 mg/l).

. With the exception of copper, the concentrations of heavy metals are lower at the outlet of the reservoir structure than at that of the reference catchment area. In the case of lead the average concentrations are respectively 5,4 et 26,3 µg/l. Lead is mainly attached to suspended matter (56 % for rue de la Classerie and 85 % for the reference catchment area). Other metals are mostly present as solutes at both sites (approximately 70 %).
. Generally, the values of pollution parameters exhibit very wide dispersion.
It thus appears that the water which has passed through the reservoir structure has low pollutant concentrations and the values of the principal pollution parameters are considerably lower than those measured at the outlet of the reference catchment area. Table III shows the reductions in pollutants obtained at the Nantes site and at other reservoir structures:

SITE	COD	SM	Pb	REF.
Car Park (Lund, Sweden)	---	95 %	40 %	[4]
Park-and-ride facility(Bordeaux-France)	79 %	36 %	86 %	[2]
Car park(Bordeaux-France)	77 %	94 %	85 %	[2]
Avenue (Paris-France)	54 %	70 %	78 %	[5]
Street (Nantes-France)	----	64 %	79 %	[5]

Table III : Reductions in pollution obtained with reservoir structures

THE QUALITY OF MATERIALS AND SOIL
Trenches were cut in 1992 and 1996 at the Nantes site and samples were taken at various levels. The measured heavy metal concentrations (Figure 3) show that micro-pollutants are mostly retained at the surface of porous asphalt. A slight increase in heavy metal concentrations at the geotextile and the upper layers of the soil underneath the reservoir structure (10 - 15 cm) were also apparent. Nevertheless, the levels measured after 4 and 8 years of operation remain below those of polluted soils.

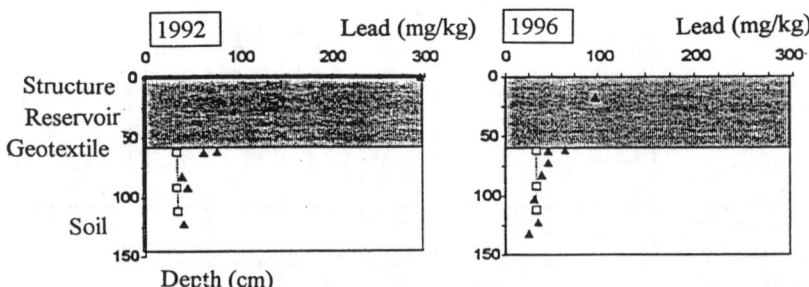

Figure 3 Vertical profiles for lead concentration (fraction <125 µm).

CLOGGING AND MAINTENANCE
A gamma ray analysis of clogged porous asphalt showed that clogging is restricted to the 2 centimetres nearest the surface. The clogging is mainly caused by sand, the percentage of clay being less than 2%. It would seem that initially sand particles are trapped for purely geometric reasons. Subsequently, finer particles bring cohesion to the clogging material. Laboratory studies have shown that a succession of cycles of

infiltration and drying through the sand with water containing clay particles leads to a progressive reduction in permeability.
In order to combat clogging it is advisable to undertake regular preventive suction cleansing of surfacings (several times per year). In the event of advanced clogging, remedial treatment requires the use of very high pressure mobile water sprays (several hundred bars) with immediate removal of the loosened material and perhaps recycling of water.
The recovery of polluted clogging materials raises the problem of what to do with them. The envisaged solution is to separate the most polluted fines from the sand particles and to recycle the latter as construction materials.

CONCLUSION

The construction of reservoir structures has increased in the last 10 years in France, initially mainly to control the risks of flooding in Bordeaux. Their use has now been extended over the whole country and they are also used in order to reduce pollutant release to receiving media during rain. Research in the areas of hydrology, the environment and mechanics has improved our knowledge of the principles of design and our understanding of the actual operation of such structures. This technique provides solutions to drainage problems which are more economical than conventional methods. Principally, it is suited to the construction of large car parking facilities which are used by large numbers of people and which take up for large areas of space in the outskirts of cities.

REFERENCES

[1] J.D. Baladès and G. Raimbault. Urbanisme et assainissement pluvial. Bull. Liaison Labo. P. et Ch., 170 (1990) pp. 47-59.
[2] J.D. Baladès, T. Guichard, M. Legret, H. Madiec. Gestion de la pollution des eaux pluviales par chaussées réservoirs en milieu urbain. T.S.M. n° 11 (1994) pp. 631-638.
[3] M. Dakhlaoui, D. Gaujous, G. Raimbault, J.P. Tabuchi. Water diffusion device design in reservoir structures. Water Sci. Tech. Vol. 32 n° 1 (1995). pp. 71-78.
[4] W. Hogland, J. Niemczynowicz and T. Wahlman. The unit superstructure during the construction period. Sci. Total. Environ., 59 (1987) 411-424.
[5] M. Legret, V. Colandini, C. Le Marc. Effects of a porous pavement with reservoir structure on the quality of runoff water and Soil. The Science of Total Environment 189/190 (1996) pp. 335-340.
[6] G. Raimbault. Structures réservoirs et topographie des aménagements urbains. Conférence Novatech 92, Lyon (1992) pp.400-409.
[7] G. Raimbault, M. Métois. Le site expérimental de structure-réservoir de Rezé. Conférence Novatech 92, Lyon (1992), pp. 213-222.
[8] J. Ranchet, F. Penaud, R. Le Grand, A. Constant, P. Oborg and B. Soudien. Comparaison d'une chaussée pavée et d'une chaussée drainante du point de vue de leur comportement hydraulique et de leur impact sur la dépollution des eaux de pluie. Bull. Liaison Labo. P. et Ch., 188 (1993) 67-72.
[9] E. Thelen, L. Fielding. Howe Porous Pavement. The Franklin Institution Press. Philadelphia. Pensylvania (1978).

DESIGN OF STORMWATER INFILTRATION TRENCHES BASED ON UNCERTAIN INFORMATION

PETER STEEN MIKKELSEN
Department of Environmental Science and Engineering
Technical University of Denmark, Building 115, 2800 Lyngby, Denmark

ABSTRACT

This paper initially discusses a selection of design procedures used in various countries now and in the past, and highlights a number of basic demands that should be fulfilled in a general design methodology. Then, it describes a new simple and easy-to-use conceptual design methodology that has been incorporated in a recent Danish design guideline. The errors introduced by resorting to a simple method are discussed in the light of the uncertainty of the involved parameters and finally, some conclusions and recommendations are given.

INTRODUCTION

Infiltration trenches, which allow water to be temporarily stored and slowly infiltrate into the surrounding soil, are typically small and expected to be inexpensive. This implies that limited resources are available for design and field-investigations, and that design methods should be simple and easy to use. On the other hand, infiltration of water into soil is a complicated problem to describe in mathematical terms, and additionally the subsurface is highly heterogeneous which makes soil parameters difficult to measure.

Formerly, soakaways were designed in a variety of ways, including *by guesswork* and *on experience*, and systematic design guidelines have only been published within the last few decades. Most current national design guidelines make use of the *rain envelope method* which is a well-known and simple method for design of detention basins. Infiltration trenches and detention basins behave much in the same manner and they may be described simplified by only two parameters; the specific storage volume ($v=V/A_r$), and the (time-invariant) specific flow restriction ($q_{FR}=Q_{FR}/A_r$), where V is the available storage volume, Q_{FR} is an average flow restriction at the outlet (the discharge capacity) and A_r is the drainage area that contributes effectively with runoff. The most significant difference is that for a detention basin the flow restriction is determined by an outlet pipe whereas for a soakaway it is determined but by the permeability of surrounding soil, the soakaway geometry and the water level in the soakaway. Thus, simplified groundwater flow theory should be used to calculate Q_{FR} for an infiltration trench, cf. e.g. (Jonasson, 1984; Stahre and Urbonas, 1990).

DESIGN OF STORMWATER INFILTRATION TRENCHES

The rain envelope method utilizes rainfall and infiltration envelope curves combined with simple mass balance considerations to find the critical storm duration and the necessary storage volume. The problem may be solved analytically by expressing IDF-curves with a mathematical equation (e.g. Mikkelsen and Jacobsen, 1993) or alternatively by automating the iterations with a computer program as suggested by Pratt and Powell (1993). However, none of these approaches cover up the general inadequacy of IDF-curves to describe storage characteristics of rainfall, including the effect of coupled rain storms. A new methodology that accounts for these problems is presented in the next section.

THE NEW DANISH DESIGN PROCEDURE

OUTLINE OF THE METHODOLOGY

The Danish tradition for design of detention basins is based on using historical rain data as input to a simple conceptual model that simulates runoff and storage of water and computes overflow statistics (the SAMBA approach). The model parameters are the necessary specific storage volume ($v=V/A_r$) of a structure, its specific discharge capacity ($q_{FR}=Q_{FR}/A_r$) and the concentration time of the catchment (t_c). Typically, t_c is of minor importance compared with the two other parameters. Results from such computations may be depicted for easy use as *storage characteristic curves* which for different return periods give v as a function of q_{FR}, e.g.

$$v = \mathscr{F}(q_{FR}) \tag{1}$$

Such curves are shown as full lines in Fig. 1 for $t_c=5$ min. As opposed to IDF-curves, these curves represent the storage characteristics of rainfall, including the effects of coupled rain storms, correctly.

Since q_{FR} is time-invariant per definition, the emptying time of a soakaway may be written $t_e=v/q_{FR}$. This implies that another relation between v and q_{FR} reads

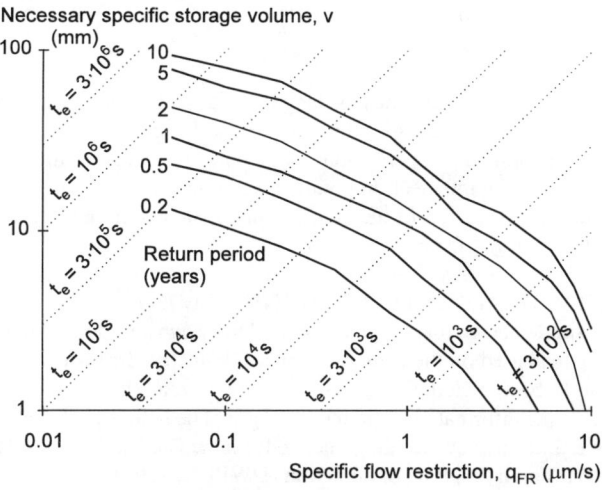

Fig. 2. The relation between v and q_{FR} as determined by the characteristics of a historical rain series ($t_c=5$ min) and by the emptying time (geometry and soil permeability) of a soakaway.

$$v = t_e \, q_{FR} \tag{2}$$

If t_e can be determined independently based on the characteristics of the soil and the trench geometry, then the design problem comes down to finding the solution to two relations with two unknown variables. This can be solved graphically as shown on Fig. 1, where curves corresponding to Eq. (2) are shown as dotted lines.

MODELLING FLOW OF WATER IN SOIL

To use Fig. 1 for practical design a methodology of calculating Q_{FR} is required. The simplest approach is to equate the discharge with a unit gradient flow (vertical flow driven by gravity) across an area equal to the wetted surface area of the trench. This corresponds to very simplified theory used for interpreting borehole permeameter data and neglects effects of unsaturated flow. Disregarding flow through the bottom area (which will eventually clog up with litter and particles) leads to the conclusion that infiltration trenches should be constructed tall and narrow rather than wide and shallow, to maximize the area of the walls, which is the only area that will be effective for infiltration over time. Assuming the water level to be on average half-way between the bottom and top of the soakaway and disregarding flow through the end walls (which are small compared to the total wall area), the average flow restriction reads

$$Q_{FR} = K \, \tfrac{1}{2} A_{side} \tag{3}$$

where K is the soil permeability and A_{side} is the area of the side walls. Based on these assumption the three parameters figuring in Eq. (2) may be written

$$v = \frac{\ell w h \phi}{A_r} \; ; \quad q_{FR} = \frac{K \ell h}{A_r} \; ; \quad t_e = \frac{w \phi}{K} \tag{4}$$

where ℓ, w and h, denotes the length, width and hight of the trench, and ϕ is the porosity of the trench stuffing. After choosing a design return period and calculating t_e from information on soil permeability (K), porosity of trench stuffing (ϕ) and a chosen trench with (w), the intersection point between the two concerned curves in Fig. 1 are found, and the dimensions of the trench (ℓ and h) are determined by using Eq. (4), after reading either v or q_{FR} from Fig. 1.

PRACTICAL USE OF THE METHODOLOGY

A single design diagram may be used to design soakaways of any type by reformulating Eq. (4) based on the actual soakaway geometry. To further ease the use of the method, v may be depicted directly as a function of t_e as shown in Fig. 2 (right) that was drawn from the information contained in Fig. 1. The latter figure is considered more easy to use and thus, the new design guideline from the Danish Water Pollution Control Committee is based on this type of depiction (DWPCC, 1994).

Some guidelines mention that soakaways should not be used, or that they should only be used when supplied with a discharge pipe, if the soil permeability is lower than a threshold value. However, it appears from Fig. 2(right) that there is no abrupt transition

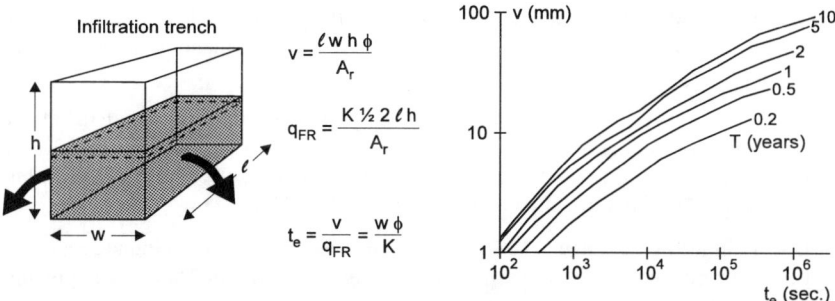

Fig. 2. Basics of the new Danish method for sizing soakaways, exemplified for a long and narrow infiltration trench. K and ϕ are the soil permeability and porosity, and w, ℓ and h are the width, length and depth.

where stormwater infiltration becomes impossible, but that the required volume may become unrealistically high when the emptying time is very large. The definition of the term *unrealistic* depends on the local rainfall pattern (including the chosen design return period), on the permeability of the soil, and on the costs of construction works. Thus, a lower threshold of soil permeability is an economic restriction and not a technical one. Design guidelines should therefore not preclude the use of stormwater infiltration in low permeability soils, but merely supply the necessary tools that allow the user to come to this conclusion under specific circumstances.

GENERAL UNCERTAINTY ASSESSMENT

Considering the large uncertainty related with determining the soil permeability at specific locations, it appears that design of infiltration systems is not very accurate compared with other stormwater management alternatives that are purely man made. Moreover, other design parameters are also uncertain, and the simplified physical model used to describe the discharge of water into soil evidently introduces systematic errors. These issues will be discussed in the following two sections.

UNCERTAINTY OF INPUT PARAMETERS

By approximating a storage characteristic curve (Eq. 1) with an equation of the form

$$v = G\, q_{FR}^{-\gamma} \tag{5}$$

and combining with Eq. (2) and (4), the design-length of an infiltration trench can be calculated with the following expression

$$\ell = \frac{A_r}{h} \left[\frac{G}{w\phi} \right]^{\frac{1}{1+\gamma}} K^{-\frac{\gamma}{1+\gamma}} \tag{6}$$

that can be used to assess the relative importance of the parameters based on first order uncertainty analysis. Mikkelsen (1995) gives such an assessment by including the following three phenomena:

1) Uncertainty related with description of the design rainfall, comprising both the effect of sampling errors and regional variation of rainfall data, amounts to 10-15% in Denmark expressed as coefficient of variation (CV) (Mikkelsen et al., 1996). This uncertainty primarily affects the parameter G in Eq. (5).

2) Uncertainties in the range 5-10% expressed as coefficient of variation are reported for determination of the impermeable drainage area (A_r) of Danish catchments (DEPA, 1992).

3) The uncertainty of the soil permeability depends on the soil type and heterogeneity, and on the measurement technique. An example based on morainic soils that are common in Denmark revealed that the uncertainty expressed as coefficient of variation varies between 15% for the best possible measurement technique (full-scale trial pit and interpretation with detailed flow theory) and 185% if only a rough soil type classification is carried out.

If only variation of soil permeability within the texture class is considered (CV=1.85) the first two uncertainty contributions do not increase the total design uncertainty considerably. In such cases a national set of storage characteristic curves will provide a sufficient information level on rainfall.

The uncertainty of K is smaller if information is available from a borehole field-test, but it is still of considerable magnitude ($CV \approx 0.40$). In this case the three uncertainty contributions are almost equally important. If a full-scale trial pit is carried out on the exact location (CV=0.15) the uncertainty of K becomes negligible, and the uncertainty connected with rainfall turns out to be critical. Thus, it may be relevant to work with storage characteristic curves determined from local rain data to decrease the total uncertainty related with design in cases where detailed information on K is available.

SIGNIFICANCE OF MODEL ERRORS
The systematic error introduced by neglecting the influence of capillarity on the shape of the plume of infiltrating water can be assessed by considering the results from two-dimensional modelling studies, e.g. Duchene et al. (1993), Herath and Mutsiake (1994) and Mikkelsen (1995). Such simulations show that the simplified flow theory (Eq. 4) may result in up to a three times underestimation of the flow from soakaways (Q_{FR}). This means that the simplified flow-theory leads to an overestimation of the required storage volume of up to 37%. This is a significant bias when at-site measurements of K are available, but the bias is negligible in situations where the soil permeability is judged from a rough texture classification or from approximate measurements.

The effect of not considering long-term changes of the hydraulic efficiency is probably the most significant systematic source of error, but this cannot be assessed without long term studies of the hydraulic behaviour of soakaways.

CONCLUSIONS

The new Danish design method for sub-surface soakaways contains a more accurate description of the rainfall dynamics than other published methods. However, the method is still very simplified compared with the physical reality. Uncertainty assessments show that in most realistic design situations the errors introduced by resorting to a simple design method are overruled by the uncertainty of the involved parameters (rainfall, drainage area and soil permeability). The largest amount of uncertainty is related with determination of the soil permeability (K) on a spatial scale corresponding to the size of an actual soakaway.

Importantly, the new methodology is more easy to use than other published design methods, and it allows for assessing the functional behaviour of infiltration trenches in a simple transparent manner. Combined with methodologies for uncertainty analysis the level of design safety can be decided depending on the accuracy of the involved parameters and the costs of data collection and construction works.

REFERENCES

Duchene, M.; McBean, E.; Thomson, N. (1993): Infiltration characteristics associated with use of trenches in stormwater management. Proc. 6th Int. Conf. on Urban Storm Drainage, Niagara Falls, Ontario, Canada, September 12-17, *II*, 1115-1120.

DWPCC (1994): Nedsivning af regnvand - dimensionering. Ingeniørforeningen i Danmark, Spildevandskomitéen. *Skrift nr. 25.* (In Danish).

Herath, S.; Musiake, K. (1994): Simulation of basin scale runoff reduction by infiltration systems. *Wat. Sci. Tech.*, *29*, (1-2), 267-275.

Jonasson, S.A. (1984): Dimensioning methods for stormwater infiltration systems. Proc. 3rd Int. Conf. on Urban Storm Drainage, Göteborg, Sweden, June 4-8, 1984, *3*, 1037-1046.

Mikkelsen, P.S.; Jacobsen, P. (1993): Stormwater infiltration design based on rainfall statistics and soil hydraulics. Proc. ASCE Int. Symp. on Engineering Hydrology, San Fancisco, California, July 25-30, 653-658.

Mikkelsen, P.S. (1995): Hydrological and pollutional aspects of urban stormwater infiltration. *Ph.D. Thesis.* Institute of Environmental Science and Engineering, Technical University of Denmark.

Mikkelsen, P.S.; Arnbjerg-Nielsen, K.; Harremoës, P. (1996): Consequences for established design practice from geographical variation of historical rainfall data. Proc. 7th Int. Conf. on Urban Storm Drainage, Hannover, Germany, September 9-13.

DEPA (1992): Bestemmelse af befæstet areal. *Spildevandsforskning fra Miljøstyrelsen*, (43). Danish Environmental Protection Agency (Miljøstyrelsen). (In Danish).

Pratt, C.; Powell, J.J.M. (1993): A new UK approach for the design of sub-surface infiltration systems. Proc. 6th Int. Conf. on Urban Storm Drainage, Niagara Falls, Ontario, Canada, September 12-17, *I*, 471-476.

Stahre, P.; Urbonas, B. (1990): Stormwater detention for drainage, water quality and CSO management. Prentice Hall, New Jersey, USA.

EVAPORATION AT THE SURFACE OF PERMEABLE PAVEMENT AND ITS IMPACTS ON THE URBAN THERMAL ENVIRONMENT

VU THANH CA
Department of Civil and Environmental Engineering
Saitama University, Urawa, Saitama 338, Japan
TAKASHI ASAEDA
Department of Environmental Science & Human Technology
Saitama University, Urawa, Saitama 338, Japan

ABSTRACT
Experiments and numerical analysis were carried out to investigate the heating characteristics of various pavement surfaces. It was found that due to small pore size in the porous ceramic pavement, this kind of pavement could keep water inside for a long time and even could absorb the water from the soil below. The consequent evaporation at this surface made its surface temperature appreciably lower than that of the non-porous and normal porous block pavement surfaces. Due to the large pore size of the material, the surface of normal porous pavement was very dry, and virtually almost no evaporation occured. The normal porous and nonporous pavement surfaces can absorb a large amount of the incoming net radiation, which increases their surface temperature and modifies the urban thermal environment in the unfavorable way.

1 INTRODUCTION

Permeable pavements were utilized originally for the drainage of urban streets. However, since they allow the exchange of water between the atmosphere and the soil below, the evaporation can occur at its surface. The evaporation, if happened, can make temperature of the permeable paved surfaces not escalate much, and subsequently the underground heat storage and sensible heat exchange between the ground surface and atmosphere are diminutive. Therefore, the utilization of porous materials for the construction of pavement may be one of the most effective methods to moderate the thermal conditions of the pavement surface.

With the purpose of understanding the heating processes at the surface of various pavements, this paper presents efforts on the study of thermal characteristics of materials of porous and traditional sealed pavements. To obtain a proper evaluation on the interaction between porous pavement and the atmosphere, field experiments were conducted with various types of alternate pavement materials. Then, a unidimensional numerical model was developed to simulate the complicated processes of heat and moisture transfer at the porous surfaces and in the underlying soil,.

2 MATERIALS AND METHODS

2.1 OBSERVATIONAL WORK AND DATA COLLECTION

Field experiments were conducted throughout the year from August 1994 to July 1995 at the Housing & Urban Development Corporation office at Kuki, 70 km north of Tokyo (36°N, 139°36'E). During the experiments, ten types of sample surfaces with horizontal sizes of 2m×2m were prepared for the observation. The layout of sample surfaces is depicted in Fig 1. Surface temperature and temperatures at various depths under the surface, ground heat flux together with meteorological conditions such as air temperature, relative humidity, wind velocity, total downward solar radiation, and downward infrared longwave radiation were automatically recorded to floppy disks. Additionally, during the extensive measurement periods of August 8-10, 1994 and February 1-2, 1995, surface reflectivity, upward longwave radiation from surfaces were also measured. Although measurements were carried out throughout the year, only the results of observations on 9-10 August, 1994 are presented in this paper, as it is intended to study the effects of various pavements on near surface atmospheric thermal environment during hot summer days.

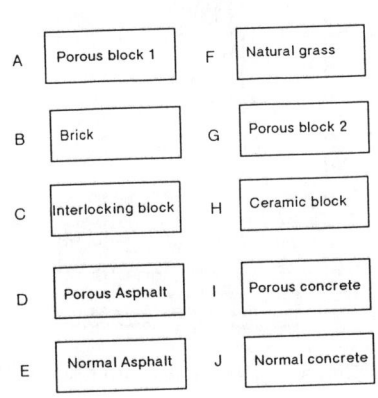

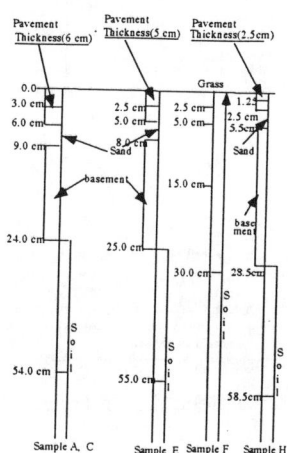

Fig. 1 Layout of the surface samples

Fig. 2 Vertical structure and temperature measurement depths of samples

Even experiments were carried out with ten pavement samples, it was recognized that many pavement samples have similar heating characteristics. Thus, only results of experiments on four samples of typical surface materials, such as porous block pavement, dark non-porous asphalt, natural grass and ceramic porous pavement are selected for this paper. The asphalt pavement sample (Sample E) was made of normal asphalt usually used for the construction of pavements and roads in urban areas. The grass surface (Sample F) is a natural surface with grass of the height of 1-3cm. The porous block (Sample A), with large inside pore size, is very permeable to water. However, it can not keep the water for a long time and the surface is dried up rapidly after a rain. On the other hand, the ceramic block (sample H) with a wide range of pore sizes is not only extremely permeable to water but also with small inside pore size, can retain water in the material for as long as several days.

Subsurface temperatures at various depths under sample surfaces were obtained by inserting thermocouples into the pavement material or into the soil at particular depths, and the measured temperatures were recorded at every 5 minutes. The measurement depths for four pavement samples are depicted in Fig. 2. Hydraulic and thermal properties of pavement materials such as porosity, hydraulic conductivity, dry density, specific heat and thermal conductivity were estimated through laboratory experiments.. Water content of pavement samples and the soil were determined at the beginning of the experiment.

2.2 OBSERVATIONAL RESULTS

Fig. 3 depicts relative humidity, wind velocity and air temperature recorded at the observational site on August 9-10, 1994. As in the figure, it was a hot summer day with air temperature reached approximately about 35°C at noon, while wind velocity and relative humidity were about 4m/s and 43%, respectively. These are typical conditions prevalent on summer days in the areas around Tokyo (Asaeda and Vu, 1993; Asaeda et al, 1996).

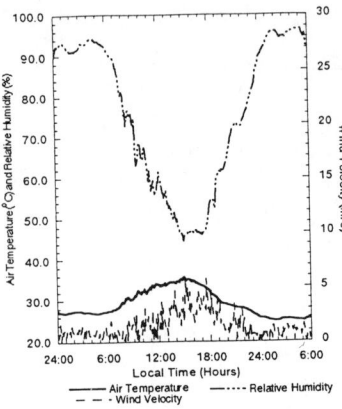

Fig. 3 Meteorological conditions on 9-10 August, 1994

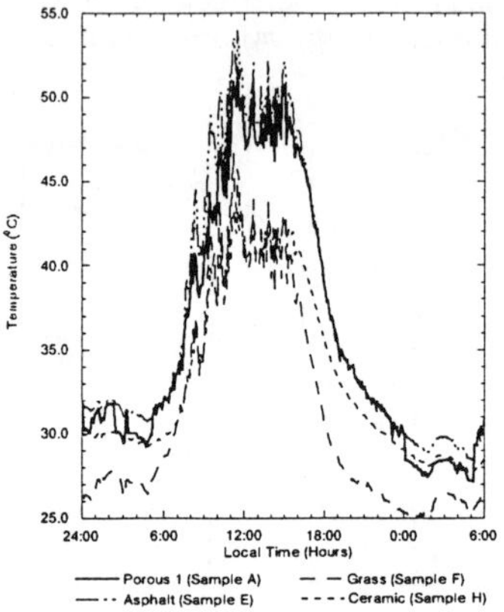

Fig. 4 Surface temperature of four samples.

Fig. 4 depicts surface temperatures of the above mentioned four samples. As in the figure, at noon, maximum surface temperature of the porous pavement sample reached almost 54°C, or nearly the same as that of the asphalt pavement. On the other hand, the surface temperature of the ceramic pavement sample reached only 43°C, which is almost the same as that of the grass surface and about 10K lower than that of the porous pavement.

Among all samples, the temperature of the natural grass surface was the lowest owing to extensive evapotranspiration from this surface. Additionally, the small conductivity of the soil prevented the conduction of heat to deep soil layers. Thus, even when the temperature at the surface was 42°C, it was only 31.5°C at 20cm deep. Temperature data at different depths under different surfaces reveal that subsurface temperature under the ceramic surface was lower than that under the porous block or asphalt surfaces but higher than that under the natural grass surface. It is due to high conductivity of the ceramic pavement material compared with that of the soil.

The difference in the heating characteristics of the porous block and the ceramic pavements arises because of the pore size inside the pavement materials. Bulky pores in the porous block reduce the capillary pressure, enabling a quick drain of the contained water. This makes the surface of the pavement very dry, and subsequently impedes the evaporation. Without evaporation, the surface is heated up like an impermeable pavement. On the other hand, by enhancing the capillary pressure through minuscule type of pores, the ceramic pavement can retain water, absorbed during a rain, for a long time. Furthermore, this pavement can even absorb a large amount of water from the underlying soil layers. Water under the surface of this pavement is evaporated, keeping the surface cooler than the surface of normal porous block pavement.

2.3 NUMERICAL ANALYSIS

A numerical model is employed for the analysis of the characteristics of heat and water transfer processes in the pavement materials. For the porous pavement and grass surfaces, the governing equations for mass and heat transfer under the surface of samples are as follows.

The equation of mass transfer inside the porous pavement or the soil is (Asaeda and Vu, 1993; Milly, 1982)

$$\left[\left(1-\frac{\rho_v}{\rho_\ell}\right)\frac{\partial\theta}{\partial\psi}+\frac{\theta_a}{\rho_\ell}\frac{\partial\rho_v}{\partial\psi}\right]\frac{\partial\psi}{\partial t}+\left[\left(1-\frac{\rho_v}{\rho_\ell}\right)\frac{\partial\theta}{\partial T}+\frac{\theta_a}{\rho_\ell}\frac{\partial\rho_v}{\partial T}\right]\frac{\partial T}{\partial t}$$
$$=\nabla\left[(K+D_{\psi v})\nabla\psi+(D_{Tv}+D_{Ta})\nabla T\right]+\frac{\partial K}{\partial z} \ . \quad (1)$$

The corresponding equation for heat transfer is

$$\left[C+L\theta_a\frac{\partial\rho_v}{\partial T}-(\rho_\ell W+\rho_v L)\frac{\partial\theta}{\partial T}\right]\frac{\partial T}{\partial t}+\left[L\theta_a\frac{\partial\rho_v}{\partial\psi}-(\rho_\ell W+\rho_v L)\frac{\partial\theta}{\partial\psi}\right]\frac{\partial\psi}{\partial t}$$
$$=\nabla\left[\lambda\nabla T+\rho_\ell(LD_{\psi v}+gTD_{Ta})\nabla\psi\right]-C_\ell q_m\nabla T \ . \quad (2)$$

where C is the total volumetric heat capacity of the soil and is estimated by

$$C = C_d + C_\ell\rho_\ell\theta + c_p\rho_v\theta_a \ ,$$

and the vapor density ρv is given by

$$\rho_v(\psi,T) = \rho_0(T)\exp\left(\psi g/RT_k\right) \ , \quad (3)$$

In Eqs. (1-3), θ is the volumetric liquid water content; θ_a is the volumetric air content; t is the time; z is the depth; C_d is the heat capacity of dry soil; c_p is the specific heat of vapor at constant pressure; Ω is the differential heat of wetting; ρ_0 is the saturated vapor density at temperature T; R is the gas constant for water vapor; T_k is the absolute temperature; q_m is the moisture flux; ρ_l is the liquid water density; $D_{\psi v}$ and D_{Tv} are the matric potential diffusivity and the temperature diffusivity of water vapor, respectively; D_{Ta} is the transport coefficient for absorbed liquid flow due to thermal gradient; T is the temperature; ψ is the matric potential; K is the hydraulic conductivity; and C_l is the heat capacity of liquid water. The evaluation of other coefficients of Eq. (1) and Eq.(2) are following Asaeda and Vu (1993).

In the impermeable pavement domain, one-dimensional heat conduction equation is applied.

The boundary conditions at the ground surface are the mass and heat fluxes. The surface mass flux is the rate of evaporation .The heat flux at the ground surface is the sum of net total radiation, sensible heat flux, and in case of porous pavement surface, latent heat flux (Asaeda and Vu, 1993).

The lower boundary conditions are constant soil temperature and matric potential at 2m depth. Initial values for temperature and matric potential were specified from the observational data.

Main drying and wetting curves of the moisture retention for the porous pavement materials, needed for the numerical model, were obtained by laboratory experiments using the well-known Richards method. The hysteresis model of Mualem (1974) is used for the determination of the scanning curves.

Comparison between computed and observed temperatures at the surface and at different depths under surface for different surfaces reveals that the model can recard the time variation of temperatures at the ground surface and various depths with acceptable accuracy. Thus, the model is used for the evaluation of the heat balance at the ground surface, and therefore, the effects of various pavements on the near surface thermal climate.

Figs. 5(a-d) depict time variation of heat fluxes at the surface As for the sign convention in the Figs. 5, downward flux is considered positive for the net radiation while upward flux is considered positive for the latent heat, sensible heat and conduction heat fluxes. Figs. 5(a-d) reveal that sensible heat flux H for the nonporous asphalt pavement was the largest during the diurnal period, clearly owing to the large net radiation R_{net} value as well as available energy ($R_{net} - G$), consequently the high surface temperature of this pavement sample, with G as the conduction heat flux. The smallest H value throughout the day was found for the natural grass. This is clearly due to the lowest temperature of the natural grass surface compared with that of other surfaces.

The heat balance for the normal porous surface is depicted in Fig. 5(a). As in the figure, the maximum value of the incoming net radiation to this surface at noon reaches about 600W/m^2. As stated previously, due to large pore size inside the pavement material, the pavement surface is very dry and almost no evaporation can occur for this surface. The maximum latent heat at noon for this surface is only about 20W/m^2. Because of the small value of the latent heat, almost all the incoming net radiation to the surface is converted to the sensible heat and ground heat fluxes. This makes the maximum sensible heat flux for this surface surpasses 430W/m^2 while the maximum ground heat flux surpasses 200W/m^2.

The heat balance for the asphalt surface, depicted in Fig. 5(b), indicates that the maximum incoming net radiation to this surface is more than 650W/m^2. This is because of the small reflectivity of this surface. With the absent of evaporation, maximum surface temperature of this pavement reached 54°C at noon (Fig. 4). This makes the maximum sensible heat exchange between this surface and the atmosphere escalating to 450W/m^2. The maximum ground heat flux for this surface reaches 300W/m^2. Measurement of upward longwave radiation from this surface indicates that the maximum value of this component of the surface heat balance reaches more than 600W/m^2. This radiation is almost absorbed by the near surface atmospheric layers Asaeda et al (1996). The sensible heat, released from the surface, and the absorbed longwave radiation heat the near surface air and contribute to the excess of the air temperature over an urban area compared with that over surrounding rural areas.

Values of latent heat L_e at the surfaces for the natural grass indicate that at noon, the largest portion of the net radiation to the surfaces is converted to latent heat. With the maximum incoming net radiation reaching the grass surface of 600W/m^2, the maximum latent heat flux is 415W/m^2 while the maximum sensible heat is only 100W/m^2 and the maximum ground heat flux is 150W/m^2 (Fig. 5c). Thus, this surface heats the air very little during the day, and does not heat the air at night.

For the ceramic pavement surface, the maximum net incoming radiation, latent heat, sensible heat, and ground heat flux are 570W/m^2, 300W/m^2, 150W/m^2 and 180W/m^2, respectively (Fig. 5d). Throughout the day, this surface behaves almost the same manner as the grass surface does. However, due to larger heat conductivity and smaller evaporation rate at the surface, this surface is heated more intensively than the grass surface is; and the heat storage under this surface is larger than that under the grass surface. Consequently, this surface heats the air more than the grass surface does both during the day and at night.

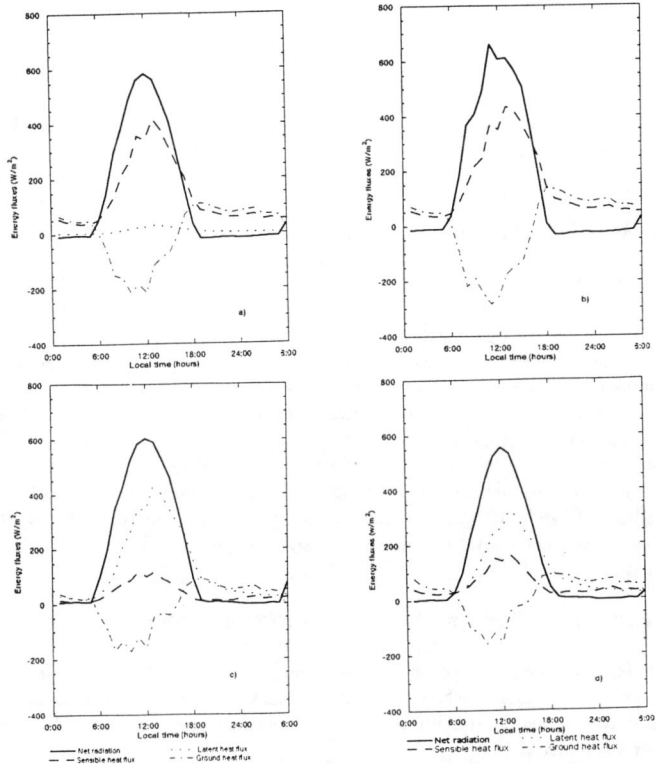

Fig. 5 Heat balance at the surface of four samples a) porous pavement (Sample A), b) asphalt pavement (Sample D), c) grass surface (Sample F) and d) ceramic pavement (Sample H)

5 CONCLUDING REMARKS

A study of the thermal transfer to the underground of the porous pavement cannot be handled properly without the coupling of moisture and heat transfer. Because of evaporation from the ceramic porous pavement surface, the surface temperature of this kind of pavement is appreciably lower than that of the non-porous and the normal porous block pavement surface. Due to the large pore size of the material, the surface of normal porous pavement is rather dry, and virtually almost no evaporation is observed at this surface. The normal porous and nonporous pavement surfaces can absorb a large amount of the incoming net radiation, which increases its surface temperature and modifies the urban thermal environment in the unfavorable way. A proper evaluation of the effects of various pavement surfaces on the near surface thermal climate can be done by the coupling of heat and mass transfer processes under the surface and at the surface to the physical processes involving the transport of heat and moisture in the atmosphere.

REFERENCES
1) Asaeda T. and T.C. Vu, *Boundary Layer Meteorol.* 15-16 (1992), 253-261.
2) Asaeda T., T.C. Vu and A. Wake, *Atmospheric Env.*, 30 (3), (1996) 413-427.
14) Milly P.C.D. *Water Res. Resear.* Vol. 18, No. 3 (1982). 489-498.
15) Mualem Y. *Water Res. Resear.* Vol. 10, No. 3 (1974). 514-520.

BRIDGE SCOUR EVALUATION PROGRAM IN THE UNITED STATES

JOHNNY L. MORRIS, P.E., AND JORGE E. PAGAN-ORTIZ
FHWA Regional Hydraulics Engineer, Atlanta, Georgia; FHWA Hydraulics Engineer,
Washington, D.C.

INTRODUCTION

The most common cause of bridge failures is scouring of bridge foundations during floods. Although the failure of the New York Thruway Bridge over Schoharie Creek in 1987 (10 lives were lost) focused national attention on scour related failures, many other bridges have been victims of scour. Examples include failures of: the I-29 crossing of the Big Sioux River in Iowa in 1962; the I-80 crossing of the John Day River in Oregon in 1964; 73 bridges destroyed by flooding in Pennsylvania, Virginia, and West Virginia in 1985; 17 bridges in New York and New England in the spring of 1987; the US 51 bridge over the Hatchie River in Tennessee in 1989 (eight people were killed); and the I-5 bridges over Arroyo Pasajero in California in 1995 (seven people were killed).

Obviously, the scour-associated bridge failures are not isolated incidents nor ones that could be ignored by highway agencies. The concern for scour related problems is evident when one realizes that approximately 86 percent of the 577,000 bridges in the National Bridge Inventory (NBI) are built over waterways, and when over 10,000 of these bridges have been determined to be scour critical--meaning that the stability of the bridge foundation has been or could be affected by the removal of bed material.

FEDERAL HIGHWAY ADMINISTRATION TECHNICAL ADVISORY

With the Schoharie Creek failure, the Federal Highway Administration (FHWA) began to review its guidance in the area of scour. In September of 1988, FHWA issued Technical Advisory (TA) T 5140.20 entitled "Scour at Bridges." On October 28, 1991, it was updated as T 5140.23 and entitled "Evaluating Scour at Bridges." This TA is very comprehensive and focuses on the development and implementation of a scour evaluation program for: designing new bridges to resist damage from scour, evaluating existing bridges for vulnerability to scour, using scour countermeasures, and improving the state-of-practice of estimating scour at bridges. It also recommends that a program be developed to evaluate every bridge over a scourable stream, existing or under design, to establish if it is vulnerable to scour.

Five Step Process
The guidance from the TA's gives a five-step procedure to follow:
1. An interdisciplinary team of hydraulic, geotechnical, and structural engineers should conduct scour evaluations.
2. New bridges should be designed to be scour safe for a superflood on the order of the magnitude of a 500-year flood.
3. All existing bridges over waterways with scourable beds should be evaluated for the risk of scour failures for such a superflood.

BRIDGE SCOUR EVALUATION PROGRAM

4. A plan of action should be developed for all bridges that are determined to be scour critical.
5. All bridges should be inspected for scour during the regular two year bridge inspection cycle.

FHWA also has two technical publications that provide technical guidance: Hydraulic Engineering Circular Number 18 entitled "Evaluating Scour at Bridges" (HEC-18) and Hydraulic Engineering Circular Number 20 entitled "Stream Stability at Highway Structures" (HEC-20). HEC-18 provides guidance in developing a scour evaluation program and analyzing bridges for scour. HEC-20 provides guidance in analyzing the effect of stream instability on bridges.

Interdisciplinary Team

In designing new and analyzing existing bridges for scour, a careful evaluation of the hydraulic, geotechnical, and structural aspects of bridge foundations is required. A team of experienced engineers is needed to make engineering judgments due to the complex nature of streams, flow patterns, soil, and structure design. The team should establish priorities for scour evaluations, determine if bridges are scour critical, and recommend countermeasures and monitoring schedules.

New Bridges

New bridges over waterways with scourable beds should be designed for scour from floods equal to or less than the 100-year flood. All bridge foundations should be checked for scour resulting from a superflood (i.e., a 500-year event or 1.7 times the 100-year event). The geotechnical analysis should assume that all stream bed material in the scour prism above the total scour line for the scour design flood has been removed and is not available for bearing or lateral support. The geotechnical analysis for the superflood should incorporate a safety factor of 1.0. HEC-18 should be followed in conducting and documenting the results of the scour evaluation studies. HEC-20 should be used in identifying stream stability problems at bridge sites.

Existing Bridges

As with the new bridges, all existing bridges over waterways with scourable beds should also be evaluated for the risk of failure from scour during the occurrence of a superflood. The 500-year flood is recommended; however, some States are now using a 100-year flood. The three basic elements of the TA relative to existing structures are to:
1. Screen all existing bridges to prioritize the order of scour evaluations on the basis of scour susceptibility.
2. Evaluate scour of all bridges with scourable beds to determine whether the bridges are scour critical.
3. Establish a plan-of-action for each identified scour critical bridge.

Target Dates

FHWA requested that all existing bridges be screened and ranked for evaluation by March 31, 1991, and the States have now essentially completed this task. In July 15, 1991, the target date for completion of the evaluation phase of the process was set for January 1, 1997. This date

applied to bridges that had been identified by the screening process as:
1. "Low risk" that cannot be coded as 4, 5, 7, 8, or 9 within Item 113, Scour Critical Bridges, of the December 1988 Recording and Coding Guide;
2. "Scour Susceptible" that are coded as 6 within Item 113; and
3. Interstate bridges with "unknown foundations."

Bridges with unknown foundations and those affected by tidal action were exempt from the target date.

The evaluation process for the scour susceptible bridges should follow the process given in HEC-18 and HEC-20. From this process, all scour critical bridges should be identified. A scour critical bridge is defined as one that is in imminent danger of scour failure based on analyzed conditions using state-of-the-art scour methodology.

PLAN OF ACTION

A plan of action for each scour critical bridge should be developed by the interdisciplinary team. It should include:
- instructions for the type and frequency of inspections to be made at the bridge site;
- monitoring the bridge's scour performance with contingency for closure;
- and/or scheduling timely design and construction of scour countermeasures.

Each of these items is covered in detail in the TA and HEC-18 and HEC-20. A brief overview of HEC-18 is given below.

HYDRAULIC ENGINEERING CIRCULAR No. 18

HEC-18 contains the state-of-the-art methodology for evaluating scour at highway bridges. The third edition of HEC-18 presents the latest advances in technology including: conversion to the metric system of units; the addition of a gradation correction factor for the pier scour equation; an equation for estimating the correction factor for the flow angle of attack with respect to a pier; an interim procedure for estimating pier scour considering the effect of debris; and updated information on scour detection equipment. Furthermore, clarification has been added for: estimating pier scour for exposed footings; pile caps located at different elevations in the flow; the effect of multiple columns skewed to the flow; scour resulting from pressure flow; and criteria for designing the foundation depth of a bridge abutment.

Chapter 1 - Introduction
This circular provides the technical guidance to implement the TA. Past bridge failures and potential ones justify the need for a bridge scour evaluation program which will assess every bridge over a scourable stream for its vulnerability to scour. Although this circular includes the state-of-the-art in scour estimation, many of the equations used need to be verified by field data. Additional research is needed to cover the various situations that must be evaluated. Scour monitoring equipment is needed to monitor scour critical bridges until they are replaced.

BRIDGE SCOUR EVALUATION PROGRAM 113

Chapter 2 - Basic Concepts and Definitions of Scour
Total scour resulting solely from a bridge consists of two components; contraction scour caused by the constriction of the waterway through the opening and local scour generated by the obstruction of piers and abutments. Other problems such as degradation, aggradation, and lateral migration in the reach of the waterway near the crossing must also be considered. Scour caused by the bridge is determined from HEC-18 procedures, while bed and bank movement in the adjacent channel reach can be evaluated using HEC-20.

Chapter 3 - Designing Bridges to Resist Scour
Hydraulic studies of bridge sites are a necessary part of the preliminary design of bridges since scour analysis must be part of the design procedure. They should address sizing of the waterway area and estimating scour for use in foundation design. Their scope should be commensurate with the importance of the highway and the consequences of failure.

Since the current technology is based on research with small models, the studies must be developed using sound engineering judgement. All results should be compared with available site and hydraulic data to arrive at a practicable foundation design.

Chapter 4 - Estimating Scour at Bridges
Before the contraction and local scour can be estimated, the engineer must: determine the rigid boundary channel hydraulics; estimate the long-term stream bed changes; adjust the rigid boundary channel hydraulics to reflect the stream bed elevation changes; and compute the bridge hydraulics. To obtain the total scour at a bridge, contraction, abutment, and pier scour must be established. A plot of the total scour depth would be a sum of the above components.

Chapter 5 - Estimating the Vulnerability of Existing Bridges to Scour
A recommended procedure for evaluating all of the bridges over scourable streams is presented (86 percent of the 577,000 bridges on the National Bridge Inventory are over waterways and are potential candidates). The short-term goal was to evaluate existing bridges with known problems. The long-term goal is to evaluate all existing bridges over scourable streams.

Chapter 6 - Inspection of Bridges for Scour
The two objectives in inspecting bridges for scour are: to record accurately the present condition of the bridge and stream, and to identify conditions that are indicative of potential problems with scour and stream stability. Should an inspector observe conditions of an immediate or potentially hazardous nature, there must be a procedure established to communicate promptly these findings to proper agency personnel.

A bridge inspection is not a scour evaluation. The factors to be considered in scour evaluation require a broader scope of study and effort than those considered in a normal bridge inspection. The purpose of a bridge inspection is to identify changed conditions which may reflect an existing or potential problem; however, a scour evaluation is an engineering assessment of what might happen in the future and the steps which can be taken to eliminate or minimize future damage.

Chapter 7 - Plan of Action for Installing Countermeasures

In developing a plan of action for a scour critical bridge, considerations may include: monitoring the site with contingency for closure; installing temporary countermeasures such as riprap with monitoring during high flows; designing permanent countermeasures, and scheduling countermeasure construction. Typical countermeasures which make bridges less vulnerable to damage or failure include riprap, guide banks (spur dikes), and channel improvements.

STATUS OF THE BRIDGE SCOUR PROGRAM

National Report

The FHWA initiated semiannual status reports on bridge scour on February 5, 1990. Through April 15, 1996, 98.7 percent of the 484,916 bridges reported over waterways had been screened into the following categories:

LOW RISK (culverts)	-	99,978 (20.6%)
LOW RISK (calc.)	-	137,386 (28.3%)
SCOUR SUSCEPTIBLE	-	131,943 (27.2%)
UNKNOWN FOUNDATIONS	-	99,252 (20.5%)
SCOUR CRITICAL	-	10,276 (2.1%)
NOT SCREENED	-	6,071 (1.3%)

This data is from the State highway agencies' April 15, 1996, status reports. The estimated cost for scour evaluations nationwide is approximately $3,000 per bridge for a total cost of $415 million for the 138,014 bridges that need to be evaluated (scour susceptible and not screened).

The FHWA has projected that several bridge owners will not complete their scour evaluations by the January 1, 1997, target date. In an effort to bring compliance to a more acceptable level, the FHWA field office personnel are reviewing the action plans for those bridge owners that are projected to be less than 90 percent completed with their scour evaluations by January 1, 1997.

The next logical step after this initial goal has been met will be to start the evaluations of bridges over tidal waterways. Guidance and procedures to obtain the hydraulic variables for a scour evaluation of bridges in tidal waterways have been expanded in the third edition of HEC-18.

Technology for determining unknown foundations is currently available through the use of nondestructive testing (NDT) methods. The National Cooperative Highway Research Program (NCHRP) sponsored a project titled "Determination of Unknown Subsurface Bridge Foundations" targeted to evaluate, develop and test NDT concepts, methods and equipment that will allow the determination of subsurface bridge foundations. The FHWA reviewed the project report which includes techniques for bridge owners to use for determining the foundation type and its depths. A FHWA Geotechnical Engineering notebook issuance will be developed which will contain a summary of the techniques contained in the NCHRP report. After January 1, 1997, bridge owners will probably be asked to apply some of these techniques to determine some types of unknown foundations and then complete a scour evaluation for those foundations.

Bridge owners have also been implementing a plan of action for installing scour countermeasures at bridges which have been determined to be scour critical. Typically, riprap is the most commonly used countermeasure for protecting bridge piers and abutments; however, the use of this countermeasure as well as others have raised concerns by environmental agencies. The FHWA is currently very active in coordinating with bridge owners and environmental agencies to create a consensus on the importance of installing a countermeasure in a timely manner at an existing bridge with a potential scour problem. Furthermore, the NCHRP is sponsoring a project titled "Countermeasures to Protect Bridge Piers from Scour" targeted to evaluate various countermeasures to protect bridge piers. Additionally, the FHWA is currently sponsoring a demonstration project, Demonstration Project 97 titled "Scour Monitoring and Instrumentation," intended for Federal, State and local engineers and managers responsible for underwater bridge inspection and identifying and monitoring bridge scour. This demonstration project presents fixed and portable instrumentation, monitoring devices, positioning, and data collection systems that serve to monitor the progress of scour at a bridge.

SUMMARY

To summarize the current scour program, Technical Advisory T 5140.20 entitled "Scour at Bridges" was issued on September 16, 1988. This TA provided recommendations and guidance on how to develop and implement a scour evaluation program. It was updated by Technical Advisory T 5140.23 issued on October 28, 1991. From these publications, the essential elements of a scour evaluation program follow:
- A screening process to identify bridges most likely to be vulnerable to damage from scour and thus need early attention.
- Engineering evaluations of existing bridges to establish which bridges are scour critical and the reporting of the results of these evaluations in Item 113 of the revised Bridge Recording and Coding Guide.
- A plan of action for monitoring and providing countermeasures for the scour critical bridges.

In November 1988, the FHWA stressed that "...immediate steps be taken to get the scour evaluation program underway in all the States...." On February 5, 1990, a semiannual reporting program was established to help stress the importance of implementing the scour evaluation program. On July 15, 1991, the target date of January 1, 1997, was established as the date by which all bridges should be evaluated which had been identified through the screening process as: (1) "low risk" that cannot be coded as 4, 5, 7, 8, or 9 within Item 113 of the December 1988 Recording and Coding Guide; (2) "scour susceptible"; and (3) Interstate bridges with "unknown foundations" and/or over tidal waterways. On November 8, 1996, the FHWA field personnel were ask to review the action plans for those bridge owners that are projected to be less than 90 percent completed with their scour evaluations by the target date.

The FHWA and State highway agencies, as well as other parties, have conducted research for many years to improve the estimating procedures for determining scour at bridges. Even so, the best available current technology represented by HEC-18 needs improvement. Ongoing and proposed research within the highway hydraulic community addresses scour needs and will result in improved methodology.

COPING WITH SCARCITY AND ABUNDANCE

Finally, the FHWA, in partnership with bridge owners will continue the scour evaluation efforts to aid in ensuring the safety of our Nation's bridges for use by the traveling public.

REFERENCES

Technical Advisory 5140.20, "Scour at Bridges," Federal Highway Administration, Washington, D.C., 1988

Technical Advisory 5140.23, "Evaluating Scour at Bridges," Federal Highway Administration, Washington, D.C. 1991.

Hydraulic Engineering Circular No. 18, "Scour at Bridges," Third Edition, Federal Highway Administration, U.S. Department of Transportation, Washington, D.C., November 1995

Hydraulic Engineering Circular No. 20, "Stream Stability at Highway Structures," Second Edition, Federal Highway Administration, U.S. Department of Transportation, Washington, D.C., November 1995

Presentation by Mr. Philip Thompson at the April 11-15, 1994, Spring Meeting of the AASHTO Task Force on Hydrology and Hydraulics in Charleston, S.C.

FINDINGS OF THE I-5 BRIDGE FAILURE

EVERETT V. RICHARDSON, J. STERLING JONES, AND JAMES C. BLODGETT

Senior Engineer, Ayres Associates, and Emeritus Professor of Civil Engineering at Colorado State University, Fort Collins, Colorado; Hydraulics Research Engineer, Turner-Fairbank Highway Research Center, Federal Highway Administration, McLean, VA; Senior Engineer, Northwest Hydraulic Consultants, Hydrologist (retired), U.S. Geological Survey, Sacramento, California

ABSTRACT

On March 10, 1995 the two I-5 bridges over Los Gatos Creek (Arroyo Pasajero) near Coalinga, California failed because of scour resulting from a large flood. Four cars and 1 truck went into the Arroyo with the loss of 7 lives. An investigation into the magnitude of the scour depths using the Level 1, 2, and 3 analysis approach recommended by the Federal Highway Administration Publications HEC 18 and 20 resulted in a determination that the minimum potential total scour depth of 25.2 ft would result in the column bents having an embedment of 15.8 ft. But more important would have exposed 8.9 ft of the cast-in-place concrete columns that have no steel reinforcement. The force of water and debris, at an angle of attack of the flow from 15 to 26 degrees, acting on the columns with a 8 to 12 ft high web wall constructed around and between them for reinforcement, caused them to fail.

INTRODUCTION

At 8:30 pm on March 10, 1995 the I-5 bridges over Arroyo Pasajero (Los Gatos Creek) near Coalinga, California failed because of scour resulting from a large flood. Four cars and 1 truck went into the Arroyo with the loss of 7 lives. The truck driver saved himself by going out of the truck's back window and grabbing a tree. At the time of the failure I-5 was closed except for local traffic. In the same storm California lost two other bridges from scour but no loss of life. This paper documents the investigation to determine why the bridge failed due to scour. The investigation used the three level analysis approach given in HEC-18 and HEC-20 (Richardson and Davis 1995 and Lagasse et al. 1995). Level 1 qualitative analysis was done by Arlo Waddoups and J. Sterling Jones, FHWA; William Lindsey, CALTRANS; James C. Blodgett and David Mueller, USGS; and E.V. Richardson, Ayres Associates. Level 2 quantitative engineering analysis was conducted by James Blodgett; E.V. Richardson; R.C. Cassano, retired CALTRAN structural engineer; and R.A. Hunrichs, USGS; Level 3 physical and computer models were conducted by J. Sterling Jones, David Bertoldi, GKY & Associates, A. Molinas, Colorado State University and Xibing Dou, Graduate Research Fellow from University of Mississippi.

Seven days after the failure, CALTRAN had in operation a temporary 2 lane bridge in the median between the washed out bridges. On March 23, CALTRAN awarded the contract to construct the replacement bridges, which were open to traffic on April 12 (33 days after the failure). The replacement bridges are 254 ft long, have reinforced cast-in-place columns embedded to elevation 385 which is 25 ft deeper than the original foundations. The abutments are spill through, with the right abutment protected by a 363 ft long guidebank and the left abutment and roadway approach protected by a 1,037 long dike.

THE BRIDGE (LEVEL 1)

The two bridges over Arroyo Pasajero transported north and south bound traffic of I-5 in the central valley of California. The bridges, built in 1967, were 122 ft long, with vertical wall abutments with wing walls and three bents spaced 27-34-34-27 ft. Each bent consisted of 6-16 inch columns spaced approximately 7.5 ft on centers. The columns were embedded 41.0 ft (elevation 410 ft) below the original ground surface (elevation 451.0) but only had steel reinforcing for 12 ft below elevation 446.7 ft. The bottom of the deck was at elevation 471. The abutments were on pile supported footings. The bottom of the footings were at elevation 446.7 and the piles were 36.7 ft long. A 1969 flood lowered the stream bed approximately 6 ft and damaged one column. In repairing the damage a web wall 8 or 12 ft high, 38 ft long and 2 ft wide was constructed around the columns to reinforce them. These dimensions were determined from discussions with CALTRAN District Engineers because no plans were available. Also, the elevation of the bottom of the wall was not documented. The most likely elevation was probably at or slightly below the new elevation of the stream bed. In addition, walls were built around the abutment footings to reinforce them.

GEOMORPHOLOGY AND LAND USE (LEVEL 1)

Los Gatos Creek is an ephemeral stream which drains from the eastern side of the coastal range onto an alluvial fan approximately 2 miles upstream of the bridges. About 1,800 ft upstream of the bridges Chino Creek (also ephemeral) joins Los Gatos Creek. At the time of the construction of the bridges Chino Creek did not join Los Gatos Creek but infiltrated into its alluvial fan. Any flow in Chino Creek that did not infiltrate into its fan was carried to Los Gatos Creek along I-5 in an interceptor ditch. Both streams have a very large sand, silt and clay sediment discharge. Sediment concentrations could be as large as 30 to 40 percent by weight. At all flows, the bed configuration would have been plane bed, standing waves or antidunes.

Los Gatos Creek normally infiltrates into its alluvial fan and disappears in a basin upstream of the California Aqueduct. The aqueduct is approximately 8 miles downstream of the bridges. During this flood the Creek deposited a large amount of debris into the basin. A local road through the basin was covered with debris over a mile wide and 10 to 12 ft thick in the middle.

Los Gatos Creek is dry most of the time and braided at low flow but flows bank to bank at high flows. It is sinuous up and downstream of the bridge. The stream bed is sand. In the reach 1.5 miles upstream of the bridge the sand has a median diameter of 0.34 mm for depths from 0.5 to 3.0 ft and 0.7 mm from 3.0 to 7.0 ft. Banks are composed of sand, silt and clay; erodible, and mostly vertical. There is a stiff, tan, clayey silt hard pan in the sand. This hard pan is not uniform, is erodible, and formed the cap of a 4 to 6 foot headcut that was found 1,000 ft downstream of the bridge after the flood. Two months after the flood following additional flows, the headcut was only 2 ft high and located 500 ft below the bridge. Original boring logs had the hard pan at elevation 404-395 ft behind the right abutment; no hard pan at the channel centerline at the bridge; and at elevation 430-422 behind the left abutment. The boring log located behind the right abutment gave slightly compact tan medium sand (457-436), compact tan slightly silty sand (436-430), tan clayey silt (430), slightly compact tan medium silty sand (430-422), slightly compact tan interlaced lenses of clayey silty sand and medium sand (422-413), very dense tan sandy gravel (413-404), compact clayey silt (404-395) and dense tan medium silty sand below 395 ft. The boring log behind the left abutment gave slightly compact tan fine sand (458-449), compact tan medium sand (449-430), stiff clayey silt (430-422), dense tan fine sand (422-412) and very dense gravely sand below 412 ft.

Differential land subsidence has occurred along Los Gatos Cr. since the 1930s due to ground water pumpage. In the time since construction (1967 to 1995) subsidence at bench mark T 1228, which is approximately 1.5 miles upstream of the bridges, is negligible. Whereas, at bench mark C 889, (about 5.3 miles downstream) the subsidence since construction is 3.8 ft. Total subsidence at these two benchmarks, from 1955 to 1995, is 1.1 and 12.6 ft, respectively.

Land use upstream of I-5 at the time of construction was mostly grazing. Now, land use up- and downstream is farming. With the advent of farming Chino Creek was channelized and has flowed directly into Los Gatos Creek instead of disappearing into its alluvial fan since at least 1971 according to the USGS 7.5 minute Guijarral Hills topographic map.

The Los Gatos Creek channel is from 300 to 400 ft wide upstream of the bridge and 150 to 250 ft wide downstream. The narrower channel downstream results from the narrow bridge opening. Also, the bridge is at an angle to the flow, which resulted in an angle-of-attack estimated in the level 1 analysis to be from 15 to 26 degrees.

FLOOD MAGNITUDE (LEVEL 2)

Maximum 24-hr precipitation amounts during the March 10 storm at precipitation stations in the Los Gatos Creek drainage area ranged from 3.6 to 6.18 inches. Based on precipitation-frequency analysis by Miller et al (1973) and a Log Pearson Type III distribution of the historical precipitation record 1912-95 for the Coalinga station, the average annual frequency for this storm is greater than 100 years. The average annual frequency of precipitation amounts recorded at Coalinga for durations of 2 and 3 days also exceeded 100 years. However, rainfall over the Chino drainage area was not as large (ranging from 1.5 to 3.0 inches).

The March 10, 1995 flood was the largest of record (1950-95) at the only long term USGS gaging station (11224500) in the Los Gatos Creek basin. The magnitude of the 1995 flood at the I-5 bridges was determined by adding the flow at the California Department of Water Resources (DWR) gage at El Dorado Avenue (11225100) near Coalinga, which is about 1.5 miles upstream, and the flow from Chino Creek. The USGS made a three section slope-area measurement at a site about 300 ft downstream of the DWR gage and upstream of the confluence with Chino Creek. Different combinations of most likely Manning n values and channel area changes resulting from scour gave discharges ranging from 16,300 cfs to 40,300 cfs. The most probable discharge was determined to be 25,000 cfs with a unit discharge of 61.4 cfs per square mile (csm). To this was added 2,300 cfs as the estimated contribution from Chino Creek.

As an independent check on the peak flow discharge at the DWR gage, peak discharges were measured by indirect methods at 6 other sites in the basin. The unit discharge at these sites varied from 158 csm to 24.2 csm, depending on the location of the measurement site and drainage area. A best fit graphical relation of peak discharge versus drainage area for these measurements indicates the peak discharge at the DWR gage at El Dorado Ave. would be about 28,000 cfs. The difference in the two calculated peak discharges is 12 percent, which is considered good agreement between the two methods.

The peak discharge at the I-5 bridges, with a drainage area of 507 square miles, was taken to be 27,300 cfs (53.8 cms). The average recurrence interval of 25,000 cfs at the El Dorado Ave. gage was calculated to be 75 years based on historic station data and 66 years based on the regional equations (Waananen and Crippen 1977).

MODEL STUDIES (LEVEL 3)

A **physical model** study was made at the FHWA Turner-Fairbank Research Center Hydraulics Laboratory to determine the effect of the web wall around the columns and the estimated angle of attack. The guidelines given in HEC-18 do not cover the I-5 pier configuration. The model study was conducted by J. Sterling Jones and David Bertoldi. Eleven tests (runs) were made with 1:20 scale. Paired tests were run with an angle of attack of 15 and 26 degrees. The model tests were made under the assumption that the columns were 18 inches (1.5 ft) in diameter and the web wall was 2 by 8 ft and 38 ft long. The as-built plans gave the column diameter as 16 inches (1.33 ft). Results of the model study are given in Table 1.

Table 1. I-5 Local Pier Scour Depths Based on Model and HEC-18 Results.

Run No.	Test Condition	Angle of Attack (degrees)	Vel Modeled (ft/s)	Adjusted Prototype Scour (ft)	Pier Width, a (ft)	Pier Length, L (ft)	K_2	HEC-18 Predicted Scour (ft)
10	No web wall	15	4.81	6.42	1.5	1.5	1.2*	6.3
		15			1.5	9.0	1.8	9.5
11	No web wall	26	4.82	6.23	1.5	1.5	1.2*	6.4
		26			1.5	9.0	2.3	12.2
3	Web wall at bed	15	4.97	7.08				NA
2	Web wall at bed	26	4.63	8.93				NA
9	Web wall above bed	15	4.86	6.67				NA
1	Web wall above bed	26	4.73	6.75				NA

$K_1 = 1.0$, $K_3 = 1.1$ for plane bed or antidune flow, flow depth modeled was 24.1 ft
*HEC-18 recommends multiplying single column depth by 1.2 when multiple columns are spaced 5 diameters or greater apart.

Model tests were conducted at incipient motion velocities for the bed material that was available for the flume. The adjusted prototype scour depths given in Table 1 were obtained by a velocity factor as follows:

$$y_{s\ adjusted} = y_{s\ from\ model} \times (V_{Actual} / V_{Modeled})^{0.43}$$

where V_{Actual} = Actual equilibrium velocity that was believed to occur after contraction scour enlarged the bridge opening.

$V_{Modeled}$ = Prototype velocity modeled by the incipient motion velocity used in the flume test.

The exponent 0.43 is the exponent of the Froude Number which contains the velocity term in the HEC-18 pier scour equation.

The most likely elevation of the piers on March 10 would have been with the web wall at or slightly above the bed. With contraction scour during the flood the web wall would have been above the bed. Therefore, the most likely local scour depth would have been 6.7 ft as given in runs 1 and 9. Note how well HEC-18 predicted the local scour depth for the condition of no web wall. With columns alone the model tests gave scour depths of 6.2 or 6.4 ft, whereas, calculated scour depths from HEC-18 equations and procedures gave 6.3 or 6.4 ft.

The FHWA **computer model** BRI-STARS (Molinas 1990 and Molinas and Wu 1995) using Yang's sediment transport equation, bed material size of 0.2 to 0.3 mm and a discharge of 27,300 cfs computed contraction scour depth of 10.0 ft.

SCOUR ANALYSIS (LEVEL 2)

Long term degradation of the channel as determined from bridge inspection reports (probably from the differential subsidence) was 10.0 ft. About 6 ft of degradation occurred in the 1969 flood, shortly after the bridge was built in 1967. Contraction scour, as determined using Laursen's live bed contraction scour equation given in HEC-18, was 8.5 ft. BRI-STARS gave a contraction scour depth of 10.0 ft. Local scour depth at the piers as determined from the model studies ranged from 6.7 to 8.9 ft (Table 1). However, the most likely local pier scour depth (6.7 ft) would result from the web wall being above the bed. Therefore, the most likely minimum total scour depth would have been 25.2 ft (10.0 + 8.5 + 6.7). Total scour depth could have been deeper than this given the uncertainty of the elevation and height of the web wall and disregarding any effect of debris on local scour.

The very large contraction scour resulted from the very wide approach channel (322 ft of active flow area) and the narrow bridge opening (122 ft). Water surface elevations measured upstream of the bridge were approximately level for 2,000 ft, showing that there was considerable ponding of water upstream of the bridge. However, some of the ponding probably results from the bridge failure.

The minimum total scour depth would have reduced the embedment of the columns from 41.0 ft to 15.8 ft. Also, the minimum total scour depth would have resulted in 8.9 ft of unreinforced column with no soil support. The lateral force of the flood waters at an angle of attack of 15 to 26 degrees on columns with 8.9 ft of their unreinforced length unsupported by the soil because of scour, and with their area increased by the web wall and debris, would cause the bridge to fail.

CONCLUSIONS

The scour that caused the failure of the twin I-5 bridges over Arroyo Pasajero (Los Gatos Creek) was the result of a large flood with from a 66 year return period based on regional frequency analysis to 75 year return period based on Log Pearson Type III analysis of historic gaging station data. The best estimate of the discharge was 27,300 cfs based on an indirect measurement in the Los Gatos Creek channel 300 ft downstream from the gaging station, which was 1.5 miles upstream from the bridges. This discharge was supported by a study using indirect flow measurements to determine the discharge per square mile at six other sites in the Los Gatos basin.

Using records of long-term degradation (10.0 ft), computation of contraction scour (8.5 ft by Laursen's equation and 10.0 by BRI-STARS) and a physical model study to determine local pier scour (6.7 or 8.9 ft), the minimum total scour depth was determined to be 25.2 ft (10+8.5+6.7). This left an imbedment of 15.8 ft, but exposed 8.9 ft of the cast in place columns without steel reinforcement. The force of the flood waters at an angle of attack of 15 to 26 degrees on columns with 8.9 ft unreinforced, with their area increased by the web wall and debris, caused the piers to fail.

Comparison of the results of the physical model study of local scour on the columns with calculations using the CSU pier scour equation given in HEC-18 showed very good agreement (6.2 or 6.4 ft in the model and 6.3 or 6.4 ft by HEC-18).

Comparison of contraction scour depths calculated by Laursen's live bed scour equation given in HEC-18 with that calculated by FHWA's computer program BRI-STARE showed good agreement (8.5 ft versus 10.0 ft).

DISCLAIMER and ACKNOWLEDGMENT

The analysis and conclusions are by the authors using data collected by the authors or obtained from the USGS, FHWA, CALTRAN, or DWR and do not represent the opinions of the latter agencies. However, the authors wish to acknowledge the cooperation of theses agencies in providing the data.

REFERENCES

Lagasse, P.F., Schall, J.D., Johnson, F., Richardson, E.V. and Chang, F., 1995. "Stream Stability at Highway Structures, 2nd Edition," HEC-20, Pub. No. FHWA-IP-90-014, Federal Highway Administration, US Department of Transportation, Washington, D.C.

Miller, J.F., Frederick, R.H. and Tracey, R.J., 1973. "Precipitation-Frequency Atlas of the Western United States," National Oceanic and Atmospheric Administration Atlas 2, Vol. XI, California.

Molinas, A., 1990. "Bridge Stream Tube Model for Alluvial River Simulation (BRI-STARS)," User's Manual, NCHRP, Project No. HR15-11, Transportation Research Board, Washington, D.C.

Molinas, A., and Wu, B., 1995. "Personal Communication."

Richardson, E.V. and Davis, S.R., 1995. "Evaluating Scour at Bridges, 3rd Edition," HEC-18, Pub. No. FHWA-IP-90-017, Federal highway Administration, U.S. Department of Transportation, Washington, D.C.

Waananen, A.O. and Crippen, J.R., 1977, "Magnitude and Frequency of Floods In California," USGS Water Resources Investigations 77-21, Menlo Park, CA.

The Need for Research on Scour at Bridge Crossings

A.C. PAROLA AND D.J. HAGERTY
Department of Civil Engineering, University of Louisville, Louisville, KY USA

D.S. MUELLER
United States Geological Survey, Kentucky District, Louisville, KY USA

B.W. MELVILLE
Department of Civil and Resources Engineering, University of Auckland
Auckland, New Zealand

G. PARKER
Department of Civil Engineering, University of Minnesota
Minneapolis, MN USA

J.S. USHER
Department of Industrial Engineering, University of Louisville
Louisville, KY USA

ABSTRACT

Scour at bridge crossings is a well-recognized cause of bridge failure. Evaluation of bridges for scour susceptibility and design of countermeasures is necessary to ensure safe bridges and minimize the disruption of traffic due to bridge closure. Current scour evaluation methods are based primarily on techniques derived from idealized small-scale laboratory physical model investigations that consider a limited number of parameters known to affect scour processes. These methods are considered as being conservative under most conditions although only limited verification has been conducted. This paper summarizes a study (Parola et al. 1997) sponsored by the National Cooperative Highway Research Program (NCHRP) to identify deficiencies in the current state-of-knowledge and practice and to develop a strategic plan for scour research.

INTRODUCTION

Each year floods in the United States inundate bridge crossings and cause scour (erosion of the streambed, streambanks and floodplain around bridge foundations) that damages or endangers bridges to the point that bridge closure is required. Costs of been estimated as more than five times the repair and replacement costs (Rhodes and Trent 1993). Scour has been the leading cause of bridge failure in the United States (Murillo 1987). Parola et al. (1996) found that over 400 bridge crossings on federal-aid routes and over 2000 non-federal aid bridges were damaged during the 1993 Midwest flooding. Evaluation of bridge crossing susceptibility to scour is essential to ensure public safety and to prevent costly disruption of traffic.

Research on scour around bridges has been conducted for well over 100 years (Engels 1894); however, estimating channel scour still remains problematic. The need to design for extreme flood events, a lack of fundamental knowledge of stream and river system development and evolutionary processes, the widely varying conditions under which such systems develop and the complexity of scour around bridge encroachments are major impediments to the formulation of highly accurate scour prediction methodologies. This paper provides a summary of a research project sponsored by the National Cooperative Highway Research Program to develop a strategic plan for scour research.

OBJECTIVES

The purpose of the NCHRP research project was to identify the research needs for bridge scour and stream stability and to develop a research plan to coordinate numerical model development, laboratory physical model studies, flood and field studies and instrumentation development. The primary goal of this project was the development of a balanced research plan that organizes and specifies the scour research and related fluvial instability research necessary to substantially improve bridge scour evaluation, inspection and design technology.

METHODOLOGY

Assessment of research needs was developed on the basis of a comprehensive review of the literature on scour and relevant fluvial geomorphology, a survey of practicing engineers and an examination of case studies of flood damaged bridges. The survey of practicing engineers was particularly helpful in determining the research priorities. The literature review provided a context from which the limitations of current scour prediction methodology could be assessed and will provide the basis for future literature reviews on scour. Examination of the conditions under which recent bridge failures have occurred was valuable in identifying parameters that have not been considered and in understanding the importance of considering the integrity of the entire highway network during extreme flood events. Project statements that outlined research investigations necessary to address the needs were developed from the assessment of research needs. The project statements were reviewed during a two day workshop by 30 individuals active in scour research or state scour evaluation programs. The comments of the workshop participants were used to modify the project statements and to sequence them into a balanced research plan.

RESEARCH NEEDS AND STRATEGIC PLAN

A combination of flood and field studies, physical model studies, and numerical model studies is required to expand the state of knowledge and improve scour prediction techniques. Figure 1 indicates the important potential contributions of each study type to the advancement of scour prediction methods. Research efforts in the past have been focused on the characterization of dominant scour processes and on the development of predictive scour equations primarily from small-scale laboratory investigations. A shift in emphasis toward flood and field studies is necessary to identify and describe scour mechanisms that have not been represented or may be difficult or impossible to accurately represent in physical model studies. Development and enhancement of multi-dimensional hydraulic and sediment transport models are required to provide the parameters necessary to evaluate complex flood conditions. Emphasis on flood and field studies and on numerical model studies does not imply reduced cost-effectiveness, usefulness or need for laboratory studies. An integration of these investigative techniques is recommended to transcend the empirical relations developed purely from small-scale laboratory studies.

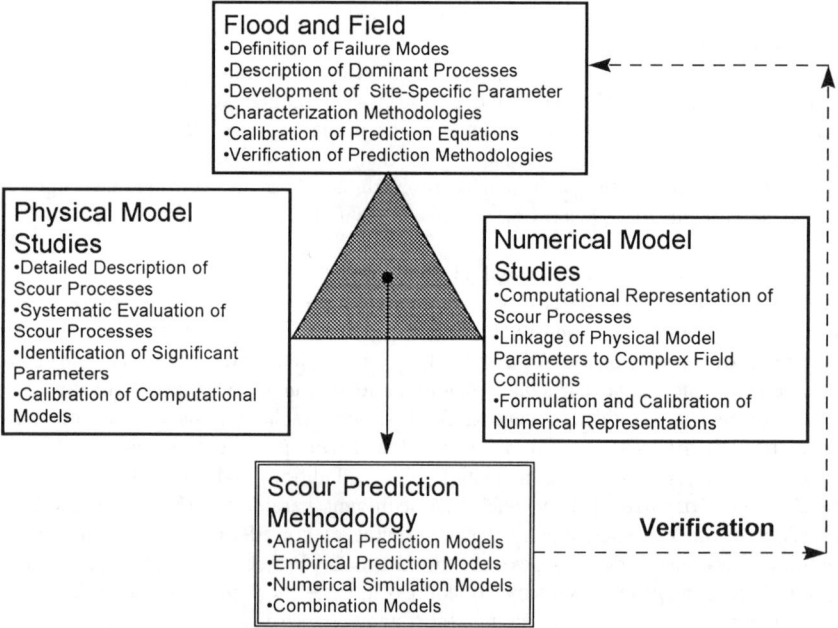

Figure 1. Advancement of Scour Prediction Methodology.

Flood and field studies provide several critical functions. First, such studies identify and describe dominant scour processes that are not amenable to study in small-scale physical models. Second, such studies characterize complex sequences and specific mechanisms of bridge failure so that prediction methodology can be targeted at actual rather than hypothetical failure modes. Knowledge of bridge failure modes and sequences of scour mechanisms is critical for cost-effective design of countermeasures. Understanding of mechanisms and sequences of scour processes is essential to develop simplified physical and numerical models that quantify scour effects. Third, investigations during floods provide verification data against which the accuracy of scour prediction methodology can be evaluated. Fourth, field studies provide a means of developing methods to characterize the erodibility of fine-grained (cohesive) soils, rock masses, and vegetal floodplain covers.

Instrumentation has been developed recently to measure flow velocity and stream topography in the complex and hazardous flood conditions characteristic of bridge crossings. This instrumentation may still be ineffective at some locations (e.g., very shallow, sediment-laden flow). Advancements in remote sensing technology will provide additional instrumentation necessary to broaden the conditions under which flow parameters can be measured. Research that adapts this instrumentation specifically for measurement of flow and scour at bridges will provide the tools necessary for collection of the relevant data.

One-dimensional and two-dimensional hydraulic models have become effective tools for evaluating complex flood conditions in terms of the parameters required for bridge scour prediction. Three-dimensional numerical models have been developed to near the level of sophistication required to represent the complex flow near bridge components such as piers. These models will be an important element of future research on flow and scour at bridges. Continued development of one-dimensional models is needed to incorporate the specific hydraulics of actual bridges, sediment transport, and unsteady flow into a single general purpose model applicable to both river and estuary conditions. Research is needed to provide estimates of velocity correction factors for flow in bends and for flow distribution through bridge openings. Although appropriate velocity information cannot be derived directly from one-dimensional models, correction factors may be developed from studies using multi-dimensional models and physical models applicable to a large number of bridge conditions. Two-dimensional and 2.5-dimensional models can be used to generate the hydraulic and sediment transport information necessary to evaluate scour at bridges except where strong flow curvature exists, such as near bridge piers and abutments. Empirical correction factors should be developed for one-dimensional models to account for complex local flow conditions. Calibration of numerical models will be highly dependent on velocity and scour measurements from physical model studies and measurements made in the field during floods.

Physical model studies have been and will continue to be a valuable and cost-effective tool for understanding complex scour processes, calibrating numerical models, and developing quantitative methods for predicting flow velocity distributions and scour depths at bridges. Modeling of dominant scour processes can be improved using information obtained in the field during floods, thereby increasing the accuracy of predictive equations derived from laboratory investigations. Field and flood studies are required at sites where fine-grained soils exhibit significant unsaturated and/or saturated scour resistance, so that mechanisms of scour in such soils may be characterized. The wide range of flow and geometric conditions that exist at actual bridge sites requires that additional research be conducted on non-cohesive materials to broaden applicability of small-scale laboratory-based methodologies.

A major source of uncertainty associated with scour prediction techniques is determining the effects of scaling the results of small-scale physical model studies to full-scale conditions. The problem of scaling is not limited to effects on flow structure and sediment transport; the impact extends to geomechanical behavior of rock and soil masses and sediment deposits. Although field data will provide insight on the feasibility of scaling laboratory data, near-prototype (larger scale) testing is necessary to provide controlled conditions from which the effects of scaling can be quantified systematically.

Many techniques and methodologies for prediction of river and coastal waterway behavior are available, but dispersed throughout the technical literature. Broad use of these techniques will be facilitated by the development of guidance documents that are easily accessible to engineers. Distillation of the literature into concise evaluation methods in manual form is necessary to provide the needed guidance to engineers conducting bridge evaluations.

Standard techniques for predicting scour depths should be developed on the basis of well described physical processes that can be characterized on a site-specific basis with standard probabilistic methods. These methods should include consideration for the non-uniformity and variability of floodplain and main channel flow, stratification of bed material and banks, and the apparent cohesion or cementation of bed materials. The variability in these parameters (and the consequent uncertainty in estimating parameter values) should be incorporated into a probabilistic description of scour.

Problem statements were written to address the identified research needs and, in particular, to remedy existing deficiencies in bridge scour prediction methods. The problem statements were developed to build on the results of previous research. These statements describe research activities that should be undertaken in a sequence that is most cost-effective, mutually supportive and avoids unnecessary duplication of effort. A series of flood and field, laboratory physical model and numerical model studies to expand the state of knowledge of scour processes and to advance scour prediction methodology were recommended in the strategic plan. Also recommended were research investigations focused on the assemblage and translation of available research results into a series of practical methods and guidelines also were recommended. A balance of benefit, cost and likelihood of success was used to prioritize research investigations. Project sequencing reflected urgency, dependency and cost.

SUMMARY

The results of this project indicate that substantial advancement in prediction methodology and countermeasure design will require an effort balanced among flood and field studies, physical model studies, and numerical model development. Improvements in evaluation of stream stability and in countermeasure design can be achieved through synthesis of research results and development of design guidance. The synthesis effort in any one area should be structured so that the methodologies are no more complex than necessary for the development of practical designs.

ACKNOWLEDGMENTS

This work was sponsored by the American Association of State Highway and Transportation Officials, in cooperation with the Federal Highway Administration, and was conducted in the National Cooperative Highway Research Program which is administered by the Transportation Research Board of the National Research Council.

DISCLAIMER

The opinions and conclusions expressed or implied in the report are those of the research agency. They are not necessarily those of the Transportation Research Board, the National Research Council, the Federal Highway Administration, the American Association of State Highway and Transportation Officials, or of the individual state participating in the National Cooperative Highway Research Program.

REFERENCES

Engels, H. (1894) "Schultz der Strompfielerfundamente gegen Untersulung." Zeitschrift fur Bauwesen, 408-416.

Murillo, J. A. (1987). "The Scourge of Scour." *Civil Engineering*, 57(7): 66-69

Parola, A. C., Hagerty, D.J., and Kamojjala, S. (1996) "NCHRP Project 12-39 (Task Final Report: Highway Infrastructure Damage Caused by the 1993 Upper Mississippi River Basin Flooding." National Cooperative Highway Research Program, Washington, D. C., 195.

Parola, A.C., Hagerty, D.J., Mueller, D.S., Melville, B.W., Parker, G., and Usher, J.S. (1997). "NCHRP Project 24-8 Final Report: Scour at Bridge Foundations: Research Needs" National Cooperative Highway Research Program, Washington, D. C.

Rhodes, J. and Trent, R. (1993). "Economics of Floods, Scour, and Bridge Failure." Proceedings, ASCE, Hydraulic Engineering '93, San Francisco, 928-933.

STREAMLINING THE PERMIT APPROVAL PROCESS FOR THE INSTALLATION OF SCOUR COUNTERMEASURES AT BRIDGES

STANLEY R. DAVIS, JORGE E. PAGAN-ORTIZ AND LINDA KELBAUGH
Maryland State Highway Administration, Baltimore, Maryland; Federal Highway Administration, Washington, D.C.; Brightwater, Inc., Ellicott City, Maryland

ABSTRACT

Bridge owners are completing their evaluation of existing bridges and are preparing to install scour countermeasures at those structures found to be scour critical or vulnerable to damage by scour. Some states have Federal regulatory agencies that provide an efficient, timely, consistent and predictable permitting process for approving the needed scour protection; other states are less fortunate and have experienced delays and inconsistent actions on permit applications. The Federal Highway Administration (FHWA) is working with the Headquarters Office of the United States Army Corps of Engineers (USACOE) to develop a more efficient, timely, consistent and predictable process in all of the USACOE's field offices. This action is essential in order to protect the traveling public.

INTRODUCTION

Imagine the surprise of the District Bridge Engineer in the Virginia Department of Transportation (VDOT) when his permit application for installing bridge scour countermeasures on a scour critical bridge was not approved. To the engineer, there is an urgent need to protect bridge foundations that are at risk of being damaged or destroyed by scour:
- Bridge owners are responsible for the stability of their bridges and for the safety of the traveling public using the bridges; in fact public safety is generally accepted as the primary responsibility of the bridge owner;
- The FHWA has required the bridge owners to assess existing bridges for vulnerability to damage from scour, and to monitor or protect those bridges found to be scour critical or vulnerable to damage from scour;
- The USACOE is obligated under Federal mandate to accept the environmental review findings of the FHWA;
- The Nationwide 3 permit application used by the District Bridge Engineer was specifically designed to simplify and streamline approval of work involved with the maintenance and repair of bridges and similar structures.

In view of all of these considerations, why was the permit application not approved?

THE NATIONAL BRIDGE SCOUR PROGRAM

Before addressing this issue, it may be useful to review the events that led to the establishment of the national bridge scour program. The collapse of the Schoharie Creek Bridge on the New York Thruway in 1987 was a catastrophic event that illustrated to the public and the engineering community the consequences of a bridge failure:

SCOUR COUNTERMEASURES APPROVAL 131

- loss of lives;
- the cost of the bridge replacement; and
- severing of an Interstate Route and high volume transportation facility, thereby creating disruption of vital local, regional and national commerce.

When the cost per day of delays and detours to truck and vehicular traffic is multiplied by the time it takes to design and build a replacement structure, the overall direct and indirect costs to the public become enormous, in most cases far exceeding the cost of replacing the bridge.

Although there have been a number of major bridge collapses before and after the Schoharie Creek bridge, none have received as much attention by the national media. Public hearings were held by the National Transportation Safety Board to determine the cause of the bridge collapse and to initiate actions to minimize the potential for any future bridge collapses. The engineering community (including the FHWA, the American Association of State Highway and Transportation Officials, the State Highway Agencies, Universities, the American Society of Civil Engineers, the Transportation Research Board, etc.) held workshops and conferences to discuss the problem of scour and how it should be addressed. FHWA took the lead in this effort and issued a directive that all existing bridges be rated for vulnerability to damage from scour. Further, the FHWA developed interim guidance on procedures to be used in (1) designing new bridges to resist scour and (2) evaluating existing bridges for vulnerability to damage from scour.

The National Bridge Inventory was updated to include more information on inspecting bridges for scour. A national effort was undertaken to study scour problems and to develop better methods of predicting the extent of scour at bridge foundations. The National Highway Institute of the FHWA has been working in cooperation with the FHWA hydraulics engineers and a consultant engineering firm to present courses on bridge scour and stream stability since 1990. To date, over 80 courses have been presented.

The available information obtained from research and field studies was collected and presented in the FHWA publication HEC-18, Evaluating Scour at Bridges. This publication is updated regularly as new information becomes available. Finally, bridge owners have been screening bridges over waterways by categories, based on their scour vulnerability, in the following categories: low risk, scour susceptible, unknown foundations and scour critical. The screening process was initiated by the FHWA on February 5, 1990. This process was followed by the scour evaluation of scour susceptible bridges, initiated by the FHWA on July 15, 1991. A target date, set by the FHWA, of January 1, 1997 was established for the completion of the evaluations of all bridges over waterways within their state. This is to include off-system bridges owned by counties, cities or other agencies.

The national scour program has been carried out primarily by engineers. While the program addressed a great variety of issues, there was no effort at the national level to involve the Federal and State Environmental and Regulatory Agencies in the program. Obtaining permits for retrofitting bridges with scour countermeasures was considered more of a routine matter to be handled by each State's Environmental Programs Division.

Most of the bridge owners are well along with their assessment of existing bridges. Most of the bridges have been rated and over 10,000 scour critical bridges have been identified. The next steps involve prioritizing the scour critical bridges and installing countermeasures at structures with the highest risk.

VIRGINIA DEPARTMENT OF TRANSPORTATION (VDOT)

The VDOT is now at the stage in its scour program where it is installing scour countermeasures. As noted earlier, the notification from the USACOE that a routine permit application to install scour countermeasures was not approved came as a surprise. It was also a signal that some of the personnel in the Federal regulatory agencies were not aware of the national scour program and its significance to the safety of the traveling public.

The VDOT has established an excellent track record over a twenty year period in working with the Federal and State agencies who review the environmental aspects of proposed highway projects. These agencies hold a monthly interagency coordination meeting with the VDOT to review each highway project. When the agencies work together as a team, the interagency meetings can serve as an efficient means for review, necessary modification and approval of highway projects.

The Federal representatives on these interagency groups need to exercise judgment in their reviews, since it is impossible to cover every detail of a highway project in environmental guidelines. One of the problems with this approach, however, is that inconsistencies can develop in how the agency's policy is interpreted by various field offices and individuals.

The VDOT initiated a training program in 1996 to orient the interagency review group as to the importance of the scour program. Several training sessions were held with presentations by the Location and Design Hydraulics Section and the Environmental Division of the DOT, their consultants and a representative of the Bridge Division from the FHWA headquarters office.

Preliminary indications are that this effort has been worthwhile. However, additional direction may be needed in order to expedite USACOE approval of countermeasures to protect existing bridges from scour. At this time, about 14 months after the permit application, the District engineer's permit is still not approved.

MARYLAND STATE HIGHWAY ADMINISTRATION (MDSHA)

In contrast to the problems faced by the VDOT, the MDSHA is fortunate to have an efficient, streamlined process for obtaining Nationwide 3 permits for scour countermeasures.

In Maryland, the USACOE representative is a former highway engineer with an appreciation of the need to provide for the safety of the traveling public. This has made the difference. Also, the Nationwide permit process was recently replaced with a process that allows the state regulatory agencies to take over more of the review responsibilities.

SCOUR COUNTERMEASURES APPROVAL 133

The following attributes of the Maryland Interagency Review Group have served to help them to operate as a team:
- an on-going interagency review process;
- open communications (no surprises); and
- a mutual trust, appreciation and respect for the roles and responsibilities of each team member.

The MDSHA made several presentations to the team on scour countermeasures. In particular, the following concepts were stressed:
- MDSHA is responsible for the safety and stability of its bridges; and
- No other representative on the team has this responsibility. The other team members have an obligation to respect this responsibility.

The evaluation of scour is not an exact science. The engineer must make a judgment as to the stability of the bridge under worst case scour conditions. In view of the consequences of a bridge collapse, the engineer's judgment can be expected to be on the conservative side. The interagency review group agreed that Maryland could process Nationwide 3 permits without the involvement of their group. For its part, the MDSHA continues to make a good faith effort to report all Nationwide 3 type actions at the interagency meetings. In addition, the MDSHA's Environmental Programs Division reviews each permit application to see how the scour countermeasures can be installed so as to limit impacts to the streams. They are concerned with the effect on submerged aquatic vegetation, mussels, and the aquatic habitat in general. Efforts are made to limit the extent of riprap coverage of the channel bottom on small streams. The MDSHA specifications require that riprap placed in the stream be clean.

The MDSHA has had reasonable success in using grout bags instead of riprap for scour protection. Grout bags have a number of advantages and disadvantages as compared with stone riprap, but on balance their performance has been good. The environmental agencies like the grout bags because they can be placed with minimal disruption to the stream and its flood plain.

SUMMARY

For its part, the FHWA is working with the headquarters office of the USACOE to develop additional guidance to expedite the approval of permits for scour protection. Perhaps one of the phrases in the Nationwide 3 permit that needs further discussion is the one advising that "minor deviations due to changes in materials or construction techniques ... are permitted". Apparently, there are different approaches by different offices and officials of the USACOE as to what constitutes a "minor deviation".

There are over 480,000 bridges over waterways in the United States and over two percent of these bridges have been found to be scour critical. It can be expected that a substantial number of these bridges will be scheduled for installation of scour countermeasures over the next few years. An efficient, timely, consistent and predictable nationwide permitting process is essential to assist the bridge owners in protecting these bridges and the traveling public.

REFERENCES

Virginia Department of Transportation Training on Bridge Scour During June 4, and June 12, 1996 for Virginia Department of Transportations Environmental Coordinators, and Environmental Resource Agencies, Respectively.

Multi-Agency Meeting Between the Federal Highway Administration, the Virginia Department of Transportation, the Maryland State Highway Administration and the National Cooperative Highway Research Program in Washington, D.C. on June 13, 1996.

Personal Communication Between Mr. Stanley R. Davis (MDSHA); Messrs. Jorge E. Pagán-Ortiz and Claude Napier (FHWA); and Messrs. David M. LeGrande, Roy T. Mills, and Steve Long (VDOT).

Evaluation of Field and Laboratory Research on Scour at Bridge Piers in the United States

DAVID S. MUELLER AND J. STERLING JONES
U.S. Geological Survey, Louisville, Kentucky, USA
and Federal Highway Administration, McLean, Virginia, USA

INTRODUCTION

The complexity of the hydrodynamics associated with flow around piers coupled with the inability of researchers to develop a generally applicable sediment transport relation have caused researchers to rely on empirical relations for the prediction of scour at bridge piers. The effect of bed material on the depth of scour is difficult to assess from most laboratory investigations because only uniform bed materials were used. Laboratory research in New Zealand concluded that as the gradation coefficient of the bed material increases the depth of scour decreases. However, there was little validation with field data. The Federal Highway Administration (FHWA) sponsored laboratory research at Colorado State University and field data collection with the U.S. Geological Survey (USGS), to evaluate the effects of bed material on the depth of scour. Correction factors, developed from these studies, to include the effect of bed material in the HEC-18 equation (Richardson and Davis, 1995) are compared.

LABORATORY RESEARCH

The FHWA sponsored an independent laboratory investigation at Colorado State University to study the effects of coarse material fractions on scour at bridge piers. On the basis of these laboratory investigations, J.S. Jones developed a K_4 correction to the HEC-18 equation to account for the effect of coarse sediments. The method was presented by Richardson and Davis (1995) in the third edition of HEC-18 with very strict criteria for its application.

Molinas and others (1996) determined from laboratory experiments that pier scour could best be characterized by using the coarse size fractions (D_{CF}), rather than the D_{50} size fraction, to determine the threshold between clear-water and live-bed conditions. They observed that if the coarse size fractions of graded bed material were transported in the approach section, the scour depths for graded bed material were about the same as for uniform bed material. If the coarse size fractions are not transported in the approach section, a correction factor, K_4, applied to the HEC-18 pier scour equation for noncohesive bed material improves the accuracy of the prediction.

$$K_4 = 2.0 + 1.25 \sqrt{\frac{D_{CF}}{D_{50}}} \left(\frac{V_o - V_i}{V_{c_{CF}} - V_i}\right)^{0.75} \ln\left(\frac{V_o - V_i}{V_{c_{CF}} - V_i}\right) \quad (1)$$

$$D_{CF} = \frac{D_{85} + 2D_{90} + 2D_{95} + D_{99}}{6} \quad (2)$$

K_4 is the correction for bed material and is always less than or equal to one; D_x is the grain size for which "x" percent of the bed material is finer; V_o is the approach velocity; V_i is the approach velocity that initiates scour around a pier and is defined as $0.4\ V_{c35}$; V_{c35} is the approach velocity that initiates motion of the grain size for which 35 percent of the bed material is finer; and V_{cCF} is the approach velocity that initiates motion of the coarse fraction in unobstructed flow.

Equation 1 was derived to as a replacement for the K_4 factor currently in the third edition of HEC-18. If it is adopted, it will probably be simplified by specifying a single fraction, say D_{90}, to represent the coarse fraction of the bed material. Equation 1 improved the HEC-18 prediction of the laboratory data for nonuniform bed material, from which it was developed, and predicted values generally erred on the conservative side.

FIELD RESEARCH

Conditions in the field can be very different from conditions in the laboratory. Laboratory conditions are uniform and usually only one variable is allowed to change at a time. Conversely, conditions in the field are nonuniform and dynamic. The USGS, in cooperation with the FHWA and State highway departments, collected 384 scour measurements at 56 bridges in 14 different States (Landers and Mueller, 1996). These data include the information necessary to study the effect of nonuniform bed material on local scour under dynamic conditions typical of the field environment.

Mueller (1996) used partial residuals and multiple linear regression to identify the key variables influencing the depth of scour at piers. The results showed the D_{95} grain size to be the dominant bed-material parameter for both clear-water and live-bed conditions. Mueller (1996) used a velocity intensity term to reflect the importance of the D_{95} grain size and to developed a K_4 correction to the HEC-18 equation for coarse bed material.

If $D_{50} < 2$ mm or $D_{95} < 20$ mm then $K_4 = 1$. If $D_{50} >= 2$ mm, $D_{95} >= 20$ mm,

and $\dfrac{V_o - V'_{D_{50}}}{V_{cD_{50}} - V'_{D_{95}}} > 0$ then,

$$K_4 = 0.4 \left(\frac{V_o - V'_{D_{50}}}{V_{cD_{50}} - V'_{D_{95}}} \right)^{0.15} \tag{3}$$

$$V'_{cD_x} = 0.645 \left(\frac{D_x}{b} \right)^{0.053} V_{cD_x} \tag{4}$$

V'_{cD_x} is the approach velocity corresponding to incipient scour in the accelerated flow region at the pier for the grain size D_x; V_{cD_x} is the critical velocity for incipient motion for the grain size D_x; B is the pier width; and V_o is the velocity of the approach flow just upstream from the bridge pier.

COMPARISON WITH FIELD DATA

The K_4 correction factors developed by Jones (Richardson and Davis, 1995), Molinas and others (1996), Mueller (1996) are evaluated using field data. The first evaluation uses the same field data used by Mueller (1996) to develop his K_4 term. A second evaluation uses independent field data from the USGS, Gao and others (1992), and Zhuravljov (1978). The proposed corrections require the use of grain sizes, which are often not available. Regression equations, developed from the field data used by Mueller (1996), are used to estimate the D_{95} and D_{35} from the D_{50}, when the D_{95} and D_{35} are not reported with the data.

$$D_{95} = 3.65 \, D_{50}^{0.954} \tag{5}$$

$$D_{35} = 0.655 \, D_{50}^{0.987} \tag{6}$$

D_{95} and D_{50} must be in ft. For the Mueller (1996) data the D_{CF} is computed as,

$$D_{CF} = \frac{D_{84} + 2 D_{95}}{3} \tag{7}$$

Using USGS data (which were used by Mueller (1996) to develop his K_4), Figure 1 shows the results of the three proposed K_4 correction factors and HEC-18 equation without a K_4 correction. As expected, the Mueller K_4 performs very well. The correction factors by Molinas and others (1996) and Jones (Richardson and Davis, 1995) do little to reduce the gross overestimation of scour depths by the HEC-18 equation.

Data from the USGS not used by Mueller (1996) in the development of the K_4 factor, selected data from Goa and others (1992), and selected data from Zhuravljov (1978) provide a more objective evaluation. Because only D_{50} was reported with these data, equations 5 and 6 were used to estimate the D_{35} and D_{95} grain sizes. D_{CF} was estimated as the D_{95} grain size. Figure 2 shows that the corrections proposed by Molinas and others (1996) and Jones (Richardson and Davis, 1995) again do very

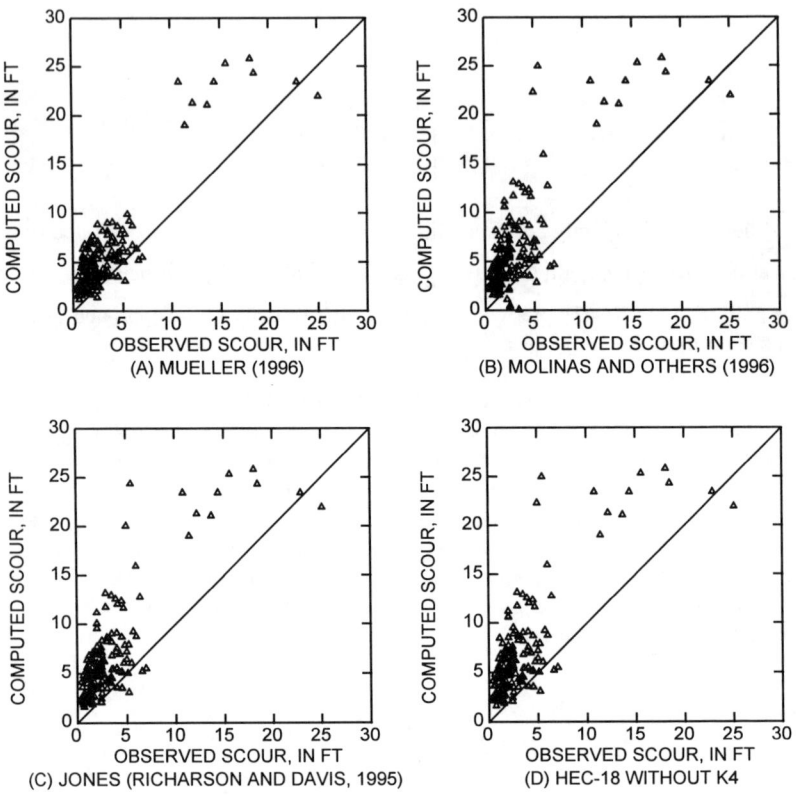

Figure 1. Evaluation of K_4 factors using data from Mueller (1996).

little to reduce the overestimation of the HEC-18 equation for coarse bed material. The correction proposed by Mueller (1996) reduces much of the overestimation but also causes underestimation of some of the data. Nearly all of the underestimations were for data from other countries. Only two observations of the independent USGS data were underestimated and they were both underestimated less than 1 ft. The underestimation of scour may be caused from estimating grain sizes using equations 5 and 6 for rivers with geology different than the rivers on which these equations are based.

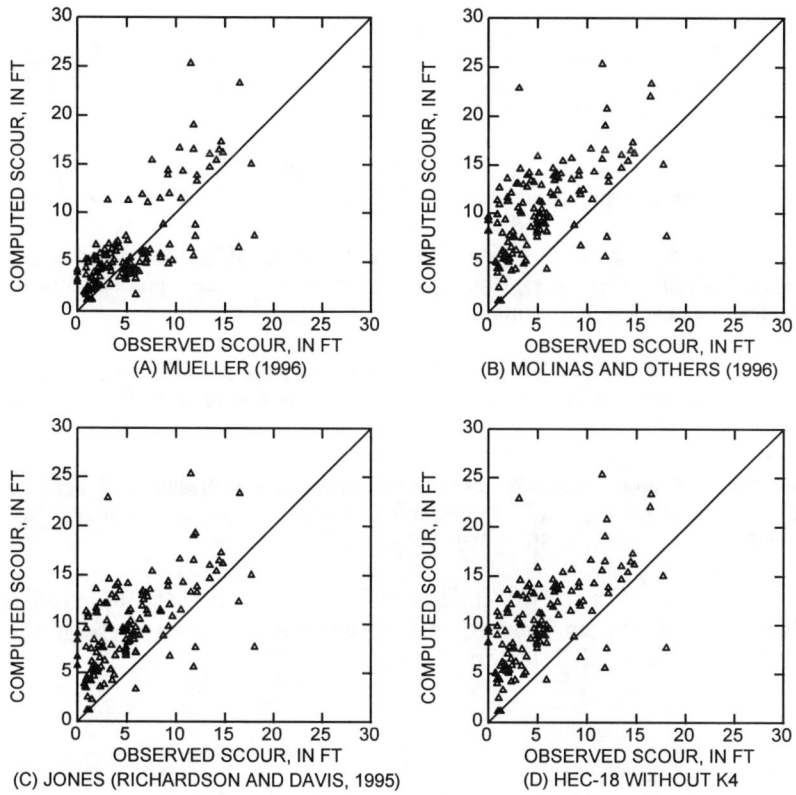

Figure 2. Evaluation of K_4 factors using independent data from the the USGS, Gao and others (1992), and Zhuravljov (1978).

CONCLUSION

The HEC-18 equation tends to overpredict the observed scour for streams with coarse bed material. The correction factor, K_4, developed from field data reduces the predicted scour in coarse bed material significantly more than the correction factors based on laboratory studies. The field-based correction factor causes some underprediction, but the only severe underprediction is associated with data collected in other countries. Further work is needed to determine the cause of the differences between laboratory- and field-based methods.

REFERENCES

Gao Dong Guang, Posada G., Lilian, and Nordin, C.F. (1992). Pier Scour Equations Used in the People's Republic of China. Colorado State University, Department of Civil Engineering, Draft Report.

Landers, M.N., and Mueller, D.S. (1996). Channel Scour at Bridges in the United States. Federal Highway Adminstration, Publication No. FHWA-RD-95-184, 140 p.

Molinas, A.A., Abdou, M., and Noshi, H.M. (1996). Effect of Sediment Gradation and Coarse Material Fraction on Bridge Scour., unpublished draft report submitted to the Federal Highway Administration.

Mueller, D.S. (1996). Local Scour at Bridge Piers in Nonuniform Sediment Under Dynamic Conditions. Ph.D. Thesis, Colorado State University, Fort Collins, Colo., 183 p.

Richardson, E.V. and Davis, S.R. (1995). Evaluating Scour at Bridges — Third Edition. U.S. Federal Highway Administration, Hydraulic Engineering Circular No. 18, Publication No. FHWA-IP-90-017, 204 p.

Zhuravljov, M.M. (1978) "New Method of Estimation of Local Scour Due to Bridge Piers and Its Substantiation." Ministry of Transport Construction, Russia, State All-Union Scientific Research Institute on Roads, Trans., 109, 20 p.

Research Needs in Geomorphology Pertaining to Bridge Scour

RICHARD L. VOIGT, JR., CARLOS M. TORO-ESCOBAR AND GARY PARKER
St. Anthony Falls Laboratory, University of Minnesota, Minneapolis USA

ABSTRACT

The failure of bridges across rivers is a well known problem facing the transportation engineer. It is typically associated with bed or bank scour, and may not be directly due to inadequate structural design. Research to date on bridge scour has tended to focus on processes in the immediate vicinity of bridge piers or abutments. These processes are, however, influenced in a fundamental way by larger geomorphic processes, which reflect both natural and human-induced change. For example, meandering rivers tend to shift, inexorably leading to a deterioration in angle of approach. Lowered base level on a stream due to, e.g., river training works can lead to upstream-migrating degradation on tributaries. The degradation itself can endanger bridge piers; the channel widening commonly associated with it can endanger abutments and approaches. This paper summarizes that part of a study conducted for the National Cooperative Highway Research Board (NCHRP 24-8) that pertains to research needs in geomorphology as they affect bridge scour problems.

INTRODUCTION

The fluvial and geomorphic processes that directly or indirectly affect bridge scour are numerous and occur at many scales. The goal of NCHRP 24-8 in regard to geomorphology was to achieve a) an overview of all the relevant processes and b) the delineation of a relatively small number of proposals for further research which are most likely to produce results of direct aid to the bridge engineer. Five topics were selected in this regard; a) channel widening at bridges induced by degradation or aggradation, b) bridge problems on alluvial fans, c) bend and confluence scour near bridges, d) impact of river basin modification on bridge scour and e) effect of channel shift on meandering streams on bridges. The order of listing is not intended to indicate any particular priority. In many cases the general nature of the geomorphic processes in question is reasonably well understood, although not necessarily in a form easily accessible to the bridge engineer. The required research effort then consists of an appropriate sifting, distillation and packaging of available information for the bridge engineer. In some cases, however, an independent research effort is warranted.

TECHNIQUES FOR PREDICTING THE RELATION BETWEEN AGGRADATION, DEGRADATION AND CHANNEL WIDENING AT BRIDGES

Channel widening is a common short-term response to river bed degradation or aggradation. (Channel narrowing usually requires substantially longer time scales.) The processes that cause a river to aggrade or degrade occur at a scale larger than the near field in the vicinity of a bridge. These processes can nevertheless have a profound effect on the bridge, both directly and indirectly through channel widening. At present a number of techniques, including numerical models such as HEC-6, are available for predicting river aggradation or degradation in response to e.g. change in base level, gravel mining and sediment overload. Only a few techniques are available, however, for quantifying concomitant channel widening, which can cause such bridge problems as alignment deterioration, abutment outflanking or destabilization, and scour around piers with shallow footings which were not originally designed to be in the main channel.

The perceived research need is the development of an empirical, field-based set of relations for predicting the expected extent of short-term channel widening in response to streambed aggradation or degradation. It is not expected that the predictions be incorporated into a numerical model of aggradation or degradation, as this would tie the methodology to a specific choice of numerical model. The relations would, however, be sufficiently quantitative in nature to be used either as predictive tools in their own right or in conjunction with any numerical model.

MANUAL FOR BRIDGE PROBLEMS ON ALLUVIAL FANS

Roadways through mountain regions such as the western part of the continental United States and Alaska must frequently be constructed on active alluvial fans. Alluvial fans are built by the successive shifting of river flows, or the successive passage of debris flows down different routes. Such fans are naturally aggrading in geomorphic time. A channel may remain in place for several decades, and then suddenly shift during a flood. Channels on alluvial fans can also undergo episodic periods of downcut between periods of sediment buildup. The importance of fan dynamics is underlined by the recent I-5 bridge failure in California.

Alluvial fans are thus inherently unstable environments for bridges. The perceived research need is the development of a manual that outlines the character of alluvial fans both in general and specifically in the context of bridge design. Such a manual would be a distillation of existing information about fan hydrology, geomorphology and dynamics. It would include case histories of bridges on fans. The style of the manual would be guided by existing manuals such as HEC-20.

BEND AND CONFLUENCE SCOUR NEAR BRIDGES

Bend and confluence scour are related phenomena, the first characteristic of meandering streams and the second characteristic of braided streams. Both are produced by secondary flow generated by streamline curvature. In meandering streams, the outside of bends tend to scour during floods and the inside fills. As a result, bed elevations on the outside of bends appear deceptively high during low flow. It is common for bridges to be placed on the outside of bends, as this often allows for the anchoring of one end against a valley wall. Correct placement of pier footings and abutments is contingent upon a recognition of the amount of bend scour that might be expected during a flood.

While braided streams are less common in the continental USA than meandering streams, they can be found in the western part of the country and abound in Alaska. Confluence scour occurs where two anabranches of a braided stream flow together. Confluence scour can lead to flow depths as much as five times the ambient values in anabranches. Experience in New Zealand suggests that bridges on braided streams are most likely to fail when a confluence forms at a pier.

While the effect of bend and confluence scour on bridges is well recognized in the technical literature, they are not specifically addressed in several commonly used manuals for bridge scour evaluation. This may be rectified by means of the preparation of a concise design manual providing quantitative methods for evaluating bend and confluence scour at bridge crossings. The manual may be a distillation of existing literature on bend and confluence scour and its effects on bridges.

IMPACT OF RIVER BASIN MODIFICATION ON BRIDGE SCOUR

While countermeasures for bridge problems must usually be implemented on a local basis, their causes often have roots in large-scale phenomena. For example, base lowering on a river due to the installation of a dam upstream can cause degradation in all the tributaries downstream of the dam. This degradation can in turn cause bridge scour problems on a basin-wide basis. While bridge engineers can rarely influence the process of basin modification, they should be able to evaluate in advance the potential effects of any given activity on the bridges in the basin. Armed with this knowledge, they can estimate the kind and amount of remedial work necessary to protect the bridges from the effects of basin modification.

Here basin modification is interpreted to include changing land use practices such as agriculture and forestry, sediment disposal from mines, gravel mining, dam construction and removal, river training, removal of riparian vegetation, construction and urbanization. The effects relevant to bridges include channel degradation (or aggradation), channel widening, change in regime from meandering to braiding, increased rates of channel shift, proclivity for bar formation and increased supply of debris.

The required study is primarily intended to be qualititative in nature, with an eye to acquainting the bridge engineer with the consequences of basin modification. The analysis should, however, include as much quantification as the generality of the topic allows.

EFFECT OF INCREMENTAL CHANNEL SHIFT ON BRIDGE SCOUR

Bridges are static structures built over river channels, which are typically prone to shift. The inevitable planform deformation and shift of alluvial rivers insures that the alignment and approach conditions at the time of bridge construction are not usually maintained indefinitely. Instead, alignment typically deteriorates in time as the channel upstream of the bridge migrates. Poor approach conditions can result in a) excess pier and abutment scour for which the bridge was not designed, b) endangerment of a bridge approach and c) worsened debris problems. Channel deformation and shift are thus major considerations for the design of bridge crossings (Fig. 1).

Channel shift is typically an incremental process. On meandering streams, for example, the problem at a bridge site often first becomes apparent two or three decades after bridge construction. Channel shift is manifested throughout a basin, and is not localized in the vicinity of a bridge. It is a natural phenomenon that occurs in the absence of specific disturbances. It may be exacerbated, however, by such basin-wide factors as land use practices, gravel mining and the removal of riparian vegetation. Remedial action such as the construction of guide banks or the installation of bank protection becomes increasingly expensive or difficult as alignment deteriorates. The bridge engineer requires a set of tools with which to evaluate the potential for channel shift and predict planform evolution into the future with and without appropriate countermeasures.

There are several modes of channel shift and deformation in alluvial rivers. For the sake of simplicity, the most immediate research need concerns the most common mode that affects bridges: incremental channel migration in meandering rivers. The product of the research effort should be a) a set of general guidelines for evaluating channel shift and its consequences, b) a numerical model for predicting shift on meandering rivers near bridges and c) a compendium of countermeasures with which to rectify the problem. The numerical model should be packaged with a straightforward interface that encourages its use by practitioners.

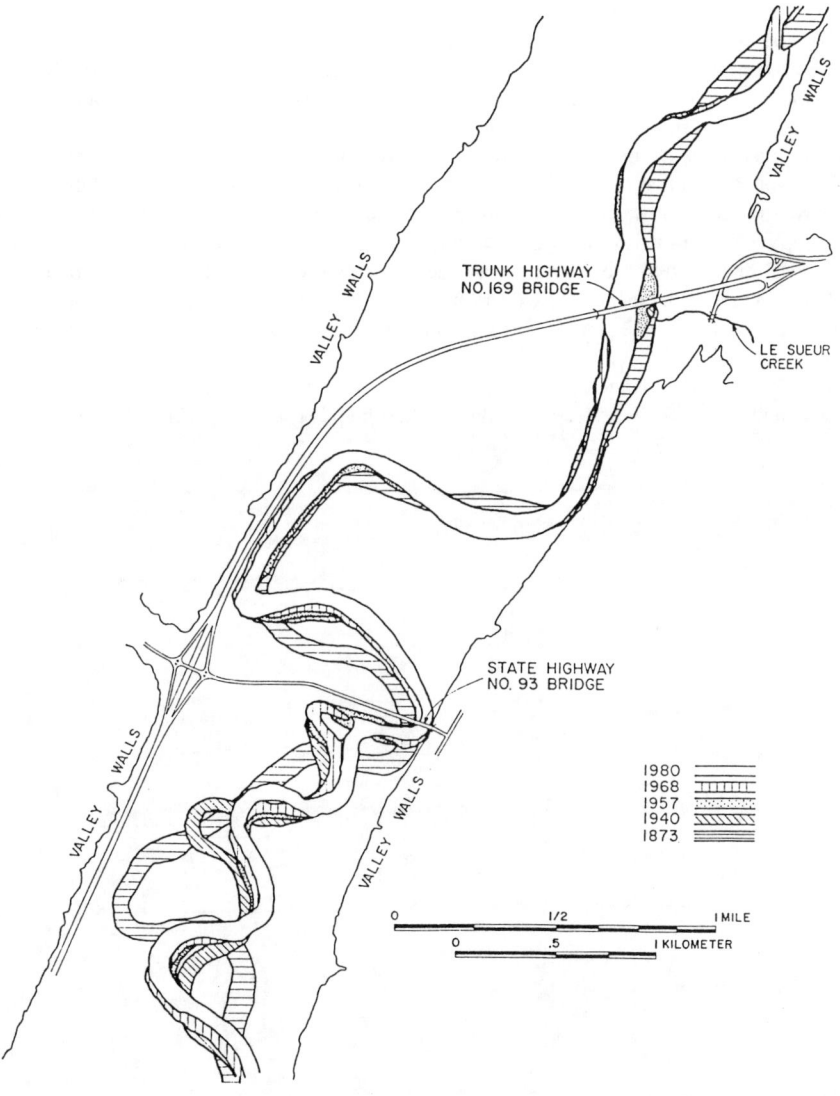

Figure 1. Channel shift on the Minnesota River near two bridges.

DISCUSSION

General summaries of geomorphic processes in a form accessible to bridge engineers are presently available from several sources. The research proposed here would result in the development of information packages in manual form that would allow for a deeper understanding of geomorphic processes as they pertain to bridges and would place that understanding on a more quantitative basis. A case in point is bend scour. It is well known that the outside of bends on meandering streams tend to scour at high flow and fill at low flow. The bed elevations on the outside of bends are thus deceptively high at low flow for the purpose of pier foundation and footing design. Several quantitative predictors are available for predicting bend scour due to floods. These methods have not, however, been broadly introduced to the community of bridge engineers. The proposed research program is designed to fill such lacunae.

Art Parola, Joe Hagerty, Bruce Melville and Andrew Simon made valuable contributions to the work reported here. Their help is gratefully acknowledged.

QUANTITATIVE TECHNIQUES FOR STREAM STABILITY ANALYSIS

P.F. LAGASSE, S.A. SCHUMM, AND L.W. ZEVENBERGEN
Senior Vice President, Senior Associate, and Hydraulic Engineer, respectively, Ayres Associates, Fort Collins, Colorado, USA

ABSTRACT
While qualitative analysis techniques provide insight on potential channel instability problems, quantitative techniques may be necessary to adequately address questions of vertical stability of the streambed, lateral (planform) stability of the stream channel, and the effects of channel instability on countermeasures. This paper surveys standard FHWA publications for guidance on these issues and presents additional techniques readily available in the technical literature.

INTRODUCTION
Many streams that highways cross or encroach upon are alluvial; that is, the streams are formed in materials that have been and can be transported by the stream. In alluvial stream systems, it is the rule rather than the exception that channels will change position and shape as a consequence of hydraulic forces exerted on the bed and banks. Consequently, many bridges, especially those on more active streams, will experience problems with scour and bank erosion during their useful life.

While alluvial channels are dynamic and subject to change, the types of change differ among rivers and between reaches (Schumm and Winkley 1994) and rates of change are highly variable and difficult to predict. For highway and bridge applications, hydraulic engineers have typically relied on a reconnaissance level of analysis and qualitative techniques to evaluate potential stream instability problems. Guidance on these techniques is currently available in such standard Federal Highway Administration (FHWA) publications as Highways in the River Environment (HIRE) (Richardson et al. 1990); Hydraulic Engineering Circular (HEC-18), Evaluating Scour at Highway Bridges (Richardson and Davis 1995); and Stream Stability at Highway Structures, HEC-20 (Lagasse et al. 1995). Qualitative (Level 1) techniques can provide insight and understanding of physical processes affecting stream channel geomorphology, and the understanding generated by such analyses assures that subsequent, more quantitative analyses (Level 2) are properly formulated.

Highway and bridge design, scour and stream stability analyses, bridge rehabilitation and countermeasure design, and channel restoration projects are all affected by changes in the morphologic characteristics of a stream. While qualitative techniques provide insight on channel processes, the application of quantitative geomorphic techniques may be necessary to evaluate the potential impact of changes in channel morphology for highway planning, design, and rehabilitation.

In general, the highway engineer needs to address three questions in regard to stream stability:

- What is the vertical stability of the streambed?
- What is the lateral (planform) stability of the stream channel?
- How will channel instability affect the design and performance of countermeasures?

This paper surveys current FHWA publications (HIRE, HEC-18, HEC-20) available to highway engineers for quantitative techniques applicable to these questions, and presents additional techniques readily available in the technical literature.

VERTICAL STABILITY

The typical effects associated with bed elevation (vertical) changes at highway bridges are erosion at abutments and the exposure and undermining of piers from degradation or scour, or a reduction in flow area from aggradation under bridges. Aggrading and degrading channels can also change planform, potentially eroding floodplain areas and highway embankments on the floodplain.

As HEC-20 suggests, the data needed for a qualitative assessment of bed elevation changes include historic streambed profiles (either cross section, longitudinal profile, or both) and long-term trends in stage-discharge relationships. Bed elevations at railroad, highway and pipeline crossings monitored over time may also be useful. In many cases, quantitative information can be obtained from cross section or streambed profile comparisons, providing an estimate of average long-term rates of degradation or aggradation (see examples in Shen and Schumm, 1981).

Both HIRE and HEC-20 provide specific quantitative procedures for estimating incipient motion and armoring characteristics of the streambed. An indication of relative channel stability can be obtained from an application of an equation for incipient motion particle size developed from the Shields diagram. Determining the critical or threshold conditions at which hydrodynamic forces are sufficient to move a sediment particle provides insight on what flow conditions might mobilize the bed and affect channel vertical stability. A simple procedure developed by the U.S. Bureau of Reclamation for determining the depth of degradation necessary to produce an armor layer sufficient to arrest vertical instability is presented in HEC-20, and HEC-18 provides specific application guidelines for determining contraction scour under live bed or clear water conditions using a modification of Laursen's (1960 and 1963) equations.

Going beyond these simple quantitative techniques to assess streambed vertical stability requires considerable expertise in sediment transport analyses. However, sediment continuity analysis and equilibrium slope concepts offer a relatively straight forward approach to more detailed vertical stability analyses (see for example, Resource Consultants & Engineers 1993). If a more rigorous analysis of channel vertical dynamics is desired, application of the U.S. Army Corps of Engineers HEC-6 computer model (USACOE 1991) or the FHWA BRI-STARS model (Molinas 1990) could be considered.

LATERAL STABILITY

The effects of lateral (planform) instability of a stream on a highway encroachment or bridge crossing are dependent on the extent of the bank erosion or channel migration and the design of the encroachment and bridge. Bank erosion can undermine piers and abutments located outside the channel and erode abutment spill slopes or breach approach fills. Migration of a bend through a bridge opening changes the direction of flow (angle of attack) through the opening, accentuating local and contraction scour. As pointed out in HEC-20, in meander bends deep pools are carved adjacent to the concave bank by the relatively high velocities on the outside of the bend. Because velocities are lower on the inside of bends, sediments are deposited in this region, forming point bars. Thus, an abutment located on the outside of a meander bend can be attacked directly or out-flanked, and growth of a point bar on the inside of a bend can significantly affect conveyance through a bridge opening.

As suggested in HIRE and HEC-20, an assessment of lateral stability can be obtained from the position of a bend or bankline at different times using aerial photographs or maps which are matched in scale and superimposed. Such comparisons can focus on a qualitative assessment of trends or, using photogrammetry techniques, can provide quantitative data. The availability of computer aided design (CAD) and software such as Descartes© enables a very precise comparison of bankline position through time. Similarly, surveyed cross sections are extremely useful, when available. For example, post-failure investigation of the U.S. Highway 51 bridge on the Hatchie River in Tennessee provided time-sequenced cross section data and an estimate of lateral migration rates that reached as high as 1.37 m/yr. In using such estimates, it must be recognized that erosion rates may fluctuate substantially from one period of years to the next.

The relative stability of an alluvial river channel can also be inferred from its planform pattern (see HIRE, HEC-20, and Brice 1982). Brice (1982) found a strong correlation between the variability of channel width through a series of river meanders and the lateral stability of the stream. A random width meandering channel with wide irregular sandy deposits (point bars) on the inside of the river bends tends to be a highly unstable river. Conversely, equiwidth streams having narrow (or nonexistent) point bars at the inside of meander bends are the most stable.

For channel relocation or restoration projects, HIRE provides empirical relationships for stable meander characteristics derived from Leopold and Wolman (1960), as well as Leopold and Maddock's (1953) hydraulic geometry relationships. In addition, HIRE establishes channel sinuosity and bed slope relationships for channel realignment which can be compared to Lane's relationship for meandering channels [$SQ^{0.25} \geq 0.0007$ where: S = slope (m/m) and Q = mean annual discharge (m^3/s)] to check for stability. Recently, Julien (1995) analytically defined the downstream hydraulic geometry of alluvial channels for particle stability under two-dimensional flows as a function of water discharge, sediment size, Shields number, and streamline deviation angle. Good agreement was found with several empirical regime equations and results were tested with laboratory and field data.

An extensive literature on meandering and channel migration is available. Numerous quantitative techniques for addressing meandering and migration issues that would be applicable to highway engineering and bridge design can be found in symposia, monographs, and publications such as Schumm (1977); Richards (1982); Elliott (1983); Knighton (1984); and Ikeda and Parker (1989). As with vertical stability, a more rigorous analysis can be considered using computer modeling of meandering processes (see for example, Chang 1988).

The stability of individual bends within the meandering pattern can also be evaluated quantitatively. For example, a maximum bend sharpness exists beyond which further significant lateral erosion is unlikely to occur. It has been shown that the maximum lateral erosion rate for a meander bend occurs when the ratio of radius of curvature (R_c) to channel width (W) is in the range of about 2 to 4 (Hickin 1975; Nanson and Hickin 1983). For values less than about 2, the erosion rate reduces sharply due to energy loss in the bend. For this condition, the rate of lateral migration is significantly reduced and a cutoff may occur. If one assumes that the planform for such a meander follows the approximate shape of a sine-generated curve (Langbein and Leopold 1966), relationships for the approximate maximum lateral erosion distance can be developed based on optimal bend shape (RCE 1993).

COUNTERMEASURES

A countermeasure is defined as a measure incorporated into a highway-stream crossing system to monitor, control, inhibit, change, delay, or minimize stream instability problems. This would include river stabilizing works over a reach of the river up- and downstream of the crossing. Countermeasures may be installed at the time of highway construction or retrofitted to resolve stability problems at existing crossings. HEC-20 provides detailed design criteria and procedures for check dams (channel vertical instability), spur fields (lateral instability), and guidebanks (scour at abutments).

Many of the qualitative and quantitative techniques for assessing vertical and lateral channel stability surveyed above are applicable to countermeasure design and evaluation. For example, qualitatively, the outside downstream portion of a meander bend can be identified as a high risk location for bankline instability, and Wolman and Leopold (1957) found that maximum depths attributable to scour along the concave margin of an unprotected bend during a flood may range from 1.75 to 2 times the depth of flow in sand-bed streams. For meandering alluvial channels, however, it would be useful to have specific quantitative techniques to assist in identification of locations where countermeasures may be necessary to ensure lateral channel stability or evaluate the performance of lateral instability countermeasures such as spurs, dikes, and revetment. Two useful techniques to address these issues are highlighted below.

In developing a quantitative technique to evaluate the potential for increased bank erosion or damage to existing bank protection associated with different flow scenarios on the Lower American River in California, Mussetter et al. (1995) considered the duration of the flows, as well as their magnitude during a specific flow event. In addition, the relative effect of the range of possible flow events was considered. This is in contrast to most design procedures for bank protection which only consider conditions for a specific discharge. To incorporate the effects of both duration and magnitude into the evaluation of the effects of various scenarios, lateral stability analyses were performed based on the concept of total energy available in the flow to perform work on the banks. Available energy, referred to as the Bank Energy Index (BEI), was computed at specific locations for each design scenario using the results of

HEC-2 hydraulic modeling and measurements of planform geometry from available mapping.

The Bank Energy Index is defined as the product of the stream power expended on the banks and the incremental time over which it was applied. Bank stream power is the product of the average main channel velocity (V_{ch}) and the shear stress acting on the bank (τ_b). For a given flood event, the total available energy at a given bank location can be determined by integrating the bank stream power over the entire hydrograph or flow duration curve:

$$BEI = \int (V_{ch}\ \tau_b) dt \qquad (1)$$

where BEI is the total available energy at a specified bank location and dt is the incremental time. The bank shear stress is computed as:

$$\tau_b = K_b\ \gamma\ d_h\ S_f \qquad (2)$$

where γ is the unit weight of water, d_h is the hydraulic depth, S_f is the energy slope, and K_b is a factor that accounts for the effect of channel curvature on the shear stress acting on the outside of a channel bend. K_b depends upon the ratio of the radius of curvature to the channel topwidth. Equations 1 and 2 can be solved for a given event by discretizing the inflow hydrographs (and stages) into a series of time steps determining the main channel hydraulic conditions and integrating the results of the calculations at each time step over the duration of the event.

From the results for the Lower American River, BEI values of about 100 based on integration of the annual flow duration curve, and 50 based on the weighted average of the various return period flood events, provided reasonable threshold values with which to identify potential lateral instability problems at unprotected sites.

A less complex quantitative technique which considers a single-event discharge and an estimate of the force on a bendway margin was developed to evaluate the performance of alternative streambank erosion protection techniques for the Army Corps of Engineers, Vicksburg District (WET 1990). Based on the concept of centripetal force, an equation for the radial stress of flow on a bendway was developed. Field investigations and computation of radial stress on banklines for 11 study areas in the Yazoo River basin in Mississippi clearly showed that rudimentary countermeasures such as used-tire revetment, sand-cement bag revetment, and cable fence spurs were generally unsuccessful in bends with even low to moderate radial stress. Stone structures including longitudinal stone dikes and stone spurs performed well in reaches of high radial stress. Isolated failures of stone structures did occur at locations with the highest radial stress.

Thus, a relatively simple quantitative analysis technique can provide specific guidelines for evaluation and design of countermeasures against lateral channel instability. Comparing bendway radial stress with structure performance helped explain structure failures and suggested that some bendways, with moderate values of radial stress, might have been protected with less intensive (and less expensive) treatment.

CONCLUSIONS

There is a wide range of qualitative and quantitative techniques currently available to highway engineers in such standard FHWA publications as HIRE, HEC-18, and HEC-20 to support stream stability analyses at highway river crossings and encroachments. As illustrated in this paper, there are additional quantitative techniques readily available in the technical literature to support highway bridge design, scour and stream stability analyses, bridge rehabilitation, and channel restoration projects. Relatively straight forward techniques such as the Bank Energy Index and calculation of radial stress on a bend can be used by highway engineers to identify reaches that will require countermeasures to ensure lateral channel stability, or to design and evaluate the performance of various countermeasure options.

REFERENCES

Brice, J.C., 1982. "Stream Channel Stability Assessment," Report No FHWA/RD/82/021, U.S. Department of Transportation Office of Research and Development, Washington, D.C., 42 p.

Chang, H.H., 1988. "Fluvial Processes in River Engineering," John Wiley & Sons, New York, NY, 432 p.

Elliott, C.M. (Editor), 1983. "River Meandering," Proceedings of the Conference on Rivers '83, American Society of Civil Engineers, New York, NY.

Hicken, E.J., 1975. "The Development of Meanders in Natural River-Channels," American Journal of Science, v. 274.

Ikeda, S. and Parker, G (Editors), 1989. "River Meandering," Water Resources Monograph 12, American Geophysical Union, Washington, D.C.

Julien, P.Y. and Wargadalam, J., 1995. "Alluvial Channel Geometry: Theory and Applications," ASCE Jour. of Hydraulic Engineering, Vol. 121, No. 4, pp. 312-325.

Knighton, D., 1981. "Fluvial Forms and Processes," Edward Arnold, London, 218 p.

Lagasse, P.F., Schall, J.D., Johnson, F., Richardson, E.V., and Chang, F., 1995. "Stream Stability at Highway Structures," Federal Highway Administration Report No. FHWA-IP-90-014, Hydraulic Engineering Circular No. 20, Second Edition, Office of Technology Applications, HTA-22, Washington, D.C., November, 144 p.

Langbein, W.B. and Leopold, L.B., 1966. "River Meanders - Theory of Minimum Variance," Professional Paper 422-H, U.S. Geological Survey, Washington, D.C.

Leopold, L.G. and Maddock, T., Jr., 1953. "The Hydraulic Geometry of Stream Channels and Some Physiographic Implications," Professional Paper 252, U.S. Geological Survey, Washington D.C.

Leopold, L.B. and Wolman, M.G., 1960. "River Meanders," Geological Society of America Bulletin, Vol. 71.

Molinas, A., 1990. "Bridge Stream Tube Model for Alluvial River Simulation" (BRI-STARS), User's Manual, National Cooperative Highway Research Program, Project No. HR15-11, Transportation Research Board, Washington, D.C.

Mussetter, R.A., Harvey, M.D., Sing, E.F., 1995. "Assessment of Dam Impacts on Stream Morphology," in Sediment Management and Erosion Control on Water Resources Projects, Fifteenth Annual U.S. Committee on Large Dams Lecture Series, San Francisco, CA.

Nanson, G.C. and Hickin, E.J., 1983. "Channel Migration and Incision on the Beatton River," ASCE Journal of Hydraulic Engineering, Vol. 109.

Resource Consultants & Engineers, Inc., 1993. "Sediment and Erosion Design Guide," for Albuquerque Metropolitan Arroyo and Flood Control Authority, Albuquerque, NM.

Richards, K., 1982. "Rivers, Form and Process in Alluvial Channels," Methuen, London, 358 p.

Richardson, E.V., Simons, D.B., and Julien, P.Y., 1990. "Highways in the River Environment," prepared for the Federal Highway Administration, Washington, D.C. by the Department of Civil Engineering, Colorado State University, Fort Collins, CO, June.

Richardson, E.V. and Davis, S.R., 1995. "Evaluating Scour at Bridges," Federal Highway Administration Report No. FHWA-IP-90-017, Hydraulic Engineering Circular No. 18, Third Edition, Office of Technology Applications, HTA-22, Washington, D.C., November, 204 p.

Schumm, S.A., 1977. The Fluvial System, John Wiley & Sons, New York, NY, 338 p.

Schumm, S.A. and Winkley, B.R., 1994. "The Variability of Large Alluvial Rivers," ASCE Press, N.Y., 467 p.

Shen, H.W. and Schumm, S.A., 1981. "Methods for Assessment of Stream-Related Hazards to Highways and Bridges," Federal Highway Administration Report FHWA/RD-80/160, Washington, D.C., p. 66-86.

U.S. Army Corps of Engineers, 1991. "Scour and Deposition in Rivers and Reservoirs," HEC-6 User's Manual, Hydrologic Engineering Center, Davis, CA.

Water Engineering & Technology, Inc., 1990. "Re-Evaluation of the Streambank Erosion Control Evaluation and Demonstration Project," for U.S. Army Corps of Engineers, Vicksburg District, Vicksburg, MS.

Wolman, M.G. and Leopold, L.B., 1957. "River Floodplains, Some Observations on Their Formation," U.S. Geol. Sur. Prof. Paper 282C, Washington, D.C., p. 87-107.

GEOMORPHIC FACTORS FOR TIDAL WATERWAY HYDRAULIC MODELING

L.W. ZEVENBERGEN, J.D. SCHALL, AND J.H. HUNT
Senior Hydraulic Engineer, Manager; Sedimentation Engineering, and Water Resource Engineer, respectively, Ayres Associates, Fort Collins, Colorado, USA

ABSTRACT

Bridge scour evaluations for highway crossings at inlets and estuaries require hydraulic modeling approaches developed specifically for the hydraulic and geomorphic characteristics of the tidal waterway. This paper reviews modeling approaches and identifies geomorphic factors which affect the selection of the appropriate model.

INTRODUCTION

Scour is the erosion of channel bed and bank material from the action of flowing water. Bridge scour is a combination of long-term degradation, contraction scour, local scour, and waterway instability. In non-tidal (riverine) environments scour occurs as a result of floods caused by upland runoff. In coastal regions, bridges are also subjected to astronomical (daily) tides and storm surges. However, the flow conditions can be different for tidal waterways. Upland runoff may have little or no effect on the flow while daily tides and storm surges can produce significant discharges and velocities.

A storm tide in coastal waters results from daily tides, wind action, and rapid barometric pressure changes. Resonance can produce a greater tidal range in portions of the waterway than on the adjacent coast. The daily and storm tidal cycle consists of a rising ocean water surface elevation that drives flow inland from the ocean (flood tide) and a falling ocean water surface elevation that releases flow from inland to the ocean (ebb tide). Hydraulic energy losses occur due to friction, expansion and contraction of the flow, and form drag caused by piers and other obstructions. Typically, the energy losses result in lower tidal ranges further inland.

Geomorphology is the science related to the shape and characteristics of the earth surface. The geomorphic characteristics of the tidal waterway, adjacent coastline, and upland basin determine the predominant type of energy loss and whether energy loss significantly affects the flow rates. The geomorphic factors affecting tidal waterways include many of the factors affecting stream stability as discussed in Hydraulic Engineering Circular No. 20 (HEC-20), Stream Stability at Highway Structures (Lagasse et al. 1995). These include channel planform (straight, meandering, braided and anabranched channels), vegetative cover, channel size, and floodplain width. In addition to these factors, bay or estuary size, upland basin size, and barrier island (or roadway) overtopping potential are also important.

HEC-18, Evaluating Scour at Bridges (Richardson and Davis 1995) includes a section on tidal hydraulic and scour processes. In HEC-18 two simplified techniques (tidal prism and orifice approaches), a moderately more complex approach (storage routing), and significantly more complex dynamic modeling (i.e. UNET, FESWMS-2DH) are discussed. This paper reviews these approaches for hydraulic modeling of tidal waterways and discusses the geomorphic factors contributing to the selection and performance of each approach.

TIDAL PRISM APPROACH

The tidal prism approach (Neill 1973) can be used when the waterway or bridge does not significantly constrict the flow, the water surface is generally level and there is little energy loss due to vegetation. As shown in Figure 1, this approach is most applicable to estuaries. The basic assumption is that the entire estuary fills and empties simultaneously with the daily tide or storm tide. While this is physically impossible, the assumption results in a simple equation for maximum discharge, $Q_{max} = 3.14 \times VOL/T$, where Q_{max} is the maximum discharge during a tidal cycle, VOL is the volume of water in the tidal prism between low and high tide, and T is the tidal period between successive high or low tides. The resulting discharge is often a conservative estimate of peak flow. Figure 2 shows the tidal stage and discharge hydrographs resulting from this approach. The maximum discharge occurs about midway between low and high tides.

In addition to waterway constriction, other geomorphic factors influence the utility of the simplified tidal prism method. If the estuary is long and the limit of tidal action is far inland, then this approach can be overly conservative because the water surface within the tidal storage zone will not be level. Another factor that can greatly influence the utility of this approach is whether flows are confined primarily to the channel or, especially during a storm surge, inundation of the floodplain occurs. When the floodplain area is inundated, the prism volume increases rapidly with respect to elevation. Because floodplain flow is shallow and vegetation causes high roughness, energy loss can be significant, making the simple prism approach overly conservative. Figure 3 shows typical estuary volume versus stage relationships. Case 1 is an estuary that has little floodplain, and Case 2 is an estuary with significant floodplain area, where the large floodplain area results in significant curvature in the storage curve. Assuming a storm tide ranges from elevation 0 to 4, the simple prism method is more applicable to Case 1, because approximately half of the prism volume is filled at elevation 2. For Case 2, flows are confined to the channel up to elevation 2 and only 20 percent of the prism volume would actually be filled at this elevation. For Case 2, the simple prism approach assumes that half of the prism volume is filled at elevation 2, resulting in 2.5 times the discharge than could actually occur based on available volume. A slight modification to the simple tidal approach would be to solve for $\Delta Q = \Delta VOL/\Delta t$ using incremental volumes from the storage curve for incremental times from the storm tide hydrograph.

ORIFICE APPROACH

The orifice approach presented in HEC-18 is applicable to constricted waterways, such as inlets to bays, where significant energy loss is confined to the inlet channel and the adjoining bay is only partially filled during a tide cycle (Figure 4). The maximum velocity is calculated as $V_{max} = C_d(2g\Delta H)^{1/2}$, where C_d is the coefficient of discharge and ΔH is the maximum differential head. The discharge coefficient includes entrance and

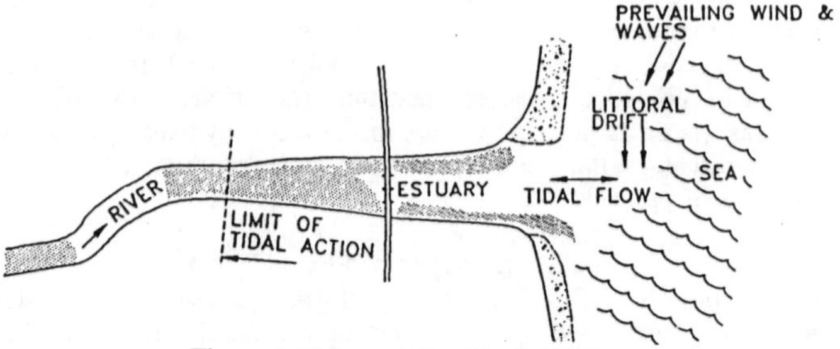

COPING WITH SCARCITY AND ABUNDANCE

Figure 1. Estuary (after Neill 1973)

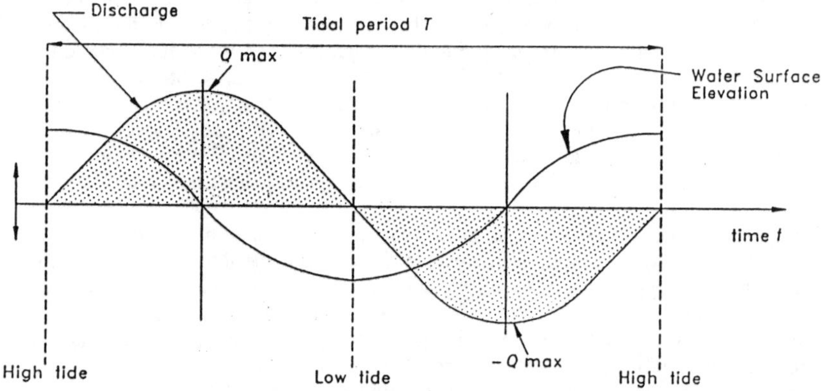

Figure 2. Simplified Tidal Hydrograph (after Neill 1973)

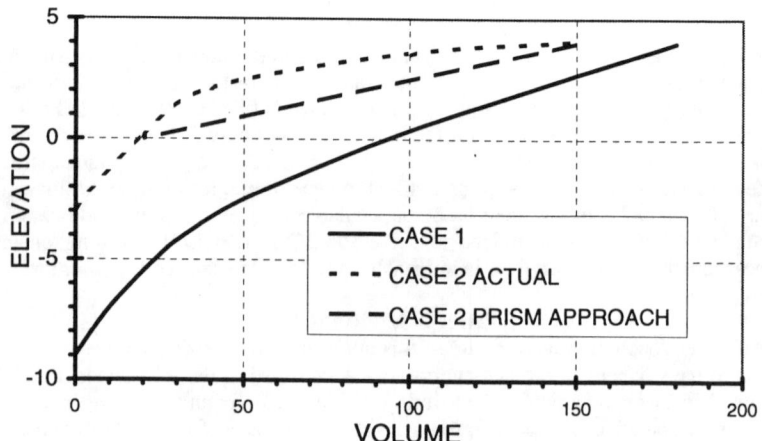

Figure 3. Typical Storage Relationships for Tidal Prism Method.

exit losses and friction loss in the channel (van de Kreeke 1967). This approach is most applicable to daily tides where the head differential can be measured. For storm tide discharge computations, the head differential would need to be estimated, which in practical application can be difficult. The discharge is then calculated from the continuity equation (Q=VA). The peak discharge occurs for the maximum head differential, which can occur at elevations other than the mid-tide. At inlets to bays, where the velocity is dependent on head differential and the discharge is the product of velocity and inlet channel area, the discharge for daily tides can increase with time if the inlet degrades. Storm surge discharges also increase as degradation occurs.

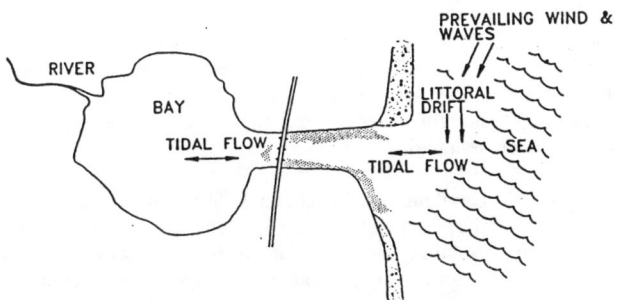

Figure 4. Bay and Inlet (after Neill 1973)

ROUTING APPROACHES

To overcome the limitations of the orifice approach (unknown head differential during a storm surge), routing methods can be applied. A method developed by Chang et al. (1994) is presented in HEC-18. The method combines the orifice equation and the storage relationship for the bay with the surge hydrograph to calculate flows through a constricted waterway such as the inlet to a bay (Figure 4). The ocean storm tide is input as a function of time, the bay storage is related to bay water surface elevation, and the discharge through the waterway is calculated from the head differential between the ocean and bay. The iterative solution technique is presented in HEC-18 using a spreadsheet or BASIC program. A similar approach is used in the ACES-Inlet model (US Army Corps of Engineers 1992). The bay is treated as a storage area and the flow in the inlet is computed for the input storm surge hydrograph. ACES-Inlet model can also compute flow in two inlets connected to a single bay.

Routing methods can include upland runoff as a separate inflow to the bay and can be modified to account for inflow over barrier islands and approach roadway embankments. If the bay is formed by a barrier island, overtopping would act as relief for the bridge. When the ebb tide drops below the overtopping elevation, unless a breach or new inlet forms, the entire bay may have to drain through the existing inlet. These conditions can also be modeled using routing methods. In the case of very large or long bays, however, the routing method can be conservative because, as with the simple prism method, the bay is assumed to fill as a level pool. For small bays, this assumption would not be overly conservative.

DYNAMIC MODELING

Dynamic modeling is recommended when complex geomorphic or hydraulic conditions make the above methods unusable or when the simplifying assumptions of the above methods are violated to such a degree that the results are overly conservative. Complex geomorphic conditions include anabranching or multiple channel inlets or estuaries, large floodplain areas constricted by main channel and relief bridges, and complex channel geometry near bridges. For estuaries with large or vegetated floodplains, where the simple tidal prism method is overly conservative due to high flow resistance, dynamic modeling is most appropriate. For large bays where an assumed level water surface is overly conservative, dynamic modeling is also most appropriate.

In a study to develop tidal hydraulic models (Ayres Associates 1994), UNET (Barkau 1996) and FESWMS-2DH (Froehlich 1996) were recommended for dynamic tidal hydraulic modeling. UNET is a 1-dimensional model and FESWMS-2DH is a 2-dimensional model. While other 1- and 2-dimensional models are also applicable for tidal hydraulic modeling, the recommended models incorporate bridge, culvert and road overtopping hydraulics. Therefore, these models were deemed most applicable for tidal bridge hydraulic and scour evaluations.

By performing hydraulic computations for channels, overbanks, bridges and culverts, and including potentially filled or inundated bay, estuary, and floodplain areas, dynamic models yield the most accurate hydraulic analysis for scour computations and countermeasure design. One-dimensional modeling is applicable for estuaries with well defined channels and for bays with single or multiple inlets. When bays are crossed by numerous causeways, especially causeways with multiple bridge openings, 2-dimensional modeling is recommended. Estuaries with multiple anabranched channels can be modeled with 1-dimensional network models (such as UNET), although the complexity may warrant the use of 2-dimensional models.

UPLAND RUNOFF

The downstream boundary condition for tidal modeling is the surge elevation hydrograph. The upstream boundary condition can incorporate upland runoff, and potentially a flood hydrograph. If the storm surge and inland flood are independent, then combining an infrequent storm surge and an infrequent riverine flood would be too unlikely a combination to consider. For example a 100-year storm surge and a 10-year riverine flood would have an approximate 1000-year recurrence interval. If the inland flood results from extreme rainfall caused by the hurricane which also produced the storm surge, combining the downstream stage hydrograph with an upstream flood hydrograph may be warranted. Care must be taken to ensure that the timing of the hydrographs is appropriate. Basin scale is a geomorphic factor influencing the decision to combine the two events in a dynamic analysis. If the basin is large, the storm surge may have receded long before the riverine flood arrives in the tidal waters, and the two events should not be combined even if both are caused by the same hurricane.

CONCLUSIONS

Geomorphic factors influence the applicability of tidal hydraulic modeling approaches. Scale is important in selecting the appropriate level of sophistication. The larger the estuary or bay, the more applicable dynamic modeling becomes. The scale of the basin also influences whether an upland flood hydrograph should be included in the analysis. For estuaries, large floodplains can result in overly conservative results from the tidal prism method. Vegetation in floodplains and tidal marshes increases flow resistance and energy loss, thereby requiring dynamic modeling. Barrier island and roadway overtopping may require the use of routing methods or dynamic modeling. Complex channel planform, such as anabranched channels, or multiple opening causeways may also require the use of 2-dimensional dynamic modeling. While tidal prism, orifice and routing methods give reasonable results in many tidal waterway analyses for bridge scour, dynamic modeling is more appropriate when these approaches are overly conservative or where there is significant geomorphic complexity. For example, in a tidal hydraulic analysis of a bridge over a large estuary with tidal marshes and wooded floodplains, the tidal prism approach resulted in a maximum velocity of 5.2 m/s, whereas dynamic 2-dimensional analysis resulted in a more realistic maximum velocity of 2.3 m/s. Cost savings of approximately $300,000 were realized from the redesigned foundations.

REFERENCES

Ayres Associates, 1994, "Development of Hydraulic Computer Models to Analyze Tidal and Coastal Stream Hydraulic Conditions at Highway Structures," Final Report, Phase I HPR552. South Carolina Department of Transportation.

Barkau, R.L., 1996. "UNET - One Dimensional Unsteady Flow Through a Full Network of Open Channels," Report CPD-66 Version 3.1, U.S. Army Corps of Engineers, Hydrologic Engineering Center, Davis, CA.

Chang, F., Veeramancheneni, R., and Davis, S., 1994. "Tidal Flow Through a Contracted Bridge Opening," Maryland State Highway Administration.

Froehlich, D.C., 1996. "Finite Element Surface-Water Modeling System: Two-Dimensional Flow in a Horizontal Plane Version 2", Users Manual.

van de Kreeke, J., 1967. "Water-Level Fluctuations and Flow in Tidal Inlets," American Society of Civil Engineers, Vol. 93, No. WW4, New York, NY.

Lagasse, P.F., Schall, J.D., Johnson, F., Richardson, E.V., and Chang, F., 1995. "Stream Stability at Highway Structures," Hydraulic Engineering Circular No. 20, Second Edition, FHWA, 144 pp.

Neill, C.R., 1973. "Guide to Bridge Hydraulics," (Editor) Roads and Transportation Association of Canada, University of Toronto Press, Toronto, Canada.

Richardson, E.V. and Davis, S.R., 1995. "Evaluating Scour at Bridges," Hydraulic Engineering Circular No. 18, Third Edition, FHWA, 204 pp.

U.S. Army Corps of Engineers, 1992. "Automated Coastal Engineering System," Technical Reference, Leenknecht, D.A., Szuwalski, A., and Sherlock, A.R., Coastal Engineering Research Center, Waterways Experiment Station, Vicksburg, MS.

Pier Scour in Resistant Material: Current Research on Erosive Power

S. P. Smith[1], G. W. Annandale[2], P. A. Johnson[3], J. S. Jones[4], and E. R. Umbrell[5]

Abstract

A new procedure for estimating the extent of pier scour into erodible rock and other resistant materials based on the Erodibility Index Method (Annandale, 1995) has been proposed by Smith and Annandale (1995). Application of the Erodibility Index Method to bridge scour requires techniques for estimating stream power at the base of bridge piers. This paper presents preliminary relationships between stream power and scour derived from ongoing experiments at the Federal Highway Administration (FHWA) Turner-Fairbank Highway Research Center.

Introduction

The Erodibility Index Method utilizes a relationship between the erosive power of water and a geomechanical classification system known as the Erodibility Index. It defines an erodibility threshold for rock and other earth materials and characterizes the erosive power of flowing water in terms of stream power. Stream power values can be compared to the erodibility threshold of the channel bed and foundation material around piers. The maximum scour depth can be predicted by determining the depth at which the stream power in the scour hole falls below the stream power required to erode the foundation material.

---

[1]Hydraulics Engineer, Colorado Department of Transportation, Staff Design Branch, 4201 E. Arkansas Ave., Denver, CO 80222

[2]Director, Water Resources Engineering, Golder Associates Inc., 200 Union Blvd., Suite 500, Lakewood, CO 80228

[3]Associate Professor, Department of Civil Engineering, Pennsylvania State University, 212 Sackett Building, University Park, PA 16802

[4]Hydraulic Research Engineer, Federal Highway Administration, Turner-Fairbank Highway Research Center, McLean, VA 22101

[5]Staff Engineer, GKY & Associates, Inc., 5411-E Backlick Road, Springfield, VA 22151

A cooperative study between the FHWA and the Colorado Department of Transportation is in progress to develop procedures for practical application of the Erodibility Index Method to pier scour prediction. As part of this study, a laboratory investigation is being conducted to produce relationships between stream power and pier scour. This paper discusses erosive power at bridge piers and presents the preliminary results of this laboratory investigation. Graphical relationships between stream power and scour depth are presented for both circular and square shaped piers.

Erosive Power at Bridge Piers

At bridge piers, highly turbulent flow conditions exist. A principal characteristic of turbulence is energy dissipation. Stream power is defined as the rate of energy dissipation per unit area and was shown by Bagnold (1966) to be the product of shear stress (τ) and velocity (V).

Smith and Annandale (1995, 1996) have studied the relationship between stream power and scour depth. Their studies indicate that stream power is greatest around a pier for the flat bed condition. This is when the turbulence is most intense. As the scour hole increases in depth, the turbulence intensity around the pier decreases with a concurrent reduction in the erosive power around the pier. The erosive power continues to decrease until it reaches equilibrium with the erosive power of the approach flow or until a material is encountered with a strength which exceeds the erosive power around the pier. The reduction in erosive power with increasing depth of scour has important implications for determining the depth of scour in a relatively resistant material, such as rock. It often happens that the strength of rock increases as a function of the distance below the surface. With the erosive power decreasing in the same direction, it is reasonable to expect that scour will cease at shallower depths in rock than would occur in cohesionless material.

Basis of Experiments

Past studies have focused on defining erosive power around piers by shear stress and velocity. Observations have indicated considerable increases in both shear stress and velocity around piers in relation to approach flow and a decreases in both as a scour hole develops. Parola (1993) observed a maximum shear stress amplification around piers of 6 to 8 times that of the approach flow. He reported Melville's (1975) observation of shear stresses as high as 4.9 times approach flow values and that Hjorth (1975) recorded ratios as high as 12. Parola (1993) recorded maximum velocities within a scour hole between approximately 1.5 and 2.0 times that of the approach flow velocity. Stream power is the product of shear stress and velocity making the above findings very significant to the basic assumptions of this investigation.

Developing a mathematical equation to compute stream power in the vicinity of a pier which accounts for the effects of pier geometry, approach flow, and scour hole geometry is extremely difficult. Direct measurement of stream power within the scour hole is presently not possible. The objective of the laboratory investigation is to develop empirical relationships between pier scour depths and stream power utilizing indirect experimental methods.

Stream Power Experiments

Three different pier models were chosen for the laboratory investigation. Models representing square, circular, and round nose piers were used. The width of each model was limited to prevent contraction scour from occurring in the flume during the experiments. The experimental runs investigating stream power at the base of the round nose piers are not yet complete and are not discussed in this paper.

Three uniform sediment samples were used as the erodible medium around the pier. Critical velocity and critical stream power was estimated for each sediment sample visually in the flume by conducting unobstructed incipient motion tests. For these unobstructed flow tests the flume was filled with water and the tailgate lowered until movement of the sediment was observed.

Experiments were run in the clear-water condition. During clear-water scour, the erosive power of the approach flow is not sufficient to mobilize the bed material and there is no inflow of sediment into the scour hole. The erosive power at the base of the pier will decrease as the scour hole develops until the erosive threshold of the bed material is reached. An equilibrium, where erosive power around the pier becomes equivalent with that of the approach flow, does not occur. At the maximum scour depth, the stream power at the base of the pier will be equal to the critical stream power of the bed material. Measurement of stream power at the base of the pier can be indirectly made by computing the critical stream power of the sediment around the model pier.

Flume flow conditions were varied for each experimental run to provide a range of approach stream power up to the incipient stream power of the three sediment samples. By observing that the stream power in the scour hole reached critical for the bed material at the maximum scour depth, a stream power ratio could be computed for each pier scour test run. The stream power ratio is a key to the Erodibility Index Method because it provides a means of estimating the stream power in the scour hole which can in turn be used to evaluate the erodibility potential of subsurface materials as the scour hole deepens. When the stream power around the pier falls below the critical for a soil or rock strata, scour deepening would cease.

For each pier scour experiment, the model pier was placed at the center of the flume and covered with sediment. The rest of the flume was lined with the same sediment and leveled. After the flume was slowly filled with water, the tailgate was lowered and the discharge increased until the desired velocity was reached. The flume was then allowed to run for a minimum of 24 hours. Main channel velocities, flow depths and discharges were recorded. At the end of each test run, the flume was drained. Each test was run until the maximum scour depth could be established. The effective stream power in the scour hole, therefore would have to be at critical when the scour hole reached its maximum depth. After each run the scour hole was photographed and the maximum depth and scour hole dimensions were measured.

In addition, an initiation of motion test was conducted for each model pier and sediment configuration. The objective of these tests was to record the velocities at which movement of the sediment around the model pier commences. This information was used to define the maximum stream power amplification around the pier. Experimental runs with approach flow conditions established just below the sediment sample's threshold of motion were also included in the testing. These experiments were run to determine equilibrium scour for each sediment and pier model.

Clear-water scour is an asymptotic process and each test would take days or weeks to actually reach maximum scour depth. A laboratory expedient was to run one long duration test for four to seven days for each sediment size. A "time to scour" relation was derived for each sediment from these long duration tests. Maximum scour depths from shorter runs were established by dividing measured scour depths by the percentage of ultimate scour that could be anticipated from the shorter duration tests.

Analysis of Data

Laboratory measurements were analyzed to study the relationship between stream power and scour depth for both square and circular shaped piers. Stream power is the product of shear stress and velocity and it is assumed that critical stream power can be determined by multiplying critical velocity by critical shear stress. Approach stream power per unit area (τV) was determined by multiplying the logarithmic velocity relation for fully developed, turbulent flow by the approach velocity. Therefore:

$$(\tau V) = \rho V^3/[5.75\log(12.27y/k_s)]^2 \qquad (1)$$

where:

ρ = water density V = approach flow velocity
y = flow depth k_s = bed roughness, defined by median particle diameter

Equation 2 (Neill, 1973) was used to compute critical velocity (V_c):

$$V_c = 1.58[(S_g-1) \; y \; D_{50}]^{0.5} \; (y/D_{50})^{0.167} \qquad (2)$$

where:

y = flow depth D_{50} = median particle diameter
S_g = specific gravity = 2.65

The calculated critical velocity values were compared with values observed during the unobstructed incipient motion tests and they correspond closely with the recorded values. Critical stream power values for each sediment sample were computed by substituting critical velocity values into equation 1.

The results of the experimental analysis are presented in Figures 1 and 2. The stream power ratio is plotted against relative scour depth. Scour depth is normalized by equilibrium scour depth. The stream power ratio is the ratio of stream power at the base of the pier and the stream power of the approach flow. The stream power at the base of the pier is assumed to be equivalent to the critical stream power of the sediment which was placed around the pier.

The data plotted in these graphs indicate a clear relationship. Stream power values are highest for shallower scour depths. The peak stream power amplification is significantly higher for the square pier. It is approximately 20 times that of the approach flow. The peak stream power amplification is approximately 14 for the circular pier. In general, stream power ratios for the square pier are slightly higher than that for the circular piers.

Conclusions

Laboratory measurements of pier scour collected at the FHWA hydraulics lab were analyzed to study the relationship between stream power and scour depth. Graphs relating stream power at bridge piers to scour depth were presented for square and circular shaped piers. This data will be useful in developing methods for practical application of the Erodibility Index Method to scour prediction.

References

Annandale, G.W., *Erodibility*, Journal of Hydraulic Research, Volume 33, No. 4, 1995, pp 471-494.

Annandale, G. W. and Smith, S. P., *Scour in Erodible Rock I: The Erodibility Index,* Proc. North American Water and Environment Congress, '96, ASCE, Anaheim, California, 1996.

Bagnold, R. A., *An Approach to Sediment Transport from General Physics*, U.S. Geological Survey, Professional Paper 422-I, 1966.

Neill, C. R., *Guide to Bridge Hydraulics*, University of Toronto Press, Toronto, Canada, 1973.

Parola, A. C., *Stability of Riprap at Bridge Piers*, Journal of Hydraulics Engineering, Vol. 119, No. 10, 1993, pp. 1080 - 1093.

Smith, S.P. and Annandale, G. W., *Preliminary Procedure to Predict Bridge Scour in Bedrock,* Proc. Of the First International Conference on Water Resources, ASCE, San Antonio, Texas, August, 1995, pp 971- 975.

Smith, S. P. And Annandale, G. W., *Scour in Erodible Rock II: Erosive Power at Bridge Piers ,* Proc. North American Water and Environment Congress, '96, ASCE, Anaheim, California, 1996.

PIER SCOUR IN RESISTANT MATERIAL

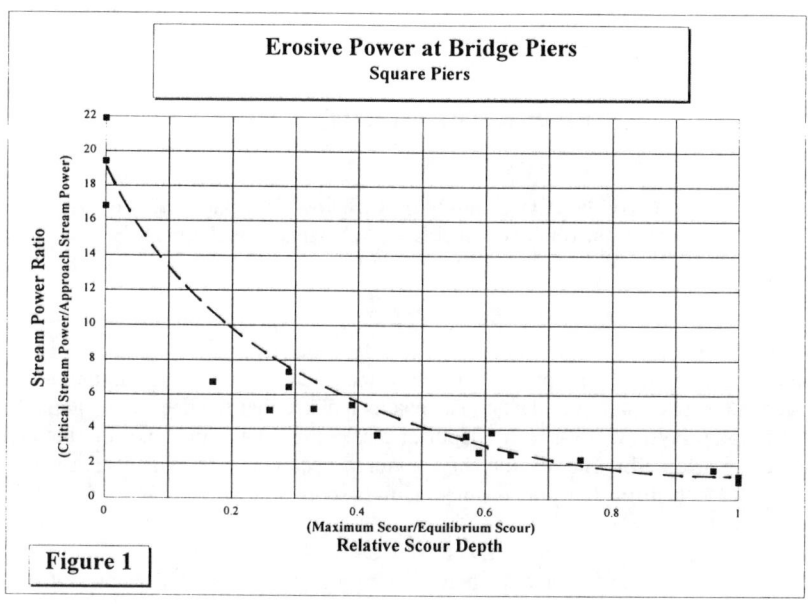

Figure 1

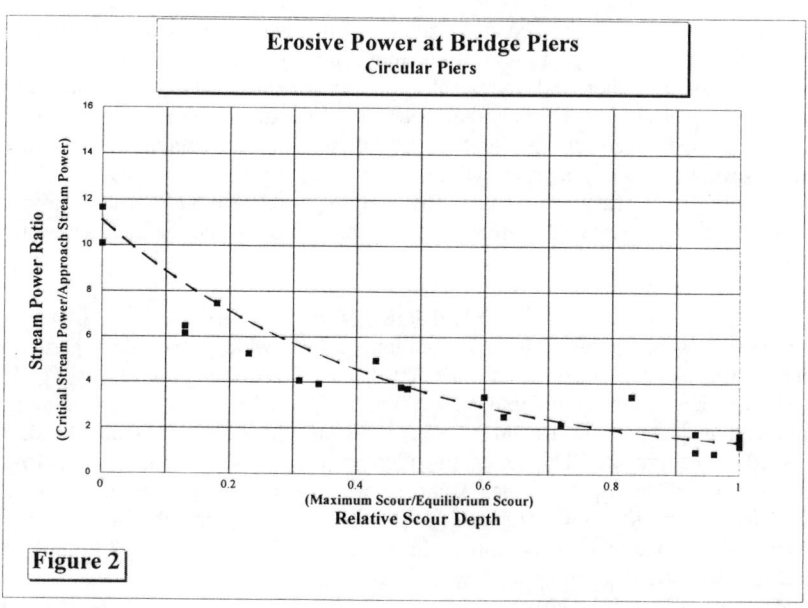

Figure 2

PIER RIPRAP PROTECTION

Melville, B.W., Lauchlan, C.S. and Coleman, S.E.
The University of Auckland, Auckland, New Zealand

ABSTRACT

A laboratory study of the stability of riprap protective layers under mobile-bed conditions is reported. Emphasis is given to the effect on riprap disintegration of the placement level of the riprap layer. The results indicate that the deeper the initial level of the riprap layer, the lesser the depth of local scour at the pier, i.e. the better the protection afforded. The stability of the riprap decreases also with increasing flow velocity. Mechanisms of riprap disintegration are discussed also.

INTRODUCTION

Posey et al (1951) and Lim and Chiew (1996) are the only known studies of bridge pier riprap protection under mobile-bed conditions. Lim and Chiew (1996) investigated riprap layers placed at the surface of the undisturbed sediment bed over a wide range of mobile-bed flows. They found that the riprap settled as bed-forms migrated past the pier and the local scour at the pier approached that for an unprotected pier at approximately the upper end of the dune regime. They concluded that riprap layers were unlikely to offer any resistance against scour when large bed features were present. An aspect which was not investigated is how placement depth of the riprap layer affects the scour. The study reported here addresses the effect of varying placement level of the riprap layer on riprap stability and the local scour at the pier.

EXPERIMENTS

Experiments were conducted in an 11.8-m long, 0.44-m wide, 0.38-m deep glass sided flume. The pier was made from a 70-mm diameter (D) clear perspex tube. Water levels were kept constant at depth, y_o = 200-mm. The bed material comprised uniform sand with median particle size, d_{50} = 0.95-mm. Riprap stones were obtained by sieving crushed road aggregate. Three sizes (d_{r50}) of riprap stones were used; 7.8-mm, 16-mm and 22-mm. The riprap stones were painted green and yellow to allow easy identification in the flow. The critical shear velocity u_{*c} for the sediments was calculated from the Shields entrainment function. Froude number ranged from 0.37 to 0.92 for the tests. The riprap layer was circular in plan (concentric with the pier) with coverage c = 4D and thickness t = $2d_{r50}$, as shown in Figure 1a. The effect of riprap placement level with respect to the original bed level was investigated by varying the placement depth Y of the layer.

Each experiment comprised two stages. First the local scour depth without riprap protection was measured. A 24-hr duration was used, which allowed the bed forms to develop fully and the live-bed scour depth at the pier to reach the equilibrium stage. Scour depths were recorded at the pier nose using a periscope which was lowered inside the pier; readings of the scour depth at the pier face were recorded at one minute intervals over a 100 minute period. The average, minimum and maximum scour depths were determined for each flowrate and these values were used to determine scour depth reduction in subsequent experiments. For the second stage, riprap was placed at the pier with the aid of a circular ring, which was removed before testing. The riprap layer was tested at four different initial levels in the sand bed for each of the flow rates. Where the riprap was placed with the top surface below the original bed level the area above was filled in with bed material. For the tests with riprap, the maximum change in level of the riprap was used as the maximum scour in the layer and compared to the unprotected maximum scour levels. Complete failure was taken to occur when the riprap could no longer protect the pier such that the scour was the same as that without protection.

FAILURE MECHANISMS

Chiew (1995) observed three failure modes for riprap protection at bridge piers under clear-water conditions:
- shear failure, where the riprap stones are entrained by the flow.
- winnowing failure, where the bed material is eroded through voids in the riprap layer.
- edge failure, where riprap stones at the periphery of the layer are undermined.

These failure mechanisms were observed also during the live-bed study reported here. An additional mode of failure due to the passage of bed forms was also observed, consistent with the findings of Lim and Chiew (1996). In general, settlement of the riprap layer is triggered by the passage of an especially large bed-form, having a deeper than normal trough elevation. If a subsequent deeper trough passes the pier, the riprap may settle further. The fluctuating bed level brought about by the passage of bed features past the pier causes the riprap stones to lose support and therefore stability. If the trough of the bed-form is deeper than the riprap layer, riprap stones are undercut and slide into the lower trough region. As a bed-form approaches the pier, high levels of turbulence and correspondingly high shear stresses are induced for short periods. Riprap stones can be plucked from the layer and transported to the lee of the pier and beyond. Once some stones are removed, the layer in that area is thinner which induces increased winnowing. Other riprap stones are exposed and can also be removed. The smaller 7.8-mm stones were most susceptible to this mechanism; they failed at relatively low values of U/U_c (U = mean flow velocity; U_c = critical value of U for bed material entrainment). Failure could be rapid where the turbulence associated with the bed forms was sufficiently high to entrain the riprap material in clusters, resulting in a swift disintegration of the layer.

Placement depth has an effect on the dominant mode of failure for the riprap layer. Figures 1b and 1c show the failure mechanisms for riprap placed flush with the bed surface and below the bed surface, respectively. For riprap layers placed flush with the bed surface the passage of bed forms causes undercutting of the layer, resulting in edge failure. Stones move forward into the approaching bed form trough and also spread out laterally. This causes a thinning of the layer allowing winnowing to occur. The riprap settles into the bed material and spreads out forming an armour layer around the pier. At this stage the layer is stable and no further lateral movement of individual riprap stones is observed with continued bed form migration through the layer, although if a deeper trough passes the pier, significant additional settlement can occur. At the nose of the pier the strong horseshoe vortex structure is capable of plucking riprap stones from the layer, allowing winnowing to occur and the riprap layer near the pier to subside into the scour region. At higher flows, a different mechanism is observed. The edge stones are unable to embed in the surrounding bed and are often transported to the scour region in front of the pier. The scour hole is filled, thickening the riprap layer near the pier and therefore reducing the effects of winnowing.

With deeper placement levels, the ability of the bed forms to undercut the riprap layer is reduced. As a consequence, winnowing against the pier face becomes the dominant mode of failure. In some cases the original riprap formation is scarcely disturbed except for a slight depression forming against the upstream pier face. Placement below the original bed also reduces the depth of subsidence of the riprap into the bed material.

EFFECT OF PLACEMENT LEVEL

Figure 2 is a plot of d_r/d_{smax} in terms of U/U_{cr} where d_r is the riprap settlement depth measured at the leading edge of the pier, as defined in Figure 1, d_{smax} is the maximum scour depth without riprap for the same mean flow velocity U, and U_{cr} is the critical mean velocity of the riprap stones, determined using the Shields entrainment function. The range of conditions tested is between 1 and 2.5 times the critical velocity of the bed material U_{cs}. The data demonstrate that the riprap settlement depth increases asymptotically with flow velocity towards the unprotected scour depth and that the deeper the initial placement of the layer, the lesser the scour experienced by the riprap at a particular flow velocity. As the flowrate increases the amount of scour reduction is reduced but the advantage of deeper placement remains. A series of curves is fitted to the data for each placement level and stone size. Each curve tends towards failure, however as the stone size increases the flow velocity required for failure also increases.

The data are plotted in Figure 3 to show the variation of scour depth (with riprap) measured from the undisturbed bed level, i.e. (d_r+Y), as a function of U/U_{cs}. The axes of the plot are the same as those used by Lim and Chiew (1996), who placed the riprap at the bed surface level for all their experiments. The present data for $Y/D = 0$, when compared to those presented by Lim and Chiew, show similar trends. Their conclusion that surface-placed riprap layers may be incapable of affording protection against scour at bridge piers where deep bed-forms occur, is supported by the new results. However, placement where Y/D > 0 significantly extends the range of flows for which protection is provided. It is noted also that filter protection of the riprap layer may affect these findings. It is intended to investigate filter layers, employed with riprap at bridge piers, as a second stage of the study reported here.

In Figure 4, the data are plotted in three groups for each of the three riprap stone sizes. It is apparent that larger riprap affords better protection at a given flow velocity. The inability of the 7.8-mm stones to sustain flows greater than around 0.65 U/U_{cr} indicates that a minimum size relationship may exist between the riprap stone size and the bed material size. For the two larger stone sizes (16-mm and 22-mm), the riprap provided better protection, the scour reduction was greater and was achieved over a wider range of flow conditions.

CONCLUSIONS
The following conclusions are drawn from this study:
1. Four destabilising mechanisms are observed for riprap protection at bridge piers under mobile-bed conditions; namely shear failure, winnowing, edge failure and bed-form destabilisation.
2. Riprap stability in live-bed conditions is highly dependent on applied flows; the higher the flow velocity, the less stable the riprap and the deeper the scour at the bridge pier. Local scour depths at bridge piers protected with riprap increase asymptotically with flow velocity towards the maximum scour depth at an unprotected pier.
3. Placement level affects the range of flows over which riprap protection without filters is effective under live bed conditions and the local scour that develops at a particular velocity; generally the deeper the placement level, the more stable the riprap and the lesser the scour at the pier.

REFERENCES
Chiew, Y.M. (1995) "Mechanics of Riprap Failure at Bridge Piers," Journal of Hydraulic Engineering, ASCE, 121(9), 635-643.

Lim, F.H. and Chiew, Y.M. (1996) "Stability of Riprap Layers Under Live-bed Conditions," draft paper to be submitted for publication (details unknown).

Posey, C.J., Appel, D.W., Chamness, E., Colorado, A. and College, M. (1951) "Investigation of Flexible Mats to reduce Scour around Bridge Piers," Highway Research Board, Report 13b, 12-22.

Figure 1: Riprap Layer Definition Sketch and Failure Mechanisms

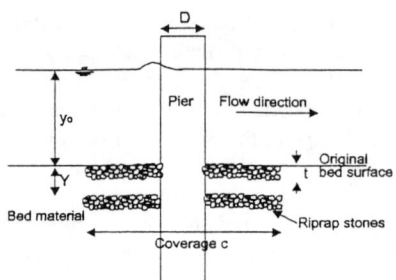

a. Definition sketch of riprap layer

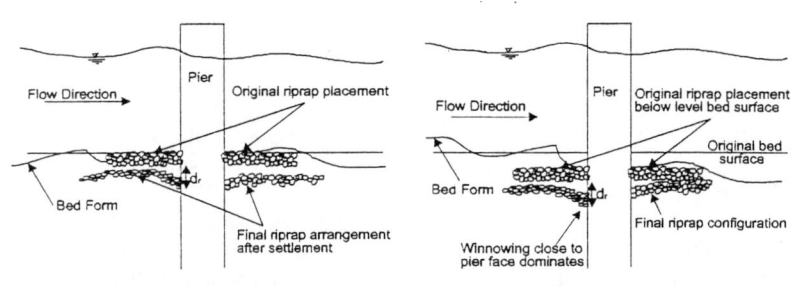

b. Riprap layer placed initially at the bed surface level

c. Riprap placed initially below the original bed surface

Figure 2: Comparison of Scour Reduction Using Riprap at Different Initial Placement Levels

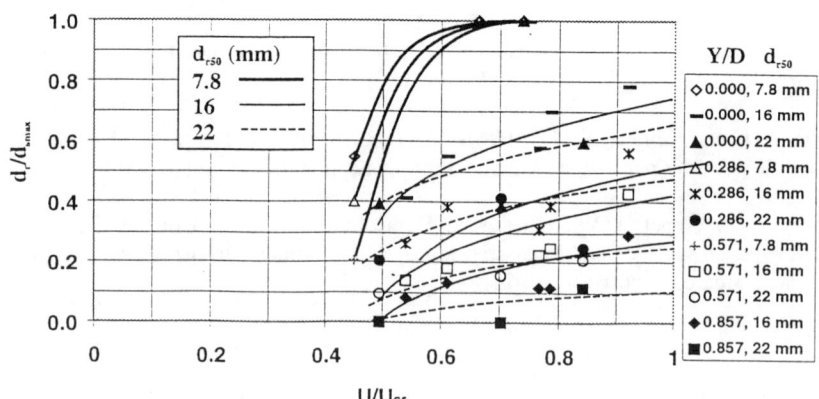

PIER RIPRAP PROTECTION 171

Figure 3: Variations in Scour Reduction with U/U_{cs}

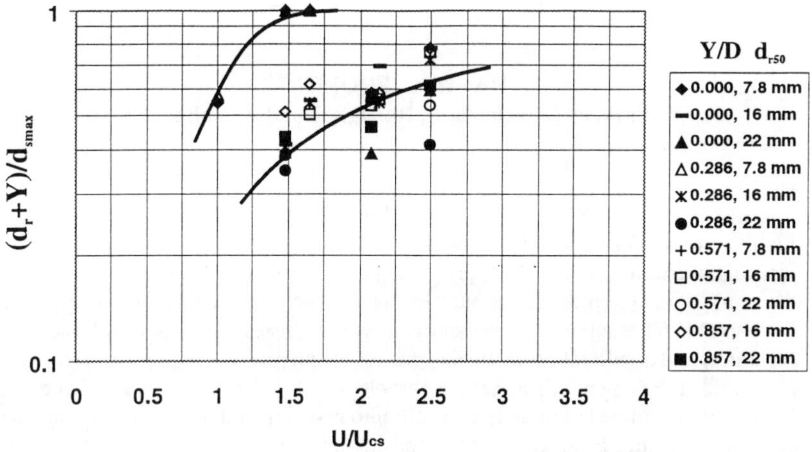

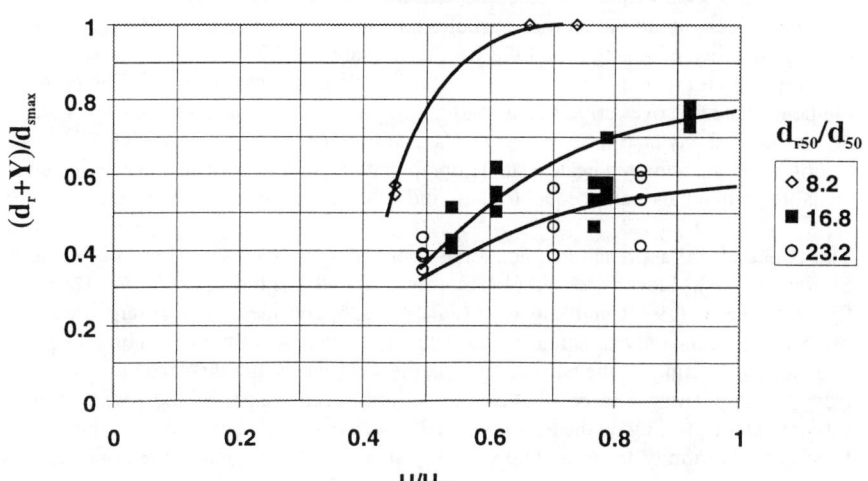

Figure 4: Effect of Riprap to Sediment Size Ratio on Scour Reduction

RIPRAP PARTICLE STABILITY
BY MOMENT ANALYSIS

DAVID C. FROEHLICH
Parsons Brinckerhoff, Raleigh, North Carolina, USA

ABSTRACT

Stability of a single rock riprap particle is evaluated based on the ratio of moments resisting overturning to moments promoting overturning of the particle. Departing from the conventional approach, the buoyant force is not subtracted from the gravitational force to obtain the submerged weight of a particle. Instead, the buoyant force is treated separately and is split into components that resist and promote overturning, similar to the conventional treatment of submerged particle weight. When gravitational and buoyant forces are considered separately and spilt into resisting and overturning components, a consistent stability factor results that tends to unity as rock specific gravity approaches one from above.

INTRODUCTION

Rock riprap is widely used to protect the beds and banks of waterways from erosion by currents. Because individual rock particles tend to be dislodged by rolling rather than sliding, stability of a single particle has been evaluated by considering overturning moments about the point of contact of the rock with an adjacent particle. The ratio of moments that resist overturning of the particle, M_R to moments that promote overturning, M_P, defines a safety factor, $SF = \sum M_R / \sum M_P$, that provides an index of particle stability. Ratios larger than unity indicate a stable riprap particle; ratios less than unity indicate an unstable particle; and ratios equal to unity indicate a neutrally stable particle.

The conventional approach to calculating overturning moments used by Stevens and Simons (1971), Stevens et al. (1976), Simons and Şentürk (1977, p. 418-446), Richardson et al. (1990), and Kawai and Julien (1996) considers the submerged weight of a rock to be the only resisting force. Submerged weight is found by subtracting the buoyant force acting on the particle (that is, the weight of water displaced by the rock) from the gravitational force (that is, the unsubmerged weight of the rock). For a submerged rock resting on the horizontal bed of a channel with zero velocity, the safety factor equals infinity for a particle with any specific gravity greater than one, and is undefined for any specific gravity less than one. However, intuition suggests that extremely dense rocks are more stable than lighter rocks of the same size and that the variation in stability should be reflected by the safety factor. The particle stability analysis that follows yields a consistent expression for the ratio of moments that tends to unity as rock specific gravity approaches one.

CHANNEL GEOMETRY

Stability calculations are made for a rock resting on the bank or bed of a channel having a trapezoidal cross section as shown in Fig. 1. Cross sections forming the ends of a length of channel are considered to remain vertical for all longitudinal slopes (a typical geometric arrangement for excavated channels lined with rock riprap.) The unit vector normal to the bank plane is given by

$$\vec{n}_b = \frac{1}{\sqrt{1 + \tan^2\alpha + \tan^2\theta}} \left[\vec{i} \tan\alpha + \vec{j} \tan\theta + \vec{k} \right] \quad (1)$$

where $\vec{i}$, $\vec{j}$, and $\vec{k}$ = unit vectors in the x, y, and z directions respectively (x and y are horizontal Cartesian coordinates, and z is the vertical coordinate); θ = lateral bank angle with horizontal, and α = longitudinal bed angle with horizontal (positive for a downslope) as shown in Fig. 1.

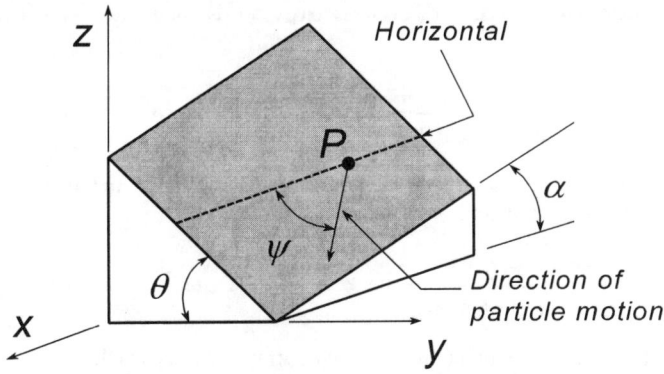

Figure 1. Section of a channel bank showing the coordinate system, bed angle α, bank angle θ, and motion angle ψ for a particle located at P.

PARTICLE FORCES

Gravitational, hydrostatic, and hydrodynamic forces creating drag and lift act on a submerged rock riprap particle. Magnitudes of the different forces and their influences on particle stability depend not only on the angles θ and α, but also on the water-surface slope ζ, and the flow direction angle λ.

The gravitational force or weight vector $\vec{W} = -\vec{k}W = \vec{W}_n + \vec{W}_t$ where

$$\vec{W}_n = \vec{W} \cdot \vec{n}_b = \frac{-W}{1 + \tan^2\alpha + \tan^2\theta} \left[\vec{i} \tan\alpha + \vec{j} \tan\theta + \vec{k} \right] \quad (2)$$

is the weight component directed normal to the bank plane, and the weight component

$$\vec{W}_t = \vec{W} - \vec{W}_n = \frac{W}{1 + \tan^2\alpha + \tan^2\theta}\left[\vec{i}\tan\alpha + \vec{j}\tan\theta - \vec{k}\left(\tan^2\alpha + \tan^2\theta\right)\right] \quad (3)$$

is directed tangent to the bank plane since $\vec{W}_n$ and $\vec{W}_t$ are orthogonal. The normal component $\vec{W}_n$ resists overturning while the vector $\vec{W}_t$ acts to displace the particle.

Bouyancy is the result of hydrostatic pressures acting on the surface of an immersed particle. Magnitude of the buoyant force B equals the weight of water displaced by the particle. Since hydrostatic forces increase with depth, the net buoyant force upon a particle is directed upwards, with a horizontal component created by a non-level water surface. The buoyant force vector is given by $\vec{B} = B[\vec{i}\sin\zeta + \vec{k}\cos\zeta] = \vec{B}_n + \vec{B}_t$ where

$$\vec{B}_n = (\vec{B}\cdot\vec{n}_b)\vec{n}_b = B\frac{\sin\zeta\tan\alpha + \cos\zeta}{1 + \tan^2\alpha + \tan^2\theta}\left[\vec{i}\tan\alpha + \vec{j}\tan\alpha + \vec{k}\right] \quad (4)$$

is the component directed normal to the bank plane, and $\vec{B}_t = \vec{B} - \vec{B}_n$ acts tangent to the bank plane. The tangential component of the buoyant force vectors is further divided as $\vec{B}_t = \vec{B}_{t1} + \vec{B}_{t2}$ where

$$\vec{B}_{t1} = \frac{B\cos\zeta}{1 + \tan^2\alpha + \tan^2\theta}\left[-\vec{i}\tan\alpha - \vec{j}\tan\theta + \vec{k}\left(\tan^2\alpha + \tan^2\theta\right)\right] \quad (5)$$

is directed up the bank slope in a direction opposite to $\vec{W}_t$, and therefore resists overturning of the particle, and

$$\vec{B}_{t2} = \frac{B\sin\zeta}{1 + \tan^2\alpha + \tan^2\theta}\left[-\vec{i}(\tan\alpha\tan\theta) + \vec{j}(1 + \tan^2\alpha) - \vec{k}\tan\theta\right] \quad (6)$$

is the remaining component that promotes overturning of the particle.

Hydrodynamic drag acts parallel to the streamline at the location of the rock particle, which makes an angle λ with horizontal, and is given by

$$\vec{D} = \frac{D}{\sqrt{1 + \tan^2\lambda + \left(\frac{\tan\lambda - \tan\alpha}{\tan\theta}\right)^2}}\left[\vec{i} + \vec{j}\left(\frac{\tan\lambda - \tan\alpha}{\tan\theta}\right) - \vec{k}\tan\lambda\right] \quad (7)$$

where D = drag force magnitude. Fluid drag is considered to promote overturning.

The hydrodynamic lift forces acts on a particle in a direction normal to the bank plane and is given by

$$\vec{L} = \frac{L}{\sqrt{1 + \tan^2\alpha + \tan^2\theta}} \left[\vec{i}\tan\alpha + \vec{j}\tan\theta + \vec{k} \right] \quad (8)$$

where L = magnitude of the lift force. Acting in the same direction as $\vec{W}_n$, fluid lift also promotes overturning of the particle.

MOMENTS AND STABILITY

The sum of moments resisting overturning of a particle is

$$\sum M_R = |\vec{W}_n|l_2 + |\vec{B}_{t1}|\cos\omega \quad (9)$$

and the sum of moments promoting overturning is

$$\sum M_P = |\vec{W}_t|l_1 + |\vec{B}_{t2}|l_1\cos\xi + |\vec{B}_n|l_2 + |\vec{D}|l_3\cos\delta + |\vec{L}|l_4 \quad (10)$$

where the moment arms l_1, l_2, l_3, and l_4 are shown in Fig. 2,

$$\cos\omega = \frac{\cos\psi \left[\sin\theta\tan\psi(1 + \tan^2\alpha + \tan^2\theta) + \tan\alpha \right]}{\sqrt{(\tan^2\theta + \tan^2\alpha)(1 + \tan^2\alpha + \tan^2\theta)}} \quad (11)$$

$$\cos\delta = \frac{\cos\psi \left[\cos\theta\tan\psi(\tan\lambda - \tan\alpha) + \tan\theta + \sin\theta\tan\psi(\tan\theta\tan\lambda) \right]}{\sqrt{(\tan\lambda - \tan\alpha)^2 + \tan^2\theta(1 + \tan^2\lambda)}} \quad (12)$$

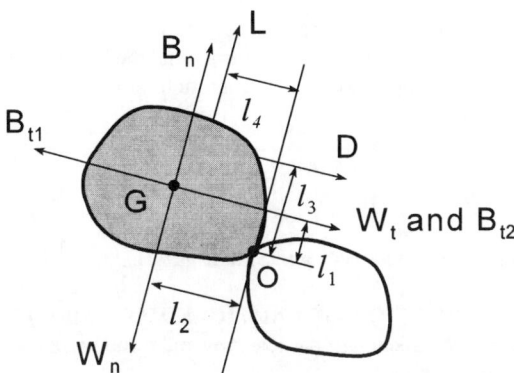

Figure 2. Illustration of forces (W_n, W_t, B_n, B_{t1}, B_{t2}, D, and L), moment arms (l_1, l_2, l_3, and l_4), center of gravity of the particle G, and point of contact O about which particle rotation can take place.

and ψ = angle of particle motion. Particle stability is given by the safety factor SF calculated as

$$SF = \frac{1 + \dfrac{\cos\omega \cos\zeta}{S_r \tan\phi}\sqrt{\tan^2\theta + \tan^2\alpha}}{\eta\left(1 + \tan^2\alpha + \tan^2\theta\right) + \dfrac{\cos\omega}{\tan\phi}\sqrt{\tan^2\theta \tan^2\alpha} + K_\zeta} \qquad (13)$$

where $\eta = M + N\cos\delta$, $M = (l_4 L)/(l_2 W)$, $N = (l_3 D)/(l_2 W)$, $\tan\phi = l_2/l_1$, and

$$K_\zeta = \frac{1}{S_r}\left[\cos\zeta(1 + \tan\zeta \tan\alpha) + \frac{\cos\xi \sin\zeta}{\tan\phi \cos\theta}\right] \qquad (14)$$

The variable η is called the stability number of a particle on a sloping bank by Stevens and Simons (1971) and Stevens et al. (1976). Calculating lift, drag, and gravitational forces as $L = c_1 \tau d^2$, $D = c_2 \tau d^2$, and $W = c_3 S_r \gamma d^3$ where c_1, c_2, and c_3 = dimensionless coefficients, and d = representative particle diameter, gives

$$\eta = \frac{S_r - 1}{S_r}\left(\frac{l_4 c_1 + l_3 c_2 \cos\delta}{l_2 c_3}\right)\tau_* \qquad (15)$$

where $\tau_* = \tau/[(S_r - 1)\gamma d]$ is dimensionless bed shear stress or Shields' parameter. For plane bed conditions (that is, with $\alpha = \theta = \delta = \xi = \omega = 0$) and small water surface slopes (that is, with $\zeta \approx 0$) the safety factor becomes $SF = 1/(\eta_o + K_{\zeta o})$ where

$$\eta_o = M + N = \frac{S_r - 1}{S_r}\left(\frac{l_4 c_1 + l_3 c_2}{l_2 c_3}\right)\tau_* \qquad (16)$$

and $K_{\zeta o} = 1/S_r$. At the condition of incipient overturning or the critical condition $SF = 1$ and the critical stability number on a horizontal bed $\eta_{co} = 1 - K_{\zeta o} = (S_r - 1)/S_r$. Dimensionless critical shear stress for a particle on a horizontal bed is then $\tau_{*co} = l_2 c_3/(l_4 c_1 + l_3 c_2)$, giving

$$\eta_o = \left(\frac{S_r - 1}{S_r}\right)\frac{\tau_*}{\tau_{*co}} \qquad (17)$$

With $M \approx N$ as suggested by Stevens and Simons (1971), $\eta = \dfrac{1}{2}(1 + \cos\delta)\eta_o$.

DIRECTION OF PARTICLE MOVEMENT

The angle ψ between the direction of particle movement and horizontal in the bank plane as shown in Fig. 1 is given by

$$\tan\psi = \frac{R\cos\theta\left(1 + \tan^2\theta\right) + \tan\alpha \cos\theta(S + Q\tan\theta)}{S\tan\theta - Q} \qquad (18)$$

where

$$R = \frac{l_1 W}{l_3 D} \left\{ \frac{\frac{B}{W}\cos\zeta[\tan\alpha - \tan\zeta(1+\tan^2\theta)] - \tan\alpha}{1+\tan^2\alpha+\tan^2\theta} \right\} - \Gamma\tan\theta \quad (19)$$

$$S = \frac{l_1 W}{l_3 D} \left\{ \frac{\frac{B}{W}\cos\zeta[\tan^2\theta + (\tan\alpha - \tan\zeta)\tan\alpha] - (\tan^2\theta + \tan^2\alpha)}{1+\tan^2\alpha+\tan^2\theta} \right\} - \Gamma(\tan\theta\tan\lambda) \quad (20)$$

$$Q = \frac{l_1 W}{l_3 D} \left\{ \frac{-\frac{B}{W}[\cos\zeta\tan\theta(1+\tan\zeta\tan\alpha)] + \tan\theta}{1+\tan^2\alpha+\tan^2\theta} \right\} - \Gamma(\tan\lambda - \tan\alpha) \quad (21)$$

and

$$\Gamma = \frac{1}{\sqrt{(\tan\lambda - \tan\alpha)^2 + \tan^2\theta(1+\tan^2\lambda)}} \quad (22)$$

With $M \approx N$ and $\eta_o \approx 2N$, the leading term in the expressions for R, S, and Q becomes $l_1 W/l_3 D = 2/(\eta_o \tan\phi)$, which is readily evaluated.

SUMMARY AND CONCLUSIONS

Stability of a single rock riprap particle was evaluated based on the ratio moments resisting overturning to moments promoting overturning. Departing from the conventional approach, the buoyant force acting on a particle was treated separately from the gravitational force when calculating moments. Gravitational, buoyant, and hydrodynamic forces were also split into components that promote and resist overturning. A consistent safety factor results that tends to unity as rock specific gravity approaches one, in contrast to the conventional moment analysis approach.

REFERENCES

Kawai, S. and Julien, P. Y. (1996). "Point bar deposits in narrow sharp bends." *Journal of Hydraulic Research*, 34(2), 205-218.

Richardson, E. V., Simons, D. B., and Julien, P. Y. (1990). "Highways in the river environment." *Publication No. FHWA-HI-90-016*, Federal Highway Administration, National Highway Institute, McLean, Virginia.

Simons, D. B., and Şentürk, F. (1977). *Sediment transport technology*. Water Resources Publications, Fort Collins, Colorado.

Stevens, M. A., and Simons, D. B. (1971). "Stability analysis for coarse granular material on slopes." Chapter 17 in *River mechanics*, H. W. Shen, ed., Water Resources Publications, Fort Collins, Colorado.

Stevens, M. A., Simons, D. B., and Lewis, G. L. (1976). "Safety factors for riprap protection." *Journal of the Hydraulics Division, ASCE*, 102(HY5), 637-655.

DESIGN RIPRAP TO PROTECT SCOUR AROUND CIRCULAR PIERS

TAE-HOON YOON
Professor, Department of
Civil Engineering,
Hanyang University, Seoul, Korea

SUNG BUM YOON
Assistant Professor, Department of
Civil & Environment Engineering,
Hanyang University, Ansan, Korea

ABSTRACT

To evaluate the accuracy and applicability of existing riprap design equations for the protection of local scour around a circular bridge pier extensive experiments were conducted. The evaluations of each design formula are made through the comparisons of calculated and observed diameters of riprap stones. Most of the design equations are based on the approach velocity of stream, and other factors such as pier width and flow depth are ignored partially or completely. Even inadequate inclusion of such effects are found. As a result, the design values are generally overestimated or underestimated in comparison with the experimental data. The equations proposed by Richardson et al. and Parola give relatively accurate results. A need for improved design formula which reflects the effects of both pier width and water depth is raised.

INTRODUCTION

To protect the local scour around a bridge pier, ripraps have been used for a long time for its easier installation and lower cost than other protection methods. A number of design criteria have been suggested to determine the size of riprap stones, but most of the existing formulae take only flow velocities into account. For example, the design formulae of Carstens(1966), Neill(1975), Breusers(1977), and Richardson et al.(1993) are based on the Isbash(1935) formula where the size of riprap stone is proportional to only velocity squared.
Quazi & Peterson(1973) proposed a formula including the effects of both velocity and water depth. Parola(1993) conducted extensive experiments and showed that the width of pier plays an important role in sizing riprap stones.
Bonasoundas(1973) presented a empirical formula which has the form of quadratic polynomial of flow velocity. Chiew's equation includes water depth and pier width.
In this study, the formulae are classified in different groups based on factors considered. Comparative analyses of the formulae were made using experimental data.

RIPRAP DESIGN FORMULAE

Most of riprap formulae are based on the studies of Isbash(1935). For flow with the horizontal bottom and without structures, Isbash proposed an equation to determine the critical flow velocity for a given size of riprap stones as

DESIGN RIPRAP TO PROTECT SCOUR

$$\frac{u_o^2}{(S_s - 1)gD_p} = 2C^2 \qquad (1)$$

where u_o is depth averaged flow velocity(m/s), S_s is specific gravity of stones, g is gravitational acceleration(m/s^2), and D_p is spherical diameter of riprap stone(m). C is a constant value and may be defined as Isbash constant. According to Eq. 1, the stability number, N_c is introduced to measure the stability of riprap stones as

$$N_c = \frac{u_o^2}{(S_s - 1)gD_p} \qquad (2)$$

Riprap design formulae are classified as 5 groups according to the type of parameters used in each formula. The first group of equations are based on Isbash formula(Eq. 1) with slightly different constants to account for the presence of bridge pier. The formulae proposed by Carstens(1966), Neill(1975), Breusers(1977) and Richardson et al.(1993) are falling in this category.
The second group of formulae are proposed by Quazi & Peterson(1973). He replaced the constant of Isbash formula by a function of relative water depth. The effect of water depth on the stability of riprap is to increase the stability of stone with increasing water depth. The third type of formula is presented by Parola(1993) based on the experiments using square piers. In his equation the Isbash constant is substituted by a function of relative pier width. The stability of stone is decreased with increasing pier width when other parameters are remained constant. Chiew's formula(1995) of fourth group is considered more appropriate in a sense that more parameters are involved. It includes water depth, velocity and pier width in terms of depth- and pier width correction factors. The fifth type of formula is proposed by Bonasoundas(1973). The size of riprap stone is determined by the quadratic polynomial of flow velocity. The types of design formulae are summarized in Table 1.

Table 1. Five types of design formulae

Group	Type	Researcher	yr.	Formula	Parameters
I	N_c=const	Carstens	1966	N_c=0.3698	S_s, u_o
		Neill	1975	N_c=0.6574	
		Breusers	1977	N_c=0.3524	
		Richardson	1993	N_c=1.285	
II	N_c=f$_1$(D$_p$/y$_o$)	Quazi & Peterson	1973	N_c=2.5(D$_p$/y$_o$)$^{-0.20}$	S_s, u_o, y_o
				N_c=0.8, 20<b/D$_p$<33	
III	N_c=f$_2$(b/D$_p$)	Parola	1993	N_c=1.0, 7<b/D$_p$<14	S_s, u_o, b
				N_c=1.2, 4<b/D$_p$< 7	
IV	D_p=f$_3$(u$_o$,y$_o$,b,b/D$_p$)	Chiew	1995	$D_p = \frac{1}{5.956\sqrt{y_o}} \left(\frac{u_o K_y K_b}{0.3\sqrt{(S_s-1)g}} \right)$	S_s, u_o, y_o, b
V	D_p=f$_3$(u$_o$)	Bonasoundas	1973	D_p=0.01(6-3.3u$_o$+4u$_o^2$)	u_o

EXPERIMENTAL PROCEDURE

Experiments were performed in a flume of 12m long, 0.45m wide, and 0.6m deep equipped with constant head tanks to analyse the effects of water depth, flow velocity and pier width on the stability of riprap stones placed around circular piers. In order to identify the individual effect of parameters on the stability of riprap stones, various size of stones, widths of bridge piers and water depths are used. Ten different grain sizes were used and their mean size ranges 2.13mm to 8.46mm. Three layers of riprap stone is placed flush with bottom elevation of channel around circular piers. The diameter of riprap coverage is 4 times as wide as pier width.

Eight flow rates, i.e., 0.04cms to 0.075cms with the increment of 0.005cms, are used for each stone size. Four pier widths (2, 3, 4, 5cm in diameter) are tested to check the effect of pier width on riprap stone movement. To determine the critical velocity for a given stone size and flow rate, the water depth is initially maintained deep, and the sluice gate at the end of channel is lowered very slowly to suppress the occurrence of unsteady flow. The critical state is obtained when a bunch of stones are removed simultaneously, and the water depth and flow velocity are measured and recorded. A set of over 200 data is obtained.

EVALUATION OF DESIGN FORMULAE

The evaluation of each formula is made by comparison of riprap size at critical state, i.e., threshold condition with that computed by design formula using flow conditions of the critical state. The results of comparison tests are described for each type of design criteria.

Group I : The stability number, N_c is constant, and the spherical diameter, D_p of riprap stone can be obtained by Eq. 2. The effects of pier width, b and approaching water depth, y_o are neglected in this type of equations. Figs. 1 and 2 are the example of this group and a clear tendency of overestimation of stone size are noted. Among the formulae in this type, Richardson equation being used by U.S. FHWA gives a relatively accurate estimation.

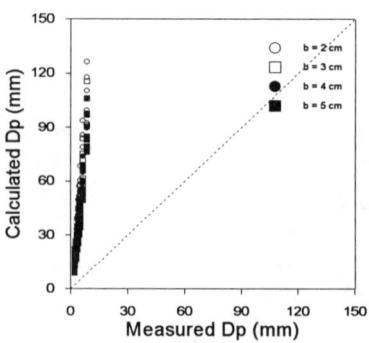

Figure 1. Comparison of measured and calculated D_p (Carstens' formula)

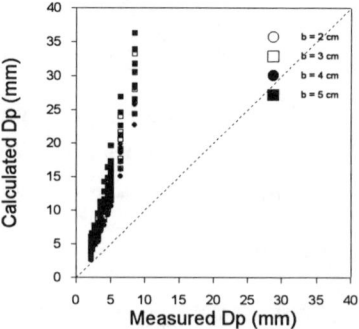

Figure 2. Comparison of measured and calculated D_p (Richardson's formula)

Group II: Quazi & Peterson equation that determines the riprap size is as follows.

$$D_p = \frac{0.85 u_o^{0.5}}{(S_s - 1)^{1.25} g^{1.25} y_o^{0.25}} \qquad (3)$$

This equation shows that the stones became more stable as the water depth is increasing. It was expected to get more accurate design values using Eq. 3 because of the inclusion of depth effect. However, the comparative study reveals more scattered prediction, especially for deep water cases, than Richardson formula where the water depth effect are ignored. This means that the effects of water depth in Quazi & Peterson formula is not represented correctly.

Group III: The pier width, b appears in Parola equation as shown in Table 1. Since his equation is given for a limited range of b/D_p, it is extended using a best fit line for the comparison purpose as

$$N_c = 1.724 \left(\frac{b}{D_p}\right)^{-0.226} \qquad (4)$$

Using Eqs. 2 and 4, the size of riprap stone can be calculated as

$$D_p = \frac{u_o^{1.63} b^{0.18}}{1.56(S_s - 1)^{0.82} g^{0.82}} \qquad (5)$$

As shown in Fig. 3, the deviations of calculated D_p from the best fit line are more or less uniform and small for the full range of data set. This means that the effect of pier width is correctly considered in this equation. It is expected that Parola's formula can give more accurate design values if the effect of water depth is included.

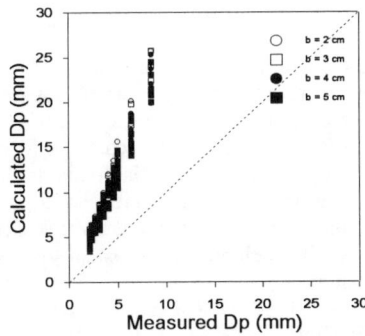
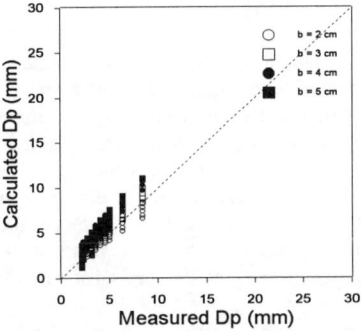

Figure 3. Comparison of measured and calculated D_p (Parola's formula)

Figure 4. Comparison of measured and calculated D_p (Chiew's formula)

Group IV: As can be seen in Fig. 4, Chiew's formula, which include both water depth and pier width, gives fairly accurate results. Considering facts that the equation is derived from scouring experimental date, and the depth-and pier width correction factors are based on the situation where the materials around the pier and in the channel are the same, the equation might need a correction if the materials around the pier is different to the bed materials.

Group V: Unlike other formulae that the riprap size is proportional u_o^2, Bonasoundas' equation, which is a second order ploynomial of velocity overestimates extremely the stone size.
Comparison was also made in terms of the ratio, r of the calculated D_p by the formulae to the measured D_p by experiment as r = D_p (computed) / D_p (measured).

Table 2. Ratios of computed D_p and messured D_p

Type	Researcher	Year	Average of r	Range of r
Group I	Carstens	1966	7.38	4.21 ~ 14.93
	Neill	1975	4.15	2.37 ~ 8.40
	Breusers	1977	7.72	4.41 ~ 15.62
	Richardson	1993	2.12	1.21 ~ 4.30
Group II	Quazi & Peterson	1973	1.32	0.42 ~ 3.75
Group III	Parola	1993	2.16	1.45 ~ 3.56
Group IV	Chiew	1995	1.16	0.61 ~ 1.90
Group V	Bonasoundas	1973	15.42	6.34 ~ 25.59

Group I shows very high ratio of r, which indicates overestimation. r values of Richardson, Quazi & Peterson and parola can be considered as in a reasonable range. Chiew's formula that include the effect of depth and pier width renders more accurate results than others. Due to improper representation of approach velocity, depth and pier width, r values by Chiew's formula turns out to be too low or too high depending on pier size.

CONCLUSIONS

To evaluate the riprap design formulae for their applicability and accuracy, a large number of experiments for different conditions were carried out. Formulae based on Isbash equation yield large riprap sizes than those by the experiments. Of those, Richardson formula being used by U.S. FHWA gives fairly accurate size of riprap. Parola formula in which pier width is properly represented is found to be relatively accurate. Chiew's equation in which water depth and pier width are incorporated is assessed to be more accurate than others, but it has drawback that the riprap size is heavily dependent on pier width resulting in too small or too large riprap size, and it calls for an improvement.

ACKNOWLEDGEMENT

This study was financially supported by Korea Science and Engineering Foundation under contract numbers(951-1201-015-2)

REFERENCES

Bonasoundas, M. (1973) Flow structure and problems at circular bridge piers, Report No.28, Oskar V. Miller Inst., Munich Tech. Univ., Munich, West Germany.
Breusers, H.N.C., Nicollet, G. & Shen, H.W. (1977) Local scour around cylindrical piers. J. Hydr. Res., Vol.15, No.3, pp.211-252.
Chiew, Y.-M. (1995) Mechanics of riprap failure at bridge piers, J. of Hydraulic Engineering, ASCE, Vol.121, No.9, pp.635-643.
Chiew, Y.-M. & Melville, B.W. (1987) Local scour around bridge piers, J. of Hydr. Res., Vol.25, No.1, pp.15-26.
Hancu, S. (1971) Sur le calcul des affouillements locaux dans la zone des piles du pont, Proceedings of the 14th Congress, International Association for Hydraulic Research, Vol.3, pp.299-306.
Isbash, S.V. (1935) Construction of dams by dumping stones in flowing water, W.S. Army Engrg. District, Eastport, Maine.
Melville, B.W. & Sutherland, A.J. (1988) Design method for local scour at bridge piers, J. of Hydraulic Engineering, ASCE, Vol.114, No.10, pp.1210-1226.
Neill, C.R. (1967) Mean velocity criterion for scour of coarse uniform bed material, Proc. 12th IAHR Congress, Fort Collins, Colo., C6.1-C6.9.
Neill, C.R. (ed) (1975) Guide to bridge hydraulics, Roads and Transportation Association of Canada, Univ. of Toronto Press, 2nd ed., 191p.
Parola, A.C. (1993) Stability of riprap at bridge piers, J. Hydraulic Engineering, ASCE, Vol.119, No.10, pp.1080-1093.
Press, W.H., Flannery, B.P., Teukolsky, S.A. & Vetterling, W.T. (1986) Numerical Recipes : The Art of Scientific Computing, Cambridge University Press.
Quazi, M.E. & Peterson, A.W. (1973) A method for bridge pier riprap design, Proc. First Canadian Hydraulics Conf., Univ. of Alberta, Edmonton, Canada, pp.96-106.
Richardson, E.V., Harrison, L.J., Richardson, J.R. & Davis, S.R. (1993) Evaluating scour at bridges, Hydraulic Engineering Circular No.18, FHWA-IP-90-017, FHWA, February, 105p.

NOTATIONS

b : width of pier (m)
C : constant
D_p : size of riprap stone (m)
g : gravity acceleration (m/s^2)
N_c : stability number
r : ratio of computed Dp and measured Dp
S_s : specific gravity
u_o : depth averaged flow velocity (m/s)
y_o : approaching water depth (m)

FAILURE BEHAVIOR OF RIPRAP LAYER AROUND BRIDGE PIERS

FOO-HOAT LIM[1] AND YEE-MENG CHIEW[2]

[1]Research Assistant and [1]Senior Lecture,
School of Civil and Structural Engineering,
Nanyang Technological University, Nanyang Avenue, Singapore 639798.

ABSTRACT

The main objective of this study is to examine the failure behavior of a riprap layer around a cylindrical bridge pier. The study shows that the inherent flexibility of a riprap layer can offer a self-healing process. It helps to reduce further erosion under a steady flow condition; and an equilibrium state is attained when the erosion ceases. When the flow velocity is increased steadily, observations show that the riprap layer will eventually fail in two modes: total disintegration and embedment. The failure occurs when the erosion power is higher than the self-healing ability. The study shows that the eventual failure mode can be determined by comparing the relative magnitude of the critical shear velocity of the bed sediment and the adjusted threshold velocity of the riprap layer.

INTRODUCTION

The formation of a scour hole around bridge piers is a common occurrence in a river. An accepted engineering method to deal with pier scour problems is to place rocks or riprap material around the pier foundation. However, field experience has shown that riprap stones often disappeared with time - with the most severe failure occurring during floods; and continual refilling is frequently needed to replenish the lost stones.

Although much effort has been channeled into the investigation of the stable size of riprap stones for design purposes, little was put into the understanding of the physics of its failure behavior. Most of the previous studies on riprap protection were aimed at riprap sizing, and designed equations have been proposed based on the threshold velocity of the riprap stone in clear-water conditions, such as Croad (1990). However, Chiew (1994) has shown that even if a riprap layer is designed based on these equations, the riprap layer can still withstand a velocity higher than the predicted threshold velocity before total failure occurs. This ability is due to the inherent flexibility of the riprap stones. The flexibility arises from the ability of the riprap stones to move relative to one another, enabling them to "self-adjust" or "self-heal" the weaknesses that form on the riprap layer. This ability to self-heal has rarely been mentioned in previous studies, but it has a significant impact on the physical process of riprap failure around a bridge pier. It dictates the fate of the riprap layer after the threshold of riprap erosion has been exceeded. The objective of this study is to examine the self-healing behavior of a riprap layer and its influence on the behavior of riprap failure.

EXPERIMENTS

The experiments were conducted in a glass-sided sediment recirculating flume that was 14 m long, 0.6 m wide and 0.6 m deep. Cylindrical piers made from clear perspex tubes with diameters, D = 25, 38, and 70 mm were used in the experiment. Four mean approach flow depths, y_o = 40, 70, 210 and 250 mm were tested. Two uniform sediments with median diameter, d_{50} = 0.26 mm and 0.96 mm were used as bed sediments and four uniform sediments with d_{50} = 1.01 mm, 2.65 mm, 5.58 mm and 9.12 mm were used as riprap stones. Figure 1 shows the definition sketch of the riprap layer. The level of the maximum degradation of the riprap layer from the undisturbed bed level was monitored using a periscope fitted inside the perspex tubes. Recordings of the displaced riprap level were made at 1 minute intervals for a total duration of 100 minutes for each experiment.

SELF-HEALING BEHAVIOR

In a recent experimental study on the stability of riprap protection around a cylindrical pier under clear-water condition, Chiew (1994) has identified three modes of erosion that destabilize a riprap layer: riprap shear; winnowing; and edge erosion. He also showed the self-healing behavior of a riprap layer to sustain a partial breakup of the layer before its eventual failure. Figure 2 shows how a riprap layer adjusts itself to the surrounding sediment bed. The movement of the stones is caused by a combined action of the three modes of erosion discussed above. Riprap shear erosion can be identified by the scour hole that forms around the pier. Winnowing erosion is identified by the general lowering of the entire riprap layer. Edge erosion is caused by flow acceleration initiated by the abrupt change of size at the edge of the layer. The finer bed materials were first eroded, resulting in the formation of a local scour hole around the edge of the riprap layer into which the riprap stones drop. This phenomenon enhances the riprap shear and winnowing erosion. Note that the winnowing and edge erosion are also shown in Fig. 2 with the general lowering of the entire riprap layer and the disintegration of the riprap stones at the edge. These processes continue and eventually an equilibrium state is reached where the scour hole becomes invariant with time.

FAILURE OF RIPRAP LAYER

At higher flow velocity, the erosion power increases to overcome the seal-healing process and a scour hole develops around the pier. At still higher flow velocity, the three modes of erosion i.e., riprap shear, winnowing and edge erosion combine to cause either a total disintegration or embedment failure of the riprap layer. The former occurs when the self-healing ability of the riprap layer cannot resist erosion, and is characterized by a complete breakup of the riprap layer. This type of failure has also been reported in Croad (1992) and Chiew (1994). Figure 3 shows a schematic diagram of the total disintegration type of failure of a riprap layer.

The second type of failure, viz., embedment failure occurs when the riprap stones are large compared with the bed sediment. This type of failure has also been reported in Croad (1993). Here, the shear velocity u_* may or may not exceed the threshold of sediment entrainment of the riprap stones $u_{*c(D50)}$, but has exceeded the threshold of sediment entrainment of the bed sediment $u_{*c(d50)}$ by a very large margin. Bed features and sediment transport start to form on the bed. The smaller bed sediments move faster than the coarser riprap stones; local erosion around individual large riprap stones cause them to self-embed into the bed. At the same time, the propagation of the bed features past the riprap layer induces a variation of the mean bed level, which causes the riprap stones to lose their stability and drop into the trough of the dune. The riprap layer disintegrates and individual stones were dispersed around both upstream and downstream of the pier. The turbulent flow field around the pier causes addition scour depth and the riprap stones around the pier drop into the trough of the dune. Figure 4 shows a schematic diagram of the embedment type of failure. If the flow condition is allowed to continue for a sufficient time, an equilibrium condition is attained. Here, the riprap stones degrade to an equilibrium level, at which the bed level at the pier fluctuates about a mean value in response to the passage of bed features, but the maximum riprap level remains invariant with time. Figure 5 shows the temporal variation of sediment bed around at the pier. Note that the maximum riprap level remains invariant with time as the bed features propagate past the pier.

Whether a riprap layer will fail as total disintegration or embedment is dependent on which of the threshold of sediment, i.e., bed sediment or riprap stones occur first. If the flow velocity has exceeded the threshold of riprap stones but lower than that of the bed sediment, the total disintegration type failure is likely to occur. Generally, this happens with a small D_{50}/d_{50} ratio. On the other hand, the embedment type failure will occur, and the D_{50}/d_{50} ratio is large.

The threshold of the bed sediment is dependent on its critical shear velocity $u_{*c(d50)}$, but that of the riprap stones around the pier involves other considerations. Chiew (1994) proposed that the threshold of scouring of riprap stones can be evaluated using

$$u_* = \frac{0.3 u_{*c(D50)}}{K(D/D_{50}) K(y_o/D)} \qquad (2)$$

where $K(D/D_{50})$ and $K(y_o/D)$ are sediment size and flow depth adjustment factor, respectively. Eq. 2 states that if the undisturbed bed shear u_* is less than the adjusted critical shear stress of the riprap stones at the pier, the riprap layer may remain intact.

However, if this shear velocity exceeds the critical shear velocity of the bed sediment $u_{*c(d50)}$, the approaching bed becomes unstable as live-bed condition commences. Under this condition, most of the riprap stones tend to settle and embed into the unstable bed rather than be entrained by the flow, resulting in the embedment mode of failure. On the other hand, if the shear velocity $u_* \leq u_{*c(d50)}$, but is greater than $0.3u_{*c(D50)}/[K(y_o/D)K(D/D_{50})]$, the tendency is for the riprap stones to be eroded on an immobile sediment bed. Here, total disintegration failure will occur.

Based on the above considerations, a limiting condition to delineate these two types of failure can be introduced by comparing the relative magnitude of $u_{*c}(d50)$ and $0.3u_{*c(D50)}/[K(y_o/D)K(D/D_{50})]$. If $u_{*c}(d50) > 0.3u_{*c(D50)}/[K(y_o/D)K(D/D_{50})]$, an embedment type of failure will occur. On the other hand, a total disintegration type failure will occur. The limiting condition can therefore be defined by the following equation

$$u_{*c(d50)} = \frac{0.3u_{*c(D50)}}{K(D/D_{50})K(y_o/D)} \quad (3)$$

Figure 6 shows this equation, in which the present experimental data and the data of Posey (1972), Croad (1990), Croad (1993) and Chiew (1994) are superimposed. The figure shows that the type failure observed in the experiments compares well with the equation proposed in this study.

CONCLUSION

The study shows that the inherent flexibility of a riprap layer has the ability to self-heal, providing an additional dimension to resist failure after the threshold of scouring of the riprap stones has been exceeded. The cause of this self-healing ability is the interaction of the bed sediment with riprap stones. An equilibrium state is attained when the erosive power of the flow is counteracted by the self-healing process, at which the scour hole becomes invariant with time. Increasing flow velocity causes the erosion power to increase, and subsequently to overcome the riprap layer's self-healing ability; and eventually the riprap layer will fail in one of the following two modes: total disintegration or embedment. The study proposed a semi-empirical equation to define the boundary of these two modes of failure.

REFERENCES

1   Chiew, Y. M. (1994), "Mechanics of Riprap Failure at Bridge Piers", Journal of Hydraulic Engineering, ASCE, Vol. 121, No. 9, pp. 635-643.
2   Croad, R. N. (1990), "Effect of Riprap on Pier Scour", Central Laboratories Report No. 90-23201, Works Consultancy Services, N. Z.
3. Croad, R. N. (1993), "Bridge Pier Scour Protection Using Riprap", Central Laboratories Report No. PR3-0071, Works Consultancy Services, N. Z.
4   Posey, C. J. (1974), "Tests of Scour Protection for Bridge Piers", Journal of Hydraulic Division, ASCE, Vol. 100, No. 12, pp. 1773-1783.

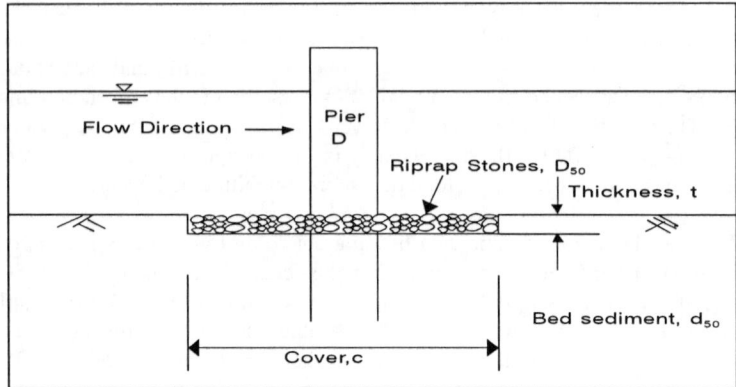

Figure 1   Definition of sketch of a riprap layer

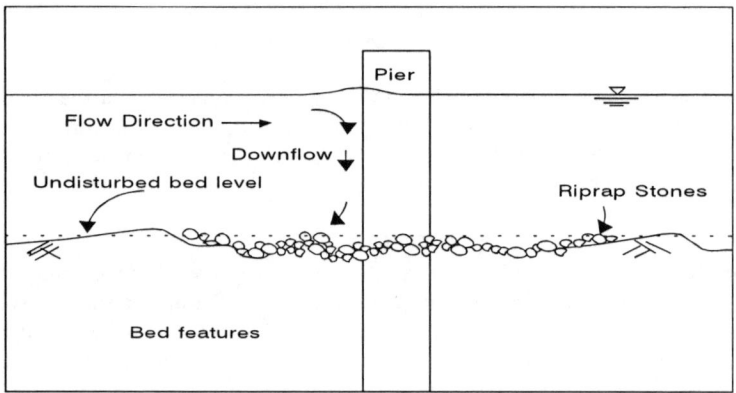

Figure 2   Self-healing behavior of a riprap layer with its surrounding sediment bed

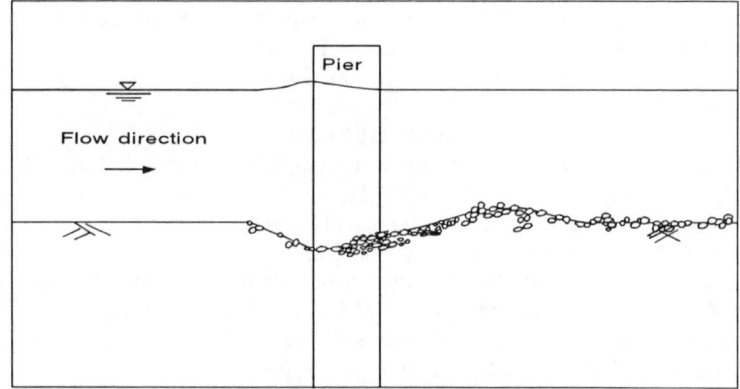

Figure 3   Schematic diagram of total disintegration failure

FAILURE BEHAVIOR OF RIPRAP LAYER

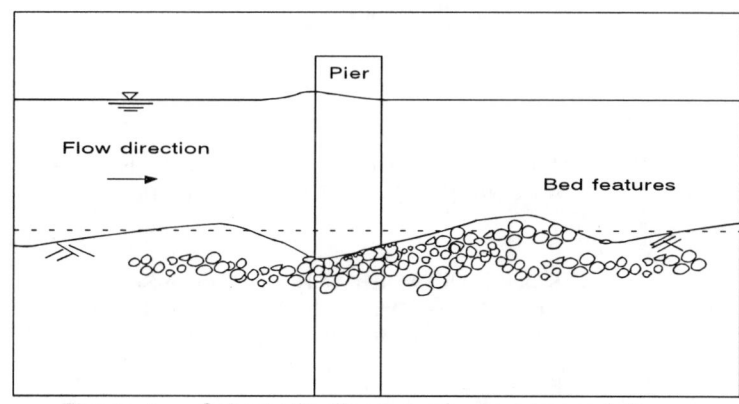

Figure 4   Schematic diagram of embedment failure

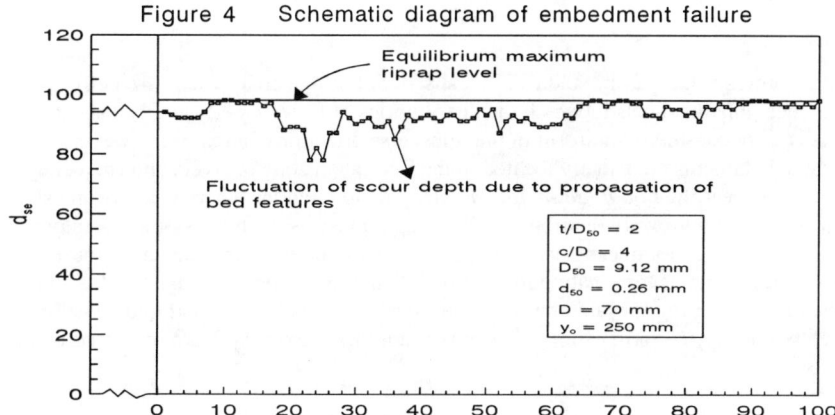

Figure 5   Temporal variation of bed level after equilibrium state in a live-bed condition

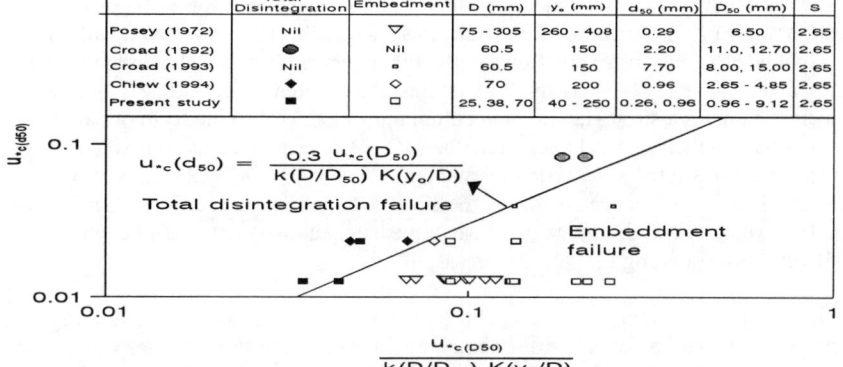

Figure 6   Classification of the behavior of failure of a riprap layer

INFLUENCE OF LATERAL MOMENTUM TRANSFER ON BRIDGE ABUTMENT SCOUR

S. KOUCHAKZADEH AND R.D. TOWNSEND
Civil Engrg. Dep., University of Ottawa
Ottawa, Ontario, Canada, K1N 6N5

ABSTRACT

Most previous laboratory studies of local scour at bridge abutments were performed in rectangular channels in which the distributions of flow velocity and bed shear stress were considered uniform in the transverse direction. In reality, however, bridge abutments are usually located in the floodplain zone of rivers and frequently terminate near the *floodplain-main channel* junction regions where velocity and shear stress are not uniformly distributed. This paper presents the results of a laboratory study performed to investigate the impact of lateral momentum transfer on local scour at abutments terminating in the floodplain of a compound channel. It is shown that, by accounting for lateral momentum transfer at small floodplain/main channel depth ratios ($y_d/H<0.3$), estimates of maximum local scour depth are increased by up to 30%.

INTRODUCTION

Watercourse cross-sectional geometry has long been recognized as an important parameter that governs the pattern of velocity distribution, boundary shear stress, momentum transfer, and secondary circulations in open channel flows (Townsend 1968; Myers, 1991). Because of the high resistance coefficients associated with the floodplain (FP) zones of rivers, flows in the FP are generally much slower than those in the main channel (MC). At small FP depths, the difference between MC and FP velocities initiates a strong lateral momentum transfer (LMT) in the form of banks of vortices having their vertical axes along the MC-FP junction regions. These vortices act as a mechanism for transferring momentum between the MC and FP zones by their continuous emergence and decay (Sellin, 1964). As a consequence of LMT, the MC flow velocity and discharge decrease immediately above the bankfull depth, while the corresponding FP values increase.

While extensive literature has been published on the subject of local scour at bridge piers and to a lesser extent at bridge abutments, since most of these studies were performed in rectangular-shaped laboratory channels, none address the issue of LMT and its impact on the local scouring process. The fact that LMT effects are not properly accounted for in past laboratory studies may be one explanation for the significant differences noted between laboratory scour data and observations in the field. Melville and Parola (1995) point to the need for further research to investigate the impact of channel shape on the local scouring process, particularly for the case when bridge abutments terminate near the MC-FP junction.

Myers and Elsawy's (1975) study of the flow interaction phenomenon quantified the impact of LMT on both the value and distribution of boundary shear stress in the MC and FP of an asymmetric compound-shaped channel. They compared observed boundary shear stress distributions obtained under *interacting* (i.e. combined MC-FP flow) conditions with those for *non-interacting* (isolated FP flow) conditions. For their shallowest FP depth, they observed 260% and 200% increases respectively in maximum and average FP shear stress. Clearly, in severely-compound flow fields, LMT significantly impacts both the value and distribution of boundary shear stress. Therefore, failure to properly account for LMT in these situations may lead to unrealistic estimation of maximum scour.

EXPERIMENTAL SET-UP

The experimental program was performed in a 12.20 m-long x 1.20 m-wide compound channel located in the Hydraulics Laboratory of the Civil Engineering Department, University of Ottawa. The channel cross-section comprises a trapezoidal-shaped MC (bottom width = 0.30 m; side slope= 0.5H:1V) located between two rectangular-shaped FPs (each 0.38 m wide), Fig. 1. The longitudinal slope of the channel is fully adjustable. The 2.43 m-long test section had a specially-designed (recessed) FP compartment to accommodate the model abutments and a sand test bed. Two types of uniform sand (median diameter, D_{50}=0.5 and 0.7 mm) were used in the tests and four different abutment shapes: namely, (*i*) vertical-wall (VW), (*ii*) semi-circular (SC), (*iii*) wing-wall (WW), and (*iv*) spill-through (ST) abutments were investigated.

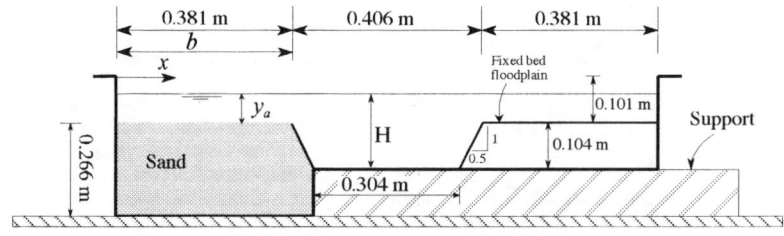

Fig. 1 - Cross-section through the channel test section

A *mini-propeller* meter, capable of detecting mean velocities as low as 0.025 m/s, was employed for measuring velocity distributions. Measurements were recorded using a data acquisition system specifically designed for this purpose. Water surface profiles along the channel were monitored via twelve manometers connected to tapping points located along the MC. To achieve *near-uniform* flow conditions, the flume tailgate was adjusted until the water surface slope closely matched the preset channel bed slope. A duration of 5 hr was selected for the tests. Long-period tests (140 hr duration) indicated that scour depths at 5 hr were approximately 60% of the equilibrium scour depth, y_{se} for the D_{50}=0.5 mm sand and 65% for the D_{50}=0.7 mm sand.

RESULTS

SHEAR VELOCITY RATIOS

Local shear velocity ratios, u_*/u_{*c} (u_* and u_{*c}= local and critical shear velocities, respectively) at the end of the 90 mm-, 150 mm-, and 210 mm-long model abutments were determined based on comparisons between observed (vertical) velocity profiles and the theoretical (*log-law*) distributions. Fig. 2 shows a sample comparison for *interacting* flow conditions, for D_{50}=0.5 mm and y_a/H=0.24 (y_a and H= FP and total flow depths, respectively). The shear velocity ratios for *non-interacting* flow conditions are presented in Fig. 3. The figure shows that u_*/u_{*c} increase as the MC-FP junction is approached.

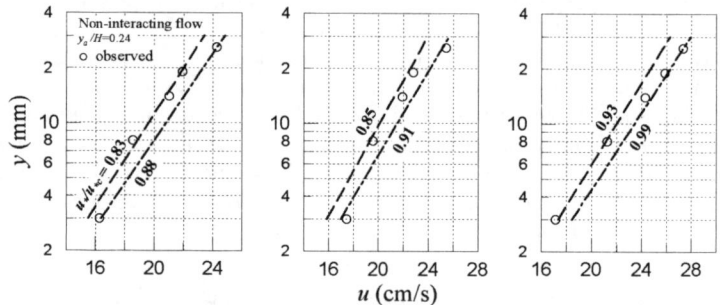

Fig. 2- Observed and theoritical vertical velocity profiles

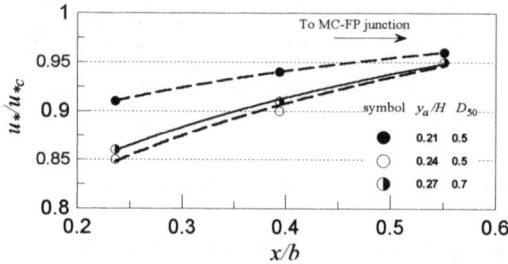

Fig. 3- u_*/u_{*c} in *non-interacting* FP flows

y_{se} FOR *INTERACTING* AND *NON-INTERACTING* CONDITIONS

Melville (1995) suggests that abutments terminating in the FP of a compound channel might be regarded as being similar to those located in a simple rectangular channel by introducing an (imaginary) vertical boundary at the MC-FP interface of the compound channel. To examine the effects of isolating the FP flow component from the MC flow on the local scouring process, our data for *interacting* conditions are compared with those for *non-interacting* conditions. The data for different abutment shapes and y_a/H =0.21 and 0.24 indicate that *non-interacting* conditions produce 15 to 30% smaller y_{se}/y_a than that for *interacting* conditions.

Isolating the MC flow from the FP flow eliminates LMT and decreases flow velocity in the region close to the MC/FP junction. This in turn decreases the shear velocity ratio in the FP. Assuming a linear relationship between y_{se} and u_*/u_{*c} in the range $0.5 < u_*/u_{*c} < 1$, the data were modified to account for the fact that $u_*/u_{*c} < 1$. The modified y_{se}/y_a for *non-interacting* conditions still show 5 to 15% smaller values than for *interacting* conditions. The differences between y_{se}/y_a for *interacting* and *non-interacting* conditions were between 10 to 35% for D_{50}=0.7 mm and y_a/H =0.27. The *non-interacting* data for D_{50}=0.7 mm, after being modified for the shear velocity ratio, are still smaller than the *interacting* data. Isolating the FP flow from the MC flow affected the 90 mm-abutments the most.

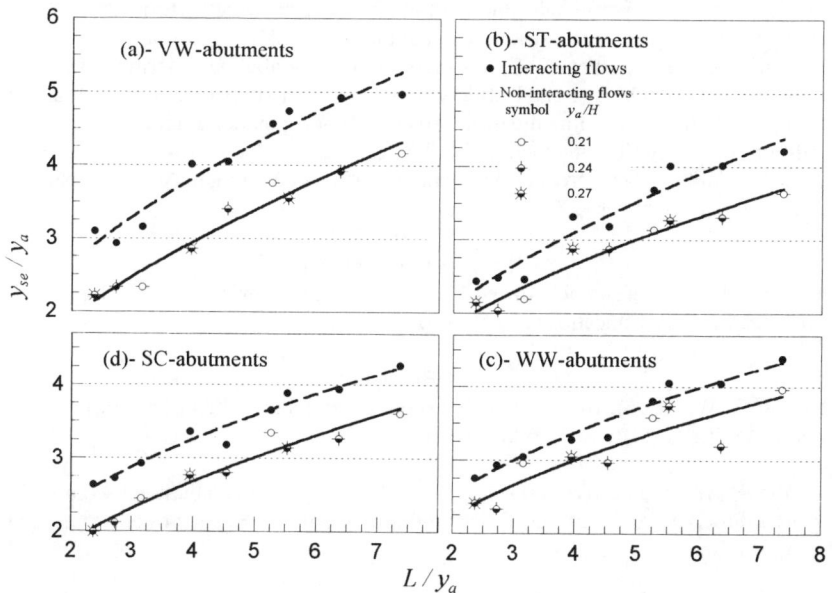

Fig. 4- Maximum local scour for *interacting* and *non-interacting* flows

The data for VW-, ST-, WW-, and SC-abutment shapes are presented in Figs. 4.a to 4.d respectively. The variations in y_{se}/y_a follow similar trends for all abutment shapes. *Non-interacting* conditions, however, decreased the scour depth at VW-abutments the most (Fig. 4-a). Also, the figures show that the relative decrease in y_{se}/y_a for non-interacting conditions is higher at shorter abutments than at longer ones. For instance, the decrease in y_{se}/y_a at 90 mm-SC abutments due to non-interacting conditions amounted to 30%, while for 210 mm-SC abutments 15% decrease in y_{se}/y_a was observed.

Comparing the *interacting* with the *non-interacting* data shows that isolating the MC flow from the FP flow reduces scour depth. Under *non-interacting* conditions the 210 mm-abutments obstruct 55% of the FP width. In such a case one would expect to observe an increase in the scour depth, because of the scour component that would normally occur as a result of the relatively high degree of flow contraction. Yet, the resulting scour depths are substantially smaller for *non-interacting* conditions than for *interacting* conditions for the 210 mm-abutments. Since traditional abutment scour equations are based on data obtained from experiments performed in rectangular channels, they predict unrealistic scour depth for severely compound flow conditions.

CONCLUSIONS

Most of the earlier experimental studies of abutment scour were performed in channels of rectangular cross-section. While this approach is appropriate for abutments located in a river's MC, it is inappropriate for abutments terminating in the FP zones, particularly if LMT effects are strong. Our study data indicate that, for the case of abutments terminating near a river's MC-FP junction regions, under conditions of strong flow interaction, LMT effects can produce a 15-30% increase in local scour depth. Therefore, design relationships for predicting maximum scour depth should account for LMT in these instances.

ACKNOWLEDGEMENTS

The first author gratefully acknowledges the scholarship provided by the Ministry of Culture and Higher Educations of I.R. of Iran.

REFERENCES

Melville, B. W. (1995). Bridge abutment scour in compound channels. J. of Hydr. Engrg., ASCE, 121(12), 863-868.

Melville, B. W., and Parola, A. (1995). The need for additional abutment scour research. Proc. of the First Inter. Conf. on Water Resources Engineering, ASCE, San Antonio, Texas, 1239-1243.

Myers, W. R. C. (1991). Influence of geometry on discharge capacity of open channels. J. of Hydr. Engrg., ASCE, 117(5), 676-680.

Myers, R. C., and Elsawy, E. M. (1975). Boundary shear in channel with floodplain. J. of the Hydr. Div., ASCE, 101(7), 933-946.

Sellin, R. H. J. (1964). A laboratory investigation into the interaction between the flow in the channel of a river and that over its flood plain. La Houille Blanche, Vol. 7, 793-801.

Townsend, D. R. (1968). An investigation of turbulence characteristics in a river model of complex cross section. Proc. Inst. of Civil Engineers (London), Vol. 40, June, 155-175.

LOCAL SCALING OF BRIDGE ABUTMENT SCOUR IN COMPOUND CHANNELS

by T. W. STURM, Assoc. Prof. and A. CHRISOCHOIDES, Grad. Res. Asst.
School of Civ. and Envir. Engrg.; Georgia Inst. of Tech.; Atlanta, GA; USA

INTRODUCTION

Clear-water scour at a bridge abutment located on the floodplain of a compound open channel is influenced by the flow distribution between floodplain and main channel in the approach and contracted bridge cross-sections. It has been suggested that the effect of the flow contraction can be accounted for by a discharge contraction ratio that depends on abutment length as well as the discharge distribution in the approach section of a compound channel (Sturm & Janjua 1994). In addition, previous equations for predicting clear-water scour depths at abutments have depended on the ratio of floodplain velocity V_1 to critical velocity V_c in the bridge approach section (Melville 1992, Sturm & Janjua 1994) with maximum clear-water scour occurring when V_1 approaches V_c. In this formulation, the independent variables for scour prediction are determined from one-dimensional numerical models.

A possible alternative to parameterizing scour depth in terms of approach velocity and discharge distribution is to relate it directly to local hydraulic conditions near the abutment face, although these conditions can only be predicted by two or three-dimensional numerical models. Possible advantages of this alternative formulation include: (1) it may provide a means of unifying scour prediction equations obtained from experiments in rectangular channels with those based on the more realistic compound channel geometry; and (2) it may offer greater sensitivity of scour predictions to changes in local depth and velocity at the location of scour initiation near the abutment face. This study explores the feasibility of relating local abutment scour to values of local hydraulic variables at the initiation of the scour process. Laboratory experiments were conducted in two different compound channel geometries with overall channel widths of 2.1 m and 4.3 m. Abutment length, discharge, and sediment size were varied, and the resulting equilibrium scour depths were measured as well as the local depth-averaged velocities at the beginning of scour. Correlation of scour depth with local abutment velocity is presented.

LOCAL SCALING OF BRIDGE ABUTMENT SCOUR 197

EXPERIMENTAL INVESTIGATION

Experiments were conducted in a 4.3 m wide by 24.4 m long flume in the Hydraulics Laboratory of the School of Civil and Environmental Engineering at Georgia Tech. Scour depths were measured as a function of discharge, sediment size, and abutment length for two compound channel cross-sections. Depth-averaged velocities at the bridge approach section and near the abutment face as well as water surface profiles were measured for the fixed-bed case.

Compound channel sections at a constant bed slope were constructed inside the flume and are shown in Fig. 1 as Cross-sections A and B. The main channel bed of Cross-section A consisted of concrete having a longitudinal slope of 0.0050. Sediment A (d_{50} = 3.3 mm) was affixed to the main channel bed and walls with varnish to provide roughness, while the same sediment was immobilized in the floodplain by mixing in a portland cement slurry for the fixed-bed experiments. In the case of Cross-section B, a lean concrete mix that utilized Sediment A as aggregate was poured to a constant slope of 0.0022. In both cross-sections, the bed surfaces resulted in fully-rough turbulent flow.

Bridge abutments were formed by a row of rectangular concrete blocks measuring 15 cm wide (in the flow direction) located at Station 32, which was 9.75 m (32 ft) downstream of the flume entrance. The abutment lengths are summarized in Table 1, as well as the ratio of abutment length L_a to floodplain width B_f. Three uniform sediments (σ_g = 1.3) were used in this research with median sediment grain sizes d_{50} as indicated in Table 1 for SED. A, B, and C.

Table 1. Experimental Abutment Lengths and Sediments

CROSS SECT.	LENGTH L_a, m	L_a/B_f	SED.	d_{50} mm
A	0.15, 0.31, 0.46	0.17, 0.33, 0.50	A	3.3
B	0.80, 1.60, 2.40	0.22, 0.44, 0.66	A, B, C	3.3, 2.7, 1.2

Both water-surface profiles and velocities were measured in the fixed-bed channels, first for uniform flow to determine normal depth, and then with the abutments in place to obtain the initial hydraulic conditions. Normal depth was set as the downstream boundary condition for all experiments with abutments. The normal depth ratio, y_{f0}/y_0 (see Fig. 1), varied between 0.13 and 0.35.

A miniature propeller current meter with a diameter of 15 mm was used to measure point velocities averaged over 60 seconds. With the abutments in

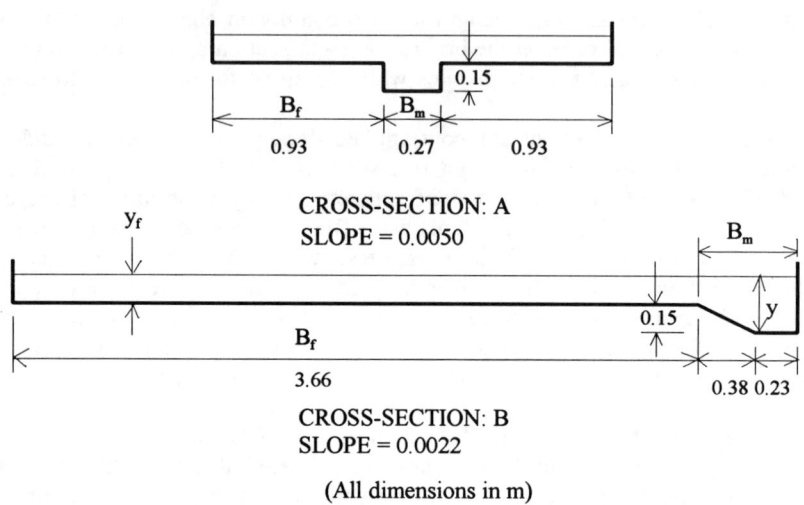

Fig. 1. Cross-sections used in scour experiments.

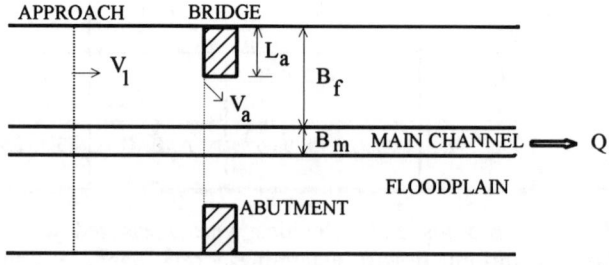

Fig. 2. Plan view of Cross-section A illustrating abutment velocity V_a.

place, detailed point velocity measurements were made across the entire channel cross-section at the bridge approach section shown in Fig. 2. Depth-averaged velocities were determined at 17 to 19 positions across the cross section. In addition, resultant velocities were measured at the upstream face of the bridge to find the maximum velocity V_a near the face of the abutment.

After the fixed-bed measurements were completed, a movable-bed section was constructed in the vicinity of the abutments. Scour measurements were made for several discharges at each of the abutment lengths given in Table 1. Scour was allowed to continue for 12 to 16 hours for Cross-section A, and for 24 to 36 hours for Cross-section B. After equilibrium had been reached, the bed elevations throughout the scour area were measured with a point gauge resulting in an uncertainty in scour depth of about ± 1.0 mm.

RESULTS AND DISCUSSION

Clear-water scour at a bridge abutment located on the floodplain of a compound open-channel occurs at an ever decreasing rate from initiation of scour until equilibrium is achieved. Theoretically, local scour is initiated when the ratio of local bed shear velocity U_* to its critical value U_{*c}, or the ratio of local flow velocity V to critical velocity V_c, exceeds unity. Furthermore, the initial rate of scour and limiting depth of scour have been shown to increase with the value of V/U_{*c} in experiments on scour by jets (Sedimentation 1977). Thus, although the local depth-averaged velocity near the abutment face is continually decreasing as the scour hole grows larger with time, it seems possible to relate the maximum depth of scour to the maximum depth-averaged velocity near the abutment face V_a at the beginning of scour as shown in Fig. 2. This velocity cannot be predicted by one-dimensional models. Biglari (1995) has applied a depth-averaged, k-ε turbulence model to predict the flow field and V_a around an abutment on the floodplain of a compound channel using Cross-section A. For Cross-section B, V_a was measured. As a test of the hypothesis that scour depends on the local depth-averaged velocity near the abutment face, a relationship for the equilibrium clear-water scour depth is sought in the form

$$d_s = f[\, y_{f0}, y_a, V_a, \rho, (\rho_s - \rho), g, d_{50} \,] \qquad (1)$$

in which d_s = equilibrium scour depth; y_{f0}= undisturbed flow depth in the floodplain set by uniform flow downstream of the bridge; y_a = floodplain flow depth at the location of V_a in the contracted section; V_a = maximum velocity near the upstream corner of the abutment face; ρ = fluid density; ρ_s = sediment density; g = gravitational acceleration; and d_{50} = median sediment diameter. Dimensional analysis of (1) results in

$$\frac{d_s}{y_{f0}} = f\left[\frac{y_a}{y_{f0}}, N_s, F_a, \frac{d_{50}}{y_a}\right] \qquad (2)$$

in which $F_a = V_a/(gy_a)^{0.5}$ = Froude number in the contracted floodplain near the abutment face; and $N_s = V_a/[(SG-1)gd_{50}]^{0.5}$ = sediment number in the contracted section as defined by Carstens (1966) with SG = specific gravity of the sediment. The value of the relative roughness (d_{50}/y_a) can be replaced by the critical value of the Froude number F_c or the critical value of the sediment number N_{sc} for the case of fully-rough turbulent flow (Pagan-Ortiz 1991) as in these experiments. For quartz sediment and water, either F_a or N_s can be considered redundant with respect to the other. If it is further assumed that scour is related to the excess of velocity (or sediment number) with respect to its critical value (Carstens 1966), a possible relationship is

$$\frac{d_s}{y_{f0}} = f\left[\frac{V_a}{V_c} - 1, \frac{y_a}{y_{f0}}\right] \qquad (3)$$

Fig. 3 illustrates the correlation of scour data in terms of the excess velocity ratio suggested by (3). The experimental scour data for Cross-section A were measured by Sadiq (1994) and reported by Sturm and Sadiq (1996), while the Cross-section B data were collected as part of this study. The values of V_a and y_a were predicted by Biglari's numerical 2D turbulence model (Biglari 1995) for Cross-section A, while they were measured for Cross-section B. The value of V_c was calculated from Keulegan's equation as a function of d_{50}/y_a with appropriate values of Shields' parameter.

The results in Fig. 3 include experimental data for six different values of L_a/B_f and for three different sediment sizes. The value of L_a/y_{f0} varied from 3 to 90 which includes both intermediate and large scale abutment lengths according to Melville's classification (1992). The variable y_a/y_{f0} in (3) is an indication of relative local water surface drawdown near the upstream corner of the abutment, but it was found not to be a significant explanatory variable for the observed scour depths in these experiments. The coefficient of determination for the least-squares-fit relationship in Fig. 3 is $r^2 = 0.88$.

CONCLUSION

Both measured and numerical predictions of velocity in the local scour region near the face of the abutment at initiation of scour explain, at least in part, the measured values of equilibrium scour depth. More work is needed to identify the influence of additional local hydraulic variables, which can only be predicted accurately by 3D numerical turbulence models, on abutment scour.

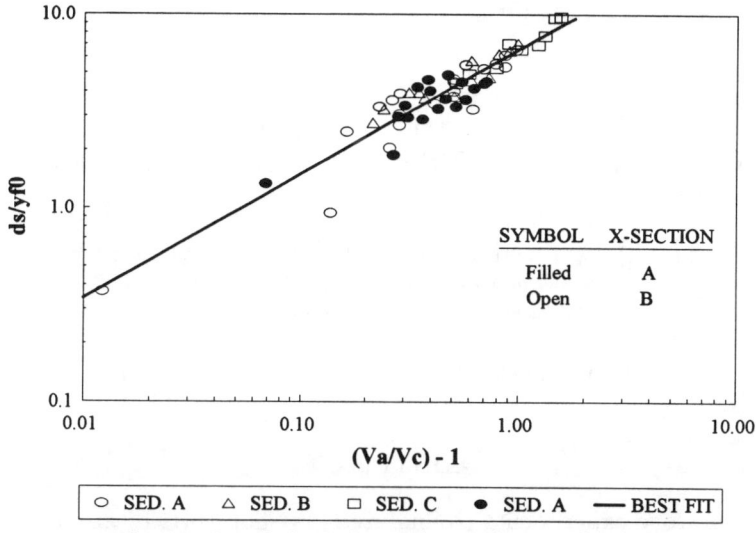

Fig. 3. Correlation of experimental scour depths.

ACKNOWLEDGMENT

This research was partially supported by FHWA Contract DTFH61-94-C-00198.

REFERENCES

Biglari, B. (1995). "Turbulence modeling of clear-water scour around bridge abutment in compound open channel." Ph. D. Thesis, School of Civil & Environmental Engrg., Georgia Institute of Technology, Atlanta, Georgia.
Carstens, M. R. (1966). "Similarity laws for localized scour." J. Hydr. Div., ASCE, 92(3), 13-36.
Melville, B. W. (1992). "Local scour at bridge abutments." J. Hydr. Engrg., ASCE, 118(4), 615-631.
Pagan-Ortiz, J. E. (1991). "Stability of rock riprap for protection at the toe of abutments located at the floodplain." FHWA-RD-91-057,Washington D.C.
Sadiq, Aftab (1994). "Clear-water scour around bridge abutments in compound channel." Ph.D. Thesis, School of Civil & Environmental Engineering, Georgia Institute of Technology, Atlanta, Georgia.
Sedimentation engrg.-ASCE manual 54 (1977). V.A. Vanoni, ed., ASCE, NY.
Sturm, T. W. and Janjua, N. S. (1994). "Clear-water scour around abutments in floodplains." J. Hydr. Engrg., ASCE, 120(8), 956-972.
Sturm, T. W. and Sadiq, Aftab (1996). "Clear-water scour around bridge abutments under backwater conditions." Transportation Research Record 1523, TRB, National Research Council, Washington, D.C., August.

VORTICITY DISTRIBUTION OF HORSESHOE VORTEX ON SCOURED BED

TOMONAO KOBAYASHI, TOSHIFUMI AIBARA,
HIROKAZU HARADA
Department of Civil Engineering
Tokyo Rika University
2641 Yamazaki, Noda-shi, Chiba 278, JAPAN

ABSTRACT

Laboratory experiments to analyze horseshoe vortices around a vertical cylinder on a scoured bed by using an image processing technique are described. The horseshoe vortices are discussed in terms of vorticity distribution. Vorticity concentration of the horseshoe vortex's counter-rotating is observed in a deep part of the scour hole, close to where the stagnation area is generated. This feature is invisible in the field of velocity distribution. Experimental results from different scour depths lead that, although a horseshoe vortex keeps its strength almost constant, its counter-rotating vorticity becomes stronger as the scour hole is deeper.

INTRODUCTION

Scouring around piers in rivers or coastal regions is a serious problem for their stability. The flow structure observed around the piers is the horseshoe vortex. The mechanism of local scouring around piers can therefore not be understood without analyzing the characteristics of this vortex. Horseshoe vortex's behavior has been studied by several researchers and some models to explain it have been proposed.

Baker (1979) investigated the horseshoe vortex and its behavior on a flat plate around a vertical cylinder, in laminar steady unidirectional flow, by using a flow visualization technique. In river flow, Melville and Raudkivi (1977) investigated the horseshoe vortex around a pier on a scoured bed, and estimated the bed shear stress with measured velocity distribution. Nakagawa and Suzuki (1971) presented a flow model around a pier on a flat bed by superposing secondary flow on a mean velocity field. Dey and Bose (1994) and Day, Bose and Sastry (1995) proposed a flow model around a pier on a scoured model.

Niedoroda (1981) pointed out that the horseshoe vortex in the wave field isn't significant in the initial stage of local scouring around a pier, in contrast to what is observed in river flow. However, the formed horseshoe vortex was the dominant flow structure around a pier on the developed scoured bed. Sumer, Fredsøe and Christiansen (1992) observed the horseshoe vortex with shedding vortices in wake on a movable sea bed by using flow visualization technique, and discussed a scour hole around a pier under waves. Kobayashi (1992) measured the velocity distribution around a pier on scoured bed under waves using a Laser-Doppler anemometer, and discussed the coherent structures in the flow. This author also simulated the flow numerically by applying discrete-vortex-method.

Most of these studies use velocity distribution to analize the horseshoe vortex behavior. However, this is useful to discuss the flow mechanism in general, *c.g.* friction or drag force. But, to analyze the horseshoe vortex itself and its behavior, vorticity distribution is a better choice, because of the vorticity involved in the vortex.

Why studies applying vorticity distribution are rare? One of the main reasons might be that the vorticity is derivative of the velocity, so for its evaluation, higher accuracy measurements of velocity in experiments are required. Recent experimental apparatus or techniques, *c.g.* image processing techniques, makes possible to easily measure velocity with high enough accuracy.

We applied an image processing technique to investigate the features of the horseshoe vortex based on its vorticity distribution. The well understanding of the horseshoe vortex's features is necessary to discuss its behavior or the local scouring around a pier in unsteady flow as well as in a wave field. This study is the first step of the detailed observations of the horseshoe vortex related to the local scouring around a pier in the unsteady flow, by applying an image processing technique and the vorticity distribution analysis of the flow.

EXPERIMENTS

EXPERIMENTAL APPARATUS

Fig.1 shows the schematic view of the experimental apparatus used in this study. A channel with 4,000 mm (L) x 200 mm (W) x 350 mm (H), made with clear acrylic, was employed. We assumed the symmetricity of the flow pattern around a vertical cylinder in the unidirectional flow, and used only a half part of a cylinder and scoured bed, that is made by hard mud, as shown in this figure. For simplicity, the scoured bed profile is assumed to be a corn-shape.

The flow in the upstream of the cylinder on the bed was visualized with dye in the vertical plane of symmetry, and observed with a video

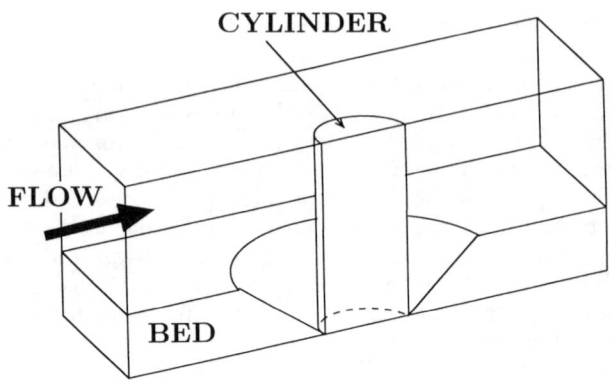

Figure 1 Schematic view of experimental apparatus.

camera. Velocity and vorticity distributions were computed with a personal computer from the video image of the flow by applying an image processing technique.

EXPERIMENTAL CONDITIONS

Four experimental cases are discussed. The experimental conditions are shown in Table 1. In all experimental cases, the observed flow field was almost stable due to the low Reynolds number.

The image processing technique we employed in this study is basic, and we couldn't measure the whole velocity field at once. Instead, velocity field at different times was obtained. This might affect the correct evaluation of the velocity or vorticity distribution. For future research, we intend to apply an advance image processing technique (*e.g.* Kobayashi and Hino; 1992) to investigate the vortex in unsteady flow with high

Table 1 Experimental Conditions.

Case	$\bar{u}$	Re	d/D	$\tan\theta$
1	24.2	4.5×10^3	0.40	0.70
2	15.6	3.0×10^3	0.40	0.70
3	25.1	4.6×10^3	0.23	0.70
4	24.4	4.7×10^3	0.0	—

$\bar{u}$: Mean velocity
Re : Reynolds Number
d : Scour Depth
D : Cylinder Diameter (150 mm)
θ : Slope Angle of Scoured bed

Reynolds numbers.

The values of the normalized scoured depth d/D and the slope angle of the scoured bed θ in Case 1 correspond to the common local scour in the equilibrium stage in rivers. Cases 1 and 2 were performed to discuss the relation between Reynolds number and the characteristics of the horseshoe vortex. The effect of the scour depth to the vortex can be discussed with Cases 1, 3 and 4.

RESULTS AND DISCUSSION

VELOCITY AND VORTICITY DISTRIBUTIONS

Fig.2 shows the velocity and vorticity distributions in Case 1. The origin is at the center of the cylinder on the unscoured bottom. x axis is along the mean flow, and z axis is on the center of the cylinder. The velocity and axes are normalized with the incoming mean velocity $\bar{u}$ and the cylinder diameter D, respectively. Vorticity with clock-wise rotation is defined positive in this figure. The velocity distribution was measured with high accuracy, and an accurate vorticity distribution was therefore evaluated. In the velocity distribution, clock-wise vortex is observed in the scour hole. This vortex is the horseshoe vortex. The vortex is also observed in the vorticity distribution as the concentration of positive vorticity. High concentration of negative vorticity is also observed at the deep part of the scour hole. This corresponds to the counter-clockwise vortex, which it is not noticeable in the velocity distribution. This important result leads that the characteristics of the vortices should be discussed in terms of vorticity distribution better than velocity distribution. The negative vorticity might be generated due to friction of the horseshoe vortex, and accumulated at the deep part of the scour hole, where the stagnation area exists.

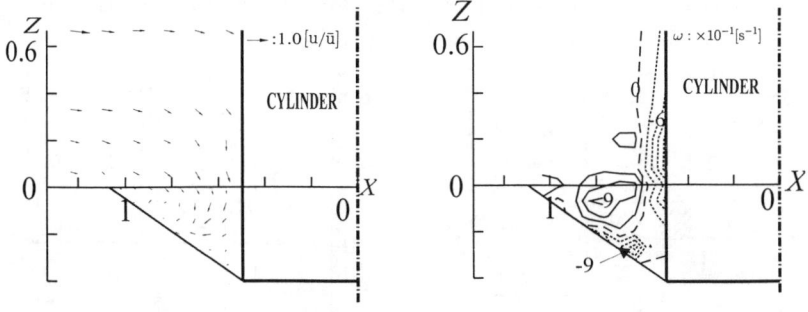

a) Velocity distribution b) Vorticity distribution

Figure 2 Velocity and vorticity distributions of a horseshoe vortex in Case 1.

HORSESHOE VORTEX AND REYNOLDS NUMBER

Fig.3 shows the vorticity distribution in Case 2, where Reynolds number is lower than that in Case 1. The strength of the vorticity distribution in Fig.3 is lower than that in Fig.2, because of the lower mean flow velocity. The tendency of the distribution in Fig.3, positive vorticity concentration related to the horseshoe vortex and the negative vorticity concentration in the deep part of the scour hole, is similar to that in Fig.2. The comparison of the vorticity distribution in Figs.2 and 3 leads that the

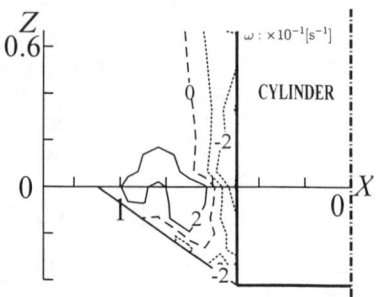

Figure 3 Vorticity distribution of a horseshoe vortex in Case 2.

influence of the Reynolds number, in the range applied in this study, is negligible in the vortex formation in front of a vertical cylinder on a scoured bed.

HORSESHOE VORTEX AND SCOUR DEPTH

Figs.4 and 5 show the vorticity distributions in Cases 3 and 4, respectively. All experimental conditions, except scour depth, in Cases 3 and 4 are almost the same as those in Case 1. The relation of the scour depth to the horseshoe vortex can be discussed by comparison of Figs.2, 4 and 5. The strong positive vorticity's concentration, that corresponds to a horseshoe vortex, is observed in front of the cylinder on the flat bed in Fig.5. Its circulation is almost independent of the scour hole depth, as found in Figs.2, 4 and 5. In Fig.4, the negative vorticity also concentrates in front of the cylinder. The negative vorticity's concentration becomes stronger in the deep scour hole in Fig.2. These figures lead to the following results; the strong horseshoe vortex appears in front of the vertical

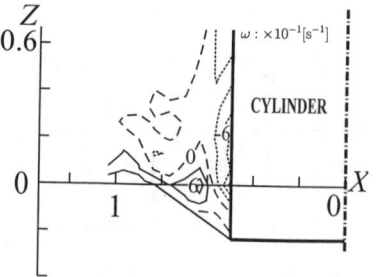 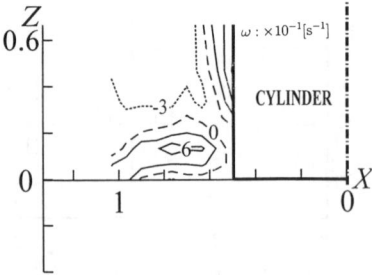

Figure 4 Vorticity distribution of a horseshoe vortex in Case 3.

Figure 5 Vorticity distribution of a horseshoe vortex in Case 4.

cylinder in the initial bed, which corresponds to the flat bed in Fig.5. The horseshoe vortex holds almost the same circulation on any scoured bottoms with different scour depth. This vortex generates friction on the bed and scour around the cylinder. As the scour hole becomes deeper, the vorticity with counter-rotation is generated and growing in the scour hole. In the deep scour hole case, the counter-rotating vorticity reduces the induced velocity or friction on the bed due to the horseshoe vortex, and the local scour reaches to the equilibrium stage.

CONCLUSIONS

The horseshoe vortex formed around a vertical cylinder on a scoured bed is investigate with its vorticity distribution evaluated from laboratory experiments. The vorticity distribution is estimated from the velocity distribution measured by an image processing technique. The formation of vortices around the cylinder is almost independent of the Reynolds number in laminar or transient flow, in the range used in this study. Vorticity concentration of the horseshoe vortex's counter-rotating is observed in the deep part of the scour hole. This flow feature is invisible in the velocity distribution. So, we conclude that to study in detail the flow structure around a cylinder in a scoured bed, the vorticity distribution of the flow is more useful than the velocity distribution. The horseshoe vortex is almost independent of the local scour depth. The counter-rotating vorticity becomes stronger as the scour hole is deeper. In the deep scour hole, the stagnation area is generated due to the existence of both the horseshoe vortex and the high concentration area of counter-rotating vorticity.

REFERENCE

Baker,C.J. (1979): The laminar horseshoe vortex. Jour. of Fluid Mech., vol.95, part 2, pp.347-367.
Dey,S. and S.K.Bose (1994): Bed shear in equilibrium scour around a circular cylinder embedded in a loose bed. Appl. Math. Modelling, vol.18, pp.265-273.
Dey,S., S.K.Bose and G.L.N.Sastry (1995): Clear water scour at circular piers: a model. Jour. of Hyd. Engrg., vel.121, no.12, pp.869-876.
Kobayashi,T. (1992): 3-D analysis of flow around a vertical cylinder on a scoured bed. Proc. of 23rd Int. Conf. on Coastal Engrg., vol.3, pp.3482-3495.
Kobayashi,T. and M.Hino (1992): Spectrum-correlation method and filters technique application for the velocity field estimation. Proc. of 6th Inter. Symp. on Flow Visualization, pp.807-811.
Melville,B.W. and A.J.Raudkivi (1977): Flow characteristics in local scour at bridge piers. Jour. Hyd. Res., vol15(4), pp.373-380.
Nakagawa,H. and K.Suzuki (1971): Flow charactersitic around a circular bridge pier. Proc. of 26th Annu. Conf. of Japan Soc. of Civil Engers., vol.2, pp.293-296. (Japanese)
Niedoroda,A.W. (1981): The descriptive physics of scour in the ocean environment. Proc. of 13th Annual Offshore Tech. Conf., pp.297-304.
Sumer, B.M., J.Fredsøe and N.Christiansen (1992): Scour around vertical pile in waves. Jour. of Waterways, Port, Coastal and Ocean Engrg., ASCE, vol.18, no.1, pp.15-31.

FLOW PATTERNS AROUND BRIDGE PIERS AND OFFSHORE STRUCTURES

KAMIL H. M. ALI, OTHMAN A. KARIM and BRIAN A. O'CONNOR
Department of Civil Engineering, University of Liverpool, L69 3BX, UK.

ABSTRACT

The ultimate objective of this research is the prediction of scour hole geometries associated with single bridge piers of various shapes and offshore platforms of complicated geometries. Accurate calculations of velocity fields and bed shear stresses were obtained using the FLUENT package. Calculated bed shear stresses can be used in appropriate sediment continuity equations for the prediction of scour hole geometries.

INTRODUCTION

Sediment transport governs or influences many situations that are of importance to mankind. Erosion and scour may sometimes undermine hydraulic structures. Lack of reliable practical knowledge of scour has left some hydraulic experts to claim that the collapse of a large bridge as a consequence of foundation erosion is a disaster waiting to happen. For example, with over 150,000 road bridges and 6000 rail bridges in the U.K., this phenomenon could pose a serious safety problem as well as a significant financial burden, Penson, S. (1996). Continuous scour at a structure can lead to its failure, thus an understanding of the scouring process is very important if steps in the design process are to be taken to protect against it. Scour is a very complex process. No single analytically derived equation is available because of the difficulties associated with the problem, such as the combined effects of complex turbulent boundary layers, time-dependent flow patterns, and sediment transport mechanism in the scour hole. An accurate prediction of the boundary shear stress is essential in the prediction of the time development of a scour hole. The authors used the FLUENT package to predict the boundary shear stress distributions for single cylinders of various shapes and for scour holes of various sizes. FLUENT was also used to predict the velocity field and boundary shear stress distributions for two different types of offshore structures under the influence of a steady current. Experiments were conducted to study the resulting scour patterns. Crushed olive stones were used as the movable bed. The FLUENT package is a general purpose computer program for modelling complicated flow fields. It incorporates techniques based on fundamental principles for simulating a wide range of fluid flow problems. FLUENT/BFC is the primary set-up package for modelling complex geometries. FLUENT uses a finite volume numerical procedure to solve the fundamental equations governing fluid flow (the Reynolds Equations). Additional turbulence model equations are also used.

RESULTS

Figure 1 shows the body-fitted co-ordinate grid used for modelling the flow round a vertical cylinder. The flow was assumed to be symmetrical and only one half of the cylinder and the flowfield was analysed. Dimensions of the cylinder and the flow parameters were based on the experiments of Yanmaz and Altinbilek (1991). The diameter of the cylinder was 67mm, water depth 0.135m. The sand particle size was 1.07mm and the specific gravity was 2.64. Bed contours were obtained at run times of 5, 20, 60, 100 and 150 minutes. Figures 2 and 3 show velocity and bed shear stress distributions for the circular cylinder and for a rigid flat bed. The maximum bed shear stress corresponds to the maximum velocity. The maximum shear stress occurs at midpoint of the circular pier extending approximately two diameters across the channel showing a 25% increase compared with the average upstream value. FLUENT solutions were obtained using the scour holes obtained by Yanmaz for t = 15, 20, 60, 100 and 150 mins. Figures 4 and 5 show the velocity vectors and bed shear stress distributions for t = 150mins. At the beginning of the experiment, t = 0mins, the bed was flat. The FLUENT results show that the initial boundary shear stress is at the side of the pier but at t = 5mins it moved to the pier symmetrical centreline at the start of the scour hole. The velocity distributions reveal that the deceleration of the velocity became more severe as the hole became deeper. In the downstream region of the hole the decrease in depth between the bed and the water surface causes the flow to converge and hence increase the mean velocity of the section, Penson, S., (1996). A steep fall in bed shear stress values occurs just after scouring action starts and as it developes further the rate of reduction becomes very slow. Negative shear stresses were obtained for those sections where reversed flow was observed. The negative sign indicates that the boundary shear stress is in the direction of the flow. The FLUENT package was also used to predict the velocity and shear stress distributions around the offshore platforms shown in Figures 6 and 7. Bed velocities measured by Butterworth (1996) were compared with those obtained using FLUENT for a flat bed. The FLUENT velocities and shear stresses for the platform in Fig. 6 (rigid, flat bed), are shown in Fig. 8. These results show that the velocities at the bed at the front piers had a small negative component in the streamwise direction. This indicates the presence of a horseshoe vortex as would be expected. The numerical velocity vectors for the pile platform are given in Figure 9.

CONCLUSIONS

1. The simulated flow in FLUENT seems to correspond well with available experimental results. The theoretical aspects of the flow dynamics, such as the horseshoe vortex, can also be interpreted from the output.
2. The FLUENT results show that the regions with the highest values of bed shear stress correspond to the region of highest floor velocity.
3. After the scouring action started, the bed shear stress falls and as the scour hole developed further the reduction became less apparent. The overall bed shear stresses decreased with the increase in scouring time and with the increase in size of scour hole.
4. The shear stresses, predicted by FLUENT, is lowest in the region of maximum erosion and is highest near the crest region.
5. The bed shear stress values, especially in the eroded part of the hole are all less than the critical value for the initiation of particle motion on a flat bed.
6. The shape of the scour hole, generated by a uniform current, around the piers of an offshore platform remains almost unchanged with time. The rate of change decelerates as time elapses. The produced scour hole can be approximated to an inverted cone with a base diameter several times that of the pier and with side slopes equal to the submerged angle of repose of the crushed olive stones.

ACKNOWLEDGEMENTS

The authors would like to express their sincere thanks to K. Owen, S. Penson and G. Butterworth, Graduate Engineers for their help in some of the numerical and experimental work described in this paper. We also thank A. M. Yanmaz for providing us with some of his experimental results.

REFERENCES

1. Yanmaz, A. M. and Altinbilek, H. D. (1991), "Study of time-dependant local scour around bridge piers, Journal of Hydraulic Engineering, Vol. 117 No. 10, ASCE.
2. Butterworth, G. (1996), "Investigation of local scour caused by currents and waves", B.Eng. Thesis, University of Liverpool.
3. Fluent Incorporated (1993), "Fluent User Guide", Version 4.2.
4. Melville, B. W. (1975), "Local scour at bridge sites", Report No. 117, University of Auckland, Auckland, New Zealand.
5. Penson, S. (1996), "Investigation of the flow around a bridge pier using the FLUENT package", M.Sc.(Eng.) Thesis, University of Liverpool.
6. Raudkivi, A. J. (1991), "Loose boundary hydraulics", Third Edition, Pergamon Press, Oxford, U.K.

(b) FLUENT

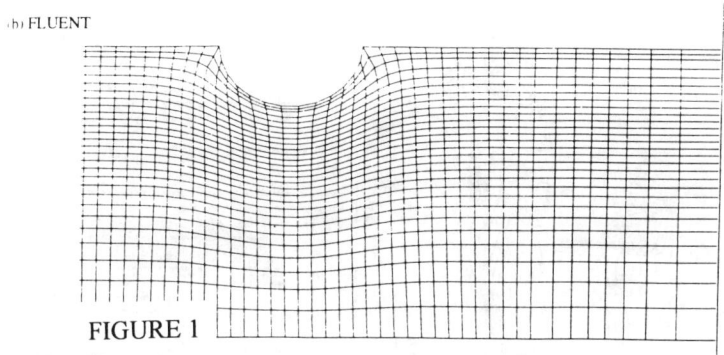

FIGURE 1

(a) FLUENT (RNG k-εModel) → 0.12 m/s

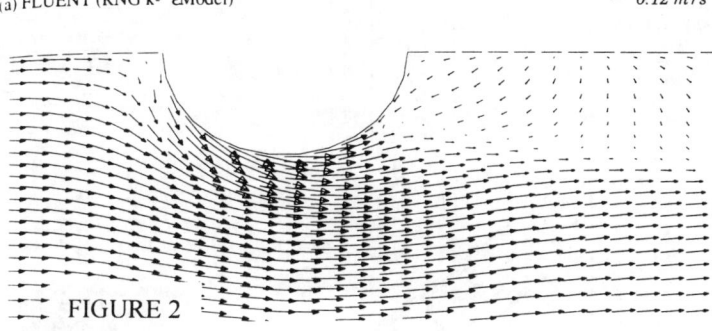

FIGURE 2

(a) FLUENT (RNG k-ε Model)

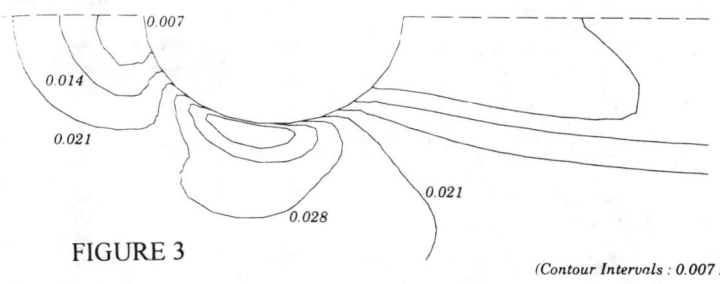

FIGURE 3

(Contour Intervals : 0.007 Pa)

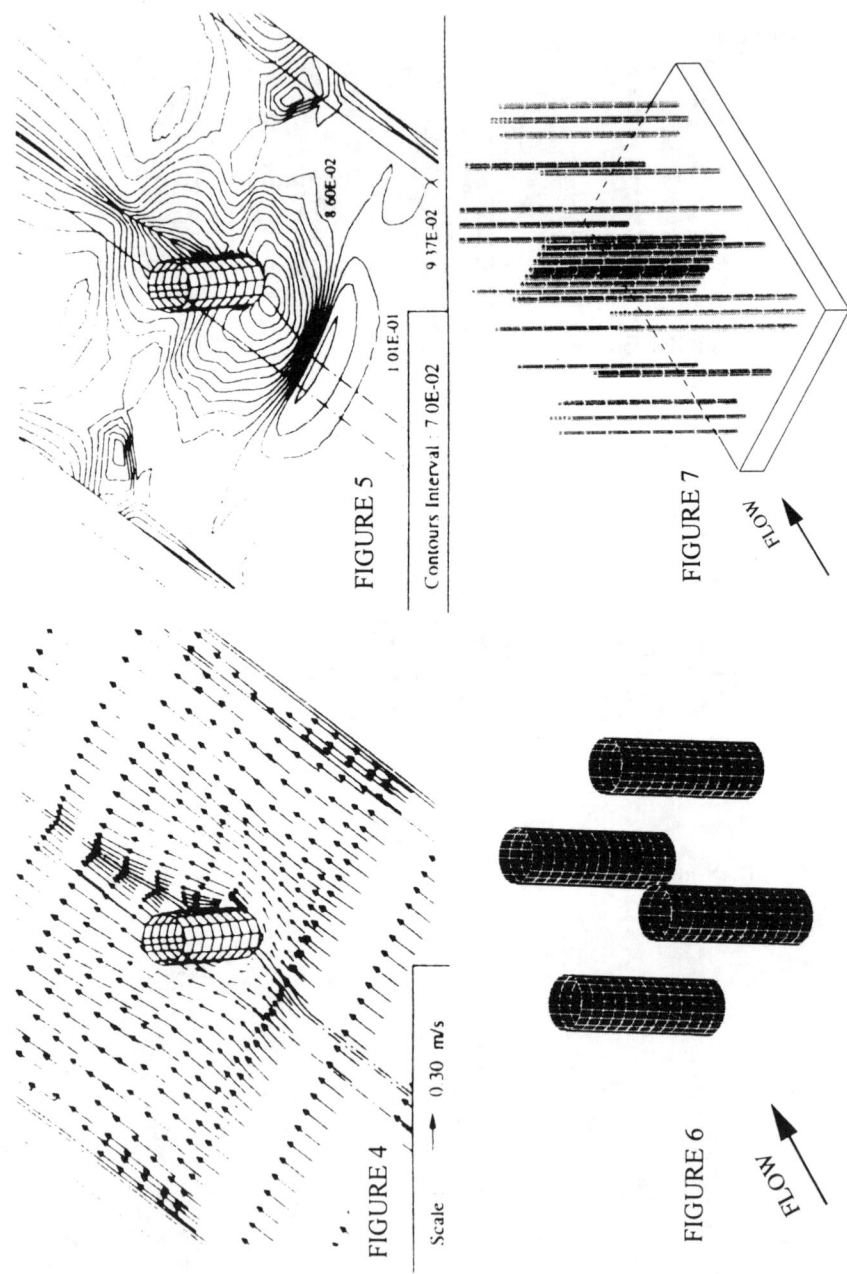

FIGURE 4

Scale: → 0.30 m/s

FIGURE 5

Contours Interval: 7.0E-02

FIGURE 6

FLOW

FIGURE 7

FLOW

FLOW PATTERNS AROUND BRIDGE PIERS 213

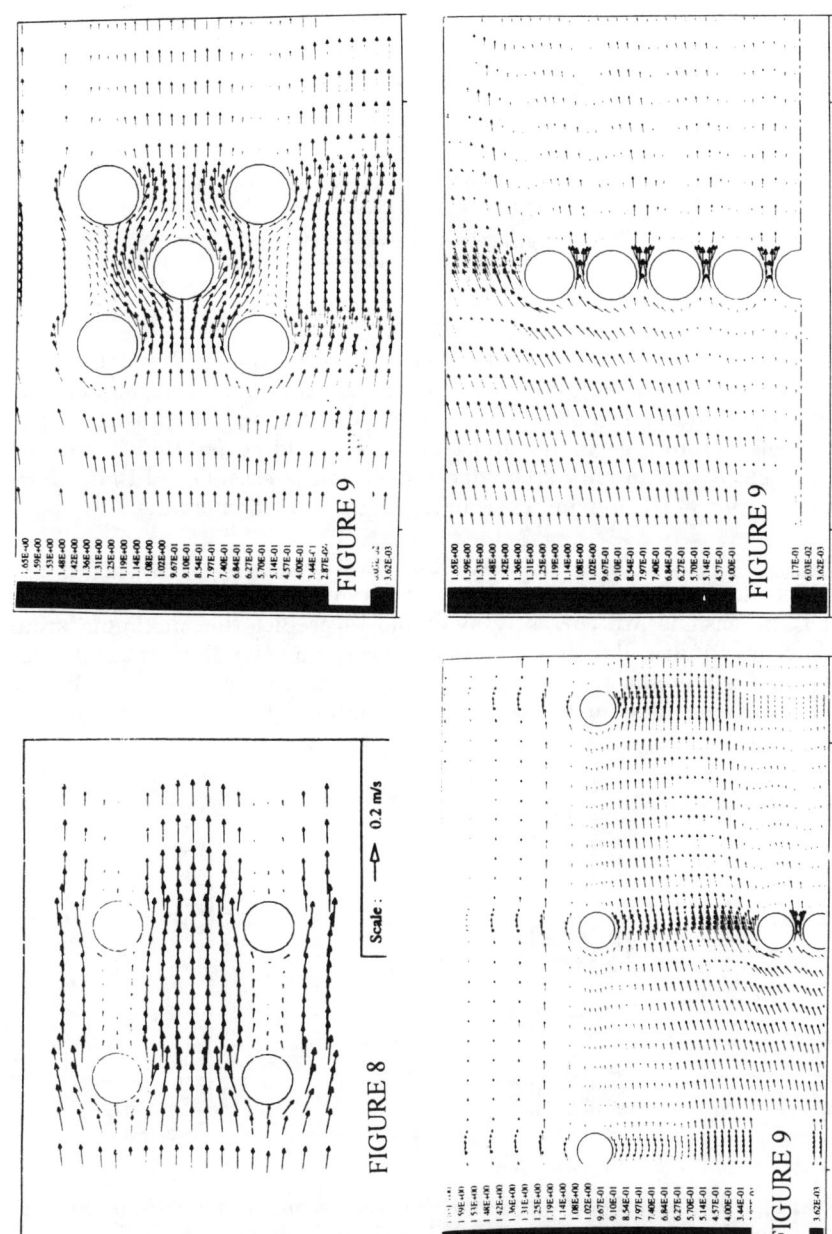

BUILDING SCOUR IN FLOODPLAINS

ALEXANDER KOHLI and WILLI H. HAGER
VAW, ETH-Zentrum
CH-8092 Zurich, Switzerland

ABSTRACT

Scour of rectangular buildings located in floodplains is analysed, based on systematic experimentation and by comparison with bridge pier and abutment scours. The effects of sediment compactness, approach Froude number, building geometry and transverse building position have been analyzed. A preliminary scour depth equation is established that allows specification of the maximum scour depth as a function of time.

INTRODUCTION

Bridge pier scour has received lots of attention during the past decade and expressions are currently available to predict the maximum scour depth of cylindrically shaped piers (Breusers and Raudkivi 1991). Bridge abutments have been analysed to a smaller degree and some results are also available (Melville 1995; Sturm and Janjua 1994).

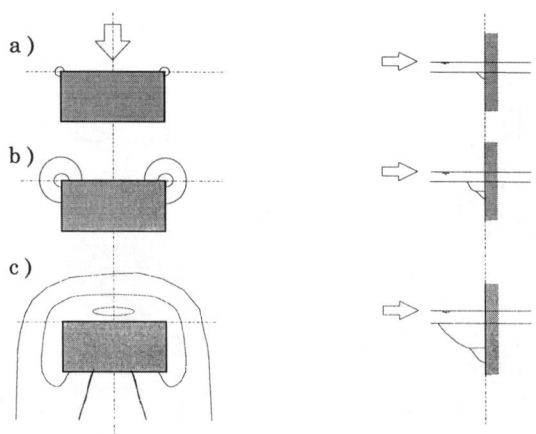

Fig. 1 *Progress of building scour* a) corner scour, b) transition scour, c) axis scour. Plan (left) and side views (right).

Buildings located in floodplains can also be seriously scoured. Rectangular-shaped objects may be regarded as a third basic case, in addition to cylindrically shaped piers, and dam-shaped abutments. Because a building is not designed for flood flows, scarce information on the scour potential is currently available. The present research project was initiated to fill this gap mainly with regard to Alpine countries, where non-cohesive material is typical, and velocities can be much in excess of, say $1ms^{-1}$.

The results of the study, which was started in late 1995, i.e. is not finished yet, can thus be applied to more general flood conditions. The purpose of this contribution is to outline main findings and a final publication will be presented later.

PHYSICAL MODEL

The experiments were conducted in a rectangular channel 1m wide, 1m deep and 12m long. The test reach is 6m long. The closed water loop has a capacity of 120 ls^{-1}. The inflow reach is 2m long and separated with a filter mattress fixed in a grid against the test reach. The transition from the inflow to the test reach is improved with an adversely sloping ramp and a 0.1m thick pervious mattress that is deeply inserted in the sandy material. The arrangement of this inflow was found to be significant for perfect approach conditions and the mattress can be regarded as flow straightener. For approach Froude numbers up to 0.50, the approach flow is perfectly plane, without any surface wave generation and resulting in an absolutely plane sand bed. For the experiments to be reported, a nearly uniform sand of average diameter d=1.3mm (±0.1mm) and density 2'650 kgm^{-3} was used.

A rectangular sharp-crested and vertical weir plate controlled the tailwater level. In all experiments, the elevation of the sand bed was adjusted to the weir crest. To inhibit sediment loss from the test reach, no sand was provided close to the weir. Further downstream, a flap gate was located to control submergence. At the beginning of an experiment, the still water level was equal to the flow depth used subsequently. On starting the experiment, the flap gate was lowered and the flow could be set up without changing the water elevation at the building. This is essential because the scour activity is extremely strong in the first instances, compared to later phases.

All experiments were conducted close to threshold conditions yet never producing a live bed scour (Ettema 1980). Such conditions are known to produce maximum scour. Typically, one experiment lasted 1 to 2 days, and maximum duration so far used was 5 days. We have never reached a so-called *end-scour* because the progression in depth has always continued, but the rate of scour evidently decreased with increasing time.

Scour Geometry

Compared to a cylindrical pier, the scour pattern of a rectangular pier is significantly different. Due to the plane building front, with sharp edges at the sides, scour development is initiated at the front edges. Shortly after the flow has started, two scour cones with an angle equal to the equilibrium angle of sand under water are formed (Fig. 1a). As the scour progresses, the two cones eventually match at the axis of the building.

The scour progresses mainly towards the building axis, and reaches at a certain time the scour depth of the edge scour (Fig. 1b,c). For even larger times, the maximum scour is thus shifted from the edges to the front axis of the building. In the following, we consider the *maximum scour depth* z as a function of time t exclusively.

Experimental Results

The effects to be discussed in the following are compactness of sand, Froude number, transverse building position in test channel, width and length of building. All data are normalized and a preliminary design equation taking into account these effects is proposed.

Dimensionless Scour Depth

The scour depth z varies with time t. A geometric measure for scour is the area of the building projected in the direction of the approach flow. Whereas the drag of a body in a flow follows essentially the product of area times velocity head, scour is mainly proportional to the approach flow depth h_o times the building width b, i.e. $Z=z/(bh_o)^{1/2}$ is a typical parameter for scour depth. As for the drag problem, Z depends on the square of the approach velocity head $V_o^2/(2g)$, where $V_o=Q/(bh_o)$ with Q=discharge and g=gravitational acceleration. Because approach Froude numbers $\mathbf{F}_o=V_o/(gh_o)^{1/2}$ were below 0.6, and a two-phase flow water-sediment is considered, free surface effects are negligible. The significant parameter corresponds to the *densimetric Froude number* $\mathbf{F}_d=V_o/(g'd)^{1/2}$. Here $g'=(\Delta\rho/\rho)g$ is reduced gravity with $\Delta\rho=\rho_s-\rho$ as density difference between sediment and fluid, and d= average grain diameter. One would thus expect that $Z \sim \mathbf{F}_d^2$.

The scour depth increases with time, and $Z \sim t$. Nondimensional time $T=(V_*/h_o)t$ was found significant, with $V_*=(g'd)^{1/2}$ as shear velocity. Note that a uniform sand of diameter d was considered.

Effect of Compactness

The effect of sand compactness is of primary interest for the test program. Fig. 2 shows tests for various conditions of sand: (1) dry loose sand, (2) dry compacted sand, (3) wet loose sand and (4) wet compacted sand. The non-dimensional time τ is introduced below. It is noted that for all four configurations the effect of compactness is so small that it may be ne-

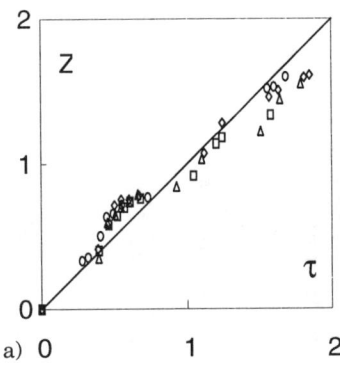 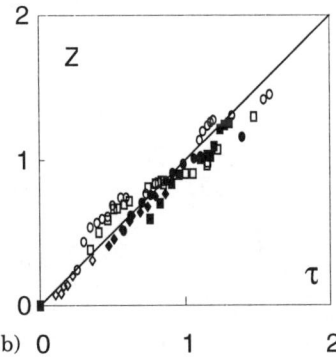

Fig. 2 a) Effect of *sand compactness* for (◇) wet compacted, (△) wet loose, (○) dry compacted and (□) dry loose sand. F_o=0.33-0.36, h_o=0.053m, (—) Eq. (1).
b) Effect of *approach flow depth* for
F_o=0.26: (◇) h_o=0.025m, (◆) h_o=0.15m.
F_o=0.35: (○) h_o=0.038m, (●)h_o=0.10m.
F_o=0.42: (□) h_o=0.035m, (■) h_o=0.077m, (—) Eq. (1).

glected. Clearly, the deviations of individual scour depth from the average curve is of the order of accuracy, and all subsequent tests were conducted with wet, relatively loose sand compacted superficially.

EFFECT OF APPROACH FLOW DEPTH
Fig 2b) relates to the effect of the approach flow depth h_o. For F_o=0.26 to 0.42, and h_o=0.025 to 0.15m the depth h_o has no additional significance when applying Eq. (1). The main experiments are thus conducted with $h_o \cong 0.05$m.

EFFECT OF APPROACH FROUDE NUMBER
The approach Froude number F_o was varied between 0.2 and 0.6, and Fig. 3a) relates to three values between 0.26 and 0.52. The effect of Froude number is seen to be quadratic, as previously assumed.

EFFECT OF TRANSVERSE POSITION
The effect of transverse building position y, measured from the channel wall to the building axis was also tested, and found insignificant (Fig. 3b). Clearly, a compensation due to constriction of flow and decrease of wall velocity leads to this result. Accordingly, buildings were normally positioned in the channel axis for obtaining a symmetrical scour pattern.

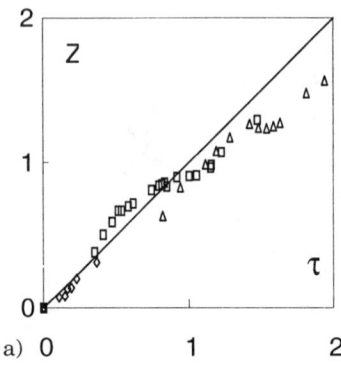 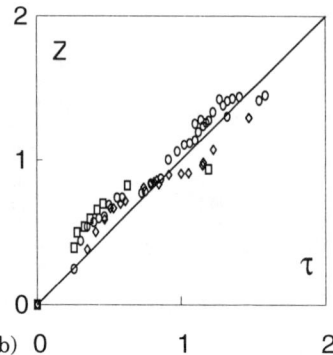

Fig. 3 a) Effect of *approach Froude number* F_o=($\diamond$) 0.26, ($\square$) 0.42, ($\triangle$) 0.52 with h_o=0.032-0.051m. Axial building with L=0.15m and b=0.2m. (—) Eq. (1).
b) Effect of *transverse position* y[mm]=($\bigcirc$) 100 (at wall), ($\square$) 150, ($\triangle$) 250, ($\diamond$) 500. Same building, F_o=0.34-0.61. (—) Eq. (1).

EFFECT OF BUILDING WIDTH
Buildings of various widths b were positioned axially in the channel. Fig. 4a) shows a minor effect of b when accounting for parameter $Z=z/(bh_o)^{1/2}$. The concept of constricted approach surface bh_o is thus verified.

EFFECT OF BUILDING LENGTH
The length L of the building in the flow direction was studied, and Fig. 4b) shows a typical result. It may be noted that the scour depth z is practically unaffected by L and wall-type buildings behave as very long buildings. The streamwise building geometry has thus no effect on the scour at the building front.

PRELIMINARY SCOUR EQUATION
The maximum scour depth z normalized to the square root of the constricted area $(bh_o)^{1/2}$ depends mainly on dimensionless time $T=(V_o/h_o)t$ and the densimetric grain Froude number $F_d=V_o/(g'd)^{1/2}$. From the experiments presented and further observations, the relative scour depth Z was related to the normalized time $\tau=0.045F_d^2 T^{1/5}=0.045F_d^2[(g'd)^{1/2}t/h_o]^{1/5}$. The preliminary scour equation thus is

$$Z = \tau. \quad (1)$$

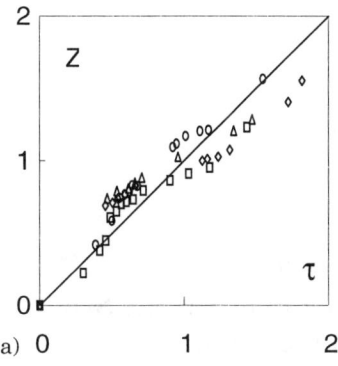 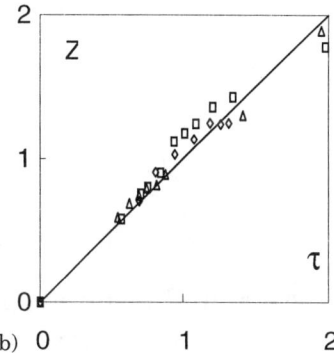

Fig. 4 a) Effect of *building width* $b[\text{mm}]=(\triangle)$ 50, $(\diamondsuit)$ 100, $(\bigcirc)$ 200, $(\square)$ 400 for $F_o=0.35$-0.40, $h_o=0.044$-0.055m. (—) Eq. (1).
b) Effect of *building length* $L[\text{mm}]=(\triangle)$ 20, $(\diamondsuit)$ 150, $(\square)$ 300 for $F_o=0.42$-0.46, $h_o=0.051$-0.055m, $b=200$mm. (—) Eq. (1).

Further effects to be included account for granulometry, angle of attack, sediment layering and building groups.

CONCLUSIONS

The maximum scour depth of buildings in floodplains is demonstrated to depend significantly on the square root of the projected area into the flow, the approach densimetric Froude number and the dimensionless time. The effects of sediment compactness, streamwise building length and transverse building position are negligible. A preliminary scour depth equation is presented.

ACKNOWLEDGMENT

This study is funded by the Swiss National Science Foundation, Grant No. 2100-042989.95/1.

REFERENCES

- Breusers, H.N.C., Raudkivi, A.J.: (1991). Scouring. *IAHR Hydraulic Structures Design Manual* **2**. Balkema: Rotterdam, The Netherlands.
- Ettema, R. (1980). Scour at bridge piers. *Report* **216**, School of Engineering, Univ. of Auckland, N.Z.
- Melville, B.W. (1995). Bridge abutment scour in compound channels. *Journal of Hydraulic Engineering* **112**(12):863-868.
- Sturm, T.W., Janjua, N.S. (1994). Clear-water scour around abutments in floodplains. *Journal of Hydraulic Engineering* **120**(8):956-972.

Physical and Computational Modeling of Bridge Scour at Oregon Inlet, North Carolina

by CONOR SHEA AND MICHAEL PORTS

Parsons Brinckerhoff Quade & Douglas, Inc., Baltimore, MD, USA

INTRODUCTION

Oregon Inlet is the northernmost inlet along the Outer Banks of North Carolina. The inlet was opened in 1846 by a severe storm and, after closing, it reopened in 1864. The inlet is very dynamic and its configuration is constantly changing. Shoals and spits grow and erode very quickly. Historically, the inlet has ranged in width from 640 m (2,100 feet) to 2,030 m (6,670 feet), but has generally maintained a main channel (gorge) depth of 9 m (30 feet). Generally, the trend is for the inlet to narrow during periods of low storm activity and widen after large storms.

The Herbert C. Bonner Bridge, which connects Bodie Island on the north side of the inlet with Hatteras (Pea) Island to the south, requires replacement. Because of the severe hydraulic conditions, design of the replacement bridge presents a unique challenge. In order to accurate estimate scour at replacement bridge piers, an approach combining extensive field reconnaissance, physical modeling of scour in a flume, and computational modeling of hydraulic flow conditions under severe tidal storm surges was required.

Prior to the construction of the Herbert C. Bonner Bridge in 1962, transport across the inlet was via ferry. As demand for transportation across the inlet increased, the ferry was unable to keep pace and the bridge was constructed to meet the demand. The bridge was opened for traffic in November, 1963. The ease of access to Hatteras and Ocracoke Island led to increased development. The bridge is now critical to the Outer Banks economic and social well being. During recent summer and fall seasons, more than 8,500 vehicles cross the bridge each day. The bridge is the only land evacuation route from Hatteras Island during hurricanes and tropical storms. Many residents of Ocracoke Island also evacuate via Bonner Bridge. The bridge also carries electrical power and telephone lines to Hatteras Island.

Almost from the time it was completed, the Bonner Bridge has experienced severe scour problems and corrosion damage. Several retrofits for scour protection have been made including the installation of armor blankets at the main navigational span, installation of support pier bents, and installation of rip-rap revetment at the south abutment. In response to concerns that the southward movement of the inlet might cause the loss of the south approach highway and abutment, a terminal groin was installed at the north end of Hatteras (Pea) Island by the State of North Carolina in 1990 to halt further migration. Because of the dynamic environment, scour at the existing bridge is monitored by sonar on a monthly basis and after major storms.

In order to reduce escalating maintenance costs associated with scour countermeasures and corrosion, the North Carolina Department of Transportation retained Parsons Brinckerhoff to design a replacement bridge that will withstand the high scour, violent storms, and harsh environment of Oregon Inlet. One of the most challenging facets of the design is to accurately assess the hydraulic conditions and potential scour at the replacement bridge that may occur under existing conditions and as the inlet configuration might changes over time. Overestimating the severity of the hydraulics conditions and scour would result in bridge piers that are both larger and deeper than necessary, and increase the construction costs of the bridge. On the other hand, underestimating the severity could lead to potential bridge failure and potential loss of life.

The hydraulic analysis and scour evaluation, which were completed in late Fall 1996, consisted of three parts: (1) Field Investigations; (2) Physical Modeling and Computational Hydraulic Modeling.

FIELD INVESTIGATIONS

Field investigation were performed to provide a better understanding of the dynamics of Oregon Inlet and provide data for calibrating the computational hydraulic model, and for physical modeling of replacement bridge piers. A survey was made of the inlet bathymetry, shoreline, and inlet shoulders. The bathymetry of the inlet was surveyed using an Acoustic Doppler Current Profiler (ADCP).

Tidal gages were installed to monitor the water surface elevations at various locations around the inlet as water surfaces fluctuated with tidal cycles. The ADCP measured depth-averaged velocities and flow direction to provide calibration data for the two-dimensional modeling..

The shoreline and configuration of shoals in Oregon Inlet change very rapidly. A survey reflects the configuration of the inlet only at the time of the survey. Initial surveys were performed in October 1995. A follow-up field investigation conducted in July 1996 found conditions had changed dramatically. The width of the inlet had decreased and an extensive shoal had risen out of the water to form an island in the middle of the inlet. By late August 1996, the island had reduced by wave action and high tides to an underwater shoal. Hence, a important part of the field investigation is the review of historic data on inlet configurations. A review was made of the extensive data collected over the years by the Corps of Engineers (U.S. Army Corps of Engineers, 1977) and the scour monitoring records kept by the North Carolina Department of Transportation. The review showed that the maximum depth at the existing bridge crossing has varied between 30 and 60 feet.

COMPUTATIONAL HYDRAULIC MODELING

Computational hydraulic modeling was performed using the Federal Highway Administration (FHWA) Finite Element Surface Water Modeling System (FESWMS-2DH) developed by David Froehlich (1996). FESWMS-2DH employs a finite element network to solve the momentum and energy equations and compute the direction and depth of flow at each node point in the finite element network. Dynamic (time-varying) conditions are simulated so that the storm surge associated with a major storm can be modeled.

A close-up of the finite element network in the vicinity of the inlet is shown in Figure 1. The grid background is the finite element network. The vector arrows represent the magnitude and direction of flow.

FESWMS-2DH was used to model the flow velocities for three design tidal storm surges: (1) the 100 Year Return Period Tidal Storm Surge; (2) the 500 Year Return Period Tidal Storm Surge; and (3) a Class IV Northeaster Tidal Storm Surge. Both flood (incoming) and ebb (outgoing) tidal storm surges were evaluated

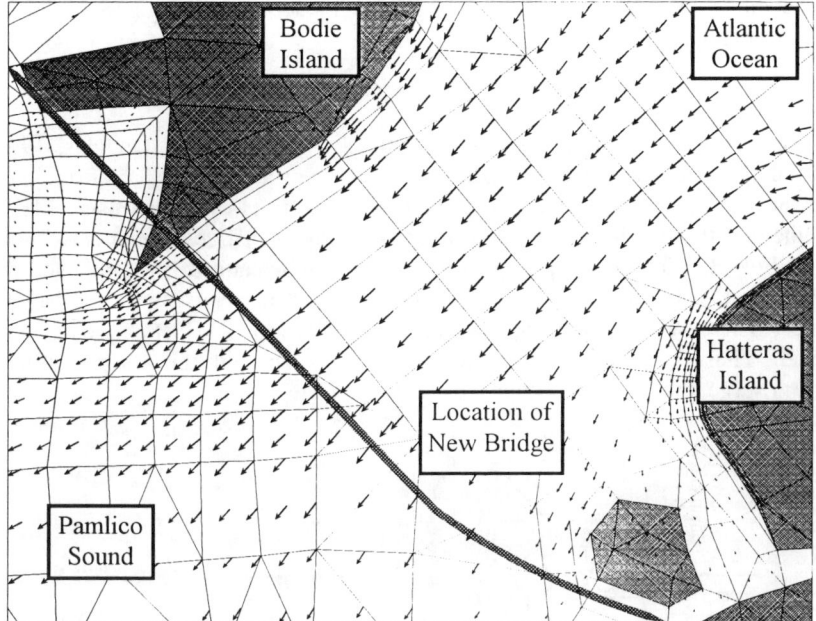

Figure 1: Portion of Finite Element Network Showing Flood Tide Entering Inlet

Because the configuration of the inlet is subject to change, it was important to evaluate the hydraulic conditions associated with potential inlet configurations that might occur over the service life of the replacement bridge. In addition to existing conditions, two alternate configurations were modeled. The alternates were:

1. a wide inlet configuration representing the widest historic inlet opening

2. a narrow inlet configuration representing the narrowest historic inlet opening.

Alternate configurations were developed using the historical records and scour monitoring data.

FESMWS-2DH results were examined to determine the most severe hydraulic conditions that might be encountered over the lifetime of the replacement bridge. The most severe conditions were produced by the existing conditions model. Results from the hydraulic modeling were used as input to scour evaluation procedures to develop estimates of maximum scour at bridge piers for the replacement bridge.

PHYSICAL MODELING

During the bridge type study (Parsons Brinckerhoff Quade & Douglas, 1995), initial scour estimates were prepared for the proposed replacement bridge piers. In the United States, scour at new highway bridges is evaluated using FHWA guidelines that are in presented *Evaluating Scour at Bridges*, commonly known as HEC-18 (Richardson and Davis, 1995). A schematic of the proposed pier configuration is shown in Figure 2. The proposed pier consists of pile cap and piles. The base of the pile cap is located at an elevation 2.0 feet above mean sea level. Thus, the pile cap will not normally in contact with the water. During a storm surge, the pile cap becomes partially inundated.

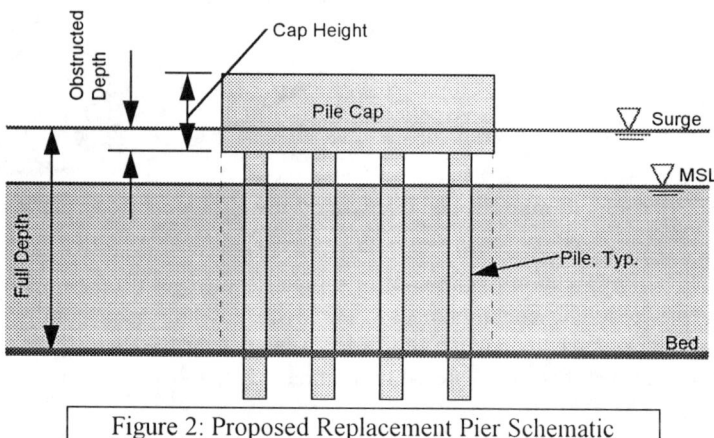

Figure 2: Proposed Replacement Pier Schematic

Under current HEC-18 guidelines, the scour generated by the pile cap is estimated by treating it as if it were a footing placed on the channel bed. This results in a scour estimate that was judged to be too conservative and unrealistic. Therefore, it was decided that physical modeling would be performed to provide better estimates of scour for the replacement bridge piers.

Physical Modeling of proposed bridge piers for the replacement bridge was performed at the FHWA Turner-Fairbank Hydraulic Laboratory in McLean, VA. A series of flume experiments measured the scour generated by physical models of the replacement bridge piers. The results showed that a pile cap located at the water surface which obstructed only a small portion of the water column, did not generate as much scour as predicted by current HEC-18 guidelines. The results from the physical modeling has been used to develop scour estimates for the replacement bridge. The scour estimates developed using the results of physical and computational model should yield substantial cost savings for the Bonner Replacement Bridge as well as having a high degree of reliability..

SUMMARY

Hydraulic conditions at Oregon Inlet are hard to evaluate. In order to produce a design for the replacement bridge that is both cost effective and resistant to scour, a three fold hydraulics and scour analysis was required. Combining field observations with both computational and physical modeling ensure an efficient, but safe design of bridge piers.

REFERENCES

Froehlich, David (1996) *Finite Element Surface Water Modeling System: Two-dimensional Flow in a Horizontal Plane, Version 2 Draft User's Manual*, U.S. Federal Highway Administration, McLean, VA, 1996.

Parsons Brinckerhoff Quade & Douglas, Inc. (1995), *Hydraulic and Scour Analysis for Bonner Bridge Replacement over Oregon Inlet, Dare County, NC*, prepared for the North Carolina Department of Transportation, 47 pages and appendix, Baltimore, MD, March, 1995.

Richardson, E.V., and S.R. Davis (1995), *Evaluating Scour at Bridges*, U.S. Department of Transportation, Federal Highway Administration, Hydraulic Engineering Circular No. 18, Publication No. FHWA-IP-90-017, Third Edition, November 1995.

U.S. Army Corps of Engineers (1977), Manteo (Shallowbag) Bay, North Carolina: Design Memorandum 1: General Design Memorandum, Phase 1: Plan Formulation, U.S. Army Corps of Engineers, Wilmington District, July, 1977.

THREE-DIMENSIONAL HYDRAULIC ANALYSIS FOR CALCULATION OF SCOUR AT BRIDGE PIERS WITH FENDER SYSTEMS

EDWARD J. KENT, PH.D., P.E.
Earth Tech
Concord, New Hampshire
and
JOHN E. RICHARDSON, Ph.D., P.E.
Flow Science
Los Alamos, New Mexico

ABSTRACT
Calculation of scour at bridge piers, which are protected by fender systems, requires evaluation of the hydraulic effects of these collision protection systems. A case study is presented involving application of the fully three-dimensional computational fluid dynamics code FLOW-3D to analyze hydraulics at a lift-span bridge, crossing a navigable waterway. The hydraulics and pier scour are analyzed for a 100-year flow condition for a lift-span pier with and without its collision protection system.

INTRODUCTION
The guidance document for the National Bridge Scour Evaluation Program, HEC-18 (Richardson et.al., 1995), recommends use of the so-called CSU equation (Richardson and Richardson, 1989) for calculation of scour at bridge piers. The key variables in this equation include velocity, depth of flow, pier width and angle of attack. A fender system can significantly affect the velocity and direction of water striking the pier structure.

This hydraulic evaluation was performed as a demonstration project using plans for a bridge which is currently under construction. In this investigation, the computational fluid dynamics code FLOW-3D (Flow Science, Inc., 1987) was used to evaluate complex hydraulic conditions for calculation of potential scour at an existing bridge over the Quinnipiac River in New Haven, CT. This bridge has a center lift span and its central lift piers are protected by a timber fender system and circular filled cofferdams. The presence of these collision protection systems make it impossible to estimate approach velocities at the lift piers using a one-dimensional model such as WSPRO (Shearman, 1990).

HYDRAULIC MODEL

To investigate the flow around piers with fender systems mathematically, one must use a computational method capable of treating transient flow phenomena involving complex geometric regions, free-surfaces, and flow through porous media (i.e., the fender system). The commercial software package, FLOW-3D, meets these requirements. FLOW-3D is a general purpose computational fluid dynamics program which uses the transient Navier-Stokes equations as governing equations for fluid motion. The solution algorithm is based on the Volume-of-Fluid (VOF) method (Hirt and Nichols, 1981).

BASIC VOLUME-FRACTION CONCEPTS

The basic idea of a volume-fraction method is to imagine the region to be modeled as subdivided into two components. For example, it may be composed of a collection of fluid and gas subregions or a collection of fluid and solid subregions. In either case, we introduce a generalized, Heavyside function (Lighthill, 1958) that assumes the value of zero in one region and unity in the other region. Then the fluid equations of motion, say the Navier-Stokes equations, are multiplied by this generalized function and integrated over a typical control volume (i.e., the cells in a Eulerian grid). Using simple approximations (Hirt and Sicilian, 1985), the resulting discrete equations are similar to those used for flow in porous media, which is not surprising since this is one way to derive those equations.

In the FLOW-3D program, two types of volume-fraction techniques are used. One of these corresponds to the original VOF method for the treatment of free surfaces and two-fluid interfaces. A second volume-fraction technique, referred to as FAVOR (Fractional Area Volume Obstacle Representation), is used in the FLOW-3D program to model complex geometries (Hirt and Sicilian, 1985). In this case, a solid volume faction (or the complement "open" fraction) is supplemented with area fractions in each of the three coordinate directions. The incorporation of these area/volume fractions into discrete control-volume flow equations automatically forces rigid, free-slip boundary conditions at all solid surfaces. This technique is particularly useful in connection with free surfaces because it allows one to more easily satisfy free-surface boundary conditions in control volumes that contain both types of boundaries.

Volume-fraction methods, as used in the FLOW-3D program, have many advantages. For instance, they require much less memory than moving grid methods, they allow arbitrary boundary displacements without grid distortion and they permit the creation and destruction of bounding surfaces without the need for complicated numerical logic.

APPLICATION TO A BRIDGE PIER WITH FENDER SYSTEM

PHYSICAL SETTING

For this study, we have chosen to model the hydraulics and calculate the potential pier scour for a bridge currently under construction over the Quinnipiac River in New Haven, CT. The bridge has a total length of 930 feet, and carries four lanes of traffic and a railroad track. Its central lift span is approximately 200 feet long. This investigation focused on one of the central lift piers which are partially shielded from flow by the ship collision protection system. Each of the lift piers is approximately 52 feet wide and 19 feet long. The lift piers are protected by a continuous timber fender system and 36-foot diameter filled cellular cofferdams. The plan view of the piers for one of the lift span towers and the fender/cofferdam system is shown schematically in Figure 1.

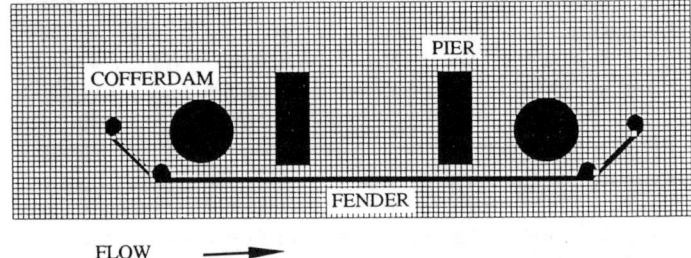

FLOW ⟶

Fig. 1. Schematic of the piers for one of the lift span towers and the fender/cofferdam system, also showing the computational mesh

THE COMPUTATIONAL MODEL

For our computational model, both the water surface elevation and the velocity upstream of the pier is fixed. Fluid is permitted to flow freely across the downstream boundary. A hydrostatic pressure distribution is also assumed to exist at the upstream boundary. Symmetry conditions are applied at the lateral boundaries. As an initial condition, the fluid is motionless. The initial condition was chosen for simplicity and is not meant to model any particular physical situation.

Figure 1 also shows the computational grid used for the base calculation. The grid contains 125 cells in the direction of flow (x), 40 cells laterally (y) and 9 cells vertically (z). Cell sizing is uniform throughout the domain although this is not a limitation of the methodology. The cell sizing is $x = 3.0$ ft, $y = 3.0$ ft and $z = 3.3$ ft.

Porous obstacles were used to model the fender system. At the location of the fender system, the fraction of each computational cell open to flow was reduced to 30% of its dimensional value. Treatment of the fender system, as a porous obstacle, allows us to model the partial flow blockage caused by the fender configuration. Solid obstacles were used to model the cofferdams and rectangular piers.

All calculations reported include the Renormalization Group (RNG) model of turbulence (Yakhot and Orszag, 1986). This model is effectively an extension of the more common k-ε model, except that it has many fewer empirical parameters and is more accurate in regions of strong shear.

NUMERICAL SIMULATION
Three separate numerical simulations were performed so that the effects of the fender system and cofferdams on calculated scour could be evaluated. The as-built configuration was modeled in the first simulation. In the second simulation, the fender system was removed but the cofferdams remained. In the final simulation, only the piers were modeled.

All computations were carried out for a period of 200 seconds, at which time the computed flow field is nearly steady. The analysis was conducted for the 100-year return period river flow of 12,100 cfs (Close, Jensen, & Miller, 1992) coupled with an upstream water surface elevation of 3.73 NVGD (MHW).

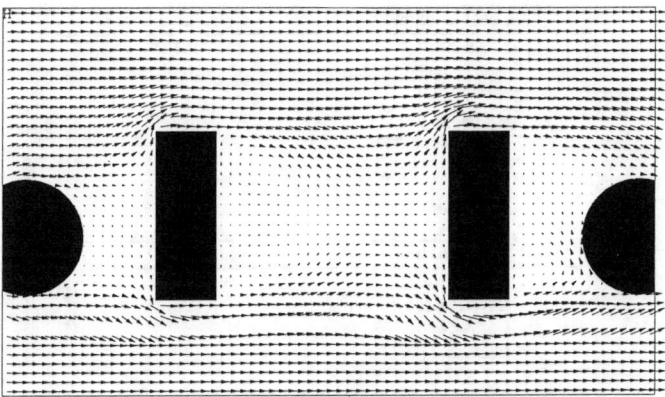

Fig. 2. Mid-depth velocities as-built configuration

Figure 2 shows the calculated flow field in the vicinity of the rectangular piers. Some circulation through the fender system is observed as well as the development of eddy structures downstream of the rectangular piers. Figure 3 shows a three-dimensional perspective of the flow around the as-built structure (note: streamlines moving through the fender system).

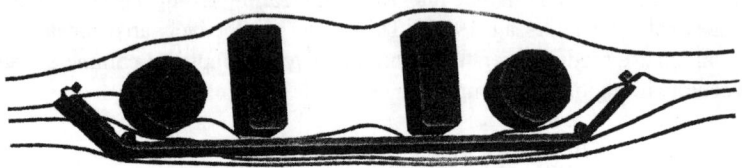

Fig. 3. Streamlines of flow around bridge pier

SCOUR CALCULATIONS

Scour was calculated for each of the pier configurations modeled. Data for the calculations was taken from a location 15 ft upstream of each lift pier at mid-depth. For simplicity, the CSU equation without modifiers was used to calculate scour. Tabulated results appear below.

Upstream Pier Q_{100}

Configuration	Calculated Scour
As-Built	9.1 ft
Some Protection (cofferdams only)	10.6 ft
No Protection	14.4 ft

Downstream Pier Q_{100}

Configuration	Calculated Scour
As-Built	10.9 ft
Some Protection (cofferdams only)	10.8 ft
No Protection	12.9 ft

The results show that the cofferdams and the fender system provide protection to both piers. When the fender system is removed, the calculated scour at the upstream pier is increased to a value similar to that calculated for the downstream pier. For this second case, the upstream pier is partially protected from the oncoming flow by the placement of the cofferdam. The downstream pier is, in turn, partially protected from the oncoming flow by the upstream pier. When the cofferdams are removed, the scour calculated at the upstream pier is maximal, as it receives no protection from the oncoming flow. The downstream pier still receives some protection from the upstream pier. The amount of protection is however reduced somewhat by the removal of the cofferdam (i.e., velocities in the vicinity of the downstream pier increase somewhat with the removal of the cofferdams).

SUMMARY COMMENTS

The results of these simulations show that the fender system, in this case, provides a certain amount of scour protection for the lift pier. Whether or not fender systems afford scour protection is site specific and depends upon the particular bridge geometry and water flow patterns. However, three-dimensional computational simulations of the type described in this paper offer a rapid and inexpensive way to investigate the scour susceptibility of pier placements complicated by fender systems and other secondary structures.

ACKNOWLEDGEMENTS

Design plans and hydrologic data used in this paper were provided by Beatrice Hunt and Nicholas Altebrando of Hardesty & Hanover, New York, New York, with the assistance of the Connecticut Department of Transportation.

REFERENCES

Close, Jensen, & Miller, P.C., 1992. *Hydraulic Calculations for Proposed Bridge Over Quinnipiac River, New Haven, CT.*
Flow Science, 1987. *FLOW-3D: Computational Modeling Power for Scientists and Engineers.* Flow Science, Inc. Report FSI-87-00-01.
Harlow, F.H. & J.E. Welch, 1965. *Numerical Calculation of Time-Dependent Viscous Incompressible Flow.* Physics of Fluids, 8:2182-2189.
Hirt, C.W. & B.D. Nicols, 1981. *Volume of Fluid (VOF) Method for Dynamics of Free Boundaries.* Jour. Comp. Phys. 39:201-225.
Hirt, C.W. & J.M. Sicilian, 1985. *A Porosity Technique for the Definition of Obstacles in Rectangular Cell Meshes.* Fourth Inter. Conf. Ship Hydrodynamics, Wash. D.C. Sept. 1985.
Lighthill, M.J., 1958. *Fourier Analysis and Generalized Functions.* Cambridge: Cambridge Uni. Press.
Yakhot, V. & S.A. Orszag, 1986. *Renormalization Group Analysis of Turbulence I. Basic Theory.* Jour. Sci. Computing 1:3.
Richardson, E.V., Harrison, L.J. & Davis, S.R. 1995. *Evaluating Scour at Bridges,* National Highway Institute, Federal Highway Administration, McClean, VA.
Shearman, J.O., 1990. *User's Manual for WSPRO - A Computer Model for Water Surface Profile Computations,* Federal Highway Administration, McClean, VA.
Richardson, E.V., Richardson, J.E., 1989. *Bridge Scour,* U.S. Interagency Sedimentation Committee, Bridge Scour Symposium, Washington, DC, January.

PREDICTION OF FLOW GENERATED BY TURBULENT OFFSET JETS

KAMIL H. M. ALI, OTHMAN A. KARIM AND BRIAN A. O'CONNOR
Department of Civil Engineering, University of Liverpool, L693BX, UK.

ABSTRACT
The paper gives predicted flowfields downstream of two-dimensional offset jets. results were obtained for flat beds as well as for scour holes of various sizes. The FLUENT computer package was used for this purpose. The effects of various turbulence models were investigated. The numerical solutions were compared with various experimental results obtained by many researchers in this field.

INTRODUCTION
Prediction of local scour holes that develop downstream of hydraulic structures plays an important role in their design. Excessive local scour can progressively undermine the foundation of the structure. Because complete protection against scour is too expensive, generally, the maximum scour depth and the upstream slope of the scour hole have to be predicted to minimize the risk of failure. The localized scour phenomenon has been the subject of extensive investigations by many researchers and numerous literature exists for scour caused by 2- and 3- dimensional turbulent jets. Most of the studies conducted on scour have been empirical because of the complexity of the physical processes (see Rouse (1939), Laursen (1952), Rajaratnam and Beltaos (1977), Ali and Lim (1986). Far less research has been conducted on offset and re-attached jets (see Ali et al (1991, 1992); Rajaratnam, N. and Subramanya, N. (1986) and Kumada et al (1973).

THE FLUENT PACKAGE
FLUENT is a general-purpose computer program for modelling complicated flow fields. It incorporates techniques based on fundamental principles for simulating a wide range of fluid flow problems (Fluent Inc. 1993). FLUENT/BFC is the primary set-up package for modelling complex geometries. FLUENT uses a finite difference numerical procedure to solve the fundamental equations governing fluid flow.

Three turbulence closure models are incorporated in FLUENT: (a) The standard $k - \epsilon$ model; (b) The Reynolds Stress Model (RSM); (c) The Renormalization

Group (RNG) model. In the present study, an evaluation is made of these models. The aim of this paper is to verify the effectiveness of the numerical procedure available in FLUENT in predicting the flow of offset jets. For that purpose velocity measurements carried out by Ali and Walley (1992) were used. The present analysis was divided into two sections:
(a)   Flat rigid bed - deep submergence (Rajaratnam and Subramanya, 1968);
(b)   Scour hole - deep submergence ((Ali and Walley (1992), t = 4770 mins.).

COMPARISON BETWEEN FLUENT'S PREDICTIONS AND RELEVANT EXPERIMENTAL RESULTS

Figure 1 shows the various flow characteristics of an offset jet. The choice of geometry setup and grid system used for the offset with a scour hole is shown in Figure 2. Figures 3 and 4 show predicted velocity profiles for one of the experiments of Rajaratnam and Subramanya (1968). It was noticed that although the general pattern of the flow was in good agreement, the predicted FLUENT results, however, underpredicted the values of the maximum velocity and underestimated the "pull" of the solid boundary. This resulted in a flatter curve of the maximum velocity profile near the impingement region (Figure 1). Location of the intersection of the dividing streamline with the offset-jet's downstream channel, i.e. the re-attachment point, X_A, (Figure 1) is usually taken at the position where the wall static pressure is a maximum (Kumada et al (1973)). Comparison was made of the predicted and measured re-attachment length ratio for the offset jet. In most cases, FLUENT's results were smaller than the measured values by 2 - 25%. The RNG and k - ϵ models gave better estimation compared to the RSM and a Higher Order Discretization scheme. To define a trajectory for the offset jet is very useful in theoretical work. The locus of the position of the maximum velocity was chosen as the reference streamline. Using the results of 6 different FLUENT runs, a least square fit was performed and the following coefficients were obtained.
C_0 = 0.9970; C_1 = -0.1949; C_2 = 0.3269;
C_3 = -1.0096; C_4 = 0.4807
The equation of profile is given by:
$$y = C_0 + C_1 x + C_2 x^2 + C_3 x^3 + C_4 x^4$$
where: $y = Y_m/h$ and $x = X/X_A$.
Figure 5 shows a comparison between the profile predicted by FLUENT and the curve obtained by Salehi (1988). The later was based on the results of many researchers. Figures 6 and 7 show experimental velocity distributions obtained by Ali and Walley (1992) in scour holes produced by horizontal jets positioned at elevations of 10 and 20cm above the original horizontal bed. The scour holes were produced after 4770 minutes (asymptotic state). Figures 8 and 9 show the corresponding velocity vectors predicted by FLUENT. Clearly, there is reasonable agreement between the experimental and the FLUENT results.

CONCLUSIONS

1. FLUENT results for simulating offset jet flow fields were compared with relevant experimental results for rigid and scoured beds. In general, the FLUENT results exhibited good qualitative agreement throughout the flow fields. In the rigid bed cases, the predicted attachment lengths were reasonably accurate to within 10% when the RNG and k - ϵ models were used. However, large errors resulted on using the RSM model. Numerical instability occurred on using the Higher Order scheme.
2. For the scoured beds produced by offset jets, reasonable agreement was obtained between FLUENT and the relevant experimental results.

REFERENCES

1. Ali, K. H. M. and Lim, S.Y. (1986), "Local scour caused by submerged wall-jets". Proc. Instn. Civ. Engrs., London, England, 81(2), pp 607-645.
2. Ali, K. H. M. and Salehi-Neyshabouri, A. A. (1991), "Localized scour downstream of a deeply-submerged horizontal jet", Proc. Instn. Civ. Engrs., Part 2, pp 1 - 18.
3. Ali, K. H. M. and Walley, P. (1992), "Local scour caused by offset jets", Internal Report, Dept. of Civil Engineering, Univ. of Liverpool.
4. Fluent Incorporated (1993), "Fluent Users Guide", Version 4.2.
5. Kumada, M., Mabuchi, I. and Oykawa, K., "(1973), "Studies in heat transfer to turbulent jets with adjacent boundaries", Bull. of JSME., Vol. 16, pp 1712 - 1722.
6. Laursen, E. M. (1952), "Observations on the nature of scour", Proc. of 5th Hyd. Conf., Bull. 34, Univ. of Iowa, Iowa, pp 179 - 197.
7. Rajaratnam, N. and Subramanya, N. (1968), "Plane turbulent re-attached wall jets", ASCE, J. of the Hyd. Div. Vol. 94, HY1, pp 95 - 112.
8. Rajaratnam, N. and Beltaos, S. (1977), "Erosion by circular turbulent wall jets "J. Hydr. Res., 15(3), pp 277 - 289.
9. Rouse, H. (1939), "Criteria for similarity in the transportation of sediment", Proc. Hyd. Conf. Studies in Engineering Bull., Univ. of Iowa, pp 39-49.
10. Salehi-Neyshaboury, A. A. (1988), "Impingement of offset jets on rigid and movable beds", Ph.D Thesis, University of Liverpool.
11. Wu, S. and Rajaratnam, N. (1995), "Free jets, submerged jumps and wall jets", J. of Hydr. Res. Vol. 33, No. 3, pp 197 - 212.

TURBULENT OFFSET JETS

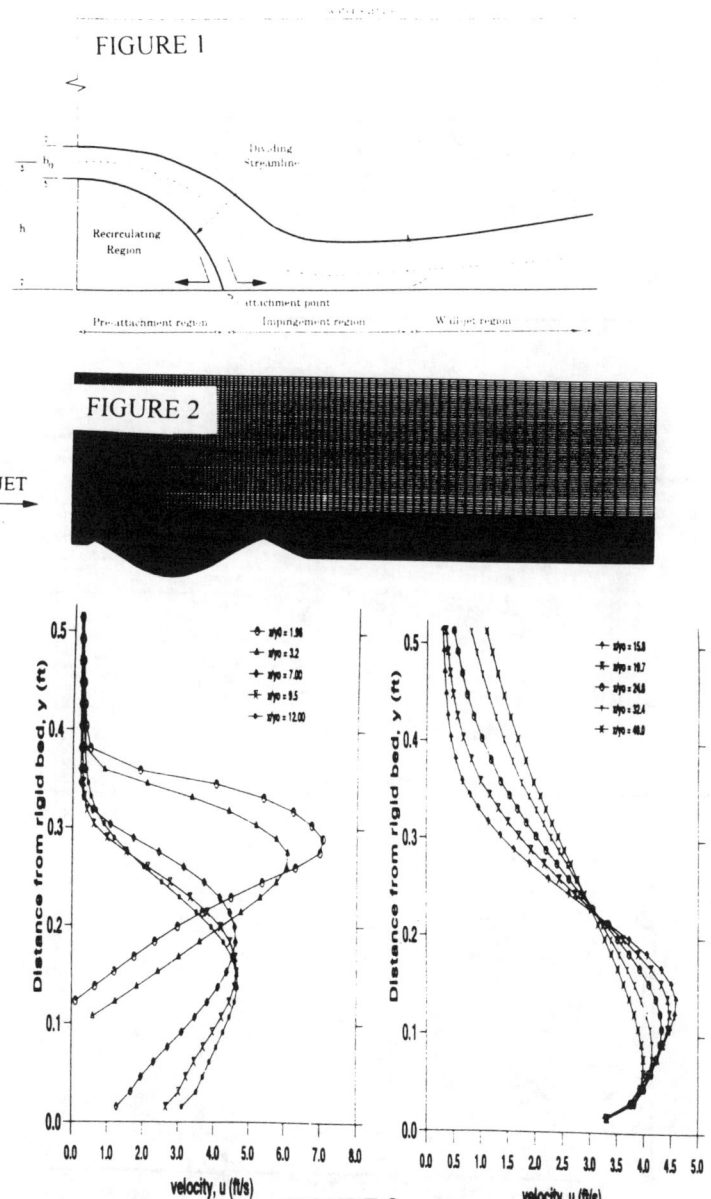

FIGURE 1

FIGURE 2

JET

FIGURE 3

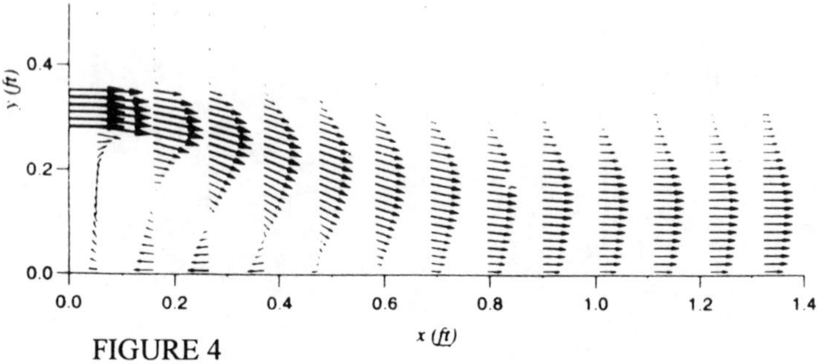

FIGURE 4

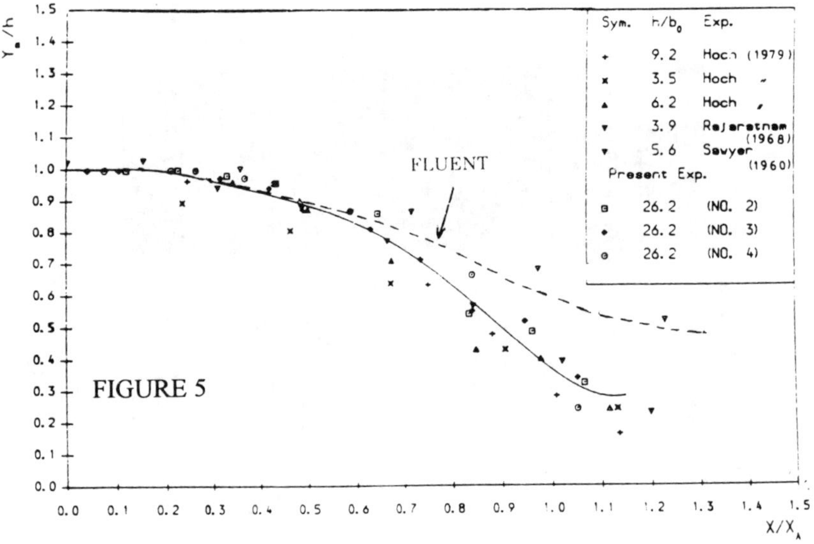

FIGURE 5

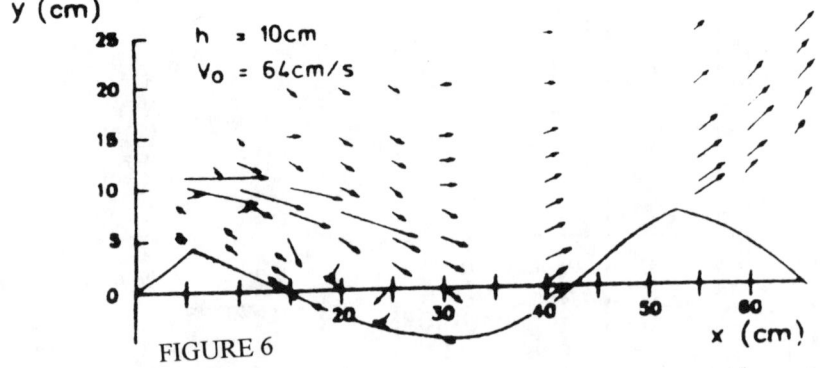

FIGURE 6

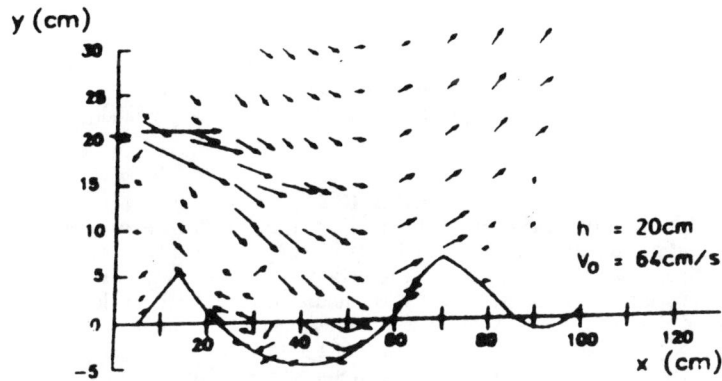

FIGURE 7

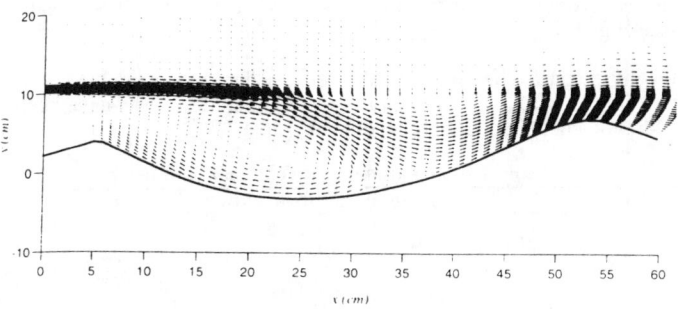

FIGURE 8

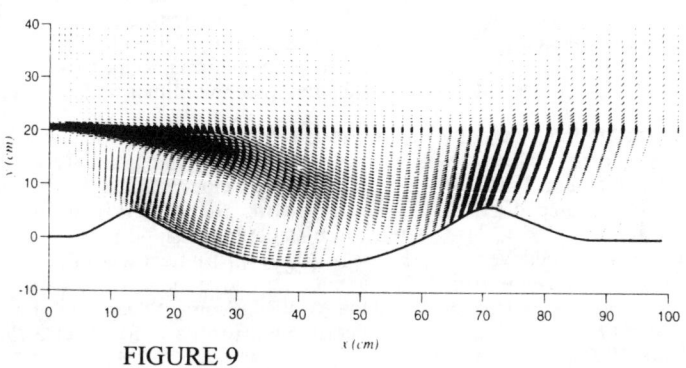

FIGURE 9

EXPERIMENTS ON FLOW UPSTREAM OF A CYLINDER

W.H. GRAF and B. YULISTIYANTO
Laboratoire de Recherches Hydrauliques,
Ecole Polytechnique Fédérale, Lausanne, Switzerland.

ABSTRACT

The flow in front of a cylinder positioned normal to the flow in an open channel has been investigated experimentally. An Acoustic Doppler Velocity Profiler (ADVP) was used to obtain instantaneously the three directions of the mean velocity in the stagnation plane. The vorticity of the flow field was calculated. Two types of flow were studied. Results of the experiments show that a horseshoe-vortex system is established; it is made up of a zone of positive vorticity forming a fully-developed vortex and a zone of negative vorticity, possibly being part of another vortex. The system develops itself in the corner at the base of the cylinder.

INTRODUCTION

If a cylinder is set in an open channel, the flow upstream will undergo a separation of the boundary layer and rolls up to form a system of vortices, known as the horseshoe vortex (see Fig. 5).

This study investigates experimentally the velocity distributions in the vertical symmetry (stagnation) plane of a flow upstream from a cylinder. Two tests have been performed, whose hydraulic characteristics are given with Table 1.

Table 1 Flow variables and channel parameters.

	cylinder		channel : uniform flow $B = 2.0$ [m]								
Test	D [m]	Re_D 10^5 [-]	Q [m^3/s]	S_o 10^{-4}[-]	n [m$^{-1/3}$s]	U_∞ [m/s]	h_∞ [m]	B/h [-]	u_{*r} [m/s]	Re_h 10^5 [-]	Fr [-]
1	0.22	1.48	0.248	6.25	0.012	0.670	0.185	10.8	0.029	1.24	0.5
2	0.22	0.95	0.149	2.80	0.012	0.430	0.173	11.6	0.021	0.74	0.33

INSTALLATION AND MEASURING EQUIPMENT

The experiments were performed in a 43.0 [m] long, 2.0 [m] wide and 1.0 [m] high tilting flume. The working section of the flume was positioned at $x = 16.0$ [m] downstream from the entrance, where a cylinder, having a diameter of $D = 22.0$ [cm] and a height of H = 0.5 [m], was installed, positioned normal to the flow. The uniform flow was fully developed and may be considered to be two-dimensional ($B/h > 5$).

An Acoustic Doppler Velocity Profiler (ADVP), developed at our laboratory (see *Lhermitte et Lemmin*, 1994), was used to measure the velocities. This instrument measures instantaneously the vertical profiles of the velocity in the three directions as well as their turbulence parameters. In this work, the ADVP-instrument using a measuring frequency of 12 Hz, operates in bistatic and tristatic mode using two and three transducers, respectively, at a time. The ADVP-instrument was installed at a fixed location below the channel bed. This obliged us to move the position of the cylinder, mounted on a moveable carriage, around the ADVP-instrument. Detailed information on the experimental installation and the experimental results is found elsewhere (see *Yulistiyanto*, 1997).

MEASUREMENTS IN THE UNIFORM APPROACH FLOW

To verify the uniformity of the approach flow, measurements were done without the cylinder, which are shown in Figs. 1 for both tests.

The friction velocity is calculated using the energy gradient, supposed equal to the slope bed for uniform flow, $S_f = S_o$, using $u_{*e} = \sqrt{g h S_f}$. Subsequently the Weisbach-Darcy coefficient is calculated using $f = 8(u_{*e}/U)^2$. The semi-empirical equation of Colebrook and White (see Graf & Altinakar, 1993, p.75), with - a = 11.55 and b = 3.86 - is then used to determine the uniform roughness, k_s.

The roughness Reynolds number, $(k_s u_*/\nu)$, can now be calculated. The calculated values are summarized in Table 2. It can be concluded that the channel bed is in the transitional regime (see Graf & Altinakar, 1993, p. 53).

Table 2 Uniform flow parameter and friction velocities.

Test	U [m/s]	u_{*e} [m/s]	f [-]	k_s [-]	$\dfrac{k_s u_{*e}}{\nu}$	u_{*r} [m/s]	u_{*cl} [m/s]	$\dfrac{u_{*cl}-u_{*r}}{u_{*r}}$	$\dfrac{u_{*e}-u_{*r}}{u_{*r}}$	B_{tr} [-]	Π [-]
1	0.67	0.034	0.0202	0.00043	14.5	0.029	0.0310	6.0%	13%	9.98	0.15
2	0.43	0.022	0.0206	0.00038	8.4	0.021	0.0207	4.5%	-9%	9.77	0.15

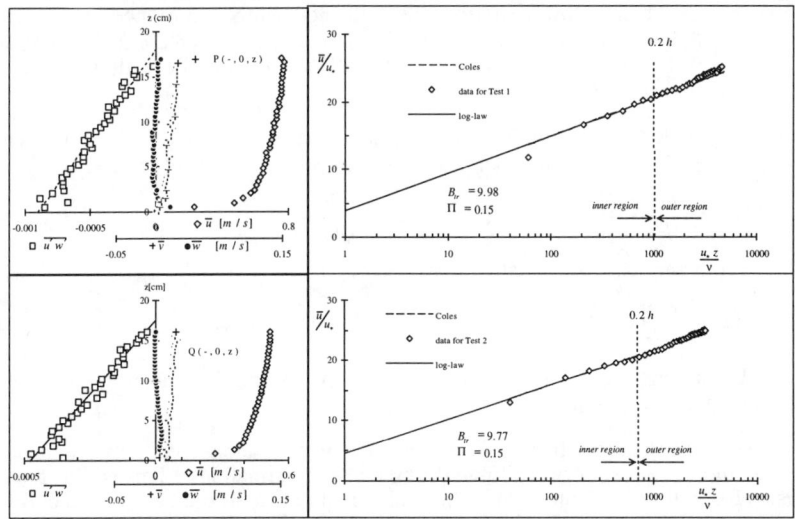

Figs.1 Mean velocities and Reynolds stress distributions in uniform flow for both tests.

The measured vertical profiles of the mean longitudinal, $\bar{u}$, vertical, $\bar{w}$, and transversal, $\bar{v}$, velocities as well as the Reynolds stress, $\overline{u'w'}$, for both tests are shown in Figs. 1.

The dimensionless longitudinal velocity profiles are also presented in Figs.1. Data in the inner region were evaluated using the logarithmic velocity distribution (see Yalin, 1992, pp. 9-10); this allows to calculate the u_{*cl}-values and B_{tr}-values, by trial and error. Data in the outer region were evaluated using the Coles' velocity defect law; this error. Data in the outer region were evaluated using the Coles' velocity defect law; this allows to calculate the Π-values. The resulting values are presented in Table 2; they compare reasonably well with values cited in the literature (see Graf & Altinakar, 1993, p. 52 and p. 58).

The measured Reynolds stress distribution allowed also to determine the friction velocity, u_{*r}. The bottom shear stress, τ_o, expressed as the friction velocity, $u_* = \sqrt{\tau_o/\rho}$, was thus obtained in three independent ways. Comparison among these values (see Table 2) is considered to be rather good.

MEASUREMENTS UPSTREAM OF THE CYLINDER

The measurements were made at 10 or 11 different stations at distances from the wall of the cylinder, $D/2 = 11.0$ [cm], in the range of $r = 1.0$ [cm] to $r = 33.0$ [cm]. Some results of the measured vertical profiles of the mean longitudinal, $\bar{u}$, vertical, $\bar{w}$, and transversal, $\bar{v}$, velocities for Test 1 (they are similar for Test 2) are shown in Fig. 2. The two-dimensional velocity distribution is shown in Figs. 3 for both tests. Each plot presents velocity distributions for a given position P (r, α, z), where r is the distance from the cylinder to P, α the radial direction at P, and z the vertical distance.

Approaching the cylinder one observes (see Fig. 2 and Figs. 3) the following: The longitudinal velocities, $\bar{u}$, decrease; their distribution becomes more uniform and close to the bed a reversed flow is noticeable with increasing importance. The vertical velocities, $\bar{w}$, being almost not existent far from the cylinder, increase in importance as downward flow, being especially strong in the lower region of the flow depth. The transversal velocities, $\bar{v}$, are of no importance.

From these velocity distributions, it is evident that a vorticity, ω_y, exists, which superposes itself upon the approach flow vorticity, ω_y^∞, to be calculated as :

$$\omega_y = \frac{\partial \bar{u}}{\partial z} - \frac{\partial \bar{w}}{\partial x} \quad ; \quad \omega_y^\infty = \frac{\partial \bar{u}}{\partial z} - 0 \tag{1}$$

Subsequently the space-averaged vorticity can be calculated as being :

$$\Omega_y = \frac{1}{A} \iint \omega_y \, dx \, dz \quad \text{with} \quad \Gamma = \iint \omega_y \, dx \, dz \tag{2}$$

where A represents the area of the region of the vorticity and Γ is the circulation, being the intensity of the vorticity.

From the velocity distributions, it is possible now to compute the vorticity components, ω_y and ω_y^∞ (used are forward finite difference approximations). This is illustrated in Figs. 4 for both tests. Note, that only a subdomain close to the base of the cylinder, $1 < r$ [cm] < 11 and $0 < z$ [cm] < 4, has here been evaluated.

On Figs. 4 it can be seen that zones of vorticities of different strength are noticeable. However, a primary vortex shows a pronounced design. The location, C, of the maximum vorticity, ω_y^{max}, is given for :

Test 1: $\omega_y^{max} = 96$ [s^{-1}]   at C ($r/D = 0.14$; $z/D = 0.028$)

Test 2: $\omega_y^{max} = 26$ [s^{-1}]   at C ($r/D = 0.25$; $z/D = 0.045$)

The horizontal and particularly the vertical components of the turbulence intensity - measured, but not shown here - have become very important.

FLOW UPSTREAM OF CYLINDER

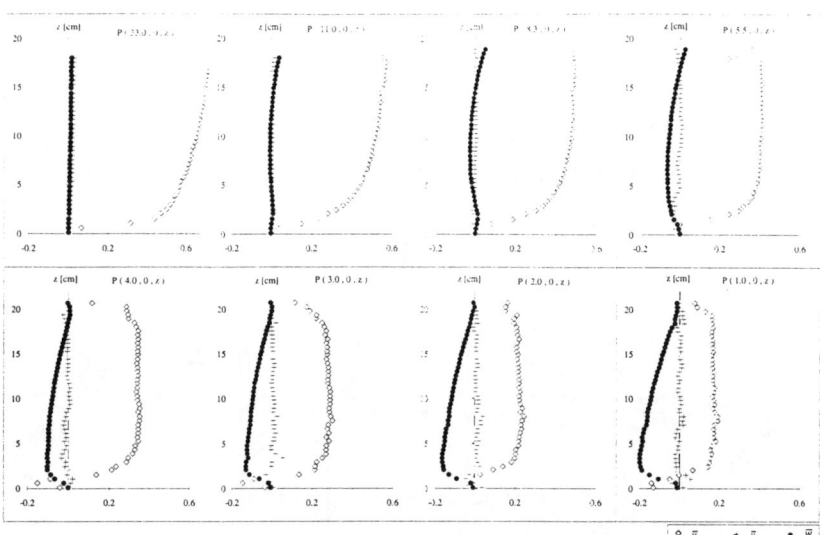

Fig. 2 Velocity profiles upstream from the cylinder for Test 1.

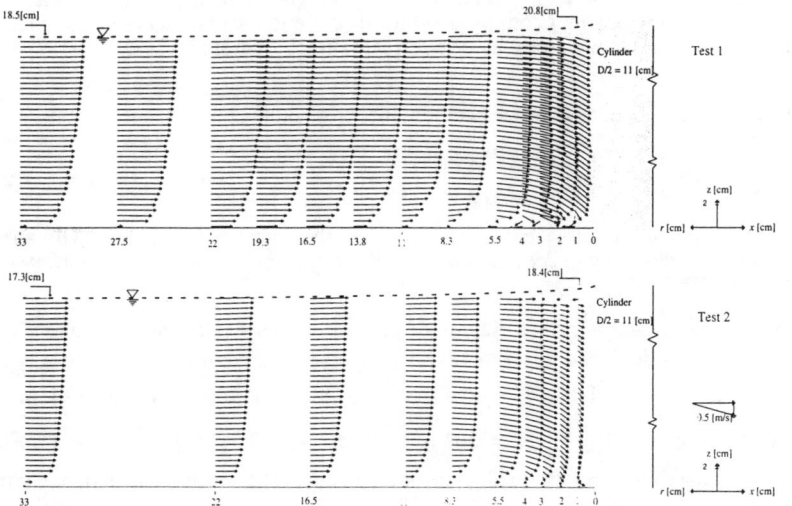

Figs. 3 Velocity vectors upstream from the cylinder for both tests.

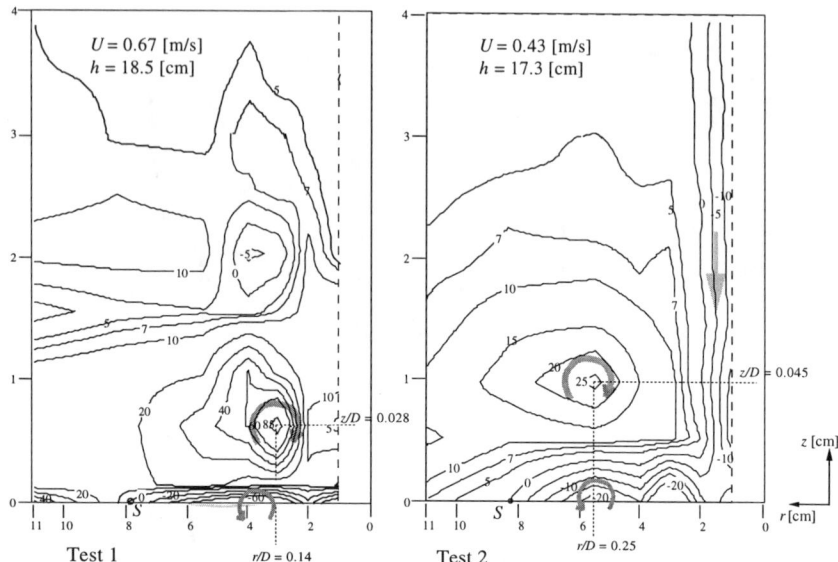

Figs. 4 System of vortices for both tests.

Underneath this clockwise vortex of positive vorticity, stretching around the cylinder, is what appears to be part of a smaller secondary vortex of negative vorticity. It is apparently fed by a current coming downwards close to the cylinder (see Fig. 3) and is subsequently washed away in the downstream direction around the cylinder (see Fig. 5).

The line of $\omega_y = 0$ represents a separation line, originating at the channel bed at:

Test 1: $S \, (\, r/D = 0.36 \, ; \, z/D = 0)$

Test 2: $S \, (\, r/D = 0.38 \, ; \, z/D = 0)$

Some of these observations are rather similar to the ones from tests done with airflow (see *Baker*, 1980, *Agui et Andreopoulos*, 1992, *Eckerle et Awad*, 1991 and *Monnier et Stanislas*, 1996).

The space-averaged vorticity, Ω_y, and circulation, Γ, has been calculated for the primary - it was delimited by $\omega_y > 7.0 \, [\text{s}^{-1}]$ - and secondary vortex, being for :

Test 1 : Ω'_y = +19.7 [s^{-1}], Γ' = 322 [cm^2/s] ; Ω''_y = -35.0 [s^{-1}], Γ'' = -17 [cm^2/s]

Test 2 : Ω'_y = +12.4 [s^{-1}], Γ' = 201 [cm^2/s] ; Ω''_y = -8.5 [s^{-1}], Γ'' = -41 [cm^2/s]

The approach flow vorticity (in absence of the cylinder) has been computed, being for :

Test 1 : Ω_y^∞ = +11.7 [s^{-1}], Γ^∞ = 150 [cm^2/s]

Test 2 : Ω_y^∞ = +9.0 [s^{-1}], Γ^∞ = 116 [cm^2/s]

This shows the to-be expected result, that the vorticity of the flow in the presence of a cylinder, Ω_y, is considerable increased.

CONCLUDING REMARKS

In the presence of a cylinder in an open-channel flow, a horseshoe-vortex system is formed at the base of the cylinder.

i) The horseshoe-vortex system consists of a fully-developed large vortex, V_1, driving a less-developed small vortex, V_2, situated underneath (see Fig. 5). It is not excluded - but not possible to see from our measurements - that other vortices do exist.

ii) The strength and location of the horseshoe-vortex system depend in our study on the approach flow velocity, U; it could be parameterized by the flow Reynolds number, Re_h. The system is stronger and closer to the base for the higher velocity.

iii) This horseshoe-vortex system plays an important role, if the bed of the channel is mobile. It is probable that the strong counter vortex does the erosion, while the primary vortex does the transportation of sediments.

iv) This study will be extended to an investigation of the flow around the cylinder (see *Yulistiyanto*, 1997)

The help in using the ADVP-instrument by Dr. U.Lemmin is appreciated.

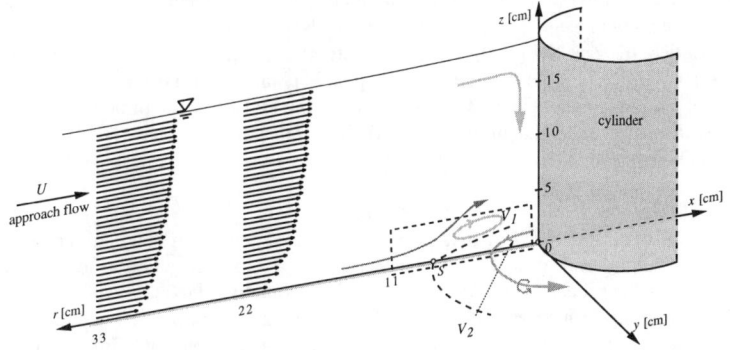

Fig. 5 Scheme of the horseshoe-vortex system, investigated.

REFERENCES

AGUI J.H et ANDREOPOULOS J. (1992) : " Experimental investigation of a three-dimensional boundary layer flow", *J. Fluids Engi.*, Vol. 114, pp. 566-576.

BAKER C.J. (1980) : " The turbulent horseshoe vortex", *J. Wind Engineering and Industrial Aerodynamics,* Vol. 6, pp. 9-23.

ECKERLE W.A. et AWAD J.K. (1991) : " Effect of freestream velocity on the three-dimensional separated flow region in front of a cylinder ", *J. Fluids Engineering*, Vol. 113, pp. 37-44.

GRAF W.H. et ALTINAKAR M.S.(1993) : *Hydraulique fluviale;* Tome 1, Presses Polytechniques et Univ. Romandes, Lausanne, CH

LHERMITTE R. et LEMMIN U. (1994) : " Open channel flow and turbulence measurement by high resolution Doppler ADVP ", *J. Atmos. Oceanic Tech.*, Vol. 11, pp. 1295-1308.

MONNIER J.C. et STANISLAS M. (1996) : " Study of a horseshoe vortex by LDV and PIV ", ERCOFTAC Bulletin, No. 30, pp19-24.

YALIN M.S. (1992) : *River Mechanics*, Pergamon Press, Oxford, GB.

YULISTIYANTO B. (1997) : *Flow around a cylinder installed in a fixed-bed open channel* , PhD dissertation 1631, Ecole Polytechnique Fédérale, Lausanne.

NUMERICAL ANALYSIS OF THREE-DIMENSIONAL TURBULENT FLOWS AROUND SUBMERGED GROINS

J. PENG, Y. KAWAHARA and N. TAMAI
Department of Civil Eng., University of Tokyo
Hongo, Bunkyo-ku, Tokyo, JAPAN

ABSTRACT

This paper presents three-dimensional numerical analysis of turbulent flows around submerged groins. A finite volume method is employed to discretize the governing equations. A k-ε model modified by Zhu-Shih is used to simulate the turbulent momentum transport. The numerical results show the flow pattern at the tip and behind the groins possess a strongly three dimensional feature. Reynolds stress distribution and the budget of the turbulent kinetic energy are discussed for the understanding of the turbulent transport processes. The bed shear stress distribution which is directly related to the transport of bed material has also been investigated.

INTRODUCTION

Groins or spur dikes along river banks have been constructed to protect bank erosion at high water depth or flood and to control river bed topography for navigation at low water depth. Recently groins have been received more attention from the standpoint of ecosystem. The hydraulic conditions such as velocity, water depth, bed shape and bed material around groins are so diverse to provide the ecosystem with suitable habitat. Hence it is important to design groins to meet the different kinds of need which do not go well with each other. However our knowledge on flood flow around groins is limited, partly because it is difficult to measure the velocity and bed topography and partly because the flow field shows highly three-dimensional nature depending on the shape of groins and river reach. Few three dimensional calculation has been applied to the flow around submerged groins, while there is a strong demand for calculation method to predict such flows. Therefore it is the objective of this study to develop a three-dimensional numerical code and to make clear the flow characteristics around groins.
In this paper, we simulate numerically turbulent flow around groins. The simulations are carried out for flood flow around five successive groins in a flat straight river with rough wall. The three dimensional flow feature behind groins and the development of the secondary flow around groins are clearly marked by the numerical results. The analysis of Reynolds stress distribution reveals the turbulent model is indispensable in the region around groins. The effect of the groin interval on the velocity and bed shear stress distribution is also examined.

NUMERICAL SIMULATION

GOVERNING EQUATIONS
For three-dimensional incompressible steady turbulent flows the governing equations are expressed as follows:

FLOWS AROUND SUBMERGED GROINS   245

$$\frac{\partial U_i}{\partial x_i} = 0 \tag{1}$$

$$U_j \frac{\partial U_i}{\partial x_j} = g_i - \frac{1}{\rho}\frac{\partial P}{\partial x_i} + \frac{\partial}{\partial x_j}\left(v\frac{\partial U_i}{\partial x_j} - \overline{u_i u_j}\right) \tag{2}$$

where U_i is the mean velocity and P pressure. Reynolds stresses $-\overline{u_i u_j}$ can be presented as:

$$-\overline{u_i u_j} = v_t\left(\frac{\partial U_i}{\partial x_j} + \frac{\partial U_j}{\partial x_i}\right) - \frac{2}{3}k\delta_{ij} \tag{3}$$

$$\text{with } v_t = c_\mu \frac{k^2}{\varepsilon} \tag{4}$$

The transportation equations for turbulent kinetic energy k and its dissipation rate ε can be written as:

$$U_j \frac{\partial k}{\partial x_j} = \frac{\partial}{\partial x_j}\left[\left(v + \frac{v_t}{\sigma_k}\right)\frac{\partial k}{\partial x_j}\right] + \Pr od - \varepsilon \tag{5}$$

$$U_j \frac{\partial \varepsilon}{\partial x_j} = \frac{\partial}{\partial x_j}\left[\left(v + \frac{v_t}{\sigma_\varepsilon}\right)\frac{\partial \varepsilon}{\partial x_j}\right] + \frac{\varepsilon}{k}\left(C_{\varepsilon 1}\Pr od - C_{\varepsilon 2}\varepsilon\right) \tag{6}$$

$$\Pr od = v_t\left(\frac{\partial U_i}{\partial x_j} + \frac{\partial U_j}{\partial x_i}\right)\frac{\partial U_i}{\partial x_j} \tag{7}$$

In the standard k-ε model, the model coefficients contained in the above equations are assumed to be constant such as: c_μ=0.09 $c_{\varepsilon 1}$=1.44 $c_{\varepsilon 2}$=1.92 σ_k=1.0 σ_ε=1.3. These constant parameters are determined from a set of experiments for simple turbulent flows. But it is well recognized that the model does not perform well for separated flows. A lot of efforts have been made to improve the predicting ability of the standard k-ε model, such as a RNG modified k-ε model derived by Speziale and Thamgam[1]. In the present paper we utilize a modified model proposed by Zhu-Shih to determine the model coefficients. In this approach a formula for c_μ is derived based on a realizability analysis of turbulent normal stresses. It is written as:

$$c_\mu = \frac{2/3}{5.5 + \eta}, \quad \text{where } \eta = \frac{Sk}{\varepsilon}, \quad S = \left(2S_{ij}S_{ij}\right)^{1/2}, \quad S_{ij} = \frac{1}{2}\left(\frac{\partial U_i}{\partial x_j} + \frac{\partial U_j}{\partial x_i}\right) \tag{8}$$

Other coefficients are the same as those in the standard k-ε model. The Zhu-Shih modified k-ε model has been found to work better than the standard k-ε model for backward facing step flow[2].

NUMERICAL METHOD
The equations are discretized by the finite volume method in a staggered grid system in which grids near the groin are packed. A consistently formulated QUICK scheme suggested by Hayase et al[3] is used for the discretization of convection-diffusion terms in the three component momentum transport equations. The SIMPLE algorithm is used to obtain the converged solution and we set the convergence criterion as the maximum normalized residual of all dependent variables less than 10^{-4}.

BOUNDARY CONDITIONS AND INITIAL CONDITIONS
Boundary conditions were specified at inlet, outlet, free surface and wall boundaries. Known distributions of velocity and turbulence quantities were imposed at the inlet and no-gradient

condition at the outlet. Free surface was treated as a symmetry plane. Along the wall boundaries including bed and side wall of river and all faces of groins the wall function approach was used to bridge the viscous sublayer in order to alleviate the fine grids requirement due to the high velocity gradient near the wall region. Initial conditions were prepared by solving the equations of continuity and momentum in the absence of viscosity effect to reduce the total computational time.

RESULTS AND DISCUSSIONS

MODEL VALIDATION
The present numerical code has been validated by comparing a numerical simulation of flow around a submerged groin with the experimental data presented by Tominaga et al [4]. The computed mean velocity is in good agreement with the experimental results[5].

FLOWS AROUND GROINS
The calculation domain under consideration of flows around five groins is schematically shown in Fig. 1. Inlet is set at upstream of the first groin and outlet at downstream away from the fifth groin. The coordinate z represents the distance from the bed and x, y the streamwise and spanwise directions respectively. Roughness height of river and groins is 5cm. Computational conditions of two cases are shown in Table 1, in which the quantity Q represents discharge, H water depth, D interval of groins and L_g length of groins.

Table 1. Computational cases (unit: m-s)

Case	1	2
Q	544.0	544.0
H	4.0	4.0
D	10.0	5.0
Lg	5.0	5.0
Grid Points	177(x) 42(y) 19(z)	128(x) 42(y) 19(z)

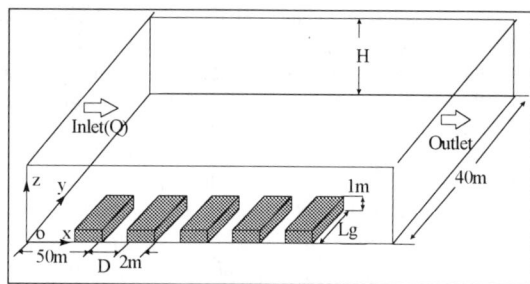

Fig.1 Schematic Diagram of flow around groins

Three dimensional flow feature around groins
Fig. 2 shows the three dimensional flow field between two groins. Velocity is normalized with U_m where U_m is the cross sectional mean velocity at inlet. For the purpose of clear view the grid size in z direction is enlarged in the figure. The top plane (z_5 =1.0m) is at the tip of groins and the bottom plane (z_1 =0.2m) is just near the bed. The region between the fourth and fifth groin is selected as the discussion area in the present section since here fully developed flows are reached. From this figure it is obvious that strongly three dimensional structure exists. Recirculation size at the back of the groin gets reduced gradually as the top surface of groin is approached. Consequently the reattachment length decreases from bottom to top planes. The location of recirculation center also varies in z direction. It moves from the tip part of groin near the bed wall toward the root of groin close the top surface plane. The reason for this is that flows over the submerged groin exert their influence on the separation flow behind groin with the recirculation flow motion suppressed gradually. In the upstream face of groin flows show upward motion because of the blockage effect of the groin.

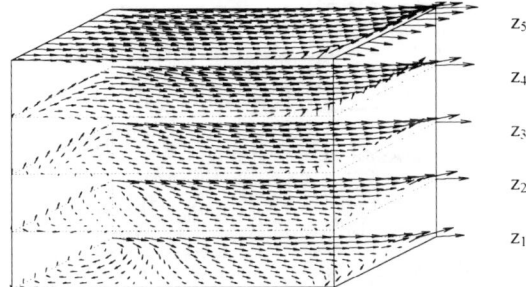

Fig. 2 Three dimensional flow field between groins.
(z_1=0.2, z_2=0.4, z_3=0.6, z_4=0.8, z_5=1.0m from the bed)

Secondary flow occurs due to the disturbance of groins. Fig. 3 shows the development of secondary flow in two cross sections. Section 1 is located in the tip area of the fourth groin and section 2 in the recirculation region behind this groin. The dimensionless V-W vector distribution in this figure is enlarged 10 times. In section 1, facing the upstream flow direction there are clockwise secondary recirculation flows above the groin. Behind the groin in the recirculation region the secondary flows go toward the bed-side corner near the bed wall and away the side back near the water surface. Separation flow, recirculation flow and secondary flow generate the streamline curvature around the groins.

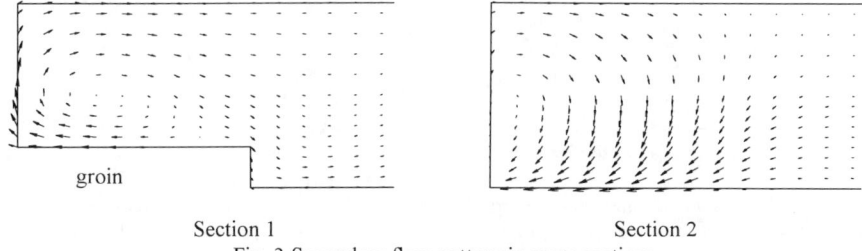

Section 1   Section 2
Fig. 3 Secondary flow pattern in cross sections

Turbulent flow characteristics

Turbulent transport processes is strongly problem dependent. The characteristics of turbulence of flows around groins are discussed based on the transport equation for the turbulent kinetic energy. Fig.4 gives the budget of turbulent kinetic energy over two cross sections (Section 1 and section 2 as mentioned above) near bed wall. The groin occupies in the y direction from y=0.0m to y=5.0m. Fig. 4(a) shows that in a small area very close to the tip range of the groin the convection and diffusion terms of turbulent energy are active while in the other range the turbulent production keeps balance with the turbulent dissipation. In the recirculation zone, shown in Fig. 4(b), the convection, diffusion, production and dissipation of turbulent energy all give some values in order. The convection of turbulent energy is negative due to the reversed flows. Again away from the groin the turbulent energy shows local equilibrium.

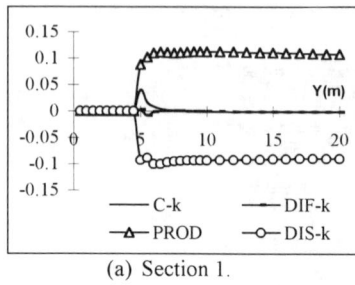
(a) Section 1.

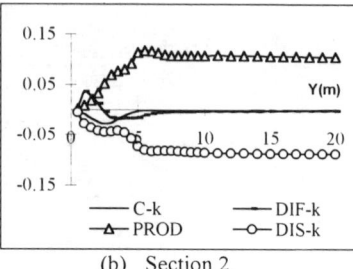
(b) Section 2.

Fig. 4 Budget of turbulent kinetic energy.

The k-ε model evaluates the influence of turbulent motion on the mean flow field through the Reynolds stresses. If the contribution of the Reynolds stresses is negligible for a certain problem then the turbulent model is unnecessary. Fig. 5 gives the $-\overline{u^2}$ component Reynolds stress gradient (normalized with $h/U_m^2 \times 10^3$ where h is the height of groin) distribution with a plane view for flows around the fourth groin. Both in the upstream and downstream of groins the Reynolds stress has a high gradient. It follows that it plays an important role in the mean velocity momentum transport in this region.

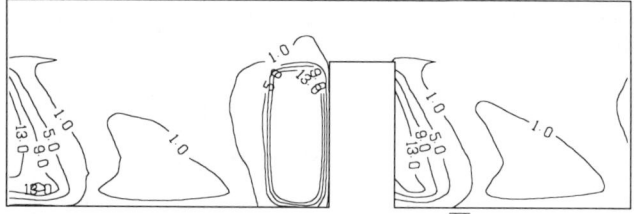

Fig. 5 Distribution of Reynolds stress gradient of $-\overline{u^2}$ in horizontal plane
(z = 0.8m above bed)

Previous discussion indicates that the turbulent motion dominates the flow behavior in the area around the groins. It is greatly necessary for a turbulent model to be used to predict the flows correctly. A relative simple turbulence model may be applicable to flows away from the groins.

The effect of groin interval

Interval between groins is an important parameter in the designing work of groin hence it is of practice interest to know the effect of groin interval on the flow field and bed shear stress distribution. Separation flow over the upstream groin may reattach the river bed and the shear stress along the back may recover the large value if the groins are spaced far apart. This is not a suitable option for the groin design since the primary function of groins is the providing of streamback protection. Too closed arrangement of groins is also not recommended because the groin system works less efficiently thus costs more. The bed shear stress distribution (expressed in $C_f \times 10^3$, $C_f = \tau_b /0.5\rho U_m^2$) shows different pattern in two cases (Fig. 6). In Case 2 when the interval of groin is small the individual bed shear stress pattern tends to overlap and merge into one which points the less efficient use of groin system. The bed shear stress takes the large value at the tip region of groins and small value between groins in both cases. The distributions of the calculated bed shear stress qualitatively agree with the field study on the river bed contour map after groins were constructed[6]. From the viewpoint of sediment transport, this pattern of bed shear stress distribution results in local scour at the tip areas of groins and deposition behind or between groins.

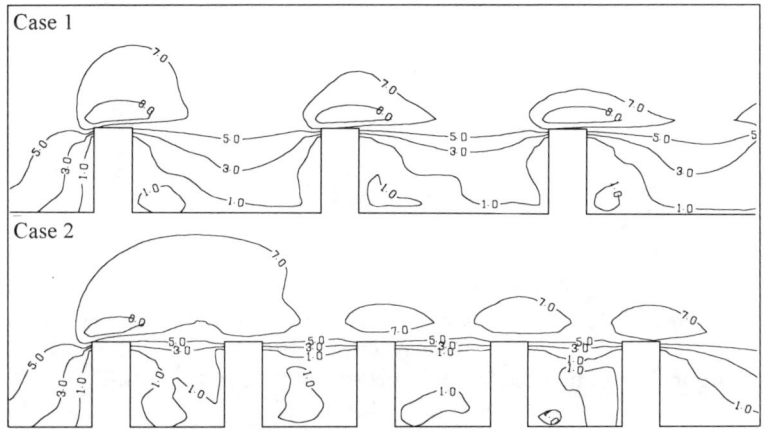

Fig. 6 Bed shear stress distribution.

CONCLUSIONS

A numerical study of three dimensional turbulent flows around groins with different groin intervals has been conducted. The three dimensional flow field and turbulent transport processes are discussed in detail which can be summarized as follows. (1) The numerical results capture strongly three dimensional flow feature behind groins. (2) Through the analysis of turbulent kinetic energy budget it is achieved that a reliable turbulent model is indispensable for flows around groins. (3) Calculated results for flow with groins in series reveal that the groin interval exerts considerable influence over the streamwise flow feature and the distribution of bed shear stress. (4) Bed shear stress takes large value at the tip area of groins where the local scour has been reported to occur in the field investigation and small value between groins connected to sediment deposition in this area.

REFERENCES

1). C. G. Speziale and S. Thamgam, Analysis of an RNG Based Turbulence Model for Separated Flows, NASA CR-189600, ICASE Rept. No.92-3.
2). J. Zhu and T. H. Shih, Calculations of Turbulent Separated Flows with Two-equation Turbulence Models, Computational Fluid Dynamics JOURNAL Vol.3. 343-354, 1994.
3). T. Hayase, J. A. C. Humphrey and R. Greif, A Consistently Formulated QUICK Scheme for Fast and Stable Convergence Using Finite-Volume Iterative Calculation Procedures, J. of Comput. Physics, 8, 108-118, 1992.
4). A. Tominaga and S. Chiba, Flow Structure around a Submerged Spur Dike, Proc. Of Annual Meeting of Japan Society of Fluid Mechanics, 317-318, 1996.
5). Y. Kawahara and J. Peng, Three-dimensional Numerical Simulation of Flood Flows around Groins, Prod. of the 2nd ACFD, Vol. 2, 539-544, 1996.
6). P. C. Klingeman, S. M. Kehe and Y. A. Owusu, Streambank Erosion Protection and Channel Scour Manipulation Using Rockfill Dikes and Gabions, Oregon State University, WRRI-98, 1984.

The Future of Irrigation

JAN VAN SCHILFGAARDE AND THOMAS J. TROUT
Director, Pacific West Area, USDA-Agricultural Research Service,
Albany, CA 94710
Agricultural Engineer, Water Management Research Lab, USDA-ARS,
Fresno, CA 93727

Irrigation has been practiced for over 6000 years. Irrigated agriculture has provided stable food supplies that enabled civilizations to grow and flourish. Today, over 220 million hectares of land are irrigated worldwide (Hoffman et al., 1990). In some countries, a large portion of the cultivated land is irrigated and large populations are heavily dependent on its production (Table 1). About 70% of irrigated land is in developing countries.

Even where the portion of the land irrigated is not large, a large portion of the high-valued food and fiber crops such as fruits and vegetables, sugar cane, and cotton, are irrigated. Irrigation reduces risks of crop failure and thus allows farmers to make the large investments necessary to produce high-value crops.

Although irrigation has been practiced for many years, much of the currently irrigated area has been developed since World War II. Rangley (1989) estimates that the total irrigated area has tripled in the last half century. This rapid expansion, motivated by the food needs of increasing populations, was made possible by advancing technology in pumps, water storage and conveyance, and from increasing public sector investment in irrigation, especially from international donor agencies.

The recent rapid expansion of irrigated area has slowed in the last decade, from over 2% per year to less than 1% per year. (Higgins et al., 1988). In a few areas, irrigated area is decreasing. Recent news stories describe environmental damages, resource depletion, social injustices, and economic failures attributed to irrigation development. Is the demise of irrigated agriculture at hand? We believe the answer is no, but it is important to evaluate past problems, propose solutions, and realistically assess the costs and benefits. The future of irrigation depends on the ability of irrigated agriculture to economically produce needed food and fiber with acceptable impacts on society and the environment.

Table 1. Countries with Major Irrigated Areas in 1986[1]

Country	Millions of Hectares Irrigated	Percent of Country's Cultivated Land Irrigated
India	55	33
China	47	48
Soviet Union	21	9
United States	19	10
Pakistan	16	77
Sub-total	**158**	
Indonesia	7.3	34
Iran	5.8	39
Mexico	5.3	21
Spain	3.3	16
Turkey	3.3	12
Thailand	3.2	16
Egypt	3.2	100
Japan	3.0	63
Italy	3.0	25
Romania	3.0	28
Total	**198**	

[1] from Hoffman et al., 1990

FACTORS AFFECTING IRRIGATION DEVELOPMENT AND SUSTAINABILITY

ECONOMICS

Irrigation development is expensive. Kortenhorst et al. (1989) using FAO data, estimated the cost for irrigation development in South Asia at $1600 per hectare and up, and in Africa, from $9500 per hectare. The on-farm cost of pressurized irrigation systems in the U.S. can easily exceed $1500 per hectare. Annual costs for irrigation labor, maintenance, and energy are also high. Thus economically viable irrigation depends upon producing significantly higher yields or higher value products than is produced without irrigation.

Irrigation can increase yields, and allow production of high-valued crops. In the United States, the average per hectare value of irrigated crops is two to three times greater than the value of non-irrigated crops (Bajwa et al, 1992). However, there are also many examples of irrigation developments, especially in developing countries, where output per hectare did not increase with irrigation (van Schilfgaarde, 1994). Reasons for poor response to irrigation include 1)growers that are unprepared or uninterested in converting to intensive, high-value agriculture, 2) undependable water supply to the irrigated farms, and 3) lack or expense of other inputs required for high yields.

High crop prices make investments such as irrigation easier to afford. However, in the recent past, world crop prices have not increased as much as most input costs, including irrigation. Several authors have predicted that increasing population will result in food shortages that will lead to dramatically increasing prices. This has not yet occurred. In fact, increasingly open international trade policies has, in some cases, created competition among producers resulting in lower prices. Many governments in the past have subsidized agriculture through artificially supporting inflated crop prices. Recent trade policy changes are reducing these subsidies.

Much of the costs of developing irrigation water supplies have been borne by governments. Although projects were often required to show positive net benefits, full costs were seldom borne by the irrigation water users. Many governments are experiencing increasing competition for their resources and are now less willing to subsidize irrigation.

WATER SUPPLY LIMITATIONS
In southern Europe, northern Africa, the Middle East, western U.S. and south and east Asia, much of the available water is being used, and the potential for further irrigation development is limited and such development would be increasingly costly. In fact, water supplies for irrigation are decreasing in some areas due to declining groundwater levels and increasing competition with urban, industrial, and environmental demands (National Research Council, 1996; CAST, 1996). Much of the potential for additional irrigation development is in more humid areas such as the sub-Saharan Africa (Kortenhorst et al., 1989) and the south-eastern U.S.

SOIL SALINIZATION

Irrigation water contains small amounts of dissolved materials. As the crop uses water, it leaves behind salts in the soil. If the salts are not periodically leached out of the plant root zone, they accumulate until they become toxic to plants. Management practices to maintain non-toxic salt levels are well established and include periodic leaching of salts out of the soil root zone and adequate drainage to carry away the leachate.

Addition of irrigation water to poorly-drained soils can result in a high groundwater table ("water logging"). High water tables restrict rooting depth and water evaporation from a high water table can result in soil salinization.

Waterlogging and salinization are major problems in many arid irrigated areas. Hundreds of thousands of hectares of irrigated land have gone out of production in central and south Asia due to salinization. Adequate drainage is often not provided with the irrigation development because of the cost. The rationale is that drainage can be added later when waterlogging and salinization become a problem. The reality is that after an irrigation development is completed, there is seldom money available to add drainage systems. Provision for and financing for drainage and drain water disposal must be an integral part of irrigation development.

TECHNOLOGICAL ADVANCEMENTS

Recent technological advances such as center pivot sprinkler systems and microirrigation help farmers apply irrigation more uniformly and efficiently. These pressurized systems also allow efficient irrigation on hilly or sandy lands that are not economical to irrigate with surface (gravity) irrigation. Much of the expansion of irrigated area in the U.S. in the last three decades has depended on sprinkler or microirrigation systems. These equipment-intensive systems require a large initial investment and good maintenance. The large majority of irrigation in the world is surface irrigation. There have not been many significant recent technological advances in surface irrigation, with the exception of laser land leveling.

Technological advances that increase irrigation efficiency can result in higher yields, lower irrigation costs, and less environmental damages. However, increased irrigation efficiency does not usually reduce water consumptively used, and thus does not make more water available for irrigation expansion or other uses. Only where irrigation drainage is unusable (ie: collects in saline groundwater or ponds) will increased efficiency result in increased water supply.

ENVIRONMENTAL CONSIDERATIONS

Water storage, diversion, and irrigation processes dramatically change the hydrology of a watershed, and impact the preexisting natural processes. Water flows in rivers are often greatly reduced in volume and drainage return flows from irrigation are nearly always lower quality than the diverted water. Societies, especially where people are well fed, are increasingly, and justifiably, concerned with these impacts. Irrigated agriculture can reduce but cannot eliminate its impacts on the environment and downstream water users. In parts of the U.S., water supply to irrigation will likely be reduced to replenish in-stream flows and drainage water from irrigated areas will likely be restricted. These trends will result in reduced water supplies and increased irrigation costs, and in some cases, will reduce irrigated area.

THE FUTURE?

Irrigated agriculture has existed for millennia, and will continue into the foreseeable future. Irrigation is critical to the basic food supply in many countries and is important in the supply of the wide variety of fruits and vegetables to which we have become accustomed. Whether in the U.S. or worldwide, the roughly 12% of agricultural land that is irrigated produces about 37% of the value of all crops. Irrigated agriculture is sustainable, up to the limit of available water supplies, when drainage and salinity management is adequate. However, even with good design and management, irrigation will impact downstream water uses, and society must be willing to accept these impacts as a cost of irrigated production.

Whether the world's irrigated area expands or diminishes in the future will depend upon the several factors listed above. The most important global factor is the demand for and price of the crops irrigated agriculture can produce. Food shortage and increased prices would provide the incentive for large investments in irrigation expansion such as were made in the middle half of this century. If yield increases through improved germplasms and crop management can keep up with the growing population, significant increase in irrigation is unlikely. Expert estimates of future food supply and demand vary widely. It is unlikely that yields can sustain the exponential increases that populations have had in this century. Irrigation is a critical input that can significantly increase yields and expand the cultivated area to help meet future food needs.

The importance of many of the other factors vary from region to region, and will result in growth in irrigation in some areas and declines in others. In arid areas where the water requirements are high and water supply is fully allocated, irrigation may decrease in response to groundwater declines, environmental demands, soil salinization, or the higher-valued water needs of urban and industrial growth. This trend is evident in the western U.S. In more humid areas such as sub-Saharan Africa and portions of South America, present irrigation development is small and the potential for supplemental irrigation is large. Future development will depend on creating irrigation systems that have acceptable environmental impacts, and that farmers can use to better their lives. The technology is available. As the demand for food increases, irrigation will be a critical component in meeting that demand.

REFERENCES

Bajwa, R.S., Crosswhite, W.M., Hostetler, J.E., and Wright, O.W. 1992. Agricultural irrigation and water use. Agricultural Information Bulletin No. 638. USDA-Economic Research Service.

Council for Agricultural Science and Technology (CAST). 1996. *Future of Irrigated Agriculture*. CAST, Ames, IA. 76 p.

Higgins, G.M., Dieleman, P.J., and Abernathy, C.L. 1988. Trends in irrigation development and their implications for hydrologists and water resources engineers. Journal of Hydrological Sciences. 33:43-59.

Hoffman, G.J., Howell, T.A., and Solomon, K.H. 1990. Introduction. In: *Management of farm irrigation systems*. Hoffman, G.J., Howell, T.A. and Solomon, K.H. (eds). Am. Soc. of Agricultural Engineers, St. Joseph, MI pp.5-10.

Kortenhorst, L.F., van Steekelenburg, P.N.G., and Sprey, L.H. 1989. Prospects and problems of irrigation development in Sahelian and sub-Saharan Africa. Irrig. And Drainage Systems. 3:13-45.

National Research Council. 1996. *A New Era for Irrigation*. National Academy Press. 203 p.

Rangeley, W.R. 1989. Influence of design on irrigation management. In: *Irrigation Theory and Practice*. Rydzewski, F.R. and Ward, C.F. (Eds). Pentech Press, London. pp. 18-25.

van Schilfgaarde, J. 1994. Irrigation - a blessing or a curse. Agricultural Water Management. 25: 203-219

PARTIAL IRRIGATION: A FUNDAMENTAL CHANGE

MARSHALL ENGLISH
Bioresource Engineering Dept., Oregon State University

ABSTRACT
Irrigation generally increases yields by 100% to 400%, and is expected to account for one half to two thirds of required increases in food production in the next few decades. A World Bank/UNDP study estimates that irrigation could be extended to an additional 110 million hectares to provide sufficient grain for 1.5 to 2.0 billion people. But this accelerating demand for irrigation will coincide with a rising concern for the environment and increasing economic competition for water. The convergence of these conflicting pressures which will force fundamental changes in irrigation practices, and those changes will probably include abandoning the most basic of traditional irrigation objectives, the objective of satisfying crop water demand.

INTRODUCTION
Recent decades have witnessed an increasing interest in the concept of partial irrigation, the deliberate under-irrigation of crops. Often referred to as deficit irrigation or supplemental irrigation, this technique has been practiced by individual farmers around the world for many years. Perhaps because it is fundamentally different from the traditional approach to irrigation it has received only peripheral attention from professionals involved in irrigation system design and management. However, as the pressures on remaining water supplies increase it will be necessary for us all to begin thinking in terms of deficit irrigation, rather than full irrigation as standard practice. Let me restate the point. Current practice treats full irrigation as the norm and deficit irrigation as a refinement to be practiced in special situations. The next two or three decades should see partial irrigation become the standard practice, with full irrigation as the special case.

FORCES FOR CHANGE
Economics will be a powerful force for change. Irrigation may increase the productivity of some lands dramatically, but it must also be recognized that agriculture is still a relatively low-value, low efficiency use of water. As such, it is often unable to compete economically for scarce water. Because cities and industries can afford to pay more for water and earn a higher economic return from water, it becomes "difficult for agriculture to prove that water used for irrigation is applied to good advantage in ensuring food security".[1]

Another of the dominant forces for change will be environmental concerns. Irrigation can profoundly alter the hydrology and other aspects of the environment of an area. Desert regions of the Pacific Northwest, for example, which once produced less than $15 per ha have increased their productivity to several thousands of dollars per ha under

---

[1] Quote taken from Web site of the World Food summit, 13-17 November, 1996, Rome.

irrigation. But at the same time, salmon runs that once proliferated have been brought to extinction in many watersheds of the region, ground water pollution from nitrates and other chemicals has made water unsafe for human consumption in some basins, and pesticide laden sediments are clogging natural streams, filling reservoirs and destroying fish habitat. Such water pollution can be regarded as a form of consumption, and these effects have brought politically powerful environmental interests into conflict with irrigated agriculture, intensifying the competition for water.

These pressures reflect a more fundamental issue, the limited supply of undeveloped or underutilized fresh water available to meet the needs of a world population that may increase by 45% within the next 30 years. Postel, et al (1996) have argued that by the year 2025 the world wide demand for fresh water will be approaching the limits of readily accessible supplies. Harald Fredericksen, in his keynote address for Sub-theme A, made a compelling argument that when climatic variability is taken into account, and when the inertia of political and social institutions are considered, much of the world is already facing a water crisis. Under such pressures, irrigated agriculture will inevitably be forced to increase the efficiency with which it uses water.

THE NATURE OF THE CHANGE

The anticipated change in irrigation practice will be a shift in the fundamental way we approach irrigation. Current objectives, reflected in standard engineering practice, are to irrigate to meet the crop demand for water. In other words, crop physiology determines the irrigation schedule. Partial irrigation represents a fundamental change in which the amount of water to be applied to a crop is based not on crop water demand but on the productivity of the water. That is, economics determine the irrigation schedule. The irrigator applies water until a point is reached at which the yield increase produced by the next increment of applied water is no longer sufficient to justify the additional use of water. This may seem like common sense, but it is a fundamental departure from current, accepted practice. In anticipation of this needed change, researchers need to devote the same level of intensity to understanding crop production functions as they have to estimating reference crop evapotranspiration.

FULL IRRIGATION

The concept of partial irrigation will be more easily understood if we first define full irrigation. I have never seen a formal definition of full irrigation, but a definition might be imputed from the irrigation guidelines suggested by the U.S. Natural Resource Conservation Service (NRCS, formerly SCS). Those guidelines stipulate an allowable soil moisture depletion, specific for each crop, at which the crop should be irrigated to avoid yield reduction. The NRCS also stipulates that when the crop is irrigated the soil profile should be completely filled in 90% of the area of the field (or 75% in the case of a low value crop). Those requirements imply the following definition of full irrigation, that *90% of the field will be provided sufficient water to insure maximum yield.* This definition is consistent with a number of recently reviewed textbooks dealing with irrigation system design, all of which stipulated, either explicitly or implicitly, that systems should be designed to deliver sufficient water to meet full crop water demand (English and Raja,

1996).

This traditional standard has a rational basis in a natural desire to achieve maximum crop yield from a given area of land. It is well established that crop yield is approximately proportional to the amount of water consumed by a crop (FAO, 1979). Thus to insure maximum yield one must avoid water stress that causes consumptive use to fall below the maximum potential rate, and that implies full irrigation. But when irrigation is expensive or when water is in short supply, maximizing the yield per unit of land is not the best objective. Under those conditions, partial irrigation is economically preferable (English, 1991).

PARTIAL IRRIGATION

That brings us to the central question of this paper; what advantage is offered by partial irrigation? Let us first consider the physical side of the question by examining the relationship between applied water and crop yield. Figure 1 shows a crop production function which relates applied water to wheat yield per unit of irrigated land (based on data from field experiments with winter wheat in the Pacific Northwest of the United States). The curve has two distinct zones characteristic of such production functions; a relatively steep zone associated with low amounts of applied water and a zone with a very low slope that converges to zero at full irrigation. At low levels of applied water there is very little waste; i.e. the crop eventually captures most of the applied water, hence the marginal productivity is high. But at higher levels of applied water the marginal productivity begins to fall off as irrigation becomes progressively less efficient. And if the amount of applied water exceeds the full irrigation requirement the yield curve begins to fall, reflecting the restricted root zone aeration, disease infestations and other problems associated with high water tables and saturated soils.

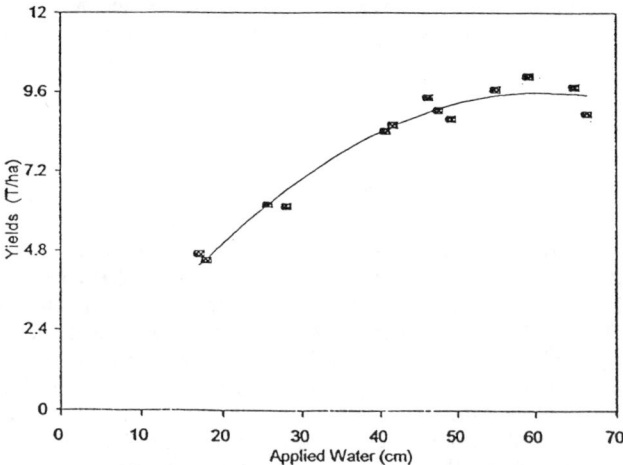

Figure 1. Yield vs. applied water for winter wheat in Oregon.

Let us take a closer look at the portion of the curve associated with small deficits, roughly the range of applied water between 49 cm and 61 cm. If the deficits are properly managed, yields in that range will be relatively insensitive to small reductions in applied water. One way to manage such deficits is by elimination of irrigations at less productive stages of growth when crops are least sensitive to stress. For example, it is estimated that wheat yields are reduced by about 11.5% for every 10% reduction in consumptive use of water when the deficits are distributed through the whole season, but if the deficits are concentrated in the vegetative stage of growth, the yield reduction would be very small, on the order of 2% (FA0 1979).

Another way to manage deficits is to reduce the amount of water applied at each irrigation. Because irrigation efficiency declines as the intensity of irrigation increases[2], the productivity of the last increments of applied water tend to be low. Consequently, elimination of the last increment will have relatively little effect on yields. It is instructive to examine this last point in terms of the spatial distribution of irrigation water and the concept of irrigation adequacy. Because the distribution of irrigation water is inherently uneven and the water holding characteristics of soils are not homogeneous, some parts of a field can be over irrigated even while other parts of the field are still under irrigated. The percentage of a field that receives at least enough water to fill the profile during an irrigation event is defined as the irrigation adequacy. (Using this definition the NRCS guidelines for full irrigation discussed earlier can be restated in terms of adequacy; i.e. full irrigation is equivalent to 90% adequacy.) Reducing the amount of water applied at each irrigation reduces the adequacy, which has the effect of reducing yield in part of the field but increasing the average application efficiency for the field as a whole.

These concepts can be illustrated by analysis of a hypothetical case. Imagine an irrigation system with a relatively high uniformity of 85% which will be used to irrigate a crop with a gross irrigation requirement of 500 mm at 90% adequacy. If that nominal adequacy were reduced to 50% the gross irrigation requirement would be reduced to 400 mm, a 20% reduction from full irrigation. At that level 50% of the field would still be over irrigated while the other half would be under irrigated. Because the effect of a deficit of this size is to reduce the adequacy the result will be a relatively modest yield reduction. In the case of Figure 1, for example, a 20 % reduction in applied water would result in a 4% reduction in yield.

We can develop some perspective about the next part of the curve, which relates excess water use to yield reduction, from a literature review published by Solomon (1985). The adverse effects of excess irrigation are influenced by the complex interplay of several factors, including in particular the depth to water table and climatic conditions, so it is difficult to describe the shape of the curve beyond full irrigation. But using data from a number of studies, Solomon developed general yield reduction curves for three degrees of sensitivity to excess water; low, medium and high sensitivity. According to his curves, an excess of 20% applied to a medium sensitivity crop will cause a yield reduction of

---

[2] There are many different definitions of efficiency relating to irrigation. The term irrigation efficiency is used here in a general sense, but is meant to encompass both water use efficiency and application efficiency.

about 4%. (Note that this order of magnitude is comparable to the effects of a 20% deficit in Figure 1.)

The above examples indicate that small deficits in the amount of water used for irrigation, if properly managed, will produce relatively small reductions in yields and that excess water use can, in some circumstances, be as much a problem as under-irrigation. It should be noted, however, that yield is not the only consideration. Some crops are sensitive to stress in ways that are not related to yield. For example, the market value of potatoes is dependent on tuber shape, which is quite sensitive to water stress at certain stages of growth. Nevertheless, in terms of major grain crops, feed crops, cotton and others of the dominant crops of the world the foregoing general discussion applies.

Let us now turn to the economic issue. We would like to know whether partial irrigation makes economic sense. We begin this discussion by defining a functional relationship between applied water and crop yield, which can be represented by the equation:

$$\text{yield} = f(w) \qquad (1)$$

Revenue earned from the crop will be equal to the yield multiplied by the crop price. If the crop price is assumed constant, the relationship between applied water and revenue will be a curve similar in shape to Figure 1, but scaled by the crop price. Let us represent such a revenue curve in a general way by the equation

$$\text{revenue} = P \cdot f(w) \qquad (2)$$

Let us also introduce a cost function that relates total cost of production to applied water, as:

$$\text{cost} = c(w) \qquad (3)$$

Net income is then the difference between revenue and cost; that is,

$$\pi = \text{Net income} = P f(w) - c(w) \qquad (4)$$

(The symbol π represents profit, a notation that is favored by economists.) We can examine the difference between partial irrigation and full irrigation using these very general equations. Conventional practice is to apply sufficient water to maximize yield which can be defined by the derivative of the yield equation, as:

$$\frac{df(w)}{dw} = 0 \qquad (5)$$

By contrast, taking a strictly economic view, the optimal level of applied water for partial irrigation is that which maximizes net income. The amount of water that will maximize net income will depend upon whether the availability of land and/or water is limiting. In the simplest case, when the amount of land to be irrigated is limiting, the optimum is defined by the equation:

$$\frac{d\pi}{dw} = P\frac{df(w)}{dw} - \frac{dc(w)}{dw} = 0 \qquad (6)$$

Let us rearrange this equation as follows:

$$\frac{df(w)}{dw} = \frac{1}{P} \frac{dc(w)}{dw} \qquad (7)$$

Comparing this equation with the equation for the yield maximizing level of water use (Equation 5) it is clear that the economic optimum must always be less than the yield maximizing level if the derivative of the cost function is positive. Since costs will generally increase with increased use of water, full irrigation will generally be sub-optimal. (An expanded treatment of this mathematical analysis has been published by English, 1991.)

English and Raja (1997) examined the economics of partial irrigation for three quite different situations; wheat production in Oregon, cotton production in California and maize production in Zimbabwe. They found that the profit maximizing level would be less than the yield maximizing level by about 15 or 16 percent in all three cases. The fact that the optima were approximately the same, a 15% deficit, in all three cases was only coincidental, but the 15% figure suggests the order of magnitude by which full irrigation may exceed the economically optimum level of water use.

FINAL THOUGHTS

If partial irrigation is more profitable, then why is it not practiced? The answer is that it *is* practiced, but only on an intuitive level. For example, irrigation recommendations are adjusted by professionals (e.g the NRCS) to account for the fact that full irrigation of an entire field is not economically justified, but the adjustments are *ad hoc*, with no explicit analytical basis for the recommendations. The nominal NRCS requirements for full irrigation, which stipulate that 90% of the field should be fully irrigated rather than 100%, is a case in point.

The fundamental objectives to be served by this paper are (1) to accelerate the shift to a new paradigm for irrigation in which any analysis would begin by determining not how much water a crop can consume but how much it can use profitably, and (2) to stimulate increased interest and research activity on the subject of crop water relationships.

REFERENCES

Postel, S.L.; G.C. Daily and P.R. Ehrlich. 1996. "Human Appropriation of Renewable Fresh Water." Science, V 271, No. 9. pp. 785-788.

English, M.J., and S.N. Raja. 1996. "Perspectives on Deficit Irrigation." Agricultural Water Management 32 (1996) 1-14.

FAO Irrigation and Drainage Paper No. 33. 1979. "Yield Response to Water." Food and Agriculture Organization of the United Nations, Rome.

English, M.J. 1990. "Deficit Irrigation; an Analytical Framework." ASCE Journal of Irrigation and Drainage Engineering. V. 1126(3), pp. 399-412.

Solomon, K.H., 1985. "Typical Crop Water Production Functions." Paper No. 85-2596. ASAE Winter Meeting, Chicago, Dec. 17-20. 1985.

AREAL EVAPOTRANSPIRATION ESTIMATION USING SEVERAL ALTERNATIVE FORMULATIONS OF THE CRAE HYPOTHESIS

LAKSHMAN NANDAGIRI
Department of Applied Mechanics & Hydraulics
Karnataka Regional Engineering College
SRINIVASNAGAR, Karnataka, India - 574 157.

ABSTRACT
Accurate estimation of ET from large irrigated areas is essential for sustainable development of land and water resources. The CRAE hypothesis offers a convenient technique for areal ET estimation using only regularly recorded metcorological data. In this study several alternative models for areal ET are derived from the fundamental CRAE hypothesis and their performances assessed using data of the Everglades Agricultural Area, Florida. Results indicate the feasibility of obtaining acceptable estimates of areal ET from irrigated land using the CRAE concept.

INTRODUCTION
The future of the irrigated agriculture in developing countries is critically dependent on efficient irrigation water management, a crucial element of which is accurate estimation of the actual evapotranspiration (ET) component. Conventional techniques for estimating ET from meterological data (e.g., Doorenbos & Pruitt, 1977) require additional information on soil/vegetation characteristics and may lead to uncertain areal estimates for diverse cropping patterns.

The Complementary Relationship Areal ET (CRAE) approach has been suggested as an alternative procedure for estimating actual ET. This approach utilises only meteorological data to produce estimates of ET from large areas (characteristic length 1 to 10 km.) and thereby avoids the need for information on the soil-vegetation complex. This is an attractive feature from the viewpoint of developing countries where not only is the data hard to come by but also vegetation and soil characteristics exhibit tremendous spatial diversity. Based upon the initial concept proposed by Bouchet (1963), the CRAE approach has been expanded by Morton (1976) and Brutsaert & Stricker (1979) and subsequently applied over a range of temporal and spatial scales to estimate ET from a variety of vegetation-soil complexes. (e.g., Ben Asher, 1981; Ali & Mawdsley, 1987; Parlange & Katul, 1992: Barr et al., 1996). No attempt has been made to apply the CRAE approach to irrigated crop lands.

Distinct Morton and Brutsaert & Stricker versions of the CRAE approach arise out differences in the choice of equations for the potential evapotranspiration (E_p) and the wet environment areal ET(E_{po}), which are components of the fundamental relationship derived by Bouchet. However, as pointed out by Brutsaert (1982), several other alternative versions can also be derived, all depending on the particular choice of the equations for E_p and E_{po}. This feature of the CRAE approach may be exploited depending on the nature of the available meteorological data set. The objective of the present study was two fold: 1) to apply the CRAE approach to an irrigated agricultural area and 2) to evaluate the relative performances of five alternative formulations of the approach using data of the Everglades Agricultural Area (EAA) located in Florida, USA.

EVAPOTRANSPIRATION MODELS

COMPLEMENTARY RELATIONSHIP AREAL ET
The basic equation of the CRAE approach is [Morton, 1983];
$$E + E_p = 2E_{po} \qquad (1)$$
where E is the actual ET from a large area unaffected by upwind boundary transitions, E_p is the potential ET from a hypothetical moist surface with radiation absorption and vapour transfer characteristics similar to those of the area and so small the the effects of the ET on the overpassing air would be negligible and Epo is the wet environment areal ET, the ET that would occur if the soil-vegetation surface were saturated and there were no limitations on the availability of water.

POTENTIAL ET (E_p)
Based on the suggestion of Brutsaert (1982), the Penman and the Pan evaporation models were considered to represent E_p in Eqn. (1). The Penman equation is
$$E_p = \frac{\Delta}{\Delta + \gamma}(R_n - G) + \frac{\gamma}{\Delta + \gamma} f(u)(e_a - e_d) \qquad (2)$$
Where Δ is the slope of saturation vapour pressure versus temperature curve at air temperature (T), γ is the psychrometric constant, R_n is net radiation, G is soil heat flux, f(u) is an empirical function of windspeed (u) and ($e_a - e_d$) is the vapour pressure deficit. Literature offers several choices for the wind function f(u) (e.g., Allen et al., 1989).

Measured pan evaporation (E_{pan}) may be converted into E_p using;
$$E_p = K_p E_{pan} \qquad (3)$$
where K_p is the pan coefficient which depends on type of pan, site and meteorological conditions. Doorenbos & Pruitt (1977) present guidelines for selecting K_p.

WET ENVIRONMENT AREAL ET (E_{po})

Priestley & Taylor (1972) suggest the following equation for estimating evaporation from wet surfaces under conditions of minimal advection.

$$E_{po} = \alpha \frac{\Delta}{\Delta + \gamma} (R_n - G) \tag{4}$$

with $\alpha = 1.26$

As an alternative to Eqn. (4), Hicks & Hess (1977) proposed a more general expression,

$$E_{po} = \frac{(R_n - G)}{a(\gamma/\Delta) + (1 + b)} \tag{5}$$

On the basis of their measurements, Hicks & Hess (1977) suggested values of $a = 0.63$ and $b = -0.15$.

FORMULATIONS OF CRAE

Substitution of E_p and E_{po} defined by Eqs. (2) to (5) into Eqn. (1) results in the following four versions of the CRAE approach.

Omitting G and combining Eqs. (1), (2) and (4) results in Eqn. (6) which is the advection - aridity model proposed and tested by Brutsaert & Stricker (1979).

$$E = (2\alpha - 1) \frac{\Delta}{\Delta + \gamma} R_n - \frac{\gamma}{\Delta + \gamma} f(u)(e_a - e_d) \tag{6}$$

Eqn. (1) combined with Eqs. (2) and (5) yields,

$$E = [2 \frac{\Delta}{(1-b)\Delta + a\gamma} - \frac{\Delta}{\Delta + \gamma}] R_n - \frac{\gamma}{\Delta + \gamma} f(u)(e_a - e_d) \tag{7}$$

Eqn. (7) was proposed by Brutsaert & Stricker (1979) but remained untested.
Eqs. (1), (3) and (4) yield,

$$E = 2\alpha \frac{\Delta}{\Delta + \gamma} R_n - K_p E_{pan} \tag{8}$$

Similarly, Eqs. (1), (3) and (5) yield,

$$E = 2\Delta \frac{R_n}{(1-b)\Delta + a\gamma} - K_p E_{pan} \tag{9}$$

Morton (1976) combined modified forms of Eqs. (2) and (4) with Eqn. (1) and derived the following model for areal ET, which was included in the present comparison.

$$E = \frac{\Delta}{\Delta + \gamma} (1.76 R_n + 2.76M) - \frac{\gamma}{\Delta + \gamma} (e_a - e_d) f_A \tag{10}$$

where M is an empirical advection term given by

$$M = (1.37 R_{nl} - 0.397 R_s)$$

in which R_{nl} is net longwave radiation and R_s is incoming solar radiation. M is subjected to the constraint; $M \geq 0$. f_A is an empirical constant that replaces the wind function $f(u)$ in Eqn. (2) and has values of 47.5 cal cm^{-2} d^{-1} mb^{-1} for $T \geq 0°C$ and 54.6 cal cm^{-2} d^{-1} mb^{-1} for $T < 0°C$.

APPLICATION

The performances of the CRAE models given by Eqs. (6) to (10) in simulating monthly mean areal ET was tested by applying them to the Everglades Agricultural Area (EAA) located in Florida, USA. Relevant meteorological data for application of the models and also measured monthly mean areal ET values for assessing the performances of the models was obtained from Shih et al (1983). The EAA, extending over an area of 190,000 ha, is devoted to intensive agricultural production, with 75% of this area planted to sugarcane and remaining to vegetable crops. The climate of the area is humid, with a May to October warm rainy season and November to April dry winter season. The original paper may be referred to for a more detailed description of the EAA. Monthly mean meteorological values of; solar radiation (1971-79), daily percentage of possible sunshine hours (1967-78), daily mean air temperature (1924-75), daily relative humidity (1958-76), daily windrun (1934-39 & 1978-79) and monthly pan evaporation (1924-75) for the EAA as given in Shih et al (1983) were used in the present study (figures in parenthesis indicate averaging period). Mean monthly areal ET (1962-71) values obtained by water balance measurements are also documented therein. Using the measured meteorological data, variables involved in the CRAE models were calculated as per procedures laid out in Allen et al (1989). The Doorenbos & Pruitt (1977) procedures for calculation of f(u) and K_p, and $\alpha = 1.26$ were adopted.

RESULTS AND DISCUSSION

Monthly mean areal ET estimated by the various CRAE formulations are compared with water balance ET in Table (1). It is evident that the seasonal trend in areal ET is replicated fairly well by all the formulations. The deviations between model predictions and measurements are presented in Table (2). Overestimation of ET in most of the months by all the models is apparent. Also overestimation or underestimation in any month is consistent between all the models, but the magnitude of deviation varies. Annual totals are all higher than the measured value and the positive deviations in annual totals ranging between 59.22 mm to 156.39 mm yield percentage errors ranging from 5.82 to 15.36. However, these errors are acceptable if one considers errors in the range of 10-20% in water balance measurements. The criteria for assessing the relative performances was taken to be the total of the absolute monthly deviations [Table (2)]. On this basis, Eqn. (8) which yielded the smallest total was the best estimator of areal ET, closely followed by Eqn. (9). Performances of Eqs. (6) & (7) did not differ much. The Morton model [Eqn. (10)] requires site specific calibration of the constant f_A. Since no such calibration was attempted in the present study, the performance of this model must be viewed in this light. Other possible sources of error in the present application are; choice of $\alpha = 1.26$, use of monthly mean data and the mismatch between the averaging periods for meteorological data and water balance measurements. A more detailed performance appraisal of the formulations presented herein must overcome these data limitations and also test the sensitivities of the models to the value of α

The CRAE hypothesis is inherently attractive owing to its meagre input data requirement and this study has shown that even pan evaporation data may be sufficient to produce acceptable estimates of areal ET.

REFERENCES

Allen, RG., Jensen, M.E., Wright J.L & Burman RD (1989). Operational estimates of reference evapotranspiration. Agron. J., 81; 650 - 662.

Ali, M.F and Mawdsley, JA (1987). Comparison of two recent models for estimating actual ET using only regularly recorded data. J. Hydrol., 93; 257-276.

Barr,AG.,Kite, GW., Granger R & Smith C (1996). Evaluating three ET methods in the SLURP macroscale hydrologic model. Submitted to Hydrologic Processes, CGU 1996 Special Issue.

Ben-Asher J (1981). Estimating ET from the Sonoita Creek Watershed near Patagonia, Arizona. Water Resour. Res., 17(4); 901-906.

Bouchet, RJ (1963). ET reele et potentielle, signification climatique. IASH Publ. 62, 134-142.

Brutsaert W (1982). Evaporation into the atmosphere. D. Reidell, Hingham, Mass., 229p.

Brutsaert W & Stricker H (1979). An advection-aridity approach to estimate actual regional ET. Water Resour. Res., 15(2); 443-450.

Doorenbos J & Pruitt W O (1977). Crop water Requirements. Irr. & Dr. Paper No. 24, FAO, UN, Rome, 179p.

Hicks BB & Hess GD (1977). On the Bowen ratio and surface temperature at sea. J.Phys. Oceanogr, 7; 141-145.

Morton FI (1976). Climatological estimates of ET.J. Hydraul. Div., Proc. ASCE, 102, HY3; 275-291.

Morton FI (1983). Operational estimates of areal ET and their significance to the science and practice of hydrology. J. Hydrol., 66; 1-76.

Parlange MB & Katul GG (1992). An advection-aridity evaporation model. Water Resour. Res., 28(1); 127-132.

Priestley CHB & Taylor RJ (1972). On the assessment of surface heat flux and evaporation using large scale parameters. Mon. Weather Rev., 100; 81-92.

Shih SF, Allen LH, Hammond LC., Jones JW, Rogers JS & Smajstrla AG (1983). Basinwide water requirement estimation in Southern Florida. Trans. of ASAE, 760-766.

EVAPOTRANSPIRATION ESTIMATION 267

Table (1) : Estimated monthly mean areal ET compared with measured ET (mm)

Month	Eqn. (6)	Eqn. (7)	Eqn. (8)	Eqn. (9)	Eqn. (10)	Water Balance Measurements
January	32.53	35.05	33.66	36.18	44.72	48
February	43.64	46.63	40.56	43.56	46.36	36
March	85.07	88.92	74.87	78.72	92.45	71
April	114.83	118.10	111.56	114.84	125.29	109
May	136.74	138.02	129.50	130.78	147.13	152
June	132.37	132.32	131.92	131.88	141.32	112
July	132.76	131.94	132.55	131.73	139.79	107
August	133.74	132.58	135.60	134.44	139.61	119
September	106.98	106.94	110.70	110.66	111.91	114
October	84.04	85.50	89.36	90.83	87.20	46
November	57.92	60.26	47.67	50.01	61.90	46
December	34.11	36.64	39.27	41.81	36.71	58
Annual	1094.72	1112.92	1077.22	1095.45	1174.39	1018

Table (2) : Deviations (mm) between measured and estimated ET

Month	Eqn. (6)	Eqn. (7)	Eqn. (8)	Eqn. (9)	Eqn. (10)
January	- 15.47	- 12.95	- 14.34	- 11.82	- 3.28
February	7.64	10.63	4.56	7.56	10.36
March	14.07	17.92	3.87	7.72	21.45
April	5.83	9.10	2.56	5.84	16.29
May	- 15.26	- 13.98	- 22.50	- 21.22	- 4.87
June	20.37	20.32	19.92	19.88	29.32
July	25.76	24.94	25.55	24.73	32.79
August	14.74	13.58	16.60	15.44	20.61
September	- 7.02	- 7.06	- 3.30	- 3.34	- 2.09
October	38.04	39.50	43.36	44.83	41.20
November	11.92	14.26	1.67	4.01	15.90
December	- 23.89	- 21.36	- 18.73	- 16.19	- 21.29
Annual Deviation	70.72	94.92	59.21	77.42	156.38
% error in ann. total	7.53	9.32	5.81	7.60	15.36
Total of absolute monthly deviations	200.01	205.6	176.96	182.58	219.45

A RADIAL MODEL OF SURFACE FLOW: APPLICATION TO LEVEL-BASIN IRRIGATION

P. GARCIA-NAVARRO [1] and E. PLAYAN, [2]
[1] Mecánica de Fluidos, CPS, Universidad de Zaragoza. 50015 Zaragoza, Spain.
[2] Estación Exp. Aula Dei (DGA-CSIC). Apdo. 202. 50080 Zaragoza, Spain.

ABSTRACT

A radial flow simulation model was applied to the estimation of infiltration and roughness parameters of point-inflow level-basin irrigation events. It was concluded that the estimation of such parameters is at least as possible for radial flow as it is for parallel flow.

INTRODUCTION

In recent decades, a large effort has been devoted to the development of unsteady flow models adapted to different types of surface irrigation systems. These models have been successfully used for many purposes, including occasional applications to the design and management of surface irrigation systems.
Numerical hydraulic models have been applied to the estimation of surface irrigation parameters, solving what has been called the inverse problem. Estimated parameters are the Manning roughness factor, n, and the parameters of a Kostiakov-Lewis infiltration equation (k, a, and f_o).
Applications of the inverse problem concept to different irrigation systems using both non-predictive and predictive techniques are reported in (Katopodes et al., 1990), Clemmens (1991) and Bautista and Wallender (1993).
In cases such as level-basins irrigated from a point source, and assuming uniformity in infiltration and roughness, the flow can be said to be radial. In two-dimensional polar coordinates, radial flow can be expressed as a one-dimensional problem.

MATHEMATICAL MODELING

TWO DIMENSIONAL CARTESIAN MODEL
The two-dimensional shallow water flow equations can be written in cartesian form as (Abbott, 1992) :

$$\frac{\partial U}{\partial t} + \frac{\partial E}{\partial x} + \frac{\partial F}{\partial y} = G \qquad (1)$$

with $U = (h, uh, vh)^T$ and where h, u and v are the depth and the velocities in the x and y directions respectively. The fluxes in the second term of the equations are,

$$\mathbf{E} = \left(uh, \quad u^2h + \frac{gh^2}{2}, \quad uvh \right)^T, \quad \mathbf{F} = \left(vh, \quad uvh, \quad v^2h + \frac{gh^2}{2} \right)^T \qquad (2)$$

The speed of the surface perturbations on still water is the wave celerity, defined as $c = \sqrt{gh}$. The right hand side of the system contains the sources and sinks of momentum arising from the bed slopes and the friction losses along the two coordinate directions,

$$\mathbf{G} = \left(-i, \quad gh(S_{ox} - S_{fx}) + D_{ix}, \quad gh(S_{oy} - S_{fy}) + D_{iy} \right)^T \qquad (3)$$

The bed and friction slopes are,

$$S_{ox} = -\frac{\partial z}{\partial x}, \quad S_{oy} = -\frac{\partial z}{\partial y}, \quad S_{fx} = \frac{n^2 u \sqrt{u^2 + v^2}}{h^{\frac{4}{3}}}, \quad S_{fy} = \frac{n^2 v \sqrt{u^2 + v^2}}{h^{\frac{4}{3}}}$$

where z is the bottom level and n is the Manning roughness coefficient.

The infiltration rate, i, can be computed using the empirical Kostiakov-Lewis equation, $i = ka\tau^{a-1} + f_0$ where τ is the opportunity time measured in minutes and k, a and f_0 are empirical parameters. The momentum transfers are estimated as:

$$D_{ix} = \frac{ui}{2}, \quad D_{iy} = \frac{vi}{2} \qquad (4)$$

POLAR MODEL

In transforming Eq. (1) from the cartesian (x, y, t) to the polar frame (r, ϕ, t), the variables (h, u, v) become the set (h, u_r, u_ϕ). The transformation is simplified by assuming pure radial flow.

The mass or continuity equation becomes

$$\frac{\partial h}{\partial t} + \frac{1}{r}\frac{\partial h r u_r}{\partial r} = -i \qquad (5)$$

and the two momentum equations combine to give a single radial momentum equation

$$\frac{\partial h u_r}{\partial t} + \frac{1}{r}\frac{\partial h r u_r^2}{\partial r} + g\frac{\partial g h^2 / 2}{\partial r} = gh\frac{\partial z}{\partial r} + g\frac{n^2 u_r |u_r|}{h^{1/3}} + \frac{u_r i}{2} \qquad (6)$$

It is possible to rewrite the system of Eqs. (20) and (21) into conservative form:

$$\frac{\partial h}{\partial t} + \frac{\partial q}{\partial r} = -\frac{q}{r} - i \qquad (7)$$

$$\frac{\partial q}{\partial t} + \frac{2q}{h}\frac{\partial q}{\partial r} + \left(gh - \frac{q^2}{h^2} \right)\frac{\partial h}{\partial r} = -\frac{q^2}{hr} + gh(S_0 - S_{fr}) + \frac{qi}{2h} \qquad (8)$$

with $q = hu_r$ and

$$S_{fr} = \frac{n^2 q |q|}{h^{10/3}} \qquad (9)$$

Standard one-dimensional numerical techniques can be adapted directly.

1D NUMERICAL SCHEME

The numerical scheme used for discretizing Eqs. (7) and (8) is the explicit McCormack in two steps predictor-corrector (McCormack 1971). It is a technique of proved efficiency for unsteady free surface flow modeling (Fennema and Chaudhry 1986, García-Navarro and Savirón 1992). Being second order accurate in space and time, it offers good resolution and has great conceptual simplicity.

In Glaister (1991) the pure radial flow equations (Eqs. 5 and 6) in the homogeneous case (no bed slope, no friction, no infiltration) are discretized using the explicit Roe scheme (Roe 1981). The geometric source terms are upwinded applying a decomposition on the basis of eigenvectors of the approximate Jacobian. This step has been avoided in the work presented here. Instead, a pointwise semi-implicit discretization has proved efficient in reducing instabilities at the advancing front. For the friction term, for instance,

$$S_{fr}\big|_i \approx \left(\frac{|q|}{K^2}\right)_i^n \left(\theta q_i^n + (1-\theta) q_i^{n+1}\right), \quad 0 \le \theta \le 1, \quad K = \frac{h^{5/3}}{n} \tag{10}$$

Boundary conditions

Having used an explicit scheme for the interior points, the method of characteristics has to be applied to specify conditions at the boundaries. A detailed description of the principles of this method may be found in several references (Abbott. 1992). The flow regime at the upstream and downstream ends determines the number of required boundary conditions. The application to a prismatic one-dimensional channel is well described in García-Navarro and Savirón (1992). The implementation for the pure radial flow equations is described in Playan and Garcia-Navarro (1997). The imposed boundary conditions consisted of a hydrograph $q=q(t)$ upstream and a zero spatial derivative of h downstream (once the advance phase has been completed).

Comparison of the Polar and Cartesian models

A two-dimensional Cartesian model (Playán et al., 1994) reproducing radial flow was used for comparison purposes. The numerical experiment simulates an irrigation event taking place on a square 1 ha (100 m x 100 m) level-basin irrigated with a constant discharge of 0.1 m³s⁻¹. Infiltration is characterized by the following parameters: $k = 0.006$ m min⁻ᵃ, $a = 0.5$, and $f_0 = 0$. Surface roughness is characterized by a Manning roughness factor $n = 0.14$. If a point inflow is located at one of the $\pi/2$ rad corners, the flow will be radial until the moment when the advancing front reaches one of the corners.

A node spacing of 1 m was used in the one-dimensional Polar model. For the two-dimensional Cartesian model, the node spacing was set to 5 m. The advance curves predicted by both models agree satisfactorily (Fig.1). The error is closely related to the grid size and sensitive to the position of the upstream node in the Polar model (r_0). This is illustrated in Fig. 2, where the final error and upstream depth (at time 40 min) are presented for values of r_0 ranging from 0.5 to 10 m. For small values of r_0 the mass balance error becomes unacceptable due to the presence of a singularity at $r = 0$. It can also be observed that the upstream depth increases with decreasing r_0, up

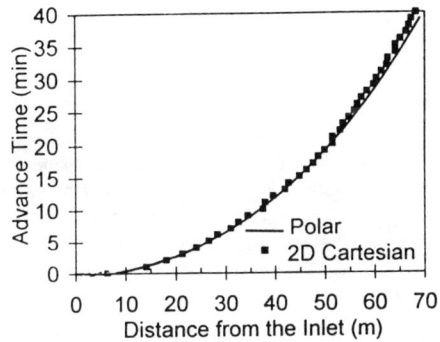

Fig. 1: Advance trajectories from the polar and cartesian models.

to a point in which some numerical stability seems to be attained. In view of these results, a 1 m mesh and $r_o = 5$ m were used in all subsequent model executions.

APPLICATION: THE INVERSE PROBLEM

In the following section, the Polar model will be used to build predictive inverse problem solvers. The Polar model can be of great use in level-basin irrigation configurations based on radial flow. In this case, it will no longer be of application after the flow reaches a field boundary that does not contain the inflow point. Ideally, this procedure could be applied to irrigation events with corner inflow at any angle ($0 < \alpha \leq 2\pi$), although $\alpha = \pi / 2$ will be the most common value.

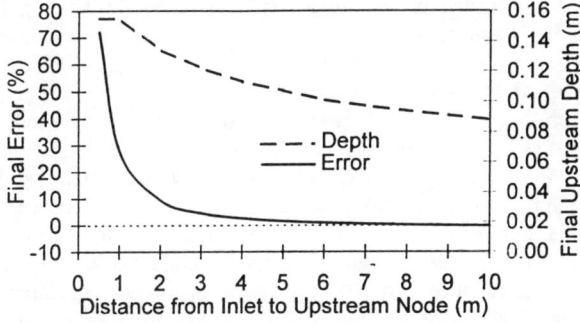

Fig. 2: Dependence of the final ($t = 40$ min) mass balance error and upstream depth on the distance from the inlet to the upstream node for the Polar model.

Objective Function

Predictive estimation techniques are based on the existence of an objective function. Such function is built with the observed and model estimated values of a variable at different locations and/or times. The objective function indicates the error caused by a poor estimation of the parameters. Estimation was limited to two parameters at a time, these being n and k. The objective function (O) will be formulated as:

$$O(n, k) = \sum_{j=1}^{J} \left[h_j^{'}(x_j, t_j) - h(x_j, t_j, n, k) \right]^2 \quad (11)$$

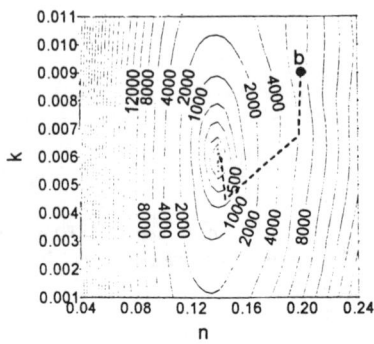

Fig. 3: Valley of minimum and selected convergence path for the optimization case n-k search. The solution point is located at the center of the plot.

where h' is the field measured flow depth and J is the total number of field observations. Both h and h' are expressed in millimeters.

Minimization Techniques

The values of n and k that minimize the objective function constitute the solution to the problem. The Polak-Ribiere minimization algorithm (as described in Press et al., 1988) was used to estimate the value of the parameters. This method performs an iterative search for the minimum in conjugate gradients. The gradient was numerically estimated using a centered difference based on a 1% perturbation. A forward difference would require less function evaluations to determine the gradient, but the centered difference proved to be more efficient in finding the solution (Playan and Garcia-Navarro, 1997).

Inverse problem: An example

Due to the exploratory nature of this research, model results were used as field data (Katopodes et al., 1990) in this numerical experiment. The solution of the problem is $n = 0.14$, $k = 0.006$. A starting guess (point b in Fig.3) characterized by $n = 0.20$ and $k = 0.009$ was supplied to the Polak-Ribiere routine. The 'field' procedure is based on flow depth measurements separated 5 m from each other and from the inlet, to a distance of 50 m. Ten flow depth measurements are collected at 5 min intervals to a final time of 30 min for a total of 60 measurements. If this experiment was to be performed in the field, the upstream quarter of a circle ($r = 5$m) should be impermeabilized to avoid infiltration. This would provide agreement between the observed and the modeled phenomena. The total discharge was 0.1 m^3s^{-1}, and node spacing was set to 1 m. The parameter f_o was set to 0. A 100 MHz Pentium® personal required 53 s of cpu time for each estimation of the objective function. Fig. 3 presents the valley of minimum corresponding to this numerical experiment. The valley was drawn from 121 evaluations of the objective function distributed evenly along the x and y axes. The corresponding value of the objective function is 0 since at this point the simulated and the 'real' measurements of flow depth are in full coincidence. The path from point b to the solution is illustrated in Fig.3. The procedure required 85 function evaluations and 6 Polak-Ribiere iterations.

SUMMARY AND CONCLUSIONS

A hydraulic simulation model suited for one-dimensional radial flow has been presented. The shallow water equations in cylindrical polar coordinates were solved using an explicit finite-difference McCormack scheme. The model has been applied to the simulation of level-basin irrigation introducing appropriate friction and infiltration terms. Pointwise semi-implicit discretization of the friction and geometrical source terms was used to reduce the instabilities at the advancing front. On coarse grids the solution is distorted, therefore making it advisable to locate the first node away from $r = 0$. A distance of 5 m proved to be satisfactory in the studied cases. Results from the Polar model were compared to those from an existing two-dimensional Cartesian model simulating radial flow.

The hydraulic model was coupled to minimization routines to build predictive level-basin inverse problem solvers. When using simulated data as input, results indicate that pairs of roughness and infiltration parameters can be identified from radial flow depth measurements. In this research uniformity has been assumed for infiltration, roughness and soil surface elevation. Estimation of the infiltration and roughness parameters is at least as possible for radial flow as it is for parallel flow.

REFERENCES

Abbott, M. B. (1992) *Computational hydraulics*. 3rd Edition. Ashgate, UK, 326 pp.

Bautista, E. and Wallender, W. W. (1993). "Identification of furrow intake parameters from advance times and rates." *J. Irrig. Drain. Engrg.*, ASCE , 119(2), 295-311.

Clemmens, A. J. (1991). "Direct solution to surface irrigation advance inverse problem." *J. Irrig. Drain. Engrg.*, ASCE, 117(4), 578-594.

Fennema, R. J. and Chaudhry, M. H. (1986). "Explicit numerical schemes for unsteady free-surface flows with shocks". *Water Resour. Res.*, 22(13), 1923-1930.

García-Navarro, P., and Savirón, J.M. (1992). "McCormack's method for the numerical simulation of one-dimensional discontinuous unsteady open channel flow." *J. of Hydraulic Research*, 308(1), 95-105.

Glaister, P. (1991). "Shallow Water Flow with Cylindrical Symmetry." *Journal of Hydraulic Research*, 298(2), 219-227.

Katopodes, N. D., Tang, J. and Clemmens, A. J. (1990). "Estimation of surface irrigation parameters." *J. Irrig. Drain. Engrg.*, ASCE, 116(5), 676-696.

McCormack, R.W. (1971). "Numerical Solution of the Interaction of a Shock Wave with a Laminar Boundary Layer." *Proceedings of the 2nd International Conference on Numerical Methods in Fluid Dynamics*, 151-163.

Playán, E. and García-Navarro, P. (1997). "Radial Flow Modeling for Estimating Level-basin Irrigation Parameters." *J. Irrig. Drain. Engrg.*, ASCE , (in press).

Playán, E., Walker, W. R. and Merkley G. P. (1994). "Two-Dimensional Simulation of Basin Irrigation. I : Theory." *J. Irrig. and Drain. Engrg.*, ASCE, 120(5), 837-856.

Press, W. H., Flannery, B. P., Teukolsky, S. A. and Vetterling, W. T. (1988). *Numerical recipes in C*. Cambridge University Press, Cambridge. 735 pp.

Roe, P. L. (1981). "Approximate Riemann solvers, parameter vectors and difference schemes." *J. Comput. Physics*, 43, 357-372.

KNOWLEDGE-BASED ADVISORY SYSTEM FOR PREDICTIVE CONTROL OF IRRIGATION MANAGEMENT STRATEGIES

MUHAMMAD ABID BODLA, Ph.D., M.ASCE
Director (Surface Water), International Waterlogging and Salinity Research Institute (IWASRI), 13-West Wood Colony, Thokar Niaz Beg, Lahore, Pakistan

ABSTRACT

Methodological development of the knowledge-based Expert System for Border Irrigation Management, ESBIM is discussed in this paper. The hybrid simulation-expert system with the embedded dynamic database provides a composite PC-based decision support system for explicit determination of management operational variables predicting optimal control of irrigation efficiencies. The applicability and generality of the integrated management support system is validated through real border irrigation case studies under widely varying operating conditions and field environments.

INTRODUCTION

Future management strategies aiming at sustainability of irrigated agriculture mainly rely on adaptive control of irrigation storage and application efficiencies. The management parameters consist of inflow rate and cutoff time and represent the only alternatives available to optimize performance of the surface irrigation systems.

Several simulation models based on conventional algorithmic programming techniques currently exist to analyze the entire surface irrigation process, however, as argued by Katapodes and Tang (1990) these models can only simulate application sequences and completely lack the ability to make management decisions. Identifying appropriate inflow rates and application times with the use of such models would require a systematic search through all possible combinations of these operational parameters for an irrigation event. Computational efforts and enormous execution time involved in this kind of exhaustive search have so far precluded the use of existing models for the desired optimal control. The model presented here incorporates heuristics and decision rules to generate solutions from study of a system's simulation responses to varying operating conditions in order to decide upon optimal operating strategies for the surface irrigation system.

THE HYBRID SIMULATION-EXPERT SYSTEM

The ESBIM model provides a three-component decision support system (DSS) for improvement of open ended graded borders with continuous flow regimes. The major components of the model are the knowledge base, the simulation interface code (SIC) communicating with the simulation model SIRMOD (Walker, 1990), the user interface and the dynamic database embedded in the system. The simulation model and expert system interact through the SIC that permits transfer of simulation results to the expert system's inference engine via its knowledge base for logical analysis of the facts and information received and allows generation of the desired optimal and alternate irrigation management strategies. The production rule scheme of knowledge representation was employed in ESBIM as it provides the desired efficient structuring and modification, and a flexible knowledge base. Finally the selected ESBIM solutions are stored in the system's database. Implementation of this advisory system is accomplished on an IBM compatible PC through the artificial intelligence *logic programming* language PROLOG (Borland Inc. 1988). Fig.1 illustrates the architecture of ESBIM.

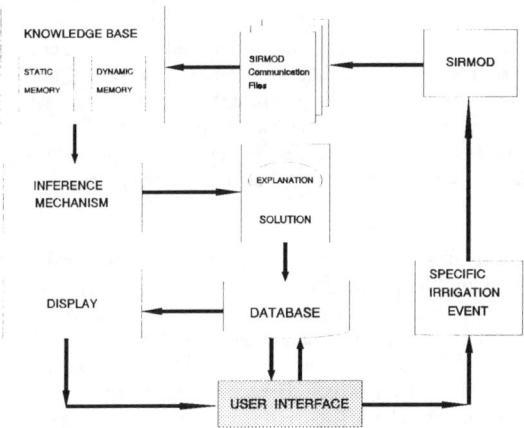

Figure 1. ESBIM Architecture

THE CONTROL ALGORITHM

The surface irrigation management problem poses a process of selecting an optimal irrigation performance level based on the interrelationships among the operating variables which consist of the field parameters (dimensions, slope, surface roughness), advance and recession trajectories, infiltration rate, and the management parameters *i.e.* unit inflow rate and cutoff time.

The control model developed for this knowledge based system utilizes results obtained through the full hydrodynamic simulations by SIRMOD to specify the performance level for a test irrigation event. Two performance categories were defined to indicate current performance level of a given irrigation event *viz.*, category I showing the storage efficiency E_s be 100 percent and category II indicating that E_s level drops below 100 percent. For each category, the ESBIM inference engine is designed to determine the extent of adjustments in the control variables: inflow rate and cutoff time to improve the irrigation performance for a given border to an optimal or alternate user defined desirable performance levels.

The Control Strategy

For a given border, application efficiency, E_a is to be maximized by careful selection of inflow rate, q_{in} and application or cutoff time, t_{co} where both of these parameters are interdependent. The first task of the control algorithm is to provide reasonable initial values of the management parameters q_{in} and t_{co}. The second task is to identify the extent and the direction of adjustments needed in these parameters. ESBIM also allows for user's choice for a flexible or rigid supply system. Finally the control algorithm serves to adjust these parameters using the following efficient control strategy.

The expert system in ESBIM invokes SIRMOD to simulate a given irrigation event. Data pertinent to current irrigation practice and the ensuing inferences are stored in the system's global database known as the working memory. The information stored relates to the input parameters i.e. infiltration parameters k, a, and f_o; the border length L; the border slope S_o; surface roughness coefficient n, the target depth of application, Z_{req}, and the current operational variables q_{in} and t_{co}, and the consequences of these current input parameters such as time of advance to the end of border t_{al}, the depths infiltrated along the length of border, the runoff volume as percentage of total volume of inflow RO, and the values of performance measures E_s and E_a under current operating conditions. The system uses the current inflow rate q_{in} and the corresponding t_{al} predicted by SIRMOD as initial estimate for the control variables q_{in} and cutoff time t_{co}.

From the simulation results, the expert system can identify the performance category for the test irrigation utilizing the production rules provided in the system's static or rule memory. The system then prompts the user to define limits to the flow regulation, if any. Depending on the performance category inferred, the system's rule base leads to fabrication of a search space consisting of a set of plausible combinations of operation variables q_{in} and t_{co} from which the system can generate optimal values for these operational parameters. The inference mechanism in the expert system uses a mixed backward-forward chaining control strategy (Bratko, 1990) for efficient function of the program.

THE DYNAMIC DATABASE INTERFACE
FOR THE EXPERT SYSTEM

The final phase of ESBIM development included design and development of a logic-based data management system to couple to the front-end of the prototype expert system developed in earlier stages.

The large number of data sets generated by the expert system can be stored into the database in a user interactive mode. These data include parameters related to individual border irrigation events and consequent management strategies resolved by the expert system's inference mechanism. With suitable development effort, a comprehensive query facility is provided. As the expert system inferencing mechanism produces the optimal and alternate operating management strategies for a given border, the results are stored in the working memory of the system. On the user's selection of the "Update" option in the "Database" pulldown menu, the Prolog interpreter is activated to store optimal and/or alternate management strategies in the database in a database file.

IRRIGATION MANAGEMENT STRATEGIES

The ESBIM database provides a pulldown menu-driven facility within the ESBIM main menu (Figure 2) to (i) store and update ESBIM generated solutions, (ii) displaying the database, and (iii) consult the database for expert system based deductive solutions.

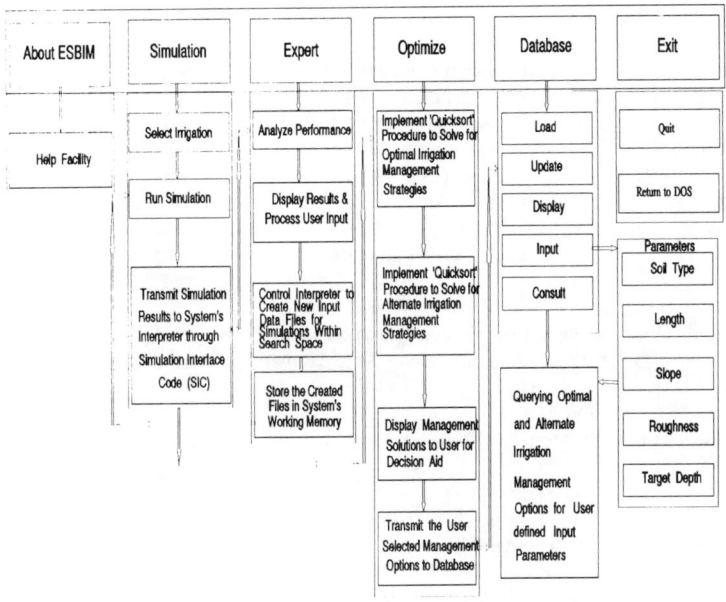

Figure 2. Principal menus and actions in ESBIM

EVALUATION OF THE EXPERT SYSTEM

Evaluation of ESBIM involved the functional and operational validation of the system using actual field data for border irrigation (Bodla, 1993). The functional validation of ESBIM prototype by applying the model for certain test cases adopted from technical literature verified the quality and robustness of the integrated simulation-expert system approach as a decision aid tool in providing the desired control for efficient irrigation management. The operational validation tested and verified efficacy of the model under real border case studies from Australia and Pakistan to establish the model's generality and reliability in an operational decision setting (Bodla, 1993).

CONCLUSIONS

The integration of hydrodynamic simulation model with Prolog based expert system along with an embedded database system in ESBIM is proven to be a very useful aid for providing the crucial irrigation management synthesis lacking in existing simulation models. The developed system can be employed as an efficient and reliable decision-aid for accurate and explicit determination of management operating parameters (inflow rate and cutoff time) resulting in complete irrigation for a given border with maximum possible improvement in system's application efficiency. ESBIM provides comprehensive design-management support to select the most appropriate operating parameters for efficient border irrigation design under local conditions through objective comparison of the series of decision alternatives produced by the system for a given border. The system is found to contain a good basis for developing a reliable and efficient feedback control for automated border irrigation systems. The system's database is shown to have successfully acquired the dynamic and deductive capabilities to furnish queried optimal and alternate irrigation management strategies for user input field parameters.

REFERENCES

Bodla, M. Abid. 1993. An integrated simulation-expert system approach for irrigation management. Ph. D dissertation, Colorado State University, Fort Collins, CO, USA.

Borland. 1988. Turbo Prolog: The Natural Language of Artificial Intelligence. Borland Int. Inc., Scotts Valley, CA, USA.

Bratko, I. 1990. Prolog Programming for Artificial Intelligence. 2nd Ed. Addison-Wesley Publishing Co., Reading, MA, USA.

Katopodes, N. and J.H. Tang. 1990. Self adaptive control of surface irrigation advance. J. Irrig. and Drain. Engrg., ASCE. 116(5): 697-713.

Walker, W.R. 1990. SIRMOD C Version; Surface Irrigation Simulation Model: User's Manual. Dept. of Agric. and Irrig. Engrg., Utah state University, Logan, UT, USA.

Limitations and Consequences of Recycling Drainage Water for Irrigation Over the Long Term

S.R. GRATTAN
University of California, Davis
Davis, CA 95658 USA

Irrigation with saline drainage water may be a viable option to certain annual-crop growers that need to reduce the volume of drainage effluent or to periodically supplement fresh water supplies. This is particularly attractive to growers on the western portion of the San Joaquin valley that must manage high saline water tables without opportunities for off-site disposal of drainage water or that are faced with reductions in fresh water supplies during periods of drought. Long-term field experiments (i.e. 6 yrs or more) conducted primarily by scientists from the USDA/ARS and University of California have demonstrated that saline drainage water applied in a cyclic manner can be used successfully as a supplemental source of irrigation water under the conditions tested. This paper reviews and synthesizes what has been learned from these studies and discusses the limits to which drainage water can be used for irrigation. The factors that affect these limits include the crop rotation and crop tolerance to salinity, the salinity and trace element concentration of both water supplies, the fraction of the season that drainage water is used for irrigation, the long-term effects on soil structure and whether the drainage water is blended or used cyclically. There are also economic, environmental and political factors that play an important role.

METHODS OF UTILIZING DRAINAGE WATER

Two methods have been proposed and tested for recycling saline drainage water. Both methods or strategies require an ample supply of good quality water that is available for irrigation at any given time during the season, particularly before and during the early part of the season. One method is referred to as the "blending strategy". This strategy involves mixing together the saline drainage water and the good quality water in certain ratios to achieve an irrigation water of suitable quality according to the crop's salt tolerance. This water is then used for each irrigation. The other method first introduced and tested by Rhoades (1984) is called the "cyclic strategy" where saline drainage water is solely used for certain crops and certain portions of their growing season. The objective of the cyclic strategy is to minimize salt stress during salt-sensitive growth stages or when salt-sensitive crops are grown.

The salinity in the crop root zone is not in steady state under the cyclic strategy but is transient allowing crops that vary in tolerance to be included in the rotation. The cyclic strategy keeps the average soil salinity lower especially in the upper, most critical portion of the profile during the early, salt-sensitive growth stage. In one theoretical study using a multi-seasonal transient state model, cyclic and blending strategies were compared where the same amount of water and salt were applied (Bradford and Letey, 1992). This study demonstrated that the cyclic strategy produced higher simulated yields of a salt-sensitive crop than did blending in a rotation with both salt-sensitive and salt-tolerant crops. However strategies did not differ when salt-tolerant or moderately salt -tolerant crops where the only crops in the simulated rotation.

In addition to allowing soil salinity to be lower at certain critical times allowing for more salt-sensitive crops to be included in the rotation, the "cyclic strategy" has many advantages over the blending method; 1) a water blending facility is not required 2) water of higher salinity can be used for periodic irrigations than if used for all irrigations and 3) greater use of the combined water supply (saline and non-saline sources) can be achieved (Grattan and Rhoades, 1990).

UPPER SALINITY LIMITS TO BLENDING

Blending saline drainage water with good quality water is often proposed as a means to expand the existing water supply. However blending does not unconditionally increase the usable water supply (Rhoades, 1984) nor is it always economically feasible (Dinar et. al., 1986). Too often growers are tempted to blend water that is too saline for use by the intended crop.

Since the purpose of blending is to increase the overall water supply to the crop, the salinity of the saline water component of the blend (i.e. drainage water) can not exceed a value where, if used directly without blending, the crop could no longer extract water and grow. If the salinity of the saline component exceeds this value, rather than increasing the overall water supply, blending will decrease the usable water supply. Therefore the upper limit depends upon the crop and the maximum salinity that crop can still transpire and grow. From a management point of view, it may be desirable to only blend with drainage water less than this theoretical upper-salinity limit.

REVIEW OF LONG-TERM CYCLIC REUSE STUDIES

Over the past decade there have been several long-term studies reported in California where saline drainage water (ECw 4 to 8 dS/m and up to 8 mg/L B) has been successfully used to periodically irrigate crops in a rotation without a significant reduction in economic yield (Ayars et al. 1993; Rhoades et al., 1988; Shennan et al., 1995). Not only have salt tolerant crops such as cotton and sugar beet been irrigated with saline drainage water (ECw of 7.4 to 7.5 dS/m) but moderately salt-sensitive crops such as tomato as well. In these studies, non-blended saline drainage water was applied to crops after they were well established with good quality water.

Furthermore these studies showed that saline drainage water can affect crop quality and in many cases positively. For example saline drainage water has increased total digestible nutrients in alfalfa (Rhoades et al., 1988 and increased soluble solids in tomato (Grattan et al., 1987) but there is not enough experience to conclude that improved quality characteristics will result when such crops are irrigated with saline water nor that these beneficial effects will justify the application of saline drainage water to crops

Despite the successes demonstrated by these investigators in using saline drainage water, recycling drainage water can not be done unconditionally. For example moderately salt-sensitive tomato could not be grown with saline drainage water for two consecutive years (Mitchell et al., unpublished data) nor could even salt-tolerant cotton be irrigated consecutively for extended years (Goyal, personal communication).

.TECHNICAL LIMITATIONS

Despite the success of these field studies, investigators emphasized concerns and potential problems using this strategy over the long-term (Ayars et al., 1993; Grattan and Rhoades, 1990: Shennan et al. 1995). Most importantly are the potential detrimental effects on soil physical quality and the accumulation of B in the crop root zone.

Soil Physical Conditions. The composition of the drainage water raises obvious concerns regarding soil-aggregate stability if this water is used for irrigation, particularly in a cyclic manner. In the San Joaquin valley the drainage water is dominated by Na-sulfate salts which if used for irrigation could build up the exchangeable sodium in the topsoil. Subsequent rains or pre-irrigations with fresh water reduce the salinity of the surface layer enough to create potentially unsuitable soil conditions (Oster et al., 1984). These unsuitable conditions could lead to aggregate instability, reduction in large pores, reduced water infiltration and soil crusting. Although investigators that conducted long-term studies in California did not indicate that water infiltration rates were reduced in plots previously irrigated with saline drainage water, several studies indicated that in certain years cotton stands were reduced in plots previously irrigated with saline water (Mitchell, personal communication; Rhoades et al., 1988; Shennan, et al., 1995).

Boron. Boron is an essential element to crops but there is a small concentration window between sufficiency and toxicity. Potential phytotoxicological concerns arise if drainage water containing more than several mg/L B is used for extended periods on the same field. Long-term reuse studies have shown that B, because of its greater adsorptive affinity than salts, is more prone to accumulate in soils (Ayars et al., 1993; Shennan et al., 1995). Therefore it requires more water to leach a certain fraction of the existing B out of the rootzone than it does to leach the same fraction of salts (Hoffman, 1980).

Toxic Trace Elements. An additional concern for reusing drainage water is the high trace element concentration in some drainage effluents. Certain trace elements, particularly Se and Mo, may accumulate in the crop and pose a hazard to the consumer (i.e. humans or animals). The dominant forms of Se and Mo in most drainage effluents are the oxyanions selenate and molybdate. While these anions are readily absorbed by plants and can accumulate in various tissues most studies have shown that these elements do not accumulate in crops to levels that are toxic to consumers because sulfate, a dominate anion in the effluent, competes in the absorption process with these oxyanions and substantially reduces the concentration in the plant tissue (Grattan et al., 1987; Shennan et al., 1995).

Appropriate management practices must be undertaken to avoid these potential problems. One possible alternative is to manage saline water table levels allowing certain crops to use water directly from the saline water table and thereby avoiding the application of drainage water to the soil surface (Ayars, personal communication).

OTHER LIMITATIONS

There are also economic, environmental and political factors that influence the long-term feasibility of recycling drainage water over the long term. Recycling drainage water can be economically efficient (Knapp and Posnikoff, 1997). However for irrigation agriculture to be sustainable requires drainage and drainage degrades water quality (van Schilfgaarde, 1990). Recycling drainage water reduces drainage volume and thus the ultimate cost to society. The presence of boron or trace elements in the effluent increases this cost.

DOES RECYCLING DRAINAGE WATER CONSERVE FRESH WATER?

Recycling drainage water conserves good quality water provided that the good quality water needed for periodic reclamation does not exceed the quantity saved during the reuse practice. Two long-term reuse studies (Ayars et al., 1993; Shennan et al., 1995) addressed this issue using reclamation formula for intermittent ponding (Hoffman, 1986). In both studies, drainage water supplied between 23 to 60% of the irrigation water requirement and they concluded that there is considerable savings in good quality water based on a salt reclamation perspective. However based on a B reclamation perspective, they concluded that it takes considerably more good quality water to reduce soil B to the original concentration than was saved during reuse. On the other hand it may not be necessary to reclaim the soil provided that reuse is intermittent. Once irrigation with only good quality water is resumed using sufficient amounts of pre-plant irrigation water, soil salinity and B will eventually reduce to satisfactory levels.

CONCLUDING REMARKS

Several long-term field studies (i.e. 6 yrs or more) conducted by the USDA/ARS and the University of California have demonstrated that saline drainage water (EC_W as high as 8 dS/m and 8 mg/L B) applied in a cyclic manner with good quality water could be used successfully as a supplemental source of irrigation water. Success of the strategy has been attributed to transient soil salinity profiles that are created under this practice allowing for more salt-sensitive crops to be included in the rotation and imposing salt-stress for shorter durations and during the portion of the season when the crop is more tolerant to salinity. However there are limits on the extent to which drainage water can be used for irrigation. Besides salinity, the major concerns with recycling drainage water over the long-term are the accumulation of B in the soil profile to toxic levels (i.e. drainage waters with over several mg/L B) and potential degradation of soil structure (i.e. using sodic drainage waters).

If proper care is taken, recycling drainage water for irrigation can be feasible and economically attractive to growers who need to reduce the volume of drainage effluent or those who face temporary shortages in good quality water supplies. Many agronomic and row crops can tolerate saline irrigation water substantially in excess of the traditional ECw-threshold values (Ayers and Westcot, 1985). This it not to say that recycling drainage water can be done successfully of the long-term. Boron and sodicity will likely limit the long-term feasibility of recycling drainage water more than does salinity. In light of the calculations conducted by Ayars et al. (1993) and Shennan et al. (1995), use of irrigation with saline drainage water containing high levels of B (e.g. 5 mg/L) does not conserve good quality water due to the volumes of good quality water needed to reclaim the soil profile. However with appropriate water management and care there is merit to intermittent and not continuous use of many drainage waters for irrigation, particularly if used in a cyclic manner.

REFERENCES

Ayars, J.E., R.B. Hutmacher, R.A. Schoneman, S.S. Vail, and T. Pflaum. 1993. Long-term use of saline water for irrigation. Irrig. Sci. 14:27-34.

Ayers, R.S. and D.W. Westcot. 1985. Water Quality for Agriculture. FAO Irrigation and Drainage Paper 29. Food and Agriculture Organization. United Nations. Rome

Bradford, S. and J. Letey. 1992. Cyclic and blending strategies for using non saline and saline waters for irrigation. Irrig. Sci. 13:123-128

Dinar, A., Letey, J. and H.J. Vaux. 1986. Optimal ratios of saline and non saline irrigation waters for crop production. Soil Sci. Soc. Am. J. 50:440-443

Grattan, S.R., Shennan, C., D.M. May, J.P. Mitchell and R.G. Burau. 1987. Use of drainage water for irrigation of melons and tomatoes. Calif. Agric. 41:27-28.

Grattan, S.R. and J.D. Rhoades. 1990. Irrigation with saline ground water and drainage water. In K.K. Tanji (ed.) Agricultural Salinity Assessment and Management. ASCE Manuals and Reports on Engineering Practices No. 71 ASCE pp. 432-449.

Hoffman, G.J. 1986. Guidelines for reclamation of salt-affected soils. Appl. Agric. Res. 1(2):65-72.

Knapp, K.C. and J. F. Posnikoff. 1997. Saline drainwater reuse in irrigated agriculture: Economics and management. Plant/Soil Conference. Calif. Chapter of ASA. Visalia, CA Jan. 15-16.

Oster, J.D., G.J. Hoffman, and F.E. Robinson. 1984. Management alternatives: crop, water, and soil. Calif. Agric. 38(10):29-32.

Rhoades, J.D. 1984. Use of saline water for irrigation. Calif. Agric 38(10):42-43.

Rhoades, J.D., F.T. Bingham, J. Letey, A.R. Dedrick, M. Bean and G.J Hoffman, Alves, W.J., Swain, R.V. Pacheco, P.G. and Lemert, R.D. 1988. Reuse of drainage water for irrigation: Results of Imperial valley study. I. Hypothesis, experimental procedures and cropping results. Hilgardia 56:1-16.

Shennan, C., S.R. Grattan, D.M. May, C.J. Hillhouse, D.P. Schachtman, M. Wander, B. Roberts, R.G. Burau, C. McNeish and L. Zelinski. 1995. Feasibility of cyclic reuse of saline drainage in a tomato-cotton rotation. J. Environ. Qual. 24(3):476-486.

van Schilfgaarde, J. 1990. Irrigated agriculture: Is it sustainable?. In K.K. Tanji (ed.) Agricultural Salinity Assessment and Management. ASCE Manuals and Reports on Engineering Practices No. 71 ASCE pp. 584-594.

Management of Drainage Water, Salt, and Selenium as Resources

by

V. CERVINKA, J. DIENER, C. FINCH, M. MARTIN, F. MENEZES,
R. MUNOZ, and D. PETERS

California Dept. of Food & Agriculture, Sacramento, Red Rock Ranch, Five Points, Westside Resource Conservation District, Five Points, USDA - Natural Resources Conservation Department, Fresno, California Dept. of Water Resources, Fresno, University of California, Davis

ABSTRACT

Society is beginning to productively utilize materials which in the past were considered waste. San Joaquin Valley growers are also beginning to manage drainage water, salt, and selenium as resources rather than disposing of them as a waste. Drainage water is being sequentially reused to irrigate trees and plants of progressively increasing salt tolerance. About 80 to 85 percent of drainage water is used to produce marketable commodities. Plants uptake selenium, and they can be harvested as "selenium enriched forage" for livestock feeding in selenium deficient areas. About 15 to 20 percent of the final volume of drainage water is discharged into a solar evaporator to evaporate water and crystallize salt. Initial research and development efforts indicate the possibility of producing marketable commodities from salt. The concept of managing drainage water, salt, and water as resources, as well as technical data from multi-year research, are presented in this paper.

INTRODUCTION

Our basic resources are sun, soil, water, air, plants, animals, and people. Societies prosper when these resources are sustained and productively utilized. Sustainable farming on irrigated land requires the continuous removal of salts imported from irrigation water. Techniques of on-farm salt leaching and removal have been practiced for many years. The real problem is how to manage the remaining salt. A common practice is to discharge this salt into rivers, lakes, seas, and desert areas -- which may lead to economic and ecological devastation in these disposal areas.

A system for the sequential reuse of drainage water reduces its volume, manages it on/near the site of origin, and uses it as a resource for the production of marketable commodities. Drainage water is not wasted but is reused several times for the production of commercial crops of a progressively increasing salt tolerance. The final volume of highly mineralized drainage water is discharged into solar evaporation facilities on farms with the intention of processing salt into a marketable commodity. This paper is based upon 10 years of experience with on-farm systems for the sequential reuse of drainage water.

CONCEPT OF SEQUENTIAL REUSE

Drainage water is reused through a biological system of crops and trees of differentiated salt tolerance. During this process, salt concentration increases as the volume of drainage water decreases. In the final stage of this sequential reuse process, the remaining drainage water is discharged into a solar evaporator or evaporation ponds for water to evaporate and for salt to crystallize. A solar evaporator is a leveled area of land with borders that is covered with black plastic liner. The volume of drainage water can be reduced to about 15 - 25 percent of the original volume and the salt concentration can reach above EC 40 (38 g/L) at the final state of the sequential reuse process. The harvested salt offers various options for its utilization: commercial salt, chemicals, construction materials or producing electricity. Research and development continues to progress on the utilization of harvested salt.

INTEGRATED MANAGEMENT

Irrigation and drainage water are managed as an integrated system. Water conservation methods are used in the irrigation of conventional farm crops. The drainage water is sequentially reused in the irrigation of moderately salt sensitive crops, moderately salt tolerant crops or trees, and lastly salt tolerant crops. Salt sensitive crops (< 1000 mg/L of salt) include vegetables, fruits, cereals, and similar farm crops. Moderately salt sensitive crops (2000 to 6000 mg/L of salt) include cotton, alfalfa, sugar beets, and other farm crops. Salt tolerant plants (> 8000 mg/L of salt) include various salt tolerant trees and halophytes. The main function of trees and halophytes is to use drainage water and reduce the volume of drainage water. High evapotranspiration characteristics are therefore desirable. Trees and halophytes also have a potential for economic return of livestock forage and/or biomass used for energy or industrial materials. Salt tolerant trees may include eucalyptus, casuarina, athel and others. Common halophytes include atriplex, saltgrass, salicornia among others.

DESIGN PARAMETERS

The bio-engineering system for sequential reuse of water and salt removal utilizes drainage water with a salt concentration of approximately 7,000 mg/L. The system concentrates salt and significantly reduces the volume of reused drainage water. This technology provides alternatives for managing salt in a relatively small area on farms called solar evaporators. It also allows for alternative management of discharging a reduced volume of drainage water into solar ponds or natural disposal sites (e.g., ocean). The crystallization of salt in solar evaporators also provides management options for salt marketing or its (short/long term) storage on farms. Major components of crystallized salt are sulfate (56.70 %), sodium (23.50 %), and chloride (8.40 %).

The management of trees and halophytes is essential for the efficient removal of salt. Trees and halophytes need to be selected for their salt tolerance, which should range from about 9,000 to 18,000 mg/L for the trees and above 25,000 mg/L for halophytes. Halophytes should preferably be perennial plants. Other required characteristics of trees and halophytes include high water demands, tolerance to frequent flooding, frost tolerance, and marketability of harvested biomass.

The relative size of each areas is as follows:

trees	10	ha
halophytes	4	
solar evaporator	2	
total	16	ha

The tree area (10 ha) has the capacity to process about 180 ML of drainage water. The farm size that can be serviced by this bio-engineering system is dependent on several factors, such as its cropping system, the quality of irrigation water, water management, soil salinity, and the use of trees for controlling groundwater conditions (water uptake from high water tables or subsurface flow).

The percentage of the infiltrated irrigation water that percolates below the root zone, known as the leaching fraction, must be sufficiently high to prevent a build-up of salts, selenium, and boron in the soil profile.

ON-FARM PROJECTS
The U. S. Bureau of Reclamation is supporting two special salt management projects in the San Joaquin Valley region: an experimental project near Mendota and a demonstration project near Five Points. The experimental project consists of 6.6 ha of trees, 2.7 ha of halophytes, and 1.3 ha of a solar evaporator. Approximately 120 ML of drainage water per year with a typical salt concentration of 7,500 mg/L is reduced. The demonstration project integrates both management of irrigation and management of drainage water. This project includes conventional farm crops (192 ha of salt sensitive crops -- such as vegetables, 52 ha of salt tolerant crops -- such as cotton or alfalfa), 5 ha of trees, 1.8 ha of halophytes, and 0.7 ha of a solar evaporator. The trees receive about 90 ML of drainage water per year with a salt concentration of about 8,000 mg/L.

The objectives of both projects include management of salt and selenium on farms, achieving sustained salt balance on irrigated land, salt harvesting, selenium reduction, and demonstration of an environmental, technical and economically feasible bio-engineering system. These projects will be completed by December 1997. A Technical Report (TR) will present an analysis of the data monitored, and will include calculations of water, salt and selenium balances. A Technology Transfer Guide (TTG) will also be prepared to facilitate future developments of drainage sequential water reuse and salt harvesting system on other farms. This TTG will include design characteristics, operational instructions, management methods for environmental considerations (wildlife safety), and system economics.

Various performance characteristics in both experimental and demonstration projects have been studied. The project data indicates these preliminary results:

1. SALT AND BORON LEACHING
A drainage system was installed on a farm, 64 ha in area, during 1995. Drainage water from this field is applied to moderately salt tolerant crops, trees, and halophytes. The salt crystallizes in the solar evaporator. Significant leaching of salt and boron from the crop root zone has been achieved as indicated in the following table:

field	salt (dS/m)		boron (mg/L)	
	1995	1996	1995	1996
A	10.0	1.8	15.3	2.3
B	13.8	3.1	13.3	3.9
C	10.9	2.0	3.5	1.7
D	2.9	1.1	2.5	0.7

2. LONG-TERM SALT BALANCE

The Mendota experimental project started in 1986. The comparison of soil profile data, analyzed in 1987 and 1995, to the depth of 2.1 m indicate that long-term salt balance has been achieved. Calculations indicate that between 300 to 700 tons of salt has been imported to the Mendota site each year in the drainage water.

3. SELENIUM REDUCTION

The typical selenium concentration in drainage water applied to the trees is about 0.3 mg/L, and this increases to about 0.7 mg/L at the final discharge into the solar evaporator. An incoming drainage water of 100 ML contains .3 kg of selenium. Some 85 ML of this water is used for the trees and halophytes; there is only .1 kg (about 30 percent) of selenium discharged into the solar evaporator.

4. NITRATE REDUCTION

This bio-engineering system for the sequential reuse of drainage water was not originally designed for the management of nitrates. However, the monitored data indicates that the amount of nitrates is significantly reduced through water reuse. To illustrate this case -- nitrates are concentrated at about 90 mg/L in the incoming drainage water, and their concentration is reduced to about 65 mg/L in the drainage water discharged into the solar evaporator. There are about 90 kg of nitrates in 100 ML of drainage water. This amount is reduced to about 9.75 kg (about 11 percent) in the 15 ML of drainage water discharged into the solar evaporator. The data indicates that the bio-engineering system for sequential water reuse can also be used for the management of nitrates.

CONCLUSION

Sequential reuse systems provide an opportunity for ecological management of drainage water and salt. The on-farm utilization of drainage water and salt encourages water conservation and good management of drainage water and salt. It benefits the ecological sustainability of farms as well as the locations of potential "waste" discharge. The comparison of two methods of drainage water and salt management, as (1) waste products and (2) resources, is illustrated in Chart 1. Drainage water and salt can be managed as resources to sustain the productivity of agriculture and the quality of environment in a way which is technically sound and economically beneficial.

DRAINAGE WATER/SALT/SELENIUM 289

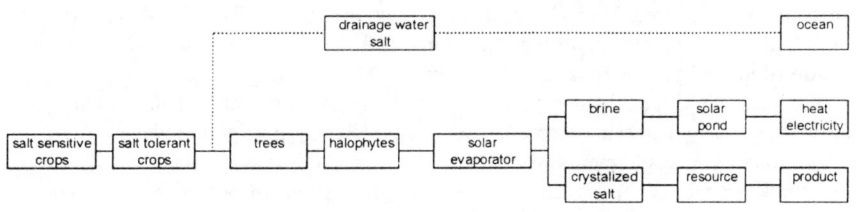

CHART 1.

REUSE SYSTEM OF IRRIGATION RETURN FLOW AND WATER QUALITY MANAGEMENT IN LOW LYING CREEK-PADDY AREA

MASAHARU KURODA* and TETSURO FUKUDA**
*Professor and **Assistant Professor, Irrigation Engineering Lab.,
Faculty of Agriculture, Kyushu University, Fukuoka 812-81 JAPAN

ABSTRACT

The operation and management of paddy irrigation system is characterized by reuse of irrigation return flow and drainage condition in low lying delta area.
The water reuse mechanism is very effective for conserving irrigation water resources but it has severe impact on water quality.
The purpose of the research is consisted of following three items: [a] Quantitative analyses on the reuse mechanism of irrigation return flow in delta area,[b] Investigation of water quality in the area, and [c] To draw conclusions on the relationship between the reuse of return flow and water quality.
A complex tank model was proposed for analyzing the water balance, and for evaluating the content ratio of reuse water being included in irrigation water.
Water quality measurements were carried for several indices such as Chemical Oxygen Demand (COD), Electric Conductivity (EC), Total Nitrogen (TN) and so on.

1. OUTLINE OF SURVEYED AREA

The typical operation and management of the irrigation system has been effectively performed utilizing the storage and buffer functions of water within creek networks in the low lying paddy area reclaimed from swamps by dikes.
As such functions exist in the area, the operation of irrigation system is characterized by the reuse mechanism of return flow of water from paddy fields to creek networks.
The water reuse mechanism is very effective for conserving irrigation water resources but it has severe impact on water quality.
The study is carried out on the typical paddy area in the Kubota district being a part of the Kase River Irrigation Project located in Kyushu, Japan.
The Kubota district is, as illustrated in Fig.1, divided into three blocks named A, B and C, from upstream side to downstream side along the dike lines from old to new ones. The net area of the paddy fields in this district is 955ha and the surface area of creek network is 89.5ha. These creeks were originally water ways in the swamp and the creek network remained like spider net inside the paddy area after reclamation.

In the block A, the natural drainage is available for every day not only at spring tide but also at neap tide. In the block B, the natural drainage is possible in usual but it is strongly limited for a few days at neap tide. In the block C, the natural drainage is impossible for several days at neap tide (Kuroda et al, 1991).

2. COMPLEX TANK MODEL FOR HYDROLOGIC ANALYSIS IN THE IRRIGATION DISTRICT

In the block A, the irrigation water mainly supplied from the main canal to creeks. In the blocks B and C, the irrigation water is supplied from both the main canal and the regulating reservoir or creeks of upper blocks through the water redistribution gates shown in Fig.1. The complex tank model was proposed to analyze the hydrologic cycle of water resources and to evaluate the irrigation operation in the district.

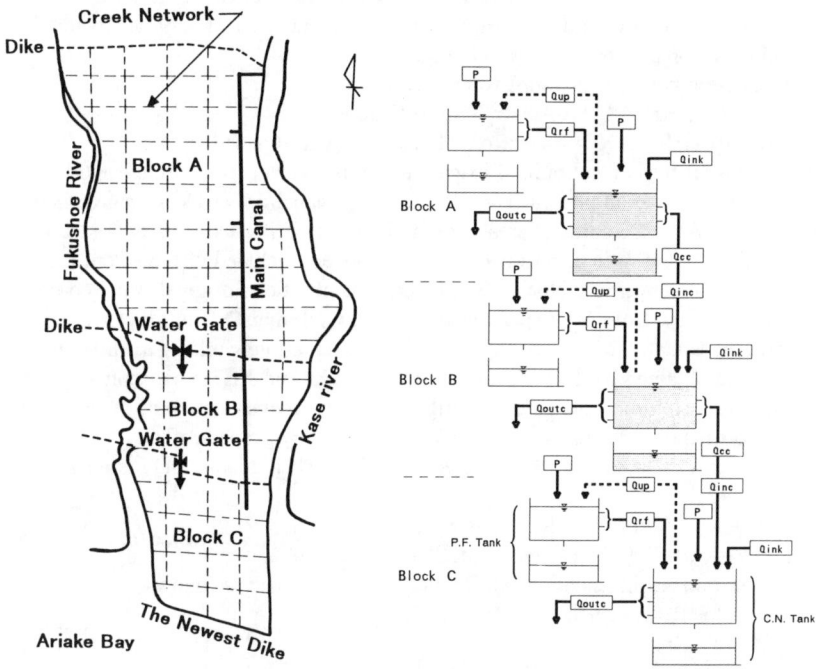

Fig. 1 Layout of the Kubota District, Kase River irrigation Project, Japan

Fig. 2 Complex Tank Model for Hydrologic process in Creek Network
P.F. Tank : Paddy Field Tank Model
C.N. Tank : Creek Network Tank Model

The construction of the complex tank model is shown in Fig.2. The complex model consists of the tanks for paddy fields (the field model) and the tanks for creek networks (the creek model) called the 2×2 type complex tank model. The complex tank models are finally arranged in series for evaluating water movement through the block A, the block B and the block C. The symbol codes in Fig.2 were defined as follows, i.e., P = effective rain fall, Qup = water application to paddy plots from creeks by pumping, Qrf = return flow from paddy plots to creeks by surface drainage and seepage. Qink = water supply from main canal to creeks, Qinc = water from the upper creek networks, Qoutc = drainage from creeks to outside of system, Qcc = water flow from upper creeks to lower creeks (Kuroda et al.,1988).

3. CONTENT RATIO OF IRRIGATION RETURN WATER (Rr) IN CREEK NETWORKS

The following analogical treatment was introduced for the quantitative estimations of irrigation return flow and for evaluating the content ratio of returned water being included in stored water in creek networks.
The following balance equations were obtained.

$$Xr_i * 10Ac * Rr_{i-1} = X_{i-1} * 10Ac * Rr_{i-1} + Qrf + Qinc * Rr_u \quad] \quad \cdots\cdots\cdots\cdots\cdots (1)$$
$$Xo_i * 10Ac * Ro_{i-1} = X_{i-1} * 10Ac * Ro_{i-1} + P * 10Ac + Qink + Qinc * Ro_u$$

in which Rr is the content ratio of irrigation return water in creek storage and Ro is the content ratio of original water from the main canal in creek storage, therefore, Rr+Ro=1. Ac is a command area of each block (ha), P is effective rain fall (mm/day). Rr_u is the content ratio of returned water component included in flow from the upper block into the concerning block. Ro_u is the content ratio of original water component included in flow from the upper block. Qinc, Qink and Qrf (m^3/day) were already denoted in Chapter 2. X is depth of water in creek network as a initial value of calculation obtained by the tank model analysis. Xr and Xo are conceptual depths of the irrigation return water and the original water in creek network, respectively. Subscripts i and i-1 means the day and the previous day, respectively.
In Equation (1), unknown terms are Xr and Xo. Then equation (1) is changed into Equation (2).

$$Xr_i = X_{i-1} + (Qrf + Qinc * Rr_u)/(10Ac * Rr_{i-1})$$
$$Xo_i = X_{i-1} + (P * 10Ac + Qink + Qinc * Ro_u)/(10Ac * Ro_{i-1}) \quad] \quad \cdots\cdots\cdots\cdots\cdots\cdots (2)$$

Using Xr_i and Xo_i in Equation (2), Rr_i and Ro_i are obtained as follows,

$$Rr_i = Xr_i * Rr_{i-1}/(Xr_i * Rr_{i-1} + Xo_i * Ro_{i-1}) \quad] \quad \cdots\cdots\cdots\cdots\cdots\cdots (3)$$
$$Ro_i = 1 - Rr_i$$

Fig.3 shows one of the example of the obtained data using Equation (3).
In fine day, the content ratio of return flow is around 40% in the block A (the upper block), 50% in the block B (the middle block) and 60% in the block C (the lower block) respectively.

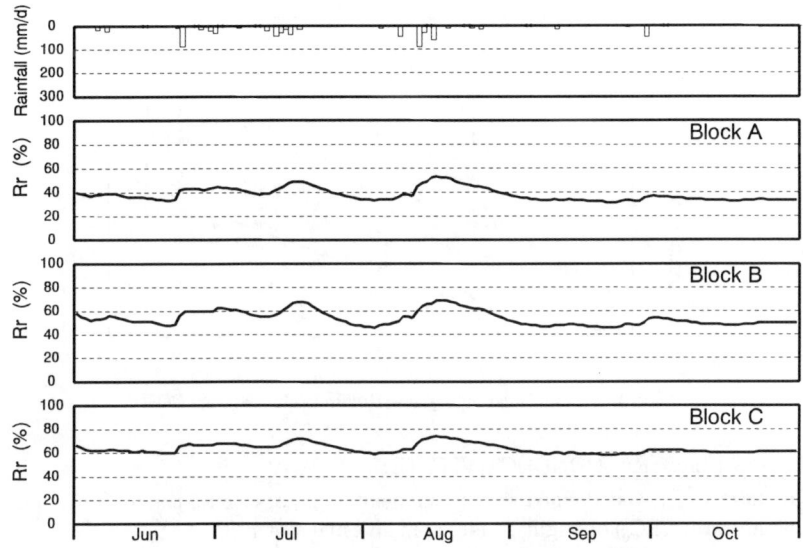

Fig. 3 Daily Changing of Content Ratio of Irrigation Return Water(Rr) in 1992

4. RELATIONS BETWEEN WATER REUSE MECHANISM AND WATER QUALITY

4.1 SURVEYING WATER QUALITY

Field measurements on water qualities were carried out from 1991 to 1995 during paddy irrigation season. Measuring points for water qualities are in canals, creeks, paddy plots, regulating reservoir and retarding basins.

The several analysis were tried to clarify the relations between the content ratio of reuse water and severe measuring items of COD, EC and TN.

4.2 ON CHEMICAL OXYGEN DEMAND (COD)

The term ΔCOD was defined as the increment value of chemical oxygen demand from the primary value (COD_0) of original water in main canal system.

The relation between ΔCOD (=COD-COD_0) and the content ratio of returned water (Rr) is shown in Fig.4. ΔCOD is increasing due to increase of the Rr values. The general aspect is given by the following function.

$$\Delta COD = 10.41 \, Rr^{1.34}$$

The primary value COD_0 is around 2~3mg/l depending climate and seasonal conditions. It is recommended, therefore, that the value of Rr should be smaller than 0.4 for keeping the restricted value of COD (less than 6.0mg/l).

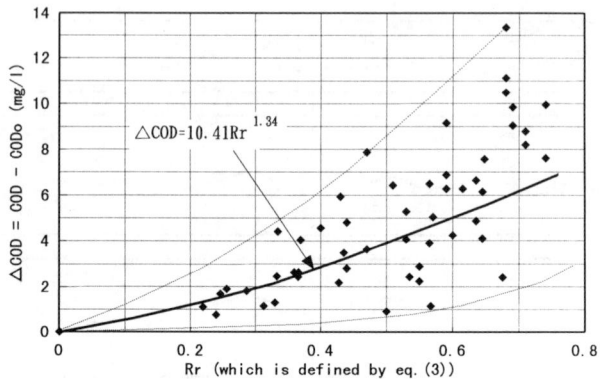

Fig.4 Content Ratio of Irrigation Return Water (Rr) and COD

4.3 ON ELECTRIC CONDUCTIVITY (EC)

The term ΔEC was defined as the increment value of electric conductivity from the primary value (EC_0) of original water in main canal system.

The relation between ΔEC (=$EC-EC_0$) and the content ratio of returned water (Rr) is shown in Fig.5. ΔEC is increasing due to increase of Rr values. The general aspect is given by the following function.

$$\Delta EC = 263.0 \, Rr^{1.80}$$

The primary value EC_0 is around $80 \sim 100 \, \mu$ S/cm depending climate and seasonal conditions. It is recommended, therefore, that the value of Rr should be smaller than 0.8 for keeping the restricted value of EC (less than 300 μ S/cm). For the EC value, the content ratio of returned water Rr is not so severe than the case of COD value in this area.

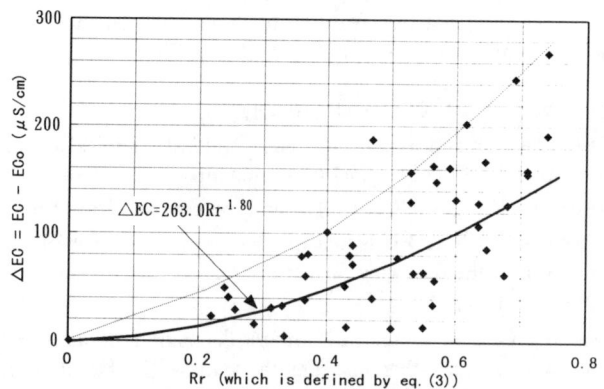

Fig. 5 Content Ratio of Irrigation Return Water (Rr) and EC

4.4 ON NITROGEN CONTENT (TN)

The term Δ TN was defined as the increment value of nitrogen content from the primary value (TN_0) of original water in main canal system.
The primary value of TN is $1.0 \sim 2.0$ mg/l in the main canal of this area.
The relation between ΔTN (=$TN-TN_0$) and the content ratio of returned water (Rr) is shown in Fig.6. ΔTN has the both symbols (+) or (-) as the results of analysis.
The (+) symbol means polluting action and the (-) symbol indicates clarification function. According to the results, it is sure that the paddy field irrigation has two faces concerning TN. One is clarification function and another is polluting action. This phenomena will be closely related to the glowing stages of paddy, climates and particularly fertilizer managements.
The general aspect on TN is given by the following function.

$$\Delta TN = 2.35 \, Rr^{1.82}$$

ΔTN is slightly changeable due to Rr values.

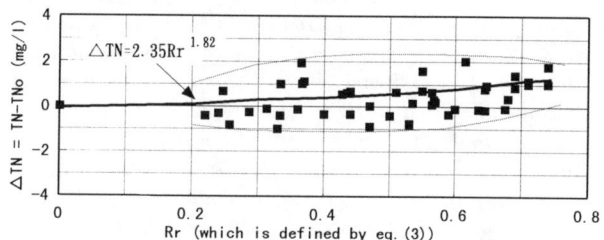

Fig. 6 Content Ratio of Irrigation Return Water (Rr) and TN

5. CONCLUSIONS

(1) Water quality indices, for example, COD and EC are gradually increased in up, middle and down-stream in order in the area of irrigation project.
(2) These facts suggest that as the reuse cycles of irrigation water is increased, the water quality index values are also increased.
(3) For the TN, it is considered that occasionally, paddy fields consume nitrogen from irrigation water and sometimes supply nitrogen into irrigation water.
(4) A recommendable operation model of irrigation system is proposed for
maintaining water quality while saving water resources. The operation of water intake from the main canal, for example, should be kept to maintain $Rr \leq 0.4$ for COD.

REFERENCES

Kuroda,M., T.Fukuda & F.Nurrochmad. 1991. Operation of irrigation system and water management in low lying paddy area with creek network. Irrigation and Drainage, Proc. of 1991 National Conference ASCE 31-37
Kuroda,M. and T.Cho. 1988. Water management and operation of irrigation system in low lying paddy area with creek network. Irrigation Engineering and Rural Planning, Japanese Soc. of IDRE 13:36-46.

WASTEWATER REUSE FOR IRRIGATION IN ARID REGIONS

N. ISMAIL [1]
M. ASCE, Arab Academy for Science & Technology, Alexandria, Egypt

ABSTRACT

Field data of effluents discharged from natural wastewater treatment plants in arid regions have been examined to assess qualifications of meeting the quality standards established for irrigation with treated sewage water. The field data are collected for the natural treatment plant of Khirbet es Samra (KS) in Jordan, which discharges its effluent into the Zerka River, which in turn runs to the King-Talal Reservoir (KTR) for further dilution and storage. Water quality parameters of wastewater treated by natural processes and measured further downstream along the river showed high organic/inorganic loads in the river, which indicates insufficient microbial oxidation and possible unauthorized waste dumping. The data taken for the plant effluent compare reasonably well with data reported on similar plants in Saudi-Arabia and the U.S.A. and show that the treated wastewater could be used for restricted irrigation. Further use for unrestricted irrigation could be permitted if operational management of oxidation ponds and receiving surface water is well maintained.

INTRODUCTION

Since the beginning of the twentieth century, there have been numerous cases of wastewater reclamation and reuse projects. In particular, during the last decade wastewater treatment and use in agriculture has received much attention especially in the arid and semiarid regions. Despite social, political, cultural and economic differences in many regions around the globe, when it comes to the fate of water resources in semiarid environments [4], many similarities actually exist (Fig. 1).
For the Arabian Peninsula, limited amounts of treated wastewater effluent are being used in Saudi Arabia, Kuwait, Qatar, Bahrain and Oman for irrigation of landscape and noncash crops. The total volume of treated effluent used in 1992 is estimated at $453,000,000$ m^3. This amount represents only 21% of the total wastewater volume, estimated at 2.2 billion m^3.

---

[1] Prof., Dept. of Construction Engineering, AAST. Formerly Director of Engineering & Technology, Wellstream Corporation, Panama City, Florida, U.S.A.

In the U.S.A. national survey on wastewater reclamation and reuse projects, 536 wastewater reuse projects were in existence in 1975. The estimated total wastewater reuse was 680 Mgal/d. Most of the wastewater reuse sites are located in the arid and semiarid western and south western states, including, California, Arizona, Colorado and Texas [5] .

QUALITY STANDARDS FOR IRRIGATION WITH SEWAGE EFFLUENT

Reclaimed wastewater regulations for specific irrigation uses are based on the expected degree of human contact with the reclaimed wastewater and the intended use of the irrigated crops. For example, the state of California requires that reclaimed wastewater used for landscape irrigation of areas with unlimited public access must be "adequately oxidized, filtered, and disinfected prior to use," with median total coliform count of no more than 2.2/100 ml . On the other hand reclaimed water quality criteria for protecting health in developing countries are often established in relation to the limited resources available for public works, and other health delivery systems may yield greater health benefits for the funds spent [9] .

The World Health Organization (WHO) has recommended that crops that will be eaten raw should be irrigated with treated wastewater only after it has undergone biological treatment and disinfection to achieve a coliform level of not more than 100/100 ml in 80 percent of the samples [9]. The criteria recommended by WHO for irrigation with reclaimed wastewater have been accepted as reasonable goals for the design of such facilities in many Mediterranean countries and in some Middle East Countries such as the existing wastewater treatment facilities of Al-Hassa area, in Saudi Arabia [1].

Adoption of more stringent regulations and treatment design alternatives (Fig. 2) are done to protect high standard of public health by preventing, at any expense, the introduction of pathogens into the human food chain [1, 3] .

JORDAN VALLEY WASTEWATER RECYCLE PROJECT

In 1985, the khirbet-es Samra (KS) natural wastewater treatment plant started to discharge its effluent to the Zerka River through Wadi Dhuleil (Fig 1).

The KS plant receives its raw waste water inflow from the capital Amman by a pipeline system at a rate of 3 x 10^6 gal/day. The Zerka River is a natural water stream located in the main urbanization center in Jordan. Domestic as well as industrial wastewater, whether treated or untreated are being disposed to the Zerka River which flows into the King Talal Reservoir (KTR). The inflow of water into the reservoir as well as the water accumulated from rainfall are being stored and used for irrigation of land in the Jordan Central Valley Zone (Fig. 1).

Existing wastewater treatment facilities at KS plant consist of a series of stabilization pond system followed by a maturation pond (Fig. 3-a). The total detention time for sewage in the pond system is 40 days. The outflow of the treated wastewater runs through a steep man-made concrete channel (Fig. 3-b) into the natural channel of Wadi Dhuleil (Fig. 3-c).

The maximum dilution of treated KS sewage effluent water by the Zerka River water ranges between 25% in winter and 60% in summer [8].

ANALYSIS OF WATER QUALITY DATA

Water quality of the Zerka River, as represented by the different physical, chemical, and biological parameters, was analyzed for several samples taken at the KS plant and downstream the discharge channel and along the river (Fig. 1).

- Dissolved Oxygen (DO) : The DO content of KS effluent at the plant exist ranged during the study period from zero to 3.3. mg/l. From site "a" downstream, it increased gradually to site "D" where concentration ranged from 6 to 12 mg/l and decreased thereafter [7].
- Chemical Oxygen Demand (COD) : There has been gradual decrease in the COD values from KS and further downstream along the river from 550 to 20 mg/l [8].
- Biological Parameters : Treated water leaving KS treatment plant contained number of fecal coliforms in the range of $10^2 - 10^3/100$ ml for short distance downstream of the concrete channel, thereafter the fecal coliforms number continued to increase up to $10^4/100$ ml [7, 8].

Natural biological wastewater treatment whether in stabilization ponds or in surface stream water depends primarily on microbial oxidation and reduction processes.
Oxidation of organic and inorganic matter requires a continuous supply of oxygen, i.e. reoxygenation of stream flow, whose rate is : K_2 is determined by the following equation :
$$K_2 \sim \frac{(D_L U)^{1/2}}{H^{3/2}}$$
where D_L = coefficient of molecular diffusion for oxygen ; m^2/day,
U = mean stream velocity , m/s, and H average water depth ; m.

In addition the mixing of these pollutants in the river flow would depend on its transverse dispersion. An accurate estimate of the transverse dispersion coefficient D_T in the Zerka slow moving river is rather difficult but an order of magnitude value can be obtained by the following equation [10] :
$$D_T = 0.2 \ HU^*$$
where U^* = shear velocity, m/s
For the discharge concrete channel values of reareation constant K_2 and the transverse dispersion coefficient D_T were estimated and found to be as large as twice the values calculated for the natural stream of the Zerka River further downstream of the KS plant exist. Turbulent mixing was the dominant mechanism of dilution in the first portion of the discharge channel (Fig. 3-b). The width of the river in the study area ranged between 2 to 3 m and the average water depth was about 0.4m.

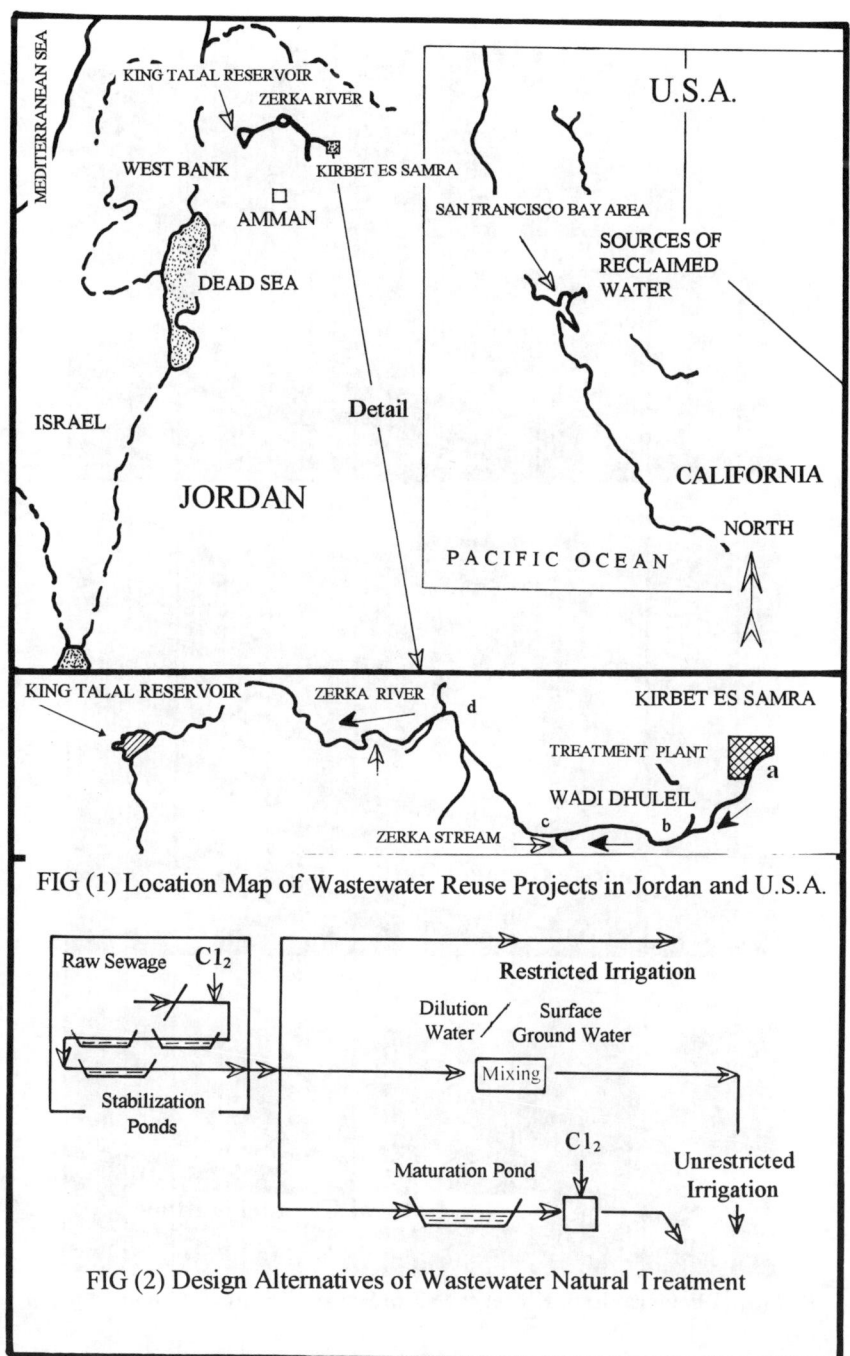

FIG (1) Location Map of Wastewater Reuse Projects in Jordan and U.S.A.

FIG (2) Design Alternatives of Wastewater Natural Treatment

FIG (3) - Photographs of Kirbet es Samra Sewage Treatment Plant - Jordan

CONCLUSIONS

For safe and unrestricted use of the effluent of KS wastewater treatment plant for irrigation, the natural treatment processes and maintenance of existing stabilization ponds should be further studied. Hydrologic mixing processes and effects of wind-induced reaeration in the discharge channel played a key role in enhancing the treatment quality of the KS wastewater outflow.

REFERENCES

1. Abdel-Rahman, W. and Shalam, A., "Reuse of Wastewater Effluent for Irrigation in Severely Arid-Regions", Water Resources Development, Vol 7, Number 4, 1991.

2. Abu-Zeid M., " Water Resources Assessment for Egypt", Water Resources Development", Vol. 8, No. 2, June 1992.

3. Davis, C., Yoloye, O. and Pawson, M., "Fostering Reclamation through Cooperation", Proc. North American Water and Environment Congress, ASCE, Anaheim, California, June 1966.

4. Grigg, N., "Water Resources Management; Principles; Regulations and Cases", McGraw-Hill Company, New York, N.Y., 1996.

5. Harnett, J. and Hall P., "San Francisco Bay Area Water Resources Study", Proc. Specialty Conference; Water Forum 81, Vol.I, ASCE, San Francisco, California, August 1981.

6. Ismail, N., Wiegel, R.L., Ryan, P. J and Tu, S.W, "Mixing of Buoyant Discharges in Coastal Waters", Proc. 21st Int. Conf. on Coastal Engineering, ASCE, M alaga, Spain, June 1988.

7. Ismail, N.M., Jawad, S. and Al-Howary, S. "Assessment of Water Quality in Zerka River",Technical Report,Department of Civil Engineering , Jordan University of Science & Technology, Irbid , Jordan, 1990.

8. Salameh, E., et al, "the Effects of Khirbet es Samra Effluent on Water Quality of Wadi Dhuleil and Zerka River", Technical Report, Water Research & Study Center, University of Jordan, Jordan, 1987.

9. Tchobanoglous, G. and Burton, F. ,"Wastewater Engineering, Treatment, Disposal and Reuse" , Mc Graw-Hill Inc., New York, N.Y., 1993 .

10. Fischer, H., List, J., Koh, R., Imberger, J. and Brooks N., "Mixing in Inland and Coastal Water", Academic Press, New York , N.Y , 1979.

Implementation of Canal Automation in Central Arizona

A.J. CLEMMENS, E. BAUTISTA, and R.J. STRAND
U.S. Water Conservation Laboratory, Phoenix, AZ, U.S.A.

ABSTRACT

The automation of irrigation distribution canals in central Arizona promises to improve water-delivery service to farmers, reduce operating costs, and improve distribution efficiency (i.e., reduce unaccounted for losses). Testing and implementation of canal automation is being carried out with two irrigation projects in Central Arizona, the Salt River Project (SRP) and the Maricopa Stanfield Irrigation and Drainage District (MSIDD). Also, a cooperative research and development agreement (CRADA) was initiated with Automata, Inc., a manufacturer of electronic sensing and control systems, for developing a canal automation product line. Implementation of canal automation within these two project will demonstrate the capabilities and limitations of this technology. The cooperation with Automata will put this technology (software and hardware) into the marketplace and allow more straightforward application to other irrigation districts.

INTRODUCTION

Declining groundwater levels in the Central Arizona Senoran Desert prompted the State of Arizona to restrict groundwater pumping and prompted irrigated agriculture districts and the U.S. Bureau of Reclamation (USBR) to construct the Central Arizona Project (CAP), which transports Colorado River water to central Arizona. Most irrigation projects in central Arizona are conjunctive use projects, using both surface and groundwater. The high cost of pumping groundwater and transporting Colorado River water to central Arizona has increased the need for more efficient use of that water. Improved measurement and control, thus, are a high priority in most central Arizona districts. Additionally, the USBR requires irrigation districts to develop and implement water conservation plans, including improved operations (e.g., canal automation).

Research on canal automation has increased significantly over the past decade. Malaterre (1995) cited 98 papers on canal regulation, with all but about 20 published within the last 10 years. Most of the earlier articles describe mechanical, analog-electrical, or ad-hoc control methods. The more recent developments in canal automation are based on digital control, which offers many advantages over traditional analog control, including convenient interface to personal computers. Any centralized control scheme essentially requires that the signals be processed digitally.

Most of the papers on canal automation deal only with closed-loop control logic, without anticipation for known demand changes. While closed-loop control is a technically complex component, by itself it is not sufficient for automation of most canals. Further, many of the failures in canal automation have resulted from incompatibility of system components (e.g., MSIDD in Clemmens et al 1994). Many districts have developed their own supervisory (manual-remote) control systems (e.g., SRP in Shipley and Juetten 1970), or use commercial systems based on programmable logic controllers (PLCs). Such systems typically do not have the capabilities to fully implement modern, digital control methods. Controllers with higher-level language programming are usually needed. Further, cooperation with companies who develop, manufacture and distribute data collection and control equipment can promote the adoption of this technology by reducing the need to conduct extensive research for every canal automation project. The purpose of this paper is to describe activities currently ongoing in central Arizona on the implementation of canal automation with two irrigation districts and on the development of an off-the-shelf canal automation product.

CANAL AUTOMATION LOGIC

Canal automation must fit within the overall operating procedure of the district, including the water ordering system, canal operations, system constraints, delivery gate control, demand flexibility allowed, etc. Different automation features are needed for different parts of the system. This may mean a combination of local-control, central-control, and intermediate, distributed-control functions. For local control, information gathered at a given site (e.g., local water level) is used to determine a control action at that site (e.g., gate movement). A Remote Terminal Unit (RTU) performs the signal processing and control functions, locally, for an individual site. For central control, information from individual RTUs are communicated to a control center. The control center processes this multi-site data and sends control signals back to the individual RTUs. For distributed control, information from one or more RTUs is passed to another RTU, where the necessary control actions are determined, without going to the control center. A control center should always be included in systems with local or distributed control so that operators can observe system behavior and collect data on long-term performance.

Control system logic is more complicated than the simple discussion above implies. Most irrigation delivery canals have a significant slope and little extra capacity for water storage. Once water flows into the canal, it will flow out of the canal at some point downstream at a later time. The time delay for flow changes to travel through the canal and the behavior of the resulting waves make the automation of canals more difficult than one would expect. Also, gate hydraulics and pool dynamic response are sometimes hard to separate, as are the interactions between neighboring pools. These problems can be minimized by separating the control of gates from the control of pool dynamics. This is done by using an <u>flow control function</u> for each check structure, which sets the gate position(s), and a <u>pool level or volume control function</u>, which determines the flow rate setting for each check (e.g., either locally or centrally).

CONTROLLER DESIGN

For closed-loop controller design, we use the approximate canal flow model developed by Schuurmans et al (1995). This model provides a simple framework for developing canal automation algorithms from two canal pool properties; the wave travel time for any segment of a canal pool flowing at normal depth, and the water surface area of the backwater upstream from each control structure. When backwater affects the entire length of a pool, the travel times through that pool are insignificant, but the flow is affected by reflection waves. Instabilities caused by these waves can be eliminated by introducing a low-pass filter, a mathematical smoothing procedure (e.g., as in ELFLO, Rogers et al 1995). We also use Schuurmans' (1996) state-feedback approach to controller design, where prior control actions are added to the state vector to account for time delays. The method can produce controllers ranging from optimal control to global tuning of local PI controllers, including Deltour's (1992) PIR (PI with delay) control.

For open-loop control, we use one of several methods including a simple time delay (e.g., Schuurmans et al 1995) and the gate-stroking method of Bautista et al (1997). This open-loop control is implemented simultaneously with but independently from the closed-loop control.

MARICOPA STANFIELD IRRIGATION AND DRAINAGE DISTRICT

The 35,000 ha Maricopa Stanfield Irrigation and Drainage District (MSIDD) is located in central Arizona and receives water from the Central Arizona Project and from groundwater. The system was designed so that all canal check structures (cross-regulators) could be controlled remotely from a personal-computer based supervisory control center. However, the district began delivering water through manual operation in 1987. Installation of canal gate remote control equipment, including motorized gates, RTUs and radio communication was not completed until 1989. The remote supervisory control system, including significant modification by the district, was completed and on line in 1991. An option was provided for automatic downstream feedback control. The scheme was tested, unsuccessfully, and abandoned (Clemmens et al 1994).

The district's WM canal (2.7 m^3/s capacity) was selected to conduct canal automation research. Data collected during 1992 provided a good test case for the evaluation of closed-loop downstream-control algorithms (Clemmens et al 1994). The location and small size of this canal allowed convenient testing of controllers without disturbing other parts of the canal system. This canal was subsequently used to develop test cases by the ASCE Task Committee on Canal Automation Algorithms (Kacerek et al 1995).

Simulation studies on the WM canal were conducted with several different closed-loop controllers and the CANAL_CAD unsteady-flow canal-automation simulation program (Holly and Parrish 1992). In general, flow rate control provided better performance than gate position control, even when flow rates were inexactly known. Also on this steep canal, accumulating requested control actions in the upstream direction provided better overall control. Third including knowledge of previous flow changes with prediction of their effects (i.e., modeling canal response delay) also greatly improved control.

Field testing on MSIDD's WM canal began in March 1995, with the assistance of researchers from Delft University of Technology. The MATLAB mathematical software package (MATLAB 1993), the unsteady simulation model MODIS (Schuurmans 1995), and MODLAB (Schuurmans 1994), a program that links MATLAB with MODIS, were used to develop and test various canal controllers. Initial tests of open-loop response verified the reasonableness of the linear model used to develop controller constants (Ellerbeck 1995). The supervisory control software used by MSIDD could not handle the automatic control functions simultaneously with their manual supervisory operations. Therefore, the automatic control function was programmed into a second generation of their software (Taylor 1995). The state-feedback control logic was programmed into this software and operated from a separate computer but through the same radio link as MSIDD's supervisory control system. Initial closed-loop tests were somewhat successful (Liem 1995). A typical test would be to start with a reasonably stable canal and make a step change in flow, for example turning off a groundwater well discharging into the canal. The control system brought the water levels back to the set points reasonably well, but encountered the following problems. First, the gates did not have reliable position sensors, making the debugging of control system problems almost impossible. Second, due to the gate width, the minimum gate movement allowed by the MSIDD control system was too large to product effective control (i.e., it caused oscillations). Flow rate control was performed centrally for there tests, but was only moderately effective. A shorter time period for local flow-rate control would have been better.

In late 1995, a Cooperative Research and Development Agreement (CRADA) was reached with Automata, Inc. to develop a canal automation product line -- plug and play canal automation -- including an entire package from the gate motor controllers to the control center software. MSIDD's WM canal was chosen as the site for development of this product line. A new gate position sensor was developed and installed at each gate for the WM canal, considering the need to measure and control gates precisely (e.g., within 1 mm). Automata's new generation RTU was chosen for use on this canal. New controller boards were built to operate the gate motors from the new RTUs. Automata obtained a multi state license for a radio frequency, to aid in more rapid application of the technology, and provided radios for this project. The RTU's were purchased by the Water Resources Laboratory of USBR, with funding from USBR's Lower-Colorado Region, Office of Water Conservation. This equipment will be installed in late 1996.

SALT RIVER PROJECT

The Salt River Project (SRP) was established in 1903 and provides water and power to agricultural and urban users within the Salt River Valley in central Arizona. SRP has a 100,000 ha service area and operates a series of seven multipurpose reservoirs on the Salt and Verde River system . A remote, manual supervisory control system was first implemented on SRP's main canals in the mid 1960's, which was replaced in 1991 with a new supervisory control center. Over the last 3 decades, a significant shift from agricultural to urban water use has taken place. In addition, SRP receives and delivers CAP water to urban and other customers through their distribution network. This has created new demands on canal operators, e.g. water transfer agreements, that have diverted their attention from normal canal operations. These new demands have prompted SRP's water operations staff to investigate the potential of canal automation to reduce the increasing burden on the operating staff. A pilot project was initiated in 1996 to study automation on the upper reach of the Arizona Canal (43 m^3/s capacity).

SRP's surface water supply comes from its reservoirs to the Granite Reef Diversion Dam, where flow enters the Arizona and South canals on opposite sides of the river. The Upper Arizona Canal is 30.6 Km long and has four pools. At that point, flow is split roughly in half to the Grand and lower Arizona canals. The two upstream pools are long and steep relative to the two lower pools. This canal was originally an earthen canal, but has since been lined with shotcrete (a concrete that is sprayed on over the existing earth surface). There is some question whether Granite Reef Reservoir has the capacity to allow automatic downstream control. The pilot project will help to answer that question.

The objective of Phase I of this canal automation pilot project is to test automatic control with an unsteady-flow simulation model. Phase I consists of 5 steps: 1) determine the dynamic response of each pool and verify relevant hydraulic relationships, 2) modify the unsteady-flow simulation model to incorporate the user-written control procedures, 3) select the controller and develop the controller constants, 4) test the control algorithm with the unsteady-flow simulation model, and 5) analysis and report.

The unsteady-flow simulation package, Mike 11 (DHI 1992) was chosen to simulate unsteady flow for the Arizona Canal, since SRP had the model set up for simulation on their entire canal network. The original Mike 11 package allows gate setting and offtake changes to be determined as a function of time from data input tables. Closed-loop control requires that gate settings be determined in real time from the observed water levels. A special user-interface was developed by DHI to allow us to write the closed-loop control functions to adjust gates from the Mike-11 computed water levels.

In the spring of 1996, several open-loop tests were performed when the district made flow changes in the Arizona canal of about 10%. These tests verified that Mike 11 was properly calibrated for this portion of the canal. The response in water level of each pool to a step change in inflow was determined by Mike 11. Each pool was modeled independently over a range of flow rates and the downstream boundary was a constant flow rate, provided by a user-written routine. This water level response was used to determine delay times, backwater surface area and reflection wave frequency for each pool, which will be used to develop a controller constants. SRP watermasters have provided seven scenarios for testing these control algorithms. This project is scheduled for completion in January 1997. Further details can be found in Clemmens et al (1997).

Phase II of the project will involve updating SRP's supervisory control system so that it can handle automatic control and demonstration of the control system to watermasters to gain their understanding and acceptance of the system. In Phase III, the control system would be tested on the actual canal in real time over an irrigation season.

REFERENCES

Bautista, E., Clemmens, A.J., and Strelkof, T.S. 1997. Inverse computation methods for open-channel flow control. . *J. Irrig. & Drain. Eng.* In press.

Clemmens, A.J., Sloan, G. and Schuurmans, J. 1994. Canal-control needs: example. *J. Irrig. & Drain. Eng.*, 120(6):1067-1085.

Clemmens, A.J., Bautista, E., Gooch, R.S., and Strand, R.J. 1997. *Salt River Project Canal Automation Pilot Project*. Intern. Work. on Reg. of Irr. Canals, Marrakesh, Morocco, Apr. Draft.

Deltour, J.-L. 1992. *Application de L'Automatique Numerique a la Regulation des Canaux*. These pour obtenir le grade de Docteur. Institut National Polytechnique de Grenoble, Grenoble, France.

DHI 1992. *Mike 11 Version 3.01* Danish Hydraulic Institute, Copenhagen, Denmark, Nov.

Ellerbeck, M.B. 1995. *Model predictive control for an irrigation canal*. Graduate thesis. Delft University of Technology. Delft, The Netherlands. August.

Holly, F.M. Jr. and Parrish, J.B. 1992. *CanalCAD dynamic Flow Simulation in Irrigation Canals with Automatic Gate Control*. Iowa Institute of Hydraulic Research, Iowa City, IA, USA. 127 p.

Kacerek, T.F., A.J. Clemmens, and F. Sanfilippo. 1995. Test cases for canal control algorithms. Proc. First Int. Conf. Water Res. Engr. Div., ASCE, San Antonio, TX. 14-18- Aug 1995.

Liem, G. 1995. *Controller Design for Irrigation Canals*. Report No. A-691, Laboratory for Measurement and Control, Delft U. of Technology, Delft, the Netherlands, 64 p.

Malaterre, P.O. 1995. Regulation of irrigation canals. *Irrigation and Drainage Sys.* 9(4):297-327.

MATLAB 1993. *MATLAB High-Performance Numeric Computation and Visualization Software*. The Math Works, Inc. Natick, MA, USA

Rogers, D.C. et al 1995. *Canal Systems Automation Manual*. Volume 2. U.S. Department of Interior, Bureau of Reclamation, Denver, CO, USA.

Schuurmans, J. 1994. *Modlab Hydrodynamic Simulation Tool*. Center for Operational Water Management, Delft University of Technology, Delft the Netherlands

Schuurmans, W. 1995. *MODIS A Model to Study the Hydraulic Performance of Controlled Irrigation Canals*. Ver. 3.2. Center for Oper. Water Mgmt., Delft U. of Tech., Delft the Netherlands.

Schuurmans, J. 1996. *Control of Open-Channels: Concepts*. PhD Dissertation, Delft Univ. of Technology, Delft, The Netherlands, Draft.

Schuurmans, J., Bosgra, O.H. and Brouwer, R. 1995. Open-channel flow model approximation for controller design. *Appl. Math Modeling* 19:525-530.

Shipley, H. and. Juetten, R.L. 1970. *Remote Electronic control - a Prerequisite for Efficient Irrigation Operations*, SRP Report, Salt River Project, July.

Taylor, K. Personal Communication. Central Arizona Irrigation and Drainage Dist., Eloy, AZ 1995.

COMPARISON OF CONSTANT LEVEL AND CONSTANT VOLUME CONTROL METHOD FOR OPEN CHANNEL FLOW

Fubo LIU[1] and Jan FEYEN[2]
[1]*Postdoctoral Researcher* & [2]*Professor, Institute for Land and Water Management Katholieke Universiteit Leuven, Vital Decosterstraat 102, 3000 Leuven, Belgium*

ABSTRACT
Two closed-loop control methods, one for maintaining a constant water level at the downstream end of each canal pool and the other for maintaining a constant water volume in each pool, are compared using a test case set by the ASCE Task Committee on Canal Automation. Based on results obtained from hydrodynamic simulations, advantages and disadvantages of the two different control strategies are compared and analyzed.

INTRODUCTION
In a system operated with closed-loop control, cross gates are regulated in response to the current status of the flow. Constant downstream level control and constant volume control are two distinctively different approaches for closed-loop control. As the term suggest, a constant downstream level control method aims at maintaining a constant water level at the downstream end of each canal pool. If changes in lateral flow are known in advance, control gates are adjusted before the changes take place in order to reduce the fluctuation of water level in the canal. The advantage of such a approach is the economy of canal embankment height. However, in such type of systems, the volume of water in each canal pool changes with the flow rate in the canal. The inflow change at the upstream end of the pool must overcompensate for the change at its downstream end in order to accomplish the required volume change, and it normally takes a longer time for the system to change from one flow condition to another. In contrast to the constant level control method, a constant volume control method maintains a constant water volume in each pool. The water surface pivots around the middle point of the pool when flow rate in the canal changes. The main advantage of constant volume control is that it enables the system to change rapidly from one flow condition to another, and no excessive time is needed to built up or deplete water storage in the system (Buyalsky et al. 1991). However, constant volume control requires level top canal at the downstream half of the canal pool.

In order to "feel" quantitatively how rapidly the constant volume approach can change the system from one flow condition to another, and under what circumstances a constant level control method enables the reduction of canal embankment height, this study compares a constant level control method and a constant volume control method using a test case suggested by the ASCE Task Committee on Canal Automation (Clemmens et al. 1996). The example canal was originally proposed for testing constant water level control algorithms. But with some minor modifications, the example canal can also be used to test constant volume control methods.

DESCRIPTION OF THE CONTROL METHODS

The closed-loop, constant downstream water level control method is formulated based on the inverse solution of the St. Venant equations. It is designed for demand oriented systems, and it is able to take into account both scheduled and unscheduled changes in offtake discharge. Detailed description of this method can be found in Liu et al. (1996). The proposed constant volume control method is based on the water volume balance in each canal pool. If all lateral offtakes are located at the downstream end of each pool, the following control equation is suggested:

$$\Delta Q_i = K_1(Q_n - Q_1) + \frac{V_{target} - V}{K_2 R} + \Delta Q_{i+1}$$

where ΔQ = required change in flow rate at the upstream end of a canal pool; i = pool index; Q_1, Q_n = respectively, current flow rate at the upstream and at the downstream end (just before the offtake) of a canal pool; V, V_{target} = respectively, current and target water volume in a canal pool; K_1 and K_2 are control coefficients; R is a time parameter, and K_2*R is the time needed to bring the water volume in a canal pool to its target value. With the calculated change of inflow rate, the required change in gate opening can be estimated.

TEST CASE

The canal has 8 pools in series which are separated by undershot gates. The system originates from a constant water level reservoir at its head, and terminates with a pump at its downstream end. In the original test canal, there is an orifice offtake at the downstream end of each canal pool. However, with constant volume control the water level at the downstream end of a pool varies with flow rate in the main canal. In the tests presented in the following, the orifice offtakes are changed to pumps, for both constant level and constant volume control. In the original example, flow rates through lateral offtakes are very small at the initial steady state. Rapid increases in offtake discharges are imposed at 2 hrs, and the inverse changes are imposed at 14 hrs. In the tests presented here, flow rates through lateral offtakes are at their maximum at the initial steady state, and the sequence of the changes in lateral flow rate is reversed. Such modifications are done for the convenience of calculating the target water volume of the canal pools. At the initial steady state, flow rate in each pool is at its maximum, and the water depth at the downstream end of each pool is close to the normal water depth. The initial water depth is the target water depth under constant level control, and the corresponding water volume is the target volume under constant volume control. The modified test canal and test scenario are described in Table 1. In addition, the bottom slope of the canal is 0.0001. There is a drop of 0.2 m at the bottom elevation of the canal after each cross gate. The Manning n coefficient is 0.02. The side slope of the canal section is 1.5.

Table 1: Test canal and test scenario

Pool number	Pool length (km)	Bottom / gate width (m)	Offtake initial flow (m³/s)	Check initial flow (m³/s)	Pool end initial depth (m)	Pool initial volume (m³)	Offtake change at 2 hrs (m³/s)	Resulting check flow (m³/s)	Offtake change at 14 hr (m³/s)	Resulting check flow (m³/s)
heading		7		13.7				2.7		13.7
1	7	7	1.7	12.0	2.1	149091	-1.5	2.5	1.5	12.0
2	3	7	1.8	10.2	2.1	62690	-1.5	2.2	1.5	10.2
3	3	7	2.7	7.5	2.1	61508	-2.5	2.0	2.5	7.5
4	4	6	0.3	7.2	1.9	63693		1.7		7.2
5	4	6	0.2	7.0	1.9	63312		1.5		7.0
6	3	5	0.8	6.2	1.7	38442	-0.5	1.2	0.5	6.2
7	2	5	1.2	5.0	1.7	25253	-1.0	1.0	1.0	5.0
8	2	5	2.3	-	1.7	24736	-2.0	-	2.0	-
end pump	-	-	2.7	-	-	-	-2.0	-	2.0	-

RESULTS AND ANALYSIS

The tests are performed using a non-linear unsteady flow simulation model which is based on the St. Venant equations and the Preissmann implicit finite difference scheme. The time interval is taken as 5 minutes, and the distance interval is 200 m. The regulation interval of control gates is 15 min. The adjustment of a gate is completed in 5 min, then the position of the gate remains constant for 10 min until the next regulation step. No restriction is imposed on the minimum gate movement. The maximum gate opening is limited by the water level upstream of the control gate at the current time step. With both the constant level and the constant volume control methods, it is possible to calculate the adjustment of control gates by measuring only two water levels in each pool, one at its upstream end, the other at its downstream end. Flow rates and water levels at the other sections required by the control calculation can be estimated based on these two water level measurements (Liu et al. 1996). However, in this study, no such estimations are included in the controller, and water levels and flows rates at all sections are given by the simulation model.

The following tests have been conducted: (1) Constant level control with unscheduled changes in lateral flow; (2) Constant level control with scheduled changes in lateral flow; (3) Constant volume control with unscheduled changes in lateral flow. Scheduled changes in lateral flow are known *in advance*, while unscheduled changes are not known. A constant volume control method never needs to know changes in lateral flow in advance, while a constant downstream level control method may include scheduled changes in later flow rate in the control calculation. For both the constant level and the constant volume control methods discussed, the controller does not need to know current lateral flow rates, although the constant level method needs to know the intended *changes* in later flow rates in the case of scheduled changes.

For the three tests conducted, the opening of each control gate, the change of water volume in each canal pool, and the change of water level at the downstream end of each canal pool are shown in Fig. 1. The corresponding water surface profile at peak flow, and the upper and the lower water surface envelop lines are plotted in Fig. 2. In test 1, since changes in lateral flow are not know in advance, the positions of control gates do not change until the water levels in the canal change (Fig. 1(a)). In test 2, changes in lateral flow are scheduled in advance, hence the openings of the control gates are adjusted before the changes in lateral flow actually take place (Fig. 1(b)). As a result of the scheduled information, the maximum deviations at the downstream end of the canal pools are reduced from about 0.5 m to about 0.2 m. Despite the large changes of flow rate in the canal, in all pools except pool 1, permanent changes of water volume from one steady state to another are only about 10% of their volume at peak flow. The change of water volume in pool 1 is about 20% of its total volume due to its extreme length (7 km). This indicates that if we want to reduce the recovery time of the system, an additional cross gate should be installed in pool 1. The time needed to change the system from one flow condition to another is about 7 to 10 hours under constant level control, either the changes in lateral flow are scheduled or unscheduled (Fig. 1(a) & Fig. 1(b)). Under constant volume control, the system can reach a new steady state within about 3 hours (Fig. 1(c)). Other tests reveal that the recovery time under constant volume control can still be considerably reduced if the regulation interval of the control gates is shorter.

The water surface profile at peak flow is the same for all the three tests. As expected, the water surface profile pivots around the middle point of each canal pool under constant volume control (Fig. 2(c)). Under constant level control, the upper water surface envelop is close to the water surface profile at peak flow when changes in lateral flow are known in advance (Fig. 2(b)). However, if the changes in lateral flow are rapid and unscheduled, the upper water surface envelop is considerable higher (Fig. 2(a)), and may even exceed the water surface envelop under constant volume control. Then the advantage of constant downstream level control is lost.

REFERENCES

Buyalski, C. P., Ehler, D. G., Falvey, H. T., Rogers, D. C. and Serfozo, E. A. (1991). Canal systems automation manual. Vol. I. U.S. Bureau of Reclamation, Denver, Color., 113 p.

Clemmens, A. J., Kacerek, T. F., Grawitz, B., and Schuurmans, W. (1996). "Recommended tests for canal control algorithms". J. Irrig. and Drain. Engrg. (Paper to be published in the special issue on canal automation).

Liu, F., Malaterre, P.O., Baume, J.P., Kosuth, P. and Feyen, J. (1996). "Development and evaluation of canal automation algorithm CLIS using test cases set by ASCE task committee." J. Irrig. and Drain. Engrg. (Paper submitted for the special issue on canal automation).

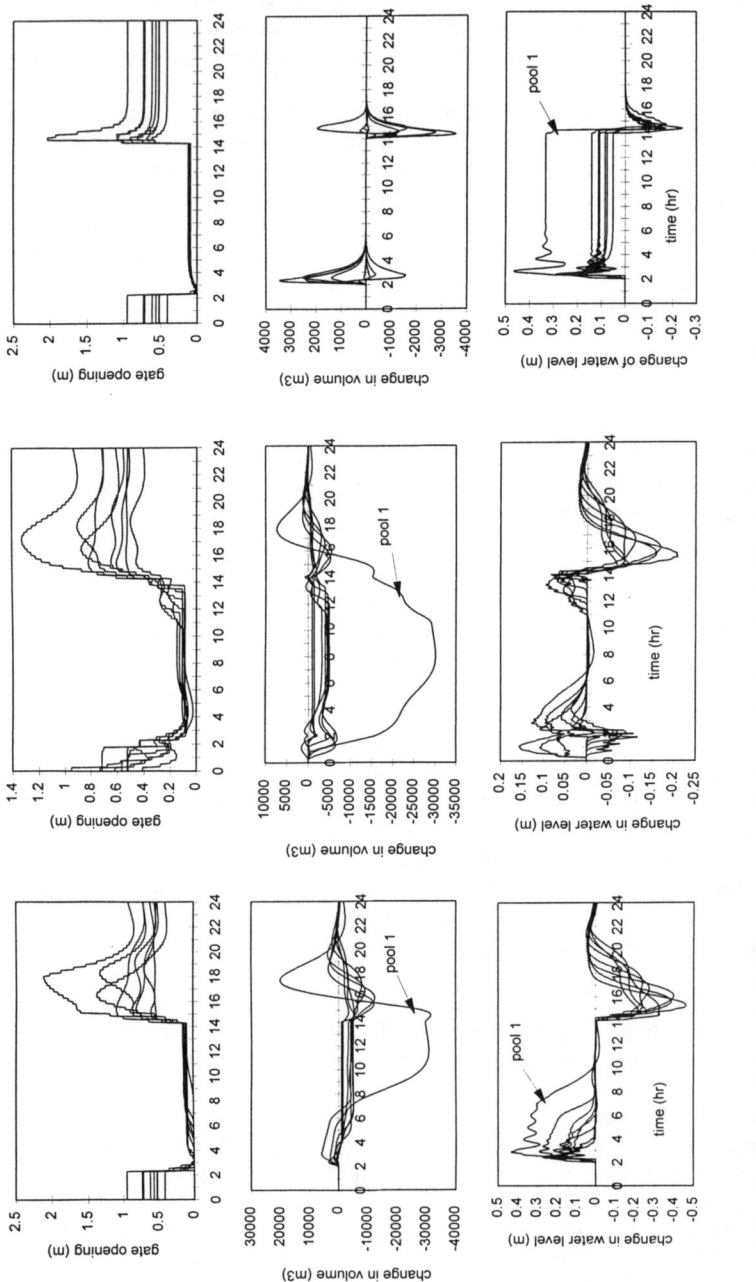

(a) constant level control, unscheduled change (b) constant level control, scheduled change (c) constant volume control

Fig 1: Response of the system under different control methods

OPEN CHANNEL FLOW METHOD COMPARISON 313

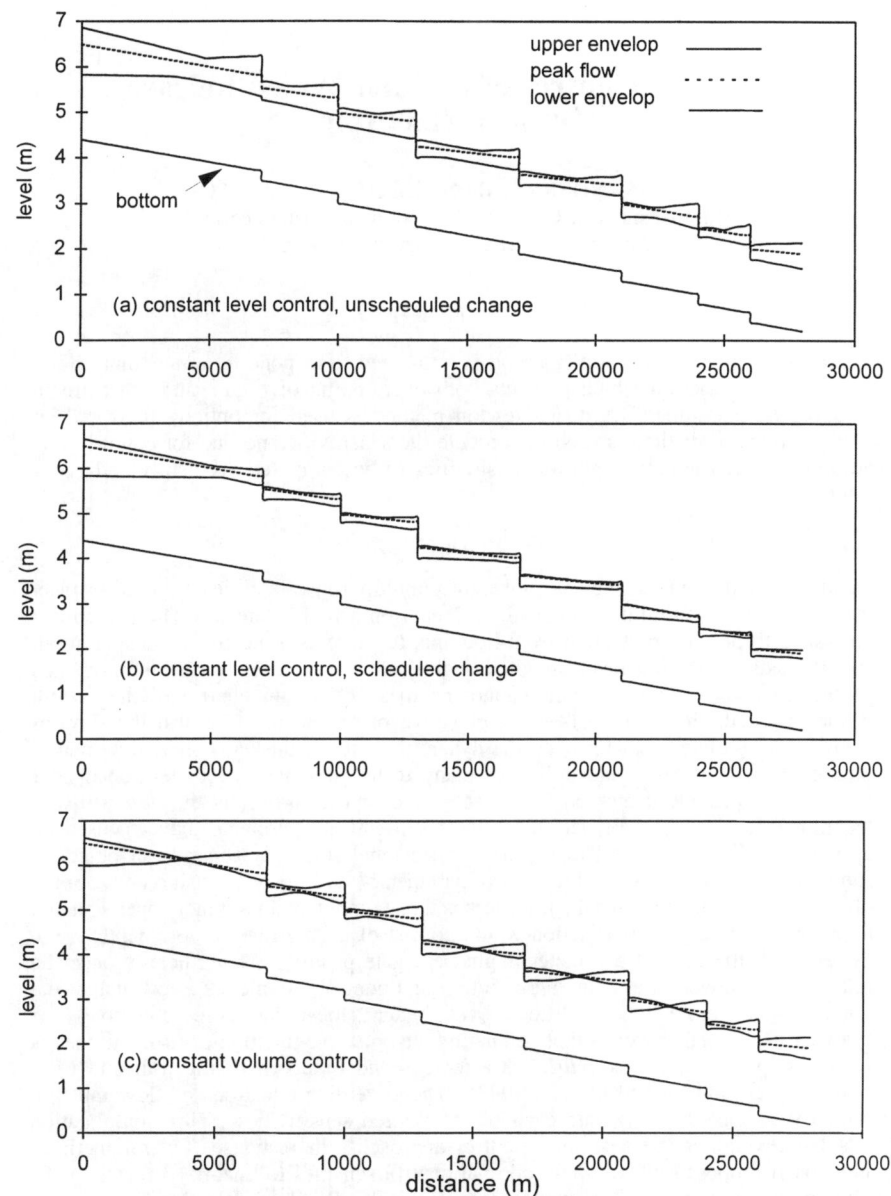

Fig. 2: water surface profile at peak flow and water sufarce envelop lines

Optimal Control of Sudden Water Release from a Reservoir

BRETT F. SANDERS AND NIKOLAOS D. KATOPODES
Department of Civil and Environmental Engineering
University of Michigan, Ann Arbor, USA

ABSTRACT

The outflow gate motion of a reservoir is optimized to respond to an incoming flood wave. A gate motion which prevents both overtopping of a dam and downstream flooding is determined. A quasi-Newton method is used for optimization, and an adjoint equation method is used to compute the sensitivities needed for control. The adjoint equation approach allows sensitivities to be computed efficiently and accurately.

INTRODUCTION

Reservoirs and open channels are inherently multipurpose facilities required to meet the needs of agricultural, municipal, and environmental interests. These facilities are asked to provide protection from flooding, to serve as a means of transportation, and to assist in the management of droughts. In many cases, all of the demands placed on open channel systems cannot be met. Optimum control of the system is then needed. It has long been an objective of engineers to control the flow of water in open channels. However, the dynamic nature of unsteady fluid flow makes control a difficult task. Nevertheless, many techniques have been developed over the years to provide a method of control. One of the earliest is the *gate stroking* technique developed by Wylie(1969). Most subsequent approaches have focused on applying traditional control theory to open channel flow. In general, an objective function is established as well as control parameters, and the relationship between the objective function and control parameters is then computed. In a typical open channel application, the objective function would consist of a flow rate or flow depth in the system, and the control parameter might be a gate position. Nevertheless, once the relationship between the objective function and control parameters is evaluated, the control parameters may be modified so as to better achieve the desired objective. The drawback of this approach is that computing the objective function/control parameters relationship, i.e., the *sensitivities*, is a tedious and time consuming trial and error process (Reddy, Dia, and Oussou 1993). The governing equations of flow must be repetitively solved to evaluate each of the desired sensitivities. This complication has led researchers to a simpler control approach. In response, several methods have been proposed which apply optimal control schemes to linearized forms of the governing equations (e.g. Sawadogo et. al., 1995; Reddy, 1991; Reddy, Dia, and Oussou, 1993). However, by linearizing the governing equations, in most cases the St. Venant Equations, the application of the control scheme is limited to a restricted class of flows. A new approach has been developed by Sanders and Katopodes (1996) which avoids the previous pitfalls of control. Repetitive solutions of the governing equations are avoided when sensitivities are computed, and the flow equations are kept in their nonlinear form. The new approach evaluates the sensitivities needed for control by solving the adjoint shallow water equations, and is termed the *adjoint equation method*.

An optimization problem associated with reservoir operation is solved to demonstrate the adjoint equation method. A one-dimensional system with a reservoir of length L_1 and a river of length L_2 is considered. The system is presented in Figure 1.

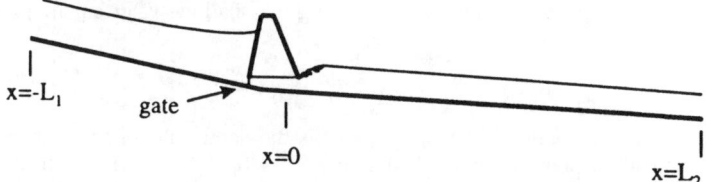

Figure 1: Sketch of the Reservoir/River System

The problem arises when a large flood wave enters a reservoir filled near to capacity. To prevent overtopping of the dam, water must be released from the reservoir. However, if too much water is released, flooding will occur downstream when the river stage rises above the levee height. The goal is to optimize the release of water from the reservoir in such a way that overtopping of the dam as well downstream flooding are avoided.

CONTROL SCHEME

OPTIMIZATION METHOD

A quasi-Newton method, known also as a variable metric method, is used to find a minimizer of an objective function J. Press et. al. (1992) provide background information on this method. Descent directions are given as

$$d_i = -A^{-1} \nabla J(x_i) \qquad (1)$$

where x_i represents the control parameters, $J(x_i)$ is the objective function, and A is the Hessian of $J(x_i)$. Subsequent steps follow as

$$x_{i+1} = x_i + \alpha_i d_i \qquad (2)$$

Using a *quasi-Newton* method, an approximation to the Hessian, $H_i = A^{-1}$ is computed instead of the true Hessian. The approximation begins with the identity matrix $H_0 = I$ and is then updated with each iteration using the BFGS (Broyden, Fletcher, Goldfarb, Shanno) algorithm. The technique guarantees movement towards the minimum with each iteration for sufficiently small α_i because H_i remains positive definite and symmetric for all i. Additionally, close to the minimum H_i will approach A^{-1}. For details on the BFGS algorithm, the reader is referred to Press et. al. (1992) or Shanno (1978). Piasecki and Katopodes (1996) additionally demonstrate the use of the BFGS algorithm in a surface water mass transport application.

DYNAMIC RESERVOIR CONTROL

Discharge from the reservoir is assumed to be controlled by a gate. A gate equation is used to relate the discharge q_{dam}, the depth h_{dam}, and the gate opening b at the dam.

$$q_{dam} = C_D \, b\sqrt{2gh_{dam}} \tag{3}$$

The parameter C_D is a discharge coefficient which is assumed to be a constant, but in general could vary with the flow conditions. The discharge given by Eq. 3 represents both the outflow from the reservoir and the inflow to the river. Moreover, the gate opening at discrete times $b(t_j)$ represents the control vector given by x in Eq. 2. Flow within the reservoir and river is evaluated by solving the shallow-water equations in one-dimension.

$$\begin{aligned} h_t + q_x &= 0 \\ q_t + \left(\left(\tfrac{q^2}{h}\right) + \tfrac{gh^2}{2}\right)_x - gh(S_o - S_f) &= 0 \end{aligned} \tag{4}$$

where $h = h(x,t)$ is the depth of flow, $q = q(x,t)$ is the depth-integrated discharge, g is the gravitational constant, S_o is the bed slope, $S_f = (q^2/(C^2 h^3))$ is the friction slope, and C is the Chezy coefficient. Subscripts x and t indicate partial differentiation. The inflow hydrograph for the reservoir is given by $q = q_o(t)$, and the downstream end of the river is assumed to be non-reflective to simulate a very long channel. Initially, a steady flow of $q = q_o(0)$ is assumed to flow through the reservoir and river. The initial depth profile in the reservoir is computed by solving for the gradually varied flow profile while the initial depth in the river is assumed to be normal. The objective function is formulated in terms of the depth of flow at a point in both the reservoir and the river.

$$J = \int_0^T \left[\int_{-L_1}^0 r_1(h) dx + \gamma_1 \int_0^{L_2} r_2(h) dx \right] dt \tag{5}$$

where

$$r_k(h) = \begin{cases} \tfrac{1}{2}\left(h(x_{kT}) - \bar{h}_k\right)^2 \delta(x - x_{kT}) & \text{if } h(x_{kT}) \geq \bar{h}_k \\ 0 & \text{if } h(x_{kT}) < \bar{h}_k \end{cases} \quad k = 1, 2$$

The depths $\bar{h}_1$ and $\bar{h}_2$ correspond to the maximum allowable depths in the reservoir and river, respectively. Exceedance of these depths implies either overtopping of the dam or flooding along the river. The points x_{1T} and x_{2T} are the *target locations*. It is at these locations that we compare the computed depths to the maximum depths given by $\bar{h}_1$ and $\bar{h}_2$. The parameter γ_1 is a scaling factor, and H is the Heaviside step function. The sensitivies needed for control are computed following the solution of the adjoint shallow water equations, which for this problem appear as (Sanders and Katopodes, 1997),

$$\begin{aligned} \phi_\tau + \left(\left(\tfrac{q^2}{h^2}\right) - gh\right)\psi_x - g(S_o - 2S_f)\psi - \tfrac{\partial r_k}{\partial h} &= 0 \\ \psi_\tau - \phi_x - 2\left(\tfrac{q}{h}\right)\psi_x + 2\tfrac{ghS_f}{q}\psi &= 0 \end{aligned} \tag{6}$$

The variables $\phi(x,t)$ and $\psi(x,t)$ are the *adjoint variables* and $\tau = T - t$. Note that the coefficients of the adjoint equations are obtained from the solution of the shallow-water equations. Additional background on adjoint equations is supplied by Marchuk

(1995) and Cacuci (1981). The initial conditions are given as $\phi(x,T) = 0$ and $\psi(x,T) = 0$, and boundaries are assumed to be non-reflective. The time T represents the time when the unsteady flow simulation ends. It also serves as the initial time for the adjoint system of equations because these equations must be integrated backwards in time to $t = 0$. Following the solution of the shallow water equations (Eqs. 4) and the adjoint shallow water equations (Eqs. 6), the sensitivities needed for control (i.e., $\nabla J(\mathbf{x}_i)$) are computed for $i = 1, \ldots, N_T$ as

$$\left(\frac{\delta J}{\delta b(t_i)}\right)_1 = -\phi(L, t_i) - \left(\frac{q(L, t_i)}{h(L, t_i)} + \sqrt{gh(L, t_i)}\right)\psi(L, t_i) \tag{7}$$

in the reservoir and

$$\left(\frac{\delta J}{\delta b(t_i)}\right)_2 = \phi(0, t_i) + \left(\frac{q(0, t_i)}{h(0, t_i)} - \sqrt{gh(0, t_i)}\right)\psi(0, t_i) \tag{8}$$

in the river. The total sensitivity of the objective function to changes in the gate position is then given by

$$\nabla J = \left(\frac{\delta J}{\delta b(t_i)}\right)_1 + \gamma_2 \left(\frac{\delta J}{\delta b(t_i)}\right)_2 \qquad i = 1, \ldots, N_T \tag{9}$$

where γ_2 is a scaling factor which allows the sensitivities in the river system to be of the same magnitude as those in the reservoir system. The optimization algorithm proceeds by taking the following steps.

1. Make an initial guess for $b(t_i)$.
2. Begin iteration loop.
   (a) Integrate shallow water equations, $t \in (0, T)$.
   (b) Integrate adjoint equations, $t \in (T, 0)$.
   (c) Compute objective function J by Eq. 5 and check for convergence.
   (d) Compute sensitivities using Eq. 9.
   (e) Update gate positions using Eq. 2.
3. End iteration loop.

RESULTS

The reservoir gate optimization problem is solved with $L_1 = 10\,km$, $L_2 = 30\,km$, $S_{o1} = 0.001$, and $S_{o2} = 0.0003$. A Chezy coefficient $C = 80$ is used for both the river and reservoir. The target location in the reservoir is placed at $x_{1T} = -1,780\,m$ with $\bar{h}_1 = 9.0\,m$ to address the overtopping concern, while the target location in the river is placed at $x_{2T} = 15,000\,m$ with $\bar{h}_2 = 2.0\,m$ to address the river flooding concern. The depth in the reservoir at the dam is initially given as $h_{dam} = 10\,m$ from which $q_o(0) = q_{dam}$ is computed using Eq. 5 with $b(0) = 0.2$. The inflow into the reservoir is then specified as,

$$q_o(t) = q_o(0) + 10 sech^2\left(0.003(t - 1000)\right) \tag{10}$$

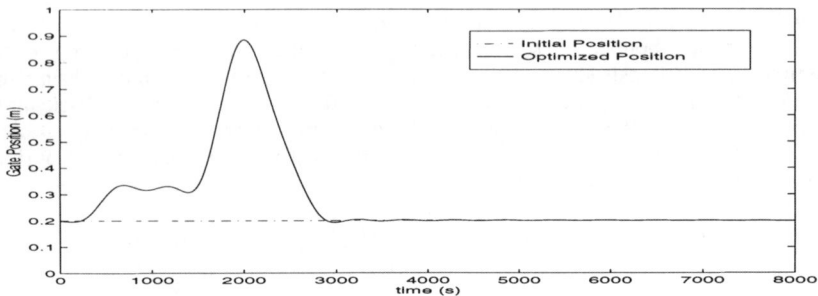

Figure 2: Gate position as a function of time.

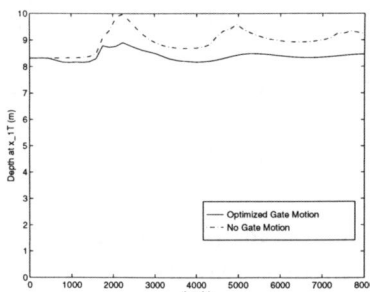

Figure 3: Depth variation at x_{1T}.

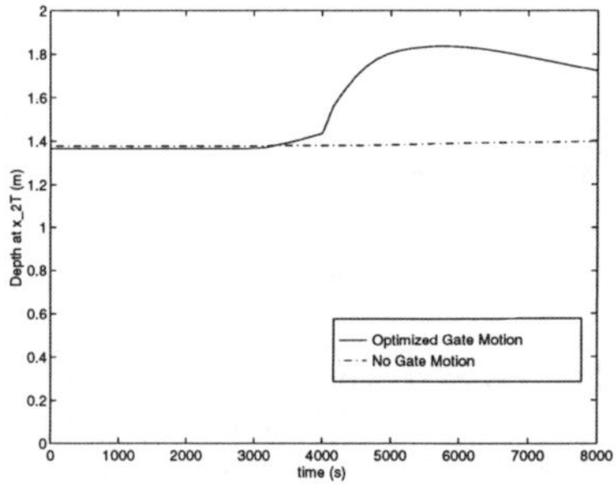

Figure 4: Depth variation at x_{2T}.

The shallow water and adjoint equations are solved using the Lax-Wendroff finite-difference scheme (Hirsch, 1988). The reservoir and river are discretized by $N_1 + 1$ and $N_2 + 1$ grid points, respectively, with $N_1 = 125$ and $N_2 = 500$. A common time step is used for both the river and reservoir, $\Delta t = 4\,s$, and a total of $N_T = 2000$ time steps are taken. Additional calculations were made with twice as many grid points to confirm convergence. Optimization was performed using a quasi-Newton method. The initial guess for the gate position was chosen as $b(t) = 0.2$ and the scaling parameters were set as $\gamma_1 = 2$ and $\gamma_2 = 32$. The scaling parameters are chosen conveniently so the objective function and sensitivities from the river and reservoir are of the same order of magnitude. Presented in Figure 2 is the optimized gate position reached after 10 iterations.

It is interesting to note that the optimized gate position given in Figure 2 is similar in shape to the reservoir's inflow hydrograph. Hence, to counter the progressive wave traveling downstream through the reservoir, the gate must be opened so that a depression wave moving upstream cancels the progressive wave at x_{1T}. The exact motion of the gate, however, is complicated and thus impossible to predict by a human operator.

To demonstrate that the depth in the reservoir meets the objective, a plot of the depth at x_{1T} is presented in Figure 3. Recalling that $\bar{h}_1 = 9.0\,m$, it is clear from Figure 3 that the objective is satisfied. To demonstrate that the depth in the river at x_{2T} meets the objective, a plot of the depth at x_{2T} is presented in Figure 4. Recalling that $\bar{h}_2 = 9.0\,m$, it clear that the objective is once again satisfied.

CONCLUSIONS

A gate motion is computed which prevents both dam overtopping and downstream flooding. The quasi-Newton method successfully finds the optimal gate position after 10 iterations, and the adjoint equation method efficiently calculates the sensitivities needed for optimization. Whereas a traditional control scheme would require 2000 solutions of the shallow-water equations to compute the sensitivities in this case because $N_T = 2000$, the adjoint equation method requires just a single solution of the shallow-water equations followed by a single solution of the adjoint equations.

REFERENCES

Cacuci, D.G., "Sensitivy Theory for Nonlinear Systems. I. Nonlinear Functional Analysis Approach," Journal of Mathematical Physics, Vol. 22, No. 12, 1981, pp. 2794-2802

Hirsch, C., "Numerical Computation of Internal and External Flows", Vol. 1, John Wiley and Sons, 1988

Marchuk, G.I., "Adjoint Equations and Analysis of Complex Systems," Kluwer Academic Publishers, 1995

Piasecki, M., and Katopodes, N.D., "Control of Contaminant Releases in Rivers and Estuaries II, Optimal Design," ASCE Journal of Hydraulic Engineering, (to appear in 1996), status: accepted.

Press, W.H., Teukolsky, S.A., Vetterling, W.T., and Flannery, B.P., "Numerical Recipes in FORTRAN," Cambridge University Press, 1992

Reddy, J.M., "Local Optimal Control of Irrigation Canals", ASCE Journal of Irrigation and Drainage Engineering, Vol. 116, No. 5, 1990, pp. 616-631

Reddy, J.M., Dia, A., and Oussou, A., "Design of Control Algorithm for Operation of Irrigation Canals", ASCE Journal of Irrigation and Drainage Engineering, Vol. 118, No. 6, 1992, pp. 852-867

Sanders, B.F., and Katopodes, N. D., "Control of Transient Open Channel Flow Using the Adjoint Equation Solution", 1997 (manuscript)

Sawadogo, S., P.O. Malaterre, P.O., and P. Kosuth, P., "Multivariable Optimal Control for On-Demand Operation of Irrigation Canals", International Journal of Systems Science, Vol. 26, No. 1, 1995, pp. 161-178

Shanno, D.F., "Conjugate Gradient Methods with Inexact Searches," Mathematics of Operations Research, Vol. 3, No. 3, 1978, pp. 244-256

Wylie, E.B., "Control of Transient Free-Surface Flow," ASCE Journal of the Hydraulics Division, Vol. 95, No. HY1, 1969, pp. 347-361

FLAP GATE FOR HYDRAULIC HEAD CONTROL

FELIX RAEMY and WILLI H. HAGER
VAW, ETH-Zentrum
CH-8092 Zurich
Switzerland

The hydraulics of a flap gate are experimentally established. The device introduced assures a constant upstream water elevation with a suitably positioned counterweight. The control of the flap gate is purely hydraulic, and design parameters are deduced from the equation of moments.

INTRODUCTION

In irrigation and sewer channels water can be stored or retained until it is used in the downstream reaches. Freeboard limitations dictate a maximum water elevation. A hydraulic device is introduced by which both water may be stored and freeboard limitation is satisfied. The flap gate control is hydraulic, and no additional regulation mechanisms are needed. If the gate clogs, the water overflows the gate, with an overflow depth much smaller than the length of the gate.

The device is based on previous suggestions by Pethick and Harrison (1981), and Kay and Ashton (1983). These studies were conducted in England and based on simplifications, such as hydrostatic pressure distribution on the flap gate. In the following, such assumptions will be removed, and conditions for the counterweight are deduced from the equation of moments.

EXPERIMENTATION

The experiments were conducted in a rectangular horizontal channel 500mm wide and 700mm high. The inlet was arranged such that approach conditions were perfect, i.e. an approach flow without both flow concentrations and surface waves. All flows were in the turbulent smooth regime. Discharges up to $50Ls^{-1}$ were run.

The flap gate was 400mm long and suspended at its top. Both the upper and lower crests were standard sharp-crested. The lower crest was always located at the channel bottom for zero discharge, i.e. when the gate was vertical. The counterweight was attached to a suspension fixed on the body of the gate, such that neither overflow nor underflow were disturbed, The counterweight mass could be moved along a pole connected to the suspension. Fig.1a) shows a definition sketch and Fig.1b) the relevant moments. The friction moment at the pivot was negligibly small.

Pressures on the flap gate were measured with a piezometer battery. In total 18 pressure taps located in the gate axis were provided. Their interdistance reduced from the top (40mm) to the bottom (10mm) of the gate.

The tailwater elevation was controlled with an overflow flap gate located at the end of the 7m long channel. In the present study free gate flow prevailed always. The discharge was measured with a V-notch weir at the channel outlet. The accuracy of discharge measurement was $\pm 1.5\%$, or $\pm 0.1Ls^{-1}$, whichever is larger. Pressure heads and water depths were measured to the next mm, due to slight head variations. Fig.2 shows the experimental arrangement.

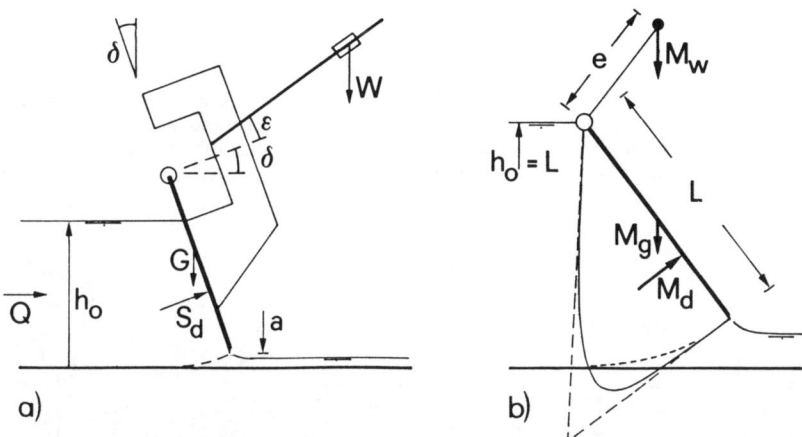

Fig.1 Automatic flap gate, definition of a) variables, b) moments.

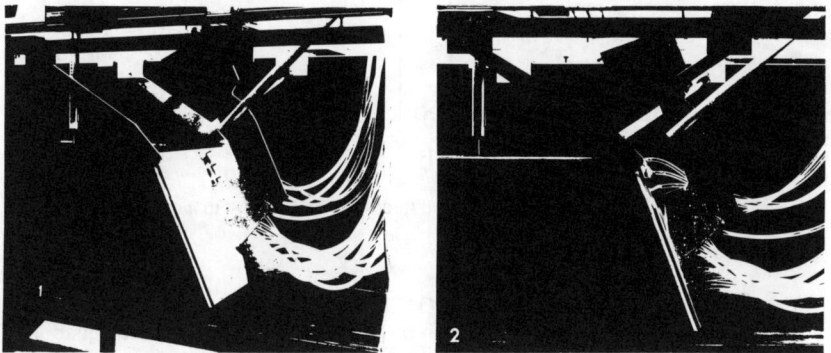

Fig.2 Photographs of model stand for D=1 and a) Q=37.1Ls^{-1}, δ=26.1°, b) Q=26.4Ls^{-1}, δ=21.6°. Note that pressure tubes have been removed, except for pressure readings.

PRESSURE DISTRIBUTION

The pressure distribution on an inclined flap gate that closes in the vertical position is non-hydrostatic. With h_o=approach flow depth, the hydrostatic pressure distribution is triangular, starting with a pressure head h_p=0 at the free surface and having a maximum h_{pM}=h_o−a at the lower crest, with a=elevation of lower crest over the channel bottom. The dynamic pressure distribution is zero (i.e. atmospheric) both at the upper and lower crests (Nago 1977). The pressure reduction close to the lower gate crest causes a significant reduction of moment on the gate suspension. The reduction increases as the gate opens and as the relative flow depth h_o/L decreases, where L=length of gate. Clearly, for a gate angle δ=0, the pressure distribution is hydrostatic, and σ=S_d/S_s=1. According to Fig.3a) the force ratio σ with S_d=dynamic force and S_s=hydrostatic force on the gate can be expressed in terms of D=L/h_o as

$$\sigma = 1 - \frac{1}{7} D \, tan\delta \; , \; \delta \leq 30° \tag{1}$$

The ratio of moments $\mu=M_d/M_s$ with M_d=dynamic moment and M_s=hydrostatic moment on the gate depends on the same parameters, given that the distances y_d and y_s from the axis of rotation to the centers of gravity of the pressure forces involve no additional parameters. With $M_s=y_s S_s$ and $M_d=y_d S_d$, or $\mu=y_d S_d/y_s S_s$, the experimental results follow (Fig.3b)

$$\mu = 1 - \frac{1}{4} D^{1/2} tan\delta \; , \; \delta \leq 30° \tag{2}$$

Note that $\sigma=\mu=1$ is assumed for "conventional" computations.

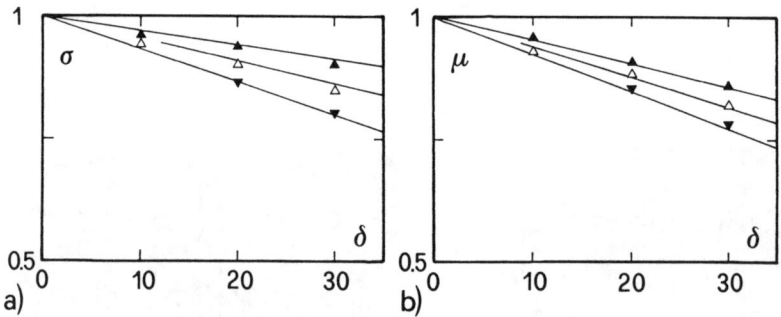

Fig.3 Correction terms a) σ relative to forces and b) μ relative to moments on flap gate for D=(▲) 1, (△) 1.56, (▼) 2.38.

EQUATION OF MOMENTS

The equilibrium condition for the gate can be formulated by applying the equation of moments. With the disturbing moment M_d, and the restoring moments M_g and M_w due to gate weight G, and counterweight W, the equilibrium equation for the design case $h_0=L$ is

$$M_d = \frac{1}{2} GL \, sin\delta + eW cos(\varepsilon+\delta) \tag{3}$$

Here e=distance from the pivot to the center of gravity of the counterweight system and ε=angle of counterweight relative to the normal on the gate (Fig.1). The static pressure force on the gate is $M_s=(1/2)\rho g cos\delta L^2 b(2L/3)$ where ρ=fluid density, g=gravitational acceleration and b=channel width. Inserting (2) for $M_d=\mu M_s$ and dividing by $cos\delta$ gives for D=1

$$\frac{1}{3} \rho g b L^3 - eW cos\varepsilon = \left(\frac{1}{2} GL - eW sin\varepsilon + \frac{1}{12} \rho g b L^3\right) tan\delta \tag{4}$$

The *equilibrium condition* requires that the gate is positioned such that $h_0=L$ for any angle δ. From Eq.(4) two conditions have to be satisfied

$$\frac{eW}{\rho gbL^3} = (3\cos\varepsilon)^{-1} \tag{5}$$

$$\sin\varepsilon = (eW)^{-1}\left[\frac{1}{2}GL + \frac{1}{12}\rho gbL^3\right] \tag{6}$$

Eq.(5) specifies the moment needed for equilibrium, and Eq.(6) determines the angle ε in terms of the three moments. Eliminating the term eW in both equations gives

$$\tan\varepsilon = \frac{(3/2)G}{\rho gbL^2} + \frac{1}{4} \tag{7}$$

If the weight of the gate is negligible compared to the counterweight then $\tan\varepsilon=0.25$, i.e. $\varepsilon=14°$ as the minimum angle. Note that the leading term in (7) is (1/4). With $(\cos\varepsilon)^{-1}=(1+\tan^2\varepsilon)^{1/2}$ the moment required from the counterweight is when neglecting terms of order G^2

$$\frac{eW}{\rho gbL^3} = \frac{1}{3}\left(\frac{17}{16} + \frac{3}{4}\frac{G}{\rho gbL^2}\right)^{1/2} \cong 0.344\left(1 + 0.35\frac{G}{\rho gbL^2}\right) \tag{8}$$

Because the relative weight of gate $G/(\rho gbL^2)$ is much smaller than unity, one has almost

$$\frac{eW}{\rho gbL^3} = 0.35 \tag{9}$$

which allows a simple estimation of the moment needed for equilibrium. The analysis also demonstrates that the effect of non-hydrostatic pressure distribution is essential. Clearly, because $|GL/eW| \ll 1$, Eq.(4) cannot be satisfied when ignoring the correction term $(1/12)\rho gbL^3$.

STABILITY OF GATE

The weight of the gate used was 46N. With a relative weight of $G/(\rho gbL^2)=0.057$ one has with $\tan\varepsilon=(3/2)0.057+0.25=0.336$ from (7), i.e. an angle $\varepsilon=18.5°$. The moment required for equilibrium is $eW/(\rho gbL^3)=0.344(1+0.35 \cdot 0.057)=0.35$ from (8), i.e. $eW=112Nm$. These two values were set at the flap gate and tested.

For zero discharge, the gate did not move from the vertical position until $h_o=39cm$ (97.5%). For a small discharge, the gate opened a little bit too much in the first instance, and closed after about 1s to the value required. For an abrupt increase of discharge up to $Q=60Ls^{-1}$ the gate followed steadily the approach flow with minor oscillations, and attained the equilibrium position. Gate instabilities that were feared prior to experimentation have not been observed. To test extreme variations of equilibrium, the gate was moved by hand out of its position. For disturbing angles of, say $-20°$ or $+20°$, oscillations about the position of equilibrium were strongly damped. It was interesting to note that the oscillations resulted mainly from surface waves in the approach channel, and much less due to the restoring moment. The gate tested was satisfactory in all hydraulic aspects. Additional tests for off-design flow with $1<D<2.4$ were also successful.

Fig.4a) shows the relative approach flow depth h_o/L for approach Froude numbers $F_o = Q/(gb^2h_o^3)^{1/2}$ up to 0.10. It is seen that the equilibrium position is satisfied to ±0.5% over the entire domain of discharge variations. Fig.4a) also shows δ as a function of F_o. The angle δ may be predicted with the momentum equation when inserting Eq.(1) for the pressure force on the gate. Kay and Ashton (1983) have presented a simplified model.

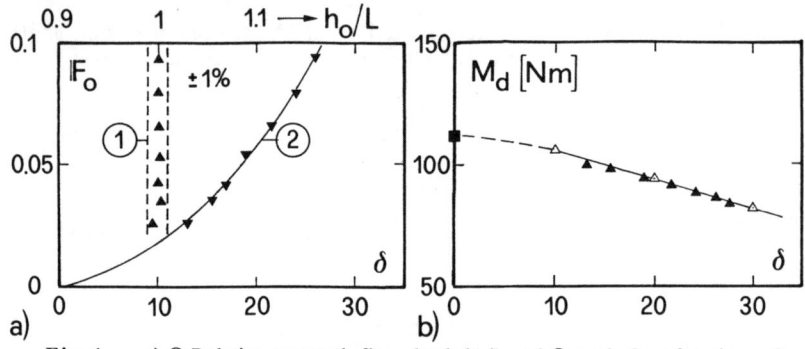

Fig.4 a) ① Relative approach flow depth h_o/L and ② angle δ as functions of approach Froude number $F_o = q/(gh_o^3)^{1/2}$, (■) computed restoring moment.
b) Verification of equation of moments, (△) disturbing moment measured due to hydrodynamic pressure on gate, (▲) restoring moments computed due to weights and counterweight, as a function of angle δ.

The equation of moments has been verified in addition by detailed computations involving all elements of the counterweight suspension (Fig.1a). Fig.4b) compares the moments so determined with the moment exerted on the flap gate due to the hydrodynamic pressure distribution. The assumptions that have led to the preceeding simplified analyses have thus been verified.

CONCLUSIONS

The hydraulics of a flap gate extended with a counterweight are explored. The hydrodynamic forces and moments of a obliquely hinged gate are experimentally determined. Then, the moment equation is applied subject to the condition that the upstream level remains constant under arbitrary gate movement. The requirements specify the angle ε, and the moment eW of the counterweight.

The analysis shows that the effect of the hydrodynamic pressure on the force is significant, and that the weight of the gate can almost be neglected. The novel device is hydraulically stable under periodic or abrupt discharge variations. It can be effectively used at locations where a constant upstream water level should be kept.

ACKNOWLEDGEMENTS

The present version of the counterweight flap gate was originally designed by Dr. Markus Schwalt, formerly PhD student at VAW. We would like to acknowledge his and Prof. Dr. D. Vischer's steady interest in this device.

REFERENCES

- Kay, M.G., Ashton, D.A. (1983). A laboratory investigation of counterbalanced flap gates for water level control. *Journal Institution Water Engineers* **37**: 506-512.
- Nago, H. (1977). Hydraulic pressure acting on shell-type gates. *Trans. Japanese Society of Civil Engineers* **9**: 170-172.
- Pethick, R.W., Harrison, A.J.M. (1981). The theoretical treatment of the hydraulics of rectangular flap gates. *19 IAHR Congress* New Delhi **B**(c): 247-254.

LOCAL SUPERVISORY AND CONTROL ALGORITHMS FOR A LABORATORY SLIDE GATE

VICTOR M. RUIZ C., RAFAEL HERRERA AND EDMUNDO PEDROZA
Instituto Mexicano de Tecnología del Agua, Paseo Cuauhnahuac 8532,
Jiutepec, Mor. 6225, Mexico. Email: vmruiz@tlaloc.imta.mx

ABSTRACT

A variety of control systems have been used in canal operation. However, some of the most common problems found in their application are not directly associated with the control algorithm itself. The main problems are the selection of the control parameters and the reliability/safety of the control system. A proportional-integral control algorithm is used for level regulation. The use of the flow rate as a control variable or the introduction of a factor considering the head variations through the gate reduces the control parameter adjustments required to maintain the control performance. A supervision system is proposed to monitor the control equipment and process. The slide gate is provided with analog and digital devices to detect abnormal operation conditions. The fault detection and diagnosis techniques used are based on process knowledge. This paper describes the initial steps in the development of a reliable safe control supervisory system for a laboratory slide gate.

INTRODUCTION

The most important goals of automatic control are to increase the reliability, availability, and safety of the processes. This must be done through a monitoring procedure. In particular, measurements must be made of the process actuators, sensors, power devices, and other control equipment to determine their status as well as of the process performance during all operating modes.

Fault monitoring should be tackled by appropriate computer processing and management actions to meet reliability and safety requirements. This includes an adequate provision for maintenance, repair, operation and an interface between man and machine.

Several methods and strategies have been developed in an attempt to attain the reliability and safety goals. Measurements indicating deviations from normal operating conditions are processed by knowledge-based expert systems, graph methods, pattern recognition, parameter identification, modeling, estimation and neural networks, to initiate a fault diagnosis (Isermann 1984, 1988, 1993). Based on the diagnosis, actions for maintenance, repair, and operation are instituted as required. These procedures require the intensive use of computers for analytical and heuristic treatment of the measured signals, process model, events and general process knowledge.

Four year ago, two modernization projects, based on remote monitoring, were completed in Mexico, "El Canal Alto" in the Yaqui Irrigation District and "Bachimba" in the Delicias Irrigation District. Both projects fell short of expectations as a result of the limited experience in the selection of the components for the remote supervisory and operation system and the lack of a supervisory system to verify the performance of the components.

The Mexican Institute of Water Technology ("IMTA") is working on the development of a control system for a laboratory slide gate. The main objective of the project is to test control algorithms, determine rules for control parameter selection, develop a supervisory system for the control system components and evaluate the equipment for the canal operation available in Mexico. The experience obtained from this project will be used to prepare the Request for Proposal that the Mexican government and farmers associations will offer to improve canal operation and monitoring.

IMTA's project is divided in two main areas, local and remote control systems. For local control, a slide gate was built and equipped. For remote control two more gates will be built, a MODBUS+ local network with Modicon PLCs will be installed and a SCADA software used. In this paper, the preparatory work done with the locally controlled slide gate at the IMTA laboratory is presented.

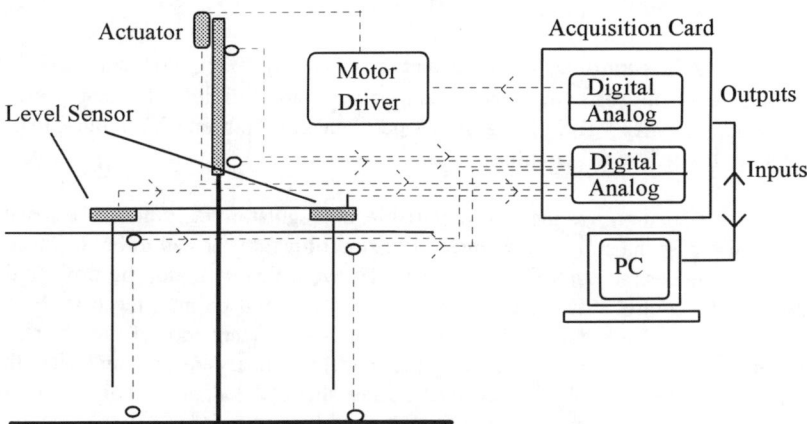

Figure 1. Slide gate with control and supervisory components.

IMTA LABORATORY SLIDE GATE

The laboratory slide gate (Fig. 1.) is equipped with a linear actuator, two pressure sensors, a potentiometer for gate position, and limit and level switches. The linear actuator has a 12 Volt DC motor, driven by a control box which allows for automatic and manual operation. Inside the box, a set of electromagnetic relays controlled by the digital outputs of the acquisition card specify the motor movement direction. Two limit switches cut the power supply to the motor when the gate is at the upper and lower positions. Two other limit switches inform the computer, through digital inputs, that the gate is fully opened or closed. Two pressure sensors are used to monitor the upstream and downstream water levels. The sensor output is relayed to the analog inputs of the acquisition card. A potentiometer connected to the actuator gear box determines the gate position. Four level-switches are connected to the digital inputs of the acquisition card to detect the upper and lower tolerated water level upstream and downstream of the gate. The acquisition card used is the Advantech PCL 812 PG. The control and supervisory algorithms are implemented using the Labtech Control Program for Windows. All software and hardware used were economical.

The gate was installed at the IMTA laboratory on a zero slope canal 60 meters long, 1m high, and 60 cm wide. The canal inflow was regulate with a manual valve. At the downstream end of the canal the level and flow were regulated by an overshot gate.

CONTROL ALGORITHM

The control algorithms used in the initial test was the Proportional-Integral (PI) regulator. The PI regulator, in a discrete and recursive form, was implemented as (Iserman, 1982):

$$\Delta u(t) = q_0\, e(t) + q_1\, e(t-1) \tag{1}$$

where $\Delta u(t)$ is the control action increment; $e(t) = Y_{ref}(t) - Y(t)$; $Y(t)$ and $Y_{ref}(t)$ are the water level upstream or downstream of the gate and its reference profile; $q_0 = K_P + K_I/2$; $q_1 = -K_P + K_I/2$; K_P is the proportional constant; and K_I is the integral constant multiplied by the sampling time.

When the gate position was the control variable, the regulator performance decreased as the head across the structure changed. The control parameter had to be readjusted to maintain the desired control performance. To avoid this problem, the flow could be used as the control variable or a correction factor could be introduced in the PI control equation. The proposed correction factor is the square root of the upstream water level over the head across the gate. The preliminary results indicated the advantages of the two solution. Further testing should be done to confirm these advantages. The PI regulator equation for the second proposal is :

$$\Delta u(t) = \sqrt{\frac{y_{upstream}}{y_{upstream} - y_{downstream}}} \left[q_0\, e(t) + q_1\, e(t-1) \right] \qquad (2)$$

The main difficulty with the proposed solutions is that the downstream water level and/or the gate positions should be measured. With many applications it is impossible to obtain a reliable measurement of the downstream level. Burt (1996) has proposed another solution using a different factor for the PI regulator based on an equation used to estimate the head variation at the gate. The proposed factor, for upstream control, is a function of the upstream water level reference profile and the gate position. In the near future, additional work will be done with the different proposals for the slide gate.

SUPERVISORY SYSTEM

In a supervisory system, strategies, fault detection and diagnosis, actions, and application, should be considered.

The strategies start with a reliability/safety analysis of the system. Based on this analysis, requirements for the design and operation of the process follow. For those requirements, the ideal of perfection, with the goal of no faults, and the principle of redundancy taking into account the possibility of faults can be applied. The automation is normally carried out at different levels. The lowest level is the regulation of the controlled variable, then monitoring and supervision, and at the top general process management.

Once the process is in operation, fault detection and diagnosis are required. Fault detection is based on signal measurement, using signal analysis or process model evolution. Then fault diagnosis may follow, using methods of pattern recognition, knowledge-based, expert systems, or other modern methods. Fault diagnosis determines the causes of failures.

Based on the results of the fault detection and diagnosis procedures, actions have to be carried out. For example, an automatic protection system may be started, or maintenance or repair tasks may be undertaken. In complex systems, the human operator plays an important role, and education and training of the process staff at different levels is important.

Finally, the application strongly depends on the components, types of processes and computer technology.

The operation control faults initially considered at the slide gate were canal overtopping and faults in sensors and actuator. The redundancy principle is applied here. The critical operation conditions related to minimum and maximum gate and level positions are monitored by multiple limit and level switches in addition to the pressure and position sensor units. A monitoring and supervisory system is placed in the computer above the PI regulator.

For fault detection and diagnosis, a knowledge based method is proposed. The status of each element of the slide gate, actuator and sensors is obtained from the data retrieved by the pressure and position sensors, and the position and water level limit switches. The analog information, its derivative (rate of variation) and the digital data are used to detect unexpected performance in the system components and determine the source of the fault. The detection and diagnosis procedures are based on a comparison between the normal evolution data of the process variables and the operation data reported by the sensors and limit switches. The main result from the fault diagnosis is a report concerning possible causes of problems. This reduces the time needed to repair the equipment and carry out other actions.

The follow up activities are specific for each fault detected and control strategy used. The actions could be to close or to open completely the gate or hold the gate at the current position and wait until a maintenance or repair procedure takes place.

Overtopping, upstream or downstream of the gate, is detected by both water level switches and pressure sensors. However, as the level switch is the most reliable device, the fault detection and diagnosis mechanisms are driven by it. If the pressure sensor does not detect the water level changes, the fault diagnosis method assumes that the sensor is defective. If the problem is upstream of the gate and an upstream control method is in use, the gate is opened completely and the computer advises the operator.

For gate position and pressure sensors faults, the measured data and their derivative evolution during operation are used with a gate model to detect any unexpected events. The action taken in those cases is to stop the gate actuator and wait for the maintenance procedure. Fault diagnosis procedures determine the source of the failure as a function of sensor values. For example, a fault in the gate position sensor may be detected as a variation in the gate position value when the gate is held constant or a lack of change in the gate position when the motor is activated.

Several procedures were developed for the detection and diagnosis of the different faults considered. All of them were implemented using LABTECH Control software which imposed many limitations. The experience obtained from these examples will be used to develop a supervisory program for a Modicon Programmed Logic Controller (PLC) or a Telesafe Remote Terminal Unit (RTU) using ladder logic and C. This equipments was chosen for its reliability, and for the MODBUS communication protocol used in canal operation in Mexico.

Further work must be done to develop a supervisory system for the operation of a canal with numerous gates, since the actions carried out to correct a fault should consider the performance of the entire canal.

CONCLUSIONS AND COMMENTS

The reliability/safety requirements for canal operation must consider the use of ad hoc control and supervisory systems to insure system performance.

The use of flow as a control variable instead of gate position reduces the control parameter adjustments required. The introduction of the head across the gate as a factor in the PI regulator equation provides advantages similar to those derived from the use of flow rate as a control variable.

Further research on control and supervisory methods should be done to improve the reliability of the control system used in canal operation. Laboratory experience should be of great benefit in the development and testing of those methods.

REFERENCES

Burt, C. (1996). *High Line Canal automatic gate control constant study*. Canal Modernization Short Course, ITRC, CalPoly, San Luis Obispo, USA.

Isermann, R. (1982). *Digital Control Systems*. Springer Verlag, Berlin, Germany.

Isermann, R. (1984). *Process fault detection based on modeling and estimation methods - a survey*. Automatica, Vol. 20, No. 4, pp. 387-404.

Isermann, R. (1988). *Process fault diagnosis based on process model knowledge*. IFAC conference on software for computer control, pp 53-65, Johannesburg, S. Africa.

Isermann, R. (1993). *Fault diagnosis of machines via parameter estimation and knowledge processong - tutorial paper*. Automatica, Vol. 29, No. 4, pp 815-835.

Accuracy of Flow Measurements in the Imperial Irrigation District

BRIAN WAHLIN, JOHN REPLOGLE, and ALBERT CLEMMENS
U.S. Water Conservation Laboratory, Phoenix, AZ, USA

ABSTRACT

The Imperial Irrigation District (IID) in southern California has an interest in water conservation measures. These measures depend on the quantification of water flows at various sites within the district. In this study, the random and systematic error components for individual flow measurements at five key sites into and exiting IID were determined. From these individual error components, an estimate of the uncertainty for the annual volume that passes through these sites was determined. This information is expected to be used to identify opportunities for further improving water management practices within the district. The 95% confidence interval for the annual volume at 4 of the 5 study sites was ±2.25%. The fifth site had a 95% confidence interval for the annual volume of ±3.50%. Very little systematic error (bias) could be identified and most of the random error is compensated because of the large number of current meterings IID performs each year.

INTRODUCTION

Water conservation programs in irrigated agriculture are generally geared toward the amount of water diverted to a particular area. The Imperial Irrigation District (IID) is at the tail end of the Colorado River, where unconsumed water flows to the Salton Sea. While some significant water conservation efforts have taken place, the volume of water flowing to the Sea is still on the order of one million acre feet per year. The purpose of this study is to estimate the random and systematic error components for individual flow measurements at five key canal and stream sites into and exiting the Imperial Irrigation District (IID). From these individual error components, an estimate of the uncertainty for the annual volume that passes through these sites were determined. These sites are: the All American Canal (AAC) at Pilot Knob, the 15.2-m (50-ft) Parshall flume at the entrance to the Coachella Canal, the Alamo River at the outlet to the Salton Sea, and the New River at the Mexican border (or boundary) and at the outlet to the Sea. The discharge at all of these sites, except for the entrance to the Coachella Canal, is measured by using current metering techniques.

ERROR EVALUATION

According to Sauer and Meyer (1992), the total uncertainty in an individual current metering can be expressed as:

$$CV_q = \sqrt{\left(\frac{CV_d^2 + CV_t^2}{N}\right) + CV_i^2 + CV_s^2 + CV_h^2 + CV_v^2 + CV_{sb}^2 + CV_{sd}^2 + CV_{sv}^2} \quad [1]$$

where CV_q is the standard error (or coefficient of variation) of an individual measurement of discharge, CV_d is the error introduced from the depth measurements, CV_t is the error caused by the pulsation of velocity, CV_i is the current meter calibration error, CV_s is the error due to the vertical distribution of the flow, CV_h is the errors caused by oblique flow, CV_v is the errors caused by the horizontal distribution of depth and velocity, CV_{sb} is the systematic error for the width measurements, CV_{sd} is the systematic error for the depth measurements, CV_{sv} is the systematic error for the velocity measurements, and N is number of vertical profiles. Details on estimating the individual error terms in Equation 1 can be found in Sauer and Meyer (1992).

At each site there are other flow measurement uncertainties besides the ones due to current metering. We assumed total error of an individual discharge measurement is composed of a random component and systematic, or bias, component as follows:

$$CV = \sqrt{CV_r^2 + CV_b^2} \quad [2]$$

where CV is the total uncertainty (coefficient of variation) for an individual discharge measurement, CV_r is the total random error of an individual discharge measurement (current metering plus other errors), and CV_b is the total bias of an individual discharge measurement (current metering plus other errors).

The total volume that passes through a given site over a time period T can be expressed by the following equation:

$$V_t = \Delta t \sum_{i=1}^{n} Q_i \quad [3]$$

where V_t is the total volume for time period T, Δt is the time increment (e.g., one day), Q_i is the average flow rate at time increment i, and n is the number of flow measurements in time period T. Assuming there is no error in the time interval, the uncertainty in the total volume measurement can be expressed as:

$$CV_V^2 = CV_b^2 + \sum_{i=1}^{n} \frac{m_i^2}{m_V^2} CV_{ri}^2 \quad [4]$$

where CV_V is the uncertainty of the total volume measurement, CV_{ri} is the random error in individual flow measurement i, m_i is the expected value of flow measurement i, and m_v is the expected value of the sum of all n flow measurements. If all the values of CV_r are equal, then the uncertainty for the total volume would becomes:

$$CV_V^2 = CV_b^2 + \frac{CV_r^2}{n} \qquad [5]$$

For this study, the uncertainty in the total annual volume that flows through a given site will be expressed in terms of a 95% confidence interval (2 standard deviations). In this case, the confidence interval for the total volume (CI_V) can be expressed in terms of the coefficient of variation as follows:

$$CI_V = \pm 2CV_V \qquad [6]$$

SUMMARY OF IID'S FLOW MEASUREMENT TECHNIQUES

Water flows into IID along the All American Canal (AAC). On the outskirts of IID, some of the water is diverted to the Coachella Irrigation District. This water is measured by using a 15.2-m (50-ft) Parshall flume at the heading of the Coachella Canal. Water also flows into IID from the New River at the boundary; however, this water is highly polluted and is not used for irrigation purposes. Drainage water is discharged into the Alamo and New Rivers which exits IID and runs into the Salton Sea. An area map of IID appears in Figure 1.

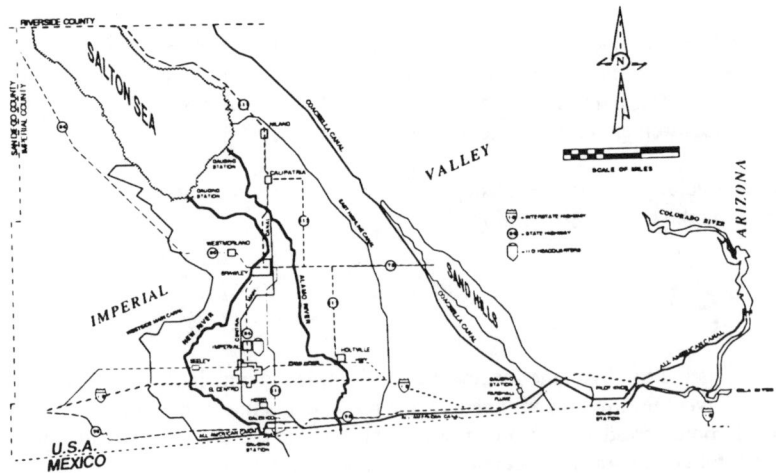

Figure 1. Area Map of the Imperial Irrigation District.

ALAMO AND NEW RIVERS AT THE OUTLET

The Alamo and New Rivers at the outlet to the Sea are current metered by IID technicians approximately once a week. All of the stream guagings are done using the 0.2/0.8 method (i.e., measuring the velocity of each vertical element at points 0.2 and 0.8 of the depth below the water surface). Prior to May 1996, IID current metered the Alamo and New Rivers at the outlet with a constant vertical spacing of 1.22 m. This led to as much as 9-11% of the total flow being in a single vertical element. After May 1996, IID used a variable vertical spacing technique in an attempt to get only 4-6% of the flow in a single vertical element as recommended by the U.S. Geological Survey (Herschy, 1985).

On days that are not current metered, IID uses stage-discharge relationships that were developed in 1979 to get an estimate of the average daily flow in these two rivers. The stage is measured on the Alamo River at the outlet using an acoustic depth sensor. The stage is measured on the New River at the outlet using a Celesco position indicator attached to a 10-cm diameter float. On both of these rivers, stage is recorded every 15 minutes, and the average daily stage is determined from these 96 readings. From the weekly current meterings, IID determines an appropriate correction to the average stage that would allow their stage-discharge relationships to work. This stage correction is then linearly distributed over the previous week. For example, if the stage correction for day 1 was -0.15 m (based on a current metering) and the stage correction for day 7 was -0.22 m (based on a current metering), then the stage correction for day 2 would be -0.16 m, etc.

The current metering site for the Alamo River at the outlet is very good. The gauging station is in an area where the river is straight and the streambed is free of large rocks, weeds, and protruding obstructions. The cross-sectional area of the river is uniform and there are no obstructions upstream to disturb the flow. There is also an acoustic velocity meter (AVM) at this site. Currently, the AVM is not used because frequent dredging of the river has disrupted the meter's calibration. However, in the past the Alamo River was current metered only once a month and the AVM was used the other 3 weeks of the month.

The gauging station for the New River at the outlet is not in a good location. The station is located just downstream from a curve in the river. As a result, there is much degradation on the west bank of the river and the cross-section of the river is highly non-uniform. Rip rap was installed on the west bank to reduce the amount of erosion, and this rip rap disturbs the flow downstream. At the gauging station, the rip rap disturbance could be detected about 4.5 m from the west bank. Also, there is an observation pier on the east bank that is located about 6 m upstream of the gauging station. The pier sticks out about 6 m into the river and disturbs the flow in the gauging station. The U.S. Geological Survey (USGS) method for estimating the error of an individual discharge (Sauer and Meyer, 1992) does not include how to estimate errors for upstream obstructions. Therefore, the CV_q obtained from the USGS method will be lower than the actual uncertainty in the current meterings.

NEW RIVER AT THE BOUNDARY

The New River at the boundary is not current metered by IID. Instead, the USGS current meters this river about every other month. The current metering station was not inspected. The USGS also uses the 0.2/0.8 method to determine the average velocity and are careful to only get about 4-6% of the total flow in a single vertical element. The stage is measured using a float-operated strip recorder. IID also uses a stage-discharge relationship for the New River at the border. This relationship was developed in 1985 and no corrections are made to it.

ALL AMERICAN CANAL AT PILOT KNOB

IID technicians also current meter the AAC at Pilot Knob once a week. There is also an AVM at this site that is used only as a check on the current meterings. The Pilot Knob current metering station is well located about 400 m downstream from seven 5.5-m wide radial gates and just before a bend in the canal. The approach channel is straight and there is no noticeable debris or vegetation to obstruct the flow. The AAC has a very uniform cross-section, and the canal bed is smooth and stable. IID uses the 0.2/0.8 method of current metering and uses constant 1.52 m vertical spacing near the bank and 3.05 m vertical spacing in the middle of the canal. About a maximum of 7.5% of the total flow can appear in a single vertical element. On days that are not current metered, IID uses the 7 submerged radial gates to estimate the discharge. There is a rating table that reports a discharge coefficient for the radial gates as a function of the gate openings. For each current metering, a shift is determined to correct the rating table so it matches the current metering. A running average of the last 4 shifts is used to calculate the average daily flow rate. The water surfaces upstream and downstream of the radial gates are measured with float-operated strip recorders.

COACHELLA CANAL

A 15.2-m (50-ft) Parshall flume measures the flow rate in the Coachella Canal. The flow into the canal is controlled by the gates at the entrance to the canal. Once a flow rate is ordered, these gates are automatically adjusted to keep the flow rate in the canal constant regardless of what is happening in the AAC. This flume was calibrated in 1982 with current meters. The calibration is 8-20% lower than standard published values (Chow, 1959). IID used the 0.6 method (i.e., measuring the velocity for each vertical element at a point 0.6 of the depth below the water surface) and only 9-11 verticals when the Parshall flume was calibrated. The small number of verticals led to about 15% of the total flow being in a single vertical element. Even though the amount of flow in a single vertical element was well above the USGS recommendation, we did not find that this introduced any additional error because the flow in the concrete-lined Coachella Canal is very uniform. The 8-20% difference between the Coachella Parshall flume and the standard tables can probably be explained by the nonstandard transition between the main channel and the Parshall flume.

ACCURACY ESTIMATIONS

The accuracy of the current meterings at each of the 5 study sites was determined using Equation 1. Detailed current meterings in which the full profile (10 points) some of the vertical elements were performed on the Alamo and New Rivers at the outlet. This verified that the 0.2/0.8 method is appropriate for these rivers and was used as a check on the USGS's method of error estimation (Sauer and Meyer, 1992). Other sources of error were also identified and the total uncertainty for the flow measurements was calculated using Equation 2. These other sources of error include: the uncertainty in stage measurements, the error in the zero determination, the uncertainty introduced for not using the proper number of verticals, the uncertainty of the AVM, and the error in integrating volume from individual flow rate measurements. Finally, the uncertainty in the total volume that passes through a given site and the 95% confidence interval were calculated using Equations 4 (or 5) and 6, respectively. These average values for 1994 through the first half of 1996 for all 5 sites are reported in Table 1. Because the New River at the outlet has such a bad current metering site, it is surprising that it actually has a lower uncertainty than the Alamo River at the outlet. However, it should be noted that the uncertainty estimate for the New River at the outlet does not include effects from the upstream obstructions because there is no way to estimate the magnitude of these obstructions (Sauer and Meyer, 1992). Therefore, the calculated value for CI_V for the New River at the outlet is a lower limit and the actual uncertainty should be somewhat higher than reported in Table 1.

Table 1. Uncertainty of Total Volume Measurements for Various Sites.

Site	CV_V	CI_V
Alamo River at the outlet	1.20%	±2.40%
New River at the outlet	1.05%	±2.10%
New River at the boundary	1.75%	±3.50%
AAC at Pilot Knob	1.00%	±2.00%
Coachella Canal	1.16%	±2.32%

CONCLUSION

The 95% confidence interval for the annual volume at 4 of the 5 study sites was approximately ±2.25%. The fifth site had a 95% confidence interval for the annual volume of ±3.50%. These results are consistent with good flow measurement methodologies. Very little systematic error (bias) could be identified and most of the random error is compensated because of the large number of current meterings that IID performs each year.

REFERENCES

Chow, V.T. 1959. Open-channel hydraulics. McGraw-Hill, New York, NY.

Herschy, R.W. 1985. Streamflow measurement. Elsevier Applied Science Publishers Ltd., New York, NY.

Sauer, V.B., and R.W. Meyer. 1992. Determination of error in individual discharge measurements. USGS Open-File Report 92-144. Norcross, GA.

CALIBRATION OF OPEN CHANNEL FLOW COMPUTER SIMULATIONS

JOHN B. PARRISH and RAM DHAN KHALSA
Iowa Institute of Hydraulic Research, University of Iowa, Iowa City, USA
Grand Junction Projects Office, USBR, Grand Junction, USA

ABSTRACT
A methodology for the field calibration of open channels for use in computer simulations is presented. The methodology's objective is to guide the practitioner in the creation of computer simulation models of open channels such that both steady state and transient outcomes can be reliably modeled, given field data for steady state conditions only. The methodology was successfully applied to the Government Highline Canal located in Grand Junction, Colorado.

INTRODUCTION
Computer modeling of open channel flow has been successfully rendered for some 20 years. There are currently a number of PC-based software packages available that solve the complete de St. Venant equations which govern, in one-dimension, the dynamic, transient behavior of open channel flow. The software used in this study was CanalCAD [1]. Successful modeling requires sufficient detail to capture the hydraulic performance of the actual channel, and then tuning to account for remaining model insufficiencies. The tuning process is accomplished by adjusting the model's roughness so that modeled water levels match their field measured values. In general, roughness tuning should only be used to refine the model's behavior, hence the final values should remain within a range of reasonable values for the channel being modeled.

Depending on the use of the computer model, the foregoing procedure is often sufficient. However, modeling of the Highline Canal required reasonable renditions of transient conditions as well as steady state conditions [2]. Since the model can only be tuned under steady state conditions, a rational approach to creation of the un-tuned model was necessary so that the model, after tuning, accounted for transient conditions as well. The need for a rational approach, as opposed to trial and error, was dictated by the Highline Canal's non-uniformity, the sheer size of the project and its necessary level of detail, and by the large number of parameters that effect the outcome.

The Highline Canal is located in Grand Junction near Colorado's Eastern border. Its source is the Colorado River. The canal consists of a series of in-line pools bounded by controlling cross-structures that maintain upstream water levels, and is populated with numerous turnouts which service adjacent farm fields. The modeled portion of the canal was approximately 74 km long, representing most of its total canal length, and had a maximum inflow of 19 cms. The canal was originally earth lined, but some canal segments are now concrete lined and membrane lined by virtue of previous renovation efforts. A unique feature of this canal was a 37 km long pool which was uncontrolled except by means of the inadvertent check-up provided by several flumes. This pool was the focus of the study, so its correct modeling was important.

The modeling objective was to provide simulated water levels that fell within +/- 45 mm of actual water levels. Actual water levels were provided for three steady state flow conditions, representing peak, post-peak, and end-of-season conditions. Additional field data was provided for one transient flow condition during which a relatively large change in canal inflow propagated downstream. Survey information included structure locations, pool invert elevations taken every 60 m, and cross-section measurements taken infrequently but at representative locations. Field data (time dependent data) included estimated inflow, estimated demands and spills, and measured water levels and flow rates at 15 key locations.

DERIVATION OF THE METHODOLOGY

The principle problem with modeling the Highline Canal's non-uniform channels was the fact that the surveyed data sets were in some respects extensive, to the point that all the collected data could not be entered into the computer, but sparse in other respects so that an accurate replication of the hydraulic attributes of the actual canal would prove difficult. It was necessary to derive a methodology whereby actual data could be interpreted and scanned for essential channel attributes that could then be utilized to construct the computer model. This approach yielded an accurate model of the Highline Canal, but perhaps more importantly, it yielded a much improved procedure for collecting the necessary data in the first place.

The actual and modeled canals were considered two systems. The modeled canal would be constructed so as to have similar attributes as required by the study, but where other attributes would vary due to the simulation software's inherent limitations and other study constraints. The principle goal in constructing an hydraulically similar computer model was creating prismatic reaches that behaved identically to the actual, non-prismatic reaches in the actual canal in terms of hydraulic grade lines, or water surface profiles.

When computing the steady-state backwater curves along a prismatic channel, the solution requires solving the following well known equation:

$$\frac{dy}{dx} = \frac{s_o - s_f}{1 - Fr^2}$$

where dy/dx is the water surface gradient, s_o the pool invert slope, s_f the energy gradient, and Fr the Froude number. This approach is not suitable for the task at hand because the actual system's channels are non-prismatic and hence s_o has no significance, and s_f cannot be determined from the data set. It became necessary to adopt a more fundamental approach and equate the energy grade lines for the two systems rather than their hydraulic grade lines. The energy line, s_f, lies above the hydraulic grade line by the kinetic energy head, $V^2/2g$. The two systems will have different velocities, V, but sensitivity studies reveal that the differences in kinetic energy heads will be on the order of 3 mm, which is of no consequence.

The energy gradient is described by Manning's equation as follows:

$$Q = \frac{c}{n} \frac{A^{5/3}}{P^{2/3}} \sqrt{s_f}, \text{ or } s_f = \frac{Q^2}{K^2}, \text{ or } s_f = n^2 \frac{Q^2}{K'^2}$$

where n is the roughness, and K' is a modified conveyance that excludes roughness. In equating energy gradients, the roughness, flow, and modified conveyance K' are also equated for the two systems. K' can be derived easily enough from the supplied data for the actual system, but direct application to the modeled system will fail given the sparsity of surveyed cross sections and the channel variability. The invert data, taken at 60 m intervals, showed considerable variability. In the long pool of interest there were a total of eight cross section surveys, yielding roughly 4.6 km of canal per survey. Thus, one must ask what information was really provided by this sparse data set?

The survey crew was instructed to collect new cross section data when cross sections "changed". Thus, cross section data provided "snapshots" of the actual channel that occurred at infrequent but descriptive intervals. Inspection of the dry canal revealed that the large variability of the invert is difficult to see, but that a change in top width where the water line was readily visible, and/or a change in "form factor" would be readily observable. The canal's dynamic behavior would be largely determined by the top width (denoted TW) and the side slope (SS) near the top width, or water surface. The underlying cross section geometry would be of no interest given equivalent conveyances K'. The top width and form factor attributes could be combined with an average invert estimated from the invert data, where the invert could be averaged at more frequent intervals than that of the cross section survey. Also, the actual channel's average invert was assumed to be parallel to its energy slope as approximated by its water surface slope.

The following constraints ensue from the foregoing observations and simplifications, where the subscripts "1" and "2" refer to the actual and modeled systems, respectively:

- $s_{f2} = s_{f1}$
- $n_2 = n_1$ and $0.020 >= n2 <= 0.025$
- $Q_2 = Q_1 =$ post-peak steady-state flow
- $K'_2 = K'_1$
- $TW_2 = TW_1$
- $SS_2 = SS_1$ near the water surface
- $s_{o2} = s_{f1}$ as approximated by the hydraulic grade line

The difference in energy line elevations at either end of a channel segment, assuming no mid-segment outflows, is as follows:

$$\Delta y = \int_0^L s_f \, dL = \int_0^L n^2 \frac{Q^2}{K'^2} \, dL \cong n^2 \frac{Q^2}{K'^2} L$$

The conveyance K' varies with length due to variance in the depth of flow, but given the previous constraint, $s_{o2} = s_{f1}$, the flow depth is constant, hence the integrand is constant in the first and second expressions, yielding a good approximation in the third and final expression.

The foregoing constraints and the final expression above where encoded in a brief program that automated the conversion of survey data to model input data. The model was tuned such that the post-peak flow and level data were honored.

Table 1. Error Statistics

	Peak	Post-Peak	End-of-Season
Average	-30 mm	3 mm	9 mm
Max	-12 mm	12 mm	82 mm
Min	-49 mm	-9 mm	-30 mm
Spread	37 mm	21 mm	113 mm

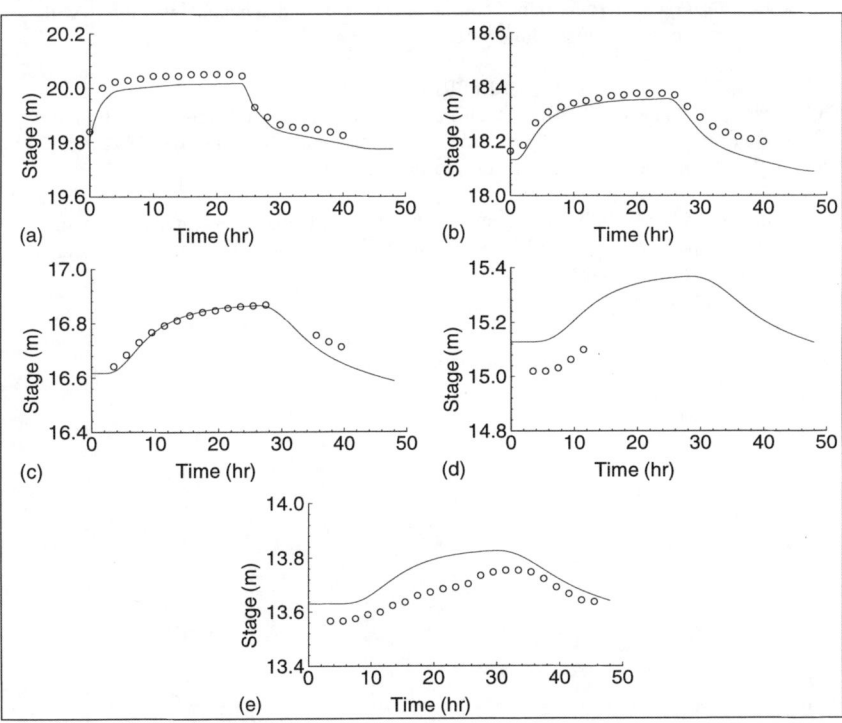

Figure 1. Stage hydrographs at five sites showing modeled (solid line) and measured (circles) transient responses.

APPLICATION TO THE HIGHLINE CANAL
The final model was used to test the deviations in the remaining steady state cases and the transient case. The error statistics for all steady state cases at the 15 water measurement sites are shown in Table 1, and the outcome of the transient test is shown in Figure 1.

CONCLUSIONS
The outcomes above suggest that the methodology derived herein was successful for the purpose of modeling water surface profiles at a variety of steady state conditions and for transient conditions. In the writer's experience, much of the error could have been reduced had a better data set been available.

The foregoing experience suggests that data sets for constructing computer models of open channels supply principally (1) the top width for an intermediate flow and (2) the side slope in the vicinity of the water surfaces for the anticipated range of flows. The lower cross-section geometry is not generally necessary. These data sub-sets should be acquired at sufficiently frequent intervals, or at a pre-determined constant increment, so that channel variability will be sufficiently incorporated into the final model.

REFERENCES
[1] F. M. Holly Jr. and J. B. Parrish. "CanalCAD: Dynamic flow simulation in irrigation canals with automatic gate control." Limited Distribution Report No. 196. Iowa Institute of Hydraulic Research, The University of Iowa, Iowa City, Iowa. 1992.

[2] J. B. Parrish and F. M. Holly Jr. "Computer simulation of the government highline canal." Limited Distribution Report No. XXX. Iowa Institute of Hydraulic Research, The University of Iowa, Iowa City, Iowa. (In Press).

Management of the Euphrates and Tigris river basins: need for a rational approach

MEHMETCIK BAYAZIT
Istanbul Technical University, Istanbul, Turkey

INTRODUCTION

The extreme case of managing water is that of the management of *multinational river basins*, where the interests of not only a large number of people but also of a number of states are involved. Although certain attempts have been made to devise legislation to determine the water rights in the case of *transboundary rivers*, no international authority exists to enforce it. As a result the problems are tried to be solved by international politics *(hydrodiplomacy)*. It is not surprising that the outcome is often far from being rational and optimal.

A striking example is that of the Euphrates and Tigris river basins. These rivers have an annual average water potential of 88 billion m^3 drained from the territories of Turkey (60%), Syria (4%) and Iraq (36%) before they join to discharge into the Persian (Arabic) Gulf (more than 90% of the waters of the Euphrates originates in Turkey). Since the breaking up of the Ottoman Empire in the beginning of the present century, the use of the waters of these two rivers has been a source of conflict among the three countries. The problems were minor until the realization of large scale river basin management projects in the last three decades, the largest of which is the *Southeastern Anatolia Project (GAP)* started by Turkey in the seventies. This project, one of the major water resources projects of the world, is expected to be completed in the first half of the 21st century and will provide 27000 GWh/year of electrical energy and irrigation water for 1.7 million ha of land. At the present time some of the major dams have been built, but irrigation projects that will consume large amounts of water (30-50% of the total flow of the Euphrates, and 10% of that of the Tigris) are yet to start. Iraq's irrigation projects are even more ambitious than those of Turkey, targeting an area of 4.0 million ha (Kolars, 1994).

Although GAP is still at the initial stage, it has created a lot of conflict between the riparian countries. The most active issue so far has been the complaints of Syria and Iraq about the reduction of the flow in Euphrates during the impoundment of the dams. This is a *short term problem*, and will not be repeated in the future. The more important issue is the *sharing* of the waters of the rivers, which will increase in significance with the development of the projects. The hostilities between the interested countries make it difficult to find solutions.

LEGAL ASPECTS

International legislation on utilization of water is based on treaties between the interested parties, conventions and general legal principles. The existing treaties have been devised for particular cases and cannot be generalized. Conventions in this area are mostly for the purpose of navigation. In 1966 International Law Association adopted the Helsinki rules on the uses of waters of international rivers. The rules deal with the *reasonable and equitable* utilization of the waters of an international drainage basin. It is set forth that the states refer the question to a *joint agency* which will formulate plans and recommendations for the most efficient use thereof in the interests of all the states. The International Law Commission of the United Nations adopted in 1994 draft articles on the law of the non-navigational uses of international water courses, which is on the agenda for the UN General Assembly in 1996. In this text it is emphasized that an international water course shall be used and developed by the states with a view to attaining *optimal utilization* thereof and benefits therefrom consistent with *adequate protection* of the watercourse. The watercourse states shall establish a *joint management mechanism* for planning the sustainable development, providing for the implementation of the plans adopted and promoting rational and optimal utilization, protection and control.

The present international legal principles consider that the upstream state may use the waters in any way it sees fit providing it does not seriously diminish their quantity or quality. Downstream states have no right of veto on reasonable upstream activities (Clark, 1995). Prior use, equitable sharing and limited sovereignty with compensation are all considered. The rules are inadequate as they do not take into account the ecology and other factors. There is a search for new acceptable rules at present.

Although some attempts have been made to set up legal rules for the utilization of international waters no international law exists today. IAHR proposed a World Water Convention to help with the establishment of a legal framework for the management of international rivers to secure their sustainable development. At present, the only solution seems to lie with a *joint basin authority* that will make decisions for the optimal development of the resources. Resources can be utilized efficiently only by cooperation between the interested countries. This can be achieved by finding a compromise for their individual demands. The real problem is the transfer of the power of the states to this authority (Carstens, 1996). This should be done by a carefully planned sequential approach starting with the technical issues.

THE EUPHRATES AND TIGRIS BASINS: PRESENT SITUATION

Turkey and Iraq established a Joint Technical Committee in 1980 to prepare a report on the problem of determining "the reasonable and appropriate amount of water that each country needs from the joint rivers", which would then be submitted to a meeting of the ministers. Syria joined the committee in 1983. The trilateral talks were stopped after 16 technical and two ministerial meetings, without producing a report (Kut, 1993). In 1987 Turkey and Syria signed a protocol for Turkey to release a minimum annual average of 500 m^3/s from the Euphrates waters. Syria and Iraq later complained that this promise was not kept during the filling of the Karakaya and Ataturk Dams. In 1989, Syria agreed to release 58% of the Euphrates waters coming from Turkey to Iraq.

Today the problem is understood as that of *sharing the waters*, especially those of the Euphrates river. Syria and Iraq argue that the waters should be allocated according to a formula which would recognize the declared needs of the riparian countries, also recognizing the historical rights of the downstream countries (Kut, 1993). Such an approach would imply that the Euphrates waters could not be used for irrigation by GAP in Turkey, since there would not be enough water available after the needs of the other countries and their acquired rights are fulfilled (Kolars, Mitchell, 1991).

Although the trilateral meetings of the Joint Technical Committee have made it possible to exchange data, to clarify the concerns and positions of the parties, and to illustrate the complexity of the problems, no definite agreements could be reached since each state considered the problem as a zero-sum game and looked for a solution that was beneficial to it, and not for an *optimal solution* that would benefit all of them. The problem of determining the share of each country will not be meaningful in the future since the available water will not be sufficient for the long-term agricultural needs of the countries. Some statesmen in the region have a pessimistic opinion such that they see the water issue as the basis of an armed conflict in the near future, imposing a security threat.

RATIONAL APPROACH

Instead of considering a transboundary watercourse as consisting of parts situated in different sovereign states, the new concept is international drainage basin as a system of surface and underground waters constituting a unitary whole (Picard, 1994). When it comes to natural resources, the state can no longer define its self-interest in terms of political boundaries, because such a position could produce outcomes that are ultimately not in the interest of the state. Just as a river basin cannot be defined on the basis of the boundaries between the districts of a country a multinational river basin should not be limited by artificial constraints such as the boundaries between the countries.

The problem of the comprehensive and rational management of the integrated water and land resources of a river basin should be studied with a *systems approach* in order to obtain *sustainable optimal solutions*. This is valid also for international basins. In the following it will be discussed how such an approach can be implemented in the case of the Euphrates and Tigris basins.

The work must be started with the collection, interpretation and evaluation of the relevant data. This is especially important in this problem where the hydrometeorological data (precipitation, evapotranspiration, runoff) as well as information on soil inventory, irrigable land potential (different figures are given by local and foreign experts), irrigation water requirements, demands and projections are insufficient, inconsistent and unreliable. All three countries will participate in the data collection and analysis, in which same guidelines for the instrumentation and methods should be adopted by all of them so that the type, format and accuracy of the data can be standardized. Satellite data must be used if necessary.

This will be followed by the generation of alternatives for development. In this stage it is necessary to consider the two basins as a whole because it is feasible to transfer water between them. The water potential of the Tigris river is larger than that of the Euphrates, in the basin of which irrigation requirements are higher (in fact, Iraq has a scheme for transferring the Tigris waters to the Euphrates). The locations of the reservoirs, irrigation areas, products to be grown, etc. should be determined considering the basin in its entirety. As an example, Turkey and Syria are considering the construction of one dam each on the Euphrates near the boundary whereas the optimal solution would be a single higher dam. Various scenarios about giving priority to hydropower or irrigation should be considered. The irrigation water requirements must be determined on the basis of improved agricultural technologies with a view to conserve the resources, thereby reducing the water needs of the countries computed according to traditional methods of agriculture. The amount and quality of return flow from irrigation should be evaluated. Salinity problems that might arise due to irrigation will have to be considered. Pricing policies must be included in the analysis in order to achieve conservation and more efficient utilization of the resources. Not only the surface waters but also the groundwater resources should be considered in the development projects. Water quality problems must be given due emphasis, especially considering the pollution of the return flow after irrigation. Reuse of the irrigation water should be considered. Effect of water quality on the fish life and ecology of the Gulf is important (Kolars, 1994). Benefits due to the control of dry and wet periods through regulation by reservoirs must be considered in evaluating the projects.

The final stage is the seeking of the optimal plan using the available optimization techniques. Hydropower-irrigation trade-offs, irrigation methods, preferred reservoir sites and water transfer between the rivers (especially important since the insufficient waters of the Euphrates river can be replenished by those of the Tigris river, which has surplus water) will be determined at this stage. Existing projects, such as GAP, and Iraq and Syrian irrigation schemes will have to be fitted in the development plan, with some modifications if necessary.

Planning of the international river basin can be performed by a team consisting of technical personnel from the three countries advised by independent international experts. An institution can be established to deal with this problem.

IMPLEMENTATION

The concept of a unified basin, not restricted by national boundaries, is not easy to accept by the countries sharing common water resources as they may have different priorities and interests relating to needs, demands, pressures and objectives (Clark, 1995). A critical question is how to eliminate or reduce obstacles to international cooperation. This is possible if the states can be made aware that benefits to each country can be increased by joint action in achieving integrated international river basin development.

The establishment of a joint technical committee with authority to prepare development plans may seem improbable, if not impossible, in the present political situation in a region where authoritarian regimes, ethnic politics and armed struggle exist (Kut, 1993). Communication at the technical level, however, is an important factor in initiating trust and cooperation. Once the joint study teams reach agreement at the technical level, the progression to the policy level becomes more feasible (Clark, 1995). Although the technical commission has no responsibility with respect to the resolution of political issues, the information it provides aids the governments in selecting and executing the optimum plan.

A step in this direction is the "Three Staged Plan for Optimum, Equitable and Reasonable Utilization of the Transboundary Watercourses of the Tigris-Euphrates Basin" proposed by Turkey in 1990. The plan envisages the joint inventory studies of the water and land resources of the region at the first stage. This would be followed by the planning for optimum utilization of resources. This plan implied that Turkey was willing to give up some of its sovereignty rights in order to achieve the optimum benefits (Kut, 1993). Unfortunately the plan was rejected by Syria and Iraq, and today it is not clear how eager Turkey is for this compromise.

In the presence of economic, ethnic and religious conflicts that lead to armed struggle in the region, it is unlikely that a rational solution can be achieved soon. However, this should not prevent the interested parties from looking for a compromise that would benefit all of them. The engineers and other specialists working in the field of water resources must continue their attempts to influence the statesmen and the public opinion. International organizations such as United Nations and World Bank should support these endeavors by providing financial and technical aid. As a result of these contributions it may be expected that an optimal plan can be realized once the political situation in the region is stabilized.

REFERENCES

Kolars, J.F., Mitchell, W.A.: The Euphrates River and the Southeast Anatolia Development Project, Southern Illinois University Press, 1991.

Kut, G.: "Burning Waters: The Hydropolitics of the Euphrates and Tigris", New Perspectives on Turkey, 9, pp. 1-17, Fall, 1953.

Kolars, J.: "Managing the Impact of Development: The Euphrates and Tigris Rivers and the Ecology of the Arabian Gulf-A Link in Forging Tri-Riparian Cooperation", Water as an Element of Cooperation and Development in the Middle East, ed. A.I. Bagis, Ankara, 1994.

Picard, E.: "Aspects of International Law of the Water Conflict in the Middle East", Water as an Element of Cooperation and Development in the Middle East, ed. A.I. Bagis, Ankara, 1994.

Clark, R.H.: "Hydrological Information and International Rivers", Time and the River, ed. G.W. Kite, Water Resources Publications, 1995.

Carstens, T.: "Hydrodiplomacy and International Water Law", IAHR Bulletin, Supplement to JHR, 34(1), 1996.

Negotiating Middle East Water Management: Lessons from Other International River Basins

DEBORAH SHMUELI and NURIT KLIOT
Department of Geography, University of Haifa, Haifa, Israel

ABSTRACT

Increasing and competing demands among countries for water, often complicated by domestic sector competition, is a major cause of international dispute. This article focuses on negotiation approaches for dealing with these conflicts, mechanisms for resolving them, and frameworks for managing shared waters. The experience gained from analysis of thirteen case studies of international river basins are applied to Arab-Israeli transboundary water issues in the Jordan Valley and the mountain aquifers.

---

[1] The research was funded by the *Israeli Institute for Water Research* during 1994-1996.
[2] The thirteen basins are the Rio Grande/Rio Bravo and Colorado River in North America; the Danube, with an in-depth focus on the Gabcikovo/Nagymaros dispute, and the Elbe in Europe; the La Plata basin in Latin America; the Senagal, Chad and Niger basins in Africa; the Tigris-Euphrates, Indus, Ganges-Brahmaputra and Mekong basins in Asia; and the Murray-Darling in Australia.

INTRODUCTION

This paper presents the case for collaborative, integrative negotiations as a preferred means for resolving water conflicts. Water has been and remains a focus of dispute, and the political, economic and environmental issues at stake are increasingly complex. It would therefore be a boon, not only to the affected parties, but to regional geopolitical stability in many areas of the world, to have mechanisms for resolution of water conflicts available and acceptable.

A research endeavor[1] recently completed involves analysis of negotiation processes and institutional frameworks for international river basin management in many parts of the world. The intent is to study what has happened in thirteen selected basins, and relate these lessons to the Middle East. [2]

A major cause of discord between and among countries has to do with increasing and competing demands for water. This is often complicated by competing sectors within the countries involved. A regional approach that transcends political boundaries and involves participation by the broad range of affected interests, offers the most promise for successful resolution of many water conflicts. However, this entails bringing the disputants into some form of negotiation or mediation. It is this search for effective approaches for dealing with differences and for a greater diversity of management options, that is the focus of this paper.

When people think of negotiation, the focus is usually on two parties in an adversarial situation. However, shared international river basins usually involve multiple parties and multiple issues. For example, the Danube River has nine cobasin states and, in addition, cobasin states receive water transfers from five non-cobasin countries. The conflicting uses for the Danube include navigation and electric power generation for the more developed countries and for the less developed Lower basin countries, drinking water, irrigation, fisheries and tourism (Galambos, 1993; Linnerooth, 1990 and 1994).

Even when there are only two nations involved, such as the U.S. and Mexico sharing the Rio Grande/Rio Bravo and Colorado basins, the range of problems gives rise to trans-national interest groups in addition to the national government policy positions (Utton, 1994; Williams, 1995). On one level, there is friction between the farmers of the two countries because of the diminished flow and increased salinity and pollution as the waters reach Mexico. However, within each country there are internal conflicts because of competition among cities for the dwindling water supply, as well as conflict between environmental lobbyists and industrial developers.

In this highly complex arena of regional water issue resolution, success depends on many factors - the political climate, the power balance among participants, the resources available, negotiating techniques and skills, creative problem solving, and solutions that benefit all parties.

NEGOTIATIONS APPROACHES
Given the diversity of water issues and the conflict characteristics, no single process of dispute resolution could be applicable to all. Dispute resolution refers to a wide spectrum of consensus approaches with which parties in conflict seek, more or less voluntarily, to reach a mutually acceptable settlement.

In many international water disputes, the classic distributive bargaining where one party "wins" the other "loses" has yielded deadlock. A Middle East example of such deadlock is the failure of the American mediator Eric Johnson's efforts between 1955 and 1958 to achieve an agreement on the allocation of Jordan River water and an integrated plan for its development. Although all parties recognized the need for cooperation over water resources, they were unwilling and unable to reach a formal agreement (Lowi, 1993, pp. 79-105).

More productive, have been collaborative approaches which offer the possibility of a viable alternative through creative solutions (Gray, 1991). Collaborative models give interested parties a stake, or at least a "say" in the agreement, and reduce post-settlement conflict by getting everyone in on the solution as it is being formed. In the Middle East in 1976, the United Nations Environmental Program Unit's (UNEP) Regional Seas program used integrative methods, to get the hostile Jordan River riparian states to collaborate on environmental issues. Despite the fact that the geopolitical situation remained hostile, this approach overcame Arab-Israeli intransigence, and resulted in the Mediterranean Action Plan (Yoffe, 1995).

Negotiation should be viewed as a search for an arrangement under which all parties would consider themselves better off than they would be without an agreement. A term sometimes used in this context is "BATNA" - best alternative to a negotiated agreement. This would be each stakeholders initial bottom line - what is available if an agreement is not reached. As more information emerges in the course of negotiations, BATNAs are often reevaluated.

A major accomplishment in itself is bringing parties to water resource disputes to the negotiating table - agreeing on the questions to be negotiated and framing these issues, agreeing as to who will be at the negotiating table, who will chair or mediate the negotiations, identifying the intended mechanisms for implementing any agreements, deciding whether or not meetings will be open to observers, and determining numerous ground rules that will organize the process. The negotiation process itself has many steps that fall into the three general categories of prenegotiation, negotiation, implementation.

Many of the negotiations studied highlight the potential of linkages in resolving long-standing stalemates. The Mexico-U.S. negotiations in all phases demonstrate this. Mexico insisted on negotiating the Rio Grande/Rio Bravo jointly with the Colorado River. Linkages with other issues occurred during the 1940 negotiations. For example, a pressing U.S. concern at that time was persuading Mexico to give adequate compensation to American investors whose oil properties had been nationalized by the Cardenas government. This concern was incorporated into the ongoing water negotiations.

If success in multilateral decision making is *defined* by satisfying some interests of all participants, it *depends* on the *power that can be brought to bear*. A graphic example is the Middle East bi-lateral water negotiations. In the Israeli-Jordanian case, Rabin and Hussein directed the negotiations between Israel and Jordan, and were personally involved in the details. Similarly, after the first round of Israeli-Palestinian negotiation, the chief Palestinian water negotiator, a schoolmate of Arafat, had direct access to him and was able to bypass his own immediate superiors when he felt that their nationalistic attitudes did not make "water sense". On the Israeli side, Rabin gave the direct orders. In both cases negotiations ended in agreement.

Once joint gains have been generated in a negotiation, the implementation issues must be dealt with, and that involves the ever-present problem of costs. The major cost questions are the total amount required, the sources of funding, and decisions on how the costs and benefits are to be allocated.

A critical success factor in multilateral negotiations is a favorable geopolitical climate. The Iron Curtain blocked basin-wide agreements in the Elbe. With unification of Germany, and the Western reorientation of the Czech Republic, basin-wide agreement is now being forged. In the realm of climate setting for negotiations, it is sometimes advisable to enlist an outside mediator to facilitate the process, especially in multi-party, multi issue, international negotiations.

Although we have emphasized the benefits of multilateral, region-wide agreements, incrementalism often plays a positive role. Dealing with one issue at a time on either a bilateral or multilateral basis can be a stepping stone toward a comprehensive settlement. In the La Plata Basin, for example, agreements have advanced from bilateral treaties to the recent multilateral accord on the Parana/Paraguay River system. However, incrementalism does not inevitably lead to basin-wide solutions. In the Ganges dispute, despite some progress at the bilateral level on flood control, no progress has been made toward regional cooperation among India, Nepal and Bangladesh.

Successful implementation sometimes requires the establishment of new institutions. "Institution", in this context, can be loosely defined as any formal or informal arrangement, convention or code of conduct relating to the agreement. It can be as simple as a rule or set of rules, as formal as a treaty, or as elaborate as an intergovernmental organization.

At this more formal end of the spectrum, there can be both success and failure. The U.S. Mexican International Boundary and Water Commission / La Comision Internacional de Limites y Aguas is generally regarded as effective. The weakness of the Niger River Commission is considered one of the major reasons for the failure of Niger basin management.

ISRAELI-ARAB WATER DISPUTE - BRIEF SYNOPSIS

When drawing lessons to be applied to the Israeli-Arab water dispute, we recognize that the situation is unique. It involves peace agreements between and among sovereign states - Israel with Egypt and with Jordan - and step-by-step agreements between Israel and the Palestinian Authority (PA). The latter is a governing body that does not exercise full sovereignty over the territory in which it is housed, let alone the territory to which it aspires.

To further complicate the situation, the agreement between Israel and Jordan, when signed on October 26, 1994, was made with the anticipation that a full and formal peace between Israel and the PA was simply a matter of time, and that agreements with Syria and Lebanon would follow.

Specifically, the Israel-Jordan treaty defined water allocation of Jordan and Yarmuk River waters, as well as Arava groundwaters, and called for joint efforts to prevent water pollution. The agreement recognized, however, that the water resources directly controlled by both countries could not meet their needs, and called for international and regional cooperative water projects.

In the "Oslo II" agreement of September 28, 1995 signed by Israel and the Palestinian representatives, the complex water issues were deferred to the final negotiations. However, a Joint Water Committee was established to manage the West Bank's water resources and to develop new supplies, as well as to stop illegal pumping of water. Other than Israel's recognition of the Palestinian claim to water rights, there was no specific reference to allocation amounts.

There has been no progress in discussions by Israel with Lebanon and Syria. Whether there will be, depends on future political developments. Of greater economic and political consequence would be an agreement between Syria and Jordan on how much each can draw from the Yarmuk. Such a situation calls for multilateral, collaborative negotiations, but the political stalemate seems to preclude it at this point.

Future water negotiations between Israel and Lebanon could revolve around the possible sale of Litani River water to Israel.

PARALLELS DRAWN, LESSONS LEARNED
1. First, there is the importance of changes in political climate. In the case of Eastern Europe, the roll-back of Soviet influence from the Danube, brought Hungary and Slovakia to litigation over the Gabcikovo\Nagymaros (GN) project. Many questions are raised by the recent change in Israel's government and the possibility of a breakdown in the peace process with the PA. Might Jordan cancel its lease arrangements with Israel over land in the Arava and south of the Yarmuk-Jordan River confluence? There is no regional compact to adjudicate such a potential dispute. In this absence, would it have been desirable for a third party to be a co-signatory to the agreement? If it is too late to consider third party backing for agreements already concluded, this might be a goal for future bi-national or multi-national accords. The instability of governing forces in all of the Arab countries and among the Palestinians, and the possibility that new governments in Israel might renounce past or future agreements, makes this question especially relevant.
2. Secondly, cost sharing is a major issue that must be addressed, especially between countries in very different stages of industrial development and environmental awareness. In the Brazil-Paraguay agreement to build the Itaipu Dam, equal sharing of costs and benefits was possible because a far wealthier Brazil provided Paraguay with a loan to cover its part of the costs. This could be of particular interest to the Israeli-Jordanian and Israeli-Palestinian future water projects.
3. A third area involves comprehensive umbrella agreements. These will be necessary in the Middle East, and should include surface, ground, wastewater, water quality, and perhaps cloud seeding to enhance precipitation, which could become an issue between cobasin states. Where water quality of surface and groundwater sources and the allocation of groundwater were not covered by the agreements, as was the case for the Danube basin and between the U.S. and Mexico, they emerged as primary points of contention and have stalled the process of joint management.
4. Following this point, water agreements must be specific and detailed, not general, or compacts "in principle". It is essential to obtain a common

agreement on a definition of water quality and how it should be tested. In addition to joint monitoring and data gathering, joint water quality laboratories should be established. Lack of consensus on the validity of data gathered was a factor that hindered resolution of the boundary dispute between Paraguay and Argentina over the Pilcomayo River.

5. The need to establish negotiations and political symmetry between economically unequal partners is highlighted by the United States-Mexican experience. For the Middle East, this means, not only that Israel must be prepared to share its advanced water technology and its superior resources, but also suggests that it may have to consider accepting regional standards that fall short of what it would desire. With pressures for rapid economic growth, the Arab countries are likely to neglect environmental consequences, as Israel has done in its developing period. If so, this could have a direct impact upon water use and quality.

6. The lessons to be learned from the NAFTA experience, raise warning signs for the Israeli-Arab economic agreements under discussion and in early stages of implementation. The planned industrial development between the Palestinians and Israelis along the Gaza Strip, for example, is advancing without the proper attention being given to the storage of industrial wastes necessary to prevent additional contamination of the joint aquifer. The U.S. Congress is now insisting on linking border economic development with environment. Such linkage in the Middle East must be structured both nationally and internationally.

7. The absence of an agreement on groundwater in the United States-Mexico case, is also an object lesson for future Israeli-Palestinian agreements. Palestinian activities in the mountain aquifers can contaminate Israeli groundwater in its coastal plain. Past Israeli policy has contributed significantly to contamination of water in Gaza. In addition, while Israel has a strong, centralized Water Authority, the Palestinians may encounter problems in centralizing their water management efforts. Governance institutions in this area are in a fluid state. The West bank is geographically separate from Gaza, and might, itself, be geographically divided by Israeli settlement salients. Moreover, Jordan and the Palestinian Entity might eventually federate. This means that agreements entered upon today must anticipate such change, and set an agreed upon general standard that would apply in all cases.

8. Finally, an independent institutional framework is necessary to implement development strategies. If an international institutional framework is to be established for the Arab-Israeli water systems, two possible arrangements could be helpful: a) membership of a powerful outside financial or political body to help secure the agreement, b) a series of successful bilateral projects as springboards for multilateral efforts.

The Middle East has much to learn and much to contribute to the principles of multi-nation water resource management. A strategy to be pursued is the linking of environmental and water issues under a single jurisdiction, in bilateral, multilateral, and eventually regional compacts.

REFERENCES

Babbitt, E., and A. McDonald, 1986, "Negotiations Analysis as a Tool in the Management of Large International Rivers," Processes of International Negotiation Project and IIASA draft paper, Department of Urban Studies and Planning, Massachusetts Institute of Technology, and American Academy of Arts and Sciences, Cambridge: MA.

Bingham, G., Wolf, A., and T. Wohlgenant, 1994. Resolving Water Disputes: Conflict and Cooperation in the United States, the Near East, and Asia, Irrigation Support Project for Asia and the Near East (ISPAN), November.

Galambos, J. 1993. "An International Environmental Conflict on the Danube: The Gabcikovo-Nagymaros Dams," IN A. Vari and P. Tamas, 1993, In Environment and Democratic Transition, Kluwer Academic Publishers, Amsterdam: NE, pp. 176-226.

Gray, B., 1991. Collaborating: Finding Common Ground for Multiparty Problems, Jossey-Bass Publishers, Los Angeles: CA.

Kliot, N., 1994. Water Resources and Conflict in the Middle East, Routledge, London: England.

Linnerooth, J., 1990. "The Danube River Basin: Negotiating Settlements to Transboundary Environmental Issues," *Natural Resources Journal,* Vol. 30, No. 3, summer, pp. 629-659.

Linnerooth-Bayer, J., 1994. "The Danube River Basin: International Cooperation for Sustainable Development." *Transboundary Resource Report,* Spring 1994.

Lowi, M., 1993. Water and Power: the politics of a scarce resource in the Jordan River Basin, Cambridge Middle East Library 31, Cambridge University Press, Cambridge: England.

Utton, A., 1994. "Water and the Arid Southwest: An International Region Under Stress," *Natural Resources Journal,* Vol. 34, Spring, pp. 1-5.

Williams, E., 1995. "The Maquiladora Industry and Environmental Degradation: A Political Analysis," paper delivered at a conference on "International Boundaries and Environmental Security: Frameworks for Regional Cooperation," Singapore, June, 1995.

Yoffe, S., 1995. "Transboundary Freshwater Resources: Environmental Negotiations Between Antagonistic States," *Freshwater Resources.*

THE JULY 1996 FLOOD ON A TRANS DMZ RIVER IN KOREA

HYOSEOP WOO
Water Resources Specialist, Korea Institute of Construction Technology, Socho-gu Umyeon-dong 142, Seoul 137-140, Korea

ABSTRACT

This article describes a historical flood that occurred on the Imjin River in South Korea during the July 26 to 28 of 1996. The Imjin River flows, from north to south, across the Demilitarized Zone(DMZ), which was set up at the end of Korean War in 1953 and since then has divided the Korean Peninsula into North Korea and South Korea. Intensities of precipitation at some places in the river basin during the flood far exceeded 200-year return period and flood stages at some reaches of the river exceeded the 100-year design flood stages. Mainly because of the tremendous magnitude of the storms but partly because of its geopolitical location, transected by the DMZ and development-restricted for the military purposes, the Imjin River basin suffered heavy losses from the flood in spite that population and industry in the basin area are relatively sparse compared to the other part of Korea.

INTRODUCTION

Two series of heavy storms during the July 26 to 28 of 1996 that had concentrated mainly on the Imjin River basin caused a historical flood on the river. This flood caused a total of 89 fatalities and property losses of more than 660 million US dollars. The number of people evacuated temporarily from their inundated houses exceeded 35,000.

During this flood, daily peak amount of precipitation at some places in the basin exceeded 400 mm and two consecutive day amount reached 700 mm, which exceeds far those of 200-year return period. A major characteristics of this flood is that the flood losses were exceptionally large because the magnitude of the storms and flood were tremendous, and partly because the storms concentrated on the DMZ and its north and south vicinities. Flood proof measures in this area have been less implemented, compared to the other part of Korea, because much parts of the area are restricted for the military purposes. Because of this reason, no major flood control dams were ever constructed in the river and even the elementary hydrological measuring stations have not been available until recent years.

THE IMJIN RIVER - A TRANS DEMILITARIZED-ZONE RIVER

As shown in Fig. 1, the Imjin River basin locates at the center of the

Korean Peninsula and across the Demilitarized Zone, which divides the Korean Peninsula into two, North Korea and South Korea. This zone, 4 km wide and 250 km long, which was set up at the end of the Korean War in 1953, is known one of the most heavily armed areas in the world. Along the narrow DMZ, hundred thousands soldiers at each side have confronted each other for more than 40 years.

Basin area of the Imjin River is 8,118 km^2, about one third of the Han River, a largest river in Korea(South), and length of the river is 255 km, about one half of the Han River. About two third of the basin area locates in North Korea, while the area locating in South Korea is only 3,009 km^2. The largest tributary of the river is the Hantan River with its basin area of 2,436 km^2, a 30 % of the total river basin area. This scale of the river surely is not considered large.

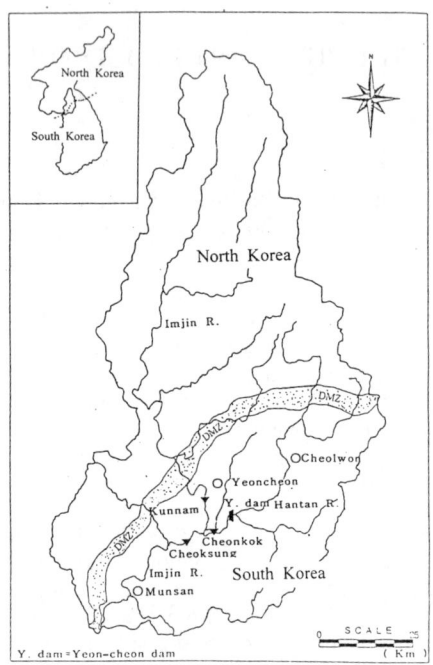

Fig. 1. The Imjin River Basin

Nevertheless, this river basin is multinational because the river flows from North Korea to South Korea, two politically independent and hostile countries. It is, however, pseudo-multinational in one sense, because the two countries were one before the Second World War and eventually they can be one nation within years like the reunified Germany.

Upland areas of the river basin are composed of high mountains of more than 1,000 m high, while lower lands are composed of relatively low hilly mountains and small and long plains along the rivers. River profile of the main stem of the river varies 0.009 in the upstream to 0.002 in the downstream. A 70 % of the river basin is covered with good forest, while the others are mainly croplands. Especially, the 4 km wide DMZ is known a world treasure of natural ecosystem. For more than 40 years since the end of the Korean War, man's access has been restricted strictly by the Armistice Agreement with a few military guards of each side being permitted to enter the "untouchable zone". Panmunjum, a famous place where the Armistice Agreement of the Korean War was signed in 1953 and after then the South (including USA) and the North military representatives have met on a regular basis, locates at the west DMZ in the river basin.

Annual precipitation of the river basin is 1,270 mm, which is close to the national average. Average annual river discharge is estimated to be about 6 billion m^3, which is equivalent to about 43 m^3/s. These values, however, were less certain because no reliable flow measurement has been made on the river since the Korean War.

THE JULY 1996 FLOOD

CHARACTERISTICS OF PRECIPITATION

The storms that caused the historic flood began in the early morning of July 26 and ceased in the morning of 28 continuing only for two full days. Table 1 shows daily precipitation amounts during the flood on major cities in the river basin. These areas received a total of from 370 to 690 mm varying places to places. Especially, City of Yeoncheon, a storm center during this flood, received a 36 % of the average annual precipitation for only one day. Total precipitation amount on this city is near 700 mm during the flood, more than one half of the annual average. As a whole, the basin area received about 40 % of the annual average for 50 hours.

station	26th	27th	28th	total	annual average	ratio to annual average(%)	
						daily maximum	consecutive day maximum
Cheolwon	225	268	34	527	1257	21	42
Dongducheon	288	244	46	578	-	-	-
Yeoncheon	162	448	77	687	1250	36	55
Munsan	206	170	67	443	1253	16	35
Cheoksung	263	329	46	638	-	-	-

Table 1. Point precipitations during the flood

unit : mm

In this study, one-day and two-day maximum precipitations for each return period on two cities, Cheolwon and Yeoncheon in the basin, were analyzed using the last 30 to 40 year hydrological records starting from 1915 with some unrecorded years in between. According to this analysis, return period of the one-day storm at Yeoncheon far exceeds 200-year, while that at Cheolwon region is equivalent to only 30-year. When that of two-day storms are considered, that at Yeoncheon also exceeds 200-year and that at Cheolwon exceeds 100-year. These results reveal that the storms during the flood surely are considered the largest ones in the 20th century in this basin.

FLOOD STAGES AND DISCHARGES

Two river stage data, collected from Kunnam gauging station on the main river reach and Cheonkok gauging station on the Hantan tributary reach, were analyzed. Figure 2 shows stage hydrographs at the two gauging stations during the flood. As shown in this figure, two hydrographs have similar shapes and peak times, which implies that two main floods, one from the main river and one from the tributary merged almost at the same time causing a greater flood on the downstream reach.

In Fig. 2, it is shown that the second peak flow on Kunnam exceeded, by 1.4 m, the 100-year design flood stage of 31.8 m, while that on Cheonkok far exceeded, by 3.2 m, the 80-year design flood stage of 31.7 m. These exceedings of the design floods caused the levee overtoppings and breaches at many places along the rivers and the failure of a small hydropower dam built across the Hantan River.

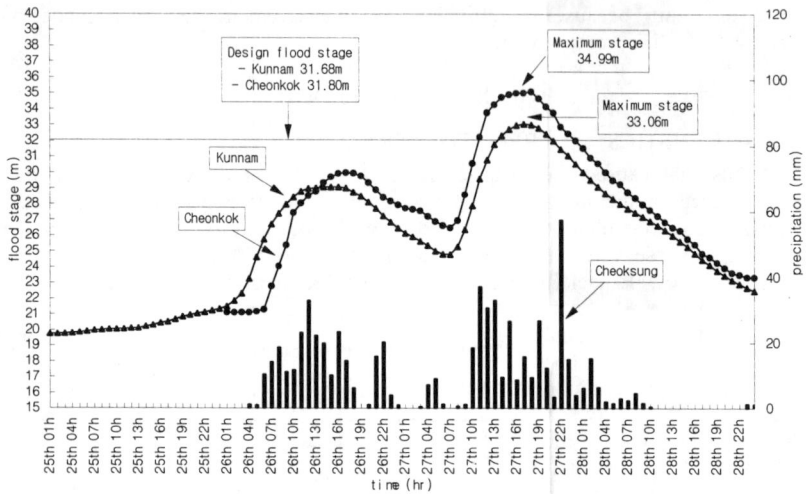

Fig. 2. Flood Stage Hydrographs during the Flood

IMPACTS OF THE FLOOD

The floods caused by the two-day storms concentrated mainly on the Imjin River and partly on the eastern vicinity of the river basin. Total fatalities were counted 89 including many military personnel and total number of the evacuated inhabitants due to the floodings were counted 35,631. About 75 % of these losses occurred in the Imjin River basin alone. These losses exceeded those that had been incurred by the August 1995 flood(Woo and Kim, to be published in Water International 1996) in the central part of the Korean Peninsula including the Han River basin and the metropolitan Seoul area. This means that the intensity and impact by the July 1996 flood on the Imjin River would be far greater than those by the August 1995 flood. It is because the population and property densities in the Imjin River basin are far smaller, with a population density of the area being one third of the national average, than those in the Han River basin.

Major flood losses included levee failures along the numerous tributaries of the river such as those at Cheolwon, Yeoncheon and Munsan and subsequent inundations of houses, roads, railways, and croplands and shutdown of water purification plants and telecommunication lines. Another characteristics of the losses due to the heavy storms is mass landslides at the mountainous areas located upstream the Hantan River, which caused most military fatalities. Among the losses incurred by this flood, only the dam failure and military damages, which appear to be unusual to general flood losses, are described below in details.

FAILURE OF THE YEONCHEON DAM

The July 1996 flood destroyed dramatically a small scale hydropower dam built across the Hantan River near City of Yeoncheon. Fig. 3 shows a picture taken one day after the dam failure. This dam, 243 m long and 23 m high, having a total reservoir capacity of about 13 million m^3 and a hydropower capacity of 6,000 kw, was built in 1985 mostly with concrete gravity and partly with earth fill at the right side of the dam. The flood caused an overtopping of the dam, starting at 9:00 AM July 27, and a breaching over the earth fill part of the dam. The earth fill part was completely washed out within an hour and subsequently a concrete part of the dam connected to the earth fill part was broken as shown in Fig. 3. The impact of this dam failure on the downstream reach was not revealed yet. Fortunately this dam was a small scale, a run-of-river type hydropower dam with a minimum reservoir capacity, which implies no major impact downstream due to the dam failure.

Fig. 3. Failure of Yeoncheon Dam

MILITARY DAMAGES

During this flood numerous landslides incurred by the heavy storms hit some military camps and facilities located near the DMZ areas. In South Korean part only, many casualties occurred mostly due to the landslides, while the number of casualties in North Korean side is unknown. There are some common causes why military camps on the mountainous areas are vulnerable to heavy storms. One is the fact that military camps are usually built on the lee side of mountains or deep valleys in order to cover from enemy's gun power, the places vulnerable to mountain slope landslides. Another is the fact that during heavy storms military trenches, dug usually along the contours of mountains, can play a role of water tank that can be collapsed and cause landslides downstream.

Antitank barriers, constructed usually across the potential access routes of enemy's tanks, were revealed to be water barriers during a large flood. An antitank barrier, in the form of a line of concrete blocks, across a stream can cause a high water level just upstream, greatly obstructing flood passage. For an extreme flood, this kind of obstruction can cause a flooding or dike breach upstream the barrier. Another odd result from this flood is that the great washing power of the flood swept away anti-personnel and -tank mines that had been buried in the DMZ. Great efforts were made in order to locate the lost mines along the numerous tributaries of the river. Unfortunately, all the tributaries in the basin flows from north to south, which can flush North Korean mines to the South Korean earth at this kind of extreme flood.

CONCLUDING REMARKS - LESSONS

The July 1996 flood on the Imjin River, a historical river since the three kingdoms of Kokuryo, Silla and Baekjae struggled each other to take the river basin area, a strategic region of the Korean Peninsula about 1,500 years ago, will be recorded truly a historical flood in every sense. The intensity of the precipitation during the flood is greatest among the 40 year record, exceeding 200-year return period and is estimated up to 500-year at some places in the basin. The intensity of the flood losses was truly large when the areas swept by the flood are relatively sparsely populated and less developed compared to the other part of Korea. This disaster, therefore, may be considered "unescapable", which means in Korea that the intensity of flood far exceeded the economically and socially acceptable upper lines. Nevertheless, this flood gave us some valuable lessons as follows:

(1) Extension of Flood Warning System

In Korea, flood warning system are being operated only in the large rivers without much attention to the smaller ones. The Imjin River has no modern flood warning system because no hydrologic data are available from North Korea occupying two third of the river basin. In order to establish the flood warning system, installation of the radar rainfall gauging system that can cover the North Korean basin is needed. Also, more hydrological measurements should be made on a regular basis, in spite that the area considered is mostly restricted in Korea.

(2) Raise of Design Standard of Dam Spillway

PMF concept for hydrologic design of dam should be considered even for a small scale dam because of the potential extent of losses incurred by dam failure. This upgrading may be further justified when global climate change and subsequent occurrence of more frequent and intense storms are to be considered.

(3) Reinforcement of Flood Countermeasure for Military Camps

Storm loss prevention tactics beside military ones are to be carefully considered when military camps and facilities are constructed on the steep mountains and valleys. Also, when antitank barriers across the mountain streams are designed, both military tactics and stream hydraulics should be carefully considered.

ACKNOWLEDGEMENT

This article is based on the report, written in Korean, "The July 1996 Imjin River Basin Flood", supervised by the author and published by Korea Institute of Construction Technology in 1996. Efforts made by the investigators, including Mr. Won Kim, Dongryul Lee, Yangsu Kim, Hyeonmin Sin, and Hyeonjun Kim who were participated in the preparation of the report, are greatly appreciated.

INTERNATIONAL RIVER BASIN MANAGEMENT
- IMPERATIVE FOR SUSTAINABLE WATER DEVELOPMENT

JELISAVETA MUŠKATIROVIĆ
"Jaroslav Černi" Institute for Development of Water Resources,
Belgrade, Yugoslavia

ABSTRACT

This paper elaborates on activities which have been implemented in the scope of a new integrated water resources management policies for sustainable development starting from various conventions and agreements to existing projects within the river basins, small watersheds and urban units. The author would like to stress the fact that at the end of this century, an adequate surrounding for the execution of adopted policies does not exist due to the unbalanced economic and political powers, frequent hostilities between different national and confessional groups, terrorism and inability and bureaucracy of international organizations for preventive actions.

The ideal approach to the integrated water resources management would require implementation of agreed policies on the whole river basins. Due to existing situation it is most appropriate to accept the basic principles of the river basin management but concentrate on particular activities which can be implemented in riparian countries. This would ensure the phased introduction of sustainable policy into the each of the participating countries. The success of each project will contribute to better understanding of interdependence between riparian states to reach the agreed goals and overcome possible conflicts. These are the imperative actions in achieving the equality between states in international river basin management.

1. INTRODUCTION

The logo for XXVII IAHR Congress, the water drop with the globe showing the continents and oceans, gives a message that each water drop preserved and wisely used is imperative for sustaining life on our planet. This message is clear although its implementation will be very demanding and long-lasting. The barriers are surrounding us. They are of economic, administrative, political, sociological, historical and cultural nature. These barriers existed for centuries and have shown themselves in various forms, depending on the level of social development, its needs and relations towards the water and other natural resources.

Today, at the and of second millennium, when it is evident that human existence is brought in the question, all preconditions for focused and coordinated activities aimed at water conservation and environmental protection are acquired. The plans and activities should aim to break barriers between individuals, urban units, regions, states as well as between people of different nationalities and confessions. The main objective of this is to treat a water as a common resource for ensuring living conditions for all forms of life for present and future, and for providing base for social and economic development.

Numerous conventions and agreements[*], which have been ratified during last decade, marked the general route, approaches and methods leading towards sustainable development.

In spite of up to date activities (adoption of general strategies, establishment of different international bodies, organization of national and international conferences and implementation of World Bank, UN supported and other projects) which give us a hope for faster improvement of human relations towards environment, the reality of present situation make us reserved and skeptical.

There is a big question if in the world, dominated by one super-power, dismembering societies, national and religious hostilities in which thousands die each day from hunger, thirst, diseases, terrorism and wars, a fair distribution of water according to demands and needs of the mankind can be achieved.

The answer on this question cannot be given - this paper only underline the importance, urgency and complexity of the task in front of our and future generations.

Everybody should contribute to the success of this task. Our generation should do the utmost to use science, technique, technology, education and legislation in order to achieve harmonious coexistence between mankind, their activities and nature. The realization of global water management policy should be undertaken step by step using wisely each positive experience which should be important chain-ring in the process of making the common future.

2. INTEGRATED WATER RESOURCES MANAGEMENT

---

[*] - Vienna (1991) International Conference-Agenda for Science for Environment and Development into the 21st Century
- Dublin (1992) International Conference on Water and Environment - Development Issues for the 21st Century
- Rio de Janeiro (1992) Earth Summit, Agenda 21

Uneven spatial and temporal variation of water resources, coupled with raising levels of surface and groundwater pollution caused by global changes and human activities are the main reasons for increased importance of providing the adequate quantities of a good quality water. Having in mind the complexity, multidisciplinary and urgency of above task it is necessary to pull together all professional and scientific potentials, governmental and non-governmental organizations and financial institutions. Special place in this activity should be played by mass media.

Strategic orientation on the integrated water resources management for sustainable development, adopted at the Summit in Rio, presents a new management approach. This is a holistic approach which integrates socio - economic interests with environmental protection, bearing in mind a sustainability of development actions. The application of this approach requires, among others: good estimation of the available water quantities, adequate water demand forecast as a function of population growth and raising life standards, as well as analysis of the environmental changes caused by human activities. This policy requires from all countries an active participation in the process of creation of approaches, methods, and activities to support sustainable development.

Huge differences amongst the rich and developing countries with abundance and scarcity of water, countries with developed environmental protection and without it, lead to the conclusion that greatest part of initial activities should be borne by major developed countries. They should be promoters of activities aiming to establish and support the implementation of a new water management approach. Their socio-economic position in the world give them the best chance to practically prove the advantages of integrated water management. This is even more so as they already have appropriate financial and technical resources as well as a high environmental expectation.

The implementation of a new approach should be structured from the lower levels: urban units, sub-basins and small regions on a national level. This does not exclude the international cooperation if the clear interests, political will and financial support exist. In this case the solutions should be compatible to the policies on the international level, i.e. the solutions should not harm any member state. This is the only way to contribute to conflict resolution, obtain public support, and agreement from political and social associations. In addition, positive experience of such approach will encourage the implementation of large scale projects in the next stage.

3. INTERNATIONAL RIVER BASIN MANAGEMENT

As holistic approach is a basis for sustainable development of water resources it is logical to consider its implementation on the natural and anthropogenic units. River basins are potentially the largest and most important water sources which provide optimal living surrounding for humans, flora and fauna. Although a river basin presents one hydrographic unit it is at the some time, considering natural resources and ecosystem, heterogeneous unit comprising of different constituents which are interconnected and interdependent.

The very rare riparian states, which have enough water from their own territory and which undertake the measures for water protection, are in a position to solve their problems alone. However, they also have an interest to contribute to water resources management to river basin level in order to save themselves from harmful effects caused by activities beyond their borders, to improve the benefits of inland navigation, tourism, natural habitats and to compensate for investments which benefit the downstream countries (erosion protection, sedimentation in the head catchment reservoirs, active flood protection measures etc.).

Greatest majority of the countries, especially developing countries, do not have enough water and furthermore the available waters are polluted. The only way of action for these countries is wise use and management of the existing water sources within the river basins. The importance of international river basins is underlined by the fact that about 47% of the total earth surface and over 40% of the world´s population belong to them.

Water management on the river basin level requires establishment of new modes of cooperation between states with clearly defined duties and obligations. This presents a corner stone of all future development policies. For developing countries and countries in transition an active support from developed ones is of particular importance in order to achieve sustainable use of natural resources.

The agreements and conventions, which promote sustainable development, are only one side of a coin. The other, still more dominant side, is expressed through the interests of economic power states, their alliances, multinational companies and political interests. This is evident from the events which occurred during last decade, which showed the inability of world´ community to actively influence processes of mass migration, terrorism and war. There is a danger that positive efforts on the global level, which have to encourage economic growth, security, social justice and environmental protection through integrated water management, should be misused in some parts of the world. Therefore, governmental , non-governmental organizations, professional and scientific association and public should harmonize their interests on national, bilateral and multilateral level. The adopted policies should exclude, as far as possible, chances for other kinds of unfair advantages.

Planning, designing and implementation of economically and socially acceptable projects which have support of two or more riparian states, can speed up and encourage processes of international river basin management. The route, which stretch from small pilot projects to multi-countries development projects provides the opportunity to learn from weakness and mistakes and bring increased levels of confidence, both within and between riparian states.

4. DANUBE BASIN MANAGEMENT-CHALLENGE FOR FOURTEEN EUROPEAN COUNTRIES

Judging by many parameters, the Danube River Basin is a very specific one and due to this it is appropriate for a pilot model for implementation of accepted policies in water domain. The advantages of integrated resources management in the upper, middle or lower part of the Danube River, are indisputable. Water demands in 14 riparian states, regardless of political systems or economic power are increasing. Quantities of domicile waters within states are significantly lower than transboundary waters; hydropower potentials are different and influences of hydraulic structures can be felt in neighboring states. Besides that, groundwater is overexploited and polluted as well as the surface waters while the biodiversity is endangered and at some places even destroyed. Particularly important are more and more pronounced spatial and temporal variations of water distribution.

These were the main reasons why the first international activities in the Danube River Basin started almost fifty years ago, i.e. 1948.

A good cooperation between Danubian countries is established in the fields of hydrology, hydrogeology, ecology, hydropower production, navigation, tourism etc. However, there is still a lack of large scale projects, which would include all Danubian countries. Existing projects are mostly result of the bilateral cooperation (HPP "Gabcikovo", HPP "Iron Gate I", HPP "Iron Gate II", HPP "Turnu Magurele", protection of natural habitats and biodiversity in the frontier areas, the protection of riparian zones due to increased water levels etc.).

These projects, bearing in mind that some of them exist over 25 years, have generated significant experience relating to the influence of transboundary projects on the river basin, as well as the questions of operation and management. Existing documentation on planning, implementation and operation of hydropower and navigation system " Iron Gate I", can provide answers on numerous issues linked to the Danube River Basin management.

FR Yugoslavia is specially interested for use and protection of the Danube River. The main reasons follows from the following facts:

- Transboundary waters of Yugoslavia are almost 10. times larger than domicile;

- Unit runoff from Yugoslav territory shows significant variation in time and space. It presents limiting factor for regional development;

- FR Yugoslavia belongs to the middle reach of the Danube River, which has the largest incremental increase of discharge (discharge increase is almost 250% from entry to exit point in Yugoslavia);
- Water quality along this river reach is heavily influenced by the upstream activities;
- FR Yugoslavia is faced with water management problems, due to formation of new states after the break-up of SFR Yugoslavia. Up to now there are no existing agreements between FR Yugoslavia, Slovenia, Croatia and Bosnia.

The interests in the water management for countries belonging to the lower part of Danube River are very similar to above mentioned. This holds true for available waters, level of environmental pollution, political and social systems, as well as economic potential.

5. CONCLUSION

The execution of a policy which incorporate the sustainable use in the management of the water resources requires considerable changes in the traditional approach to water management. This is reason for gathering all major policy makers on the task of creation of better life on this planet. However, it should be underlined that numerous factors exit, which may directly or indirectly slow-down its realization.

Although the present situation does not encourage implementation of integrated water management, it does not mean that the world should wait. Slowly and wisely, multidisciplinary projects should start on the level of urban units, sub-basin and river basins and in a latter stage they can be put together on the international river basin level.

The flexibility of such concept is particularly pronounced due to great differences between parameters which affect the water regime, impossibility of exact estimation of further global changes as well as consequences of human activities. In order to ensure implementation of integrated approach it is necessary to define basic principles, while the routes of their achievement should be functions of interests, possibilities, available support and confidence.

Planning and realization of large, complex, multidisciplinary and interdisciplinary projects require adequate cooperation in research, surveillance, analysis, effect estimation and application of existing and new technologies. It is important to pull together scientific and professional resources in the area of sustainable use and protection of water resources and environment. Success of these projects on the basis of positive effects and consensus building is imperative for quicker and more comprehensive engagement of varies countries in implementation of the integrated water resources management approach within international river basins.

River Basin Management:
Battleground for Economic Growth

CHRISTOPHER D. UNGATE
Tennessee Valley Authority, Chattanooga, Tennessee USA

Dams, levees, and related structures are the principal approach employed to manage rivers in the twentieth century. Dams allowed humans to "correct" nature's errors by supplying water where it is scarce and diverting water from areas prone to flood. These human interventions have reduced flood risk and provided drinking water to key population centers; provided dependable waterways to move commodities on inland rivers; created agricultural economies in arid lands; and produced hydropower to electrify rural areas. The adverse and sometimes unanticipated environmental, ecological, and social effects of many of these structures exposes the incomplete understanding of the complexity of river systems that existed before the dam-building era.

No where has reliance on dams to manage rivers for economic benefits been as strong as in the U.S. There are about two million dams of all sizes in the United States alone.[1] Of this number, there are about 75,000 that are of sufficient size to be tracked in the national dam safety database.[2] There are an average of 24 dams in each county (a local government jurisdiction of varying size below the state level) and over 6,000 dams in the state of Texas alone. Calculations have even been made estimating that the effect of U.S. dams on the speed of the Earth's rotation.[3]

---

[1]Graf, Walter L., "Landscapes, Commodities, and Ecosystems: The Relationship Between Policy and Science for American Rivers", in Sustaining Our Water Resources, National Academy Press, 1993.

[2]Water Control Infrastructure, National Inventory of Dams, 1992, Federal Emergency Management Agency Report No. 245, October 1993.

[3]U.S. Water News, "Construction of water reservoirs alters Earth's spin," March 1996.

Over 40 percent of U.S. dams are privately owned and the remainder are owned and operated primarily by local jurisdictions. Only about 2,000 dams are owned by the federal government. However, the small number of dams owned by the federal government belies the significant impact of federal government dams on the national economy. The U.S. Department of Agriculture's Soil Conservation Service (now the Natural Resource Conservation Service) was responsible for planning, designing, and/or constructing about 25,000 dams during the 1950s through the 1970s--the hey-day of dam building in the U.S.--primarily for the purposes of flood control and water supply. The U.S. Army Corps of Engineers owns and operates about 550 dams that provide much of the flood control and navigation benefits on major rivers of the U.S.--principally the Mississippi-Missouri-Ohio river system. The Bureau of Reclamation owns and operates over 350 dams for water supply and flood control that are the primary life support for irrigated agriculture in the western U.S. Both the Corps and the Bureau, together with the Tennessee Valley Authority (TVA), produce hydropower as a secondary purpose of their operations for navigation, flood control, and water supply. This "by-product" represents over 60 percent of the installed hydropower capacity in the U.S. and between 10 and 20 percent of the electric energy used in the U.S. in any year.

The large number of conflicts over the operation of dams in the U.S. today is a symptom of the significance of the economic benefits of dams, the lack of mitigation of their environmental and social impacts, and the inadequacy of the management approaches for major rivers on which dams are situated. The operation of federal dams on several rivers, such as the Missouri, Appalachicola, and Alabama, is the subject of litigation among states and other groups, or is specified by court order, such as on the Columbia. Legislative interventions on other rivers (in California's Central Valley, for example) attempt to address the conflict among operating priorities of dams and the impacts of their operations. Conflicts between environmental and recreation interests and the owners of privately-owned and non-federal publicly-owned dams that produce hydropower are resolved through a six-year regulatory process managed by the U.S. Federal Energy Regulatory Commission (FERC).

INTEGRATED APPROACH: CONCEPTS AND REALITIES

The lack of an integrated U.S. river management policy belies its origins. A holistic, interconnected systems approach was the underpinning of early American river perspectives.[4] The origin of the concept of river basins as

---

[4]Walter L. Graf, *supra.* note 1.

single, integrated environmental systems is attributed by many to American geographer George Perkins Marsh and American geologist John Wesley Powell.[5] The idea of integrated, multi-purpose river basin management was developed during the Progressive Era (during the administrations of Presidents Theodore Roosevelt, William Howard Taft, and Woodrow Wilson). Elements of the integrated approach included coordination of the goals and functions of federal water agencies, comprehensive water quality and quantity planning, cost-sharing tied to the benefits of projects, linking of land use to water use, and the comprehensive evaluation of all issues from a basinwide perspective.[6] However, Progressive Era proposals focused on the utilitarian uses of water--navigation, flood control, irrigation, and hydropower--and did not include broader concepts of ecosystems and watershed protection.

Progressive Era river basin management proposals were never adopted by the U.S. Congress. Instead, the Corps, the Bureau, and the Federal Power Commission (now the FERC) were authorized to construct or license individual water resource projects aligned to their missions of navigation and flood control, irrigation, and electric power, respectively. Progressive Era proposals were resurrected during the New Deal, which sought to use them as a vehicle for creating jobs and economic growth. New Deal proposals were rejected as well (with the exception of the Tennessee Valley Authority (TVA)) because the Congress disapproved of central planning for water resources.[7]

The primary role of engineers and of politics in water resource development explains the success of mission-oriented dam construction programs of federal agencies from the New Deal era through the 1960s, and the limited acceptance of the integrated approach.[8] Engineers were thought to be able to solve society's water resource problems in a detached and non-political way. In reality, federal engineers were the agents of Congressional committees and local interests in designing and constructing water resource projects located in or benefitting almost every local jurisdiction in the country. The focus of these projects were the utilitarian goals of navigation, flood control, water

---

[5]Daniel Schaffer, "Managing the Tennessee River: Principles, Practice, and Change," The Public Historian, Vol. 12, No. 2 (Spring 1990).

[6]Robert W. Adler, "Addressing Barriers to Watershed Protection," Environmental Law, Vol. 25, 1995, pp. 1005-6.

[7]Ibid., p. 1008.

[8]Ibid., p. 1013-4.

supply, irrigation and hydropower--all for economic development, and did not address environmental or recreational impacts or benefits which now are generally accepted as essential in a balanced economy.

With the growth of the environmental movement and the introduction of President Lyndon Johnson's Great Society programs, Congress passed the Water Resources Planning Act of 1965. This bold, but flawed, piece of legislation declared it to be "the policy of the Congress to encourage conservation, development, and utilization of water and related land resources of the United States on a comprehensive and coordinated basis" by all levels of government and affected interests--calling for an integrated approach by all river basin stakeholders.[9] Implementation of this policy was compromised because the law did not supersede any existing laws underlying mission-oriented dam building programs, but focused only on planning. The Water Resources Planning Act was never repealed, but the planning agencies implementing it were disbanded in 1981 during the Reagan Administration.

TVA, as "the great American experiment" of the comprehensive, multi-purpose river management approach, focused its attention during the 1930s and 1040s on building dams for navigation, flood control, and hydropower to restore the economic health of an impoverished region. In this it was no different than other mission-oriented federal agencies. For the next three decades, TVA focused principally on building thermal electric generating plants--especially nuclear plants--to support the continued growth of the Tennessee Valley economy. During the late 1970s and 1980s, a new generation of TVA environmental scientists as well as engineers began to address the environmental and recreation impacts of the TVA reservoir system, prompted by adverse reaction to the last of TVA's dam building efforts and by the agency's original mandate to promote "the proper use, conservation, and development of the natural resources of the Tennessee River drainage basin" for the "orderly and proper physical, economic, and social development" of the region.[10] The ultimate success of these integrated resource management initiatives is dependent upon the successful transition into deregulated electricity markets by TVA's massive power program, which now overshadows TVA's original river management function in size.[11]

---

[9]Water Resources Planning Act of 1965.

[10]Tennessee Valley Authority Act of 1933.

[11]Palmer-Bellevue, "The Ties That Bind," 1995, p. 1-14.

FORCES OF CHANGE

The massive size of dams creates impressions of permanence, of resistance to change, and of control of natural forces for the common good. Certainly the sheer number of dams that have been built and the huge cost of operating, maintaining, modifying and/or removing them makes them the centerpiece of river basin management in the U.S. for decades, if not centuries, to come.

This perception of permanence does not carry over to the owners of dams, how they operate them, or how society decides to control that operation to protect and promote the public good. Significant conflicts over the environmental and recreation impacts of dams built for other purposes have yet to be resolved. These conflicts expose the complex interactions of numerous government agencies operating at national, regional, and local levels--a system that has been described as "similar to a marbled cake, with several levels of government intermingled in an irregular pattern."[12] Public criticism of the size, complexity, and cost of government programs, of which water resource management is a good example, has severely constrained the resources allocated to manage dams and other natural resource programs.

The desire to both improve the efficiency of government programs and the legislative effort to deregulate electric utilities has created an interest in privatizing federal hydropower facilities. Legislation that would sell all federal dams on several river systems was proposed in the 104th Congress; however, a substantive public discussion of how this action fits into a comprehensive U.S. river basin management strategy has yet to be held. Meanwhile, a populist watershed management movement is addressing sources of pollution not regulated or effectively controlled by government, and is giving voice to public concerns over how water resources are managed. National conferences such as Watershed '93 and Watershed '96 exhibited the abundance and creativity of these efforts.

Adler identifies five key issues that must be resolved before watershed and river basin management programs can resolve these issues and address public concerns over the management of water resources:[13]
- Scale: Should programs proceed at the scale of whole river basins that encompass whole ecological regions or smaller watersheds involving fewer jurisdictions?
- Boundaries: Related to the issue of scale is the question: Should pro-

---

[12]Robert W. Adler, *supra* note 6, at p. 991.

[13]Robert W. Adler, *supra* note 6, at pp. 1088-1104.

grams organize around political, geographical, or ecological boundaries?
- Control: How should power over programs be allocated among levels of government and between government and constituent groups?
- Mission: Should the purposes of programs be procedural (ensuring that all stakeholders are involved) or substantive (setting clear, enforceable mandates)?
- Consistency: To what degree should programs be consistent on a regional or national scale?

On rivers in which large dams have been built that store water from one season or year to another, the scale and boundary of the organization that manages them logically should be established around river basin boundaries to assure safe operations and the achievement of their benefits. However, these river basin boundaries cross the many boundaries of cities, counties, states, and neighboring countries, creating conflict between the managing organization and these jurisdictions. The history of dis-integrated river basin management in the U.S. is a product of these conflicts. It is easy to understand why American geologist John Wesley Powell recommended that new states formed during the Progressive Era be organized around watershed boundaries.[14]

The degree of success of the lone experiment of comprehensive, multi-purpose river basin management, the TVA, has been attributed to its broad mission and its success in mustering regional support for its activities.[15] Its river management boundaries were drawn around the Tennessee River basin. Control over the river is shared with the states, who carry the principal responsibility for pollution control and fish and wildlife management. The flexibility of its enabling statute has allowed it to adapt to the emerging environmental and recreational interests of its stakeholders. Perhaps the simple fact that TVA is the designated decision maker for river flow management in the Tennessee River basin provides a forum for resolving conflicts that is essential to successful implementation of an integrated river management system. There are many way to address Adler's issues of scale, boundary, control, mission, and consistency, but the TVA model is instructive for the degree of its success in resolving jurisdictional conflicts within a large river basin.

---

[14]Robert W. Adler, *supra* note 6, at p. 1005.

[15]Robert W. Adler, *supra* note 6, at pp. 1062-3.

Slovenian water management strategy
- towards EU water policy

FRANCI STEINMAN, Prof. Dr. Ing.
PRIMOŽ BANOVEC, Assist. Dipl. Ing.
University of Ljubljana, Faculty of Civil Engineering and Geodesy,
Ljubljana, Slovenia

ABSTRACT

In the following article the basic characteristics of Slovenian water environment, ways of managing these natural resources, experiences of introducing various organizational forms as well as guidelines for the future are presented. In the framework of current continental legislation, state Appropriation Principle regarding waters has been adopted along with the introduction of concession system. With regard to technical-technological views, activities for further accommodation to current standards (directives) and the foreseen EU water policy are well under way.

Organizational structure and water management strategy were strongly influenced by the change of socio-political organization, accomplished by the proclamation of Slovenian independence in 1991 as well as by the transition to market system. In the period of 1991-1996, Slovenia, with its legislation, organization and laid-down guidelines, managed to approach the EU so that Association Agreement [1] was signed in 1996. According to gross domestic product Slovenia already reaches some EU countries and at the same time it exceeds other countries in the period of transition.

In the past 5 years, fundamental expert starting-points have been prepared (strategy, national plan), at the same time, however, first experiences with concessionaires, particularly foreign ones, have been acquired. In addition, we have stated that water management is connected with a series of difficulties such as lack of capital, skilled staff (equivalent to foreign concessionaires) as well as organizational and management predicaments, etc. Cooperation in various international projects has shown, however, that similar difficulties are also characteristic of other countries in the period of transition.

INTRODUCTION

The Republic of Slovenia on its specific geostrategic position presented always a bridge of different cultures, economies, political systems and even geographic properties. This passage between various features is regarded as continuous so that the characteristics of this country are, to a certain extent, also specific to the EU countries and other socialist countries in the period of transition.

After the independence of Slovenia in 1991, water management was influenced by the transition to market economy via following movements:
- from socialized towards private ownership,
- from a manufacturing towards a services economy,
- from large towards small enterprises,
- from markets of former Yugoslavia towards Western markets,
- from a supply side economy towards a demand side economy.

The efficiency of the transition in Slovenia is best illustrated by the growth in GDP (Figure 1) [2].

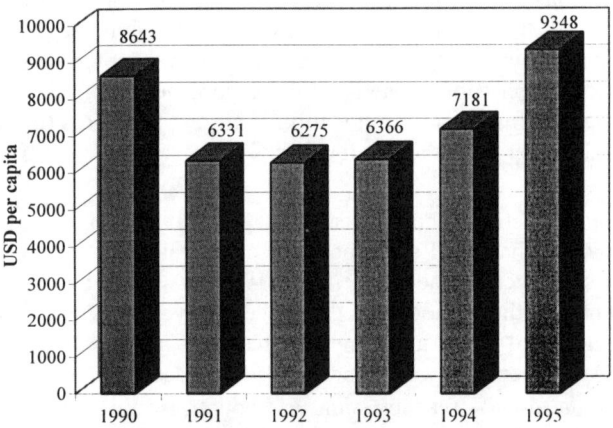

Figure 1: Growth in GDP per capita in 1990 - 1995

GEOGRAPHIC CHARACTERISTICS

Although relatively small as a state, Slovenia represents a wide diversity of landscapes, which implies a large variety of precipitation, between 800 mm/year (Pannonian plain) and 4000 mm/year (The Alps). A calcareous formation (Karst) with caves and underground rivers covers 44% of the national territory while almost 43 % of territory is exposed to intensive erosion processes.

The entire catchment areas of water flows comprise 43,274 km², therefore, in the territory of Slovenia (20.256 km²), from the viewpoint of land use, we can exert influence only on 47 % of this surface. The total water outflow is 34 billion m³ per year, from which 57 % of outflow is generated in Slovenia. The influence of the existing accumulations capable of retaining only 0,3% of the entire water outflow is insignificant as to the management of water quantities but at the same time strong on the quality of the waters.

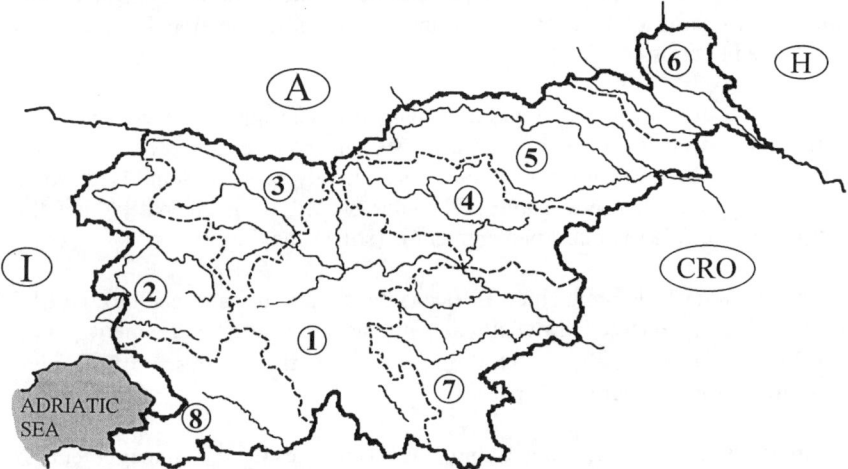

Figure 2: The administrative water management units (1-8) are based on larger catchment areas.

The population (1.99 million) is concentrated mainly in the valleys and alongside larger waterflows, so the interests of water management are closely intertwined with the needs of urbanization, agriculture, traffic, etc.

ORGANIZATIONAL AND STATUS ASPECTS OF THE WATER MANAGEMENT

The organizational structure and the way of water management has been always strongly affected by the current socio-political organization. Former Yugoslav self-management system differed from other socialistic, centrally planned economies in some ways but shared some common attitudes and practices. Until 1991 there were three characteristic periods:

- Until 1970 water management was performed entirely by the state through its Administration for water management which also financed and supervised all the works.

- Later on, until 1981, the process of decentralization was under way, transferring projects and the realization to the administrative bodies of regional catchment areas. The state was in charge of legislation as well as of administrative and state inspectional supervision, whereas Administration for water management supervised its financial and technical part.
- In the following decade Administration for water management was abolished; only project service for water resources planning was retained. The state was merely in charge of legislation, administrative and inspectional supervision (restricted to emissions).

Since 1991 transition to free market system has been made and new national institutions as well local authorities have been introduced. However, a certain experience in decentralized management has already been present. With some acts adopted since then and some still under procedure in the parliament the position and the role of water management is (still) changing.

The general aim of Slovenia since 1991 has been a gradual approach to the EU. Efficient changes enabled Slovenia to sign the association agreement [1] in 1996. Further development is also directed towards the progressive fulfilment of conditions for full membership.

The most important act concerning the water resources management is Environmental Protection Act, adopted in 1993, where the Appropriation principle was used for all natural sources. All water was declared to be state owned. A system of permits for all uses or pollution was established and a new legislation on Public Services and concessions was adopted. Possible forms of organization of those services and system of concessions are practically the same as those existing in other market economies.

In order to enable an effective and well-organized water management, two sector documents were prepared in 1991-1992: Water Management Strategy [3] and National Programme of Water Management [4]. They were based primarily on the trends of developed countries, considering the principles of directives of the EU, EPA USA documents, the water policy of comparable European countries and the results of interstate projects on water management [6] as well as on adopted state obligations (Agenda 21, etc.). So the groundwork for the preparation of a new Water Act (to be passed in 1997) has been laid.

THE SITUATION OF THE WATER MANAGEMENT

The characteristics of the water management in the period of transition are as follows:
- The process of reorganization - the fundamental acts are in the process of passing, therefore the institutions which are to manage water resources have not been completely formed.
- Subordinate economic position - due to the process of independence, priority was given to sectors essential for the functioning of state (army, police, foreign affairs, etc.). Therefore the means earmarked for water management were drastically reduced [3] (Figure 3).

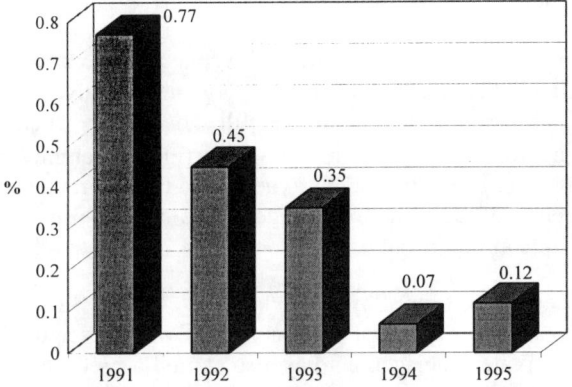

Figure 3: The means (% state budget) earmarked for the water management.

- Subordinate social position - investments in water management give long term economical results. In unstable economical and political circumstances, mainly personal consumption and services (e.g. transportation) were on the increase.
- Lack of knowledge and experiences for the operation in new market conditions and particularly in the field of concessions [5].

HOW TO GO FORWARD

Comparisons with other countries in the period of transition show some similarities [6,7]:
- instability of economy and political structures,
- changing of legislation and institutions,
- lack of means for water management,
- lack of managerial know-how; in distinction from disposable technical-technological knowledge.

In such circumstances according to present economical and political possibilities a flexible accommodation of measures would be most appropriate. Since operating conditions (legislation, authorities) have not been clearly defined, the system of water management is becoming poorly manageable, unclear and ineffective.

The transfer of foreign knowledge and technologies takes place not only in the framework of various international projects but also through direct commitment of foreign enterprises in the liberalized market. However, a direct transfer of foreign experiences in water management is not possible, therefore various pilot projects are under way at present helping the principles of comprehensive water management [8] to be put into effect.

The adopted directives of the EU (of which Slovenia wants to be a full member) create great demands that are, financially and organizationally, a heavy burden to bear also for many member countries [9]. To achieve such demanding objectives will be feasible only through long-term programmes in complex stages. There have still been discussions whether to take a broad approach , i.e. by gradual increase of demand of arrangements, or to use best available technology principle at new arrangements.

Until 2000, a new long-term Physical Planning Act of Slovenia is to be obtained, based on the development programmes of various sectors harmonized on the state level. Water management has also started a preliminary work [10] which requires primarily the laying of the foundations of the water policy.

REFERENCES

1. Association agreement between the European Communnities and the Member States on the one part, and the Republic of Slovenia on the other part, 1996,
2. Statistical Office of the Republic of Slovenia (different publications),
3. Steinman F. et all, 1991, Water Management Strategy,
4. Rožič et all, 1992, National Programme of Water Management,
5. F. Steinman, P. Banovec, 1996, Does the current situation in the water managment enable consignation of concessions? ISBN 961-6167-08-1
6. C. Whitehead, M. Noe, 1995, Strategic action plan for the Danube catchment,
7. Somlyódy L., 1994, Managing Water Quality in Central and Eastern Europe,
8. Van Rooy R.T.J.C., 1995, Towards comprehensive water management in The Netherlands (2) bottlenecks, European Water Pollution Control 5/6,
9. D. Stoller, 1996, Costs of water treatment: on the treshold of bearable, Abwassergebühren: an der Grenze des Erträglichen (VDI Nachrichten),
10. Steinman F. et all, 1996, Water Management Contents in the Physical Planning Act of the Republic of Slovenia.

The Okavango River Basin

DANIEL P. MILLER
Stanley Consultants, Muscatine, Iowa

ABSTRACT
The Okavango River, in Angola, Namibia, and Botswana is important for its wilderness environment and resources for human development. The river terminates in an inland delta, providing perennial wetland habitat within the Kalahari desert. This is one of the few remaining regions in southern Africa capable of sustaining populations of large mammals and reptiles. Researchers are aware both that human development threatens the present ecosystem and that, even outside of human influence, this river is prone to rapid, significant ecological changes.

THE RIVER
The Okavango river basin stretches over more than 245,000 square kilometers in Angola, Namibia, and Botswana. Average annual rainfall over the basin varies from 2,000 mm in the Angolan highlands to 400 mm over northeast Namibia and northwest Botswana. Mean annual discharge at the Botswana border is about 10 billion (10×10^9) cubic meters. One hundred kilometers into Botswana, the river terminates in an alluvial fan known as the Okavango Delta. This ribbon of water and swamps provide a unique environment for plant- and wildlife amidst an otherwise parched landscape.

The Okavango begins as the Rio Cubango in the highlands of Angola, where the 151,000 square kilometer watershed receives from 1000 to 1400 millimeters of precipitation per year. About six percent of this precipitation runs into the Cubango and constitutes the majority of the ultimate flow; for as the river proceeds in a southeasterly course the climate becomes increasingly arid. At its delta terminus, precipitation averages about 400 mm per year, evaporation 2000.

Precipitation here is seasonal. The winter months, May through November, are very dry. Heavy rains in December, January, February, and March make for wide fluctuations in the river flow, from a low of about 120 cms in November to as high as1000 cms at the end of the rainy season.

In its first 600 km, between Huambo and the Namibian border, the river flows through mountainous terrain in channels cut deep into the surrounding topography. Along the 420 km stretch that forms the Angola-Namibia border, the slope flattens out and the floodplain grows progressively wider. Sedge grasses, papyrus, and phragmite reeds grow along the riverbanks. The floodplains support grasslands, and, as the river progresses eastward, increasing riparian woodlands and intermittent swamps. Near the eastern end of the border reach, the Cubango is joined by the Rio Cuito, flowing from the southern slopes of Mt. Sacangombe. By now the riparian swamps have become several hectares in area and support a wide variety of wildlife. The water is a fertile breeding ground for fish and reptiles. The 5 to 15 km wide floodplain forests and meadows are home to birds and mammals. Outside the floodplain lies desert, barren during the winter but bursting with life during the short rainy season. In very wet years, the river may spread beyond the basin boundaries, occasionally reaching the Etosha Pans in northern Namibia.

The combined rivers flow south, across the western end of the Caprivi Strip, and enter Botswana near Mohembo in the northwest. For a hundred kilometers the river, here called the Okavango, meanders through a 15 km wide panhandle floodplain before reaching the Gumare fault at the head of the delta.

THE DELTA

The delta is an alluvial fan in a graben region between two faults at the western end of the Great East African Rift. The 150 km between the Gumare fault at the upper end and the Kunyere fault at the lower end of the delta has filled with eolian Kalahari sands to a very level grade. In places the sand is more than 300 meters deep. Local topography varies by less than two meters. There is about 65 meters of fall in the 250 km from Mohembo to Maun, at the southern end of the delta.

At the top of the delta the Okavango splits into the Thaoge and Nqoga channels. One hundred years ago, the Thaoge was the primary channel where the river flanked the delta region to the west, and flowed into Lake Ngami. The first European explorers reported the lake to be 80 to 100 km long and up to 2m deep. Over time the channel became clogged and flows were diverted eastward. Today, the Thaoge river is completely plugged near Gumare and the channel beyond is essentially abandoned. The Nqoga has become the main distributary of the delta, dispersing 70 percent of Okavango inflow among the Jao, Maunchura, and Mborogo rivers and adjacent peat swamps. Lake Ngami is completely dry most of the year.

From the upper delta channels to the Khwai, Santantadibe, Boro, Shashe, and Kunyere rivers, flowing out of the southern end of the delta, there is no continuous channel. Rather the water spills through porous reed embankments, into flanking perennial swamps. High summer flows cause the swamps to swell, ultimately to about twice their perennial area. The southern rivers do not begin to flow until the water overcomes the sill created by the Kunyere fault. These rivers ultimately dissipate in the Makgadikgadi Pans in central Botswana.

The discontinuous flow between the incoming and outgoing rivers has two significant effects:

- First, the water is filtered as it enters the swamp through the reed embankments. The lower part of the embankment, in particular, filters much of the suspended sediment from the river. The resulting aggradation, combined with rampant vegetation eventually chokes the channel, and a new course is found. The reed filters, meanwhile, play a large role in maintaining the high quality of the delta water.

- Second, the flood surges are extremely attenuated as they pass through the swamp. As the water rises the swamps expand across the wide, flat delta. Were the swamps not there, the passage of the flood wave between Mohembo and Maun would take about 10 days, through the swamps it takes about four months. This absorption of the flood flows has the effect of delta waters reaching their greatest expanse when the surrounding countryside is encountering the dry season, thus providing habitat for migrating wildlife during the winter. The wildlife population in the delta is some 10 times larger in July and August than in January and February.

The constriction and closure of a river channel is accompanied by death and desiccation of the flanking swamps. Eventually, the peat, which can be as much as 5 meters thick, begins to burn. The underground fires may last more than a decade, burning deeper and deeper below the surface as the water recedes and the peat dries. When the fires have completely burned out the peat will have lost 98 percent of its volume and the surface subsides to the original sand level. The newly exposed soil is rich in nutrients and soon supports a diverse plant and wildlife community. The area may become re-inundated soon after it subsides, or, if the new channel is separated by higher ground, the area may remain dry or become part of the seasonal swamps. Thus the 18,000 square kilometer delta is in constant rotation between perennial swamp, seasonal swamp, and perennial dry ground as swamps are formed, die out, burn, and become rejuvenated. A cycle of channel initiation, development into a main water course, ultimate clogging, desiccation, burning of the adjoining swamp, and return to dryland vegetation takes on the order of 150 to 200 years.

The vast perennial and seasonal swamps play an important role in the distribution of water from the upper channels to the departing channels of the lower delta, but their effects are difficult to predict. The reed banks of the channels stand a bit higher than the flanking swamps, and the water level in the channel is typically slightly higher than in the swamps. When the seasonal flood reaches the Nqoga, Thaoge and Maunchura rivers it quickly spills over into the surrounding swamps. The swamps expand widely through the flat terrain. As the water spreads across the delta, the velocity is nearly zero and it is difficult to delineate a clear flow path. Imperceptible changes in vegetation can have a significant impact on the apportionment of the exiting flow among the lower rivers. The out-flowing Boro River, for example, has experienced six significantly different flow regimes in the last 54 years.

HUMAN POPULATION

Human habitation in the Okavango basin since the early nineteenth century has been limited by disease vectors, mostly malaria and sleeping sickness, and by recurrent droughts. An epidemic in 1896 destroyed most of the host animals for the tsetse fly, so, between 1900 and 1930, human populations grew. In the mid twentieth century, the tsetse fly returned and the human population began to decline. In the 1980's a cordon fence was constructed on the southern perimeter of the delta to separate settled areas from wildlife in the delta. The primary purpose of the fence is to prevent the spread of hoof and mouth disease and sleeping sickness from wildlife to cattle. The fence also prevents competition between domestic livestock and wildlife over pasturage.

Currently about 100,000 people in Botswana, and another 115,000 in Namibia live within the Okavango basin and draw upon its resources for their livelihood. Angolan populations in the Cubango and Cuito basins have been suppressed by civil war but are expected to grow, since this is one of the nation's best agricultural regions. Sustenance farming is the main livelihood of the rural communities in all three countries. The river provides a source of water, building materials, and, in the river-sustained grasslands, feed for livestock. In the delta, malapo farming is practiced, whereby a crop is cultivated in the floodplain after the annual flood subsides. The nutrients and moisture deposited by the flood result in good yields.

Fishing is prevalent and highly productive along the entire river, and serves as an important source of protein. Commercial and sport fishing are notable economic activities within the delta. Sustenance hunting is practiced in the narrow woodlands along the river and on the outskirts of the delta. There is some trophy hunting licensed in restricted areas of the delta, but the number of hunters and count of animals taken is relatively small. Tourism is a very important industry in the delta. The current number of visitors, from all over the world, is estimated between 25,000 and 35,000 per year and expected to rise.

INTERNATIONAL COOPERATION

There is realization among all three riparian nations that demands for further development of Okavango's resources will steadily increase and that careful planning is necessary to avoid irreparable damage and to ensure equitable utilization. Some of the pressing concerns are:
- With peace at hand in Angola, demands for water will grow, both for domestic consumption and irrigation.
- With Namibian water resources substantially committed, that nation is looking to the Okavango for supply, possibly through connection with the East-West National Carrier pipeline. Irrigation is also possible, especially within the Caprivi Strip.

- In communities south of the delta, ground water supplies are questionable and an increasing population is at risk of drought. There have been efforts to channel water out of the delta for storage in reservoirs and for use in irrigation, but these efforts have met with staunch protests from environmentalists, both local and international. The need, practicality, and effects of such efforts are hotly debated throughout Botswana.

The three nations entered into an agreement in October 1994 establishing the Okavango River Commission (OKACOM) to jointly plan and manage the basin's resources. The Southern Africa Development Community (SADC), the United Nations, the United States, Great Britain, The Netherlands, Germany, Sweden, Australia, and South Africa have participated in research in the region. The International Union for the Conservation of Nature (IUCN), Greenpeace, and National Geographic are closely involved. It is widely recognized that the delta is unique in southern Africa for its capability of sustaining such a large, diverse plant, wildlife, and human community.

REFERENCES
1. R.H Hughes and J.S. Hughes, A Directory of African Wetlands, IUCN, 1992
2. Okavango Research Group, University of the Witwatersrand, Johannesburg, Papers Published in the Period 1986-1993,
3. Directorate of Environmental Affairs Ministry of Environment and Tourism, Republic of Namibia, Namibia's Environmental Assessment Policy, Jan 1995
4. University of Botswana, Development of the Okavango Research Centre, Project Memorandum, June 1995
5. N. Ellery & T. S. McCarthy, Principles for the Sustainable Utilization of the Okavango Delta Ecosystem, Republic of South Africa, January 1994
6. Else Skjonsberg & Yvonne Merafe, The Okavango Fisheries Socio - Economic Study, Gaborone, Oslo, August 1987
7. National Planning Commission, Namibia: Population and Development Planning, Republic of Namibia, Windhoek, March 1994
8. Government of Botswana, Ministry of Local Government and Lands, Department of Town and Regional Planning, Programme for the Planning of Resource Utilisation in the Okavango Delta Region, Draft Final Report Volume II, Programme Phase, Swedeplan, Gaborone
9. J. van der Heiden, The Okavango Delta: Current state of planning and conservation, in Wetlands Conservation Conference for Southern Africa, IUCN, Switzerland, 1992
10. Kalahari Conservation Society, Government of Botswana, Ministry of Local Government and Lands, Department of Town and Regional Planning, Ecological Zoning Okavango Delta, Final Report Volume I - Main Report, Snowy Mountains Engineering Corporation, Australia, April 1989
11. IUCN, The IUCN Review of the Southern Okavango Integrated Water Development Project, IUCN, Switzerland, 1993

386 COPING WITH SCARCITY AND ABUNDANCE

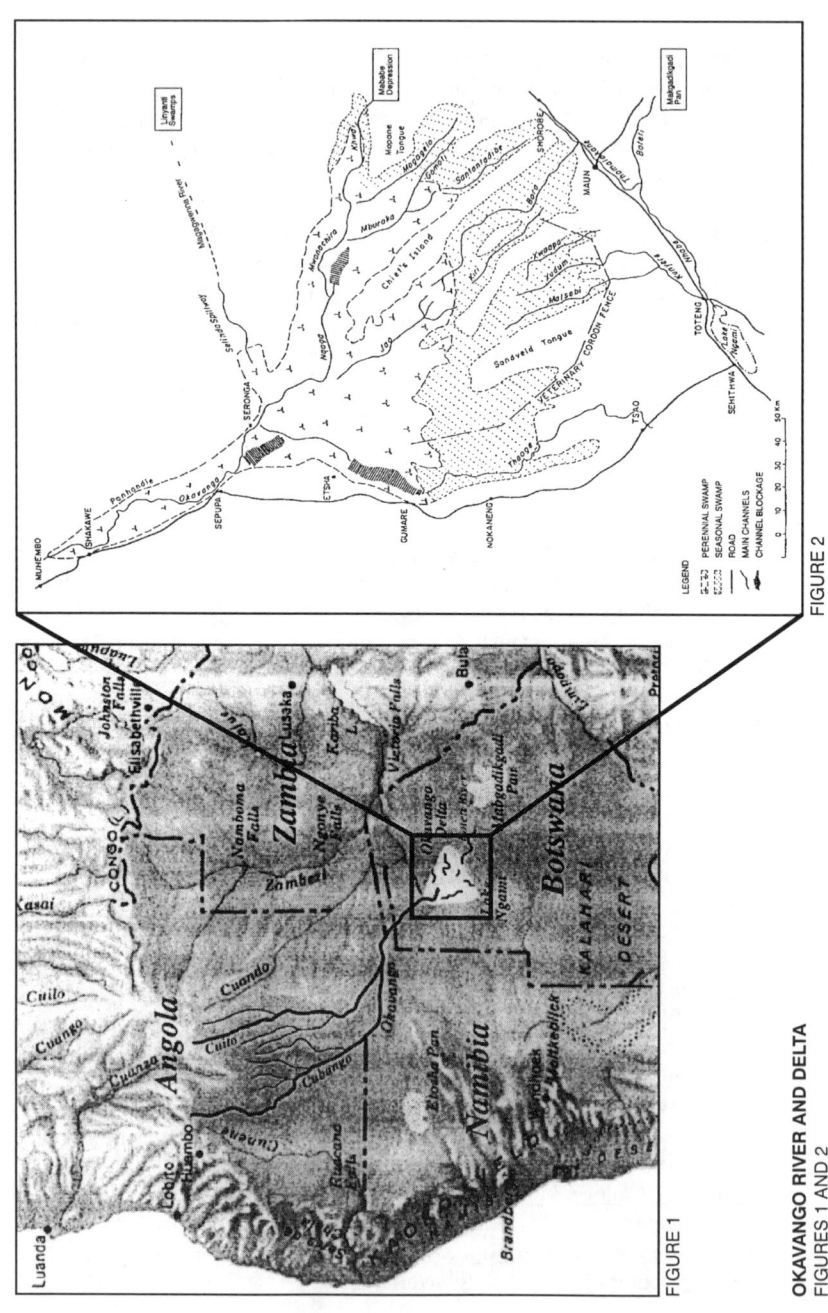

FIGURE 2

FIGURE 1

OKAVANGO RIVER AND DELTA
FIGURES 1 AND 2

MANAGING EXTREMES: THE IJC EXPERIENCE

LISA BOURGET AND MURRAY CLAMEN
Engineering Advisers, International Joint Commission
Washington, D.C. and Ottawa, Canada

ABSTRACT

The International Joint Commission was established by the Boundary Waters Treaty of 1909 to approve and oversee the construction and operation of structures affecting transboundary waters between the U.S. and Canada. The long-term relationships and binational cooperation fostered through such oversight provides a broad foundation of trust, understanding, and preparedness for water management in times of flood and drought. This cooperation is illustrated by three case studies: high lake levels on Lake Superior and Lake Ontario, and drought conditions on Osoyoos Lake.

INTRODUCTION

Nearly 300 streams form or cross the 8000 kilometre boundary between Canada and the United States. The use of these waters can cause disputes as well as issues of mutual concern. These disputes are often exacerbated during periods of flood and drought. With the signing of the Boundary Waters Treaty ("Treaty") of 1909, the United States and Canada established the International Joint Commission ("Commission" or "IJC") to oversee issues concerning boundary and transboundary waters shared by the two countries.

BACKGROUND

The Boundary Waters Treaty and the IJC

The Treaty sets out a number of provisions which, among other things, have had far reaching implications for water management between the two countries. It specifies certain provisions related to uses, obstructions or diversions of boundary and transboundary waters, should these operations affect the natural level or flow of the boundary waters in the other country. In addition, the Treaty also provides for questions or differences arising between the two governments to be referred to the IJC for examination and report. Such studies (termed References) are advisory in nature yet generally have tended to carry considerable influence and persuasive power due to the way they are developed and issued.

The IJC is made up of six commissioners, three from the United States, appointed by the president with Senate advice and consent; and three from Canada, appointed by the Governor General in Council. Permanent offices are maintained in Ottawa and Washington. A third office in Windsor, Ontario, assists the IJC with its responsibilities under the Great Lakes Water Quality Agreement, signed by the U.S. and Canada in 1972. The Commission also relies on the services of government and public experts from both countries to conduct its studies and to oversee operation of projects it approves.

Applications and Orders of Approval
An entity interested in constructing and operating a facility with transboundary effects applies through its respective national government to the Commission for approval. The IJC then conducts detailed technical investigations and holds public hearings to identify what impacts these facilities could have on all interests and to consider the merit of each application based on a variety of viewpoints and technical information.

If the IJC approves the application, its consent (called Order of Approval) typically includes conditions and criteria governing the construction and operation of the facilities. In most cases, the IJC also requires that an international board of experts be established to develop regulation plans and to supervise the operation of these facilities in order to ensure that conditions and criteria in the Orders are met. Over the years, changing conditions have required periodic revisions to the Orders and the regulation plans.

CASE STUDIES

This paper will examine three examples where the IJC has approved structures in boundary waters, focussing on the management of those structures during periods of flood or drought. Two of these examples are in the Great Lakes basin, where precipitation over several years is the main natural factor causing extreme high or low lake levels. The third example is located in the west, where drought is often a concern.

LAKE SUPERIOR REGULATION
In 1914 the IJC approved applications from Algoma Steel Corporation in Canada and the Michigan Northern Power Company in the United States to divert some of the water of the St. Marys River at the outlet of Lake Superior for hydropower generation. The Order established the International Lake Superior Board of Control to oversee operation of the facilities. The construction of a 16-gate Lake Superior Compensating Works structure was a key requirement in the Commission's 1914 Order. This facility was built to offset, or compensate for, the increased outflow capacity of the St. Marys River that resulted from the hydropower developments.

The IJC has updated the 1914 Order over the years to meet the changing conditions and requirements in the system. In a 1979 revision the Commission included a provision for extreme water supplies which indicates that, during extreme supply conditions, the IJC will indicate the appropriate outflows from Lake Superior, taking into account upstream and downstream interests. During 1986, at the time of the highest-recorded water levels on all the Great Lakes except Lake Ontario, the IJC made use of this provision for the first time by directing that Lake Superior outflows be below those specified by the regulation plan (Plan 1977) to help alleviate high water level problems on Lakes Michigan-Huron and downstream. The maximum effect was to raise Lake Superior by 4.4 inches (11.2 cm) and to reduce Lakes Michigan-Huron, St. Clair and Erie by 3.0 inches (7.6 cm), 1.8 inches (4.6 cm) and 1.3 inches (3.3 cm) respectively (Figure 1).

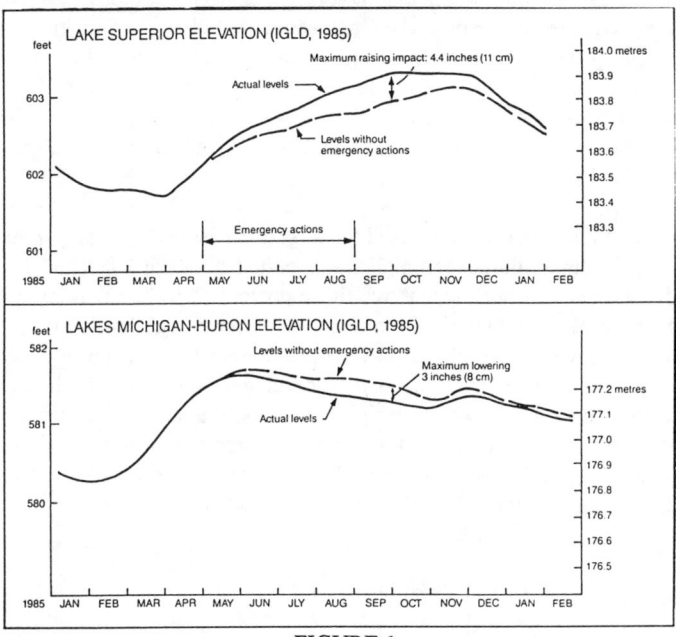

FIGURE 1

The decision to provide downstream interests some degree of protection from high water without causing harm upstream on Lake Superior, required a high degree of cooperation and coordination between the Commission and its Board at a time of heightened public interest in Great Lake water levels. Considerable information and opportunities to comment were provided to all concerned interests on this important international action. Because of the continuing oversight of the Commission and its well established protocols, the situation was managed very effectively and demonstrated the benefits of the Commission process in emergency situations on one of the largest lakes in the world.

LAKE ONTARIO-ST. LAWRENCE RIVER REGULATION

In 1952 the IJC issued an Order of Approval to the applications from Canada and the

United States to construct hydropower facilities in the international reach of the St. Lawrence River at Cornwall, Ontario and Massena, New York. The Moses-Saunders power dam that crosses the river between Cornwall and Massena is the principal regulatory structure. The order established the International St. Lawrence River Board of Control to ensure compliance with the provisions of the Orders by the operators of these works.

Recognizing that future water supplies to Lake Ontario would at times be higher or lower than those experienced in the past, the Commission included an emergency provision in the order. This provision, Criterion (k), specifies that in the event that supplies exceed supplies of the past, the works shall be operated to provide all possible relief to the riparian owners upstream and downstream. In the event that supplies less than supplies the supplies of the past occur, the works should be operated to provide all possible relief to navigation and power interests.

This important criterion has been utilized on several occasions to manage extreme water supplies. For example, during the high water period between 1985 and 1987 the IJC increased Lake Ontario outflows above those prescribed by the regulation plan (Plan 1958D) to prevent Lake Ontario from rising to extreme high levels due to continued extreme high inflows to the lake from the upper Great Lakes (Figure 2). Nature also helped because the very mild weather and favourable ice conditions in the St. Lawrence River that winter helped to make these high flows possible. During that time, water level conditions in the Montreal area and downstream were monitored closely so as not to aggravate the existing high level conditions there. Figure 2 also shows Lake Ontario water levels with and without regulation during a dry period in the 1960s.

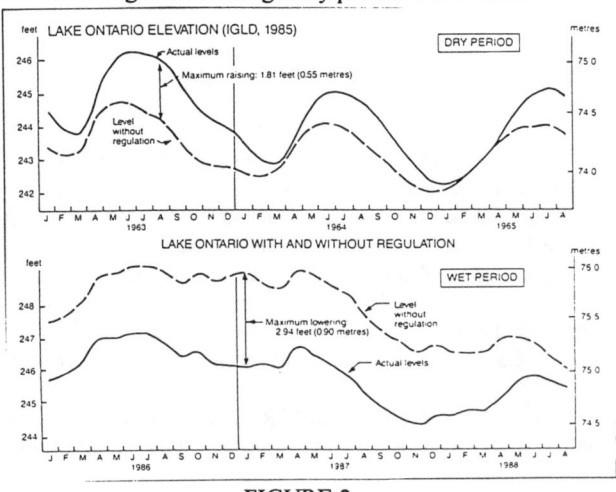

FIGURE 2

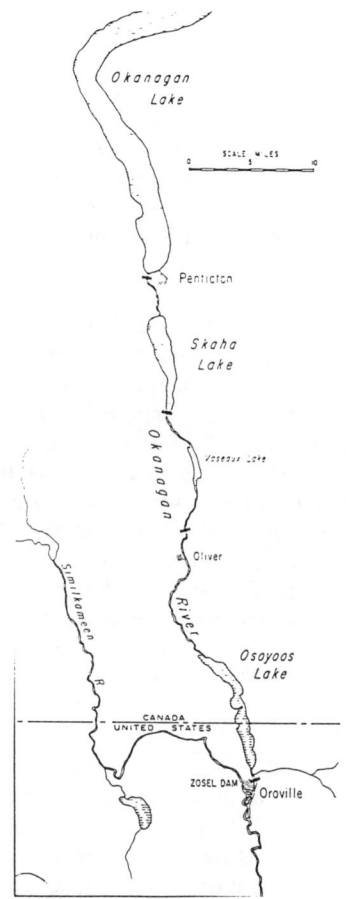

What is particularly important to appreciate is that various interests such as commercial navigation, hydropower, shoreline riparians, recreation and the environment co-exist in a challenging relationship under normal supplies. Under extreme supplies, these relationships are strained even further. It is the Commission's established mechanisms for reviewing binational agreed-upon data, and balancing all interests in this intensely used and highly populated system, that is essential for effective international water management.

OSOYOOS LAKE REGULATION

Osoyoos Lake spans the border between Washington and British Columbia (Figure 3). Zosel Dam, which raises water levels on the lake, was rebuilt in accordance with Orders of Approval issued by the IJC in 1982, which also specifies management and operating requirements. The IJC established a new International Osoyoos Lake Board of Control to supervise the operation of the new structure.

FIGURE 3

Permissible levels on Osoyoos Lake are specified, in part, by whether or not drought conditions exist. Drought conditions are determined by examining the forecast April-July flow volume of the Similkameen River, which has its confluence just downstream of Zosel dam and can back water through the dam and across the border; the forecast April-July net inflow to Okanagan Lake, located upstream of Osoyoos Lake in Canada; and the forecast midsummer elevation of Okanagan Lake. Under drought conditions, more water may be stored than under non-drought conditions. However, raising the level of Osoyoos Lake can be problematic for those living along its shores, mostly Canadians. Therefore, in 1993 and 1994, Washington and B.C. signed a Memorandum of Understanding such that, under drought conditions, the state would limit the amount by which it raised the lake and the province would supply an amount of water equivalent to the unused storage from their upstream reservoirs upon demand by the state.

Although this arrangement worked well for 1993 and 1994 (1995 and 1996 were not drought years), it is a year-by-year agreement. The Board investigated the probability that there will be sufficient storage in Okanagan Lake to trade flow for a reduction in the Osoyoos Lake elevation during drought years. It determined that historically, approximately one in three droughts along the Similkameen River is *not* accompanied by a drought in Okanagan Lake, thus permitting the trade-off. However, under the remaining two in three droughts, drought will occur simultaneously on the Similkameen River and on Okanagan Lake simultaneously and there will not be sufficient water in Okanagan Lake to permit the trade-off.

The Board is completing its investigation of hydrologic conditions on Osoyoos and Okanagan Lakes, and on the Similkameen River, and is examining possible impacts of climate change/variability. Currently, the IJC Order of Approval specifies the permissible range of Osoyoos Lake elevations during drought and non-drought years, and year-by-year arrangement between the state and province are possible within that range of elevations. The work done by the Board, and the relationships established among experts across the border, may offer insight into other possible alternatives to water management of these boundary waters. If any proposed alternatives are consistent with the current IJC Order of Approval, additional informal arrangements may facilitate management of the waters to everyone's benefit. Otherwise, the IJC might consider amending its Order of Approval to formalize new management guidelines that take advantage of the latest knowledge and technology, balance the interests of the diverse users of the water system, and exemplify the cooperative principles of the Boundary Waters Treaty.

CONCLUSIONS

The IJC has been instrumental in managing transboundary waters, and its unique mechanisms have been highlighted perhaps most in times of flood and drought. The deliberate distancing from national interests and the long-term management responsibilities, typically handled by IJC-appointed boards, has allowed professionals to build long-term relationships across the border based on professional respect and understanding. Furthermore, the distancing from national interests also facilitates coordinated investigation of various issues, such as climate change/variability. The nature of the management activities means that coordinated, consistent data is freely available on both sides of the border, and that this data is fully understood and utilized. These long term transboundary relationships, understanding, and data availability provide thenecessary foundation for better water management during times of crisis, such as flood and drought.

Comprehensive Assessment of the Freshwater Resources of the World

PAUL H. KIRSHEN
Stockholm Environment Institute, Boston, MA USA, also
Tufts University, Medford, MA USA
and
KENNETH M. STRZEPEK
University of Colorado, Boulder, CO USA

ABSTRACT
At its Second Session in 1994, the United Nations Commission on Sustainable Development (CSD) called for a comprehensive assessment of current and future freshwater resources, needs and problems. The Stockholm Environment Institute (SEI) was commissioned by the Swedish Government to assume the Swedish responsibilities for preparing the assessment together with United Nations organizations. The report will be organized into four chapters:

1. A statement explaining the need for such an assessment;
2. A description of the availability, quality and variability of freshwater resources of the world and the use to which they are put at present;
3. An investigation of current and future water needs and the problems that must be faced at the global, regional and national levels;
4. Strategies and options for the sustainable development of freshwater resources of the world.

INTRODUCTION
At its Second Session in 1994, the United Nations Commission on Sustainable Development (CSD) called for a comprehensive assessment of current and future freshwater resources, needs and problems. The Stockholm Environment Institute (SEI) was commissioned by the Swedish Government to assume the Swedish responsibilities for preparing the assessment together with United Nations organizations. A Steering Committee for the Comprehensive Freshwater Assessment (referred to below as CFWA) was established with representatives from SEI and relevant United Nations Organizations (the Administrative Committee on

Coordination Subcommittee on Water Resources, FAO, UNEP, WMO, UNESCO, WHO, UNDP, UNIDO, and the World Bank). The CFWA will submit a report for submission to the CSD and ultimately to the UN General Assembly. The report will be organized into four chapters:

1. A statement explaining the need for such an assessment;
2. A description of the availability, quality and variability of freshwater resources of the world and the use to which they are put at present;
3. An investigation of current and future water needs and the problems that must be faced at the global, regional and national levels;
4. Strategies and options for the sustainable development of freshwater resources of the world.

Since at the time of the preparation of this paper the report is still undergoing preparation, this paper only presents some of the draft findings of the CFWA. These findings may change as the report undergoes review. The report itself is based upon a series of commissioned background papers.

MOTIVATION

Chapter 1 of the CFWA presents the following summary to motivate the global assessment. "The following numbers are sufficient to indicate the trend towards freshwater supply vulnerability. Between 1950 and 1995, world population increased from 2.5 billion to 5.6 billion and is expected to reach 8.3 billion by 2025, an increase of 2.7 billion during the next 30 years. In 1995 54% of the readily available water was already appropriated. This means that by 2025 the traditional development process will be approaching a 90% appropriation level. This supply vulnerability situation will be exacerbated by a probable decrease in water quantity availability due to the continuous degradation of water quality which is taking place in industrial and urban areas and from non-point pollution from the agricultural sector, all over the world. The degree of vulnerability will vary from region to region depending on economic and environmental circumstances. Poverty alleviation in some regions will therefore be very challenging.

An additional challenge will be the increase in food production needed to feed the 2.7 billion additional people that are projected to inhabit the earth between 1995 and 2025, because this will have to come from irrigated agriculture which at present uses about 70% of the total amount of freshwater used. Thus, the availability of sufficient amounts of water of adequate quality may become an issue of paramount importance for global food security exacerbated by competition from the urban areas and industry.

Furthermore water is intimately linked with human health. The World Health Organization has estimated that 80 per cent of all diseases and over one third of deaths in developing countries are water related. Industrial and household wastes, pesticides, nutrients, fertilizers, airborne pollution, toxic chemicals and sewage, which are present in water resources worldwide all take their toll. The economic cost of water pollution in terms of human health is high. According to the World Bank, a recent outbreak of cholera in one region killed 1,000 people and cost $1000 million in lost income from agricultural export and tourism.

Finally the protection and health of the environment, coastal zones and oceans are intimately linked with the quantity and quality management of inland freshwater resources, and at present there is great concern about this issue in most regions of the world " (Keating, 1996).

Chapter 1 also notes that water resources issues have been receiving global attention since the 1970's which has continued through the UNCED Conference in Rio de Janeiro. There have been many expressions of alarm over the apparently slow rate of progress towards the sustainable use and development of water resources. Most recently, the Committee on Natural Resources of the Economic and Social Council of the United Nations "noted with alarm that some 80 countries, comprising 40 per cent of the world's population, are already suffering from serious water shortages and that, in many cases, the scarcity of water resources has become the limiting factor to economic and social development". It further noted that "ever-increasing water pollution has become a major problem throughout the world, including coastal zones". The UN Commission on Sustainable Development, at its second session in 1994, echoed these concerns and noted that, in many countries, a rapid deterioration of water quality, serious water shortages and reduced availability of freshwater is severely affecting human health, the ecosystem and economic development (Keating, 1996).

PRESENT SITUATION

As summarized by Davis (1996), ninety seven and one-half percent of all water on earth is salt water, primarily in the oceans. The remaining 2.5 percent is fresh water, almost all of which is stored in the icecaps of Antarctica and Greenland and as groundwater. The most accessible water resource for human and ecosystem use is the freshwater available in lakes, reservoirs and river channels. This amounts to only 0.26 percent of the total freshwater in storage or 0.007 percent of all water on earth. Only this small amount is renewable and actually available for use on a sustainable basis. Soil moisture, while it does not contribute directly to runoff, is important for supporting rainfed agriculture.

Fresh water in rivers and streams is replenished by the rains and snows. The runoff from all of the globe's rivers averages 42,600 cubic kilometers per year. However there are huge continental, regional, local and seasonal variations. For example the arid and semi-arid zones of the world, which constitute 40 percent of the landmass, produce only 2% of global runoff. By contrast 16 percent of all global runoff comes from one source, the Amazon River. Regional variability means that Australia has only 0.8 percent of global runoff to meet all of its requirements and Europe has 6.8 percent. Africa is better supplied with 9.5 percent of the total, but 32 percent of that comes from the Congo-Zaire River basin alone, while other areas have no runoff at all. Seasonal variability makes much of the water supply unavailable when it is needed. Typically rivers have low flows when the demand for irrigation and other uses is the highest. Unless reservoir storage has been constructed, only a small part of the resource can be used. Extreme variability, such as periods of floods or very low flows, not only affect the amount available but also causes major natural disasters with significant economic and social consequences.

Building upon this and other background information, Chapter 2 then provides more quantitative and qualitative information on present water withdrawals, water scarcity, groundwater over utilization and pollution, water supply and sanitation, water pollution, soil erosion and salinisation, climate change, and population stress. It also shows the global distribution of present water stress by nation using some of the indices developed for Chapter 3.

SCENARIOS OF 2025 WATER USE

Chapter 3 discusses the driving forces behind future demands for water and considers the possible implications of these forces on future water resources demands. This is done in the context of several scenarios of future water demands. One of these scenarios is the Conventional Development Scenario (CDS) developed by the Boston Center of SEI(Raskin et al, 1996). The CDS is neither a prediction of what *will* happen nor a statement of what *should* happen. It describes the direction we are headed and the problems we may encounter -- if current patterns and driving forces are played out.

The guiding principles of the CDS are *evolution, convergence, and integration.* Demographic, socio-economic and technological patterns gradually evolve without significant surprises, radical technological innovations, or fundamental policy changes. Developing and transitional regions are assumed to converge gradually toward OECD economic and water practices. Ultimately, in the CDS, the world becomes progressively more integrated both economically and culturally.

The conventional development paradigm assumes that the engines for economic growth and wealth allocation are unregulated markets, private investment, and competition; population increases at mid-range projections with a continuation of rapid urbanization; industrialization progressively absorbs nations and regions on the periphery of the marketed world economy; human motives are dominated by the value of possessive individualism with material wealth the basis for the "good life"; and the nation-state survives as the central unit of governance.

The CDS first develops future demands for water withdrawals for 10 global regions for each of three sectors of water use; household and commercial, industrial, and irrigation. These are then disaggregated to the country level. In order to evaluate the consequences of this scenario, the CDS background document (Raskin et al.,1996) develops several indices of water resources stress and presents maps of the world with nations sorted with these indices. The indices include:

- *use-to-resource ratio*, the annual water withdrawals divided by annual renewable water resources, provides an overall gauge of the average pressure on available resources and the threat to aquatic ecosystems.
- *the coefficient of variation of precipitation (COV)*, the standard deviation of annual precipitation divided by the mean annual precipitation, measures the degree of variability in hydrological patterns and the sensitivity of rainfed agriculture to variations in precipitation; the higher the COV the more variable the precipitation.
- *import dependence*, the percentage of a national water supply that flows from external sources, measures the geopolitical security of national water resources with higher percentages reflecting greater vulnerability; availability is dependent on developments in upstream riverine countries and the maintenance of international allocation arrangements; also, downstream nations are subject to degradation of water quality.
- *storage-to-flow ratio*, national reservoir storage capacity divided by average annual water supply, measures the capacity of water resources infrastructure to cope with water fluctuations; higher ratios imply more resilience against floods and droughts.
- *average income* (GDP per capita), a proxy for a nation's capacity to cope with water problems and uncertainties, and to deliver basic water services to its citizens

These measures are selected for their relevance to the issue of water vulnerability, and the availability of data both now and for the scenario simulations. The use-to-resource measure reflects the physical pressure on water resources, *on average*. The next three variables together -- COV, import dependence and storage-to-flow ratio -- represent different aspects of water resources *reliability*. Finally, average income recognizes that the level of vulnerability depends on economic *coping capacity*.

POLICY OPTIONS

Chapter 4 will present the CFWA recommendations for policy options and as of the time of preparation of this paper are still being finalized. As Keating (1996), notes, however, "since it will take time to change many unsustainable development patterns, urgent and decisive action must begin now. Experience has shown us that the consequences of inaction, in terms of human suffering, social disruptions, foregone economic opportunities and the cost of undoing the harm caused to the resource and the environment will usually outweigh the human and financial resources needed in to engage into a sustainable development path. Many of the problems are of a local and regional nature, and action is primarily a national (and regional) responsibility. Nevertheless, it would be illusory to believe that anything short of a global commitment would provide the means to sustainability. Because some of the water crises could be very severe, the whole world has a stake in resolving the problems.

ACKNOWLEDGMENTS

The CFWA is the product of the CFWA Steering Committee of which we are members. The Committee is acknowledged for their hard work and dedication.

REFERENCES

Davis, D., Global Water Assessment, Chapter 2, Draft, Stockholm Environment Institute-Stockholm Center, February, 1996.

Keating, M., CFWA Draft, Stockholm Environment Institute-Stockholm Center, October, 1996.

Raskin, P., Gleick, P., Kirshen, P., Pontius, R., and Strzepek, K., Water Futures: Assessment of Long-Range Patterns and Problems, Stockholm Environment Institute-Boston Center, August, 16, 1996.

Intranetted Management of Water Resources

M.B. ABBOTT AND S. SHIPTON
IHE, Delft, The Netherlands

SUMMARY

The future management of water assets necessitates the active participation of a wide range of stakeholders. The first operational tools necessary to facilitate this participation by using intranets have been extended further with the main purpose of enhancing the transparency of the positions adopted by the participating individuals and organisations.

1. INTRODUCTION

We take it as given that future management systems will be intranetted to a large number of interested individuals and organisations who will then manage the corresponding water assets collectively. This intranet-mediated collective management of water resources currently proceeds through negotiation processes that are increasingly supported by networks for transporting and distributing knowledge and data and the means for processing this knowledge and data into negotiating positions. The resulting process, which can be regarded as a kind of language game, has become an important subject of research in hydroinformatics and one that is closely associated with the elaboration of new tools for supporting the transmission and intentional processing of the knowledge and data necessary for supporting the negotiation process. Since the choice of tools and the social arrangements that the use of these tools necessitates are so closely interconnected, research of this kind is of the *sociotechnical* kind that is now becoming so widespread in hydroinformatics. They are thus of a kind where many, if not most, technical aspects are subject to social considerations and where the necessary new social structures can only be realised by introducing new technical means.

2. THE SOCIOTECHNICAL IMPLEMENTATION OF THE COVENANT

From the side of current society, the implementation of the covenant that gives mankind the responsibility to protect its natural environment is usually followed through four 'moments' (see, originally, Callon, 1986). These are the moments of (1) *problematisation* in which an interested individual or group of individuals takes up the interests of certain creatures and their habitats by posing the difficulties that these creatures experience as problems that are susceptible to a sociotechnical solution, (2) *interessement*, in which actors within the group are locked into this group, *qua* group, through their individual perception of a common interest, (3) *enrolment*, in which these actors are assigned roles and these roles are coordinated, and (4) *mobilisation*, through which the group-perceived interests of the represented creatures are actively pursued. This is a continuous cyclic process and one which must be supported through all the moments of all its cycles by suitable hydroinformatics tools. The first two moments are currently being increasingly realised using Internet, while the second two are better realised using intranets, albeit still normally situated within the Internet environment.

Moments 3 and 4 also introduce a new element, of *social learning*, in which participants in the interest group raise the level of their knowledge and understanding of the creatures and habitats that they claim to represent by using computer-based tools for processing the most relevant Internet-distributed knowledge and data. A considerable further market then appears for such tools for promoting social learning.

The increasing mobilisation of large and politically influential groups of persons into environmental interest groups and the provision of tools capable of processing knowledge and data to suit the interests of these groups leads already to a demand to participate actively in the decision making process by pressing an own group interest within a politically-established and institutionalised framework. *Group Delphic* methods that depend upon the interaction of such engaged, knowledgeable and politically influential groups are thus pushed increasingly to the fore. (See, as an example of a pre-Internet and indeed pre-computer precursor, Webler *et al*, 1995). The decision process itself then becomes distributed through wide-area networks and increasingly takes the form of a partially-network-supported negotiation with a much larger number and range of stakeholders than hitherto. It then becomes imperative to provide network- and computed-based common platforms, environments and languages for these participating stakeholders if any kind of working consensus is to be obtained within society as a whole.

3. EXPERIENCE AND FIRST EXPERIMENTS WITH INTERNETTED SYSTEMS: EAGLE AND CASCADE 1

The first major operational intranetted *environmental impact assessment support* and *decision support systems* (EIASS and DSS) within the ambit of the present work was the EAGLE system for the on-line monitoring and management of dredging works for the multi-billion dollar road and rain link across the Sound between Denmark and Sweden, a project that is to be completed in the year 2001. This system, which has been operational since mid-1995, is used to describe the impact of dredging spills upon marine organisms in the Sound. Such organisms as eelgrass, which is of primary concern to fishing interests, mussels, which are a source of income to mussel farmers, eider ducks, which are another major concern, and others besides are represented by a number on interest groups, both on the Danish and the Swedish side. The interests of the link operator and contractors are represented by a special body, called the *Øresundskonsortiet* (*Sound Consortium*), while other users include a number of Danish and Swedish government environmental agencies. The widely varying knowledge and data needs of such a heterogenous collection of end users are accommodated through another body again, which is a *Control Centre*, staffed by experts from the Danish Hydraulic Institute and the Danish Water Quality Institute. The operation of the EAGLE is governed by statutes agreed between the Danish and Swedish sides based upon legislation enacted by the parliaments of the two countries. The system, which has been described more fully by Larsen and Nielsen (1996) integrates a complete range of advanced numerical models, on-line measuring networks and transits, remote sensing facilities, GIS facilities, network hardware and software and flexible user interface facilities. It is run as an intranet with quite extensive firewalls, macro-virus detectors and other such security software.

Experience with systems of this kind led in 1995 to the construction of two experimental prototypes for studying means to enhance specifically the *transparency* of the assessment and decision-making processes. It was observed, in effect, that the assumption and reasoning of one group of users had to be made transparent to other groups of users if a consensus was to evolve through the use of such a system. Since these prototypes were such that knowledge and data 'cascaded' through then, they were called *Cascade 1* and *Cascade 2* (see Shipton, *née* Šimić, 1996). Both were constructed on the basis of the long-established Battelle methodology (Bureau of Reclamation, 1972). In this methodology, environmental quality is measured by an index that is normalised to the range [0,1]. A *value function* then relates the various levels of parameter estimates to levels of environmental quality as perceived by the specific user group. This function is generated by a code from the elicited intentions of the user. When graphed, each such function provides a visual representation of the assumptions of its user group. Within the ambit of the study of signs, or *semiotics*, as applied in hydroinformatics, this reduced to the application of a specific class of *sign vehicles*.

The value functions are the elements which are most likely to be changed during applications, either because of advances in domain knowledge or simply in order to reflect different and often labile opinions about the parameters' environmental qualities, with most of the value functions being built upon subjective, and variously-interested, opinions. With a value function code being the only link to the specific parameter in the application, however, all that is then necessary is to replace the existing function by running a code, which now appears only as an attribute of the corresponding parameter. Of course, the value function has to be prepared for use by the individual user or user group (either as pairs of EQ values or as the corresponding graph). If the value function is of a new type, on the other hand, this level of generalisation will not adequately accommodate the change. However, by coding the entire hierarchy (with codes being assigned to the parameters, components and categories) a structure has been created which can maintain a stable environment for any application. Moreover, each parameter's attributes are passed to the application controls after they have been retrieved from the tables. This makes the application responsive to any changes which do not change the structure. As an example, if the change in a parameter's name is recorded in the table, the change will be reflected all through the application (e.g. in the menu items and notebook pages). Parameters and their attributes are retrieved when the application is started and stored in memory in the form of records. The rigorous application of object-orientated methodologies is of course essential for these purposes.

Instead of designing and implementing templates for each parameter separately (for a system having typically some 80 parameters), it is desirable to reduce the problem to the level of implementing parameter classes. Six parameter types were accordingly introduced into this prototype, taking account of their group-perceived appearance and behaviour in the application (where a parameter is here the application's object). From the software-implementation point of view, of course, it is not important what the parameter really is, but it is necessary to specify the numbers and types of inputs that have to be handled. Accordingly, a parameter evaluation template has been introduced for each type. In order to ensure that the template's instance represents the parameter's template, values specific to the parameter have then to be passed as attributes. As an example, the same template is called to evaluate both of such physically disparate types as *pest species* and *dissolved oxygen*.

4. THE OBJECTIVES AND REALISATION OF CASCADE 2

Experience with a considerable number of DSSs, and with the EAGLE system in particular, had shown that an adequate coupling with a well-supported, proprietary GIS system was a *conditio sine qua non* for the successful application of almost any such system in this area. Almost all thinking about the natural and human-social environments is a thinking about distributions in space. Further to this again, of course, all interventions in these environments proceed also in time, so that temporal descriptions and resolutions of events have also to be incorporated. Further to this, the need was expressed by several end-users to introduce improved facilities for making sensitivity analyses at a number of (often alternative) levels, to enhance the presentation capabilities and to incorporate more advanced and specifically context-sensitive help facilities.

Usually, however, it was observed that there was a need to consider only a finite number of different locations and time frames. Indeed, in respect to time, it was common that only short and long term impacts were considered. In order to accommodate even these extensions into the first prototype, however, it became necessary to introduce an evaluation matrix for each parameter. Moreover, the data had then to be handled for each geographical cell both during the evaluation of the impact and during the presentation of the results. Because the amount of data increased substantially, the original way of handling the intervention archive was no longer adequate. Changes were accordingly made both in the databases and in the template designs. These changes were introduced into the second experimental prototype with a view to exploring the feasibility and modalities of introducing spatial and temporal resolutions that could later be integrated with GIS presentations. No changes were introduced in this second experimental prototype in the domain knowledge representation, so that it was not necessary to restructure the database used to store this knowledge. It was however decided to restructure the tables which were used to store procedural data, taking into account that all data entering the system during an evaluation should be accessible, (i.e. input data should be accessible for each evaluation cell defined for the evaluation). This proved, in practice, to be critical for presenting the individuals user's customised aggregation process. In order to keep its data organisation as orderly as possible, a separate directory was created for each new project that was considered. Tables created during the project definition then came to form a procedural database that was also stored in the directory.

Thus in *Cascade 2* the main table was used to store the evaluation codes of the evaluation cells. These evaluation codes were subsequently used to identify all the links in the data aggregation process. Besides the project table, tables had then to be created to store the evaluation input data. In order to facilitate subsequent referencing, this table was created already during the project definition: creating another, alternative, view then led to the creation of another evaluation table. Tables for the alternatives and references were identical in structure, thus again facilitating the subsequent processes of comparison. A work table was created to store most of the dynamic data (and specifically the evaluation data for the parameter which was under immediate consideration). Such a flexible structure with dynamic paths to the tables (in which, again, a path is dependent upon a current project, i.e. the project's directory) was greatly facilitated by the possibility to introduce local aliases in the tool that was used in order to construct this prototype. In a similar way, it also became possible in this system to pass run-time information about the directory in which a project's database was stored to the data-sensitive controls of the system. The parameter templates had also to be changed in order to provide more efficient ways to introduce data. It was also decided for the purposes of this prototype to separate the evaluation of the reference from the evaluation of the alternatives. Evaluation was done first for the reference and then for each of the created alternatives. For each parameter, a full evaluation matrix was made available as a two-layered notebook. A number of tools were available to speed up this process and to help the evaluator in keeping the relevant information available. At the level of the parameters, the evaluation setup could be changed by excluding some of the frames from the evaluation. Although this option was available only for the reference, with the resulting structure every change made in the setup could be applied for every project alternative as well.

The new approach necessary to the evaluation process used in Cascade 2 was realised by defining a number of *evaluation modes*. Five modes were introduced in this prototype but only two of these were numerical and could therefore introduce evaluation-cell feature into the aggregation procedure. In each parameter's evaluation matrix, different modes could be present for the different time-allocation cells. For each parameter, all modes were counted during the aggregation and reported in the presentation module.

REFERENCES

Abbott, M.B. and Shipton, S., 1996, Promoting distributed social learning and collaborative decision making through networking, or: Group Delphi in Cyberspace; http://www.ihe.nl/hi

Bureau of Reclamation, 1972, *Environmental evaluation system for water resources planning*, U.S. Department of the Interior, Washington, D.C.

Callon, M., 1986, Some elements of a sociology of translation; domestication of the scallops and fishermen of St Brieuc Bay. In *Power, Action and Beliefs: a New Sociology of Knowledge*, J. Law (ed), Routledge, London.

Larsen, L.C., and Nielsen, A.H., 1996, EAGLE - An Environmental Information and Decision Support System; http://www.dhi.dk

Šimić, S., 1996, *A decision support system for environmental assessment*, M.Sc. thesis, IHE Delft, No.261, March.

Webler, I., Kastenholz, H., and Renn, O., 1995, Public participation in impact assessment: a social learning perspective. *Environmental Impact Assessment Review*, 15 (5), pp 443-464.

AQUARIUS: A GENERAL MODEL FOR EFFICIENT WATER ALLOCATION IN RIVER BASINS

GUSTAVO E. DIAZ [1] and THOMAS C. BROWN [2]
1. Dept. Civil Engr., Colorado State University, Fort Collins, CO 80523
2. Rocky Mountain Experiment Station, USFS, Fort Collins, CO 80524
gdiaz@lamar.colostate.edu   tcbrown@lamar.colostate.edu

ABSTRACT

This paper introduces AQUARIUS, a state-of-art computer model devoted to the *temporal* and *spatial* allocation of water among competing uses in a river basin. Version 96 of the model is driven by an economic efficiency operational criterion that calls for the reallocation of stream flows until the net marginal returns in all water uses are equal; that is, until a Pareto optimal arrangement is reached. This is achieved by systematically examining—using nonlinear optimization—the feasibility of reallocating unused or marginally valuable storage and releases in favor of alternative uses. Because water systems components can be interpreted as objects of a flow network in which they interact, the model considers each component of a water system as an equivalent node or link in the programming environment using an object-oriented programing language (C++). Future versions will allow simulation of existing allocation rules and priorities.

INTRODUCTION

To efficiently manage today's increasingly complex water systems, new modeling tools are needed —tools dedicated not only to the continued support of the traditional uses, but also to the growing societal recognition of nontraditional water uses for a host of environmental and recreational objectives. Traditional water uses —for which most existing water management systems were designed and are typically operated— include flood control, hydropower, irrigation, and urban water supply. Nontraditional uses include preservation of biological and geomorphological integrity of a river, as well as provision of opportunities for water-based recreational activities such as fishing and boating.

This paper introduces AQUARIUS, a state-of-art computer model devoted to the *temporal* and *spatial* allocation of flows among *competing* water uses in a basin. We envision AQUARIUS as an analysis framework rather than a single dedicated model for water allocation. In this first version of the model, Version 96, we adopted an economic criterion for determining water allocation in a river basin. This was done primarily because economic demands have traditionally played a key role in water allocation decisions, and also in light of the greater accessibility of economic value estimates for some nontraditional water uses. Documentation of the model can be found in Díaz and Brown (in press).

THE EFFICIENT ALLOCATION OF WATER

AQUARIUS V.96 is driven by an *economic efficiency* operational criterion that calls for the reallocation of stream flows until the marginal returns in all water uses are equal, i.e., until a Pareto optimal arrangement is reached. Each traditional use (e.g., hydropower) and nontraditional use (e.g., water recreation) is represented by a *demand curve*, also known as marginal benefit function. An example of this type of curve is shown in Figure 1 for instream recreation in the Bitterroot River in Montana, involving dry fly fishing, floaters and shoreline recreationists.

Because—in contrast to Figure 1—nonlinear demand curves are more the rule than the exception, AQUARIUS uses exponential models of the form $P = a\, e^{Q/b}$, with $a > 0$ and $b \leq 0$, to represent convex (to the origin) demand functions. Under this multipurpose modeling approach, the model systematically examines the feasibility of reallocating unused or marginally valuable storage and releases in favor of alternative or competitive uses, identifying trade-offs between the various water user groups.

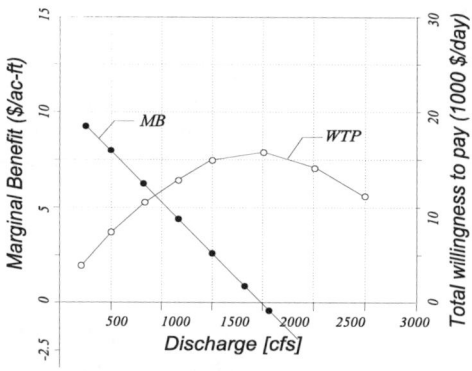

Fig.1 Marginal and total recreation value as a function of instream flow (Duffield et al., 1992)

For a water use for which the level of allocation has been *pre-determined* but for which the demand function (in economic terms) is practically impossible to define, the analyst can experiment with surrogate demand curves until the required level of water allocation for that particular uses is satisfied. Indirectly, this approach indicates the societal willingness to pay for incremental increases of flow for nontraditional water uses, or perhaps, the level of economic subsidy required to sustain that activity in open competition with other uses in the basin.

SOLUTION OF THE WATER ALLOCATION PROBLEM

In the model, water allocation throughout a river system and for an entire planning horizon is based on a global objective which is to *maximize the sum of all economic benefits* stemming from the instream and offstream use of water —as expressed by their willingness to pay— subject to the operational constraints of the system such as: reservoir storage limits, firm water supply levels, max/min instream flows, max/min diversions, seasonality of water demands, etc. Given demand functions for the various water uses j, the global benefit function (B) to be maximized over the various time periods i is

$$B = \sum_{i=1}^{np} \sum_{j=1}^{nu} \int_0^{a_{ij}} f_{ij}(x_{ij}) \, dx_{ij}$$

np : total number of time periods
nu : total number of water uses

where x is the level of output in the demand function $f(x)$ and a denotes the level of allocation. It should be remembered that B is maximized when a_{ij} are set such that the marginal prices are equal for all i,j. In other words, total benefits are maximized when levels of consumption are such that the marginal benefits for each use across all uses and time periods are equal (provided that an unconstrained solution to the allocation problem is found). B can, of course, only be maximized over the j uses for which marginal benefit functions are specified. If relevant uses are omitted because their benefit functions cannot be specified, the model can still represent them by adding the necessary physical constraints to the formulation.

The water allocation problem postulated above requires consideration of a complex nonlinear objective function. A variety of approaches exist in the literature for dealing with the solution to this type of problem, none of which is uniquely superior or universally proven. The solution technique implemented in AQUARIUS V.96 takes advantage of the special case of the general nonlinear programming problem that occurs when the objective function is reduced to a quadratic form and all the constraints are linear. The method entails approximating the original nonlinear objective function by a quadratic form using Taylor Series expansion and solving the problem as a quadratic programming problem. A succession of these approximations is performed —following a technique knows as Sequential Quadratic Programming— until the solution of the quadratic problem reaches the optimal solution. This method of solution is an extension of the work reported earlier by Díaz and Fontane (1989).

MODELING RIVER BASINS

One of the unique characteristics of AQUARIUS is its implementation using an Object-Oriented Programming (OOP) language (C++). This modeling approach implements the concept of a system as articulated in systems engineering. Water systems are ideal candidates to be modeled under an OOP framework, where each system component (each reservoir, demand area, diversion point, etc.) is conceptualized as an equivalent object in the programming environment.

CREATING A FLOW NETWORK

The user interacts with the model through the so-called network-worksheet screen (Figure 2), which allows the analyst to readily represent the water system of interest using the inherent capability of the object-oriented paradigm for graphical representation. The model provides four elements for user interaction: (i) the network worksheet (NWS), (ii) the menus, (iii) the water system components (WSC) palette, and (iv) the object tools palette.

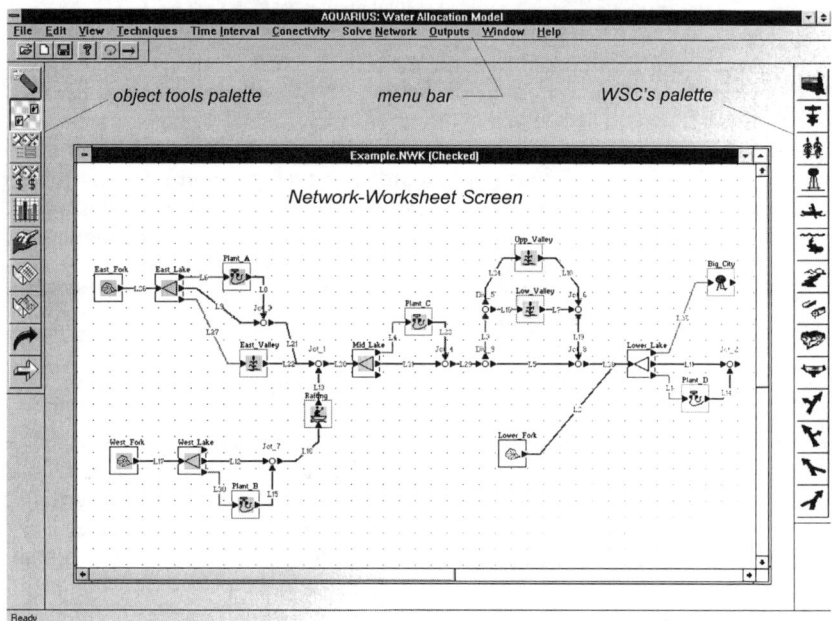

Fig.2 AQUARIUS Network-Worksheet Screen

In the NWS each system component corresponds to a graphical node or link (object) of the flow network. These components are represented by *icons*, based on a pictorial representation of the object. By *dragging and dropping* these icons from the menu, the model creates instances of the objects on the screen. In this manner, one by one, all the necessary system components are created. WSCs can be repositioned anywhere in the NWS or be removed from it. Once nodes (e.g., reservoirs, demand areas) are placed, they can be linked by means of natural river reaches and conveyance structures, which are also objects available from the WSC palette. This operation is carried out by simply left-clicking on the outgoing terminal of a node, and next into the incoming terminal of the other node. This procedure facilitates the assembly or alteration of water systems by simply "wiring up" their system components in the NWS. The creation and alteration of flow networks is further facilitated by copying and inserting an object or whole portions of an existing network onto the same or a new NWS. The *Copy/Paste* procedure not only creates new instances of the object(s), but also duplicates their data structure (creating clones of the original objects).

ENTERING INPUT DATA

The input data to the model have been divided into two basic groups: *physical* and *economic* data. The physical data include the information customarily associated with the dimensions and operational characteristics of the system components, such as maximum capacity of a reservoir, percent of return flow from an offstream demand area, and efficiency of a powerplant. The economic data consist mainly of the demand functions of the various water uses competing for water. The input data entered for any system component remain part of the object, even after the network is saved on a storage disk. When the network is reloaded, all data saved from the previous session are retrieved in exactly the same form.

SOME OPERATIONAL ASPECTS

AQUARIUS was conceived to simulate the allocation of water using any time interval of analysis: daily, weekly, monthly. Future versions of the model will also support time intervals of non-uniform lengths. Under the latter scheme, we can, for example, think of a year long operation horizon subdivided as: the first 7 days (short-term), the following 3 weeks (medium-term) and the remaining 11 months (long-term). This partition of the operation horizon into intervals of different length may coincide with the way inflows to the river system are forecasted.

The user can use AQUARIUS in a full deterministic optimization mode (for general planning purposes), or in a quasi-simulation mode, with restricted foresight capabilities. For that purpose, the model distinguishes between the *period of analysis*, used to specify the length of the whole segment of time for which the model will simulate (and optimize) the operation of the system, and the *optimization horizon*, used to specify how far ahead into the future the model should look to build the optimal operational policies. Formulating the water allocation problem entirely within the domain of the objective function provides the user with the capability to redirect the water allocation process in any desirable direction in *real time*, directly from the screen, as the optimization progresses. This unique feature of the model provides the analyst with an expeditious and innovative mode of exploring several *what if...* scenarios.

River flow networks contain a myriad of state, decision and economic variables that the user may need to consider for the analysis. AQUARIUS facilitates the interpretation and analysis of all that information through readily accessible graphical and tabular output display formats. Because of space limitations in this paper, a case study has not been included. The software runs on a personal computer under a Microsoft Windows 95 or Windows NT operating system.

WORK IN PROGRESS

As discussed earlier, water allocations achieved using Version 96 of the model maximize economic efficiency. Under the dominant water allocation doctrine in the western United States —the *prior-appropriation doctrine*— water is allocated following a time-based priority rule, whereby the water available to satisfy a new application is reduced by the sum of all prior established rights. A time-based allocation in a heavily appropriated river can become *inefficient* as values change if, as is commonly the case with water, institutional barriers or market failures impede voluntary transfers of rights from lower-valued to higher-valued uses. Thus, actual water allocations in the West may be quite different from the economically efficient allocations achieved using V.96.

In order for AQUARIUS to predict allocations under current or alternative priorities and operating rules, a simulation facility is presently being added to the model. An attractive tool for achieving such simulation is a *network-flow* model. Simulation models based on network-flow programming constitute a "hybrid" formulation that combines some advantageous features of simulation with some optimizing capability. The object-connectivity capabilities of AQUARIUS are fully exploited to couple data requirements of the network flow algorithm to the topology of the river basin to be modeled. The network-flow solver —to be available in V.97— will be an alternative "engine" to the fully-dynamic optimization algorithm already implemented in V.96.

Regardless of the process by which the existing allocation of water in a basin became established, it may be helpful to compare the *current allocation* with an *efficient allocation*. Such a comparison may indicate promising opportunities for private water trades, or, where such trades are hampered or precluded by institutional barriers, may indicate areas where institutional reforms can allow a more efficient water allocation to occur. Also, where water developments are publicly financed, the comparison may indicate directions that the public entity should consider to increase the efficiency of the public project.

LITERATURE CITED

Díaz, G.E., and D.G. Fontane, 1989. Hydropower optimization via sequential quadratic programming. ASCE, Journal of Water Resources Planning and Mgmt..Vol.115(6).

Díaz, G.E., and T.C. Brown, in press. AQUARIUS: An object-oriented model for the efficient allocation of water in river basins. General Technical Report, Rocky Mountain Forest and Range Experiment Station, U.S. Forest Service, Fort Collins, CO.

Duffield, J.W., C.J. Neher and T.C. Brown, 1992. Recreation benefits of instream flows: Application to Montana's Big Hole and Bitterroot Rivers. Water Resources Research. Vol.28, No.9, pp.2169-2181.

Implementation of Object-Oriented Programming for Dynamic Simulation of Open Channel Flow

John A. Hinckley, Jr. and Shaw L. Yu
Department of Civil Engineering
University of Virginia, Charlottesville, VA, USA
and
Perry J. LaPotin
U.S. Army Cold Regions Research and Engineering Laboratory,
Hanover, NH, USA

INTRODUCTION

Computer models have been developed to improve the understanding and management of environmental systems for the betterment of society. Models have been created to simulate open channel flow to predict the aerial extent of flooding. The beneficial aspect of these models is their ability to yield reliable flood forecasts by solving sophisticated channel routing equations. However, these models are not easily understood by people with limited computer experience. People are unable to understand the basic structure of these models because the equations solved by the models are embedded in numerous lines of C or FORTRAN code. Consequently, the effective use of these models is limited to skilled computer programmers.

Complicated computer code limits water management operations when senior flood control experts are absent from their posts during flooding events. When a senior flood control expert is absent, their substitute is responsible for issuing flood forecasts. Often the substitute is less familiar with the model being used which can delay the issuance of critical flood forecasts. Consequently, a computer model with a high level of user interactivity is desirable to provide quick and accurate flood forecasts.

Object-oriented programming (OOP) is a computer language designed with the intent to make computer code easier to understand (Cox, 1986). This language uses specific objects to symbolize "varying elements of a system." Furthermore, these objects can be connected to symbolize "differential and integral relationships" (McKim et al., 1993).

In addition to being more user friendly than other traditional programming languages, OOP provides an improved framework for observing the behavior of environmental systems. For example, each object, which symbolizes an element of an environmental system can be animated and observed during the entirety of a simulation. Therefore, every component of an object-oriented program can be easily viewed and understood.

OVERVIEW OF MODELING WITH STELLA®II

An understanding of the objects in the STELLA®II programming environment is prerequisite for understanding the contents of this paper. Therefore, the following section will begin with a brief description of those objects. The following object descriptions were taken from Hinckley (1996).

Rectangles represent *levels (integral equations)*. Levels accumulate or decrement depending on the rate-values that are connected to them (i.e. they are assigned an initial condition, and then allowed to integrate the differential equations symbolized by *rates* which are described later in this section). The rectangles are referred to as the 'state variables' for the system since they have the capacity to change *states* through time and space. The term 'steady state' is used to describe a state variable invariant in time (and/or space).

Open circles are referred to as *converters* and function to convert inputs to outputs. The inputs may be equations or logical statements (open circles) or numerical relationships (circles containing tildes). Converters do not accumulate but change instantaneously over the simulation run.

A *cloud* represents *sources* or *sinks*. If an arrow points into the cloud it must be a sink. Conversely, an arrow pointing away from a cloud implies that the cloud must be a source. In Figure 1, the cloud represents a source. The double-lined arrow is referred to as a *pipeline* and resembles a section of pipe. The pipeline is used to simulate the movement of a fluid from one location to another. For example, in Figure 1 the pipeline simulates the movement of water from the cloud to the level. Valves represent *rates (differentials)*. The object is meant to symbolize a plumber's valve that opens or closes depending on physical conditions. When the valve opens, water will 'flow' from the cloud (the source) into the level. The combination of a cloud, pipeline, and valve is referred to as a 'flow'.

Solid arrows are referred to as *connectors* or information flows, and function to depict the causal linkages among the objects (variables) in the model. Connectors have no numerical value (McKim et. al., 1993; Hinckley, 1996; Richmond and Peterson, 1994).

DYNAMIC SIMULATION OF OPEN CHANNEL FLOW 413

The structural diagram in Figure 1 can be used as a basic rainfall model. For example, the cloud represents an actual cloud. The flow represents the movement of water from the cloud to the Earth's surface. The rainfall rate is set within the converter. The level represents the storage on the Earth's surface.

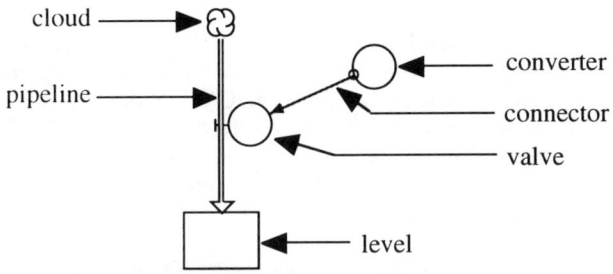

Figure 1. Basic STELLA®II diagram.

MODEL DEVELOPMENT

The objects shown in Figure 2 collectively form a "boxcar structure". The boxcar structure, which serves as the backbone of the OOP model, is a simplified channel routing structure. The volume of water stored in two consecutive reaches correspond to the levels entitled "storage_1" and "storage_2". The inflow to the uppermost reach (reach one) corresponds to the flow entitled "inflow_1". "Outflow_1" is both the outflow from reach one and the inflow to reach two. "Outflow_2" is the outflow from reach two. The converters entitled "outflow_calc_1" and "outflow_calc_2" calculate the outflow from reaches one and two. The connectors show that "outflow_1" is calculated with the value of "inflow_1" and "storage_1". Likewise, "outflow_2" is calculated with the value of "outflow_1" and "storage_2". The connectors also indicate that the values of outflow calculated by "outflow_calc_1" and "outflow_calc_2" are embedded in "ouflow_1" and "outflow_2" respectively. The objects shown below and many other ancillary converters (which are not shown for simplification) solve the St. Venant equations.

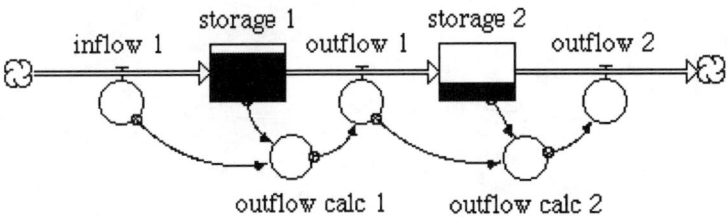

Figure 2. Simplified OOP representation of an open channel.

Please note the shading of the two levels in the Figure 2. During simulations, the shaded area of the level represents the volume of storage in a reach. Furthermore, the shaded area will increase and decrease over time as the calculated value of the volume of water fluctuates. Hence, Figure 2 indicates that at a given time, more water is stored in reach one than reach two.

Two steps were taken to check the accuracy of the model. The first step was made to ensure proper encoding of the St. Venant equations in the OOP environment. In this step, the output from the OOP model was compared with the output from an similar model (Schroeter 1990) which also solves the St. Venant equations according to the forward finite difference solution described in Veissman and Lewis (1996). Both model outputs, generated with the same inflow hydrograph, had excellent agreement.

The second step involved testing the model with historic flow data from the Grand River Basin, the largest river system in Southwestern Ontario, Canada. The OOP model simulated the flow of a 56.7 km stretch of the Grand River (from Brantford to York) having an average gradient of 0.34 m/km and draining an area of 825 km^2. The model was tested with data from flooding events which occurred in May, 1974 and April, 1979 (Schroeter and Epp 1988).

Figure 3 shows the results from the May, 1974 event comparison. According to Schroeter and Epp (1988), this rainfall event is the largest on record for the Grand River Basin. The figure shows close agreement between the magnitude and timing of the peak flow. But, before the peak, the observed flow is greater than the simulated flow. This disparity may be attributed to lateral inflow, which is not currently incorporated into the model. Despite the slight disagreement between observed and simulated flows before the peak, there is excellent agreement between the observed and simulated flows following the peak discharge.

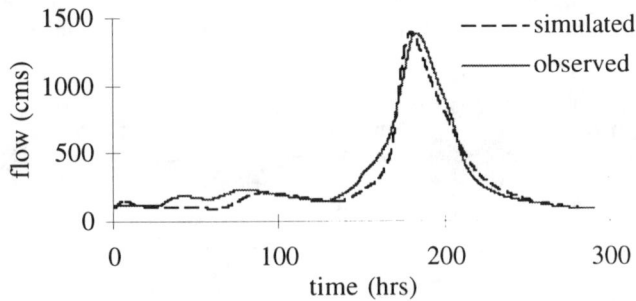

Figure 3. Observed and simulated hydrographs for the May, 1974 event.

The April, 1979 event differs from the May, 1974 event in that principle discharge resulted from rainfall and snowmelt. The results from the April, 1979 are comparable to those shown in Figure 3 (Schroeter and Epp, 1988).

RESULTS AND DISCUSSION

An object-oriented open channel flow model has been designed and successfully tested. Comparison of OOP model outputs with those of an existing finite difference model suggests that the St. Venant equations were successfully encoded with the OOP environment. Further testing of the OOP model with historic data yielded accurate estimates of open channel flow. A noteworthy aspect of the research is the model's accuracy in simulating the May, 1974 record breaking flood. The level of agreement between the observed and simulated hydrographs in Figure 4 indicates that the model is robust within the operating framework of the Grand River Basin event.

The described simulations advance computational modeling by providing accurate and easier to use techniques for solving the St. Venant equations. By animating the objects, users can actually view the behavior of each model component over time. Therefore, the user does not have to peruse through numerous lines of source code to understand or de-bug a simulation. Users can easily change equations or parameter values embedded in the model by simply clicking on the object of interest.

Future developments within the OOP model will involve adapting elements for the upper St. Francis River of southeast Missouri. While being tested with ten years of historic data from the St. Francis, the model will be expanded to incorporate lateral inflow and backwater effects. Future development of the model will also include the incorporation of a Digital Elevation Model (DEM) to better estimate the aerial extent of flooding.

REFERENCES

Cox, G.W.(1990) Ecosystem simulation using STELLA, *Laboratory manual of general ecology*, Wm. C. Brown Publ. Co.

Hinckley, J.(1996) Object-GAWSER, Object-oriented guelph all-weather storm event runoff model, phase I: training manual, Special report 96-4, U.S. Army Corps of Engineers, Cold Regions Research and Engineering Laboratory, Hanover, NH.

McKim, H.L., E.A. Cassell and P.J. LaPotin (1993) Water resource modeling using remote sensing and object-oriented simulation. *Hydrological Processes*, **7**:153-165.

Richmond, B. and S. Peterson (1994) STELLA II. Technical documentation. High Performance Systems, Hanover, New Hampshire.

Schroeter and Associates (1990) FLOOD/PC. Technical documentation, Version 2.0. Guelph, Ontario, Canada.

Schroeter, H.O. and R.P. Epp (1988) Muskingum-Cunge: a practical alternative to the HYMO VSC method for channel routing. *Canadian Water Resources Journal*, **13**(4): 68-79.

Veissman, W. and G.L. Lewis (1996) *Introduction to Hydrology*. New York: Harper Collins Publishers, 4th edition.

The Caspian Sea Storm Surge Coastal Protection

M.GALANT, A.ZVEGINTSEV, P.LYSSENKO
VOLGA Ltd. Consulting Engineers, Moscow Russia

ABSTRACT
Storm surge phenomenon was studied for the North-West part of the Caspian sea. The combination of deterministic and probabilistic numerical methods allowed to outline the approach to the storm surge forecasting for condition of significant mean sea level variations.

INTRODUCTION
The Caspian Sea is the world largest natural lake with pronounced long-period mean sea level (MSL) variations. At present the sea is undergoing the phase of sea level rise. At the same time due to natural climatic conditions the Caspian sea is actively experiencing the phenomenon of storm surges (especially in its northern shallow part where the maximum sea storm surges were historically registered as 4.0-4.5 meters. The ongoing MSL rise which is about 1.5 m since 1977 and periodical storm surges significantly endanger the habitat and economic activity in the northern (N) coastal area of the Caspian sea. Due to very small surface slopes (by order of 10^{-4}) the flooding during the large storm surges may cover the coastal zone up to 20 - 50 km wide. Thus timely forecasting of large storm surges and protection of industrial and infrastructure objects is of primary importance for the entire N and NW region of the Caspian sea.

The general methods for analytical prediction of the storm surges are well known and were numerously tested and verified. Nevertheless for particular conditions of the Caspian sea (especially its N part) interaction of atmosphere, shallow sea and the coastal area (including protective and other structures) have several specific features which had to be studied to make the rational decisions on engineering tasks. The main problems may be summarized as follows:

1. What kind of relationship does exist between the storm surge height and MSL?
2. How to assess the probability of the large storm surges and time dependent scenario of their development?
3. What type of influence does coastal topography have on local sea level variations during the storm period in comparison with the neighbouring open sea zone?

The investigation of these problems is the main contents of the present paper.

NUMERICAL MODELLING

Interaction of atmosphere, sea and flooded and non-flooded coast was studied on the bases of MIKE21 modelling system developed by Danish Hydraulic Institute (DHI). It is one of the most world widely spread modelling systems for this type of problems. Three nested hydrodynamic models were developed with a grid size of 9610 m (model No.1 - entire Caspian sea 1250 x 700 km), 1470 m (model No.2 - NW part) and 372 m (model No.3 - 30 x 30 km area around the town of Lagan' for which engineering aspects were specifically tested) (see Fig.1). Data, initial and boundary conditions transfer between the models were determined by internal MIKE21 options.

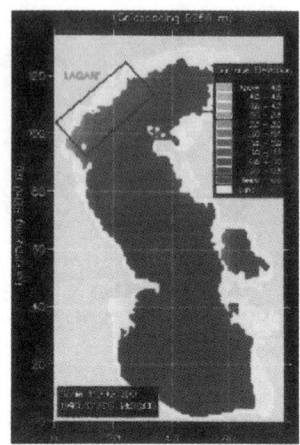

Fig.1 The Caspian sea computational grid system.

Models verification was carried out by spatial and point data of historical records (atmospheric pressure fields, wind speed and direction, time series of water level variations, information of coast flooding boundaries) for 5 storm surge events including the disastrous one of November 1952. The sea level variations in time and space were quite reasonably reproduced by the computer model. The accuracy of the obtained results is sufficient for necessary assessments and practical outcomes of the study (see Fig. 2 and 3).

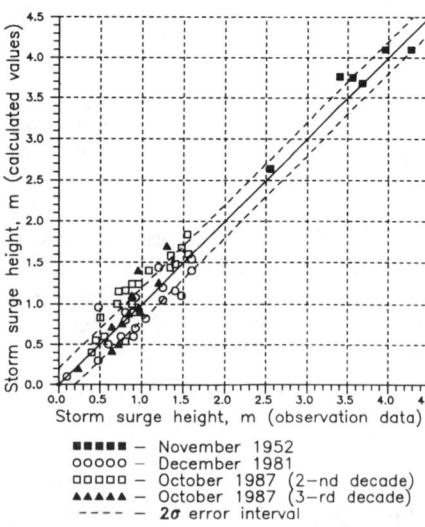

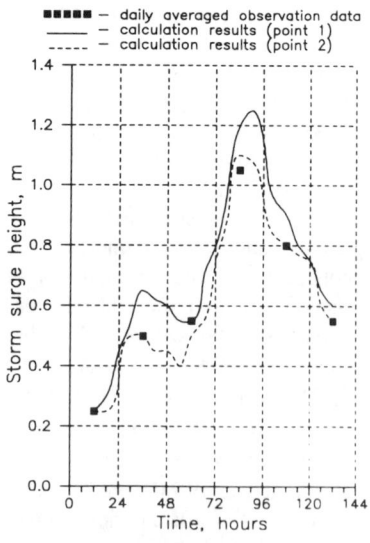

Fig.2 Comparison between daily averaged observed and calculated storm surge heights.

Fig.3 Example of operational storm surge forecast (26/10-01/11 1987)

The route mean square deviation σ=0.21 m (Fig.2) is practically of the same order as possible error level connected with 6-hourly periods of water level recordings.

STORM SURGE HEIGHT AND MEAN SEA LEVEL

The relationship between storm surge height and MSL was determined analytically for unique meteorological conditions of November 1952 and MSL elevation varied within the range of -29.0 ÷ -24.0 meters (Baltic System). Results for the shallow northern part of the Caspian sea are shown in Fig.4 (for deep water Middle and S Caspian sea parts the variation of storm surge height ξ versus MSL is not traced). Decreasing of the storm surge heights ξ along with MSL rise ∇_0 (sea depth) for similar meteorological conditions and bathymetry is directly coming out of simplified momentum equation ($\partial/\partial t\cong 0$, $Fr=u^2/gh<<1.0$):

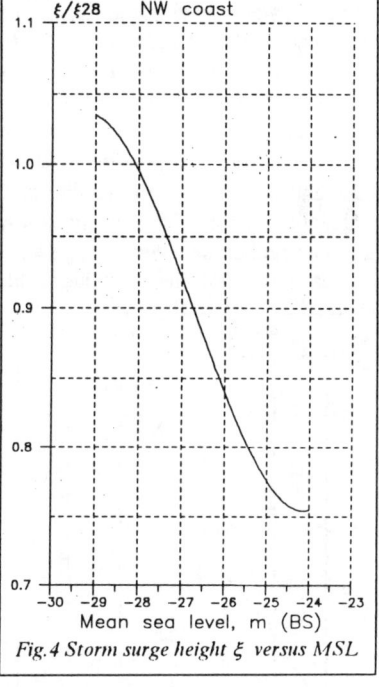

Fig.4 Storm surge height ξ versus MSL

$$\frac{\partial \xi}{\partial x_i} \cong \frac{\tau_{ai} - \tau_{wi}}{\rho g h} = \frac{k_a ww_i - k_w uu_i}{\rho g \left(h_0 + \delta h + \xi\right)} \quad (1)$$

where τ_{ai}, τ_{wi}, w_i, u_i, x_i- direction components of wind shear stress on the air-water border τ_a, sea bed shear stress τ_w, wind speed w and depth average current speed u, k_a and k_w - coefficients, $h_o(x)$ - water depth for "normal" MSL elevation (∇_0=-28.0 m), δh - MSL level deviation from its "normal" value, ρg - specific weight of water.

Equation (1) can be analytically solved only for several not-interesting cases. However it is obvious that with increasing δh the slopes $d\xi/dx_i$ and ξ values are decreasing for constant τ_a and τ_w. Moreover, the increasing of ξ up to the order of h_o and δh values decreases the $d\xi/dx_i$ rate during each storm surge event. This ξ and $d\xi/dx_i$ feedback additionally limits the value of large storm surges for the shallow N Caspian sea. This result (see Fig.4) has a great practical outcome. It confirms that the design crest elevation of proposed protective structures on the N coast of the Caspian Sea is rising relatively slower than the MSL.

LARGE STORM SURGES PROBABILITY ASSESSMENT

The basic concept was adopted as follows: probability of any storm surge is equal to probability of meteorological conditions which may cause this storm surge. Data collection for a 10-year period was performed for statistical analysis assuming the average wind fields distribution over the entire sea area instead of local point wind speed and direction measurements. Besides information on 11 extreme events was collected and studied in detail for the period of 1952-1987 (atmospheric pressure fields for each event with 5-6 days duration and half-a-day time interval). It was estimated that the typical storm surge prone synoptic situation for NW Caspian sea coast coincides with the stable anticyclone with the center located to the NE from the sea and covering the sea by its peripheral part. In these cases there were registered E and SE winds with 10-12 m/s·day time gradients and 27-30 m/s mean-hourly maximum values. Multiple simulations have showed that due to sea water mass inertia maximum water level rise gradients are achieved only 10-15 hours and maximum level setup - not earlier than 24-30 hours after the storm start.

Basing on the data of historical records and simulation results a "typical" variation pattern was developed with a single parameter as a maximum wind speed w_{max} (fig.5). It is based on a 6-day time interval - so called "natural synoptic period" and reflects the worst possible level setup conditions for a given maximum wind speed w_{max} and direction φ. Probability of the extreme storm surges is determined by the probability characteristics of the storm winds (wind speed and direction) and also by non-linear relations "wind - surge height" which depend on coast line orientation and sea depth. General theory of this problem is beyond the contents of this paper.

For NW Caspian sea coast where the most storm surge prone wind direction ($\varphi=135^0$, SE) coincides with its most probable direction according to the "wind rose" it was derived

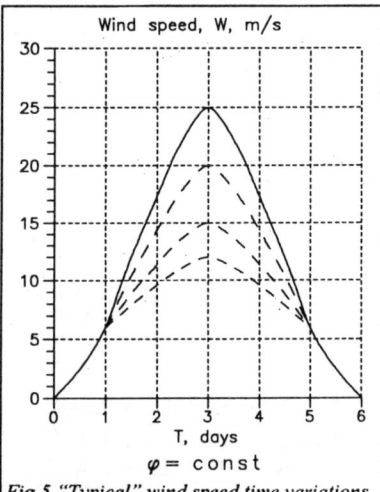

Fig.5 *"Typical" wind speed time variations with different* w_{max}

$$\xi = \xi_1 \left(1 + \frac{ln\,T - ln|ln P_0|}{ln\,N_0} \right)^m, \qquad (2)$$

where: P_0 - probability of the situation when the storm surge height is not exceeding the ξ value during the period T (years);
$N_0 = 11$-15 mean yearly number of extreme winds ($w_{max\geq}10m/s$, $\varphi=135^0$, SE);
ξ_1 - storm surge height with average frequency - once in a year ($P_0 = 1/e = 0.36788$)
For N Caspian sea $\xi_1 = 0.6$-1.0 m and $m = 0.9$-1.1 for different coastal points.

ξ_1 value is usually well known from hydrographic data statistics, while for $\xi \geq \xi_1$ it does not produce so reliable results. Calculations according (2) allow to determine the storm surge height of any probability for a selected time period T_0. Verification of (2) against data of historical records have not revealed disagreements greater than ± 5 per cent. Statistical calculations for conditions of disastrous event of 1952 have demonstrated that this event had a yearly probability value $P_0 = 0.9999$. It did not seem possible earlier to derive this result from statistics of direct measurements because the period of historical instrumental records at any point of the N Caspian sea level stations does not exceed 100 years.

Equation (2) is the basis for flood risk assessment and for solution of engineering tasks on coastal protection of the NW Caspian sea.

STORM SURGE WAVE TRANSFORMATION

Phenomenon was studied within the framework of project on coastal protection of town of Lagan'. Computer simulations were carried out on a model No.3 for MSL=-26.0 m and typical storm wind variations (Fig.4). The coastal topography and protective structures are given in Fig.6. The plot is combined with results of water level time series in different points of coastal area under consideration. The most important conclusions coming out of simulation results are as follows:

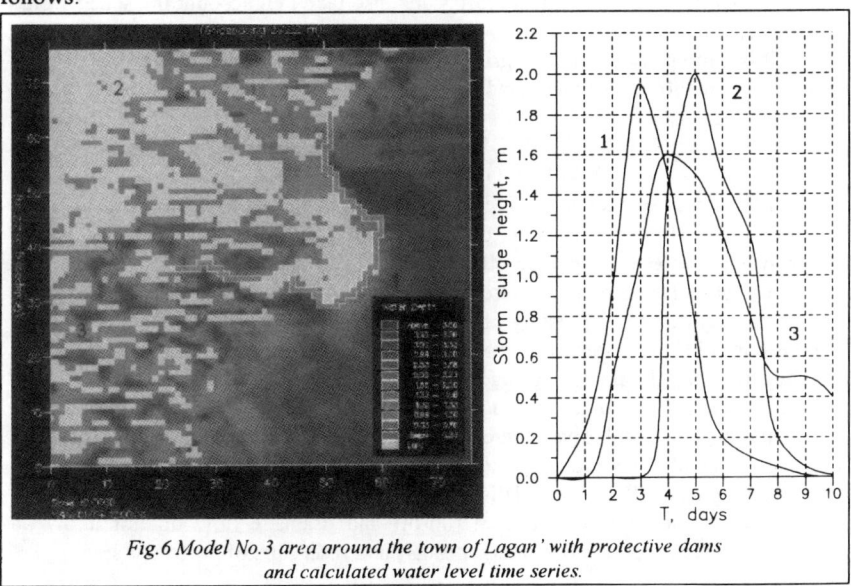

Fig.6 Model No.3 area around the town of Lagan' with protective dams and calculated water level time series.

1. Time variations of water levels in the open sea and coastal points are essentially different. Thus critical errors may arise if operational forecast output or design of protective structures are based only on storm sea level variations in the open sea.

2. For semi-circle protective structures layout (main alternative, Fig.6) the flooding of the territory is possible from the opposite side to the sea with a certain time lag. Only additional dikes, which position and crest elevations were determined out of simulation results can provide reliable flood protection.

The first conclusion is of principle importance. The physical reasons of the water level time lag in the sea and in the coastal area may be qualitatively traced out of (1). Assume $\tau_a = kww_i$ has a synchronous variation in the entire area and positive value during the synoptic period. The positive X-axis direction is towards the coast line. Then for the sea level rise phase ($\partial \xi / \partial t > 0$) $\tau_a - \tau_w \leq 0$, $\partial \xi / \partial x_i \leq 0$ for the sea points and $\tau_a - \tau_w \approx -\tau_w < 0$, $\partial \xi / \partial x_i < 0$ for the coastal zone. For the moment of extreme water level in the sea ($\partial \xi / \partial t \approx 0$) for the points of flooded coast $\tau_a - \tau_w \leq 0$ and always $\partial \xi / \partial x_i > 0$. It means that for positive flow direction ($\partial \xi / \partial x_i < 0$) the wind drag force compensates the land surface hydraulic friction, and for negative direction ($\partial \xi / \partial x_i > 0$) - increases its action. In the studied case (see Fig.6) these effects developed the time lag in water level variations on the coast and in the sea in the range of 1-3 days.

The practical conclusion is quite clear: correct forecasting of water level variations in time and space on a flat (without strictly pronounced surface slope) flooded coast is impossible without modelling of the vast flooded zones. Topographical descretisation of the flood-prone area must correspond to the plan dimensions of the local topographic forms and also take into account the space varied hydraulic resistance factor.

It is quite possible that the grid size of 376.0 meters applied for model No.3 is not even optimal for the studied conditions.

INTERACTION OF THE SEA AND INTERNAL WATERS OF PROTECTED AREA

To assess the technical efficiency of protective structures of the town of Lagan' a 1-D hydrodynamic model of internal water streams, reservoirs, floodplains and navigation outlet was setup as MIKE11 (DHI) application.

The boundary conditions were transfered from MIKE21 simulations. The main item of the studies was the testing of design alternatives for navigation canal with or without control sluice gates to be closed during the storm event. The general answer was negative as floodplain storage capacity appeared to be not suffiicient to accumulate the inflow volume through the canal during the storm surge period. The easy coming natural solution of erecting a dike around the navigation canal was suggested as an alterntive to construction of concrete control structure with sluice or submersible gates. MIKE11 simulation results proved that multiple culverts that are generally used for diversion of local run-off and drainage flow do not allow an impermissible water level rise in internal canals and reservoirs of protected area.

CONCLUSION

Operational and statistical storm surge forecasting necessarily requires solution of the above described basic problems accounting for natural conditions of the flooded territory and the sea specific climatic features. This might be successfully gained on the way of application of aero-hydrodynamic numerical model supported by reliable hydrometeorological statistics. The studies performed may be considered as a true example of this approach, however it should not be taken as a guide for all possible real cases as "truth is always something real".

A New Model for Reliability-Based Optimal Design of Water Distribution Networks

CHENGCHAO XU AND IAN C. GOULTER
Central Queensland University, Rockhampton Qld. 4702 Australia

INTRODUCTION

Over last few decades, considerable effort has been devoted to the development of reliability-based optimization algorithms for the design of water distribution networks. Summaries of this work are given in Mays (1989) and Goulter (1995). This paper presents a new approach for reliability-based optimization of water distribution networks. The approach is capable of recognizing the uncertainty in nodal demands and the pipe capacity as well as the effects of mechanical failure of system components. The main contribution of model lies in the use of a new efficient algorithm, based on the concept of the first order reliability method (FORM), to compute approximate values of the hydraulic reliability of water distribution networks. A probabilistic hydraulic model is used in the model to account for uncertainty in nodal demands and the pipe capacity. The model also incorporates a strategy for identifying the critical nodes upon which the reliability constraints are imposed in the cost minimizing step. The computational efficiency of the optimization is enhanced by deriving the first order derivatives analytically using a sensitivity analysis based technique.

MODEL FORMULATION

Reliability-based optimization for the design of redundant water distribution networks can be mathematically formulated in the following general form:

Min $F(\mathbf{d})$ (1)

subject to

$\beta_j(\mathbf{X},\mathbf{d}) \geq \Phi^{-1}(R_j)$ $j=1,2,...N$ (2)

and

$\mathbf{d}^L \leq \mathbf{d} \leq \mathbf{d}^U$ (3)

where $\mathbf{d}$ = vector of design variables which may include pipe and water tank sizes and pumping capacities; $F(\mathbf{d})$ = objective function to be minimized; $\mathbf{X}$ = vector of basic random variables, namely, the random nodal demands and random values of pipe roughness; $\mathbf{d}^L$ and $\mathbf{d}^U$ = lower bounds and upper bounds respectively for the design variables; $\beta_j(\mathbf{X}, \mathbf{d})$ and R_j = system reliability index and specified reliability requirement respectively for system operating configuration j with uncertainty in nodal demands and the pipe capacity; N = number of reliability constraints imposed on the optimal design; and $\Phi^{-1}(\bullet)$ = inverse standard normal cumulative distribution function.

A major feature of this formulation is that the hydraulic requirements are considered implicitly through the reliability constraints which are evaluated using a probabilistic hydraulic model. The reliability aspects of component failures are also considered implicitly through the inclusion of the reliability constraints for the various failure conditions. However, it is not computationally feasible or even appropriate for even a medium sized system to include all possible pipe failures. In order to reduce the computational effort to reasonable levels, only a limited number of the critical failure events are imposed as constraints. These critical events may be identified by ad hoc heuristics, e.g., a failure of the pipe near to the sources is in general more critical than the one further distant from the sources. Engineering judgement and experience of the designer play a very important role in identifying these critical failures (Cullinane et al., 1992). The reliability of a given network configuration may also be approximated by the reliability of the most critical node within the configuration (Morgan and Goulter, 1985). In order to alleviate computational difficulties associated with finding to find the most critical node of the system, the most critical node may also be assumed to be the node which is determined as being most vulnerable by a stochastic hydraulic analysis using a mean values first order second moment (MVFOSM) method (Xu and Goulter, 1996). With such approximations, each reliability constraint can be assessed in a single FORM computational procedure (Xu and Goulter, 1997).

RELIABILITY EVALUATION AND SENSITIVITY ANALYSIS

In this paper, failure of a network is based on the criterion of insufficient head. For a given network operating configuration, the most critical node is first identified by the MVFOSM and its reliability is then evaluated by FORM. The first step of FORM is to construct the limit state function $G(\mathbf{Y})=g(\mathbf{X})=H_L(\mathbf{X})-H_L^{min}=0$, where $\mathbf{X}$ = vector of basic random variables in the original space; $\mathbf{Y}$ = corresponding standard normal variables in the transformed space, i.e., $\mathbf{Y}=(\mathbf{X}-\overline{X})/\sigma$ for independent normal X, where σ and $\overline{X}$ = standard deviation and mean respectively of random variable X; H_L = random nodal head at the critical node L which is an implicit nonlinear function of the basic random variables $\mathbf{X}$; H_L^{min} = corresponding minimum head requirement. The "most probable failure point" (MLFP), i.e., the combination of conditions at which failure is most likely to occur at the critical node, is then determined by constructing iteratively a sequence of points $\mathbf{Y}_1$, $\mathbf{Y}_2$,... such that (Liu and Der Kiureghian, 1991):

$$Y_{k+1} = \left(Y_k^T \mathbf{a}_k + \frac{G(Y_k)}{|\nabla G(Y_k)|} \right) \mathbf{a}_k \qquad (4)$$

where $\nabla G(Y_k)$ = gradient vector of the performance function evaluated at the kth iteration and $\mathbf{a}_k$ = unit vector normal to the performance surface in the direction away from the origin and $\mathbf{a}_k = -\frac{\nabla G(Y_k)}{|\nabla G(Y_k)|}$. The gradient of the performance function can be efficiently computed by the chain rule of differentiation, i.e.,

$$\nabla G(Y) = \frac{\partial G}{\partial Y} = \frac{\partial G}{\partial H_L} \frac{\partial H_L}{\partial X} \frac{\partial X}{\partial Y} = \frac{\partial H_L}{\partial X} J_\sigma^T \qquad (5)$$

where $J_\sigma = (\sigma_1, \sigma_2, \ldots \sigma_k)^T$, and the partial derivatives of nodal heads at the critical node with respect to the changes of basic random variables are computed from the probabilistic hydraulic equation, i.e., $\left[\frac{\partial H_L}{\partial X}\right]^T = -J_L^{-1} J_X$, where J_L^{-1} = Lth row of the inverse of the Jacobian matrix J of system equations; J_X = sensitivity matrix with respect to changes in the basic random variables.

This process usually converges to the MLFP, Y^*, in a few iterations. The left hand side of Eq. 2 is then given by $\Omega = |Y^*|$, where Ω = reliability index for the critical node under the given conditions.

An important feature of FORM is its facility for computing reliability sensitivity measures with respect to any design parameters. It can be shown that the sensitivity of reliability index, Ω, with respect to pipe diameter, **d**, is given by:

$$\frac{\partial \Omega}{\partial \mathbf{d}} = \frac{1}{|\nabla G(Y^*, \mathbf{d})|} \frac{\partial H_L}{\partial \mathbf{d}} \qquad (6)$$

Similarly, the partial derivative of the nodal head with respect to the design variables can be estimated by $\left[\frac{\partial H_L}{\partial \mathbf{d}}\right]^T = - J_L^{-1} J_\mathbf{d}$, where $J_\mathbf{d}$ = sensitivity matrix with respect to the change of pipe sizes evaluated at the MPFP, Y^*.

SOLUTION TECHNIQUE

Since costs of pipe networks are in general nonlinear functions of the decision variables and the reliability index, $\beta_j(X, \mathbf{d})$, is an implicit function of the stochastic nodal demands and pipe capacity, X, and the decision variables, $\mathbf{d}$, the reliability-based optimization formulated above is a nonlinear programming problem. In this paper, the model was solved by integrating the FORM with the GRG2 optimization program (Lasdon and Waren, 1986) as shown in Figure 1. Beginning with an initial design, the FORM routines, which employ a network solver and sensitivity analysis routines to obtain the values of performance function and its gradient, are called to evaluate the reliability index for different operating scenarios. If the reliability requirements are met, the optimality conditions are checked. If the optimal solution has not been achieved or the solution violates the reliability constraints, a sensitivity analysis of the reliability with respect to the design variables is performed and the reduced gradient of objective function evaluated to determine the search direction for a new solution. This process is repeated until all constraints and optimality conditions are met.

NUMERICAL EXAMPLES

The proposed reliability-based optimization model was applied to the example network shown in Figure 2. The pipe cost for this example is given by $CT_i = 0.39 \times L_i \times d_i^{1.51}$, where CT_i = total cost for pipe i ($\$10^6$); L_i = total pipe length (km) and d_i = pipe diameter (m). The minimum pressure at each demand node is specified as 50 m. All nodal demands and the values of pipe Hazen-Williams (HW) coefficients are assumed to be independent normal random variables with expected values equal to 75 m^3/h (except for node 9 which has demand of 225 m^3/h) and 110 respectively. The coefficients of variations for nodal demands and pipe coefficients are 0.3 and 0.1 respectively. All pipe lengths are assumed to be 1.0 km. Reliability constraints are imposed on the normal pipe configuration and degraded configurations corresponding to the failure of one component. This results in a total of 13 reliability constraints for this problem. The reliability requirements for the normal operating situations change from 0.9 to 0.99 and for failed conditions from 0.6 to 0.95 respectively. The solutions for each case using an initial pipe diameter of 40 cm for all pipes are shown in Table 1. As expected, larger pipes are selected to fulfil the reliability requirements as the bounds on reliability increase.

CONCLUSION

This paper has presented a new optimization model for reliability-based design of water distribution networks. The inherent uncertainty in the nodal demands and the values of pipe coefficients as well as the impacts of component failures are able to be considered in the model. The proposed model is very flexible and permits the incorporation of the experience of the designer in identifying the critical failure events which need to be included into the optimization. The computational efficiency for the solution of nonlinear optimization is improved by deriving the first order derivative of the cost and reliability analytically.

REFERENCES

Cullinane, M.J., Lansey, K.E. and Mays, L.W. (1992). "Optimization-availability-based design of water distribution networks." Journal of Hydraulic Engineering, ASCE, 118(3), 420-441.

Goulter, I.C. (1995). "Analytical and simulation models for reliability analysis in water distribution systems." In "Improving efficiency and reliability in water distribution systems" Ed. by E. Cabrera and A.F. Vela, Kluwer Academic Publishers, London, 235-266.

Lasdon, L.S. and Waren, A.D. (1986). "GRG2 User's Guide." Department of General Business and Administration, The University of Texas at Austin, Texas, U.S..

Liu, P. L. and Der Kiureghian, A. (1991). "Optimization algorithms for structural reliability." Structural Safety, 9(3), 161-177.

Mays, L.W. (Ed) (1989). Reliability analysis of water distribution systems. ASCE, New York.

Morgan, D. and Goulter, I. C. (1985). "Optimal urban water distribution design." Water Resources Research, 21(5), 642-652.

Xu, C. and Goulter, I.C. (1996). "Uncertainty analysis of water distribution networks." In "Stochastic Hydraulics'96" Ed. by K.S. Tickle et al., A.A. Balkema, Rotterdam, 609-616.

Xu, C. and Goulter, I. C. (1997) "Reliability assessment of water distribution networks using the first order reliability method." Proceeding of 27th IAHR Congress, San Francisco.

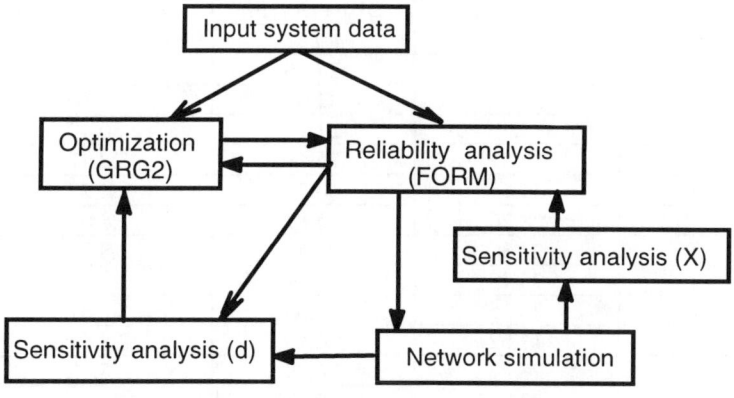

Figure 1 Schematic of optimization solution

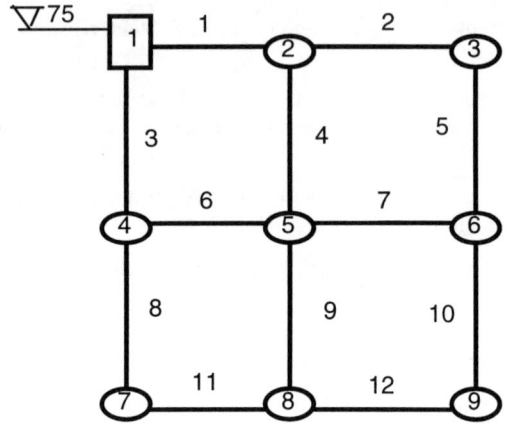

Figure 2 Example network

Table 1 Optimization results

Pipe	Case 1** R_1=0.9 R_2=0.6 Diameter (cm)	Case 2** R_1=0.9 R_2=0.8 Diameter (cm)	Case 3** R_1=0.99 R_2=0.95 Diameter (cm)
1	39.16	40.57	46.20
2	20.07	27.46	28.94
3	39.95	41.05	41.50
4	35.40	31.12	30.48
5	15.82	26.48	28.47
6	34.15	30.75	34.27
7	28.26	25.87	28.03
8	21.37	27.47	28.89
9	25.19	25.72	28.01
10	28.15	26.63	28.81
11	17.52	26.49	28.44
12	25.52	26.29	28.42
Cost ($)	688,919	755,530	835,145
CPU (s)*	47.6	31.6	18.3

NB: R_1: Required reliability under normal operating conditions
R_2: Required reliability under failure of one system component
* On VAX 6000 Mainframe at Central Queensland University

Reliability Assessment of Water Distribution Networks Using the First Order Reliability Method

CHENGCHAO XU AND IAN C. GOULTER
Central Queensland University, Rockhampton Qld. 4702, Australia

INTRODUCTION

In recent years, considerable effort has been devoted to quantitative reliability assessment for water distribution networks. As a result of these efforts, a variety of reliability measures and evaluation algorithms have been developed. Goulter (1995) has recently provided a 'state of the art' review of these reliability measures and associated evaluation techniques with an extensive list of references. Two distinct types of failures termed as "demand variation failure" and "mechanical failure" are identified as needing consideration in the derivation of measures for reliability. Demand variation failures consider situations where the demands imposed on the system exceed the system capacity, whereas mechanical failures consider situations associated with the failures of system components, e.g., main bursts, blockage of valves, loss of pumping stations etc, which reduce the hydraulic capacity of the network and lead to operational failure of water supply. This paper focuses on the hydraulic reliability aspects of demand variation failures. However, the results can be integrated with the mechanical reliability to generate overall reliability measures for water distribution networks.

The key factors contributing to demand variation failure are increases in nodal demands and decreases in pipe capacity. The impacts of the uncertainties arising from the random nature of each of these two factors on the hydraulic performance of water distribution systems are modelled by a probabilistic hydraulic model which uses an algorithm based on the first order reliability method (FORM) to estimate the hydraulic reliability of the water distribution network. The algorithm itself involves repetitive solution of hydraulic equations and a sensitivity analysis to derive the gradient of the performance function used to identify the failure points.

RELIABILITY OF WATER DISTRIBUTION NETWORKS

There is currently no consensus on the definition of reliability for water distribution networks. In this paper, failure of network performance is defined on the basis of insufficient delivery pressure with the assumption that the nodal demands are always met. Accordingly, the nodal reliability is defined as the probability that the nodal demand is met at a pressure greater than or equal to the prescribed minimum pressure head.

Due to the uncertainty in the nodal demands and the values of pipe roughness, the hydraulic head at each individual node is also probabilistic. Let H_i^{min} be the minimum head specified for demand node i. The system performance with respect to the pressure requirement at node i can be formulated as follows:

$$Z_i = H_i - H_i^{min} \qquad (1)$$

where Z_i the performance function for node i and H_i is the random nodal head at node i. Note that H_i is an implicit function of the random nodal demands, pipe roughness and other network parameters, i.e.,

$$\mathbf{F(H,X)} = 0 \qquad (2)$$

where $\mathbf{F}(\bullet)$ is a vector of functions representing the mass balance at each node; $X=(x_1,...,x_K)^T=(X_D, X_C)^T$ is a vector of basic random variables, namely the probabilistic nodal demands, X_D, and the random values of pipe coefficients, X_C; H is a vector of random nodal heads and K is the number of random variables. The performance function in Eq. (1) can be rewritten in terms of the basic random variables X as follows:

$$Z_i = H_i(\mathbf{X}) - H_i^{min} = G_i(\mathbf{X}) \qquad (3)$$

By definition, the failure state is described by $G_i(\mathbf{X})<0$. The hypersurface defined by $G_i(\mathbf{X})=0$, which separates the system performance space into 'failure' and 'safety' region, is referred to as the limit state surface or the failure surface in structural reliability analysis (Hasover and Lind, 1974).

For a prescribed minimum head at node i, the nodal reliability is given by

$$R_i = 1.0 - P_i^F \qquad (4)$$

where P_i^F = probability of supply failure at node i, i.e.,

$$P_i^F = \int_\Omega f_\mathbf{X}(\mathbf{X}) d\mathbf{X} \qquad (5)$$

where $f_\mathbf{X}(\mathbf{X})$ is the joint probability density function of X and Ω is the failure region defined by $G_i(\mathbf{X})<0$. The solution of this multiple integral is, in general, very complicated. One possible way to obtain an approximate solution is to use a Monte-Carlo simulation technique. However Monte Carlo simulation requires a large number of trials to ensure results of reasonable accuracy. Alternatively, the mean value first order second moment method (MVFOSM) (Yen et al., 1986) can be used to compute the first two moments of the random nodal heads, e.g., Xu and Goulter (1996). However, such an approach is only suitable for situations with small variations in the values of the random variables. In situations involving large variations in the values of the random variables, the effect of nonlinearity in the hydraulic model becomes significant and the accuracy of the MVFOSM models deteriorates dramatically. Other approaches include point estimation methods details of which can be found in a recent review paper by Tung (1996). This paper describes a first order reliability method (FORM) approach to determine the nodal hydraulic reliability P_i^F.

FIRST ORDER RELIABILITY METHOD (FORM)

FORM was originally developed for structural reliability analysis (Hasover and Lind, 1974) but has recently been applied to the water resources engineering, e.g., Yen et al. (1986). FORM attempts to reduce the error of the probability estimate arising from the nonlinearity of performance functions by expanding the Taylor series at some point $X^* = (x_1^*, x_2^*, \ldots x_K^*)^T$ on the failure surface defined by $Z_i = G_i(X^*) = 0$ rather than at the mean values, $\overline{X}$, i.e.,

$$Z_i = G_i(\mathbf{X}^*) + \mathbf{A}^T (\mathbf{X} - \mathbf{X}^*) = G_i(\mathbf{X}^*) + \sum_{j=1}^{K} a_j (x_j - x_j^*) \tag{7}$$

where a_j is the partial derivative of G_i with respect to the jth random variable and can obtained from $\mathbf{A}^T = -\mathbf{J}_i^{-1} \mathbf{J}_x$ in which both the ith row of the inverse of the Jacobian matrix, $\mathbf{J}_i^{-1}$, and the sensitivity matrix, $\mathbf{J}_x$, are evaluated at the failure point X^*. Since X^* is on the failure surface, the first term on the right hand of Eq. (7) is equal to zero. For the independent random variables $\mathbf{X}$, the mean value and variance of the performance function can be obtained by:

$$\overline{Z}_i = \sum_{j=1}^{K} a_j (\overline{x}_j - x_j^*) \tag{8}$$

$$\sigma_{Z_i}^2 = \sum_{j=1}^{K} a_j^2 \sigma_{x_j}^2 \tag{9}$$

where σ_x is the standard deviation of random variable x. Assuming that the performance function for each individual node follows a normal distribution, the probability of nodal failure can approximated by:

$$P_i^F = 1.0 - \Phi(\beta) \tag{10}$$

where $\Phi(\bullet)$ is the cumulative distribution function for a standard normal variable and $\beta = \overline{Z}_i / \sigma_{Z_i}$.

The major task in reliability analysis using FORM is to locate the most likely failure point X^* on the failure surface. There are many algorithms available in the literature to determine this most probable failure point, e.g., Yen et al. (1986). For problems with non-linear performance functions such as those considered in this paper, there is no closed form solution and iterative schemes have to be employed. In this paper, the iterative algorithm developed by Rackwitz and Fiessler (1978) incorporating the modification made by Liu and Der Kiureghian (1991) to improve the convergence is implemented. The algorithm starts from a point which is not necessarily on the failure surface and progresses towards the most likely failure point through a series of "improved" intermediate points identified iteratively using the formulae in Eqs. (11)-(13):

$$\alpha_j^n = \frac{a_j^n \sigma_{x_j}}{\left[\sum_{k=1}^{K}\left(a_k^n \sigma_{x_k}\right)^2\right]^{1/2}} \qquad j=1,2,...,K \qquad (11)$$

where the superscript n denotes the current iteration number and a_j^n represents the values a_j at iteration n.

$$\beta_i^{n+1} = \frac{G_i(\mathbf{X}^n) + \sum_{j=1}^{K} a_j^n (\bar{x}_j - x_j^n)}{\sum_{j=1}^{K} \alpha_j^n a_j^n \sigma_{x_j}} \qquad j=1,2,...K \qquad (12)$$

and

$$x_j^{n+1} = \bar{x}_j - \alpha_j^n \beta_i^{n+1} \sigma_{x_j} \qquad j=1,2,...,K \qquad (13)$$

In implementing the iterative algorithms of Eqs. (11)-(13) for water distribution networks, the performance function $G_i(\mathbf{X}^n)$ at each iteration is evaluated by network simulation. The gradient of performance function with respect to the basic random variables is then computed efficiently using the inverse of Jacobian matrix available from the network simulation (Xu and Goulter, 1996). Overall computational efficiency is further enhanced by using a very fast algorithm based on the bifactorization technique, taking into accounts the sparse structure of the Jacobian matrix of water distribution networks, to solve the linearised system during simulation.

In the modified algorithm proposed by Liu and Der Kiureghian (1991), a merit function is used to monitor the convergence process and the new point X^{n+1} is refined at each step and accepted only if it results in a smaller value of merit function.

It should be noted that FORM can be applied to the case in which the random variables are non-normal and dependent on each other by tranforming the non-normal dependent variables into equivalent normal independent variables at the failure point. Further details of this process can be found in Yen et al. (1986).

NUMERICAL EXAMPLES

An example network shown in Figure 1 is used to illustrate the method. Details of network data can be found in Xu and Goulter (1996). All the nodal demands and the values of pipe coefficients are modelled by independent normal random variables. For given levels of uncertainty, e.g., coefficients of variation (CoV) for nodal demands and values of the pipe roughness coefficients set to 0.3 and 0.1 respectively such as used in this case, the proposed procedure can be applied to estimation of the probability of node failure and the corresponding nodal reliability. For example, consider node 12 for which the minimum nodal pressure head is specified as 41.6 metres. The performance function for this node can be formulated as:

$$Z_{12} = H_{12} - 41.6 \tag{14}$$

The iterative scheme in this case converges to $\beta=1.9987$ after 4 iterations, which gives a probability of nodal failure of 0.0229 or equivalently a nodal reliability of 0.9771. The reliabilities at other nodes can be evaluated similarly.

To compare the accuracy of the FORM and assess its improvement over the simple MVFOSM, a Monte-Carlo simulation study was also undertaken to provide the 'benchmark' solution. Figure 2 shows the results of reliability analysis for node 12. For the cases in which the coefficient of demand variation is less 0.3, all three methods give comparable results. However, as the demand variation increases above 0.3, the difference between the FORM, MVFOSM methods and Monte-Carlo simulation gradually increases. A similar trend is observed for increases in the variation of the values of pipe rougness coefficients. Figure 2 also shows that in all cases FORM offers improvements on the accuracy of nodal reliability over the MVFOSM method particularly for cases involving the large variations in the random nodal demands.

CONCLUSION

A framework for hydraulic reliability analysis of water distribution networks based on the first order reliability method (FORM) is presented. The method is capable of taking into consideration the stochastic nature of demands and random values of pipe roughness coefficients. An efficient algorithm using simulation and sensitivity analysis techniques is developed to identify the failure point. It is shown that the proposed method is able to accurately determine the hydraulic reliability for a wide range of nodal demands and pipe roughnesses.

REFERENCES

Goulter, I. (1995). "Analytical and simulation models for reliability analysis in water distribution systems." In "Improving efficiency and reliability in water distribution systems" Ed. by E. Cabrera and A.F. Vela, Kluwer Academic Publishers, London, 235-266.
Hasover, A.M. and Lind, N.C. (1974). "An exact and invariant first order reliability format." Journal of Engineering Mechanics Division, ASCE, 100(EM1), 111-121.
Liu, P. L. and Der Kiureghian, A. (1991). "Optimization algorithms for structural reliability." Structural Safety, 9(3), 161-177.
Rackwitz, R. and Fiessler, B. (1978). "Structural reliability under combined random load sequences." Computer and Structures, 9, 489-494.
Tung, Y.K. (1996). "Uncertainty analysis in water resources engineering." In "Stochastic Hydraulics' 96" Ed. by K.S. Tickle et al., A.A. Balkema, Rotterdam, 29-46.
Yen, B.C., Cheng, S.T. and Melching, C.S. (1986). "First-order reliability analysis." In "Stochastic and risk analysis in hydraulic engineering" Ed. by B.C. Yen, Water Resources Publication, Littleton, Colo., 1-36

Xu, C. and Goulter, I. C. (1996). "Uncertainty analysis of water distribution networks." In "Stochastic Hydraulics' 96" Ed. by K.S. Tickle et al., A.A. Balkema, Rotterdam, 609-616.

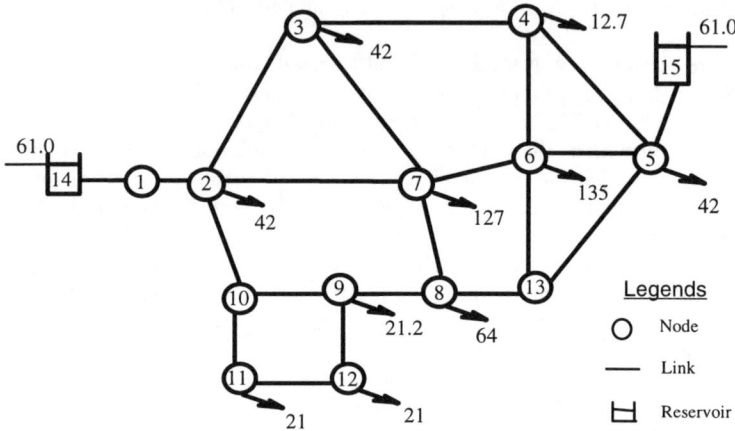

Figure 1 Schematic of example network

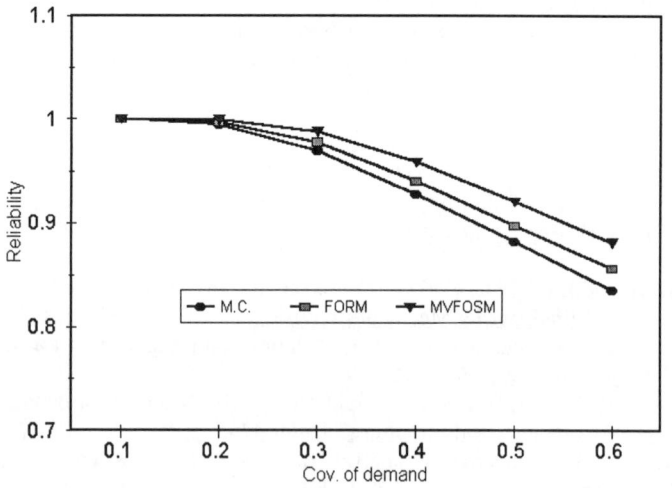

Figure 2   Reliability vs demand variation (node 12)

Parameter Dimension Estimation for
Water Distribution Networks

CHENGCHAO XU and KEVIN S. TICKLE
Central Queensland University, Rockhampton, Qld. 4702 Australia

INTRODUCTION

Mathematical models of water distribution networks require many parameters before they can be used to make predictions of system behaviour under various conditions. Of several types of parameters, the pipe coefficients are the most difficult to obtain since they are not directly measurable. In practice, the pipe coefficients are usually inferred from the measurement data of their dependent variables such as pipe flows and nodal heads taken at some discrete points within a network in a process of model calibration, or inverse modeling. Since, in general, the number of field measurements available is not enough to exactly identify each individual pipe coefficient within a system, it is common practice to group the pipes to reduce the number of effective parameters to be estimated. This is usually achieved by assigning the same value to the pipes in a group based on prior information on characteristics and conditions of pipes and the experience of network modeler. The grouped parameters are then estimated by applying nonlinear regression analysis. Criterion for such an analysis is usually based on the minimization of a norm of the difference between the calculated values from the model and the measurement data. Several algorithms have been developed for parameter estimation of water distribution networks, e.g., Lansey and Basnet (1991).

In these approaches, the measurement data was usually considered as without any uncertainty. However, measurement uncertainty will effect parameter estimation and thus model prediction error. Also, an increase in parameter dimension will usually improve the goodness of fitting in terms of the system modeling error, but will increase the parameter uncertainty and vice versa (Yeh, 1986).

The determination of trade-offs between the parameter dimension and system modelling error in water network calibration has recently been investigated by Malick et al. (1994). In their approach, the first order second moment analysis of uncertainty was used to evaluate the propagation of measurement errors in field data to parameter estimates and model predictions. The covariance matrices for parameter estimates and model predictions are calculated for different combinations of parameter dimensions and grouping. The optimal parameter dimension and the best pipe grouping are then determined based on the relative magnitudes in the modeling errors and the uncertainty in parameter estimates (or model prediction). The calculation of parameter covariance matrix in this approach involves the pseudo inverse of parameter sensitivity matrices.

In the case where the parameter sensitivity matrix is singular or ill-conditioned, the procedure proposed by Malick et al. (1994) may fail to provide a satisfactory solution because the inverse of parameter sensitivity matrix would be unable to be determined. In this paper, an alternative approach based on singular value decomposition (SVD) is proposed to overcome this problem. SVD has been found to be able to provide a useful solution even if the parameter sensitivity matrix is rank deficient, e.g., Menke (1989) and Wasantha Lal (1995). The SVD method provides not only the necessary information to determine the parameter dimension but also the information to calculate the covariance and correlation matrices, which are very useful for further parameter re-grouping.

OUTLINE OF THE METHOD

For a given set of measurement data, the parameter calibration is formulated as a nonlinear regression problem specifically to minimise the sum of squared differences between the computed and observed values. An iterative scheme based on the Newton-Raphson (NR) technique is used in the minimisation procedure. At each NR iteration, network simulations are performed to evaluate the values for the variables corresponding to the measurement data and sensitivity analyses are undertaken to determine sensitivity matrix the parameter estimates (Xu and Goulter, 1996). Once the output errors and parameter sensitivity matrix are known, the required corrections to the parameter estimates needed to minimize the output error can be obtained by solving the following linear system of equations:

$$A \Delta X = e \qquad (1)$$

where ΔX and e are vectors of corrections to the parameter estimates and output errors respectively, i.e. $e_i = H_i - h_i$, where H_i is the observed nodal head and h_i is the calculated value from model simulation; A is m×n parameter sensitivity matrix whose element (i,j) is given by $\partial h_i / \partial x_j$; m and n are the number of measurements and parameters to be estimated respectively. The estimated parameters used for next iteration are updated using the equation:

$$X^{k+1} = X^k + \alpha \Delta X \qquad (2)$$

where k denotes the iteration number and α is a correction factor used to improve the convergence of the NR procedure and is set to 0.6 in this study. The iterative procedure terminates when the corrections to the parameter estimates are within the specified tolerance.

If the parameter sensitivity matrix A is full rank, Eq. (1) can be solved by any variant of Gaussian elimination for an even system and by the pseudo inverse for an overdeterminated system. However, in the case that **A** is ill-conditioned or rank deficient, these methods will fail to provide a satisfactory solution. The SVD method on other hand is able to provide a useful solution for all conditions: under, even, over or mixed determined systems (Menke, 1989). The method factorises a m×n matrix into the product of three matrices:

$$\mathbf{A} = \mathbf{U}\Lambda\mathbf{V}^T \tag{3}$$

where U and V are respectively n×n and m×m matrices of orthogonal eigen vectors that span the data and parameter spaces, i.e., $U^TU=I$ and $VV^T=I$; Λ is an n×n diagonal matrix whose elements are called the singular values. They are usually arranged in order of decreasing size ($\omega_1, \omega_2,...\omega_n \geq 0$).

The SVD reveals many important properties of the matrix **A**. If at least one of the ω_i is equal to zero or close to zero, the sensitivity matrix is considered as singular or ill-conditioned. The number (L) of nonzero singular values is equal to the rank of the parameter sensitivity matrix, i.e., the maximum number of independent parameter groups that can be identified from the measurement data available. In the case of rank deficiency, the model parameters can be regrouped to reduce the dimension to an appropriate level and calibration is repeated for the new parameter groups. Alternatively, the calibration is continued by simply replacing $1/\omega_i$ by zero if ω_i falls below a small cut-off value in the Eqs. (4) and (5). The results so obtained solve a problem which mininizes the sum of squares of modeling errors while calculating L, the maximum number of independent parameter groups (Menke, 1989). The solution to Eq. (1) in this approach can be written as

$$\Delta \mathbf{X} = \mathbf{V}\Lambda^{-1}\left(\mathbf{U}^T\mathbf{e}\right) \tag{4}$$

For uncorrelated measurement data, the covariance of the estimated model parameters is given by (Menke, 1989)

$$Cov(\mathbf{X}) = \Gamma = \mathbf{V}\frac{\sigma^2}{\Lambda^2}\mathbf{V}^T \tag{5}$$

where σ is an m×m diagonal variance matrix of the measurement errors. The elements of diagonal of Γ are the estimates of the uncertainty (variance) in the parameters.

It can be seen from Eqs. (4) and (5) that both the parameter estimation and the covariance of the estimated parameters are very sensitive to the smallest singular value. The choice of the cut-off for the singular value, i.e., determination of the parameter dimension, is therefore crucial. An increase in the cut-off threshold value (i.e. an effective reduction in parameter dimension) will reduce the parameter uncertainty but it may increase the modeling error. The precise cut-off value is usually chosen by a trial and error process in order to balance the modeling error and parameter uncertainty. Plots of the sizes of the singular values against their index numbers can be useful in this process. However, a more appropriate approach is to relate the cut-off value to the condition number of the nonsingular matrix which provides upper bounds for the amplification of the measurement error in the solution of Eq. (1). The condition number of the matrix **A** is defined by $\kappa(\mathbf{A})=\omega_1/\omega_{min}$, i.e., the maximum singular value divided by the minimum singular value. To avoid the undesirable magnifications of measurement error at small cut-off singular values, the cut-off value of $0.001\omega_1$ as suggested by Wasantha Lal (1995), is used in this study.

The correlation among parameters can be obtained by $\rho_{ij} = \Gamma_{ij} / \sqrt{\Gamma_{jj}\Gamma_{ii}}$. If there are pairs of parameters that are highly dependent, the calibration can be simplified by grouping these parameters. The correlation matrix derived from Eq. (5) can thus be used to isolate the group of parameters as demonstrated by Wasantha Lal (1995).

ILLUSTRATIVE EXAMPLE

The example network shown in Figure 1 is used to illustrate the method proposed in this paper for the determination of parameter dimension and parameter estimation of water distribution networks. All network model parameters are known except for the nine pipe coefficients. The measurement data is generated by a network simulation with coefficients of 120 and 100 for pipes 1-5 and 6-9 respectively. A Gaussian noise of N(0, 0.1) is added to the measurements to simulate the measurement error in practice. Calibration starts with all 110 pipe coefficients.

<u>Case 1</u> Twelve nodal head measurements which are taken at nodes 2,3,5 and 7 for three demand conditions (1, 0.7, 1.5) are used to calibrate the nine pipe coefficients. Although measurements are taken at 4 nodes, they are measured at three different conditions. One may think that these 12 measurements are sufficient for calibration of pipe coefficients. This perception is however wrong. Figure 2 shows the singular values against their index numbers, which clearly indicates that the maximum parameter dimension is 4. Since the system equation is rank deficient, the use of other calibration methods involving the solution of the inverse of a set of system equations is bound to fail. However, SVD provides a reasonable solution to this problem as evidenced in Table 1.

<u>Case 2</u> All 6 nodal heads at the demand nodes are measured under normal demand conditions. The system is indeterminate since m (6) is less than n (9). SVD, however, is still very useful in this case by providing the information on the parameter structure which can possibly be identified from the available measurements. The SVD solution for this case indicates the maximum parameter dimension is six and the parameter correlation matrix indicates that the parameters for the pipes with flows converging to the same demand node, e.g., (2,3), (5,6) and (8,9), are highly dependent and should be grouped together. These parameter groups can be used as a starting point to further examine the parameter behaviour. Figure 3 illustrates the variation of modeling error and parameter uncertainty with parameter dimension. The optimal parameter dimension can be determined from the shapes of

these two curves and in this case it is about 4 and calibration results are also given in Table 1.

CONCLUSION

The ill-conditioned problem is commonly encountered in the inverse modeling of water distribution networks. In general, an ill-conditioned problem is extremely hard to notice a priori, even for a small system. This paper has presented a new approach based on the SVD method which is not only able to detect the occurrence of ill-conditioned situations but also provides a useful solution for the problem. It has been shown that SVD is particularly useful in identifying the parameter dimension and grouping in the calibration of water distribution networks.

REFERENCES

Lansey, K. E. and Basnet, C. (1991). Parameter estimation for water distribution networks. Journal of Water Resources Planning and Management, ASCE, 117(1), 126-143.

Malick, K. N., Lansey, K. E. and Tickle, K. S. (1994). Determining optimal parameter dimensions for water distribution network models. In "Water policy and management: solving the problems" Eds D.G. Fontane & H.N. Tuvel, ASCE, New York, 758-761.

Menke, W. (1989). Geophysical data analysis: discrete inverse theory. Academic Press, Inc. New York.

Wasantha Lal, A.M. (1995). Calibration of riverbed roughness. Journal of Hydraulic Engineering, ASCE, 121(9), 664-671.

Xu, C. and Goulter, I.C. (1996). Uncertainty analysis of water distribution network. In 'Stochastic Hydraulics'96" Ed. K.S. Tickle et al., A.A. Balkema, Rotterdam, 609-616.

Yeh, W.-W. G. (1986). Review of parameter identification procedure in groundwater hydrology: the inverse problem. Water Resources Research, 22(2), 95-108.

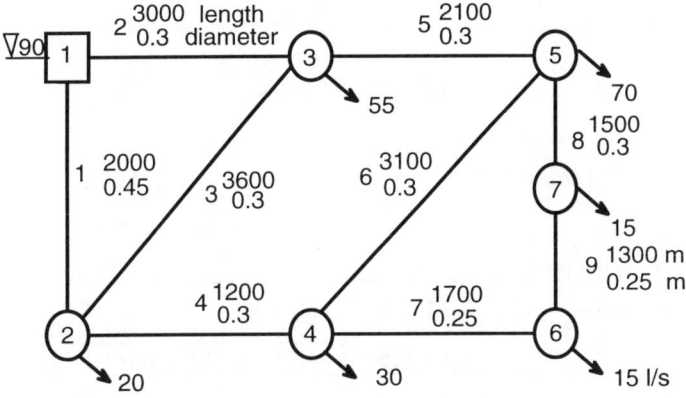

Figure 1 Schematic of example network

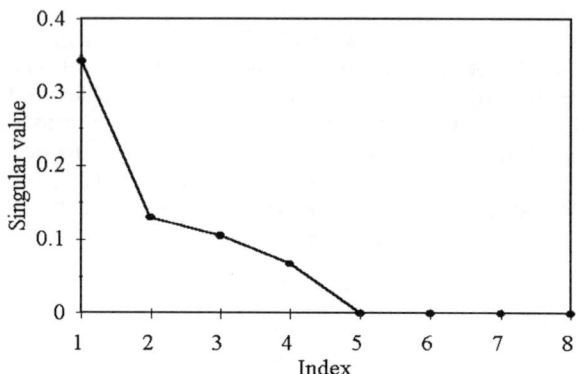

Figure 2 Singular values vs index number

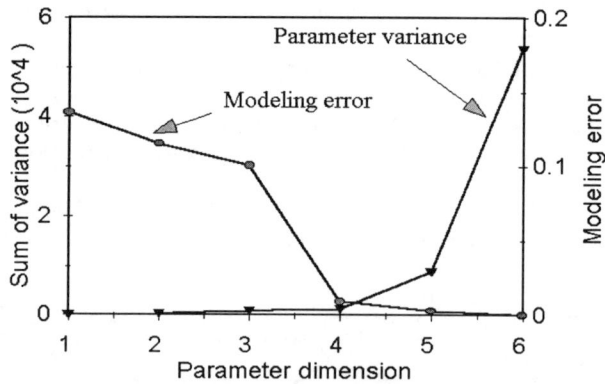

Figure 3 Parameter dimension vs model error and trace of Γ

Table 1 Calibration results

Pipe	1	2	3	4	5	6	7	8	9
Case 1	121.6	117.9	111.9	126.0	109.5	105.3	106.7	109.8	109.6
Case 2	122.2	113.5	113.5	126.9	110.1	110.1	110.1	110.1	110.1

WATER QUALITY MODELING AND RISK ANALYSIS OF THE KEELUNG RIVER, TAIWAN

Jan-Tai Kuo*, Albert Y. Kuo**, and Wen-Shiang Chung*

*Dept. of Civil Engineering and Hydraulic Research Laboratory, National Taiwan University, Taipei 106, Taiwan.
**School of Marine Science/ Virginia Institute of Marine Science, The College of William and Mary, Gloucester Point, Virginia 23062-1346, U.S.A.

ABSTRACT
A time-dependent, laterally averaged, two-dimensional hydrodynamic and water quality model, HEM-2D, was applied to the tidal reach of the Keelung River in northern Taiwan to model the distribution of DO, CBOD, and nitrogen series (organic nitrogen, NO_3-N, and NH_3-N). After calibration and verification, the model was used to investigate the influence of a channel regulation project on the water quality conditions. The model was also used to predict future water quality conditions as the result of the Tanshui River clean-up project which is being implemented. Furthermore, considering the uncertainty of model parameters and other input data, risk analysis techniques, mean value first-order second moment method and Monte Carlo simulation, were employed to calculate the probability that DO, CBOD, and NH_3-N concentrations will exceed river water quality standards.

INTRODUCTION
The Keelung River has the longest tidal excursion and worst pollution condition among the three tributaries of the Tansui River in northern Taiwan. The channel regulation project of the Keelung River (Fig. 1), including a cutoff for two meanders, began in 1991 and was completed in 1993. The channel regulation causes changes of physiographic, hydraulic, and water quality characteristics in the river. The Tansui River clean-up project is currently underway and will increase the percentage of household connections of sewage to sanitary sewers of a separate sewer system. The clean-up project also includes interception systems and wastewater treatment plants which will significantly reduce the pollutant loadings into the river. At the same time, due to the interception of a large amount of wastewater to the ocean outfall system, the fresh water flow in the river will decrease. This could affect the extent of water quality improvement, especially during the low flow condition. The purposes of this study are to investigate the influence of channel regulation on river water quality, and to predict the river water quality in future at various phases of the Tansui River clean-up project.

The water quality modeling study also concerns the uncertainty due to model structure, model parameters, and other input parameters (such as waste loading, boundary conditions, and flow rate). In this study, risk analysis was carried out to estimate the exceedance probability of violation of water quality standards in each river segment.

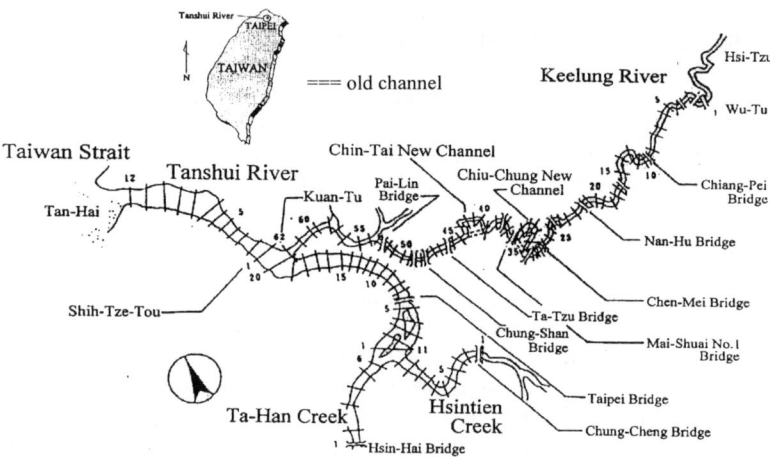

Fig. 1   The Keelung River and the Tanshui River system with it's segmentation

HYDRODYNAMIC AND WATER QUALITY MODELS

The vertical two-dimensional, time-dependent hydrodynamic and water quality model, Hydrodynamic Eutrophication Model-2D (HEM-2D), developed by Park and Kuo(1993) was used for this study. The hydrodynamic model and water duality model are linked internally. The hydrodynamic model generates outputs for flow velocity, surface elevation, and salinity that are used in the water quality model to calculate water quality constituent distributions. The water quality model is based on the mass-balance of a dissolved or suspended substance in the water column:

$$\frac{\partial (CB)}{\partial t} + \frac{\partial (CBu)}{\partial X} + \frac{\partial (CBw)}{\partial Z} = \frac{\partial}{\partial X}(K_X B \frac{\partial C}{\partial X}) + \frac{\partial}{\partial Z}(K_Z B \frac{\partial C}{\partial Z}) + BS_i + BS_e \quad (1)$$

where t = time; u, w = flow velocities in X and Z directions; respectively; C = concentration; B = river width; K_X, K_Z = dispersion coefficients in X and Z directions, respectively; S_i = internal sources and sinks; and S_e = external sources and sinks. Eight water quality constituents, including phytoplankton, organic nitrogen, ammonia nitrogen, nitrite-nitrate nitrogen, organic phosphorus, inorganic(ortho) phosphorus, carbonaceous biochemical oxygen demand (CBOD), and dissolved oxygen (DO) are simulated in HEM-2D. In this study, CBOD ($CBOD_5$), DO, and nitrogen cycle (3 constituents) were chosen to be simulated and investigated.

MODEL CALIBRATION AND VERIFICATION

The upstream boundary for the model is located beyond the limit of tide at Chiang-Pei Bridge (Fig. 1) and downstream boundary is at the river mouth. The tidal reach of the Keelung River was divided into 35 segments with $\Delta X = 1.0$ km. The number of segments decreased to 30 after channel regulation. In the vertical direction, the depth was divided into 6 layers with $\Delta Z = 1.0$ m, except the surface layer which is 1.5m. The field data for this study were taken from 1993 to 1994 (Kuo et al., 1995), including two measurements before channel regulation. Hydrodynamic model calibration and verification, including those for mean tidal range, surface elevation, longitudinal velocity, turbulent mixing, and salinity distribution had been carried out in the first phase of this study (Chen, 1994). The hydrodynamic model parameters were taken from the result of that study.

The water quality model was calibrated by simulating the water quality distribution on June 24, 1994. A spin-up time of 12 tidal cycles for hydrodynamics was needed. Field data collected on June 2, 1994 were used to specify the initial conditions of water quality. The daily varying boundary conditions between June 2, 1994 and June 24, 1994 were obtained from data interpolation and input to the model. The simulated results for the range and mean values of water quality on June 24, 1994 were compared with field data to adjust the kinetic coefficients. Field data on Sep. 9, 1993 and Oct. 3, 1993 before channel regulation were used for water quality model verification (Fig. 2). Field data on May 6, 1994 and May 29, 1994 were selected for model verification after channel regulation. The calibration and verification results are favorable.

UNCERTAINTY AND RISK ANALYSES

In this study, the Mean value First-Order Second Moment method (MFOSM) and Monte Carlo simulation(MC) were chosen to investigate uncertainty of model prediction and risk of river water quality condition in violation of water quality standards.

In MFOSM, for a performance variable Z, which is a function of input variables $X_1, X_2, X_3 \cdots, X_n$, that is, $Z = G(X_1, X_2, X_3 \cdots, X_n) = G(\mathbf{X})$, using Taylor series expansion about the mean values of n input random variables, and neglecting higher order terms, Z can be expressed as:

$$Z \cong G(\overline{\mathbf{X}}) + \sum_{i=1}^{n} \left[\frac{\partial G}{\partial X_i} \right]_{\mathbf{X}=\overline{\mathbf{X}}} \cdot (X_i - \overline{X}_i) \qquad (2)$$

If X_i are uncorrelated, the mean and variance of Z, E(Z) and Var(Z), can be linearized as equations (3) and (4) (Yen et al., 1986):

$$E(Z) = G(\overline{X}_1, \overline{X}_2, \overline{X}_3 \cdots, \overline{X}_n) = G(\overline{\mathbf{X}}) \qquad (3)$$

$$\text{Var}(Z) = \sum_{i=1}^{n} \left(\frac{\partial G}{\partial X_i}\right)^2_{X=\overline{X}_i} \cdot Var(X_i) \qquad (4)$$

Where $(\partial G / \partial X_i)_{X=\overline{X}_i}$ is the sensitivity coefficient, and $(\partial G / \partial X_i)^2_{X=\overline{X}_i} \cdot Var(X_i)$ can be given in the form of :

$$S_i = \left(\frac{\Delta Z}{\Delta X_i}\right)^2 \cdot Var(X_i) \qquad (5)$$

where S_i is the contribution to the overall variance of Z from input variable X_i.

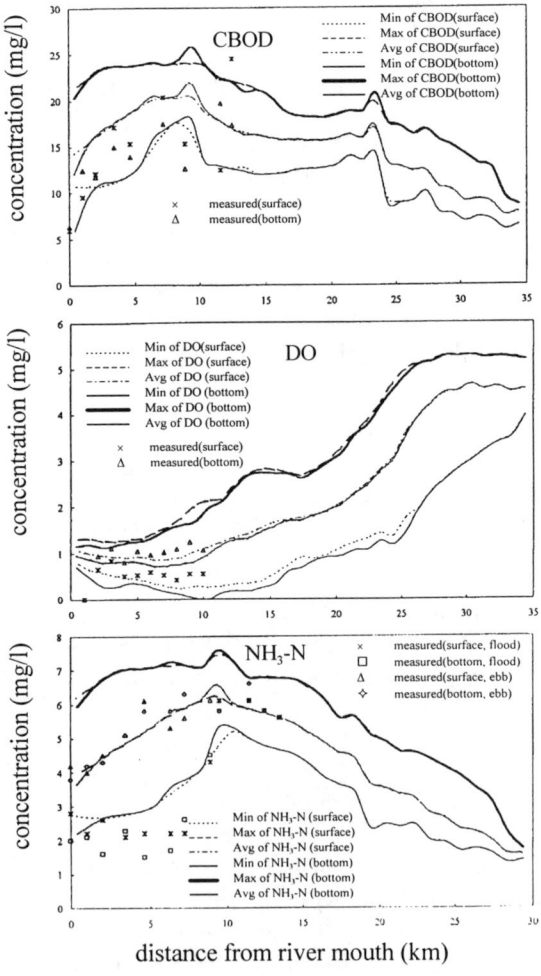

Fig. 2   Model verification for October 3, 1993

WATER QUALITY MODELING AND RISK ANALYSIS 445

MODEL APPLICATIONS

INFLUENCE OF CHANNEL REGULATION

Comparing with the old river channel, the channel length is shortened by about 5 km as a result of the channel regulation project. The simulation results for the Keelung River after meander cutoff include DO, CBOD, and NH_3-N for daily maximum, average, and minimum concentrations under the condition of Q_{75} (Q_{75} is the flow that is equaled or exceeded 75% of time). The results show that in general DO concentrations increase after meander cutoff, especially in the section of new channel and upstream of it, but the change of daily minimum DO concentration is insignificant. However, CBOD concentration changes significantly around the new channel; it increases in some sections and decreases in some other sections. The trend for the change of NH_3-N concentration is similar.

PREDICTION ON FUTURE WATER QUALITY

The Tansui River clean-up project will decrease waste loadings year by year due to the construction of sewage interception systems, as well as collection and treatment systems. The project includes building of the large Pa-Li Primary Treatment Plant with ocean outfall and some other smaller secondary treatment plants. At the same time, the lateral freshwater inflows into the Keelung River will decrease significantly and affect river water quality during low flow. This study considers the conditions for years 1993, 2001, and 2011 with progress of the clean-up project. Simulation results show that although river water quality will improve with time, the daily minimum DO concentration in the river will not meet the water quality standard (DO>1 ppm) in all segments under a low freshwater flow condition.

RISK ANALYSIS

The exceedence probability of water quality constituents in violation of river water quality standard for segments of interest were calculated. First, it is necessary to simulate the variance of all input parameters such as diffusion coefficient, Manning coefficient, CBOD decay rate, nitrification rate, sediment oxygen demand, waste loading, boundary condition, Q_{75}, and temperature. The calibrated parameters are used as estimated mean values. The exceedance probability of DO, CBOD, and NH_3-N concentrations in year 2011 under Q_{75} in reaches of interest can be shown (the case of DO is shown in Fig. 3) with a comparison between MFOSM and MC. The results indicate that the mean estimates from the two methods are close to each other. For the case of DO, mean values of DO concentrations are higher by MFOSM, except in a few segments. The results also show that values of coefficient of variation from MC are lower than those from MFOSM. From calculated results of exceedance probability of constituent concentrations in violation of river water quality standards, it can be seen that in most reaches water quality meets standards(DO >1 ppm, CBOD <6 ppm, NH_3-N <5 ppm). With these standards, the results also show high risk of not meeting standards for daily minimum DO and daily maximum CBOD and NH_3-N concentrations. The estimate of mean values of model parameters for both MFOSM and MC is the important input variable in determining the results of risk analysis.

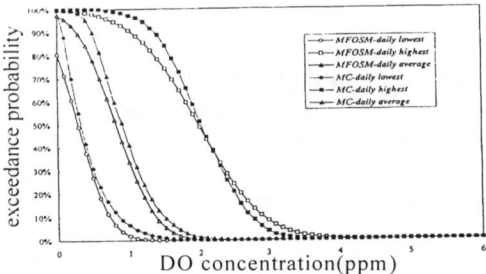

Fig. 3 Exceedance probability of various DO concentration at Pai-Lin Bridge

CONCLUSIONS

The HEM-2D model was employed for a water quality study and risk analyses in the Keelung River. Favorable model calibration and verification results were obtained. Modeling results show that DO concentration increases after channel regulation, but the change of daily minimum DO concentration is insignificant. CBOD and NH_3-N concentrations change significantly around the new channel. For the simulation and prediction of water quality at three stages of the clean-up project(years 1993, 2001, and 2011), DO concentrations will increase significantly with time, and both CBOD and NH_3-N concentrations will decrease significantly. The mean value first-order second moment method and Monte Carlo simulation were adopted for risk analysis on water quality in the Keelung River. Modeling results show high risk for violation of river water quality standards for daily minimum DO and daily maximum CBOD and NH_3-N concentrations. It also shows that under Q_{75} low flow condition, water quality is strongly influenced by the upstream freshwater discharge.

ACKNOWLEDGEMENT

This study was supported by the Taipei Municipal Government. We would like to express our thanks to Prof. S.L. Lo of the Graduate Institute of Environmental Engineering, National Taiwan University for his contributions to the field survey and suggestions to this study.

REFERENCES

Chen, C.W. (1994). *Simulation on the Influence of Salinity Distribution in Keelung River after Channel Regulation*. Master Thesis, Graduate Institute of Environmental Eng., National Taiwan University, Taipei, Taiwan (in Chinese).

Kuo, J.T. and S.L. Lo and A.Y. Kuo(1995). *A Study on the Influence of Channel Regulation and Monitoring System in Keelung River, vol. III : The Influence of Channel Regulation on Water Quality in Keelung River*. Hydraulic Research Laboratory, National Taiwan University, Technical Report, Taipei, Taiwan (in Chinese).

Park, K. and A.Y. Kuo (1993). *A Vertical Two-Dimensional Model of Estuarine Hydrodynamics and Water Quality*. Spec. Rep. Appl. Mar. Sci. and Ocean Eng. No.321, Virginia Inst. Mar. Sci, Glouceser Point, VA.

Yen, B.C., S.T. Cheng, and C.S. Melching (1986). First Order Reliability Analysis, in: *Stochastic and Risk Analysis in Hydraulic Engineering*. B.C. Yen (ed.), Water Resource Publications, Littleton, Colorado. pp. 1-36.

Reliability Analysis for Water Quality Management in the Han River

Kun-Yeun Han*, Sang-Ho Kim**
* Professor of Civil Eng., Kyungpook National Univ., Daegu 702-701, KOREA
** Research Assistant, Kyungpook National Univ., Daegu 702-701, KOREA

ABSTRACT

A reliability analysis for water quality management in a river is studied using the AFOSM and MFOSM methods. The variations of discharge and water quality in the headwater, tributaries, and reaction coefficients are considered. The computed results obtained with the AFOSM and MFOSM methods are compared with those from the Monte Carlo method in the QUAL2EU model in terms of mean, standard deviation, and coefficient of variation. The results of the AFOSM and MFOSM methods agree well with those of the Monte Carlo method. Reliabilities of not violating existing water quality standards at several locations in the Han River are computed and presented using AFOSM, MFOSM, and Monte Carlo.

INTRODUCTION

As the country has rapidly grown and industrialized, the Han River has served as the principal source of water supply in nearby cities. In this study, stochastic water quality analysis as well as deterministic analysis are considered for the basin-wide water quality management in the country. In water-quality modeling, modelers should take into account the errors included in the model parameter. They should not account only for deterministic conditions of the model parameters, since any variation in these parameters may cause large errors. Deterministic models, being no more than simplified descriptions of the natural system, have some errors associated with them. Much research has been conducted to quantify uncertainties in model projections.

METHODOLOGY

In this study, a stochastic water quality model is developed considering the variability in the model parameters. These parameters are considered as variables represented by probability distributions instead of constant values. The AFOSM (Advanced First-Order Second-Moment) and MFOSM (Mean Value First-Order Second-Moment) methods are used in this analysis. The QUAL2EU model is modified and expanded to include AFOSM and MFOSM algorithms. Six subroutines are modified and eight subroutines are added to QUAL2EU.

AFOSM METHOD

The basic concept of this method is to linearize the performance function at a failure point, x_i^*. The first order expansion of the performance variable for uncorrelated variables at a failure point is

$$Z = g(x_i^*) + \sum_{i=1}^{m} (X_i - x_i^*)\frac{\partial g}{\partial x_i} \tag{1}$$

The expected value and standard deviation of Z are

$$E[Z] = \sum_{i=1}^{m} C_i (\overline{X_i} - x_i^*) \tag{2}$$

$$\sigma_Z = \sum_{i=1}^{m} a_i C_i \sigma_i \tag{3}$$

where $C_i = \frac{\partial g}{\partial X_i}$ (4)

$$a_i = \frac{C_j \sigma_j}{[\sum_{j=1}^{m}(C_j \sigma_j)^2]^{1/2}} \tag{5}$$

and $g(x)$ is a functional representation of the performance function, Z. If all uncertain variables are normally distributed are suitably transformed to a normal distribution, the risk P_f can be estimated by

$$P_f = 1 - \psi(\beta) = 1 - \psi(\frac{E(z)}{\sigma_z}) \tag{6}$$

where β is the reliability index, $\psi(\beta)$ is the standard normal CDF evaluated at β. The Rackwitz algorithm is used to find x^* and β.

MFOSM METHOD

In this method, the performance variable can be obtained by Taylor series expansion about the mean values of the uncertain variables.

$$Z = g(\overline{x_i}) + \sum_{i=1}^{m} (X_i - \overline{x_i})\frac{\partial g}{\partial x_i} \tag{7}$$

The expected value and the standard deviation for statistically independent variables are

$$E[z] = \bar{z} = g(\bar{x}_i) \tag{8}$$

$$\sigma_z = [\sum_{i=1}^{m}(C_i \sigma_i)^2]^{1/2} \tag{9}$$

APPLICATION TO THE HAN RIVER

DETERMINISTIC ANALYSIS

The QUAL2E model is applied to assess the water quality variations in reach of the Han River shown in Fig. 1. The analysis is performed in the reach from Paldang Dam to Hangang Bridge considering the effect of the Jamsil submerged weir. A sensitivity analysis is made to determine the significant reaction coefficients and the optimization technique by BFGS (Broyden-Fletcher-Goldfarb-Shanno) method is used to estimate the significant reaction coefficients.
The calibration and verification are performed by using observed water quality data. Fig. 2 shows the calibration results. The calculated pollutant concentrations for BOD, DO, T-N, and T-P show good agreement with the observed values.

RELIABILITY ANALYSIS
The suggested reliability model is applied to the same reach as in the deterministic study. Variations of discharge and water quality in the headwater, tributaries, and reaction coefficients are considered. Table 1 shows the uncertain parameters used in this study. These parameters are assumed to be normally distributed, and uncorrelated. The results computed using AFOSM and MFOSM are compared with the Monte Carlo method in the QUAL2EU model in terms of mean, standard deviation, and coefficient of variation. 2000 iterations are used in the Monte Carlo simulation. The AFOSM, MFOSM, and Monte Carlo results are also compared as probability distributions of constituent concentrations for specified locations. Table 2 shows the simulation results of this study. The computed values by the AFOSM and MFOSM methods are in good agreement with those of the Monte Carlo method. Risk analysis is performed to evaluate the probability of not violating existing water quality standards at some locations in the Han River in terms of BOD and DO using AFOSM, MFOSM, and Monte Carlo.

Figure 3 shows the result of reliability analysis with respect to DO and BOD downstream of Jamsil Weir. From the decision making point of view, the performance of AFOSM, MFOSM, and Monte Carlo seem to be equivalent, however, AFOSM and Monte Carlo seem to be adequate for more sophisticated

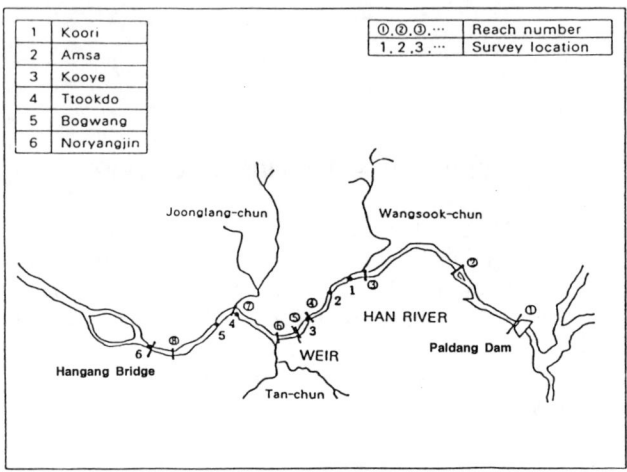

Fig. 1 Location map of the Han River

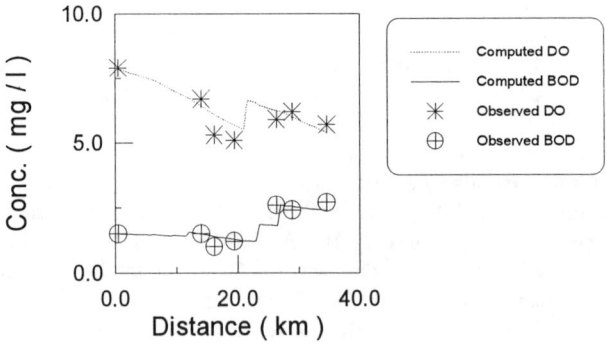

Fig. 2 Calibration results of the deterministic analysis

Table 1 Estimated uncertainties of input parameters

No. of Uncertain Parameter	Group Name	Component	Distribution Type	C.O.V. (%)
1	RXNC	BOD Decay Rate	Normal	20.00
2	RXNC	BOD Settling Rate	Normal	20.00
3	RXNC	SOD Rate	Normal	20.00
4	FFHW	Headwater Flow	Normal	5.00
5	FFHW	HWTR-DO	Normal	10.00
6	FFHW	HWTR-BOD	Normal	20.00
7	FFPL	Point Load Flow	Normal	5.00
8	FFPL	PTLD-DO	Normal	10.00
9	FFPL	PTLD-BOD	Normal	20.00

simulations. Stochastic models using AFOSM, MFOSM, and Monte Carlo can make a great contribution for basin-wide water quality management in Korea.

CONCLUSION

A reliability model for the prediction of water quality variations in a river is studied. The AFOSM and MFOSM algorithms are applied to the QUAL2EU model. The results computed using AFOSM and MFOSM are compared with those from the Monte Carlo method in the QUAL2EU model in terms of mean, standard deviation, and coefficient of variation. The AFOSM, MFOSM, and Monte Carlo methods are also compared for estimated constituent concentration exceedance probability distributions for specified locations. The results of AFOSM and MFOSM methods are in good agreement with those of the Monte Carlo method.

Risk analysis was presented using AFOSM, MFOSM, and Monte Carlo to compute the probability of not violating existing water quality standards at some locations in the Han River in terms of BOD and DO.

REFERENCES

Ang, A.H. and Tang, W.H. (1984). *Probability concepts in engineering planning and design, Volume II: Decision, Risk and Reliability,* John Wiley & Sons.
Han, Kun-Yeun et al. (1993). "Water quality management using deterministic and stochastic models in the Nakdong River." paper no. D-7-1, *Proc. of IAHR,* pp. 203-210, Tokyo, Japan.
Mays, L.W. and Tung, Y.K. (1992). *Hydrosystems engineering and management,* McGraw-Hill.
Rackwitz, R. and Fiessler, B. (1977). "Non-normal distributions in structural reliability." *SFB 96, Report 29,* Technical University of Munich, pp. 1-22.

Table 2 Simulation results of AFOSM, MFOSM methods downstream of Jamsil Weir

Nonexceedance Probability of DO (%)				Exceedance Probability of BOD (%)			
Conc.(mg/l)	AFOSM	MFOSM	Monte Carlo	Conc.(mg/l)	AFOSM	MFOSM	Monte Carlo
5.60	0.06	0.06	0.10	0.60	99.84	99.79	99.75
5.80	0.39	0.39	0.60	0.70	99.29	99.16	99.15
6.00	1.80	1.80	2.15	0.80	97.53	97.28	96.75
6.20	6.23	6.22	6.80	0.90	93.06	92.72	91.75
6.40	16.50	16.47	17.50	1.00	84.13	83.84	83.00
6.60	33.98	33.93	34.95	1.10	69.99	69.85	68.80
6.80	55.88	55.84	55.90	1.20	52.11	52.10	52.20
7.00	76.05	76.06	73.70	1.30	33.99	33.90	35.85
7.20	92.03	89.78	87.45	1.40	19.16	18.87	20.60
7.40	97.54	96.64	94.30	1.50	9.26	8.85	11.45
7.60	99.15	99.16	98.65	1.60	3.83	3.45	5.45
7.80	99.90	99.84	99.65	1.70	1.35	1.11	2.30
8.00	99.98	99.98	99.75	1.80	0.41	0.30	0.80

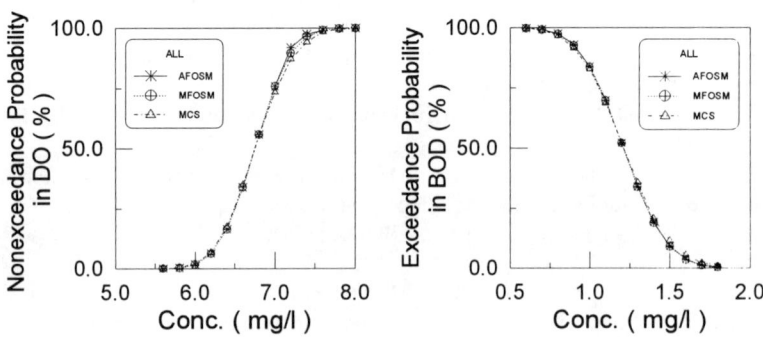

Fig. 3 Comparison of computed probabilities of DO and BOD downstream of Jamsil Weir

UNCERTAINTY IN THE DESIGN
OF STREAM CHANNEL RESTORATIONS

PEGGY A. JOHNSON
Department of Civil Engineering, Pennsylvania State University
University Park, PA 16802

MASSIMO RINALDI
Department of Civil Engineering, University of Florence
Florence, Italy

ABSTRACT

Stream channel restorations are generally undertaken for a variety of reasons, including to improve ecological habitat, recreate natural alluvial features, restore the environmental value of engineered artificial waterways or degraded streams, alleviate flooding, and stabilize unstable channels, particularly near highways and highway structures. Uncertainty in the restoration design can result in an unstable channel that is subject to erosion and/or depostion and which can threaten the safety of bridges, culverts, and roadways, as well as creating a poor aquatic habitat.

INTRODUCTION

In the 1960's and 1970's, many stream channels were modified for flood control by straightening, widening, and clearing. In the years that followed, channels in many parts of the country were directly and indirectly modified as a result of urbanization. As a result, wetlands were drained, aquatic habitats were destroyed, and significant erosional problems developed. Erosion of stream channels in the vicinity of structures, such as bridges and highways can decrease the safety of those structures.

The response to these problems has been to restore stream channels to a more natural state and in doing so improve the aquatic habitat, the riparian habitat, and the stability of the stream. Stream channel restoration covers a wide variety of modifications including relocating channels, decreasing the channel slope using weirs, changing the planform of the stream by creating meanders, stabilizing banks, providing roughness in the bed and banks for ecological diversity, and changing the cross-sectional geometry to improve the stability.

Guidance for many of these restorations is minimal at this time. In addition, definitions of channel stability and the evaluation of other parameters involved in restoration design are vague and sometimes subjectively determined. Uncertainty in the resulting restoration design can result in an unstable channel that is subject to erosion and/or depostion and which can threaten the safety of bridges, culverts, and roadways, as well as creating a poor aquatic habitat.

CHANNEL RESTORATION DESIGN

There are a variety of objectives and procedures for restoring a stream channel to a more natural state. Design procedures to rehabilitate or restabilize unstable streams require a cross sectional geometry and a planform, typically based on the bankfull depth or discharge. The stream to be restored may be reconverted exactly to its original course prior the channelization, when it is known, or to a channel form similar to the original or to adjacent reaches or streams in the fluvial system.

UNCERTAINTY IN RESTORATION DESIGN

Uncertainty in the equations and design techniques that lead to a stream channel restoration design include random variability in the parameters, uncertainty in the models used to estimate design parameters, and uncertainty in evaluating parameters in the field.

BANKFULL ELEVATION

Probably the most important and elusive parameter in stream channel restoration is bankfull elevation. Gordon et al. (1992) describe four examples for which bankfull elevation is not well defined: (1) where the bank tops are not at the same elevation, (2) where stream reaches are unstable, (3) where an obvious break between the stream banks and the floodplain does not exist, and (4) where complexities such as benches and terraces exist. In these cases, the interpretation of bankfull becomes observer dependent. Since bankfull discharge is dependent on bankfull depth, differences in bankfull depth can result in significant differences in estimations of bankfull discharges. The variability in bankfull discharge resulting from the wide variety of definitions can be quantified in terms of the coefficient of variation Ω. Coefficients of variation for data collected by Williams (1978) using 16 methods of estimating bankfull ranged from 0.20 to 1.84 (Johnson and Heil, 1996).

MEANDER DIMENSIONS

The meander wavelength, amplitude, and radius of curvature for a restoration design are often estimated from the Leopold and Wolman equations (1957; 1960). Rinaldi and Johnson (1996) examined the uncertainty in these equations in terms of bias for a set of 18 streams in central Maryland. They measured the meander wavelength, amplitude, and radius of curvature in the field and compared the measured parameters to those calculated from the Leopold and Wolman equations. They found that the average biases (the ratio of the predicted parameter to observed parameter) for the 18 sites were 2.7, 2.2, and 2.5 for the meander wavelength, amplitude, and radius of curvature, respectively, with coefficients of variation of 0.25, 0.54, and 0.37, respectively. The study demonstrated the problems with using the Leopold and Wolman equations for small eastern streams in the design of channel restorations. There were essentially two types of problems: (1) using data out of the range of the data used to calibrate the equations and (2) using dissimilar data in that the meanders in small eastern streams are typically not freely meandering, with meander characteristics often controlled by vegetation.

BANKFULL DISCHARGE

The bankfull discharge is often estimated from the 1.5-year discharge. This is a reasonable estimate, but can differ significantly from the actual bankfull discharge. The 1.5-year discharge can be estimated in several ways; however, streams that are to be restored are often relatively small, ungaged streams, so Manning's equation is often used.

Uncertainty analyses of Manning's equation have been conducted (Ang and Tang, 1984; Mays and Tung 1992). The results showed that the uncertainty in the resulting discharge is primarily a result of the uncertainty in the roughness coefficient n. The roughness coefficient is typically determined by comparing the stream channel to pictures or tables. It can also be calculated from Manning's equations if all other parameters are known. In a study conducted by the U.S. Army Corps of Engineers (Hydraulic Engineering Center 1986), a group of 77 experienced engineers were asked to estimate the value of Manning's n for 10 stream reaches for the 100-year storm event. Each participant was shown pictures of each stream reach. The stream reaches were highly varied and included channels from the southwest, east, and midwest, as well as concrete channels. They could base their estimate on experience and tables and pictures. The average n ranged from 0.02 for the concrete channel to 0.062 for the southwestern channel. The coefficients of variation ranged from 0.20 to 0.35, with the higher coefficients of variation generally corresponding to the higher average n; however, this correlation is not necessarily a very strong one. The data were tested for lognormal, uniform, and normal distributions, and found to be lognormally distributed. No test statistics were provided.

A first-order analysis of Manning's equation was conducted by Mays and Tung (1992) to determine the uncertainty in flow rate computed from Manning's equation. Cross-sectional area and wetted perimeter were assumed to be deterministic variables. Uncertainties in Manning's roughness and channel slope were assumed to be 0.2 and 0.3, respectively, but provided no indication of how those values were determined. In another example, the coefficients of variation were given as 0.053 and 0.068, respectively, again with no explanation of these values.

SLOPE
The slope of the designed restoration project is determined from a map according to the layout of the meander pattern. Johnson (1996) conducted a simple experiment in which a group of 18 students were asked to determine the channel slope on a five-foot interval contour map. The resulting coefficient of variation was surprisingly high, at 0.25. This is approximately the same variation as expected in Manning's n; therefore, even though this was a relatively small sample, the brief experiment shows that the uncertainty in estimating channel slope from a topographic map is significant and should probably not be omitted from an uncertainty or reliability analysis. The uncertainty in estimating the slope may be somewhat greater than that found from the map due to an inability to determine the precise location of a thalweg on a map of any scale.

CROSS SECTIONAL SHAPE
The dimensions of the channel cross section are based on the bankfull depth, the geometry selected (e.g., triangular or trapezoidal), and the bankfull discharge. Since Manning's equation is typically used to determine the bankfull channel dimensions, the same uncertainty in determining n and S are encountered as discussed above. However, in this case, the uncertainty in the bankfull discharge is also included.

EXAMPLE

The Hammond Branch in central Maryland is a small stream that had been straightened in several reaches to accommodate the construction of highways in the area. The result has been to destabilize the stream and cause excessive down cutting in some areas. The mean bankfull discharge for this stream, calculated from Manning's equation, is $Q_b = 5.16$ m^3/s.

The uncertainty in the restoration design for Hammond Branch consists of uncertainty in the estimate of the bankfull discharge, the meander characteristics, and the bankfull channel width. A first-order analysis can be used to assess the uncertainty in this estimate; however, the cross-sectional area and perimeter are correlated, so that a simple first-order analysis would be inappropriate. The correlation between the area and perimeter is assumed to be $\rho = 0.6$ and the correlations between the other parameters in the equation are assumed to be insignificant.

The coefficients of variation for n, S, A, and P in Manning's equation are assumed to be 0.15, 0.25, 0.4, and 0.4, respectively, based on findings by Johnson (1996), Hydraulic Engineering Center (1986), Williams (1978), and Johnson and Heil (1996). The uncertainty in the bankfull discharge using a first-order analysis yielded $\Omega_Q = 0.79$.

The uncertainty in the meander characteristics can be assessed based on the study by Rinaldi and Johnson (1996). The Leopold and Wolman equations for meander wavelength and amplitude can be multiplied by a correction factor N that adjusts the equations for generally overpredicting. For the wavelength, $N_\lambda = 1/2.67$ and for the amplitude, $N_a = 1/2.22$. Based on field surveys (Rinaldi and Johnson 1996), the coefficients of variation for the Leopold and Wolman equations were found to be 0.25 and 0.54 for wavelength and amplitude, respectively. Therefore, the uncertainty in λ and a are equal to the coefficients of variation, assuming that the uncertainty in the design bankfull width W_b is not accounted for.

The uncertainty in W_b must be accounted for separately since the process of obtaining this parameter is dependent on S which, in turn, is dependent on λ and a. From Manning's, W_b is a function of Q_b, S, y_b, and n. Using a first-order analysis and assuming that the correlations between the remaining variables are not significant, the uncertainty in W_b is:

$$\Omega_{W_b} = [(0.79)^2 + (0.15)^2 + (0.25)^2]^{1/2} = 0.84$$

REDUCING UNCERTAINTY IN RESTORATION DESIGN

The uncertainty in the example restoration design may result in failure of all or portions of the restored channel. The uncertainty can be reduced by adding structures, such as riprap walls and weirs, to the bed and banks to stabilize them. It is also possible to reduce the uncertainty by addressing the sediment transport requirements of the stream. A simple method to establish the sediment transport characteristics indirectly is to determine the stream power or shear stress in excess of the stream power or shear stress required to move the sediment on the channel bed. If there is significant excess shear stress at flow depths within the banks, then the flow through the channel will likely erode the bed and possibly the banks.

Another way to decrease the uncertainty in the design of the restoration project is to use meander equations developed specifically for the region in which the stream lies. A large portion of the uncertainty in design resulted from using equations that are based on large, freely meandering streams. Although it is time-consuming to conduct the field surveys that are required to establish such equations, the result will be less time and money spent on maintaining the channel.

REFERENCES

Ang, A.H., and Tang, W.H., *Probability Concepts in Engineering Plainning and Design, Volume Ii - Decision, Risk, and Reliability*, John Wiley and Sons, Inc., New York, NY, 562 pp., 1984.

Gordon, N.D., McMahon, T.A., and Finlayson, B.L., *Stream Hydrology, an Introduction for Ecologists*, Wiley and Sons, New York, NY, 526 pp., 1992.

Hey, R.D., European progress on river renaturalization, *Proceedings, North American Water and Environment Congress, ASCE*, Anaheim, California, June, 1996.

Hydraulic Engineering Center, *Accuracy of Computed Water Surface Profiles*, U.S. Army Corps of Engineers, Davis, California, 1986.

Johnson, P.A., 1996. Uncertainty of hydraulic parameters. *Journal of Hydraulic Engineering*, ASCE, 122(2), 112-115, 1996.

Johnson, P.A., and Heil, T.M., Uncertainty in estimating bankfull conditions, *Water Resources Bulletin* (in press).

Leopold, L.B., and Wolman, M.G., River channel patterns: braided, menadering and straight, *U.S. Geological Survey, Professional Paper* 282-B, 39-85, 1957.

Leopold, L.B., and Wolman, M.G., River meanders, *Bulletin of the Geological Society of America*, 71, 769-794, 1960.

Mays, L.W., and Tung, Y.K., *Hydrosysterms Engineering and Management*, McGraw-Hill, Inc., 530 p., 1992.

Rinaldi, M., and Johnson, P.A., Characterization of stream meanders for stream restoration, *Journal of Hydraulic Engineering*, ASCE, in press.

Williams, G.P., Bankfull discharge of rivers, *Water Resources Research*, 14(6), 1141-1154, 1978.

Risk-Based Design of Roadway Crossing Structures Considering Intangible Factors

YEOU-KOUNG TUNG
Department of Civil & Structural Engineering
Hong Kong University of Science & Technology
Clear Water Bay, Kowloon, Hong Kong

A. MAINARD WACKER
Wyoming Highway Department
Cheyenne, Wyoming USA

VICTOR HASFURTHER
Department of Civil Engineering
University of Wyoming
Laramie, Wyoming

INTRODUCTION

The purpose of a highway is to serve the public. The commonly used design parameter is the design flood frequency which is arbitrary and is based on practices that have "evolved" over the years. During the late 1970's and early 1980's, the concept of risk-based (least total expected cost, LTEC) analysis was developed for determining the design flood frequency for various hydraulic structures. The risk-based analysis yields a structural design associated with the minimum total amount of structural installation cost and expected future flood related damage cost. The corresponding design frequency is herein called the LTEC design frequency. However, there exist many intangible factors that might affect the final adoption of a design flood frequency. In this pa-per, two types of design frequencies, namely, the LTEC and extended-LTEC design frequencies, are studied. The LTEC design frequency is the conventional, solely economically-based design frequency using tangible factors whereas the extended-LTEC design frequency considers other intangible factors in the determination of flood design frequency for highway drainage structures.

In this study, the risk-based design is a conventional one which considers the

ROADWAY CROSSING STRUCTURES DESIGN   459

inherent hydrologic uncertainty in calculating the expected economic losses. The de-sign frequency is a decision variable rather than a pre-selected design parameter as in the traditional design procedure. The flood related damages are primarily consisted of damages to buildings, crops, embankment and pavement, and traffic-related damage. In the analysis, the annual expected flood damage cost should be the incremental damage as the result of the presence of the roadway crossing structures.

The concept of risk-based design has been recognized for many years. Since the pioneering work of Pritchett (1964), several studies on the evaluation of economic risks and optimal design were performed for box culverts, pipe culverts, and bridges (e.g., Young et al., 1974; Tseng et al., 1975; Corry et al., 1980; Schneider & Wilson, 1980; Tung & Mays, 1982). An overall view of the risk-based design of highway drainage structures was given by Tung and Bao (1990).

INTANGIBLE FACTORS AFFECTING THE DESIGN FREQUENCY
In addition to economic costs of the project, there are other intangible aspects that might be equally, if not more, important in the determination of an appropriate design frequency for highway drainage structures . In the following, we briefly discuss some intangible factors that could affect the selection of appropriate design frequency for highway drainage structures.

HYDRAULIC AND ENVIRONMENTAL EFFECTS
Construction of drainage structures for highway crossings results in changes in hydraulic characteristics of rivers. Stream responses to changes may be confined locally or may be more wide spread. It is difficult to quantitatively and definitively relate the design frequency to the potential hydraulic effects on a stream system. Heuristically, increases in the design frequency represents less encroachment on the floodplain and less disturbance to the natural hydraulic characteristics of flow.

Impact of roadway crossings on the environment primarily arises from the potential increase in sediment concentration as the result of change in hydraulic characteristics. Highway drainage structures designed with a lower return period are more susceptible to overtopping by major floods rendering large quantities of embankment material being eroded and carried into the stream. There could be other changes in stream systems induced by the roadway crossing that might have some impact, for better or worse, on the aquatic ecosystem.

PUBLIC SERVICE
This term broadly covers the general serviceability performance of roadways to the public. It primarily relates to the notion that traffic interruption should only occur

due to extraordinary circumstances such as from very large floods. Simplistically, the public serviceability of a highway at a roadway crossing can be measured by its ability to provide continuous service to the public without being interrupted by flooding. Clearly this measure is closely related to the design frequency used for drainage structures. The larger the design frequency for a drainage structure, the less frequent will the traffic be interrupted by flooding which naturally would have a higher serviceability and cause less inconvenience, both tangible and intangible, to the traveling public.

LEGAL LITIGATION
State highway departments must design roadway crossings with extreme care to best serve the general public. Even if a highway engineer carefully practices drainage design for roadway crossings with all the legal ramifications in mind, litigation may still result for the perceived negligent design of a roadway drainage structure. Intuitively, use of a larger return period would result in less chance of being involved in litigation for a transportation department. Conversely, attempting to avoid litigation by designing for very large return periods at all drainage sites would generally be very costly for the public.

OTHER FACTORS
In addition to the four intangible factors mentioned above, the following intangible factors may also be added to the list: potential loss of life, national defense highway, and impact on local economy and environments.

MULTIPLE-ATTRIBUTE DECISION-MAKING (MADM)
All the intangible factors mentioned above are non-commensurable and, most of them, are in conflict with the economical consideration of drainage structure design. Use of a multiple-attribute approach enhances more realistic decision-making and the design frequency so determined will be more acceptable in practice and defensible during litigation or negotiation with others.

METHOD
In this study, a simple method called the "simple additive weighing (SAW) technique" was employed to consider intangibles for determining the extended-LTEC design frequency. This technique involves an analysis of an information matrix consisting of a decision-maker's subjective evaluation of their preference to each of the attributes involved for the alternatives under consideration. The relative merit of each alternative is judged on the basis of its final rating computed as

ROADWAY CROSSING STRUCTURES DESIGN 461

$$F_i = \frac{\sum_{j=1}^{N} R_{ij} W_j}{\sum_{j=1}^{N} W_j}, \text{ for } i = 1, 2, \ldots, M \tag{1}$$

in which F_i is the final rating for alternative i; R_{ij} is the rating for alternative i with respect to attribute j; W_j is the weight for attribute j representing the relative importance of attribute j; N and M are, respectively, the total number of alternatives and attributes.

ISSUES
Several important issues need to be considered for implementing the proposed technique:
(a) Who is (or are) the decision-maker(s)?
(b) What are the attributes to be considered?
(c) How to design a procedure for opinion survey?
(d) How to synthesize and analyze survey results?

AN APPLICATION
The methodological framework was developed to determine an appropriate design flood frequency for three types of roadway crossing structures (i.e., bridges, box culverts and commercial pipe culverts) for sites typical to Wyoming. Information on numerous sites representative of Wyoming as well as structure configurations and costs were provided by the Wyoming Department of Transportation (WDT). The flood frequency relationship used in the analysis was the regional regression equations developed by Druse et al. (1988). Optimization model was developed for determining the optimal layout of drainage structures and their corresponding LTEC design frequency.

In this study, along with the tangible cost factors, three intangible factors (i.e., maintenance frequency, litigation potential, and public service) were quantitatively considered. To remove as much subjectivity as possible in the decision-making process, a national survey was conducted. In the survey, relative importance and desirability of the various tangible and three intangible factors were rated by engineers holding different positions of responsibility in many federal, state, and county transportation agencies. The responses from the survey were examined and analyzed extensively (Tung et al., 1993) from which models that quantitatively simulate the decision-maker's preference rating were developed. As expected, results from the national survey results clearly show that different degrees of variability exist in the preference ratings from the different survey respondents. To incorporate this inherent variability from the survey responses into the decision-making, fuzzy set theory in conjunction with the SAW method was applied.

For illustration, Table 1 contains the site characteristics for an example bridge under consideration in an urban area located in the mountainous region of Wyoming. The resulting LTEC design frequencies corresponding to the site condition are tabulated in the upper part of Table 2. In the lower portion of Table 2, the ratings of various factors and the mean extended-LTEC design frequencies, along with the values plus and minus one standard deviation, for each intangible factor, as compared with the total cost factor, are tabulated. For each LTEC design frequencies, the mean flood frequency was estimated from the regional flood frequency relationship. The standard deviation was used to provide an engineer with estimates reflecting the margin of error to be expected.

REFERENCES
1. Corry, M.L., J.S. Jones and P.L. Thompson. 1980. "The Design of Encroachments on Flood Plains Using Risk Analysis," Hydraulic Engineering Circular, No. 17, U.S. Dept. of Transp., FHWA, Washington, D.C.
2. Druse, S.A., H.W. Lowham, M.E. Cooley, and A.M. Wacker, 1988. Streamflow in Wyoming, Wyoming Highway Department.
3. Pritchett, H.D., 1964. "Application of the Principles of Engineering Economy to the Selection of Highway Culverts," Stanford University, Report EEP-13.
4. Scheider, V.R. and K.V. Wilson. 1980. "Hydraulic Design of Bridges with Risk Analysis," Report, FHWA-TW-80-226, U.S. Dept. of Transp., FHWA, Washington, D.C.
5. Tseng, M.T., A.J. Knepp and R.A. Schmalz, 1975. "Evaluation of Flood Risk Factors in the Design of Highway Stream Crossings," Report No. FHWA-RD-75-54, U.S. Dept. of Transp., FHWA, Washington, D.C..
6. Tung, Y.K. and L.W. Mays, 1982. "Optimal risk-based hydraulic design of bridges," J. of Water Resour. Plan. and Mgmt., ASCE, 108(WR2):191-203.
7. Tung, Y.K. and Bao, Y., 1990. "On the optimal risk-based design of highway drainage structures," J. of Stochastic Hydrology and Hydraulics, 4(4):311-324.
8. Tung, Y.K., Hasfurther, V., Wacker, A.M., Bao, Y., and Zhao, B. 1993. "Least total expected cost (LTEC) analysis for selecting a defensible design flood frequency for highway drainage structures in Wyoming," Report, Wyoming Water Resour. Center, Univ. Of Wyoming, Laramie, Wyoming.
9. Young, G.K. and M.R. Childrey. 1974. "Impact of Economic Risks on Box Culvert Design," An application to 22 Virginia sites, prepared for FHWA, Report.

ROADWAY CROSSING STRUCTURES DESIGN 463

Table 1. Site Characteristics of Example Bridge Design Problem

LOCATION: Urban Area	**REGION:** Mountainous Region
DRAINAGE STRUCTURAL TYPE: Bridges	
CHANNEL CHARACTERISTICS:	
Bankfull channel width = 90.0 ft	Bankfull channel depth = 4.5 ft
Transverse floodplain slope on the left = .0300	Transverse floodplain slope on the right = .0300
Longitudinal channel slope = .0120	Manning roughness on floodplain = .0400
Manning roughness in main channel = .0350	
BASIN CHARACTERISTICS:	
Drainage area = 350.0 sq. miles	Annual average precipitation = 12.00 inches
Geographic factor = 1.1	

FLOOD FREQUENCY BY BASIN CHARACTERISTIC METHOD:

2-YR	5-YR	10-YR	25-YR	50-YR	100-YR	200-YR
972.2	1676.8	2235.9	3157.6	4030.0	4823.7	5900.5

ROADWAY CHARACTERISTICS:	
No. of lanes = 2	Design width of pier = 3.00 ft
Bridge deck width = 40.00 ft	Design embankment height = 16.00 ft
Pavement thickness = 14.50 in	Embankment soil types = Low cohesive
ECONOMIC PARAMETERS:	
Interest rate = .0420	Design project life = 50.0 years
No. of buildings = 5	Building values = 100.000 ($1000)
Building elevation above channel bottom at the drainage structure site = 9.00 ft	
TRAFFIC CHARACTERISTICS:	
Detour length = 5.00 miles	Rate of repair = 60.20 cu. yd./hr
Mobilization time = 15.20 hrs	Average vehicle speed on detour = 30.00 mph
Accident ratio = 265.70 per million veh-miles	Avg. daily traffic count = 1000.00
UNIT COST CHARACTERISTICS:	
Cost adjustment factor = 1.32	Mobilization cost ($) = 450.00
Unit embankment cost ($/cu. yd.) = 3.00	Unit bridge cost ($/sq. ft.) = 50.00
Unit pavement cost ($/cu. yd.) = 40.00	Unit cost of occupant ($/person) = 20.00
Unit damage cost ($/accident) = 6250.60	

Table 2. Results of Example Application of Risk-Based Design

LTEC DESIGN FREQUENCY & DESIGN DISCHARGE			
	- STDEV	MEAN	+STDEV
	15.7 YRS	25.3 YRS	41.4 YRS
	2652.1 CFS	3173.2 CFS	3787.7 CFS
INTERVALS OF EXTENDED-LTEC DESIGN RETURN PERIODS AND DESIGN DISCHARGES:			
	- STDEV	MEAN	+STDEV
MAINT. FREQ.	40.9 YRS	48.4 YRS	86.5 YRS
	4670.7 CFS	4799.9 CFS	5323.2 CFS
LITIG. POTEN.	42.1 YRS	53.2 YRS	104.3 YRS
	4692.8 CFS	4875.2 CFS	5543.2 CFS
PUBLIC SRVC.	39.0 YRS	39.0 YRS	39.0 YRS
	4637.1 CFS	4637.1 CFS	4637.1 CFS

CASE STUDY: LESSON AND PROBABILISTIC DETERMINATION ON ERODED EARTH SPILLWAY REHABILITATION

S. SAMUEL LIN[1], JOSEPH S. HAUGH[2] AND L. LYNN CLEMENTS[3]
[1], [2] Virginia Dam Safety Program, Richmond, Virginia and
[3] Rapidan Service Authority, Ruckersville, Virginia, U.S.A.

Abstract A valuable lesson is learned from a dam's severely eroded earth spillway due to a rare flood occurring in Virginia in 1995. Erosion is an inevitable consequence of an earth spillway because of soil's limited resistance to the attack imposed by a significant flood. However, the spillway structural stability must be maintained at its spillway design flood (SDF). RCC (roller compacted concrete) is recommended to rehabilitate the damaged emergency spillway instead of earthen channel spillway due to its unreliable stability and durability, particularly considering that the spillway is constructed on a very steep slope. This scheme has been chosen for the least cost of rehabilitation on the basis of risked-based economic analysis resulting from the comparison of alternatives and based on the remainder of the dam's service life.

INTRODUCTION

The White Oak dam was constructed in 1965 as a flood control and water supply structure in Madison County, Virginia. The drainage area is 3,240 acres of rural, mountainous terrain. The lake normally covers 50 acres and has a flood storage of 610 acre-feet. The earthfill dam consists of a 63-foot high earthen embankment with a 75-foot wide earthen/rock side open channel emergency spillway and a reinforced concrete principal spillway consisting of a 17X12 feet riser and 350 feet of 36-inch diameter outlet pipe. The dam is classified in the significant hazard category due to life-in-jeopardy of a downstream dwelling in case of dam failure. The emergency spillway, about 50 feet away from the dam constructed with a 15% slope outlet channel, was designed for safe passage of 0.7PMF (Probable Maximum Flood) exceeding the required SDF of 0.5PMF.

On June 27, 1995, a heavy, sustained storm blanketed the county and dumped more than 24 inches of rain within six hours inducing about a 500-year frequency flood at the dam site. The emergency spillway handled more than10 feet depth of flow at the control section at a velocity exceeding 13 fps and a flow of about 10,320 cfs, about

10% lower than the SDF. This extreme flow undermined the toe of the spillway and blew out approximately 300 feet of the 75-foot wide channel to a maximum depth of 30 feet and width of 150 feet below the control section, and extended the width of the channel by another 40 feet for the last 100 feet. A total of 10,500 CY of silts, clays and foundation rocks were eroded from the spillway and adjacent boundaries. This high flow inundated the downstream areas causing damages to the stream channel, bridge, and farm lands and contributing to the downstream sediment load of the floods. Fortunately, the spillway's control section is solid granite and 30 feet in length survived the assault with some resulting 3-foot deep erosion ruts. The only earth remaining in the spillway was upstream of the control section and some of the channel walls. The investigation and condition of the spillway, ruined by dynamic effects of tremendous flow has been explored by Clements et al. (1996).

LESSON FROM DISASTER

Devastating scouring of the spillway could extend to the remaining original control section of the spillway or the remaining portion of the right abutment. This could be brought on by the occurrence of an equal or a greater magnitude flood event. In the year and a half following the flood, the exposed bedrock of the spillway has become weaker due to weathering and started fracturing. The deterioration of both bank slopes of the spillway continues. An issue was raised regarding the spillway's present condition: it should be considered as severely damaged or even failed. Scour damage is expected for an earth spillway once it is activated. The design capacity of an emergency spillway is considered adequate in hydraulic performance if the dam stays safely and the reservoir is not drained due to the spillway's breach at the SDF. However, the spillway is not considered functioning properly due to its structural stability failure resulting from severe damages.

The earthen channel spillway prototype has met the test of a flood close to the SDF for its hydraulic performance but not for its channel stability. The physical condition of the remaining spillway is uncertain if an equilibrium of the extensive erosion is established and scouring ceases with sacrifice of channel materials when a similar flood recurs through this spillway. If the June 1995 flood was the SDF, then the worst condition could have been the remaining portion of the right abutment of the dam being washed away, and consequently could trigger dam breach starting from the toe of the dam. The present spillway cannot function within the condition of safely passing the SDF without endangering the dam's embankment. Therefore, its condition is not considered as safe as originally designed. This is the primary reason why this damaged spillway needs to be rehabilitated. (Lin, 1996)

ALTERNATIVES OF SPILLWAY REHABILITATION

The Virginia Dam Safety Regulations, Sec. 5.6 life of the impounding structure states "Components of the impounding structure, the impoundment, the outlet works, drain system and appurtenances will be durable in keeping with the design and planned life of the impounding structure." (DCR, 1989) The best description of the performance standard for the original spillway should be that the spillway must be capable of passing the design storm and withstanding the flow velocities of the design storm without a stability failure. Particularly, the failure of an earth spillway which can cause either the lake to be drained or the dam's failure as a chain reaction should be prevented. Spillway rehabilitation planning involves many choices among feasible alternatives. It is necessary to consider relative economy with the premise of structural safety in order to make a rational choice among the possible alternatives and thus to establish the most favorable overall plan. Specifically, a recommended choice is properly influenced both by the comparison in terms of dollars and by durability that is intangible. Engineers researched alternatives such as (1) approximate earthen/rock replacement, (2) earthen foundation with rock or concrete armoring, (3) earthen foundation with concrete armoring, (4) reinforced concrete replacement, (5) RCC stepped spillway, and (6) combination of earth and RCC repair.

AN ACCEPTABLE SCHEME SELECTION

The research of all these schemes includes structural design, hydraulics, existing geology (Ganoe, 1995), available material, site constraints and construction costs. (Street & Kintzer) The following two analytical steps are adopted in conjunction to determine the most acceptable scheme for the spillway's rehabilitation.

STEP I: DURABILITY AND COST COMPARISON

Under full flow conditions, most earth emergency spillways are expected to erode. The idea is to save the dam. However, the cohesive strength of any type of soil is not strong enough to withstand the hydraulic loadings of a major flood's supercritical flow at the slope of 15% at a depth of about 10 feet. Especially, this 500-yr flood created the flow velocity exceeding 25 fps at the toe of the dam considerably exceeding the permissible velocity for a well-vegetated earth spillway, 6fps (SCS, 1985). The destructive situation of the original spillway which has endangered the stability of the right abutment of the dam should not be allowed to recur. The failure of the dam could even cause life-in-jeopardy. The original spillway contained rock, which although not as strong as solid bedrock, did contain some structure and was more resistant to erosion than compacted earthfill. In addition, the available cohesive soil in the area is scarce. Alternative (1) earthen/rock replacement is not acceptable according to the regulations' safety standard. On the same level of structural stability, the costs of the alternatives (2), (3) & (4) are significantly higher than alternative (5), the RCC stepped spillway. (Clements et al., 1996) The proposed RCC stepped spillway and its associated stilling basin can confine the major flood within a limited topographic length to a flow velocity not causing any major damage at the downstream channel. Alternative (6), the combination of earth and RCC scheme presents no danger to the integrity of the dam if it were to fail, is the least cost compared with other feasible alternatives. However, the tradeoff between cost and spillway protection for alternatives (5) & (6) needs to be further studied.

STEP II: QUANTITATIVE DECISION ANALYSIS

The devastating losses of the White Oak dam would include not only the dam and its appurtenant structures but also the loss of resources beneficial to the reservoir's service areas and extreme environmental impacts. Flood control benefit, life loss and social influence are not considered in this case due to a relatively minor benefit to dam failure damages and difficulty in obtaining an objective estimate. Risk is termed as the probability of a structure's failure combined with its consequences. Risk costs are the costs in risk of loss in project benefits, property losses, and environmental impacts on account of 500-yr flood recurrence uncertainty. None of alternatives (5) RCC and (6) combination shown in Table 1 is an apparent solution to the decision of a repair method for the least total cost in the long term without investigation of possible losses relevant to their stability protection. These alternatives can be thoroughly examined in terms of their rehabilitation costs and risk (damage) costs at one time loss.

Engineering design and decision making under uncertainty properly require probability considerations. From the principle of multiplying probabilities, the probability, **J**, that at least one event which equals or exceeds the t_p-yr event ($P = 1/t_p$) will occur in any series of N-yr is:
$$J = 1 - (1-P)^N \quad \quad \quad (1)$$
Both the probability of failure of the spillway and the associated consequences of failure are needed for a risk-based economical analysis. The construction and damage risk costs must be estimated for comparisons of different schemes for the same magnitude 500-yr (t_p) flood within 120 years (N), the remaining service life of the dam. The probability of this flood recurrence within 120-yr is 21% (J). The results of risk-based economic analysis of alternatives (1), (5) & (6) are shown in Table 1 in which the earthen channel scheme is also included only for comparison. Even though the total expected costs between the RCC and combination options are close, the damage prevention factor far outweighs the initial cost's consideration. The selection is the RCC stepped spillway scheme based on structural integrity, geological harmony, site constraints, hydraulics, and risk economic analysis. The hydraulic merits include using a RCC stepped spillway associated with a stilling basin which can dissipate the energy of the SDF to a velocity causing no major downstream damage.

DISCUSSION AND CONCLUSION

1. Any damaged earth spillway should be repaired to prevent deteriorating conditions caused by floods during which it is in use. The condition of the existing eroded spillway with a huge scoured stilling basin could function as an energy dissipator for a certain magnitude flood. However, some situations such as extreme hydrologic events are uncertain during the remaining service life of the dam, and progressive deterioration of the present spillway. Therefore, restoring this spillway to a satisfactory condition is necessary and also required by the regulations.
2. The high speed supercritical flow could cause severe erosion on the earthen spillway bottom. The use of compacted earthfill is not satisfactory for an open

channel spillway which uses a control section associated with an exit channel slope that exceeds about 4%. (SCS, 1985) In the situation of a slope of 15 percent, excessive supercritical flow could occur. Even surface capping or grouting of the compacted earthfill material would not be a satisfactory solution for the high velocities that would occur on the steep slope. Any infrastructure is expected to meet the minimum required safety standards. There is no compromise of public safety and cost acceptable for infrastructures such as dams.
3. Furthermore, from a dam safety standpoint, it is important to prevent the potential of dam breach. Tremendous scouring of the compacted earthfill emergency spillway could extend to the remaining original control section of the spillway or the remaining portion of the right abutment. This could be brought on by the occurrence of an equal or a greater magnitude flood event. In contrast, in the perspectives of hydraulic performance and material strength, the RCC alternative is durable and survivable for major floods in the long run of the dam operation.
4. Restoring the spillway to its original, pre-disaster condition is not only unattainable but also unacceptable from the safety standpoint of structural integrity. A rehabilitation scheme of the spillway must assure the spillway's durability, survivability and stability or reliability on the economic basis. The RCC stepped spillway scheme is recognized as the best of the eligible alternatives.
5. Life value is not involved in the risk analysis for this case. A life loss converted to be a money unit of cost can be very helpful in a quantitative decision making process to determine an appropriate SDF on a same dam hazard class basis. (DCR, 1989)

SUGGESTION AND RECOMMENDATION

The best way for us to learn from the lessons in natural disasters is to explore humans' carelessness and/or mistakes and recognize that our memory and remedial action are the best opportunity to avoid the same mistake recurring. The following thoughts are proposed to achieve the improvement of an earth channel spillway's safety:
1. The spillway stability of concrete or masonry materials is required to meet engineering standards. For an earth spillway, sufficient engineering analyses should also be required to assure not only its hydraulic performance but also its structural stability at the SDF, such as alignment/layout (particularly a channel slope), foundation investigation, tailwater depth, etc. The scour analyses using NRCS's DAMS2 can predict potential scour locations and depths for proper erosion control measures, such as vegetating or armoring the spillway channel and/or even constructing a stilling basin can protect the outlet channel from or alleviate erosion. Any existing earthen channel spillway with a slope greater than 4% which could create a supercritical flow should be hydraulically and geologically reviewed for its structural stability against the SDF.
2. A conceptual and qualitative model should be developed as a driving force to define earthen channel stability with respect to erosion. The following potential factors are proposed for regulators as an engineering guideline to justify the stability of the spillway's performance against the SDF: (a) loss of the control section of the spillway due to headcutting; (b) loss of the integrity and configuration of the spillway; (c)

ERODED SPILLWAY REHABILITATION

continuous scouring when the SDF occurs and migrates laterally to the dam abutment or toe, or causes both side slopes of the spillway to become unstable; (d) consideration of the economic factor as an index (for instance, rehabilitation cost is higher than the original construction cost at the present value).

3. The service life of a dam for long-term purposes is full of risks because of hydrologic, hydraulic, or seismic impact, geotechnical problems or structural deterioration or improper operation procedures. All such impact processes or deficiencies in dam performance are subject to variation due to random and nonrandom causes. Nonrandom variation can be controlled by monitoring, inspecting, repairing, and effective operation and maintenance plans. Random variation cannot be eliminated. For instance, the hydrologic process is fraught with uncertainties, but integrated, proactive risk-assessment methods can help owners and engineers make better, informed decisions. Through risk assessment for relevant uncertainties, appropriate remedial measures to correct deficiencies of a dam or its appurtenant structures (particularly spillway) can reduce the risks to life and property due to flooding resulting from the potential of reservoir draining or dam failure.

Table 1 White Oak Dam Spillway Rehabilitation: Items for Risk Analysis

ALTERNATIVES ITEMS	RCC (Alt. 5)	EARTHEN (Alt. 1)	COMBINATION (Alt. 6)
Rehabilitation	Capital	Cost,	C_c ($)
Construction	900,000	800,000	700,000
Loss Items	Potential Damage Costs, C_d ($)		
Spillway Damage	0	800,000	700,000
Dam Failure	0	2,000,000	0
Stream, Wetlands & Farmlands Damage	0	100,000	50,000
Route 657 Bridge Loss	0	250,000	250,000
Loss of Water Supply	0	1,000,000	0
Stream Pollution	0	200,000	100,000
Loss in Risk	Total Risk Cost, C_r ($)		
Total Expected Loss ($J \times C_d$)	0	934,500	231,000
TOTAL COST ($C_t = C_c + C_r$)	900,000	1,734,500	931,000

Note: J = probability of the damaging 500-yr flood event during the remaining 120-yr service life of the dam, or 21%.

REFERENCES
Clements, L. Lynn, S. S. Lin, B. L. Kintzer, B. T. Street and B. W. Ganoe (1996). "Case Study: Learning from Earthen Spillway Erosion and Rehabilitation", ASDSO 1996 Dam Safety Conference, September 8-11, Seattle, Washington.
Ganoe, Brian W. (1995). Emergency Spillway Damage Assessment on White Oak Dam, Natural Resources Conservation Service (NRCS), Richmond, Virginia.
Lin, S. Samuel (1996). Sept. 20, 1996 White Oak Dam Field Trip Report, Virginia Dam Safety Program, Richmond, Virginia.
Street, Trent and Barry Kintzer (1996). Design Report on White Oak Dam, NRCS, Richmond, Virginia.
U.S. Department of Agriculture, Soil Conservation Service (SCS) (1985). Earth Dams and Reservoirs, TR60, Washington, D.C.
Virginia Department of Conservation and Recreation (DCR) (1989). Virginia Dam Safety Regulations, Richmond, Virginia.

Reservoir Reliability Design Under Interannual Climatic and Hydrologic Variability

OSCAR MESA, GERMÁN POVEDA, LUIS CARVAJAL, and JOSÉ SALAZAR
Universidad Nacional de Colombia, Facultad de Minas, Medellín, Colombia

ABSTRACT

The hydro-climatology of tropical South America and Colombia is highly coupled with low-frequency large-scale oceanic and atmospheric phenomena occurring over the Pacific and the Atlantic oceans. In particular, ENSO's signal modulates climatic and hydrologic conditions on time scales ranging from seasons to decades. With some regional differences in timing and amplitude, tropical South America exhibit negative anomalies in rainfall and stream flows associated with the warm phase of ENSO (El Niño), and positive anomalies with the cold phase (La Niña). We illustrate such a dependence for the hydro-climatology of Colombia. Reservoir design and operation acquire a new dimension under long-term cycles and quasi-periodicity such as those inherent to ENSO forcing. New models to estimate reservoir reliability based on regime dependant processes or double well potential undergoing stochastic fluctuations are considered. Two regimes are used to represent the two phases of ENSO and better reproduce the long range properties of the record associated with reservoir performance.

INTRODUCTION

Annual rainfall over Colombia and tropical South America on inter-annual time scales is strongly influenced by low-frequency large scale ocean-atmosphere phenomena. In particular, El Niño-Southern Oscillation (ENSO) is a major forcing mechanism of the climatic and hydrological anomalies (*Poveda and Mesa*, 1995, 1996). Such low frequency aperiodic oscillations are fundamental in reservoir design and operation. The 1991-92 El Niño event caused losses about US$ 1,000 million to the economy of Colombia, due to prolonged electricity shortages, forced by the drought. For long time, hydrologists have recognized that a sequence of low flows is much more critical than a very low discharge. This has been the motivation of the long range modeling efforts. The realization of the climatic control of this low frequency oscillation is nevertheless quite new and various fundamental questions deriving from that relation are in order. In the first place, planning and real time operation of reservoirs can benefit significantly from predictions based on global physical models at climatic scales (seasons to years). This implies that traditional methodologies based on the coupling of Markovian models for the hydrology and stochastic optimization for the operation need revision. Secondly, reservoir design requires reconsideration, in view of the dynamical basis of the hydrologic variability. Traditional stochastic hydrologic models have a covariance structure with too short a range to represent the aperiodic recurrence present in the streamflow records. Otherwise, long memory models lack physical grounds and more dependance structure than necessary. The identification of the climatic cycles opens a new and precise way for hydrologic modeling. In particular, the role of the scale of fluctuation is fundamental.

ENSO AND THE HYDRO-CLIMATOLOGY OF COLOMBIA

El Niño is the anomalous warming of sea surface temperatures at the eastern and central tropical Pacific. Important components of this anomaly in the global climate are the deepening of the oceanic thermocline in the eastern Pacific, and the weakening of the predominant surface easterly trade winds. During El Niño the center of convection shifts from the western to the central Pacific Ocean, due to a large scale perturbation in the Walker cell. This zonal circulation is driven by differences in sea surface temperature and convection along the equatorial Pacific Ocean. The accompanying Southern Oscillation produces a pressure gradient between the western and the eastern equatorial Pacific, it is quantified with the Southern Oscillation Index (SOI). This index is the standardized difference between the Tahiti and Darwin sea level pressure. Negative anomalies of the SOI are associated with warm events (El Niño) and with a weakened Walker cell, and positive anomalies are associated with cold events (La Niña) and with a stronger Walker cell. ENSO is an aperiodic oscillation with an average recurrence of about four years, varying from two to ten years (*Trenberth*, 1991). ENSO has profound influences over the whole global circulation, affecting weather and climate and producing tremendous socioeconomic effects worldwide, including floods, droughts, shortages in food production and hydro-power generation, and epidemics *(Glantz et al.,* 1991; *Ropelewsky and Halpert,* 1996). Global rainfall anomalies are probably more active players than a passive spectators of the climatic changes triggered by both phases of ENSO.

Overall, there is a coherent behavior of hydrological anomalies in tropical South America during extreme phases of ENSO (see *Poveda and Mesa,* 1996). With some regional differences in timing and amplitude, the region exhibits negative anomalies in rainfall and stream flows associated with the warm phase of ENSO (El Niño), and conversely, positive anomalies with the cold phase (La Niña). As an example, Figure 1 shows the time evolution of the smoothed anomalies in monthly values of the Southern Oscillation Index (SOI) and the series of the Nare river discharges at Santa Rita, Colombia. The simultaneous correlation coefficient of the smoothed series is 0.72. This correlation is statistically significant up to the 99.99% level. The number of degrees of freedom in the F-test was reduced using the scale of fluctuation of the processes (*Vanmarcke,* 1983). This suggests that ENSO may explain up to 50% of the variance of the Colombian hydrology beyond the annual cycle, if we assume a linear relationship. Of course the relationship ENSO-hydrology is highly nonlinear, and therefore we are missing some features of this type of dependence.

Correlations between climatic variables over the Pacific Ocean and the Colombian rainfall and runoff series are larger when climatic variables lead the Colombian hydrology by 3-4 months. This fact provides good possibilities to develop predictive models. Among various alternatives, we have used the following techniques: neural networks (*Mesa et al.*, 1994b), singular spectral analysis (*Carvajal et al.*, 1994), linear inverse modeling *(Poveda and Penland,* 1994). Certainly, use of such techniques contributes to improve hydrological forecasting, with tremendous practical consequences.

Existence of persistence, cycles and trends have been central to the interest of hydroclimatologists for decades. The whole issue of the Hurst effect in geophysical records (*Mesa and Poveda*, 1993) and the persistence of those processes gains a new interpretation under the presence of long-periodic forcing and global warming.

REGIME DEPENDANT PROCESSES AND NON-LINEAR STOCHASTIC DIFFERENTIAL EQUATIONS

Linear autoregressive models have been used traditionally in streamflow simulation, despite of their long recognized shortcomings. In this section we apply non-linear stochastic models to incorporate ENSO influence on the Colombian hydrology. We consider: (1) Regime dependent Autoregressive (RAR) modeling that combines different linear autoregressive components selectively according to the state of either an external time series or the process itself, and (2) Non-linear Stochastic Differential Equations with a potential formed by two stable wells separated by un unstable state. The two stable states represent the two phases of ENSO. We test the models to preserve short and long-term properties of the inflows.

REGIME DEPENDANT AUTOREGRESSIVE (RAR) MODEL. These models have been introduced by *Tong* (1983). RAR and Instrumental RAR were applied to various rivers, to illustrate we use monthly river discharges of the Riogrande (RG8, Antioquia, Colombia). See *Salazar et al.*, 1994b for details. To evaluate model performance, short-term and long-term properties of simulated traces were compared with those of the

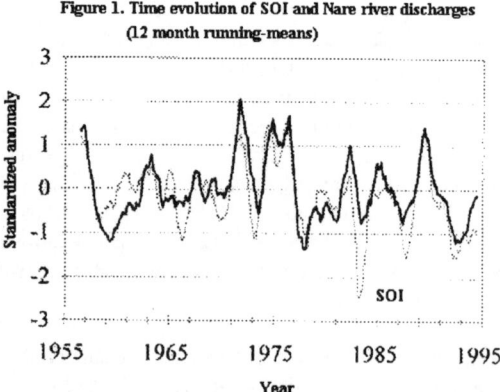

Figure 1. Time evolution of SOI and Nare river discharges (12 month running-means)

inflows: power spectra, extreme events, and storage-related characteristics. They were also compared with results obtained with an AR(1) log-transformed model. Given the strong influence that ENSO exerts on the hydrology of Colombia, the Southern Oscillation Index (SOI) was used as the external variable Y. Results for the Instrumental RAR(2, 1) are given by

$$X_t = 0.9 + 0.57\sqrt{\frac{1.05}{1.15}}(X_{t-1} - 0.84) + \sqrt{1.05(1 - 0.57^2)}e_t \quad \text{if} \quad Y_{t-2} \geq 0.4$$

$$X_t = -0.3 + 0.53\sqrt{\frac{0.6}{0.61}}(X_{t-1} + 0.28) + \sqrt{0.61(1 - 0.53^2)}e_t \quad \text{if} \quad Y_{t-2} < 0.4 \quad ,$$

with $e(t)$ being a white noise process with zero mean and unit variance.

NON-LINEAR STOCHASTIC DIFFERENTIAL EQUATIONS. The time evolution of inflows are represented by the equation

$$\frac{dx}{dt} = -x^3 + \lambda x^2 - \rho x + v + F(t) \quad .$$

This equation represents the evolution of a non-linear system within a double-well potential (Figure 2) driven by stochastic noise (*Gardiner*, 1985; *Demaree and Nicolis*, 1990). Parameters λ, ρ, and ν are defined in terms of the three equilibria, such that $\lambda=a+b+c$, $\rho=bc+ca+ab$, and $v=abc$. The two stable states (a and c) represent the climatic conditions forced by the warm and cold phases of ENSO. Parameters a and c were estimated as the modes of the empirical bimodal distribution of the discharges. To determine b (unstable equilibrium) and q^2 (noise variance) we imposed preservation of the mean and variance of the historical record. See *Salazar et al.*, 1994a for details.

One considerations has to be made here. The marginal probability distribution function of the process has very fast decaying tails when the double-well potential is a fourth order polynomial. This is not the case for river discharges records. For this reason we assumed a second order polynomial (normal distribution) for the tails (before a and after c) while using the same fourth order double well-potential between a and c.

RESULTS

Table 1 presents statistics of long-term and storage related characteristics for the historical record and for (log-transformed) auto-regressive, non-linear stochastic differential equation (NSDE), and regime-dependent autoregressive models applied to Riogrande river in Colombia. Results refer to a zero-mean threshold for 2,000 generated data, whereas the record has only 612 data. Notice that results are standardized in terms of monthly means and standard deviations. Comparing with observed statistics, NSDE and RAR produce better results than LTAR for maximum run length and area. Whereas mean values are similar. This advantage of the nonlinear over the autoregressive models is very important for reservoir design and reliability analysis. The rescaled adjusted range analysis confirms this superiority (observed 69.6 against 57.3 for ARTL, 62.6 for NSDE, 61.8 for RAR and 60.8 for Inst. RAR). Rescaled range values for the models are the mean of 10 simulations of 612 long series.

Observed marginal distributions of different rivers in Colombia show a tendency to bimodality as implied from the models presented here. This is in agreement with the analysis of the physical mechanisms associated with the two phases of ENSO.

Table 1. Storage and long-term characteristics of historical and simulated traces

	Historical	ARLT	NSDE	RAR	Inst. RAR
Maximum of Series	4.1	6.2	3.4	3.8	4.4
Minimum of Series	-2.2	-2.2	-2.5	-3.0	-2.8
Total Number of Runs	83	267	270	279	269
Mean run Length (+)	3.1	3.2	3.0	3.4	3.4
Max Run Length (+)	25	19	21	25	25
Area of Run of Max Length(+)	31.4	19.2	36.9	27.2	30.9
Mean Run Area (+)	2.8	2.8	3.1	2.9	2.8
Max Run Area (+)	41.7	19.2	36.9	27.2	32.0
Length of Run of Max Area(+)	24	19	21	25	16
Mean Run Length (-)	4.2	4.3	4.4	3.8	4.0
Max Run Length (-)	29	23	23	24	36
Area of Run of Max Length (-)	-25.9	-18.3	-21.2	-20.2	-29.4
Mean Run Area (-)	-2.8	-2.9	-2.9	-2.8	-2.9
Max Run Area (-)	-25.9	-21.4	-21.2	-23.5	-29.5
Length of Run of Max Area (-)	29	21	23	24	36

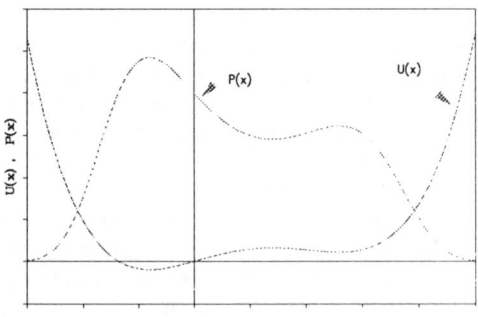

FIGURE 2. Probability Density Function, $P(x)$, and Potential, $U(x)$.

References

Carvajal, L. F., O. J. Mesa, J. E. Salazar, and G. Poveda, 1994: Singular spectral analysis to study hydrological time series in Colombia (in Spanish), *Proc. XVI Latin-American Congress on Hydraulics and Hydrology*, IAHR, Santiago, Chile, Vol. 3, 97-108.

Demarée, G. R., and C. Nicolis C.: 1990, Onset of sahelian drought viewed as a fluctuation-induced transition. *Q.J.R. Meteorol. Soc.* , 116, 221-238.

Gardiner, C. W.: 1983, *Handbook of Stochastic Methods*. Springer Verlag, Berlin.

Glantz, M, R. Katz, and N. Nicholls, Eds., 1991: *Teleconnections Linking Worldwide Climate Anomalies*, Cambridge University Press, 535 pp.

Mesa, O. J., and G. Poveda, 1993: The Hurst effect: The scale of fluctuation approach. *Water Res. Res.*, **29**, 3 995 - 4 002.

Mesa, O. J., L. F. Carvajal, J. E. Salazar, and G. Poveda, 1994: Hydrologic prediction using neural networks (In Spanish), *Proc. XVI Latin-American Congress on Hydraulics and Hydrology*, IAHR, Santiago, Chile, Vol. 3, 385-396.

Poveda, G., and O. J. Mesa, 1993: Methodologies to predict the Colombian hydrology considering the ENSO event (in Spanish), *Revista Atmósfera*, Colombian Meteorological Society, Bogotá, **16**, 26-39.

Poveda, G., and O. J. Mesa, 1995: The Relationship between ENSO and the hydrology of tropical South America. The case of Colombia. *Proc. Fifteenth Annual American Geophysical Union Hydrology Days*, Fort Collins, CO, Hydrology Days Publications, 227-236.

Poveda, G. and O. J. Mesa, 1996: Feedbacks between hydrological processes in tropical South America and large scale oceanic-atmospheric phenomena. Submitted to *J. Climate*.

Poveda, G., and C. Penland, 1994: Prediction of mean monthly river discharges in Colombia using linear inverse modeling (in Spanish), *Proc. XVI Latin-American Congress on Hydraulics and Hydrology* IAHR, Santiago, Chile, Vol. 4, 119-129.

Ropelewsky, C. F., and M. S. Halpern, 1996: Quantifying Southern Oscillation-precipitation relationships. *J. Climate*, 9, 1043-1059.

Salazar, J. E., O. J. Mesa, G. Poveda and L. F. Carvajal, 1994a: Application of a non-linear continuous model to study hydrological time series (in Spanish), *Proc. XVI Latin-American Congress on Hydraulics and Hydrology*, IAHR, Santiago, Chile, Vol. 4, 169-180.

Salazar, J. E., O. J. Mesa, G. Poveda, and L. F. Carvajal, 1994b: Modeling the effect of ENSO on Colombian hydrology using regime-dependent auto-regressive processes (in Spanish). *Proc. XVI Latin-American Congress on Hydraulics and Hydrology*, IAHR, Santiago, Chile, Vol. 4, 181-191.

Trenberth, K., 1991: General characteristics of El Niño-Southern Oscillation. *Teleconnections Linking Worldwide Climate Anomalies*, R. M. Glantz, R. Katz, and N. Nicholls, Eds., Cambridge University Press, 13-42.

Vanmarcke, E., 1988: *Random Fields: Analysis and Synthesis*, MIT Press, 382 pp.

Acknowledgments. Support from COLCIENCIAS and EPM of Colombia is gratefully acknowledged.

STATISTICS OF MAXIMUM FLOWS

EDUARDO VARAS
Professor Dept. Hydraulic and Environmental Eng., Pontificia Universidad Catolica de Chile, Casilla 306, Correo 22, Santiago,Chile

ABSTRACT

Sample statistics and probability weighted moments of floods recorded in several gaging stations located in Chile, England, France and USA are compared. Results show a remarkable similarity between probability weighted moment (PWM) ratio diagrams, even though the hydrologic conditions of the four regions are different. Relations between PWM of a certain order and PWM of the preceding order are well defined and show a small dispersion. This indicates that it is possible to estimate accurately a superior order PWM as a function of the PWM of smaller order. Consequently, it is possible to get a good estimate of the first 5 PWM as a function of the mean flood and an estimate of the second order moment, and thus calculate the parameters of the probability models which are used to represent floods.

INTRODUCTION

The reliability of flood estimates depends on the length of record, consistency of the observed floods and accuracy of measurements. Unfortunately the suddenness and violence of the phenomenon, hinders the possibility of having abundant and precise observations, since when large flows occur, gaging stations tend to have operational problems. For this reason, flood estimates calculated on the basis of a single station record tend to be of limited reliability. Furthermore observed records are usually short (20 to 50 years) compared with the return periods of interest in design (100 to 1000 years). This is particularly relevant in cases of developing regions where records are scarce and short.

This paper compares sample statistics of maximum daily records of 75 gaging stations located in the United Kingdom, France, California and Chile, to examine if there is any similarity between the adimensional data collected in regions that are hydrologically different.

GENERAL DESCRIPTION OF THE REGIONS

CHILE
Selected stations have at least 20 years of record and represent 804 station-years of information. They are located in the watersheds of Rapel, Mataquito, Maule and Itata

rivers in central Chile. The relief of the area is characterized by two mountain ranges and a central valley. Climate is mediterranean with a wet winter period (June-August) where most precipitation falls. Annual rainfall varies in average from 500 mm to 2500 mm with a marked increase towards the south. The area above 2000 m usually receives precipitation as snow. Specific yield of the basins increases towards the south with values in the range of 10 to 90 l/s/km^2.

The average maximum daily flows in the region is 393 m^3/s, with values varying between 44 and 1102 m^3/s. The average coefficient of variation is 0.64 with values in the range of 0.41 to 0.92. Average flood per unit area is 477 l/s/km with values in the range 124 to 1638 l/s/km. The coefficient of skew is positive except in one station. The average value is 1.42 and the maximum value is 10.0. Only four stations have coefficients of skew lower than 0.3, since observations are highly skewed.

UNITED KINGDOM
Central UK is represented by 20 gaging stations with 20 years each. (Cunnane, 1989). Most floods are caused by frontal storms occurring in winter and summer, although others are caused by convective storms and snowmelt. The average maximum daily flood is 56 m^3/s, varying between 6 and 561 m^3/s. The average coefficient of variation is 0.55 with values in the range of 0.33 to 1.14. Average flood per unit area is 76 l/s/km with values in the range 31 to 149l/s/km. The coefficient of skew is almost always positive. The average value is 1.36 and the maximum value is 3.89.

FRANCE
The sample includes 17 stations located in France which have on average 52 years of record each. The shortest record is 14 years long and the largest has 95 years of record. In all they represent 884 station-years. The average maximum daily flood is 597 m^3/s, varying between13 and 3456 m^3/s. The average coefficient of variation is 0.46 with values in the range of 0.21 to 0.69. Average flood per unit area is 235 l/s/km with values in the range 23 to 733 l/s/km. The coefficient of skew is almost always positive. The average value is 0.94 with values in the range -0.65 to 3.49.

USA
Six stations in central California representing 239 station-years were available. Record lengths vary between 25 and 54 years. The average peak flood is 51 m^3/s, varying between 10 and 76 m^3/s. The average coefficient of variation is 1.06 with values in the range of 0.73 to 1.70. Average flood per unit area is 56 l/s/km with values in the range 18 to 147l/s/km. The coefficient of skew is always positive. The average value is 1.56 and the maximum value is 3.32.

PROBABILITY WEIGHTED MOMENTS
Probability weighted moments (PWM) are defined as the expected value of three terms: the random variable (x) to the l power, the probability distribution function to

the j power (F) and the complement of this function to the k power. The l,j,k order PWM is thus calculated with the following expression:

$$M_{l,j,k} = E(x^l F^j (1-F)^k) = \int_0^1 x^l F^j (1-F)^k dF \qquad (1)$$

Conventional moments represent a special case of PWM, with null j and k exponents. Greenwood et al. (1979) have recommend this method to estimate the parameters of several distribution probability models due to the advantages that the method has in cases of short records.

Landwehr et al. (1979a,b) recommend the following expression to calculate the PWM. These equations give biased estimates of k moments as a function of sample size (n), of the values in ascending order (x_i) of the rank number (i) of each value. According to the authors biased estimates are prefered.

$$M_k = \frac{1}{n} \sum_{i=1}^{n} x_i ((n-i+0.35)/n)^k \qquad (2)$$

It is possible to obtain similar expressions for PWM calculated with null k exponent and non null j exponents or to calculate j-PWM as a function of k-PWM. (Cunnanne, 1989)

STATISTICS AND MOMENTS
STATISTICS
Using the selected samples for each region, both statistics (mean, variance, coefficients of skew and kurtosis) and the first 5 PWM were calculated. Results obtained are summorized in Tables 1 and 2. The first table indicates for each region average and the range of values, for basin size, floods, standard deviations of floods, and skew and kurtosis coefficients. A large variation of sample characteristics can be appreciated. Figure 1 shows a scatter plot of the coefficient of skew as a function of the coefficient of variation for each station . No clear association is indicated. Values in different regions present large variations and no defined tendency. Other moment ratio diagrams present considerable scatter. Relations between kurtosis as a function of skew are better defined, although they also show large deviations.

MOMENTS
Table 2 present the average characteristics of the first five PWM for each of the regions considered in the study. The first PWM is the mean flood in each location and the rest of the PWM are presented in a standarized form, being each of them divided by the first moment. The similarity between the values for each region is remarkable, although regions differ appreciably. Second standarized moment is 0.66 for all stations and mean values for each region vary from 0.62 to 0.76.

Figures 2 and 3 show also that there is a distinct and clear relation between moments of higher order as a function of the PWM of lower order. No distinction can be observed between samples of the different regions and all locations define the same tendency. It must be noted that the USA stations follow the same pattern although the values represent peak flows and not maximum daily flows.

CONCLUSIONS

The main conclusions that can be derived from the results presented are:
a) No clear relations can be obtained between the statistics of maximum flows in different hydrologic regions or in any one region. Individual locations present a large scatter and it is not possible to obtain a good estimate of higher order statistics as a function of lower order statistics.
b) There is a remarkable similarity between standarized PWM for all locations considered.
c) Relations between higher order PWM and the corresponding lower order PWM are clearly defined and present a small variation. Consequently higher order moments can be accurately predicted from lower order PWM.
d) The similar behaviour that the different regions exhibit show the convenience of comparative hydrology where information of regions with more records can be used to supplement scarce information in other regions.
e) Although the consistency and similarity between PWM can be expected given their definitions, the similarity of results indicate that higher order moments can be accurately calculated using regional relationships and thus one can estimate parameters of probability models with three or more parameters with small samples.

ACKNOWLEDGEMENTS

This paper represents part of the work done in the Research Project 1950981 financed by Fondo Nacional de Investigación Científica y Tecnológica, Chile.

REFERENCES

CUNNANE, C. (1989) Statistical Distributions for Flood Frequency Analysis. **Operational Hydrology Rep No. 33**. World Meteorological Organization.
GREENWOOD, J.A., LANDWEHR, J.M. y WALLIS, J.R.(1979) Probability weighted moments: definition and relation to parameters of several distributions expressable in inverse form. **Water Resources Res, vol 15,** 5,1049-1054.
LANDWEHR, J.M., MATALAS, N.C. y WALLIS, J.R. (1979a) Estimation of parameters and quantiles of Wakeby distributions: 1. Known lower bounds. **Water Resources Res, vol 15**, n6, 1361-1372.
LANDWEHR, J.M., MATALAS, N.C. y WALLIS, J.R. (1979b) Estimation of parameters and quantiles of Wakeby distributions: 2. Unknown lower bounds. **Water Resources Res, vol 15**, n6, 1373-1379.
VARAS, E (1996) Estadígrafos de caudales máximos, **XVII Congreso Latinoamericano de Hidráulica,** Octubre 21-25, Guayaquil, Ecuador, 111-120.

Table 1. Statistics for Each Region

Region	Rec	Mo	Desv	Var	Skew	Kurt	Max	Area
UK								
Avg	20	56	26	.55	1.36	3.83	125	755
Dev	0	125	53	.21	1.38	6.24	248	1644
Max	20	561	238	1.14	3.89	16.4	1107	7490
Min	20	6	2	.34	-.62	-1.4	9	74
France								
Avg	52	598	233	.46	.94	2.3	1166	12489
Dev	22	895	332	.12	1.04	5.22	1619	28232
Max	95	3456	1275	.69	3.49	18.61	6300	110000
Min	14	13	6	.22	-.65	-1.56	33	30
Chile								
Avg	29	394	244	.64	1.42	1.54	1002	1376
Dev	11	268	163	.13	1.93	3.32	639	1610
Max	67	1101	568	.93	10.0	12.5	2648	6350
Min	15	44	41	.41	-.16	-1.61	166	163
USA								
Avg	39	51	53	1.1	1.55	3.06	234	56
Dev	12	23	28	.36	.9	4.94	157	46
Max	54	76	92	1.7	3.3	13.0	515	147
Min	25	10	11	.73	.76	-.09	40	18

Table 2 : Probability Weighted Moments

Region	Mo	mj1	mj2	mj3	mj4
UK					
Avg	56	0.64	0.48	0.39	0.34
Dev	125	0.03	0.04	0.04	0.04
Max	561	0.71	0.57	0.49	0.44
Min	6	0.60	0.44	0.34	0.28
France					
Avg	598	0.62	0.46	0.37	0.31
Dev	895	0.03	0.03	0.03	0.03
Max	3456	0.67	0.52	0.43	0.35
Min	13	0.57	0.40	0.31	0.26
Chile					
Avg	394	0.67	0.52	0.43	0.36
Dev	268	0.03	0.04	0.04	0.05
Max	1101	0.73	0.59	0.50	0.46
Min	44	0.62	0.46	0.36	0.24
USA					
Avg	51	0.76	0.63	0.54	0.48
Dev	23	0.05	0.07	0.07	0.08
Max	76	0.84	0.74	0.66	0.61
Min	10	0.70	0.55	0.55	0.39

STATISTICS OF MAXIMUM FLOWS 481

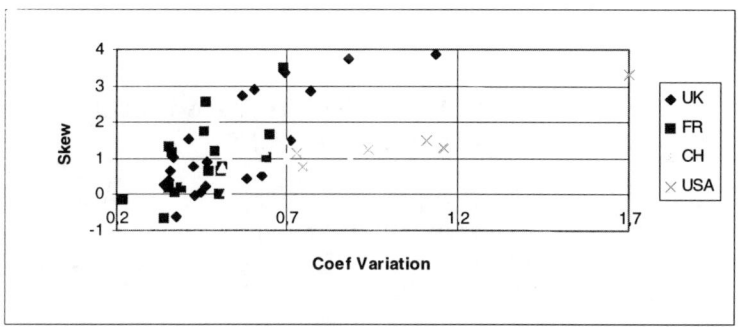

Figure 1: Skew vs Coefficient of Variation

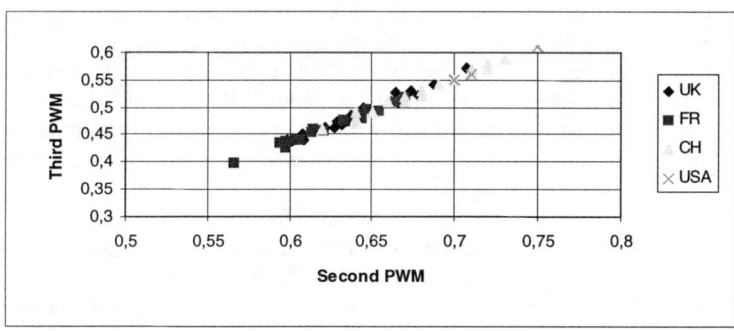

Figure 2: PWM Diagram : Third PWM vs Second PWM

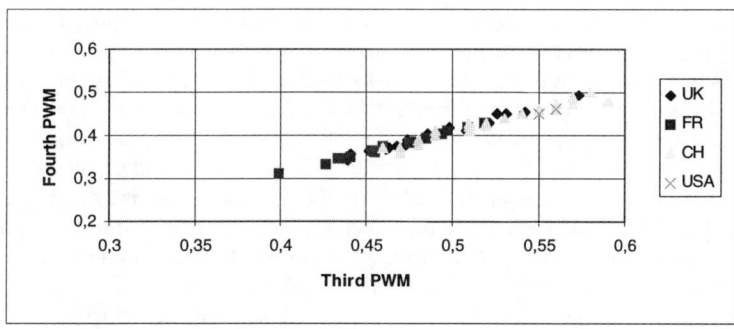

Figure 3 : PWM Diagram : Fourth PWM vs Third PWM

RISK ESTIMATION OF MONTHLY RAINFALL IN SEMIARID REGIONS

BONIFACIO FERNÁNDEZ
Professor, Hydraulics and Environmental Engineering Dept. Pontificia Universidad Católica de Chile. Casilla 306, Correo 22, Santiago, Chile.

ABSTRACT
This paper develops an algorithm for estimating the risk of occurrence of events defined as a run in statistically independent time series where periodicity is important. The procedure uses the basic ideas of return period and risk, as they are defined in hydrology, in order to be applied to meteorological droughts using monthly precipitation time series as water supply. Estimation of the risk related to the occurrence of a run of several consecutive months without precipitation in the semiarid zone of Chile, in the south of the Atacama Desert, are used as examples.

INTRODUCTION
The purpose of this paper is to present calculation algorithms developed for estimating the risk of occurrence of meteorological droughts defined on the basis of monthly precipitation series. These are relatively complex events characterized as a succession of periods of scarcity on the basis of non-seasonal although statistically independent time series. Use is made of the definition previously adopted for droughts as proposed by Yevjevich (1967), according to which, given one series of supply and another series of demand, droughts are defined as a succession of consecutive values in which the supply never manages to satisfy the demand.
The estimation of return periods and risks of drought involves a certain amount of difficulty, in that a succession of values must be taken into account rather than a single value such as is the case with floods. In addition, in this case addressed herein the variables that define the supply are periodical in nature. Fernández and Salas (1994, 1996) have made a general analysis of the problem of estimating the return period for droughts and have proposed a method which sets out approaches to this problem under various conditions. This paper discusses an extension of these procedures to cover the case of non-seasonality i.e., consideration of situations in which the probability of the variable taking on a given value are dependent on time, which occurs when considering monthly hydrological variables. More specifically, the method is applied to the case of monthly rainfall, which may in addition be considered as being statistically independent. Extension to statistically dependent process is in progress.

RISK OF COMPLEX EVENTS

Normally in hydrology, it is customary to define return periods by relating them with the probability of occurrence of simple, independent, non-periodic events, such as is ordinarily the situation with maximum annual floods (Lloyd, 1970; Kite, 1977). The immediate practical application of this development is its use in estimating the risk of failure of hydraulic works caused by one or more floods of a size greater than the design flood and occurring during the service life of the facility. In order to extend these calculation methods to complex events, return periods are defined as the anticipated amount of time until the event in question first occurs (Benjamin and Cornell, 1970; Stedinger et al., 1993), which for purposes of estimating risks of failure does not necessarily require imposing the condition that the series begins immediately at the time of the last occurrence of the event (Fernández and Salas, 1994). Making use of the procedure proposed by Fernández and Salas (1996), return period, T, is estimated as follows for the case of complex events:

$$T = \sum_{n=1}^{\infty} n f_n \qquad (1)$$

where f_n is the function introduced by Schwager (1983) to calculate the probabilities of occurrence of successions defined in the following manner:

f_n = the probability that the event in question will first occur at the instant n.

In addition, the function S_n is defined as follows

S_n = the probability that the event will occur at the instant n or before.

For events defined as a succession, it is assumed that their occurrence is verified at the instant at which they are brought to completion. If L is the service life of a project and the event is the design condition, then S_L is the risk of failure of the system.

RISKS AND RETURN PERIODS APPLIED TO PERIODIC EVENTS

Let it be assumed that the results of a chronological series are observed for each instant t = 1, 2, 3, ..., n, ..., in which values of $X_1, X_2, X_3, ..., X_n, ...$ are obtained corresponding to the periodic values of period ω. These periodic series are also described as $\{X_{v,t}\}$, where v=1, 2, ..., N, to indicate time in years, and t=1, 2, ..., ω, to denote periods or seasons.

In this paper, the event of interest consists of a succession of r consecutive months at a time, a condition which may be thought of as being representative of a meteorological drought of r months' duration $R(R_1, R_2, ..., R_r)$. The calculation algorithm additionally requires defining the function $q_n(l)$ as the probability that at any instant n the succession is brought to completion whose last l values satisfy the condition of the last l values of the set R, i.e.:

$q_n(l)$ = Prob. (X_n coincides with R_r, X_{n-1} coincides with R_{r-1},..., X_{n-l} coincides with R_{r-l})

If the series X is made up of statistically independent variables and if the succession defining the event of interest indicates the minimum value separating the normal state of the drought at each element, $R(P_1,P_2,P_3,...P_r,)$, the above function may be calculated as follows:

$$q_n(l) = \text{Prob}(X_n \leq P_r)\text{Prob}(X_{n-1} \leq P_{r-1})\cdots\text{Prob}(X_{n-l+1} \leq P_{r-l+1}) \tag{2}$$

To calculate f_n, use is made of the recursive algorithm proposed by Schwager (1983) and previously used by Fernández and Salas (1996) for non-seasonal situations. In the periodic case, the initial conditions are similar, i.e.:

$$f_n = S_n = 0 \text{ for } n<r \tag{3}$$

as well as:

$$f_n = S_n = q_n(l) \text{ for } n=r \tag{4}$$

and finally, for instants greater than the length of the succession, it is evident that:

$$S_n = S_{n-1} + f_n \text{ for } n>r \tag{5}$$

The probability of the succession occurring for the first time at instant n corresponds to the probability that a succession of length r in n, i.e., $q_n(r)$, should be brought to completion, taking into consideration that the event has not occurred at any time before then. For such purposes, it is necessary to consider the probabilities that the event has not actually occurred prior to $n-r$ and that it is not completed due to effects of symmetry in the definition of R, which leads to the following:

$$f_n = q_n(r)\left\{1 - S_{n-r} - \sum_{i=1}^{r-1}\frac{f_{n-i}}{q_{n-i}(r-i)}\right\} \tag{6}$$

where both $q_n(r)$ and $q_{n-i}(r-i)$ are calculated by means of equation (2).

This algorithm enables the functions f_n and S_n to be calculated for n=1,2,3,...,n,... for subsequent use in obtaining the return period by means of equation (1). In addition, it is apparent that S_L corresponds to the risk of failure for a service life of L units of time since it represents the probability of the event in question occurring at least once during the service life of the facility, whose failure is conditioned to the occurrence of the succession R once at the least.

It should be noted that when work with periodic variables is involved, both f_n and S_n depend on the timing of the initial instant. This situation may be clearly appreciated in the examples.

RISK OF MONTHLY RAINFALL DROUGHTS

It is of interest to estimate the risk associated with events consisting of various consecutive months without any appreciable precipitation or with no precipitation at all at various meteorological stations in the semiarid zone of Chile. This zone is located between the 27th and 34th parallels, South Latitude, immediately to the south of the Atacama Desert and north of Santiago. The major characteristics of the meteorological stations selected are shown in Table 1, and the location of each in the study area is shown on Figure 1.

Table 1. Characteristics of the meteorological stations selected for this study.

Name	Series	Location			Annual Precip. (mm)	
		Latitude	Longitude	Elevation	Average	St. Dev.
Copiapó	1952-94	27ª21'	70ª21'	380	13.9	16.5
Vallenar	1950-94	28ª34'	70ª47'	469	32.8	34.9
La Serena	1915-94	29ª54'	71ª15'	132	96.1	62.9
Ovalle	1954-94	30ª36'	71ª12'	220	116.9	76.9
Illapel	1953-94	31ª36'	71ª11'	310	181.3	106.1
San Felipe	1950-94	32ª45'	70ª44'	636	213.6	116.7
Santiago	1915-94	33ª27'	70ª42'	520	328.7	140.3

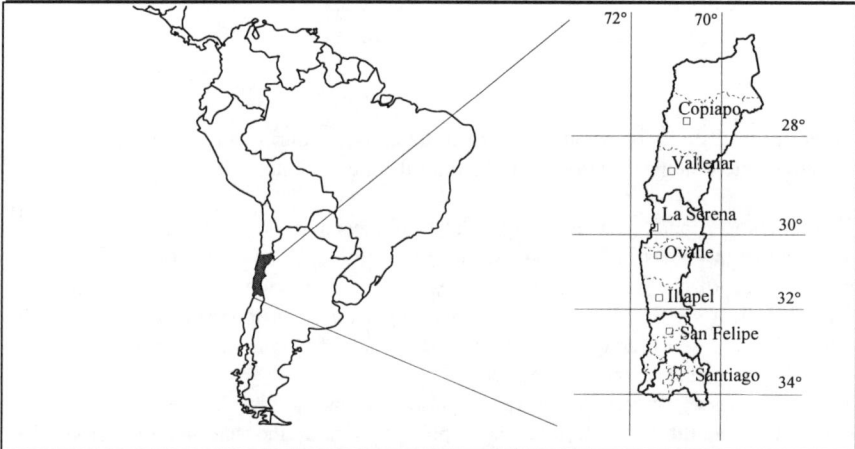

Figure 1. Location of meteorological gagging stations considered for this study in the semiarid zone of Chile.

This semiarid zone is characterized as having a scarce amount of annual precipitation and a dry season usually extending from October to April of each year. The amount of annual precipitation increases when moving from north to south, from an average of little more than 10 mm per year in Copiapó to somewhat more than 300 per year in Santiago. The probabilities of the amount of precipitation occurring in any given month being null or negligible are indicated in Figure 2 for each of the meteorological stations selected.

It is likewise of interest to evaluate the return period for occurrences of six consecutive months at a time without precipitation or, more properly speaking, with negligible precipitation only. Stated in another way, the event of interest is characterized by a succession $R(P_1<e, P_2<e, P_3<e, P_4<e, P_5<e, P_6<e)$ in which e is the amount standing for a negligible amount of precipitation. It is assumed that the series of interest begins with January. The function $q_n(l)$, where $l=1,2,3,...,6$, is also dependent on the timing of the initial month and the season in question. Due to the annual periodicity, $\omega=12$, this function is repetitive, starting with July of the second year.

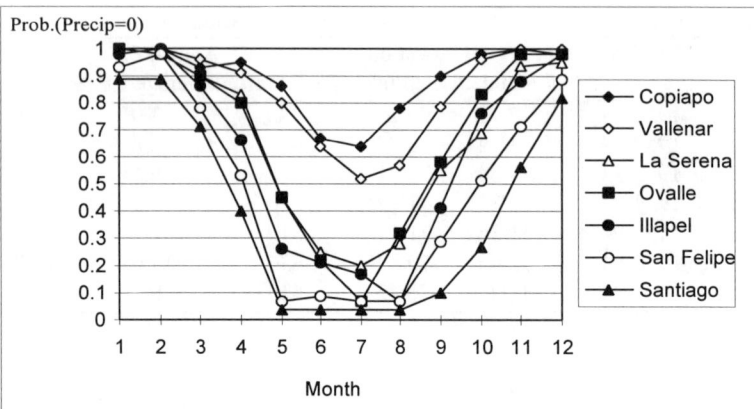

Figure 2. Probability of having null precipitation occur during any given month at the selected meteorological stations in the semiarid zone of Chile.

The functions f_n and S_n have been calculated by means of the proposed algorithm such as to arrive at values of n that are sufficiently large such as to be able to assure the convergence of the sum which estimates the return period. The values of both functions clearly reflect the effects of periodicity, as may be appreciated in Figure 3, in which the values of f_n and S_n are shown as a function of n, in months, for the conditions occurring at the La Serena gagging station.

Table 2 shows the values for the return period exhibiting consistent events during six consecutive months at a time without precipitation at the selected meteorological stations when the series is started with January. Additionally shown is the risk of this occurring during the first five years of a project at each of the localities where these meteorological stations are situated.

Table 2. Return periods for six consecutive dry months and their risk of occurrence in a five-year period for different localities of the semiarid zone of Chile.

Location	Return Period months (years)	Risk during 5 Years (probability of occurrence)
Copiapó	9.4 (0.78)	1.000
Vallenar	10.2 (0.85)	1.000
La Serena	16.7 (1.39)	0.999
Ovalle	15.5 (1.29)	1.000
Illapel	19.0 (1.58)	0.994
San Felipe	35.3 (2.94)	0.845
Santiago	82.7 (6.89)	0.481

ACKNOWLEDGMENTS

This paper represents part of the work done in the Research Project Fondecyt 1950983.

financed by Fondo Nacional de Desarrollo Científico y Tecnológico de Chile - Fondecyt. In addition, support from the USIA Affiliation Program between Colorado State University and Pontificia Universidad Católica de Chile is also acknowledged.

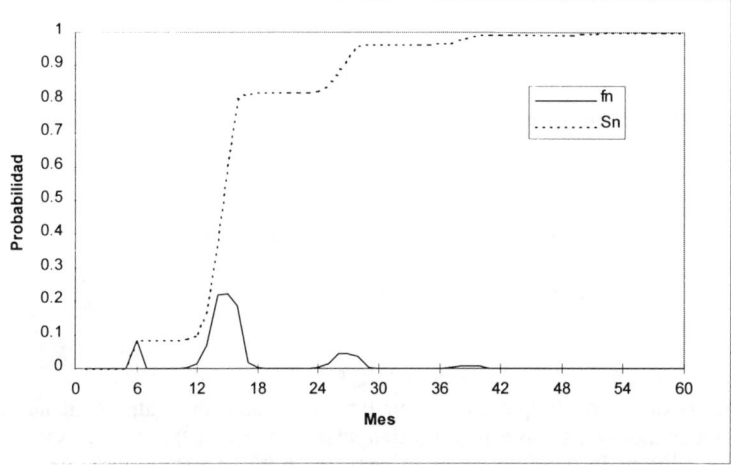

Figure 3.   Periodic functions f_n and S_n for six months at La Serena.

REFERENCES

Fernández, B. 1995. Return Period and Risk of Low Flows. Hydra 2000, XXVIth IAHR Congress, London. J. Gardiner, editor. Vol. 4:42-47.
Fernández, B. y J.D. Salas. 1994. Return Period and Risk of Hydrological Events. XVI Congreso Latinoamericano de Hidráulica, Vol. 3:221-232. Noviembre. Santiago, Chile. In Spanish.
Fernández, B. y J.D. Salas. 1996. Return Period and Risk of Hydrological Events. 1 Mathematical formulation. Considered for publication in Journal of Hydrologic Engineering.
Lloyd, E.H. 1970. Return Period in the Presence of Persistence. Journal of Hydrology, 10-3:291-298.
Schwager, S.J. 1983. Run Probabilities in Sequences of Markov Dependent Trials. JASA, 78:168-175.
Stedinger, J.R., R.M. Vogel, and E. Foufoula-Georgiou. 1993. Frequency Analysis of Extreme Events, Chapter 18 in Handbook of Hydrology. D.R. Maidment, Ed. MacGrawHill, N.Y.
Yevjevich, V. 1967. An Objective Approach to Definition and Investigation of Continental Droughts. Hydrology Paper 23. Colorado State University, Fort Collins, Colorado.

Poisson Pulse Queueing Model for Residential Water Demands

STEVEN G. BUCHBERGER and TRENT G. SCHADE
Department of Civil and Environmental Engineering
University of Cincinnati
Cincinnati, Ohio 45221-0071 USA

ABSTRACT

The intensity, duration, frequency and volume of residential water demands are modeled as a nonhomogeneous Poisson rectangular pulse (NPRP) process. Assuming plug flow conditions in the distribution system, the NPRP premise leads directly to expressions for the probability distribution and moments of the busy servers, flow rate, Reynolds number, travel time, and disinfectant concentration at any point along the supply network. Field measurements show good agreement between modeled and observed water demands at 18 single family homes along a dead-end main.

INTRODUCTION

Recent amendments to the Safe Drinking Water Act have led to a proliferation of models for predicting changes in water quality between the points of treatment and consumption (Clark et al. 1993, Rossman et al. 1994). Modeling water quality in distribution systems is difficult for many notable reasons, including complex dynamic hydraulic conditions. Flow patterns change continuously over time and across space in response to random demands imposed by many consumers dispersed throughout the service area. The intertwined issues of random flows and uncertain water quality are especially acute in peripheral dead-end regions of the network. Here travel times are long, stagnant conditions are common, chlorine residuals may be low and individual consumers measurably influence flow. Taken collectively, these factors often stymie water quality models applied to dead-end mains of a water distribution system.

This paper highlights some features of a parsimonious stochastic model of residential water use. The model is based on the premise that the intensity, duration, frequency and volume of water demands follow a nonhomogeneous Poisson rectangular pulse (NPRP) process. Focusing on dead-end mains, the NPRP premise leads directly to expressions for the mean, variance, and probability distribution of busy servers, flow rates, travel times, and disinfectant concentrations at any point along the supply line.

NPRP MODEL CONCEPT

WATER PULSE

The frequency of residential water use is assumed to follow a Poisson arrival process with a time dependent rate parameter. When a water use occurs, it is approximated as a rectangular pulse of random duration and random intensity as illustrated in Figure 1. It is unlikely that more that one pulse will start at the same instant. However, owing to the random duration of each water pulse, it is possible that two or more pulses with different starting times will overlap for a limited period. When this occurs, the total water use at the residence is the sum of the individual intensities from the coincident pulses.

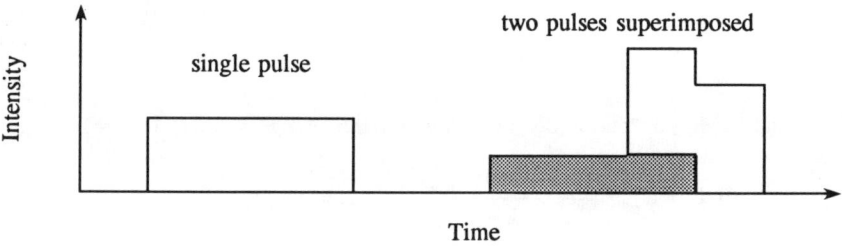

Figure 1. Poisson rectangular pulse process.

DEAD-END MAIN

Dead-end mains are a sequence of links in the distribution network having terminal point(s) that do not rejoin the looped portion of the system as shown in Figure 2. Since water moves through a dead-end main in the downstream direction only, the principle of mass conservation alone is sufficient to estimate flow rates. Dead-end mains are surprisingly common. Utility surveys show dead-end mains account for about 25 percent of the total infrastructure in a typical distribution system and tend to service a disproportionately high percentage of the residential consumer base.

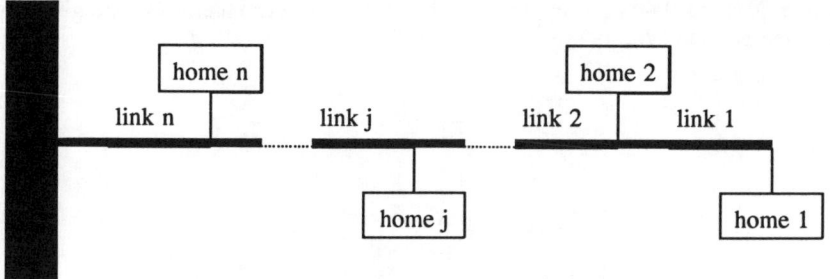

Figure 2. Definition sketch of dead-end main.

NPRP MODEL RESULTS

Owing to space limits, only the homogeneous case is presented here. Details on the nonhomogeneous case can be found in Buchberger and Wu (1995) and Wu (1996).

BUSY USERS

Noting the analogy between demands on a telephone exchange (Erlang, 1917-18) and demands on a water system, the approach taken here exploits a rich base of work intimately connected to queueing theory. Suppose customers arrive at residence j according to a homogeneous Poisson process with parameter λ_j and receive service at rate μ_j. The service rate can be viewed as the inverse of the mean duration of a pulse. Let K_j be the number of busy water servers in residence j under steady state conditions. The total number of busy servers in the block of n residences is

$$K_n^* = \sum_{j=1}^{n} K_j \qquad (1)$$

Note that K_n^* has a Poisson distribution with parameter $\rho_n^* = \Sigma(\lambda_j/\mu_j)$, j=1,...,n,

$$P[K_n^* = k] = \frac{(\rho_n^*)^k \exp(-\rho_n^*)}{k!} \qquad k = 0,1,2,... \qquad (2)$$

A typical single family residence has $\rho_1^* \approx 0.05$ (Buchberger and Wells, 1996). From Equation (2), water in a dead-end main is stagnant with probability $P[0]=\exp(-\rho_n^*)$.

FLOW RATES

The intensity of water use at a busy server in any residence is assumed to be an independent and identically distributed positive continuous random variable with mean α and variance β^2. The flow rate through link n (see Figure 2) resulting from water demands occurring downstream at K_n^* busy servers is

$$Q_n^* = \sum_{j=1}^{n} Q_j \qquad (3)$$

where Q_j is the flow into residence j. The mean and variance of Q_n^* are given by (Buchberger and Wu, 1995)

$$E[Q_n^*] = \sum_{j=1}^{n} \rho_j \alpha_j \qquad (4a)$$

$$Var[Q_n^*] = \sum_{j=1}^{n} \rho_j (\alpha_j^2 + \beta_j^2) \qquad (4b)$$

where $\rho_j = \lambda_j/\mu_j$. Equation (4a) is a generalized version of the expression given by Linaweaver et al (1966) for the expected demand on a water system serving n homes.

TRAVEL TIMES

Travel time through a dead-end link depends on three factors, namely, link volume, pulse volume, and customer arrival rate. The arrival rate experienced by link n due to downstream customers demanding water is given by

$$\lambda_n^* = \sum_{j=1}^{n} \lambda_j \qquad (5)$$

The total volume of water $V_n^*(t)$ swept through link n during time $\{0,t\}$ is a compound Poisson process

$$V_n^*(t) = \sum_{k=0}^{N_n^*(t)} [X_n^*]_k \qquad t \geq 0 \qquad (6)$$

where $N_n^*(t)$ arises from a Poisson distribution with parameter $(\lambda_n^* \, t)$ and represents the number of water pulses generated by the block of n residences during interval $\{0,t\}$. X_n^* is the volume of a single water pulse and is assumed to be an independent and identically distributed random variable with mean ϕ_n^* and variance $(\psi_n^*)^2$.

Assuming plug flow, the cumulative distribution of water travel time through a link can be related to the probability of downstream demands as follows

$$P[T_n \leq t] = P[V_n^*(t) \geq L_n] \qquad (7)$$

where T_n is the travel time through link n and L_n is the total volume of link n. Equation (7) indicates that the travel time through link n is governed by how long it takes cumulative water demands to displace the link volume. The kth moment of travel time follows directly from the distribution function

$$E[T_n^k] = \int_0^{\infty} k t^{k-1} P[T_n > t] dt \qquad (8)$$

It can be shown that the mean and variance of T_n are given by

$$E[T_n] = \frac{L_n}{\lambda_n^* \phi_n^*} \qquad (9a)$$

$$\text{Var}[T_n] \sim \frac{L_n}{(\lambda_n^*)^2 \phi_n^*} \left[1 + \left(\frac{\psi_n^*}{\phi_n^*}\right)^2 \right] \qquad (9b)$$

Equation (9a) for the mean travel time is exact, irrespective of the probability distribution of pulse volumes. It can be shown that the denominator term ($\lambda\phi$) is equivalent to the mean flow (see Eq 4a). Hence, Equation (9a) gives $T_n = L_n/Q_n$ which is identical to the well-known deterministic result for mean residence time in a closed reactor. Equation (9b) is exact for exponential pulse volumes; in other cases it is an excellent approximation provided $L_n/\phi_n^* > 30$, a condition often satisfied in practice.

RESIDUAL CONCENTRATIONS

Assuming first-order reaction kinetics, the concentration of chlorine in a parcel of water residing for an elapsed time t in link n of the dead-end main is

$$C_n(t) = C_0 \exp[-K_n t] \tag{10}$$

where C_0 is the chlorine concentration when the parcel entered link n and K_n is the decay constant. Using Eq (10), the probability distribution of chlorine concentration at the end of link n can be expressed in terms of travel time through the link,

$$P[C_n(t) \le c] = P\left[T_n \ge t = \frac{1}{K_n}\ln\left(\frac{C_0}{c}\right)\right] \tag{11}$$

Denoting $\Lambda_n = \lambda_n^*/K_n$ and then assuming exponential pulse volumes, the mean and variance of the dimensionless disinfectant concentration are given by (Wu, 1996)

$$E\left[\frac{C_n(t)}{C_0}\right] = \exp\left[\frac{-L_n/\phi_n^*}{1+\Lambda_n}\right] \tag{12a}$$

$$\mathrm{Var}\left[\frac{C_n(t)}{C_0}\right] = \exp\left[\frac{-2L_n/\phi_n^*}{2+\Lambda_n}\right] - \exp\left[\frac{-2L_n/\phi_n^*}{1+\Lambda_n}\right] \tag{12b}$$

Note that Eq (12a) gives a slightly higher result than the estimate obtained by simply substituting the mean travel time from Equation (9a) into Equation (10).

FIELD VERIFICATION

To verify the NPRP hypothesis, residential water demands have been monitored over a four year period at 55 instrumented homes serviced by dead-end mains in Milford, Ohio about 20 miles northeast of Cincinnati. In a recent field test nearly 59,000 water demands were recorded around the clock for a 30 day period at 18 single family homes on a dead-end loop (Schade, 1996). Results, shown in Figure 3, demonstrate that the NPRP model provides an excellent fit to the hourly variation in the probability distribution of busy servers at the study site.

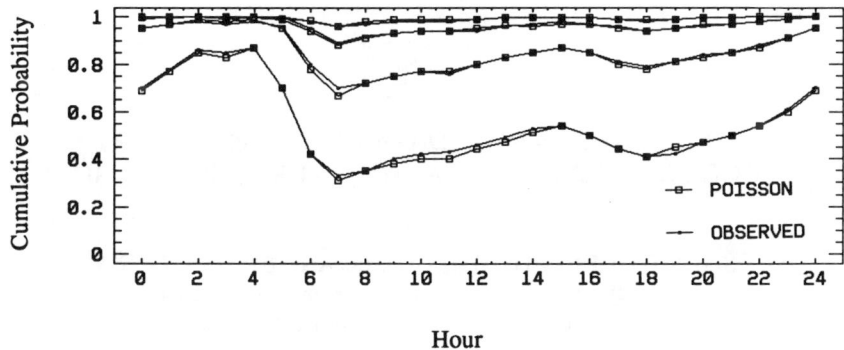

Figure 3. Hourly distribution of busy servers
(k=0,1,2,3 from bottom to top)

SUMMARY

The premise that residential water use occurs as a nonhomogeneous Poisson rectangular pulse process leads to a reasonable, rigorous and robust framework for solving contemporary problems that arise with water distribution systems. Several results from the NPRP model are outlined. The NPRP approach offers a stochastic complement to deterministic models which are indispensable tools for designing and operating distribution systems to meet water demands and quality standards.

ACKNOWLEDGMENTS

This work was supported by awards from NSF (BCS-9257608) and AWWARF (294).

REFERENCES

Buchberger, S.G. and L. Wu (1995) "Model for instantaneous residential water demands", **ASCE J Hydraulic Eng**, 121(3): 232-246.

Buchberger, S.G. and G. Wells (1996) "Intensity, duration, and frequency of residential water demands", **ASCE J Wat Res Plan & Mngt**, 122(1): 11-19.

Clark, R.M., W.M. Grayman, R.M. Males and H.F. Hess (1993) "Modeling contaminant propagation in drinking-water distribution system", **ASCE J Env Eng**, 119(2): 349-364.

Erlang, A.K. (1917-18) "Solutions of some problems in the theory of probabilities of significance in automatic telephone exchanges" **Post Office Electric Engr J**, 10, 189-197.

Linaweaver, F.P. Jr., J.C. Geyer and J.B. Wolff (1966) **Residential Water Use**, Johns Hopkins University, Baltimore, Maryland.

Rossman, L.A., R.M. Clark, and W.M. Grayman (1994) "Modeling chlorine residuals in drinking-water distribution systems", **ASCE J Env Eng**, 120(4):803-820.

Schade, T.G. (1996) "Water demand and travel time in a residential dead-end loop", **MS Thesis**, University of Cincinnati, 64 pages.

Wu, L. (1996) "Stochastic model of flow in the periphery of a water distribution system", **PhD Dissertation**, University of Cincinnati, 207 pages.

AN EVALUATION OF DATA NEEDS TO SUPPORT FLOOD FREQUENCY ESTIMATION AT REGULATED SITES

S. ROCKY DURRANS, SAŠA TOMIĆ AND STEPHAN J. NIX
Department of Civil and Environmental Engineering
The University of Alabama, Tuscaloosa, Alabama, USA

ABSTRACT

A basic input for the design of hydraulic structures and flood protection facilities, as well as for nonstructural approaches to flood protection, is provided through flood frequency analysis. A tremendous amount of effort has been devoted to the development and improvement of methods of flood frequency analysis during the 1900s. Unfortunately, however, the majority of this work has ignored the fact that most streams in the world have become increasingly regulated by the actions of man. Most research on flood frequency analysis continues to focus on traditional methods that were never intended for application to sites where the effects of regulation are present. It is the focus of our ongoing research to study the problem of flood frequency estimation at regulated sites, and to make an assessment of the data needs and estimation approaches that are the most suitable for resolution of this problem. This paper reports some of our initial findings, and outlines the strategy of our future research.

INTRODUCTION

A basic input for the design of hydraulic structures and flood protection facilities, as well as for nonstructural approaches to flood protection, is provided through flood frequency analysis. A tremendous amount of effort has been devoted to the development and improvement of methods of flood frequency analysis during the 1900s. Unfortunately, however, the majority of this work has ignored the fact that most streams in the world have become increasingly regulated by the actions of man. Most research on flood frequency analysis continues to focus on traditional methods that were never intended for application to sites where the effects of regulation are present.

The primary focus of our ongoing research is to study the problem of flood frequency estimation at locations immediately downstream of regulating structures, such as dams. There are, at present, no generally accepted methods by which flood frequency estimates can be made in such instances. It is therefore not surprising to find that a number of different and rather *ad hoc* procedures are being used by different organizations. Some agencies, such as the Tennessee Valley Authority (M. Goranflo, personal communication, 1996) are using rather standard statistical techniques, and effectively ignore the lack of homogeneity of the basic data. Others have been seduced into believing that continuous rainfall-runoff modeling is the answer. Those of this latter persuasion employ such methods to synthesize streamflow hydrographs, and then subject the annual peaks in those hydrographs to frequency analyses. This is certainly an attractive option (which explains its seductiveness), but there is again a problem with homogeneity of the annual peaks. Rainfall-runoff modeling also exhibits a "loss of variance" problem (Kirby, 1975; Thomas, 1982; Muzik, 1994). We note that these problems with continuous simulation are not inherent to the ideology of continuous simulation itself, but rather arise as a consequence of inadequacies in current mathematical models to represent the entire range of watershed response.

In our research, we are pursuing the regulated flood frequency problem through an integrated deterministic-stochastic modeling approach (Laurenson, 1974; Durrans, 1994,1995). The stochastic components of the approach permit us to realistically handle the random elements associated with streamflow conditions, reservoir pool levels, etc., while the deterministic components allow us to preserve such physical constraints as conservation of mass and, if necessary, momentum and energy. Ostensibly, the presence of the deterministic components should also permit flood frequency analyses for sites downstream of regulating structures to be accomplished in prescriptive rather than just descriptive ways. Thus, there is a potential for improving the decision-making process as it relates to reservoir operation policies.

The following section of this paper presents, in a very generic way, the form of the estimation strategy that is being employed in our research. The particulars of exactly how the generic method should be implemented are the main topics of our efforts. Issues of special relevance are the data needs to support the estimation process, as well as the way(s) in which the estimation should actually be accomplished. These issues are addressed in subsequent sections of this paper.

MODELING APPROACH

The various characteristics of a flood hydrograph, such as the peak discharge and volume, downstream of a regulating dam are clearly dependent on conditions existing in the reservoir during the passage of the flood event. For example, the reservoir pool level existing at the beginning of the flood event has a very significant effect, as do settings of the reservoir outlet gates. For instance, in an arid region where a reservoir may have been drawn down to nearly a completely empty state during the dry season, the onset of a flood may do no more than refill the reservoir and contribute to little or no outflow.

If the initial pool level of a reservoir were always the same, and if the outlet gates were always operated in exactly the same way when floods occur, then a series of annual flood peaks (say) downstream of the dam could be considered to be homogeneous in the sense that they all arose from the same underlying population distribution (assuming, of course, that the inflow flood peaks are also homogeneous). The real world is not so kind, however. Reservoir pool levels change as a consequence of antecedent conditions (even if floods were to always occur at the same time of year), and political pressures and public fears conspire to cause reservoir gates to be operated in different ways during flood events. The net effect of this is that the characteristics of the regulated flood frequency distribution must be conditioned on the initial reservoir pool level and the outlet gate operations. The conditional distributions for the various combinations of pool level and gate setting can then be integrated into the final regulated flood frequency curve using the theorem of total probability. Mathematically, this can be expressed as

$$F_Y(y) = \int_\Omega F_{Y|\Lambda}(y|\lambda) f_\Lambda(\lambda) d\lambda \qquad (1)$$

Here, y is a random vector of regulated flood characteristics and λ is a random vector of reservoir conditions. $F_Y(y)$ is the desired cumulative distribution of regulated flood characteristics, $F_{Y|\Lambda}(y|\lambda)$ is the cumulative distribution of regulated flood characteristics conditioned on a particular set λ of reservoir conditions, and $f_\Lambda(\lambda)$ is the joint density function of reservoir conditions. The symbol Ω denotes the space of feasible reservoir conditions. The conditional cumulative distribution function $F_{Y|\Lambda}(y|\lambda)$ is related to the joint distribution of upstream (unregulated) flood characteristics and reservoir conditions, which is denoted as $F_{X\Lambda}(x,\lambda)$:

$$F_{Y|\Lambda}(y|\lambda) = G[F_{X\Lambda}(x,\lambda)] \qquad (2)$$

The mapping of $F_{X\Lambda}(x,\lambda)$ into $F_{Y|\Lambda}(y|\lambda)$, symbolized by $G[\ ]$, can be accomplished using Monte Carlo simulation. Such simulations involve routing of random inflow flood hydrographs through the reservoir, and essentially represent the solution of a stochastic differential equation (Soong, 1973).

Equations (1) and (2) are very generic and can be applied to problems other than the regulated flood frequency problem considered here. For instance, they can be applied for regionalization of flood frequency behavior; i.e. for transfer of information from one site to another along the same stream. They can also be applied to derive the distribution of runoff characteristics on the basis of distributions of rainfall and basin characteristics.

The dimensions of the random vectors in equations (1) and (2) are a major concern. From the viewpoint of keeping the estimation procedure from becoming overly complex, it is desirable to make the dimensions small. However, the physics of a particular application may operate in an opposite direction, and may indicate that the dimensions should be relatively large. To illustrate, consider an application to a small reservoir with an uncontrolled outlet, where the distribution of reservoir conditions would be univariate and would involve only the pool level at the beginning of a flood event. Complex regulating structures with several controllable outlets would require the use of multivariate distributions.

The major question being addressed in this research is that of how complex (i.e. how highly dimensional) must be the distributions of reservoir conditions and unregulated flood characteristics. In essence, we are applying equations (1) and (2) in an inverse way to seek answers to questions of the nature: *If data of type 'x' were available for flood frequency estimation purposes, would they significantly impact the results and thus imply that their collection is needed? Also, if data of type 'x' are important, how much of those data are needed?* To answer these questions, we are applying equations (1) and (2) in a systematic way of gradually increasing their complexities (dimensions) to study the impacts on the estimation of the regulated flood frequency distribution. This will allow us not only to answer the questions posed in this research, but it will also allow us to assign priorities to future data collection efforts.

DATA AVAILABILITY

Naturally, an assessment of data needs for resolution of the regulated flood frequency problem should begin with a survey of the data types and quantities that are currently collected at regulating structures. In our current research, we are focusing on the characteristics of reservoirs in the Southeastern United States. While we have been in contact with one Western water agency (namely, the U.S. Bureau of Reclamation), our survey has been limited mainly to utility companies and agencies in the Southeast. The reason for our focus on the Southeast is the wet and humid nature of the climate, which causes most reservoirs to be nearly full most of the time. This tends to simplify the problem, and it thus provides a rational starting point for this research. Future efforts will consider the more complicated problems associated with reservoirs in arid climates.

In accomplishing the survey, 41 individuals were contacted in more than 20 different agencies and sub-agencies. These included the Corps of Engineers, the Bureau of Reclamation, the U.S. Geological Survey, and the Tennessee Valley Authority. Utility companies contacted included Southern Company Services, Alabama Power Company, and Georgia Power. Each of these agencies collects and/or distributes at least some data for reservoir inflows, outflows, and pool levels. Most of these data are stored digitally, but they are not always collected automatically. These data are usually reported at a frequency of one day.

An issue that we believe is significant is that, while most agencies collect data on individual outlet gate settings and discharges, these data are usually maintained only in handwritten, hard-copy form and thus are not readily available for use. Only in a few cases are these data maintained in a digital form. It appears, for the most part, that they are stored in cardboard boxes in out-of-the-way places where accessing them would be a nightmare at best. The significance of these data lie mainly in the physics of the reservoir routing problem, and also in the fact that a desired outflow discharge from a major structure can be obtained in any of a number of ways (there is a unique relationship between gate settings and the resulting discharge, but the converse is not necessarily true).

An additional finding in our research is that the Federal Emergency Management Agency (FEMA) does not appear to have a comprehensive national database that is indicative of *where* flood damages have historically occurred (B. Blanton, personal communication, 1996). We find this significant also, as we believe that most flood damages in the U.S. (and in most of the rest of the world, for that matter) are occurring, or have occurred, on regulated streams and rivers. We base this argument on the observation that most streams and rivers are now regulated in one form or another, and on the fact that individuals have a higher propensity to occupy and use lands in supposedly protected areas than in areas which are not protected. The existence of concrete evidence, in the form of a database, that defensible flood estimates on regulated watercourses are more important than those for natural watercourses could be what it takes to convince the research community that their efforts have been largely misdirected.

FUTURE RESEARCH

With an assessment of current data collection efforts essentially complete, this research can now turn to the problem of systematically evaluating and prioritizing these and other possible data collection efforts. As noted earlier, this will be accomplished through the use of computer modeling and Monte Carlo simulation to solve equations (1) and (2), and by systematically increasing the complexities of the representations of various types of reservoirs and their inflow hydrographs. We intend to begin with simple reservoirs with uncontrolled outlets, to progress from there to reservoirs with a single controllable outlet, and to ultimately address reservoirs with multiple outlets whose gate settings can be changed during the passage of a flood event.

In the long term, future research will need to address the problem of estimation of regulated flood frequencies at locations some distance downstream of a regulating structure. Because of the attenuating effects of floodplain storage and other factors, these flood distributions are different from that immediately downstream of the regulating structure. Again, however, equations (1) and (2) represent a framework within which the problem may be solved; the only difference is that channel routing, as opposed to reservoir routing, is necessary. This problem of information transfer along a river reach is fundamental to the notion of regionalization of flood frequency information, and essentially provides a physically-based framework which can be used to assess the suitabilities of the implied structures in current regionalization schemes.

SUMMARY

This paper has presented a brief summary of our progress to date on the problem of estimation of flood frequency characteristics at locations immediately downstream of regulating structures. Progress to date has centered on a survey and evaluation of current data collection efforts, and has led to the conclusion that while some types of data are regularly collected and easily available, others are not. This suggests that there is a need for some rethinking of the types of databases that are routinely maintained. Future research will further establish the types and quantities of data that are necessary for resolution of the regulated flood frequency problem, and will also result in a prioritization of those data collection needs.

ACKNOWLEDGEMENTS

The research reported in this paper has been supported by the U.S. Geological Survey, through the Alabama Water Resources Research Institute, which support is gratefully acknowledged.

REFERENCES

Durrans, S.R., Total probability methods for problems in flood frequency estimation, *Selected Proceedings*, International Conference on Statistical and Bayesian Methods in Hydrologic Sciences, UNESCO Publishing, Paris, in press, 1995.

Durrans, S.R., Integrated deterministic-stochastic approach to flood frequency analysis, *Proceedings*, 14th Annual AGU Hydrology Days Conference, Colorado State University, Fort Collins, Colorado, 1994.

Kirby, W., Model smoothing effect diminishes simulated flood peak variance, American Geophyscial Union, *EOS* 56(6):361, 1975.

Laurenson, E.M., Modeling of stochastic-deterministic hydrologic systems, *Water Resources Research* 10(5):955-961, 1974.

Muzik, I., Understanding flood probabilities, in *Stochastic and Statistical Methods in Hydrology and Environmental Engineering, Vol. 1*, K.W. Hipel, ed., Kluwer Academic Publishers, Dordrecht, The Netherlands, 1994.

Soong, T.T., *Random Differential Equations in Science and Engineering*, Academic Press, New York, 1973.

Thomas, W.O., Jr., An evaluation of flood frequency estimates based on rainfall/ runoff modeling, *Water Resources Bulletin* 18(2):221-230, 1982.

PROTECTION OF THE ENVIRONMENT AGAINST FLOODS - A STATISTICAL PROBLEM ?

REINHARD POHL
Dresden University of Technology, 01062 Dresden, Germany, Corporate Member of IAHR

ABSTRACT

Very often the understanding of processes in nature and environment can be improved by considering from a probabilistic point of view. In the present paper, statistical methods of hydraulic design are proposed. Three examples are presented: a stilling basin, flood protection at a river and freeboard at a dam. It was the aim to make calculations at a reasonable effort. According to the opinion of the author, the results obtained are more convincing than in the calculation of individual values resulting from definite load cases. In future, it will be necessary to lay down limiting values for permissible probabilities of failure.

INTRODUCTION

Of course, in the dimensioning of hydraulic structures nothing should be left to chance. Because using the actual methods and tools of engineering work, in nearly all cases the appropriate dimensioning of hydraulic structures is possible.
However, there always remains a certain element of risk which results from the fact that input data can scatter or show a trend which might be influenced by global changes mainly in case of field data. Reaction of the public to risks taken over involuntarily is very different and includes supersession and overrating. For this reason, in the following it is tried to describe comparatively simple approaches and procedures by which the hydraulic- hydrological safety of hydraulic structures can be determined.

ACTUAL DESIGN PRACTICE AND PROBABILISTIC CONSIDERATIONS

The basic concept of hydraulic design generally accepted and frequently practised in Germany at present is to select a definite case (or several ones, if necessary) for which the calculation is made. In the sense of the hydraulic-hydrological consideration such a case is characterized by a definite discharge Q (e.g. HQ_{50}, HQ_{100} for weirs and river training measures respectively, HQ_{1000} for dams) under special flow conditions. Comparing the result (for example that of water level) with

a limiting or critical value, which possibly can include a safety factor results in the acceptance or refusal. By an appropriate adaptation of the influenceable input values or parameters, within repeated calculations the hydraulic structure can be dimensioned to fit the demands of safety as well as of economy.
The discharge Q which in the following is to be considered as the essential random quantity can be described by a flood hydrograph Q = Q(t) for instance in the form

$$Q = Q_s \cdot \left[\frac{t}{t_A} \cdot e^{(1-t/t_A)} \right]^{n_f} \quad \ldots\ldots\ldots\ldots\ldots\ldots\ldots\ldots\ldots\ldots\ldots\ldots\ldots\ldots\ldots\ldots\ldots\ldots\ldots(1)$$

as given by *Sinniger and Hager 1984* with the peak discharge Q_S, the time to peak t_A and the shape factor n_f. Although for the dimensioning often only the peak value is used, all three values mentioned are random quantities. The series of peak values is obtained by selection of the maximum values from subintervals (years) of the period of observation.
The design flood BHQ can for instance be found by extrapolation of the peak values (e.g. BHQ = HQ_{1000}) by adapting the distribution function of best fit (e.g. LogNormal, Gumbel, Pearson III, Fig. 1).

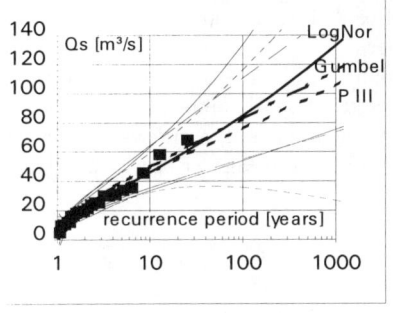

Fig. 1 : Gauge Hartha/ Saxony (Würschnitz) Peak discharge vs. recurrence period

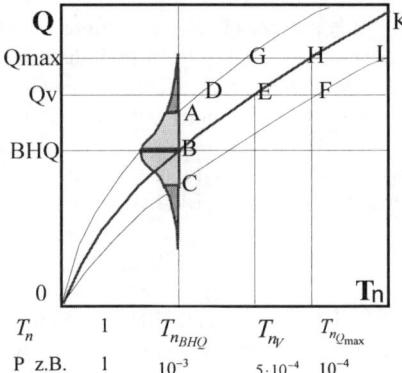

Fig. 2 : Peak flood discharge vs. recurrence period (without discharge volume)

In the common hydraulic dimensioning with quantile values of the peak discharge it is ensured that the discharge capacity is equal to this value, however, nothing can be said about the probability of hydraulic-hydrologic failure of the system as a whole, which is normally less.
Related to Figure 2, the previous dimensioning corresponds only to one calculation for the point B, whereas failure only occurs for $Q > Q_v$ i.e. beyond point E. When uncertainty in the extrapolation (i.e. due to too short data series) is to be taken into consideration by means of a transverse distribution between A and C (see *Kirnbauer 1981*), in the case of the present Figure an incomplete result could turn out without any failure ($P_v = 0\%$), because the discharge Q belonging to the upper limit of the confidence interval in the point A is less than Q_v.

A possible failure should be calculable by taking into consideration the discharge values between E and H analogously to the chosen. The only theoretically possible values beyond the maximum possible discharge Q_{max} (between H and K) can be replaced by Q_{max} which corresponds to the probably maximum flood or to a discharge determined by envelope curves.

Considering the design case (Q ≥ BHQ), the values of the section B - H of the curve must be used. For arbitrary years, the section 0 - H must be used.

In the previous considerations it was assumed that the peak discharge of the flood is decisive solely for dimensioning. In these cases, an exact correlation can be established between the distribution of the discharge value Q_S and the distribution of the resulting value.

When the random quantities t_A and n_f are introduced this is somewhat complicated as from the distributions for the hydrograph parameters Q_s, t_A, n_f at least theoretically, the joint (trivariate) density of distribution $f(Q_s, t_A, n_f)$ can be derived and the probability of failure would correspond to the integral over the failure space B:

$$P_{\ddot{u}} = \iiint_B f(Q_s, t_A, n_f) \cdot dQ_s \cdot dt_A \cdot dn_f \quad \dots\dots(2)$$

Within the present paper, the „failure" should be defined as a non-desired hydraulic state, which does not necessarily induce damages.

Even when the marginal distributions are simplified, because of the transformation equations which cannot be solved explicitly (flood routing and freeboard calculations) an exact solution is hardly or not possible and, as it was mentioned above, it is necessary to use a simulation solution by means of statistical testing (Monte-Carlo-Method, MCM).

FIELDS OF APPLICATION IN HYDRAULIC ENGINEERING

DIMENSIONING A STILLING BASIN: Using the energy conservation law and the momentum conservation law, a clear correlation can be established between the discharge and required length of the stilling basin as well as the required downstream depth. From the condition $h_{u\ exist} > h_{u\ requ}$ (sufficient downstream water depth) and $L_{exist} > L_2$ (sufficient length of the stilling basin) quantile values for Q can be found, for which the system works correctly and vice versa.

It seems to be useful to determine the probability of failure following the classic definition of probability

$$P_V = \frac{\text{number of cases of failure}}{\text{number of all cases}} \quad \dots\dots(3)$$

The example introduced in Figure 3 deals with an uncontrolled weir with a following, 2-dimensional stilling basin with a deepened bottom and two different downstream key curves. There are plotted 100 discharge values on the abscissa following their Gumbel-distribution according to Fig.1. The test value graphs indicate firstly whether the existing downstream water depth h_u and the deepening δ of the stilling basin (minus 5%) is greater than the conjugate water depth h_2 (see

Preissler, Bollrich 1996) and secondly whether the existing length of the stilling basin L_{exist} is greater than the required length L_2. Consequently, negative test values stand for the above defined hydraulic failure of the stilling basin. From the present example the failure probability can be obtained by counting the discharge points on the Q-axis where the test function is negative.

For the assumed input data, the probability of failure is of $P_{V(h)} = P_{V(h \cup L)} = 0,13$ at downstream open channel flow (dashed line). For the backwater discharge (due to culvert) beginning from about 34 m³/s (solid line) the analogous value is $P_{V(h)} = 0,14$ and $P_{V(h \cup L)} = 0,15$ with the failure due to a too short stilling basin being of $P_{V(L)} = 0,01$.

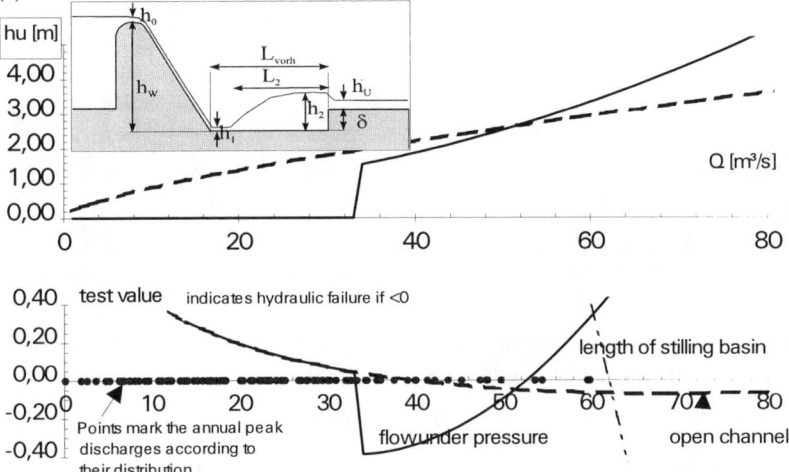

Fig. 3 : Downstream depth vs. discharge (top) and operation test values (bottom) of the 2-dimensional stilling basin (example) - negative check values mean underdimensioning

FLOOD PROTECTION AT RIVERS: For the hydraulic dimensioning of the cross-section of rivers the distribution of the water levels can be used. However, this can only be assigned directly to the distribution of the peak discharge Q_S if there are special conditions. In the case of unsteady flow, bed movement and eventual waves respectively a situation as in Figure 4 results, where the probability of inundation is marked by P_3 and P_2 (without waves).

HYDRAULIC-HYDROLOGIC SAFETY OF DAMS: According to the common practice in several countries the expected value of HQ_{1000} is used for dam design (checked by the PMF) that brings up the problems which have been discussed in connection with Fig. 2. It has been proposed to apply a probabilistic approach based on the comparison of water levels to overcome these problems (*Pohl 1995, 1996*): At the beginning of a flood event a random initial reservoir level h_0 (see Fig. 5) is to be found, depending on previous floods or droughts, and on the reservoir operation.

The density function f_1 describes this value. Any large flood inflow described by the hydrograph (eq.1) with random parameters (Q_S, t_A, n_f) will normally produce a rise of water level until the maximum level H_{max} in the uncontrolled storage is reached. The density function f_2 describes the random position of H_{max}. The third random variable is the wave generating wind causing wave run-up and wind set-up, in so far as it simultaneously appears with the flood. These wind effects require a freeboard height h_f (density function f_3). The above defined "failure" occurs when:

$$h_o + h + h_f = H_{max} + h_f > b = \text{crest level} \quad \ldots (4)$$

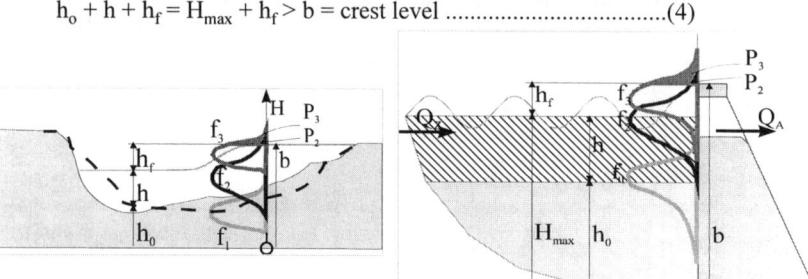

Fig. 4: Distribution of the maximum water levels at a cross-section of a river

Fig. 5 : distribution of three main load values: initial water level h_0 ; additional elevation h due to flood ; required freeboard height h_f

The probability of overtopping after n repetitions of the annual flood can be calculated with eq.(3). The theoretical solution for the probability of overtopping is

$$P_{\ddot{u}} = P(h_o + h + h_f > b) = \iiint_B f(h_o, h, h_f) \cdot dh_o \cdot dh \cdot dh_f \quad \ldots (5)$$

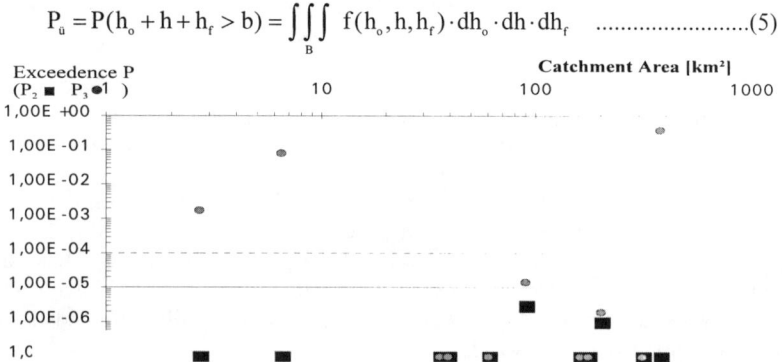

Fig. 6 : Probability of exceedance for the direct overflow (P_2) for at least one time annually and the freeboard excess due to wind waves (P_3)

Because of difficulties with the exact solution a Monte-Carlo simulation (*Pohl 1996*) was used.
The quantile values of the distributions cut by the crest level b on the vertical axis correspond to the failure of the dam caused by overtopping (see Fig.5). Referred to the density function f_2 (dark area) this means, that a direct overflow occurs with a

coherent nappe: $P_2 = P(h_0 + h > b)$. The lighter grey area below the density function f_3 marks the less dangerous case of overtopping of individual waves, however, this case represents a hydraulic failure, too: $P_3 = P(h_0 + h + h_f > b)$. In Figure 6, the results of the above procedure for a selection of 11 Saxon dams are shown.

It can be recognized that the probability P_2 of direct overflow is always smaller than 10^{-5} and thus it seems to be sufficiently safe. In case of three barrages, the probability of overflow of individual waves is more than 10^{-3}, therefore, it is recommended to examine these cases in more detail in order to check if remedial measures are necessary. For the calculation, various input values were used which were not explained in detail in this paper. The results may vary in dependence on these input data and on additional assumptions, so that the results presented in Fig.6 are to be understood as examples.

CONCLUSIONS

By means of the approaches presented it is possible to assign probabilities to the correct hydraulic operation or to the failure of hydraulic structures or systems. These results may be used as decision tools for the dimensioning and/or the assessment of existing hydraulic structures. If it is necessary for the decision, costs may be assigned to the probabilities by means of cost and damage functions respectively.

Evaluating the results, there is a problem concerning the partial lack of acknowledged limiting (or critical) values of reliable probabilities of failure. Of course, such values depend on the hazard potential in case of failure. Mainly in case of danger to human life, probabilities of failure from endangerings taken over involuntarily are controversially accepted by the public, whereas they are largely accepted in insurance mathematics. Searching for allowable probabilities perhaps one could find them in considerations on the utilization of nuclear power where $P < 10^{-6}$ seems to be acceptable (*Otway and Ermann 1970*). However, this value refers to the death of one person per year which is not to be expected automatically in the hydraulic-hydrological failure defined above. As a basis for discussion on finding a decision, therefore, for example for dams, the limiting values $P_2 < 10^{-5}$ or $P_3 < 10^{-4}$ (solid and dashed line in Fig. 6) could be used. For river engineering measures and stilling basin dimensioning measures, greater probabilities are permissible and economically acceptable.

REFERENCES

KIRNBAUER, R.:Ermittlung von Bemessungshochwässern im Wasserbau.- In: Wiener Mitt., Bd. 42, TU Wien, 1981, MARTIN, H.: Ansätze und Methoden zur Ermittlung der Überflutungssicherheit.- In: Wiss. Z. TU Dresden, 37(1984)4, S. 243-253, OTWAY, H.J. ; ERMANN, R.C.: Reactor siting and design from a risk viewpoint.- In: Nuclear Eng. No.13, Elsevier publ. 1970, risk of overtopping of dams.- In: Proc. Research and Development in the Field of Dams.SwissNCoLD, Sept. 1995, pp 733-739, PREIBLER, G; BOLLRICH, G.: Technische Hydromechanik.- Berlin: Verl. Bauwesen, 1996. SINNIGER, R.; HAGER, W. H.: Retentionsvorgänge in Speicherseen.- In: Schweizer Ingenieur und Architekt.- 104(1984)26.- S. 535-539

ON THE UNCERTAINTY OF THE RISK OF FAILURE OF HYDRAULIC STRUCTURES

JOSE D. SALAS
Department of Civil Engineering
Colorado State University
Fort Collins, Colorado 80523, USA

and

JUN H. HEO
Department of Civil Engineering
Yonsei University, Seoul, S. Korea

ABSTRACT

The uncertainty of the risk of failure of flood related hydraulic structures is determined by estimating the variance of the risk of failure assuming that the underlying model is Gumbel and the method of moments estimation. It is shown that the coefficient of variation of the risk of failure may be significant even for long sample sizes.

INTRODUCTION

A typical design of many flood related structures is based on specifying the return period T (in years) for which the structure will be designed, then the design flood magnitude is obtained from the frequency distribution of the annual floods. The specified return period depends on the type of hydraulic structure and for this purpose tables are available in design manuals and books (Viessman et al, 1989). Likewise, when a flood related structure has been already built and after some years of operation, one wishes to re-evaluate or re-assess the performance of the structure, the problem in this case is to determine the return period T or the exceedance probability p=1/T given the known capacity of the referred structure. Furthermore, whether in design or evaluation stage, one may be interested in determining the risk of failure or the reliability of the structure.

The frequency distribution of annual floods is usually determined by analyzing the available flood records, by selecting a given probability distribution function (PDF), and by estimating the parameters of the PDF (Stedinger et al, 1993). Since the available flood sample is of limited size, the parameters and quantiles of the distribution are uncertain.

Quantifying such uncertainty has been of much interest in literature (Kite, 1988; Chowdhury and Stedinger, 1991). Likewise, if the problem is to determine the return period for a given flood magnitude, because of the same reasons, the return period becomes uncertain and consequently the corresponding risk of failure or reliability become uncertain quantities as well. While quantifying the uncertainty of the T-yr flood has been extensively studied in literature, however, this is not the case with the uncertainty of the risk of failure or reliability. The purpose of this paper is to propose a procedure for quantifying the uncertainty of the return period, risk of failure, and reliability.

RETURN PERIOD AND RISK OF FAILURE

Return period and risk of failure are related quantities. In Civil Engineering practice, the return period has been defined as the average number of trials to the first occurrence of an event of magnitude greater than a predefined critical event (Benjamin and Cornell, 1970). If the events are independent and the exceedance probability p of the critical event remains constant over trials, the return period T can be determined by $T=1/p$, and it is measured in units of trials. Vogel (1987) reviewed the concept of return period under several statistical conditions. Also, Loaiciga and Mariño (1991) discussed some important issues related to the concept of return period, particularly the assumption of stationarity of the underlying probability distribution function. Literature abounds on the general topics related to flood frequency analysis and the determination of return periods or exceedance probabilities of hydrologic events. A good bit of the literature addresses the cases where the hydrologic events are assumed to be independent and identically distributed (iid), (see for example, NERC, 1975; Kite, 1988; Stedinger et al, 1993).

The uncertainty of the return period for extreme rainfall and flood events has been recognized by some investigators. Davis et al (1972) using Bayesian approaches studied the distribution of the return period of maximum point rainfall when uncertainty arises in the parameters of the Poisson and exponential distributions. Likewise, in the context of designing coastal protection works and channel improvement, Watt and Wilson (1978) presented a method in which the uncertainty associated with return period was taken into account. However, no explicit consideration was given to the associated uncertainty of the risk of failure.

The risk of failure of flood related hydraulic structures such as culverts and bridges, is directly related to the return period T or exceedance probability p of the design flood. In these cases the risk of failure is typically defined as the probability that the number of floods greater than the design flood in a n-year period is greater or equal to one. Thus, assuming that annual floods are independent it may be shown (Yen, 1970) that the risk of failure is given by $R_n = 1 - (1-p)^n = 1 - (1-1/T)^n$. Although the failure of hydraulic structures may result from forcing factors other than hydrologic (Kite, 1977; Chow et al, 1988), the emphasis here is on risk of failure resulting from extreme hydrologic events. Regarding methods for evaluating the risk of failure of hydraulic structures there is also plenty of literature available (see for instance, Yen and Ang, 1971; Tung and Mays, 1981; Lee and Mays, 1986; Plate and Ihringer, 1986; and Yen et al, 1986; Yen and Tung, 1994).

UNCERTAINTY OF THE RISK OF FAILURE

In engineering practice, it is common to specify a return period and derive a corresponding design size. So, traditionally, return period $T=1/p$ is specified and the design size ξ_q is given through an assumed distribution $F(.)$ as the solution in ξ in equation $F(\xi) = q = 1-p$. Next we describe a method to account for the uncertainty in estimates of risk of failure of hydraulic structures, and in particular existing structures. We will assume for illustrative purposes the Gumbel distribution as the underlying probability model of annual flood peaks which are assumed iid,

$$F(x) = \exp\{-\exp[-\frac{x-x_0}{\alpha}]\} \quad (1)$$

in which x_0 and α are the location and the scale parameters, respectively. Given a sample $X_1, ..., X_N$, in which N is the sample size, the parameters x_0 and α can be estimated, and let $\hat{x}_0$ and $\hat{\alpha}$ denote such estimators. Thus, for a specified flood design peak ξ_q, one gets the corresponding non-exceedance probability q from Eq.(1). Such q depends on the unknown parameters x_0 and α and so an estimate of q is given by

$$\hat{q} = \exp\{-\exp[-\frac{\xi_q - \hat{x}_0}{\hat{\alpha}}]\} \quad (2)$$

In such case, one can also estimate the risk of failure of a hydraulic structure in a n-year period as

$$\hat{R}_n = 1 - (1-\hat{p})^n = 1 - \hat{q}^n \quad (3)$$

We are interested in assessing the uncertainty in this estimate of failure.

A first order approximation of the expected value of $\hat{R}_n$ is

$$E(\hat{R}_n) \approx 1 - [E(\hat{q})]^n$$

and $E(\hat{q})$ in turn, can be approximated by

$$E(\hat{q}) \approx \exp\{-\exp[-\frac{\xi_q - E(\hat{x}_0)}{E(\hat{\alpha})}]\} \quad (4)$$

If $\hat{x}_0$ and $\hat{\alpha}$ are unbiased, or nearly so, then $E(\hat{q}) \approx q$ and

$$E(\hat{R}_n) \approx 1 - q^n = 1 - \left[\exp\{-\exp[-\frac{\xi_q - x_0}{\alpha}]\}\right]^n \quad (5)$$

Likewise, a first order approximation of the variance of $\hat{R}_n$ can be derived from

$$Var(\hat{R}_n) \approx \left(\frac{\partial \hat{R}_n}{\partial \hat{x}_0}\right)^2_\mu Var(\hat{x}_0) + 2\left(\frac{\partial \hat{R}_n}{\partial \hat{x}_0}\right)_\mu \left(\frac{\partial \hat{R}_n}{\partial \hat{\alpha}}\right)_\mu Cov(\hat{x}_0, \hat{\alpha})$$

RISK OF FAILURE OF HYDRAULIC STRUCTURES 509

$$+ \left(\frac{\partial \hat{R}_n}{\partial \hat{\alpha}} \right)^2_\mu Var(\hat{\alpha}) \qquad (6)$$

where $(\ )_\mu$ implies the random variables in the derivatives are replaced by their corresponding expected values. It may be shown that

$$\frac{\partial \hat{R}_n}{\partial \hat{x}_0} = \frac{n}{\hat{\alpha}} \exp[-\hat{y} - n e^{-\hat{y}}] = \frac{n}{\hat{\alpha}} (-\ln \hat{q}) \hat{q}^n \qquad (7)$$

where $\hat{y} = \frac{\xi_q - \hat{x}_0}{\hat{\alpha}}$. Similarly,

$$\frac{\partial \hat{R}_n}{\partial \hat{\alpha}} = \frac{n}{\hat{\alpha}} (-\ln \hat{q}) \ln(-\ln \hat{q}) \hat{q}^n \qquad (8)$$

Furthermore, the variances and covariance in Eq.(6) depend on the selected estimation procedure. Again, for illustration we will consider the method of moments. It may be shown (NERC, 1975)

$$Var(\hat{x}_0) = 1.168 \frac{\alpha^2}{N}$$

$$Var(\hat{\alpha}) = 1.10 \frac{\alpha^2}{N}$$

$$Cov(\hat{x}_0, \hat{\alpha}) = 0.096 \frac{\alpha^2}{N}$$

Thus, from the previous equations after simplification one can get

$$Var(\hat{R}_n) \approx \frac{n^2}{N} (q)^{2n} (-\ln q)^2 [1.168 + .192 z + 1.1 z^2] \qquad (9)$$

where $z = -\ln[-\ln q]$. Equation (9) shows that the variance of the risk of failure is a function of the design life n, the non-exceedance probability q (or return period T), and the sample size N.

Table 1 gives the expected value and the standard deviation of the risk of failure, determined respectively from Eqs.(5) and (9), as a function of the non-exceedance probability q, the design life n, and the sample size N. It shows that the uncertainty of the risk of failure evaluated by its standard deviation can be quite significant. As expected, its significance is more important for small sample sizes. For example, for N=10 and n=10 the coefficient of variation of the risk of failure may become greater than one. However, even for sample sizes that are generally considered as long samples (say N=50 or N=100), the coefficient of variation of the risk of failure may be greater than 0.5.

SUMMARY
An approach for estimating the uncertainty of the risk of failure of hydraulic structures has been developed by using the Gumbel distribution and the method of moments of parameter estimation. The expected value and the variance of the risk of failure have been derived explicitly. It has been shown that the variance of the risk of failure is a function of the non-exceedance probability q (or return period T), the design life n, and the data

sample size N. Current work includes extending the approach to other distributions and methods of estimation, and evaluating the applicability of the approach in actual decision making.

Table 1. Expected Value and Standard Deviation of the Risk of Failure as a Function of n, T, and N

N	q	T	n=10		n=50	
			$E(R_n)$	$\sigma(R_n)$	$E(R_n)$	$\sigma(R_n)$
10	0.90	10	0.651	0.311	0.995	0.023
	0.98	50	0.183	0.226	0.636	0.503
	0.99	100	0.096	0.145	0.395	0.484
50	0.90	10	0.651	0.139	0.995	0.010
	0.98	50	0.183	0.101	0.636	0.225
	0.99	100	0.096	0.065	0.395	0.216
100	0.90	10	0.651	0.098	0.995	0.007
	0.98	50	0.183	0.071	0.636	0.156
	0.99	100	0.096	0.046	0.395	0.153

ACKNOWLEDGEMENTS

The support of the NSF Grant CMS-9625685 on "Risk and Uncertainty Analysis Under Extreme Hydrologic Events" is gratefully acknowledged.

REFERENCES

Benjamin, J.R. and Cornell, C.A., 1970. Probability, Statistics and Decision for Civil Engineers, McGraw Hill Book Company, New York.

Chow, V.T., Maidment, D.R., and Mays L.W., 1988. Applied Hydrology, McGraw Hill Book Co., New York.

Chowdhury, J.U. and Stedinger, J.R., 1991. Confidence Interval for Design Floods with Estimated Skew Coefficient, ASCE Jour. Hydr. Div., 117(HY1), 811-831.

Davis, D., Dusckstein, L., Kisiel, C., and Fogel, M., 1972. Uncertainty in the Return Period of Maximum Events: A Bayesian Approach, Proc. Inter. Symp. on Uncertainties in Hydrologic and Water Resources Systems, Tucson, Arizona, p. 853-862.

Kite, G.W., 1988. Frequency and Risk Analyses in Water Resources, Water Resources

Publications, Littleton, Colorado, 257 p.

Lee, H.L. and Mays, L.W., 1986. Hydraulic Uncertainties in Flood Levee Capacity, ASCE Jour. Hydraul. Eng., 112(10), 928-934.

Loaiciga, H.A. and Mariño, M.A., 1991. Recurrence Interval of Geophysical Events, ASCE Jour. Water Resour. Plann. & Manag., 117(3), 367-382.

NERC (Natural Environment Research Council), 1975. Flood Studies Report, Vol.1, Hydrological Studies, Whitefriars Press Ltd., London.

Plate, E.J. and Ihringer, J., 1986. Failure Probability of Flood Levees on a Tidal River, in Stochastic and Risk Analysis in Hydraulic Engineering (B.C. Yen Editor), Water Resources Publications, Littleton, Colorado.

Stedinger, J.R., Vogel, R.M., and Foufoula-Georgiou, E., 1993. Frequency Analysis of Extreme Events, Chapter 18 in Handbook of Hydrology, D.R. Maidment, Editor, McGraw Hill Book, Co., New York.

Tung, Y.K. and Mays, L.W., 1981. Risk Models for Levee Design, Water Resour. Res., 17(4), 833-841.

Viessman, W., Lewis, G. L. and Knapp, J. W., 1989. Introduction to Hydrology, 3rd Edition, Harper and Row Pub., New York.

Vogel, R.M., 1987. Reliability Indices for Water Suypply Systems, ASCE Jour. Water Resour. Plann. & Manag., 113(4), 563-579.

Watt, W.E. and Wilson, K.C., 1978. An Approach to Optimal Design of Hydraulic Structures, in Reliability in Water Resources Management (E.A. McBean et al, Editors), Water Resources Publications, Littleton, Colorado, p. 75-90.

Yen, B.C. and Ang, A.H.S., 1971. Risk Analysis in Design of Hydraulic Projects, in Stochastic Hydraulics (C.L. Chiu, Editor), Proc. First Intern. Symp., University of Pittsburgh, Pittesburgh, Pensylvania, p. 694-701.

Yen, B.C., Cheng, S.T., and Melching, C.S., 1986. First Order Reliability Analysis, in Stochastic and Risk Analysis in Hydraulic Engineering, Water Resources Publications, Littleton, Colorado, p. 1-36.

Yen, B.C. and Tung, Y.K., 1994. Reliability and Uncertainty Analyses in Hydraulic Design, American Society of Civil Engineers, N. York, 291 pages.

Long-Term Memory in Hydrologic Series?

A. R. RAO AND D. BHATTACHARYA
School of Civil Engineering
Purdue University
West Lafayette, IN 47907-1284, USA

ABSTRACT

The objective of this study is to investigate the behavior of the Hurst coefficient when the short term memory is considered. It is shown that after accounting for short-term memory in monthly hydrologic time series, the Hurst exponent is close to 0.5, indicating no evidence of long-term memory. Thus, short-term memory models like the autoregressive-moving average models are adequate to model hydrologic time series.

1. Introduction

The Fractional Gaussian Noise (FGN) model, first introduced into hydrology by Mandelbrot (1969a) is used for modeling hydrologic time series. The FGN model is also suggested as a model for persistence in hydrologic time series (Mandelbrot and Wallis, 1969a, 1969b and 1969c). In this paper, Mandelbrot's and Andrew Lo's (1991) versions of the range statistic, first proposed by Harold Edwin Hurst (1951), are used to investigate the Rescaled-Range characteristics of hydrologic time series.

Mandelbrot and Wallis (1969c) divided the range by the standard deviation of the sample and called it the "rescaled range" or "R/S" statistic. The rescaled range statistic is not designed to distinguish between short-term and long-term memory. Usually short term memory is present in hydrologic time series. Therefore, any empirical investigation of long-term memory in hydrologic time series analysis must first account for short-term memory in a series. This can be accomplished by using Lo's (1991) modified rescaled range statistic, which filters the short-term memory in a time series. In this paper, the Hurst exponent is computed by using the rescaled and modified rescaled ranges respectively. Comparison of these two Hurst exponents reveal the effects of short term memory on the Hurst exponent.

2. The Rescaled Range Statistic

To estimate the Hurst exponent, the rescaled range statistic has been used extensively since its formulation by Mandelbrot. For a time series of n

observations, $X_1 \ldots X_n$, the rescaled range statistic, denoted by $Q(n)$, is defined as range $R(n)$ divided by the standard deviation $S(n)$.

If the standard deviation of the entire series is replaced by the standard deviation of the partial series then we have an estimate of rescaled range statistic $\hat{Q}(n)$. The relationship between rescaled range statistic $\hat{Q}(n)$ and the number of observations n is given by equation (1) (Mandelbrot & Wallis 1969c),

$$\hat{Q}(n) = \frac{R(n)}{S(n)} = C n^h \quad 0.5 < h < 1.0 \qquad (1)$$

where C is a constant and h is the Hurst exponent. For an independent and identically distributed (i.i.d) sequence, h is equal to 0.5. However, for a series in which observations have statistically significant correlation at small lags, the Hurst exponent is larger than 0.5. This property of the Hurst exponent to deviate from 0.5 is used as an indicator of long-term memory in the series. The slope of the $R(n) / S(n)$ versus n plot is an estimate of the Hurst exponent H and varies between 0.5 and 0.95 for hydrologic data.

3. The Modified Rescaled Range

In order to filter out the effect of short-term memory in a series, the rescaled range statistic $\hat{Q}(n)$ was modified by Lo (1991) so that its statistical behavior is invariant over a general class of short-memory processes. This is accomplished by using the modified rescaled range statistic $\hat{Q}(n,q)$ which is defined by equation (2).

$$\hat{Q}(n,q) = \frac{1}{\hat{\sigma}_n(q)} \left[\max_{1 \leq k \leq n} \sum_{j=1}^{k} (X_j - \bar{X}_n) - \min_{\leq k \leq n} \sum_{j=1}^{k} (X_j - \bar{X}_n) \right] \qquad (2)$$

where:

$$\hat{\sigma}_n^2(q) = \frac{1}{n} \sum_{j=1}^{n} (X_j - \bar{X}_n)^2 + \frac{2}{n} \sum_{j=1}^{q} \omega_j(q) \sum_{i=j+1}^{n} (X_i - \bar{X}_n)(X_{i-j} - \bar{X}_n) \qquad (3a)$$

$$\hat{\sigma}_n^2(q) = \hat{\sigma}_x^2 + 2 \sum_{j=1}^{q} \omega_j(q) \hat{\gamma}_j \qquad (3b)$$

where $\hat{\sigma}_x^2$ and $\hat{\gamma}_j$ are the sample variance and autocovariance estimators of X.

The weights $\omega_j(q)$ in eqs. (3a) and (3b) are defined in eq. 4,

$$\omega_j(q) \equiv 1 - \frac{j}{q+1}; \quad q < n, \tag{4}$$

The modified rescaled range statistic $\hat{Q}(n,q)$ differs from rescaled range statistic $\hat{Q}(n)$ in the denominator. In $\hat{Q}(n)$, the denominator is the square root of a consistent estimator of the variance of X_t, $t=1,...,n$. If X_t has statistically significant autocorrelation, the estimator $\hat{\sigma}_n(q)$ includes the autocovariance up to lag q. The weights $\omega_j(q)$ are chosen such that they yield a positive $\hat{\sigma}^2(q)$. The truncation lag q is chosen by considering the correlogram or the spectral density of the data at hand. The modified Hurst exponent, denoted by H^1, is estimated by using a plot of the modified rescaled range statistic, $R(n)/\hat{\sigma}_n(q)$ against the number of observations n on a logarithmic scale.

4. Data Used

The Hurst exponent and the modified Hurst exponent were estimated for hydrologic time series from the Midwestern United States. In order to remove the annual periodicity in the monthly data, the monthly mean was subtracted from the observations and divided by the monthly standard deviation. The resulting series is called the standardized series and is used for further analysis. Data used in the study are presented in Table 1, where skewness coefficient and lag one autocorrelation coefficient are of the standardized series.

5. Results.

Plots of rescaled range $\hat{Q}(n)$ and the modified rescaled range $\hat{Q}(n,q)$ against the number of observations n of the series on a logarithmic scale for the average monthly flow in the Wabash River at W. Lafayette, IN are shown in Figure 1 and 2 respectively. In Figure 2, $\hat{Q}(n,q)$ is computed by using autocovariance of the series up to lag $q = n/5$ where n is the number of observations. The mean of the estimates are shown as circles along with the maximum and minimum of the estimates corresponding to n. The least squares fit for mean estimates is shown as a straight line along with its slope.

Table 1. Summary of Data Used and Results

Times Series	Period of Record	Mean	Std. Dev.	Skew. Coeff.	Lag One Auto.Corr. Coeff.	H	H^1 (q=n/10)	H^1 (q=n/5)
Avg. monthly flow in Wabash River at W.Lafayette, IN	1924-1993	6618 cfs	6060 cfs	1.888	0.387	0.683	0.496	0.516
Avg. monthly flow in Mississippi River at Clinton, IA	1874-1993	47936 cfs	30354 cfs	1.213	0.645	0.719	0.463	0.479
Maximum monthly flow in the Wisconsin River at Merrill, WI	1903-1991	4853 cfs	4231 cfs	1.707	0.443	0.717	0.497	0.553
Average monthly temp. at Algoona, IA	1893-1992	57°F	21.8°F	0.057	0.248	0.683	0.527	0.514
Monthly precip. at Urbana, IL	1903-1992	3.12 in	1.96 in	1.009	0.037	0.638	0.559	0.524

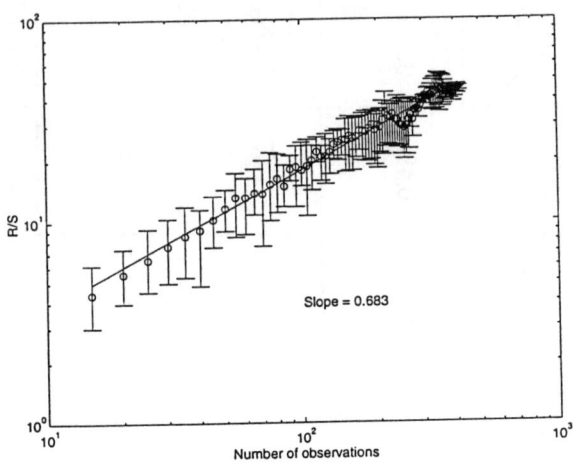

Figure 1. R/S analysis of the standardized average monthly flows in the Wabash River at W. Lafayette, IN

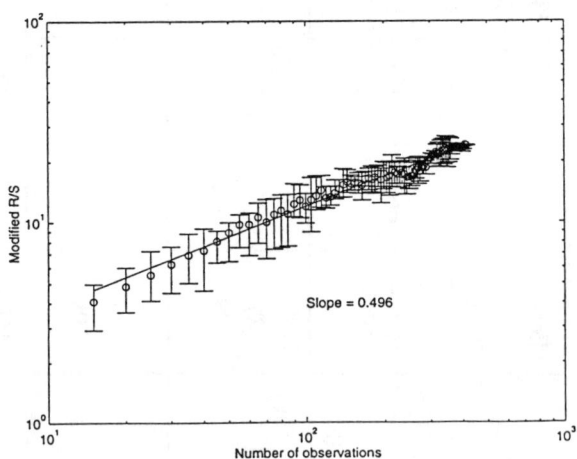

Figure 2. Modified R/S analysis for the same data as in Figure 1. $\hat{Q}(n,q)$ is computed using a truncation lag of $q = n/5$.

The Hurst and modified Hurst exponent for all the series are given in Table 1. For the average monthly streamflow, maximum monthly streamflow, average monthly temperature series the Hurst exponent is always greater than 0.5. The modified Hurst exponent H^1, on the other hand, is always close to 0.5. Since the modified Hurst exponent H^1 is computed using the modified rescaled statistic which filters the short-term memory in a series, it can be inferred that $R(n)/\hat{\sigma}_n(q) \sim \sqrt{n}$. However, for the standardized precipitation series the difference between the Hurst and modified Hurst exponents is smaller compared to the other series (Table 1). This can be attributed to the fact that the autocorrelation in standardized precipitation series is very weak. These results lead to the inference that upon filtering of short-term memory the Hurst exponent is close to 0.5.

The effect of variation of the truncation lag "q" on the modified Hurst exponent was also considered. For the average and maximum monthly streamflow series, an increase in the truncation lag from $q = n/10$ to $q = n/5$ leads to an increase or decrease in the modified Hurst exponent H^1 which can be explained by the correlation present in the series between these lags.

6. Conclusions

When the modified rescaled range statistic, which accounts for short-term memory, is used, little evidence is found to support the long-term memory hypothesis in hydrologic time series. The modified Hurst exponent is always close to 0.5, which is the case for i.i.d sequences.

References

1. Hurst, H. (1951), 'Long term storage capacity of reservoirs', *Transactions of the American Society of Civil Engineers* **116**, 770-799.
2. Lo, A. (1991), 'Long-term memory in stock market prices', *Econometrica* **59**(5), 1279-1313.
3. Mandelbrot, B. & Wallis, J. (1969a), 'Computer experiments with fractional gaussian noises, part 1, 2, 3', *Water Resources Research* **5**, 228-267.
4. Mandelbrot, B. & Wallis, J. (1969b), 'Robustness of the rescaled range r/s in the measurement of noncyclic long run statistical dependence', *Water Resources Research* **5**, 967-988.
5. Mandelbrot, B. & Wallis, J. (1969c), 'Some long run properties of geophysical records', *Water Resources Research* **5**, 321-340.

Precipitation Variability and Curve Numbers

JOSEPH A. VAN MULLEM
Natural Resources Conservation Service, Bozeman, Montana

ABSTRACT
Rainfall records from three widely separated experimental watersheds were used with the Green-Ampt infiltration model to predict runoff curve numbers. The average curve numbers did not vary by location for fixed infiltration parameters. Curve numbers were found to be dependent on both storm duration and rainfall intensity.

INTRODUCTION
One of the most common methods used to convert storm rainfall into storm runoff is the curve number method. The curve number is a single parameter which relates the volume of storm runoff to the volume of storm rainfall for a given soil, land use, cover, and condition. Tables listing curve numbers for various types of land use, soils, and cover are provided in several Natural Resources Conservation Service (NRCS, formerly the Soil Conservation Service) publications (USDA 1986, USDA 1989).

The curve number for a watershed may be found from the rainfall and runoff records for the watershed. The curve number for a single soil and cover may be found from small watersheds or plots with a single soil and cover condition. Because of the wide variability of curve numbers between individual storm events, it takes ten years or more of hydrologic data to calculate an average curve number for a watershed.

Much of the variability of curve numbers can be related to characteristics of the storm, and to the antecedent conditions of the watershed. For watersheds where surface runoff dominates, rainfall intensity and duration are important storm characteristics.

The curve numbers found in NRCS tables were developed from research areas in various locations in the United States. However, most of these areas are in agricultural areas in the eastern and midwestern parts of the country. Curve numbers are applied uniformly to all parts of the country, and to many other parts of the world. Although a few curve numbers have been developed for special crops and cover types, most are applied without regard to location. So fallow land in Maine is assumed to have the same curve number as fallow land of the same soil type in Arizona, and rangeland in Montana the same as rangeland in Florida.

Pierson et al (1995) used three different infiltration models and attempted to determine curve numbers from rainfall simulation data at sites in Idaho and Hawaii. They showed that the distribution of rainfall may have an influence on curve number values. Infiltration models compute greatly different runoff volumes from the same storm rainfall with different temporal distributions (Van Mullem 1991a). Hydraulic conductivity in the Green-Ampt model has been related to curve numbers for a number of temporal distributions used as design storms (Van Mullem 1991b).

PURPOSE

The purpose of this paper is to determine if rainfall in widely distant locations has enough variability to result in different curve numbers. The Green-Ampt model will be used to calculate runoff and the curve number. If differences exist, their identification may enable us to regionalize curve numbers. This would result in more accurate modeling, and more accurate estimates of peak discharges in ungaged watersheds.

METHODS

The NRCS runoff equation is:

$$Q = (P - 0.2S)^2 / (P + 0.8S), \text{ for } P > 0.2S \tag{1}$$

Where Q is storm runoff, P is storm rainfall, and S is the maximum potential retention. The curve number is a dimensionless index which varies from 0 to 100. It is related to the parameter S by:

$$CN = 25000 / (254 + S) \tag{2}$$

where S is in mm.

Equation (1) may be solved for S using the quadratic formula (Hawkins 1973) as:

$$S = 5 [P + 2Q - (4Q^2 + 5PQ)^{1/2}] \tag{3}$$

The curve number for a single event may be found by solving for S from equation (3) and then solving for CN with equation (2). The watershed curve number may be found as the median of the curve numbers of the annual maximum runoff events (USDA 1986).

The Green-Ampt infiltration equation is:

$$f = K (1 + n y_f / F) \tag{4}$$

where f is the infiltration rate at any time, K is the effective hydraulic conductivity, n is the initial soil moisture deficit, y_f is the capillary suction head, and F is the accumulated infiltration. The Green-Ampt model is physically based, simple to use, and is an increasingly popular method of finding the infiltration component of rainfall.

Sites were selected from USDA, Agricultural Research Service locations. Three sites were chosen which each had over twenty years of record, and were located far enough apart so that their precipitation characteristics might be different. The sites were at Coshocton, Ohio; Hastings, Nebraska; and Safford, Arizona. A small watershed with precipitation and runoff data was selected at each site.

The GetPQ program (Dripchak and Hawkins 1993) was used to identify the individual storm events. The largest non-winter runoff event was selected for each year. From these, the largest half were selected. So the selected storms were those which resulted in the 2-year or greater storm runoff event. The data from Coshocton included 41 years of record. To test the variability between different periods of record, the Coshocton record was split into two periods.

The Green-Ampt model was then used to calculate the runoff for each of the selected storms over a range of hydraulic conductivity. No attempt was made to fit the model parameters to each site. Saturated hydraulic conductivity (K_{sat}) was varied from 3.8 to 20.3 mm/hr to represent typical values. The effective hydraulic conductivity was taken as one half of K_{sat}. The capillary suction head was assumed to be a function of K_{sat} according to the equation (Van Mullem 1991b):

$$y_f = 614 \ (K_{sat} + .51 \ )^{-.493} \ (mm) \tag{5}$$

The initial moisture deficit was set at 0.2 and the interception storage and depression storage at 2.5 and 1.25 mm respectively.

The rainfall and computed runoff from each storm was then used to compute an equivalent curve number. The median curve number for each watershed and each value of Ksat was then found and compared between watersheds. Differences should be caused by differences in precipitation, since all other parameters are fixed.

Table 1 shows some of the differences between the rainfall events at these sites. The total storm rainfall at Safford, Arizona is only half of what it is at the other sites. The storm intensities for the 10- and 30-minute durations are significantly higher at Hastings, Nebraska. Total storm duration is shortest at Safford and is significantly longer at Coshocton.

Table 1. Average Rainfall Characteristics

Location	No of storms	Total rainfall (mm)	Storm duration (hrs)	Intensity, 10 min (mm/hr)	Intensity, 30 min (mm/hr)
Hastings	10	53.7	2.94	86.6	56.4
Safford	11	25.7	2.20	60.5	35.6
Coshocton	10	52.8	3.50	60.5	35.3

RESULTS

Table 2 lists the median and mean curve number for each site and for each value of Ksat. The standard deviation about the mean is also shown. Two sets of data are shown for Coshocton as a result of the split sample. The Coshocton means are lower than at the other sites. The standard deviation is also greater at Coshocton. The differences between any of the means are not statistically significant at the 10 percent level. The median values are less affected by extreme low events and are plotted in Figure 1. The median curve numbers at Hastings are slightly higher than at the other locations.

Table 2. Calculated Curve Numbers.

Site	Ksat mm/hr	Median Curve No.	Mean Curve No.	Std Dev
Coshocton 1	3.81	89.37	87.90	4.31
	7.62	86.84	84.58	5.58
	20.3	81.62	78.28	7.81
Coshocton 2	3.81	89.62	87.74	4.69
	7.62	86.21	84.08	6.57
	20.3	81.16	76.88	10.07
Hastings	3.81	91.55	90.44	2.64
	7.62	89.18	87.76	3.61
	20.1	85.04	82.67	5.48
Safford	3.81	90.66	89.65	3.88
	7.62	87.35	86.50	4.84
	20.1	82.22	82.21	3.61

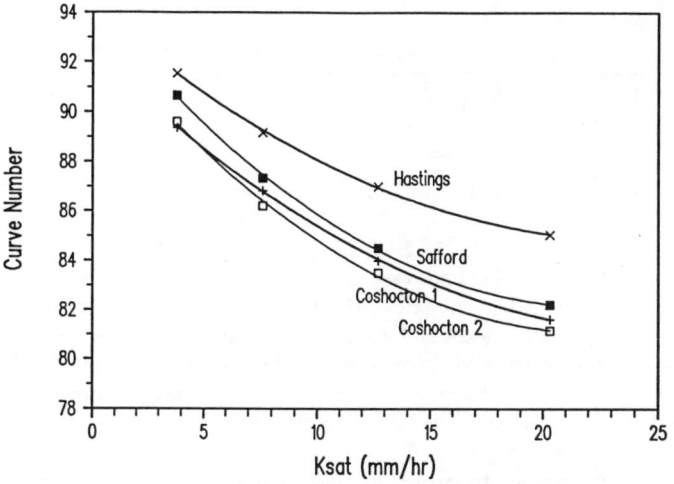

Figure 1. Variation of Curve Number with K_{sat} for the three locations studied.

DISCUSSION

The lack of variability between curve numbers at these locations is suprising, considering the differences between the precipitation events shown in Table 1 for these locations. Curve numbers generally decrease with duration and increase with storm intensity. Therefore there are some compensating effects for the locations studied here. Using the curve numbers calculated from a K_{sat} of 7.6 mm/hr, curve number is plotted against duration in Figure 2 and against maximum 10 minute intensity in Figure 3. The linear regression equation relating CN to duration (D) hrs and 10 minute intensity (I) in/hr for these sites is:

$$CN = 86.22 + 1.23\ I - 1.32\ D. \qquad (6)$$

R^2 was 0.65 and both variables are significant at the one percent level.

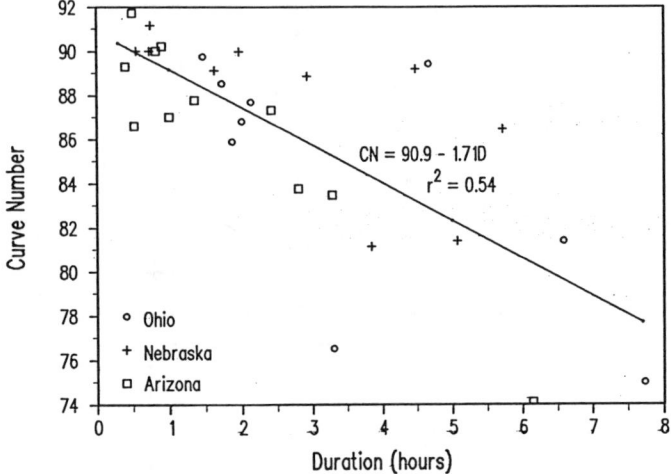

Figure 2. Variation of Curve Number with Storm Duration

CONCLUSIONS

Calculated runoff curve numbers did not vary as a result of precipitation variability between three locations in the United States. This supports the practice of using curve numbers developed at one location for a soil and cover type at other locations. Relationships between the calculated curve number and storm duration and intensity were found to be significant at the one percent level. This indicates that storm characteristics are important to curve number determination.

Further studies will be necessary to assure that curve numbers are applicable without regard to location. The effect of duration and intensity may also have application to the hypothetical storm distributions used in design.

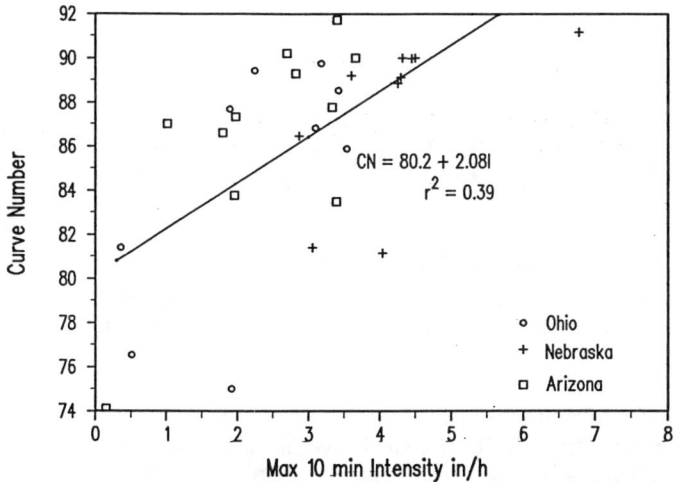

Figure 3. Curve Number variability with maximum 10-minute storm intensity.

REFERENCES

Hawkins, R. H. 1973. Improved prediction of storm runoff in mountain watersheds. J. Irrig. and Drain. ASCE 99(4) 519-523.

Dripchak, M. M. and Hawkins, R. H. 1993. GetPQ Ver. 1.5, Data Reduction Program Series, Water Resources Program, Univ. of Ariz., Tucson, Arizona

Pierson, F.B., R.H. Hawkins, K.P. Cooley and S.S. Van Vactor. 1995. Experiences in estimating curve numbers from rainfall simulation data. Proceedings, Watershed Management Symposium, August 14-16, 1995. San Antonio, Texas. ASCE.

Van Mullem, J.A., 1991a. Precipitation distributions and Green-Ampt runoff. J. Irrig. and Drain. ASCE 117(6) 944-959.

Van Mullem, J.A., 1991b. Green-Ampt and the curve number. ASAE Paper 91-2610. Chicago Ill. Dec 17-20, 1991.

USDA, Soil Conservation Service. 1986. National Engineering Handbook, Hydrology Section 4. Washington, DC.

USDA, Soil Conservation Service. 1989. Urban Hydrology, TR55. Washington, DC.

Errors and Variability of Reservoir Yield Estimation as a Function of the Coefficient of Variation of Annual Inflows

JOSÉ NILSON BEZERRA CAMPOS
Prof. Department of Hydraulic and Environmental Engineering (DEHA)
Universidade Federal do Ceará, Campus do Pici, 60.451-970, Fortaleza, Brazil
E-mail nilson@ufc.br

FRANCISCO DE ASSIS DE SOUZA FILHO
Prof. Centro de Tecnologia - Universidade de Fortaleza
R. Bento Albuquerque, 2010, Fortaleza, 60.190-080, Brazil
E-mail assis@feq.unifor.br

JOSÉ CARLOS DE ARAÚJO
Prof. Department of Civil Engineering (DECIV) - Mining School
Universidade Federal de Ouro Preto, Ouro Preto, 35.400-000, Brazil

ABSTRACT

An analysis of the precision of reservoir yield evaluation as a function of the coefficient of variation of annual inflows (Cv) is performed. Two aspects are involved: the size of inflow time series required for a given precision in the estimation of the population mean of annual inflow; and the variance of reservoir yield as a function of Cv and of the size of the time series. The value of Cv analyzed range from 0.2 - characteristic of Northeast of the USA rivers - to 1,4 - characteristic of some rivers in Brazil's semi-arid Northeast. The results showed that uncertainty and errors affect markedly rivers with high coefficient of variation. For instance, if a 30-year time series of annual inflow is good in estimating the mean of the population of annual inflows for Northeast of USA, to reach the same precision in the Northeast of Brazil it would be necessary about 690 years of data. Besides, assuming a known population mean value, the reservoir yield for Cv = 0.2 is about 2.8 times greater than that for Cv = 1.4. Thus, the higher the Cv, the lower the reservoir yield and the less accurate its preview.

INTRODUCTION

A water resources planner is concerned, in quantitative point of view, to provide water reliable for the development of a given region. In many places of the world, surface reservoirs play the most important part in this process. The operation of these water systems depends on the future streamflow series, not known by the planner. To take decisions about reservoir operation rules, the planner relies on the knowledge of past events. To make the best the planner must use the best the available data.

Assuming that inflow probability distribution function (p.d.f.) can be described by two parameters, mean (μ) and standard deviation (σ), two problems are analyzed in this paper: 1) errors in estimating μ from the recorded data and time series size necessary for a given precision; and 2) assuming that population parameters are known, the variance of reservoir yield as a function of the Cv (= σ/μ). The difference between the two cases is: 1) in the first case one can theoretically decrease the estimation error increasing the observed data - as long as time goes on; 2) the variance is a characteristic of the process uncontrolled by the planner - the variability involved in the process should be incorporated in the planning of operation rules and in regulamentation of water rights.

UNCERTAINTY ON THE POPULATION MEAN OF ANNUAL INFLOW

In this context, uncertainty is defined as a state condition of unreliable knowledge (Reckhow and Chapra, 1983). Taking the confidence interval for the mean of the time series observed inflows and using the t-test, one has:

$$\mu - \bar{x} = t \cdot \left(s / \sqrt{n}\right) \quad (1)$$

where $\bar{x}$ and s are respectively the mean and standard deviation of observed inflows, t the Student variate and n the sample size. Defining $(\mu - \bar{x})/\mu$ as the relative error **d** for the sample mean ($\bar{x}$) as an estimation of the population mean (μ), then for a given **d** equation (1) can be rewritten

$$n = k \cdot Cv^2 \quad (2)$$

where n is the required time series size to estimate the population mean (μ) with a given precision **d**, once the coefficient of the population is Cv and k is a constant that depends on desired precision. Using equation (2), Campos (1996) showed that, if a 30-year series is good enough to estimate the mean of the annual inflows in Northeast of USA, in Northeast of Brazil, to reach the same precision it would be necessary 691 years of data. This shows that, even though it is possible to reduce uncertainty by increasing the number of observed values, the hydrologists of Brazil's Northeast must deal with this uncertainty for a long time.

VARIABILITY OF YIELD FOR A GIVEN HORIZON TIME

The second issue of the paper is studying the variability of reservoir yield assuming that the uncertainty is reduced to zero. A third step, not analyzed here, is lump the two cases: uncertainty on the population parameter and natural variability of probability distribution of the yields. To reach that a computer experiment using synthetic generation of inflows an simulation of reservoir operation was performed.

To estimate the reservoir yield, it was used the mutually exclusive rule as proposed by Moran (1954) and the dimentionless form equation of reservoir budget as proposed by Campos (1987). The reservoir yield function can be written in the form:

$$fm = \phi(Cv, fe, fk) \quad (3)$$

where fm is the dimensionless yield for a given reliability (M/µ); M the reservoir yield; fk the dimensionless reservoir capacity (K/µ); K the reservoir storage capacity; fe the dimensionless evaporation factor (fe = $3\alpha^{1/3}E_v/\mu$) with α the reservoir shape parameter and E_v the yearly evaporation from the lake. The paper studies a particular case with fk equal to one, fe equal to 0.10, the water supply reliability equal to 90% and the Cv ranging from 0.2 to 1.4. For each Cv value, reservoir simulations were performed for horizon planning ranging from 30 to 1000 years. For each horizon planning 50 samples were generated.

RESULTS

The simulations mentioned above showed that, the higher the Cv, the lower the reservoir yield Q (see Figure 1). For given conditions of fk = 1.0, fe = 0.10 and µ = 100 v.u. (volume units), Q decreases from 78 v.u. (Cv = 0.2) to 28 v.u. (Cv = 1.4), i.e., for the same conditions, the yield for a river with Cv = 0.2 is almost three times higher than that for Cv = 1.4, which is a considerable disadvantage for semi-arid regions, such as Brazil's Northeast.

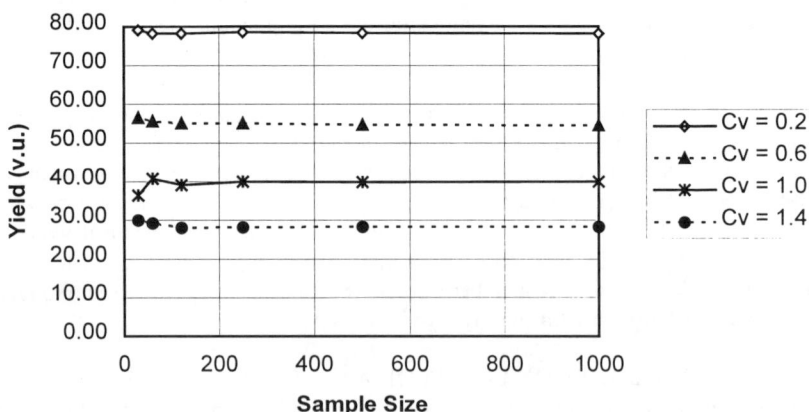

Figure 1: Reservoir Yield as a function of Cv

In fact, table 1 presents several examples of rivers with different coefficient of variation of annual inflows, ranging from 0.19 in King river, Australia, up to 1.76 in Banabuiú river, in Brazilian Semi-arid. Vogel et al. (1995) investigated 160 different places in USA and found average Cv of 0.25.

Table 1 - Coefficient of variation of annual inflows for rivers in the world

River	Country	Watershed area (km²)	Cv	Source
Mekong	Laos	299.000	0.17	Mc Mahon
King	Australia	451	0.19	Mc Mahon
Sieber/Herzberg	Germany	82	0.25	Billib
Lava Tudo	Brazil	1.147	0.36	(-)
Pelotas - Passo Socorro	Brazil	8.365	0.37	(-)
Warrangaba	Australia	8.750	1.11	Mc Mahon
Diamantina	Australia	115.000	1.19	Mc Mahon
Jaguaribe/Orós	Brazil	21.000	1.24	(-)
Banabuiú/Sen. Pompeu	Brazil	5.200	1.76	(-)

The investigation of reservoir yield error produced the data of Figure 2. The error E ($E_i = |Q_i - \overline{Q}|/\overline{Q}$) decreases with Cv and, for a given Cv, it decreases with the sample size. It means that, for a given accuracy of yield preview (E = 5%, e.g.) it is necessary 76 years of data for a place with Cv = 0.2, whereas, for Cv = 1.4, it would be necessary about 758 years, almost ten times more. During the simulation process, 50 values of yield Q were generated for different values of Cv. Figure 3 shows the variability of Q for Cv = 1.4 and Cv = 0.2. Observe the difference of behavior for low and high values of Cv. In order to verify if the generated values of reservoir yield behave according to the normal distribution function, values of $(Q - \overline{Q}) / S_Q$ were plotted against its sample frequency in Figure 4. The results show that the histogram fits properly the theoretical curve. In the previously mentioned formulas, $\overline{Q}$ is the mean and S_Q the sample standard deviation.

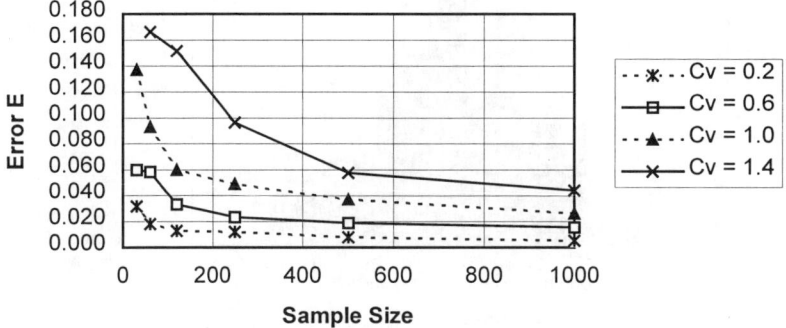

Figure 2. Error against reservoir horizon planning (sample size)

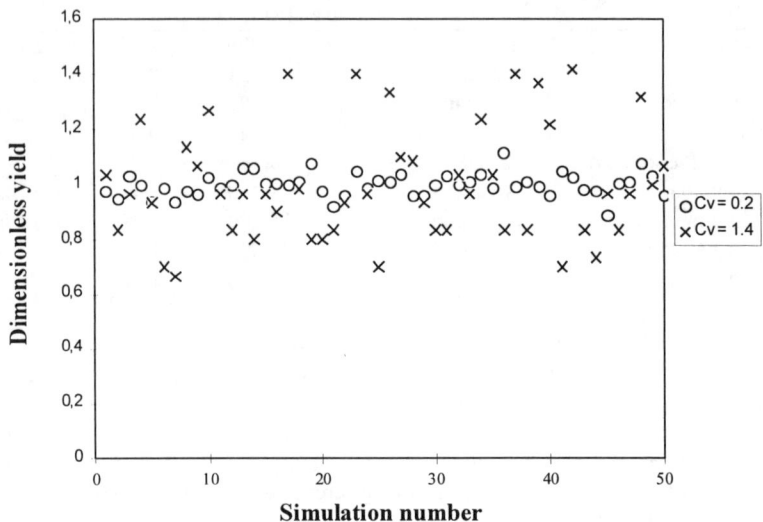

Figure 3. Dispersion diagram for reservoir yield

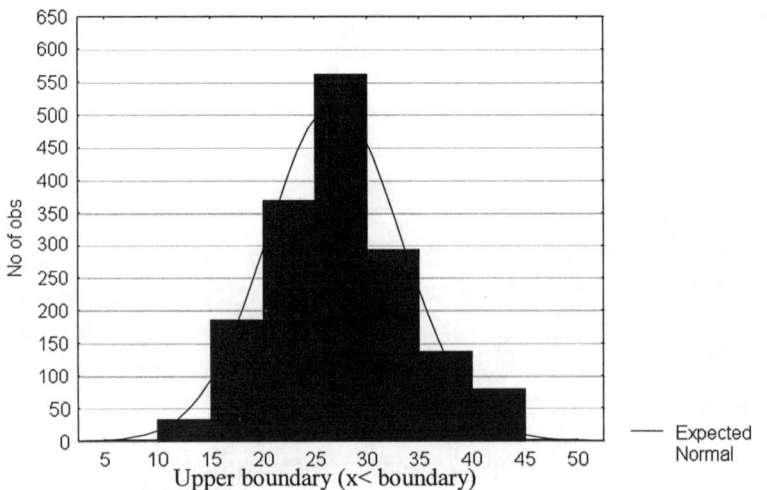

Figure 4. Fitting of errors to normal p.d.f.

CONCLUSIONS

From the results obtained in this paper, it can be concluded that the reservoir yield is very sensitive to Cv. For instance, for rivers like the ones in Northeastern United States (Cv ≅ 0.25) yield is about 2.5 times higher than for rivers in Brazilian semi-arid region (Cv ≅ 1.4) for given conditions. Error of generated yields decreases with lower values of Cv and, for a given Cv, it decreases with longer reservoir horizon. Consequently, high Cv regions, such as Brazil's Northeast, present greater uncertainty levels than low Cv regions. As it is not convenient to increase so much the reservoir planning horizon, the uncertainty of the reservoir yield must be accurately evaluated and included in the planning process and in regulamentation of water rights.

REFERENCES

BILLIB, M. (1988) "Wasserwirtschaftliche Plannung in Semiariden Gebieten" IWHLW, Universität Hannover, Hannover, Germany.

CAMPOS, José Nilson B. (1987) "A Procedure for Reservoir Sizing on Intermittent Rivers under High Evaporation Rate." Diss. Colorado State University. Fort Collins, Co.

CAMPOS, José Nilson B. (1996). "O Processo de estocagem de águas em reservatórios: o papel da variabilidade dos deflúvios" (in Portuguese). In: III SIMPÓSIO DE RECURSOS HÍDRICOS DO NORDESTE. ASSOCIAÇÃO BRASILEIRA DE RECURSOS HÍDRICOS. Salvador, Brazil.

KENNETH, H. RECKHOW and CHAPRA, Steven C. (1983) Engineering approaches for lake management vol. 1: Data analysis and Empirical modeling. BUTTER WORTH PUBLISHERS Maryland.

Mc MAHON, T.A. and RUSSEL, G.M. (1978) "Reservoir Capacity and Yield" Elsevier Scientific Company. New York.

MORAN, P.A.P. (1954) "A Probability Theory of Dams and Storage Systems". Australian Journal of Applied Science vol. 6.

VOGEL, R. M., NEIL, M. F. and RALPH, A. B. (1995) "Storage-Reliability-Resilience-Yield Relations for Northeastern United States" Journal of Water Resources Planing and Management, set/oct: p.365-374.

MULTIVARIABLE MARIMA MODELING OF WATER RESOURCES TIME SERIES: APPLICATIONS TO RIVER NILE TEN-DAY DISCHARGES

AHMED H. EL-SAYED[1] and MOHAMED ERRIH[2]
[1]Asst. Prof., Water and Water Structures Eng. Dept., Faculty of Eng., Zagazig Univ., Egypt
[2]Asst. Prof., Institute of Hydraulics, Univ. of Sciences and Tech., P.O. Box 1505 Oran, Algeria

ABSTRACT
Recent developments with respect to transfer function-noise models are used to model and forecast ten-day natural inflows of the River Nile at Aswan. The input series are the natural ten-day discharges of the Blue Nile at Khartoum, the White Nile at Tamaniat, and of the river Atbara at mouth. Formal modeling procedures suggest a three-stage iterative process, namely: model identification, parameter estimation and diagnostic checks. An example of modeling multivariate streamflow series is concluded. An alternative modeling procedure is used and compared. The results obtained suggest that **MARIMA** models reproduce quite well the main statistical characteristics of the time series analyzed. It is concluded that the transfer function-noise model produces better forecasts for the natural ten-day inflows at Aswan. It is also assumed that the same conclusions apply for most water resources time series.

INTRODUCTION
Management and planning studies of water resources systems require the analysis of various types of hydrologic variables. For instance, sizing the needed capacity of a reservoir may be made based on the analysis of streamflows at one or more sites. Likewise, operational studies of reservoir systems may require forecasting rainfall and runoff. In general, investigations of water resources may involve data generation and/or forecasting, not only of hydrologic variables (inputs) but other variables, related to the use of water such as irrigation, hydropower and urban water supply (outputs). Sets of mutually related series defined at several points along a line, over an area or across space, or sets of mutually related series of various kinds defined at a point, are usually called multiple, multi-point or in general multivariate time series (Salas *et al.*, 1985). Multivariate time series may also be viewed as single or multiple series at a given site, having statistically distinguishable properties at various seasons of the year (Matalas, 1977). Some examples of multivariate time series in water resources are: series of annual or seasonal precipitation at several gaging stations in a region; streamflow series at various points along a river or at various rivers; different water quality variables at a particular river cross-section; series of precipitation, evaporation, inflows and outflows of given reservoir system; and series of different kinds of water demands such as for irrigation, hydropower, industry and urban water consumption.
The aim of the present paper is to analyse and discuss the technique known as **MARIMA** (Multivariate Autoregressive Integrated Moving Average Model) for modeling multivariate time series of inputs and outputs of water resources systems.

MODEL IDENTIFICATION

Identification of a transfer function component is usually more involved than that of an **ARIMA** (Autoregressive Integrated Moving Average Model) model. For multiple inputs situation, we may follow a procedure similar to that suggested in Box and Jenkins (1976) with some modifications. A generalized form of the **TF** (Transfer Function) model may be written in the form (Montgomery et al., 1990):

$$Y_t = \sum_{j=1}^{m} \frac{w_j(B)}{d_j(B)} X_{j,t-b_j} \qquad (1)$$

where B : backward shift operator, such that $BX_t = X_{t-1}$; and each input $X_{j,t}$ has a delay b_j and has a transfer function representation-type operator $\delta_j(B)$ and a moving-average-type operator $\omega_j(B)$, such as:

$$\omega(B) = \omega_0 - \omega_1 B - \omega_2 B^2 - \ldots - \omega_s B^s;$$

$$\delta(B) = 1 - \delta_1 B - \delta_2 B^2 - \ldots - \delta_r B^r$$

If the original series are nonstationary, the differencing may be required to produce stationary. If we denote $y_t = Y_t - Y_{t-1}$ and $x_{j,t} = X_{j,t} - X_{j,t-1}$ then

$$y_t = \sum_{j=1}^{m} \frac{\omega_j(B)}{\delta_j(B)} x_{j,t-b_j} \qquad (2)$$

The ten-day discharges of the Blue Nile at Khartoum, the White Nile at Tamaniat, the river Atbara at mouth, and of the main Nile at Aswan will be used to illustrated the complete process of transfer function analysis (Figure 1). The inflows of the White Nile at Tamaniat are the inflows of the main Nile at Tamaniat minus the inflows of the Blue Nile at Khartoum. The input series (X_t) are the ten-day inflows of the Blue Nile at Khartoum ($X_{1,t}$), the White Nile at Tamaniat ($X_{2,t}$), and of the Atbara river at mouth ($X_{3,t}$). The output series (Y_t) is the ten-day inflows of the main Nile at Aswan.

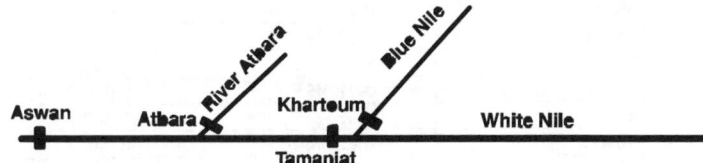

Figure 1 The stations at the river Nile and its tributaries

The first step in the modeling process is to build an **ARIMA** model for each of the three input series and to obtain three prewhitened input series (i.e. residuals), such, $e_{1,t}$, $e_{2,t}$ and $e_{3,t}$ for the Blue Nile at Khartoum, the White Nile at Tamaniat and the river Atbara at mouth, respectively. In this application, only the inflows of the concurrent period 1960-1981 are used as an identification period for the model and inflows for the period 1982 as a validation period. Visual inspection of four series reveals nonstationarity and seasonal behaviour. The sample autocorrelation functions also confirm this observation (El-Sayed, 1991).

The standard univariate modeling techniques indicate that the Blue Nile inflows at Khartoum can be modeled by an **ARIMA (1,0,1) (1,1,1) 36** process which is then used to prewhiten the output inflows at Aswan:

$(1 - 0.764B)(1 + 0.115B^{36})(1 - B^{36})X_{1,t} = (1 + 0.285B)(1 + 0.813B^{36})e_{1,t}$ (3)

where B^{36} is the seasonal backward shift operator, such that $B^{36}X_t = X_t - X_{t-36}$.
The second input of the White Nile at Tamaniat can be modeled by an ARIMA (1,1,1) (1,0,1) 36 process:

$(1 - 0.093B)(1 - B)(1 - 0.939B^{36})X_{2,t} = (1 + 0.941B)(1 + 0.831B^{36})e_{2,t}$ (4)

which in turn is used to prewhiten the output. Similarly, a univariate model for Atbara inflows is the ARIMA (1,1,1) (1,0,1) 36 process:

$(1 - 0.621B)(1 - B)(1 - 0.987B^{36})X_{3,t} = (1 + 0.995B)(1 + 0.828B^{36})e_{3,t}$ (5)

which is also used to prewhiten the output. Table 1 presents the cross correlation between each of the three prewhitened inputs and the prewhitened outputs.
The cross correlations exhibit significant values at lags 0 and 1 for the first input, at lag 0 for the second input and at lag 0, 1 and 2 for the third input series. This is consistent with a tentative model identification of r=2 (r and s refer to the degree of $\delta(B)$ function in the TF model), s=0, and b=0 (b is the absolute delay before the input series begins to influence the output series) for the transfer function relating the Blue Nile and Aswan inflows, r=1, s=0, and b=0 for the transfer function relating the White Nile and Aswan inflows, and r=3, s=0, and b=0 for the transfer function relating the river Atbara and Aswan inflows. A common practice in Box-Jenkins modeling is to try a few different models and make a choice amongst them at the diagnostic stage (Box and Jenkins, 1976). For the three input series we shall choose a three models with (r,s,b)=(1,1,0).

Table 1 Cross correlations between the prewhitened input and output series

Lag	CROSS CORRELATIONS		
	Blue Nile at Khartoum and River Nile at Aswan	White Nile at Tamaniat and River Nile at Aswan	River Atbara at mouth and River Nile at Aswan
0	0.359532	0.239462	0.150516
1	0.315558	0.039345	0.185592
2	0.099222	0.019516	0.138102
3	0.102651	0.011865	-0.029649
4	-0.067901	0.055237	0.124553
5	-0.050966	0.021744	0.074012
6	0.053065	-0.030041	-0.080497
7	-0.045588	-0.022449	0.009132
8	0.011983	-0.031646	0.021507
9	0.051630	-0.036413	0.026397
10	-0.033576	-0.035596	0.038300
11	0.027123	-0.085628	0.008259
12	-0.063813	-0.062279	0.017812

MODEL ESTIMATION

The preliminary parameter estimates for the transfer function models relating the three inputs and the output are:

$(1 - 0.2458B)y_t = (0.6414 - 0.1788B)x_{1,t};$ (6)

$(1 - 0.1982B)y_t = (0.6630 - 0.1973B)x_{2,t};$ (7)

$(1 - 0.2673B)y_t = (-0.040 - 0.5753B)x_{3,t};$ (8)

where $x_{1,t}$, $x_{2,t}$, $x_{3,t}$, and y_t are the seasonally differenced series of the ten-day inflows of the Blue Nile at Khartoum ($X_{1,t}$), White Nile at Tamaniat ($X_{2,t}$), Atbara river ($X_{3,t}$), and Aswan (Y_t), respectively.
The noise series (n_t) is estimated by combining the three models as:

$$n_t = y_t - \frac{0.6414 - 0.1788B}{1 - 0.2458B} x_{1,t} - \frac{0.663 - 0.1973B}{1 - 0.1982B} x_{2,t} + \frac{0.040 + 0.5753B}{1 - 0.2673B} x_{3,t} \qquad (9)$$

The combined effect of the three inputs have been modeled as the sum of the individual inputs. This is probably reasonable, since $x_{1,t}$, $x_{2,t}$ and $x_{3,t}$, are essentially uncorrelated (Montgomery *et al.*, 1990). The autocorrelation and partial autocorrelation functions of the estimated noise series (n_t) indicate that (n_t) can be modeled as seasonal **ARIMA** (1,0,1) (0,0,1) 36 process:

$$(1 - 0.9915B)n_t = (1 + 0.9980B)(1 + 0.7441B^{36})a_t \qquad (10)$$

where a_t is the series of residuals after filling the TF model.
Therefore, the combined transfer function model plus noise is:

$$y_t = \frac{(\omega_{10} - \omega_{11}B)}{(1 - \delta_{11}B)} x_{1,t} + \frac{(\omega_{20} - \omega_{21}B)}{(1 - \delta_{21}B)} x_{2,t} + \frac{(\omega_{30} - \omega_{31}B)}{(1 - \delta_{31}B)} x_{3,t} + \frac{(1 - \theta_1 B)(1 - \theta_2 B^{36})}{(1 - \phi_1 B)} a_t \qquad (11)$$

where θ_1 is the moving average coefficient; θ_2 is the seasonal moving average coefficient; and ϕ_1 is the autoregressive coefficient.
The **TF** model has been identified and the parameters have been estimated to yield

$$y_t = \frac{(0.5752 - 0.3144B)}{(1 - 0.1679B)} x_{1,t} + \frac{(0.7347 - 0.2424B)}{(1 - 0.0956B)} x_{2,t} +$$

$$+ \frac{(0.1471 - 0.3983B)}{(1 - 0.354B)} x_{3,t} + \frac{(1 + 0.9767B)(1 + 0.7591B^{36})}{(1 - 0.9913B)} a_t \qquad (12)$$

DIAGNOSTIC CHECKING

It is common in **ARIMA** modeling to identify more than one model form, estimate the parameters for each model, and then do a careful diagnostic check to test the validity of the model. The same is true in TF modeling. Of particular interest in this case are two items: (i) the final residual series designated a_t and (ii) the relationship between this a_t series and the prewhitened input series, which has been designated e_t (Makridakis *et al.*, 1983).
In the process of directly estimating the transfer function parameters, an assumption is made that the prewhitened input series e_t is dependent of the random noise component a_t. Thus, an important part of the diagnostic process is to indicate this assumption.
The cross-correlation function (**CCF**) between the final residuals a_t and the three prewhitened input series $e_{1,t}$, $e_{2,t}$, and $e_{3,t}$ are essentially zero. Therefore, we can conclude that the **TF** model expressed in Equation 10 satisfies the assumption of independence between tha a_t and the e_t's series.

FORECASTING

Once a suitable time series model has been obtained, we may use the model for various purposes. A major application of a time series model is forecasting. The forecasting version of the Program **RATS** (Thomas, 1990) is used to forecast the ten-day inflows of the river Nile in both the historical and the validation periods.
Column 2 of Table 2 shows the forecasts obtained from the **TF** model, Equation 10, in the validation period, year 1982. Column 1 shows the original values of the ten-day inflows at Aswan in the same period. For the comparison purposes, we have generated forecasts for the ten-day inflows at Aswan using a multiplicative seasonal **ARIMA** model (El-Sayed, 1991). Forecasts obtained from this model in the validation period are shown in Column 3 of Table 2. We note

Table 2 Forecasts of the river Nile inflows (millions cubic meters) at Aswan

Year	Month	Decade	ASWAN (1)	TF model (2)	ARIMA model (3)
1982	January	1	137.66	115.46	108.68
		2	147.43	132.79	108.96
		3	120.00	128.63	137.66
	February	1	240.19	108.00	110.42
		2	93.44	87.34	128.70
		3	75.74	60.33	109.01
	March	1	31.20	35.02	65.79
		2	2.71	11.65	25.23
		3	0.00	8.81	1.79
	May	1	5.13	9.21	3.53
		2	25.15	21.66	18.19
		3	40.06	30.54	27.91
	April	1	56.55	33.38	39.58
		2	70.16	32.72	53.22
		3	28.66	42.21	86.10
	June	1	52.99	53.99	65.78
		2	75.79	77.12	81.18
		3	116.17	87.38	104.77
	July	1	89.86	93.16	127.52
		2	186.29	198.67	207.65
		3	394.89	331.21	364.30
	August	1	479.18	497.24	509.90
		2	726.25	674.14	643.56
		3	729.84	788.17	845.27
	September	1	918.17	807.43	832.93
		2	726.13	693.10	760.62
		3	502.84	506.30	548.28
	October	1	358.78	423.80	501.61
		2	257.48	308.42	318.85
		3	309.64	294.92	205.34
	November	1	354.35	257.87	210.48
		2	172.12	196.48	230.70
		3	95.04	125.40	174.84
	December	1	73.83	99.35	99.14
		2	118.36	135.49	100.37
		3	116.84	103.45	96.40

immediately that the transfer function model produces superior forecasts in the validation period. The forecasts results indicate how the forecasted inflows are close to the actual ones and hence the adequacy of the specified **TF** model for Aswan inflows.

SUMMARY AND CONCLUSION

Using multivariate approaches in modeling of water resources time series has been important subject in water resources literature the past 30 years. Suggested approaches used in the early stages were based on multiple regression analysis as well as principal component analysis.

Following the last lead presented by the well-known book of Box and Jenkins (1970), most recent approaches in the water resources literature suggest the three-stage iterative model building, namely: model identification, parameter estimation and diagnostic checks. Model identification consists in determining the structural form of the model, whether the model involves constant or periodic parameters, and finding the order of the model. Several tools have been used which may help in such a task. They include: cross-correlation analysis, autocorrelation, and partial-correlation analysis. Once the model is defined, model parameters are estimated using Gauss-Newton algorithm. The residuals of each component of the model are recorverd. The model diagnostic checks are necessary in order to see whether the selected model is appropriate and whether it is better than other component models.

An application of multivariate modeling approach has been included in this paper. The main objectives have been: to present the concepts and the estimation procedure in some detail, and to compare with an alternative modeling approach. Generated samples are used for comparing the historical and generated inflows of the two approaches.

The following conclusions are drawn from the results of the foregoing application:

(1) The transfer function-noise model provides an effective mean of forecasting ten-day inflows to the Aswan High dam reservoir based on Blue Nile, White Nile, and Atbara river;

(2) The most recent statistical processes involved are used to identify, estimate and verify a reasonable model;

(3) The model diagnostic checking stage indicates the **TF** model with the three covariate series is better than a seasonal **ARIMA** model, and

(4) A forecasting sample of both the historical and the validation periods suggests also that the transfer function-noise model provids better forecasts than a particular **ARIMA** model.

REFERENCES

Box, G.E.P., and Jenkins, G.M., *'Time Series Analysis Forecasting and Control'*, San Francisco, California, Holden Day, 1976

El-Sayed, A.H., *'Hydrological Parameters and Their Effect on River Supply'*, Ph.D. Thesis, Faculty of Eng. Zagazig Univ., 1991

Makridakis, S., Wheelwright, S.C. and Mc Gee, V.E., *'Forecasting: Methods and Applications'*, 2nd Ed. New York, John Wiley and Sons, 1983

Matalas, N.C., 'Generation of Multivariate Synthetic Flows', In: *Mathematic Models for Surface Water Hydrology*, T.A. Cirinai *et al.* (Editors), John Wiley and Sons, London, 1977

Montgemory, D.C., Johnson, L.A., and Gardiner, J.S., *'Forecasting and Time Series Analysis'*, 2nd Ed., Mac Graw-Hill, Inc., New-York, 1990

Salas, J.D., Tablios, G.Q., and Bartolini, P., 'Approachs to Multivariate Modeling of Water Resources Time Series', *Water Resources Bulletin*, Vol. 21, N°4 and 5, 1985

Thomas, A.D., *'RATS: User's Manual'*, Version 3.10, 1990

Uncertainty Analysis of a Water Resources System in a Competition and Privatization Environment

RICARDO SMITH, ISAAC DYNER, SANTIAGO MONTOYA & CARLOS FRANCO
National University of Colombia, Water Resources Graduate Program,
Medellin, Colombia.

ABSTRACT

In several countries the public ownership of the utilities has been questioned. Procedures to sell public owned utilities or to incentivate private participation has been implemented. In Colombian new laws related to the electricity sector were enforced in order to guarantee competition in the electricity generation sector. With this changes Colombian government is looking to diminish the size of the public sector, to gain efficiency and to have additional resources for other sectors. The roll of the central government in the electricity generation system has moved from a long term enforced central planning to a follow up of the electricity sector evolution. In this sense the Colombian central government needs a tool that will allow to follow the evolution of the utilities capacity in the new competition environment, so that if it is not appropriate the government can take corrective measures. A model that represents the evolution of the Colombian electric sector, based in System Dynamics is presented. Some uncertainties are included in the model such as financial, fuel prices, regulatory laws, and others. Model results are presented with some conclusions and recommendations.

INTRODUCTION

Energy sector in less developed countries nowadays face liberalization and market privatization processes. Problems relative to balance government regulation and free market economy arise in this new environment. The uncertainties associated with the modeling of a water resources systems evolution in a privatization and competition environment include:

VULNERABILITY TO HYDROLOGY. 78 % of Colombian electricity generation capacity is from hydropower plants. The generation capacity is highly influenced by the system hydrology conditions. These conditions are highly uncertain. Uncertainty increase due to the significative impact (negative impact) of El Niño Southern Oscillation (ENSO) phenomena over Colombia. Electricity price levels are highly related to hydrology conditions.

RESERVOIR CAPACITY REGULATION. The hydrologic dependence of the Colombian electricity generation system is greater because of the system low capacity regulation. Only 10% of the hydropower system has seasonal regulation. Most of the hydropower projects are of the run-of-the-river type. This low regulation capacity compromise the electricity generation capacity during ENSO events. The government is trying to reduce this vulnerability of the system by incentivating the building of thermopower plants.

REGULATORY UNCERTAINTY. The market interchange of electricity is done under a central government regulatory and environmental laws framework. Government regulations looks to guarantee a free market economy for the electricity interchanges. Environmental regulation looks to keep pollution at an acceptable level. These laws are adjusted from time to time (regulatory uncertainty) according whit what the central government think is most appropriate. For example recently a price incentive was established to incentivate private investors to build new capacity. This price is added to the electricity generation selling price. This price is expected to be a temporary measure.

GAS AVAILABILITY AND PRICES. Most of the new investment in electricity capacity generation is in gas fueled thermopower plants . The gas price is uncertain and it depends on the country gas reserves (availability) and on the government gas policies. Gas supply is done by a public monopoly, so prices could be manipulated by the central government. Recently gas prices are close to 2.0 US$/MBTU, a low price (30% of the real cost) set by the government to encourage investments in gas fueled technologies and private participation.

DSM PROGRAMS. Demand side management (DSM) could be implemented creating uncertainty about electricity demand growth and future capacity needs. Recently Colombian government has implemented a efficient electricity bulbs program. If this program succeed it is expected to diminish electricity demand by 7.8% in the next 13 years

RATE OF RETURN AND COMPANY STRUCTURE. The main uncertainty in the market for the private investor are the expectatives of the investment rate of return. In a highly uncertain market investors trends to expect higher rates of returns. In a competition environment publicly owned companies will use higher rate of returns for investment decisions. The higher the uncertainty the higher the investment rate of return.

INVESTMENT RECOVER PERIOD. In a highly uncertain risky market, the economic agent wants to recover the investment as soon as possible. Under actual economic conditions, hydropower plants has long construction periods and low rate of returns. Gas fueled thermopower plants has short construction periods (2 years), low investment recovery periods and higher rate of return. It is then expected that the hydro-thermal composition of Colombian capacity generation system to change over the next years. New investments will be mainly made in gas fueled thermopower plants because of financial conditions.

Financial capability of the capacity generation companies has to be considered in the modeling process. In a competition environment a company can invest only if it is in an adequate financial situation. Otherwise the financial institutions (Banks) will not lend the necessary money for the investments. It could happen that in a favorable investment situation a capacity generation company will not make any investment because to its financial situation. This situation could worsen if company selling prices to the users (tariffs) are not appropriate (subsidized). This is a common situation in several publicly owned companies, and its solution is uncertain.

TRANSMISSION UNCERTAINTY. The generators use the National Interconnected Grid to sell his energy to the consumers. The system capacity generation expansion requires a well planned growth of the transmission grid system. Investors will always build capacity in the region that will give them the maximum return of its investments. In his investment decision investors has to take into account charges for the use of the national transmission grid. This charges will depend on the region installed capacity and on the transmission grid capacity in that region. Transmission charges will be greater in a given region when the installed capacity is close to the transmission capacity.

SYSTEM MARGINAL PRICE (SMP). In Colombia, generated electricity is hourly dispatched using a pool system that sets the system marginal price (SMP) which gives the profits to the different companies. The SMP has high fluctuations and is highly uncertain. Because in the actual system hydro-thermal composition low regulation hydropower projects are dominant, the SMP is highly dependent on hydrologic conditions. The generators could agree to manipulate the SMP to convince potential buyers to sign long term contracts whit higher prices in order to avoid volatile prices.

Highly uncertain SMP or low SMP because of a high hydrology condition could discourage new investments. For example over the last year Colombia has an over the average hydrology condition that combined with inframarginal public generation projects has produced a very low SMP. The government has then introduced a regulated capacity price to be added to the SMP to encourage investments in generation capacity.

UNCERTAINTIES NOT CONSIDERED. Four uncertainties are not considered in this version of the model. The first one are related to the general macroeconomics government policies; second, the uncertainty of exchange rates US$/COL$; third, local political pressures over the new investors; and four, efficient technologies future developments, e.g. building OCGT plants as efficients as CCGT plants.

PROPOSED MODEL - DESCRIPTION AND RESULTS

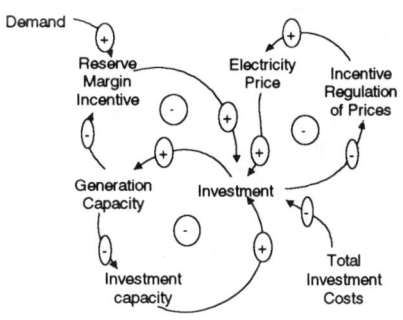

Figure 1. Casual diagram considering competence environment.

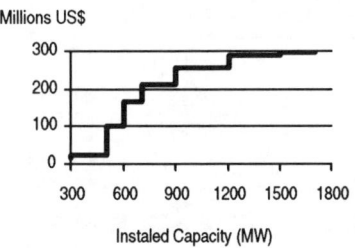

Figure 2. New net reinforcement cost (*Adapted from Corredor et al, 1995*).

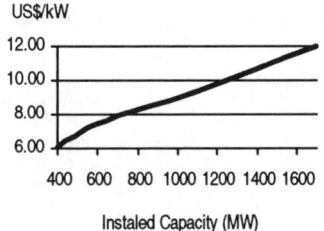

Figure 3. Change in the transmission charges by incremental grid use (*Adapted from Corredor et al, 1995*).

A model based in System Dynamics has been developed to model the evolution of Colombian electricity sector generation capacity in the new competition environment. The developed model includes several of the uncertainties mentioned above. The model aim is to be used as a policy analysis model for the Colombian government (Ministry of Energy).

The model causal diagram could be seen in Figure 1.

The model assumes that the agents in the market face three types of uncertainties: Expected price (US$/kWh), discounted investment (US$/kWh) and the investment financial capability (binary variable). Any company can invest in new capacity if the price paid exceeds the discounted cost of the new investment, depending on the company financial capability. The system expand all the capacity modeling the price that the generator can expect based on his expectatives of the system evolution.

The price depends on the available capacity for each type of generation technology The water availability for the reservoirs (generation capacity) can be proposed as a mean income of water per year (%). A diminished hydrology can push both high-level SMP and long-term contract prices due to unavailability of energy at the reservoirs, then new investors can be encouraged to built more capacity, looking for less hydrology vulnerable technologies such as OCGT and CCGT plants.

Transmission cost uncertainty is modeled using functions that represent the incremental cost of install new plants in a specific region (see Figures 2 and 3). The transmission grid cost can be modified adding the cost obtained using the described functions. New signals to the mean energy investment cost are send, and the agents will re-evaluate his investments.

Financial uncertainty can be solved by modeling the cash flow for every agent. Those companies that can't invest in new capacity are not considered, so the expansion program is done by private investors (without financial limitations), and current companies that could make profits in the system.

The central government has to invest, depending on the future evolution of the system.

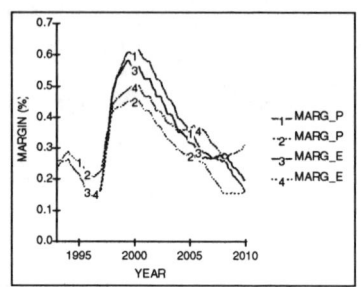

Figure 4. Reserve margin for power and energy

The demand uncertainty can be analyzed with a different model designed to study the effect of DSM programs in Colombia (Franco, 1996). This model was build in a System Dynamics platform, and it models programs like efficient bulb replacement and natural gas residential use (heating and cooking). The resulting demand can be linked to the electricity sector evolution model under uncertainty. New demand projections can mean a new scenarios to the liberalization process. Figures 4 to 6 shows some model results. Figure 4 shows the Colombian system reserve margin evolution for power and energy under different assumptions. Figure 5 shows the evolution of the companies electricity market share and figure 6 the evolution of generation capacity by technology. These figures and others results shows that new investments create a great reserve power margin, increasing with the change in electricity consumption. Private investors take

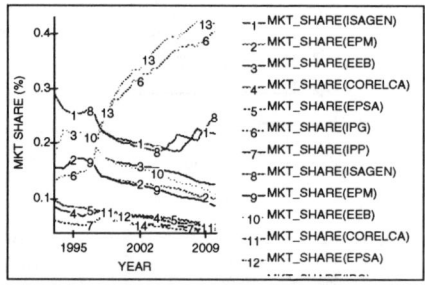

Figure 5. Market share by company

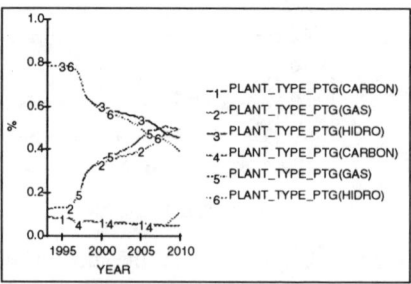

Figure 6. Generation capacity evolution by technology.

control over the system. Future investments are made in cheaply, efficient, short-term recovery period gas plants, reaching 50% of the system capacity around the 2005 year.

SUMMARY

A electricity generation evolution model based on System Dynamics, as a tool to evaluate scenarios and policy analysis, can deal with the uncertainties created by the liberalization process and competition environment in the Colombian electricity sector.

Feedback capabilities allowed by the methodology can help to stablish synergies produced by elements like average inflow rate to reservoirs, rate of return expectations, cash flow evolution, regulatory laws, and others. The model results are helpful looking to diminish the government policies influence over the new electricity market system. Results show how regulatory policies, with nowadays price incentives, encourage capacity investments during the next five years. After year 2000 this effect, joined to the planned reduction of price incentives, diminish the capacity investment levels which reduce again the power reserve margin of the system. Dynamic regulation must send steady signals to new investors, but has to look for mechanisms to prevents this investment cycles. The risk level to blackouts will be less in the future with the building of OCGT plants in the system, and slowly reducing dependence on the hydraulic system and its vulnerability to low hydrology events.

References

BUNN D.W.. LARSEN, E.R.,1992. "Sensitivity of reserve margin to factors influencing investment behaviour in the electricity market of England and Wales". Energy Policy, may 1992, Pages 420-429.
BUNN D.W., 1994. Evaluating the effects of privitizing electricity. Journal Oper. Res. Soc., Vol. 45, No. 4, pages 367-375
CORREDOR A., Pablo H.; GÓMEZ V., Juan Diego; VILLEGAS R., Andrés. "Transmission charges by grid use: A expansion planning decision element". Energética, No. 15, Water resources graduate program, National University of Colombia, 1995.
DYNER R., Isaac. System Dynamics Plataforms for Integrated Energy Analysis. University of London. London Business School. June 1996.
FRANCO, Carlos J. "A national desintegrated model to analyze the effect of rational use comsumption policies". Thesis submitted to the National University of Colombia for the degree of Master Sciences. National University of colombia, September, 1996.
FORD, Andrew; YOUNGBLOOD, Annette. Technical Documentation of the Electric Utility Policy an Planning Analysis Model, Version 4. Los Alamos National Laboratory. Los Alamos, New México. Apr. 1982.
SMITH, Ricardo A. "Generation Systems Planning Elements within Competence Environment". II International Forum about Energy Planning. Santa Fé de Bogotá, September 13th-15th, 1995.

ns
WATER RESOURCE SYSTEM OPERATION UNDER HYDROLOGIC UNCERTAINTY: THE MAE KLONG RIVER BASIN, THAILAND

TAWATCHAI TINGSANCHALI
and
VEERAKCUDDY RAJASEKARAM
School of Civil Engineering, Asian Institute of Technology,
P.O Box 4, Klongluang, Pathumthani 12120, Thailand

ABSTRACT

The stochastic variability of hydrological and climatological variables results in considerable uncertainty in system operation of the Mae Klong River basin, Thailand. With further development of irrigation command area and additional requirement of water for domestic and industrial use in the future, measures to improve the system operational reliability by modifying the present operation ruling levels of reservoirs, attains considerable importance among others. The system operational reliability has been evaluated by using the water balance function which expresses the monthly available water in excess of demands as a function of stochastic variables and deterministic variables such as the operation ruling levels of reservoirs for each month. A non-linear optimization on a combined function of monthly water balances has been carried out to determine the optimal reservoir operation ruling levels while the system operation reliability is maximized. The monthly observed data of hydrological and climatological variables from 1965 to 1994 has been used for the analysis.

INTRODUCTION

Water resources engineering systems involve natural processes or phenomena which have random output, hence the resulting operation involve certain degree of risk which could be studied in the probabilistic platform by describing the random variables involved by appropriate probability distribution functions. Moreover, failures are the joint occurrences of excessive load and weak resistance (structural failure) or excessive water demand and inadequate supply (operational failure) of the system [1]. Provided a water resources system is safe against structural failures, it is important for a designer to look into the system safety against operational failures by computing the reliability of system operation, duly considering the uncertainties that arise from the stochastic behavior of hydrological and climatological variables. The present study focuses on computing the operational reliability of a water resources system, namely, the Mae Klong River Basin system in Thailand, and to optimize the system operation to maximize reliability and hence an assured benefit from the system.

SYSTEM OPERATION UNDER UNCERTAINTY

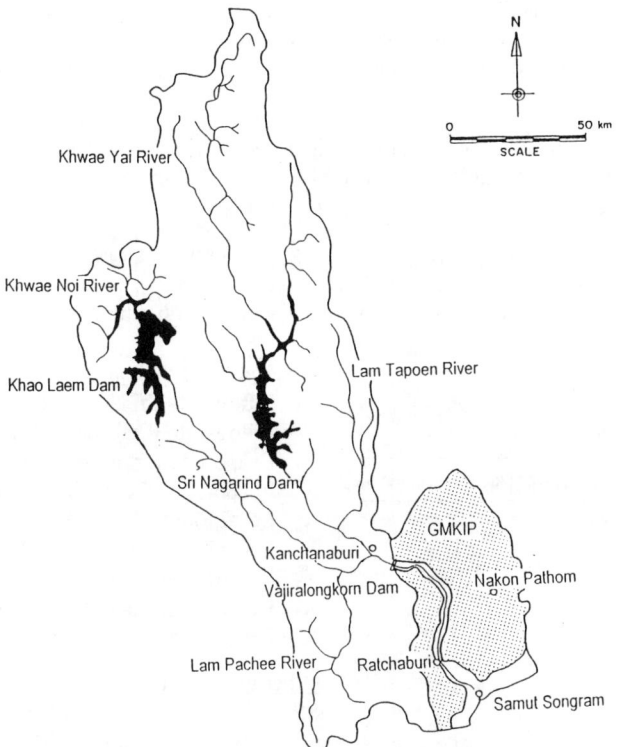

Figure 1 The Mae Klong River Basin

DESCRIPTION OF STUDY AREA

The Mae Klong River basin is located in the western part of Thailand (Figure 1), with a total drainage area of 30,840 sq.km, and an average annual yield of 12,700 MCM. There are two main tributaries, namely, the Khwae Yai river and the Khwae Noi river, and two sub-tributaries, namely the Lam Tapoen river and the Lam Pachee river. The main stream called the Mae Klong river starts from the confluence of the major tributaries and ends at the Gulf of Thailand. The Greater Mae Klong Irrigation Project (GMKIP) is located on both banks of the Mae Klong River and gets water diverted from the Vajiralongkorn barrage located at the head end of this river. At present, there are two major reservoirs in the basin, namely, the Sri Nagarind Dam, located across the Khwae Yai River with a total storage capacity of 17,745 MCM and the Khao Laem Dam, located across the Khwae Noi river having a total storage capacity of 8,860 MCM.

The water requirement of the GMKIP (235,300 ha, at present) is fulfilled, and the full extent of 483,300 ha will be completed shortly. GMKIP has a variety of crops such as rice, sugar cane, upland crops, vegetables and fish ponds. In addition, there are many small scale irrigation and domestic water supply projects in the basin. Augmenting flow in the adjacent basin, namely, the Tha Chin basin and meeting the salinity control water requirement (50 m³/s during dry season) at the tail end of the Mae Klong River are existing downstream demands of the basin. The optimal operation of reservoirs to meet these demands along with a new demand of 50 m³/s to supplement the water requirement of the Bangkok Metropolitan Authority is carried out in the present study.

RELIABILITY OF SYSTEM OPERATION

The stochastic variables associated with the present study have been illustrated along with the schematic representation of the Mae Klong River basin in Figure 2.

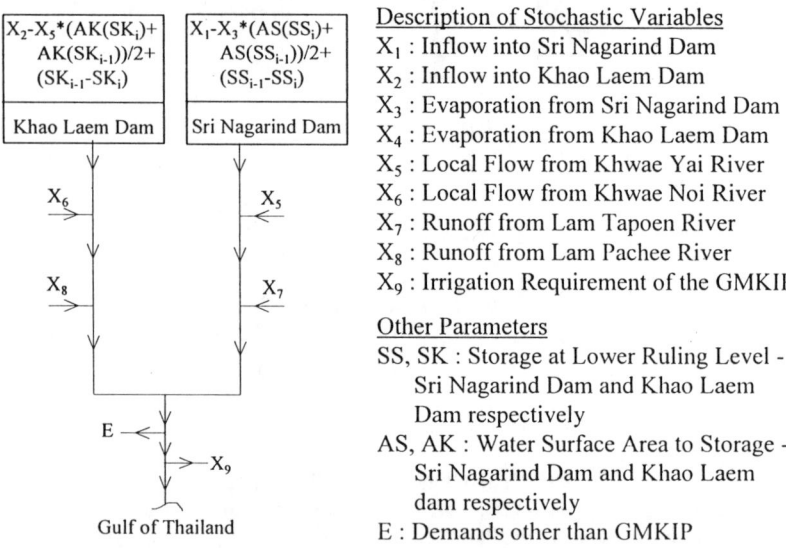

Description of Stochastic Variables
X_1 : Inflow into Sri Nagarind Dam
X_2 : Inflow into Khao Laem Dam
X_3 : Evaporation from Sri Nagarind Dam
X_4 : Evaporation from Khao Laem Dam
X_5 : Local Flow from Khwae Yai River
X_6 : Local Flow from Khwae Noi River
X_7 : Runoff from Lam Tapoen River
X_8 : Runoff from Lam Pachee River
X_9 : Irrigation Requirement of the GMKIP

Other Parameters
SS, SK : Storage at Lower Ruling Level - Sri Nagarind Dam and Khao Laem Dam respectively
AS, AK : Water Surface Area to Storage - Sri Nagarind Dam and Khao Laem dam respectively
E : Demands other than GMKIP

Figure 2 Schematic of the Mae Klong River Basin for Reliability Analysis

The system water balance function (i.e., the 'performance function' or 'state function' of system operation [2]), G(X) for month i is formulated as follows:

$$G(X) = X_1 + (SS_{i-1} - SS_i) - X_3[AS(SS_i) + AS(SS_{i-1})]/2$$
$$+ X_2 + (SK_{i-1} - SK_i) - X_4[AK(SK_i) + AK(SK_{i-1})]/2$$
$$+ X_5 + X_6 + X_7 + X_8 - E - X_9 = \sum_{n=1}^{9} C_n X_n + C_{10} \quad \text{...............(1)}$$

Using Shinozuka's approach [3] to determine the reliability of operation, the limit state equation, i.e., G(X)=0 is represented as a hyper-surface in the coordinate system of the corresponding normalized random variables denoted by $X_i' = (X_i - \mu_{Xi})/\sigma_{Xi}$, where μ and σ respectively denote the mean and standard deviation of the stochastic variable concerned. When all the stochastic variables involved in the performance function follow normal distribution, the reliability r is determined from the equation, $r = \phi(\beta)$, where ϕ stands for the standard normal probability distribution and β, known as the reliability index, is found as the shortest distance from the origin of the normalized variables to the hyper-surface [3]. Moreover, the point on the hyper-surface corresponding to the shortest distance is termed as the most probable failure point.

The variables X_1 through X_9 are correlated among themselves as they all emerge from the same hydrologic conditions of the Mae Klong River basin for the month concerned. Therefore, a set of (statistically) un-correlated variables Y_1 through Y_9 are formulated [2] and the performance function $G(X)$ is transformed as,

$$F(Y) = \sum_{n=1}^{9} D_n Y_n + D_{10} \quad \quad (2)$$

in which the coefficients D_1 through D_{10} are expressed as follows:

$$D_i = \sum_{n=1}^{9} C_n \sigma_n t_{n,i} \; ; i = 1,..,9$$

$$D_{10} = C_{10} + \sum_{n=1}^{9} C_n (\mu_X)_n \quad \quad (3)$$

where $t_{i,j}$ represents the elements of the orthogonal transformation matrix $[T]$ that satisfy the relationship $[T]^t[M][T]=[\lambda]$ in which $[M]$ represents the matrix of correlation coefficient of the stochastic variables X_1 through X_9 and $[\lambda]$ represents the diagonal matrix of eigen values of $[M]$. The expression for β based on the transformed performance function (Equation 2) is derived as,

$$\beta = \frac{D_{10}}{\sqrt{\sum_{n=1}^{9} D_n^2 \lambda_n}} \quad \quad (4)$$

The stochastic variables involved in this study have been proved for their normalcy using the Chi-Square test, with a confidence level of 90%. Having found β, and hence the monthly reliability, values of $\{X_1,..,X_9\}$ which satisfy the performance equation (i.e., $G(X) = 0$), are determined and are known as the values at the most probable failure point. The values at the most probable failure points are indicated with an asterisk superscript in the subsequent derivations.

RELIABILITY-BASED OPTIMAL SYSTEM OPERATION
In order to improve the reliability of operation of the Mae Klong River Basin system, it is proposed to improve the lower and upper operation ruling levels of the main reservoirs, namely, the Sri Nagarind dam and the Khao Laem dam. The monthly performance function $G(X)$ (Equation 1) expresses the amount of water left at the end of monthly operation, after meeting the requirements and losses in the month concerned. Since the storage corresponding to a lower operation ruling level of a month appears in the subsequent month as well, the 'water balance product function' is defined as,

$$P = G(X^*)_1 . G(X^*)_2 ... G(X^*)_{12} \quad \quad (5)$$

Simplifying $G(X^*)_i$ as $G_i = t_i + S_{i-1} - S_i$ where t_i represents other terms than storage of operation ruling levels, and $S_i = SS_i + SK_i$, the following non-linear model is used to optimize the lower operation ruling levels:

$$\underset{S_1,\ldots,S_{12}}{Max} P = (t_1 + S_{12} - S_1).(t_2 + S_1 - S_2)\ldots(t_{12} + S_{11} - S_{12}) \quad \ldots\ldots(6)$$

subject to the constraints,

$[S_i]_{min} < S_i < [S_i]_{max}$; $i = 1,..12$ (storage limit constraints)

$t_i + S_{i-1} - S_i > 0$; $i = 1,..12$ (monthly water availability constraints)

The model is solved using the method of Lagrange's multipliers, and the optimal solution is found to coincide with the global maximum of the water balance product function. The values of S_i's are divided among the reservoirs in the ratio of their active storages. Moreover, the non-linearities involved in the evaporation terms have been handled by repeating computations until unique results are obtained. A similar consideration with E and X_9 (i.e., water demand terms) excluded from G(X) has been used to optimize the upper ruling levels.

RESULTS AND DISCUSSION

The reliability index (Equation 4) is considered as the indicator for system operation, in contrast to the conventional simulation analysis in which the system operation study is carried out using a lengthy record of hydrologic variables to analyze the impact of their variability. The denominator of the expression for β ultimately consists of the standard deviations of the stochastic variables, and the eigen values of the matrix of correlation coefficients among the stochastic variables, while, the numerator consists of the available storage release from reservoirs based on the lower operation ruling levels. As such, the reliability index β (and hence, the reliability), takes into account of the variability of all the stochastic variables involved as well as the availability of water for operation. Moreover, the monthly reliabilities determined are on the conservative side for the following reasons: (i). the reservoir water levels are always maintained above the lowest operation ruling levels, hence the available storage release according to these ruling levels is on the safe side, and (ii). the reliability index is estimated based on the (shortest) distance to the most probable failure point [3], and the occurrence of stochastic events need not to corresponds to this condition in reality, hence the actual reliability could be greater than the one computed.

Because of the individual variability of stochastic variables in each month, the monthly reliabilities computed are different and the lowest is found in May, though the water balances for different months are made equal after optimization of lower operation ruling levels.

Table 1 - Monthly Reliabilities with Present and Proposed Ruling Levels

Month	Jan	Feb	Mar	Apr	May	Jun	Jul	Aug	Sep	Oct	Nov	Dec
Present	5.4×10^{-4}	6.8×10^{-4}	6.3×10^{-4}	7.3×10^{-2}	0.54	0.77	0.97	1.0	1.0	0.98	0.49	3.2×10^{-3}
Proposed	0.95	0.94	0.84	0.68	0.63	0.79	0.75	0.74	0.73	0.72	0.66	0.87

SYSTEM OPERATION UNDER UNCERTAINTY

Table 1 illustrates the reliabilities computed with the present and proposed operation ruling levels and shows remarkable improvement in the overall system reliability which is taken as the lowest of all months. The overall system reliability has improved to 0.63 from very low value. Figure 3 compares the present and proposed operation ruling levels for the two major reservoirs.

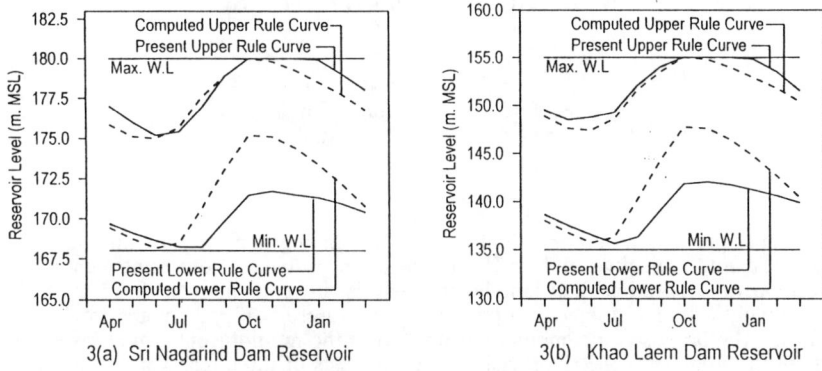

3(a) Sri Nagarind Dam Reservoir 3(b) Khao Laem Dam Reservoir

Figure 3 Operation Rule Curves of Reservoirs

Since the monthly operational reliabilities have been improved with the proposed operation ruling levels, the yield from various crops and hence the net-benefit from them could be assured with better reliability compared to the present condition.

CONCLUSIONS

A methodology to determine the operational reliability of a water resources system with multiple components has been demonstrated with specific example to the Mae Klong River basin system in Thailand. The system performance has been mathematically expressed using the stochastic variables such as inflows, evaporation, water requirement etc., and deterministic variables such as the operation ruling levels of reservoirs, and the system reliability has been evaluated. Since the monthly storage releases from reservoirs are found to be the main factors influencing the reliability among others, an optimization to yield a balanced system of storage releases has been carried out and hence the new operation ruling levels for reservoirs have been determined. The new set of ruling levels is found to yield better reliability of system operation compared to the present one and hence the yield from the system is assured.

REFERENCES

1. YEN, B.C and ANG, A.H-S (1971) Risk Analysis in Design of Hydraulic Projects, in Stochastic Hydraulics, Proceedings of the First International Symposium, CHIU (ed.), University of Pittsburgh, Pennsylvania, U.S.A
2. ANG, A.H-S and TANG, W.H (1984) Probability Concepts in Engineering Planning and Design, Vol. II, John Wiley and Sons, New York, U.S.A
3. SHINOZUKA, M (1983) Basic Analysis of Structural Safety, Journal of Structural Division, ASCE, Vol. 109, No. 3, pp. 721 - 740.

STOCHASTIC MULTIOBJECTIVE OPTIMIZATION OF MULTIRESERVOIR SYSTEMS

Y. C. WANG
Deparment of Water Resources Engineering
Newjec Engineering Consulting Corporation, Tokyo, Japan

ABSTRACT

This paper presents an optimization method for finding noninferior solutions of a stochastic multiobjective multireservoir operation problem. Constraint technique, decomposition iteration, and simulation analysis are conjunctively used to deal, respectively, with multiobjective optimization, multireservoir system, and stochastic inflows. The proposed methodology is applied to the multiobjective multireservoir system in the upper Tone river basin, which consists of three reservoirs in parallel and is operated primarily for hydropower, water supply, and flood control. 49 noninferior solutions of the multiobjective multireservoir system are obtained, from which the decision maker can find the most satisfactory operating policy.

INTRODUCTION

Water resource systems tend to include multiple reservoirs which are operated to serve several objectives such as hydropower, flood control, and water supply for municipal, industrial, and agricultural uses. Moreover, the stochasticity of inflows into reservoirs is one of basic characteristics of water resources systems. It is multireservoir, multiobjective, and stochastic inflows that represent three difficult points in water resource systems analysis. Literature review indicates that, as far as stochastic inflows are taken into account, the choice of reservoir optimization techniques is mainly restricted to Stochastic Dynamic Programming (SDP) and Stochastic Linear Programming (SLP). SDP is preferred to SLP for reservoir operation problems, because SDP has much smaller computational effort (including computer time and memory) than SLP. However, SDP is usually not feasible in computation when it is applied to complex large-scale water resources systems, because computational effort increases exponentially with the numbers of reservoirs, objectives, and random variables. This phenomenon was thought of as the 'curse of dimensionality'. Some methods have been presented to alleviate the curse of dimensionality, but they are effective only for cases where state variables are assumed to be deterministic. Several attempts have been made to solve large-scale stochastic problems (Yeh, 1985), but only a single objective is taken into account.

This paper presents a multiobjective optimization method for finding noninferior solutions or satisfactory operating policies of multireservoir systems with stochastic inflows. Constraint technique, decomposition iteration, and simulation analysis are conjunctively used to deal, respectively, with the vector optimization, large-scale multireservoir systems, and the stochasticity of inflows. The detailed description of mathematical models are given in the following section. It should be pointed out that the models are developed primarily for multireservoir systems comprising reservoirs

MULTIRESERVOIR SYSTEMS OPTIMIZATION 549

in parallel. The methodology presented in this paper was applied to the multiobjective multireservoir system in the upper Tone river basin in order to test its effectiveness.

METHODOLOGY

Consider a system consisting of multiple reservoirs in parallel with stochastic inflows and multiple objectives (three objectives in this paper, that is, hydropower, water supply, and flood control). A mathematical model for identifying the optimal operating policy of the system is written as

$$\text{Max}\left\{ E\left[\sum_{i=1}^{m} HP_{it}(S_{it},R_{it}) \right], \sum_{i=1}^{m} WS_{it}(S_{it},R_{it}), \sum_{i=1}^{m} FC_{it}(S_{it},R_{it}) \right\} \quad (1)$$

s.t.
$$R_{it}=S_{it}+Q_{it}-S_{i,t+1} \quad (2)$$
$$S_i^{min} \leq S_{it} \leq S_i^{max} \quad (3)$$
$$R_i^{min} \leq R_{it} \leq R_i^{max} \quad (4)$$
$$HP_i^{min} \leq HP_{it} \leq HP_i^{max} \quad (5)$$
$$(i=1,2,...,m; \ t=1,2,...,T)$$

where R_{it} =the release from reservoir i during period t; S_{it}=the initial storage volume of reservoir i at the beginning of period t; Q_{it}=the inflow into reservoir i during period t; HP_{it}=the hydropower generated by the hydropower plant corresponding to reservoir i; WS_{it}=the firm water supply of reservoir i during period t; FC_{it}=the reliability-based flood control capacity during period t; E is expectation with respect to stochastic inflows; Superscripts *min* and *max* mean, respectively, lower bound and upper bound of physical parameters; m is the total number of reservoirs under consideration; T is the total number of time periods within one year.

It is noted that expectation is not taken on items 2 and 3 of objective function (1), since the firm water supply WS_{it} and the reliability-based flood control capacity FC_{it} are defined within the context of reliability.

Vector optimization is transformed into scalar one, using constraint technique. The scalar optimization model corresponding to expressions (1) to (5), with hydropower as objective and water supply and flood control capacity as constraints, is written as

$$\text{Max}\left\{ E\left[\sum_{i=1}^{m} HP_{it}(S_{it},R_{it}) \right] \right\} \quad (6)$$

s.t.
$$R_{it}=S_{it}+Q_{it}-S_{i,t+1} \quad (7)$$
$$S_i^{min} \leq S_{it} \leq S_i^{max}-FC_{it} \quad (8)$$
$$\max\{WS_{it},R_i^{min}\} \leq R_{it} \leq R_i^{max} \quad (9)$$
$$HP_i^{min} \leq HP_{it} \leq HP_i^{max} \quad (10)$$
$$\sum_{i=1}^{m} WS_{it} \geq WS_t \quad (11)$$
$$\sum_{i=1}^{m} FC_{it} \geq FC_t \quad (12)$$
$$(i=1,2,...,m; \ t=1,2,...,T)$$

where objective function (6) is to maximize the total expected hydropower; Equations (8) and (9) are modified forms of equations (3) and (4); Equations (11) and (12) are new, representing objective constraints; WS_t and FC_t are, respectively, the total firm water supply and the total flood control capacity of reservoirs.

In theory, the model comprising (6) to (12) can be solved by means of some optimization techniques such as Stochastic Dynamic Programming (SDP). But in practice, it is impossible to solve it directly because of the so-called curse of dimensionality. In accordance with the curse of dimensionality, decomposition iteration technique is devised to break down the large-scale system consisting of m reservoirs in parallel into m subsystems containing only a single reservoir, which can be solved by SDP. In particular, decomposition iteration technique is to optimize the operating policy of one reservoir (or subsystem) while the operating policies of other reservoirs are kept unchanged. One iteration is finished after each reservoir is optimized once. Iterations continue until a constant operating policy for each reservoir is reached. Obviously, an initial operating policy for each reservoir must be given in advance of iteration.

Assume that reservoir i (i =1, 2 ,..., m) is optimized. An optimization model for reservoir i is written as

$$\text{Max} \sum_{t=1}^{T} E[HP_{it}(S_{it}, R_{it})] \quad (13)$$

s.t.

$$R_{it} = S_{it} + Q_{it} - S_{i,t+1} \quad (14)$$

$$S_i^{min} \leq S_{it} \leq S_i^{max} - FC_{it} \quad (15)$$

$$\text{Max}\{WS_{it}, R_i^{min}\} \leq R_{it} \leq R_i^{max} \quad (16)$$

$$HP_i^{min} \leq HP_{it} \leq HP_i^{max} \quad (17)$$

$$WS_{it} \geq WS_t - \sum_{\substack{j=1 \\ j \neq i}}^{m} WS_{jt} \quad (18)$$

$$FC_{it} \geq FC_t - \sum_{\substack{j=1 \\ j \neq i}}^{m} FC_{jt} \quad (19)$$

$(t = 1, 2, ..., T)$

The model comprising expressions (13) to (19) is different from the model comprising expressions (6) to (12) in two aspects. First, the objective function is reduced from m reservoirs in expression (6) to one reservoir in expression (13). Next, expressions (11) to (12) are modified so that WS_{jt} and FC_{jt} ($j = 1,2,...,i-1,i+1,...,m$) are transferred to the right-hand side of inequalities, as shown in expressions (18) and (19), since WS_{jt} and FC_{jt} are known after the operating policy for reservoir j are found. It is expressions (18) and (19) that demonstrate that reservoirs are associated with rather than independent of each other.

The stochastic single-reservoir optimization model comprising expressions (13) to (19) can be solved by SDP. In order to avoid solving the large simultaneous set of equations in SDP for determining the probability of the release from a reservoir, synthetic inflows are employed to simulate the future operation of reservoirs. Since the firm water supply and the flood control capacity are determined by the simulation rather than by solving the large simultaneous set of equations in SDP, it can be expected that computational time is considerably reduced.

APPLICATION

The methodology described previously was applied to the multiobjective multireservoir system in the upper Tone river basin. The Tone river is the biggest river in Japan with a watershed area of 16,840 km². The upper Tone river basin refers to the river basin upstream of Kurihashi gage station as shown in Fig.1 and its watershed area amounts to half of the total watershed area of the Tone river basin.

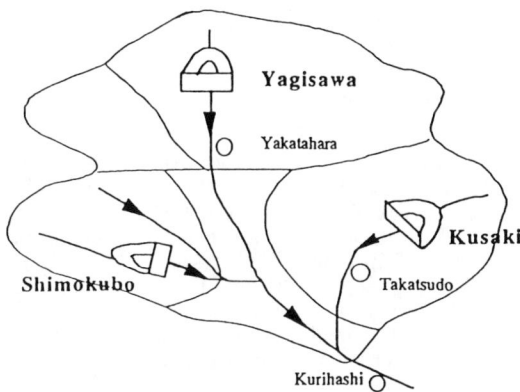

Figure 1. Geographic Distribution of Three Reservoirs

The system comprises three reservoirs: Yagisawa, Shimokubo, and Kusaki reservoirs. The main physical parameters of the reservoirs are given in Table 1.

Table 1. Main Physical Parameters of Three Reservoirs

Physical parameter	Reservoir		
	Yagisawa	Shimokubo	Kusaki
Reservoir storage capacity ($10^6 m^3$)	204.3	130.0	60.5
Normal storage capacity ($10^6 m^3$)	182.2	130.0	60.5
Dead storage capacity ($10^6 m^3$)	28.5	10.0	10.0
Flood control capacity ($10^6 m^3$)	22.1	35.0	20.0
Annual firm water supply [a] (m^3/s)	6.5	3.3	5.0
Hydropower generation capacity (MW)	240.0	15.0	56.5

[a] The reliability of annual firm water supply is 90%

The historical monthly inflow data are used to determine the distribution of monthly inflows. With the Kolmogrov-Smirnov goodness-of-fit test, the lognormal is accepted as the proper distribution for all months (Wang, 1992). Thomas-Fiering model is employed to generate 300 years' inflow for each reservoir. The statistical parameters of historical and synthetic inflows into Yagisawa are listed in Table 2.

Table 2. Statistical Parameters of Historical and Synthetic Inflows into Yagisawa

	Mean (m^3/s)		Standard Deviation(m^3/s)		Lag One Correlation	
	Historical	Synthetic	Historical	Synthetic	Historical	Synthetic
Jan.	4.18	4.18	2.61	2.62	0.7847	0.7854
Feb.	3.71	3.71	2.05	2.06	0.1521	0.1514
Mar.	5.85	5.85	4.06	4.05	0.4759	0.4728
Apr.	32.17	32.17	9.08	9.08	−0.2935	−0.2945
May	45.44	45.44	12.32	12.32	0.5968	0.5962
Jun.	25.04	25.04	11.88	11.82	0.1237	0.1212
Jul.	19.00	19.00	9.01	8.99	0.3937	0.3938
Aug.	10.30	10.30	6.52	6.51	0.4681	0.4684
Sep.	7.85	7.85	4.17	4.15	0.4339	0.4344
Oct.	8.97	8.97	6.24	6.20	0.2836	0.2841
Nov.	9.63	9.63	4.74	4.71	0.3565	0.3565
Dec.	6.50	6.50	4.16	4.14	0.3722	0.3770

One month is taken as one period. The seasonal change of water supply is considered in such a way that the monthly firm water supply WS_t in month t is derived from multiplication of total annual firm water supply, WS, by a constant coefficient. The reliabilities of the annual firm water supply, WS, and the total flood control capacity, FC, are given the values 90% and 95%, respectively.

The sum of the design values of the flood control capacities (see Table 1) of the three reservoirs is $77.10 \times 10^6 m^3$. Assume that the changing range of the total flood control capacity is from $77.10 \times 10^6 m^3$ to $77.10 \times 10^6 \times 150\% = 115.65 \times 10^6 m^3$. Within the range, the constrained parameter FC is given seven values as listed in Table 3. The sum of the actual annual firm water supplies (see Table 1) of the three reservoirs is 14.80 m³/s. Let the changing range of the total firm water supply be from 8.3 m³/s to 22.2 m³/s. Within this range, the constrained parameter WS is given seven values as listed in Table 3. The optimal operating policy obtained by employing SDP to solve each single-reservoir is used as the initial operating policy of each reservoir.

With the values of the constrained parameters FC and WS as well as the corresponding initial operating policies, the total expected annual hydropower, HP, is obtained by using the algorithm described previously. The total expected annual hydropower, HP, corresponding to different combination of FC and WS, are listed in Table 3 and shown in Fig.2.

Table 3. Optimal Solution Results for Three Reservoirs

No.	I[a]	II	III	No.	I	II	III
1	77.10	8.30	359.5	26	98.65	20.40	345.6
2	77.10	10.80	358.5	27	98.65	21.40	342.2
3	77.10	14.80	356.5	28	98.65	22.20	336.5
4	77.10	18.40	353.5	29	105.65	8.30	352.6
5	77.10	20.40	350.5	30	105.65	10.80	351.6
6	77.10	21.40	347.8	31	105.65	14.80	349.3
7	77.10	22.20	342.1	32	105.65	18.40	346.2
8	84.65	8.30	357.3	33	105.65	20.40	343.0
9	84.65	10.80	356.3	34	105.65	21.40	340.1
10	84.65	14.80	354.3	35	105.65	22.20	334.4
11	84.65	18.40	351.8	36	110.65	8.30	350.7
12	84.65	20.40	348.8	37	110.65	10.80	349.9
13	84.65	21.40	345.4	38	110.65	14.80	347.5
14	84.65	22.20	339.7	39	110.65	18.40	343.7
15	91.65	8.30	356.0	40	110.65	20.40	340.4
16	91.65	10.80	355.0	41	110.65	21.40	337.4
17	91.65	14.80	353.0	42	110.65	22.20	331.6
18	91.65	18.40	350.5	43	115.65	8.30	349.2
19	91.65	20.40	347.3	44	115.65	10.80	348.1
20	91.65	21.40	343.9	45	115.65	14.80	345.8
21	91.65	22.20	338.0	46	115.65	18.40	341.1
22	98.65	8.30	354.7	47	115.65	20.40	337.9
23	98.65	10.80	353.7	48	115.65	21.40	334.5
24	98.65	14.80	351.4	49	115.65	22.20	328.7
25	98.65	18.40	348.9				

[a] I = Flood control capacity, FC ($10^6 m^3$)
II = Annual firm water supply, WS (m³/s)
III = Annual expected hydropower, HP(mwh)

MULTIRESERVOIR SYSTEMS OPTIMIZATION 553

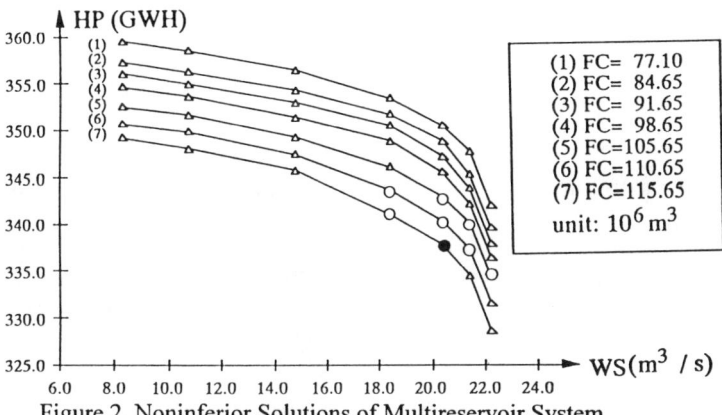

Figure 2. Noninferior Solutions of Multireservoir System

The noninferior solutions in Table 3 or Fig.2 are very important decision making information for the decision maker. Which of the noninferior solutions should be selected as the most satisfactory alternative depends on the decision maker's preference. For example, among the three objectives, the greater priorities may be given to water supply and flood control because drought and flood are two quite intranquil factors in Tone river basin. As a result, the preferred alternatives may probably be in the right lower hand part of the set of the curves in Fig.2. In fact, the detailed decision analysis indicates that the 8 points with marks ○ and ● in Fig.2 are the preferred alternatives, of which the most satisfactory solution is the point with mark ●.

CONCLUSIONS

A methodology for stochastic multiobjective optimization of multireservoir systems is presented in this paper, and its effectiveness is tested by applying it to the multiobjective multireservoir system in the upper Tone river basin. Conclusions derived from this paper are summarized here. First, constraint technique is effective in the sense that the trade-offs among objectives are explicitly considered and noninferior solution are found and displayed as shown in Fig.2. Secondly, decomposition iteration consists in breaking down a multireservoir system into multiple single-reservoir subsystems. Since state variables in decomposition iteration increase linearly rather than exponentially with the number of reservoirs, computational space is saved. Thirdly, simulation analysis is used to solve for the firm water supply and the reliability-based flood control capacity without solving the large simultaneous set of equations in SDP. Thus, computational time is substantially reduced. Finally, a conjunctive use of constraint technique, decomposition iteration, and simulation analysis alleviates the curse of dimensionality considerably. Since this method is effective for solving the three-objective three-reservoir system, it is promising to apply it to the problem containing more objectives and reservoirs.

REFERENCES

Wang, Y.C., N. Tamai, and Y. Kawahara (1992). "Comparison of stochastic dynamic- and linear- programming for reservoir operation." Proc. of the 4th Symposium on Water Resources, 351-356.

Yeh, W. W-G. (1985). "Reservoir management and operations models: A state-of-the-art review." Water Resour. Res., 21(12), 1797-1818.

Coping with uncertainty in the economical optimization of a dike design

P.H.A.J.M. VAN GELDER, J.K. VRIJLING, K.A.H. SLIJKHUIS
Delft University of Technology, Faculty of Civil Engineering
Delft, The Netherlands

ABSTRACT
A design philosophy for dikes based on an economical cost model is described and examined on its sensitivity with respect to statistical- and model uncertainty. The philosophy is applied on an example of a dike along the Dutch coast.

1 INTRODUCTION
The Netherlands is a unique country by the fact that it can continue to exist thanks to its sea- and river dikes. Without these water retaining structures 2/3 of The Netherlands would be inundated quite regularly and most of its population should have to move elsewhere. In the past dikes were designed by building them as high as the highest known water level at that particular location. More recently other design philosophies have been developed. The approach where a dike has to withstand a certain water level with a fixed probability per year has become quite common. Probabilities of exceedances for water levels once per 10,000 years are adopted for the Dutch sea dikes and once per 1,250 years for the Dutch river dikes [Van Gelder, 1996a]. Other design philosophies have been suggested but didn't become popular so far. However, the approach in which the dike design is determined by an economical optimization [Van Dantzig, 1956] has much advantages in comparison with other design philosophies. This approach will be explained in section 2 of this paper and applied to a location along the Dutch coast. It will be shown how the approach can deal with statistical- and model uncertainties in sections 3.1 and 3.2 respectively. In section 4 the influence of these uncertainties will be examined on the economical optimal dike height. Finally the conclusions are drawn in section 5.

2 ECONOMICAL OPTIMIZATION OF THE DIKE HEIGHT
Taking account of the cost of dike building, of the material losses when a dike-break occurs, and of the frequency distribution of different sea levels, the optimal dike height can be determined by economical optimization [Van Dantzig, 1956]. Assume that H_0 is the current dike level and that we want to determine the amount X by which the dikes must be heightened to the height H (see figure 1).

ECONOMICAL OPTIMIZATION OF DIKE DESIGN

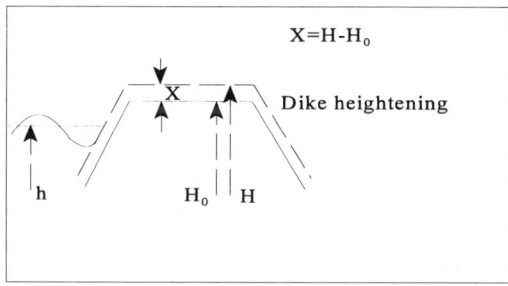

Let h at any moment denote the sea level along the dikes, then no loss is incurred as long as h≤H; if h>H then we assume a loss with an amount of W including migration costs of the population and cattle, privation of production, damage to houses, buildings, industry etc.

Figure 1: Schematic overview

The probability distribution of the sea level h will be denoted by F(h) and will be subject of discussion in sections 3.1 and 3.2. With a dike height of H, each year the expected loss is given by (1-F(H))W. If we assume a constant rate of interest δ the expected value of all future losses are given by:

$$R = F(H)W\Sigma_{t=0}^{\infty}(1+\delta)^{-t} = F(H)W/\delta$$

The total costs of heightening a dike will be assumed linear:

$$I = I_0 + I'X$$

where I_0 is the initial cost and I' the subsequent cost of heightening per meter.
The economical optimal dike height follows by minimizing the expression R+I over the variable H (or X).

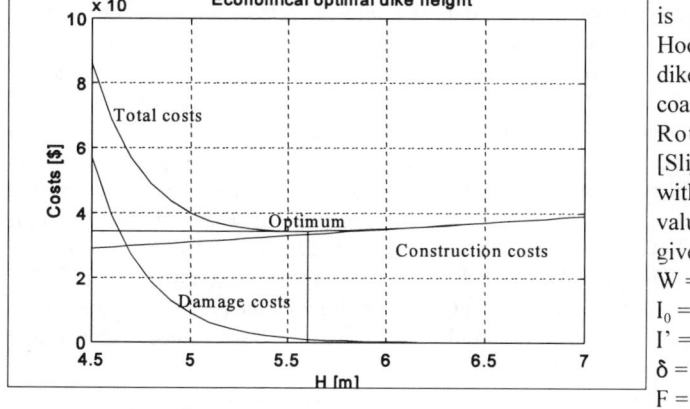

The design method is applied to the Hook of Holland dike along the Dutch coast near the port of Rotterdam in [Slijkhuis, 1996] with the following values for the above given variables:
W = $ 24.2 x 10^9
I_0 = $ 110 x 10^6
I' = $ 40 x 10^6
δ = 1.5%
F = Gumbel

Figure 2: The economical optimization lines

The economical optimal dike height follows to be H_{opt} = 5.60m with a corresponding probability of failure of $F(H_{opt})$ = 6.43 x 10^{-6} per year.

3 UNCERTAINTIES

In the design of dikes attention should be paid to the statistical - and model uncertainties of the sea levels. The analysis of estimating sea level exceedance probabilities has always been a controversy in the literature. The procedure that is traditionally followed:
(1) observe a historical record of sea levels
(2) pick a probability density function that seems reasonable
(3) estimate the parameters of the pdf from the historical records
(4) make inferences about theoccurrencee of extreme sea levels
In step (1) one usually has to deal with a limited amount of data. Although the sea levels along the Dutch coast are measured quite intensively for more than a century now and also flood historical data is available [Van Gelder, 1996b], the amount of data to predict sea levels with return periods of 1000-10000 years is very limited. For the Hook of Holland dike a data set of year maxima of sea levels in the period 1887 - 1995 is available . This dataset is corrected for the sea level rise (of 20 cm per 100 years along the Dutch coast) and since it consists of year maxima, it can be considered a homogeneous dataset.
In steps (2) and (3) there exist a number of uncertainties under which:
- Statistical- or parameter uncertainty which is associated with the estimation of the parameters of the model of the stochastic process due to limited data.
- Model uncertainty which is associated with the uncertainty that a particular probabilistic model of the stochastic process may not be the true or best model.

3.1 PARAMETER UNCERTAINTY

A parameter θ of a distribution function is estimated from the data. It is therefore a function f of the data and because the data is a random variable, the parameter itself is also a random variable. The parameter uncertainty can be described by the distribution function of the parameter. In [Slijkhuis, 1996] an overview is given of the analytical and numerical derivation of parameter uncertainties for certain probability models (Exponential, Gumbel and Log-normal). One of her conclusions was that a bootstrapping method is a fairly easy tool to calculate the parameter uncertainty numerically. Other methods to model parameter uncertainties like Bayesian methods can be very well applied too ([Van Gelder, 1996c]). Bootstrapping methods are described in for example [Efron et.al., 1993]. Given a dataset $x=(x_1,x_2,...,x_n)$, we can generate a bootstrap sample x^* which is a random sample of size n drawn with replacement from the dataset x. The following bootstrap algorithm has been used for estimating the parameter uncertainty:
1. Select B independent bootstrap samples x^{*1}, x^{*2}, ..., x^{*B}, each consisting of n data values drawn with replacement from x.
2. Evaluate the bootstrap evaluation corresponding to each bootstrap sample;

$\theta^*(b)=f(x^{*b})$ for b=1,2,...,B.

ECONOMICAL OPTIMIZATION OF DIKE DESIGN

3. Determine the parameter uncertainty by the empirical distribution function of θ*.

The algorithm has been applied to the location parameter A and scale parameter B of the Gumbel distribution with an MaxLik-fit to the Hook of Holland data. The following distributions and approximating normal distributions were obtained (figure 3):

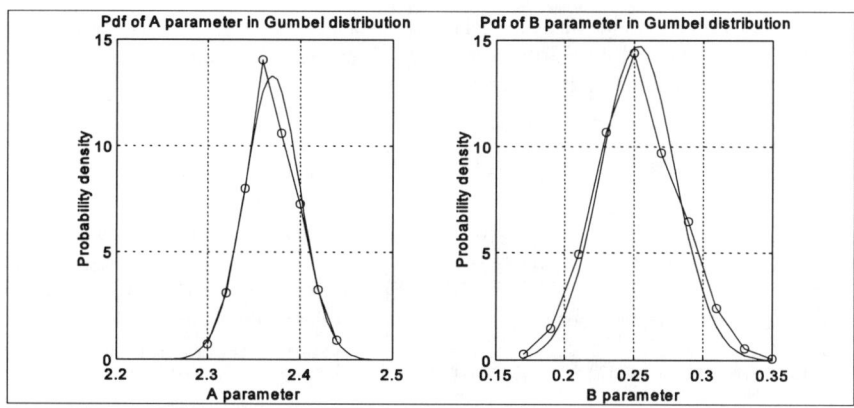

Figure 3: Parameter uncertainties in the Gumbel distribution

The influence of the parameter uncertainty on the frequency line for the sea levels year maxima is examined in the next figure. The 5 lines correspond with values for the location and scale parameters as given in table 1.

Line	A	B
1	μ	μ-2σ
2	μ-2σ	μ
3	μ	μ
4	μ+2σ	μ
5	μ	μ+2σ

Table 1

Note the high sensitivity of the frequency lines.

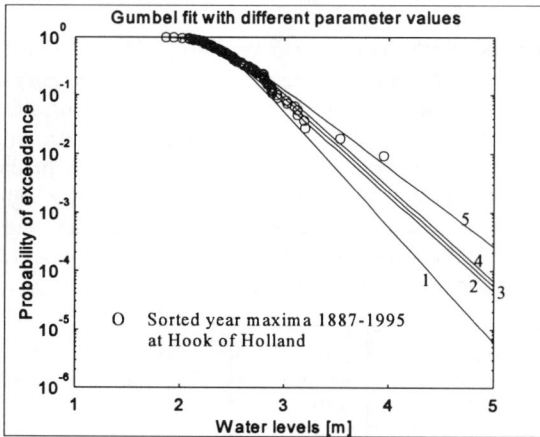

Figure 4: Influence of parameter uncertainty on Gumbel fit

If we are involved in calculating the expected frequency line for the sea levels h, then the inferences we make on h should reflect the uncertainty in the parameters θ. In the Bayesian terminology we are interested in the so-called predictive function:

$$F(h) = \int_\theta F(h|\theta) f(\theta) d\theta$$

where F(h|θ) is the probabilistic model of sea levels, conditional upon the parameters θ and F(h) is the predictive distribution of the sea levels, now parameter free. In popular words: "the uncertainty in the θ parameters have to be integrated out".

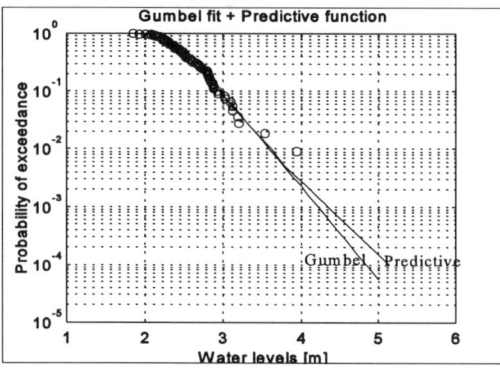

Figure 5: Integrating out the uncertainty

The predictive distribution can be interpreted as being the distribution F(h|θ) weighted by f(θ). *In making inferences on sea levels it is important to use the predictive function for h,*(fig. 5) as opposed to the probabilistic model for h with some estimator for the parameter set θ, i.e. f(q|θ*). This is because using point estimators for uncertain parameters underestimates the variance in sea levels.

3.2 MODEL UNCERTAINTY

In most cases, the probability model of a physical magnitude cannot be clearly estimated from statistical techniques. It is also seldom possible to derive the exact model theoretically. The correct probability model of a physical magnitude may never be known. Instead of assuming a specific probability model for prediction, it is much better to assume that each model could be potentially correct. Predictions may be made from any probability model. Yet, the model that reveals a wide scatter in the probability plot will have a large standard deviation around its predicted value. Consequently, the prediction from this model should be given less weight relative to those models that exhibit less scatter. Weighting factors can be calculated by techniques described in for example [Pericchi et.al., 1983].For the Hook of Holland data, the following models were fitted and the corresponding weighting factors were calculated:

LogNormal	0.05
LogPearson III	0.46
Gumbel	0.49

Table 2: Weighting factors

4 THE INFLUENCE OF UNCERTAINTY ON THE ECONOMICAL OPTIMIZATION
Rather than performing the economical optimization with the probability distribution $F(h|\theta)$, we will perform it with the predictive distribution, where the parameter uncertainty is "integrated out".

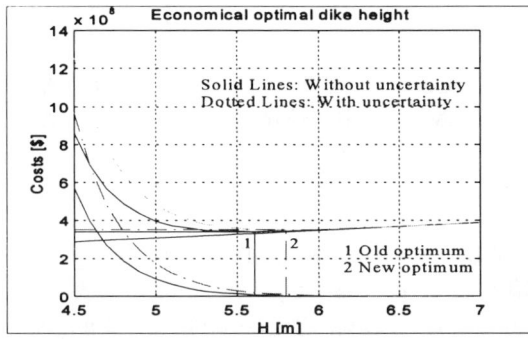

An increase in the optimal dike height from 5.60m to 5.80 m is the result when parameter uncertainty in the probabilistic model is taken into account. The corresponding costs almost remain the same however. The influence of model uncertainty on the optimal dike height is rather small, since the Gumbel model already gave a high weighting factor.

Figure 6: Economics with and without parameter uncertainty

5 CONCLUSIONS
In this paper we have studied the influence of statistical- and model uncertainty on an economical design philosophy for dikes. Techniques like Bootstrapping, Bayesian prediction and Model weighting are very suitable for this analysis. An application of the dike height at Hook of Holland is presented where the influence of statistical- and model uncertainties on the economical optimal dike height is in the order of 20cm. The uncertainty in other variables like W, I_0, I' and δ is examined in [Slijkhuis, 1996].

6 REFERENCES
[1993] Efron, B., Tibshirani, R.J., *An Introduction to the Bootstrap.* Chapman & Hall.
[1983] Pericchi, L.R., and Rodriguez-Iturbe, I., *On some problems in Bayesian model choice in hydrology*, The Statistician, Vol.32.
[1996] Slijkhuis, K.A.H., *The influence of statistical uncertainty on the design height of dikes [IN DUTCH].* Msc-thesis, Delft University of Technology, October 1996.
[1956] Van Dantzig, D., *Economic decision problems for flood prevention.* Econometrica, 24, pp.276-287, New Haven.
[1996a] Van Gelder, P.H.A.J.M., Roos, A., and Tonneijck, M.R., *On the probabilistic design of dikes in the Netherlands.* Applications of Statistics and Probability, Volume 3, pp.1505-1508, A.A. Balkema.
[1996b] Van Gelder, P.H.A.J.M., *How to deal with wave statistical and model uncertainties in the design of vertical breakwaters.* Probabilistic tools for vertical breakwater design, MAS3-CT95-0041, Workshop Grenoble, 24-25 October 1996.
[1996c] Van Gelder, P.H.A.J.M., *A new statistical model for extreme water levels along the Dutch coast.* Stochastic Hydraulics, pp. 243-250, Mackay.

SENSITIVITY AND UNCERTAINTY ANALYSIS OF A REGIONAL SIMULATION MODEL FOR THE NATURAL SYSTEM IN SOUTH FLORIDA

A. M. WASANTHA LAL, JAYANTHA OBEYSEKERA, RANDY VAN ZEE
Senior Engineer, Director and Supervising Professional, Hydrologic Systems Modeling Division, South Florida Water Management District.

ABSTRACT

The sensitivity and uncertainty estimates of a regional model to a selected set of lumped and distributed parameters is determined using the first order method, the Rosenblueth method and the Latin Hypercube sampling (LHS) method. Evapotranspiration (ET) crop coefficients is identified in the study as the parameter most sensitive to the output, followed by roughness. Singular value decomposition is used to determine possible grouping of parameters.

INTRODUCTION

Parameter uncertainty, input uncertainty and the algorithm uncertainty are responsible for the total model uncertainty. Regional hydrologic models for South Florida use large numbers of distributed parameters such as ET crop coefficients and flow roughness which change with local natural conditions, and cannot be determined with precision (Trimble, 1995). Model outputs also depend on topographic elevations, river widths, detention depths and other features which cannot be determined through direct computations. The study is aimed at determining the sensitivity of selected model outputs to some of the parameters. Model sensitivity analysis provides an opportunity to study the behavior of physical systems to varying natural and man-made conditions.

METHODS OF ANALYSIS

The simplest method for sensitivity analysis is the first order method in which the following definition of the sensitivity matrix $\mathbf{A} = [a_{i,j}]$ is used.

$$a_{ji} = \frac{\partial y_j}{\partial x_i} \approx \frac{y_j(x_i + \Delta x_i) - y_j}{\Delta x_i} \quad \text{for each} \quad i = 1, 2, \cdots, n, \; j = 1, 2, \cdots, m, \tag{1}$$

in which, y_j, $j = 1, 2, \ldots, m$ are the m observations; x_i, $i = 1, 2, \ldots, n$ are the n parameters. Singular value decomposition of matrix $\mathbf{A}$ can be used to identify groups of parameters sensitive to groups of observations (Lal, 1995). The output variance can be computed approximately using

$$Var(y_i) = \sum_{j=1}^{m} a_{ij}^2 \, \sigma_j^2 + 2 \sum_{j<k}^{m} a_{ij} a_{ik} \, \sigma_j \sigma_k \, \rho_{jk} \quad i = 1, 2, \ldots n \tag{2}$$

in which, σ_j^2 = variance of parameter j; $\rho_{j,k}$ = correlation between parameters j and k.

The second method used is the factorial design method in which more than one parameter is changed at a time to generate parameter sets for the model runs (Box, et al., 1978). The output is used to determine sensitivity of parameters as well as parameter interactions. The same combinations of parameters are used in the Rosenblueth method to compute output variances (Binley, 1991). Using the method, the N th moment in the output is expressed as

$$E[(y)^N] = \frac{1}{2^m}[(q_{+++\ldots m})(y_{+++\ldots m})^N + (q_{-++\ldots m})(y_{-++\ldots m})^N + (q_{---\ldots m})(y_{---\ldots m})^N] \tag{3}$$

in which, m = number of parameters; $(y_{---\ldots m})$ gives the N th moment when each of the m parameters are set at mean plus or mean minus one standard deviation according to a + and - signs; $q = 1 \pm \rho_{i,j} \cdots$ as described by Binley, (1991).

When using the LHS method (Sing, 1990), probability distributions are assumed for each parameter, and each distribution is divided into a number of intervals. A Gaussian distribution is assumed in the study. Parameters are randomly selected such that each interval appears in exactly one model run. The model outputs generated are statistically analyzed.

APPLICATION TO SOUTH FLORIDA

Figure 1 shows the area of South Florida studied using the Natural System Model (NSM 4.2). A total of 27 parameters were selected for the study by creating regional parameters for ET and roughness. Overall detention depth and the river width are also considered as parameters. All parameters or groups are varied as percentages of base values. Effect of ground elevation was included in the study by considering unit displacements in several regions. A total of 58 flow, stage and hydro-period observations are used to understand the influence of the parameters.

The first order method is used to estimate the output uncertainty assuming both zero and maximum correlations among parameters to obtain upper and lower bounds of variance. The actual correlations among parameters are not available for the study. When using the Rosenblueth method, parameters of the same type such as ET and roughness were grouped to reduce the 27 independent parameters to $N = 4$ groups, which still requires $2^N = 16$ computer runs. The number of runs made for the LHS method is limited to 200 because each run takes approximately 1 Hr to run in a Sparc 10. Two ranges of uncertainties of input parameters is assumed. The extreme range is based on the assumption that it is physically impossible for the parameters to exceed a certain range. The calibration range is computed considering that any values outside the range would offset the calibration, and become easily noticable. The extreme ranges used for ET, roughness and detention depth are 12%, 43% and 30% of parameter base values. Trimble (1995) used a similar approach, and applied some of the methods to the South Florida Water Management Model (SFWMM).

RESULTS AND DISCUSSION

The sensitivity matrix for average conditions in Fig. 2 shows the relative importance of ET over other parameters. ET and roughness effects are somewhat spread over the region because many land use types are also spread. Effects of the topographic elevation changes are localized because only small areas were considered. Table 2 shows upper and lower bounds of output uncertainty of selected gages which correspond to zero and maximum correlations. The contributions of individual parameters are also shown in the table. Sensitivity of parameters depend on many state variables too. Water level for example is more sensitive to param-

REGIONAL SIMULATION MODEL 563

Table 1: Output uncertainty resulting from assumed parameter uncertainties. Parameters: EFrmN, EFrmS = ET crop coeff. of fresh marsh in the North and the South; ESgrN, ESgrS = ET crop. coeff. of saw grass in the North and the South; MSagrN, MSagrS = Mannings roughness of saw grass in the North and the South.

Observations: QTami, QShSl = discharge across Tamiami Trail and Shark Slough, in ac ft / day; S-10, S3A-28 = stages of gages S-10 and S3A-28 in ft.; T1-7 = hydroperiod at gage 1-7 in days/year.

	EFrmN	ESgrS	ESgrS	MSagrN	MSagrS	Avg.	ϵ_{uncorr}	ϵ_{corr}
Std. err.	12%	12%	12%	43%	43%			
QTami	56	135	173	64	70	3633	445	1700
QShSl	23	46	60	23	22	1300	231	881
S-10	0.08	0.02	0.00	0.03	0.00	15.0	0.68	1.04
S3A-28	0.01	0.03	0.04	0.02	0.00	8.4	0.13	0.48
T1-7	15	3	0	3	0	339	18	41

eters when it is below the ground than above.

Singular value decomposition of the sensitivity matrix normalized using averages of parameter values and the standard deviations of observations show that the most significant parameter group among the selected consist of almost all the ET coefficients in the central region which remain wet most of the time. This group affects all the stages in the area and the flows. The second parameter group has ground elevation of the central sawgrass region, which mainly affects the local water stages. The singular values for the problem do not decay rapidly as in many small scale models (Lal, 1995) showing that the model actually has a large parameter dimension.

The factorial design method gives more detailed information of sensitivities. For average flow across Tamiami Trail for example,

$$Q_{tam} = 3647 - 7007ET - 784Mann - 56.2Wid + 13.6ET \times Mann + 26.1ET \times Wid + \ldots \quad (4)$$

in which, ET, Mann, Wid = changes of all the values of ET, Manning's roughness and river widths, considered as fractions of base values. The equation shows that interaction terms are relatively negligible, and that sensitivity to one parameter is only slightly affected by another parameter. Equation 4 shows that the average discharge of 3647 Ac ft/day reduces by 7007 Ac ft/day when all the ET values are increased by 100%, assuming a linear model. First order and LHS methods give approximately the same results. The LHS method can also provide estimates of statistical significances and standard deviations.

CONCLUSIONS

First order methods are simple and fast, but relatively inaccurate. They can provide upper and lower bounds of uncertainty bands when the parameter correlations are not known. Rosenblueth and factorial methods are more accurate, but require a large number or computer runs when the number of parameters is large. Factorial method can detect parameter dependencies as well. LHS method require a large number of computer runs, but this number does not increase exponentially. LHS method gives realistic estimates of parameter uncertainties and significances assuming the parameters to be independent. Since it is not easy to find a single method that can replace all three methods used for sensitivity and uncertainty analysis, it is beneficial to use all three methods at the screening stage, specially when there are many parameters.

REFERENCES

Binley, A. M., and Beven, K. J, (1991). "Changing responses in hydrology: assessing the uncertainty in physically based predictions", *Water Resources Research*, 27(6), 1253-1261.

Box, G. E. P., Hunter, G. W., Hunter, J. S., (1978). *Statistics for experiments*, John Wiley & Sons, New York.

Sing, M. G. (1990). *Reprints from Systems and Control Encyclopedia*, Pregamon Press, New York, 4230-4236.

Trimble, P. (1995). "An evaluation of the certainty of system performence measures generated by the South Florida Water Management Model", Masters Thesis, Florida Atlantic University, Boca Raton, FL.

Wasantha Lal, A. M. (1995). "Calibration of Riverbed Roughness", *J. Hydr. Engg.*, ASCE, 121(9), 664-670.

ACKNOWLEDGEMENTS

The authors wish to thank Marie Pietrucha and Beheen Trimble of the Hydrologic Systems Modeling Division of the South Florida Water Management District for assistance during different phases of the study.

REGIONAL SIMULATION MODEL 565

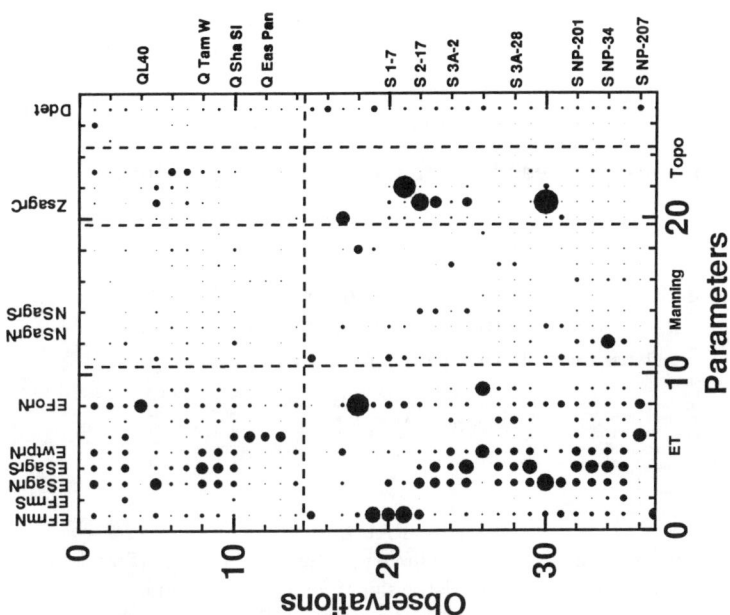

Figure 2: The sensitivity matrix.

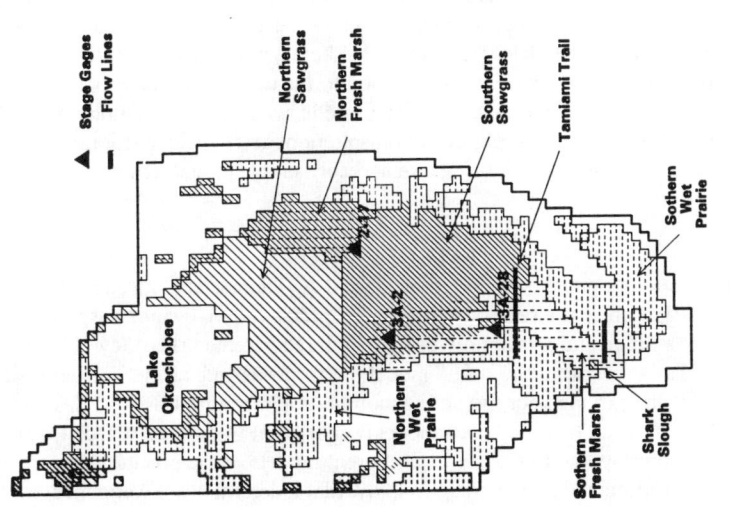

Figure 1: A map of South Florida.

A Simple Extension to the 'abc' Watershed Model

RICHARD M. VOGEL, ANTIGONI ZAFIRAKOU-KOULOURIS, AND
THADDEUS GREEN
Department of Civil and Environmental Engineering
Tufts University
Medford, MA, 02155

ABSTRACT

Fiering (1967) introduced a very simple, three parameter rainfall-runoff model termed the *"abc"* model which describes the relationships among precipitation, evapotranspiration, groundwater storage and streamflow for a catchment. This model has been shown by others to be useful for modeling monthly and annual streamflows. The weakest feature of this watershed model involves the algorithm for computing actual infiltration and actual evapotranspiration. We provide an extension to the *"abc"* model which captures the observed relationship between actual evaporation, potential evaporation and soil moisture. An application to the upper Colorado river basin demonstrates that this simple extension can provide a very realistic simulation of the relationships among precipitation, evapotranspiration, groundwater storage and streamflow, at the annual level, using a remarkably simple model structure with only four model parameters.

INTRODUCTION

The *"abc"* model, originally conceived by Harold A. Thomas and introduced by Fiering [1967], models the relationships among precipitation, evapotranspiration, groundwater storage and streamflow using only three parameters. Alley (1984) documents that a variant of the *"abc"* model performs as well as the Thornthwaite-Mather and Palmer models for predicting monthly soil moisture levels. Salas and Smith [1981], showed that the *"abc"* model is equivalent to various ARMA(p,q) stochastic streamflow models, assuming different models of the precipitation process. It is both comforting and enlightening to know that a deterministic watershed model is equivalent to a wide class of stochastic models, because stochastic and deterministic approaches are often seen as separate and independent. This study only considers the use of the "*abc*" model for use in modeling long-term (annual) streamflow. The main objective of this study is to introduce and evaluate a simple extension to the '*abc*' model which captures the observed relationship between long-term actual evaporation, potential evaporation and soil moisture storage.

THE '*abc*' WATERSHED MODEL

The "*abc*" model, as originally reported by Fiering [1967], is a rainfall-runoff model based on a nonseasonal, lumped watershed system, where precipitation, infiltration, evapotranspiration and groundwater storage contribute to the annual runoff. Uniform rainfall is assumed over the basin. If precipitation at time *t* is represented by P_t, then infiltration is $I_t=aP_t$ and actual evapotranspiration is $E_t=bP_t$, where *a* and *b* represent the fraction of rainfall which infiltrates and evaporates, respectively. The remaining component of rainfall $P_t-I_t-E_t=(1-a-b)P_t$ results in surface runoff to the stream channel. Groundwater storage at time t, is defined as G_t and the groundwater outflow to the stream channel or baseflow, is defined as a fixed fraction cG_{t-1} of groundwater storage in the previous period. Finally, the streamflow Q_t, is given as the combination of surface and groundwater inputs in addition to model error

$$Q_t = (1-a-b)P_t + cG_{t-1} + \varepsilon_t \qquad (1)$$

Groundwater storage G_t is derived by continuity as previous groundwater storage G_{t-1} less groundwater outflow plus infiltration and model error

$$G_t = (1-c)G_{t-1} + aP_t + v_t \qquad (2)$$

Since the parameters represent percentages they have upper and lower limits $0 \le a,b,c \le 1$, and since infiltration and evapotranspiration combined cannot exceed total precipitation, $0 \le a+b \le 1$. Salas and Smith [1981] combine (1) and (2) to show that if P_t is an independent normal variable, Q_t can be interpreted as an ARMA(1,1) process.

AN EXTENSION TO THE "*abc*" MODEL

One disadvantage of the "*abc*" model is that it only captures the influence of climate on hydrology using precipitation, ignoring the important influence of temperature and soil moisture on actual evapotranspiration. The "*abc*" model assumes that actual evapotranspiration $E_t=bP_t$ depends soley on available moisture, whereas in reality, actual long-term evapotranspiration depends upon both available moisture and potential evapotranspiration. A number of empirical relationships have been introduced which express observed relationships between long term actual evaporation, potential evaporation and precipitation (See Kuhnel et al. for a review). A relationship introduced by Turc (1954), which was modified slightly by Pike (1964), termed the Turc-Pike relation is

$$E_t = \frac{P_t}{\sqrt{1+\left(\frac{P_t}{PE_t}\right)^2}} \tag{3}$$

where PE_t is potential evaporation. Studies by Budyko (1974) for over one thousand catchments in the USSR provide further empirical support for the Turc-Pike relationship. The Turc-Pike relationship is illustrated in Figure 1 for a wide range of P_t and PE_t values. The Turk-Pike relationship is able to reproduce the long-term water balance over a very broad range of climates, ranging from a desert climate, where $P_t/PE_t<0.5$ to a humid climate, where $P_t/PE_t>>1$.

Actual evaporation, E_t, is modeled in the 'abc' model using $E_t=bP_t$, whereas the Turk-Pike relation documents that the percentage of rainfall which evaporates, b, is a nonlinear function of both precipitation P_t, and potential evaporation, PE_t. The relationship $E_t =bP_t$ only holds for an extremely arid climate in which case $E_t \cong P_t$ and b=1. In an extremely humid climate E_t is roughly constant so that $E_t \cong PE_t$. These two limiting climates are represented by the left and right-most extremes of the Turk-Pike relationship, respectively. For example, in a desert climate where soil moisture storage is extremely limited, actual evaporation is approximately equal to moisture supply hence $E_t \cong P_t$. Similarly, in a humid environment in which soil moisture is unlimited, actual evaporation proceeds at its potential rate $E_t \cong PE_t$. Therefore, if one employs the 'abc' model to model long-term hydrology without modification by equation (3), it is only likely to provide a good approximation in an arid environment and the correction in (3) will become more and more important as the ratio P_t/PE_t increases.

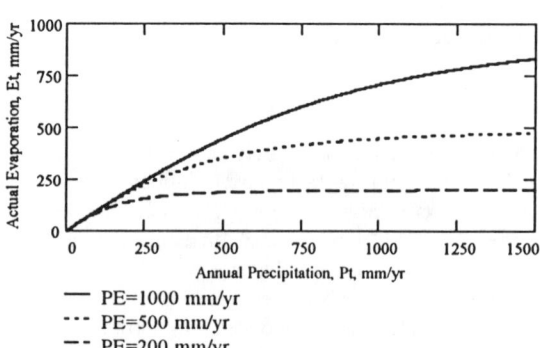

Figure 1 - Relationship Between Long-Term Actual Evaporation, E_t, Potential Evaporation PE and Precipitation P_t, given by the Turk-Pike (1954, 1964) Relationship in Equation (4)

Estimation of Annual Potential Evaporation

Potential evaporation PE_t can be measured using an evaporation pan or a lysimeter and it can be modeled using an energy budget such as the Penman equation, or a simpler temperature-based method (see Fennessey and Vogel, 1996, for references). To remain consistent with the modeling philosophy of the 'abc' model we chose to model PE_t using a temperature-based method. Fennessey and Vogel (1996) document that Penman estimates of annual potential evaporation are linearly related to annual average temperature at 34 NOAA First Order observatories in northeastern USA. In this initial study, we model PE_t in (3) using

$$PE_t = d + e\, T_t \qquad (4)$$

Use of equations (1)-(5) leads to what we term the 'acde' model.

APPLICATION TO THE UPPER COLORADO RIVER BASIN

In this section, we evaluate the ability of the 'abc' and 'acde' models to reproduce the relationships between rainfall, evapotranspiration, groundwater and streamflow for the upper Colorado river basin. The upper Colorado river basin is a 296,000 square kilometer catchment. Revelle and Waggoner (1983) summarize values of annual precipitation, temperature and virgin streamflow for the upper Colorado river basin over the period 1931-1976. Virgin streamflow is the measured flow at Lee Ferry, Arizona, plus estimated depletions within the upper basin, evaporation from reservoirs, and changes in reservoir storage. Values of annual rainfall and temperature are areally weighted averages of the records of individual weather stations. See Revelle and Waggoner (1983) for further details. Average values of P_t and Q_t are 321 mm/year and 62 mm/year respectively, hence actual annual evapotranpiration is 81% of the annual precipitation.

The parameters of the 'abc' and 'acde' models are estimated using ordinary least squares procedures and summarized in Table 1 along with values of Z, defined as the sum of the squared differences between the the observed and modeled streamflows over the period 1931-1976. A comparison of the observed and estimated streamflows is provided in Figure 2 along with a comparison with the multivariate statistical model developed by Revelle and Waggoner (1983) $Q_t = 9274 + 52P_t - 2400T_t$, where now Q_t is in $10^6 m^3$, P_t is in mm and T_t is in °C.

Table 1 - Comparison of Goodness-of-Fit and Parameter Estimates

Model	Model Parameter Estimates					Z
	a	b	c	d	e	
'abc'	0.00518	0.8312	1.0			4783
'acde'	0.0205		1.0	0.7222	0.02657	2723
Revelle and Waggoner (1983)						2904

Figure 2 - Comparison of Simulated and Observed Streamflows

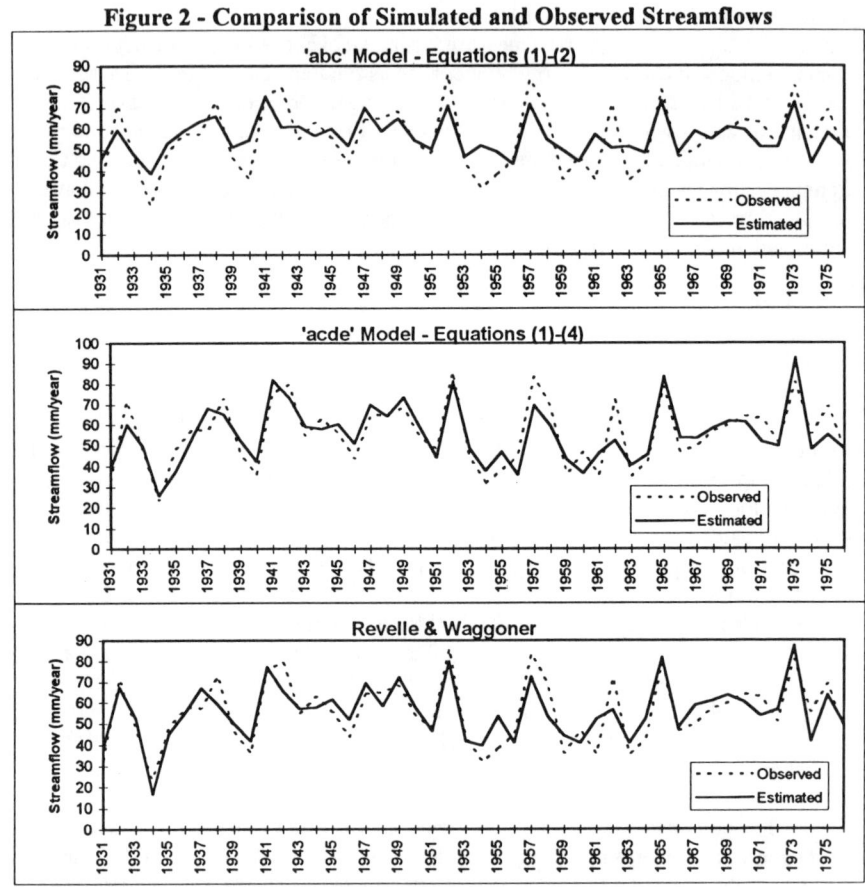

CONCLUSIONS

We document that the 'abc' model introduced by Fiering (1967), which assumes that evapotranspiration is proportional to moisture supply, will probably only provide a good approximation to the annual hydrology in a very arid region where actual evapotranspiration is approximately equal to the moisture supply. In other regions, actual long-term evapotranspiration is a nonlinear function of both moisture supply and potential evaporation. Using the upper Colorado river basin as an example, we document that use of equations (1)-(4) can provide an accurate representation of the year-to-year hydroclimatology of a complex river basin with only four model parameters. This simple model seems to apportion the various stores of water including evapotranspiration, groundwater storage and streamflow, consistently, over the period 1931-1976. Research is currently underway to evaluate further extensions to the model, including the addition of seasonality, snowmelt and spatial scaling properties of the model parameters. Future work will also address the important issue of model parameter estimation because Kuczera (1981) documented that use of generalized least squares procedures can lead to significant improvements over ordinary least squares in terms of '*abc*' model parameter consistency.

ACKNOWLEDGEMENTS

Although the research described in this article has been funded by the United States Environmental Protection Agency through grant number R 824992-01-0 to Tufts University, it has not been subjected to the Agency's required peer and policy review and therefore does not necessarily reflect the views of the Agency and no endorsement should be inferred.

REFERENCES

Alley, W.M., On the Treatment of Evapotranspiration, Soil Moisture Accounting, and Aquifer Recharge in Monthly Water Balance Models, *Water Resources Research*, Vol. 20, No. 8, pp. 1137-1149. 1984.

Budyko, M.I., *Klimat I zhizn, Gidrometeor. Izdat.*, Leningrad. Translation by D.H. Miller as *Climate and Life*, Academic Press, New York, 1974.

Fiering, M.B., *Streamflow Synthesis*, Harvard University Press, Cambridge, Mass., 1967.

Kuczera, G., On the Relationship Between the Reliability of Parameter Estimates and Hydrologic Time Series Data Used in Calibration, *Water Resources Research*, Vol. 18, No. 1, pp. 146-154, 1982.

Kuhnel, V., J.C.I. Dooge, J.P.J. O'Kane, and R.J. Romanowicz, Partial Analysis Applied to Scale Problems in Surface Moisture Fluxes, *Surveys in Geophysics*, Vol. 12, pp. 221-247, 1991.

Pike, J.G., The Estimation of Annual Runoff from Meteorological Data in a Tropical Climate, *Journal of Hydrology*, Vol. 2, pp. 116-123, 1964.

Revelle, R.R. and P.E. Waggoner, The Effects of Carbon Dioxide-Induced Climate Change on Water Supplies in the Western U.S., in *Changing Climate*, pp. 419-432, National Academy Press, Washington, D.C., 1983.

Salas, J.D., and Smith, R.A., Physical Basis of Stochastic Models of Annual Flows, *Water Resources Research*, Vol. 17, No. 2, pp. 482-430, April 1981.

Turc, L, 'Le bilan d'eau des sols, Relation entre la precipitation, l'evaporation et l'ecoulement', *Ann. Agron.*, Vol. 5, pp. 491-569, 1954.

Vogel, R.M. and N.M. Fennessey, Regional Models of Potential Evaporation and Reference Evapotranspiration for the Northeast USA, *Journal of Hydrology*, Vol. 184, pp. 337-354, 1996.

THE EFFECT OF CLIMATE CHANGE ON DANUBE FLOW REGIME

by Dr.B. GAUZER and Prof.Dr.Ö. STAROSOLSZKY
Water Resources Research Centre (VITUKI), Budapest, Hungary

ABSTRACT

Water management of the Danube River, largest in Central-Europe depends on the changes in the flow regime due to climate change. The consequences of the climate change over the upstream catchment may have an impact of the water supply and use along the riparian countries.

Climate change scenarios are available for Central Europe as part of the global climate modeling. The change of monthly temperature and precipitation patterns can be derived and the generated data sets can be used for precipitation-runoff models. The simulation of the daily flow for the streamgauging stations can be bused upon observed daily discharge time series of different selected periods.

This method was used for the study of the simulated river flows for selected hydrological observing stations and for the evaluation of the simulated time series for three stations (Wasserburg-Inn, Pfelling-Danube and Nagymaros-Danube). The effect of the temperature and precipitation changes causes shift in the time of floods and in the ratio of the flows due to snowmelt. The study was supported by the PECO Project of the European Union Commission.

DATA AVAILABLE
The actual simulation included the system of River Danube and its nine tributaries, represented by altogether 28 cross-sections, 13 on the Danube and 15 on the 9 main tributaries. Observations of discharge at 12-hour time steps have been used.

To estimate snowmelt and rainfall-runoff processes data from 32 meteorological stations over the catchment have been used, namely observations of precipitation and air temperature at 12-hour time steps including daily minima and maxima of air temperatures. As a first step time series of active precipitation and soil frost are estimated for each meteorological observation site. Consequently areal means of active precipitation and soil frost are calculated for each of the drainage basins belonging to individual cross-sections, serving as input for the GAPI (Gamma Antecedent Precipitation Index) rainfall-runoff model. Runoff time series produced by GAPI are routed along the river with the use of DLCM (Discrete Linear Cascade Model). The above procedure has been carried out with the set of optimized parameters used in the operational forecasting practice.

METHODS OF SIMULATION
For the simulation of flow time series a model set developed in VITUKI with one component from the National Meteorological Service was used. It contains four models:
- Quantitative precipitation model (of the National Meteorological Service);
- Snowmelt processes model;
- Rainfall-runoff model (GAPI);
- Coupled structural stochastic model based on the unsteady flow equations (DLCM), including a stochastic submodel.

By these models the daily flow time series for the Nagymaros streamflow gauging station were calculated for each climate scenarios - among others -, for a dry year selected from the data base of the Hungarian Hydrological Forecasting Service. The calculations resulted in runoff data for other stream-gauging stations (e.g. for Pfelling on Danube and Wasserburg on Inn), but in our study the flow time series of Nagymaros were mainly discussed.

Flow conditions of River Danube at Nagymaros station are analyzed in connection with the changing of daily precipitations and air temperatures in the drainage basin, with a special emphasis on snowmelt induced runoff. Runoff models capable to follow the given processes are utilized for the simulation of runoff.

GENERAL DESCRIPTION OF THE SCENARIOS
For the scenarios a relatively dry period, from 1. September 1990 to 31. August 1991 have been chosen. The period starts and ends with snow free conditions.

Four scenarios for daily sums of precipitation and air temperature have been analyzed:
(0) The basic scenario is derived from observed precipitation and air temperature values.

The further scenarios are for 2050 under the IS92a emissions scenario (established by the Intergovernmental Panel on Climate Change):
(1) The first scenario is derived from the UK Meteorological Office high resolution equilibrium experiment (UKHI);
(2) The second scenario is derived from the UK Meteorological Office transient experiment (UKTR);
(3) The third scenario is derived from the Canadian Climate Centre high resolution equilibrium experiment (XCCC).

The scenarios are described by monthly changes of precipitation and temperature, gridded at a resolution of 0.5°x0.5°; values for precipitation are percentages, while those for temperature are absolute. The hydrological regime of the Hungarian section of River Danube is mostly determined by the processes which take place in the Upper Danube basin in Bavaria and Austria and in the drainage basin of the tributary Inn. Drainage basins of tributaries Traun and Enns in Austria have secondary importance. Changes of monthly areal average values were created for these three tributaries, the behaviour of the Slovakian and Hungarian tributaries was assumed to be same as for basins of Traun and Enns. The monthly changes are applied to the daily values of observed data, that is the temporal structure of the daily precipitation data and the coefficient of variation for air temperatures are assumed to remain constant.

The simulation of runoff was carried out for the above scenarios, and the results received have been analyzed. The following questions have been tackled:
- seasonal and monthly distribution of runoff;
- changes in the winter low-flow period;
- changes in the peak discharges of floods during the winter;
- changes in the peak discharges of the spring snowmelt induced floods.

RESULTS OF THE RUNOFF SIMULATIONS
Due to the small amount of precipitation only insignificant flood waves passed on the Danube during winter and spring. Significant snow resources appeared only in the drainage basin of River Inn which resulted flood waves in June. The subsequent flood in August was mostly induced by liquid precipitation.

As a result of 15-20 percent increase of precipitation and the rise of air temperature given in the UKHI scenario 31-67 percent increase of effective precipitation was received in the basin of the Upper Danube during winter. Decreasing effective precipitation during May was the consequence of smaller snow accumulation. Increasing effective precipitation during summer was the consequence of the more intensive snowmelt in the high mountainous area.

At the UKTR scenario the more significant increase of precipitation and rise of air temperature resulted more significant increasing of effective precipitation in the period between October and December.

In spite of the smaller increase of air temperature, effective precipitation resulting from the XCCC scenario is in most cases similar to the values resulting from the UKHI scenario.

The drainage basin of River Inn has shown a similar pattern. The basin is more rainy in comparison with the Upper Danube. Because of the higher elevation the snow accumulation starts in October, and 4.9°C rise of air temperature caused 600 percent increase of effective precipitation given in the UKTR scenario in December. Decrease of effective precipitation during summer was significant at the UKHI and UKTR scenarios.

Increase of runoff was significant in the period from December to February for the UKHI scenario, from November to February for the UKTR and XCCC scenarios in the basin of the Upper Danube. Changes were more significant in January. The UKTR scenario resulted significantly lower peak discharges of the August flood to the observed ones (1016 m³/s, 1401 m³/s), as a consequence of the significant (32 percent) decrease of effective precipitation.

In the basin of River Inn changes were significant for the UKHI and UKTR scenarios in the period between November and May. Peak discharges of the bi-modal flood in June decreased for the UKHI and UKTR scenarios, while peak discharge of the first flood in June increased significantly for the XCCC scenarios (1337 m³/s, 2116 m³/s), as a consequence of the more intensive snow accumulation and melting.

Monthly and seasonal volumes of runoff for the cross-section Danube-Nagymaros are listed in Tables 1-3. The results received were in concordance with the effective precipitation volumes of the main contributing catchments.

Scen.	IX.	X.	XI.	XII.	I.	II.	III.	IV.	V.	VI.	VII.	VIII.
Obs.	1526	1438	1675	1234	1273	805	1204	1124	1683	2760	2890	3040
UKHI	1557	1531	1827	1482	1614	1114	1340	1247	1839	2696	2656	3043
UKTR	1499	1575	2108	1662	1829	1161	1419	1335	1932	2497	2890	2603
XCCC	1623	1539	1928	1510	1478	1065	1113	1124	1829	3305	2706	3151

Table 1. Monthly volumes of runoff for cross section Danube-Nagymaros (m³/s)

Scen.	IX.-XI.	XII.-II.	III.-V.	VI.-VIII.
Obs.	1546	1104	1337	2897
UKHI	1638	1403	1474	2798
UKTR	1727	1551	1562	2663
XCCC	1697	1351	1388	3054

Table 2. Seasonal Volumes of runoff for cross section Danube-Nagymaros (m³/s)

Scen.	Q_{min}	Q_{max}	Q	σ
Obs.	688	6750	1728 (100%)	986
UKHI	779	6640	1835 (106%)	870
UKTR	782	5252	1882 (109%)	797
XCCC	796	6920	1869 (108%)	1025

Table 3. Statistical characteristics of runoff for cross section Danube-Nagymaros (m³/s)

Changes were significant for the UKHI and UKTR scenarios in the period between November and April, and between November and February for the XCCC scenario, but water levels remained in the low flow range.

The bi-modal flood in June was partly induced by snowmelt. Peak discharges of the first flood slightly decreased for the UKHI and UKTR scenarios, due to the higher air temperature during winter, and increased significantly for the XCCC scenarios as a consequence of the stronger snowmelt. Peak discharges of the second flood in June decreased because of the earlier snowmelt.

The August flood was induced mostly by liquid precipitation, and significant part of the flood originated from catchments outside of the two main drainage basins of runoff generation. Significant changes appeared only for the UKTR scenario due to the decreasing effective precipitation (6778 m^3/s, 5189 m^3/s).

CONCLUSIONS

From the observed and simulated flow conclusions on the water regime can be derived. Changes of discharges during September and October were not significant as a consequence of the major role of the subsurface inflow during this period. Since minimum discharges appeared during winter, the effect of rise of temperatures were significant. Numbers of days when discharges were less than 800, 900, and 1000 m^3/s at Nagymaros are listed in Table 4. It can be concluded that during this period, due to the fact, that minimum discharges appeared during February, the investigated scenarios have a significant advantageous impact on the safety power plant operation (Thermal Poser Station Százhalombatta, Nuclear Power Station Paks).

Scen.	800 m^3/s	900 m^3/s	1000 m^3/s
Obs.	19	27	48
UKHI	-	-	11
UKTR	-	-	6
XCCC	-	19	34

Table 4. Duration of different discharge values Danube-Nagymaros (m^3/s)

A slight rise of discharges has been detected during winter as a consequence of the temperature rise. As a consequence of the dry winter period, changes remained insignificant during spring.

The number of days with the rise of discharges appeared at 296, 280, and 279 consequently.

The first peak of the bi-modal flood in June shows, that the impact of the higher temperature values during winter was compensated by the higher intensity of process of melting and the higher initial flow preceding the flood wave in case of UKHI and UKTR scenarios and over-compensated for the XCCC scenario. Decrease of peak discharges for the second flood was the consequence of the more intensive process of melting.

Since the August flood originated from liquid precipitation, peak discharges were strongly related to the changes of precipitation values.

The decrease of effective precipitation during summer was partly compensated by the increase of subsurface inflow as a consequence of higher effective precipitation during winter and spring.

The rise of the annual average runoff was most significant for the UKTR scenario, due to the most expressed increase of precipitation. The change of standard deviation was influenced by the peak discharge values.

Concluding the present analysis, it is necessary to mention that the simple technique used to transform monthly changes into daily values and also the short data series used may limit the possibility to derive general conclusions from the results received.

* * *

The study of other periods under similar scenarios may offer a wider scope about the potential flow regime changes. The results of the study will be forwarded to the report to be submitted according to the contract EV5V-CT93-0293 with the European Union.

REFERENCES

Starosolszky, Ö. et al. (1994): Climate change impacts on the water resources of Hungary including the upper Danube basin. IIASA. Contract No 94-115. Budapest-Laxenburg.

Gauzer B., and Starosolszky, Ö. (1996): Effect of potential air temperature and precipitation change on the flow regime of the Danube. Időjárás. Journal of the Hungarian Meteorological Service, Vol. 100. No. 1-3. January-September 1996. Budapest.

WMO-UNEP (1996): Climate change 1995. Impacts, adaptations and mitigation of climate change. Contribution of WG-II to the Second Assessment Report of IPCC. Cambridge University Press.

RELIABILITY OF OPERATION OF THE VOLGA WATER-RESOURCE SYSTEM UNDER GLOBAL CLIMATE CHANGE CONDITIONS

A. L. VELIKANOV
Water Problems Institute, Russian Academy of Sciences
Moscow, Russia

ABSTRACT

The paper presents results of the investigation of operation of the Volga River water-resource system for two different scenarios of climate change - one involving a decrease in the Volga River runoff in the foreseeable future and another involving a runoff increase. Probabilistic estimates of possible consequences of changes in the hydrological regime for the volume and reliability of the yield of the water-resource system, including ecological water releases to the lower Volga, are given.

INTRODUCTION

The problem of global climate change under the human impact remains one of the most important problems of today, and it is far from its solution. However, if the change manifests itself fairly distinctly, large water-resource systems (WRS) will be one of the first to respond to this event. Climate changes will inevitably result in adequate changes in hydrological characteristics and in water demands. This situation can lead to a situation in which WRS will not be able to fulfill their main functions. For this reason, even under conditions of great uncertainty of climatic scenarios, it appears necessary to evaluate the behavior of WRS under climate change conditions. When solving this problem, two principal questions arise: what model of river runoff is to be adopted for a scenario and how can changes in the operation of WRS be assessed under new conditions? To construct hydrological characteristics, we use long-term samples from climatic parameters of the scenario under consideration. The best way of assessing the results of operation of a water-resource system under various climatic changes is the use of the system yield probability curves. Proceeding from the above, investigations were carried out to assess the reliability of operation of the Volga water-resource system for two scenarios of global climate change in the foreseeable future.

THE VOLGA WATER-RESOURCE SYSTEM

The Volga River basin is one of the principal economic regions of Russia. Here about a fourth of the former USSR's population resides and a fourth of industrial and agricultural output of the former USSR are produced. Half of the fish catch from inland water bodies of the country and 90% of sturgeon catch are accounted for by the Volga-Caspian sea basin. The main water users in the Volga basin are: hydropower production, water transport, fishery, irrigated farming, municipal and industrial water supply.

In order to solve the problem of the water supply of national economy, the runoff of the Volga and Kama rivers was regulated by a chain of 11 reservoirs with a total volume of about 190 cukm, (their total useful volume being 90 cukm). The installed capacity of the hydropower stations of the cascade is 11.7 million kW, the annual power production amount to 36 billion kwh. The cascade reservoirs created a navigable route with a guaranteed depth of 4 m. They gave water to 2.5 million ha of irrigated land and supplied water to industrial enterprises and for municipal needs. The consumptive water use in the basin now makes up 12-14 cukm/yr.

As a result of the WRS operation in the Volga River basin, its natural hydrological regime was largely modified. Under natural conditions, about 62% of the annual flow passed in April-June through the Kuibyshev site, whereas the remaining 38% of the flow passed during the July-March dry period. Under runoff regulation conditions, only 44% of the flow passed during the flood, whereas 56% of the flow passed during the dry period. In low-flow years this difference grows. Winter water discharges grew 2-3 times in the dam lower pools; substantial fluctuations of diurnal discharges and levels were observed. All this created grave ecological problems, which can be aggravated under the impact of man-induced climate changes.

HYDROLOGICAL CHARACTERISTICS

At the present-day level of hydrological knowledge and available hydrometeorological data, it is practically impossible to give reliable quantitative estimates of temporal and spatial characteristics of water resources of large WRS for the distant future taking into considerations the possible climate changes. Probabilistic-stochastic approaches, based on the hypothesis of the stationarity of the processes cannot be unreservedly recommended either. Therefore it is proposed to use different approaches to the estimation of future water resources.

One of the possible approaches is the combined analysis of variation of regional hydrometeorological and runoff characteristics. Six individual basins with typical flow formation conditions and their contribution to the Volga River runoff were chosen. The combined analysis of differential integral curves of long-term variations in water discharge, air temperature, and precipitation for the parallel observation period of 1891-1988 was used for these basins - both for mean annual values and separately for every season: winter, spring, summer and autumn. Periods were distinguished in the observation series for the past years, which could serve as analog for future climatic and prognostic hydrological scenarios for the Volga River from the viewpoint of their combinations of air temperature, precipitation and ruxnoff. Thus, according to climatologists' forecasts, in 2000-2005, in the Volga basin, an increase in winter temperatures by 1.5-2.0°C should be expected, if the global temperature rises by 1°C. For this climate scenario, the Volga basin will find itself in the zone of precipitation decrease. The analog of the above conditions is the 1971-1977 period, when the increase in the mean annual temperature occurred due to a winter temperature rise by 1°C. At the same time, a decrease in annual precipitation was observed, which led to a decrease in the annual Volga River flow at the Volgograd site by 10-15 cukm for this period, which corresponds to the forecast of the State Hydrological Institute. This fact allows us to adopt the hydrological series of 1971-1977 as a model of the flow series for 2000-2005 and use it in the water-resource calculations. For the second scenario of climate changes by 2020-2050, the forecast of temperature, precipitation, and runoff will be different. If the global temperature rises by 2°C (due to winter temperatures increase in the basin by 3-4°C), the precipitation will be larger too, which will entail an increase in the Volga annual runoff by 23 cukm. Such a combination of the dynamics of air temperature, precipitation, and runoff corresponds to the 1980-1988 period.

We can also use a selection for the future scenarios of climate changes for some years, representing the corresponding changes in hydrometeorological elements under conditions of high-water, lower-water and mid-water years. While forming hydrological scenarios for large WRS for prognostic changes in the hydrological characteristics, we can also use a water balance model. Such an approach was tested for the Volga River basin, when the water balance elements were calculated separately for cold and warm seasons of the year for nine sectors of the basin for a year (1953/1954); the water discharge in this year was close to the mean annual value.

SIMULATION MODEL

Simulation modeling has become a most widely used technique in water-resource management. It allows us to analyse WRS operation under variable natural and economic conditions. A simulation model, developed at the Water Problems Institute of the Russian Academy of Sciences is based on the Out-of-Kilter algorithm, which has become universal due to its simplicity and convenience. The direct use of simulation models based on this algorithm for concrete WRS is generally impossible. Therefore, some modifications were implemented, allowing us to take into account the specificity of hydropower stations, organization of special ecological releases in the outlet site of the Volgograd system, maintenance of navigable depths and power releases (in summer and winter, respectively). The WRS of the Volga River basin is represented in the simulation model as a grid with nodes as reservoirs, water intakes, and sites of inflow of large tributaries, and with links as river reaches and canals. A certain priority is assigned to every water user and to the desirable water reserve in the reservoir. This allows us to

OPERATION UNDER CLIMATE CHANGE 581

reduce the task of water distribution according to priorities to an optimization problem of minimizing of costs of overflows within the grid during the calculated time interval. In accordance with the special features of the analysed problem, we provided a possibility to use 22 intervals within a year and to include into the scheme 50 nodes and 100 links.

Computer simulation of this model resulted in determination of the values of water releases and reservoir levels in the WRS. As a result, it was possible to develop a special program allowing us to obtain power indices of the operation of the WRS of the Volga-Kama chain hydropower stations. The input information for the calculation within intervals are initial level marks of reservoirs, time of the beginning of floods, information on ice formation, evaporation, winter runoff coefficient, water use, sanitary releases, lateral inflow, etc. The results of the calculations within the interval are final reservoir level mark, value of the head, capacity and energy of hydropower stations, inflow, discharge through the hydropower station turbines and release facilities, total discharge in the lower pool, the value of the decrease (filling) of reservoir storage, losses for evaporation, ice formation, etc. The volumes of available inflow of waters, which passed through power stations turbines, volumes of releases, volumes of the decrease and filling of the reservoir storage, the average seasonal capacity of the hydropower station, and gross hydropower production are calculated for the seasons of the year.

After the calculations are completed, the mean annual values of the above parameters for a long-term series, as well as values of the probability of the firm capacity of hydropower stations and the entire chain for a number of regular discharge years, for a number of calculation intervals and for the volume of unproduced hydropower are calculated. The probability of water releases to the lower pool of the Volgograd Hydropower Station in the flood period, and monthly water discharges in it during the winter period (January-February and December) are also calculated. The latter indicate the ecological situation in the Volga basin during the freezing period in the lower Volga. The developed mathematical model allowed us to make a series of computations for the various scenarios of expected climate changes in the future (in the 2000-2005 period).

INVESTIGATIONS RESULTS

Our investigations have shown that the characteristics of the yield of the Volga WRS largely depend on man-induced climate changes and result from changes in the annual flow values, as well as changes in the flow interannual distribution. The response of the characteristics of the systems yield under natural flow variation conditions can be evaluated using probability curves. These curves make it possible to assess not only the change in the firm yield of a WRS, but also to determine its reliability for various global climate change scenarios. In our investigations, we constructed probability curves for various indices of operation of the Volga water-resource system. These indices are: total annual power production of the hydropower stations, their firm winter capacity, volume of spring fishery releases to the lower reaches of the Volga, which guarantee the natural reproduction of fish populations in the northern Caspian Sea, higher winter discharges in the lower pool of the Volgograd Hydropower Plant, which are ecologically most dangerous, and requisite depths in unregulated sites of the water-resource system. The results indicate that, in the near future, at the beginning of the 21st century, under climate warming conditions (years 2000-2005), we can expect an decrease in the mean annual power production of the hydropower stations of the Volga-Kama reservoir chain by 5-7%, which will lead to the increase in the consumption of organic fuel by 1 million ton/yr. A certain redistribution of power production within a year can occur: the winter power production will be lower, and this will improve ecological conditions in winter. However, the situation with spring ecological water releases can become more serious than now, which will require a reconsideration of the reservoir chain operation rules and the priorities of water distribution aimed at increasing ecological releases. Our analysis has shown that the maintenance of required reliability of satisfying the demands for spring ecological releases will entail a decrease in hydropower production of the chain stations by 2.0-2.5 billion kwh/yr.

In the distant future, beyond the year of 2025, we can expect an increase in the mean annual and safe water yield of the Volga WRS without infringement of ecological demands on the WRS operational regime and the reliability of ecological releases can even increase. On the whole, we may assert that the Volga water-resource system will function rather reliably owing to its design parameters under predicted global climate change conditions and will provide a firm yield within allowable limits. Our investigations have also shown, that the regulatory capacity of the reservoir system will ensure the attenuation of negative effects of hydrological changes, caused by climate variations. But this will require more thorough investigations of the development of WRS operation rules under nonstationary conditions.

NETWORK FLOW MODEL APPLIED TO GEUM RIVER BASIN

G. B. YEON, Prof., Dept. of Civil Eng., Chung Cheong College, Cheong-Ju, Korea
S. B. SHIM, Prof., Dept. of Civil Eng., Chung Buk University, Cheong-Ju, Korea
K. H. YOON, Project Manager, Div. of Water Resources Eng., KICT, Seoul, Korea

ABSTRACT

The purpose of this paper is to construct a network model for the optimal allocation of limited water resources to the nodal system with given priorities. The solution technique for the model is based on the Out of Kilter Algorithm (OKA).
For the verification and application of the theoretical methodology and computer programs, the Geum river system is selected. Using release of Daecheong dam and water demand in Geum river basin, optimal allocation of water resources is accomplished for 4 cases (case 1 - case 4) which consider priority numbers in the demand nodes. The results of the application show that the model can reasonably represent the physical system, and water shortage at the demand nodes with high priority numbers is reduced. Its system solution was verified with that by revised simplex algorithm.

1. INTRODUCTION

In complex river basin consisting of natural river and artificial channels, it is very important to improve largely the benefits of system by allocating effectively limited water resources when water demands exceed water supply during low water level or drought.
Especially, water demands are increasing geometrically on account of the development of industrial and welfare facilities. When the allocating problems of water come up in order to supply entirely water to each water demand point, there will needs honest principles to allocate the water. If the allocating methods only follow from upstream to downstream, it may give rise to the difficult situations in the justice of water allocating management.
In order to supply appropriately water demand to each demand points during low water level or drought the optimal allocation system model of water resources which consider the importance of the use of water following the priorities of demand points have to be developed.
Network models can construct simply the complex river basin system as a node and link and can allocate consistently the limited water resources through the whole basin.

OKA (Out of Kilter Algorithm) is used as the solution method of network flow since it was developed as the solution of the least cost flow about network model. The Texas Water Development Board (1972) used the network flow analysis in order to evaluate the development alternative of complicate system for the plan of extension of the Texas Water system. Sigvaldason (1976) used the network flow analysis to evaluate the alternative policies for the conflicting purpose of 48 multipurpose reservoirs in the basin of Ontario Trench river. Graham (1986) used the network flow analysis for predicting allocation and use of increased runoff from simulated silvicultural activities on the Rio Grande National Forest. Chung (1987) had combined simulating model and out of kilter algorithm, so that he had large results in flexibility and computation. Yeon (1994) constructed the network model based on out of kilter algorithm to allocate optimally the limited water resources considering a priority number.

So, the purpose of this paper is to construct a network model for the optimal allocation of limited water resources to the nodal system with given priorities. The solution technique for the model is based on the Out of kilter Algorithm. For the verification and application of the theoretical methodology and computer programs the Geum river system is selected. Its system solution was verified with that by the revised simplex algorithm.

The limits of this study and application are following. All input data such as release, the variance of water demand and the construction of reservoir system are deterministic. When a network is constructed, it assumes that inflows, demands and losses rise in nodes.

2. CONSTRUCTION OF OPTIMAL ALLOCATION SYSTEM

NETWORK CONSTRUCTION OF RIVER BASIN SYSTEM

In order to solve the allocation problems of limited water resources the complex river basin system can express as mathematical structures which consist of node and link in point view of system engineering. Nodes in the network represent reservoirs, demand points, canal diversions, and river confluences. Links joining the nodes depict river reaches, canals, and pipelines. The nodes and links comprising the physical system are actually only a subset of the total network required to perform the simulation. In addition to these "real" components, the model has to plus "artificial" nodes and links as shown in Figure 1 for a simple system. These additional components are necessary in order to ensure that the network is fully circulating, as required by the solution algorithm, and to represent inflows, reservoir storage, spill, demands, channel losses, and return flows accurately. Especially, each link (i,j) in the network is represented by three parameters; lower bound L_{ij}, upper bound U_{ij}, and cost per unit flow C_{ij}. These parameters need to construct a capacitated flow which can be solved by OKA.

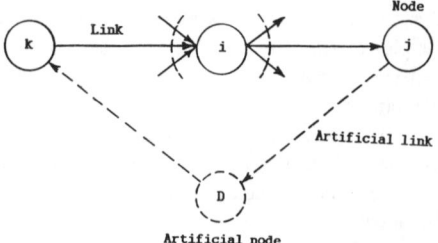

Figure 1. Simple capacitated network flow configuration

3. APPLICATION AND DISCUSSION

APPLICATION BASIN AND ITS PRESENT CONDITIONS

The application basin was Cheongju city, Daejeon city and Sintanjin, Gongju, Buyeo, Cheongang, Yiksan among chief demand points using Daecheong reservoir as water supply source. And Cheonan, Gunsan, Jeonju regions where supply water by inter-basin transfer are selected also as application basin.

In the whole condition for the water supply of application basins (KOWACO, 1989), it forecast that water shortage will raise on account of supplying water to an application basin and to neighbor basins (Sapgyocheon, Mangyongang basin) by 2001 year. Figure 2 was configured schematically the water supply condition in the application area using node-link concept. It consists of 23 nodes and 30 links including non-storage from Daecheong reservoir to each demand points.

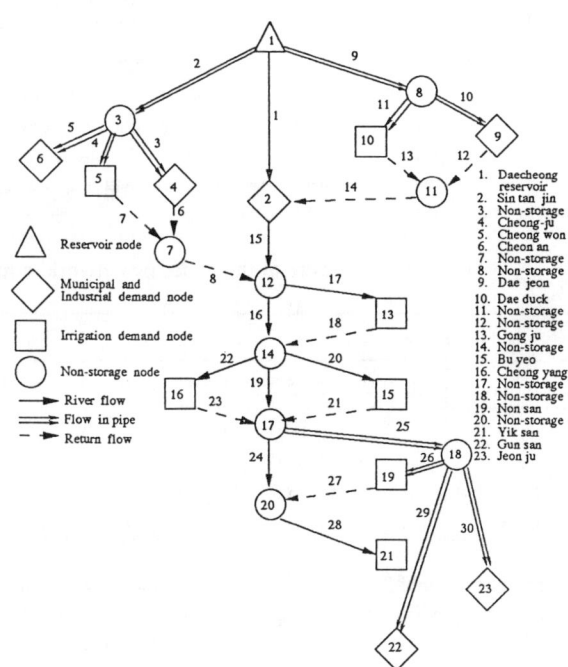

Figure 2. The configuration of the Geum river basin for network system

INPUT DATA AND ANALYSIS

The network model has been used Daecheong reservoirs' release in 1982 that was a comparative drought year since it operated. And the standard water demand in each demand points have used the 2001 years' demand.

The priority number necessary to operate network model should be decided after considering economic, political system analysis of each demand points and experts' opinions. In this paper 4 cases (case 1 - case 4) which consider priority numbers in the demand nodes have been decided by researcher after considering various conditions for each demand points (Table 1). This priority number can design to integral number between 1 and 99. Once water demand for each demand points are decided in a standard year, and upper and lower bound of each links could be decided following water demand.

Table 1. Priority number related to the nodes for Network model operation

Node No.	2	4	5	6	9	10	13	15	16	19	21	22	23
Case 1	30	30	30	30	30	30	30	30	30	30	30	30	30
Case 2	30	26	30	30	26	30	30	30	30	30	30	26	30
Case 3	28	26	30	30	26	30	28	28	30	28	30	26	30
Case 4	28	26	29	30	26	29	28	28	30	28	30	26	30

ANALYSIS AND DISCUSSION OF RESULTS

When water demands increase largely than supply capacity of water resources in the river basin, the release data of Daecheong reservoir and 2001 years' water demand of each demand points were used in order to apply and verify the network model.

Table 2 shows water shortage for demand water of each points when network model operate based on above conditions.

Table 2. Shortages at the nodes by network model operation

Unit : MCM

Node No.	2	4	5	6	9	10	13	15	16	19	21	22	23	SUM
Demand	26	117	78	53	314	19	61	49	20	151	111	91	160	1250
Case 1	0	0	0	0	21	5	0	0	5	92	44	25	44	236
Case 2	3	0	18	0	0	5	0	20	6	93	45	0	44	234
Case 3	0	0	53	15	0	8	0	0	8	38	64	0	44	230
Case 4	0	0	39	15	0	8	0	0	13	38	73	0	44	230

In above Table 2, case 1 is the result when network model was operated in order to analysis the condition setting equivalent priority number 30. Node No. 9, No. 10, No16, No. 19, No. 21, No. 22, No. 23 raise Water shortage 236 MCM.

Figure 3 indicate water shortage ratios (%) for demand water of each point, it shows that Node No. 9, No. 10, No16, No. 19, No. 21, No. 22, No. 23 suffer water shortage ratios (%) for demand water of each point 7%, 26%, 25%, 61%, 40%, 27%, 28%.

Figure 4 indicate monthly system losses, it means flow in spill link and happen when the link in network system is in maximum condition. System loss 66, 38, 54, 77, 94, 43, 26, 13, and 88 MCM in case 1 - case 4 happened on Jan., Feb., Mar., Apr., May, Sep., Oct., Nov., and Dec. This system loss happen when releases from a reservoir exceed water demand, it spill at each demand point.

In case 2, because Node No. 4 (Cheongju), No. 9 (Daejeon), and No. 22 (Gunsan) are big cities, and its priority were adjusted from "30" to "26" in order to supply preferentially water to these cities (Table 1). So Node No. 4, No. 9, and No. 22 could eliminate water shortage. In case 3, priorities number were adjusted from "30" to "28" to consider the demand condition of Node No. 2 (Sintanjin), No. 13 (Gongju), and No. 15 (Buyeo), No. 19 (Nonsan) which required the domestic, industrial and agricultural water. So Node No. 2, No. 13, and No. 15, No. 19 could almost reduce the shortage.

NETWORK FLOW MODEL 587

In case 4, priorities number of Node No. 5 (Cheongwon), No. 10 (Daeduck) which had raised large shortage also were adjusted to consider the demand conditions.

So Node No. 5 could reduce shortage from 53 MCM to 39 MCM.

Using release Daecheong dam and water demand in Geum river basin, optimal allocation of water resources is accomplished for 4 cases (case 1 - case 4) which consider priority numbers in the demand nodes. So water shortage could reduce from 236 MCM to 230 MCM by adjusting priority numbers as case 1- case 4. Above is that water shortage could not be solved fundamentally only by adjusting priority numbers of demand points, but limit water resources could be allocated optimally following priority numbers because the sensitiveness of allocation could be observed.

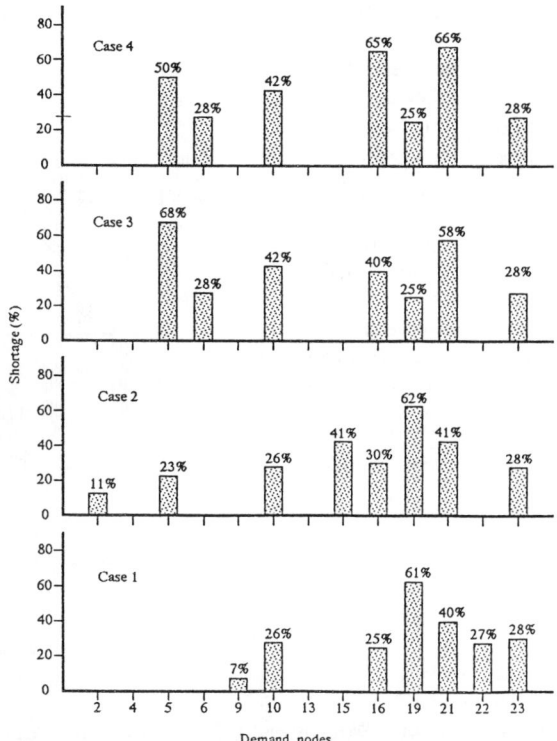

Figure 3. Shortage ratios (%) at the nodes by network model operation

OPTIMAL ALLOCATION AND COMPARISON BY SIMPLEX ALGORITHM

To verify the applied results of network model the optimal allocation by revised simplex algorithm has performed and compared with that of network model. The results of network model using OKA and revised simplex algorithms almost agreed when the same applicable basin and input data had used to give the same conditions. So the validity of network model using OKA could be verified when it was used for the optimal allocation of water resources.

4. CONCLUSION

A network model for the optimal allocation of limited water resources to the nodal system with given priority has been constructed. The solution technique for the model is based on the Out of Kilter Algorithm (OKA). For the verification and application of the theoretical methodology and computer programs, the Geum river system is selected. Using release Daecheong dam and water demand in Geum river basin, optimal allocation of water resources is accomplished for 4 cases (case 1 - case 4) which consider priority

numbers in the demand nodes. So water shortage could reduce from 236 MCM to 230 MCM by adjusting priority numbers as case 1- case 4.
water shortage could not be solved fundamentally only by adjusting priority numbers of demand points, but limit water resources could be allocated optimally following priority numbers because the sensitiveness of allocation could be observed. Its system solution was verified with that by revised simplex algorithm.

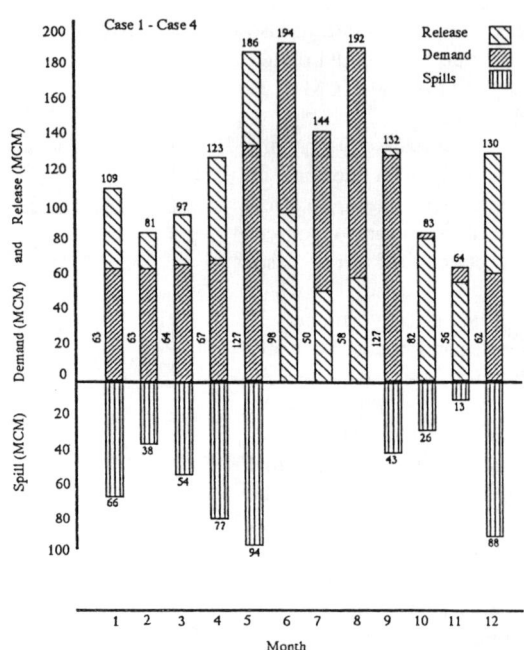

Figure 4. Monthly spill by network model operation

REFERENCES

Chung, F. I., M. C. Archer, and J. J. DeVries, Network flow algorithm applied to California aqueduct simulation, J. of Water Resources Plng. And Mgmt., ASCE, 115(2), 131-147, 1989.

Graham, L. P., et al., Allocation of augmented water supply under a priority water rights system, W. R. R., 22(7), 1083-1094, 1986.

KOWACO, Report for water demand following development western sea part, Korea Water Resources Corporation, 1989.

LINDO System, IBM PC for the interactive and discrete optimizer, Microsoft Corporation, 1992.

Sigvaldason, O. T., A simulation model for operating a multipurpose multi-reservoir system, W. R. R., 12(2), 2763-2780, 1976.

Texas Water Development Board, Economic optimization and simulation techniques for management of regional water resources systems, Austin, Texas, 1972.

Yeon, G. B. And S. B. Shim, Optimal allocation of water resources based on the network model, J. of Korea Water Resources Association, 26(5), 1994.

RESERVOIR MANAGEMENT DURING DROUGHT

by David Stephenson
University of the Witwatersrand, Johannesburg, South Africa

INTRODUCTION

The concept of variable draft from reservoirs can increase the total yield, and postpone additional water schemes, thereby saving costs. The economic loss due to water restrictions in times of drought must be balanced against the cost of assured water. To minimise the impact of rationing, the reservoir release rule should be optimized.

Alternative operating procedures or objectives have been proposed for optimizing the yield of reservoirs, e.g.

    Maximum total yield (Chen, 1972)
    Minimum economic loss (Stephenson, 1996)
    Continuous hedging (Shih and ReVelle, 1994)
    Proportional risk (Basson et al, 1994)
    Sharing (Dudley, 1990)
    Capacity allocation (Lund and Reed, 1995)

A separate problem is to restrict the consumption to the desired release. This can be achieved by pressure, control, limited time of availability, high tariffs, fines, appeals, or mechanical control valves or orifices.

CASE STUDY

The Min-Der reservoir in Taiwan is undersized for meeting the full demand from it. The possibility of additional water supply is limited so it is necessary to manage the water in the reservoir to meet demands in the best possible way.

The Min-Der reservoir on the Houlung river has a catchment area of 536 km^2, a reservoir capacity of 14 million m^3 and mean annual runoff of 906 million cubic metres due to the 1992 mm p.a. rainfall which occurs largely during typhoons. The average annual supply rate from the reservoir is 28 million cubic metres.

An operating rule was derived using linear programming and this indicated how much the draft should be dropped if the reservoir level was low at the beginning of dry seasons in particular. The probability of operating at different levels is also indicated. As the reservoir is over-utilized, the full demand can rarely be met and irrigation requirements should be curtailed for at least one or two seasons each year.

METHODOLOGY

The procedure in deciding on an operating rule for the reservoir is as follows:

HYDROLOGY

Long term time series of monthly inflows into the reservoir were synthesized. River flow records are only available since 1970, and were therefore of limited use in obtaining 100 year or other extreme flows. Time series could be generated synthetically by stochastic means or deterministically. A physically based computer RAFLER (Stephenson and Paling, 1992) was used with monthly raindata for surrounding raingauges. Acceptable agreement was obtained with the existing streamflow records giving confidence in the longer term projections.

The monthly streamflow series was analyzed to obtain risk of running the reservoir dam to different levels with various combinations of starting storage and draft, over four seasons.

PROBABILITY MATRIX

A matrix relating probable end to starting storage and draft was then set up for solution by linear programming. The probability equations were based on queuing theory (Langbein, 1958). Each draft was associated with a cost coefficient for the optimization exercise. The coefficients were functions of season, level of draft and priority of restriction (the lowest effects being restricted first). The Quatro-pro package was used to optimize the variables.

ECONOMIC ANALYSIS

The economic costs of various levels of water restriction were obtained by questionnaires sent out to various consumers. Fig. 4 shows the resulting composition of the objective function (economic cost versus level of supply of water) which is made up of irrigation, domestic and industrial consumption.

The resulting optimum draft levels associated with various reservoir levels and seasons are graphed in Fig. 5. Once the desirable level of supply was established, a method of controlling water usage to the desired supply level was developed. This was based on use of tariffs to control water use. The water consumption can be controlled at the desired supply level by imposing high water tariffs. Hopefully the consumer will balance water payments against economic loss due to restriction so that the tariff could be established, but if the consumer is unaware of his water consumption economic production relationship use of questionnaires or past experience may be necessary to set tariffs.

SIMULATIONS

After the optimization exercise to determine drafts, the system was simulated over 100 years of synthesyzed inflow to study the water level and draft variations. The simulation model was also able to accommodate alternative draft rules. For example, Rule 1 did not allow zero draft, so if the reservoir ran low it could even run to below zero storage numerically. Rule 2 accepted zero draft if the reservoir ran low to avoid running out of water. Operating rule 3 was the old draft rule not based on increasing cost of rationing and the reservoir ran dry frequently.

Costs, amounts of water used and number of storages less than or equal to zero for all three operating rules can be found in Table 1.

Table 1: Comparison of Average Annual Cost, Average Annual Water Used and Cumulative Storages Less Than or Equal To Zero for the Three Operating Rules

Operating rule	Average Cost (M NTD/annum)	Average water used (M m^3/annum)	Storage months $<=0$
1. Optimum value	42 567	31.9	3
2. With zero draft	63 866	32.3	0
3. Max draft	159 580	41.9	15

The reservoir ran dry for one period of three months using Operating rule 1 in the 100 years of simulation. Operating Rule 2 resulted in no storages less than or equal to 0. The Old Operating Rule resulted in 15 months with storage less than or equal to 0 in 11 diferent years. This is for 1996 draft levels and the solution can be expected to deteriorate as the demand increases in the future.

An example of the results is given for the inflow hydrology for 1951 to 1960. Figures 6 and 7 illustrate the storages and drafts in the reservoir resulting from Operating Rules 1 and 2. Operating Rule 2 had 3 zero drafts in 1953 and Operating Rule 1 did not have any zero drafts or storages.

CONCLUSIONS

The average yield, and life of reservoirs can be prolonged by developing a management procedure. Draft should be restricted progressively during drought. In this way, the possibility of running dry is minimized.

Various users should be ranked in order of economic importance and restricted successively.

Water consumption can be controlled by use of tiered tariffs.

REFERENCES

Basson, M.S., Allen, R.B., Pegram, G.G.S. and Van Rooyen, J.A., 1994. Probabilistic Management of Water Resources and Hydropower Systems. Water Ress. Pubs. Fort Collins, Colorado.
Chen, L.-S., 1972. Rule curve for the operation of an irrigation reservoir. Joint Comm. on Rural Reconstruction, Taipei.
Dudley, N., 1990. Alternative Institutional Arrangements for water supply probabilities and transfers. Proc. Seminar on Transferability of Water Entitlement. Univ. New England, Armidale.
Langbein, W.B., 1958. Queuing theory and water storage. ASCE, J. Hydr. Div. HY5, Oct, 1811.1-24.
Lund, J.R. and Reed, R.U., 1995. Drought water rationing and transferable rations. ASCE, J.Water Ress. Planning Management, 121(6), Nov/Dec, 429-437.
Shih, J.S. and ReVelle, C., 1994. Water Supply operations during drought, continuous hedging rule. ASCE, J. Water Ress. Planning Management, 120(5), Sept/Oct, 613-629.
Stephenson, D., 1996. Drought management as an alternative to new water schemes. Water SA, Oct.
Stephenson, D. and Paling, W.A.J. 1992. An hydraulic based model for simulating monthly runoff and erosion. Water SA, 18(1) 43-52.

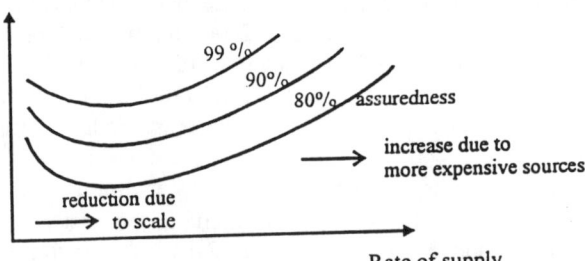

Fig. 1 : Effect of assuredness on cost of water

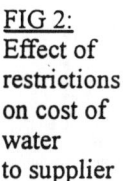

FIG 2:
Effect of
restrictions
on cost of
water
to supplier

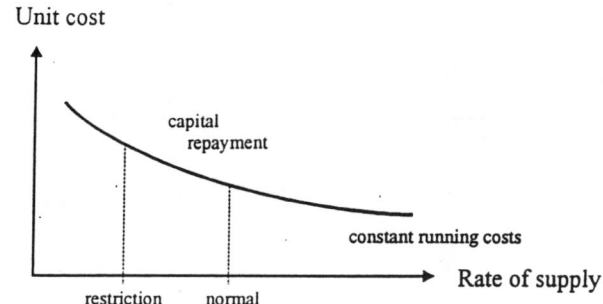

FIG 3:
Effect of
tariff on
consumption

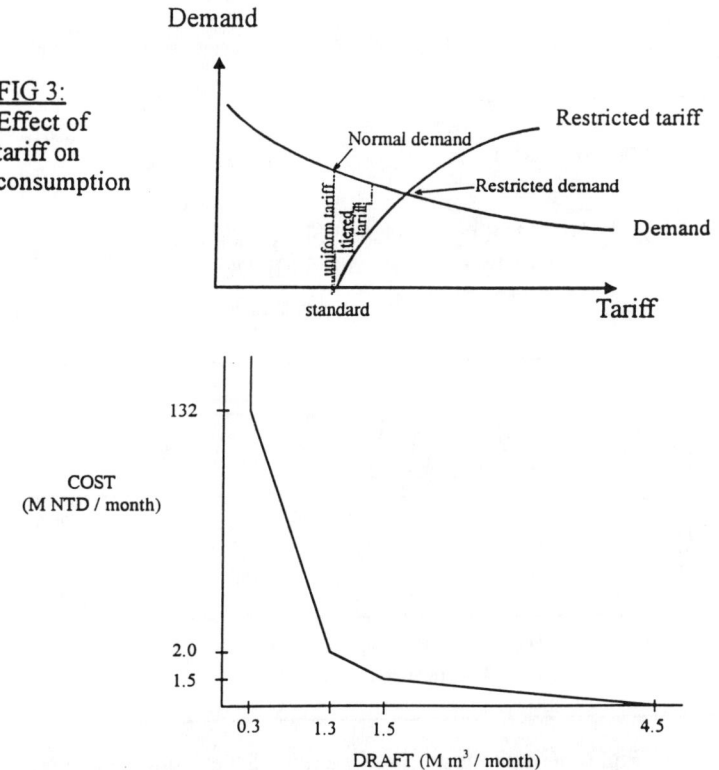

Fig. 4 : Average monthly economic loss function (not to scale)

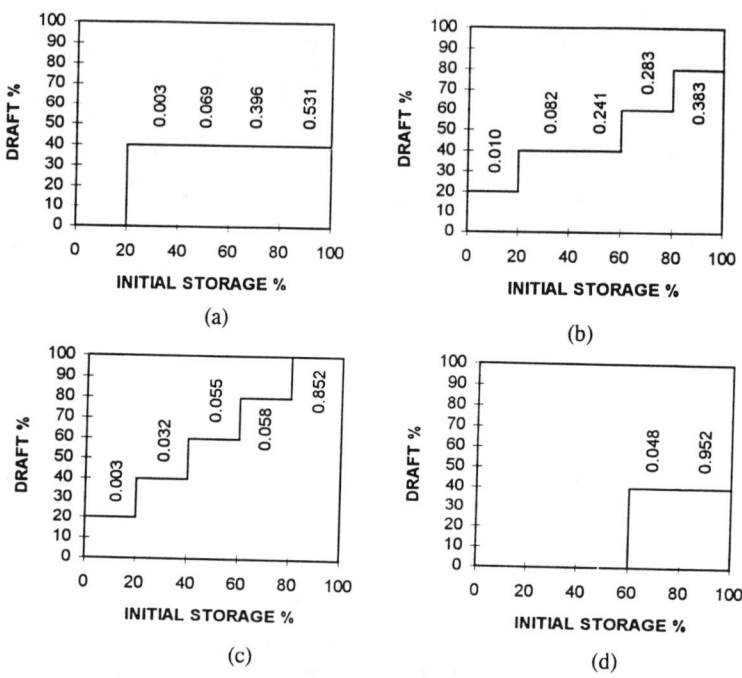

Fig. 5 : Graphical representation of Operating Rule 1 as defined by Quattro Pro and LINDO: (a) season 1; (b) season 2: (c) season 3; (d) season 4.

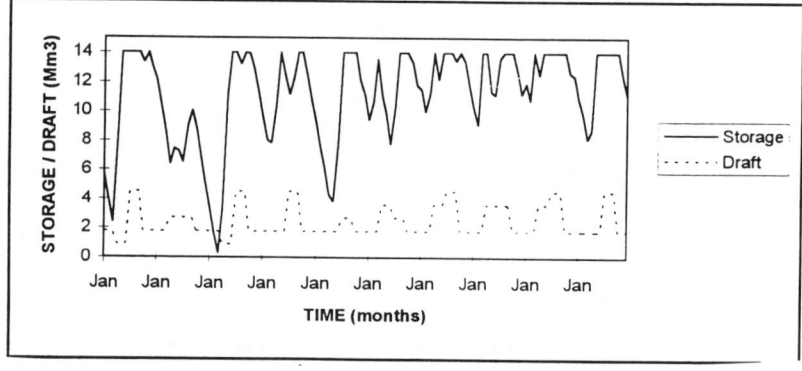

Fig. 6 : Graph illustrating Operating Rule 1 storage and draft for the hydrological years 1951 - 1960

Stochastic Modeling of Hydrological Droughts

HSIEH WEN SHEN
Professor, Department of Civil and Environmental Engineering,
University of California, Berkeley, California, U.S.A.

JENQ TZONG SHIAU
Graduate Student, Department of Civil and Environmental Engineering,
University of California, Berkeley, California, U.S.A.

ABSTRACT

The far-reaching impacts of droughts are often spread on larger areas and tend to persist over longer time. Understanding the characteristics of droughts is an essential and important component for design, planning and management of a water resources system. In this study, streamflows are under investigated and hydrological droughts are considered as the period during which streamflows below monthly median. As drought events defined by theory of runs, two important properties of droughts, distributions of drought duration and severity are fitted from data. Then, the occurrences of drought events can be modeled as the alternating renewal process. The return period of drought events is also determined based on the derived severity distribution and mean interarrival time of droughts.

INTRODUCTION

The absence of a precise and universally accepted definition of drought causes the confusion about whether a drought exists or not [Wilhite et al., 1985]. The *theory of runs* [Yevjevich, 1967] offers as an objective method to define the onset, termination, and severity of droughts clearly. However, the occurrences of droughts are not included in the runs analysis.

As truncation level separates the streamflows into drought and wet states which occurred alternatively, the drought phenomena can be modeled as the *alternative renewal process* [Kendall et al., 1992; Loaiciga et al., 1996]. Therefore, the interarrival time between drought events is the sum of drought and wet duration. This relationship can lead to several useful distributions, such as the arrival time of n-th drought events, the expected number of drought events occurred by time t, ... etc..

It becomes standard practices to know the *return period* of drought events, defined as the droughts with certain severity of greater, in design, planning and management of specific water resources system. From derived severity distribution and mean interarrival time of droughts, the return period of drought events with certain severity or greater is easily derived.

DEFINITION OF HYDROLOGICAL DROUGHTS

The *hydrological drought* is considered as the period during which the streamflows below certain threshold. This threshold is called truncation level in *theory of runs* [Yevjevich, 1967], the definition of runs is illustrated in Figure 1. According to runs analysis, the negative runs are referred as the drought events. Two important drought characteristics, duration (DL) and severity (DS) are under investigated first.

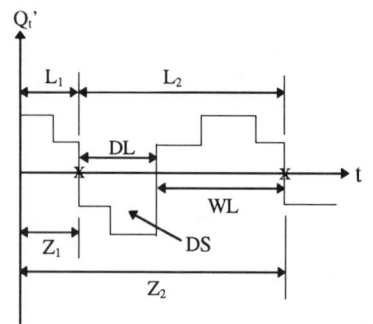

Figure 1. Definitions of runs and alternating renewal process

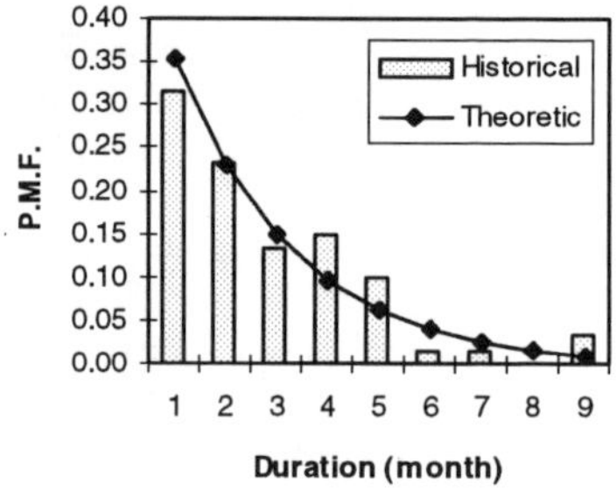

Figure 2. Distribution of drought duration

A 28-year monthly streamflow series is used as historical data in this study. The truncation level is considered as the median flow of each month, which implying that drought state is streamflows below the median and wet state is streamflows above the median. Due to the high correlation of monthly streamflows, this flow series is assumed as the first order Markov chain [Jackson, 1975]. Therefore, the probability mass function of drought duration (DL) and wet duration (WL) are:

$$f_{DL}(dl) = P_{DW} P_{DD}^{dl-1} \quad , \quad dl = 1, 2, 3, \ldots \quad (1)$$

$$f_{WL}(wl) = P_{WD} P_{WW}^{wl-1} \quad , \quad wl = 1, 2, 3, \ldots \quad (2)$$

where $P_{DW} = P(Q_t' \geq 0 \mid Q_{t-1}' < 0)$
$P_{DD} = P(Q_t' < 0 \mid Q_{t-1}' < 0)$
$P_{WD} = P(Q_t' < 0 \mid Q_{t-1}' \geq 0)$
$P_{WW} = P(Q_t' \geq 0 \mid Q_{t-1}' \geq 0)$
Q_t' : truncated flow at time t

These transition probabilities estimated from historical data are P_{DW}=0.353, P_{DD}=0.647, P_{WD}=0.357, and P_{WW}=0.643. As the median flow selected as truncation level, P_{DW} and P_{DD} should be equal to P_{WD} and P_{WW}, respectively, this indicates that the distribution of drought and wet duration are identical. The minor differences exist in estimations due to the limited samples of historical data. These theoretic distributions are well fitted by comparing with the historical data, both shown in

Figure 2 and 3.

The drought severity can be fitted as gamma distribution [Mathier et al., 1992], the probability density function is:

$$f_{DS}(ds) = \frac{ds^{\alpha-1}}{\beta^{\alpha}\Gamma(\alpha)} e^{-ds/\beta}, \quad ds > 0 \qquad (3)$$

The parameters estimated from historical data are $\alpha=0.771$ and $\beta=128.152$. Shown in Figure 4, the theoretic distributions is also fitted well with the observed data.

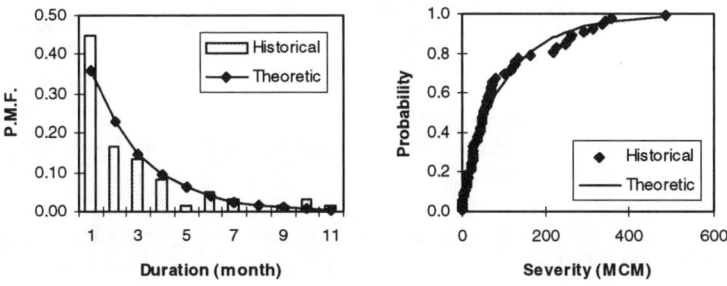

Figure 3. Distribution of wet duration **Figure 4.** Distribution of drought severity

THE ALTERNATING RENEWAL PROCESS

As drought events defined by theory of runs, the truncated flow series is composed of two events: drought and wet events which are occurred alternatively. Therefore, the streamflow series can be thought as a sequence of alternative drought and wet periods and modeled by the *alternating renewal process* [Kendall et al., 1992; Loaiciga et al., 1996]. The definitions of alternating renewal process are also illustrated in Figure 1.

The time between two successive drought events, called interarrival time L, is equal to the sum of drought and wet duration, namely:

$$L = DL + WL \qquad (4)$$

Given the drought and wet duration distributions as Eq.(1) and (2), and under the assumption of independence of DL and WL, by using of generating function method, the probability mass function of interarrival time L is given by the following expression:

$$f_L(l) = \frac{P_{DW}P_{WD}}{P_{DD} - P_{WW}}(P_{DD}^{l-1} - P_{WW}^{l-1}), \quad l = 2, 3, 4, \ldots \qquad (5)$$

Figure 5 shows this theoretic distribution and observed interarrival time.

Based on the relationship of Eq.(4) and distributions of drought and wet duration, Eq.(1) and (2), the mean interarrival time of drought events, E(L), becomes:

$$E(L) = \frac{1}{P_{DW}} + \frac{1}{P_{WD}} \qquad (6)$$

The mean interarrival time of droughts estimated from historical data is 5.60 months, while computed from above equation is 5.63 months.

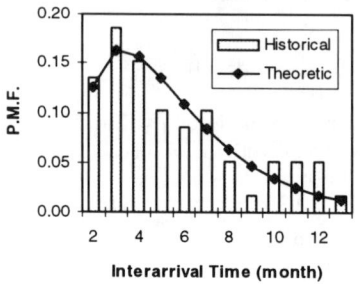

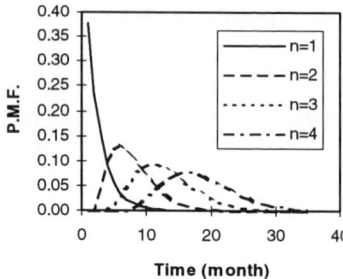

Figure 5. Distribution of drought interarrival time

Figure 6. Distribution of arrival time of n-th droughts

Let L_1 be the arrival time of the first drought and for $n \geq 2$, let L_n be the interarrival time between (n-1)-th and n-th drought events. Then, the arrival time of n-th drought event, denoted by Z_n, is equal to:

$$Z_n = \sum_{i=1}^{n} L_i \quad , \quad n = 1, 2, 3, \ldots \qquad (7)$$

Generally, since L_1 is not an interarrival time, it may have the distribution different from L_n, $n \geq 2$ (Eq.(5)). Based on the fact that before the first drought event coming is wet state, hence, L_1 is assumed that distributed as the wet duration, Eq.(2). Under the assumption of independence of L_n, $n \geq 1$, the probability mass function of Z_n can be derived by generating function method:

$$f_{Z_1}(t) = P_{WD} P_{WW}^{t-1} \quad , \quad t = 1, 2, 3, \ldots$$

$$f_{Z_n}(t) = P_{DW}^{n-1} P_{WD}^{n} \sum_{j=0}^{t-2n+1} \frac{(n+j-2)!}{j!(n-2)!} \frac{(t-n-j)!}{(t-2n-j+1)!(n-1)!} P_{DD}^{j} P_{WW}^{t-2n-j+1} \quad , \quad n \geq 2 \, , \, t \geq 3 \qquad (8)$$

This distribution for different n is shown in Figure 6.

Let N(t) denote the number of drought events occurred by time t. According to elementary renewal theorem, the number of drought events per unit time is equal to the reciprocal of mean drought interarrival time for the long run [Ross, 1983; Wolff, 1989], namely:

$$\lim_{t \to \infty} \frac{N(t)}{t} = \frac{1}{E(L)} \qquad (9)$$

Estimated from historical data, t=336 months, N(t)=60, E(L)=5.60 months, therefore, N(t)/t is perfect agreement with 1/E(L), which equal to 0.179 and 0.178, respectively.

According to Figure 1, $N(t) \geq n$ if and only if $Z_n \leq t$, therefore, $P(N(t) \geq n) = P(Z_n \leq t)$. Hence, $P(N(t)=n) = P(N(t) \geq n) - P(N(t) \geq n+1) = P(Z_n \leq t) - P(Z_{n+1} \leq t)$. According to the

STOCHASTAIC MODELING OF DROUGHTS 599

distribution of Z_n, Eq.(8), P(N(t)=n) has the following epression:

$$P(N(t)=1) = 1 - P_{WW}^t - \sum_{i=3}^{t} P_{DW} P_{WD}^2 \sum_{j=0}^{i-3} \frac{(i-j-2)!}{(i-j-3)!} P_{DD}^j P_{WW}^{i-j-3} \quad , \quad t = 1, 2, \ldots$$

$$P(N(t)=n) = \sum_{i=2n-1}^{t} P_{DW}^{n-1} P_{WD}^n \sum_{j=0}^{i-2n+1} \frac{(n+j-2)!}{j!(n-2)!} \frac{(i-n-j)!}{(i-2n-j+1)!(n-1)!} P_{DD}^j P_{WW}^{i-2n-j+1} \quad (10)$$

$$- \sum_{i=2n+1}^{t} P_{DW}^n P_{WD}^{n+1} \sum_{j=0}^{i-2n-1} \frac{(n+j-1)!}{j!(n-1)!} \frac{(i-n-j-1)!}{(i-2n-j-1)!n!} P_{DD}^j P_{WW}^{i-2n-j-1} \quad , \quad n \geq 2 \; , \; t \geq 3$$

The distribution of N(t) for different time and n is shown in Figure 7.

The expected number of droughts occurred by time t, E[N(t)], can also be determined:

$$E[N(t)] = (1 - P_{WW}^t) +$$

$$\sum_{n=2}^{[(t+1)/2]} \sum_{i=2n-1}^{t} P_{DW}^{n-1} P_{WD}^n \sum_{j=0}^{i-2n+1} \frac{(n+j-2)!}{j!(n-2)!} \frac{(i-n-j)!}{(i-2n-j+1)!(n-1)!} P_{DD}^j P_{WW}^{i-2n-j+1} \quad (11)$$

where [(t+1)/2] : the largest integer of (t+1)/2.
The expected number of droughts occurred within period of historical data (28 years) by above equation is 59.9, while there are 60 drought events observed in this period.

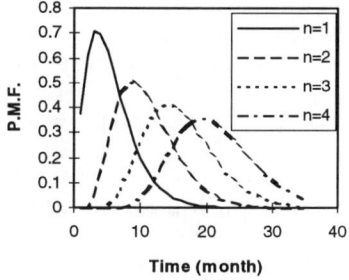

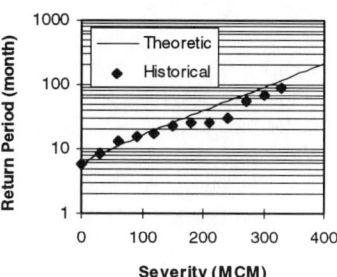

Figure 7. Distribution of N(t)   **Figure 8.** Return period of droughts

RETURN PERIOD OF DROUGHTS

The *return period* or *recurrence interval* of a hydrological event is defined as the average elapsed time between occurrences of an event with a certain magnitude or greater [Haan, 1977]. In this study, the return period of drought events is considered as the mean interarrival time of a drought event with certain severity or greater.

The *renewal rate* of drought events is the number of drought events per unit time. For sufficient long time, the renewal rate is equal to the reciprocal of mean drought interarrival time (see Eq.(9)). The drought events with severity equal or greater than specific value, ds, have probability 1-F_{DS}(ds), where F_{DS}(ds) is the cumulative distribution function of drought severity. Then the renewal rate of such drought events is equal to [1-F_{DS}(ds)] / E(L) for sufficient long time [Stedinger et al., 1993].

Hence, the return period of drought events with severity equal or greater than ds is:

$$E(T_{ds}) = \frac{E(L)}{1 - F_{DS}(ds)} \tag{12}$$

Figure 8 illustrates this theoretic relationship of return period and drought severity, the observed return period from historical data also shows consistent trend.

SUMMARY AND CONCLUSIONS

1. Defined by the theory of runs, hydrological droughts are considered as the period during which the streamflows below the median in this study. Drought characteristics, such as drought duration, Eq.(1), and severity, Eq.(3), are fitted from the historical data.
2. Based on the phenomena that drought and wet duration are occurred alternatively, the occurrence of drought events can be modeled as the alternative renewal process. The distribution of drought interarrival time, Eq.(5), the arrival time of n-th drought events, Eq.(8), the distribution of N(t), Eq.(10), and the distribution of expected number of drought events, Eq.(11), are derived.
3. The return period, defined as the mean interarrival time of drought events with certain severity or greater, are determined from the relationship of the distribution of severity and mean interarrival time of droughts, Eq.(12).

REFERENCES

Haan, Charles T., 1977, *Statistical Methods in Hydrology*, Iowa State University Press, Ames, Iowa, pp.293~311.

Jackson, Barbara Bund, 1975, Markov Mixture Models for Drought Lengths, *Water Resources Research*, 11(1), pp.64~74.

Kendall, D. R., and J. A. Dracup, 1992, On the Generation of Drought Events Using an Alternating Renewal-Reward Model, *Stochastic Hydrology and Hydraulics*, 6(1), pp.58-68.

Loaiciga H. A., and R. B. Leipnik, 1996, Stochastic Renewal Model of Low-Flow Streamflow Sequences, *Stochastic Hydrology and Hydraulics*, 10(1), pp.65~85.

Mathier, L., L. Perreault, B. Bobe, and F. Ashkar, 1992, The Use of Geometric and Gamma-Related Distributions for Frequency Analysis of Water Deficit, *Stochastic Hydrology and Hydraulics*, 6(4), pp.239~254.

Ross, Sheldon M., 1983, *Stochastic Processes*, Wiley, New York, pp.107~114.

Wilhite, Donald A., and Michael H. Glantz, 1985, Understanding the Drought Phenomenon : The Role of Definitions, *Water International*, 10, pp.111~120.

Wolff, Ronald W., 1989, *Stochastic Modeling and the Theory of Queues*, Prentice Hall, Englewood Cliffs, New Jersey, pp.53~68.

Yevjevich, Vujica, 1967, An Objective Approach to Definitions and Investigations of Continental Hydrologic Droughts, Colorado State University, *Hydrology Papers No.23*, pp.4~5.

Comparison of Mean Areal Precipitation Estimates Derived from NEXRAD Radar vs. Rain Gage Networks

BRYCE FINNERTY and DENNIS JOHNSON
Office of Hydrology, National Weather Service, Silver Spring, MD U.S.A.

ABSTRACT

Historically, hydrologic forecasts and calibration of the National Weather Service (NWS) hydrologic models are typically prepared using inputs of 6-hour mean areal precipitation (MAP) estimates derived from rain gage networks. However, operational hydrologic forecasts can now be prepared using 1- or 6-hour MAPs derived from high resolution gridded precipitation estimates from the NWS Next Generation Radar system, NEXRAD. An initial analysis of 7 months (May 1993 - December 1993) of 1- and 6-hour operational MAP time series for nine basins revealed differences between the gage and radar MAPs in the long term accumulations, individual storm totals, and timing of events. The 7-month radar MAP accumulations were 10-25 percent less than the gage MAP accumulations. The 6-hour radar and gage MAPs had similar estimates of the timing of events, but the 1-hour radar and gage MAPs showed more discrepancies in the timing of events. The radar MAPs captured more of the variability in the precipitation fields than the gage MAPS at the 1-hour time step. However, the variability of the radar and gage MAPs was nearly equal at the 6-hour time step.

INTRODUCTION

New challenges in operational hydrologic forecasting have materialized because of the installation of an advanced national system of radars called Weather Surveillance Radar 1988-Doppler (WSR88-D), or NEXRAD (Hudlow, 1988). NEXRAD produces high resolution gridded precipitation estimates at a 4x4 km^2 spatial scale and a 1-hour time scale. These NEXRAD data provide spatial and temporal estimates of highly variable precipitation fields that are not possible from point measurements of precipitation from rain gage networks.

Hydrologic forecasting at the NWS generally uses a lumped modeling approach in which hydrologic model parameters are averaged over a river basin. The Sacramento Soil Moisture Accounting (SAC-SMA) model (Burnash et al., 1973), which is the most commonly used hydrologic model in the NWS, uses a mean areal precipitation (MAP) input and outputs discharge. The SAC-SMA model parameters are calibrated using 6-hour model input MAPs derived from historical point gage precipitation measurements. These parameters are then used operationally with 6-hour gage MAPs to produce river forecasts with a 6-hour time step. River forecasters now have the option of using 1- or 6-hour radar-derived mean areal precipitation (referred to as MAPXs) for operational forecasting. However, the model parameters have not been recalibrated to the MAPX data because only a short 7-month period exists where historical gage and discharge data overlap with the NWS archive of radar data. Therefore, a detailed analysis of model input MAPs is required to understand how these model inputs will impact model performance. The objective of this paper is to analyze the differences in operational methods of estimating 1- and 6-hour MAPs derived from radar and gage networks.

ANALYSIS

The study area is nine basins in the region near the Oklahoma-Arkansas-Missouri state boundaries. This region was analyzed because of its dense gage network, six overlapping radar umbrellas, and one of the longest available periods of archived NEXRAD radar products in the United States to date. MAP and MAPX time series were derived for the period from May 7, 1993 through December 31, 1993. This is the period in which the historical gage and discharge data overlap with archived radar products.

The radar data used for the study are the 4x4 km^2 resolution Stage III data. Stage III is a merged radar-gage precipitation field design to provide the spatial resolution of radar data while preserving the precipitation accumulations measured by gages. It assumes the gage data as "ground truth" and scales the radar accumulation estimates to match the gages (Shedd and Smith, 1991). The influence of the gages on the Stage III data diminish with distance from the gage. The gage MAPs were calculated using both the hourly and daily gage accumulations obtained from the National Climatic Data Center's 3200 precipitation data.

Summary statistics were generated for the 1- and 6-hour case, for all nine basins, and for the entire 7-month duration of the MAP and MAPX time series. Statistics include the mean, standard deviation, and coefficient of variation, all of which are conditioned upon the occurrence of precipitation. The conditional statistics provide information about the ability of radar and gage networks to detect precipitation, as well as their ability to estimate precipitation rates. Cumulative sums for each basin and the percent bias of accumulations from radar vs. gage networks were also tabulated.

RESULTS

Tables 1 and 2 show that for the 7-month period, the Stage III MAPX cumulative totals have a negative bias ranging from -10 to -25 percent as compared to gage MAPs. This under catch of the MAPXs exists for both the 1- and 6-hour cases. The fact that the 6-hour conditional mean of the MAPX is larger than MAP for some basins is attributed to the spatial and temporal averaging of point gage measurements to time increments and areas where it is not raining. This produces rain for time intervals where no rain was occurring, reduces the conditional mean, and hides the actual cumulative bias that exist at the 6-hour case. MAPX is approximately 25 percent greater at detecting the variability in the precipitation at the 1-hour time step than the gage MAPS. However, the normalized mean (coefficient of variation) shows that the variability of the MAPs and MAPXs is nearly equal at the 6-hour time step. This indicates that the radar's ability to capture the variability in precipitation at a 1-hour time step is largely lost when averaging to a 6-hour time step.

Figure 1 illustrates how the cumulative sum of both the MAP and MAPX time series behaves over the 7-month duration for the TIFM7 basin. Only this basin is shown because all nine basins exhibited similar behavior. The figure shows how the gages and radar have different estimates of the volume and timing of individual precipitation events. Figure 1 also shows a plot of the cumulative differences in the MAP and MAPX time series, which highlights the timing and storm-by-storm catch of the two rainfall measurements. Both methods appear to be similar in their estimation of

Table 1 - Conditional Summary Statistics for 9 Basins
Using 1-hour MAP and MAPX Time Series

Basin ID	SUM (mm)		HOURLY MEAN (mm)		HOURLY STD. DV. (mm)		COEF. of VARIATION		% BIAS of SUMS
	MAP3	MAPX	MAP3	MAPX	MAP3	MAPX	MAP3	MAPX	
JOPM7	1207	1040	2.07	1.51	3.26	3.26	1.575	2.160	-13.9
TIFM7	1104	996	1.48	1.32	2.69	2.66	1.820	2.020	-9.8
WTTO2	977	798	1.27	1.04	2.68	2.20	2.107	2.111	-18.3
KNSO2	948	837	1.50	1.41	2.67	2.83	1.782	2.000	-11.7
ELDO2	1056	930	1.51	1.34	2.55	2.90	1.680	2.176	-11.9
TALO2	933	839	1.47	1.33	2.77	3.28	1.880	2.460	-10.1
TENO2	979	784	1.62	1.18	2.85	2.90	1.755	2.460	-19.9
VBLA4	1055	798	1.73	1.07	2.99	2.29	1.730	2.128	-24.4
MLBA4	864	726	1.523	1.11	2.37	2.48	1.547	2.240	-15.9

%BIAS = [(MAPX-MAP)/MAP] x 100

Table 2 - Conditional Summary Statistics for 9 Basins
Using 6-hour MAP and MAPX Time Series

Basin ID	SUM (mm)		6-HOUR MEAN (mm)		6-HOUR STD. DV. (mm)		COEF. of VARIATION		% BIAS of SUMS
	MAP	MAPX	MAP	MAPX	MAP	MAPX	MAP	MAPX	
JOPM7	1207	1040	5.51	5.10	9.29	9.52	1.69	1.87	-13.9
TIFM7	1104	996	4.30	4.70	7.90	8.55	1.84	1.82	-9.8
WTTO2	977	798	3.60	3.82	6.94	6.89	1.93	1.80	-18.3
KNSO2	948	837	4.35	4.65	7.75	8.32	1.78	1.79	-11.7
ELDO2	1056	930	4.19	4.72	7.68	8.79	1.83	1.86	-11.9
TALO2	933	839	4.15	4.49	8.34	9.34	2.01	2.08	-10.1
TENO2	979	784	4.45	4.15	8.64	9.02	1.94	2.17	-19.9
VBLA4	1055	798	4.67	3.80	7.69	6.20	1.65	1.63	-24.4
MLBA4	864	726	4.23	3.65	6.58	6.77	1.56	1.85	-15.9

event timing, but a spike appears in the cumulative difference plot when their timing is off. MAPXs have a lower accumulation even in the summer, which may be due to averaging point gage measurements over areas where it is not raining. Or, the radar may simply be underestimating precipitation rates. Radar generally performs better than gage networks during high intensity convective storms because of the radar's high resolution spatial coverage (Seo and Smith, 1996; Smith et al., 1996).

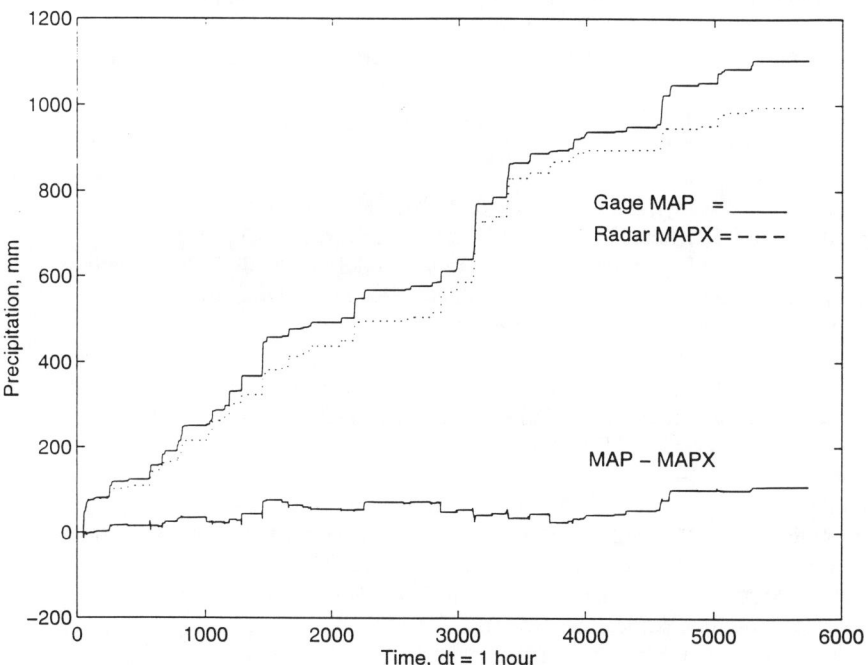

Figure 1: 7-month cumulative sum of 1-hour MAP and MAPX time series for the TIFM7 basin. Time=0 is May 7, 1993 and time=5736 is December 31, 1993. The radar MAPX has a lower cumulative sum of precipitation for the period analyzed. The bottom of the figure shows a plot of the difference between gage and radar mean areal precipitation estimates (MAP - MAPX).

CONCLUSIONS

The most significant result of this preliminary study is that MAPX has a negative bias in the 7-month accumulations for all nine basins analyzed. Any bias in the hydrologic model inputs may impact operational flood forecasts, soil moisture accounting, and long-term water supply forecasts. The 1-hour MAPX detects more precipitation variability than the gage MAPs, but this effect was minimized when averaging to a 6-hour time step. There are clear differences in each method's ability to estimate the timing and volume of individual events; however, no clear trend exists for all events. This indicates that hydrologic forecasting can benefit from having both measurement systems available to the forecasters, from which they can choose. Further analysis is required to evaluate what impact these differences between MAP and MAPX model inputs may have on operational hydrologic forecasting.

REFERENCES

Burnash, R.J.C., Ferral, R.L., and McGuire, R.A., 1973. A generalized streamflow simulation system - conceptual modeling for digital computers. U.S. Department of Commerce, National Weather Service and State of California, Department of Water Resources.

Seo, D.J., and Smith, J.A., 1996. Characterization of climatological variability of mean areal rainfall through fractional coverage. Water Resources Research, 33(7): 2087-2095.

Shedd, R.C., and Smith, J.A., 1991. Interactive precipitation processing for the modernized National Weather Service. Preprints, Seventh International Conference on Interactive Information and Processing Systems for Meteorology, Oceanography, and Hydrology, New Orleans, Louisiana, American Meteorological Society, pp. 320-323.

Smith, J.A., D.J. Seo, M.L. Baeck, and M.D. Hudlow, 1996. An intercomparison study of NEXRAD precipitation estimates. Water Resources Research, 32(7): 2035-2045.

MANAGING WATER SCARCITY THROUGH MAN-MADE RIVERS

SAAD A. ALGHARIANI
Alfateh University, Tripoli, Libya

ABSTRACT

Poor surface and groundwater resources distribution on the national, regional and global levels has been creating severe water shortages in several parts of the world. The construction of Man-Made Rivers (MMR) to achieve a more desirable distribution of available water supplies is both technically feasible and economically rational. The Libyan experience with its Great MMR Project is briefly described and its results are introduced as an example to support the above conclusions. Similar future projects may be constructed to serve other water scarcity zones such as the countries of the Middle East, North Africa and the Southwestern USA.

INTRODUCTION

In several parts of the world, the potentially available water resources from both surface and groundwater basins are neither evenly distributed in space and time nor equally accessible in the right quantity and quality. To ameliorate the resulting water shortage situations in these areas several solutions have been suggested and implemented. One of these solutions has been exemplified in large scale mass transfer of surplus water from one water aboundance basin to another water deficit basin or from one water abundance region of a country to another water scarcity region. Examples of past experience with interbasin water transfers include the California State Water Project and the Colorado River transfers in USA, the Karakum Canal in the former Soviet Union and the Triple Canal and its complete link canal system in Pakistan. All these examples and similar other projects executed or still under construction at the present time have been limited to interbasin transfer of surface water within the confines of the national political boundaries of the countries concerned. More recently there have been serious calls for mass water transfer from water abundance regions in sovereign nations to water scarcity regions in other sovereign nations by agreement [1]. The Middle East peace Pipeline proposed by the Turkish government is the most recent example. Increasing water scarcities and escalating water demand in several arid zone areas such as North Africa, the Middle East and the Southwestern USA may provide the political impetus and the socioeconomic rationale for the realization of such transboundary water transfer projects.

In addition to surface water huge unexploited groundwater basins of acceptable water quality do exist in many parts of the world. Some countries in the arid zones of North Africa and the Middle East are entirely dependent on the exploitation of the aquifers in these basins for their growth and development. In cases of uneven national or regional distribution of these aquifers large amounts of water can be economically pumped and collected from wellfields to be transferred to where it is needed. This paper discusses some aspects of the potential successes and expected limitations of water transfer projects through Man-Made Rivers as a viable option of remedial solutions for present and future water shortages in the dry areas. The Libyan experience with its newly constructed Man-Made River (MMR) project is given as an illustrative example for the discussion. Similar transboundary projects are desperately needed for the peaceful growth and prosperity of many other water stressed countries in the arid zone. These projects can contribute to the principles of equity, environmental integrity and economic efficiency which are the basic objectives of development sustainability.

NATURAL AND MAN-MADE RIVERS

Natural rivers and water courses represent the major drainage network responsible for the collection, transfer and redistribution of the more than 40,000 cubic kilometers annual runoff produced by precipitation on the earth's land surface. Their global distribution and hydrological regimes are determined by several geographical and climatic factors beyond man's control and more often, unsuitable to serve his needs. River technology has been developed since ancient times with the specific purpose of maximizing the beneficial uses of available surface water supplies and minimizing any undesirable consequences of their erratic nature. These objectives have been achieved to a considerable extent through the construction of barrages, dams, diversion channels and other hydraulic structures of flow regulation and flood control. With the recent advances in hydraulic engineering sciences, and the continuous demand for more water supplies, river technology has been refined to the degree where it is possible to achieve complete control and redesign of entire river basins according to any hydrological needs dictated by the socioeconomic and environmental concerns of a region. It is even arguable whether natural rivers and water courses are still considered natural any longer. The available technology can be used to divert and reroute natural rivers and streams and even to construct Man-Made Rivers (MMR) in the form of large scale interbasin water transfer systems to redistribute available water supplies from both surface and groundwater sources among water abundance and water scarcity zones according to their actual needs. It is intended here to extend the name of a Man-Made River (MMR) to all water transfer projects of a magnitude comparable to a natural river or a water course, regardless of whether the water supply source is a surface or a groundwater basin.

THE LIBYAN (MMR) PROJECT AS A CASE STUDY
PROJECT CONCEPTION

Libya is located in the heart of the North African arid zone. As indicated in Figure,1, 93% of the country is covered by the North African Sahara desert. Renewable water resources are less than 200 cubic meters per person per year. Most of the population and important economic activities are concentrated in the northwestern and northeastern plains of the Jefara and Benghazi respectively (Figure 1). The strategic coastal areas surrounding the Gulf of Sirte in the middle is sparsely populated and represents a demographic and geopolitical vacuum separating the two most economically and sociopolitically important parts of the country. In addition, the most important oil fields in North Africa lie behind this region and their oil production is exported through the several seaports scattered along the coastal areas of the Gulf. Water shortages in this region impose severe restrictions to its growth and development.

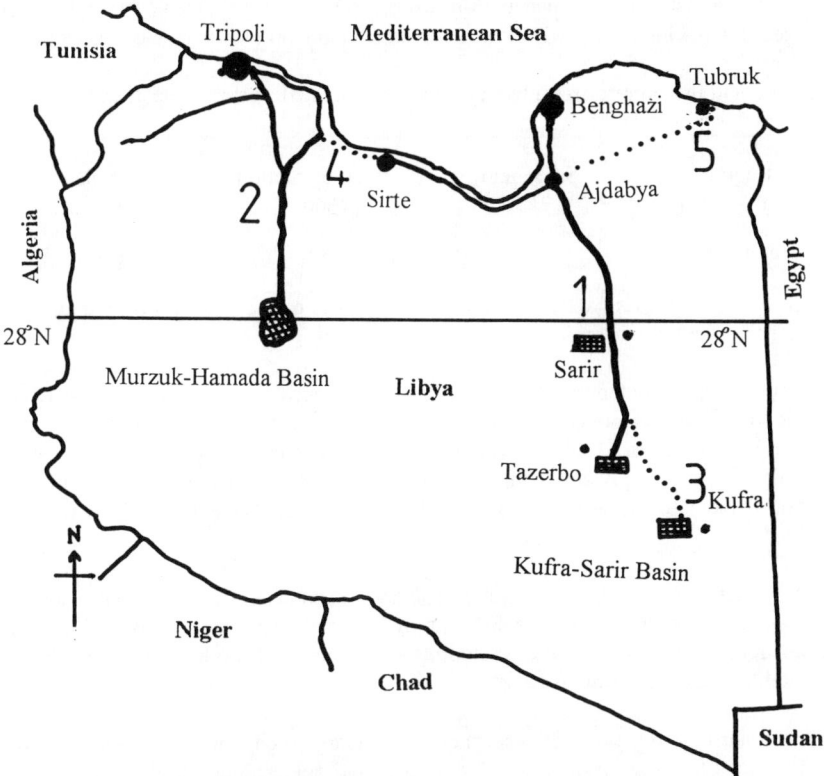

Figure 1. Stages of the Libyan Man-Made River and its feeding groundwater basins.

Deeper in the southern regions there exist two huge regional groundwater basins of enormous storage volume of excellent waer quality (Figure 1). The Kufra-Sarir basin complex lies in the southeasern part opposite to the Gulf of Sirte and the northeastern Benghazi plain. The Murzuk-Hamada basin complex lies in the southwestern part opposite to the northwestern Jefara plain

Increasing water demands of the coastal regions during the past 20 years led to severe pumping and overdraft of the local groundwater aquifers of limited extent and annual recharge. These aquifers have been exposed in several locations to unacceptable levels of piezometric declines and seawater intrusions with disastrous environmental and socioeconomic impacts. To remedy the resulting situation three options were available: expansion in seawater desalting, moving people and related economic activities to the groundwater basins of the southern desert regions or transferring the large amounts of water available in the aquifers of these basins to the water deficit regions in the coastal plains and the Gulf of Sirte. Detailed technical, economic, financial, environmental and sociopolitical investigations of these options revealed that water transfer and redistribution to correct the unbalanced national water resources budget indicated in Table 1 is the most viable option. Thus the idea of the

Table 1. Within country groundwater balance estimated during 1985 (million cubic meters per year MCM/Y)

Region	Safe yield	Annual extraction	Water balance
Above 28°N	745	1600	-855
Below 28°N	4610*	750	3860

* believed to be a fossil water available on the basis of a drawdown of 1 meter per year for 50 years of economic pumping.

huge and complex water transfer and redistribution network of the Man-Made River (MMR) project was conceived. The intention was to develop a comprehensive national network connecting all local networks for different water uses and balancing the national water supply and demand. Before embarking on this project the following issues pertinent to all water transfer projects in general have been discussed and resolved:

1. The present and future water requirements of the southern regions are fully met and safeguarded. In other words the transferred or exported water from the basins in these regions represents a surplus after including all the needs of the local activities in the reasonable foreseeable future.

2. The water requirements of the northern regions to which water is transferred have been reduced to the minimum possible amount through tapping all alternative water resources which are cheaper on the long range than the cost of the transferred water.

In addition, all effective savings in existing uses have been implemented without impairing economic production efficiency.

3. It has not been proven yet that the southern exporting aquifers are hydraulically connected with any recharge water sources in the area. Thus it is considered that the transferred water from these aquifers is not naturally replaced once it is withdrawn. Therefore its development has been undertaken with the full understanding that it will be depleted within a limited period of time depending on the storage volume which can be economically transferred. Within this period the project should generate economic returns sufficient to develop other water resources, such as desalination, to replace the transferred water supplies in the importing regions.

4. It is recognized that a water transfer project, once made, becomes essential to the welfare, if not the existence, of the people it serves. Thus the project must be continued in service or substituted by another source of water. Otherwise, all the economic and social activities based on the project cannot be sustained in the future. The imported water is a new element added to the physical environment of the northern regions. The addition of this water certainly enhances the economic development and population growth in these regions. If the transferred water were to be discontinued due to aquifer exhaustion or any other reason, the human activity in the northern regions would experience catastrophic curtailment unless other alternative supplies are secured.

5. It has been realized that the project will have profound socioeconomic and environmental impacts, both during its construction period and throughout its operating life time. Whatever the economic, social and environmental costs which may result in the southern and northern regions, these costs must be borne by the national or regional authorities once the project is constructed and operating. Such costs should be mitigated when recognized in advance. They must be compensated for by water users through an efficient and effective water pricing system.

PROJECT DESCRIPTION
In view of the previously mentioned demographic and water resources situation the (MMR) project has been planned and designed to be implemented through five consecutive stages as shown in Figure 1.

Stage one is to construct two large scale wellfields in the Sarir and Tazerbo areas of the Kufra-Sarir groundwater basin. Each wellfield has been designed to supply 350 million cubic meters of water per year (MCM/Y) to be transferred through a separate prestressed renforced concrete pipeline of 4 meters diameter. The two pipelines discharge a combined constant flow of 700 MCM/Y in a huge balancing circular reservoir of 4 MCM capacity located near the city of Ajdabya on the Mediterranean coast. The Sarir wellfield pipeline branches westward along the coast towards the city of Sirte. The Tazerbo wellfield pipeline branches eastward towards the city of Benghazi and its surrounding plains. The transferred water by both branches is intended to supply the coastal areas along their routes with fresh water to support their immediate and future urban, industrial and agricultural development. 99% of this stage was completed and it has been operating since 1991.

Stage two is to construct and operate the largest wellfield in the world comprising more than 484 wells spread over an area of 160 x 120 kilometers located in the Murzuk-Hamada groundwater basin. This wellfield has been designed and developed to yield the world's largest ever single unphased abstraction of 900 MCM/Y which will be transferred to the northwestern water deficit regions through two separate 4 meters diameter pipelines similar to those used in stage one. One of these lines will transfer up to 700 MCM/Y to the coastal areas and the Jefara plain to sustain growth and development in this region and to correct the negative environmental impacts of declining water tables and deteriorating water quality. This pipeline has been put under operation since September 1,1996. The other pipeline will transfer 175 MCM/Y to the scattered communities along the northwestern mountain range and hills.

Stage three is to develop a third wellfield in the Kufra-Sarir basin near the Kufra oasis to supply a further 560 MCM/Y through a pipeline connection with stage one.

Stage four is to connect the western Sirte pipeline branch of stage one with the eastern pipeline branch of stage two and pump a further 350 MCM/Y to the northwestern region of the coastal areas and the Jefara plain.

Stage five is to extend the eastern branch of stage one to the city of Tubruk to provide around 200 MCM/Y for the municipal and industrial water uses along its route.

After its completion the project will transfer and redistribute a total of more than 2000 MCM/Y of water through probably the largest and most complicated man-made water distribution network of its kind in the dry areas of North Africa and the Middle East. It is impossible in the limited space of this article to provide a full description of the civil and hydromechanical engineering technologies used in the design, construction and operation of the different stages of this project. It is sufficient to mention that the latest innovations in engineering techniques and material science developed in the west have been employed and tested to the fullest possible extent. In addition to its major utilitarian function as a hydraulic system for water transfer and redistribution the project may be considered in a sense as a testing ground for modern techniques in pipe manufacturing, wellfield construction, corrosion control, communication and control systems and large scale operation and management. the experience to be gained from this project in these areas will certainly be of valuable use in future similar projects.

PROJECT ECONOMICS AND SUSTAINABILITY

Whenever large scale mass water transfers are considered the financial resources available for investment in these projects and the expected cost of the transferred water are of prime concern. It is essential to compare the average unit cost of transferred water with other potentially available alternative supplies. The economics of large scale water transfer projects are relatively very complicated and uncertain especially in developing countries that are dependent on internatinonal markets for the procurement of most of the material equipment and technical expertise.

The economic analysis performed during project conception estimated the average unit cost of transferred water at about 0.25 US Dollars per cubic meter, which is highly competitive with other alternatives such as desalination estimated at 2.5-5.0 US Dollars per cubic meter. Actual economic studies performed after the completion of stage one[2] revealed that the average unit cost of water to the users gate with the cost of capital set at 7% interest is 0.83 US Dollars per cubic meter and the cost to turnout end reservoirs only is 0.55 US Dollars per cubic meter. With free capital (0% interest) the average unit cost of water to the users gate is 0.34 US Dollars per cubic meter and the cost to turnout end reservoir only is 0.21 US Dollars per cubic meter. Several factors contributed to these seemingly high costs. The project has been designed, implemented and supervised by foreign international consulting firms and contractors. Foreign bidders usually ask up to 100% increases in real value of their contracts in Libya. They attribute their inflated bids to hard uncertain working conditions and high risk expectancy. It is possible that the average unit cost of water would not be more than 0.3 US Dollars at worst if the project were implemented in some other parts of the world or by local Libyan firms and expertise. But even with these seemingly high cost, the project is still viable under Libyan conditions where there is no other option except sea water desalination which may cost up to 5.0 US Dollars per cubic meter under the present working conditions in the country. The project provides many other quantifiable and non quantifiable benefits which may be of over-riding importance in terms of national socioeconomic and environmental policy. Such benefits include income redistribution, employment in agriculture and other sectors, water and food security, enhanced skills of the workforce, improved health and environmental benefits.

As to the question of sustainability there is no clear and definite answer yet. The present studies are based on the assumption of 50 years of continuous operation. The water sources aquifers of the Kufra-Sarir basin alone can provide up to 40,000 MCM/Y with a one meter per year lowering of the water table over the whole reservoir extent of more than 500,000 square kilometers [3]. Thus the question of sustainability seems to depend on water production costs and managerial skills rather than on available water supplies which are apparently sustainable for hundreds of years even in the absence of natural recharge of the aquifers. Sustainability can be assured if the transferred water is utilized in such a way as to provide the national economy with the means and strength that enable it to develop alternative water supplies when the (MMR) sources become uneconomical to pump and transfer or are exhausted altogether.

CONCLUDING REMARKS

Water transfer and redistribution through Man-Made Rivers can be both technically feasible and economically viable, especially under severe water stress conditions similar to Libya. The Libyan (MMR) project offers an excellent pioneering example to future potential prospectors in North Africa, the Middle East and other parts of the world.

It may sound like a science fiction to speak about linking continental rivers, bringing water from the ice caps to the arid regions of the globe, implementing the North American Water and Power Alliance (NWAPA) project and diverting the River Congo to irrigate the Sahara desert. The technology is already available. What is needed for extending the MMR concept to serve further areas through the design, financing and construction of a global network of water canals and Man-Made Rivers is the political impetus and socioeconomic rationale that an expected severe future water scarcity can provide. The Martian (Canali) studied by Schiaparelli, 1877 and Lowel, 1894 may turn out to be no more than the extrapolated foreshadowings of how the earth will look like in the future. The California Aqueduct, the Libyan (MMR), the Karakum Canal and other contemplated similar projects may seem to be the modest beginnings of a faster drive towards more gigantic future projects for which the young generations of hydrologists and hydraulic engineers can lay the foundation and pave the way.

REFERENCES

1. Kuffner, U. "Water Transfer and Distribution Schemes, "Water International, Urbana, ILL, USA, Vol.1. 18, No.1 March 1993, pp.30-34.

2. Brown and Root (Overseas) Limited, "Estimate of Cost of Water, "Report No. (F111321A), the Management and Implementation Authority of the GMMR Project, Benghazi, Libya, January 1990, pp. 1-25.

3. Pallas, P., "Water Resources of the Socialist People's Libyan Arab Jamahiriya,"the Geology of Libya. Vol. II, Academic Press, London, England, 1980, pp. 585-592.

INFLUENCE OF OVER-BASIN DIVERSION RESERVOIR ON WATER MANAGEMENT

YOUN-JAN LIN,   CHANG-SHIAN CHEN,   EDWARD S.C. YOUNG

Department of Civil Engineering	Department of Hydraulic Engineering	Engineer, HoShin Engineering
Ming Hsin Institute of Tech. & Com.	Feng Chia University	Constants Corp., Ltd.
Hsin-Chu, Taiwan, ROC	Taichung Taiwan, ROC	Taipei, Taiwan, ROC

ABSTRACT

Data from preliminary study of Chan-Min reservoir is used to investigate the water quality and quantity impacts of this over-basin diversion reservoir. Time series analysis was employed to synthesize possible discharge series. Simulations with various operation policies, based on design flow and shortage index, were performed to analyze the influence of over-basin diversion on water quality and quantity of a river. Results are obtained from four different criteria : of the current operation rules and curves, of altered operation rules and curves, change of datum stage of diversion weir, and maintaining required base-flow. When required base-flow is set at 14.62 cms, the design flow Q_{75} can be maintained as before the construction of the reservoir. However, from the aspect of water resource distribution, this required base-flow induces the greatest impact on the utilization of water resources.

INTRODUCTION

All rivers of Taiwan are short and steep. It is estimated about 75% of the annual rain-fall goes into the ocean. Yet the limited rain-fall is usually distributed unevenly both in time and space. The construction of over-basin diversion reservoir, built to diverse the ample water resource from other basin for storage, can help. The purpose of this research explores what impacts the operation rules, established before the diversion, have over the water quality and quantity after operations. Synthetic flows generated from stochastic hydrologic models are used with various evaluation schemes to analyze the water quantity and quality impacts from over-basin diversion reservoir.

METHODOLOGY

The basic theory and methodology used includes :
1. SYNTHETIC INFLOW   ARMA model was used for hydrological time series analysis.

2. OPERATION STRATEGY FOR OVER-BASIN RESERVOIR:

(1) Adjusting Operation Rules: Most reservoir operations are based on the upper limit curve, lower limit curve, and inactive pool level. The usual methodology consists of having the existing operation rules either raised or lowered, ordinarily with 5%, 10%, and 20% ratios. The raise would results in higher water elevation in the reservoir, while the lowering decreases the water elevation.

(2) Altering Datum Stage of Diversion:. Maintaining the reservoir operation rules, raising the datum stage of diversion lowers the water intake into the reservoir but the flow downstream would increase.

(3) Maintaining Channel Base-flow:. If the river flow is higher than the based-flow, then the forced based-flow is fully released. If, however, the flow is below force base-flow, the river flow is completely released.

3. IMPACT INDICES

We use the shortage indices to indicate the efficient usage of water quantity and Q_{75} to express the influence on water quality.

(1) Water Quantity Impact Indices (SI_1, SI_2) Shortage indices are used to indicate the utilization and distribution of water resources. According to US Army Corps of Engineers, water shortage can be defined by Shortage Index (SI_1):

$$SI_1 = \frac{100}{N} \sum (\frac{AS}{AD})^2$$

where AS stands for annual shortage, AD is the annual demand, and N is the years analyzed. SI_1 is suitable for stable discharge water resource system, but lacks the ability to show the seriousness of shortage.

To amend for the above, another shortage index (SI_2), which is better suited for planning and actual operation and includes the characteristics of both SI_1 and Water deficit index, widely used in Japan. It is expressed:

$$SI_2 = \frac{100}{N} \sum (\frac{\%-day}{100*365})^2$$

where %-day is the shortage index; 100*365 represents 100% supply for 365 days annually, which corresponds to annual target supply; and N is the years of computation.

(2) Water Quality Impact Index (Q_{75}) According to the 8th ordinate of details of execution under Water Pollution Prevention Law of ROC, the water quality for all rivers uses 75% of the historical annual flow as the design flow. Therefore, when regarding to water quality, the Q_{75} represents the water quality of simulation results.

EXAMPLE SIMULATIONS

Data was originated from preliminary study of Chan-Min reservoir, which is an off-channel reservoir with Shuang-Shi-Teuei diversion weir as its inlet. It is planned for public water use for suburb of Tai-Chung City. The watershed area is 188.39m^2, with effective storage of 79,104,000m^3. The diversion weir has a direct annual runoff of 1,701,110 K.Ton with the maximum intake of 40 cms. Historical data include both 10-day (the traditional reservoir operation period in Taiwan) flow data at the dam site and 10-day flow data at the site of diversion weir for 31 years.

1. MODEL DESCRIPTION : From historical data, flows at the dam site is far less than that at the weir site. So the inflows from the weir constitute the major portion of inflow at the dam. This study uses operation tables from the TPWCB for verification. This operation is a simple method to satisfy the public water demand, no particular rule curves or operation rules followed. Records from 1959 to 1989 are chosen for the simulation. The reservoir 10-day operation formula is :

$$CMS(I+1) = CMI(I) + [SOT(1) - SOT(I)*0.1] -$$
$$EV(I) + CMS(I) - CMOP(I) - DS(I)$$

where $CMS(I+1)$ = final storage, $CMS(I+1)$ = initial storage,
$CMI(I)$ = inflow , $SOT(I)$ = water intake from diversion weir,
$SOT(I)*0.1$ = intake loss, $EV(I)$: Evapotransporation
$CMOP(I)$: public water demand , $DS(I)$: direct runofff

The shortage formula is : $PL(I) = CMTP(I) - CMOP(I)$
where $PL(I)$ is the shortage for public demand
$CMTP(I)$ is the planned water supply for public demand
$CMOP(I)$ is the actual supply for public demand.

The worse and the second worse drought occurring years are used to verify our computation.

2. FLOW SYNTHETICS AND ANALYSIS : The 10-day flow records of Shuang-Shi-Teuei diversion weir were synthesized using the SCA package and ARMA (1,3) series into a set of 100-year flows. Through consistency and statistical characteristics examination (not shown here), it is concluded that the synthetic flow are acceptable.

3. SIMULATION OF OVER-BASIN RESERVOIR : The planned water usage of 750 K.Ton/day at the year 2021 is used in combination with previously described reservoir operations to simulate the 100-year synthetic flow operation. The results are used as the basis of the impact on water quality and quantity of over-basin reservoir.

   (1) Impact on Discharge Accumulative Curve by Changing the Reservoir Operation Rule Curve With changing operation rules, we used the rule curves in wet season, then raised and lowered rule curves in dry seasons by 5%, 10%, 15%, 20% respectively to obtain eight different rule curves. The effects were insignificant. Under the condition of fixed rule curves, our approaches are : (A) when the storage is above the upper limit of rule curve, the water supply is 100%; (B) when the

storage is between the upper and lower limits, the water supplies are at 90% and 70%; (C) when the storage is below the lower limit of rule curve, the water supplies are at 70% and 50% (the original is at 60%). The simulation results are similar to those with altered rule curves. Therefore, the alteration of rule curves have little effect on discharge.

Regarding to the change of datum stage of diversion, the diversion weir datum stage, which was regulated by TPWCB, is taken as major reference. We reserved the diversion datum stage during the rainy season (Jun. to Sep.) and changed that at drought season (Jan. to Mar., Oct. to Dec.) into a fixed value between 8.54 to 23.35 cms. Without changing the rule curves, the reservoir operation is simulated. It shows that altering diversion datum stage of the weir would cause partial increase on flow, which would results in a better reservoir operation. But since the diversion weir does not have enough ability to adjust the water resources, the result shows great influence caused by the seasons and they can not maintain the Q_{75} design flow before the reservoir is built.

For maintaining channel base-flow, without altering the rule curves and diversion weir datum stage, the reservoir is forced to release fixed base-flow downstream. Maintenance of base-flow is within the range of 0 to 20.62 cms. Results are shown in Figure 1. It clearly indicates the storage capacity of the reservoir by storing the flows in the rainy seasons and releases at drought seasons. Not only is the present reservoir operation improved drastic, the original river discharges as well. When base-flow is within the range of 16.62 to 20.62 cms, all the operations clearly show that Q_{75} are better than that before the dam construction. This is a clear evidence that maintaining base-flow operation provides effectiveness for the water quality downstream.

(2) Impact of Resources Distribution by Altering Reservoir Operation Strategies.   It is found that both maintaining base-flow and altering datum stage of diversion weir would clearly increase shortage index (as shown in Figure 2, 3, 4). In order to satisfy the water quality standard downstream, there must be some deficit at public water demands.

From Figure 2 and 3, regarding base flow maintenance, the SI_1 raised from 1.56 to 13.87, which could be seen as a raise of average shortage rate (ASR) from 1.25 to 3.72. SI_2 is raised from 0.14 to 1.45, with the ratio of product of shortage rate and time of continuous shortage to the state of non-shortage within a year (ASRTNR) average increase from 0.38 to 1.20. Regarding to diversion weir datum stage changes, SI_1 is raised from 0.87 to 5.07, ASR from 0.93 to 2.25. SI_2 is raised from 0.08 to 0.49, ASRTNR from 0.274 to 0.701. Though forced base flow could increase the flows downstream and improve the water quality, the impacts on public water usage are serious.

If combining the effects of forced base flow and datum stage change on Q_{50}, Q_{75}, and Q_{90}, it is evident that both would have clear improvement on Q_{50} and Q_{75}, while Q_{90} shows improvements before 14.62 cms and declines after that. This is because at Q_{90} the reservoir storage could not satisfy the downstream base flow demand during the drought season. Therefore, during the drought seasons, the storage capacity of the reservoir is not capable to meet the downstream base flow demands. With insufficient release, flow could not be maintained at a fix values and starts to decline.

From above, to improve the water quality of natural river, the adjusting effect of maintaining base-flow is better than that of datum stage. But in term of water shortage, the side effect of base-flows maintenance is worse than that of datum stage change. Furthermore, of the above three methods, only the operations of forced base-flow can achieve water quality standards downstream. From Figure 1, when the base-flow is greater than 14.62 cms, Q_{75} of operation would come close to the Q_{75} of original natural river. Otherwise, the water quality of the river will be worse than that before the construction of the dam. When the base-flow is 14.62 cms, $SI_1 = 6.96$ (ASR of 2.64), and $SI_2 = 0.74$ (ASRTNR of 0.86).

CONCLUSTIONS AND SUGGESTIONS

Since the long term synthetic flow series deducted from stochastic hydrological model incorporates the various possibilities of the system, the results of the implementation of synthetic flow series on the impact of water quality and quantity for over-basin reservoir are very valuable.

1. Three approaches are used to evaluate the impact on water quality and quantity for over-basin reservoir. They are alteration of the rule curves, changing the diversion weir datum stage, and maintaining base-flow. Only maintaining base-flow could maintain the water quality to Q_{75} of the natural river.

2. To maintain the Q_{75} of the natural river, the forced base-flow from the reservoir must be 14.62 cms. But the SI_1 would rise from 1.56 of the original rule curve operation to 6.96 (ASR from 1.25 to 2.64 within the next 100 year), SI_2 would rise from the original 0.14 to 0.74 (SRTNR from 0.38 to 0.86 within the next 100 year).

3. Under the competition between resource utilization and environmental protection, concepts from Figure 3 to 4 could be used to implement benefit analysis to find the point of balance.

4. For different reservoir operation strategies, the discharge accumulative curves display different impact, as shown in Figure 1. Further study of their impacts on water quality and environment would be beneficial to the water quality control strategy.

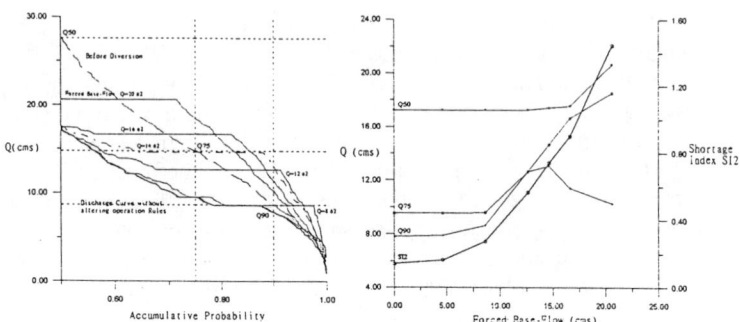

Figure 1. Accumulative Probability of discharge   Figure 2. Design Flow, Forced Base-Flow and SI_2

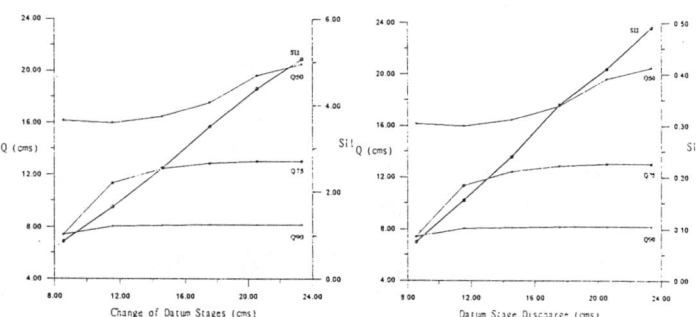

Figure 3. Design Flow, Weir Discharge Change, and SI_1

Figure 4. Design Flow, Weir Discharge Change, and SI_2

REFERENCE

1. Thomas, H. A., and M. B. Fiering, (1962) "Mathematical Synthesis of Stream Flow sequences for the analysis of river basin by simulation", in Design of Water Resource System, edited by A. Maasset al., Harvard University Press, Cambridge, Mass.
2. Maass, A., M. M. Hufschmidt, R. Dorfman, H. A. Thomas, Jr., S.A. Margin, and G.M. Fair, (1962) "Design of Water-Resource System", Harvard University Press, Cambridge, Mass.
3. Rowe, W.D., and Y.Y. Haimes, (1981) "Methodology and Myth", Risk/Benefit Analysis in Water Resource Planning and Management. Plenum Press.

FLOOD RISK ANALYSIS IN THE UPPER ADRIATIC SEA DUE TO SEA LEVEL RISE AND LAND SUBSIDENCE

G. GAMBOLATI[1], P. TEATINI[1], L. TOMASI[1],
M. GONELLA[2], C.S. YU[3], C. DECOUTTERE[3]
[1] Department of Mathematical Models for Applied Sciences
University of Padova - Via Belzoni 7 - 35131 Padova, ITALY
[2] MED Ingegneria S.r.l. - Via della Paglia 35 - 44100 Ferrara, ITALY
[3] Laboratory of Hydraulics - Catholic University of Leuven
De Croylaan 2 - B-3001 Heverlee, BELGIUM

ABSTRACT

The low lying coastal areas in the Upper Adriatic Sea that may be potentially flooded have been detected by an integrated modeling approach that predicts the mean sea level rise due to global climate change and storm events and the natural and anthropic land subsidence. A G.I.S. has been applied to manage the results from the various models and to perform a risk analysis of inundation of those areas in the next century.

INTRODUCTION

The shoreline of the Upper Adriatic Sea comprised between the cities of Cattolica and Monfalcone (Italy) is characterized by locations (such as Venice and the Romagna Riviera) of great tourist interest and areas (such as the northernmost lagoons and the river Po delta) with a precarious hydrogeological setting (Figure 1).

The risk analysis of the potentially flooded areas in the Upper Adriatic and the study of the expected coastline evolution in the next century represent the final outcome of the CENAS project ("Study on the coastline evolution of the eastern Po plain due to sea level change caused by climate variation and to natural and anthropic subsidence") funded by EU Environment Research Programme 1990-1994.

The study has been performed by the use of an integrated approach accounting for the variations of both the ground and the sea level in the next 100 years. Land may subside due to sediment natural compaction and subsurface fluid (water and gas) withdrawal. The mean sea level may rise

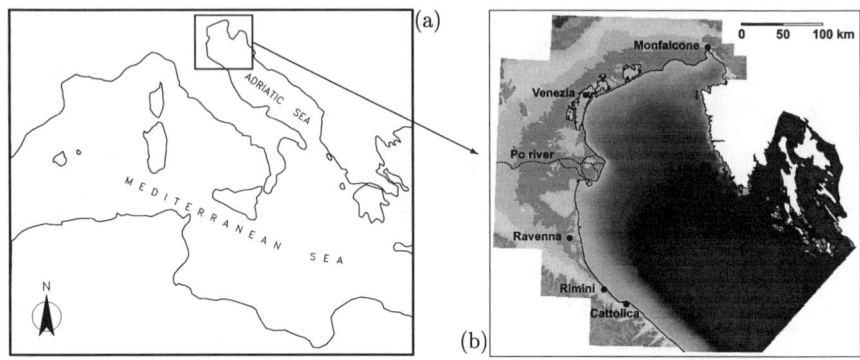

Figure 1: (a) Adriatic Sea and (b) D.E.M. of the study area.

due to global climate change, tide, storm surge and wave set-up of particularly intensive meteo-marine events. These different vertical effects have been predicted by various models, and handled with a G.I.S. (Geographical Information System) tool.

MODELING ANALYSES
The prediction of each individual event has been obtained with appropriate mathematical numerical simulations.

LAND SUBSIDENCE MODELS
The prediction of natural land subsidence, assessed in 0.5 mm/y at Venice, 2÷2.5 mm/y in the Ravenna area and 4÷5 mm/y in the Po river delta (Figure 2), has been obtained with 1-D non-linear finite element models of soil compaction driven by unsteady groundwater flow in the accreting sedimentary basin underlying the Upper Adriatic Sea during the last 10^6 years[1, 2].

Land settlement due to water withdrawal from the multiaquifer system underlying the Romagna region (extending south of the Po river delta) has been predicted by coupling a 3-D finite difference flow model with a 1-D finite element consolidation model. Both models have been calibrated using the piezometric decline and subsidence records of the last 50 years (of the order of some tens of meters and centimeters, respectively), and have been applied with some realistic programme of water pumping. In the most critical areas the prediction indicates an anthropic land settlement of the order of 1 m.

FLOOD RISK ANALYSIS

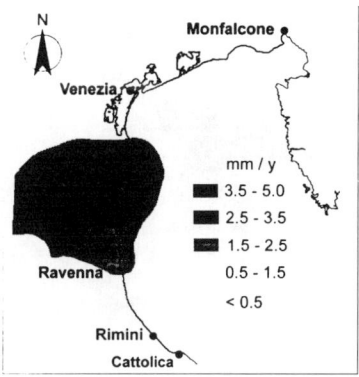

Figure 2: Present and projected natural land subsidence in the eastern part of the Po river plain.

Land subsidence caused by gas production has been estimated by a 3-D non-linear poro-elastic finite element model. The model has been applied to the Angela-Angelina gas field, located offshore in front of Ravenna at a depth between 3000÷4000 m, and has predicted a final settlement of about 15 cm of the coastal area which lies above the reservoir. The modeling approach is quite similar to the one used to simulate the land subsidence over the major Dosso degli Angeli gas field located north of Ravenna[3, 4].

HYDRODYNAMIC AND WAVE MODELS

The forecast of the mean sea level rise due to storm effects has been made by taking into account tides, storm surges and wave set-up.

The Adriatic Sea model to simulate tides and storm surges has been built from a 2-D depth averaged model[5], with a resolution of 6×6 km. The model is mainly driven by the boundary forcing at the southern opening made from seven tidal components and by the atmospheric forcing over the whole basin. Storm surges are simulated by taking into account the effects of both the atmospheric pressure gradients and the winds blowing over the sea surface. Surface winds have been generated using a Duun-Christensen[6] type formula with the coefficients calibrated from the experience. For the air-sea drag, a relationship suggested by Wu[7] has been used.

The wave climate during an extreme event has been calculated from wind data in the Adriatic Sea using the WAM model on a 12×12 km grid[8]. The WAM model is a third generation wave model established by the WAM group in 1988, and is based on the principle that a sea state is a superposition of a large number of sinusoidal components each having a different frequency and running in a different direction. The wave spectrum is found by solving the energy balance equation. The wind acts as a source term that generates the waves. Other source terms are the non linear wave-wave interaction and the dissipation due to bottom friction. Shoaling and depth refraction have been taken into account, while current refraction has not been included since currents in the Adriatic Sea are known to be generally

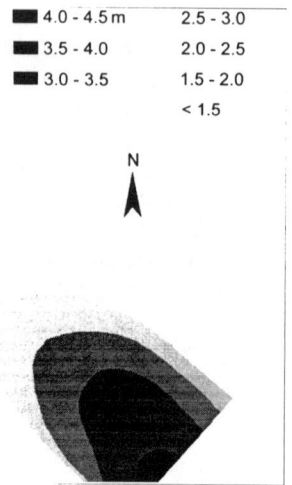 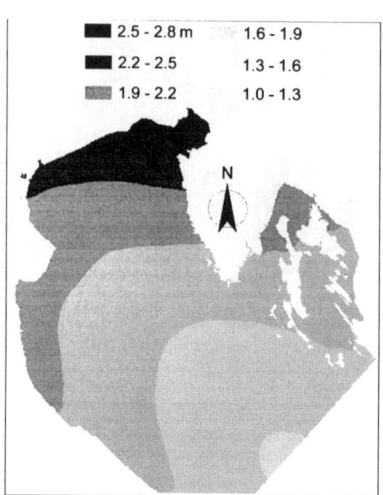

Figure 3: Output (significant wave height H_s) of the WAM model applied to the Upper Adriatic Sea (January 31, 1986).

Figure 4: Tide-surge levels for a storm with a 100 year return period simulated by the hydrodynamic model.

small. Finally, the wave set-up has been determined by using the significant wave height H_s, the wave period and the directional spreading factor of the radiation stress.

The hydrodynamic and wave models have been calibrated over seven selected historical storms occurred in the period 1986-1992. As an example, Figure 3 gives the results from the WAM model for the storm of January 31, 1986. The forcing factors of the most severe event, i.e. the one that has produced the highest surge level along the Ravenna coast, and is characterized as a "scirocco" type storm with the wind blowing from the south - south-east direction at a moderate speed, a long fetch and a duration of several days, have been increased so as to obtain expected extreme levels with 1, 10 and 100 year return periods. These levels turn out to be equal to 1.5, 2.0 and 2.5 m, respectively, along the Ravenna shoreline. The map of the tide-surge levels with a 100 year return period is provided in Figure 4.

FLOODED AREAS AND RISK ANALYSIS

The outcome of the numerical analyses and the construction of ad hoc maps have been managed with the G.I.S. known as GRASS developed by the USACERL. In particular GRASS has been quite effective in detecting the potentially flooded coastal lowlands and in producing risk maps with a

FLOOD RISK ANALYSIS 625

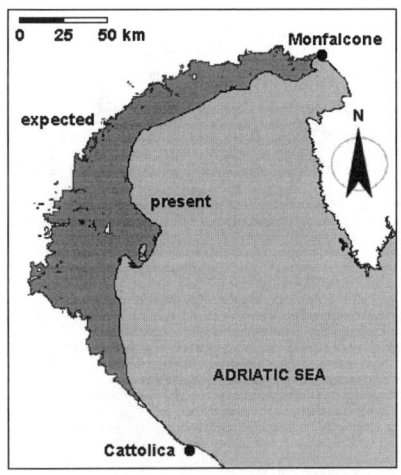
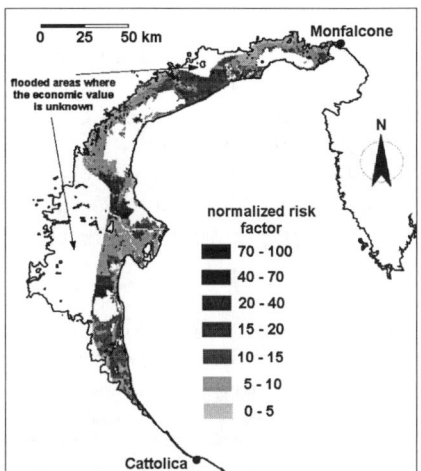

Figure 5: Potentially flooded lowland in the year 2100 with a return period of 100 years.

Figure 6: Risk factor related to the flooded coastal area shown in Figure 5.

space resolution of 200 m. A factor risk R has been defined and associated with each potentially flooded area. R depends primarily on the product between three factors, the hazard, the vulnerability and the economic value.

The hazard H and the vulnerability V represent the probability for the flood to occur with a return period T and the percentage of damage caused to the area by the inundation, respectively. The vulnerability maps are, on a first approximation, the maps of the water elevation on the flooded areas, and have been obtained for the selected return periods by using both the ground and the mean sea level maps. By the D.E.M. (Digital Elevation Model) of the Italian National Geological Survey (with a 1 m vertical resolution on a regular grid of 10" longitude times 7.5" latitude), as improved with data from CTR (Technical Regional Maps) of the Emilia-Romagna region, the ground elevation maps up to the year 2100 have been built by decreasing the present ground height by the amount of natural and anthropic land subsidence predicted by the corresponding models. The maps of the mean sea level rise have been constructed by combining the simulated storm effects with the eustatic rise caused by climate variation[9] (estimated by IPCC92 at 40 cm for the next 100 years). Finally, the economic value E map has been obtained using GRASS and the map of land use provided by the Italian Environmental Ministry, A.R.S. Service. Figure 5 shows the lowland which is potentially flooded with a return period of 100 years, while Figure 6 provides the map of the related risk factor.

CONCLUSION

An integrated modeling approach (consisting of groundwater flow model, compaction model, poro-elastic model, tidal-storm surge model and wave model) coupled with a G.I.S. has been applied to and used for the evaluation of the potentially flooded lowlands along the Upper Adriatic Sea shoreline and the analysis of the risk connected to the inundation in the next century. The results show that the procedure discussed above represents a very powerful tool for the control and effective management of low lying coastal areas.

Acknowledgment This research was supported by the EC Environment Research Programme (contract: EV5V-CT94-0498, Climatology and Natural Hazards).

REFERENCES

(1) G. Gambolati, Equation for one-dimensional vertical flow of groundwater. 1. The rigorous theory, *Water Resour. Res.*, **9** (3), 721-733, 1973.

(2) G. Gambolati, G. Giunta and P. Teatini, Numerical simulations of non-linear groundwater flow in accreting sedimentary basins, in *XI Conference on Computational Methods in Water Resources*, edited by A. Aldama et al., 23-32, Comp. Mech. Publications, Southampton UK, 1996.

(3) G. Gambolati, M. Putti e P. Teatini, Land subsidence, in *Hydrology of Disasters*, edited by V.P. Singh, Chapter 9, 231-268, Kluwer Academic, Dordrecht (The Netherlands), 1996.

(4) G. Gambolati, P. Teatini and W. Bertoni, Numerical prediction of land subsidence over Dosso degli Angeli gas field, Ravenna, Italy, in *Current Research and Case Studies of Land Subsidence*, Proceedings of the Joseph F. Poland Symposium, edited by J. Borchers and C.D. Elifrits, Star Publ. Co., Belmont (California), 1997. In press.

(5) C.S. Yu, M. Fettweis, J. Monbaliu and J. Berlamont, Storm surge simulations in the Adriatic Sea, Scientific report, EC/Environment CENAS, K.U.Leuven, Belgium, 1995.

(6) J.T. Duun-Christensen, The representation of the surface pressure field in a two dimensional hydrodynamic numeric model for the North Sea, the Skagerrak and the Kattegat, *Dt. Hydrogr. Z.*, **28**, 97-116, 1975.

(7) J. Wu, Wind stress coefficient over sea surface from breeze to hurricane, *J. Geophys. Res.*, **87** (C12), 9704-9706, 1982.

(8) C. Decouttere, C.S. Yu and J. Berlamont, Storm wave simulation in the Adriatic Sea, Scientific report, EC/Environment CENAS, K.U.Leuven, Belgium, 1996. In preparation.

(9) S.C.B. Rafer, T.M.L. Wigley and R.A. Warrik, Global sea-level rise: past and future, in *Sea Level Rise and Coastal Subsidence*, J.D. Millinan and B.U. Haq (eds). Kluwer Academic, 11-45, 1996.

SIMULATION OF FLOODS IN PING RIVER USING IIS DISTRIBUTED HYDROLOGICAL MODEL (IISDHM)

R. JHA, S. HERATH and K. MUSIAKE
IIS, University of Tokyo, 7-22-1, Roppongi, Minato-ku, Tokyo-106, Japan

ABSTRACT

Newly developed IIS distributed hydrological model (IISDHM) is used to simulate floods. Major three components of the model, unsaturated flow, overland flow and river flow are modeled by using three dimensional Richard's equation, two dimensional diffusive form of St. Venant equation and one dimensional full dynamic form of St. Venant equation for river network respectively. The model is applied to Ping river basin, a tributary of Chao Phraya river basin having a catchment area of 6300 sq. Km and hourly simulation is performed for a period of four months (Jul. to Oct., 1992). The simulated peaks and observed peaks are found to match well. GIS of the catchment consisting of topography, soil distribution, landuse and river network has been developed at 1 km grid size and an interactive software is developed to input the data.

1. INTRODUCTION

Thailand is a tropical and agriculture county. It is difficult to simulate the flood in tropical countries without considering soil moisture condition. The same amount of rainfall generates less runoff in the beginning of the rainy season than late rainy season. The water balance at P.1 station for 1992 from July to November have shown in Fig. 1. The Fig. shows that losses in July is 96% where as the losses in October and Nov. are reduced to 86% and 56% respectively only. The main crop is rice during rainy season and plantation time is July. Farmers prepare their lands (puddling) for rice plantation and Puddling needs more than 100mm of water.

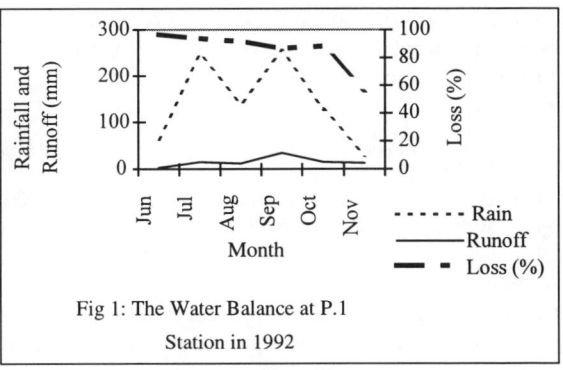

Fig 1: The Water Balance at P.1 Station in 1992

After the transplantation, rice needs a certain depth of ponding until some days before harvesting. In the tropical countries the evapotranspiration is also quite high. The annual potential evapotranspiration in Chaing Mai region of Thailand is more than 1611mm, however, the average rainfall is 1250 mm per year only. A considerable amount of water is loosed by evaporation. A good simulation cannot be done without considering all the factor mentioned above. To take into account of all these factors, a physically based distributed model has been used to simulate the flood in Ping river basin considering the full dynamic routing for channel networks.

2. MODEL DESCRIPTION

The dynamic modeling of one dimensional unsteady flow in open channels is usually based on the numerical solution of the well known St. Venant equation (1871). Diffusion version of St. Venant equation have been used by Moussa and Bocqullion (1996), and Akan and Yen (1981), however, Fread (1985) and Amein and Cunge (1970) have developed the algorithms to use full dynamic equation to simulate the floods. Some researchers have proposed extremely simplified model to predict the flood such as, storage routing (Goodrich, 1931), Muskingum routing (McCarthy, 1938), Kainin and Miljukov routing (Fread, 1985).

The major process performed in the model for each time step are described as;
(a) when a grid receives the rainfall, it is added to the previous storage of the grid. The major losses such as infiltration is calculated on the basis of hydraulic conductivity of the soil and soil moisture status of the top layer, and evapotranspiration is modeled by Kristensen and Jensen model (Kristensen and Jensen, 1975). A water balance of the grid done and excessive rainfall is calculated. This process is done for each grid in the catchment. (b) Excessive rainfall in all grids are routed as overland flow. (c) The river grids receive water from overland routing and river routing is done to get the river discharge and depth at different points of the river. The main components are described below:

2.1 SUBSURFACE FLOW

Soil moisture distribution in the unsaturated subsurface zone is calculated by solving three dimensional Richard's equation. The X and Y components of the Richard's equation are solved explicitly for previous time step. The governing equation is written as,

$$C(\psi)\frac{\partial \psi}{\partial t} = \frac{\partial}{\partial z}[k(\psi)\frac{\partial \psi}{\partial z} + k(\psi)] + \frac{\partial}{\partial x}[k(\psi)\frac{\partial \psi}{\partial x}] + \frac{\partial}{\partial y}[k(\psi)\frac{\partial \psi}{\partial y}] - S \quad (2.1.1)$$

Where $C(\psi)$ is specific moisture capacity function; $K(\psi)$ is unsaturated hydraulic conductivity and S is source or sink term.

2.2 OVERLAND FLOW

Overland flow may generate either, when the rainfall rate exceeds the infiltration rate and ponding occur or when ground water level rises upto the ground surface. Overland flow is simulated in each grid square by solving the two dimensional diffusive wave approximation of St. Venant equation, which is written as:

SIMULATION OF FLOODS

$$\frac{\partial h}{\partial t} + \frac{\partial (Uh)}{\partial x} + \frac{\partial (Vh)}{\partial y} = q \quad (2.2.1)$$

$$\frac{\partial h}{\partial x} = S_{ox} - S_{fx} \quad \text{in } X \text{ direction, and} \quad \frac{\partial h}{\partial x} = S_{oy} - S_{fy} \quad \text{in } Y \text{ direction} \quad (2.2.2)$$

Where h is water depth; U and V are velocities in X and Y directions respectively; q = excessive rainfall; S_{ox} and S_{oy} are slopes in X and Y directions; S_{fx} and S_{fy} are energy slopes in X and Y directions. Applying Manning equation, i.e.

$$U = \frac{1}{N_x} h^{0.667} S_{fx}^{0.5} \quad \text{and} \quad V = \frac{1}{N_y} h^{0.667} S_{fy}^{0.5} \quad (2.2.3)$$

where N_x and N_y are roughness coefficient in X and Y directions.
When we put Equation (2.2.3) in Equation (2.2.1), the Equation (2.2.1) becomes non linear. It is difficult to find a stable solution for set of two dimensional non linear equations. To simplify the problem, we make following simplifications:

The Equation (2.2.1) can be written as; $\frac{\partial h}{\partial t} + U\frac{\partial h}{\partial x} + V\frac{\partial h}{\partial y} = q \quad (2.2.4)$

Where U and V are calculated in the previous time step. The implicit finite difference form equations are written for all nodes and solved by successive over relaxation method (SOR) using Chebysher acceleration.

2.3 RIVER ROUTING

The dynamic form of St. Venant equation are used to simulate the river network. The continuity equation is as follows;

$$\frac{\partial A}{\partial t} + \frac{\partial Q}{\partial x} = q_L \quad (2.3.1)$$

Where q_L is lateral flow per unit length and per unit time.
The dynamic form of momentum equation is written as:

$$\frac{\partial Q}{\partial t} + \frac{\partial}{\partial x}\left(\frac{Q^2}{A}\right) + gA\left(\frac{\partial h}{\partial x} + S_f\right) = 0 \quad (2.3.2)$$

Where Q is discharge; A is area of cross section, and S_f is water level gradient in X direction. T is time step, h is elevation of water level.
The Weighted four point finite difference approximation (Chow et al. 1988) of Eq. 2.3.1 and 2.3.2 are written for each point of river network. This produce gives (2N-2) non-linear equations. Where N is number of points in river network. For each branch there is two boundary conditions (upstream and down stream). If the down stream is connected to junction, then d/s boundary condition is junction condition. At each junction the number of junction conditions are same as number of channels joining at that junction, and they are (1) the water elevations are same and (2) the mass balance at the junction. Now there are 2N nonlinear equations and 2N unknowns are solved for each time step Newton Raphon method (Fread, 1985).

2.4 DATA PREPARATION AND INPUT

In this model, the preprocessing and post-processing of the data are done in ARC/INFO, a GIS software. The model can directly read the output from GRID sub module of ARC/INFO, such as topography, land use soil type, rainfall code, meteorological station code, catchment boundary, etc. The time series data such as rainfall, potential evaporation are prepared in any spread sheet software. An interactive software has been developed to read the data input directly from the screen and write in a particular file and format, which will be the input to the model.

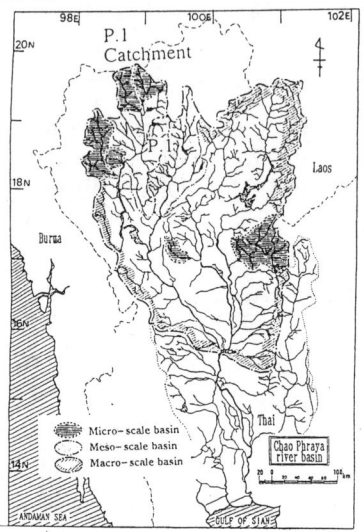

Figure 2 : Location map

3. MODEL APPLICATION

The above described model was applied to Ping river basin in Thailand. Ping river basin is a sub-catchment of the Chao Phraya river basin, which is the largest river basin in Thailand. Ping catchment is located at north-west side of Thailand (Fig. 2), between latitude $18^0 45\,'N$ to $19^0 45\,'N$ and longitude $99^0 45\,'E$ to $100^0 15\,'E$ with a catchment area of 6300 sq. km. The DEM was down loaded from United State Geological Survey (USGS) ftp site (http: //edcwww.cr.usgs.gov /landdaac/ 30asdcwdem/ 30asdcw dem.html). The resolution of the DEM was 30 seconds, , which is equal to 922 m in the study area. It was resampled to 1000 m and used to delineate the catchment boundary and to generate rivers using ARC/INFO (Fig. 3). The comparison between generated rivers and rivers from Digital Chart of the World (DCW) is shown in Figure 3. The

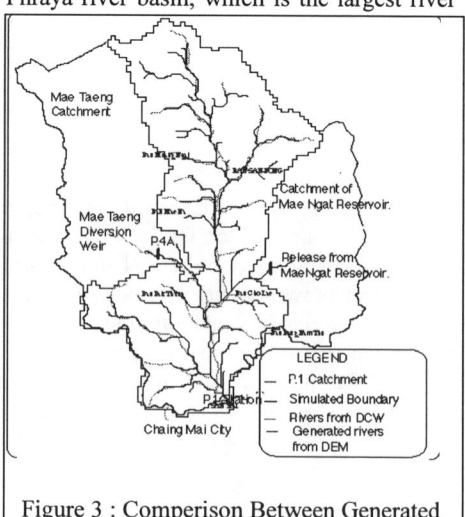

Figure 3 : Comperison Between Generated Rivers and DCW Rivers

comparison shows good agreement between two river. A network of 17 channels are constructed for river routing. Daily rainfall of 12 rain gauge stations are available, out of the 12 stations, hourly rainfall data of 3 stations are available. on the basis of hourly rainfall pattern, the daily rainfall is distributed to hourly rainfall data and used in the computation. Daily pan evaporation of one station is known which is used as potential evapotranspiration of whole catchment. Mae Ngat reservoir, which has a catchment area of 1281 sq. km. and Mae Taeng irrigation weir (catchment area 1903 sq. km.) are located north-eastern and north-west part of the catchment respectively (shown in Figure 3). The daily release from Mae Ngat reservoir and Mae Taeng weir are used as boundary conditions of the upstream of the these two rivers. The actual simulated area is 3116 sq. km.

4. RESULTS AND DISCUSSION

On the basis of experience and trial simulation, the surface storage and leaf area index for different crops in different months have been decided. A comparison of hydrographs by dynamic and observed at Chaing Mai (P.1) gauging station (shown in location map, Fig. 2) is presented in Fig. 4. from Jul. 1992 to Oct. 1992. The figure shows that the peaks of the simulated and observed hydrographs match reasonably well. The water balance during four months has been presented in Figure 5. The surface runoff is 10% only, whereas, the actual evapotranspiraion is 35% and other losses such as increase in soil moisture, surface storage and ground water replenishment is 54% during this period.

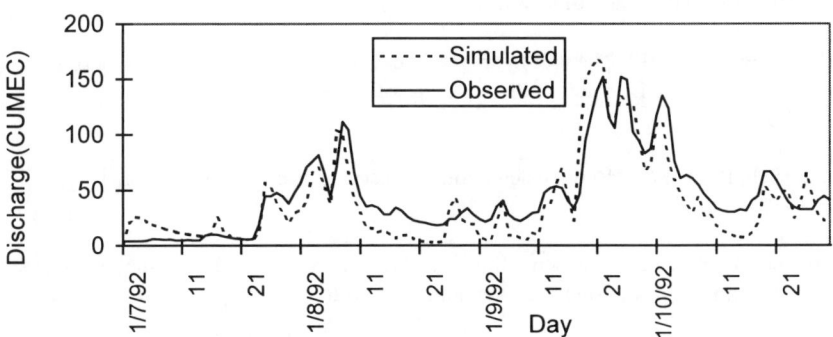

Figure 4: The Comparison between Simulated and observed Hydrograph at P.1 Station

The data requirements for IISDHM are substantial. The experience in Thailand indicates that the main part of data already exists with different Government and non-government organization such as departments of irrigation, meteorological, mines and universities. The total computational time taken for simulating 3116 sq. km. catchment at 1 km mesh and 1 hour time resolution for four months was about 7.0 hours in a DEC-ALPHA vt-300 computer running Windows NT. The computational power available currently shows that application of distributed catchment model to forecast the flood is possible operationally. The IISDHM has been successfully applied in the Ping river basin, Thailand. However, to check the universal applicability of the model, this model should be applied in different countries.

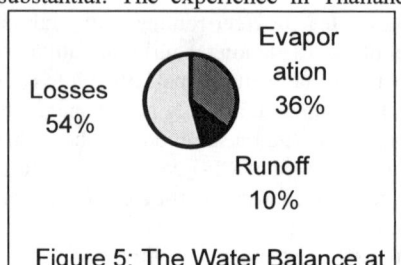

Figure 5: The Water Balance at P.1 Station

REFERENCES

Amein, M. And Fang, C.S. 1970, Implicit flood routing in Natural Channels, J. Hyd Div., ASCE, Vol.96, no. HY12, pp 2481-2500, 1970

Chow, V.T., Maidment, D.R. and Larry, W. Mays, 1988, Applied Hydrology, McGraw-Hull International Edition.

Fread , D. L., 1885, Channel Routing in M. G. Aderson and T.P. Burt (Eds), Hydrological Forecasting, Wiley, New York

McCarthy, G.T. 1938, The unit hydrograph and flood routing in Conference of the North Atlantic Division of the US corps of Engineers, New London, Connecticut

Goodrich, R. D. 1931. Rapid calculation of reservoir discharge. Civ. Eng, 1, pp.417-418

Kristensen, K.J. and Jensen, S.E. (1975), A Model for Estimating Actual Evaporation from Potential Evaporation, Nordic Hydrology, Vol. 6. pp. 70-88.

Regional Model of Glaciers Runoff and its Application for Hydrological Forecasts

V.G.KONOVALOV
(Central Asian Research Hydrometeorological Institute,
Observatorskaya Str. 72, 700052 Tashkent, Uzbekistan Fax: 7-3712-33-20-25)

INTRODUCTION

The report considers theoretical and applied aspects of regional modelling, computation and forecasting of annual and long-term regime of total melting and glaciers runoff in Central Asian conditions. The integrity of N_{gl} glaciers ($N_{gl} \geq 2$) as a whole which located in certain river basin is considered as a unique glacial area for description and calculation of processes V_m total glaciers melting and W_{gl} glaciers runoff. Depending on the size of a basin, the total number of glaciers is divided into k of their groups, which are included certain quantity of n_j quasi-homogeneous glaciers ($n_j \geq 1$): $N_{gl} = n_1 + n_2 + n_3 + ... + n_k$. At the determination of annual and long-term regime of glacial areas are taken into account all the main peculiarities and regularities which were ascertained at the separate glaciers. As the methodical base to compute long-term series of characteristics of glaciers hydrological regime are used mathematical and physic and statistical models of accumulation and melting processes of snow and ice which describe the annual cycle of glaciers regime. Special features of climate, relief and glaciers in the each river basin are considered at the stage of data preparation for computations. To make the method for computation of the mass balance and glaciers runoff the most universal it is envisaged that the input information includes as a rule the data of network measurements of the air temperature, cloudiness, precipitation and morphometric parameters of glaciers which are contained in the Inventory of Glaciers.

GENERAL APPROACH FOR ONE-DIMENSIONAL CASE

To develop the practical method of $V_m(t)$ calculation let us present this variable as one dimensional function of the altitude z. Then

$$v(t)_z = \int_{z_e}^{z_b} M(z,t) \cdot S(z)dz,$$

where z_b is the altitude a.s.l. of glaciers beginning, z_e is the altitude a.s.l. of the glacier snout, s(z) - is the function of glacier's area distribution depending on the area altitude. After having applied the generalized theorem on the average to the integral of the product of the two functions M(z) and s(z) we'll have

$$v_m(t)_z = M(\tilde{z},t)S,$$

where $\tilde{z}$ - is the certain altitude in the z_b - z_e interval, S - is the total glacier's area. The total melt volume for a number of glaciers in a basin can be presented as the particular sums related to the homogeneous groups of glaciers. In the each k-th group consisted of n_k glaciers

$$v_m(t)_{n_k} = n_k(1/n_k) \cdot \sum_{i=1}^{n_k} M(\tilde{z}_i,t)S_i .$$

Let us divide the interval z_b - z_e into separate sections, where M(z) function is preset to be linear. In this case the expression for calculation of the total melt volume of N_{gl} glaciers has the form:

$$V_m(t) = n_1 \sum_{j=1}^{m} [C_m - a\gamma \cdot \text{Mo}(\tilde{z}_{1j})]\tilde{S}_{1j} + n_2 \sum_{j=1}^{m} [C_m - a\gamma \cdot \text{Mo}(\tilde{z}_{2j})]\tilde{S}_{2j} + \ldots$$

$$+ n_k \sum_{i=1}^{m} [(C_m - a\gamma \cdot \text{Mo}(\tilde{z}_{kj})]\tilde{S}_{kj}$$

$$\text{Mo}(\tilde{z}_k) = \sum_{i=1}^{n_k} \tilde{z}_i p_i ,$$

(p_i - is the weighing factor)
where m is the number of sections in z_b - z_e interval. This number is taken to be five in order to determination of $V_m(t)$ include the calculation of the main components of the total melt volume and the runoff from the glaciers.

REGIONAL MODEL FOR GLACIERS RUNOFF

$$V_M(t)_m = M_c(\bar{z}_{im},t)S_{im} + M(\bar{z}_i,t)S_i + M(\bar{z}_f,t)S_f +$$
$$+ M(\bar{z}_{ws},t)S_{ws} + M(\bar{z}_{ss},t)S_{ss}$$

$$M_c = M \cdot f(h_c).$$

Here M_c is the intensity of ice melt under the moraine cover (*im*), *i* is the bare ice, *f* is the old firn, *ws* is the winter snow, *ss* is the summer snow, $f(h_c)$ is the function of extinction of ice melting under the moraine cover with the thickness h_c. To obtain the data on the total melt volumes V_m and ice-melt runoff W_{gl} it is necessary to summarize the applied components:

$$V_m = \sum_{d_{bp}}^{d_{ep}} V_M(t)_m,$$

$$W_{gl} = N_{gl} \sum_{d_{bi}}^{d_{ei}} [V_{im}(t) + V_i(t) + V_f(t)],$$

where d_{bi} and d_{ep} - are the dates of the beginning and the end of the calculation period, d_{bi} and d_{ei} - are the dates of the beginning and the end of the icemelt period.

The **REGMOD** model includes some methods and set of PC programs which realize the description of total melting and glaciers runoff processes in the coordinates z, t. Those are the following.
a) Local and regional formulae of of melting intensity of snow, bare ice and ice under moraine cover,
b) The method of calculation of moments of beginning and end of ice melting period and glaciers runoff formation,
c) The model of snow boundary movement on glaciers surface during the ablation period,
d) Space and temporal variability of the main meteorological elements.

MULTIDIMENSIONAL EXTRAPOLATION OF METEOROLOGICAL ELEMENTS IN THE REGMOD MODEL

Further improvement of the **REGMOD** model was achieved by means of elaboration of principally new method of multidimensional extrapolation of

precipitation q and air temperature T. At present computation of annual course of precipitation or air temperature for a point with coordinates z, v, λ is of two staged procedure:
(a). Horizontal extrapolation of a basic meteorological station data from the point $x_0(z_0, v_0, \lambda_0)$ to the point $x_i(z_0, v_i, \lambda_i)$ within the same elevation level with a mean altitude z_0 ; namely, at z_0.~ const.
(b).Extrapolation of $x_i(z_0, v_i, \lambda_i)$ to the given elevation z using analytical approximation of the x-element vertical profile in the point with coordinates z_0, v_i, λ_i. For numerical description of vertical profile $x_i(z_0, v_i, \lambda_i)$ a set of $x_{ik}(z_{ik}, v_{ik}, \lambda_{ik})$ values was used, which are estimated by k-numerical models of spatial distribution of the given meteorological element (k is the number of elevation levels).

The proposed method has the following advantages:
(i) makes it possible to extrapolate precipitation and air temperature data of one or several basic stations to the whole mountainous territory of Central Asia;
(ii) improves extrapolation quality due to averaging of computational results for several basic meteorological stations;
(iii) makes it possible to get numerical estimates of annual course of T and q in high alpine basins where there are no meteorological stations or their number is not sufficient to set up local relationship of $T(z)$ and $q(z)$.

The method is designed for application in regional models with distributed parameters which describe glaciological and hydrological processes, and in simulation methods of runoff forecasts. Long-term series of mean annual T values and annual q sums at several basic meteorological stations were taken as initial data for horizontal extrapolation. The point-element x_i having coordinates z_0, v_i, λ_i was theof object extrapolation; it characterized one of the areas in a glacial region or a group of glaciers from their integrity within a basin.

The presented methodics of spatial and temporal extrapolation of air temperature and precipitation was realized as a separate procedure in the improved version of **REGMOD** model. By this model were made the computations of long-term series of seasonal and monthly volumes of total melting, snow and glacial runoff, annual mass balance in the ablation areas in the River Pjandge basin. The total area of glaciation in this basin is 6962 km^2 . It turned out that in the River Pyandge basin the values of snow and glacier feeding which were averaged for 1935-1980 period are quite correspondent with the estimates of other authors.

RUNOFF FORECASTING FOR RIVERS WITH ICE - AND SNOW MELT FEEDING

The analysis in general form of the possibility of long-term forecast of the glacier runoff showed that the sum of the winter and spring precipitation is the most applicable predictor. The following conditions which provide the efficient utilisation of this predictor were formulated:
- the relative area of glaciers must not exceed 10% of the whole area of catchment,
- to forecast the ice and snow melt (or snow-rain) components of the total runoff are used the separate dependencies,
- the sum of winter-spring precipitation is the common predictor to forecast ice - and snow melt feeding of rivers.

To facilitate the application of the above mentioned principles was received the list of Central Asian catchments where the first condition is satisfied. Further we need to calculate long-term serie of W_{gl} and to use these data for obtaining the dependence $W_{gl} = f(q)$ - where q is the sum of winter-spring precipitation. Then computed serie of ice melt volumes is used to separate snow melt feeding W_{sn} from the total runoff and to construct the dependence $W_{sn} = f(q)$. Finally the long-term forecasts of the water content of vegetation period in the river basins with ice - and snow melt feeding are fulfilled by the separate forecasting of these main components of total runoff. Local dependencies were derived for forecasts of W_{sn} and W_{gl} in Zeravshan river basin confirms the complete contrast of influence of winter-spring precipitation on the process of snow - and ice melt components formation in a basin.

The possibilities of the long-term forecasting of total melting and glaciers runoff based on the quasiperiodic properties of time series of V_m and W_{gl} were examined also. The practical solution in the each concrete case depends on the form of autocorrelation function of these long-term series. In particular the stable repetition (each six years) of the peaks of the positive and negative values of correlation was revealed for the Malaya Almaatinka and Pskem river basins. As the results, the autoregression equations of the first and second orders which have rather satisfactory characteristics of the forecasting value were obtained to forecast the glacier runoff in these river basins with earliness of four and six years. The test of equations using the independent data showed good results. Besides, the equations of multiple linear regression are widely used to forecast W_{sn} and W_{gl}.

Solution of problem of long-range forecast of water inflow to the Nurek reservoir at the Vakhsh river is the most important for economy development of Tadjikistan, Turkmenistan and Uzbekistan in condition of extremely restricted water resources in Central Asia. Application of the **REGMOD** model for Vakhsh river basin make it possible to improve quality of hydrological forecasts essentially and to meet needs of numerous users with information about water resources. Probabilities of monthly runoff forecasts for Vakhsh river are presented in the Table 1.

Table 1 Quality estimations of computations and forecasts of water inflow into the Nurek reservoir (Tadjikistan)

Months, season	Computations		Forecasts		
	S/σ	P%	Term of forecasts, months	S/σ	P%
April	0.7	72	1	0.7	72
May	0.7	70	2	0.8	69
June	0.6	79	3	0.8	66
July	0.6	80	4	0.6	72
August	0.6	86	5	0.8	66
September	0.7	82	6	0.7	80
April-September	0.3	100	6	0.3	92

Notes: S/σ - is relative root mean square error, P% - probability. Computations and forecasts of runoff are started at April,1.

MAIN RESULTS

(a). Methodics, PC program and information base for computation of long-term series of the hydrological regime of glaciers integrity and their groups in Central Asia were elaborated. (b). The computations of long-term series of monthly and seasonal volumes of snow and ice total melting, snow and glacier's runoff, values of annual mass balances in the ablation areas were realized for the Zeravshan, Vaksh and Pyandge river basins. (c). An universal methodics of spatial and temporal extrapolation of monthly precipitation sums and mean monthly air temperature has been developed for the Central Asia territory in the ranges of $35\text{-}45^\circ N$ and $67\text{-}81^\circ E$. The method is a part of a model with distributed parameters used for determination of hydrological regime and mass balance of glaciers integrity and groups of them in the river basins of Central Asia. Quality estimate of the method has shown quite satisfactory convergence of the computed and actual air temperature at meteorological stations. (d) Mathematical model describing input of liquid and solid precipitation on the watershed surface and their transformation into runoff is used in Hydrometeorological Service of Uzbekistan for discharge forecasting during of April-September and separate months of this season. Basic component of the model is estimation of glaciers runoff, which present considerable part of the Vakhsh and Pyandge rivers water yield.

Coupling 1-D and 2-D models for flood management

A. PAQUIER, B. SIGRIST
Cemagref, Lyon, France

ABSTRACT

Coupling 1-D model for main water channel and 2-D model for flood plain is a suitable method to simulate flooding process. It is a supplementary way to cope with global management of rivers. A method of coupling is described and part of the first phase of validation is shown. We also mentioned one case of application of the coupled model to a river of France.

INTRODUCTION

For many years, flood management has been based on sending waters downstream as rapidly as possible in order to avoid too high levels. This method generally leads to increased flood damages downstream. At present, it is preferred to slow down water by spreading it in areas where it causes no damage (meadows, marshes, etc.). So it is even more necessary to simulate precisely floods. For this purpose, engineers use 2-D simulations instead of 1-D simulations. The main disadvantage of 2-D computations is the complexity of computational grid ; particularly, in order to represent the discharge in the main water channel accurately enough, small cells are required to take into account the rapidly changing topography. This is particularly true if the same model is used to cope with global management of the river which requires simulation of a large range of discharges. For numerical stability, such small grid must be extended to areas close to the main water channel and to both directions (longitudinal and transversal). A consequence is both an increase of the number of cells and a decrease of the time step.
Thus, we propose to keep the classical 1-D modelling of the main water channel and to use 2-D model only for the flood plain. The aim of such coupling of 1-D and 2-D models is to increase the flexibility provided to the user in order to build a model closer to the reality more easily.
In this paper, we describe the way the coupling was performed. Then, we give two simple cases used for validation and one case of application to a river located in the South-East of France.

METHOD OF COUPLING

For river modelling, 2 directions can always be clearly defined :
1. the direction of the main flow (main direction) which will be the axis of the 1-D model
2. the lateral direction locally normal to the previous one in main water channel (classical cross sections of a 1-D model). This cross direction can be adapted to be more or less perpendicular to the flow in the flood plain for large discharges.

On that basis, a grid can be defined [1]. The code for the grid generation will provide cross sections for the 1-D model and a somewhat regular grid adapted to topography for the 2-D model.

1-D and 2-D equations to be solved are shallow water equations. They consist in one equation of mass conservation and 1 (1-D) or 2 (2-D) equations of conservation of the quantity of movement. Coupling will thus consist in transferring mass and quantity of movement from the 1-D model to the 2-D model and conversely. When using a grid as defined here above, 2 kinds of transfers occur :
1. Transfer through an edge parallel to the main direction : 1-D model for main water channel and 2-D model for the flood plain.
2. Transfer through an edge parallel to the cross direction : " 1-D model upstream, 2-D model downstream" or " 1-D model downstream, 2-D model upstream".

Of course, when modelling a whole river, the two kinds of transfers are likely to occur. Here below, we will only consider transfer of type 1. In such a case, transfer is considered through the top of the banks of the minor bed. If there are embankments where shallow water equations are no longer valid, it is possible to compute mass and quantity of movement to be transferred by using weir-type relations. More generally, the problem will be the following one : computing normal discharge (equivalent to mass transfer) and flux of the quantity of movement through one edge knowing, on one side, 1-D values which are cross sectional area and velocity in the main direction and, on the other side, 2-D values which are water depth and velocities in the two horizontal directions.

Solvers for shallow water equations were an explicit second-order Godunov-type finite difference scheme for 1-D and an explicit second-order Godunov-type finite volume scheme for 2-D [2] which may be considered as an extension of the 1-D scheme. The two schemes are thus very close which simplifies coupling. However, in 1-D equations, the terms of transfer appear in the second member : lateral discharge for the equation of mass conservation and the corresponding quantity of movement for the other equation. Thus, it is easy to change 1-D solver without changing the method of coupling.

In our method, the part of the 1-D model between 2 successive cross sections is assimilated to a 2-D cell where the variables are defined in the following way:
1. Bottom level : half the sum of the levels of left and right banks.
2. Water depth : computed from the bottom level and the water level of the 1-D model.
3. Velocity on main direction : velocity of 1-D model.
4. Velocity on cross direction : zero.

Then, a 2-D computation is performed on the whole modelled areas. This computation includes 3 steps :
1. Approximate values of the variables at the middle of the edges and at an intermediate time are computed from an estimate of the gradients of the variables and from an estimate of the variation of the variables on half a time step.
2. Fluxes through the edges are computed generally by solving Riemann problems but weir-type relations can also be used.
3. From fluxes and estimates of second member, final values at the centre of the cells are computed.
For the minor bed, only the resulting values of fluxes through the edges are kept to be used as lateral inflows in the 1-D computation which is performed later. Both transfers (from 1-D to 2-D and from 2-D to 1-D) should occur once at every time step. However, as the numerical schemes considered use an intermediate time step, it is also possible to transfer twice every time step ; this method gives more numerical stability in very unsteady phases (for instance, beginning of flooding).

CASES OF VALIDATION

The aim of validation consists in checking the determination of the transfer terms, the 1-D and 2-D models having been validated previously [4], [5]. Due to the somewhat complex conditions when a coupled model is used, we decided to validate the model by comparison to 2-D results.

1. WAVE ON A FLAT HORIZONTAL AREA

For that case, the distinction between 1-D and 2-D areas is completely artificial. The computational grid for the coupled model is constituted by a 1-D strip of four cells surrounded by three 2-D strips of four cells on each side of the 1-D strip. The boundary conditions are reflecting walls all around the computational area. The initial conditions are rest with a water depth of 2 metres in the 1-D area and 1 metre in the 2-D area. First, we checked that the symmetry of the problem is kept during the computations and that the 4 cells of any strip gave identical results. Secondly, we checked that the celerity of the wave is close to the theoretical velocity of 3.8 m/s (no bottom friction introduced). Third, comparison with 2-D results on the same grid (thus theoretically providing identical results) was performed. Result is shown on figure 1 at the centre of the cell just beside the central cell ; for all the cells, maximum difference is less than 1cm.

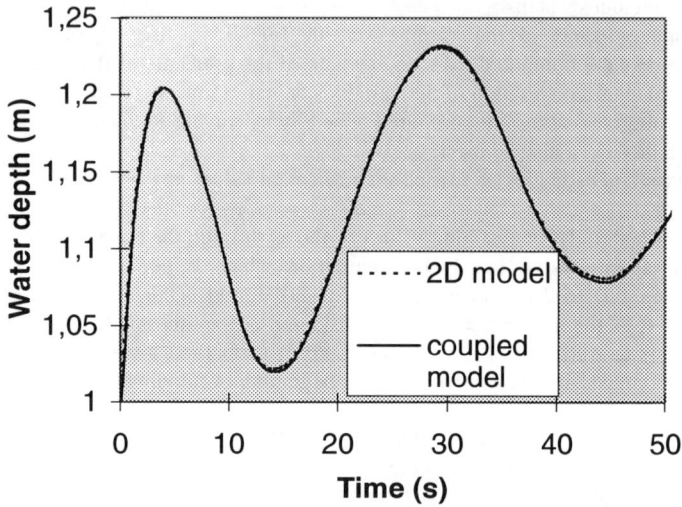

Fig. 1 - Wave on a flat horizontal area - Water depth at 14 m from centre.

2. JUNCTION WITH FLOODING

The computational grid is shown on figure 2.

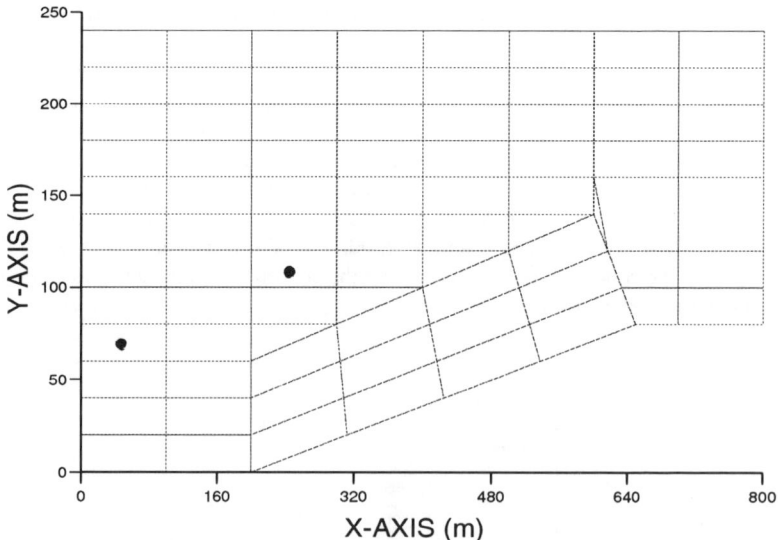

Fig. 2 Computational grid for the junction with flooding.

The two upstream 1-D strips are joining in the downstream 1-D channel, 2 cross sections of a triangle 5 m deep for 20 m wide becoming a cross section of a trapeze 5 m deep, 20 m wide at the bottom and 40 m wide at the top. No slope is considered. The initial conditions are rest with a water depth of 5 m in the 1-D channels. A triangular discharge hydrograph with a peak of 150 m3/s at 50 s and return to 0 at 100 s is introduced in each channel. The rise of water level causes the flooding of the 2-D flat horizontal area. Figure 3 shows the variation of water levels at 2 different locations (points on figure 2) for the coupled model and for a 2-D computation with same grid on the 2-D part and 10 cells in the width of the 1-D parts. Note that similar results are obtained although the confluence in 1-D requires further hypothesis (see [3] for details of 1-D solving at confluence).

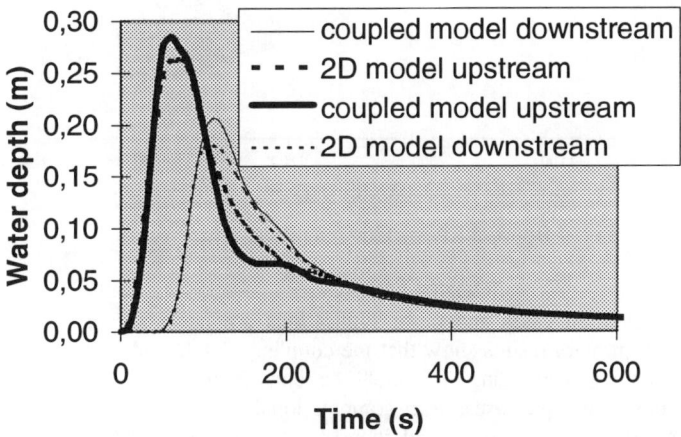

Fig. 3 Junction with flooding - Water depths in the flood area.

APPLICATION

The coupled model has been applied to a part of the valley of the Bourbre river downstream the town of L'Isle d'Abeau near Lyon (France). The computational grid (3711 cells) is shown on Figure 4. It includes a confluence and varied topography for both 1-D and 2-D areas. A flood of approximate return period of 100 years was simulated. A full 2-D computation with several cells in the cross direction of every main water channel (total of 4223 cells) was also performed. Results are generally similar (a few centimetres of difference for maximum water levels) for a ratio of computational time of about 5.

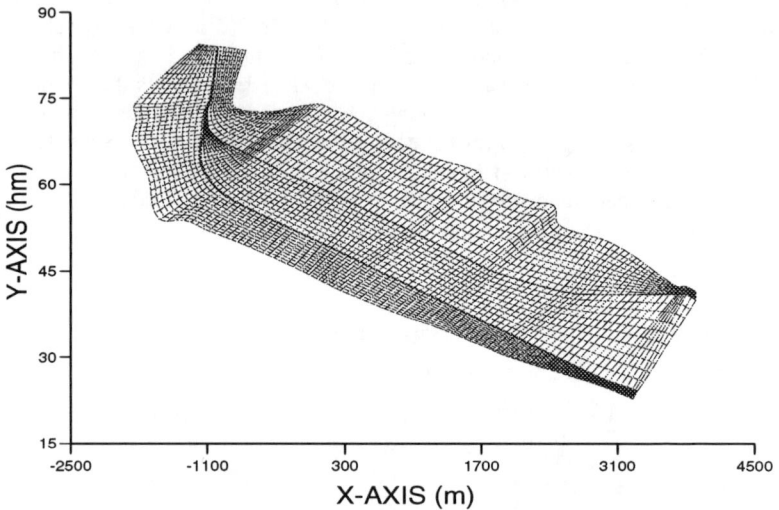

Fig. 4 Bourbre River - Computational grid.

CONCLUSIONS

Although more validation cases and more applications are necessary to definitely secure the method, primary results show that the coupling of 1-D and 2-D models is convenient for simulating flooding. It should be considered as a supplementary possibility for engineering purposes, with computational time requirements several times lower than full 2-D computations and, in most cases, similar accuracy.

REFERENCES

[1] P. Farissier, P. Givone (1993), *Mapping and management of flood plains*, Application of Geographic Information Systems in Hydrology and Water Resources Management. IAHS publication N° 211, IAHS Press, Wallingford, UK, 485-490.

[2] A. Paquier (1994), *New methods for modelling dam-break wave*, Specialty Conference "Modelling of flood propagation over initially dry areas", Milano, Italy, ASCE, 229-240.

[3] A. Paquier (1995), *Dam-break in a valley with tributary,* Hydra 2000, Proceedings of the XXVIth IARH Congress (London), 272-277.

[4] A. Paquier (1996), *Validity of a 1-D model for simulating dam-break wave,* Proceedings of Hydroinformatics 96, Zürich, Balkema, 409-416.

[5] A. Paquier, P. Farissier (1996), *Use of a 2-D model for simulating the flooding of a plain,* Proceedings of Hydroinformatics 96, Zürich, Balkema, 129-136.

FORECASTING DROUGHT RISKS FOR A WATER SUPPLY STORAGE SYSTEM USING BOOTSTRAP POSITION ANALYSIS

GARY TASKER AND PAUL DUNNE
U. S. Geological Survey, Reston, VA and West Trenton, NJ, USA

ABSTRACT

Forecasting the likelihood of drought conditions is an integral part of managing a water supply storage and delivery system. Position analysis uses a large number of possible flow sequences as inputs to a simulation of a water supply storage and delivery system. For a given set of operating rules and water use requirements, water managers can use such a model to forecast the likelihood of specified outcomes such as reservoir levels falling below a specified level or streamflows falling below statutory passing flows a few months ahead conditioned on the current reservoir levels and streamflows. The large number of possible flow sequences are generated using a stochastic streamflow model with a random resampling of innovations. The advantages of this resampling scheme, called bootstrap position analysis, are that it does not rely on the unverifiable assumption of normality and it allows incorporation of long-range weather forecasts into the analysis.

INTRODUCTION

Forecasting drought risks for a surface-water supply system is a difficult and necessary task for many water supply managers. Water management decisions concerning proper storage levels in reservoirs and water use limitations made at the beginning of a possible drought can be critical if below normal streamflows persist over a period of a few months. Position analysis (Hirsch, 1978) is a tool that water managers can use to forecast risks associated with a specific operating plan for the basin over a period of a few months. It can aid the water manager in deciding which plan of operation to implement by providing a means to evaluate and rank each proposed plan of operation in terms of future drought risks. Position analysis relies on the generation of a large number of possible monthly flow traces (a few months in length) which have been initialized with the current reservoir storages and current streamflows. The large number of flow traces are used to incorporate into the analysis the broad range of meteorological conditions that may occur but cannot be accurately forecast.

In this paper a stochastic model of streamflows is used to generate flow traces at multiple sites. The stochastic model generates synthetic streamflows by separating the flows into carryover components to model the serial correlation and random components called innovations. Hirsch (1981) generated traces for a single site using a random number generator to produce normally distributed innovations. However, the assumption of normality of innovations may not exploit all the information in the sample. A new approach is presented herein in which the innovations of the multi-site stochastic model of streamflows are generated by drawing at random, with replacement, from the historical sample of innovations. Efron (1979) calls this method of randomly resampling with replacement a bootstrap sample. The advantage of bootstrapping is that it does not rely on the unverifiable assumption of normality.

The bootstrap runoff traces for a position analysis model may be used as inputs to a model simulation of a water supply storage and delivery system for drought management. For a given set of operating rules and water use requirements for a system, water managers can use such a model to forecast the likelihood of specified outcomes such as reservoir levels falling below a specified level or streamflows falling below statutory passing flows a few months ahead conditioned on the current reservoir levels and streamflows. Thus the model can be used to determine the effectiveness of specified changes in operating rules or drought restrictions in reducing drought risks.

BOOTSTRAP RUNOFF TRACE GENERATOR

Data input to the position analysis forecasting model are multiple runoff traces generated by a stochastic model of monthly runoff based on the historic streamflow records for gaged sites in the area. The stochastic runoff model used for this model is the log-transform autoregressive moving-average (LT-ARMA(1,1)) cyclic model described in Hirsch (1981). Denote the runoff in year i (i=0,1,2,....)and month j (j=1,2,...,12) as X_{12i+j}, and let Y_{12i+j}=log(X_{12i+j}), and define the variable

$$Z_{12i+j} = \frac{\left(Y_{12i+j} - \overline{Y}_j\right)}{S_j} \tag{1}$$

where $\overline{Y}_j$ and S_j are the sample mean and standard deviation, respectively, of the logarithms of the observed runoff values for month j. Therefore, Z_{12i+j} represents the standardized deviation for year i and month j from the log-transformed mean runoff for month j. The serial dependence is modeled as a periodic moving average model of the form:

$$Z_{12i+j} = \phi Z_{12i+j-1} + E_{12i+j} - \theta_j E_{12i+j-1} \tag{2}$$

where $E_{(.)}$'s are independent random variables with mean of zero. The thirteen parameters, ϕ and θ_j, are estimated by the method described in Hirsch (1979). The first and third terms on the right-hand side of equation 2 can be thought of as the carryover components of the deviation from mean monthly log-transformed runoff due to antecedent moisture and delayed runoff in a basin while $E_{(.)}$ is the innovation or random component of the deviation from mean monthly log-transformed runoff due to meteorological conditions in month j of year i. Breaking up the standardized runoff values into carryover components and innovations allows one to generate synthetic runoff by random resampling from the sample innovations. A position analysis trace 12 months long can be generated by setting the starting value of Z equal

FORECASTING DROUGHT RISKS   647

to the value of the log-transformed standardized runoff for the present month and sequentially computing Z's using a series of 12 innovations randomly selected from the sample innovations. To preserve the spatial covariance in the traces that are generated for multiple sites, the same randomly selected year is chosen for each site to begin the sequence and the next eleven innovations are chosen in their historical order. Thus the innovations for multiple sites stay together in the bootstrap sequence and the lag-zero cross correlations of the innovations are preserved.

Selecting the innovations at random in position analysis has another advantage aside from being able to produce many equally likely possible traces. The bootstrap model allows the user to specify the probabilities of having a normal sequence, a wetter than normal sequence, and a drier than normal sequence that coincide with the forecasts of the National Weather Service. Suppose that a 90-day forecast says that it is 5% more likely that conditions will be drier than normal. Then the bootstrap selection process can be modified to make it 5% more likely to select a relatively negative 3-month sequence of innovations than a relatively positive sequence of innovations, which would make it more likely for a "dry" position analysis trace to be generated.

AN APPLICATION TO A RESERVOIR SYSTEM IN NEW JERSEY

The reservoirs and pumping stations of the Raritan River Basin Reservoir System, located in central New Jersey, and its interconnections to the Delaware-Raritan Canal Water Supply System supply potable water to nearby communities. This combined system includes Spruce Run Reservoir, Round Valley Reservoir, Hamden Pumping Station, and the Delaware-Raritan Canal. A detailed description of the system can be found in Dunne and Tasker (1996).

The basin model is a continuity accounting model consisting of a series of interconnected nodes. At each node the monthly inflow volume, outflow volume, and change in storage is determined and recorded for each month. Figure 1 is a flow diagram showing the nodes (as boxes) and flow paths as lines. The model is driven by the natural inflows into each node which are the natural runoff traces generated by the bootstrap position analysis model. A set of operating rules control the reservoir and canal releases and pumpage to meet passing flow and withdrawal demands.The basin model runs with a set of default operating rules for releases, pumpage, and diversions which the user may change in order to see the effects of alternative sets of rules and evaluate their usefulness.

The following is a brief description of a complex set of operating rules for the basin. The model user may modify these rules, the reservoir capacities, passing flow requirements, or maximum pumping capacities before running the model to test alternative methods for management of the resources.Releases from Round Valley Reservoir (capacity 55 BG) to South Branch are made to meet Stanton passing flow requirements that cannot be met by unregulated flows or by releases from Spruce Run. Releases to North Branch are made to meet withdrawal demands and passing flow requirements at Manville and Bound Brook that cannot be met by unregulated flows, by releases from Spruce Run reservoir (capacity 11 BG)

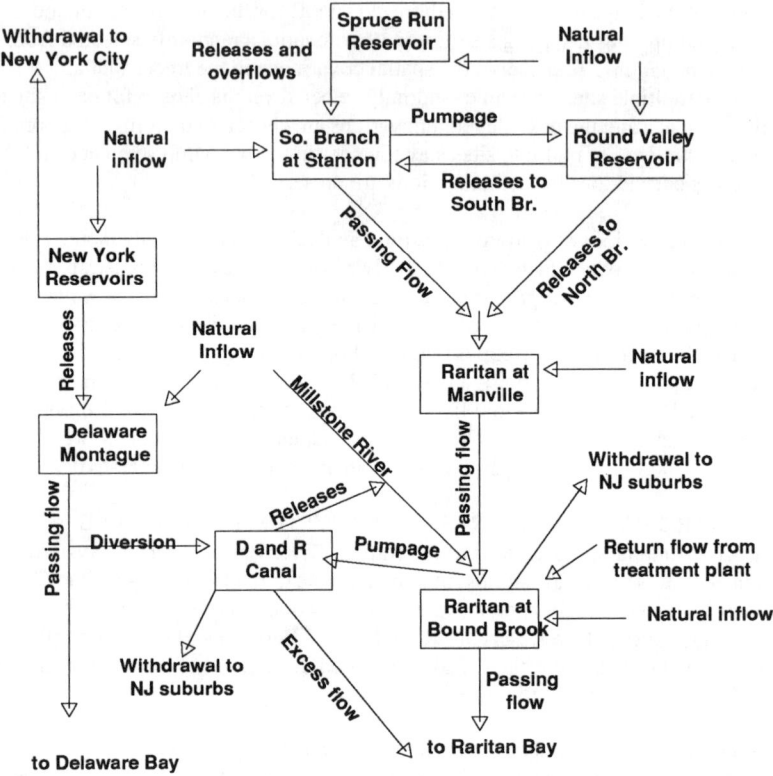

Figure 1. Flow diagram for basin model of Raritan Water Supply System. Boxes represent model nodes and arrows represent inflows and outflows.

or by diversions from the Delaware-Raritan Canal. Inflow into Round Valley reservoir comes from natural runoff supplemented by pumped water from the Hamden pumping station above the Stanton gage. Pumping is only done when minimum passing flows at Stanton are met and when storage in Round Valley is below target levels. The model user sets the target storage levels for each month.

Suppose, for example, that it is currently the end of April, all reservoirs are 80 percent full, April runoff is known, and the long range weather forecast is for 5 percent drier than normal conditions. A water manager wishes to test the effects on Round Valley reservoir storage of changing the recreation minimum levels of Spruce Run storage. In rule 1, recreation minimum storages are set at 3 BG. In rule 2, the recreation minimum for Spruce Run is raised to 7.5 BG which will mean greater recreation use of the reservoir. Ten of the 300 model generated traces of Round Valley storage for rule 2 are

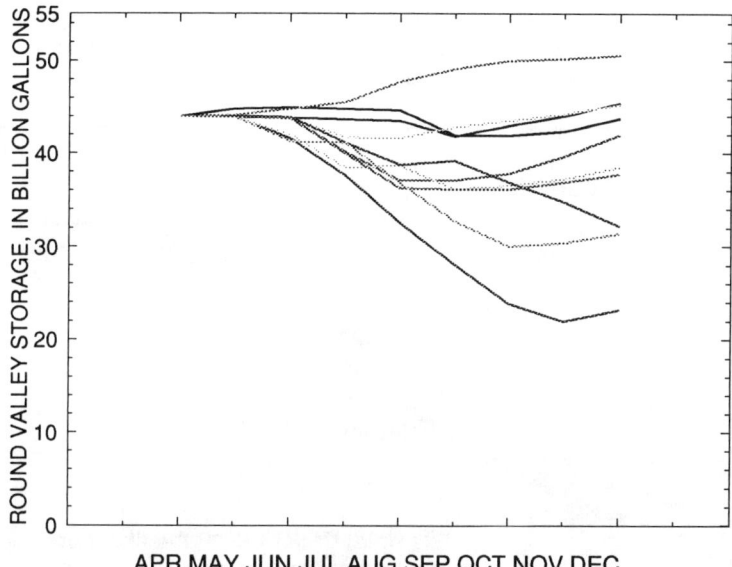

Figure 2. Ten of 300 bootstrap position analysis traces for storage in Round Valley Reservoir based on initial May storages and flows and operating rule 2.

shown in figure 2. A frequency analysis of all traces for rule 1 and 2 for the end of November (figure 3) show the increased likelihood under rule 2 of storages in Round Valley at the end of November falling below a specified level. An analyst may wish to use this information to compare the expected increase in recreation benefits for Spruce Run to increase in pumping costs to refill Round Valley and loss in recreation benefits for Round Valley.

SUMMARY AND CONCLUSIONS

A method of random resampling of innovations in stochastic models is used to generate a large number of traces of natural monthly runoff to be used in a position analysis model for a water-supply storage and delivery system. Resampling the innovations eliminates the need to make parametric assumptions about the distribution of the innovations and simplifies parameter estimation. One only assumes that the best estimate of the population of the innovations is the sample itself. Lag-one cross correlation of the innovations in a multi-site application is preserved by having the innovations for multiple sites stay together in a contemporaneous resampling scheme. The effects of long-range weather predictions can be included by setting the probabilities that relatively large negative or large positive innovations will be randomly drawn. The position analysis model (Hirsch, 1978) is a powerful forecasting tool that allows for the quantification of risks associated with different management options. It provides the water resources manager with a means to rank alternative operating rules and evaluate risk-cost trade-off.

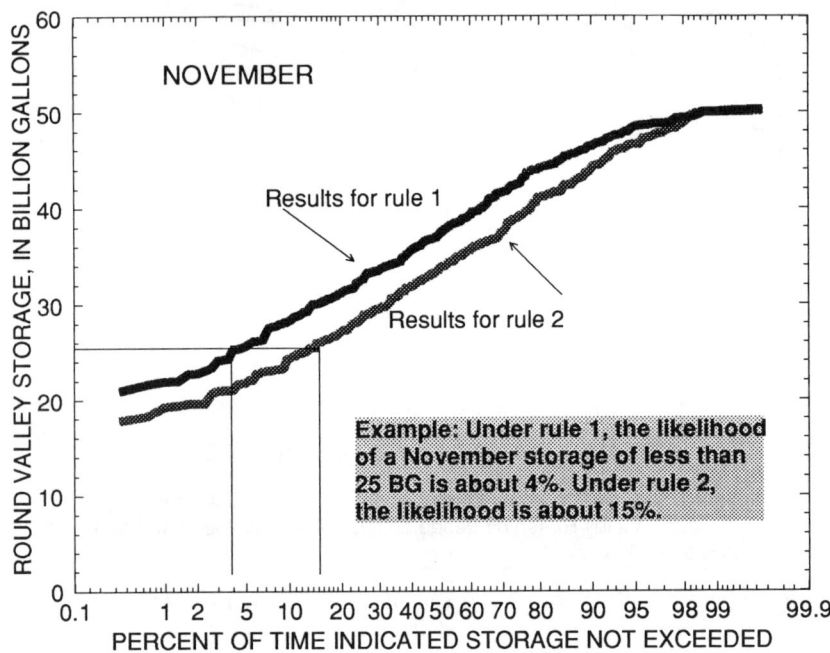

Figure 3. Storage frequency curves for Round Valley Reservoir at end of November based on 300 bootstrap position analysis traces, assumed initial May storages and flows, and assumed operating rules.

REFERENCES

Dunne, P., and Tasker, G., 1996, Computer Model of the Raritan River Basin Water-Supply System in central New Jersey: U.S.G.S. Open-file Report 96-360, 62p.

Efron, B., 1979, Bootstrap methods: Another look at the jackknife: *The Annals of Statistics,* v. 7, no. 1, pp 1-26.

Hirsch, R.M., 1978, Risk Analysis for a Water-Supply System--Occoquan Reservoir, Fairfax and Prince William Counties, Virginia: *Hydrological Sciences Bulletin* v. 23, no. 4, pp 476-505.

Hirsch, R.M., 1979, Synthetic Hydrology and Water Supply Reliability: *Water Resources Research,* v. 15, no. 6, pp 1603-1615.

Hirsch, R.M., 1981, Stochastic hydrologic model for drought management: *J. Water Resources Planning and Management Division, ASCE,* v. 107, no. WR2, pp 303-313.

FLOOD ATTENUATION EFFECTS OF NATURAL FLOOD BASINS IN THE SACRAMENTO VALLEY, CALIFORNIA

J.C. VICK AND P.B. WILLIAMS
Philip Williams & Associates, Ltd., San Francisco, U.S.A.

ABSTRACT

Flood control works in the Sacramento River Valley have eliminated most of the valley's flood basin storage. Under natural conditions, basin storage and release reduced valley outflow (relative to inflow) from November through March and increased valley outflow (relative to inflow) from March through July. Elimination of the basins has resulted in increased winter and reduced spring and summer valley outflows (relative to inflow).

INTRODUCTION

During the last five years, there has been an increasing focus on ecological restoration based on the reestablishment of natural hydrologic and geomorphic processes. Defining pre-disturbance geomorphic and hydrologic patterns is a key component of such a restoration effort. From these historic patterns, natural geomorphic, hydrologic, and ecologic processes can be better understood and restoration goals, targets, and methods can be clearly defined.

The Sacramento River and Sacramento-San Joaquin Delta are now the focus of several restoration efforts. The reestablishment of natural hydrologic patterns is a major factor of the restoration. Hydrologic patterns in the Sacramento Valley and Delta have been severely altered by human intervention. One major intervention in this system was the completion of the Sacramento River Flood Control Project, which eliminated flood basin storage. Clearly defining the historic hydrologic patterns of Sacramento Valley outflow (a major component of Delta inflow) is critical to understanding an restoring the natural values of this system.

ENVIRONMENTAL SETTING

The Sacramento River is one of the largest rivers in California. It drains the northern portion of the Central Valley and discharges into the Sacramento-San Joaquin Delta. Under natural conditions, outflow from the Sacramento River provides 73% of the

Delta's inflow (CDWR, 1994). Prior to the construction of modern flood control works, the Sacramento River was bordered by natural flood basins (Figure 1). These basins were inundated approximately annually by flows overtopping the natural levees that separated them from the river or flowing through levee crevasses. These basins stored up to four million acre-feet of water (USCOE, 1916), slowly releasing the stored flows back to the mainstem river through small sloughs.

The Sacramento River Flood Control Project, authorized by the Flood Control Act of 1917, eliminated most if the valleys flood basin storage capacity. This project, in conjunction with private reclamation efforts, eliminated overflow into the Colusa, American, and Sacramento Basins and reduced the Yolo and Sutter Basins to leveed bypass channels. Only the Butte Basin is still subject to widespread inundation.

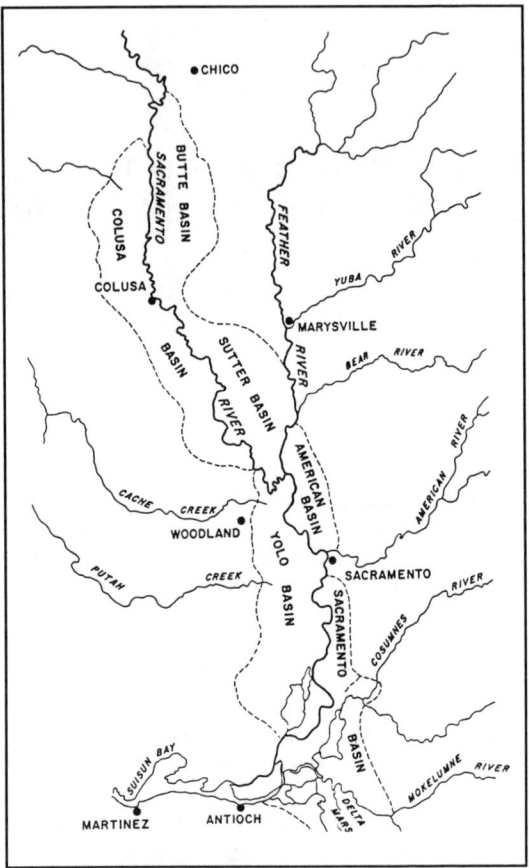

Figure 1. Flood Basins of the Sacramento Valley (Gilbert, 1917)

COMPARISON OF ANNUAL HYDROGRAPHS

To determine the hydrologic impacts of the elimination of flood basin storage, we compared Sacramento Valley inflow and outflow hydrographs precipitation for periods before and after construction of major flood control works. The inflow hydrographs were based on gauge data at the Feather River at Oroville, American River at Fair Oaks, and Sacramento River at Red Bluff. These stations contribute 71% of the Sacramento Valley's total natural runoff (Roos, 1973). Monthly precipitation was based on Roos's (1973) eight-station average, which includes stations located at Davis, Chico, Rocklin, Marysville, Nevada City, Redding, Red Bluff, and Yreka.

The valley outflow at the mouth of the Sacramento River (including the Yolo Basin), was based on estimates of discharge for the years 1879-1885 completed by W.H. Hall (1886) and by the California Department of Water Resources (CDWR) for the years 1922-1943 (CDWR, 1995). Hall's data represent Sacramento Valley outflow during the period prior to completion of the Sacramento River Flood Control Project (i.e., when the basins still functioned to store flood waters). The CDWR data represent outflows during and after construction of the flood control project but prior to the construction of major storage reservoirs in the basin.

To compare different types of water years, we chose the years with median, maximum, and minimum total annual discharges from the 1879-1885 period and then years from the 1923-1942 data that had similar total annual discharges for comparison.

RESULTS AND DISCUSSION

The temporal flow distribution hydrographs clearly depict the flood basin's effects on seasonal flow patterns and flood peak attenuation (Figures 2 and 3). The basins affected seasonal discharge patterns by reducing fall and winter outflow and increasing spring and summer outflow. In the median and wet years, flood basin storage reduced outflow from November through February, as the basins filled. During the median year, discharge from the basins increased valley outflow (relative to inflow) from March through July. In the wet year, the increased discharge extended from April through August. The flood basins did not reduce winter outflow or augment spring or summer outflow during the dry year.

In addition to altering seasonal outflow patterns, the flood basins affected peak discharge timing in the median and wet years. In the median year, inflow peaked in March. Outflow also peaked in March, but high discharges were extended through April. In the wet year, flood peak attenuation by the basins is more pronounced. The inflows peaked in March, whereas outflow did not peak until May.

After completion of the flood control project, effects of the flood basins on flood peak timing and seasonal discharge patterns were eliminated. In both the median and wet years, the temporal flow distribution patterns of the outflow hydrograph nearly exactly correspond with the inflow hydrograph. In the median year, February and March flow were increased (relative to inflow) and May through September flows were reduced (relative to inflow). In the wet year, January through May flows were increased (relative to inflow) and July and August flows were reduced (relative to inflow).

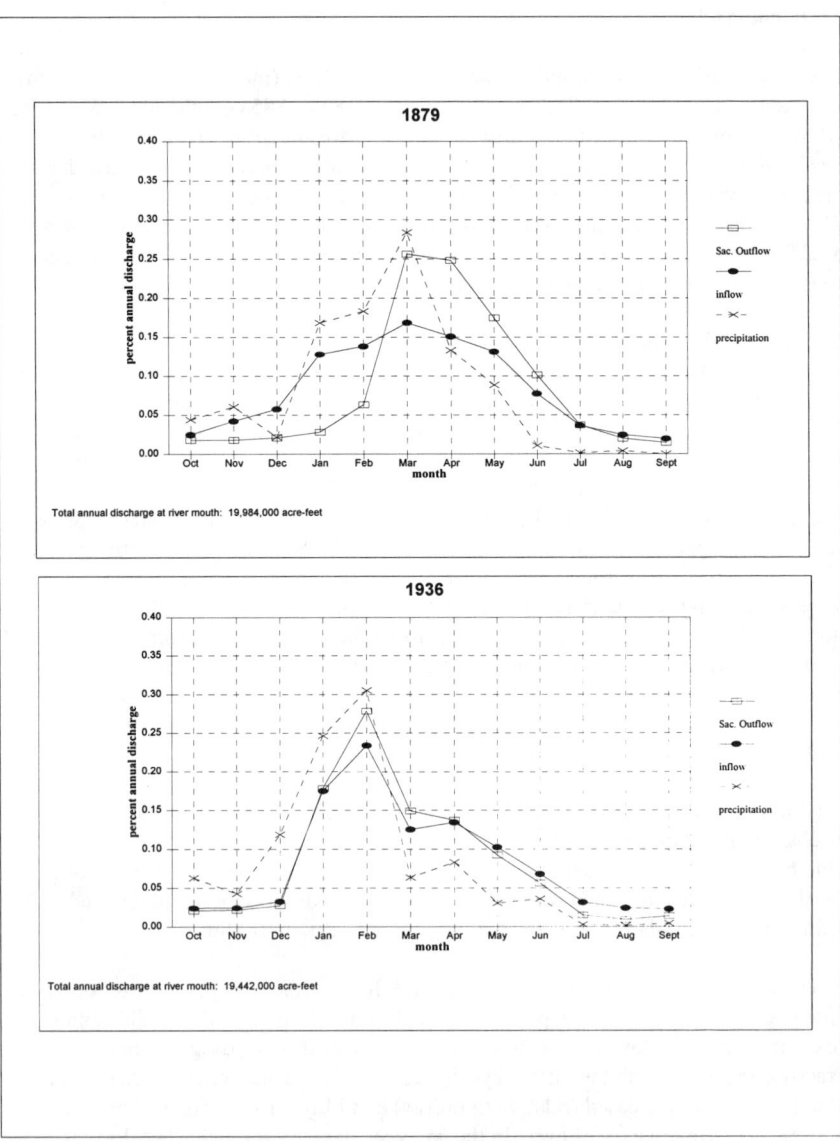

Figure 2. Sacramento Valley Inflow, Outflow, and Precipitation Hydrographs for 1879 and 1936. These represent median years before (1879) and after (1936) construction of the Sacramento River Flood Control Project.

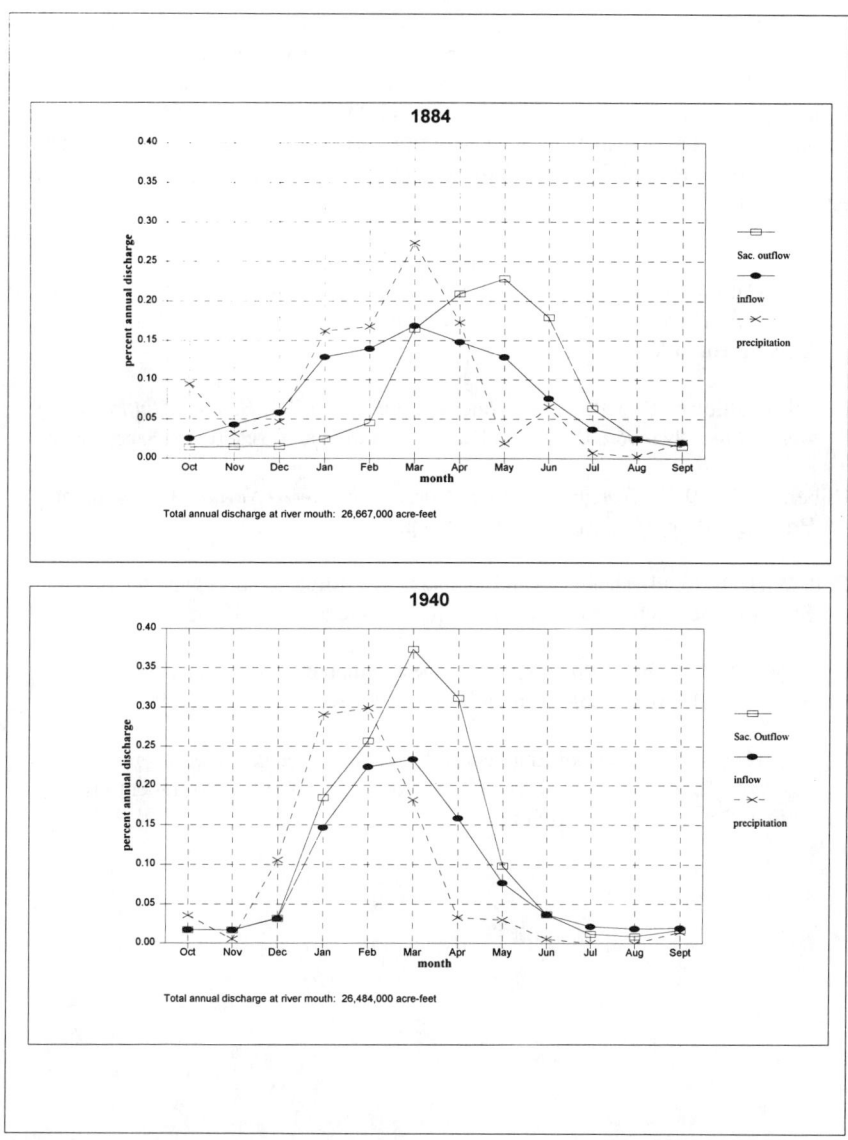

Figure 3. Sacramento Valley Inflow, Outflow, and Precipitation Hydrographs for 1884 and 1940. These represent wet years before (1884) and after (1940) construction of the Sacramento River Flood Control Project.

CONCLUSIONS

The hydrologic impacts of eliminating flood basin storage are clearly depicted by comparison of the pre-flood control and post-flood control valley inflow and outflow hydrographs. In their natural state, the flood basins stored high fall and winter flows (reducing basin outflow from November through March) and released these flows in the spring and summer (increasing basin outflow from March through July). This redistribution of annual flows was important to ecological processes in the river and estuary system.

REFERENCES

CDWR (California Department of Water Resources). 1994. *California Central valley Unimpaired Flow Data: October 1920 through September 1992*. California Department of Water Resources, Sacramento. 54pp.

CDWR (California Department of Water Resources). 1995. *Historic Outflow Data: Model Run 1995c06a*. California Department of Water Resources, Sacramento.

Gilbert, G.K. 1917. *Hydraulic-Mining Debris in the Sierra Nevada*. Government Printing Office, Washington, D.C., 154pp.

Hall, W.H. 1886. Physical Data and Statistics of California: tables and memoranda. State Engineering Department of California, Sacramento. 451pp.

Roos, M. 1973. Drought Probabiliyt Study Sacramento River Basin (Progress Report). California Department of Water Resources, Sacramento. 90pp.

USCOE (U.S. Army Corps of Engineers). 1916. *Supplemental Report on Flood Control of the Sacramento River (House Report No. 616)*, Government Printing Office, Washington D.C.169pp.

# Hydrodynamic approaches to design balancing ponds

M.J.J. PIROTTON
University of Liège, Institute of Civil Engineering,
Dept. of Applied Hydrodynamics and Hydraulic Constructions
6, Quai Banning, B.4000 Liège, Belgium

ABSTRACT
This paper aims to summarize the global hydraulic studies carried out in order to prevent the floodings of a small river located in the western part of Belgium. A first stage is devoted to compute the hydrograph resulting from extreme rainfalls on the catchment, using a distributed physically based model. Its numerical propagation on the studied site reflects the actual situation in the vicinity of the river site. Due to natural reservoirs, the flood peak is dampened and its base extended, i.e. the flood subsides. This peak reduction is first fitted to the reality by unsteady computations that handle the lateral exchanges between the planned floodplains and the main river. Besides, they led to optimise the location and the size of the planned lateral reservoirs as well as to forecast their impacts on the surroundings. This paper summarizes simplified approaches that saved much computational effort.

INTRODUCTION
« L'Espierres » is a small river flowing through the border between France and Belgium. This tributary is parralel to a navigable waterway for small ships along 8 kilometres before joining the Scheldt river. Because of the significant difference of water quality, mixing is not permitted between both flows. When floods occur in the Espierres river, the resulting flood stage implies a complete cross-section filling. Gradually, water spreads on the floodplains and flows into the bordering canal. Two basins were planned in Belgium in order to protect the near by sites of the river. The simulations have to reflect the ability of the lateral weirs to diverge the prescribed volumes and to forecast the smoothing of the hydrograph propagating along this 8 km long section.

HYDROLOGICAL PHYSICALLY BASED MODEL
We focus here on the thin water layer propagating on natural tridimensional slopes in order to compute the lateral discharge for each river element. This additional inflow that propagates in the main drainage path is then computed in a final stage with the specific software further described.

It is commonly supposed that a dynamic equilibrium exists between friction and gravity components in the flow direction. Resulting from the scale difference between the spatial dimension and water layer thickness, we assume biunivoque relations between the speed components and the water depth.

The resulting non-linear equation has been entitled kinematic wave approximation and was proposed for the first time in the hydrology field by Woolhizer and Ligett.

$$\frac{\partial h}{\partial t} + \frac{\partial(a_i h^{m+1})}{\partial x_i} + B(h) = 0 \qquad \text{on the domain D} \qquad (1)$$

with the following notations :
- $h(x_i,t)$: the water depth
- $B(h)$: the general function including the infiltration speed into the soil, the intensity of rainfalls and other effects of exchanges and losses
- a_i : functions including topographical data, surface and flow characteristics

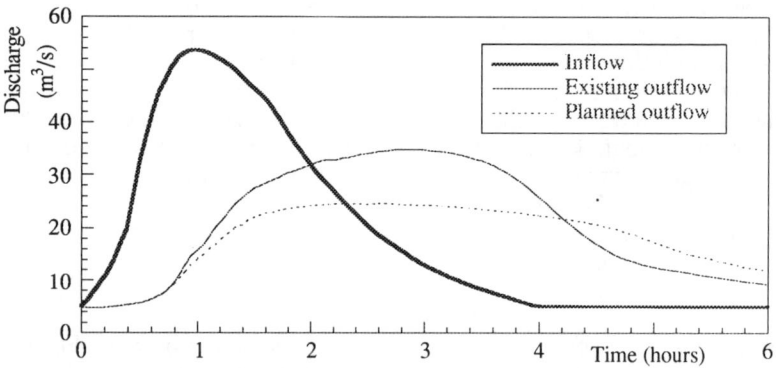

Figure 1 - Flood subsidence between the entrance and the outlet

The analysis of theoretical solutions assesses that hydraulic jumps arise on complex topographies at the scale of the thin water layer. The physical relevance of those kinematic shocks suggests the unique formulation of the weak solution of the problem. In order to prevent the occurrence of multi-peaked solutions highlighted by significant errors in the global volume balance (classical Galerkin method), a shock capturing finite element approach is used which introduces discontinuous test functions (see [1]).

The package works on any digital terrain model. It handles spatial and temporal variations of rainfall and soil properties. Partially dried elements deal with the effects of infiltration. Discretising the natural 5,000 ha catchment slopes submitted to significant rainfalls, we compute the ensuing hydrograph that propagates in the main drainage path. The temporal evolution of the upstream imposed discharge is illustrated on figure 1.

OPEN-CHANNEL FLOW

As far as the problem is handled in a steady way, unrealistic friction factors issue from a reasonable fitting between computed water profiles and measured values. This stage suggests two major remarks. Firstly, local overflowings are so important that we have to face the understanding of the flow in compound channels, the importance of which is frequently reported in the literature. Secondly, the computed inflow could be locally overestimated. Since the global balance confirms the accuracy of the hydrological approach, it seems essential to account transient influences that smoothen the hydrograph along its propagation. The smoothing would not result from the roughness properties but much more from the exchanges between the main path and lateral areas.

The sketches on the figure 2 highlight two potential circumstances that affect the suitable reasoning to achieve about lateral exchanges.

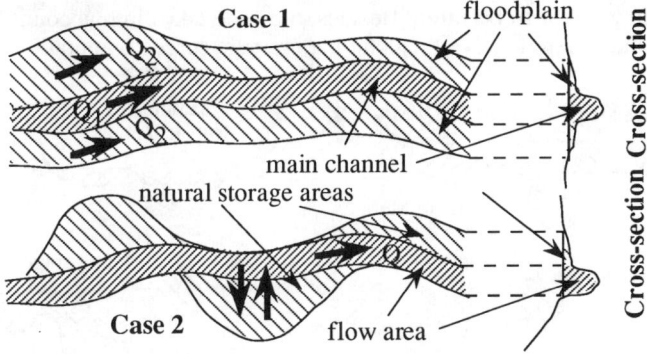

Figure 2 - Hydraulic influences between the main channel and the floodplain

In the first case, the typical shape of the cross-section on the whole is spatially uniform enough to conceive that each sub-section is hydraulically homogeneous, inducing two main streams. Several recent papers explain why common methods based on conveyance considerations lead to substantial errors, as well as why the right approach has so far to be found.

In the second event, the open water course is locally disturbed, especially on the floodplain, by topographic unevenness and manmade structures. All these hazards prevent the discharge to develop freely on the floodplain. It acts as lateral storage areas, with hydraulic death zones where water movements occur only in the transverse direction. The occurrence of culverts in the Espierres river in order to flow beyond embankment fills for highways or railways and the variable height of river banks imply that this last event has to be preferred.

Dealing with the complete set of over-depth integrated quasi-bidimensional Navier-Stokes equations, the river software reproduces all transient flows occurring in nets of natural rivers. They result from lateral exchanges with floodplains through a suitable statement of the term of lateral inflow or outflow.

The coexistence of several flow rates with shocks and bores in ramified nets of variable cross section arms requires the development of suitable capturing methods. The implemented scheme belongs to the class of implicit Petrov-Galerkin methods with dissymetric continuous test functions. The degree of decentering is optimized according to the local flow conditions. This shock capturing technique performs reliable simulations of unsteady sharp transitions, as verified for dam-break flood-wave simulations and jumps propagations.

Concerning the reservoir flood routing, only storage phenomena are considered. The influence of impulsive motion of the inflow is neglected and water surface is assumed to be horizontal. As illustrated on figure 4, the interface between both sub-sections acts as an imaginary solid sharp wall. Thus, we extend the available stage-discharge curve of lateral diversion weirs to this situation with an efficiency factor to be fitted. Besides, the code takes into account backwater effects of the different culverts, including pressure surge computations.

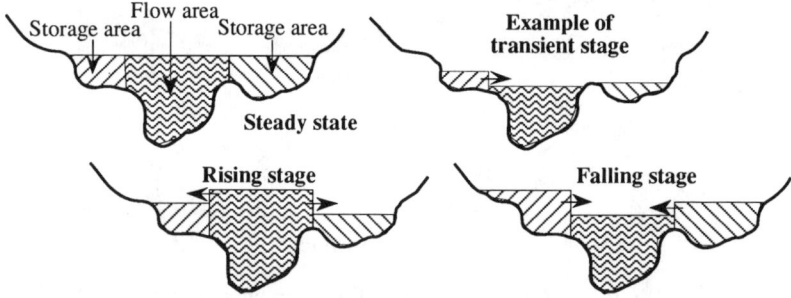

Figure 3 - Illustration of the transient exchanges

EXISTING SITUATION AND FORECAST IMPROVEMENTS

An excellent fit between computed values and gauged data is obtained with a physical relevant roughness coefficient that corresponds to referenced values for the same conditions. Figure 4 suggests the significant smoothing of discharge along the propagation, due to the natural storage in the surrounding fields.

After many computations, reservoirs were designed to lie on a respective maximum area of 6 ha and 11 ha. The figure 5 shows the location of the basins and the planned lowering of the maximum water profile. The weir sill is 200 m long for the first basin and 500 m long for the second one. In order to maintain wetted areas behind the weirs, a slight difference of level is maintained between the bottom of the reservoirs and the sills, with a delayed emptying by gates.

DESIGN BALANCING PONDS 661

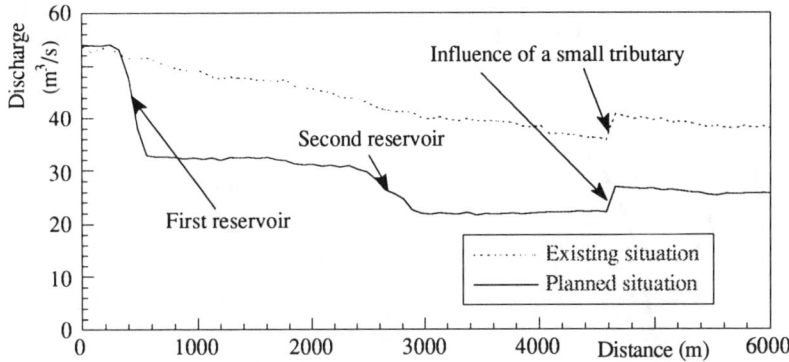

Figure 4 - Maximum discharge at each cross-section

The sole roughness of the main channel cannot induce any subsidience of the flood during its propagation, as suggest in figure 4 the constant levels of maximum discharges linking the characteristic stairs. No lowering occurs between both reservoirs where the flow is maintained in the main channel.

Figure 5 highlights the fundamental difference of results between a steady analysis and transients computations since the prospected modifications bring no significant effects upstream of the reservoirs.

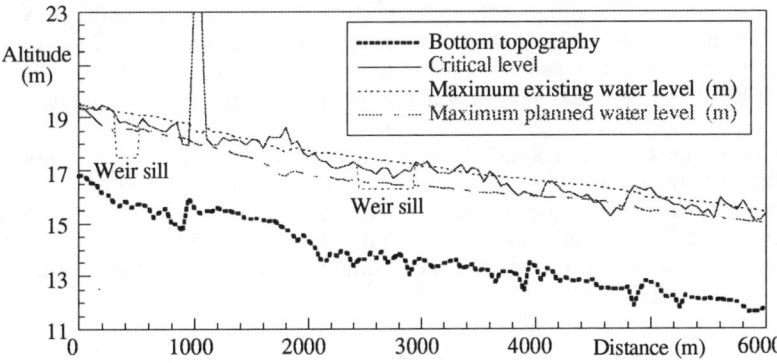

Figure 5 - Maximum water level at each cross-section

Figures 6 explains the gradual work of the fisrt weir for every 50 m spaced section of the sill. The computation proves that the downstream part is more efficient, inducing more lateral exchanges, as often theoretically demonstrated.

Computations assessed that discrepancies of the efficiency factor for this kind of diversion had minor effects on the global exchange balance, due to the characteristic time of the hydrograph. A middle value was selected in accordance with the works of Sinniger et al [2] for classical lateral thin crested weirs.

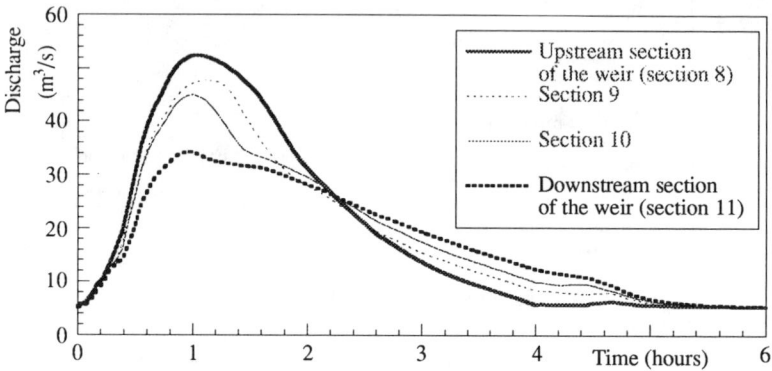

Figure 6 - Temporal evolution of discharge along the upstream weir

CONCLUSION

This article proved that we have to evolve towards transient computations in order to reproduce current floodings situations, to ensure a fit design and an adequate management. In several actual cases, the smoothing of the initial hydrograph can be mainly explained by lateral exchanges between the main channel and the floodplain. Natural as well as manmade obstacles prevent in practice a secondary flood to freely develop on floodplains. A first approach consists therefore in considering these areas as death zones. They maintain their water level by lateral exchanges which imply only transverse discharges.

The computation for a small belgian river assessed that they are correctly modelized by lateral weirs laws since the dynamics of these diverted flows can be neglected. With the assurance of reliable simulations gained by comparison with gauged measurements, simulations were then devoted to forecast the lowering of the water profile after the buildings of flooding basins in order to set up an economical design.

REFERENCES

[1] PIROTTON M.J.J., *A Global Approach of the Unsteady Surface Flows Computations Including Shocks, by Finite Elements*, "Numerical Methods in Engineering Simulation", editors : M. Cerrolaza, C. Gajardo, C.A. Brebbia, Computational Mechanics Publications, pp. 45-52, Mérida, Venezuela, 1996.

[2] SINNIGER R.O. ET HAGER W.H., *Constructions Hydrauliques, Ecoulements stationnaires*, Presses Polytechniques Romandes, Lausanne.

THE THREAT OF FLOODING AND THE PROBLEM OF PROTECTION OF TERRITORIES IN THE DELTA OF THE URAL RIVER IN VIEW OF THE RISING OF THE CASPIAN SEA LEVEL

V.F.POLONSKY, L.P.OSTROUMOVA,[1] Y.G.VIKULOV, R.R.MULIKOV[2]
1-State Oceanography Institute, 6 Kropotkinsky Per., Moscow, Russia
2-Department of Ecology and Bioresources, 10-a Abay Ave., Atyrau, Kazakhstan

ABSTRACT
Researchers investigated the processes of distribution of backwater from the sea to the mouth portion of the Ural River, wind-induced surges, hydromorphological characteristics of the delta. They also developed methods to estimate changes in the water level regime in the mouth portion of the Ural River in the presence of fluctuations in the level of the Caspian Sea, assessed the threat of flooding of various regions and objects in the delta of the Ural River, provided a hydrological substantiation for protective measures taking into account possible changes in the background level of the Caspian Sea.

INTRODUCTION
The range of fluctuations in the level of the closed water body, Caspian Sea, amounted to approximately 5 m within the recent millennium, with its most acute rises and drops taking several decades, featuring the rate of change of 10 to 30 cm per year. The periods of a relatively stable condition of the sea level appeared to be more prolonged. In the presence of significant fluctuations of the Caspian Sea, the territories adjacent to the sea edge and the deltas of rivers flowing into it are flooded and dry out from time to time. Especially complicated are these processes in river deltas, because flooding of residential areas here may occur not only from the sea, but also tens of kilometers from it - from the river - where reaches backwater from the sea. Owing to the sea level rise which occurred in 1979-95 - by 2.4 m, to the -26.6 m BS mark in the Ural River delta - there emerged an acute problem of protecting the large industrial center of Kazakhstan - the city of Atyrau - as well as the territories and objects adjacent to it. A scientific substantiation of protective measures shall include complex research and development of methods of calculation of levels in the mouth portion of the Ural River in the presence of a complicated interaction of the background sea level, discharge of the Ural River, wind-induced surges, taking into account morphological particulars of the delta and the mouth off-shore zone. In this case, besides the data of hydrometeorological stations, to assess discharge distribution in the delta braiding system, adjust level surfaces along the delta area, and investigate regularities of surge penetration into the delta, researchers used the findings of special expeditions arranged in 1989-93. Besides, level recorders were placed on a large scale along the area of the Ural River delta, and measurements of the braiding network of the delta were made. These data permitted to develop empirical methods to calculate level changes in the delta of river and sea origin, as well as to parametrize and verify the joint hydrodynamic model of the Caspian Sea, the mouth off-shore zone, and the delta of the Ural River.

GENERAL DESCRIPTION OF PROCESSES IN THE URAL RIVER DELTA

The modern stage in the forming of the Ural River delta began after the sea transgression in the late 18th century when, owing to the flooding of the relict delta, the river began flowing into the sea along the single course./1/ Later on, a ramified delta was formed by the early 30s of the 20th century in the presence of a relatively high sea level (with -25, -26 m BS marks), which had 10÷12 water courses flowing into the sea (Fig.1). While redistributing between Zolotoy and Yaik branch systems in the delta after a sharp drop of sea level, the overwhelming portion of the Ural River discharge (about 90%) flew into the sea, by the 70s, through the Ural Caspian Canal (UCC) built in 1932. This discharge has dispersed in the lower parts of the UCC among 20 fish passes. Practically all of the rest of the discharge flew into the sea through the Fish Canal built in 1965. Taking into account the newly built canals, a new hydrographic network has formed in the Ural River delta during the recent 60 years, with the masses of reed plants surrounding it on the former expanses of the mouth off-shore zone. Beginning with 1979, backwater from the sea goes up the mouth portion of the Ural River. When the sea (average) background level rose 2 m, water levels in Atyrau increased 1.7 m in low-water periods, and 0.5 ÷1 m, during freshets. Besides, the mouth portion of the Ural River is affected by sea surges causing additional level rises in the river. As a result, there increases the threat of water overflowing river bed banks and flooding the adjacent territories. In the upper part of the delta and in the vicinity of Atyrau, level rises from low-water periods to freshets are stipulated mainly by the river discharge. Here, this rise amounts to 2 ÷2.5 m at the peak of a high freshet. Surge-induced level rises comprise a relatively small addition to freshet levels here. In the seaward delta zone, on the contrary, the main flooding factor is surge-induced level rises, which may reach 1.5÷2 m. The change of discharge levels from low-water period to freshet amounts in this zone to about 0.5 m. Thus, to assess the threat of flooding of various delta regions and the city of Atyrau from the river, it is necessary to make estimates for level surfaces along the mouth portion of the river in the presence of different combinations of water discharges in the river and threatening surge-induced level rises in the mouth off-shore zone. This task can be solved using both field data and a numerical hydrodynamic model of the Ural River delta. However, the threat of flooding from the sea cannot be assessed without a numerical hydrodynamic model of the mouth off-shore zone of the Ural River.

EMPIRICAL METHODS OF CALCULATION OF DISCHARGE LEVELS AND SURGE-INDUCED RISES OF THE LEVEL IN THE MOUTH PORTION OF THE URAL RIVER

During the present-day rise of the background level of the Caspian Sea (H_s) (from -29.0 m abs. in 1978 to -26.6 m abs. in 1995), hydrologic stations (HS) in the Ural River delta - Peshnoy, Dzhambul, Rakusha, Atyrau - have accumulated the data on the rise of the water level in these points owing to the rise of Hs level. To assess the changes in discharge (background) levels without surge-induced fluctuations of the sea level, researchers used dependences describing the levels at these stations (H_i) as a function of total discharges of the Ural River (Q_0) arriving in the mouth portion at Atyrau HS section. They compared coordinates of these dependences for different years in the presence of different background Caspian Sea levels. As a result, empirical dependences were obtained, describing the changes in the water level at investigated stations with the breakdown by water discharges (Q_0) as a function of changes in the background sea level $H_i=f(Q_0,H_s)$ in the range of observed changes. In the absence of acute man-made alterations to the delta morphology and its hydrographic network, these dependences can be used to make an approximate forecast of discharge levels in the mouth portion of the Ural River, if there is a further increase in the level of the Caspian Sea. If the bottom of the adjacent off-shore zone feature a smooth gradient, discharge levels are thinning at the depth of approximately 2÷3 m.

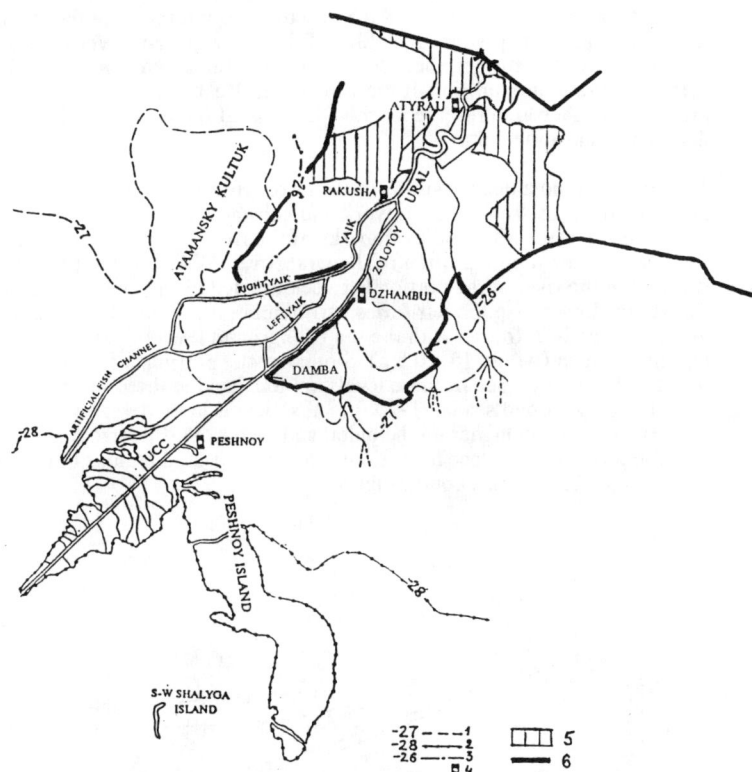

Fig.1. Scheme of the Ural River delta: 1 - Sea edge of the Ural River delta. early 1990s; 2 - Sea edge of the Ural River delta, early 1980s; 3 - Sea edge of the Ural River delta, early 1930s; 4 - Hydrometric stations; 5 - Regions of greatest threat of flooding from the river; 6 - Antisurge circular dike.

We can suggest that if each of the portions adjacent to the mentioned HS are flooded 2-m high, relations $H_i=f(Q_0,H_S)$ will become asymptotic to the $H_i=H_S$ line. Then, having extrapolated smoothly empirical relations $H_i=f(Q_0,H_S)$, we can obtain estimated forecast dependences. For example, for Atyrau, such dependence for an applicable range of -29.0 to -25.0 m BS will look like:

$$H_A = A \bullet H_S^2 + B \bullet H_S + C, \quad (1)$$

where: $A = -D_1 \bullet 0.017855 \bullet (Q \bullet 10^{-3})^2 + D_2 \bullet 0.029912 \bullet (Q \bullet 10^{-3}) + D_3 \bullet 0.037101$; $B = -D_4 \bullet 0.287138 \bullet LOG(Q) + D_5 \bullet 2.080098$; $C = D_6 \bullet 0.029875 \bullet Q^{0.703831}$.
Here and below, Dj - dimensional factors.

If we take into account the rate of change in the level of the Caspian Sea (about 15 cm per year), the recommended method of forecasting the levels at the mouth portion of the river may be reliable with an advance period of up to 10 years. This is enough to carry out protective measures in time. To a larger extent, the accuracy of such forecasts is determined by the correctness of a forecast of the sea and discharge level. Fig.2 presents predicted profiles of the water surface. When the sea background level rises, even in freshet, there forms a backwater-specific profile of the water surface at the mouth portion of the Ural River. Relatively great discharge gradients of the water surface are preserved if there is a rise of the background sea level higher than the head of the delta.

To reveal the most dangerous, as to water overflow and terrain flooding, individual portions of the hydrographic network and substantiate recommendations for their protection, a comparison of level marks of the river-bed banks and adjacent territories was made at the mouth portion of the Ural River. With H_s less than -26.0 m BS, level surfaces in the river at the portion between the delta head and the sea edge even in freshet are lower than the territories surrounding the delta. Higher up the delta head, discharge levels in freshet with the sea background level of -26.5 and -26.0 m BS do not differ much (within 10÷20 cm). Similarly, the portions of a possible flooding from the river in the presence of these levels are localized in these same places (Fig.1). With the background sea level -25.0 m abs., levels in the river (with $Q_0=2,000$ m^3/s) will rise 0.3 ÷0.5 m higher up the delta, and water overflow from the river will become possible any place here, except for the elevated regions of the city of Atyrau within 24.0÷23.5 m abs. contour lines.

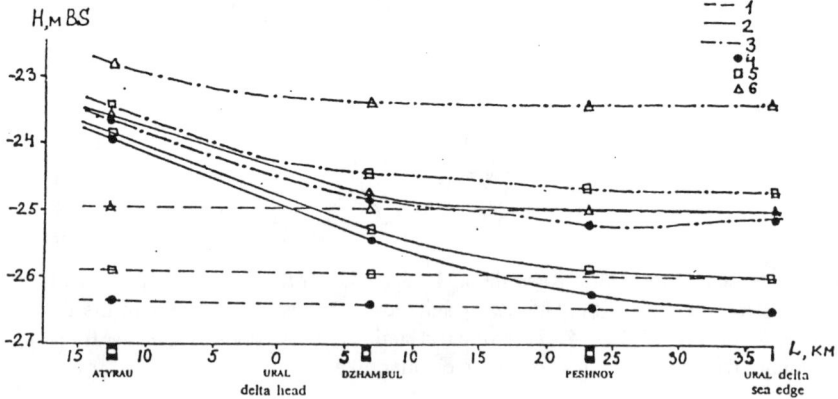

Fig. 2. Forecasted Water surface profiles at the Ural River mouth portion: 1 - Water surface profiles with the total discharge in the head oif the Ural River delta $Q_0=200$ m^3/s; 2 - Water surface prfiles with the total discharge in the head of the Ural River delta $Q_0=2,000$ m^3/s; 3 - Water surface profiles at the maximum level during the storm.

While investigating the regularities of surge penetration into the mouth portion of the river, one may use such characteristic as the surge factor, C_{sg}:

$$C_{sg}=H_{sg}/H_{sg},\qquad(2)$$

where: H_{sg} - surge height in a reference point at the mouth off-shore zone used to assess the surge fading at the mouth portion of the river, H_{sgj} - surge height at a point of the mouth portion of the river.

It has been found that when river discharge values drop, surge factors at the mouth portion of the river increase. The intensity of surge dying down in the lower (off-shore) zone of the Ural River delta rises sharply with increased water discharges in the Ural River in freshet. /2/ Surges penetrate freely into the delta in the presence of water discharge values specific for low-water periods and extend higher up the Ural River as far as Atyrau, practically without fading. In the presence of higher discharge values, the extent of surge-induced level fluctuations is sharply attenuated and amounts above Atyrau to within 10% of the surge heights in the UCC mouth.

Revealed dependences of surge attenuation at the mouth portion of the Ural River can be used to estimate the total water levels for specified discharges of the Ural River and surge heights in reference points at the mouth off-shore zone (mouth of the UCC) in the presence of sea background levels of about -27.0 m BS. These heights can be estimated from wind parameters in Peshnoy residential settlement using our suggested dependence:

$$H_{sg} = -D - \bullet 0.1231 \bullet I^2 + D_8 \bullet 0.81713 \bullet I - D_9 27,244022, \quad (3)$$

For a particular surge, the (I) value proportional to the momentum conveyed by the wind to the water shall be determined through a summation of square projection values of average, between observation periods, wind speeds to the effective direction for the period of calculation:

$$I = \sum_{i=1}^{i=k} (V_i^2 \bullet Cos\alpha_i), \quad (4)$$

where: V_i - wind speed for an individual period, m/s; α_i - angle between the observed and the efficient wind directions. The summation is done using the data of time-specific wind observations beginning with period $i=1$ as of the moment of the beginning of the surge and ending with period $i=k$ as of the moment of reaching the maximum speed of the surge wind during the surge period. The efficient wind direction is assumed to be the one coinciding with UCC center line in its mouth portion.

Surge observation series at the off-shore zone of the Ural River mouth are not uniform and are small statistically. However, the wind observation series in Peshnoy is enough for statistical assessment. Hence, dependence (3) was used to switch over from I values of a definite probability to Hsg values of the same probability.

USE OF NUMERICAL HYDRODYNAMIC MODELLING TO ASSESS THE FLOODING THREAT IN THE URAL RIVER DELTA

In the presence of fluctuations in the sea background level, both the long-term and the operative forecasting of the level in the mouth portion of the Ural River make it expedient to use hydrodynamic modelling. For this purpose, we applied the joint numerical hydrodynamic model of the delta, off-shore zone of the Ural River mouth, and the Caspian Sea developed by V.F.Polonsky, S.Y.Kasyanov, and Y.G.Fillipov. Variant calculations were made to estimate distribution of flooding levels and contours in the Ural River delta for various combinations of Caspian Sea background level ranging between -26.5 and -25.0 m BS, surge and wind settings, and water discharges in the river of various probability. Results of calculations showed a significant protective role of the reed-plant barrier in front of the Ural River delta. Its presence reduces surge-induced levels in the delta by 20 to 50 cm. It was shown that in the absence of protective structures, presence of sea background level of -26.5 m, and surges of less than 5% probability ($H_{sg} > 1.5$ m over the sea background level), a serious threat of flooding from the sea is incipient for residential settlements on the left shore of Zolotoy branch below the head of the died out Zolotyonok arm. With the sea background level of -26.0 m, permanently flooded will be settlements on the left shore of Zolotoy branch. The territory of Damba residential settlement will be permanently below the sea level, therefore it is expedient to evacuate it. Besides, in surge periods with heights featuring less than 5% probability, the threat of flooding will also be incipient for residential settlements Yerkenkala, Taskala, and such important economic objects as the settler of the city sewage and the oil refinery. With the sea background level of -25.0 m, all residential settlements on the left shore of Zolotoy branch below the head of Zarosly arm, including Dzhambul and Taskala, will be subject to permanent flooding from the sea. The sea will move directly to the settler of the city sewage and the oil refinery. With the absence of protective measures, even in average surge and freshet periods, the whole of the delta will be flooded, except for the elevated hills of Atyrau City. In a surge period, the lower parts of the city terrain may appear to be two meters below the water surface.

As one of protection measures, an opportunity of distributing the discharge along a number of older, died out at present, delta courses was examined, or of discharging the freshet waters at the head of the delta using natural terrain depressions around Atyrau. However, these measures were found to be irrational, because with existing water surface gradients, and taking into account the distance from the place of discharge to the sea, the effect of water level reduction in freshets may amount to only 0.5 -1 m in Atyrau. Besides, erection of water discharge canals is rather complicated and labor consuming. Moreover, if they are implemented, there will be a number of negative consequences for the environment. Therefore, a main protective measure is considered to be erection of dikes both along the river, and against the sea. A possible flooding from the river in freshet along the whole of its mouth portion can be easily prevented by raising the river bed banks to the height of estimated marks of the total discharge and surge levels (Fig.2). With the sea level range being -26.0 m BS and below, a long circular dike (Fig.1) will be efficient to protect from surges not only the city of Atyrau and residential settlements along the Ural River, but also agricultural facilities. If the sea level reaches the mark of -25.0 m BS, a minor circular dike will be efficient to protect only the city of Atyrau with its adjacent important economic objects. In both cases, the height of protective dikes for most of their length ought to amount, according to our estimates, to within 2 m. However, at the most threatening parts adjacent to concave dike contours, especially those facing the strongest south-west storms, their height must be up to 3 m. Such most dangerous point in the first option of protective dike (Fig.1) is the cone east of Damba and Dzhambul residential settlements.

REFERENCES *(in Russian)*

1. Polonsky, V.F, Lupachyov, Y.V., Skriptunov, N.A. Hydromorphological Processes in River Mouths and Methods of Their Estimation (Forecasts). St.Petersburg, Gidrometeoizdat Publishers, 1992, 384 p.

2. Polonsky, V.F., Ostroumova, L.P. Hydrological Processes at the Mouth Portion of the Ural River in the Presence of the Caspian Sea Level Rise. Herald of the National Academy of Sciences of Republic of Kazakhstan, No. 3, Almaty, NANRK Publishers, 1995, p.p. 11-23.

NEW DIMENSION TO SPILLWAY GATE OPERATION FOR REDUCING THE FLOOD INTENSITY

Sunil Kute and M.J.Deodhar

ABSTRACT

The incoming flood approaching the reservoir that has already reached its Full Reservoir Level has always been enforcing the operation of spillway to its critical stage. Under such circumstances, the dam authorities have no other option than to let out total incoming flood over the spillway to avoid enchrochment on free board.

In the present paper, a hypothesis is put-forth which suggests to take more water from the reservoir in advance till approaching floods are increasing. Once the flood hydrograph reaches the point of contraflexture, the spillway gates are operated with desired closing rate. This results not only in reducing the flood intensity to the desired one but also assures the full reservoir storage.

The control of discharge achieved has been designated with a new term M.J.D.(Monitered Joint Discharge)

INTRODUCTION

1.1 Normally the spillway of a dam is operated when the reservoir is full and the inflow occurs. Initially the spillway gates are operated partially and the discharge is let down keeping the reservoir water level at F.R.L.(Full Reservoir Level). Under these conditions, the inflow is equal to outflow. As the inflow increases the gates are opened more and more and finally the gates are fully opened to discharge the maximum flood. This corresponds to no flood attenuation and intensity of incoming and outgoing floods is the same.

1.2 As inflow starts incoming into the reservoir, initially the discharge may be let down more than the incoming flood. This more outflow will be at the cost of reservoir storage. This will result in lowering the water level in the reservoir. As the inflow increases after a critical stage the gates may be operated in such a way that the some part of the inflow will be absorbed in the reservoir to compensate the volume drawn initially from the reservoir storage. The part of inflow which is thus, utilised to meet the requirement of the reservoir sotorage will result in increasing the lowered water level in the reservoir. This will lead in reducing the flood intensity of higher discharge.

HYPOTHESIS

2.1 To implement this concept, the inflow should be known which can be calculated from

1. Outflow from spillway

2. Changed storage of reservoir worked out from change in water level of reservoir.

2.2 The release of higher outflow than inflow can be implemented only upto the point of contraflexture of inflow hydrograph. This point of contraflexture can be located if inflow hydrograph is studied at a regular interval. The change in the slope of inflow flood hydrograph will indicate the exact location of point of contraflexture. The moment the point of contraflexture is defined as the flood hydrograph tends to change its slope beyond this point, the increased release of outflow is reduced suitably to compensate the withdrawal from reservoir storage. Monitering of controlled increase of outflow upto the point of contraflexture and reducing thenafter, to compensate the reservoir storage is designated as "Monitered Joint Discharge(M.J.D.)". The M.J.D. can be corelated with desired factor of safety in reducing the flood intesity over the spillway.

2.3 The M.J.D. and corresponding gate operation has to be controlled at the spillway site where outflow discharge and change in reservoir water level is readily available.

CASE STUDY

3.1 The above hypothesis was applied to a specific case for an existing dam. The following data was considered.

a] Outflow from spillway(Fig. 1)
b] Storage capacity of the reservoir (Fig. 2)
c] Full Reservoir Level which was treated as initial water level

3.2 As the inflow started incoming into the reservoir, the spillway gates were operated partially to let down more volume of water than the inflow at the cost of reservoir storage. Due to this, water level dropped by 8.0 mm when outflow discharge was 1300 m³/sec. after 1.66 hours. From this data which was readily observed at spillway site, the inflow was worked out to be 600 m³/sec. This procedure was continued at an interval of 1.66 hours to get the nature of inflow hydrograph. Simultaneously, the inflow hydrograph was plotted to study its slope. It was noticed that the point of contraflexture occured at 16.66 hours. Thenafter the outflow was reduced steadily allowing the water level in reservoir to rise upto F.R.L. The spillway gates were operated in such a way that water level was maintained at F.R.L. The procedure followed is shown graphically in fig.3. It can be seen that the intensity of flow hydrograph has been reduced from 20000 m³/sec to 18000 m³/sec.

DISCUSSIONS

4.1 Application of the hypothesis presented here,reveals that

there is definitely flood attenuation , thus reducing the
intensity of the flood. However, this has some constraints

1   A close watch has to be kept on
    i]   Rate of Reduction in water level of the reservoir
    ii] Assessing the point of contraflexture of inflow
         hydrograph.
    iii] Acurate observation of the water level in reservoir
         and subsequently the change in reservoir storage
         is obligatory.

2   If the catchment is fan shaped, the inflow hydrograph
    rising curve will have a steep slope. In this case
    increase in outflow than inflow will pose a difficult
    task.

CONCLUSIONS

5.1 Implementation of the hypothesis can be concluded as

1. Flood intensity can be reduced by controlling increase
   in outflow upto the stage where inflow hydrograph
   reaches its point of contraflexture.

2. The greater factor of safety in functioning of
   spillway can be achieved.

3 Higher flood than the designed flood can be
  discharged safely.

4. Flood intensity on the down stream of the dam can be
   reduced.

5. Monitered Joint Discharge (M.J.D.) corelated with
   designed factor of the safety can be established in
   case of each dam.

------o------

COPING WITH SCARCITY AND ABUNDANCE

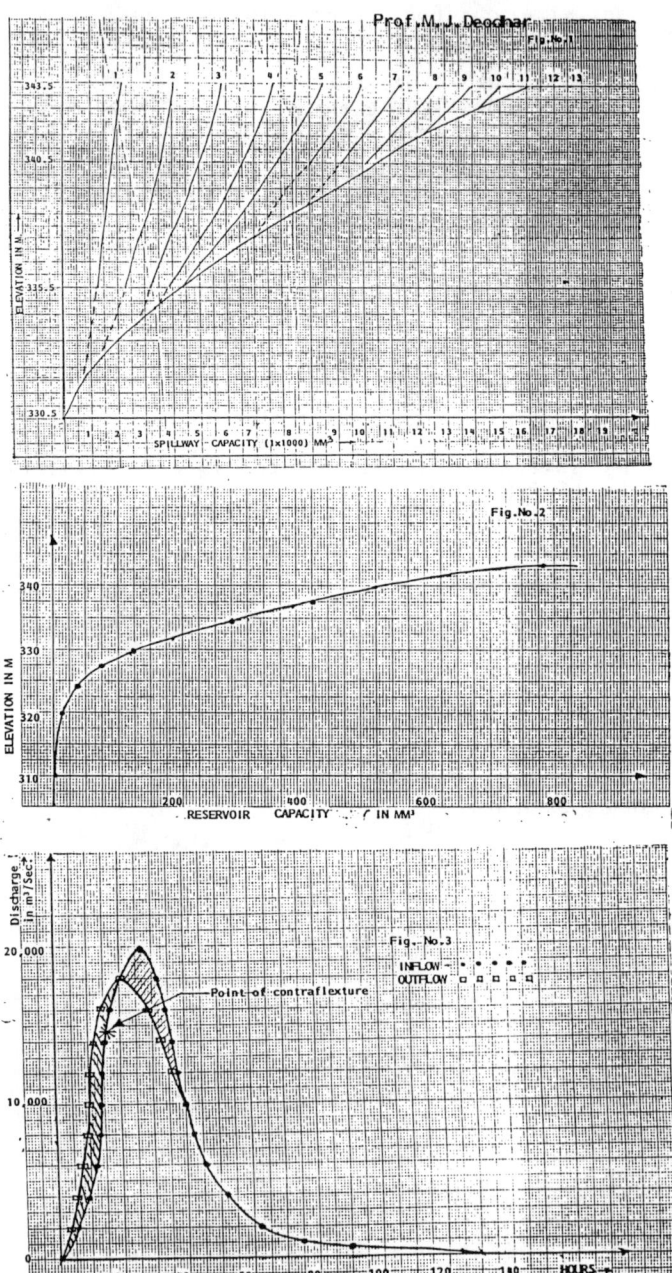

A FLEXIBLE IRRIGATION WATER SUPPLY
WHY AND HOW

PAUL R. CROSS, P.E.
General Manager
Lake Chelan Reclamation District
Manson, WA 98831
USA

WHY

A flexible irrigation water supply is important first and foremost to the farmer. Flexible supplies mean that a farmer can grow crops efficiently and economically to compete in a world wide market place. Irrigation water supply systems should be designed and operated to meet the needs of the farmer not the convenience of the operator. Flexible water supplies mean that the educated farmer can choose almost any means of on-farm irrigation practice and be nearly equally efficient. With a flexible supply efficiency depends upon the complexity of the on-farm system, the level of care and the amount of labor the farmer chooses to add to the effort.

HISTORICAL PERSPECTIVE

Irrigation water supplies and distribution systems have been historically designed around four fixed and rigid parameters. These parameters are total acres served, an average instantaneous flow rate per acre, an expected duration per irrigation in whole day increments to be delivered by the purveyor and an equal rotational frequency. These parameters led to supplies that could be both easily and simply controlled. Demand was predictable and continuous. Diversions from streams or reservoirs could be steady and ordered well in advance if artificially controlled.

Moderately good irrigation efficiency under these systems was possible but in most cases severely constraining to the farmer. These constraints sometimes allowed only certain crops to be grown efficiently. Other crops were not sustainable under the rotational constraints when coupled with the allowable rates and existing soil conditions. In most cases irrigation efficiency could only be obtained by supplying a significant labor effort to manage the water supply provided to the on-farm system. In years past when labor costs were relatively modest and additional labor supplies readily available the consequences of a constrained system were tolerable. As time went by labor became more expensive and less available, water supplies become fully appropriated or sometimes over-appropriated, farm sizes increased and water efficiency declined.

ECONOMICS

Farmers today compete in a world wide market place. They need to economically and sustainably grow food. The world economy is also becoming highly competitive. Farmers are now required to concentrate on both total yield and crop quality to both stay competitive and to market their product. Flexible water supplies can improve yields a number of ways including improving germination conditions and allowing for proper timing of pre-irrigations. This can lead to more vibrant crops and longer growing seasons. That in turn can lead to the ability to do multi-cropping in certain areas for certain combinations of crops. In tree fruit production a flexible water supply can be used for secondary functions such as frost control and fruit cooling. Frost control during the development stages of certain fruit crops is essential to crop production. Evaporative fruit cooling during hot summer days is used to prevent sunburn of certain fruit crops and to improve the quality and marketability of the fruit.

ON-FARM SYSTEMS

A flexible water supply allows the farmer the opportunity to choose an on-farm system that best meets the needs of the desired crop, the cost and availability of labor and other local economic or social situations. All major methods of on-farm irrigation systems can be nearly equally efficient if they can be managed properly (Hanson, 1995). For certain crops both yield and crop quality require that select on-farm systems be used. Other choices are constrained by the capital investment and payback on the on-farm system. The rotational nature of some crops may preclude the farmer from investing in a permanent irrigation system that is capital intensive. Economic conditions such as labor costs and availability, the cost of water, the cost of electricity or other support mechanisms may impact the cost and sophistication of the on-farm system.

Water supplied under a rigidly controlled rotational system can reduce the efficiency of even the most sophisticated, well designed on-farm system (Replogle, 1987). On-farm management can be handcuffed from responding to changing crop irrigation requirements. Flexible supplies can allow furrow and border strip systems to select optimum instantaneous rates to match infiltration. Flexibility to manipulate the rate and duration of flows in the course of the irrigation can insure uniformity of applications. Systems designed to apply water uniformly such as sprinkler systems still need flexibility in rate when irregular field shapes require irrigation laterals to be different lengths. Flexibility in duration helps to avoid overfilling the soil profile when a partial set does the job. Flexibility in frequency can give the grower the opportunity to avoid certain weather conditions such as rain or wind that may change the water requirement or impact the uniformity of the application. Even trickle or drip irrigation systems can be mismanaged without flexibility in rate, frequency and duration. These systems can be run too long and result in deep percolation or can be run too short and result in deficit irrigation or crop stress.

IRRIGATION WATER MANAGEMENT

Water use efficiency and runoff or deep percolation are interrelated. Excess water applications lead to rising water tables, increased demands on drainage systems and most typically degradation of receiving water bodies or ground waters. Waters that run off the land or percolate through the root zone to the ground water are typically reduced in quality by sediments, nutrients or herbicides. It becomes vitally important to manage water applications to match target crop demands. Irrigation water management starts with the on-farm system. The on-farm system must be designed to apply water at a rate less than or equal to the infiltration rate of the soil. Infiltration rates can vary over time. A flexible supply can be utilized to adjust the rate of application to match the infiltration rate changes and to avoid surface runoff. The avoidance of runoff can mean less erosion and significantly less contributions of sediments and nutrients to receiving water bodies.

Once the instantaneous rate of irrigation is adjusted correctly the irrigation water manager must next insure a uniform application across the field. This is typically achieved by having control over the duration of the irrigations. The manager must finally match the volume of water applied to the desired volume needed. The desired volume must consider the current moisture levels in the soil, the water holding capacity of the soil, the projected crop irrigation requirements and the frequency that irrigations can be applied. Applied volumes that take the soil moisture over the water holding capacity of the soil will result in either deep percolation or saturated soil conditions. Saturated soils may not lead to water quality problems but they do result in poor conditions for the growth of most crops.

Deep percolation generally results in the leaching of nutrients, fertilizers and herbicides out of the root zone. This is not only a problem environmentally and economically but also typically hurts crop production and sometimes crop quality. Applying volumes less than the soil water holding capacity can be done but care should be made to quantify the actual crop water requirements. If irrigations are less than the crop water requirements then deficit irrigation occurs. Care must be taken to avoid drying out the soil below the maximum allowable depletion of the crop being grown. Soil moistures below that level will stress the crop and result in reduced yields, reduced crop quality or both.

WATER QUALITY CONCERNS

When runoff or deep percolation does occur drainage waters and ground or surface receiving waters can be negatively impacted. Nutrients, fertilizers and herbicides meant to remain in the root zone of the crop can be carried away from their target areas. Runoff can carry sediments off the field further degrading surface water receiving bodies.

In many areas the sustainability of agriculture is threatened by environmental concerns. As regulators continue to tie water quantity and water quality issues together water efficiency must improve and total water supplies diverted from instream flows must decrease. Non-point source pollution becomes a factor in other social issues such as the care of endangered species and competition among various water user interests for what is in the public interest. Other benefits include reduced drainage, lower water tables and less salinity.

HOW

Truly on-demand systems are not economically practical in most situations (Clemmens, 1987). A practical compromise is a system utilizing a limited rate arranged system. Limited rate arranged systems can be designed and managed to provide the flexibility in frequency, rate and duration that is needed. Storage is a critical component in the system to absorb the large daily variations in demand. Canal automation provides more sophisticated methods of meeting rapid variations in demand as well as in-canal storage (Buyalski, 1991). In most cases it appears that storage reservoirs in combination with canal automation is the most logical approach. Other papers in these two sessions of the proceedings will discuss the "how" in more detail. These papers will discuss levels of modernization typical of most irrigation districts throughout the western United States, design concepts for pipelines and canal reservoirs and case studies of limited rate arranged systems showing their successes and failures in providing the desired levels of flexibility.

REFERENCES

Hanson, R., et al 1995. Field Performance of Microirrigation Systems, Proceeds Microirrigation Conference 1995.

Replogle, J., 1987. Irrigation Water Management with Rotational Scheduling Policies, Proceeds Planning, Operation, Rehabilitation and Automation of Irrigation Water Delivery Systems 1987.

Clemmens, A. J., 1987. Delivery System Schedules and Required Capacities, Proceeds Planning, Operation, Rehabilitation and Automation of Irrigation Water Delivery Systems 1987.

Buyalski, C.P., et al 1991. Canal Systems Automation Manual, Vol. 1, U. S. Bureau of Reclamation, Denver, CO

Irrigation District Modernization for the Western U.S.

C. M. BURT, S.W. STYLES, M. FIDELL, AND E. REIFSNIDER

Director, Irrigation Training and Research Center (ITRC), Cal Poly, San Luis Obispo, CA, USA; Project Manager, ITRC; Graduate student, Cal Poly; former head of the Water Conservation Office, USBR Mid-Pacific Region, Sacramento, CA, USA

OVERVIEW

Many irrigation districts throughout the western U.S. have been actively engaged in modernization efforts. In most cases, the modernization improves the level of water delivery service (flexibility and reliability) provided to farmers. In many cases, the impetus and/or funding has come from external sources. These external sources include persistent droughts, an opportunity to sell water which is conserved and transferred, the need to increase in-stream flow rates by reducing diversions from rivers, and the need to improve downstream water quality by decreasing the drainage outflows.

The Irrigation Training and Research Center (ITRC), located within the BioResource and Agricultural Engineering Department at California Polytechnic State University in San Luis Obispo, has been actively involved in assisting irrigation districts for many years. This assistance has taken many forms. One has been the offering of numerous short courses for district personnel on topics such as SCADA (Supervisory Control and Data Acquisition), flow measurement, canal automation, and canal modeling. These courses have typically been funded by organizations such as the US Bureau of Reclamation (USBR), the California Energy Commission, California Dept. of Water Resources, or individual districts. ITRC also provides troubleshooting and brainstorming consulting services to many water districts. Regardless of the initial motivation for modernization which a district may have had, ITRC consistently works with the districts to achieve better water delivery service to farmers while also meeting or exceeding the original goals.

STATUS AND NEEDS SURVEY

In 1995, ITRC gathered data from 61 agricultural districts within the Mid-Pacific Region of the USBR by interviewing irrigation district personnel and studying their Water Conservation Plans to complete a "Status and Needs

Survey" (Burt et. al., 1996). These districts cover 902,000 hectares, comprising about 90% of the irrigated acreage in Mid-Pacific Region districts. Data were analyzed to determine general demographic information, the degree of water delivery flexibility provided to farmers, and the extent of existing and planned district modernization. The interview process defined needs for direct technical assistance and training.

This paper summarizes some of the results of the Status and Needs Survey, and gives a brief overview of what is being done by some of the districts.

FLEXIBILITY INDICES
Urban homeowners are accustomed to receiving water from the tap "on demand" (i.e., without providing advance notice), with unlimited flexibility in frequency (when), duration (how long), and flow rate. In the Mid-Pacific Region, agricultural water users (i.e., farmers) receive water with a high degree of equity (not measured in this study) and with much more flexibility than most of their counterparts in other areas of the world. Nevertheless, the flexibility of water deliveries in the Mid-Pacific Region does not compare with the "demand" flexibility provided to homeowners.

<u>Frequency Flexibility</u>
835,500 hectares have policies which allow farmers to receive water on an unlimited frequency schedule (Table 1), as long as they order water in advance. For farmers who have an unlimited frequency schedule, the mean advance notice time was 26 hours, and the mean number of times a farmer cannot get water on his requested day is once per season.

63,800 ha (7% of the total area) use a form of rotation schedule. Of these, 56,100 ha use a fixed rotation with trading turns between farmers, and 7700 ha use a modified rotation schedule. None of the districts surveyed use a strict fixed rotation (no trading turns) or a fixed rotation during peak water use periods.

Table 1. Analysis of Districts with Various Frequency Policies* (n=61)

Type of Schedule	Total Hectares	% Total	Number of Districts
Fixed Rotation (with trading turns)	56,100	6	1
Modified Rotation	7,700	1	1**
Unlimited Frequency	835,510	93	60

* "Frequency" pertains to a farmer choosing the day he receives water.
** One district had unlimited frequency on most of the district area, but had a modified rotation on other areas.

Flow Rate Flexibility

Only one district responded that farmers could not receive different flow rates for each irrigation - although this district allows farmers to receive several different flow rates throughout the season. The remaining districts have policies allowing farmers to receive different flow rates at each irrigation.

Similarly, 56 districts have no restrictions on changing a flow rate *during* an irrigation event; the average advance notice before changing flow rates during an irrigation is 13 hours. Three districts do not allow any flow rate changes during an irrigation. Seventeen districts have a policy of no advance notice required before a flow rate change.

Duration Flexibility

Thirty-four districts have policies allowing farmers to receive water for any arranged duration. The remaining districts allow delivery durations of 12 hours, 24 hours, or other fixed increments. The advance notice required before farmers can shut off the water ranged from 0 to 24 hours, and averaged 6 hours; seven districts do not require advance notice to shut off.

Duration flexibility is important for all forms of on-farm irrigation, but it is very difficult for irrigation districts to allow farmers to shut water off unannounced or at odd times - canals and pipelines with conventional control hardware can overflow if this happens. Farmers would like more duration flexibility to reduce over-irrigation, and avoid unnecessarily high bills and deep percolation of water and nutrients. Drip and microirrigation systems are easily automated to provide the correct amount of water to replace evapotranspiration (ET) plus losses due to non-uniformity, so they are ideally suited for management with unlimited duration flexibility. Since soil infiltration rates change through the season with surface irrigation, farmers rarely know exactly when they will complete an irrigation. Since an irrigation could be finished at any hour of the day or night, farmers can prevent overirrigation if they can shut off their water with no advance notice.

Farmers want a high degree of flexibility in irrigation delivery duration; ideally farmers operate their own turnouts. If the district requires that a district employee operate the turnouts, the farmer's ability to automate an on-farm irrigation system disappears.

Many water conveyance systems, delivery canals and pipelines are not designed with adequate control systems to permit farmers to operate turnouts. Often, when one farmer makes a flow rate change, the ditchrider must move along the complete length of the supply canal or pipe to readjust the flows of other open turnouts. On average, district personnel must be present to open and close farm turnouts nearly 50% of the time. On average, district personnel operate gates within one hour of the prescribed time. When there is not enough flow to match a water order, 22 districts pro-rate the order and 27 districts postpone the water.

Most irrigation districts have areas of their distribution system with limited capacity. When farmers request water orders, district personnel must check the pipeline/canal capacity to ensure there is enough capacity to supply that order without adversely affecting other users.

Flexibility Index (District Level)
The above mentioned aspects of district delivery policies regarding frequency, flow rate and duration were indexed to quantify the "extent" of flexibility within each district. Each parameter (frequency, flow rate and duration) has a rating from 0 - 5, with 5 as the most flexible score. The sum of these individual indices gives the "Flexibility Index". A flexibility index of 15 is the highest score possible.

The Flexibility Index defined in Table 2 was developed as a performance index that can be used in future studies. The average indices for frequency, flow rate, and duration were 3.3, 4.3, and 4.0. The average total flexibility index (i.e., the sum of the frequency, flow rate, and duration indices) was 11.6 out of a possible 15. Overall, the flexibility indices were high - all districts had flexibility ratings greater than 10. The overwhelming majority of districts (54) had flexibility ratings less than 13; one district received a perfect score of "15".

Table 2. Definition of the Flexibility Index

Points	Condition
	FREQUENCY
1	Always a fixed rotation
2	Fixed rotation with trading, or limited frequency, or fixed rotation during peak season only
3	24 hours or more advance notice required before delivery is made
4	Less than 24 hours advance notice required before delivery
5	Farmer does not need to notify district before delivery
	FLOW RATE
1	Same flow rate must always be delivered
2	Several flow rates are allowed during the season
3	A different flow rate is available each irrigation, with up to 2 changes per irrigation allowed
4	Flow rate can be changed any time, provided advance notice is given to the district
5	Flow rates can be different and changed by the farmer without giving advance notice to the district
	DURATION
1	District assigns a fixed duration of irrigation
2	District assigns a fixed duration, but allows some flexibility
3	Farmers must select a duration with a 24 hour increment
4	Farmers can choose any duration, but must give notice before changing
5	Farmers can have any duration, with no advance notice required before changing

ON-FARM IRRIGATION METHODS

Recognizing the types and acreage using different irrigation methods helps in understanding the degrees of supply flexibility required by farmers. Farmers vary in their need for technical and educational support depending on their irrigation method; drip systems require frequent, flexible water deliveries. Over half (53%) the total acreage represented by the Survey used surface irrigation methods (i.e., furrow, border strip, or basin). Sprinkler and drip irrigation represented 19% and 13% of the total irrigated acreage, and is expected to increase. The remaining acreage was irrigated rice (6%) or used combination irrigation methods (i.e., hand-move sprinkler and drip on row-crops) (Table 3).

Table 3. On-farm Irrigation Methods Used Within District Service Areas (n=61)

Irrigation Method	Hectares	Percent of Total
Furrow	325,700	38
Border Strip or Basin	130,300	15
Hand Move or Side Roll Sprinklers	89,900	11
Center Pivot or Linear Move	1,200	<1
Permanent Sprinklers (trees or vines)	24,000	3
Rice	49,200	6
Drip on Row Crops	7,500	1
Microspray or Drip (trees or vines)	98,600	12
Solid Set Sprinklers on Row/Field Crops	34,800	4
Combination	82,900	10
TOTAL	844,100	100

WATER PRICING

The majority of interviewed districts (45 districts representing 666,100 ha) charge for water on a volumetric basis. The mean price is $398/ha-m ($47.80/AF). Twelve districts representing 225,100 ha use a fixed pricing structure; seven districts charge different prices depending on the crop type.

DELIVERED WATER

The water supply allotted to the districts is highly variable, by both district and by year. Districts that experience wide fluctuations in water supply almost always see ground-water recharge as a major concern, and their policies may emphasize recharge during wet years rather than flexible deliveries during average or dry years. Districts had .76 m average gross water available for deliveries the last ten years, including both surface and groundwater supplies.

ITRC/USBR ASSISTANCE

The Status and Needs Assessment determined what types of structures, communication systems, flow measurement devices, conveyance facilities, etc. were in use at the present, and what the districts plan to invest in for future improvements. As a result of those findings, ITRC and USBR embarked on an aggressive program of technical assistance to districts. This program is offered to districts on a cost sharing basis, and only upon the request of the districts.

Key components of this assistance include:
- Rapid appraisals of the district modernization needs. A 1-2 day survey is conducted with district personnel, and recommendations are then given.
- Improvement of flow measurement and flow control techniques. This includes assistance with the selection and design of structures, as well as training.
- Development of RFQs (Request for Qualifications) and RFPs (Request for Proposals) for SCADA systems. These documents provide detail of the required hardware and software for the district. The development of these documents is an important learning opportunity for district personnel, as they must develop a master plan for modernization in order to properly specify the SCADA needs.
- Design of improvements for drainage or irrigation water recirculation and storage facilities, to reduce surface discharges and to increase delivery flexibility.
- Technical assistance in selecting proper structures for flow control and water level control (upstream or downstream), or for improvements in delivery through pipelines.
- Development of improved PI (Proportional Integral) algorithms for upstream control, and demonstration of the Begemann gate for upstream water level control.

The response by the districts to this technical assistance has been strongly enthusiastic.

REFERENCE

Burt, C.M., K. O'Connor, S. Styles, M. Lehmkuhl, C. Tienken, and R. Walker. 1996. Status and Needs Assessment: Survey of Irrigation Districts - USBR Mid-Pacific Region. Irrigation Training and Research Center (ITRC). Cal Poly. San Luis Obispo, CA 93407.

EXPERIENCE IN OPERATING A LARGE LIMITED RATE ARRANGED SYSTEM - WESTLANDS WATER DISTRICT

STEPHEN H. OTTEMOELLER
Director of Resources, Westlands Water District, Fresno, California, USA

ABSTRACT
The benefits of flexible agricultural water deliveries are frequently discussed in terms of potential cost and water savings. Westlands Water District, a 600,000 acre (240,000 ha) agricultural district in the San Joaquin Valley of California, has operated a limited rate arranged distribution system for thirty years. The economic and water efficiency benefits of their system are readily apparent, but there are also some lessons which can be learned from Westlands' operations experience and, perhaps, incorporated into future new distribution system or rehabilitation designs.

THE DISTRICT
Westlands Water District is located on the west side of the San Joaquin Valley in central California. The District averages 15 miles (24 km) in width and stretches 70 miles (112 km) from the town of Mendota on the north to Kettleman City on the south. The San Luis Canal/California Aqueduct (SLC), a canal owned jointly by the State of California and the federal government, runs the length of the District such that about two-thirds of the district is served by gravity and about one-third of the District is served by pumped water. The land to the east of the SLC generally slopes gently from an elevation of about 320 feet (100 m) above mean sea level to about 160 to 200 ft (50 to 65 m). The land to the west of the SLC generally has steeper slopes, rising to over 500 feet (150 m), and the fields are less likely to be leveled and more likely to be served by sprinkler or drip irrigation systems. Some of the land west of the SLC is served by gravity from the Coalinga Canal (CC) which branches from the SLC via the Pleasant Valley Pumping Plant.

The District's water supply is distributed from the SLC through a distribution system which consists of a closed, buried pipeline network designed to convey irrigation water to 160- to 320-acre (64- to 128-ha) fields. The distribution system includes 1,034 miles (1,650 km) of pipe varying in size from 10 to 96 inches (250 mm to 2,240 mm). The gravity portion of the system consists of 36 separate laterals, each with a head works component at the SLC that includes small recirculation pumps to provide adequate pressure to lands adjacent to the canal. The pumped portion of the system consists of

27 laterals, each of which includes at least one automated pumping plant at the canal and some of which include multiple lifts.

Water is delivered year-round to 568,000 acres (227,200 ha) of irrigable land. Based on a 1990 survey of District farmers, the various types of on-farm irrigation systems in the District included furrow (38%), border strip (5%), combination sprinkler/furrow (38%), sprinkler only (16%), and drip/trickle (3%). Since 1990, the planting of several thousand acres of trees has increased the percentage of drip systems. The two predominant crops in Westlands are cotton (272,000 acres (108,800 ha) in 1996) and tomatoes, primarily processing (92,600 acres (37,000 ha) in 1996). Other significant crops, varying in acreage from 22,000 acres (8,800 ha) to 10,000 acres (4,000 ha), include garlic, wheat, cantaloupes, beans, lettuce, almonds, onions and alfalfa. The significance of the current crop mix is that it is different than the crop mix used to design the capacity of the distribution.

SYSTEM CAPACITY & OPERATIONS RULES

The basic design flow capacity of the distribution system is one cubic foot per second (cfs) (28 liters per second (LPs)) for each 80 acres (32 ha). This design flow rate was calculated using an average annual and maximum monthly delivery requirement assuming a cropping pattern that existed at the time of the design in 1965. It also assumed that monthly distribution would be uniform and irrigation sets would be 24 hours. The typical field size served by one delivery is 160 acres (64 ha). The design flow rate at the delivery point for each field is 4 cfs (128 LPs)for a 160 acre (64 ha) field and 7 cfs (196 LPs) for a 320 acre (128 ha) field.

Each field delivery typically consists of a 10 inch (250 mm) or 12 inch (300 mm) vertical pipe with a propeller meter that records flow rate and cumulative volume. Delivery pressure varies considerably, but the system is designed to deliver a minimum of five feet (1.6 m) of head. Flow is typically controlled by butterfly valves immediately downstream of the meter. Because pressure can vary depending on the activity of other farmers on a lateral, the actual flow can vary after a farmer has set the flow.

As noted earlier, there are a variety of different types of on-farm irrigation systems used in Westlands. Connections to the distribution system range from a simple drop pipe which discharges into a head ditch or is connected to gated pipe to complex on-farm distribution systems and drip systems. Sprinkler operations typically require a booster pump except in a few portions of the system where pipeline pressure is sufficient.

Westlands' distribution system is operated on a Limited Rate Arranged demand schedule. Farmers are required to notify the district office at least 24 hours in advance of any initiation or change of irrigation activity. Activities which require notice are turning a delivery on or off, changing the flow rate or changing the field to which water is being delivered. An "on" order must include information regarding the starting date, the requested flow rate, and the expected daily duration of flow if less than 24 hours.

Typically, flow changes occur in the morning, but considerable flexibility is allowed. Farmers operate their own delivery valves and set the desired flow rate. While advance notice is required by District operating rules before turning off a delivery, the nature of the system has allowed the District to be flexible regarding this rule.

Each day the district determines, based on the water orders from the farmers, what flow rate to request from the state Department of Water Resources (DWR) which operates the SLC. The District orders a flow rate for each lateral and canal pool and is required to be accurate within plus or minus five percent. To assist in making an accurate order to the DWR, District personnel read the meters at the head works of each lateral and compare the prior day's actual flow rates to the ordered flows. It is rare that actual flow rate is equal to the ordered flow, so judgments must be made to arrive at an appropriate rate to order from the DWR. Because the flow capacity of the SLC downstream of Westlands is 7,000 cfs (200,000 LPs) and significant deliveries are occurring to other DWR customers on a daily basis, the impact of inaccuracies in the order to the DWR is not a significant operational issue. Actual deliveries are accounted for on a monthly basis using meter readings at the head works. The effect of this accounting method is to allow greater flexibility to District farmers than would be the case if Westlands were at the end of a canal.

The District has a contract with the United States for an annual water supply of 1,150,000 acre feet (138,100 ha-m). Due to variations in the weather and other environmental and operational limitations, the actual supply varies. The contract supply, and any other supply available to the District, is allocated to farmers. Farmers are billed monthly for their water use based on the amount used in the prior month. It is, therefore, up to each farmer to determine how best to use his total water supply during the year. Water users are allowed to order and take water as long as they have an available water supply. Intra-district water transfers are allowed and occur frequently.

OPERATIONAL ISSUES

Unless a distribution system is designed and operated as a true demand system, there will always be a need to deal with the limitations of the system. While engineers and system designers may have developed a very complete and workable system taking into account economic and physical limitations, farmers are always looking for the best way to maximize their profits. To the extent there are aspects of a delivery system that limit a farmer's discretion to make decisions, there will always be pressures to push the limits of their discretion. Westlands operating rules are generally intended to afford water users the maximum discretion possible without compromising the integrity of the system or adversely impacting other farmers.

In order to allow such discretion, the District must rely on the responsibility of the water users to follow the rules. District staff directly involved in the operations of the system include six Customer Accounting Technicians in the main office and an Operations Supervisor, three Senior Operators and 13 Operators in the field. Each

Operator is responsible for an average of 45,000 acres (18,000 ha) and about 250 delivery locations. It is, therefore, impossible for the Operators to be "policemen" in the field. While there are necessarily strong penalties for certain violations of District operations rules, the District has been very successful relying on the integrity and responsibility of its farmers.

LIMITED DESIGN CAPACITY
The most significant operational limitation of the District's distribution system is related to the basic design criteria of 1 cfs per 80 acres. While it was always recognized that individual irrigation events would require a higher flow rate, as evidenced by the design criteria for individual deliveries of about twice this rate, cropping patterns have changed such that there is a greater likelihood that there will be competing demands for the full capacity of a lateral. In addition, advances in irrigation technology and efforts to attain better distribution uniformity have resulted in the need for higher flow rates of short duration. To deal with the physical limitations of the system and the competing demands for use of the delivery capacity of the distribution system, the District has established a basic entitlement to delivery capacity of 1 cfs per 80 acres. This entitlement is used in various ways depending on the circumstances.

The problem of limited lateral capacity is less likely to manifest itself on gravity laterals than on pumped laterals. Gravity laterals can typically delivery more than the design flow rate of the pipelines and problems tend to occur at the low end of the lateral or at isolated sublaterals. For this reason, District personnel do not contact water users on the basis of water orders even if the total orders for a gravity lateral exceeds design capacity. When isolated problems occur, water users contact the District and an evaluation is made of the specific problem. Typically the problem is caused by a few water users taking a significant flow volume and impacting the ability of other farmers to take their desired flow rate. In such cases, the first effort is to resolve the problem by contacting the involved water users and requesting that they reduce their delivery rates. If this approach fails, the District will determine a fair distribution of the available capacity and require that water users limit their delivery rates to the calculated rate. It is rare that deliveries on gravity laterals have to be limited to the 1 cfs per 80 acres.

Pumped laterals are automated systems which are controlled by water levels in terminal tanks or regulating reservoirs. If the capacity of the pumping plant is exceeded by deliveries on the lateral, the system will shut down to protect the pumps. Therefore. daily orders for each pumped lateral are checked against the design flow rate. If orders exceed capacity, water users are contacted and the pumped system is balanced for each pumping plant reach. In balancing the lateral, the 1 cfs per 80 acres is applied to the number of irrigable acres farmed by each water user in the design service area of the lateral who has ordered water. In most cases, water users will voluntarily comply with the District's determination of allowable flow rates. If necessary, the District will dispatch an operator to verify that water users are complying and to make adjustments where required.

At times, some water users will try to increase their deliveries after they believe the operators have left for the day and are not available to enforce the delivery limits. Unfortunately, the effect of such changes is usually that the system goes into an overdraft condition. The overdraft conditions causes an alarm to be transmitted by radio to a central monitoring location and an operator is dispatched to try to stop the overdraft from reaching a point where the pumps automatically shut down. If the operator is unsuccessful, the entire lateral is without service until the system can be restored, usually several hours. While most water users are aware of the consequences of exceeding their allotted flow rate, such problems still tend to occur during the more intense summer irrigation season. District rules allow Operators to turn off and lock the meters of any water users who repeatedly cause such problems.

INCOMPLETE DISTRIBUTION SYSTEM

The inherent capacity limitations of the distribution system are exacerbated by the fact that approximately 15% of the District still does not have a completed system. The areas without a permanent distribution system are primarily areas served by pumped systems. While there are farmer-owned pipeline systems which divert water directly out of the SLC, they are generally not as energy efficient and often are more limiting in terms of delivery capacity. The result is that water users frequently request that they be allowed to take water from a portion of the completed system to serve lands not designed to be served by that lateral. To accommodate those demands, the District has established rules for determining how much of a lateral's capacity is available for delivery outside the design service area. The rule is generally based on the 1 cfs per 80 acre "entitlement" for lands within the service area and "excess capacity" can be delivered outside the design service area of a lateral.

IMPACTS OF FEDERAL RECLAMATION LAW

Another demand for delivery of water outside the design service area of a lateral or sublateral is related to the land ownership changes that have occurred since the distribution system was designed and constructed. Following the federal Reclamation Reform Act of 1982, the nature of land holdings in the District changed considerably. In many cases, the distribution system was constructed to serve 320 acre (128 ha) parcels. As Reclamation Law ownership and leasing rules changed, it has become rare that a 320 acre (128 ha) field is owned and operated by a single entity. The Reclamation Reform Act also included provisions that relate the cost of water to the status of the land which receives the water.

In order to properly account and bill for water use by each farming entity, it became necessary to install new meters in situations where a former 320 acre (128 ha) unit is now farmed as two 160 acre (64 ha) units. In many cases, farmers elected to install a new delivery adjacent to his field, with service from a lateral or sublateral not designed to serve that field, rather than construct one-half mile (0.8 km) of pipeline. Such decisions were made with the knowledge that taking water outside the design service of a lateral was subject to the needs of water users within the design service area. As in the case where the distribution system is incomplete, the entitlement to the capacity of a lateral is limited to 1 cfs per 80 acres currently being irrigated in the design service area of the lateral. While this situation does not present a significant problem in most cases, it does put added pressure in areas where capacity limitations are already a problem.

SEASONAL APPLICATION EFFICIENCY

Within the limitations and operation rules described above, the flexibility afforded to Westlands farmers by the distribution system contributes significantly to the District's high seasonal application efficiency (SAE). Given the significant water supply deficiencies that can occur in Westlands, it is critical that farmers be allowed to make the necessary decisions and take the necessary actions to maximize the efficient use of water. The District-wide SAE averaged 83 percent during the period 1978 through 1990.

POSSIBLE SYSTEM IMPROVEMENTS

Because the most significant limitation of Westlands' distribution is the rate of delivery or flow stream to individual fields, any improvements to the system to allow greater flexibility to the farmer would require physical modifications. While the existing system has considerable flexibility compared to other systems, farmers are currently unable to deliver large enough flow streams to achieve the labor savings that might be achieved through daytime-only irrigation. Dr. John L. Merriam has conducted a preliminary review of one of the District's laterals to identify opportunities for improvements.

Dr. Merriam's concept for improvement includes the installation of one or more small reservoirs along the alignment of existing laterals. The reservoirs would allow the delivery at high flow stream rates to the lower portions of the lateral that can not currently be achieved because of pipe sizes and overall lateral capacity limits at the head works of the lateral. Additional components of the improved system would include high volume tailwater return systems and groundwater wells for additional conjunctive use. Given the value of water in Westlands and the potential cost of dealing with excess deep percolation in some areas of the District, it is likely that the cost of such improvements could be well justified.

DESIGN AND CONGESTION CONSIDERATIONS FOR FLEXIBLE IRRIGATION SUPPLY SYSTEMS
PART I - CONCEPT

JOHN L. MERRIAM and STUART STYLES
BioResource and Agricultural Engineering Department
California Polytechnic State University
San Luis Obispo, California, 93407 USA

ABSTRACT

The CONCEPT (PART I) and the APPLICATION (PART II) present with illustrations the need for and how to obtain a flexible supply system to facilitate the optimum on-farm use of an irrigation supply. A precise and descriptive list of pertinent terms in included. The acceptance of the practical need for and advantages of daytime irrigation and the reasonable range of congestion made practical by using a flexible supply system is developed.

INTRODUCTION

The basic objective of creating a flexible irrigation supply system is to develop a system that has negligible restraint on the total on-farm cultural management procedures. This state-of-the-art system is capital intensive to reduce labor and other expenses and increase crop production and net return. It is concerned with the project supply system and the farm operations considering them as one financial unit. The project engineer-planner must consider the on-farm operations and needs as well as those of the project. "He must think like an educated farmer."

For the educated farmer to optimize his total operations, restraints on the on-farm irrigation aspects must be reduced to their economical limits. He must be able at the point of application to control an assured supply as to frequency, rate, and duration-- this is scheduling.

Scheduling is often thought of by the engineer as delivering water on the day the crop needs it with an adequate volume in a 24 hour or some other fixed period. However, for on-farm irrigation management, control of the rate and duration is probably more important. "It is not just the volume of water delivered, but the way it is delivered to be usable."

For surface methods, the soil intake rate and method affect the duration and the stream size. The mechanical methods utilize fairly small streams and rather continuous use, but duration control is essential. Only by controlling the supply at the outlet can the farmer optimize his management capabilities for all methods.

A Limited Rate Arranged-demand schedule is needed which permits the grower on the arranged day to control the rate (up to a limit) and duration. Essentially, a Demand schedule on an arranged day system is necessary for optimization. (ASCE 1984)

The system, to be flexible, needs a high level of automation. A variable source (reservoir capacity) and distribution system that can deliver large streams of water at a stable condition as farm and system flow rates vary is essential. Use of a semi-closed float valve pipeline gives the best control (Merriam 1987).

The concept is equally adaptable to developed and under-developed countries and to new or upgraded systems. An organized group to operate the system under a Limited Rate Arranged-demand schedule: an irrigation district, a mutual water company, or a Water Users Association is essential. The emphasis of the concept is that the farmer is the client and the owner--"The farm and the project are one financial unit."

The farmer oriented concept of a flexible supply system almost always result in a great reduction in night time irrigation for surface methods and a desire for a large enough stream to effectively utilize the irrigation labor. In Coachella Valley, California, there are over 250 farmer built reservoirs usually serving 40 or 80 ac (16 or 32 ha) fields within the 80,000 acre (32,000 ha) project. They convert the small 3 cfs (85 lps) stream with a 24 hour duration to the management desired larger stream and effective duration to save labor and increase efficiency (see Figure 1).

Figure 1
Photo of reservoirs in citrus in Coachella Valley, CA, USA

TERMINOLOGY

A new, essentially a replacement terminology is needed for developing countries. It is equally well adapted to developed countries. Its consistent use assists in comprehension. The distribution system to the farms will almost invariably and economically be low pressure pipelines with the many advantages of pipelines[1] over ditches (Van Bentum 1994).

The term watercourse is widely used to describe the earth ditches and the maintenance problem of weeds and silt which need to be cleaned out by the farmer in most areas.

---

[1] "The studies made by the Liaison and Coordination Unit disclosed that underground cement concrete pipe systems should take the place of open channel lined or unlined, for the delivery system in the command, as this alternative allows the farmer to remit water in the field free from transit losses, and the O and M problems, and last but not the least, involves practically half the cost of the normal open channel system -- and that it takes less than half the time to complete the job." Technical Series No. 15, USAID Liaison and Coordination Unit, Water Resources Dept., Bhopal, Government of Madhya Pradesh, India. (1992).

IRRIGATION SUPPLY SYSTEMS - CONCEPT

The term must **NOT** be used in flexible supply lexicon. It is replaced by <u>farm distribution pipeline</u> or <u>distributors</u> on which the farm outlets are placed within <u>pipeline distribution areas</u>. The great difference in the capabilities of the two as to right-of-way, maintenance procedures, and delivery capability (upstream vs. downstream control) prevent interchangeable use. Though their location and basic objectives are similar, their utilization is not.

The farm distribution pipelines will be supplied by a <u>conveyance line</u> or a <u>lateral</u> operating under downstream control. It almost always will be a semi-closed (or low pressure closed) pipeline. The lateral will respond directly to on-farm downstream demands.

The laterals will be supplied by <u>branch</u> or main pipelines or canals, or a reservoir. These must be able to act automatically as a flexible supply. They can be upstream or downstream controlled as long as they can supply a farmer initiated demand automatically from in-canal storage, operation procedure, operational spillage or reservoir storage. To automate the pipelines, <u>semi-closed</u> (Merriam 1987) Harris float valve controlled pipelines are essential. The float valves responding to downstream variable demands provide a stable minimum and maximum low pressure in the lines for engineering purposes. Importantly, they also provide a stable pressure and flow rate at the farm turnout as flow rates are changed which a closed pipeline cannot do.

In most cases, the main and branch canals will receive water from rivers or primary reservoirs at fairly stable rates and operate with manual controls or under some degree of automatic upstream or downstream controls as is prevalent on most current projects. The <u>flexible supply system</u> design provides a way to convert the fairly steady canal flows to the farmer needed variable flow which may become a no flow condition at night. The economical balance of costs between canal sizing, in-canal storage, and canal operation and the essential reservoir capacity for night and operational storage for the delivery system, will almost always involve a <u>service area reservoir</u>.

Conceptually a <u>service area</u> is a group of farms supplied at one point from a main or branch, or under less common conditions from a lateral. Below this point automation is essential. However, at this point manual canal turnout controls are practical with a planned operation program using the <u>service area reservoir</u> to convert steady canal and turnout flows into variable flows at the farm.

The <u>service area reservoir</u> (Merriam 1997) is most effectively located near the center of the service area although it may be at the canal turnout. With a basic consideration of the flexible supply concept being to overcome the need for night time irrigation required by steady canal flows, the minimum storage capacity is the unused night time flow. This is impractically small as it requires rigid control of the farmers' supply completely in contrast to the flexible supply concept. A 24 hour service area water requirement capacity is adequate but requires close coordination with the canal operation. A somewhat larger storage capacity simplifies canal operation procedure and precision. It may often be the best choice particularly in retrofit programs as current canal controls may then be adequate. An economical balance as to reservoir capacity costs and canal capital and operation costs is desirable considering current and future prices. However the cost of increasing reservoir capacity is usually very small and excess capacity is usually very practical plus some intangible benefits such as recreation and fishing.

The irrigated area below the service area reservoir, usually about half or a little more of the total, is best served by a semi-closed pipeline system providing a stable farm turnout condition regardless of flow rate or fluctuating reservoir level. A minimum head on the pipe is 3 to 6 feet (1-2 m). The reservoir storage below this is dead storage and the operating range is above it. It is usually in the range of 6 to 10 feet (2 to 3 m). The area needed is seldom as much as 1% of the service area.

Excavation for balance cut and fill for the reservoir embankment often provides below ground dead storage. The total dead storage may be appreciable and so provide silt storage and make fish culture practical.

The following conceptual presentation of the design of a flexible supply system is facilitated by the use of the terms: irrigated farm (or field), unit farm area, unit farm stream.

The irrigated farm is the farm or field that will be irrigated on the arranged day. Importantly, it requires an arranged day regardless of size.

The unit farm area is the selected representative-for-design numerical area (hectares or acres) that can be irrigated in one day (8-14 hours) by an irrigator having a large flexible unit farm stream and good equipment. In small land ownership locations, its size is selected in the upper range of ownership sizes, currently at about 1 to 3 ha (3 to 8 ac). The size can be varied in different parts of large projects. In the USA land subdivision units control practical unit area sizes to 20 or 40 acres (8. or 16. ha) for surface irrigation methods.

The unit farm stream is selected as the probable near maximum stream size needed to apply an irrigation using good equipment in one or more sets in the daytime to a unit farm area at good efficiency. Usually this concerns the initial stream for a furrow irrigated field since it has the slowest intake rate and its duration must be adequately considered. Border-strip and basin easily conform. Furrow Advance Ratios (Time of Advance/Time of Infiltration at the lower end) should be between 0.3 and 1.0 for good distribution uniformity (Merriam 1988). The stream size must be carefully selected based on evaluation of actual (if possible) field condition and soil infiltration range of values. It must be done with consideration of present and future conditions. Its size relates to application efficiency and hours spent irrigating (a farmer cost) and capital investment as to pipe capacities needed (a project cost). "The farm and the project are one financial unit."

The decision on the size of this unit farm stream appreciably affects the cost of the distribution lines and the laterals, and less so on the branch canals or pipelines and negligibly so on the main canal--the same volume of water is delivered in a day. Changing pipe sizes from 8" to 10" and 12" (200mm to 250 and 300mm) increases flow by two and three times. However costs increase only about 15% and 40% and project costs perhaps only 2% to 5% with corresponding small increases in water rate charges. They have major impact on farm management and irrigation labor. It is seldom truly economical to be very restrictive in this choice.

For farmer utilization of upgraded irrigation systems and methods, some simple mathematical processes are convenient for management. Evapotranspiration of a crop is given in inches(or mm)/day, rainfall is 0.01 inch or mm/day, soil moisture deficiency is in percent of root zone depth, inches or mm. Ordered or applied water

IRRIGATION SUPPLY SYSTEMS - CONCEPT   693

needs to be easily convertible into comparable depth units. In the British system a flow rate of 1.0 cfs for 1.0 hr applies 1.0 inch on 1.0 acre. A comparable metric flow unit applying 1.0 cm. on 1.0 ha in 1.0 hr. is 1.0 basic stream which is 27.78 lps. This is a very practical sized stream, and easily visualized by an irrigator. Its use rather than lps is easier and facilitates upgraded management. In developing countries it is far easier for a farmer to comprehend and arrange for 1, 1-1/2 or 2 streams than to request 28, 37, or 56 lps. It is also true in all other countries.

Congestion is the fundamental expression of how flexible a system is: how much reserve capacity does it have, how much of the time is it in use. For this presentation based on experience that farmers do not often willingly irrigate at night, capacities are based on daytime only use though some night use may actually be arranged.

$$\text{Congestion (\%)} = \frac{\text{number of farms}}{\text{number unit streams} \times \text{irrigation cycle days}} \times 100$$

If there were 10 farms, 1 stream, and the cycle (frequency) was 10 days, the congestion would be 100%, -- used every day and no reserve except at night.

If two streams were available, it could be used only 50% of the time, lots of reserve. This flexibility might be obtained by instead of using 8" (200mm) diameter pipe all the way, by using 10" (250mm) for the upper half (or better two thirds) and 8" (200mm) for the lower half. The increase in the distribution pipeline cost would be about 7% for double capacity. For the times when only one stream was being used or the second stream was taken near the upper end, the actual arranged stream could be double or nearly so taken for a shorter time. This could be a major labor saving for the farmer. It is part of the value of the flexible supply system.

Acceptable Congestion is a variable. It is a matter of judgment within an acceptable range. If the distributors have a lower congestion and there is lots of reserve time, the laterals can have a little more congestion. The total number of irrigated farms requiring a day of use, not the area, enters into the consideration. A whole pipeline distribution area in a developing country having ten farms requiring two streams for 50% congestion, may be only one field in the USA utilizing a large stream for only one day.

The larger the number of farms on a distributor, or pipeline distributors on a lateral, or laterals on a branch or main canal, the more nearly the operation will approach an average use rate. Congestion can be increased with more intensive use, but reserve for unusual weather or other conditions must be retained. Consideration may be given to variations in cropping pattern and amount of land fallow at peak periods over the life of the project. "Don't limit the future by what is built now."

Table 1 contains a reasonable range of congestion values as a basis for judgment.

Table 1
Practical Congestion Ranges, percent

distributors	lower lateral	upper lateral	branch	main
50 - 60	60 - 70	65 - 80	70 - 85	85

The following PART II, Illustrative Design section illustrates the use of the table with decreasing relative capacity needed as more days of use (streams) are required. An early paper by Clements in France in 1965 developed the concept applying it to large canal flows as related to area being serviced by sprinklers. Clemmens developed a similar concept, but it also was related to area and flow rates for continuous canal flows rather than needed days of availability. (Clemmens 1987) The needed days from the congestion concept is subsequently related to pipeline capacity for part time flow, not to steady canal flows. All concepts are related to probability of flow becoming more nearly average with increasing number of users or increasing area.

REFERENCES

ASCE (1984). Recommended Irrigation Schedule Terminology. Conference Proceedings, ASCE I&D Div. On-Farm Irrigation Committee, R. Walker, Ch. July 1984, Flagstaff, AZ, USA.

Clemmens, A.J. (1987) Delivery System Schedules and Required Capacities. Symposium Proceeding. Planning, Operation, Rehabilitation and Automation of Irrigation Delivery Systems. D. D. Zimbelman, Ed., ASCE I&D Div. July 1987.

Merriam, J.L. (1987). Design of Semi-closed Pipeline Systems. Symposium Proceedings, Planning, Operation, Rehabilitation and Automation of Irrigation Water Delivery Systems. D. D. Zimbelman, Ed. ASCE I&D Div., July 1987.

Merriam, J.L. (1988). Simple Furrow Advance Ratio Evaluation Technique For Upgrading Management. Conference Proceedings of ASCE I&D Div. July 1988.

Merriam, J.L., S. Styles. (1997). Incorporation of Reservoirs into Irrigation Supply Systems to Simplify Flexible Operations. Proceedings of the XXVII Congress of the Association for Hydraulic Research / American Society of Civil Engineers, Water Resources Division. August 1977, San Francisco, CA USA.

Van Bentum, Robert, I. Smout (1994). Buried Pipelines for Surface Irrigation. Water Engineering Development Center, Loughborough University of Technology, Leicestershire. LE 11 3TU, England.

DESIGN AND CONGESTION CONSIDERATIONS FOR FLEXIBLE IRRIGATION SUPPLY SYSTEMS
PART II - APPLICATION

JOHN L. MERRIAM and STUART STYLES
BioResource and Agricultural Engineering Department
California Polytechnic State University
San Luis Obispo, California, 93407 USA

ILLUSTRATIVE DESIGN

The design following the concepts in Part I is based on an operation plan using a Limited Rate Arranged-demand schedule though it may be less effectively used with other schedules. This schedule requires the irrigator to apply in advance for a day upon which he is permitted the use of the system for the arranged irrigation farm. An assured minimum and a maximum rate (limit) is set in the design procedure by the pipeline or turnout capacity. The farmer used rate seldom is the maximum limiting rate. The required arranged schedule prevents overloading. The reasonable congestion limit provides assurance of availability under most conditions. Implicit in this design procedure is that the farm will be usually irrigated during the daytime. This condition is a very high priority among the educated (experienced) farmers using a flexible supply system. During the arranged day, the irrigator can take water as he wishes as to rate and duration--demand up to the system limiting rate, and at night if so arranged.

In the design procedure a base map at a workable scale showing topography, ownership or subdivision boundaries, and soil is needed (a GIS is helpful). A tentative irrigation layout of distributors, laterals, branches (and main) is superimposed. Under some conditions, it may be reasonable to modify field boundaries to facilitate irrigation methods. On one project, the top and bottom property lines were relocated to approximate a contour so reasonable cross slope irrigation grades would not leave odd shaped pieces of land. Long narrow repeatedly subdivided by inheritance fields were consolidated into shorter wider units prior to designing the irrigation layout.

Pipeline distribution areas should be small enough to require only one or two streams. Three is generally undesirable as creating more flow variation, and difficulty in metering if that is desired. However one Project in Egypt having 49 very small farms will serve up to eight in a day from three sub distributors using some half streams and half day arrangements.

The unit farm area value is selected with consideration of present and future farm boundaries. It is usually a bit larger than a majority of the farms or fields presently used. If the actual ownership is much larger, two days can be arranged and considered when reviewing the congestion. Two small areas could be allocated half days. There is considerable leeway and judgment at this point of laying out the distributor and lateral pipelines.

The size of the unit farm stream is not pertinent at this stage. The procedure is to select the number of streams (days of use) needed at various locations which can later be converted to flow rates and pipe sizes.

The irrigation cycle length is related to crop, climate and soil variations. Moderate variations in the design cycle length (frequency) with an acceptable effect upon congestion are of little consequence. The magnitude of the irrigation cycle must be representative of the actual conditions under peak use plus a little reserve. Its precise value is not important, but a design value must be selected. It is a key value in considering congestion and it must be a practical whole number.

On the map at each distributor turnout from the lateral, the required number of streams needed for the distributor for acceptable distribution congestion should be noted--1 or 2, seldom 3.

Moving up the lateral, the cumulative number of streams should be noted for each lateral reach. These items can be tabulated along with other helpful information. This is illustrated in Table 2 for a 10 day irrigation cycle and 10 distributors on a lateral. It utilizes an arbitrary percent reduction related to probability (practically essentially linear) to decrease the number of stream on the lateral to conform to an acceptable congestion value.

Table 2
Streams and Congestion for Lateral No.

Distributor No.	1	2	3	4	5	6	7	8	9	10
Number of Farms	10	14	6	16	5	12	18	11	15	13
Number of streams	2	2	1	3	1	2	3	2	3	2
Distributor congestion %	50	70	60	53	50	60	60	55	50	65
Total of farms	10	24	30	46	51	63	81	92	107	120
Total of streams	2	4	5	8	9	11	14	16	19	21
Reduction percent	0	0	5	5	10	10	15	15	20	25
Adjusted no. of streams	2	4	5	7	8	10	12	14	15	16
Lateral congestion %	50	60	60	66	64	63	67	66	71	75

The use of the simple linear reduction percent procedure selected to obtain a chosen congestion at the inlet and starting with zero at the lower end, is a representation of probability. It should be supplemented with site affected judgment. At distributor 4, the number of streams was reduced from 8 to 7 even though the indicated percent reduction of 5% would not justify it. The rounded values of percent are intentionally shown to emphasize that precision is not justified when answers are in whole numbers, but judgment should be used. The follow-up question--what is the increase in the annual project cost per unit of water delivered to use a pipeline for 8 streams with a congestion of 58% over one conveying 7 for this reach at a congestion of 66%.

The next design step is to select the unit farm stream size. Its value is representative of a large area so it will not have an exact value, but it must be practical. Whether it is 80, 100 or 110 lps is not of concern. With a flexible schedule, adequate education, and pipelines, the application efficiency can be at least 75%. Select a value using judgment. It must be adequate to cover the unit farm in one day, usually daytime only. Thinking like a farmer, select the largest one since it will save time and labor, and then thinking like a cost-minded engineer, select the smallest. As a design engineer select the economical one considering the farm and the project as one financial unit, and then make it a bit larger -- "Don't limit the future by what is built now."

On the map for each reach, note the flow rate (number of streams x unit stream flow rate). Then for the available gradient note a pipe size for each reach. Hydraulic design then follows with consideration of minimum head on farm turnouts, minor losses, Harris float valve losses (Merriam 1987), needed pressure at the inlet, etc. (Van Bentum 1994).

The Harris float valves have two functions in a semi-closed pipeline system. First is to break line pressure into steps in conformance with pipe and joint capabilities to resist pressure usually about 5. to 6. m. This will permit lower cost low pressure pipe. Concrete 1.0 m length non-reinforced tongue and groove mortar joint irrigation class pipe (ASAE S261) frequently proves to be the most economical with a maximum head of about 6. m. In developing countries it usually does not require foreign exchange. The second function is pertinent to flexible system operation: maintain a stable flow condition at the farm turnout as flow conditions are changed in the pipeline system creating an unstable pressure condition. Many times the distributor can be stabilized from the float valve stand on the lateral.

As the several lateral lines join the branch (or main) pipeline, the process is repeated with higher values of congestion as the many more farms tend to approach a more nearly average flow. In table 2 a distributor having ten farms and two streams, would average one stream a day but could have two or zero on some days. A lateral having a 120 farms and 16 streams would average 12 farms a day but would probably range for 8 to 16, never go to zero and restricted to 16 maximum. The arranged schedule provides adequate restraint on the number of users.

A representative service area on most projects will range from perhaps 100 to a 1000 hectare (250 to 2500 acres). However the projects in Egypt are 20. to 50. ha each having its own pump and reservoir. A branch or lateral pipeline would take off from an automated main canal flowing continuously at a fairly stable average rate varied by seasons. The essential on-farm flexibility of rate and duration would be created by some canal operations but mostly by a service area reservoir (Merriam 1997). It would be located near the center of the service area. It would be capable of storing at least the overnight unused flow for the service area in conjunction with appreciable canal operation. A full day's storage requires moderate canal coordination. Some additional storage greatly reduces the need for precise canal operation permitting mismatches to be adjusted a day later and costs very little more. It could be compatible with manual canal operation and simple operations to upgrade an old canal and generally would be the most practical size.

Ground slope and topography have appreciable impact on location and reservoir design. The pipelines serving the area below the reservoir must have an adequate operating head of about 2. m (6. ft.) so there often needs to be this much dead storage above the ground. Generally, the reservoir embankment is excavated from within the reservoir creating below ground dead storage and an opportunity for fish culture.

As an alternate to gravity operation, pumping should be considered as is done in Egypt and from the Imperial Irrigation District's large 17,000 acre (6800 ha) partially flexible supply project. (Dimmitt 1992) Much less total reservoir capacity is needed as dead storage becomes small. Water can be pumped in or out depending on topography and design.

System capacity for a flexible daytime only operation relative to a 24 hour or rotational system needs to be increased. The increased costs invariably are more than compensated for by on-farm benefits of: reduced and more convenient labor; increased yields; more efficient irrigation and water conservation; reduced potential drainage and salinity; reduced inter-farmer (top ender - low ender) conflicts. Project benefits are appreciable. The Orange Cove Irrigation District, California, USA, reduced its field crew by one half. (Chandler 1990).

The distributors need to have doubled capacity for daytime only usage relative to 24 hour operation. This would require for example, increasing from 200mm to 250mm diameter pipe for the entire length. (Increasing to 300mm would triple capacity and 350mm increases flow 4.5 times.)

If relative pipe costs (concrete) were 200mm (1.00), 250mm (1.13), 300mm (1.41), 350mm (1.78), to change from a small stream for 24 hour to daytime only would cost 13% more for pipe, with twice the flow rate. To make possible the use of two large daytime only streams with low congestion would increase pipe cost about 45%. Lateral costs would increase less.

If the annual costs of the pipelines were as much as one fourth of the project annual costs, the 45% increase to obtain an optimum flexible supply would only be (1/4 x 45%) about 11%, easily compensated by the many benefits.

Branch pipelines (or canals) can function either with upstream or downstream control depending on whether a service area reservoir is used or not. For smaller service areas, the branch can be replaced by a pipe lateral with downstream control if the main canal can provide through operation or in-canal storage the needed storage: for the rejected night time flow; for the lesser changes caused by on-farm irrigation operations such as initial and cutback furrows flows; for early or late turn-on and turnoffs; for different set sizes, etc. The Orange Cove irrigation District (Chandler, et al. 1990) converted to a flexible supply operation by nearly doubling the capacity of the old project controlled distributor, and permitting the farmers to operate their own turnout valves on an arranged day under a Limited Rate Arranged-demand schedule. There has been tremendous support of the program by the farmers who ultimately pay all the costs.

Where main canals have inadequate capabilities to handle the major changes resulting from the flexible schedules, additional storage capacity must be developed. The service area reservoir is usually the most practical procedure. With such a reservoir the portion above the reservoir can operate under upstream control. It may be a pipeline or a canal if off-takes have adequate head to function as at Gadigaltar (Merriam 1990) which has five laterals and seven distributors taking off from the canal. Flow rates are set at least twice a day into it from the supply reservoir. The daytime flow is a little more than the arranged withdrawals to be sure that there is always excess to the needs. The excess goes on into the service area reservoir. The evening set assures the reservoir will be adequately full by the next morning to satisfy the arranged needs in the lower portion. The main canal or supply source must accept the moderate flow changes.

Where inadequate head exists in a ditch, an open system pipeline (upstream control) can be used. The open stand overflow is built high enough that the laterals taking off have adequate head. In the case of the Pima-Maricopa Flexible Irrigation Project (Lindstrom 1997), Harris float valves are set with the valve base at the required overflow elevation and the floats set to shut off flow if the reservoir becomes full. With this combination of open and semi-closed pipelines, the system can be operated with either upstream control with a fixed preset flow rate changed twice a day so the main canal can be easily operated, or with downstream control with the canal taking moderate fluctuation in demand.

With the branch operated in upstream control mode in conjunction with a service area reservoir, its capacity need only be a bit larger than the average 24 hour flow rate. This is one of the benefits derived from having a service area reservoir. If it is operated in downstream control mode, with storage at the upper end of the canal, it must have full capacity to supply all the service area demands in the daytime. Nearly twice as large capacity as with a service area reservoir.

REFERENCES

ASCE S261. Design and Installation of Nonreinforced Concrete Irrigation Pipeline Systems. American Society of Agricultural Engineers Standards.

Chandler, J.C., R.M. Moss, J.L. Merriam. 1990. Automated Low Pressure Pipe for Flexible Deliveries: Orange Cove Irrigation District, California. Symposium Proceedings, R4, XIV International Congress. International Commission on Irrigation and Drainage, Rio de Janeiro, Brazil. April 1990.

Dimmitt, A.K., K.I, McLoughlin, F.Z. Kumand, and D.G. Welch. 1992. Planning for Water Conservation Through Irrigation System Modernization and Rehabilitation. Proceedings ASCE I&D Div. August 1992.

Lindstrom, S. (1997). The Pima-Maricopa Flexible Irrigation Supply Project. Proceedings of the XXVII Congress of the International Association for Hydraulic Research / American Society of Civil Engineers, Water Resources Division. San Francisco, California. August 1977.

Merriam, J.L. (1987). Design of Semi-Closed Pipeline Systems, Symposium Proceeding, Planning, Operation, Rehabilitation and Automation of Irrigation Water Delivery Systems, ASCE I&D July 1987, Portland, OR, USA.

Merriam, J.L. (1990). Gadigaltar Tank Irrigation Pilot Project. Water Resources Dept., Bhopal, Madhya Pradesh, India. 235 Chaplin Lane, San Luis Obispo, CA 93405, USA.

Merriam, J.L., S. Styles. (1997). Incorporation of Reservoirs into Irrigation Supply Systems to Simplify Flexible Operations. Proceedings of the XXVII Congress of the Association for Hydraulic Research / American Society of Civil Engineers, Water Resources Division. August 1977, San Francisco, CA USA.

Van Bentum, R., I. Smout. 1994. Buried Pipelines for Surface Irrigation. Water Engineering Development Center, Loughborough University of Technology, Leicestershire. LE 11 3TU, England.

USE OF RESERVOIRS AND LARGE CAPACITY DISTRIBUTION SYSTEMS TO SIMPLIFY FLEXIBLE OPERATIONS
PART I - SYSTEM CAPACITY

JOHN L. MERRIAM and STUART STYLES
BioResource and Agricultural Engineering Department
California Polytechnic State University
San Luis Obispo, CA 93407, USA

ABSTRACT

This two part paper and the accompanying two part paper "Design and Congestion Consideration for Flexible Irrigation Supply Systems" present the need for and a way to obtain a flexible water supply essential for effective on-farm water use. A supply flexible in frequency, rate, and duration under control of the irrigation at the point of application assists total on-farm management. This results in increased crop production, reduced and more convenient irrigation labor, more effective water use, and reduced deep percolation and runoff loss.

The needed large system capacity to obtain a reasonable level of congestion to assure water when needed (frequency and rate), and the use of reservoirs to reduce the need for night time irrigation (rate and duration) while stabilizing canal flows, is presented and illustrated with case studies.

INTRODUCTION

With the increasing acceptance worldwide of the value at the farm level of a flexible water supply, upgrading of the supply systems to larger, variable flow and reasonable congestion capabilities is being recognized as essential. The on-farm problems created by use of a rotation schedule which permits a canal to operate continuously at a constant flow rate (an engineer's dream but a farmer's restraint) are beginning to be considered in planning. The value for surface methods of large flows and daytime only sets with half or less as much labor conveniently and more effectively used, is considered in the economics of the projects. The convenience of irrigating when and with the flow rate desired and as modified has value to a farmer for which he is willing to pay a higher water charge. "It is not just the volume of water delivered but also the way it is delivered to make it usable, that is important." The farmer control of the water permits appreciable reduction in drainage and salinity problems caused by excess water application. (Styles 1977) A Limited Rate Arranged-demand schedule (ASCE 1984) is the desired practical schedule. With it on the arranged day, the irrigator can take and vary the desired flow up to the system limit when and as long as needed to infiltrate the desired depth.

To obtain flexibility which permits reduced on-farm water use, increased yields, and reduced and more economical labor, requires flow rate changes of appreciable magnitude. The supply and distribution systems must be able to transmit the variable flows. Daytime use only at least doubles flow rate requirements. A reserve capacity to permit choice of frequency at the lower end of a line may double capacity again, but which in the upper portion with averaging the many users may need only a very small extra capacity. The night time non-used flow of an area left in a canal becomes the next day's flow further down stream, or it can simply be placed in a re-regulating reservoir.

Using the modified flow further downstream involves what may become a complicated cannel operation problem. On long canals it becomes almost an impossible problem without putting many restraints on the on-farm use conditions or constructing re-regulating reservoirs. With the use of the non-restraining schedules, fluctuation storage becomes essential although operational spillage may be practical where the spillage is reused.

The essential storage for fluctuating flows may be obtained from initial supply reservoirs, in-canal storage, in-the-canal reservoirs, beside-the-canal reservoirs filled and/or emptied by pumping of gravity, service area reservoirs which are emphasized in this paper for new or rehabilitated projects, and of course by combinations supplemented by canal operations.

More than minimum storage and conveyance capacity is almost invariably economical because of reduced operation constraints at the farm and project level. For a twelve hour delivery schedule with just twelve hour storage, rigidity and precision remain essential. Twenty four hour storage still has rigidity for each day's total flow, while more capacity permits one day's discrepancies to be compensated for on the next day. This greatly simplifies operation precision usually at very little increase in cost of unit volume of water delivered.

The number and size of the farms, the desired duration and maximum flow rate and congestion (Merriam 1997) to an irrigation unit field has impact on the size of the storage capacity needed, and great impact on the interconnecting conveyance and delivery capacity. The desired delivery capability to provide an application of 125 to 150 mm (5.0 to 6.0 inches) per work day (say about 10 hours) is controlled by soil intake rate limitations. If a unit field for one irrigation set in the USA is about 8.0 ha (20 ac) for a 6.0 hour furrow infiltration time and 4.0 hour advance time resulting in a set time of 10.0 hours applying 125 mm (5.0"), the flow average rate would be

.125 m x 8.0 ha x 10,000/10 hours = 1000 m^3/hr = 280 lps (10 cfs) per 8.0 ha field

If basins were used, a larger flow rate of 1500 m^3/hr for 6.7 hours might be desirable at the farm level but might be uneconomically large for the project and so require a compromise between the farm and project benefits. The actual flows taken should seldom be the maximum design limiting flows which should be appreciably larger than the average ones to remove farm restraints--"The farm and the project are one financial unit."

In a developing country where 1.0 ha (2.5 acres) might be an irrigation unit area and the duration remaining the same at 10 hours being related to soil and method, the average flow rate would be about

.125 m x 1.0 ha x 10,000/10 hours = 125 m^3/hr = 35.0 lps (1.3 cfs) per 1.0 ha field

This average flow might be too small a stream to be very efficient and a larger stream might be made available. Two streams combined for a half day still covering two farms may be practical.

The design (not the illustrated average) unit stream size as well as the irrigation unit field area must be carefully determined and be adequate for future conditions and any practical method. Basins are capable of utilizing large streams, so an average size stream would require longer labor. Furrows are most efficient if large initial streams are used and then cut back or return flow systems installed to recover runoff. Border strips are quite compatible with an average rate. Several sets smaller in area and shorter in duration can be made with the flexible streams to make up a full workday on a unit field area. For border strips and furrows the first irrigations applied usually have a larger than average intake rate which must be considered when selecting the design unit stream.

Sprinkler and microirrigation methods generally use small streams for many days but daily durations need to be controlled to apply the desired depth which changes with crops, season and soils and so require appreciable flexibility in duration.

These design unit stream sizes should be determined by actual irrigation evaluations as the stream size must not be a limiting condition for farm operation. It must be appreciably larger than the average size. "The farm and the project are one financial unit." "Don't limit the future by what is built now", especially for pipeline systems.

PIPE CAPACITY DESIGN ILLUSTRATIONS

It is to be emphasized that while construction of reservoirs in addition to major storage reservoirs, has been common knowledge for a long time, their use as a tool to facilitate upgraded on-farm use of water and labor and resulting improved on-farm operation management has not. The resulting flexibility in supply in conjunction with a competent distribution system can accomplish the basic on-farm objective of overcoming common management restrictions. An educated manager can then perform at the optimum level. (Merriam 1987a; Merriam 1991)

For illustration of reservoir and pipeline capacities assume ET is 8.0 mm/day (.33 ipd) (for actual design the peak rate for several consecutive days should be considered) on a 200 ha area served by a lateral, and 60% application efficiency. Then for this illustration the daily volume (average) would be

$$.008 \text{ m pd} \times 200 \text{ ha}/60\% = 2.67 \text{ ha m/day}$$

and the frequency of application for a Management Allowed Deficiency (Merriam 1966) of 75 to 100 mm would range from

$$7.5 \text{ mm}/8.0 \text{ mm pd} = 10 \text{ days} \quad \text{to} \quad 100 \text{ mm}/8.0 \text{ mm pd} = 13 \text{ days}$$

The average 24 hour supply flow rate to the 200 ha service area would be

$$2.67 \text{ ha m} \times 10,000/24 \text{ hours} = 1,110 \text{ m}^3/\text{hr} = 310 \text{ lps (10.9 cfs) steady flow}$$

For a 12 hour flow, the rate would be twice as great, and even larger for 10 or 9 hour durations required by different soils, but the needed daily volume remains the same.

For a 1.0 ha farm for a 12 hour set in a 10 day irrigation cycle, the average flow rate would be 31 lps (1.1 cfs), which might be too small to be practical. For a 10 hour set and a one-third increase, the practical flow rate limit might be 45 lps (1.6 cfs) which could be cutback. For a 13 day cycle and a 12 hour set, an average flow rate is 40 lps so the design limit might be 55 lps.

The representative average number of 8.0 ha unit farms per day for a 10 day cycle in a 200 ha service area in the USA would be

$$(200 \text{ ha}/8.0 \text{ ha}) / 10 \text{ days} = 2.5 \text{ farms/day}$$

and in a developing country with 1.0 ha unit farms

$$(200 \text{ ha}/1.0 \text{ ha}) / 10 \text{ days} = 20 \text{ farms/day}$$

The supply and conveyance capacity and the needed reservoir capacity to operate this area with a flexible schedule and acceptable congestion is related to probability and an economically acceptable congestion--what degree of assurance of delivering water on the date first requested under a limited rate arranged schedule. For example, on the arranged date for up to a limiting flow rate of one-third larger than the 10.0 hour average, say 360 or 45 lps, varied as desired for as long as needed on that day, what capacity will be needed. These limiting conditions must be determined with great care to not appreciable restrict on-farm operations.

For this illustration, in the USA the "average" would be either two or three streams. With a flexible schedule it might be four or one or even zero. With three streams used as a design limit to determine the "limiting" pipeline capacity rate, with congestion percent defined as (Merriam 1997)

$$\text{congestion \%} = \frac{\text{number of farms}}{\text{number of unit streams x days in cycle}} \times 100$$

a congestion figure would be 25/30 = 83%, which is rather restrictive. With four streams it would be 25/40 = 62% which would be acceptable with twenty-five 8.0 ha field units or farms which then could be covered in six days out of a ten day cycle.

For a developing country with many small units to be covered in say 10 days, a higher level of congestion would cause negligible problems with the 200 farms. An 80% congestion level resulting in irrigating up to 25 farms per day would probably be acceptable. With 1000 farms the usual daily request would approach nearer to the average.

With the USA example it is very likely that with only 25 unit area farms, on some days that one or four stream(s) would be requested. However, with the developing country situation with 200 farms it would be highly improbable with an average 20 farms that 10 or 30 would request water on any day. However, holidays and rains do have great impact. The acceptable level of congestion affects the needed reservoir storage and system capacity. The canal turnout peak capacity for the USA example is either three or four streams of 250 lps average. The peaking capacity should be perhaps 30% greater so four 375 lps streams (1500 lps) would be acceptable for flexibility. Allowing for the probability that not all four farms would simultaneously take peak flow, a flow of 1,300 lps could be acceptable, but would it really save much on meter charges over the 1,500 lps capacity?

For the developing country condition, an illustration for a rigid rotation schedule is presented in Table I for a branch pipeline taking off from a canal to deliver water to a lateral and distributor lines for the 200 1.0 ha farms. This table shows for 100% congestion (20 farms per day) and then for 80% congestions (averaging 20 farms per day but allowing up to 25) in a ten day irrigation cycle, the flow rates and relative pipe diameters needed at the beginning and at the midpoint for a rigid schedule: fixed rate - 31 lps, fixed duration - 12 hours, daytime, fixed frequency - 10 days.

The 80% congestion allows some flexibility in frequency but not for rate and duration. This is commonly done in the United States with 24 hour duration's. This Restricted Arranged schedule is not a good one but is simple for illustration purposes. Night time unused flow will be absorbed in the canal or a reservoir.

Pipe capacities and relative sizes shown in Table I for rigid rate and duration schedules, show that to provide for 80% congestions (flexibility in frequency), capacity is increased from 620 lps to 775 lps, a 25% increase. Pipe diameter is increased from 100 to 108, only an 8% increase. If pipe costs are about comparable to diameter, a 10% increase in pipe costs may be anticipated. If pipe costs alone are about 20% of total project costs, an appreciable degree of flexibility in frequency can be obtained for only about 2% increase in project costs.

Table I

(a) 100% congestion, cover all 200 farms in rotation in ten days with 20 31 lps streams		
flow at canal	620 lps	relative pipe diameter 100
flow at midpoint	310 lps	relative pipe diameter 77
(b) 80% congestion, coverable in eight days out of ten with up to 25 streams		
flow at canal	775 lps	relative pipe diameter 108
flow at midpoint	340 lps	relative pipe diameter 83

Table II

80% congestion, up to 25 40 lps streams, coverable in eight days, durations variable. This is a flexible schedule.		
(a) no reservoir, pipe size decreases with length		
flow at canal	1000 lps	relative pipe diameter 119
flow at midpoint	500 lps	relative pipe diameter 91
(b) mid area reservoir, 24 hour flow, pipe size decreases only in lower half		
flow at canal	500 lps	relative pipe diameter 91
flow at midpoint	500 lps	relative pipe diameter 91

In Table II a flexible Limited Rate Arranged schedule is illustrated without and with a mid area service reservoir, with 80% congestion allowing appreciable flexibility in frequency, with a 30% increase over average flow rate, and with durations as needed to match soil intake rates and farm conditions -- a very good schedule. Arranged scheduling will limit deliveries to only 25 farms on any one day even though more might be requested, or permit night time irrigation.

Table II relating to a satisfactory flexible schedule of adequate frequency, rate and duration shows that without a service area reservoir with all flexibility supplied by the main canal, that for a 60% increase in off take capacity only a 19% increase in initial pipe diameter would be required.

However, by construction a service reservoir (Table II (b)) almost all of the project flow variables can be absorbed by the reservoir. This will also permit the main canal to have very steady flows and appreciably reduce its cost. This alternate, by using the upper portion of the service area main 24 hours per day rather than 12 hours as in Table II(a), reduces the capacity to 500 lps (50%) and the pipe diameter to 76% for its entire length in the upper half.

These are the essential benefits of service area reservoirs. They make possible the many on-farm benefits from flexibility without appreciably changing the usual canal operations. This makes upgrading of many existing systems possible and greatly reduce canal operation problems, costs and operational spillage. The Cost of the reservoir is compensated for by reduced pipe costs and the many on-farm benefits. "The farm and the project are one financial unit."

REFERENCES

ASCE. (1984). Recommended Irrigation Schedule Terminology. Conference Proc. ASCE I&D Div. On-farm Irrigation Committee, R. Walker, Ch. July 1984. Flagstaff, AZ, USA.

Merriam, J.L. (1966). A Management Control Concept for Determining the Economical Depth and Frequency of Irrigation. Transactions, ASAE, vol. 9, no. 4.

Merriam, J.L. (1987a). Reservoirs Help On-farm Operation and Automation. Conference Proceedings, ASCE I&D Div. July 1987. Portland, OR, USA.

Merriam, J.L. (1987b). Design of Semi-closed Pipeline Systems. Symposium Proceedings, Planning, Operation, Rehabilitation and Automation of Irrigation Water Delivery Systems, D. D. Zimbelman, Ed., ASCE I&D Div. July 1987. Portland, OR, USA.

Merriam, J.L. (1991). Flexible Supply Schedule Missing Link for Effective Surface Irrigation and Automation. Conference Proceedings, ASCE, I&D Div. July 1991. Honolulu, HI, USA.

Merriam, J.L., S. Styles. (1997). Design and Congestion Considerations for Flexible Irrigation Supply Systems, Part I Concept, Part II Application. Congress Proceedings, IAHR/ASCE Water Resources Engineering Division. August 1997. San Francisco, CA, USA.

Styles, S. (1997). Alleviation of Surface and Subsurface Drainage Problems by Flexible Delivery Schedules. Congress Proceedings, IAHR/ASCE Water Resources Engineering Division. August 1977. San Francisco, CA, USA.

USE OF RESERVOIRS AND LARGE CAPACITY DISTRIBUTION SYSTEMS TO SIMPLIFY FLEXIBLE OPERATIONS
PART II - RESERVOIR CAPACITY AND CASE STUDIES

JOHN L. MERRIAM and STUART STYLES
BioResource and Agricultural Engineering Department
California Polytechnic State University
San Luis Obispo, CA 93407, USA

RESERVOIR CAPACITY DESIGN ILLUSTRATIONS

ELEMENTAL ON-FARM RESERVOIR CAPACITY
One day storage plus several hours operational reserve is minimal for practical flexible on-farm operation. Overnight (12-16 hr) storage can be used with appreciable on-farm scheduling restrictions on startup time or supply canal management and larger capacity facilities. The small cost saving is seldom economical. It is essential to develop an operation plan that included the flexible on-farm needs of variable rates, durations, startup times, and labor.

Equation for 24 hr. storage which should be increased by about one third for practical on-farm operations, inexact flow rates, starting time variations, etc. is

$$\text{cycle (days)} \times \frac{\text{ET mmpd}}{10 \times \text{eff. \%}} \times \text{area ha} = \text{ha m}$$

for a USA illustration for one 8.0 ha (20. ac.) field

$$10. \times \frac{9.0 \text{ mmpd}}{10 \times 60\% \text{ eff.}} \times 8.0 \text{ ha} = 1.2 \text{ ha m}$$

which should be increased to about .16 ha m to facilitate operations. If a reservoir is proposed for 3.5 m height with a 2.0m working range in the top portion, and 1.5 m dead storage for minimum head of the pipeline, the needed average area would be:

$$1.6 \text{ ha m}/2.0 \text{ m} = 0.8 \text{ ha}$$

This would be adequate at 60% congestion to flexibly serve a service area of a group of six 8.0 ha fields during the ten day cycle with the reservoir being filled (or nearly so) previous to use.

If the supply stream can be varied from in-canal storage day to night, a smaller storage can be used but with appreciable operation needed in the supply system. For just 24 hour storage the supply from the canal will have to be varied precisely daily in conformance with anticipated demands and "arranged" supply with possibly a second refining adjustment. However, if the reservoir is increased 30% or more, storage is obtained which nearly eliminates operation precision for a small annual cost per ha m for the water regulated per year. "Don't limit the future with what is built now."

If in a developing country the service area of a distributor were 40 to 80 ha (a USA farm unit), a similar reservoir could be used. It would be placed near the center of the service area as illustrated in Table IIb rather than at the upper end.

USING THE PROJECT SUPPLY STORAGE RESERVOIR
Where a lateral pipeline (an automated level top canal may be cheaper on very nearly level ground) can be closely connected to the project supply reservoir (or equivalent), flexibility can be obtained by having adequate capacity in the lateral pipeline without a supplemental reservoir as in Table IIa but the canal will have large day to night and daily fluctuations. Such capacity would need to be large enough in the upper initial reaches to supply all anticipated streams. From the PART I introductory illustration in the USA (see Table IIa) this would be four 375 lps streams reducing to three and then two near the lower end and one in the last reach. It would be used essentially only in the daytime. It would utilize the existing storage capacity or canal operations so the fluctuating offtakes would be satisfied. If a closed or semi-closed pipeline were used, it would be a fully automated system as used on the Orange Cove Irrigation District, CA. (Chandler 1990)

USING A SERVICE AREA RESERVOIR
If the PART I 200 ha USA illustrations were not near the storage reservoir but took off from a main or branch canal on which it was desired to maintain a nearly stable rate, a service area reservoir of 20 ha m or more capacity would be located near the center of the service area as presented in Table II(b). With this location, for a conceptual illustration the basic condition (which must be modified to fit the actual site conditions and a flow rate and duration related to soil intake rates) illustrates a 12 hour set for daytime only irrigation. With this assumption, the stream run in the upper portion would be the 1-1/4 stream full flow rate needed for a 24 hour run desired to stabilize canal flow rather than the 2-1/2 stream needed for 12 hrs. The night-time non-used flow at this same rate then flows on into the service area reservoir to provide the flow needed the next day in the lower area. In practice the probability is that in the USA example, three and occasionally four 375 lps streams could be needed at times in the total area. This would increase the required design capacity to two streams for 24 hours rather than the 1-1/4 average flow rate. This would be limited by arrangement to only two each in the upper and lower portions. An operation program should be made for maximum conditions, and then a bit more added --"Don't limit the future with what is built now." With many outlets for the developing country illustration, the incremental increase would be less.

Additionally, since the variable farm turnout flow rates in practice are usually taken for less than 12 hours, the rate must be larger than the average 12 hr flow rate illustrated though the needed volume remains about the same. This practical condition requires that a larger than average flow be planned from the canal during the day-time period to satisfy the upper portion farm needs which are taken for less than 12 hours, plus some operational reserve. For the few hours increment of time difference, this larger flow will bypass on into the service area reservoir. For the rest of the 24 hours, the smaller night-time flow which is not reduced by off-takes in the upper portion may be controlled by a valve on a semi-closed pipeline system where it outlets into the reservoir. If an open pipeline system (upstream control) is used in the upper portion, the reduced flow is set at the canal. The use of the arranged schedule permits these two daily flow rates to be anticipated and taken from the canal and the reservoir absorbs the inevitable small mismatches. If the lower area is about 55% of the total area rather than half, the day and night flows can be more nearly equal.

If the reservoir has just minimum capacity, precision of control is needed. Extra capacity makes it possible for approximate flow rates to be used, some extra in the day time, some less at night, and discrepancies absorbed in storage and made up in the next day's order. Make it bigger for easier operation.

With these operations, the conveyance capacity in the upper half in Table II (b) must be larger than average, appreciably so for the 2.5 USA farms, but very little for 20 farms in a developing country. This does require reservoir capacity measurements to calculate the two or three needed daily flow changes, or it can be automated. The needed 12-hour canal flow changes are small. They may be done with in-canal storage. The use of daily farm orders under the arranged schedule will be needed as usual. However, the actual farm off take rates will be as needed and may not conform closely with the farm water orders as made though the volume will be.

If the service area lateral is a closed or semi-closed pipeline, operations can be automated. If the upper area line is closed at the outlet into the reservoir during the day, or a moderate outflow is set and the line has adequate capacity, upper area farm off-takes can be made as desired and the inflow into the pipeline from the canal equates exactly and automatically. The canal supplies this fluctuation from planned scheduled flow or in-canal storage. With pipeline off-takes, a constant head at the canal is not essential and in-canal storage fluctuations are acceptable. At night all flow would go into the reservoir at the desired rate to balance canal conditions with a float valve preventing the reservoir from overflowing. It is emphasized that flows changes, if set, are made at the outlet end of the semi-closed pipeline, downstream control. If open pipelines or canals are used in the upper portion, controls are set at the canal, upstream control.

The area below the reservoir starts with a nearly full reservoir. A little space is left for mismatches and variable startup times for the reservoir the next morning. Off-takes through closed or semi-closed pipeline are taken by the farmers as desired and about as scheduled. They are automatically supplied from the reservoir storage. Most operations will require only morning and evening reservoir adjustments. All farm turnouts adjustments are made by the farmer at his valve with the reservoir absorbing all modifications in the lower area as excess flow (operational spillage) does in the upper area.

This technique may be the simplest to upgrade existing systems as very little change is needed in the canal system. It may be done in small units as each service area is independent. This is being done in Egypt.

CASE STUDIES
MULTIPLE PROJECT STORAGE RESERVOIR WITH CANAL OPERATION
Orange Cove, California USA. The 11,200 ha irrigation district is the first one on the over 400 km long Friant-Kern canal serving many irrigation districts. The Orange Cove I.D. serves land from numerous pipeline laterals by gravity below the canal and by pumping to above the canal. It is upgrading its system from a Fixed Rate and Duration schedule by increasing pipeline capacity by two thirds and constructing terminal reservoirs at the upper end of all uphill pumped pipelines. The District obtains flexibility in the lower portions by main canal fluctuations, and in the upper portion by varying the pumping rate, and reservoir fluctuations. The canal operations are absorbed by in-canal storage with the daily pass through volume being stable. The new Limited Rate Arranged-demand schedule permits the farmer on his arranged day to start and stop and modify his flow rate. The flow is metered. The farmer control is

greatly appreciated by them and the District work has been reduced by over half. (Chandler 1990)

Mardan SCARP, Pakistan. This small 25 ha Pilot Project with 12 farmers withdraws water from a large canal so the small fluctuating offtakes have negligible effect. This small area has been converted from a rotation schedule to a larger flow pipeline system with control at each farm outlet of the frequency, rate, and duration under a Limited Rate Arranged-demand schedule. Impressive results have been obtained from upgraded on-farm management, not from increased water use since the previous supply provided essentially adequate volume. Irrigation labor is reduced and made more convenient which is greatly appreciated especially the elimination of night time irrigation and the top ender-low ender problem. Yields of the various crops have been increased 30-50% because management can now control the water to provide better germination and growth, and match the needs of the several small areas of different crops. "It is not just the volume of water delivered, but the way it is delivered to make it useful." (Merriam 1991). This Demonstration Project showed that by improved irrigation, the drainage project that was installed could have been superseded by an upgraded irrigation project. (Styles, 1997).

ON-FARM RESERVOIRS
Coachella Valley, California USA. The irrigation district covers about 32,000 ha and delivers water to each 16 ha unit with a maximum fixed flow rate of 100 lps in 24 hour units. Farm holdings are typically 16 or 32 ha with several larger units. Over 250 on-farm owner built reservoirs have been constructed to convert the rigid flow rate and duration to a flexible operation. A large part of the justification is to reduce and make more convenient irrigation labor by supplying larger flow rates and day time use. Better water use efficiency and less drainage also results (see Figure 1).

Figure 1
Photo of reservoirs in citrus in Coachella Valley, CA, USA

Egypt. A variation of the above program is being applied to groups of many small (.2 to 2 ha) holdings and many farmers (50 to 100). A service area reservoir is constructed about two thirds the way along the distribution pipeline. Water is pumped up from a canal at a nearly steady rate to stabilize canal levels. The reservoir permits variable farmer offtake rates and durations during the day as arranged, and permits the canal flow to be adequately stable. A Water Users Association operates the Limited Rate Arranged-demand schedule with a congestion of 50% to 60% with daytime only usage though night time use can be arranged. Fish culture will be combined in the reservoir with the irrigation program.

SERVICE AREA RESERVOIRS
Gadigaltar, Madhya Pradesh, India. An 1150 ha semi-closed pipeline system has 101 Harris float valves, over 65 km of low pressure concrete pipe, over 500 farmers in 67 groups each with its own offtake and meter, and each farmer within the group having a personally controlled off take valve. The system is operated by a Water User Association using a Limited Rated Arranged-demand schedule with 50%-70% congestion. This is facilitated by an 8 ha m service area reservoir near mid-area. The upper area off take pipelines are supplied from a 5.4 km long sloping canal operating under upstream control with the excess operational spillage and night time flow being reregulated in the service area reservoir. The lower area below the reservoir is supplied through branch and lateral pipelines receiving their supply automatically from the reservoir via a 1.3 km long level top canal maintained constantly full by and AVIO float controlled gate. (Merriam 1990)

Imperial Irrigation District, California USA. This very large 189,000 ha district has established a 6800 ha Pilot Project Neyrtec Area involving seven sloping ditch laterals each about 13 km long. A "lateral interception" ditch was built across the lower end of the laterals and the operational spillage was concentrated in a 25 ha m service area reservoir. It is pumped from this reservoir to supply water to another large "lower" area.

This project which has only a limited objective of permitting the farmer to control duration of flow at his turnout, can ultimately increase the system capacity to also provide some degree of flexibility of rate. The frequency and 24 hour duration rate are currently arranged, but can be modified on the Pilot Project. The service area reservoir makes this important project possible. (Dimmitt 1992)

REFERENCES

Chandler, J.C., R.M. Moss and J.L Merriam. 1990. Automated Low Pressure Pipe for Flexible Deliveries, Orange Cove Irrigation District, Orange Cove, CA, USA. ICID 14th Congress Symposium Proc. R4.

Dimmitt, A.K., K.I. McLaughlin, F.Z. Koumand, and D.G. Welch. 1992. Planning for Water Conservation through Irrigation System Modernization and Rehabilitation. American Society of Civil Engineers. National Conference Proceedings, Irrigation and Drainage Division, pg. 294-299.

Merriam, John L. 1990. Gadigaltar Tank Irrigation Pilot Project, Khargone, Madhya Pradesh, India. Private printing ISPAN 1811 N. Kent St., Rm 1001. Arlington, VA 22209, or John L. Merriam, 235 Chaplin Lane, San Luis Obispo, CA 93405.

Merriam, John L. 1991. Flexible Irrigation Supply Pilot Projects, Part I Principle, Part II Case Studies. Proc. July 1991. American Society of Civil Engineers, Irrigation and Drainage Division Specialty Conference. Honolulu, HI, USA.

Styles, S. 1997. Alleviation of Surface and Subsurface Drainage Problems by Flexible Delivery Schedules. XXVII Congress Proceedings IAHR/ASCE Water Resources Engineering Div. August 1997. San Francisco, CA, USA.

Pima-Maricopa *Flexible* Irrigation Project

SHANE LINDSTROM
Water Management Engineer
Gila River Indian Community
Sacaton, AZ

INTRODUCTION

The Gila River Indian Community (GRIC) of central Arizona is planning a new irrigation project. The Pima-Maricopa Irrigation Project (P-MIP) will deliver irrigation water to 146,330 acres. Included in the P-MIP will be the primary, secondary and tertiary delivery systems. Early in the planning process, the advantages of a flexible delivery system were identified. Estimating economic advantages has been difficult as very little actual data is available.

To quantify economic benefits of a flexible irrigation water delivery system and coordinate the primary canal system operations, a 700-acre Demonstration Project (DP) has been proposed and designed by the GRIC. The DP will be used to test and establish irrigation and operational techniques. Knowledge gained from the DP will be applied to refine the design and operation of the complete P-MIP. As a demonstration area, it will be used to educate water users in irrigation practices and system capabilities.

It is projected to reduce on-farm development and operational costs by delivering through a pipeline a farmer-controlled, adjustable rate to the point of need, and simplify operations of the primary (main canal) delivery system.

DESIGN CRITERIA

Major features of the design include:
♦ Water delivered to the point of need
♦ Daytime only use
♦ Point of delivery (farm level) controlled rate and duration
♦ An adjustable, *stable* flow rate up to 12 cfs at a farm turnout
♦ A minimum of four feet of head available at every farm turnout
♦ Limited Rate Arranged Delivery Schedule
♦ Low pressure pipelines with Harris Float Valve control

LAYOUT

The proposed flexible irrigation water delivery system combines both the conveyance of water (supply system), and the distribution of water to the point of need. The distribution system is traditionally part of the on-farm design and installed by the farmer. By combining both aspects, the overall development costs for the total project will be reduced and sustainable agriculture will be assured for the Gila River Indian Community.

Service areas have been defined with a *Main line* coordinated with a Service Area reservoir conveying water from the primary canal to a series of *Lateral lines*. *Distributor lines*, serving about 120 acres, branch off and are aligned along the head of all fields. A riser extends to the surface every 330 feet. Additional farmer-installed risers can be constructed wherever needed depending on the appropriate irrigation method (level basin, furrow, border-strip, sprinkler, etc.). Constructing additional on-farm head ditches to bring the water to the point of need will not be necessary. Small farmers can farm, garden, or pasture without the high cost of head ditch development. Gated pipe will be practical. Large farmers can concentrate on adapting and funding modern systems that use water and labor efficiently.

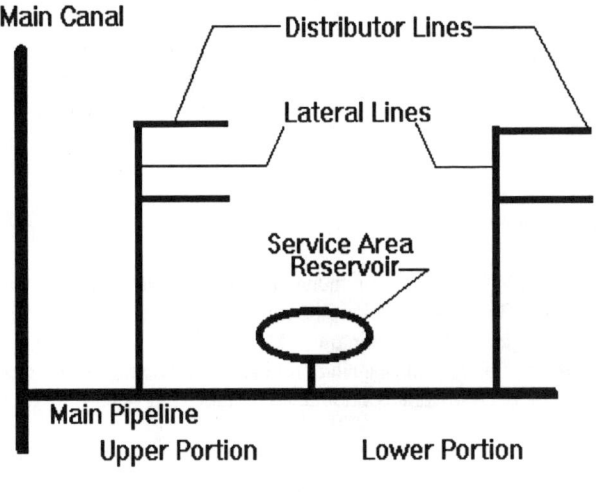

Figure 1 - Schematic Layout

OPERATION, CONTROL AND AUTOMATION

This design uniquely combines features of both upstream and downstream operations. The on-farm operations, which are paramount to this design, will use downstream operational concepts, while the supply system (primary canal and upper main pipelines) will use the conventional upstream operation.

Additional problems associated with deliveries to small 10 to 20 acre units are also handled by the delivery system. Forcing farmers to conform to the system (i.e., using rotational delivery concepts) eliminates the goal of flexibility.

In simple terms, this flexible design attempts to handle all problems associated with an irrigation water delivery system and not force either the primary canal or individual farmers to be affected by operational constraints. This is accomplished by semi-closed pipelines with Harris Float Valves, a reregulating service area reservoir, and a suitable delivery schedule.

SEMI-CLOSED PIPELINES WITH HARRIS FLOAT VALVES

The design incorporates extensive use of semi-closed pipelines. Semi-closed pipelines combine several of the best features of closed and open pipelines. The very desirable ability of the closed system to respond at the inlet end to changes in flow rates from a downstream location (a farm turnout) is retained. This permits a valve adjustment at a downstream turnout changing the flow rate to adjust the intake rate to match immediately, because the pipeline functions as a communication and a conveyance system. This essentially automates the operation of the system, which is a necessity for modern (and future) irrigation systems (Merriam 1987a).

Automation is too often confused with elaborate structures and electronic controls. This suggests a high level of operational management and hardware to ensure the system performs adequately. The simplicity of automation in this design ensures a very high level of reliability that does not and cannot exist in electronic circuitry. It is proven, simple, reliable technology.

The capability of an open pipeline to operate at low pressures is retained, permitting the use of less expensive pipe and equipment, and providing practical conditions at farm turnouts where low pressure is desired to control erosion and enhance the precision of flow settings.

The drawbacks in closed systems of large pressure variations with changing flow rates, and in open pipelines the problems associated with air entrainment and the inability to act as a communication system, are eliminated.

The fundamental concept for the semi-closed system control involves a water level sensing capability on the downstream side of the controlling valve. This involves a Harris Float Valve (HFV). As flow changes downstream, the water level in the standpipe that contains the HFV changes. The float activated valve then operates, permitting downstream passage of water at the same rate (Merriam 1987b).

REREGULATING SERVICE AREA RESERVOIR

In this design, a regulatory reservoir is placed near the middle of the service area. The service area is further designated as an upper (upstream from the reservoir) and a lower (downstream from the reservoir) areas. One purpose of the reservoir will be to provide storage to automatically accept varied flows from the upper service area resulting from farmers starting and stopping irrigations, or not taking water at night. Almost all irrigation can be done in the daylight. The turnout flow from the primary canal will be kept nearly constant (upstream operation) to allow near steady-state operation of the primary canal. The distributor lines will only convey water from the main line and a lateral when a farmer-controlled valve is opened (downstream operation). The rejected flow from the upper portion is routed automatically into the reservoir (see Figure 1). Another purpose of the reservoir is to supply water to the downstream service area. Downstream operations will then be fully realized in the lower services area. Water will stay in the reservoir until a downstream demand exists. The reservoir is strategically located to allow gravity flow in and out, sized to hold a two-day supply of water for the lower service area and located off-line enabling it to be bypassed to reduce siltation (Merriam 1997a).

Another major benefit of the regulatory reservoir is that it will reduce the size of the upper potion of the main pipeline. Instead of having to size the main pipeline to service the entire area, it can now be sized to service only the upper area in the daytime with the night time flow passing through to the reservoir. A similarly sized pipe services the downstream service area below the reservoir.

DELIVERY SYSTEM SCHEDULE

There are a great variety of delivery schedules used. The main components of the delivery schedule are: the frequency in which the water can be delivered, the rate of flow delivered, and the duration of these deliveries. The chosen schedule for this design is the Limited Rate Arranged. This was chosen as an economical alternative to a demand schedule. With this schedule, a water user must arrange for the day of irrigation. This allows the irrigation district to manage the overall system by controlling who receives water on which day, but not controlling the hours of delivery within that day.

On the arranged day, the water user will have complete control over the water. The duration, within that 24-hour period is up to the farmer. The irrigator can start and stop irrigations based on *his* time and labor considerations and adjust the rate up to a maximum of 12 cfs.

CONGESTION

Distributor lines are designed to serve a maximum of 120 acres. To ensure turnout stability, each will have a Harris Float valve and a meter and only one user at a time can irrigate from a distributor line. It is estimated that with a flow rate of 12 cfs, 20 acres can be irrigated during a day. Daytime only irrigation will be strongly encouraged. This would allow a minimum irrigation interval of six days within a typical ten to twelve day cycle (60% congestion). The mains and lateral are able to serve the entire service area in seven to eight days (75% congestion). This low congestion value provides a high degree of assurance of having water when desired - nearly a demand system (Merriam 1997b). With this irrigation interval, congestion will be negligible. If a farming operation controls a whole 120 acre block, congestion would be virtually nonexistent. Unauthorized use of water should be a minor problem. If a farmer is assured of timely water deliveries, there will not be a need to take it without authorization.

ON-FARM OPERATIONS

The goal of this system is to remove restraints imposed on the farmer by his delivery system. The restraints on frequency, rate, and duration affect the farmer in a multitude of ways, compromising his overall management of his farm. Restrictions cannot be placed on on-farm water use practices without placing limitations on the type of irrigation system employed, the type of crops that can be grown (Replogle, 1984). The restraints are particularly noticeable to the farmer on the amount, convenience and cost of labor.

Since the rate and duration are controlled at the farm level, all irrigation methods, both surface and pressurized are possible. While this system was designed and sized for surface irrigation (the predominate form of irrigation on the GRIC), conversion of acreage to pressurized systems could be easily accomplished.

SURFACE AND SUBSURFACE RUNOFF AND DRAINAGE

Current storm runoff patterns will not be affected by buried pipelines as they would with ditches. Road and farm bridges will not be necessary. Farm equipment will have unlimited access to fields. These features will obviously save a considerable amount on the total cost of the project, and will be documented by the Project.

Runoff from irrigated fields has long been a problem. Expensive solutions exist to capture this water, or where applicable, high development costs can eliminate it. The logical alternative to disposal of this water is to reduce the quantity by controlling the source. Since the flow rate can be adjusted *during the irrigation*, runoff can be greatly reduced by cutting back the inflow once runoff becomes excessive.

The drainage and poor water quality problems that arise can be avoided rather than overcome (Merriam and Burt, 1988). The beneficial environmental aspects of this design will be difficult to quantify, but cannot be overlooked.

AGRICULTURAL EDUCATION CENTER

An Agricultural Education Center will be established as an integral part of the Demonstration Project. It will be a 10 to a 20-acre site where education and not crop production is the objective. Current cultural practices use 24 hour a day operations. To assure Project success, it is imperative that area growers and Community members are educated in how to maximize the benefits from this system (daytime irrigations only).

The surface methods of furrows, border-strips, and basins are well adapted and can be effectively used with adequate training of irrigators. It may be practical to have an irrigator training program to certify an irrigator who would be the only person permitted to take water from the system without the use of a zanjero. This will ensure participation of the area growers.

SUMMARY
- Equitable to large and small agricultural water users
- Point of delivery (farm level) controlled rate and duration
- Adjustable, stable flow rate up to a maximum of 12
- Reduces on-farm development cost
- Reduces operation and maintenance costs of primary delivery system
- Virtually fully automated
- Minimum of four feet of head (pressure) available at every turnout
- Cost effective
- Proven, simple, effective technology
- Adaptable to ALL irrigation methods
- Reduced, effective, and convenient labor
- Efficient and uniform water application

As of January 1997, the GRIC is still pursuing National Environmental Policy Act compliances for the Pima-Maricopa Irrigation Project and funding has not been authorized for the construction of the Demonstration Project.

REFERENCES

Merriam, John L., 1987a. Pipelines for Flexible Deliveries. 1987b, Design of Semi-closed Pipelines System. Planning, Operation, Rehabilitation and Automation of Irrigation Water Delivery Systems. ASCE I&D Symposium Proceedings, July 1987. D.D. Zimbelman, Editor, Portland, OR.

Merriam, John L. and Stuart Styles, 1997a. Incorporation of Reservoirs Into Irrigation Supply Systems to Simplify Flexible Operations. 1997b, Design and Congestion Considerations for Flexible Irrigation Supply Systems. Water for a Changing Community. IAHR/ASCE Congress Proceedings, August 1997, San Francisco, CA.

Replogle, J. A., 1984. Some Environmental, Engineering and Social Impacts of Water Delivery Schedules. Proceedings, Twelfth Congress, ICID, Ft. Collins, CO.

Merriam, John L. and Burt, Charles M., 1988. Alleviation of Surface and Subsurface Drainage Disposal Problems by Improved Delivery Scheduling. USCID Meeting, September 1988, San Diego, CA.

ALLEVIATION OF SURFACE AND SUBSURFACE DRAINAGE PROBLEMS BY FLEXIBLE DELIVERY SCHEDULES

STUART STYLES
BioResources and Agricultural Engineering Department
California Polytechnic State University, San Luis Obispo, CA 93407 USA

ABSTRACT

It is shown that all irrigation methods have higher potential efficiencies and uniformities than are commonly obtained in the field. When the manager has control of the water supply rate and duration (i.e., flexible scheduling), higher values can usually be economically attained. Such control can alleviate or eliminate many subsurface drainage and salinity problems. Surface runoff can be more easily controlled by the manager and can be essentially eliminated by return flow systems.

INTRODUCTION

The practical application of irrigation water to satisfy crop transpiration requirements means that some excess be applied by the manager. This includes extra water applied to compensate for:

- the application of some excess to be sure of applying enough (Efficiency);
- non-uniform application (Distribution Uniformity - DU);
- maintaining a salt balance in the soil (Leaching Requirement - LR),
- less than perfect management, and
- more convenient labor operations

These practical excesses show up as surface runoff and deep percolation in the field. In addition, there are excesses that show up due to poor system design and management. To alleviate these excesses requires management control and good systems. Good systems need knowledgeable designers who can "think like an educated farmer." For management to exercise control, the irrigation water supply must be adjustable. This means that a flexible schedule is essential. The most practical schedule is a Limited Rate Arranged-demand (ASCE 1984).

At the farm level, the Limited Rate Arranged schedule provides flexibility in the frequency, rate, and duration under control of the irrigator. The frequency (day of irrigation) must be arranged by the irrigator with the supplying project so that there is some degree of control of the maximum project supply rate. The magnitude of the restriction on frequency is related to the degree of congestion built into the supply system (Merriam 1997a). On the arranged day, the irrigator arranges for the rate and duration. This may vary from there being no restriction on the rate (up to the system limit) and duration (i.e., a demand schedule on the arranged day), to such restrictions as a fixed rate and duration. A fixed rate and duration are potentially very restrictive on management and as such are undesirable at the farm level.

The system limited rate does not greatly affect management of the irrigations but does affect the irrigation labor needs and economics. For the most desirable on-farm labor utilization, the stream should be large enough that an entire field can be irrigated in one daylight period (Merriam 1997b).

The history of irrigation shows that projects have traditionally delivered water from steadily flowing canals in a continuous manner because engineers place primary importance on their needs, and usually unknowingly neglect the farmer's need of a variable supply. The results have been drainage and salinity problems for the farmer. In arid and semi-arid countries most of the needed subsurface drainage can be alleviated by appropriate irrigation.

METHODS AND MANAGEMENT
Mechanical Pressurized Methods

For sprinklers and micro-irrigation systems, good distribution uniformity is usually designed into the system and prolonged by proper maintenance. In general, deep percolation results from improper duration. To alleviate this requires a schedule permitting the irrigator to turn off the water when enough water has been applied.

The definition for Distribution Uniformity ratio was originally presented by the ASCE I&D Division, On-Farm Irrigation Committee (ASCE 1978) as:

$$DU = \frac{\text{av. of low quarter depth of water infiltrated}}{\text{av. depth of water infiltrated}}$$

With pressurized systems, the differences in water applied (and assumed infiltrated) at several points is the result of the pressure variation, nozzles, and elevation changes. This results in essentially a normal distribution pattern which remains constant over time and is the result of the system design. Assume a representative value of DU = 0.70 is used (see Table 1) and an irrigation event applies 2 inches of water. This means the low quarter receives 30% less than the average (1.4 inches) and the high quarter 30% more (2.6 inches). The difference, 60%, is nearly doubled the low quarter infiltrated depth (2.6 vs 1.4 inches) and if the desired water infiltrated was 1.4 inches the water applied in excess of 1.4 inches becomes deep percolation and must be taken care of with either natural or artificial drainage. Good design (and maintenance) of pressurized irrigation systems is essential to minimize deep percolation.

When over-irrigation occurs by running too long (duration), excess water applied will go to deep percolation. A sprinkler set running 12 hours when 10 hours was adequate causes an additional 20% of the water to percolate below the rootzone. A flexible schedule without a duration restriction on management is essential to permit control of the deep percolation loss.

In response to a lack of flexible duration from the irrigation district delivery, there has been a significant observed increase in the number of irrigation wells that have been utilized by growers who have pressurized systems. This has occured even where less expensive irrigation district water was available.

Furrow Irrigation
Furrows are sloping channels down which water is run, infiltrating laterally and vertically. The key to proper furrow irrigation is the management of the duration and flow rate down the furrow. Some helpful definitions include:

- the duration needed to reach the lower end is the Time of advance (T_{adv})
- the duration water is available at the lower end is the Time at lower end (T_l)
- the water application time Time of application (T_a)
- soil moisture deficit (SMD) is the water depleted from the rootzone
- In this paper the advance ratio (AR) = T_{adv} / T_l (Merriam 1988). Others have alternatively defined the AR as T_{adv} / T_a (Burt 1995).

Water should remain running (or ponded) at the lower end long enough to infiltrate the desired amount of water. The desired amount of water is defined as the soil moisture deficicit (SMD). There should always be runoff (or ponding) during the irrigation in order to adequately irrigate the lower end of the field. Ending an irrigation prior to reaching the end of the furrow will underirrigate the lower end of the field and eventually lead to salt build up problems for most soils.

Within the restraints of soils and topography which affect all surface methods, furrows are the most adaptable to achieving good DU and efficiency. The time of advance (T_{adv}) is readily modified by growers by changing the stream size, furrow shape, slope, surface conditions, and length of the field. The desired time of application (T_a) can be varied by changing the furrow shape, spacing, surface condition, soil moisture deficiency, and stream size (wetted perimeter). A number of these conditions are modifiable by varying the frequency, stream size, or duration of irrigation (Merriam 1978a).

The impact of irrigation labor convenience and the length of daylight are important practical aspects of surface irrigation. This means that the duration of a set and the flow rate available has an economical impact. Where practical, daytime only irrigation is advantageous. The ability to control the desirable stream size, advance time, and irrigation time to create a practical application time, is of great importance. A stream adequately large enough to set all the furrows in a field at one time and to be modified where cutback streams are used with no change in the number of furrows running, can significantly save labor.

A tailwater recovery system (TRS) is very practical when a farm turnout structure can automatically maintain a stable rate to the farm system. There are various techniques available to accomplish this. Two options used by growers include:

- using a Harris float valve to a pipeline, or
- a Neyrtic float gate to a level top ditches

These systems require that they be used in conjunction with a project supply that is flexible in rate. A TRS at the lower end of a field can return runoff from the furrows (or border-strips) into the farm side of the turnout structure. As the return flow intermittently varies, the system automatically cuts back the supply from the project by the same amount and all the return flow goes back onto the field with the field delivery rate remaining constant. This combination permits a large enough initial stream to be set in all furrows to have a good AR and convenient labor while having no runoff from the field. No cutback is needed and as the intake rate decreases the flow rate to the field is automatically cut back to just match the intake requirement.

Some growers utilize an alternative approach where the TRS returns water to the head ditch with a constant flow rate. This has been successfully done in the Imperial Irrigation District in Southern California. This requires that a large TRS pond be used to store a portion of the irrigation runoff and to return the the water for a set time at a constant flow rate.

By having a flexible stream size, the manager can control the AR. The AR is related to how the essential excess furrow irrigation water is distributed between deep percolation and runoff. This management tool is described in detail by Merriam (1988).

- A slow but good AR of 1/1 will have about 16% Deep Percolation and about 12% runoff and a DU of about 0.80.
- A medium AR of 1/2 will have about 8% Deep Percolation and about 22% runoff and a DU of about 0.87.
- A rapid AR (perhaps too rapid for good economies) of 1/4 will have about 5% Deep Percolation but about 35% runoff with a DU of 0.93.

All of these variations could be made in the same field merely by changing the initial stream size with its resulting change in application time and advance time. This management tool is made possible by a flexible supply. The reduction of the drainage excess is obvious. The deep percolation for the AR = 1/1 is 16%. For the AR = 1/2, it is 8%, only half as much. This is the emphasis of this paper.

The manager has the tools to control where the excess water goes and how much IF he has control of a flexible supply. He can have more control with furrows than any other method.

Border-Strip
With the use of the border strip method of having water flow down a level across strip between border ridges it is a complicated procedure to have good DU and efficiency. The recession curve on a border strip is site specific and always about the same shape for most irrigations. To have good uniformity, the advance curve must be essentially parallel which can be done by modifying the flow rate. The water also needs to be shut off-on time. The optimum shut off time usually occurs about 2/3 to 3/4 down the field. A flexible flow rate and duration water supply is needed.

The further complication of correlating the SMD with the time of irrigation and the length of the strip to have high efficiency are discussed in several papers (Merriam 1978b), (Merriam 1987), (Burt 1995).

Measured Distribution Uniformity
The following Table 1 is condensed from Table 1 in a paper by Hanson, et al (1996). This table reports the DU values obtained from the many evaluations made in California by the Mobile Labs. These typically are from commercial farms with a water supply that is often rigid duration, fixed flow schedules.

Table 1 Average DU

Method	Sample Size	DU%	Std. Dev.
Hand move/Solid Set sprinklers	164	62	15 c
Continuous move Sprinklers	57	75	10 a
Undertree sprinklers	28	79	16 ab
Micro-irrigation (permanent crops)	458	73	15 a
Micro-irrigation (row crops)	23	63	16 c
Furrow	157	81	14 b
Border	72	84	14 b

Standard Deviation values with the same letter are statistically similar. Level of confidence = 95%

The purpose of this paper is to emphasize that in order for knowledgeable irrigation managers to best utilize their skill, they must have a flexible supply to reduce and allocate on-farm water losses and also to use the water more effectively. Evaluations which have been made usually show that surface systems are not used at their potential. The reason usually found is that the proper stream size is rarely available and 24 hour durations force compromises. This results in low efficiencies. The moderate flow rates and 24 hour durations are not conducive to effective labor use and daytime only irrigation is not practical so they are labor intensive.

The DU values of the mechanized pressure methods are not appreciably affected by changes in available flow rates since they are designed and installed for their optimum rates and pressures. Duration does not affect their DU.

However with the moving water surface methods, flow rates and durations have an appreciable effect upon DU, efficiency and labor. With upgraded supply systems as advocated in this paper, surface irrigation operation can be appreciably improved. The surface DU's tabulated above are higher than those for the pressurized methods and can become higher. The standard deviation shown indicates the potential is above 90%.

If flexible supply systems and schedules are made available, surface irrigation methods where applicable can compete with the most efficient pressurized systems in water, labor, and energy conservation. Flexible systems and schedules are beneficial to upgrading all irrigation methods, but are especially beneficial to surface methods.

REFERENCES

ASCE (1978). On-Farm Irrigation Committee, E. G. Kruse, Ch. Describing Irrigation Efficiency and Uniformity. Journal ASCE, I&D Div., Vol. 104, No. IR 1. March 1978.

ASCE (1984). On-Farm Irrigation Committee, R. E. Walker, Ch. Recommended Irrigation Schedule Terminology. Conference Proceedings, ASCE, I&D Div. July 1984. Flagstaff, Arizona, USA.

Burt, Charles M. (1995). The Surface Irrigation Manual, Waterman Industries, Exeter, California, USA.

Hanson, B., W. Bowers, B. Davidoff, D. Kasapligil, A. Carvajal and W. Bendixen (1995). Field Performance of Microirrigation Systems. Proceedings of the ASAE 5th Intl. Microirrigation Congress. April 1995. Orlando, Florida, USA.

Merriam, John L. and Jack Keller (1978a). Farm Irrigation Systems Evaluation: A Guide for Management. Utah State University, Logan, Utah.

Merriam, John L. (1978b). Border -Strips Irrigation Design--Practical Approach. ASAE paper No. 78-2008. ASAE Summer Conference. June 1978. Logan, Utah.

Merriam, John L. (1988). Simple Furrow Advance Ratio Evaluation Technique for Upgrading Management. Conference Proceedings, ASCE I&D Div. July 1988. Lincoln, Nebraska, USA.

Merriam, John L. and Stuart Styles (1997a). Design and Congestion Considerations for Flexible Irrigation Supply Systems. Congress Proceedings, XXVII, IAHR/ASCE, Water Resource Div. August 1997. San Francisco, California, USA.

Merriam, John L. and Stuart Styles (1997b). Use of Reservoirs and Large Capacity Distribution Systems to Simplify Flexible Operations. Congress Proceedings, XXVII, IAHR/ASCE, Water Resources Div. August 1997. San Francisco, California, USA.

ENTRANCE FLOW AND THE ACHIEVEMENT OF UNIFORM FULLY-DEVELOPED OPEN CHANNEL FLOW

HECTOR R. BRAVO AND JOHN W. MEINECKE
University of Wisconsin–Milwaukee, Milwaukee, USA

INTRODUCTION

Advances in measurement techniques have led to re–examination of simple and complex open channel flows. The simplest flow that exhibits 2D turbulent structures is uniform, fully developed flow in a wide channel. The attainment of such a flow is a non–trivial problem. The flow development length is about 240 hydraulic radii, starting from a fairly uniform entrance flow.

A flume recently installed at UWM has small rectangular head tank, because of space limitations. The tank width, length, and depth are equal to 1, 1.6, and 1.6 flume widths, respectively. The head tank provides a downward enlargement of 0.933 flume widths and originally had neither a transition nor baffling. The resulting entrance flow was non–uniform, highly turbulent, and showed considerable surface waviness.

A test program was completed to study the effect of different head tank improvements on the entrance flow mean and turbulence characteristics. This paper summarizes the tested configurations, the measurement techniques including LDV, the detailed measurements, and a simple method for quantitative analysis. A 2D flow was verified to exist near the downstream end of the flume, thus setting the stage for basic studies of open channel turbulence.

BACKGROUND AND THEORETICAL CONSIDERATIONS

Nezu and Rodi (1986) discussed the conditions required to achieve fully developed, 2D open channel flow. The location of the test section, y, should satisfy $y/(4R) \geq 60$, with R=hydraulic radius. The aspect ratio b/h (width/ flow depth) should be ≥ 6 to achieve 2D flow in the central region ($|x|/h < (b/h-4)/2$).

Analyses of boundary layer development in an open channel start with an ideal, uniform entrance condition. If the entrance condition is non–uniform, then the development length may be larger and exceed the flume length.

Building a small head tank that produces good entrance flow is difficult. It is necessary to calm the high–velocity inflow from a pipe having a small fraction of the flume cross–sectional area. Rouse (1961) described the use of transitions, lattice screens, and diffusers in head tanks. Stilling that cannot be accomplished by tank size must be made by baffling. Baines and Peterson (1951) investigated flow through screens, including their capacity to modify the velocity distribution and the turbulence characteristics. Rouse (1961) recommended using a tandem of lattice screens with a solidity ratio S (closed area/total area) no grater that 0.5, spaced at least eight to ten times the mesh size.

FLUME AND HEAD TANK

The tiltable flume used is 18.288 m long, 0.762 m wide, and 0.508 m deep. The length, width, and depth of the head tank are 1.219 m, 0.762 m, and 1.219 m, respectively. The head tank is 0.711 m deeper than the flume, as shown in Fig. 1. Water is supplied as a submerged jet issuing from a vertical, 0.254 m–diameter pipe. The pipe was initially located near the center of the tank; it was then moved upstream to allow the addition of a vertical transition and a series of identical lattice screens.

Four head tank configurations were tested. Case 1 is the original configuration. Case 2 included the addition of a vertical transition of circular longitudinal section. Case 3 included the addition of a series of two lattice screens standing on the bottom of the tank, without a transition. Case 4 included both the vertical transition and a series of three lattice screens. The wooden screens have an intermediate solidity ratio S = 0.47, and are spaced eight times the mesh size.

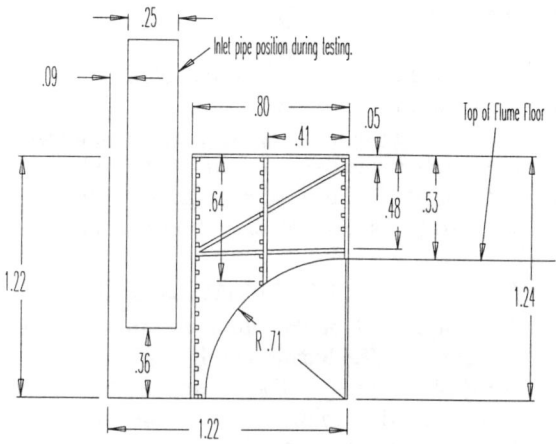

Figure 1. Longitudinal section along centerline of head tank.

MEASUREMENT TECHNIQUES

Mean velocity was measured using a two component electromagnetic velocity meter and a PC–based data acquisition system. Measurements of mean and turbulence characteristics were obtained using a one component, 10 mW TSI laser–Doppler velocimeter (LDV) operated in backscatter mode. The data acquisition software was set to choose 10 K velocity realizations at each measurement point. The optics component of the LDV system are mounted on a motor–driven, two–component traversing system. The uniform measurement grid used had horizontal and vertical spacings of 0.076 m and 0.030 m, respectively. The vertical spacing was halved for LDV measurements near the head tank. The measurements of fully developed flow used a vertical step of h/50.

INTERPRETATION OF RESULTS

The following experimental conditions were used: h=0.152 m, b/h=5, R=0.109 m, mean bulk velocity U_m=0.386 m/s, Froude number F=0.315, Reynolds number R=1.5 x 10^5, slope S=2.9 x 10^{-4}, Q=0.0448 m^3/s, friction velocity U_*=$(gRS)^{1/2}$=0.0176 m/s, and screen Reynolds number R_s=1.7 x 10^4.

3D plots were prepared of all the measured distributions. Quantitative analysis was simply done by comparing the moments of the measured distributions, taking as a reference the moments of a theoretical uniform entrance flow. The transverse mean position of the latter is x/b = 0.5. The mean square displacement for a uniformly distributed, continuous variable is σ_x/b = 1/12. The mean square displacement for the discrete measurement grid used is σ_x/b = 1/15.91. Results obtained at y/R = 7 are shown in Figs. 2 (mean velocity distribution) and 3 (turbulence intensity), and Table 1. Case 1, produced flow concentration near the mean position (σ_x/b = 1/16.5 < 1/15.91) and high average turbulence intensity (u'/U = 0.148). High (and fairly uniformly distributed) values of u'/U indicate strong surface waviness. In case 2 the transition practically eliminated flow separation and reduced surface waviness (u'/U = 0.121). Peak velocities were reduced, but the velocity varied linearly across the flow, with the mean position shifting to the right (x/b = 0.513). In case 3 the lattice screens reduced the average turbulence intensity (u'/U = 0.123) by the mechanisms of jet diffusion and decay, reduced peak velocities, but the flow remained concentrated near the mean position (σ_x/b = 1/17.0). The lack of flow guidance reduced the effectiveness of the lattice screens. In case 4 the simultaneous addition of a transition and lattice screens produced a flow with mean position and mean square displacement closest to the desired uniform velocity case (x/b = 0.499, σ_x/b = 1/16.1), and a lower turbulence intensity (u'/U = 0.098).

Table 1 shows (between parentheses) results obtained at y/R = 21 and Fig. 4 shows a sample velocity distribution. The mean position is slightly to the right, the mean square displacement is practically the desired one (σ_x/b = 1/15.9) and the average turbulence intensity is the lowest of all cases (u'/U = 0.063).

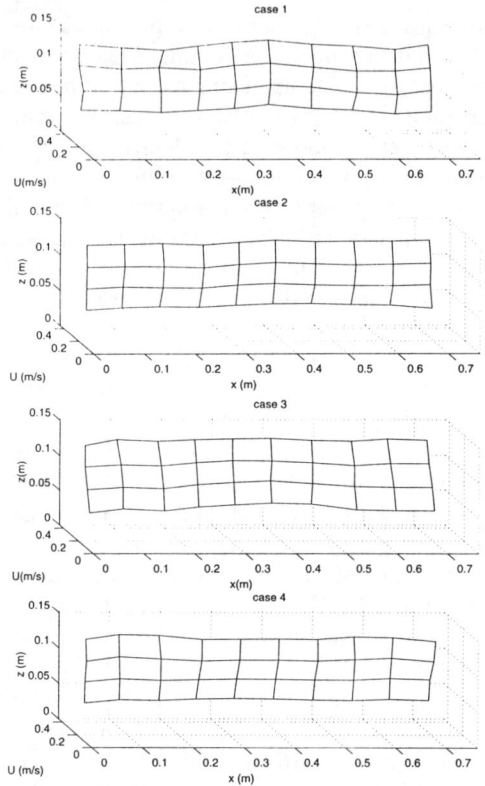

Figure 2. Mean velocity distributions measured at y/R=7.

Table 1. Velocity distribution and turbulence intensity at y/R = 7 (and y/R=21).

	mean velocity		
	mean position x/b	mean square displacement σ_x/b	u'/U
case 1 (w/o transition, w/o screens)	0.495 (0.499)	1/16.5 (1/16.0)	0.148 (0.092)
case 2 (w/ transition only)	0.513 (0.512)	1/16.3 (1/16.4)	0.121 (0.083)
case 3 (w/ screens only)	0.506 (0.499)	1/17.0 (1/16.5)	0.123 (0.082)
case 4 (w/ transition and screens)	0.499 (0.505)	1/16.1 (1/15.9)	0.098 (0.063)
uniform velocity distribution	0.5000	1/15.91	

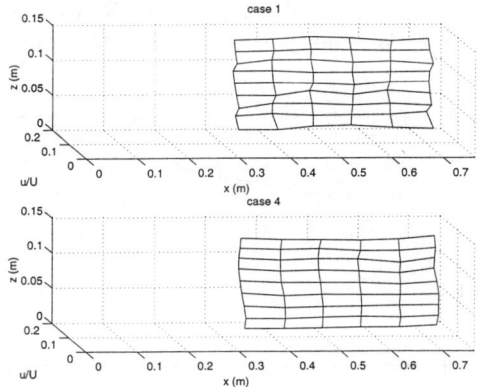

Figure 3. Turbulence intensity distributions measured at y/R=7.

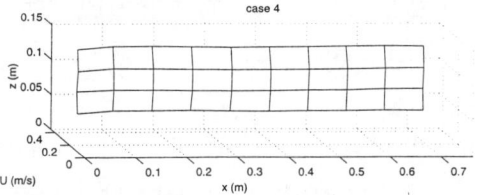

Figure 4. Velocity distribution measured at y/R=21.

Fig. 5 shows mean and turbulence characteristics measured at y/R=160 at the flume center, plotted using the following dimensionless variables: $z^+=zU_*/\nu$, $U^+=U/U_*$, z/h, and u'/U_*. Both profiles show good agreement with data and laws presented by Nezu and Rodi (1986); those laws are the solid lines in Fig. 5. Mean velocity follows the law of the wall and shows a wake effect. Practically identical profiles were obtained at other locations in the central region.

CONCLUSIONS

The effects of the addition to a head tank of a vertical transition and/or a series of lattice screens was investigated. A simple method was implemented for quantitative analysis, based on the moments of the measured distributions.

The transition eliminated flow separation, reduced peak velocities and waviness, but the velocity varied linearly across the flow. A series of lattice screens reduced peak velocities and the average turbulence intensity, but the flow remained concentrated near the mean position. The simultaneous addition of a transition and lattice screens permitted to obtain a practically uniform entrance flow. Flow measurements obtained at a distance y/R=160 show good agreement with experimental data and laws presented by Nezu and Rodi (1986).

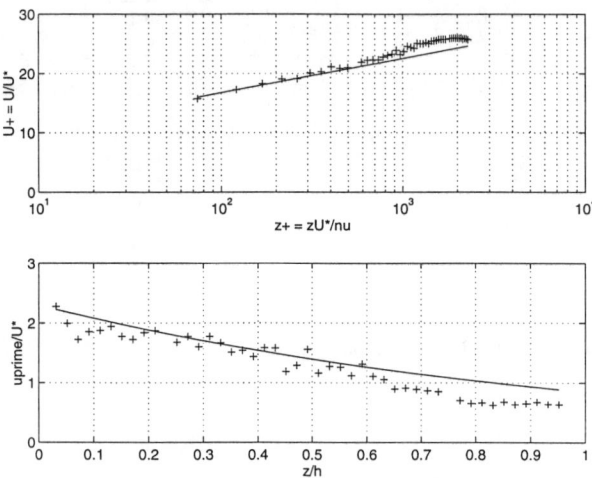

Figure 5. Velocity and turbulence intensity distributions measured at y/R=160.

ACKNOWLEDGMENTS

John Meinecke built the tank improvements and collaborated in the setup of the LDV and data acquisition systems. Measurements were completed by John Meinecke and Yunhan Zheng. Hector Bravo directed the tests and wrote the manuscript. Kwang Lee, Dept. Chair, was a staunch supporter of the project.

REFERENCES

Baines, W.D. and Peterson, E.G., 1951. "An investigation of flow through screens," Trans. ASME, Vol. 73, p. 467–480.

Nezu, I. and Rodi, W., 1986. "Open–channel flow measurements with a laser Doppler anemometer," J. of Hydraulic Eng., Vol. 112, No. 5, p. 335–355.

Rouse, H., 1961. "Laboratory instruction in the mechanics of fluids," The Univ. of Iowa, Studies in Engineering, Bulletin 41.

POSITIVE FRONT OF DAMBREAK WAVE

GUIDO LAUBER and WILLI H. HAGER
VAW, ETH-Zentrum
CH-8092 Zurich, Switzerland

The positive wave front is determined for a dambreak wave in an initially dry downstream bed. The channel geometry is rectangular and turbulent flow on a smooth boundary is considered. The theoretical results are compared with experiments for bottom slopes up to 50%.

INTRODUCTION
Dambreak waves have received particular attention in the past decades, mainly since the advents of numerical modelling. Classic approaches were provided by De Saint-Venant, Boussinesq, Ritter, Massau and Schoklitsch, and a historical resume on this period is given by Hager and Chervet (1996). A detailed review of experimental contributions on the dambreak wave is also available (Hager and Lauber 1996). From the latter state-of-the art, only scarse contributions to the front development were found. In particular, the effect of bottom slope has hardly been considered. The present project aims to answer some relevant questions:
- How fast does the wave front propagate?,
- What is the time-location of the wave front?,
- What is the effect of bottom slope?,
- Can the Froude similarity be applied?
- Does a front with a constant velocity of propagation exist?, and
- Can wave fronts be predicted with a simple approach?

EXPERIMENTS
Dambreak waves were reproduced in a channel 500mm wide, 700mm deep and 14m long. 4m downstream from the upper end, a vertical gate driven by pressurized air was inserted. Accelerations up to 4g could be generated, but effects of opening time on the dambreak wave disappear for an acceleration larger than 2g. The right channel wall was of glass to allow for video visualization, the bottom and the left wall were of black PVC. The estimated surface roughness height was $k_s=0.005$mm. Flows were essentially all in the turbulent smooth regime. Detailed descriptions of the test stand and the observational procedure are available (Lauber and Hager 1995).

In a preliminary study, the minimum depth at the dam location was determined to $h_o=300$mm, to inhibit effects of viscosity. Comparisons of waves for $h_o=300$, and 500mm indicated perfect agreement when scaled according to Froude similarity. For locations up to x=10m (downstream end of channel), i.e. $x/h_o=30$, the viscosity effect is negligible.

The present paper refers to the leading wave front of a dambreak wave issued into a dry prismatic channel of variable slope. The conventional Ritter solution (e.g.

Chaudhry 1993) is known to overestimate the front velocity, and effects of bottom slope up to 50% have not been covered so far. Further, results involving the entire wave profile, and the velocity distribution are available by 1997 (Lauber 1997) and will be presented separately. It should be noted that the wave front is an essential feature of the dambreak wave.

From detailed observations close to the breach section [x/h_o]<1, two positive waves may be identified (Fig.1): (1) *Initial wave* with a propagation velocity $v_I=[(10/9)gh_o]^{1/2}$, and (2) *Dynamic wave* with $v_d=2(gh_o)^{1/2}$, according to Ritter (Liggett 1994). The initial wave results essentially from orifice flow and its velocity is much smaller than v_d. After a short time, the dynamic wave overtakes the initial wave, and it is this phase that is considered subsequently. The time origin of the dynamic wave is $T=(g/h_o)^{1/2}t=2^{1/2}$ and this corresponds to the time of free fall of the surface particle at the gate until reaching the channel bottom. As shown by Martin (1990), the inital phase can be considered as a free fall of the fluid mass. In this initial phase, the shallow water conditions are invalidated and effects of streamline curvature are dominant. This phase is excluded hereafter.

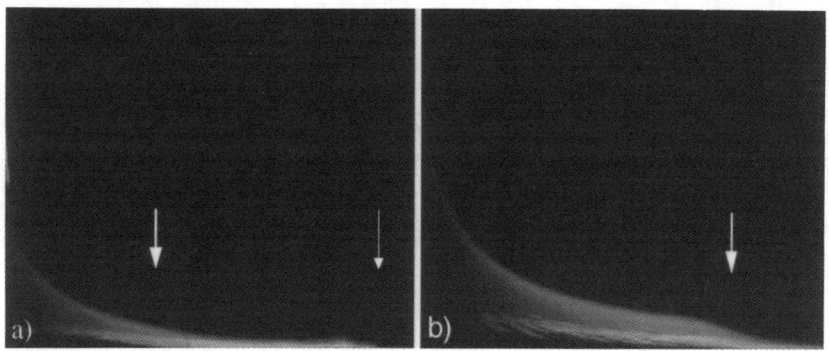

Fig.1 Free surface downstream of dam section (left end) at T=1x0.114, b) 2x0.114. Initial wave (small arrow) and dynamic wave (larger arrow) are indicated.

WAVE FRONT VELOCITY

Except for small relative times $T<2^{1/2}$, the De Saint-Venant equations may be used for modeling dambreak flows. In the characteristic form, they read for the positive wave in a rectangular channel (Liggett 1994)

$$\frac{dx}{dt} = v + c , \qquad (1)$$

$$\frac{d(v+2c)}{dt} = g(S_o - S_f) , \qquad (2)$$

where x=streamwise coordinate along the channel, with origin x=0 at the dam section, t=time, with origin t=0 at the time of rupture, v=cross-sectional velocity, $c=(gh)^{1/2}$=shallow water propagation velocity, g=gravitational acceleration, and $S_f=(v^2/2g)(f/4R_h)$=friction slope, where f=resistance coefficient, and R_h=hydraulic radius. Usually, the flow is shallow and R_h=h=flow depth.

POSITIVE FRONT OF DAMBREAK WAVE 731

At the positive wave front (subscript F), Eq.(2) yields $(v_F-2c_0)[T=2^{1/2}]=0$ because the front flow depth is $h_F=0$, and the initial velocity in the reservoir is $v(0)=0$. Integration of Eq.(2) subject to the initial condition $V_F(T=2^{1/2})=2$ in an initially dry channel gives

$$V_F = 2 + (S_0-S_f)(T-2^{1/2}). \qquad (3)$$

All capital letters are non-dimensionalized, respectively, by the characteristic length h_o, the characteristic time $(h_o/g)^{1/2}$ and the characteristic velocity $(gh_o)^{1/2}$.

From experimental results (Lauber 1997) the wave front region (or tip region) has a nearly constant velocity $V=V_F$. The friction slope S_f can thus be written as $S_f=V_F^2 f_m/(8\alpha)$ where $h_m=\alpha h_o$ is the average height of the front region and f_m the corresponding friction coefficient. Inserting in (3) and solving the quadratic equation for V_F gives

$$V_F = \frac{4}{\tau}\frac{\alpha}{f_m}\left[\left(1 + \frac{f_m}{\alpha}\tau(1+\frac{S_0}{2}\tau)\right)^{1/2} - 1\right], \qquad (4)$$

where $\tau=(T-2^{1/2})$ is relative time and $j=S_0\alpha/f_m$ the slope effect.

Two cases are of particular relevance:
- $\tau\ll 1$, for which $V_F=2+0(\tau)$, i.e. no effect of bottom and friction slopes. This is the front velocity predicted by Ritter;
- $\tau\gg 1$, for which to order τ^{-1}

$$V_F = 2(2j)^{1/2}\left[1 + \frac{1-(2j)^{1/2}}{S_0\tau}\right]. \qquad (5)$$

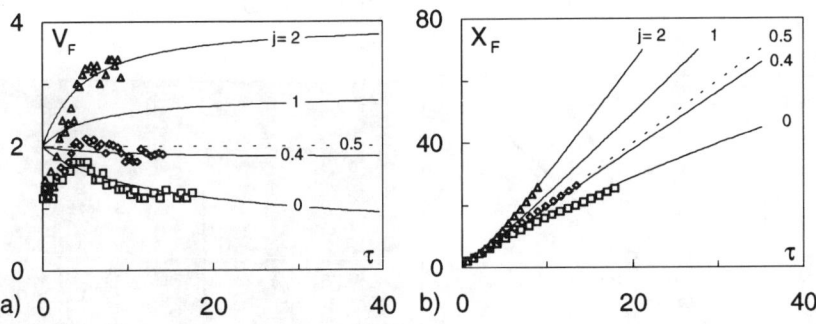

Fig.2 a) Non-dimensional front velocity V_F for various bottom slopes S_0 with $(f_m/\alpha)=1/4$. b) Nondimensional front location $X_F(\tau)$ with $\tau=T-2^{1/2}$. $S_0=(\square)$ 0, ($\Diamond$) 0.1, (Δ) 0.5, (- - -) equilibrium slope.

Asymptotically, $V_F=2(2j)^{1/2}$ is the extreme velocity equal to the *uniform velocity* ($S_o=S_f$). After a long time, the wave front velocity is thus equal to the uniform velocity. For $j=1/2$, the front velocity is always constant and equal to $V_F(j=1/2)=2$. Fig.2a) compares Eq.(4) with observations, and agreement is noted except for small τ with the transition from the initial to the dynamic waves. The coefficient f_m/α was adjusted to $f_m/\alpha=1/4$, independent of S_o and h_o. For $j>0.25$, pseude-uniform conditions are practically reached ($\pm 5\%$) if $\tau>40$. Our channel was much too short for development of those conditions, therefore.

WAVE FRONT LOCATION

The location of wave front $X_F(T)$ can be determined with Eq.(1) by setting $c_F=0$, i.e. $dx_F/dt=v_F$. Using Eq.(4) for the wave front velocity yields a singularity in setting the initial condition when $j>0$. Therefore, the modified Eq.(5) was used by imposing the condition $V_F(\tau=0)=2$, i.e.

$$V_F = 2(2j)^{1/2}\left[1 + \frac{1-(2j)^{1/2}}{S_o\tau+(2j)^{1/2}}\right], \quad j>0.1 \tag{6}$$

$$V_F = \frac{4}{\bar\tau}[(1+\bar\tau)^{1/2}-1], \quad j = 0 \tag{7}$$

where $\bar\tau=(f_m/\alpha)\tau$. Integration subject to the initial condition $X_F(\tau=0)=0$ gives

$$X_F = \int V_F\, d\tau = 2(2j)^{1/2}\left[\tau + \frac{1-(2j)^{1/2}}{S_o} ln\left(1+\frac{S_o\tau}{(2j)^{1/2}}\right)\right], \quad j > 0.1 \tag{8}$$

$$X_F = 4\frac{\alpha}{f_m}\left[2(1+\bar\tau)^{1/2}-2 + ln\left(\frac{4((1+\bar\tau)^{1/2}-1)}{\bar\tau((1+\bar\tau)^{1/2}+1)}\right)\right], \quad j = 0. \tag{9}$$

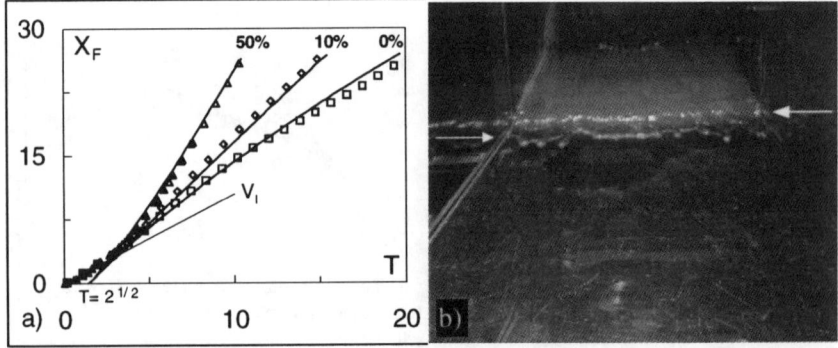

Fig.3 a) Wave front location X_F as a function of relative time T, notation Fig.2, b) Dambreak wave formation with initial wave (front arrow) and dynamic wave (back arrow). Wave passing after short time.

Fig.2b) compares (8) and (9) with observations, and agreement is noted again. For the equilibrium condition j=1/2 the front velocity is constant from the beginning of the dambreak and $X_F=2\tau$. For j<1/2, the front velocity decreases and it increases for j>1/2.

CONCLUSIONS

The positive front of a dambreak wave propagating on a smooth sloping and rectangular prismatic channel is determined. Based on experiments in which scale effects are excluded, two wave types may be identified: (1) The *initial wave* for relative time $T=(g/h_o)^{1/2}t<2^{1/2}$ with a propagation velocity $v/(gh_o)^{1/2}=1.05$, independent of slope and roughness, and (2) the *dynamic wave* for $T \geq 2^{1/2}$ where effects of bottom slope and boundary roughness are significant. Using the characteristic form of the De Saint-Venant equations the front velocity may be determined according to Eq.(4). Integrating again, the front location X_F as a function of T is predicted, according to Eqs.(8) and (9). These expressions are compared with experiments and agreement is noted. The characteristics of the wave fronts are significant features of the dambreak wave as they determine the wave arrival and front velocity. It is also shown that uniform flow is asymptotically reached at the wave front.

ACKNOWLEDGEMENTS

This project was supported by the Swiss National Science Foundation, grant number 2100-039216.93.

REFERENCES

- Chaudhry, M.H. (1993). *Open channel flow*. Prentice Hall: Englewood Cliffs.
- Hager, W.H., Chervet, A. (1996). History of dambreak wave. *Wasser Energie Luft* **88**(3/4): 49-54 (in German).
- Hager, W.H., Lauber, G. (1996). Experiments to dambreak problem. *Schweizer Ingenieur und Architekt* **114**(24): 515-524 (in German).
- Lauber, G. (1997). Experiments on dambreak waves. *PhD Thesis*. Swiss Federal Institute of Technology ETH: Zurich, Switzerland (to be published).
- Lauber, G., Hager, W.H. (1995). Optical sensing of extremely unsteady water flows. *Wasser Energie Luft* **87**(11/12): 275-278 (in German).
- Liggett, J.A. (1994). *Fluid mechanics*. McGraw-Hill: New York.
- Martin, H. (1990). Rapidly varied unsteady channel flows, in *Technical Hydrodynamics* 2, G. Bollrich, ed. VEB Verlag für Bauwesen: Berlin (in German).

ANALYSIS OF FLOW FIELD IN THE MEANDERING CHANNEL

Kouichi OZAWA
Ritsumeikan University Kyoto Japan

Nobuyuki TAMAI
Tokyo University Tokyo Japan

ABSTRACT

It is known that the bed form of a meandering channel relates to especially the secondary flow. In this study primary and secondary flow distributions are obtained considering the convective acceleration terms changing to the longitudinal direction in momentum equations. Depth averaged solutions were derived by a perturbation method. The calculated values for theoretical three-dimensional distributions are compared with the experimental results obtained for the meandering channel. The results are fairly consistent between theoretical and experimental values.

1. INTRODUCTION

The vertical distribution of the secondary flow was studied analytically in the case of a constant curvature before time. In that case it was considered that the flow condition is constant to the longitudinal direction . And also it was supposed that the secondary flow was fully grown. Rozovskii[1] developped the analysis of the secondary one in these flow fields and ascertained actually his theory by experiments or field s observations. He distinguished to each case of laminar and turbulent flow. For the former using a quadratic distribution of a primary flow he obtained sixth order equation regarding non dimensional vertical co-ordinate ξ . And for the latter using a logarithmic distribution and a eddy viscosity of quadratic equation he obtained the equation containing a logarithmic expression.Engelund[2] also obtained the sixth order equation of ξ . He used a vertical mean eddy viscosity determined by supposing that the logarithmic distribution is approximately consistent to the quadratic one.Kikkawa,Ikeda and Kitagawa[3] obtained the distribute equation that has a finite value at the bed. They used a flow function of the secondary flow and calculated the variation of the bed height. Johannesson and Parker[4] expressed the vertical distribution by the sixth order equation of ξ .They used same eddy viscosity as Engelund and the perturvation solution of first order.

In the case of a meandering channel that the curvature is not constant,the vertical distribution of the secondary flow was obtained by arranging the result in the case of a constant curvature. Ikeda and Nishimura[5] introduced the coefficient considering the fact that the secondary flow does not grow fully and the phase lag to the plan of the meandering channel. And Johannesson and Parker[6] tried to obtain rigorously the coefficient and the phase lag.

2. ANALYSIS OF THE FLOW

In this study it is supposed that the depth of the flow is fully shallow compared with the width of the channel. And the vertical averaged velocities,obtained by the perturbation method,are refered . The vertical distribution of the secondary flow in a constant and the quadratic distribution of the primary flow are used as first approximate values. Integrating the momentum equation,considered the convective acceleration,more precise solution is obtained.

(1)VERTICAL AVERAGED VELOCITIES ETC.

ANALYSIS OF FLOW FIELD IN MEANDERING CHANNEL 735

For the vertical averaged velocities etc.the second order solution,obtained by the perturbation method,is used. In this case the fundamental equations are continuity one,longitudinal and transverse momentum ones for a shallow water depth[7),8)].
Using the parameter of the perturbation expansion ε (=$B_0/2/R$),they are as follows

$u(=\bar{u}_s/V)=u_0+\varepsilon u_1+\varepsilon^2 u_2$ (1)
$v(=\bar{u}_n/V)=v_0+\varepsilon v_1+\varepsilon^2 v_2$ (2)
$h(=h_a/H_0)=h_0+\varepsilon h_1+\varepsilon^2 h_2$ (3)

In which B_0 and $R(=L/2\pi\theta_0)$ are respectively channel width,minimum radius of curvature and which $u_0,u_1,\cdots,h_2$ are functions of $s(=s_c/R),n(=n_a/B_0/2)$(refer Fig.1).
The boundary conditions in this case are as follows.

a)at side walls $v_0+\varepsilon v_1+\varepsilon^2 v_2=0$
b)at upstream end $u_0+\varepsilon u_1+\varepsilon^2 u_2=u_{i0}+\varepsilon nu_{i1}+\varepsilon^2(n^2-1/3)u_{i2}$
c)at downstream end $(1/2)\int_{-1}^{1}(h_0+\varepsilon h_1+\varepsilon^2 h_2)dn=1$.

In which $u_{i0}=1,u_{i1},u_{i2}$ are determined using experimental data by the minimum square method.
The differences of zero,first and second order perturbation solutions are respectively uniform,linear and quadratic change(refer Fig.2).The bed configuration is expressed by following form(refer Fig.3).

$\eta(=h_b/H_0)=\eta_0+\varepsilon\eta_1+\varepsilon^2\eta_2$ (4)

In which η_0 etc. are determined like as shown later(3.(1)).

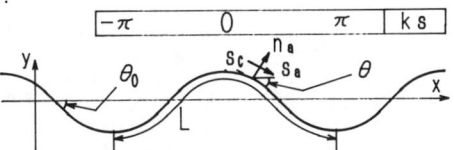

Fig.1 Sine-generated curve

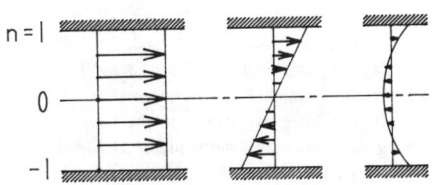

Fig.2 Transverse distribution of primary flow velocity(0,1,2nd order model)

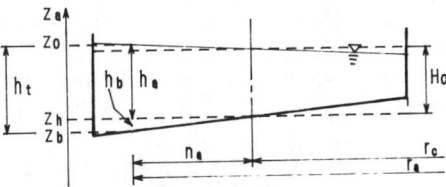

Fig.3 Transverse section

(2)FIRST APPROXIMATE SOLUTION OF PRIMARY AND SECONDARY FLOW
As for the vertical distribution of the main flow,a quadratic equation of ξ is used.In this case vertically averaged eddy viscosity is used as by Engelund and also by Johannesson & Parker were done.

$f_1(\xi)=O_1+O_2\xi+O_3\xi^2$ (5)

In which $O_1=M/M_1, O_2=1/M_1, O_3=-1/(2M_1), M=M_1-1/3, M_1=(\alpha C)/\sqrt{g}, \alpha$ =a coefficient regarding eddy viscosity,g=a gravity acceleration,C=a coefficient of Chezy formula, $\xi=z_a/h_:$. As for the vertical distribution of the secondary flow,the sixth order equation of ξ is used.

$\hat{u}_n=P_1+P_2\xi+P_3\xi^2+P_4\xi^3+P_5\xi^4+P_6\xi^5+P_7\xi^6$ (6)

In which $\hat{u}_n=u_n/V,V$=section averaged velocity of co-ordinates(s,n),$P_1,P_2,\cdots P_7$ are function of (s,n),their details are expressed in appendix.

(3)ANALYTICAL EQUATION
As for the expression of the three dimensional distribution of the primary flow,Eq.(5) of the vertical distribution and Eq.(1) of vertically averaged velocity are combined. That is

$\hat{u}_s=(u_0+\varepsilon u_1+\varepsilon^2 u_2)f_1(\xi)$ (7)

In which $\hat{u}_s=u_s/V$. As for the secondary flow,comparing Eq.(6) with $(v_0+\varepsilon v_1+\varepsilon^2 v_2)f_1(\xi)$ plus Eq.(6),they are almost equivalent except the section of nearly zero curvature. Therefore for the first approximation of the secondary flow, Eq.(6) is used. For the water depth($h_:=h_a+h_b$) the second or-

der solution of the vertically averaged equation,Eq.(3) and Eq.(4) are used. For the bed slope zero order solution(uniform flow) is used.
The longitudinal and transverse momentum equations are expressed by Eq.(8),(9).

$$\nu_t \frac{\partial^2 u_s}{\partial z_a^2} = u_s \frac{\partial u_s}{\partial s_a} + u_n \frac{\partial u_s}{\partial n_a} + \frac{u_s \cdot u_n}{r_a} + g \frac{\partial}{\partial s_a}(h_a + z_h) \quad (8)$$

$$\nu_t \frac{\partial^2 u_n}{\partial z_a^2} = u_s \frac{\partial u_n}{\partial s_a} + u_n \frac{\partial u_n}{\partial n_a} - \frac{u_s^2}{r_a} + g \frac{\partial h_a}{\partial n_a} \quad (9)$$

The first approximate equations etc. of the primary and secondary flows are substituted to the right side of these equations. And integrating these vertical distributions of the main and secondary flow are obtained. One of the boundary conditions is the shear stress=0 at the water surface,namely, $\nu_t \partial u_s/\partial z_a = 0, \nu_t \partial u_n/\partial z_a = 0$. The boundary values in the case of integrating these equations vertically ,A,B,C and D in Fig.4, are decided in order to consist with the experimental results.
The results for the primary and secondary flows are as follows.

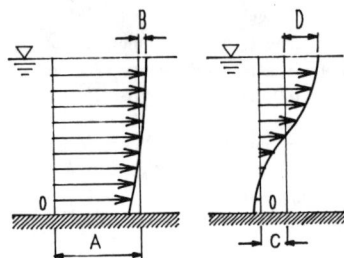

Fig.4 Vertical distribution model of primary and secondary flow velocity

$$\hat{u}_s = O_7 + O_8 \xi + O_9 \xi^2 + O_{10} \xi^3 + O_{11} \xi^4 + O_{12} \xi^5 + O_{13} \xi^6 + O_{14} \xi^7 + O_{15} \xi^8 + O_{16} \xi^9 + O_{17} \xi^{10} \quad (10)$$

$$\hat{u}_n = P_8 + P_9 \xi + P_{10} \xi^2 + P_{11} \xi^3 + P_{12} \xi^4 + P_{13} \xi^5 + P_{14} \xi^6 + P_{15} \xi^7 + P_{16} \xi^8 + P_{17} \xi^9 + P_{18} \xi^{10} + P_{19} \xi^{11} + P_{20} \xi^{12} + P_{21} \xi^{13} + P_{22} \xi^{14} \quad (11)$$

In which $O_7, O_8 \cdots, P_8, P_9, \cdots, P_{22}$ are the functions of co-ordinates(s,n). They are expressed in appendix.

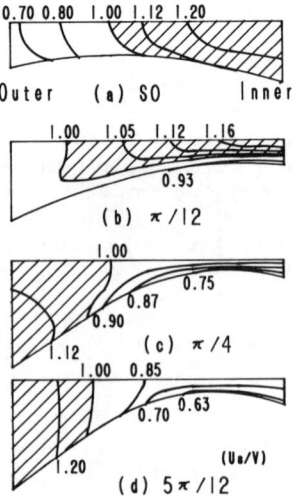

Fig.5 Contour of primary flow velocity

(calculated)

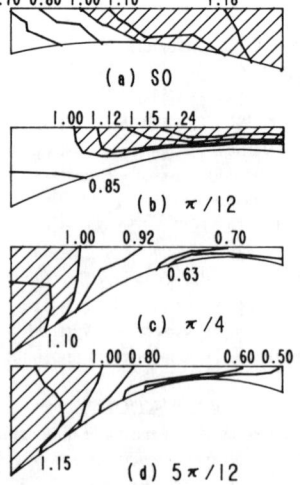

Fig.6 Contour of primary flow velocity

(experiment)

ANALYSIS OF FLOW FIELD IN MEANDERING CHANNEL

3. COMPARISON OF CALCULATED VALUES WITH EXPERIMENTAL RESULTS

(1)PRIMARY FLOW

The contours of velocities in the sections are expressed in Fig.(5),(6). These are respectively calculated and experimental results. Experimental conditions are as follows.Channel width,30cm,meandering length,2.485m,maximum deviation angle,45°,water discharge,1.335l/s and mean water depth is 1.94cm. The plan of (a)~(d) are expressed in Fig.7. In this case,the coefficients of the bed form, $\eta_0=0$, $\eta_1=n(a_0 \sinks + a_1 \cosks)$, $\eta_2=(n^2-1/3)(a_2+a_3\sin2ks+a_4\cos2ks)$ are $a_0=0.747, a_1=3.83, a_2=12.86, a_3=9.53$ and $a_4=3.33$. These are obtained by comparing with experimental results. The calculated and experimental values of contours are on the whole consistent. Namely,the lines of mean velocity($u_s/V=1.00$)are nearly same in form and position. And more high parts than mean values(shadows in Figures) are situated in same sides.

(2)VELOCITY VECTOR

The velocity vectors combined by the primary flow with the secondary one at each point of a section are expressed in Fig.7 and 8. These are respectively the calculated results and experimental ones[9].

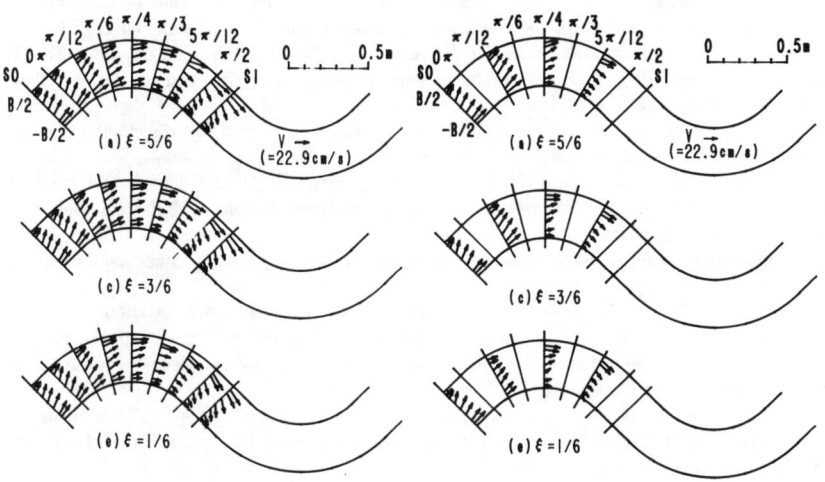

Fig.7 Distribution of velocity vector (calculated)

Fig.8 Distribution of velocity vector (experiment)

Each (a)~(e) corresponds to near the water surface~near the bed by step $h_t/3$. The analysis of the vertical distribution of the secondary flow in the meandering channel was based on the flow which is not change to the longitudinal direction. But the experimental data obtained by the meandering channel which the curvature changes continuously are different from the data in the case of a constant curvature. In the results of the experiments of the movable bed,a few vertical distribution which becomes zero,when the distribution is integrated,is seen(3 cases in 28 examples). The vertical distribution obtained by considering also the convective acceleration in this study is fairly consistent with the experimental results. Many plots of the vertical distribution in this case are situated in one side(both calculated,experimental results,about 70 percentages). It is able to say that the estimation of the vertical distribution of the secondary flow has been fairly improved than before. As for the example in the case of which the vertical integration of the distribution becomes to zero,it is seen at the outer side of the section(5 π /12) in Fig.7. The velocity vector near the water surface faces towards outer side,and its magnitude is large. On the other hand the vector near the bed do towards the center of the channel,and also the magnitude is large. Therefore it is considered that in the downstream region of this position the concentration of the flow to the outer bank side at near the water surface occurs.

There are following characteristics about the magnitude of the velocity vector. It changes fairly to the transeverse direction(for example,in (a)5 π /12 in Fig.7,the magnitude in the right side is about 50～60% comparing one in the left side).And it changes a little to the vertical direction(for example,in any section in Fig.7,the magnitude at the position of (e) is about 80～90% comparing with one at (a)).

4. CONCLUSION

The vertical distribution of the secondary flow in the meandering channel has been sought hitherto considering the phase lag or coefficient for growth of the secondary flow for the case of a constant curvature. But the experimental data are scarcely consistent with those of the constant curvature. In order to obtain more precise distribution the convective acceleration of momentum equations is considered to perticipate in the calculation. For the velocity distribution of the primary flow also the same device is practiced. The obtained results are fairly consistent with experimental ones. The flow characteristics obtained are as follows.

(1) From the point of which the curvature is zero to the maximum curvature,the direction of the secondary flow is on the whole the outer wall sides at any vertical location in water.

(2) In the downstream over the position of the maximum curvature the direction of the secondary flow changes,the velocity vector composing with the primary flow faces to the following outer wall side in the meandering channel.

(3) The magnitude of the secondary flow at near the inner bank is generally small for a sectional shape with deep scour at the outer bank.

(4) The position where the magnitude of the sencondary flow becomes to the maximum value is a little downstream over the turning point of curvature,and in the central of the width the vertical distribution of here diminishes a little from the water surface to the bed,but has a same direction,namely,the outer bank side.

(5) The vertical distribution of the primary flow is comparatively uniform. As for the distribution on the plane the value at the inner side in the entrance section is large. But at the same side of the maximum curvature the value becomes small.

APPENDIX

O_1 =M/M_1,O_2=1/M_1,O_3=−1/(2M_1),M=M_1−1/3,M_1= α C/$\sqrt{g}$, α =0.066,P_1=−HNR·cos ks(OP1−1/3),P_2=HNR·cosks(OP2−1),P_3=−HNR·cosks(OP3−1/2),P_4=−HNR·cosks·OP4,P_5=−HNR·cosks·OP5,P_6=−HNR·cosks·OP6,P_7=−HNR·cosks·OP7,HNR=K_1·RAL·DEL(H1G+EET1)RCRA1,K_1= ε ^{2},RAL= M_1/α ^{2},DEL=H_0/R,H1G=h_0+ ε h_1,EET1= η $_0$+ ε η $_1$,RCRA1=1−ncosks,OP1=O_1 ^{2}/3+5O_1O_2/12+3 (O_2 ^{2}+2O_1O_3)/20,+7O_2O_3/30+2O_3 ^{2}/21,OP2=O_1 ^{2}+O_1O_2+(O_2 ^{2}+2O_1O_2)/3+O_2O_3/2+O_3 ^{2}/5,OP3= O_1 ^{2}/2,OP4=O_1O_2/3,OP5=(O_2 ^{2}+2O_1O_3)/12,OP6=O_2O_3/10,OP7=O_3 ^{2}/30,O_7=U2G+K_2·RAL·DEL· (H2G+EET2)(O7A/3+O7B·5/24+O7C·3/20+O7D·7/60+O7E·2/21+O7F·9/112+O7G·5/72+O7H·11/18 0+O7I·3/55), O_8=−K_2·RAL·DEL·(H2G+EET2)(O7A+O7B/2+O7C/3+O7D/4+O7E/5+O7F/6+O7G/7 +O7H/8+O7I/9),O_9=K_2·RAL·DEL·(H2G+EET2)·O7A /2,O_{10}=K_2·RAL·DEL·(H2G+EET2)·O7B/6,O $_{11}$=K_2·RAL·DEL· (H2G+EET2)· O7C/12, O_{12}=K_2· RAL·DEL·(H2G+EET2)·O7D/20,O_{13}=K_2·RAL ·DEL·(H2G+EET2)·O7E/30,O_{14}=K_2·RAL·DEL·(H2G+EET2)·O7F/42,O_{15}=K_2·RAL·DEL·(H2G+EE

T2)·O7G/56,O_{16}=K_2·RAL·DEL· (H2G+EET2)·O7H/72 ,O_{17}=K_2·RAL·DEL·(H2G+EET2)·O7I/90,U
2G=u_0+ ε u_1+ ε^2u_2,H2G=h_0+ ε h_1+ ε^2h_2,K_2= ε^2,EET2= η_0+ ε η_1+ ε^2 η_2,U1G=u_0+ ε u_1,GR
H=RCRA(F_r^{-2} ∂ H2G/ ∂ s-j/2),F_r^{-2}=gh_0/V^2,RCRA=1- ε ncosks+ ε^2n^2cos2ks, P_6 =K_3·V2G+(H2
G+EET2)(-U2G· RCRA· A33·C2S-A34/ ε ·HNR/K_1· cosks·C2N+cosks· RCRA·U2G^2· C2A-PRE/
6)-C1/2, P_9=(H2G+EET2)(-U2G·RCRA·A33·C1S-A34/ ε ·HNR/K_1·cosks· C1N+cosks·RCRA·U2G
^{2}·C1A-PRE),P_{10}=UNS·S2A+UNN·N2A+UNAC·AC2A+(H2G+EET2)PRE/2,P_{11}=UNS·S2B+UNN·
N2B+UNAC·AC2B,P_{12}=UNS·S2C+UNN·N2C+UNAC·AC2C,P_{13}=UNS·S2D+UNN·N2D+ UNAC·
AC2D,P_{14}=UNS·S2E+UNN·N2E+UNAC·AC2E,P_{15}=UNS·S2F+UNN·N2F,P_{16}=UNS·S2G+UNN·N
2G, P_{17}=UNS·S2H+UNN·N2H, P_{18}=UNS·S2I+UNN·N2I,P_{19}=UNN·N2J,P_{20}=UNN·N2K,P_{21}=UN
N·N2L,P_{22}=UNN·N2M,V2G=v_0+ ε v_1+ ε^2v_2,FR2E= ε (F_r^2-1)/2,UNS=K_2(RAL·DEL)^2(H2G+EE
T2)U2G·RCRA·A33, UNN=K_2(RAL·DEL)^2(H2G+EET2)A34/ ε ·HNR/K_1·cosks,UNAC=-K_2·RAL·
DEL(H2G+EET2)U2G^2·RCRA·cosks

REFERENCES
1) Rozovskii,I.L.,:Flow of Water in Bends of Open Channels , Academy of Sciences of Ukranian SSR,Kiev,1957,Translated by Prushansky,Y.,The Israel Programfor Scientific Translations,1961.
2) Engelund,F.:Flow and bed topography in channel bends,Proc.ASCE.J.Hyd.Div.Vol.100,HY11, pp.1631-1647,1974.
3) Kikkawa,H.,Ikeda,S. and Kitagawa,A.:Variation of bed profile with in curced open channel ,Proc. JSCE,Vol.251,pp.65-75,1976.(in Japanese)
4) Johannesson,H. and Parker,G.:Velocity Redistribution in Meandering River , Proc. ASCE.J.Hyd. Div.Vol.115 ,pp.1019-1039,1989.
5) Ikeda,S. and Nishimura,T.:Three-dimensional flow and bed topography in sand-silt meandering rivers,Proc.JSCE,Vol.369/ II -5,pp.99-108,1986.(in Japanese)
6) Johannesson and Parker:Secondary flow in mildly sinuous channel, Proc.ASCE. J. Hyd. Div.Vol. 115,pp.289~308,1989.
7) Ikeuchi,K. and Tamai, N.: Evolution of the depth-averaged flow field in meandering channels, Proc.,JSCE,Vol.334,pp.89-101,1983.(in Japanese)
8) Ali.H.N.,:Introduction to Perturbation Techniques,John Wiley & Sons,1981.
9) Tamai,N. and Ali,A.M. :Depth-averaged flow fields in meandering channels with alluvial equilibrium bed ,Proc.Hydraulic Engineering,JSCE,Vol.,pp.685-690,1985.

Fig.1 Sine-generated curve
Fig.2 Transverse distribution of primary flow velocity(0,1,2nd order model)
Fig.3 Transverse section
Fig.4 Vertical distribution model of primary and secondary flow velocity
Fig.5 Contour of primary flow velocity(calculated)
Fig.6 Contour of primary flow velocity(experiment)
Fig.7 Distribution of velocity vector(calculated)
Fig.8 Distribution of velocity vector(experiment)

UNSTEADY FLOW CHARACTERISTICS IN A COMPOUND CHANNEL

JAYARATNE B.L.[*], N. TAMAI[*], Y. KAWAHARA[*], K. KAN[**]

[*] Dept. of Civil Engineering, University of Tokyo, Japan
[**] Dept. of Civil Engineering, Shibaura Institute of Technology, Japan

ABSTRACT

Unsteady flow characteristics were investigated experimentally recording detailed velocity measurements at three sections in a compound channel during repeated passages of a hydrograph. Extensive water depths measurements were recorded at eleven sections from upstream to downstream. Water depth variations, velocity distributions, discharge variations are presented. Flow depth and water surface slopes in the main channel and the flood plain are different to each other and it strongly effects on the flow mechanism. Flood flow propagation is also discussed.

INTRODUCTION

The knowledge of hydraulic characteristics of unsteady flow in open channels is essential for understanding the river flow behaviour as flows in natural rivers are often unsteady. Flow structures in compound channels are complicated and three dimensional. Strong interaction between main channel and flood plain flows is one of the most important feature of compound channel flow.

Kawahara & Tamai (1989) explained the 3-D flow fields and mechanism of lateral momentum transfer between main channel and flood plain using an algebraic stress model along with the k-ε model. Tominaga & Nezu (1991) revealed the three dimensional turbulent structures of compound open channel flows associated with secondary currents. Studies on unsteady open-channel is rare due to the difficulties in measurements. Tominaga et al. (1994) conducted experiments of unsteady flow in a compound channel and pointed out that unsteady flow characteristics of compound channels are different from that of single cross section rectangular channels. Jayaratne et al (1995,1996) discussed the unsteady flow behaviour when vegetation exists on the flood plain and effects of unsteadiness on temporal water depth variation. In this study, unsteady flow characteristics were examined experimentally in a compound open channel at three different sections from upstream to downstream in order to enhance the accuracy of the measured data which can be compared with numerical simulation.

EXPERIMENTAL PROCEDURE

The experiments were conducted in a 25m long, 1m wide, tilting flume. Rectangular flood plain made of thin steel plates has been set on one side of the flume. The width was 60cm and the height of 5.25cm. Both flood plain and main channel were smooth. The channel slope was set as 0.001. Depth and velocity measurements were recorded at the three sections 5m apart being first section 8m from the flume entrance which guarantees the fully developed flow. Unsteady flow hydrographs were generated by controlling the opening of a valve in the pipe system using relevant software.

Primary velocity components (u) were measured simultaneously at the same locations of these three different sections using 3mm diameter micropropellers. The linear relation between velocity and output voltage should be same throughout the measurements. However, when the submerged depth of the sensor of the propeller is very low, it may violate this linearity. Those data were omitted in data processing. Data were recorded at least every 4cm distance in horizontal direction. But close to the interface, it was gradually reduced to 0.5cm to obtain the more detailed velocity distribution. This distance was kept 0.5cm in vertical direction. The sampling rate was 10Hz and the ten readings per second were averaged to render mean velocities. An I-type electromagnetic velocity meter (SFT-200-05) was used to measure the longitudinal and transverse velocity components (u and w) at interface, mid of the main channel and mid of the floodplain.

The water depths were measured using servo meters at the same points where the velocity measurements were taken in the transverse direction. Depth data were recorded at 11 sections being 1m distance apart. First and middle, last sections are same for both velocity and depth measurements. Three servo meters were used simultaneously, keeping first one as reference and other two meters being moved.

There were many measuring points and the flume is an outdoor facility, the experiments should be finished as early as possible before weather condition (specially wind) causes undesirable effects on the measurements. Therefore velocities at one point were measured by passing the hydrograph only once. The signals from the measuring instruments were recorded with a personal computer for later treatment.

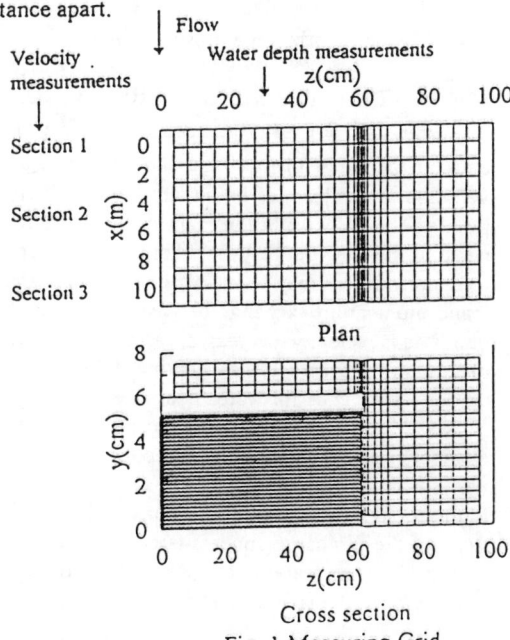

Fig. 1 Measuring Grid.

RESULTS & DISCUSSIONS

WATER DEPTH & WATER SURFACE PROFILES:

Figure 2 presents the water depths measured at mid of the main channel and mid of the flood plain. The peak of the depth on the flood plain occurs a little later than that in the main channel indicating a slight time lag. Fig. 3 shows water surface profile at t=205s when depth at section 2 is maximum. It can be clearly seen that slope of the water surface in the main channel and the flood plain at a given distance is different to each other due to the different wave velocity and water depth. It substantially affects the velocity variations in the main channel and the flood plain as water surface slope directly links with the driving force.

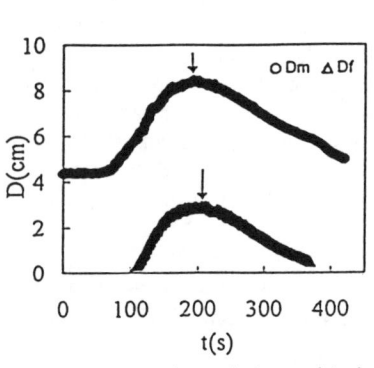

Fig. 2 Water depths variations with time

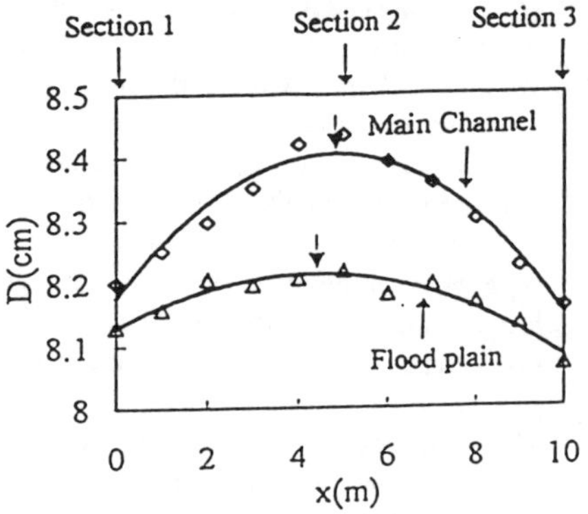

Fig. 3 Water surface profile at t=205s

MEAN VELOCITY & DISCHARGE VARIATIONS:

The time variation of cross sectional mean primary velocity in the main channel U_m and on the flood plain U_f at the section 2 are presented in Fig. 4. The averaged velocities are calculated by integrating the primary velocity recorded at the measuring points. The velocity in the main channel reaches its maximum much earlier than the peak of the flow depth. After that, velocity decreases slightly until water depth starts to decrease. This may be mainly due to the unsteadiness (effect of water surface slope) and momentum exchange between main channel and flood plain flow. When water depth falls velocity decreases rapidly. The velocity on the flood plain attains its maximum little earlier than depth reaches its peak. However, there is no significant velocity difference till water depth begins to fall. Velocity in the main channel again shows a little increment for a short period from the time of flow confined to main channel in the falling stage as effect of lateral mixing is over.

Figure 5 shows discharge variations with time for main channel and flood plain flow. Peak of the discharge appears later than peak of the velocity. But it is slightly earlier than peak of the water depth. According to the experimental results, this phenomenon is valid for both main channel and flood plain flows.

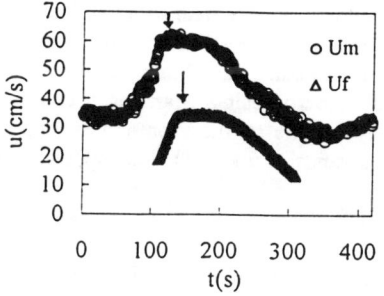
Fig. 4 Mean velocity variations with time

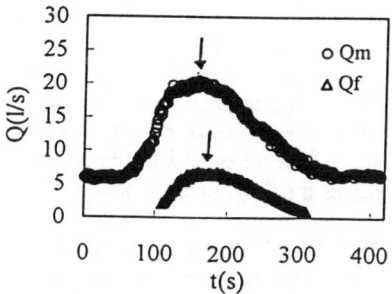
Fig. 5 Discharge variations with time

DEPTH AVERAGE VELOCITIES:
Depth averaged primary velocities across the section 2 at given time instances are shown in Fig. 6. Time instances were selected for both rising and falling stages of water depths 6.5cm and 7.5cm. According to Fig. 4, peaks of the mean velocities in the main channel and the flood plain occur at t=115s and t=143s respectively. It means, at this time interval, velocity in the main channel decreases while it increases on the flood plain. The rising stage of D=6.5cm at t=123s lies within above two time instances. When the time is at t=148s (D=7.5cm, rising), velocity in the main channel decreases slowly whereas it has reached to its peak on the flood plain. Therefore, velocity on the flood plain at t=123s is considerably lower compared to that at t=148s while it is higher in the main channel. Exhibiting this result, transverse velocity profile at t=123s and t=148s crosses at each other close to the interface.

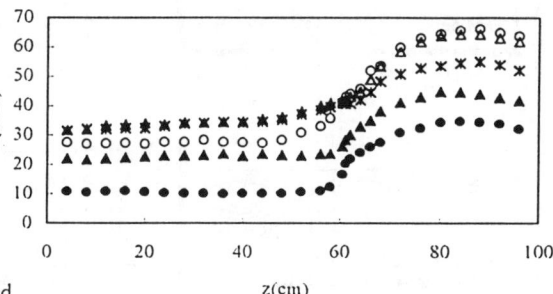

Fig. 6 Depth averaged velocity variations across the section 2.

○ D=6.5cm t=123s, rising
△ D=7.5cm t=148s, rising
✗ D=8.45cm t=205s, peak
▲ D=7.5cm t=272s, falling
● D=6.5cm t=326s, falling

Cross-sectional averaged velocities are calculated taking into account of longitudinal water surface slope [d(D)/dx] in order to understand the above mentioned cross velocity phenomena. The calculated results are compared with the measured data in Table 1.

Table 1: Calculated & Measured Velocities

Time (s)	Depth (cm)	u(cm/s) -Calculated		u(cm/s) -Measured	
		Um	Uf	Um	Uf
123	6.5	64.35	20.12	62.42	26.42
148	7.5	59.72	30.93	57.34	32.21
205	8.45	52.65	30.31	52.01	32.14
272	7.5	41.28	21.24	38.37	22.26
326	6.5	35.11	12.89	28.46	14.10

It reveals that this cross velocity phenomenon (velocity in the main channel decreases while it increases on the flood plain) occurs mainly by the effect of water surface slope on driving force. However, mean velocity in the main channel integrated by point velocities is slightly lower than the calculated velocity using Manning's equation whereas it is opposite on the flood plain. It may be due to the momentum transfer between main channel and flood plain which reduces the main channel velocity and increases the flood plain velocity.

PRIMARY VELOCITY VARIATIONS:

The difference in velocity at three cross sections is found to be small. Fig.7 shows the primary velocity contours at different water depths. At the beginning of the hydrograph, flow is steady with water depth 4.37cm. First contour diagram represents water depth 4.5cm just few seconds after the hydrograph reached to its unsteady region. Velocity contours at water depths 6.5cm and 7.5cm (both rising and falling stages) and peak of the hydrograph are given here. When the flow is confined to the main channel, maximum velocity appears close to the free surface whereas it is below the free surface for compound channel flow due to the interaction between the main channel and the flood plain flows. As a result of these secondary currents, velocities close to the interface are decelerated and its effect extends even at the mid of the main channel and hence velocity shows its maximum about 4cm away from the mid of the main channel. It is clear that velocity in the rising stage is higher than that of the falling stage at same water depths.

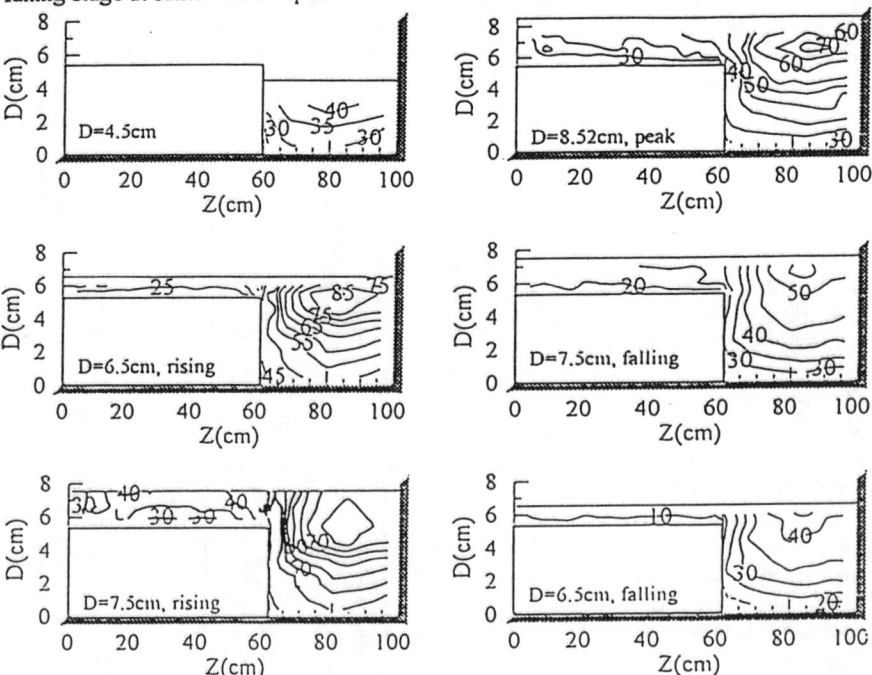

Fig. 7 Primary velocity contours at section 1.

FLOW ROUTING:

Water depth variations show some similar shapes with a time lag in Fig.8. In rising stage, difference of water depth between sections is higher as a result of higher temporal water depth variation than that of the falling stage. Time lags between peaks of the hydrographs are approximately equal to the calculated time lags using Kleitz-Seddon's kinematic wave theory. Fig. 9 represents the velocities measured at 3cm above the channel bed and mid of the main channel for the section 1, 2 and 3. It can be seen that the maximum velocity gradually reduces with the flow propagation from upstream to downstream. First section attains its maximum first and subsequently second and third sections reach their maximums respectively.

Table 2 Time lags between peaks of hydrographs

Branch of channel	Calculated		Measured	
	Sections 1-2	Sections 2-3	Sections 1-2	Sections 2-3
Main channel	4.9	5.0	5.5	6.6
Flood plain	9.3	9.5	10.6	9.9

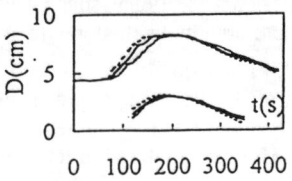
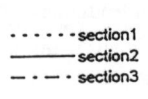
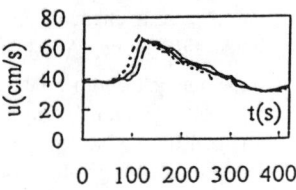

Fig. 8 Water depth variations

Fig. 9 Velocity variations

CONCLUSIONS

The velocity reaches its maximum first and subsequently discharge and water depth attain their peaks. However these time lags are mainly significant only for the main channel flow. Velocity in the main channel starts to decrease in the second phase of the rising stage by the effects of water surface slope and flow mixing between main channel and flood plain flows. Maximum velocity reduces gradually with flow propagation from upstream to downstream.

ACKNOWLEDGEMENT: The authors gratefully acknowledge that this study is partially supported through Grant-in Aid for Scientific Research A(1), No.08305017 by Ministry of Education, Science and Culture.

REFERENCES

1. Jayaratne, B.L., H.Tu, N.Tamai and K.Kan, (1995): Unsteady flow characteristics in compound channels with vegetated flood plains, XXVI IAHR Cong., London Vol. 1, pp.379-384.
2. Jayaratne, B.L., N. Tamai, Y. Kawahara and H. Tu, (1996): Velocity variations of unsteady flow in compound channels with or without vegetation on flood plain, 10th Cong., APD-IAHR, Malaysia.
3. Kawahara,Y. and N.Tamai, (1989): Mechanism of lateral momentum transfer in compound open channel flows, Proc. XXIII IAHR.,Canada, Vol. B, pp.B463-B470.
4. Tominaga, A. and I. Nezu, (1991): Turbulent structure in compound open-channel flows, J. Hydr. Eng. ASCE Vol. 117(1), pp.21-41.
5. Tominaga,A., J.Liu, M.Nagao and I.Nezu, (1995): Hydraulic characteristics of unsteady flow in open channels with flood plains, XXVI IAHR Cong., London Vol. 1, pp.373-378

The Mixing Mechanism In Turbulent Two-Stage Meandering Channel Flows

P. RAMESHWARAN AND B. B. WILLETTS
Department of Engineering, University of Aberdeen, Aberdeen, AB24 2UE, Scotland

ABSTRACT

The paper reports the findings of small-scale experiments on overbank flow in two-stage meandering channel systems. Eight geometric parameters were investigated, namely main channel sinuosity (r_{mc}), aspect ratio (AR_{mc}), cross-sectional shape and bank side slope (SS_{mc}), flood plain roughness (k_{fp}) and longitudinal slope (S_{fp}), meander belt width relative to flood plain width (MBW/FW) and flood bank sinuosity (r_{fp}). Observations were made of the mixing mechanism of the flood plain and main channel flow which influences energy losses. The effect of some of these parameters is presented through stage-discharge relationships. Energy losses are analysed in term of Darcy-Weisbach resistance coefficients for the case involving straight flood banks. Scale effects in such flows are investigated using data from the UK Flood Channel Facility (UK-FCF). The paper presents a possible approach to estimation of the Darcy-Weisbach resistance coefficient for two-stage meandering channels in the roughness dominated flow region.

INTRODUCTION

In recent years, two-stage meandering channels have been intensively studied in order to understand the mixing mechanism of the flood plain and main channel flows and to develop a accurate prediction method for the conveyance capacity (Sellin et al., 1993; Ervine et al., 1993). Such channels are sustainable and environmentally friendly, maintaining fully functional ecological systems in the main channel. The resistance behaviour of two-stage river systems when flooded is of great interest because it dictates the stage or the depth of flow for a particular flood discharge. Estimating the system conveyance is very difficult because it is associated with three-dimensional mixing of flood plain and main channel flows, and the associated vorticity. The main sources of energy dissipation in two-stage meandering channels are bed roughness, geometric shape losses, expansion-contraction losses and interaction losses (Ervine et al., 1993).

Recently, James and Wark (1992) and Greenhill and Sellin (1993) produced methods of estimating conveyance in two-stage meandering channels. These methods were based mainly on the UK-FCF Series B programme which varied only a limited number of the influential parameters. Therefore, during the past two and half years small scale physical model studies have been carried out in an extension of that programme specifically designed to incorporate some of the neglected parameters.

This paper presents the influence of some geometric parameters on stage discharge relationship and resistance behaviour, and a possible approach to estimating flow resistance in the roughness dominated region of two-stage meandering channels flows.

EXPERIMENTS

The experiments were performed in a 11 m long and 1.2 m wide tilting flume with a recirculating hydraulic circuit. The main channels, of uniform sinuosity, were constructed in a flat flood plain as shown in Figure 1. The main channel top width was always 200 mm and the bank slope always 60°. In some experiments the main channel was trapezoidal in cross-section with maximum bank-full depth 52.5 mm, and in others the main channel was 'natural'. The natural bed was allowed to form naturally by placing mobile material with an initially flat surface 52.5 mm below the flood plain and exposing it to bankfull flow until topography change had become very slow. The resulting flow-moulded surface was stabilised to prevent further evolution. During formation of this natural channel, material was fed at the upstream end of the channel at the equilibrium transport rate. The topography of the natural channel is shown in Figure 2. The investigations were carried out for smooth and two different rough flood plain surfaces. For each of the two different rough surfaces, namely Rough A and Rough B, a regular array of surface mounted 5 mm spheres was used as shown in Figures 1 and 2. Calibration experiments were performed in a 0.3 m wide flume and the calculated equivalent roughness heights for Rough A and Rough B are 1.33 and 5.01 mm respectively.

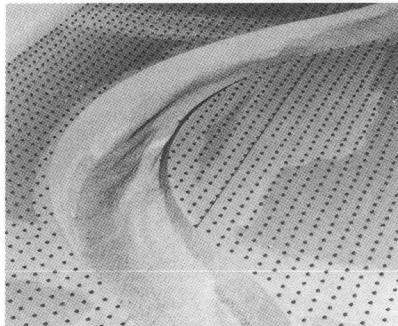

Figure 1. General view of a experimental two-stage meandering channel.

Figure 2. Photograph showing natural channel bed.

Discharge was measured by Rotameter at low flow (≤ 3.5 ls^{-1}) and by orifice plate and manometer at higher flows. Flow depths were measured with a point gauge linked to a digital output and reading to the nearest 0.5 mm. A tailgate was used to eliminate drawdown and elongate the area of uniform flow.

FLOW RESISTANCE

In a fully turbulent flow region, the velocity is given by the so-called Prandtl-von Kármán universal velocity distribution law. The logarithmic equations take the following form for rough boundaries.

$$\frac{u}{u_*} = \frac{1}{\kappa} ln \frac{y}{k_s} + A_r \tag{1}$$

where u is the time averaged velocity, $u_* \left(= \sqrt{gRS}\right)$ is the shear velocity, κ is the von Kármán constant ($\kappa = 0.41$), y is the distance from the solid boundary, k_s is the Nikuradse equivalent sand roughness, R is the hydraulic radius, S is the hydraulic gradient and A_r is a constant ($A_r = 8.5$). Equation (1) can be integrated over the channel cross-section to provide flow resistance equations. Keulegan (1938) derived the following equation for cross-sectional averaged velocity, U, of turbulent flow as a function of the shape of the channel cross-section.

$$\frac{U}{u_*} = \frac{1}{\kappa} ln \frac{R}{k_s} + A_r - \frac{1}{\kappa}(1-\beta) - \bar{\varepsilon}\frac{U}{u_*} \tag{2}$$

Here, β is the shape function of the channel and $\bar{\varepsilon}$ is resistance associated with the free surface. For the sake of simplicity, Keulegan (1938) suggested $\beta = 0.1$ and $\bar{\varepsilon} = 0$ irrespective of shape, since the variation of these quantities with shape is relatively small. Thus, using the Darcy-Weisbach equation and equation (2), the resistance coefficient relationship for a two dimensional open channel can be expressed as follows.

$$\frac{1}{\sqrt{f}} = 2 log \frac{R}{k_s} + 2.23 \tag{3}$$

where $f\left(= 8gRS/U^2\right)$ is the Darcy-Weisbach resistance coefficient and g is gravitational acceleration.

RESULTS AND DISCUSSION

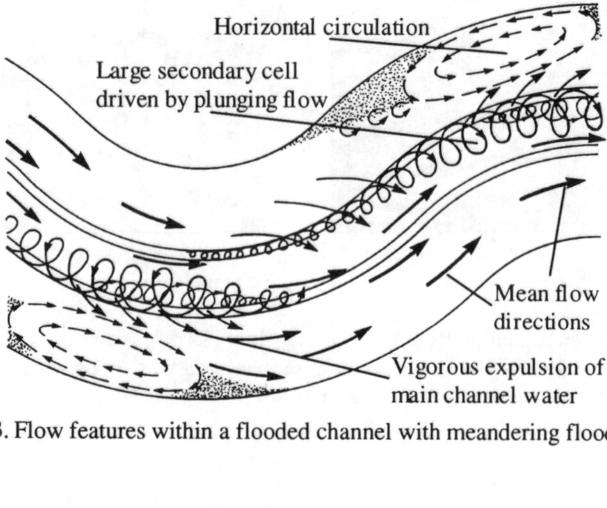

Figure 3. Flow features within a flooded channel with meandering flood banks.

The flow resistance and energy dissipation are largely governed by the momentum exchange associated with mixing. Figure 3 shows flow structure in a two-stage meandering overbank flow between meandering flood banks. The vigorous exchange of water between flood plain and main channel drives large secondary flow cells in the main channel. Large secondary cells grow along the outer bank upstream of each bend apex and decay very rapidly downstream of the bend apex. Expulsion of flow from the main channel to the flood plain occurs in the bed apex region. A small secondary circulation occurs in the bend apex region underneath the large cell with the opposite sense. A horizontal circulation occurs on the flood plain just after the bend apex and blocks the flood plain contribution to conveyance. Similar flood plain behaviour was observed by Shiono et al. (1994). The flow features within a flooded meander channel with straight flood banks are described by Sellin et al. (1993).

Table 1. Summary of the reported experiments.

Test No.	r_{mc}	AR_{mc}	SS_{mc}	Cross-Section	Roughness	MBW/FW	S_{fp} ($\times 10^{-3}$)	r_{fp}
1	1.77	4.49	60°	Trapezoidal	Smooth	0.86	1.000	1.00
2	1.37	4.49	60°	Trapezoidal	Smooth	0.85	1.000	1.00
3	1.11	4.49	60°	Trapezoidal	Smooth	0.58	1.000	1.00
4	1.11	4.57	60°	Natural	Smooth	0.58	1.000	1.00
5	1.11	4.57	60°	Natural	Rough B	0.58	1.333	1.00
6	1.11	4.57	60°	Natural	Rough B	0.73	1.333	1.00
7	1.11	4.57	60°	Natural	Rough B	0.73	1.297	1.03
8	1.11	4.57	60°	Natural	Rough B	1.00	0.900	1.11

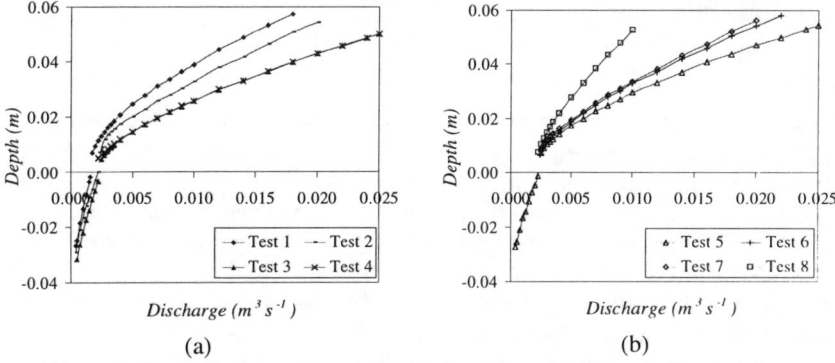

Figure 4. Stage-discharge curves for (a) Smooth and (b) Rough flood plains.

The essential parameters of the reported tests are listed in Table 1. Stage-discharge curves are shown in Figures 4a-4b. Discharge increases steadily with depths up to bankfull. In all cases, when the flow first goes overbank, significant depth increases occur because the wetted perimeter increases much more than the flow area, so the retarding effect of the boundary on the flow increases suddenly. Thereafter, discharge rises increasingly rapidly with depth. Sinuosity and flood plain roughness have pronounced influence on the stage-discharge behaviour and the associated energy dissipation. A meandering flood bank also has a significant influence on flow.

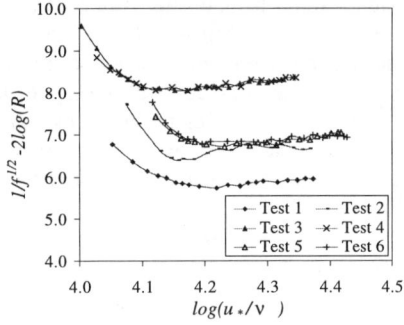

Figure 5. Quantity $1/\sqrt{f} - 2\log(R)$ as a function of $\log(u_*/\nu)$.

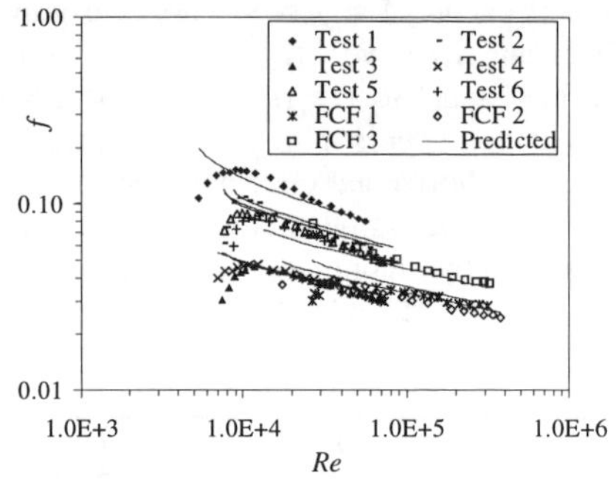

Figure 6. Darcy-Weisbach resistance coefficient vs. Reynolds number.

Table 2. Reported UK-FCF experiments (Sellin et al., 1993).

FCF	r_{mc}	AR_{mc}	SS_{mc}	Cross-Section	Roughness	MBW/FW	S_{fp} ($\times 10^{-3}$)	r_{fp}
1	1.37	9.14	45°	Trapezoidal	Smooth	0.61	0.996	1.00
2	1.37	14.60	45°	Natural	Smooth	0.61	0.996	1.00
3	2.04	14.60	45°	Natural	Smooth	0.86	1.021	1.00

To ascertain whether the mean velocity of flow in a two-stage channel flow is affected by the viscosity, $1/\sqrt{f} - 2\log R$, $(= 2.23 - 2\log k_s)$, is plotted as a function of $\log(u_*/\nu)$ in Figure 5. Each experiment produces a line consisting of a curved portion and an almost straight line parallel to the abscissa. The curved line region is dependent on viscosity and roughness while the other region, the roughness dominated flow region, is independent of viscosity. The differences between the curves arise from geometric differences in the configuration. It is postulated that for overbank flow, the meandering inner-channel behaves like added roughness in the roughness dominated flow region and the effective roughness k_s can be specified in terms of the geometric parameters and surface roughness as follows:

$$\frac{k_s}{h_{mc}} = \frac{k_{fp}}{h_{mc}} + \Phi\left(AR, SS_{mc}, \frac{MBW}{\lambda}, \frac{MBW}{FW}, \frac{k_{mc}}{k_{fp}}, Re'\right) \qquad (4)$$

where $h_{mc} = A_{bf}/B$ and $Re' = 4U_{bf}B/\nu$.
Here, λ is the meander wave length, k_{mc} is the main channel roughness height, A_{bf} is the bankfull area of main channel, B is the top width of the main channel, U_{mc} is the

calculated bankfull mean velocity of the main channel and ν is the kinematic viscosity. By this means a 'whole cross-section effective roughness' can be specified which enables an overall value of f to be calculated from equation 3.

Experimental data, together with data of Macleod and Ervine (1996) and from the UK-FCF (Sellin et al., 1993; See Table 2) have been used to define function Φ by regression analysis using equations (3) and (4). Figure 6 compares the experimental results with predictions of f made using the Φ so obtained. The influence of the variables is effectively conveyed by the function Φ in the roughness dominated zone, as evidenced by the fit of points to the lines. The function Φ is:

$$\Phi = (AR)^{-0.82}(SS_{mc})^{-1.15}\left(\frac{MBW}{\lambda}\right)^{1.84}\left(\frac{MBW}{FW}\right)^{2.58}\left(\frac{k_{mc}}{k_{fp}}\right)^{-0.02}(Re')^{-0.77} \quad (5)$$

CONCLUSIONS

Secondary circulation in the main channel associated with mixing between main channel and flood plain flows, influences energy dissipation. An increase of main channel sinuosity or flood plain roughness changes the circulation and increases the energy dissipation rate significantly. An approach to the prediction of a Darcy-Weisbach resistance function for meandering channels with a straight flood banks has been presented for the roughness dominated flow region. The proposed procedure takes into account geometric parameters and scale effects. The resistance relationship obtained is a reasonable predictor in the roughness dominated region.

ACKNOWLEDGEMENTS

The authors gratefully acknowledge the financial support of EPSRC. They are grateful for the benefit of helpful discussions with Dr D.A. Ervine, Professor R.H.J Sellin and their colleagues.

REFERENCES

Ervine, D.A., Willetts, B.B., Sellin, R.H.J. and Lorena, M. (1993) Factors affecting conveyance in meandering compound flows, *J. Hydraul. Engrg.*, 119, 1383-1399.
Greenhill, R.K. and Sellin, R.H.J. (1993) Development of a simple method to predict discharges in compound meandering channels, *Proc. Instn. Civ. Engrs Wat., Marit. & Energy*, 101, 37-44.
James, C.S. and Wark, J.B. (1992) Conveyance estimation for meandering channels. *Report SR 329*, Hydraulics Research, Wallingford.
Keulegan, G.H. (1939) Laws of turbulent flow in open channels. *J. Res.*, 21, 707-741.
Macleod, B. and Ervine, D.A. (1996) Personal Communication.
Sellin, R.H.J., Ervine, D.A. and Willetts, B.B. (1993) Behaviour of meandering two-stage channels, *Proc. Instn. Civ. Engrs Wat., Marit. & Energy*, 101, 99-111.
Shiono, K., Muto, Y., Imamoto, H. and Ishigaki, T. (1994) Flow structure in meandering channels. *Proceedings of MAFF Conference*, Loughborough University, 4-6 July, 9.3.15-9.3.26.

Estimation of flow resistance in ice covered channels

Dong Zengnan, Mao Zeyu, Wang Yongtian
Department of Hydraulic Engineering
Tsinghua University, Beijing 100084, China
Mu Gaofeng, Wang Jing
Xinjiang Institute of Water Resources and Hydroelectric Research, China

ABSTRACT

This paper presents an improved method for determining roughness coefficient of the underside of ice cover based upon direct measurement of velocity distribution in the flow under ice cover. It is assumed that velocity distribution has logarithmic character in the regions near the channel bottom and ice cover. Results of this presented analysis have indicated that measured velocity profiles can be used to evaluate resistance coefficients of ice-covered channels with more reliable accuracy.

1 INTRODUCTION

Formation of ice cover on flowing waters greatly alters flow characteristics of the rivers or channels. From engineering point of view, estimation of conveying capacity of ice-covered rivers is very often required in river management involving the operation of reservoirs when releases based upon these estimations are required to avoid damages from flooding. It is of special significance for the management of river that experiences frigid winters and flooding risk exists at high discharges, and determination of the height of protective dikes becomes very important.

Studies of flow with ice cover are comparatively limited in comparison to studies of free surface flow. The additional boundary formed by the undersurface of the ice cover can vary from very smooth to extremely rough, and the measurement of the size and distribution of roughness at the undersurface is not normally possible. In the past, generally speaking two methods have been proposed to determine the roughness coefficient of the underside of ice cover[1,2,3,4]: reach analysis from water surface slope and cross-section analysis from velocity distribution. This paper presents an refined method of determining roughness coefficient of the underside of ice cover based on direct measurements of velocity distribution in the flow under ice cover.

2 APPROACH TO ANALYSIS

It is generally assumed that maximum velocity divides the flow depth into two regions. In both regions the velocity distribution may be expressed by Prandtl-Von Karman semi-logarithmic velocity distribution equation (Fig.1):

$$u = u_* [\frac{1}{K} \ln \frac{y}{k_s} + C] \tag{1}$$

where: u-velocity at a given depth; K-Karman constant usually assumed to be 0.4; u_* -shear velocity; y-depth measured from the rough boundary (ice, bottom); K_s -linear dimension of roughness. Eq.(1) may be presented in the following simplified form

$$u = a \ln(y) + b \tag{2}$$

Measured velocities under ice cover at various depth may be plotted in the coordinates u versus ln(y). They should concentrate along a straight line whose equation may be determined using least mean square fit method. The dimensional value a in Eq.(2) denotes the slope of the line, and dimensional constant b the intersection of the line with the velocity axis. The correlation coefficient is calculated for each case to estimate the difference between measured velocity values and the theoretical velocity distribution.

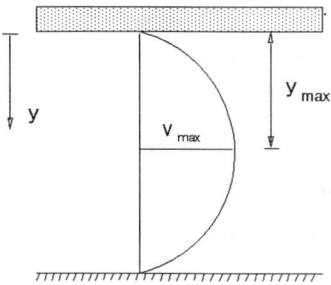

Figure 1 Division of flow depth into ice and bed zones

With the known parameters a and b in Eq.(2), the maximum velocity u_{max} for flow depth y_{max} can be calculated as

$$u_{max} = a \ln(y_{max}) + b \qquad (3)$$

For any given y_0, the following value is subsequently calculated:

$$\int_{y_0}^{y_{max}} u \, dy = y_{max}[a \ln(y_{max}) + b] - y_0[a \ln(y_0) + b] - a(y_{max} - y_0) \qquad (4)$$

The average velocity over the depth ($y_{max} - y_0$) can therefore be calculated

$$V = \frac{Q}{A} = \frac{u_{max} y_{max}}{y_{max} - y_0} - \frac{y_0 u_0}{y_{max} - y_0} - a \qquad (5)$$

By taking y_0 as zero, the average velocity for the subarea above the maximum velocity plane

$$V_i = U_{max} - a \qquad (6)$$

The boundary shear velocity in a prismatic channel can be approximated as

$$V_* = \sqrt{g R S_f} \qquad (7)$$

where: R - hydraulic radius; S_f -energy gradient. From V_* and the average velocity of the zone under consideration, the Darcy-Weisbach friction factor for the underside of ice cover in the station where velocity distribution was measured can be calculated from

$$f_i = 8 \left(\frac{V_*}{V_i}\right)^2 \qquad (8)$$

A corresponding value of Manning's n was then computed from

$$n_i = \sqrt{\frac{f_i}{8g}} R_i^{1/6} = \frac{1}{\sqrt{8g}} \sqrt{f_i} R_i^{1/6} \qquad (9)$$

3 FIELD MEASUREMENT

Field measurements were obtained from a 15 km long approaching channel of hydroelectric power plant in Xinjiang of China. Measurements were taken in the period from October 18 1995 to April 24, 1996. The channel bed consisted of highly heterogeneous bed material. Channel width varied from 8 to 12 m, and flow depth at the points where measurements were taken varied from 1 to 2.97 m. The Manning's resistance coefficient of the channel bed is about 0.0275. During the study period, a continuous ice cover was present in the reach from December 14, 1995 to March 6, 1996. Velocity profiles were measured at three stations, with or without ice cover are shown, starting from the end of the channel and located approximately 5 km apart, by means of propeller current meter placed on a vertical rigid rod. The local velocity was measured at 8 or more points for each vertical profile. A minimum of 3 profiles were measured at each cross section. Examples of measured velocity distribution are shown in Fig.2.

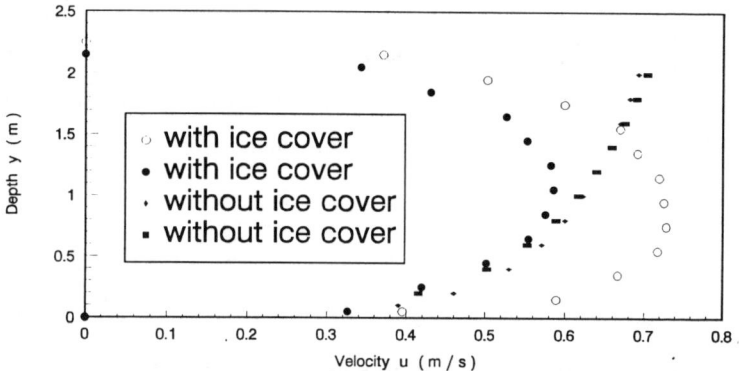

Fig. 2 Velocity distribution over flow depth (with or without ice cover)

Numerous velocity distributions measured under ice cover in various places of the approaching channel were used for the determination of roughness coefficients. The input data are measured velocities over the depth of flow from the underside of ice cover to the position of maximum velocity at y_{max}. Examples of input data and calculated values for the approaching channel with an ice cover are given in the following table,

y_{max} (m)	u_{max}' (m/s)	a	b	u_{max} (m/s)	V (m/s)	n_i
1.10	0.585	0.111	0.587	0.598	0.487	0.0287
1.00	0.578	0.196	0.613	0.613	0.417	0.0282
1.20	0.561	0.111	0.548	0.568	0.457	0.0309
1.50	0.748	0.144	0.696	0.754	0.610	0.031

where u_{max}' - maximum velocity measured at the depth y_{max} ; u_{max} - maximum velocity calculated at the depth y_{max}. Analysis of the field data for the study reach led to the conclusion that the semi-logarithmic velocity distribution approach generally appeared to give more consistent and acceptable results.

4 CONCLUSIONS

Roughness of the underside of ice cover is a very important factor which controls the flow with ice cover. This roughness is very often represented by Manning coefficient. Determination of this coefficient in nonuniform flow is only possible by means of measured velocity distribution under ice cover. In this paper a refined method is presented for determining roughness coefficient of the undersurface of ice cover based upon direct measurements of velocity distribution in the flow under ice cover. This approach appears to have led to improved representation of the velocity distribution in natural channels.

REFERENCE

1 Vedula, S. and Achanta, R. R. (1985), 'Bed shear from velocity profiles: A new approach', ASCE, Journal of hydraulic engineering, Vol. 111, No. 1.

2 Majewski, W., Baginska, M. and Walczak, P. (1986), 'Determination of roughness coefficient for ice covered rivers by means of direct measurements of velocity distribution', Symposium on measuring techniques in hydraulic research, Delft.

3 Larsen, P.A. (1969), 'Head losses caused by an ice cover on open channels', Journal of the Boston Society of Civil Engineers, No. 1.

4 Hendrikson, F. M. and Davar, K. S. (1986), 'Estimation of resistance to flow in ice covered channels using binary velocity distributions', IAHR symposium on ice, Iowa city, Vol.1.

TURBULENT STRUCTURE OF OPEN AND ICE-COVERED FLOW IN A CHANNEL

D.S. KUZNETSOV*, E. I. DEBOL'SKAYA **
*Moscow State University, **Water Problems Institute of Russian Academy of Sciences, Moscow, Russia

One of the main problem connected with calculations of a current structure in any flows is the parametrization of turbulent characteristics. There are many dependencies for the velocity vertical profile for open and ice-covered flows but there are not many for turbulent momentum transfer coefficient (eddy viscosity) $A(z)$ though it is the most important characteristic for example during calculations of the transport process of impurities. The objective of this paper is to obtain a parametric dependence of $A(z)$ on the base of a numerical simulation.

MODEL

Equations used in the model describe the steady flows when the gradient of pressure and turbulent viscosity are balanced:

$$\frac{\partial}{\partial z}\left(A(z)\frac{\partial u}{\partial z}\right) = -gi \tag{1}$$

$$\alpha \frac{\partial}{\partial z}\left(A(z)\frac{\partial b}{\partial z}\right) - \frac{b^2}{A(z)} + A(z)\left(\frac{\partial u}{\partial z}\right)^2 = 0 \tag{2}$$

$$A(z) = l\sqrt{b} \tag{3}$$

$$l = c^{1/4}\kappa \left|\frac{\partial u/\partial z}{\partial^2 u/\partial z^2}\right| \tag{4}$$

where $A(z)$ is the momentum transfer coefficient (eddy viscosity), $u(z)$ - a velocity of the flow, i - a slope of the surface level, $l = l(z)$ - a scale of turbulence, $b = b(z)$ - a turbulent kinetic energy, $\kappa = 0.4$ - the von Karman constant, $\alpha = 0.32$, $\gamma = 0.09$, $c = 0.4$. The axis x lies in the bottom surface, the axis z is directed vertically upwards. The origin of coordinates is placed at the bottom.

Boundary conditions for $A(z)$ at the bottom and at the ice surface can be received from the expressions:

$$A\Big|_{z=z_b} = A_b = \frac{\tau_b}{\rho_w \frac{\partial u}{\partial z}\Big|_{z=z_b}} = \frac{u_{*b}^2}{\frac{\partial u}{\partial z}\Big|_{z=z_b}}, \quad A\Big|_{z=z_i} = A_i = \frac{\tau_i}{\rho_w \frac{\partial u}{\partial z}\Big|_{z=H-z_i}} = \frac{u_{*i}^2}{\frac{\partial u}{\partial z}\Big|_{z=H-z_i}},$$

where τ_b, τ_i - boundary shear stresses, u_{*b}, u_{*i} - dynamic velocities, z_b, z_i - dimensions of roughness juts of the bottom and ice surfaces respectively. From the logarithmic speed distribution law in layers adjacent to z_b, z_i it follows:

$$\frac{\partial u}{\partial z}\Big|_{z=z_b} = \frac{u_{*b}}{\kappa z_b}, \quad \frac{\partial u}{\partial z}\Big|_{z=H-z_i} = \frac{u_{*i}}{\kappa z_i}.$$

Taking into account that $z_b << z_0$, $z_i << H - z_0$ and (1) is valid, we can write:

$$A_b = u_{*b}\kappa z_b = \kappa z_b \sqrt{giz_0}, \quad A_i = u_{*i}\kappa z_i = \kappa z_i \sqrt{gi(H-z_0)}$$

where z_0 is a vertical coordinate of the surface of zero shear stress or maximum of velocity, H - a depth. To obtain the boundary value of the turbulent energy let the scale of turbulence at the bottom and ice surfaces be equal to dimensions of the roughness juts. Then theKolmogorov's hypothesis (3) gives:

$$b\Big|_{z=z_b} = b_b = (A_b/l_b)^2 = \kappa^2 giz_0, \quad b\Big|_{z=z_i} = b_i = (A_i/l_i)^2 = \kappa^2 gi(H-z_0).$$

At the bottom and ice surfaces the velocity of the flow is assumed to be zero:
$$u\Big|_{z=z_b} = 0, \quad u\Big|_{z=z_i} = 0.$$

Boundary conditions at the upper surface of the open flow are [2]:

$$A\frac{\partial u}{\partial z}\Big|_{z=H} = \frac{\tau_W}{\rho_W}, \quad A\Big|_{z=H} = 4.3 \cdot 10^{-4} W^2, \quad b\Big|_{z=H} = 2 \cdot 10^{-4}\beta W^2,$$

where $\tau_W = 10^{-3}\beta\rho_a W|W|$ - is the shear stress at upper surface of water due to the wind influence, ρ_a - the air density, ρ_W - the water density; the coefficient β can take two different values depending on the degree of wave development which is determined from the wind velocity W and the flow depth H : for $H > \mu W$ $\beta = 0.23$, for $H < \mu W$ $\beta = 0.75$, where $\mu = 7.58$ s - is a dimensional and the wind velocity W is expressed in m/s everywhere.

To determine the coordinate z_0 in an ice-covered flow it is enough to know the bottom-ice roughness relation. Really, as it follows from empirical data the average velocity of the near bottom flow section is equal to that of the top section (near the ice surface). The interface between these sections coincides with plane of zero shear stress, i.e. with z_0-plane. Taking into account the Chezy formula for the average velocity and Shtricler formula:

$$n_{b,i} = 0.15 \cdot z_{b,i}^{1/6}/\sqrt{g}, \tag{5}$$

we can receive

$$z_0 = H/(\tilde{k}+1), \tag{6}$$

where $\tilde{k} = \left(\frac{n_i}{n_b}\right)^{3/2} = \left(\frac{z_i}{z_b}\right)^{1/4}$, n_b and n_i are Manning's roughness coefficients of bottom and ice surfaces respectively.

The position of the point z_0 for the open flow is determined from the ratio of the shear stress at the top boundary τ_W to the bottom friction stress without wind $\tau_b^0 = \rho_W giH$, i.e. from the parameter:

$$k = \frac{\tau_W}{\rho_W giH}, \qquad (7)$$

where ρ_W is the water density. To obtain the dependence of z_0 on k it is necessary to notice that for the gradiently-viscous regime of the flow described by the equation (1), the following relation is valid:

$$\tau_b = \tau_W + \rho_W giH \qquad (8)$$

Taking into consideration (7) and (8) it can be obtained:
$\tau_b / \tau_W = 1 + 1/k$.

From linearity of distribution of the turbulent friction stress with depth it follows that the point z_0 position at any wind (fair and contrary) will be determined by expression $z_0 = H(k+1)$. At the fair wind $k>0$ and $z_0 > H$.

From equation (1) - (4) the differential equation of the first order about $A(z)$ with the point of singularity $z = z_0$ was obtained. The solution of this equation may be written in the following form:

$$A = \left[\kappa c^{1/4} \int_{z_b}^{z} \frac{\sqrt{b} dz}{z - z_0} + \frac{A(z_b)}{z_b - z_0} \right] (z - z_0) \text{ at } z < z_0, \quad A = \left[\kappa c^{1/4} \int_{z}^{H} \frac{\sqrt{b} dz}{z - z_0} + \frac{A(H)}{H - z_0} \right] (z - z_0) \text{ at } z > z_0.$$

To calculate these integrals, it is necessary to know the value of the turbulent energy b which was obtaned from equation (2) numerically by the run method

PARAMETRIZATION

In case when the current is influenced by a favorable wind the following dependence for $A(z)$ was proposed based on numerical experiments results:

$$A(z) = a z \sqrt{gi(H-z)} + \frac{\eta W^2}{H} \left(\frac{z^3}{gi} \right)^{1/2}, \qquad (9)$$

where $a=0.45$, $\eta = 2 \cdot 10^{-6}$ are non dimensional constants, W - wind velocity.

In case when the current is influenced by a head wind, the point z_0 lies within the flow and it is necessary to divide the cross section of the flow into two parts. For the bottom part, when $z < z_0$, the dependence of $A(z)$ can be expressed in the form:

$$A(z) = a z \sqrt{gi(z_0 - z)}. \qquad (10)$$

For the top part, when $z > z_0$, it is possible to write:

$$A(z) = GW^2 (z - z_0) \left[\frac{4(H-z)}{(H-|z_0|)^2} + \frac{1}{H - z_0} \right] \qquad (11)$$

where $G = 4.3 \cdot 10^{-4}$ s - the dimensional constant taken from empirical dependence of turbulent exchange factor on wind velocity of the surface "water-air" [2].

In Fig. 1, vertical distributions of $A(z)$ are given in case of the effect of various wind velocity on the flow. These distributions were received due to numerical calculations of equation (1)-(4) and from formulas (9),(10),(11).
In case of ice-covered flow the following dependence for $A(z)$ was proposed:

$$A(z) = az\sqrt{gi(z_0 - z)} \quad \text{if } z < z_0 \tag{12}$$

$$A(z) = a(H - z)\sqrt{gi(z - z_0)} \text{ if } z > z_0. \tag{13}$$

Fig.2 shows the curve $A(z)$ plotted by results of the numerical calculations and by formulas (12), (13). It is necessary to remark that the zero value of $A(z)$ at $z = z_0$ is reached both from formulas (12),(13), but this effect is absent on our curves because we used the averaging of $A(z)$ over three nearest points.
The substitution of expressions (12) and (13) in expression for velocity, obtained by integrating the equation (1) separately in each layer, allows to receive a parameteric dependence for ice-covered current velocity $u(z, i, H, z_b, z_i)$:

$$u(z) = u_0 + \frac{2\sqrt{gi\,z_0}}{a}\left(\sqrt{1-y} - \ln\frac{\sqrt{1-y}+1}{\sqrt{y}}\right) \quad \text{at } z < z_0 \tag{14}$$

$$u(z) = u_0 + \frac{2\sqrt{gi\,z_0}}{a}\left(\sqrt{y-1} + \frac{1}{2}\ln\left|\frac{\sqrt{y-1}-r}{\sqrt{y-1}+r}\right|\right) \text{at } z > z_0, \tag{15}$$

where $y = z/z_0, y_b = z_b/z_0, r = (d_i/d_b)^{1/8}, u_0 = \frac{2\sqrt{gi z_0}}{a}\left(\ln\frac{\sqrt{1-y_b}+1}{\sqrt{y_b}} - \sqrt{1-y_b}\right)$.

Fig.3 shows a distribution of velocity, constructed according to field measurements, numerical experiments and formulas (14) and (15): fig.3(a) - the comparison with Larsen's experiments [1], fig.3(b-d) - with our measurements in the river Moskva.
In the case of a free flow which is influenced by a head wind the integration gives for the bottom layer ($z < z_0$) expression (14) and for the top layer ($z > z_0$):

$$u_t(z) = u_0 + \frac{gi}{\varphi}\ln\left(\frac{\tilde{y}-1}{\tilde{y}-y}\right), \text{ where } \varphi = 4G\frac{W^2}{(H-|z_0|)^2}, \tilde{y} = \left(H + \frac{(H-|z_0|)^2}{4(H-z_0)}\right)/z_0.$$

The model of the open and ice-covered current permits to receive the distribution of turbulent characteristics of the flow along the vertical.

REFERENCES

2. Larsen P.A. (1969), Journal of the Hydraulics Division, ASCE, 96 (HY3), pp.703-724
3. Pierson, W.J. (1955),Wind waves, J. Adv. Geophys., 2, 93

ACKNOWLEDGEMENT

This research has been supported by grant 96-05-65151 from the Russian Foundation for Basic Research.

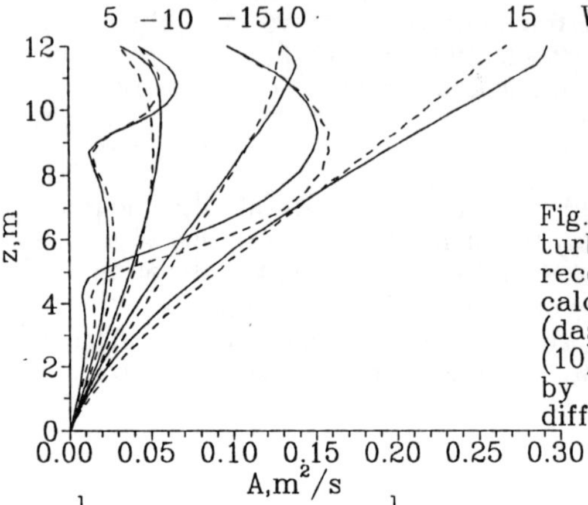

Fig.1 Vertical distributions of turbulent exchange factor A(z) received through numerical calculations of equations (1)–(4) (dashed lines) and formulas (9), (10),(11) (solid lines), caused by the influence of wind having different velocities upon the flow.

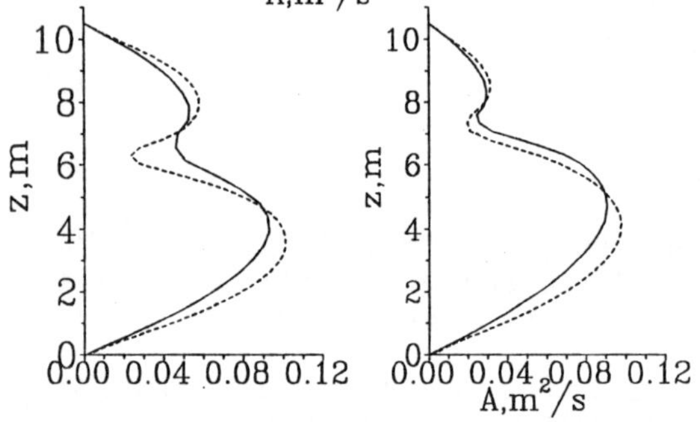

Fig2. Vertical distributions of turbulent exchange factor A(z) plotted by results of numerical calculations (dashed lines), by formulas (12),(13) (solid lines), using Larsen's experimental data.

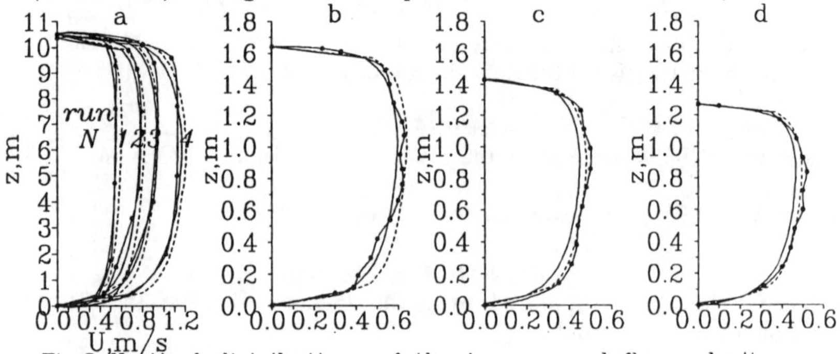

Fig.3 Vertical distributions of the ice–covered flow velocity constructed by field measurements (solid lines with centered symbols), numerical experiments (dashed lines) and formulas (14),(15) (solid lines): a – comparison with Larsen's experiments (1969), b,c,d – with our measurements in the r.Moskva

FLUID MIXING AND BOUNDARY SHEAR STRESS IN COMPOUND MEANDERING CHANNEL

ISHIGAKI,T.,Y.MUTO, N.TAKEO and H.IMAMOTO

Ujigawa Hydraulics Laboratory, Disaster Prevention Research Institute
Kyoto Univ., Kyoto, Japan

ABSTRACT
Flow structure in a compound meandering channel is investigated here with experimental results by flow visualization and bed shear measurement. Fluid mixing between in-bank and over-bank flow and its effect on boundary shear stress are discussed. The results show that the mixing is different in shallow cases of flooding depth and deeper cases. This difference can be explained by depth dependence of the flow structure.

INTRODUCTION
Since Toebes & Sooky (1967) examined a stage-discharge curve in a compound meandering channel, the flow structure has been investigated by some researchers (e.g., Imamoto & Ishigaki; 1983; Arnold, 1987; Ervine, 1993; Shiono et al., 1994). From the results it is found that over-bank flow is running straight when flooding depth is deep and the structure of secondary flow is different from that in a simple meandering channel.

Fluid mixing between main channel and flood plain flow is strong as observed in straight channel by many researchers, however, the mixing is stronger in meandering channel because the in-bank and over-bank flow are crossing each other and secondary flow is produced by the centrifugal force. This paper deals with three dimensional structure of the flow in a compound meandering channel. Velocity, vorticity and divergence on horizontal plains were obtained from visualized pass lines by a tracer method. Boundary shear stress was measured by a film sensor (Gust & Wearherly, 1985).

EXPERIMENTAL METHODS
In a straight channel, of which length is 20 m, width B=1.2m and slope I=1/800, flood plains are set as shown in Fig.1. The main channel is composed with straight and arc channels. The width, b, is 15 cm and the flood plain height, h, is 5 cm. Sinuosity s (=L/Lw , L: curved channel length, Lw: meandering wavelength) is 1.37.

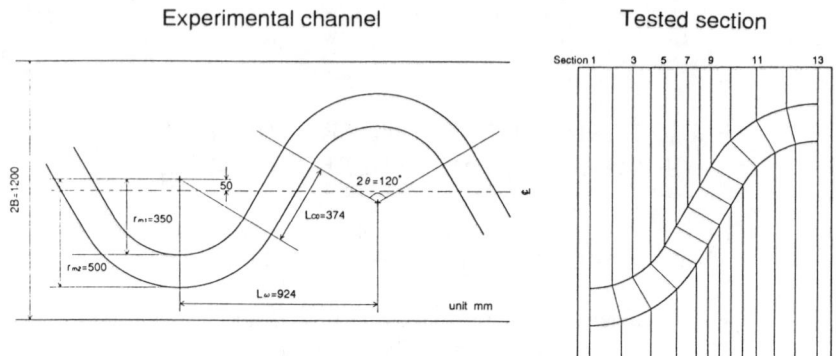

Fig.1 Experimental channel and tested section.

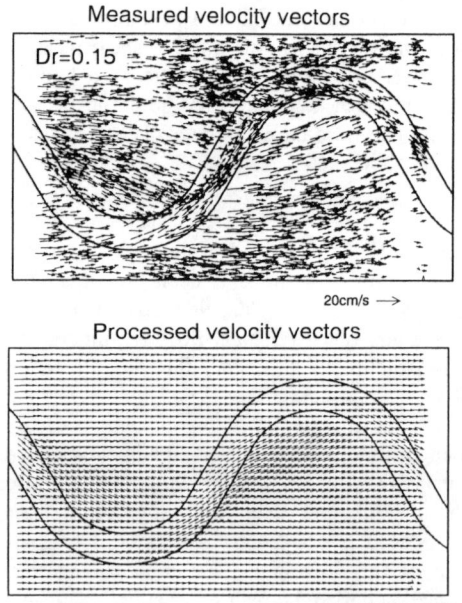

Fig.2 Measured (upper) and processed (lower) velocity vectors. (Dr=0.15)

Seven waves are set in the channel and forth and fifth waves from the upper end are tested sections. Twelve sections in half wavelength are used for measurements. Section 1 is at the apex, and sections 5 and 9 are upper and lower end of the straight channel as shown in Fig.1. Flow structure in a compound meandering channel shows different aspect as increasing water depth, thus we investigate the structure with a parameter Dr named as a relative depth, which is the ratio of main channel and flood plain depth, H and hf=H-h. In experiments, H is 5.0 to 10.0 cm, the discharge Q is 1.43 to 19.05 l/s, Reynolds number Re is 1900 to 13600, Froude number Fr is 0.34 to 0.54.

Velocity vectors, vorticity and divergence are figured out from the data obtained by flow visualization. A tracer method is used and velocity vectors are digitized with length of pass line of tracer. Tracers are waterproof paper of 5 mm square and neutral buoyant plastic bead of 0.45 mm diameter. Vertical component of vorticity and divergence are calculated by following equations.

$$\zeta = \frac{\partial u}{\partial y} - \frac{\partial v}{\partial x} \qquad div = \frac{\partial u}{\partial x} + \frac{\partial v}{\partial y}$$

Longitudinal component of velocity u and lateral component v are interpolated at nodes of 3 or 1.5 cm square grid by using measured velocity vectors as shown in Fig2.

Boundary shear stress is measured by heated thin-film gauge (WTG-50A: Micro-Measurement) with CTA circuit (DANTEC 55D01). Relation between output voltage and boundary shear stress is calibrated in a straight channel flow of large aspect ratio. In the calibration boundary shear stress τ is equal to ρgRI, where ρ= density of water, g= gravity acceleration, R= hydraulic radius and I= energy slope.

FLUID MIXING BETWEEN INBANK AND OVERBANK FLOW

Distribution of velocity is complex as shown in Fig.2. At first we discuss on the lateral distribution of velocity averaged along the longitudinal axis over one wavelength in Fig.3. It shows that the distributions in small Dr cases are different from those in large Dr cases. There are high speed flows at the center and both sides of the channel in large Dr cases. From the results we select the data in two cases, Dr=0.15 and 0.50, to investigate flow structure.

Fig.4 shows high and low speed parts of velocity on the water surface. In Dr=0.15, high speed fluid in the main channel is flooding and spreading over the flood plain. On the contrary, in Dr=0.50, the flooding flows are moved to both sides of the channel and over-bank flow runs straight around the channel center. Details of flow structure are shown in Fig.5. It shows the distributions of vorticity on the vertical axis and two dimensional divergence on the water surface. Clockwise rotation is positive in the figure of vorticity and up-welling part of fluid is positive in the figure of divergence. In Dr=0.15 case, maximum value of vorticity is observed in the wake region of apex. This area spreads out in the main channel, and the flooding flow is moved out. This explains that the positive area of divergence spreads along the concave bank. The results for the Dr=0.5 case are quite different from the above observation. A pair of counter rotating vorticities goes along the center of the channel, and the vortex around the apex is stretched and moved toward the both sides of the channel.

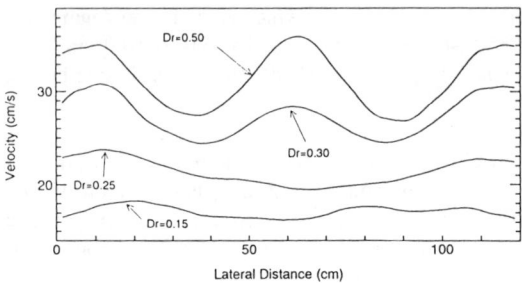

Fig.3 Lateral distribution of longitudinal mean velocity on the water surface.

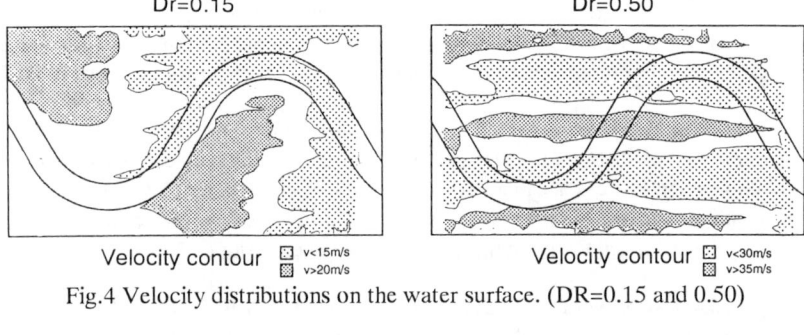

Fig.4 Velocity distributions on the water surface. (DR=0.15 and 0.50)

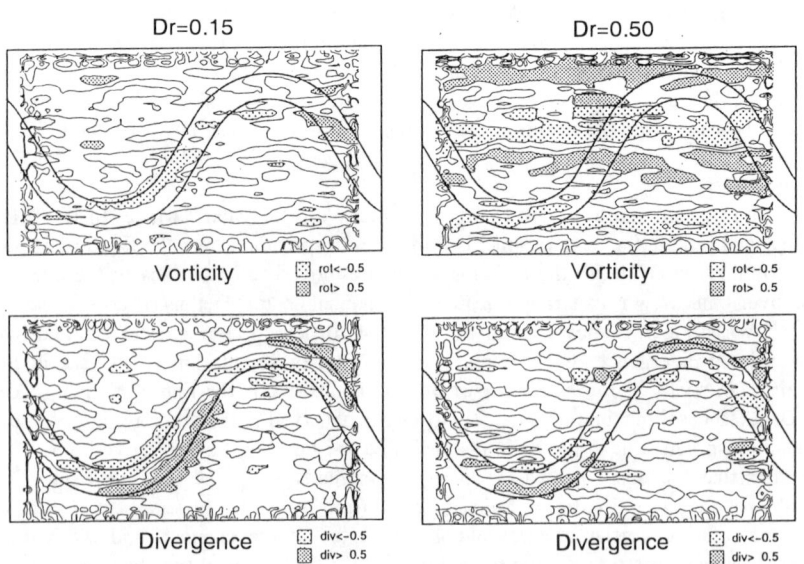

Fig.5 Distributions of vorticity on the vertical axis and two dimensional divergence on the water surface. (DR=0.15 and 0.50)

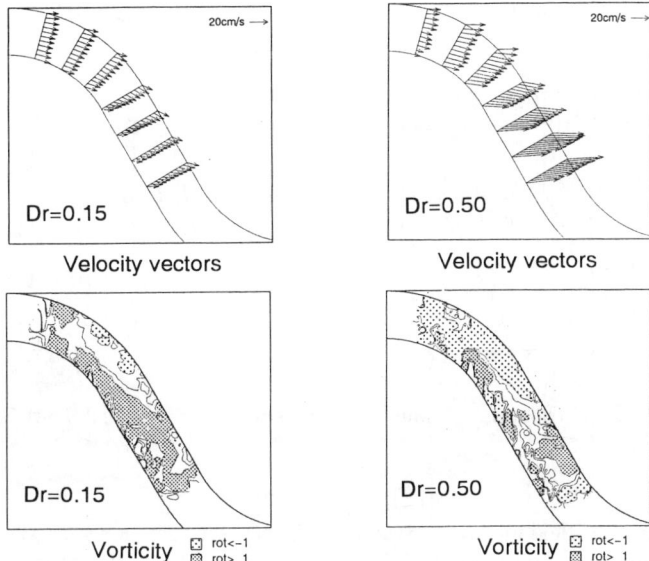

Fig.6 Distribution of velocity and vorticity on the boundary between in-bank and over-bank flows. (Dr=0.15 and 0.50)

From the former results it is considered that fluid mixing between in-bank and over-bank flow changes as flooding depth. Flow structure in a horizontal plain just above the flood plain (h < z < h+3mm) was visualized by using neutral buoyant tracers with light sheet of 3mm thick, and velocity vectors were digitized with length of pass line. Fig.6 shows the distributions of velocity and vorticity calculated with visualized data. This figure also shows the difference of the flow structure in Dr=0.15 and 0.50. That is, the over-bank flow in Dr=0.50 is crossing straight over the in-bank flow, and they have the different structures. This structure can also be explained by the precise data of LDA measurements by Shiono et al. (1994).

BOUNDARY SHEAR STRESS

Boundary shear stress was measured at each 3 or 6 cm pitches along the lateral 6 sections as shown in Fig.7. The data is normalized by mean shear stress $\tau = \rho gRI$. It is found that the distributions on the flood plain are closely related with the flow structure shown in the previous figures. In the Dr=0.15 case, the maximum value is observed around the flooding area of in-bank flow. Whereas the maximum value for Dr=0.50 is observed around the channel center and both sides where the flow speed is high as shown in Fig.3 and 4. At the downstream side of section 7 an distinct peak value is observed. It is considered that this is because high speed flow at the center rushes at the edge of bank. These results indicate that maximum shear stress is acting on different place as flooding depth is increasing, and protecting the bank over a long distance from erosion is necessary.

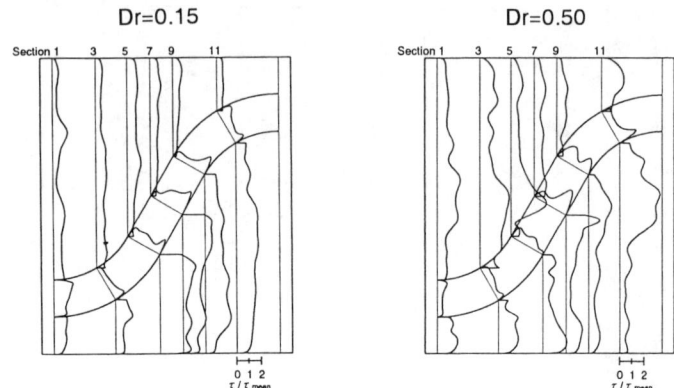

Fig.7 Distributions of boundary shear stress. (Dr=0.15 and 0.50)

CONCLUTIONS

To investigate the fluid mixing between in-bank and over-bank flows, distributions of velocity, vorticity and divergence on the water surface and boundary shear stress were measured in a compound meandering channel, of which sinuosity of main channel was 1.37 in a range of relative depth Dr=0 to 0.50. In these hydraulic conditions the mixing process changes remarkably because of changing the flow structure. For small Dr cases, in-bank flow affects the whole structure of flow, but not for large Dr cases. For large Dr cases, over-bank flow runs straight and the flooding fluid from the main channel is moved to the both sides of the channel. Such difference of flow structure is also recognized in boundary shear stress distributions. As a result, the maximum boundary shear stress acts on a wide area because its acting position changes as flooding depth increases.

REFERENCES

Arnold,U., Zur bilddaten- und modellgestutzten Bestimmung der Schadstoffausbreitung in naturnahen Fliessgewassern, Mitteilungen, IWW, RWTH Aachen, Nr. 52. 1987.

Ervine,DA, Willetts,BB, Sellin,RHJ and Lorena,M, Factors affecting on conveyance in meandering compound flows, J.Hydr.Eng., Vol119, No.12, pp.1383-1399, 1993.

Gust,G. and Weatherly,GL, Velocities, Turbulence, and Skin Friction in a Deep-Sea Logarithmic Layer, J.Geophys.Res., 90, pp.4779*4792, 1985.

Imamoto,H. and Ishigaki,T., Characteristics of Flow in a Curved Open Channel with Flood Plains, Proc. 27th Conf. on Hydraulics, JSCE, pp.67-72,1983 (in Japanese).

Shino,K. Muto,Y., Imamoto,H. and Ishigaki,T., Flow structure in meandering channels for overbank, MAFF, Conf. River and Coastal Engineers, 1994.

Toebes,GH and Sooky, AA, Hydraulics of Meandering River with Floodplains, Proc. ASCE, Vol.93, No. WW2,, pp.213-236, 1967.

A simple model of gravel-bed roughness

V. I. NIKORA, D. G. GORING AND B. J. F. BIGGS
National Institute of Water and Atmospheric Research
Christchurch, New Zealand

ABSTRACT
Statistical properties of gravel-bed roughness are investigated using field and laboratory data. The probability distribution of bed elevations appeared to be near-Gaussian. The empirical structure function of bed elevations consisted of two regions: a scaling region at small spatial lags and a saturation region at large lags. Thus, the structure function of bed elevations could be parametrised by introducing five parameters which fully describe roughness properties. We hypothesised that the scaling exponents are universal which reduces the number of roughness parameters to characteristic longitudinal, transverse and vertical scales. These can be determined from empirical longitudinal and transverse structure functions.

INTRODUCTION

The roughness of gravel-bed rivers is a fundamental quantity of river hydraulics since it determines the properties of the average flow, turbulence, flow resistance, and bed particle motion. Also, the bed roughness is one of the key factors in understanding near-bed processes important for benthic ecological systems. However, an appropriate quantitative description of gravel-bed roughness is still a problem and new efforts should be applied to clarify this problem.

Traditionally, river engineers evaluate bed roughness by characteristic particle diameters (most often it is the size of the median or intermediate axis of the particle), that, is the diameter with n-percentage finer, where n is usually 50, 65, 84 or 90%. According to this approach (*characteristic particle size approach*), the equivalent sand roughness height k_s can be roughly estimated as $3.5d_{84}$ or $6.8d_{50}$. Due to its simplicity the characteristic particle size approach is widely used in river hydraulics. However, the bed roughness can be characterised by a single size percentile, say d_{84}, only if we assume that the shape of the particle size distribution, bed arrangement, particle shape and orientation, packing, spacing and sorting, shape and density of particle clusters, etc., are universal, i.e., hold at all sites and flows. However, these assumptions have not been properly tested and are probably invalid in natural rivers.

Another way to describe the bed roughness is to consider it as a random field of bottom elevations $Z(x,y,t)$ (origin of Z is lower than channel bed surface, at the arbitrary level), where x and y are the longitudinal (the main flow direction) and transversal coordinates, and t is time (*random field approach*, Nikora, 1982). A complete picture of an arbitrary random field is provided by the m-dimensional probability density when $m \to \infty$, that contains full information about the field. This approach takes into account all possible factors influencing bed roughness and therefore is preferable to the approach based on the use of the single particle size. Quantitative description of gravel beds by means of m-dimensional probability functions is theoretically possible, but in practice it meets great difficulties. In this situation an acceptable compromise is the use of moment functions (correlations, spectra, structure functions, etc.) for a quantitative evaluation of the gravel-bed roughness. If the bed elevation field is stationary, homogeneous and Gaussian the moment functions of the second order will provide full information about the roughness. In general, the spectrum of the bed elevation field $Z(x,y,t)$ is a function of spatial coordinates x and y, time t, projections K_x and K_y of the wave vector $\mathbf{K}$, and the time frequency ω. For the simplest case of the "frozen" bed we can accept the hypothesis of the local spatial homogeneity that reduces the roughness description to the two-dimensional spectrum $S(K_x,K_y)$, the correlation function $R(\Delta x,\Delta y)$ and the structure function $D(\Delta x,\Delta y)$ (Δx and Δy are spatial lags). If the field $Z(x,y)$ is non-Gaussian, moment functions of the higher orders should also be considered.

The aims of this paper are: (1) to investigate statistical properties of the field $Z(x,y)$, and (2) to develop a simple model of gravel-bed roughness using the random field approach.

DATA

We use in our analysis our own laboratory and field data as well as published field data of Furbish (1987).

1. Field measurements were carried out on 3 reaches of Canterbury (New Zealand) gravel-bed rivers with width 7-8 m, water discharge 0.45-2.01 m³/s, slope 0.00276-0.0154 and bed particle size d_{50}=46-97 mm. The following characteristics were measured: longitudinal bed profiles, three-axis bed particle sizes (the longest a, the intermediate b and the shortest c) along the profiles, water surface slope, channel cross-sections, water discharge and bulk particle

size distribution by Wolman method. A simple profiler was designed and manufactured to measure gravel-bed profiles in the field. The profiler consists of a rigid frame containing 200 sliding rods with 10 mm distance between centers of adjoining rods. This allowed us to measure gravel bed profiles 2 m long with spatial resolution 10 mm. Longitudinal profiles were measured at the following positions: 25, 50 and 75% of the width from the stream bank. To reduce statistical uncertainty two to three profiles were measured in most cases with distance between profiles slightly larger than the typical size of bed particles. In no cases were particle clusters observed. Hence, we were able to investigate grain ("skin") roughness in its pure form.

2. Laboratory measurements were carried out in two outdoor flumes with gravel beds created manually using bed material from the Waimakariri River. Particle sizes varied from 20 to 60 mm (flume 1) and from 40 to 120 mm (flume 2). Longitudinal bed profiles 6-7 m long were measured with sampling interval 9.6 mm using an automatic profiler PV-7 (Delft Hydraulics Laboratory).

3. The transverse structure of gravel-bed roughness was assessed using 10 transverse profiles published by Furbish (1987). They were measured in different gravel-and-cobble reaches on North Boulder Creek, a mountain stream in the Colorado Front Range. For our analysis we digitised the above profiles with sampling interval 10 mm. Unfortunately, no information on particle size was presented, by Furbish (1987).

STATISTICAL PROPERTIES OF GRAVEL-BED PROFILES

1. To evaluate the shape of bed elevation probability distribution we used skewness and kurtosis coefficients. In all cases these coefficients appeared to be close to zero (Table 1). This allows us to propose that the distribution of bed elevations is near-Gaussian.

Table 1. Skewness Sk and kurtosis Ku coefficients for gravel-bed profiles.

Data	Sk	rms of Sk	Range of Sk	Ku	rms of Ku	Range of Ku
Longitud. Profiles-field	0.51	0.38	-0.17-1.03	-0.10	0.80	-1.20-1.45
Transverse profiles-field	0.16	0.37	-0.37-0.66	-0.10	0.29	-0.54-0.45
Longitud. Profiles-lab	-0.21	0.25	-0.42-0.22	0.31	0.42	-0.31-0.46

Here: Sk and Ku are the mean values of skewness and kurtosis coefficients, respectively; rms is the standard deviation of Sk (or Ku).

2. The functions $S(K_x, K_y)$, $R(\Delta x, \Delta y)$ and $D(\Delta x, \Delta y)$ contain the same information about the bed elevation field since they relate to each other by Fourier transformation which is a linear operation. The only difference is that the

former presents it in the wave number domain while the latter is in the space domain. In this paper we preferred to use the structure function $D(\Delta x, \Delta y)$ since it can be applied to describe non-uniform bed profiles also. The structure function $D(\Delta x, \Delta y)$ of bed elevation $Z(x,y)$ is defined as an average squared increment $\{Z(x+\Delta x, y+\Delta y) - Z(x,y)\}$. In practice the 2-D structure function can be estimated from:

$$D(\Delta x, \Delta y) = \frac{1}{(N-n)(M-m)} \sum_{i=1}^{N-n} \sum_{j}^{M-m} \{Z(x_i + n\delta x, y_j + m\delta y) - Z(x_i, y_j)\}^2 \quad (1)$$

where $\Delta x = n\delta x$, $\Delta y = m\delta y$, δx and δy are the sampling intervals, N and M are the total numbers of measuring points of bed elevations in directions x and y, respectively. We have used relationship (1) to calculate longitudinal $D(\Delta x, \Delta y = 0)$ and transverse $D(\Delta x = 0, \Delta y)$ structure functions for all gravel-bed profiles described above. In all cases empirical structure functions consist of two main regions: at small spatial lags structure functions behave like power functions $D(\Delta x) \propto \Delta x^{2H_x}$ and $D(\Delta y) \propto \Delta y^{2H_y}$ (region 1), while at sufficiently large lags they become constant (region 2) (Figure 1). Exponents H_x and H_y appeared to be the same within each family of profiles (field longitudinal profiles: H_x=0.67-0.70; laboratory longitudinal profiles: H_x=0.46-0.54; field transverse profiles: H_y=0.73-0.78) but differed among them. Spatial scales Δx_o and Δy_o dividing regions 1 and 2 (the procedure is shown in Figure 1) can be defined as horizontal characteristic scales of bed roughness. The third, vertical, scale ΔZ_o can be obtained from region 2 as $\Delta Z_o = \sqrt{D(\Delta x = \infty)/2} = \sqrt{D(\Delta y = \infty)/2} = \sigma_z$, where σ_z is the standard deviation of bed elevation. This follows from relationship (Monin and Yaglom, 1975):

$$D(\Delta x, \Delta y) = 2\{\sigma_z^2 - R(\Delta x, \Delta y)\} \quad (2)$$

where $R(\Delta x, \Delta y)$ is the correlation function of bed elevation. When $\Delta x, \Delta y \to \infty$, $R(\Delta x, \Delta y) \to 0$ and we have $D(\infty) = 2\sigma_z^2$.

MODEL OF GRAVEL-BED ROUGHNESS

From the above it follows that the distribution of gravel bed elevation deviates somehow from the normal law (while the flatness of the distribution corresponds to that of the normal law the skewness though close to zero is definitely positive, Table 1). However, as a first approximation, for practical purposes we can assume normality of gravel-bed elevation and use moment functions of the second order to describe its structure. As it was shown above, the structure function is a convenient tool to describe three-dimensional roughness. Unfortunately, we could investigate only cross-sections $\Delta x = 0$ and $\Delta y = 0$ of the structure function. Thus we are unable to define $D(\Delta x, \Delta y)$ at all possible Δx and Δy. However, from the physical considerations and from the measurements, the main anisotropy axis of the field, $Z(x,y)$, should coincide with our axis x and y. It follows that if the shape of $D(\Delta x, \Delta y)$ is universal, the cross-sections $D(\Delta x, \Delta y = 0)$ and $D(\Delta x = 0, \Delta y)$ will describe the bed elevation structure function completely. Taking into account the above experimental facts sections $D(\Delta x, \Delta y = 0)$ and $D(\Delta x = 0, \Delta y)$ can be successfully parametrised in the form (Figures 1 and 2):

$$D(\Delta x, \Delta y = 0) / D(\Delta x = \infty, \Delta y = 0) = (\Delta x / \Delta x_o)^{2H_x} \text{, when} \quad \Delta x < \Delta x_o \quad (3)$$
$$D(\Delta x, \Delta y = 0) / D(\Delta x = \infty, \Delta y = 0) = 1 \text{,} \quad \text{when} \quad \Delta x > \Delta x_o \quad (4)$$

$$D(\Delta x = 0, \Delta y) / D(\Delta x = 0, \Delta y = \infty) = (\Delta y / \Delta y_o)^{2H_y} \text{, when} \quad \Delta y < \Delta y_o \quad (5)$$
$$D(\Delta x = 0, \Delta y) / D(\Delta x = 0, \Delta y = \infty) = 1 \text{,} \quad \text{when} \quad \Delta y > \Delta y_o \quad (6)$$

From (3)-(6) the gravel-bed roughness properties can be completely described by five parameters: three characteristic linear scales Δx_o, Δy_o, and ΔZ_o plus two scaling exponents H_x and H_y. The scaling exponents H_x and H_y have the sense of fractal Hurst exponents (Feder, 1988) which characterise self-affine properties of bed profiles. They appear to be the same for the investigated river reaches (Figure 2) and this hints at their universality. If it is so, the number of roughness parameters reduces to three linear scales.

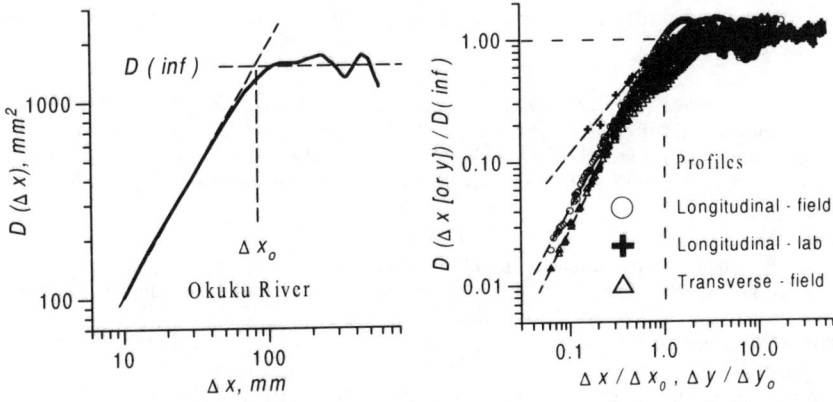

Fig. 1. Example of $D(\Delta x, \Delta y = 0)$. *Fig. 2.* Structure function parametrization.

DISCUSSION

Several questions arise from the above considerations. The first one is why we have difference between scaling exponents for natural gravel-beds and manually created beds. From fractal theory (Feder, 1988) values $H_x=0.67$-0.70 and $H_y=0.73$-0.78 for natural rivers reflect positive persistence of bed elevations while $H_x=0.49$-0.54 for laboratory flumes correspond to the classical case of the trace of Brownian motion (which is random by nature). The possible reply is that rivers try to arrange bed particles in a special, optimal, way using various mechanisms (particle arrangement, sorting, orientation, etc.). This causes special organisation of particles on the natural beds that produces the Hurst exponent sufficiently larger than 0.5. The same reason can be applied to explain possible universality of scaling exponents H_x and H_y. In the case of manually created beds particles are packed randomly and as a result H_x appeared to be close to 0.5. Another important question is how characteristic scales that follow from our model relate to traditional d_n-description of gravel-beds? Our measurements showed the existence of some correlation (not strong) between longitudinal Δx_o and vertical ΔZ_o scales and the shortest and intermediate sizes of particles, respectively. However, to clarify this problem as well as to test independency of Δx_o, Δy_o, and ΔZ_o (they also can be correlated) additional data are needed.

CONCLUSIONS

1. Statistical properties of gravel-bed elevations were examined and a simple model of bed roughness was suggested. The model offers several advantages over conventional d_n-approach since it takes into account directly all bed properties.
2. Additional efforts should be applied to test the model and to examine possible relations between model parameters and traditional particle size approach. Also, the role of bed particle clusters should be clarified. Our research is proceeding on these problems.

Acknowledgments. The research was conducted under contracts CO1614 and CO1519 from the Foundation for Research Science and Technology (New Zealand). The authors are grateful to A. Stokes for design and performance of the bed profiler as well as to S. Brown, P. Mason, C. Kilroy, and K. Eberle for assistance with field measurements and preliminary data processing.

REFERENCES

Feder, J. (1988). *Fractals.* Plenum Press, New York.
Furbish, D.J. (1987). Conditions for geometric similarity of coarse stream-bed roughness. *Mathematical Geology,* 19(4), 291-307.
Monin, A.S., and Yaglom, A. M. (1975). *Statistical Fluid Mechanics.* Vol. 2, The MIT Press, Cambridge.
Nikora, V.I. (1982). Statistical properties of river bed microforms. In: *Problems of Land Hydrology.* Hydrometeoizdat, Leningrad (USSR), 143-151 (in Russian).

Analytical model for hydraulic roughness of submerged vegetation

D. KLOPSTRA, H.J. BARNEVELD, J.M. VAN NOORTWIJK[*],
E.H. VAN VELZEN[**]
[*] HKV_{CONSULTANTS}, Lelystad, The Netherlands
[**] Ministry of Transport, Public Works and Water Management; Institute for Inland Water Management and Waste Water Treatment (RIZA), Arnhem, The Netherlands

ABSTRACT
A new, analytical, physically based, model of the vertical flow velocity profile and the hydraulic roughness of submerged vegetation has been developed. For the vegetation layer and the surface layer, different turbulence models have been applied. Model simulations correspond well with results from flume experiments reported in literature. Because the analytical model includes only one empirical relation, it potentially has a wide range of applicability. Due to lack of data, the model is not yet validated for field conditions.

INTRODUCTION
As a consequence of nature rehabilitation, which has become an important aspect of river management world wide, vegetation characteristics of flood plains are expected to change in the future. In order to assess the effects on river functions such as the safe conveyance of flood waves, it is important to know how the hydraulic roughness of the flood plains will be affected. In this paper, results from studies on hydraulic roughness of vegetation reported in literature are used for development and verification of a physically based model of the vertical flow velocity profile and hydraulic roughness of submerged tall vegetation such as reeds. The emphasis is on an analytical expression, which can be easily incorporated in numerical hydraulic software packages.

VELOCITY PROFILE OF SUBMERGED VEGETATION
The velocity profile of submerged vegetation (illustrated in figure 1) is treated separately for the vegetation layer and the surface layer. The two profiles will be smoothly matched through boundary conditions at the interface.

VEGETATION LAYER
The momentum equation, assuming uniform and steady flow, reads:

$$\frac{\partial \tau(z)}{\partial z} = F_D(z) - \rho \cdot g \cdot i \qquad (1)$$

with: τ = shear stress (kg/ms^2), ρ = density of water (kg/m^3), z = vertical co-ordinate (m), g = acceleration due to gravity (m/s^2), i = energy gradient (-) and in which the drag-force $F_D(z)$ on the vegetation is defined by:

$$F_D(z) = m \cdot D \cdot C_D \cdot \frac{1}{2} \cdot \rho \cdot u(z)^2 \qquad (2)$$

with: m = vegetation elements per m^2 (m^{-2}), D = diameter stem vegetation element (m), C_D = drag coefficient (-), $u(z)$ = flow velocity at level z (m/s).

The turbulent shear stress can be described by the concept of Boussinesq:

$$\tau(z) = \varepsilon \cdot \frac{\partial u(z)}{\partial z} \qquad (3)$$

with: ε = turbulent viscosity (kg/ms) = $\rho \cdot v_t$ and v_t = eddy viscosity (m^2/s).

In conformity with the turbulence models described in e.g. Rodi (1980), v_t is assumed to be characterised by the product of a velocity scale and a length scale of large scale turbulence, which is responsible for the vertical transport of momentum. In conformity with Tsujimoto and Kitamura (1990), the characteristic velocity scale is assumed to be represented by the flow velocity $u(z)$. The characteristic length scale α is assumed to be independent of z. The turbulent shear stress (3) then reads:

$$\tau(z) = \rho \cdot \alpha \cdot u(z) \cdot \frac{\partial u(z)}{\partial z} \qquad (4)$$

The momentum Equation (1) now transforms to:

$$u(z) \cdot \frac{\partial^2 u(z)}{\partial z^2} + \left(\frac{\partial u(z)}{\partial z}\right)^2 = \frac{m \cdot D \cdot C_D \cdot u(z)^2}{2\alpha} - \frac{g \cdot i}{\alpha} \qquad (5)$$

which has the following analytical solution:

$$u(z) = \sqrt{C_1 \cdot e^{-\sqrt{2 \cdot A} \cdot z} + C_2 \cdot e^{\sqrt{2 \cdot A} \cdot z} + u_{s0}^2} \qquad (0 < z < k) \qquad (6)$$

with: k = vegetation height (m) and

$$A = \frac{m \cdot D \cdot C_D}{2\alpha} \qquad (7)$$

$$u_{s0} = \sqrt{\frac{2 \cdot g \cdot i}{C_D \cdot m \cdot D}} \qquad (8)$$

u_{s0} is the characteristic constant flow velocity in non-submerged vegetation, which also follows directly from (5) with all velocity gradients set on zero. Constants C_1 and C_2 in (6) follow from boundary conditions. At the bed ($z=0$) the bottom shear stress is neglected and the flow velocity is assumed to be equal to u_{s0}. At the top of the vegetation layer the boundary condition is determined by the shear stress:

$$\tau(k) = \rho \cdot g \cdot (h-k) \cdot i \qquad (9)$$

with: h = water depth (m)

With these boundary conditions, the following values for C_1 and C_2 are derived:

$$C_1 = \frac{-2 \cdot g \cdot i \cdot (h-k)}{\alpha \cdot \sqrt{2} \cdot A \cdot (e^{k \cdot \sqrt{2 \cdot A}} + e^{-k \cdot \sqrt{2 \cdot A}})} \tag{10}$$

$$C_2 = -C_1 \tag{11}$$

The velocity profile for the vegetation layer is now established. The only unknown parameter is the characteristic length scale α.

SURFACE LAYER

For the surface layer, Prandtl's mixing length concept is adopted resulting in the well known logarithmic flow velocity profile. The virtual bed of such a profile does not coincide with the top of the vegetation but appears to lie at a distance h_s under that level. The flow velocity profile thus can be written as:

$$u(z) = \frac{1}{\kappa} \cdot u_* \cdot \ln\left(\frac{z-(k-h_s)}{z_0}\right) \qquad (k < z < h) \tag{12}$$

with: κ = Von Kármán's constant (-)
h_s = distance between top of vegetation and virtual bed of surface layer (m),
z_0 = length scale for bed roughness of the surface layer (m)
u_* = virtual bed shear stress for the surface layer: $u_* = \sqrt{g \cdot (h-(k-h_s)) \cdot i}$

h_s and z_0 follow from the continuity condition that both actual value and gradient of the flow velocity of the vegetation and the surface layer should be equal at the interface ($z=k$). These conditions result in the following values for h_s and z_0:

$$h_s = g \cdot \frac{1 + \sqrt{1 + \frac{4 \cdot E^2 \cdot \kappa^2 \cdot (h-k)}{g}}}{2 \cdot E^2 \cdot \kappa^2} \tag{13}$$

$$z_0 = h_s \cdot e^{-F} \tag{14}$$

with:

$$E = \frac{\sqrt{2 \cdot A} \cdot C_3 \cdot e^{k \cdot \sqrt{2 \cdot A}}}{2 \cdot \sqrt{C_3 \cdot e^{k \cdot \sqrt{2 \cdot A}} + u_{v0}^2}} \tag{15}$$

$$F = \frac{\kappa \cdot \sqrt{C_3 \cdot e^{k \cdot \sqrt{2 \cdot A}} + u_{v0}^2}}{\sqrt{g \cdot (h-(k-h_s))}} \tag{16}$$

$$C_3 = C_2/i \tag{17}$$

$$u_{v0} = u_{s0}/\sqrt{i} \tag{18}$$

HYDRAULIC ROUGHNESS OF SUBMERGED VEGETATION

From the average flow velocity in the vertical U, which follows from the integrals of (6) and (12), the hydraulic roughness expressed as the value of Chézy (m$^{1/2}$/s) can be obtained via $C=U/\sqrt{(h.i)}$. To integrate (6) analytically, the first term under the square root sign is neglected in comparison to the second term. The impact of this simplification is only noticeable at extremely low vegetation densities for which the model is not developed. The following value of Chézy is then obtained:

$$C = \frac{1}{h^{1/2}} \cdot \left\{ \begin{array}{l} \frac{2}{\sqrt{2 \cdot A}} \cdot (\sqrt{C_3 \cdot e^{k \cdot \sqrt{2 \cdot A}} + u_{v0}^2} - \sqrt{C_3 + u_{v0}^2}) + \\ \frac{u_{v0}}{\sqrt{2 \cdot A}} \cdot \ln\left(\frac{(\sqrt{C_3 \cdot e^{k \cdot \sqrt{2 \cdot A}} + u_{v0}^2} - u_{v0}) \cdot (\sqrt{C_3 + u_{v0}^2} + u_{v0})}{(\sqrt{C_3 \cdot e^{k \cdot \sqrt{2 \cdot A}} + u_{v0}^2} + u_{v0}) \cdot (\sqrt{C_3 + u_{v0}^2} - u_{v0})}\right) + \\ \frac{\sqrt{g \cdot (h - (k - h_s))}}{\kappa} \cdot \left((h - (k - h_s)) \cdot \ln\left(\frac{h - (k - h_s)}{z_0}\right) - h_s \cdot \ln\left(\frac{h_s}{z_0}\right) - (h - k) \right) \end{array} \right\} \quad (19)$$

The hydraulic roughness thus can be calculated analytically when vegetation characteristics (m, D, C_D, k), water depth h and characteristic length scale of large scale turbulence α are known. The only unknown parameter α will be analysed next.

MODEL VERIFICATION

The performance of the analytical model is assessed in two successive steps:
1. Comparison with measured flow velocity profiles from flume experiments by varying the characteristic length scale α in such a way that the shape of the measured velocity profile is represented;
2. Comparison with measured hydraulic roughness values from flume experiments. Experimental results used for this are summarised in Table 1. Equal drag coefficients have been applied for all verification tests: C_D=1.4 for cylinders/reed and C_D=2.0 for strips. These are average values from the papers of Table 1. Figure 1 shows two typical results of the first verification step, for which in total 23 measured flow velocity profiles have been used. This verification step shows that values of α can be selected in such a way that calculated and measured flow velocity profiles are in good agreement. This means that the assumption that α is independent of z does not have to be withdrawn. To make the analytical model generally applicable, α has been correlated to hydraulic and vegetation characteristics, with the following best-fit result (Figure 2):

$$\alpha = 0.0793 \cdot k \cdot \ln\frac{h}{k} - 0.00090 \quad \text{and} \quad \alpha \geq 0.001 \quad (20)$$

With this relation for α, the performance of the analytical model is tested against hydraulic roughness values which follow from the flume experiments of Table 1. The results of this second verification step are shown in Figure 3. Taking into consideration that constant drag coefficients are applied, the performance of the analytical model is good.

	vegetation characteristics			
Paper	shape	m (m^{-2})	D (m)	k (m)
Tsujimoto and Kitamura (1990)*	cylinders	2,500	0.0015	0.0459
Shimizu and Tsujimoto (1994)*	cylinders R	10,000	0.0010	0.041
	A	2,500	0.0015	0.046
Starosolsky (1983)	reed	220	0.0046	0.15|0.25
Tsujimoto, Okada and Kontani (1993)*	cylinders sphere on top	10,000	0.00062 0.003	0.065
Nalluri and Judy (1989)	cylinders 1C	400	0.006	0.15
	2C	200	0.006	0.15
	strips 5B	833	0.005	0.16
	6B	833	0.005	0.16
Kouwen et al (1969)	strips	1,000	0.005	0.10
* including flow velocity profiles				

Table 1: Flume experiments used for model verification.

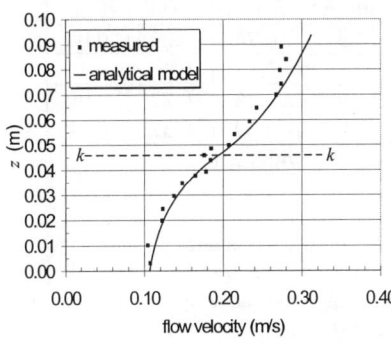
a. Run A31 (Shimizu & Tsujimoto, 1994)

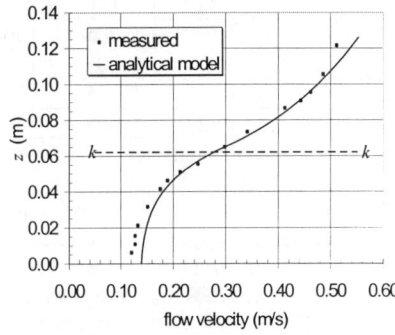
b. Run BZ11 (Tsujimoto et al., 1993)

Figure 1 Measured and calculated flow velocity profiles

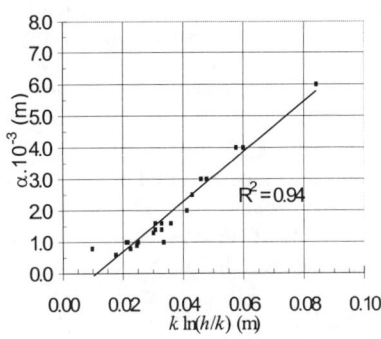

Figure 2 Relation for α

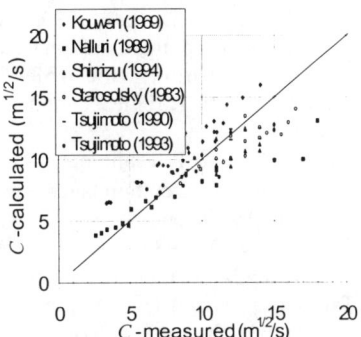

Figure 3 Measured and calculated C- values

DISCUSSION

The new analytical model for the vertical velocity profile and the hydraulic roughness of submerged vegetation appears to be an important step towards a generally applicable analytical model (HKV_{CONSULTANTS}, 1996). Hydraulic roughness values calculated with the model correspond well with results from flume experiments. However, the model can not yet be validated for field situations due to lack of data. Model results for field situations (see Table 2) show that under certain conditions (i) α exceeds the values for which the relation for α was fitted, (ii) the calculated virtual bed level for the surface layer is below the actual bed level (i.e. $h_s > k$) and (iii) the length scale for the bed roughness z_0 is of the order of magnitude of h_s. This, in combination with the resulting low Chézy-values, illustrates the need for additional research on the validity of the modelling concepts, as well as the relation for α. This should be combined with a profound field measurement program (or large scale flume experiments) so as to validate the study results.

Test	Input parameters				Calculated parameters			
	h (m)	k (m)	D (m)	m (m^{-2})	C (m$^{1/2}$/s)	h_s (m)	z_0 (m)	α (m)
1	5.0	0.5	0.005	100	17.5	0.74	0.26	0.09
2	5.0	2.0	0.005	100	8.7	1.14	0.46	0.14
3	5.0	0.5	0.005	500	16.9	0.46	0.22	0.09
4	5.0	2.0	0.005	500	7.4	0.69	0.37	0.14

Table 2 Results analytical model for field situations with reed.

REFERENCES

HKV_{CONSULTANTS}, 1996: Analytical model of hydraulic resistance of submerged vegetation (In Dutch). Commissioned by Rijkswaterstaat/RIZA, The Netherlands.

Kouwen, N., T.E. Unny and H.M. Hill, 1969: Flow retardance in vegetated channels. ASCE. Journal of the irrigation and drainage division. Proceedings of the American Society of Civil Engineers. June 1969.

Ministry of Transport, Public Works and Water Management, 1996: State of the art resistance of vegetation (to be published); RIZA, Arnhem, The Netherlands.

Nalluri, C. and N.D. Judy, 1989: Factors affecting roughness coefficient in vegetated channels. Proc. Int. Conf. on Channel Flow and Catchment Runoff. University of Virginia, Charlottesville; p 589-598, VA.

Rodi, W., 1980: Turbulence models and their applications. State-of-the-art paper. IAHR.

Shimizu, Y. and T. Tsujimoto, 1994: Numerical analysis of turbulent open-channel flow over a vegetation layer using a k-ε turbulence model. Journal of hydroscience and hydraulic engineering. Vol. 11, no 2. January 1994.

Starosolsky, Ö., 1983: The role of reeds in the shaping of currents. Proceedings of the 20th International Association for Hydraulic Research. Moscow.

Tsujimoto, T. and T. Kitamura, 1990: Velocity profile of flow in vegetate bed channels. Progressive Report June 1990. Hydr. Lab. Kanazawa University.

Tsujimoto, T, Okada, T and K. Kontani, 1993: Turbulent structure of open-channel flow over flexible vegetation. Progressive Report. December 1993. Hydr. Lab., Kanazawa University.

1-D or 2-D Models for River Hydraulic Studies?

R. WALTON, J.B. BRADLEY and T.R. GRINDELAND
VP, CEO, and Senior Engineer, WEST Consultants, Inc., Seattle, WA, USA

ABSTRACT

Model selection is usually straightforward if site conditions are known and the purpose and tolerances of the study understood. However, there are occasions when the choice is not so clear. At two sites in Washington State, we have applied both one- and two-dimensional models to examine hydraulic aspect of bridge scour and bank stabilization. Comparisons of the model applications illustrate their differences and provide insight into the basis for model selection. We also we consider whether 'neglected' physical processes can be estimated from the results of one-dimensional models, and how useful these calculations might be.

INTRODUCTION

Most hydraulic modelers would select a one-dimensional model, such as HEC-2 or HEC-RAS, to simulate steady flows in narrow, uniform streams, and two-dimensional models, such as RMA-2 or FESWMS, to simulate steady flows in wide, shallow rivers. It is clear that there are times when a one-dimensional model is quite adequate and gives reasonable results, and other occasions when multi-dimensional models are required. But there are applications where the choice is not immediately clear. Two-dimensional models are usually selected when there is a need to either resolve the lateral distribution of a system variable (such as head, velocity, or a constituent) or the system is considered to be too complex to be studied using a one-dimensional model.

As computers become more powerful, and graphically-based modeling systems become more 'user friendly', hydraulic modelers have increasingly more sophisticated analytical tools available. However, the modeler, often less well-trained in 'modeling' but better trained with 'computers', is faced with the difficult task of deciding what type of model is more appropriate for a given application, as both types 'appear' easy to use. During the course of evaluating scour and bank stabilization at many bridges throughout the Pacific Northwest, the selection of an appropriate model has usually been clear. However, at two sites in Washington State, we have applied both types of models, and can consider what we gained by using two-dimensional models.

SAUK RIVER MODEL

On the Sauk River in Skagit County, Washington, flow in a secondary channel is attacking the left bank immediately upstream of a major highway crossing, endangering the bridge and requiring quick implementation of protective measures. A range of options to stabilize the left bank upstream of the bridge were developed, and 'soft' solutions quickly dropped because of the very high velocities (up to 5-6 m/sec) attacking the left abutment during large floods. To evaluate various 'hard' engineering alternatives, the two-dimensional, finite-element, hydrodynamic model, RMA-2 (run under FastTABS), was used to model the 100-year flow in the bridge reach. Using the usual amount of survey data for bridge scour and bank protection assessments, the FastTABS system was used to quickly develop a hydrodynamic model to simulate the circulation patterns and near-bank velocities for a number of bank stabilization options, including spurs, groins, and bank guides.

The one-dimensional model, HEC-2, was also run, originally to develop downstream starting heads, but then extended through the bridge reach. Figure 1 shows the steady-state water surface elevations and velocity vectors from the two-dimension model for existing bridge conditions. Figure 2 compares the along-river water surface elevations, and the lateral velocity distribution at a section through the secondary flow channel using a conveyance-based post-processing of the HEC-2 model results (Walton and Bradley, 1995). The results show that while along-river stages compare well, the one-dimensional model does not reproduce the velocities in the secondary flow channel. As the aim of the study was to consider various mitigation alternatives to control the velocities in the secondary flow channel, it seems clear that the selection and use of the two-dimensional model was appropriate. In addition, the two-dimensional model (Figure 1) shows that there is superelevation of about 0.4 m, which is important when considering the vertical extent of bank protection.

CISPUS RIVER MODEL

Major flooding during February 1996 undermined the central pier of the Tom Music Bridge over the Cispus River in Lewis County, Washington. The bridge dropped six feet, and significant erosion was seen upstream of the right abutment, including channel migration to the right and bar formation at the left bank. The two-dimensional finite-element model, FESWMS, was applied in the modeling system SMS to evaluate the hydraulic and scour performance of a proposed replacement bridge, and to estimate riprap sizes to protect the banks from further erosion and migration.

The one-dimensional model, HEC-RAS, was also run, first to compare the hydraulics of two replacement bridge configurations, and second to set downstream stages for the two-dimensional model. Figure 3 shows the steady-state water surface elevations and velocity vectors from the two-dimension model for existing bridge conditions. Figure 4 compares the along-river stages and the lateral velocity distributions at a section upstream of the bridge across the major bar that has formed along the left bank. There is good agreement in the longitudinal distribution of stages, and fairly good agreement in current speeds except that components of velocity parallel to the cross section can be seen in the two-dimensional model but not in the one-dimensional model. Upstream of the bridge, the two-dimensional model shows superelevation at the right bank of about 0.3 m. This is important because high water marks from the February 1996 flood indicated that the flow (considered to exceed the 200-year event) came close to, but did not overtop a road running parallel to the left bank.

COMPARISON OF STUDIES

The purpose of the Sauk River study was to evaluate various bank protection measures along the left bank upstream of the bridge. To do this, the two-dimensional model proved very useful as some of the alternatives, such as bank guides and spurs, significantly changed the flow characteristics in the secondary flow channel. These differences, which were crucial in evaluating the alternatives, would have been extremely difficult to configure with a one-dimensional model.

The purposes of the Cispus River study were (1) to estimate contraction scour at the proposed replacement bridge, and (2) to size riprap to protect both abutments and the upstream right bank. Contraction scour is calculated using velocity and depth information in the bridge opening and about one bridge length upstream. When the results from the one- and two-dimensional models were compared, they were quite similar, and given the relatively large uncertainty in calculating contraction scour, the values from either model would have been appropriate. The same was true for calculating riprap sizes.

Superelevation, on the order of 0.3-0.4 m, was seen in the results of the two-dimensional models. If determining water surface elevations had been the purpose of the study, an initial assessment might have been that the two-dimensional model provided information not available from the one-dimensional model results. However, superelevation can be estimated analytically using (Woodward and Posey, 1941):

$$\Delta y = C v^2 W / g r$$

where Δy is the superelevation measured from the centerline elevation, C is a correction factor (close to 1 in natural streams), v is the average velocity, W is the channel width, g is the acceleration due to gravity, and r is the radius of curvature. Using the one-dimensional model results, the analysis gave estimates of 0.4 m for the Sauk River in the secondary flow channel, and 0.3 m for the approach to the Cispus River bridge. The results compare well for both systems.

DISCUSSION

Over the past 10 years, there have been significant improvements in numerical models. Much of this has been to improve the ease with which models can be set up through graphical interfaces. In river hydraulics, recent examples are the replacement of HEC-2 with HEC-RAS, and the use of RMA-2 and FESWMS under FastTABS and then SMS. However, some of these 'improvements' can be misleading. There is now often the tendency to select a two-dimensional model simply because two dimensions must be better than one! This is often accompanied by an increase in cost. But are two-dimensional models always better or is it sometimes lack of experience? In addition, there may be no data, or only post-event indications such as 'estimated' flows or high-water marks, for model calibration.

The Sauk and Cispus River models were both very easy to set up in the graphical interfaces, with all the tools available for editing and data creation. However, both two-dimensional models proved to be extremely difficult to run for sites with high Froude numbers (approaching 0.8) and complex geometries with locally very steep gradients. One difficult aspect was proceeding from a 'cold start' in which the water surface was initially horizontal to a steady-state solution in which the downstream water surface elevation was 4-5 m lower. At both sites, the one- and two-dimensional models gave similar velocities and depths if either scour or at-bank riprap was being estimated. The main advantage of the two-dimensional models was their ability to evaluate complex situations like flow interactions with 'hard' structures such as bank guides and spurs on near-bank velocities, and the formation of eddies at large changes in channel width. Even lateral differences in water surface elevations can be roughly estimated using analytical methods, and the use of local conveyance calculations to estimate lateral velocity distributions (also a post-processing of one-dimensional model results) generally give reasonable distributions for use in scour and riprap calculations.

Two-dimensional models are very useful tools, but one-dimensional models, properly used with additional analytical calculations, can often give very reasonable engineering answers. More than ever, it is important to not only understand the site conditions, the purpose of the study, and the order of accuracy implied in the final result, but also how and under what conditions these models work. If all this is understood, it should be easier to select the right model for the job.

REFERENCES

Walton, R. and J.B. Bradley, (August 1995), "HEC-2 Modifications for Bridge Scour Analyses", ASCE, Water Resource Engineering Division Specialty Conference, San Antonio, TX.

Woodward, S.M. and C.J. Posey, (1941), Hydraulics of Steady Flow in Open Channels, Wiley, N.Y.

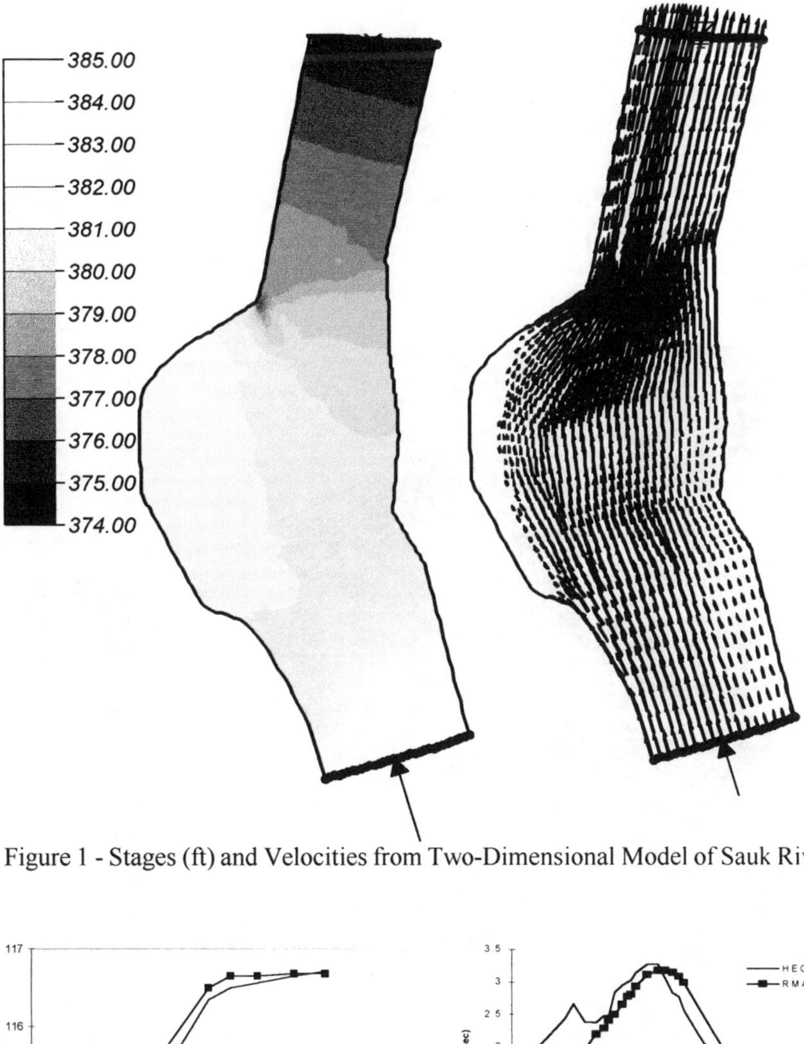

Figure 1 - Stages (ft) and Velocities from Two-Dimensional Model of Sauk River

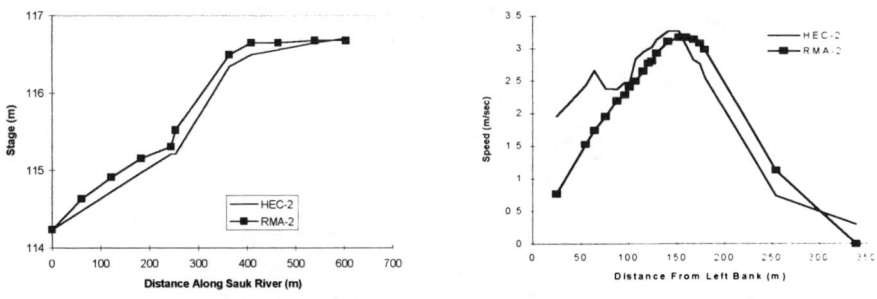

Figure 2 - Comparison of Stages and Speeds for Sauk River Models

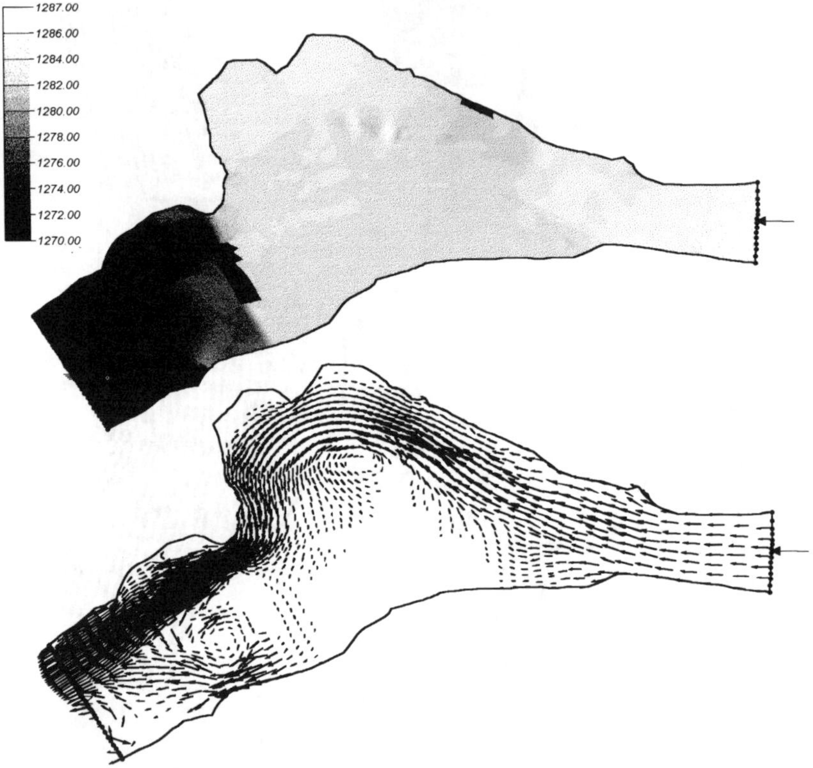

Figure 3 - Stages (ft) and Velocities from Two-Dimensional Model of Cispus River

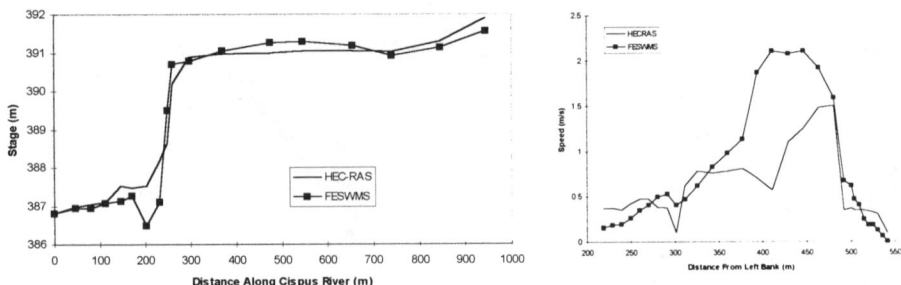

Figure 4 - Comparison of Stages and Speeds for Cispus River Models

Steady Flow over an Obstacle with Contraction and Sill

ADRIAN W.K. LAW
School of Civil and Structural Engineering
Nanyang Technological University
Nanyang Avenue, Singapore 639798

ABSTRACT

This paper investigates the aspect of one-dimensional modeling of the centrifugal effect on flow over an obstacle with simultaneous geometric changes in both the bottom and sidewall boundaries. Experiments were performed to provide laboratory measurements on the water surface profile along the obstacle. The measurements were compared to the results of a numerical analysis based on the control volume approach by Fenton (1996).

INTRODUCTION

The St. Venant long wave equation holds a central role in practical engineering analysis for open channel flow. Its one-dimensional simplicity allows direct program implementation for desktop computing. In the derivation of the equation to its simple form, hydrostatic pressure distribution has been assumed. The assumption is satisfactory in most cases. However, in some rapidly varying flows such as transcritical flows, i.e. where the flow regime transits from subcritical to supercritcal, it is well known that the hydrostatic assumption is not valid. In this case, the centrifugal dynamic pressure due to streamline curvature needs to be accounted properly to obtain physically meaningfully results.

There were many attempts in the past to incorporate this centrifugal effect into one-dimensional flow profile simulation. To retain the one-dimensional simplicity, approximations are required to express the three-dimensional streamline movement in terms of some bulk parameters such as the average velocity and the average water surface elevation across a cross-section of the flow. Several formulations are available. Dressler (1978) used a curvilinear coordinate system that follows the bottom boundary to formulate the problem. Through the curvilinear coordinate system, the centrifugal pressure is inherently incorporated. The curved-flow Dressler equation was verified experimentally in a two-dimensional spillway flow in

Sivakumaran and Yevjevich (1987). However, the equation does not appear to allow the flow regime to change from subcritical to supercritical state (Fenton, 1996). Naghdi and Vongsarnpigoon (1986) used the approach of directed Corserect surface, and obtained an equation for a two-dimensional fluid sheet in Cartesian coordinate after linear approximations. The resulting equation was shown to be capable of describing the flow over a two-dimensional obstacle, including a possible undular hydraulic jump downstream. Recently, Fenton presented an alternative approach based on control volume in Cartesian coordinate.

A number of verification experiments were performed in the past that allow the examination of the applicability of the different formulations. Nearly all experiments of this type utilize two-dimensional bed curvature. However, for most open channel flows, it is common to have continuous geometric changes in both the bottom and sidewall boundaries. In order to yield an improved long wave equation that incorporates the centrifugal effect for practical usage, it is necessary to consider this simultaneous changes of flow boundaries in both directions. The objective of this paper is to evaluate the centrifugal correction through an experimental study that involves flow over an obstacle with a combined bottom sill and side contraction. The results should represent a more general assessment on the correction approach than based on only changes in bottom curvature.

ANALYSIS

Among the different approaches described previously in the introduction, Fenton (1996) formally incorporated the possible variations in the channel width. The following describes his approach and the simplifications that can be applied to the present case.

Fenton proposed a 1-D simplified long wave equation based on the consideration of momentum in a control volume approach:

$$\frac{\partial Q}{\partial t} + \frac{\partial}{\partial x}\left(\frac{Q^2}{A}\right) + \frac{1}{\rho}\int_A \frac{\partial p}{\partial x} dA + gAS_f = 0 \qquad (1)$$

whereby the centrifugal pressure is included to the first order as a constant body force over a cross-section as

$$\frac{1}{\rho}\frac{\partial p}{\partial z} = -g - \overline{\alpha}\frac{Q^2}{A^2} \qquad (2)$$

where $\overline{\alpha} = \dfrac{\kappa}{\cos\theta}$ with κ being the streamline curvature and θ the local angle of inclination. Fenton suggested to approximate $\overline{\alpha}$ as

$$\overline{\alpha} = \omega_o \frac{h_s''}{1+h_s'^2} + \omega_1 \frac{H''}{1+h_s'^2} \quad ; \quad \frac{\partial \overline{\alpha}}{\partial x} = \omega_o \frac{h_s'''}{1+h_s'^2} + \omega_1 \frac{H'''}{1+h_s'^2} \tag{3}$$

where H is the mean water depth over the cross section, h_s is the bottom elevation, and ω_o and ω_1 are both constants to be evaluated that represent the contribution to the average "vertical" streamline curvature from the channel bottom and the water surface respectively. For transcritical flow over a sill, he reported that the formulation suffers parasitic numerical instability which reduces the flow depth to zero at some point past the top of the sill in a two-dimensional setting.

Here the main interest is to examine this approach in the presence of simultaneous changes in both the bottom and sidewall geometry. For steady flow in a rectangular channel with changing width and bottom elevation, Equation (1) can be simplified to

$$Q^2 \frac{\partial}{\partial x}\left(\frac{1}{A}\right) + \frac{1}{\rho}\int_A \frac{\partial p}{\partial x} dA + gAS_f = 0 \tag{4}$$

thus isolating the effect of the centrifugal pressure. Integrating (2) and then substituting into (1) and after some manipulation, we get

$$\frac{\partial \overline{\eta}}{\partial x}\left(gA + \overline{\alpha}\frac{Q^2}{A}\right) - \left(1+2\overline{\alpha d}\right)\frac{Q^2}{A}\frac{\partial A}{\partial x} + \overline{d}\frac{Q^2}{A}\frac{\partial \overline{\alpha}}{\partial x} = 0 \tag{5}$$

where $\overline{\eta}$ is the mean water surface elevation, and $\overline{d}$ is the depth of the centroid of the flow cross-section below the mean water surface. For rectangular section $\overline{d}$ is equal to $H/2$. The mean water surface elevation can be expressed as $\overline{\eta} = H + h_s$. Note that this expression for $\overline{\eta}$ is slightly different that used in Fenton (1996). Equation (5) can be further manipulated to yield the following third order differential equation:

$$\frac{d^3H}{dx^3} - \frac{2}{H}\left(S_o + \frac{HB'}{B}\right)\frac{d^2H}{dx^2} - \left(\frac{1+h_s'^2}{\omega_1}\right)\left(\frac{2}{H^2}\right)\left(\frac{1-Fr^2}{Fr^2}\right)\frac{dH}{dx} - \left(\frac{\omega_0}{\omega_1}\right)\left(\frac{2h_s''B'}{B} - h_s'''\right)$$
$$+ \frac{2}{H}\left(\left(\frac{\omega_0}{\omega_1}\right)S_o h_s'' + \left(\frac{1+h_s'^2}{\omega_1}\right)\frac{B'}{B}\right) - \frac{2}{H^2}\left(\frac{1+h_s'^2}{\omega_1}\right)\left(\frac{S_o - S_f}{Fr^2}\right) = 0 \tag{6}$$

where S_o is the bed slope = $-h_s'$. Equation (6) becomes the governing equation for one-dimensional steady flow computation with simultaneous changes in bottom and sidewall geometry. The values of ω_0 and ω_1 remain unclear. As discussed in Fenton

(1996), logically ω_0 represents the combined effect of the contribution to the curvature from the bed as well as the effect of the bed in determining the surface elevation, and should have a value of close to 1.0, whereas ω_1 represents the contribution to the curvature from the surface and should have a value close to 0.5.

EXPERIMENTS

Laboratory experiments were performed in a 16 m long recirculating flume with glass wall and constant head supply upstream. An obstacle consisting of both contraction and sill is constructed with dimensions shown below:

$$b = b_o\left(1 - 0.4\cos^2\left(\frac{x}{3h_{so}}\right)\right) \quad ; \quad |x| \le 1.5\pi h_{so} \quad (7a)$$

$$h_s = h_{so}\left(0.05 + \cos^2\left(\frac{x-S}{3h_{so}}\right)\right) \quad ; \quad |x - S| \le 1.5\pi h_{so} \quad (7b)$$

where x is zero at the maximum contraction, and h_{so} = 60 mm. With the obstacle, the width of the channel is reduced to a minimum of 94 mm, 60% of the upstream width b_o=156 mm, while the bottom elevation is elevated to a maximum h_{smax} of 63 mm. The distance between the maximum contraction and the maximum height of the sill is 200 mm. Note that h_{so} is not equal to the maximum height of the sill due to a correction of 0.05 h_{so} for the bottom mounting plate of the obstacle. Also it should be noted that the reduction in width was achieved with symmetric protrusion from both channel sidewalls. The obstacle was made of plexi glass to minimize friction losses.

A total of seven experiments is performed. In each experiment, water surface elevation at the middle of the channel section was accurately measured over the obstacle, with the longitudinal resolution of the point measurements being 2 cm. The variation of the water surface elevation across a section was also examined but it is typically small compared to the layer thickness, and was in general less than 2 mm.

In designing the obstacle, the radius of bed curvature is made sufficiently large. The minimum radius of bed curvature is 0.27 m, larger than the typical flow depth over the obstacle which is in the order of 0.1 m. In addition, to minimize the effect of separation from the side wall downstream of the maximum contraction, the rate of change in width was made sufficiently small to prevent a sudden expansion. The minimum radius of curvature along the side wall is 0.54 m. The contraction is staggered ahead of the sill also for the consideration of preventing stronger separation from the side wall with supercritical flow.

RESULTS

Numerical computation was implemented on (6) based on the fourth order

Runge-Kutta numerical scheme. The friction slope S_f was simulated using the Manning's equation with a roughness of $n = 0.01$ for the glass walls. Similar to Fenton, the numerical results suffer strong parasitic numerical instability. In most cases, the computed water depth becomes zero at some point past the critical control.

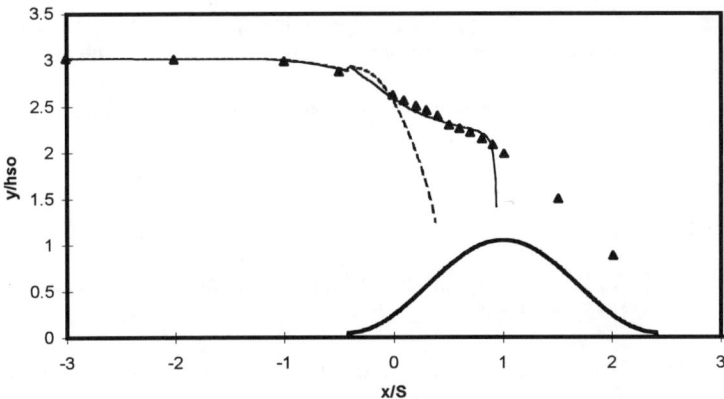

Figure 1. Comparison between the numerical results and the laboratory measurements (▲). The dotted line represents results for $\omega_o = 1.0$ and $\omega_l = 0.5$. The solid line is for case of $\omega_o = 0.2394$ and $\omega_l = 0.01$.

The comparison of numerical and experimental results is shown in Figure 1 for one of the tests with $Q = 0.00962$ m^3/s. Two different sets of values for ω_o and ω_l (first set: $\omega_o = 1.0$ and $\omega_l = 0.5$; second set: $\omega_o = 0.2394$ and $\omega_l = 0.01$) are presented. The first set approximates what is anticipated in a logical manner. However, the numerical results deviate significantly from the laboratory measurements even before the control point. The second set is what yields the best agreement between the computed results and the measurements up to slightly beyond the control point. The smaller values in the second set suggests that the contribution to the pressure distribution from the centrifugal effect is significantly less than what is anticipated. It also indicates that the contribution from the curvature of the surface elevation is minimal compared to the bottom elevation. However, the results are preliminary at this stage due to the presence of the parasitic instability. It should be noted that the present profile of the bottom geometry exhibits a discontinuity of h_s and h_s'' at the beginning of the sill. This discontinuity will have some effect on the numerical results based on the fourth order Runge Kutta scheme. However, it does not appear that the effect is significant enough to override the preliminary observations obtained above.

SUMMARY

The laboratory measurements provide the necessary information for the verification of the available one-dimensional simulation approaches, in the case of steady flow over an obstacle with simultaneous changes in sidewall and bottom geometry. In this paper, we have examined the comparison between these measurements with the Fenton's approach. The preliminary observation is that the centrifugal effect is significantly less than what is anticipated, and that the curvature effect from the surface elevation is minimal compared to the bottom elevation. However, the observation is preliminary at this stage due to the strong parasitic instability in the numerical computation.

REFERENCES

Dressler, R.F., 1978. "New nonlinear shallow-flow equations with curvature." J. Hydr. Res, Vol.16, No.3, pp.205-222.

Fenton, John D. 1996. "Channel flow over curved boundaries and a new hydraulic theory." Preprint, 10th Congress IAHR/APD, Langkawi, Malaysia.

Law, Adrian W.K., 1985. "Single layer flow over an obstacle consisting of a contraction and a sill." Report No. UCB/HEL-85/06 Hydraulic Engineering Laboratory, University of California at Berkeley.

Naghdi, P.M. and Vongsarnpigoon, L. 1986. "The downstream flow beyond an obstacle." J. Fluid Mech., Vol. 162, pp. 223-236.

Sivakumaran, N.S., Tingsanchali, T., and Hosking, R.J. 1983. "Steady shallow flow over curved beds." J. Fluid Mech., Vol. 128, pp.469-487.

Sivakumaran, N.S. and Yevjevich, V. 1987. "Experimental verification of the Dressler curved-flow equations." J. Hydr. Res., Vol. 25, No.3, pp.373-391.

LATERAL VELOCITY VARIATIONS IN A COMPOUND CHANNEL
- A Practical Approach -

TU H., S. TAKAKI and H. TAKAMATU
*Dept. of Water Res. & Hydr. Engrg., Pacific Consultants Co. Ltd.
Dai-ichi Seimei Buldg., Nishi Shinjyuku, Tokyo 163-07, Japan*

ABSTRACT
Presented in this paper is a simple approach for evaluating the lateral distribution of the depth-averaged velocities in a compound channel. Results obtained by the proposed method agree reasonably well with those from experiments and numerical simulations.

1 INTRODUCTION
In Japan, many large and small rivers flow through, or nearby, population centers where there is a severe shortage of land for commercial, residential and industrial uses. The local residents and municipalities naturally turn to the rivers' floodplains, if any, as their precious recreation open spaces. Proper planning and design, however, are prerequisites for reclaiming floodplains. Though the topic has since the sixties (Sellin 1964) drawn the interests of many researchers (Tu et al.1995), it may still take many years of researches and other combined efforts before all the problems could be solved. Before that, however, practicing engineers need some approximations and practical tools to carry out their immediate engineering design work.

One example is how to determine the lateral distribution of the depth-averaged velocity in a compound channel. The present study, based on past research results, proposes a simple, practical approach which is verified with both numerical simulation results and experimental data. The proposed method can be easily used once are given the geometry, the roughness coefficients, the discharge and the slope.

2 PREVIOUS INVESTIGATIONS
Traditionally, compound channels were treated by arbitrarily dividing the compound cross section into relatively large homogeneous and easier to analyze sub-areas. In Table 1 are summarized some of these methods for the determination of a compound channel's discharge. Among these, the single-section method takes the cross section as a whole in calculating the hydraulic radius, leading to an over-estimation of the flow discharge, as first observed by Sellin in 1964 (see also Lambert and Sellin 1996).

Of the divided-channel methods, in both case ① b) and case ① c) one does not consider the interfaces as part of the wetted perimeter, assuming that there exists no shear stress at the interfaces as indicated by the dotted lines. Therefore, they may be seen as two simple examples of the Zero-shear stress interface method, ②. On the other hand, in case ① a), the interfaces ci, fj need to be counted into the wetted perimeters of the sub-divided sections. This may be explained as the follows: the smaller the floodplain depth, the more important would be the flow interactions at interfaces ci, fj. Also for this very reason, interfaces ci, fj may be neglected in case ① b). In the present paper, it is based on the apparent-shear stress concept that will be studied the lateral variations of the depth-averaged velocities.

Table 1 A Summary of Divided-Channel Methods

Method	Details	Note
Single-section method	$Q = AU = A\dfrac{1}{n}R^{2/3}I^{1/2}$	The discharge would be over-estimated (Sellin 1964)
① traditional method — Divided-channel method	a) ci, fj are also taken as wetted perimeter	when h_f is relatively small (Posey 1967)
	b) ci, fj are not taken as wetted perimeter	when $h_f \geq 0.5 h_m$ (Posey 1967)
	c) ck, fk are not taken as wetted perimeter	
② zero-shear interface method	After determining the so-called zero-shear stress interface, divide the floodplain from the main channel	It is difficult to determine the location of the zero-shear stress interface
③ Apparent shear method	Use a semi-empirical relation for the shear in the interface between the main channel and the floodplain	Currently the most widely used method

It is also to be noted that the method discussed in this paper is meant as a practical tool for practicing engineers. To obtain the detailed flow structures in compound channels, such efforts as model study or numerical simulations will be necessary (see for example, Krishnappan and Lau 1986, Keller and Rodi 1988, Knight and Shiono1989, etc.).

3 A PRACTICAL APPROACH

Consider part of a compound channel from its center line (CL), as in Fig.1. The other part may be treated in a similar fashion. One has: the main-channel and floodplain depths, h_m and h_f; their respective width, B_m and B_f (note that B_m here is half of the total main channel width); roughness coefficient, n_m and n_f; hydraulic radius, R_m and R_f; sub-sectional mean velocity without the interaction, U_{mo} and U_{fo}, or with interaction, U_m and U_f; slope, I; lateral coordinate, z, with the intersection of the floodplain and the main channel as its origin (Fig.1); and the depth-averaged velocity, U(z).

Note that $b_e = b_1 + b_2$ is the width of the region where the main channel flow interacts with the floodplain flow. Rajaratnam and Ahmadi (1981) proposed that:

$$b_1 = \frac{2}{\sqrt{3}} b_m \qquad b_m = 3.78 h_{mf} \quad (1)$$

$$b_2 = 2.5 b_f \qquad b_f = 2.5 h_f \quad (2)$$

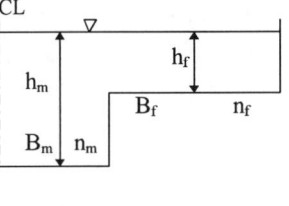

The sub-sectional mean velocities without the interaction are given by the Manning relation as:

$$U_{mo} = \frac{1}{n_m} R_m^{2/3} I^{1/2}$$

$$U_{fo} = \frac{1}{n_f} R_f^{2/3} I^{1/2}$$

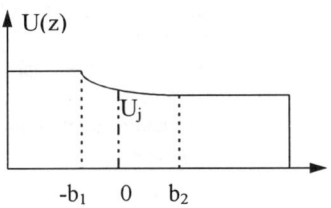

Fig.1 Definition Sketch

However, due to the interaction between the main-channel and floodplain flows, there exists an apparent shear. Taking this added shear into the momentum equation, one may derive (see for example, Tamai 1992) the true sub-sectional mean velocities, as:

$$U_m = U_{mo} \sqrt{1 - (b_e / B_m)(1 - \lambda)} \qquad (3)$$

$$U_f = U_{fo} \sqrt{1 + (b_e / B_f)(R_m / R_f)(1 - \lambda)} \qquad (4)$$

where U_{mo}, U_{fo} and b_e, etc., are as defined above; while λ, an attenuation factor of the bed shear stress in the interaction region, is proposed by Nicollet and Uan (1979):

when $R_f/R_m < 0.3$

$$\lambda = \left[\frac{1-\omega}{2}\cos(\frac{\pi}{0.3}\frac{R_f}{R_m}) + \frac{1+\omega}{2}\right]^2 \quad (5)$$

when $R_f/R_m \geq 0.3$

$$\lambda = \omega^2 \quad (6)$$

with

$$\omega^2 = 0.81(\frac{C_f}{C_m})^{1/3}(\frac{R_m}{R_f})^{1/18} \quad (7)$$

or, in case of large aspect ratio $\quad \omega = 0.9(n_m/n_f)^{1/6}$

Further, based on experimental results, Rajaratnam and Ahmadi (1981) suggested an empirical relation (not shown here due to space limit) for the longitudinal velocity's variations in both the transverse and vertical directions. By integrating that equation in the vertical direction, we obtained the depth-averaged velocity as (see Fig.1):

at $-b_1 \leq z \leq 0$:
$$U(z) = \left[1 - 0.75(\frac{z+b_1}{b_m})^2\right](U_m - U_j) + U_j \quad (8)$$

at $0 < z \leq b_2$:
$$U(z) = (U_j - U_f)\exp\left[-0.693(\frac{z}{b_f})^2\right] + U_f \quad (9)$$

with $\quad U_j = 0.27(U_m - U_f) + U_f \quad (10)$

Equations 8, 9 and 10, together with Eqs.1 to 7 may be used for calculating the lateral variations of the depth-averaged velocity in a compound channel. The procedures are the follows: first prepare the basic data; then determine the width of the interaction zone with Eqs.1 and 2; subsequently obtain the sub-sectional mean velocities using Eqs.3 and 4; and finally calculate the depth-averaged velocity with Eqs.8, 9 and 10. All this can be done conveniently with spread sheets (e.g., Excel).

4 VERIFICATIONS

In this section we shall use the above procedures to evaluate the depth-averaged velocity's lateral variations, and compare the obtained results with those from either experiments or numerical computations. Shown here are three test cases: the first uses experimental data by the Public Works Research Institute (1981); the second and third take the simulation results of the Lower Ara River, Japan.

Compared in Fig.2 are the results from the proposed approach and those from tests in a 3m wide, asymmetrical compound channel. The agreement is rather good. In Fig.3 are shown the lateral velocity variations in a straight compound channel, of the lower Ara River. Again the agreement is satisfactory. The last case is for a cross section in a meandering reach (Fig.4, the radius being $r_c = 3300m$), also of the lower Ara River. According to the US Army Corps of Engineers (see Maynord 1995), the velocity in the outer bank of channel bends is given by: $U(z)_{outer}/U(z) = c_v = 1.74 - 0.52 \, log(r_c/B)$, and $c_v = 1$ if $r_c/B > 26$, where $U(z)$ is the equivalent in a straight channel, and B is the total channel width. It is seen that this relation should be used (Fig.4). However, results (not

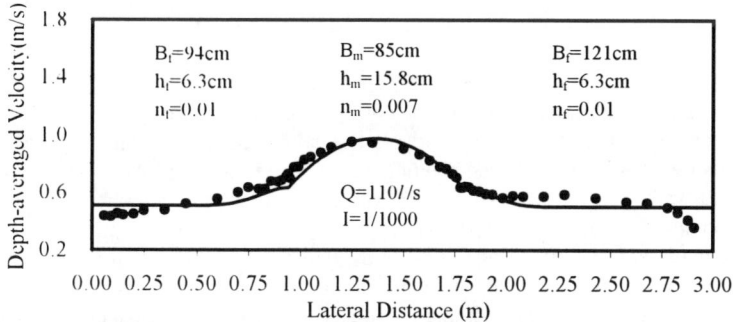

Fig.2 Verification of the Proposed Method (—), Case 1: with Experimental Data (•)

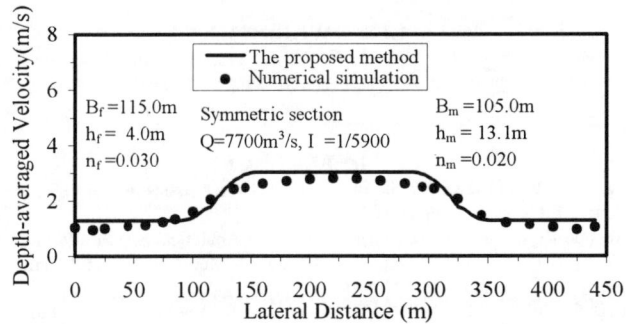

Fig.3 Verification of the Proposed Method, Case 2: with Numerical Simulation Results; in a Straight Reach of the Ara River

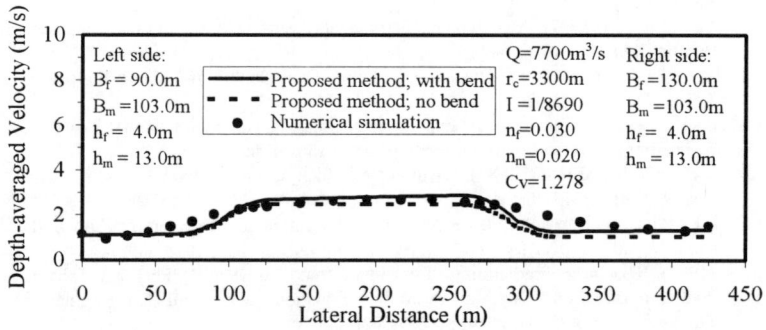

Fig.4 Verification of the Proposed Method, Case 3: with Numerical Simulation Results; in a Meandering Reach of the Ara River

shown here) from succeeding river bends, during large floods, have shown that both the outer and the inner bank velocities should be modified by the Corps' formula.

5 CONCLUSIONS

Floodplains are important in Japan, not only for discharging floods, but also in providing recreation areas for the local population, who recently are claiming riverfronts rich in nature. However, with tree planting, for example, the floodplain roughness, and thus the water elevation, would increase. In order to quickly evaluate the corresponding hydraulic effects, engineers need, among others, a practical tool to estimate the lateral variations of the depth-averaged velocity in a compound channel, without resorting to numerical simulation which would take much more time and money. This paper proposes a simple approach, which is verified using experimental and numerical computation results. The method has been proposed as a tool for the lower Ara River, which the local municipalities may find helpful in their river planning and administration.

ACKNOWLEDGMENT

Our gratitude goes to Mr. T. Watanabe for his help in processing the data, and to the Technology Research Center for Riverfront Development, as well as the Ministry of Construction's Administration Office of the Lower Ara River, for their kind support in this investigation.

6 REFERENCES

Keller, R. J. and W. Rodi (1988), "Prediction of Flow Characteristics in Main Channel/Floodplain Flows", J. Hydr. Res., 26(4), pp.425-441.

Knight, D. W. and K. Shiono (1989), "Two Dimensional Analytical Solution for a Compound Channel", Proc. 3rd Int. Symp. on Refined Flow Modeling and Turbulence Measurements", Tokyo, Japan, pp.503-510.

Krishnappan B. G. and Y. L. Lau (1986), "Turbulence Modelling of Floodplain Flows", J. Hydr. Eng., ASCE, 112(HY4), pp.251-266.

Lambert, M. F. and R. H. J. Sellin (1996), "Discharge Prediction in Straight Compound Channels Using the Mixing Length Concept", J. Hydr. Res., No.3, pp.381-194.

Maynord, S. T. (1995), "Gabion-Mattress Channel -Protection Design", J. Hydr. Engr., Proc. of ASCE, Vol.121, No.1, pp.519-522.

Nicollet, G. and M. Uan (1979), "Écoulements Permanents à Surface Libre en Lits Composés", La Houille Blanche, No.1, pp.19-30.

Posey, C. J. (1967), "Computation of Discharge Including Over-Bank Flow", Civil Engineering, ASCE, April, pp.62-63.

Public Works Research Institute (1981), "Experimental Data of Compound Channel Flows", Lab. of River Engrg., PWRI Report No.1666, Japan, p.138 (in Japanese).

Rajaratnam, N. & R. M. Ahmadi (1981), "Hydraulics of Channels with Flood Plains",J. Hydr. Res., Vol.19, No.1, pp.43-60.

Sellin R. H. J. (1964), "A Laboratory Investigation into the Interaction between the Flow in the Channel and that over Its Flood Plain", La Houille Blanche, 110, p.689.

Tamai, N. (1992), "Discharge Prediction for Flow in a Compound Channel: Part II: Discharge Prediction Based on a Depth-averaged Flow Equation", Australian Civil Engineering Transactions, Vol.CE34, No.4, pp.295-302.

Tu H., N. Tamai and Y. Kawahara (1995), "An Experimental Investigation of Unsteady Compound-Channel Flows - with vegetation or with embayment -", J. of Fac. of Engrg., the Univ. of Tokyo, Vol.XLIII, No.2, pp.173-200.

Comparison of Water Surface Profiles from Physical and Numerical Models in Mixed Regime Flow

MICHAEL E. MULVIHILL, M ASCE
Hydraulic Engineer, US Army Corps of Engineers
Los Angeles, CA, USA
Professor of Civil Engineering, Loyola Marymount University
Los Angeles, CA, USA

SCOTT E. STONESTREET, M ASCE
Hydraulic Engineer, US Army Corps of Engineers
Los Angeles, CA, USA

ABSTRACT

Results of numerical and physical model studies of flood control channels in Los Angeles County are presented with an emphasis on comparing the computed water surface profiles to the average water surface profiles from physical model flumes. These studies result from ongoing design work by the US Army Engineer District, Los Angeles and by the US Army Waterways Experiment Station in Vicksburg, MS.

DESCRIPTION OF STUDY

The US Army Engineer District, Los Angeles, has completed a major review study of the Los Angeles County Drainage Area (LACDA) to determine the adequacy of the existing flood control system. Elements of this study have been discussed by US Army (1992), Mulvihill (1992), Hite (1993), Stonestreet (1994) and Mulvihill (1995). The watershed area includes rugged mountains, foothills, valleys, and densely urbanized coastal plains and comprises over 3779 sq km (1459 sq mi). As a result of the study, the Corps of Engineers, along with the Los Angeles County Department of Public Works, plans to increase the capacity of two mainstem flood control channels, the lower Los Angeles River and the Rio Hondo Channel, and modify a number of highway, railroad, and utility bridges which cross these flood control channels.

DESCRIPTION OF FLOW IN FLOOD CONTROL CHANNELS

The LACDA flood control channels are fixed boundary, artificial trapezoidal channels lined with concrete and/or grouted stone. A mixed flow regime exists in the channels whereby supercritical flow occurs in the open channel sections away from bridge obstructions with average Froude numbers ranging from 1.0 to 2.6. However, bridge piers obstruct the flow at many locations forcing flow to critical depth within the pier obstructions. At these locations, the bridge piers act as hydraulic control points and subcritical flow occurs upstream of the bridge for a short distance. Given the relatively low Froude numbers of the supercritical flow, the hydraulic jumps tend to occur as undular jumps consisting of several standing waves. In addition, channel transitions and local inflow from storm drains may act as hydraulic control points in the channels.

Other parameters of concern which impact the water surface profile include superelevated water surfaces, air entrainment, and complex geometry created by highly skewed bridge crossings and a series of bicycle ramps which pass under the bridges on the channel side slopes.

PHYSICAL MODEL STUDY

Due to the uncertainties inherent in using traditional one-dimensional numerical models to predict water depths and energy losses in the vicinity of bridge piers, the results of a physical model study were used to develop a more reliable design. A special concern was the impact of the unstable flow regime (i.e. flow near critical depth) on flow depths. By using physical models, the three-dimensional flow patterns could be observed and geometric modifications could be readily tested. This model study consisting of five model flumes, was constructed and tested at the US Army Waterways Experiment Station in Vicksburg, MS.

Depth data was collected in the model flumes using point gages. At each sampling cross-section, a total of five to seven sample points was used in order to obtain the lateral profile of the water surface. Both the average and maximum depths were obtained during model testing. By computing the average of the average depths for a given cross-section, a one-dimensional, average depth was obtained which could be compared to computed depths. It was assumed that this approach sufficiently smoothed out irregularities in the raw model data, such as shock waves, and negated superelevation effects which have been observed to be significant.

ONE-DIMENSIONAL NUMERICAL MODELING

The numerical analysis was conducted at the US Army Engineer District in Los Angeles. The District's one-dimensional numerical model was used to calculate the water surface profiles for the channels. This computer model utilizes the standard step method and Manning's roughness coefficients to evaluate friction losses. The model has the capability to include expansion and contraction losses at channel transitions. Energy losses at bridges are evaluated by the Koch and Carstenjen momentum method for rapid flow and Yarnell's equation for tranquil flow. Bridge pier debris loading may be entered by the program user. At channel junctions, water surface elevations are evaluated by a momentum-based subroutine.

COMPARISON OF MODEL RESULTS

Figure 1 illustrates the hydraulic plan and profile of a portion of the Rio Hondo flood control channel. This reach of the channel has a concrete base width of 30.5 m (100 ft) with one vertical to 2.25 horizontal grouted stone side slopes. The channel invert slope averages about 0.00233 with notable short, steep reaches immediately upstream of the bridges which have an average slope of about 0.01134. The flow regime for this section of channel is supercritical.

Figure 2 shows the hydraulic plan and profile of a portion of the lower Los Angeles River channel. This concrete channel has a base width of 91.4 m (300 ft) and one vertical to 2.25 horizontal side slopes. The channel invert slope averages about 0.00151 with a short, steep reach of invert having a slope of 0.00668. The flow regime for this portion of the river is mixed with subcritical flow occurring upstream of the bridge obstructions and supercritical flow occurring downstream.

These figures provide a comparison of average water surface elevations between the numerical and physical models. This comparison shows that:
- for rapid flow (figure 1) the two water surfaces are essentially the same,
- for tranquil flow upstream of bridges (figure 2) the location of hydraulic jumps computed by the numerical model appears to be conservative,
- in each case the computed supercritical depths downstream of bridges appear to be lower than the physical model depths,
- overall, depths compare closely for non-skewed bridges; whereas, depths computed numerically for skewed bridges are not as accurate, but nonetheless appear to be conservative.

As anticipated, the physical model results varied from those predicted by the numerical model, especially in the vicinity of bridge pier obstructions. The physical model clearly showed the wave action and general flow conditions at the bridge piers.

CONCLUSION

Many investigators have studied the accuracy or appropriateness of traditional, one-dimensionally-based, standard step computations for estimating flow depths in artificial channels with subcritical flow. In many cases, these studies have been for computations which were based on Manning's coefficient for energy losses. A close review of plotted water surface profiles for mixed flow regimes, as shown herein, as well as many additional profiles which have been reviewed by the authors, indicates that these computations provide a relatively accurate estimate of the average depth for a given discharge in flood control channels. An overall trend of conservatism is observed at locations of subcritical flow in mixed regime systems. In contrast, somewhat less accuracy is noted for reaches of supercritical flow specifically where complex bridge pier and channel geometry affect the flow depth.

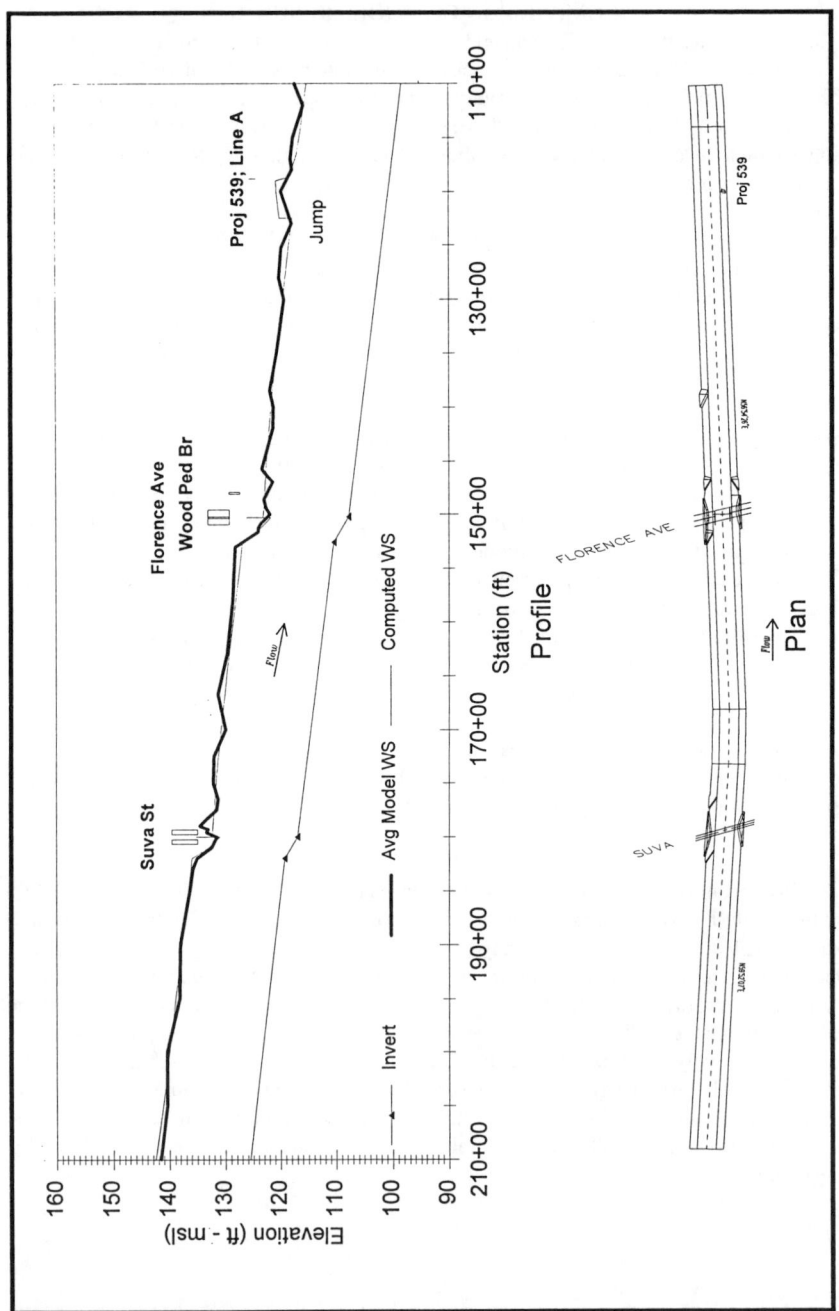

Figure 1. Plan & profile of Rio Hondo Flood Control Channel

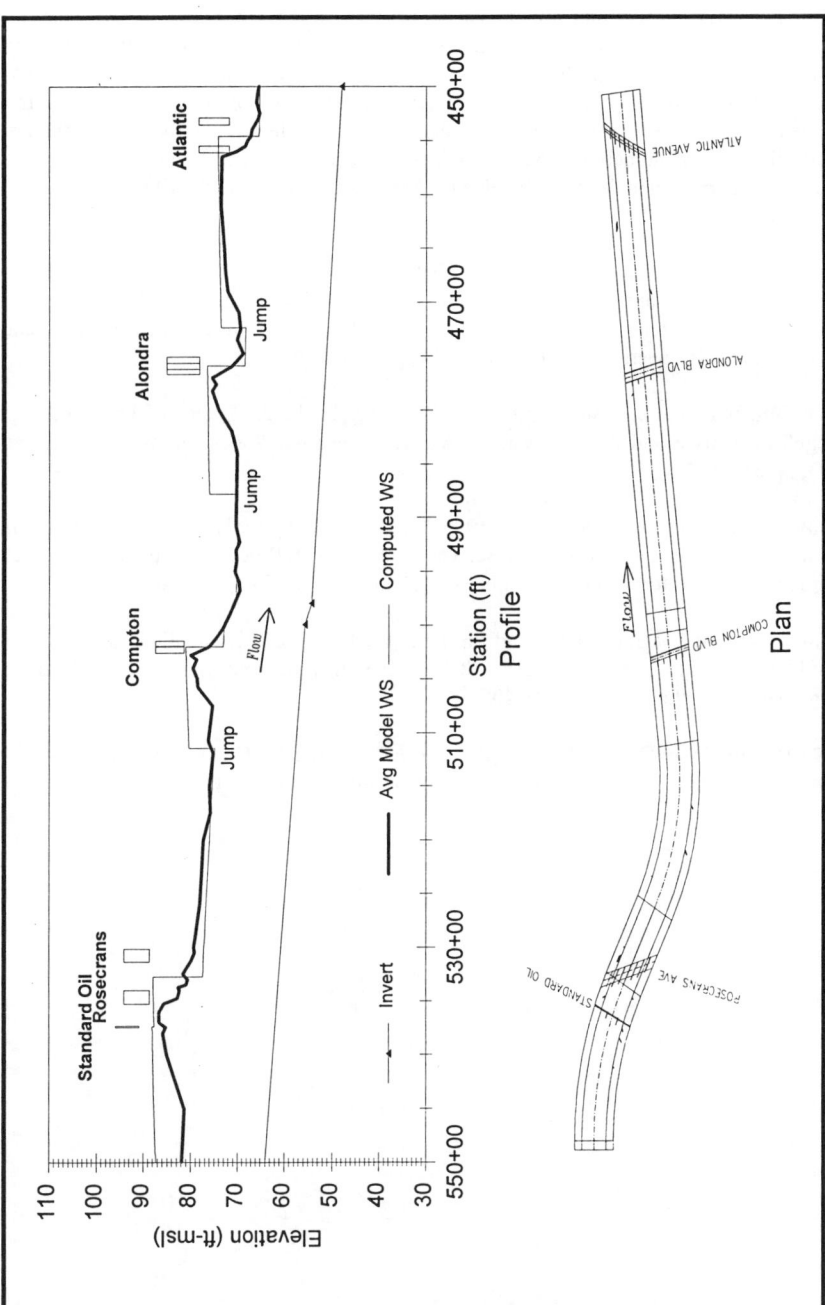

Figure 2. Plan & profile of Los Angeles River channel

Physical models provide an explicit means for determining the full three-dimensional picture of the water surface. Of special note is the amount of wave activity present in rapid flow which is not accounted for in traditional one-dimensional numerical models. It has been observed that many perturbations exist in artificial flood control channels which may create adverse wave affects, specifically near bridges. It is hoped that the state-of-the-art multi-dimensional numerical modeling will continue to advance and produce stable, reliable numerical models for high velocity flow in artificial channels with complex geometry.

REFERENCES

Hite, J.E. Jr, S.E. Stonestreet and M.E. Mulvihill. 1993. " Model Study of Rio Hondo Flood Control Channel Los Angeles, California." ASCE Proceedings National Conference on Hydraulic Engineering, San Francisco CA, pp 1695-1700.

Mulvihill, M.E. and S.E. Stonestreet. 1992. "Revised Hydraulic Design of The Los Angeles County Flood Control System." ASCE Proceedings Water Forum '92, Baltimore, MD, pp 612-617.

Mulvihill, M.E. and S.E. Stonestreet. 1995. "A Methodology for Evaluating the Hydraulic Performance of Bridges Over Fixed Bed Channels." ASCE Proceedings International Conference on Water Resources Engineering, San Antonio, TX, pp 1282-1286.

Stonestreet, S.E., M.E. Mulvihill and J.E. Hite. 1994. "Revised Hydraulic Design of the Rio Hondo Flood Control Channel." ASCE Proceedings National Conference on Hydraulic Engineering, Buffalo, NY, pp 401-405.

US Army Engineer District, Los Angeles. 1992. "Los Angeles County drainage area review, final feasibility report and environmental impact statement."

SPUR DIKE EFFECTS ON THE RIVER NILE MORPHOLOGY AFTER HIGH ASWAN DAM

By
Prof. M.M.SOLIMAN, ENG. K.M.ATTIA, Prof. KOTB,
Prof. A.M.TALAAT, & Dr.A.F.AHMED
Cairo, Egypt

INTRODUCTION

Prior to the High Aswan Dam (HAD) construction, the principal method in training the Nile River were spurs and embankments to confine water in a narrower channel during floods. These structures were made from dumped stone with their top elevation roughly equal to flood level. Spurs were for the most part located along the concave bank of the river bends. The spurs were built with a long sloping nose. This sloping nose resulted in the development of a diving helicoidal current that caused extensive bank erosion downstream and between the spurs. Also a deep scour hole generated downstream the spur, resulted sometimes in slumping the main body of the spur into the scour hole and destroying the structure. There were about 80 large spurs in the Nile River. After the construction of the dam, these spurs are too large in terms of the reduced flows. Therefore, smaller spurs are used. The most common functions of these spurs are to protect the river banks from erosion. No much attention is paid to the spur's dimension, their spacing, and optimum use in terms of the channel improvements, regulation and modification.

The main objectives of this study are to investigate the effect of spur dikes on the Nile bend channels, with special emphasis on water levels, and velocity components. The investigation is carried out through 2-D mathematical model.

THE MATHEMATICAL MODEL

A 2-D mathematical model (TRUSOLA) is used to simulate a Nile bend channel located about 3 km downstream of Naga Hammadi barrage. The model is developed by Delft Hydraulics Laboratory (DHL.,1991).

This model is selected to suit the conditions in the Nile where shallow flow depths are present. The shallow water assumption is used to model the flow, and the vertical momentum equation is reduced to hydrostatic pressure relation. The vertical water motion is derived from the horizontal flow field, using the continuity equation. The program followed the boussinesq approximation. Different lengths and spacing are used to simulate the effect of spur dikes on water levels and velocity components. About fifty runs were carried out.

ORTHOGONALITY AND GRID SPACING

For the representation of the bed topography of rivers and flow velocities, it is convenient to use orthogonal curvilinear grids. The advantage of using an orthogonal grid is that the partial differential equations, which describe the two dimensional flow and, hence, the finite difference approximation, becomes substantially simpler than if a general non orthogonal curvilinear grid is applied,(Olesen,1992). Thus, the truncation errors will be smaller and accuracy is much better in an orthogonal grid. Although the system is more flexible, there are still restrictions with respect to the orthogonality and the change in spacing(Wijbenga,1985 a).

For the study reach downstream Naga Hammadi an orthogonal curvilinear grid is generated with a total number of 65 and 27 in both the flow and transverse directions respectively. The grid size in the flow direction is designed as 20 m starting from the bend centerline (spur dike position). This area is located around node number 31 which is considered the centerline of the channel ber ' as well as peak point for the curved part of the channel. The size of the grid increased further upstream and downstream the centerline node. Similar procedure is used for the grid size in the transverse direction, (see Fig. 1)

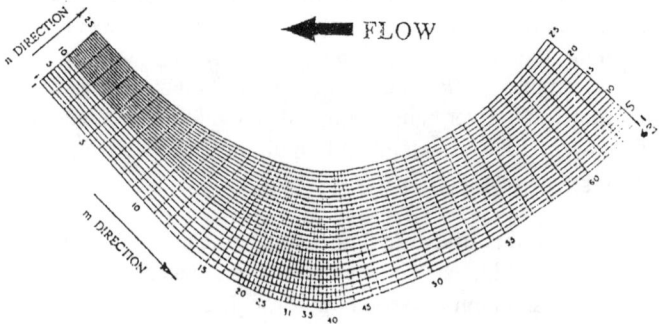

Figure 1: Layout of the designed grid

THE STUDY CASES

A number of selected cases were simulated. A basic case representing a channel bend without any spur dike is simulated in order to have an insight of the bend morphological feature such as water levels and velocity components (U and V). This run is used as a reference for comparison with spur dike cases. Six runs were conducted using single spur dikes of different lengths. For all runs the spur is fixed at the centerline node, 31. The aim of these runs is to study the effect of single spur dike installation on the channel bend. Eighteen runs were conducted using double spur dikes. Twenty-one runs were carried out to simulate the effect of triple dikes.
The simulated lengths (L) varied between 30m and 200m and spacing between dikes varied from 1 to 7-L..

ANALYSIS OF RESULTS

The *total heading up* for some runs are summarized in table 1. One can deduce that single spur dike has similar effect in terms of the total heading up as double spur dikes. Furthermore, analogous results have been ensued when using triple spurs. This may be explained as when using spur of length 50 m in double or triple arrangements and distance apart equal to L, the main flow separates at the tip of the first spur dike and a zone of dead water filled the area between the spurs creating a pseudo wall. The spacing of L=50 m is too short to allow the formation of an eddy zone of recirculating water (Attia,1996). The flow in this case can be nominated as quasi smooth flow and the double and triple arrangements act similar to the single dike.

Spur lengths of 100m and 160m restrict about 20% and 32% of the channel width respectively. In case of double and triple arrangements and spacing L, an insignificant increase or decrease in the total heading up was occurred. However, the spur lengths show fair effect on the total heading up and this can be noticed in Table 1. Slight. effect has been shown in the case of single and double arrangements for 2L spacing, however the total heading up increases again in case of triple spur arrangements. This result may be explained that the spacing of 2L creates a wake interference, because the spurs are placed so close to each other that the wake and vortex at each spur are interfering with those developed at the following spur, resulting in intense and complex vorticity and turbulence mixing. In such a flow, the length of the spurs is relatively unimportant, but the spacing is obviously of major importance. The spacing of 3.5L shows slight effect when using spur length of 50m between ,double ,and triple arrangements. As the length of the spur increased to reach 100 m long, the effect on the heading up also increased. A similar trends are complied when using spur length of 160 m. The spacing of 3.5 L gives significant increase in total heading up especially in case of triple spur dikes.

Table 1: Heading up due to river bends with spur dikes

Spur length (L) in meters	Single	Double			Triple		
		L	2L	3.5L	L	2L	3.5L
30	5.5						
50	5.57	5.56	5.57	5.58	5.57	5.59	5.63
100	5.79	5.75	5.77	5.84	5.78	5.86	6.05
160	6.18	6.1	6.13	6.38	6.18	6.47	7.23

Note: heading up due to bend without spurs = 5.46 cm. (for more detail refer to Atia, 1996).

One of the major principals for judging effects and changes is to improve the conditions of *flow patterns* in the river or at least to minimize the interference with the existing flow regime of the river. The determination of the principal velocities (magnitude and direction) in the vicinity of new structures will also provide a basis for evaluating the need for protection against resulting scour and erosion.

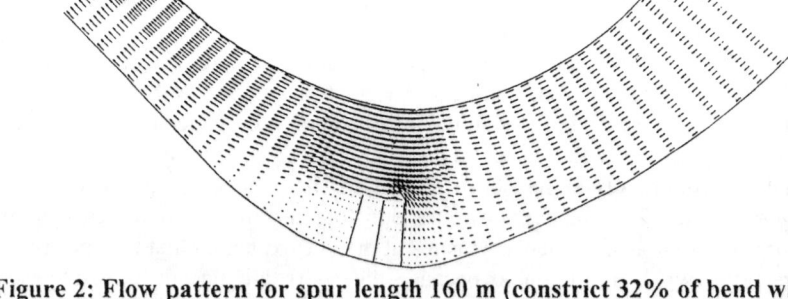

Figure 2: Flow pattern for spur length 160 m (constrict 32% of bend width)
The direction and magnitude of the flow velocities for the entire runs confirmed that the flow separated at the tip of the first spur. The lengths and directions of the arrows indicate the magnitude and direction of the flow velocities. An almost dead water zone was found to occupy the area among the spurs where the velocity vectors dwindled. These observations refer to a pocket of material deposited between the spurs particularly for spur lengths of 30, 50, and 80 m

with spacing of 2L and 4L. The spacing of 7L between spurs was found to be very large that each spur acted as a single one.

The eddy zones (recirculating water) appear very clearly in the cases that have spur length of 160 m (see Fig. 2). Outside from the eddy zones the main flow was disturbed from its usual approach pattern. In fact the boundaries of the eddy zones and the flow within it followed a complex three-dimensional pattern. The extent of the eddy zones is not steady, for the boundary oscillates irregularly.

The spacing of 2L and 4L between the spurs gave similar observation and made the group act similar to continuous protection or training of the river. However, from the economical point of view the spacing of 4L is preferable due to the increase in the length of the covered areas by similar number of spurs.

The longitudinal velocity component U increased due to the effect of single spur dike construction. The increase in the velocity was found to be proportional to the spur length. The same results were observed for the double and triple spurs. The findings were attributed to the blockage effect. However, the spacing between spurs showed reverse relations.

An increase in the transverse velocity component V was also found due to the construction of single spurs. Accordingly the resultant velocity increased due to the construction of single spur dikes in channel bends.

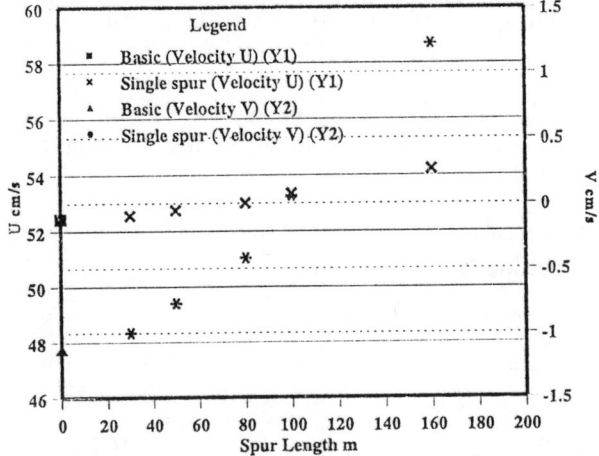

Figure 3 : Effect of spur dike length on U & V components, for single spur

The lowest value of V was observed for the double spur arrangement whereas the highest value occurred by the single spur construction. However for double and triple spur dike construction no general trend was found.

CONCLUSION

- Spur dikes would cause appreciable increase in the heading up if it constricted more than 6% of the channel width.

A pocket of deposited material occurs between the spurs for all tests with spacing less than 4L. This pocket disappeared for a spacing of 7L.

Flow separation occurred at the tip of the first spur for the entire tests. Eddy formation only appeared for spur lengths greater than 18% of the channel width.

The longitudinal velocity U increased with the increase of the spur lengths that caused a decrease in the channel width. The transverse velocity V increased in case of single spurs, whereas for double and triple spurs no conclusive trend was found.

REFERENCES

-Attia M.K., "Spur dike effects on straight and channel bends" Ph. D. Thesis, Ain Shams University, Cairo Egypt, 1996.

-Delft Hydr. Lab. (DHL), "A simulation program for hydrodynamic flow and transport"; Technical manual for (2-D TRUSOLA model), The Netherlands 1991.

-Olesen,W2-D "Math. modelling of morphological processes in the Brahmaputra River" Int. Conf. on Protection and Develop. of the Nile & other Major Rivers, Cairo, Egypt 1992 .

-Wijbenga, J.H., "Determination of flow patterns in rivers with curvilinear co-ordinates "; Publication 352 , Delft Hydraulics, 1985a.

Two-Dimensional Floodwave Analysis Resulting from Breached Levee

Kun-Yeun Han*, Jae-Hong Park**, Jong-Tae Lee***
* Professor of Civil Eng., Kyungpook National Univ., Daegu 702-701, KOREA
** Research Assistant, Kyungpook National Univ., Daegu 702-701, KOREA
*** Professor of Civil Eng. Kyonggi Univ., Seoul 120-702, KOREA

ABSTRACT

A hydrodynamic model of one and two-dimensional combined analysis for floodwave propagation from breached levee in protected lowland is studied. Overflow through the breached levee is treated as internal boundary condition in a channel. The model is applied to the actual levee-break case and two dimensional velocity distributions and inundated depths are presented to demonstrate the simulation results. The computed results have good agreements with observed data in terms of inundated depth, flood arrival time, and flooded area.

INTRODUCTION

The flood damage due to levee breach has become greater because of the increased residential areas and industrial facilities near the river bank. For past several years, we have experienced several flood disaters in major rivers. The flood hazard has been raised serious social problem together with lives and property damages in the country.

Compared to sophiscated two-dimensional finite difference and finite element model, the diffusion hydrodynamic model has advantages in its versatility to simulate floodwave over complex topography. It does not require a specific algorithm to handle the moving boundary. And it considers effectively the complex phenomena such as street flow, flow through subdivision and flow around building, or obstruction in urban areas.

Combined model of one and two-dimensional floodwave analysis is applied to the actual levee breach in the downstream of the Han River. One-dimensional unsteady flow analysis is performed in the reach considering the levee-break flow. Overflow through the broken levee is calculated considering breach width, breach duration time, and submergence effect. The computed results of two-dimensional model such as inundated depth, velocity vector, flood arrival time, and flooded area are discussed.

FLOOD INUNDATION MODEL

The basic expression of flood routing in main channel is descibed by the following one dimensional Saint-Venant equation and in which for the flowrate through broken-levee, q_L term and the momentum effect of lateral outflow through broken levee, L term is introduced.

$$\frac{\partial Q}{\partial x} + \frac{\partial (A+A_0)}{\partial t} - q_L = 0 \tag{1}$$

$$\frac{\partial Q}{\partial t} + \frac{\partial (Q^2/A)}{\partial x} + gA(\frac{\partial h}{\partial x} + S_f + S_e) + L = 0 \tag{2}$$

where A is the active cross sectional area, A_0 is the inactive cross sectional area, x is the longitudinal distance along the channel, t is the time, S_f is the friction slope, and S_e is the expansion contraction slope.

Substitution of finite difference approximation defined by the weighted four-point method yield two nonlinear equation with respect to unknown h and Q. Newton-Raphson method is applied to a system of nonlinear equations.

FLOODWAVE IN INUNDATED AREA

The two-dimensional diffusion wave is written as

$$\frac{\partial}{\partial x} F_x \frac{\partial h}{\partial x} + \frac{\partial}{\partial y} F_y \frac{\partial h}{\partial y} = \frac{\partial h}{\partial t} \tag{3}$$

Consider a unit cell, which width is b_x, b_y and surface area is $A_{x,y}$. For the numerical analysis of equation (3), the integrated finite difference version of the nodal domain integration method is used.

The calculation is repeated until the difference between h_c and h_i is smaller than the specified tolerance (10^{-4}m). The flowrates across the adjacent element is summed, then the flow depth is computed by Eqn. (4) ~ (6).

$$\Delta Q_c^k = Q^k]_{dir1} + Q^k]_{dir2} + Q^k]_{dir3} + Q^k]_{dir4} \tag{4}$$

$$\Delta h_c^k = \frac{\Delta Q_c^k (\Delta t)}{b_x b_y} \tag{5}$$

$$h_c^{k+1} = h_c^k + \Delta h_c^k \tag{6}$$

INTERFACE MODEL

For the derivation of the breach hydrograph which represents the amount of the flowrates into the inland, the interface model is developed considering levee breach mode and submergence effect. The effective parameters to the breach hydrograph are water level of channel and inland, breach width, failure duration time and final breached levee height.

The flow direction can be recersed depending on the difference of levels between in and outside of the levee. Thus, the higher level of the two is named $\hat{h}$ and lower one is $\tilde{h}$ for generalized form, and this can be controlled by directional signal α (Lee and Han, 1989).

(i) when $\hat{h} > h_b$

$$q_L = \frac{\alpha \beta}{\Delta x_i} [C_1 b_t (\hat{h} - h_b)^{1.5} + C_2 (\hat{h} - h_b)^{2.5}] \qquad (7)$$

(ii) when $\hat{h} < h_b$, $\tilde{h} < h_b$ and $\hat{h} = \tilde{h}$

$$q_L = 0 \qquad (8)$$

where β is the submergence correction factor, C_1, C_2 are the discharge coefficients for rectangular and trapezoidal breach respectively, h_b is instantaneous elevation of breach bottom.

Two dimensional model in this study partly evolved from the diffusion hydrodynamic model (Hromadka and Yen, 1987). The original routing algorithm is revised and expanded to improve numerical stability, check mass balance error, and broaden to deal with interface algorithm between channel and inundation area.

APPLICATION

Ilsan levee, which was originally constructed during 1933~35, then repaired several times thereafter, is located at approximately 1.2 km downstream from the Hangju Bridge in the Han River as shown in Fig. 1.

The torrential rain, amount of 488 mm, fell in the Han River basin for 40-hour period from September 9 to September 11, 1990. At September 12, 01:30 AM, the piping hole, 5 cm in diameter, was first found through the levee. The size of the hole was increased to 1 m at 02:40 AM. At 03:30 AM, the reach of 30 m was breached, and the escaping warning was announced to the neighboring residents. At 2:00 PM, the breach width was developed up to approximately 200 m, and the inundated zone was enlarged its range to neighbouring areas.

Finally, nearly 5,451 ha of resident areas and farmlands, 2,100 houses, and 7,374 residents were suffered from the disaster. The total amount of damages

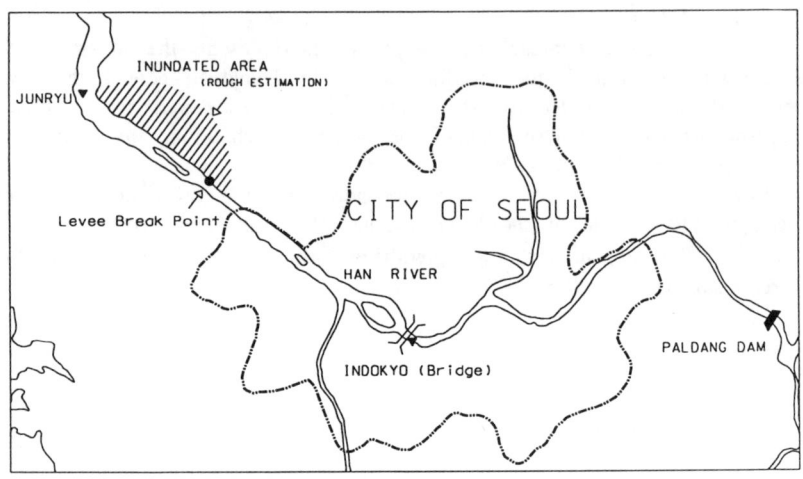

Fig. 1 Location map of study area

was estimated at approximately six hundred billion dollars.
Available field data included ; (1) geometric data from aerial photographs and field survey; (2) land use conditions to estimate the roughness coefficients; (3) surveyed flooded areas; (4) flow depth and flood arrival time at 15 flood gage stations.
To compute the accurate inundated flowrates, the floodwave analysis in channel is performed in the 38 km reach from Indokyo to Junryu. A total of 40 cross sections located at unequal interval is used and average slope is 1/12,000. Manning roughness coefficient, n, varies from about 0.023 to 0.028. The discharge rate varies from the low flow of about 300 $m^3/\sec$ to the flood discharge of over 32,000 $m^3/\sec$ in the upstream boundary.
The height of breached levee was 5 m. The levee was gradually breached up to 200 m width in 3 hours. The boundary conditions at upstream and downstream are the flood stage hydrographs at Indokyo and Junryu respectively.
Based on the breach outflow hydrograph, two-dimensional floodwave analysis is performed in the inundated area. The surface topography is discretized into uniform square grid element. Each element is assigned a location on the grid system with elevation, and Manning roughness coefficient. It is descretized into 765 elements and its size is 250 m x 250 m. A Manning's n value for each element varies from 0.025 to 0.040 depending on land use conditions, vegetations, and flow obstructions.

The flowrates through breached levee has reached a peak value of 3,551 $m^3/\sec$, at the elapsed time of 4.5 hours. A total duration was approximately 21hours. The submergence correction factor was varied from 1.0 to 0.1 as the inunated depth increases. At 12:00 PM, the overflow through broken levee was ceased. The calculated values are matched with the observed data (Kyonggi Province, 1990).
A flow velocity vector at some specified time level is illustrated in Fig. 2 and 3. A time-lapse simulation of the floodwave propagation is also shown in Fig 4. The computed inundation area coincides with the observed data. Mass conservation errors are maintained smaller than 10^{-3} %.
Comparisons between the computed and observed flow depth and flood arrival time at some stations are also presented. The computed results have good agreements with the observed data (Seoul-Sinmoon; Joongang-Ilbo; Chosun-Ilbo; Kyunggi-Ilbo, 1990).

CONCLUSION

Floodwave analysis through broken levee in protected lowland offers a engineer to predict the flooded areas and take measures against flood hazards. The weighted four point method is used in one-dimensional analysis, whereas diffusion hydrodynamic method in two-dimensional analysis. Two models are combined by introducing interface model. One-dimensional flood wave analysis in channel is performed considering that overflow through breached levee is treated as internal boundary condition.
The model is applied to the Ilsan levee-break, which occurred on September 12 ~ 13, 1990 in the downstream of the Han River. One-dimensional unsteady flow analysis has been executed in a 34-km reach from Indogyo to Junryou considering the flowrate through the broken levee. The velocity distributions and inundated depths have been presented to demonstrate the simulation results. The computed results have good agreements with observed data in terms of inundated depth, flood arrival time, and flooded areas.

REFERENCES

Fread, D.L. (1985). "Channel Routing", in Anderson, M.G., and Burt, T.P. (eds), *Hydrological Forecasting*, John Wiley & Sons, New York, pp. 437-503.
Hromadka, T.V., and Yen, C.C. (1987). "Diffusion Hydrodynamic Model." *Water Resources Investigation 87-4137*, US Geological Survey.
Kyonggi Province.(1990). *Observed flood data resulting from Ilsan levee-break.*
Lee, J.T., and Han, K.Y. (1989). "A dynamic levee breach model and its applications to hypothetical and actual flood." *Proc. of 23rd Congress of International Association for Hydraulic Research*, IAHR, pp. 441-448.
Seoul-Sinmoon, Joongang-Ilbo, Chosun-Ilbo and Kyunggi-Ilbo. (1990). *Data from daily newspapers.*

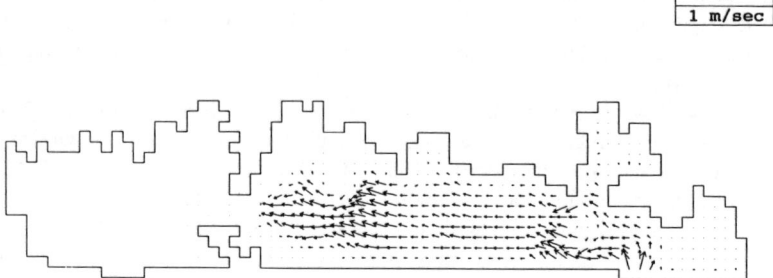

Fig. 2 Two-dimensional velocity distribution (At 09:00, Sept. 12)

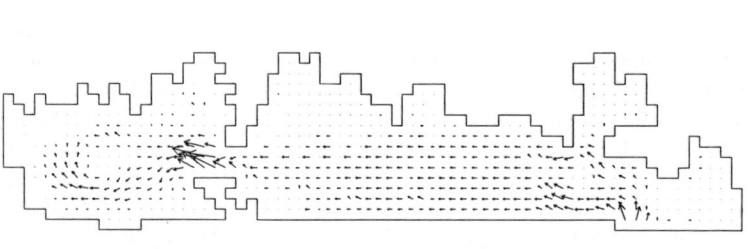

Fig. 3 Two-dimensional velocity distribution (At 14:00, Sept. 12)

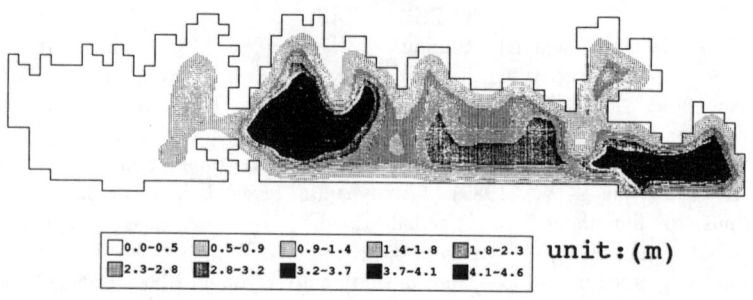

Fig. 4 Inundation map from breached levee (At 14:00, Sept. 12)

Two-Dimensional Large Eddy Simulation for Shallow Recirculating Flow

M. NASSIRI, S. BABARUTSI and V. H. CHU

Department of Civil Engineering and Applied Mechanics
McGill University, Montreal, Quebec H3A 2K6, Canada

ABSTRACT

Shallow recirculating flows in an open-channel expansion are computed using a 2-D LES (Two-Dimensional Large Eddy Simulation) method. The Smagorinsky model of the sub-grid scale is adapted in conjunction with a model for the bed-generated turbulence. Calculations are conducted for recirculating flows of different depths. The results obtained for the mean velocities, the contaminant concentrations, and the Reynolds stresses are compared with available laboratory data. The values of the Smagorinsky parameter varying from 0.17 to 0.2 are selected to optimize the model performance.

INTRODUCTION

Turbulent flows in inland and coastal waters are often shallow in the sense that the horizontal length scale of the motion is large compared with the depth of the motion. The large-scale motion, being confined in the small depth between the free surface and the channel bed, is essentially 2-D (two-dimensional). The computation of this 2-D motion can be made either by a Reynolds averaging method (Rastogi and Rodi, 1978; Babarutsi and Chu, 1991; Babarutsi, Nassiri and Chu, 1996) or by a LES (Large Eddy Simulation) method (Madsen et al., 1988; Nadaoka and Yagi, 1993). Since the required computation time for the 2-D LES is the same as the Reynolds averaging method, the 2-D LES method is preferred for its ability to produce both the mean and the unsteady fluctuating flow field. Highly unsteady turbulent flows can be simulated by the 2-D LES. The method can also be used for verification of a Reynolds averaging model.

The 2-D LES method is not the same as the original LES method for three-dimensional flow calculations. The Smagorinsky parameter C_S in three-dimensional simulation is known to have a value varying from $C_S \simeq 0.06$ to 0.21 (Moin, 1982; Kobayashi and Togashi, 1993). However, in 2-D LES, Madsen et al. (1988) have

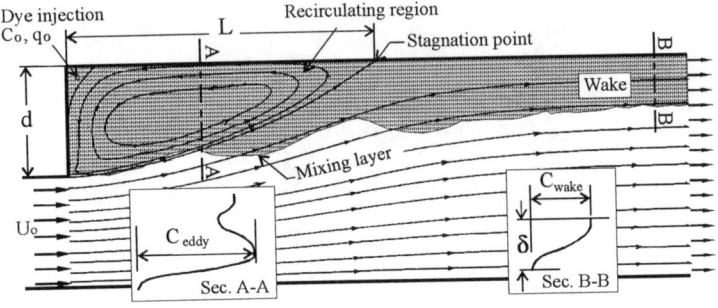

Figure 1: Top-view of the recirculating open-channel flow. The width of the channel is 60 cm. The expansion width d is 30 cm. Dye of concentration C_o was injected at a steady rate q_o into the corner of the recirculating zone.

proposed the value of C_S in the range of 0.4 to 0.8 to calculate the near-shore tidal current, . Nadaoka and Yagi (1993) in their simulation of river flow with vegetations, have obtained reasonable results without even using the subgrid-scale modelling. Their method is equivalent to 2-D LES using zero Smagorinsky parameter.

The existing uncertainty with the Smagorinsky parameter is the main difficulty for the application of the 2-D LES method. In this paper, a wide range of C_S value is used to calculate recirculating flows using the 2-D LES method. The results are compared with available experimental data. The objective of the investigation is to obtain a Smagorinsky parameter suitable for shallow flow simulation.

FORMULATION OF 2-D LES

In the 2-D LES of the shallow open-channel flow, the large-scale turbulent motion is calculated by the following time-dependent depth-averaged equations:

$$\frac{\partial \zeta}{\partial t} + \frac{\partial h U_k}{\partial x_k} = 0 \tag{1}$$

$$\frac{\partial U_i}{\partial t} + \frac{\partial U_i U_k}{\partial x_k} = -g\frac{\partial \zeta}{\partial x_i} - \frac{c_f}{2h} U_i \sqrt{U_k U_k} + \frac{\partial}{\partial x_k}[(\nu_s + \nu_{sg})\frac{\partial U_i}{\partial x_k}] \tag{2}$$

in which $(U_i; i = 1, 2)$ are the time-dependent depth-averaged velocities in the x_i-direction; ζ is the water surface elevation. The momentum exchange due to the small-scale turbulent motion is included in the equations by the turbulent

Test	h (cm)	U_o (m/s)	q_o (ml/s)	C_o (gr/ml)	$C_o q_o$ (gr/s)	$\dfrac{L}{d}$	$\dfrac{\delta}{d}$	$\dfrac{C_e}{C}$	$\dfrac{C_w}{C}$
T1	1.45	0.228	1.184	0.930	1.101	2.73	0.50	19.5	10.0
T2	2.50	0.146	1.184	0.930	1.101	4.66	0.75	9.5	5.5
T3	3.50	0.243	2.900	1.054	3.057	5.46	0.90	9.0	5.0

Table 1: Test conditions of the shallow recirculating flow experiments by Babarutsi and Chu (1991).

viscosity ν_s which is assumed to be proportional to the friction velocity, $U_* = \sqrt{\frac{1}{2} c_f U_k U_k}$, and the water depth, h, by the relation

$$\nu_s = c_v U_* h \qquad (3)$$

where c_f is the friction coefficient. The empirical constant, c_v, is 0.08 (Rastogi and Rodi, 1978). The viscosity for the subgrid scale, ν_{sg}, is given by the Smagorinski model:

$$\nu_{sg} = (C_S \Delta)^2 [\frac{\partial U_i}{\partial x_j}(\frac{\partial U_i}{\partial x_j} + \frac{\partial U_j}{\partial x_i})]^{1/2} \qquad (4)$$

where C_S is the Smagorinsky parameter, and Δ the sub-grid length scale. The contaminant concentration distribution is determined by the mass-transport equation

$$\frac{\partial C}{\partial t} + \frac{\partial U_k C}{\partial x_k} = \frac{\partial}{\partial x_k}[\frac{\nu_s + \nu_{sg}}{\sigma_t}\frac{\partial C}{\partial x_k}] + S_s \qquad (5)$$

where C is the time-dependent depth-averaged concentration, S_s the source/sink term, and $\sigma_t = 1.0$ the Prandtl-Schmidt number.

RECIRCULATING FLOW

The 2-D LES method is used to compute the recirculating flows downstream of a sudden open-channel expansion as shown in Figure 1. The computations are conducted for three water depths h = 1.45 cm, 2.50 cm, 3.50 corresponding to tests T1, T2, and T3 of the experiments by Babarutsi and Chu (1991). Table 1 summarizes the conditions of the tests. Numerical results are obtained for the reattachment length, L, the recirculating flow rate, q, the wake width δ, the peak concentration within the eddy, C_{eddy}, and the peak concentration in the wake C_{wake}.

For the solution of the equations, an explicit time integration is employed on a staggered grid using control-volume finite difference method. The velocity field is determined by central differencing scheme. The solution of the mass-transport equation is by the Lagrangian Second Moment Method (Nassiri and Babarutsi, 1996). For computational stability, the Courant number $Cr = U_o(\Delta t)/(\Delta x)$ is

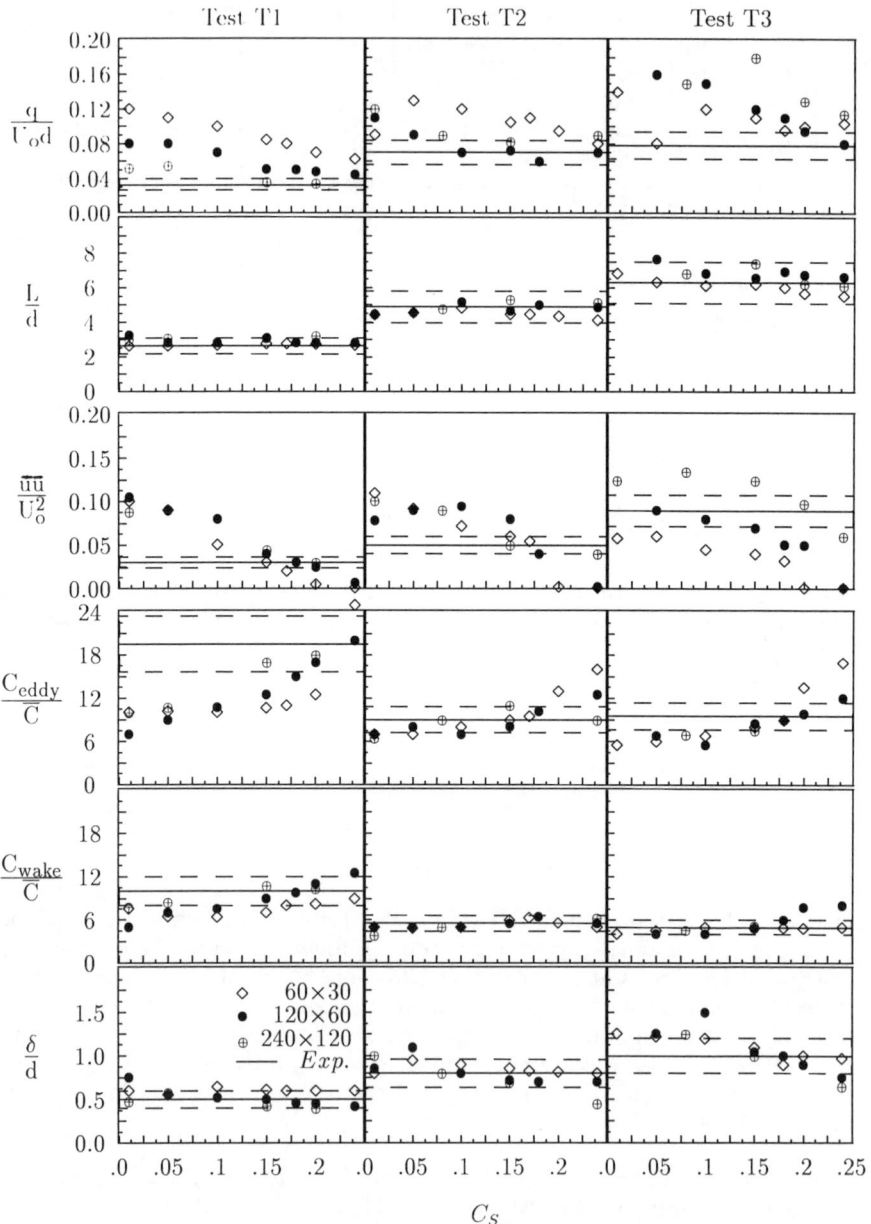

Figure 2: Effect of the Smagorinsky parameter C_S on the recirculating flow.

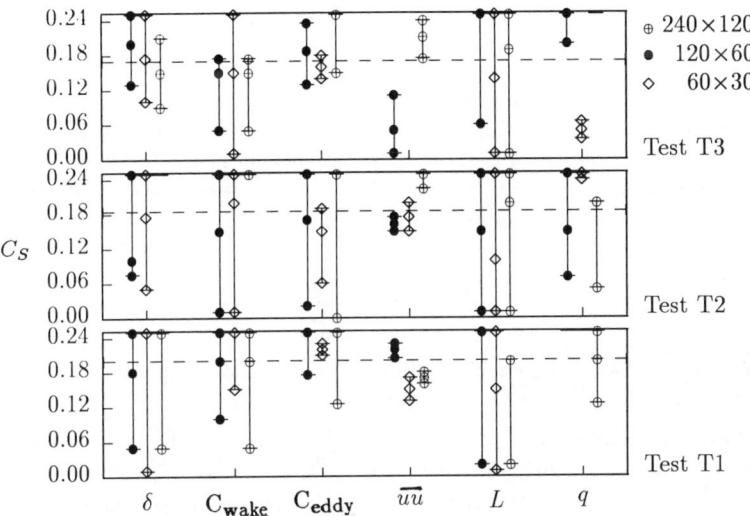

Figure 3 : Range of C_S values when the flow features, δ, C_{wake}, C_{eddy}, $\overline{uu}$, L and q, are predicted within the $\pm 20\%$ range of the measurement.

kept below 0.05.

The calculations are conducted using the Smagorinsky parameter C_S in the range of 0.01 to 0.25 and 60×30, 120×60, and 40×120 grids. The normalized results obtained for q, L, $\overline{uu}$, C_{eddy}, C_{wake} and δ are plotted in Figure 2 compared with the experimental measurements. The solid lines indicate the measured values while the dashed lines the $\pm$ 20% range. The symbols ∘, •, and ⊕ denote the numerical results obtained by the 60×30, 120×60 and 240×120 grids, respectively. Some of the flow features, such as the reattachment length, L, the wake width, δ, and the concentration in the wake, C_{wake}, are relatively independent of the C_S value and the grid size. Other featues such as the recirculating flow rate, q, the concentration in the eddy, C_{eddy}, and the Reynolds stress, $\overline{uu}$, are sensitive to the value of C_S and the grid size.

An attempt is made to find the value of C_S that would produce acceptable results. Figure 3 shows the range of C_S value in which the computational results fall within $\pm 20\%$ of the experimental measurements. The dash lines in the figure are selected so that all features of the flow are reproduced within $\pm 20\%$ of the measurements. The recommended Smagorinsky parameter for tests T1, T2, and T3 is $C_S = 0.20$, 0.19, and 0.17, respectively.

CONCLUSIONS

A 2-D LES method is implemented for the depth-averaged calculations of shallow recirculating flows in open-channel. Computations are conducted for a range of Smagorinsky parameter, C_S, varying from 0.01 to 0.25 and for three different grids (60×30, 120×60, and 40×120). The recirculating flows are predicted within ±20% of the measurements when the value of the Smagorinsky parameter, C_S, is within the range of 0.17 to 0.2.

REFERENCES

[1] Babarutsi, S. and Chu, V. H. (1991) "A Two-length-scale model for Quasi-two-dimensional Turbulent Shear Flows," *Proc. 24th Congress of IAHR*, Vol. C, pp. 51-60.

[2] Babarutsi, S. and Chu, V. H. (1991) "Dye-concentration Distribution in Shallow Recirculating Flows," *J. Hydr. Eng.*,ASCE, Vol. 117, No. 5, pp. 643-659.

[3] Babarutsi, S., Nassiri, M., and Chu, V. H. (1996) "Computation of Shallow Recirculating Flows Dominated by Friction," *J. Hydr. Eng.*, ASCE, Vol. 122, No. 7, pp. 367-372.

[4] Nassiri, M. and Babarutsi S. (1997) "Computation of Dye-Concentration in Shallow Recirculating Flow," to appear in *J. Hydr. Eng.*,

[5] Kobayashi, T. and Togashi, S. (1993) "Comparison of turbulence models applied to backward-facing step flow by LES data base," *Eng. Turbulence Modelling and Experiment 2*, Elsevier, pp. 133-142.

[6] Madsen, P., Rugbjerg., M. and Warren, I. R. (1988). "Subgrid modeling in depth integrated flows", *Proc. 21st Int. Conf. on Coastal Engineering*, ASCE, pp. 505-511.

[7] Moin P. and Kim, J. (1982), "Numerical investigation of turbulent channel flow", *J. Fluid Mech.*, Vol. 118, pp. 341-377.

[8] Nadaoka, K. and Yagi, H. (1993d) "A Turbulent Model for Shallow Water and its Application to Large-eddy Computation for Longshore Currents", *J. Hydr., Coast. and Envir. Eng.*, JSCE, No. 473, 25-34 (in Japanese).

[9] Rastogi, A. K. and Rodi, W. (1978) "Prediction of heat and mass transfer in open channels", *J. Hydr. Eng.* ASCE, Vol. 104, pp. 397-420.

[10] Smagorinsky (1963) "General Circulation Experiments with the Primitive Equations, The Basic Experiment," *Mont. Weather Rev.*, Vol. 91, pp. 99-104.

Numerical Analysis of Horizontal Vortices in Compound Open Channel Flows by the Two-layered Flow Model

by I. KIMURA*, T. HOSODA**, Y. MURAMOTO** and R. YASUNAGA***
* Dept. of Civil Engrg, Wakayama National College of Tech.,
77 Nada-noshima, Gobo, Wakayama 644, Japan
** Dept. of Civil Engrg, Kyoto Univ., Yoshida Sakyo-ku, Kyoto 606, Japan
*** Tokyo Electric Power Co. Ltd., Uchisaiwai, Chiyoda-ku, Tokyo 100, Japan

ABSTRACT

In compound open channel flows, it is important to clarify the horizontal vortices due to the shear instability at the junction between the main channel and the flood plain. The numerical method based on the two-layered flow model is developed to simulate the 2-D and 3-D flow structure caused by horizontal vortices in compound open channel flows with a rectangular main channel. Numerical simulation is done under the hydraulic conditions of the laboratory tests by Ikeda et al (1995). The model performance is examined through the comparison between the numerical results and the experimental ones.

INTRODUCTION

It is well known that horizontal vortices can be seen along the junction between the deep main channel and the shallow flood plain(s) due to the shear instability in compound open channel flows. These vortices affects the resistance to flow and sediment transport in rivers during flood. Laboratory tests have been done on above phenomena. Fukuoka and Fujita (1989) studied the resistance to flow in compound open channels with the smooth and rough flood plains. Ikeda et al.(1995) elucidated the instability of horizontal vortices in terms of the shear instability and Karman's vortex street stability. They also clarified the 3-D flow structure which is caused by horizontal vortices using the 2-D laser Doppler anemometer. On the other hand, numerical methods with turbulence models have been used for the analysis of the steady secondary circulation in lateral sections (Naot et al., 1993). 2-D shallow water flow equations, which have been used for the vortex formation below an abrupt expansion (Hosoda and Kimura, 1993) and the flow with vegetation (Nadaoka and Yagi, 1993), are not applicable in compound open channel flows with a rectangular main channel because the strong 3-D structure exists at the junction between the main channel and the flood plain, where depth changes discontinuously. 3-D numerical model, which requires long CPU time and large computer memory, is not practical for the engineering purpose.

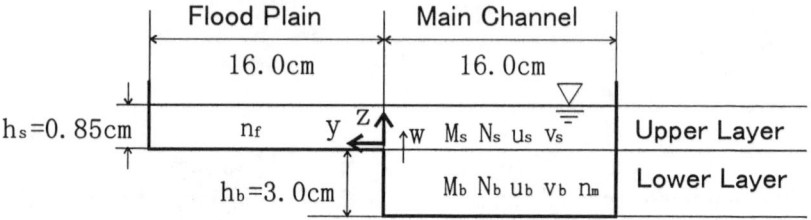

Fig.1 The cross-sectional geometry and the coordinate system

From the practical point of view, we propose the numerical method based on the two-layered flow model to analyze the compound open channel flows efficiently. The model can reproduce not only the horizontal vortices but also the vertical flow structure caused by the vortices to some extent. The horizontal and vertical flow pattern are reproduced numerically under laboratory test (by Ikeda et al.) conditions. The model performances are examined through the comparison between the numerical results and the experimental ones.

BASIC EQUATIONS AND NUMERICAL METHOD

Basic equations of the two-layered flow model are derived by integrating the 3-D flow equations over the control volume in the upper and lower layers shown in Fig.1 in the same way as the TLDAM method by Liu (1991). The continuity and momentum equations in both layers in the main channel and in the flood plain are described below:

[Continuity Equation in the Upper Layer in the Main Channel]
$$\frac{\partial h_s}{\partial t} + \frac{\partial M_s}{\partial x} + \frac{\partial N_s}{\partial y} - w_i = 0 \tag{1}$$

[Continuity Equation in the Lower Layer in the Main Channel]
$$\frac{\partial M_b}{\partial x} + \frac{\partial N_b}{\partial y} + w_i = 0 \tag{2}$$

[Continuity Equation in the Flood Plain]
$$\frac{\partial h_s}{\partial t} + \frac{\partial M_s}{\partial x} + \frac{\partial N_s}{\partial y} = 0 \tag{3}$$

[Momentum Equations in the Upper Layer in the Main Channel]
$$\frac{\partial M_s}{\partial t} + \frac{\partial \beta u_s M_s}{\partial x} + \frac{\partial \beta v_s M_s}{\partial y} - (wu)_i + gh_s \frac{\partial h_s}{\partial x} = gh_s \sin\theta$$
$$+ \frac{\partial(-\overline{u'^2})_s h_s}{\partial x} + \frac{\partial(-\overline{u'v'})_s h_s}{\partial y} - (-\overline{u'w'})_i + \frac{\partial}{\partial x}\left[\frac{\tau(v)_{xxs}}{\rho} h_s\right] + \frac{\partial}{\partial y}\left[\frac{\tau(v)_{xys}}{\rho} h_s\right] \tag{4}$$

$$\frac{\partial N_s}{\partial t} + \frac{\partial \beta u_s N_s}{\partial x} + \frac{\partial \beta v_s N_s}{\partial y} - (wv)_i + gh_s \frac{\partial h_s}{\partial y}$$
$$= \frac{\partial(-\overline{v'u'})_s h_s}{\partial x} + \frac{\partial(-\overline{v'^2})_s h_s}{\partial y} - (-\overline{v'w'})_i + \frac{\partial}{\partial x}\left[\frac{\tau(v)_{yxs}}{\rho} h_s\right] + \frac{\partial}{\partial y}\left[\frac{\tau(v)_{yys}}{\rho} h_s\right] \tag{5}$$

[Momentum Equations in the Lower Layer in the Main Channel]

$$\frac{\partial M_b}{\partial t} + \frac{\partial \beta u_b M_b}{\partial x} + \frac{\partial \beta v_b M_b}{\partial y} + (wu)_i + gh_b \frac{\partial h_s}{\partial x} = gh_b \sin\theta - \frac{gn_m^2}{(h_b + h_s)^{1/3}} u_b \sqrt{u_b^2 + v_b^2}$$

$$+ \frac{\partial(-\overline{u'^2})_b h_b}{\partial x} + \frac{\partial(-\overline{u'v'})_b h_b}{\partial y} - (-\overline{u'w'})_i + \frac{\partial}{\partial x}\left[\frac{\tau(\nu)_{xxb}}{\rho} h_b\right] + \frac{\partial}{\partial y}\left[\frac{\tau(\nu)_{xyb}}{\rho} h_b\right]$$ (6)

$$\frac{\partial N_b}{\partial t} + \frac{\partial \beta u_b N_b}{\partial x} + \frac{\partial \beta v_b N_b}{\partial y} + (wv)_i + gh_b \frac{\partial h_s}{\partial y} = -\frac{gn_m^2}{(h_b + h_s)^{1/3}} v_b \sqrt{u_b^2 + v_b^2}$$

$$+ \frac{\partial(-\overline{v'u'})_b h_b}{\partial x} + \frac{\partial(-\overline{v'^2})_b h_b}{\partial y} - (-\overline{v'w'})_i + \frac{\partial}{\partial x}\left[\frac{\tau(\nu)_{yxb}}{\rho} h_b\right] + \frac{\partial}{\partial y}\left[\frac{\tau(\nu)_{yyb}}{\rho} h_b\right]$$ (7)

[Momentum Equations in the Flood Plain]

$$\frac{\partial M_s}{\partial t} + \frac{\partial \beta u_s M_s}{\partial x} + \frac{\partial \beta v_s M_s}{\partial y} + gh_s \frac{\partial h_s}{\partial x} = gh_s \sin\theta - \frac{gn_f^2}{h_s^{1/3}} u_s \sqrt{u_s^2 + v_s^2}$$

$$+ \frac{\partial(-\overline{u'^2})_s h_s}{\partial x} + \frac{\partial(-\overline{u'v'})_s h_s}{\partial y} + \frac{\partial}{\partial x}\left[\frac{\tau(\nu)_{xxs}}{\rho} h_s\right] + \frac{\partial}{\partial y}\left[\frac{\tau(\nu)_{xys}}{\rho} h_s\right]$$ (8)

$$\frac{\partial N_s}{\partial t} + \frac{\partial \beta u_s N_s}{\partial x} + \frac{\partial \beta v_s N_s}{\partial y} + gh_s \frac{\partial h_s}{\partial y} = -\frac{gn_f^2}{h_s^{1/3}} v_s \sqrt{u_s^2 + v_s^2}$$

$$+ \frac{\partial(-\overline{v'u'})_s h_s}{\partial x} + \frac{\partial(-\overline{v'^2})_s h_s}{\partial y} + \frac{\partial}{\partial x}\left[\frac{\tau(\nu)_{yxs}}{\rho} h_s\right] + \frac{\partial}{\partial y}\left[\frac{\tau(\nu)_{yys}}{\rho} h_s\right]$$ (9)

where subscript "s" and "b" designate the value in the upper and lower layer, respectively; subscript "i" denote the value at the interface between two layers ; t = the time; h = the depth; u and v = x (streamwise) and y (lateral) components of layer-averaged velocity; w = the velocity components in z (vertical) direction; M and N = the discharge flux defined as hu and hv, respectively; $-\overline{u_i' u_j'}$ = the layer-averaged Reynolds stress tensors (u_1'=u', u_2'=v'); $\tau(\nu)_{ij}$ = the layer averaged molecular viscosity tensors; n_m and n_f = Manning roughness coefficient in the main channel and in the flood plain, respectively; θ =bed slope; ρ =density of water; g=acceleration due to gravity.

The simple model presented in Eq.10 was used to evaluate the layer-averaged Reynolds stress tensors.

$$-\overline{u_i' u_j'} = D_h \left(\frac{\partial u_i}{\partial x_j} + \frac{\partial u_j}{\partial x_i}\right) - \frac{2}{3} k \delta_{ij}$$ (10)

where u* = the local friction velocity ($\equiv \sqrt{gn^2(u^2+v^2)/h^{1/3}}$), k = depth-averaged turbulent kinetic energy evaluated by the empirical formula proposed by Nezu and

Table 1 Hydraulic conditions used in the numerical simulation

n_s	n_b	$\nu\,(m^2/s)$	α	$\Delta t\,(sec)$	bed slope	Re	Fr
0.011	0.011	0.1×10^{-6}	0.1	0.0005	1/1000	14500	0.517

Re : Reynolds number in main channel , Fr : Froude number in main channel , Δt : time increment

Nakagawa (1993), who proposed the universal expression, Eq.11, for turbulent kinetic energy distribution :

$$\frac{k}{u_*^2} = 4.78 \exp\left(-2\frac{z}{h}\right) \tag{11}$$

The depth-averaged turbulent kinetic energy is $2.07u_*^2$ when Eq.11 is integrated from the bottom to the surface. B in Eq.10 is the function as described below used by Tominaga (1987) to reproduce the log-law velocity distribution near the side wall.

$$B = \begin{cases} 4.0\frac{h'}{h_w}\left(1.0 - \frac{h'}{h_w}\right) &, \quad h' \leq h_w/2 \\ 1.0 &, \quad h' > h_w/2 \end{cases} \tag{12}$$

where in the upper layer, h' = the distance from the side wall of the flood plain; h_w = the depth in the flood plain, whereas in the lower layer, h' = the distance from the side wall of the main channel; h_w = the thickness of the lower layer. In the upper layer in the main channel, the eddy viscosity coefficient D_h near the junction is evaluated by Eq.13 to introduce the effect of flow in the flood plain.

$$D_h\big|_{y=y'} = \begin{cases} \alpha[(h_s+h_b)u_*]_{y=y'} B\left(\frac{y'}{3h_{s0}}\right) + \alpha[h_s u_*]_{y=+0}\left\{1-\left(\frac{y'}{3h_{s0}}\right)\right\} &, \quad -3h_{s0} \leq y' \leq 0 \\ \alpha[(h_s+h_b)u_*]_{y=y'} &, \quad y' < -3h_{s0} \end{cases} \tag{13}$$

where y=+0 denotes the depth definition point which is the closest to y=0 (the junction between the main channel and the flood plain). It is assumed in Eq.13 that the D_h near the junction can be interpolated by D_h of the flood plain at the junction and of the main channel.

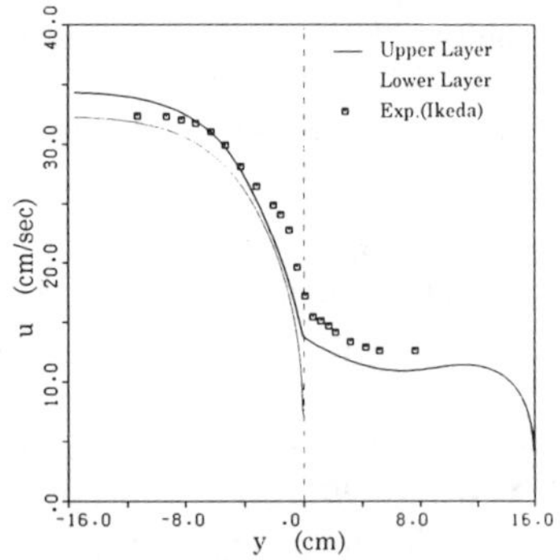

The initial and boundary conditions of the calculation are
- Initial conditions
  Initial velocities in the main channel and in the flood plain are evaluated by the Manning law, respectively.
- Boundary conditions
  The streamwise length of the calculation area is 8m. The periodic boundary conditions are used in the upstream and downstream ends of the calculation area. The non-slip condition is used for the velocity adjacent to the side-wall.

Fig.2 Time-averaged velocity distributions

Numerical simulation with the basic equations Eqs.1 - 9 is done under the hydraulic conditions of the laboratory test by Ikeda et al (1995). The hydraulic conditions are listed in Table 1. The finite volume method with the QUICK scheme for the convective inertia term is adopted as the numerical method. The Adams-Bashforth method with 2nd order accuracy is used for the time integration.

CONSIDERATION OF NUMERICAL RESULTS

The time averaged velocity distributions along the lateral section is compared with the

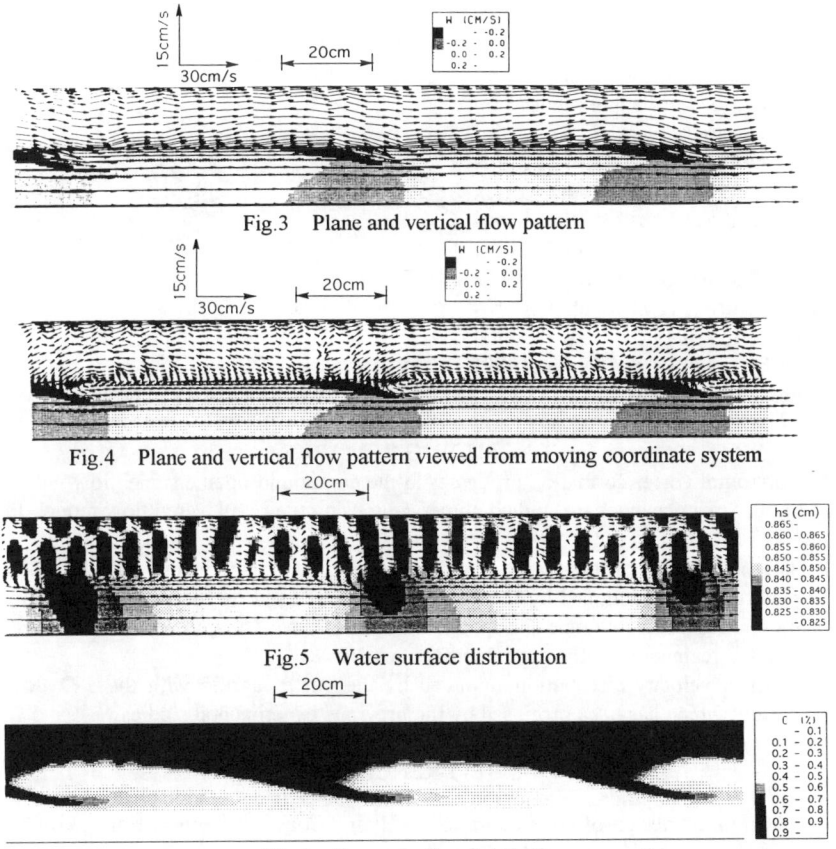

Fig.3 Plane and vertical flow pattern

Fig.4 Plane and vertical flow pattern viewed from moving coordinate system

Fig.5 Water surface distribution

Fig.6 Concentration distribution

experimental results (Fig.2). The calculated results are in good agreement with the experimental ones.

Fig.3 shows the velocity vectors in the upper layer and Fig.4 shows the same one viewed from the coordinate system which moves at the same speed as the vortex convection (16cm/s). Fig.5 shows the water surface distribution. These figures indicate that the horizontal vortex is reproduced numerically and the time and spatial intervals of vortices are about 5.0 sec and about 0.9 m, respectively. On the other hand, those of the experimental results are 0.73 m and 3.9 sec, respectively. The time and spatial intervals of the calculated vortices are longer than the experimental ones. Fig.6 shows the mass concentration distribution obtained by the 14 sec calculation after the initial conditions (c=0 in the main channel, c=1.0 in the flood plain, where c= concentration). It is pointed out that the mass transportation in the compound open channel flows is dominated by the horizontal vortices.

In Fig.3 and Fig.4, the vertical velocity distributions are superimposed on the plane velocity vectors. These figures show that the vertical velocity is negative in upstream of the vortex center, whereas positive in downstream of the vortex center. These vertical flow pattern suggests the 3-D flow structure (Fig.7) which was proposed in the previous experimental studies by Fukuoka and Fujita (1989) and Ikeda et al.(1995).

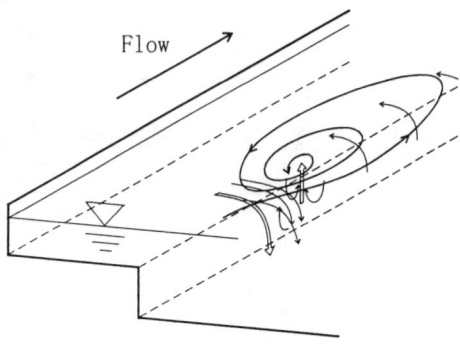

Fig 7 3-D flow pattern induced by the vortex

CONCLUDING REMARKS

The horizontal vortex formation processes in the compound open channel flow with a rectangular main channel are studied numerically using the two-layered flow model. In summary:

1) The horizontal vortices at the junction between the main channel and the flood plain can be reproduced by the numerical model based on the two-layered flow model. However, the time and spatial intervals of the calculated vortices are not in good agreement with the experimental ones.
2) Vertical velocity distributions induced by the vortex agrees with the 3-D flow pattern which has been proposed by the previous experimental studies.

REFERENCES

Fukuoka, S. and Fujita, K. "Prediction of flow resistance in compound channels and its application to design of river courses." J. Hydr., Coast. and Envir. Engng, JSCE, No.411/ II -12, 63-72 (in Japanease).

Hosoda,T. and Kimura,I.(1993). "Vortex formation with free surface variations in the shear layer of plane-2D open channel flows." Proc. of 9th Symp. on Turbulent Shear Flows, Kyoto, Japan, Vol.1, P112, 1-4.

Ikeda,S., Murayama,N. and Kuga,T (1995). "Stability of horizontal vortices in compound open channel flow and their 3-D structure." J. Hydr., Cast. and Envir. Engng, JSCE, No.509/ II -30, 143-154 (in Japanease).

Liu, B. Y. (1991). "Study on sediment transport and bed evolution in compound channels." Ph.D Thesis, Kyoto University.

Nadaoka,K and Yagi,H.(1993). "Horizontal large-eddy computation of river flow with transverse shear by the SDS & 2DH model." J. Hydr., Coast. and Envir. Engng, JSCE, No.473/ II -24, 35-44 (in Japanease).

Naot,D. Nezu,I. and Nakagawa,H.(1993). "Hydrodynamic behavior of compound rectangular open channels." J. Hydr. Engng, ASCE, Vol.119, No.3, 390-408.

Nezu,I. and Nakagawa,H.(1993). *Turbulence in open channel flows*. IAHR Monograph, Blkema, Rotterdam, 53-56.

Tominaga, A.(1987). "Tree dimensional turbulence structure of flows in a straight open channel." PhD Thesis, Kyoto University, Japan (in Japanese).

2-D MODELS FOR FLOWS IN THE RIVER WITH SUBMERGED GROINS

SHIROU AYA[1], ICHIRO FUJITA[2], NOBUYUKI MIYAWAKI[3]
[1]Department of Civil Engineering, Osaka Institute of Technology
Ohmiya, Asahi-ku, Osaka 535, Japan
[2]Department of Civil Engineering, Gifu University
Yanagido, Gifu 501-11, Japan
[3]Kyushu Branch, CTI Engineering Co. Ltd.,
2-1-10 Watanabedori, Chuo-ku, Fukuoka 810, Japan

ABSTRACT

The 2-D mathematical models for river flows in a channel with submerged groins have been developed in the generalized curvilinear coordinate system to investigate the flow behavior in a complicated channel. The effect of a groin was described as a counter force of the fluid force acting on a submerged groin. They were applied to the flood flows in the Yodo River and the velocity distributions were successfully compared with those observed by using PIV of the video images of flows.

INTRODUCTION

The river improvement works in the Yodo River system have more than 100 years history. Their purpose had been the maintenance of the navigation channel, but it was changed to flood protection and water utilization, and the Yodo River has a two/three-stage compound channel at present (Fig.1). A lot of groins were constructed for the river management on the middle stage of the channel until 1950's, but the last improvement works in 1970's dredged the low flow channel and deconstructed almost of groins, and the water level has been kept about 70cm higher than before. However, the lower reach of the Yodo River is the habitat of the protected fish, Acheilognathus longipinnis, and 24 groins have still been reserved between 11.8km and 14.6km, and they work as submerged groins at present. This paper will present the mathematical models for the hydraulic analysis of such complicated waters as the Yodo River, that is, the 2-D models for flows in the river channel with submerged groins in the curvilinear coordinate system, and the results of its application to the Yodo River will also be examined with those obtained by the field observation by using the Particle Image Velocimetry (PIV) for the surface velocity distribution of the river flows.

MATHEMATICAL MODELS

MODELING A SUBMERGED GROIN

The effect of a submerged groin in the river channel is modeled as a counter force acting on water for a fluid force acting on a submerged groin. Referring to the fluid force acting on a wing, the drag D and the lift L acting on a submerged groin can be described in the Cartesian coordinate system as follows (Fig. 2).

$$D = \frac{1}{2}\rho C_D A_G \sin \alpha U |U| \qquad (1)$$

$$L = \frac{1}{2}\rho C_L A_G \sin \alpha U |U| \text{SIGN}(\cos \alpha) \qquad (2)$$

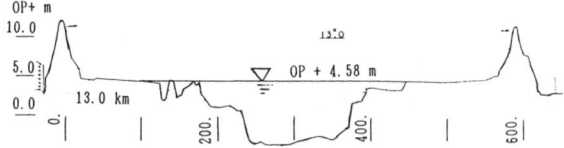

Fig. 1 The cross-section of the Yodo River at 13.0km.

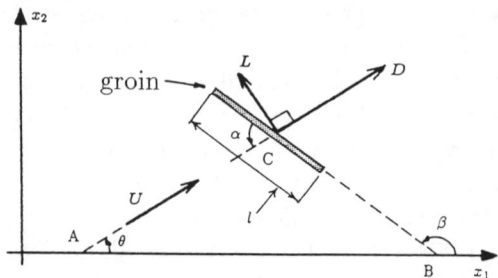

Fig. 2 Forces acting on a submerged groin.

where A_G is the projected area of a groin to the flow direction and equal to $l_G h_G$, l_G the groin length, h_G the groin height, C_D the drag coefficient, C_L the lift coefficient, α the attack angle, ρ the density, and SIGN(a) signature function. Denoting the angle between the groin and the x_1 directions by β, and that between the flow and the x_1 directions by θ, α is described as follows.

$$\alpha \equiv \theta + \pi - \beta \qquad 0 \leq \alpha \leq \pi \tag{3}$$

Using Eqs. (1) and (2), E_i the x_i component of the counter force for the sum of the drag and the lift is formulated as:

$$\begin{pmatrix} \overline{E_1} \\ \overline{E_2} \end{pmatrix} = \begin{pmatrix} \cos(-\theta) & \sin(-\theta) \\ -\sin(-\theta) & \cos(-\theta) \end{pmatrix} \begin{pmatrix} D \\ L \end{pmatrix} \tag{4}$$

Therefore, the external force terms $\overline{F_i}$ in the 2-D momentum equations can be described as:

$$\overline{F_i} = -\frac{1}{hS^x}\frac{\overline{E_i}}{\rho}\delta(x_1 - x_{1k})\delta(x_2 - x_{2l}) \tag{5}$$

where S^x is the area in the x_1-x_2 plane of the control volume and equal to $dx_1 dx_2$, h the depth, δ Dirac's delta function used to indicate the existence of a groin, and x_i the Cartesian coordinate system.

2-D MODELS IN THE CURVILINEAR COORDINATE SYSTEM

Actual river has complicated channel geometry. Therefore mathematical models are required to have not only high accuracy but also high resolution for channel geometry, and this paper uses the depth-averaged 2-D models in the generalized curvilinear coordinate system. The models are derived by the transformation of the 2-D depth-averaged models in the Cartesian coordinate system into those in the generalized curvilinear coordinate system by use of the relationship:

$$U^i = \frac{\partial \xi_i}{\partial x_j} U_j \qquad (i, j = 1, 2) \tag{6}$$

where U^i is the ξ_i contravariant component of the depth-averaged velocity vector, U_i x_i component of the depth-averaged velocity, ξ_i the generalized curvilinear coordinate. And the final forms of the mathematical models are obtained as follows.
Continuity equation:

$$\frac{\partial}{\partial t}\left(\frac{\zeta}{J}\right) + \frac{\partial}{\partial \xi_j}\left(\frac{hU^j}{J}\right) = 0 \tag{7}$$

Momentum equation in the ξ_1 direction:

$$\frac{\partial U^1}{\partial t} = +J\frac{\partial x_1}{\partial \xi_2}\left[\frac{\partial}{\partial \xi_j}\left(U^j\frac{\partial x_2}{\partial \xi_i}U^i\right)\right] - J\frac{\partial x_2}{\partial \xi_2}\left[\frac{\partial}{\partial \xi_j}\left(U^j\frac{\partial x_1}{\partial \xi_i}U^i\right)\right]$$
$$+U^1\left(\frac{\partial U^j}{\partial \xi_j}\right) + \frac{J^2}{h}\left[\frac{\partial x_2}{\partial \xi_2}\left(\frac{\partial}{\partial \xi_j}\left(\frac{h\,\overline{\tau^{1j}}}{J\,\rho}\right)\right) - \frac{\partial x_1}{\partial \xi_2}\left(\frac{\partial}{\partial \xi_j}\left(\frac{h\,\overline{\tau^{2j}}}{J\,\rho}\right)\right)\right]$$
$$-gJ^2\left(G_{22}\frac{\partial \zeta}{\partial \xi_1} - G_{12}\frac{\partial \zeta}{\partial \xi_2}\right) - \frac{1}{h}\frac{\tau^{1b}}{\rho} + \overline{F^1} \tag{8}$$

Momentum equation in the ξ_2 direction:

$$\frac{\partial U^2}{\partial t} = -J\frac{\partial x_1}{\partial \xi_1}\left[\frac{\partial}{\partial \xi_j}\left(U^j\frac{\partial x_2}{\partial \xi_i}U^i\right)\right] + J\frac{\partial x_2}{\partial \xi_1}\left[\frac{\partial}{\partial \xi_j}\left(U^j\frac{\partial x_1}{\partial \xi_i}U^i\right)\right]$$
$$+U^2\left(\frac{\partial U^j}{\partial \xi_j}\right) - \frac{J^2}{h}\left[\frac{\partial x_2}{\partial \xi_1}\left(\frac{\partial}{\partial \xi_j}\left(\frac{h\,\overline{\tau^{1j}}}{J\,\rho}\right)\right) - \frac{\partial x_1}{\partial \xi_1}\left(\frac{\partial}{\partial \xi_j}\left(\frac{h\,\overline{\tau^{2j}}}{J\,\rho}\right)\right)\right]$$
$$-gJ^2\left(-G_{21}\frac{\partial \zeta}{\partial \xi_1} + G_{11}\frac{\partial \zeta}{\partial \xi_2}\right) - \frac{1}{h}\frac{\tau^{2b}}{\rho} + \overline{F^2} \tag{9}$$

where $\overline{\tau^{ij}}$ is the turbulent and dispersion stress and the gradient type transport can be usable, ζ water surface elevation, J the transformation Jacobian, and G_{ij} the metric tensor. $\overline{F^i}$ is the external force term caused by a groin, and using Eq.5, it is described as:

$$\overline{F^i} = \frac{\partial \xi_i}{\partial x_j}\overline{F_j} \tag{10}$$

APPLICATION TO THE YODO RIVER

SIMULATION CONDITIONS

The 2-D models described above were applied to the lower reach of the Yodo River, and two sets of the numerical experiment were conducted: a) the long area test, and b) the short area test: to examine the effects of the submerged groins, and its simulation area was from 12.8km to 13.4km. The grid size was about 5m x 2.5m. C_D was selected at 3.8, and C_L 0.3 after Fukuoka et al (1995). The eddy viscosity was estimated at 1 m^2/s. The selected flow was the flood in July, 1993, and its peak discharge was estimated at 3800 m^3/s.

PIV FOR THE SURFACE VELOCITY DISTRIBUTION
PIV uses the surface patterns in brightness caused by drifting woods, foams, and small unevenness of the water surface as the tracer in image processing, and its details and advantage were already presented in

Three dimensional modeling of flow and transport mechanisms in meandering two-stage channel flows

J.Russell Manson and Gareth Pender
Bucknell University, Lewisburg, U.S.A. and Glasgow
University, Glasgow, U.K.

ABSTRACT
Results are presented of a computational study of turbulent free surface flow in a meandering two-stage channel flowing in overbank mode. The three dimensional Reynolds Averaged Navier-Stokes equations were solved using an operator splitting technique with a linear constitutive law employed for the turbulent stress-strain relationship. The computational results are compared qualitatively with experimental measurements. Comparisons are drawn for the mean flow field, the turbulence kinetic energy and the water surface elevation. The important mechanisms are highlighted as well as the models ability and inability to capture them.

Introduction
Overbank flows in two-stage meandering channels remain a considerable challenge to analysis by theoretical models. Advancements made in the field of experimental data collection have greatly increased our understanding of the flow physics. The translation of this understanding into reliable predictive procedures has been only partially successful. The aim of the present paper is to contribute to the debate on modeling these flows and to present some results of the authors' simulations. The reader is refered to Sellin, Ervine and Willetts (1993) for a description of the important flow mechanisms in meandering two-stage channels. A significant structure is the strong shear layer at the cross-over region. Flow expelled from the main channel, at this cross-over region, will carry with it scalars (turbulence, sediment, pollutant) into a wake region on the floodplain downstream of the cross-over. The computational grid used to model these mechanisms requires significant computer resources (storage and speed).

Description of the Present Problem
The test case geometry is taken from Kiely (1989) is depicted in figure 1(a). It is basically a single meander with main channel flow depth 8 cm and flood plain depth 3 cm. The total width of the channel is 120 cm. and the width of the main channel is 20 cm. The bed slope is set at 0.001 and the flow rate is 14.1 litres/sec. The sinuosity is 1.25, the aspect ratio of the main channel is 4 and the depth ratio is 0.38. Some of the numerical predictions were compared with other experimental data with similar relative depths and sinuosities to Kiely (1989). The other results used for qualitative comparison were Schroder, Stein and Rouve (1991) and one of the S.E.R.C. Series B experiments.

Computational Model,Grid and Boundary Conditions
Unfortunately, space does not permit a adequate description of the computational model however the reader is referred to Pender and Manson (1994) for a fuller description.

Basically the fully three dimensional Reynolds averaged Navier-Stokes equations are solved by a fractional step projection method with appropriate boundary conditions. A rigid lid assumption is used to represent the free surface and a relatively coarse grid was adopted, figure 1(b), which was cartesian and so had to be chosen to closely approximate the real geometry in a staircase fashion as also tried by McGuirk and Palma (1992). The inflow velocity profile adopted was a very simple approximation to the experimental values. Turbulence parameters at the inlet were also order of magnitude approximations. Boundary conditions at the inlet were specified as steady and the model was run until steady conditions prevailed throughout the computational region. The present study uses a grid 61 x 49 x 17. The computer runs took place on an IBM 3090-VF at the University of Glasgow, U.K.

Primary Velocity Field

The velocity profiles in the lateral direction at several flow depths are shown in figure 2(a) for sections 25, 31, 35 and 39. These correspond to the first bend apex, two sections in the cross-over region and the second bend apex respectively. The bankfull level is at 5.0 cm. Kiely's results (1989) are shown in figure 2(b) for the equivalent locations. Some of the mechanisms appear to be reproduced. In particular, figure 2 shows: (i) The velocities in the main channel are lower than on the floodplain; (ii) the velocities on the floodplain increase with depth above the floodplain; (iii) there is a transfer of fluid from the main channel to the floodplain; and (iv) for the downstream section in the cross-over region and at the downstream bend apex the velocity profile has a minimum mid-way between the main channel and the edge of the floodplain. This is evidence of the wake effect observed by Kiely (1989). To aid in the visualization of the results a streamline plot was constructed. Figure 3 shows that some streamlines which begin in the main channel continue down through the main channel while others move out onto the floodplain. This corroborates the observations of Sellin, Ervine and Willetts (1993).

Secondary Velocity Field

The secondary velocity field was compared qualitively with the results of Schroder, Stein and Rouve (1991). The comparison was disappointing however since the numerical model results give no clear indication of the same secondary flow structure as experiment. A reverse flow was observed at the channel bed but a downward plunging flow was not reproduced at the channel centre. Kiely (1989) also found it difficult to isolate the secondary flow structure owing to the dominating effect of the flow explusion onto the floodplain. The vigorous explusion of main channel water onto the right flood plain was captured in fact the flow structure was dominated by this mechanism. Inadequate grid resolution is almost certainly the reason for the numerical model's inability to resolve this feature.

Water Surface Elevation

The total pressure in the surface cell is separated into a uniformly sloping component due to the gravitational force and a deviation from this constant slope. Figure 4(a) shows the deviation from the uniform slope. Figure 4(b) shows a contour plot of depth from one of the S.E.R.C. Series B experiments. Qualitatively, these results are very encouraging. The computations show, in agreement with experiment, regions of maximum water level within the main channel immediately prior to the floodplain just after the bend apex. These are followed on the floodplain by regions of minimum. This provides,for the first time, mathematical evidence that the expansion contraction flow, postulated by Ervine and Ellis (1987) occurs in such systems. This phenomenon has previously been verified by experiment but had not been verified numerically. This is an important capability of the model as it will be possible to evaluate energy loss coefficients from the results. This will enable the computation of coefficients for simpler conceptual models of meandering two stage channel flow, Wark (1993). At either end of the meander away from the main channel curvature, the surface smoothly returns to a longitudinally uniform slope.

Turbulence Characteristics

A direct comparison is not possible owing to the way Kiely (1989) has defined turbulence intensity. However utilising the fact that the normalised kinetic energy is proportional to the square of the turbulence intensity some observations can be made. Figure 5 shows Kiely's measured turbulence intensity and the model prediction for the normalised turbulent kinetic energy both at 1.0 cm above the floodplain bed. A favourable comparison between the two plots is observed. Note that only the normalised turbulent kinetic energy contours above 0.01 are plotted which would correspond to turbulence intensities of 10% (0.1). Turbulence is generated most noticeably at the cross-over region where the velocity gradients are strongest. The turbulence generated here is carried on to the floodplain with the mean flow expelled from the main channel. This leads to higher turbulent mixing in this region. This is the wake effect observed by Kiely (1989).

Summary of Work and Conclusions

The three dimensional simulation was successful in capturing the following: a vigorous explusion of water from the main channel onto the floodplain; trends in the water elevation surface; the wake effect, i.e. a minimum velocity on the floodplain midway between the main channel and the floodplain walls, and higher turbulence levels in the wake. The three dimensional simulation did not convincingly capture the the secondary motions in the cross-over region probably due to grid coarseness. However, the authors' view is that research efforts should initially concentrate on obtaining accurate grid independent solutions for a wide variety of geometries and flow rates with a linear two equation model of turbulence before (if ever) moving on to Reynolds Stress Transport Models. Although it can be argued theoretically that the R.S.T.M. captures more flow physics, in practical terms these features may not be significant. Sellin, Ervine and Willetts (1992) suggest, when considering both low and high sinuosities, that, '...... *The co-flowing lateral shear stress (zero mass transfer), so influential in the straight channel case, is insignificant at both the sinuousities here. The mechanisms arising from the cross-flow and driven by horizontal shear layers are much more important* ....'

References

Keily,G. (1989), 'An Experimental Study of Overbank Flow in Straight and Meandering Compound Channels', **Thesis submitted to University College, Cork, Ireland towards Degree of Doctor of Philosophy**

Sellin,R.H.J., Ervine,D.A. and Willetts,B.B. (1993), 'Behaviour of Meandering Two-stage Channels', **Proc.Inst.Civil Engineers (U.K.) - Water, Maritime and Energy,** Vol.101, Mar, pp 37.

Pender, G. Manson,J.R. (1994), 'Developments in Three Dimensional Numerical Modelling of River Flows', in **River Flood Hydraulics**, Editors:White and Watts, proceedings of the 2nd international conference, 21-25 March, York, England.

Schroder,M., Stein,C.J. and Rouve,G. (1991), 'Application of the 3D-LDV technique on physical model of meandering channel with vegetated flood plain', 4th International Conference on Laser Anemometry (Advs. Apps.) , August, Cleveland, Ohio.

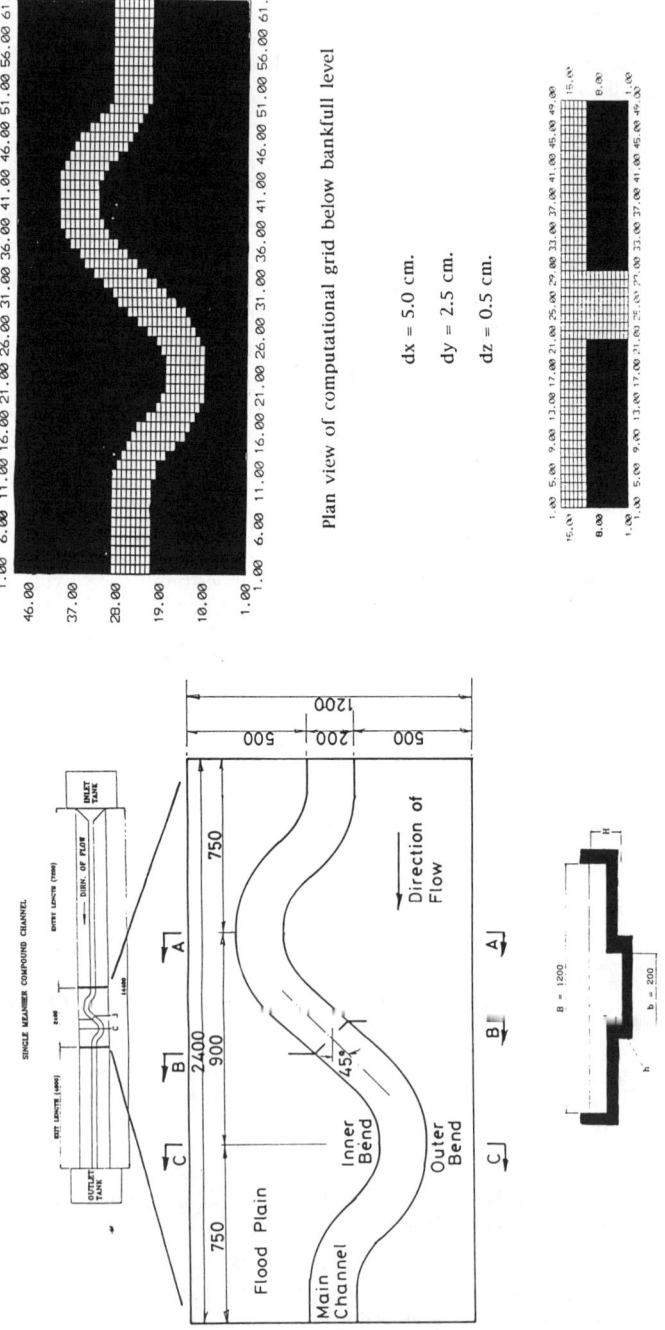

Figure 1(b) Computational grid

Figure 1(a) Test case geometry (from Kiely (1989))

MEANDERING TWO-STAGE CHANNEL FLOWS 839

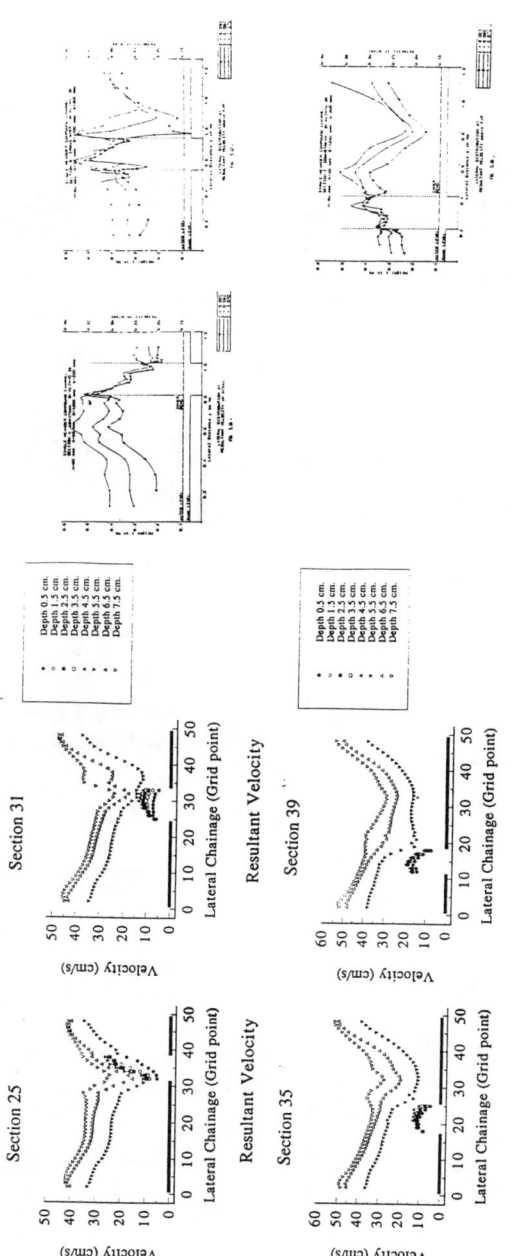

Figure 2(b) Measured velocities (after Kiely (1989))

Figure 2(a) Numerical predictions for primary velocity at different depths

840 COPING WITH SCARCITY AND ABUNDANCE

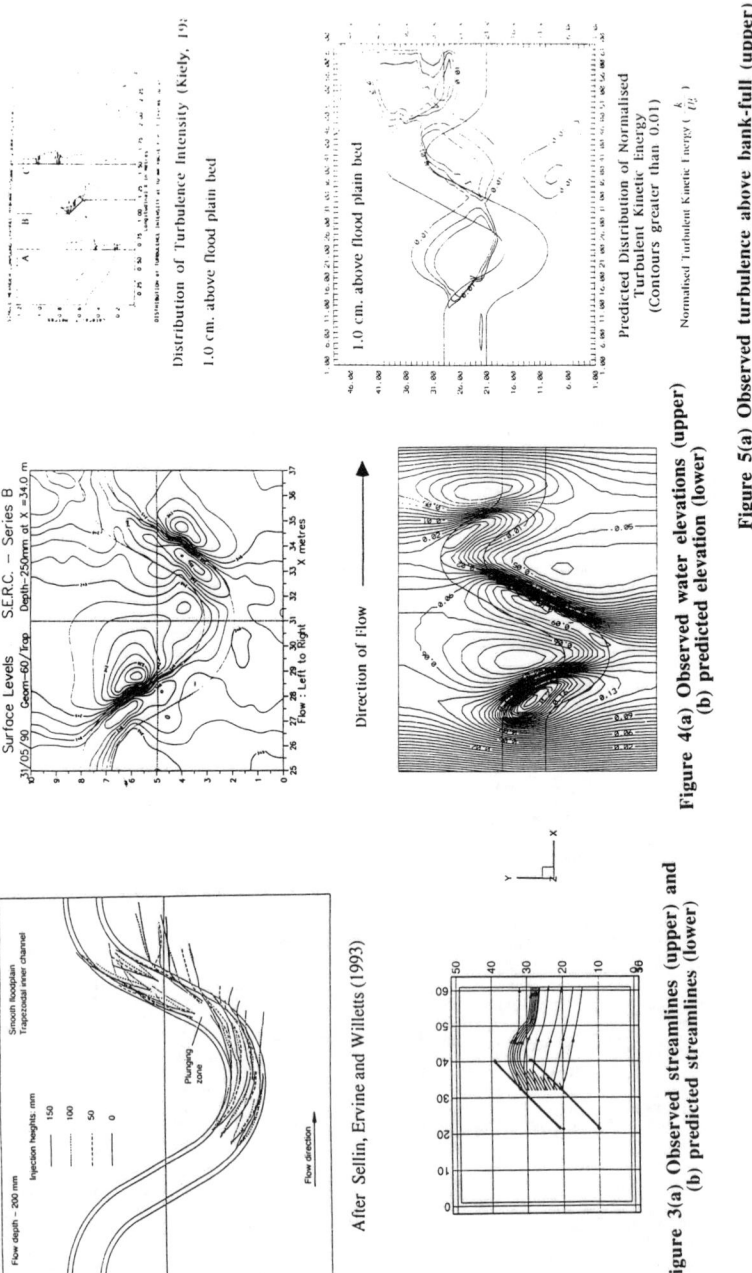

Figure 3(a) Observed streamlines (upper) and (b) predicted streamlines (lower)

Figure 4(a) Observed water elevations (upper) (b) predicted elevation (lower)

Figure 5(a) Observed turbulence above bank-full (upper) (b) predicted turbulence above bank-full (lower)

MODELING FOR STUDY OF THE HYDROLOGIC PROCESSES IN WATERSHED

ADILSON PINHEIRO, PHILIPPE MAISON AND BERNARD CAUSSADE
INP - ENSEEIHT - Institut de Mécanique des Fluides de Toulouse
UMR CNRS-INP/UPS 5502
Avenue du Professeur Camille Soula - 31400 - Toulouse - France
Tél. (33) 5 61 28 58 02 - Fax (33) 5 61 28 58 99 - email: caussade@imft.fr

ABSTRACT
In this paper, a physical approach is presented in order to model the water flow in the soil and the transfer in the soil-plant-atmosphere system of a distributed conceptual hydrologic model. It is based on the resolution of Richards' Equation applied to the unsaturated zone of the soil. Groundwater flow is simulated by an exponential law. Water fluxes in the soil-vegetation-atmosphere system take into account soil surface changes (vegetated soil, bare soil, etc.). Those could be supplied by remote-sensing.
This approach has been validated by means of comparison with data measured at both experimental plot scale and representative and experimental catchment scale, where physical characteristics as porosity, soil granulometric composition and soil use are known.

INTRODUCTION
One of the main challenges in modern hydrology consists in making realistic prediction of the soil heterogeneity impact on the watershed hydrologic regime. Moreover this problem necessitates connection with meteorological models in order to predict climate variations and their impact. The study of these problems entails a better comprehension of interactions between climate, soil and vegetation and, other hand of conjoined effects with topography in flow generation at local or higher scales.
Concerning modeling, a conceptual point of view allows to solve some difficulties because of the simplification of the physical concept used to represent the terrestrial water cycle. Nevertheless, in this case, some variables are often neglected or simply implicitly taken into account. This is more particularly the case for soil flow and for soil-plant-atmosphere transfer. Results are besides global.
A hydrological processes representation on watershed is realistically realized with physically based models (Beven, 1989). Natural or anthropic changes of watershed physical characteristics and spatial and temporal variability of input and output are taken into account by such kind of models. In these models, internal mechanisms are based on mass, energy and momentum conservation laws. Moreover, soil-plant-

atmosphere transfer is driven by models usually called SVATS (Soil-Vegetation-Atmosphere Transfer Scheme).

In this paper, we present a physically based approach of water flow in the soil and transfer in the soil-plant-atmosphere system in the distributed conceptual model CEQUEAU (Morin et al., 1981). Our aim is to model physically the vertical water flow in the soil and so to take into account soil characteristics which are measurable parameters. We are also interested in the representation of soil-plant-atmosphere balance. These representation of hydrological phenomena are based on the Hutson and Wagenet model (1989).

The CEQUEAU model leans on a spatial representation of the watershed in squares called *whole-squares*, which are physically homogeneous by averaging. These squares are then divided in partial squares with regards to the flow direction. This second division governs water transfer on the hydrographic network which is realized in between partial squares.

This model consists of two functions. First, the production function, which represents the different hydrological processes from the arrival of atmospheric water on the ground to the moment at which this water joins the watercourse. The second one is the transfer function which governs in time and in space the formation and evolution of water flow up to the watershed outlet. The production function leans on a fictitious reservoir concept.

UNSATURATED FLOW

Water flow in the unsaturated zone of the soil, at the scale of whole-square, is governed by Richards' equation:

$$\frac{\partial h}{\partial t} C(\theta) = \frac{\partial}{\partial z}\left[K(\theta)\frac{\partial H}{\partial z}\right] - U(z,t)$$

where θ is the volumetric water content, h is the soil water pressure head, $C(\theta)$ is the differential water capacity which is the slope of the $\theta(h)$ curve, $K(\theta)$ is the hydraulic conductivity, H is the hydraulic head and U is a sink term representing water lost per unit time by transpiration.

The Crank-Nicholson implicit method is used to solve this equation. The symmetric tridiagonal matrix obtained is then solved by using the Thomas algorithm. This resolution provides the values of h, and then flux and flux density.

The retentivity function is composed of two parts: the first one is described by an exponential law (Campbell, 1974)

$$h(\theta) = a(\theta/\theta_s)^{-b}$$

and the second one is represented by a parabolic equation (Hutson et Cass, 1987)

$$h(\theta) = \frac{a(1-\theta/\theta_s)^{1/2}(\theta_c/\theta_s)^{-b}}{(1-\theta_c/\theta_s)^{1/2}}$$

where θ_s is the volumetric water content at saturation and a and b are constants. The constant a is sometimes regarded as an air-entry value. The point of intersection of these two curves is: $h_c = a[2b/(1+2b)]^{-b}$ and $\theta_c = 2b\theta_s/(1+2b)$.
The hydraulic conductivity at the water content θ is calculated according to Campbell (1974):

$$K(\theta) = K_s(\theta/\theta_s)^{2b+2+p}$$

where K_s is the hydraulic conductivity at saturation and p is a pore interaction.
The equations governing root growing are those from Tillotson et al. (1980). The presence of plants allows the calculation of crop cover fraction by a sigmoidal empirical equation. The uptake water rate is calculated according to Nimah et Hanks (1973):

$$U_i^{j-1/2} = \left[h_{root} + z_i(1-R_c) - h_i^{j-1/2} - s_i^{j-1/2}\right]\left[RDF_i^{j-1/2}K_i^{j-1/2}/\Delta_x\Delta_z\right]$$

where h_{root} is the root water potential at the soil surface, $1-R_c$ is the root water resistance term, s_i is the osmotic potential, RDF_i is the proportion of total active roots at the node i, z_i is the depth at the node i, Δx is the distance into the soil from the measurement point of h and s, and j is the time reference.

EVAPOTRANSPIRATION

The daily potential evapotranspiration (ETP) is calculated from the Thornthwaite formula. This value is divided into two parts: a potential evaporation (EP) which is produced by bare soil and a potential transpiration (TP) coming from vegetated soil. This sharing varies as the crop cover index. Actual transpiration (TR) is calculated in terms of plant water absorption.

OVERLAND AND GROUNDWATER FLOW

Overland and groundwater flows are modeled using a linear reservoir expression. In this case, the propagation coefficient may vary as aquifer hydrodynamic characteristics such as transmissivity, hydraulic conductivity and specific storativity.

APPLICATIONS AND RESULTS

The Experimental Field of Poucharramet has been chosen to validate the model. This field is located in the south west of France and is composed of 4 drained plots of about 1.1 ha each. On this site, the data basis were obtained by Guiresse (1996) from the Laboratoire de Pédologie of ENSAT concerning drained discharges, by Koreta (1996) about water content vertical transects and by Météorologie de France regarding climatologic data. Simulations have been realized between 1992 and 1994. Each year, from the 16th week, maize is cultivated on these plots
The model provides results at both local and global scale. Figure 1 presents vertical transect of water content on the 11/11/92. We notice that the model gives an effective simulation of the vertical heterogeneities of the soil. Figure 2 shows the comparison between measured and calculated hydrographs in drains. We observe that the model seems to simulate correctly the measured values. A comparison of results realized with our physically-based model and with CEQUEAU is presented on figure 3.

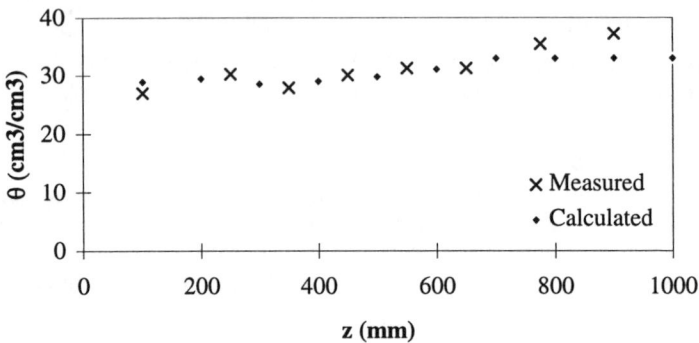

Figure 1 - Measured and calculated humidity vertical transects on the plot 2.

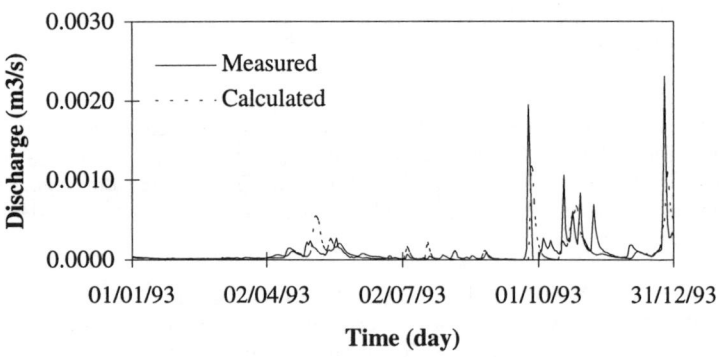

Figure 2 - Measured and calculated hydrographs on the plot 1.

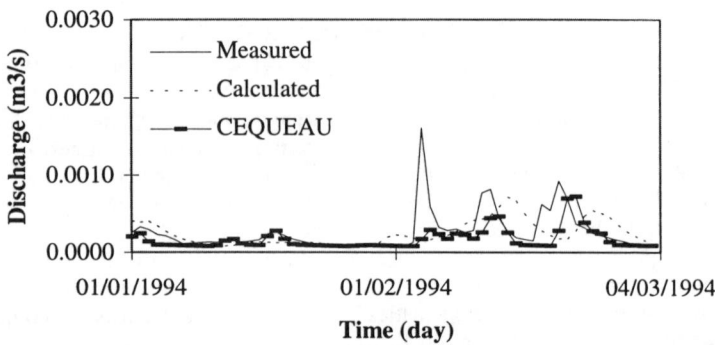

Figure 3 - Comparison between physically-based and CEQUEAU models

Moreover, the model allows to analyze the influence of different hydrological processes in a watershed. In this case, a study of the plants influence on evapotranspiration has been done. This study is presented on figure 4. We note that actual transpiration follows the development of maize culture.

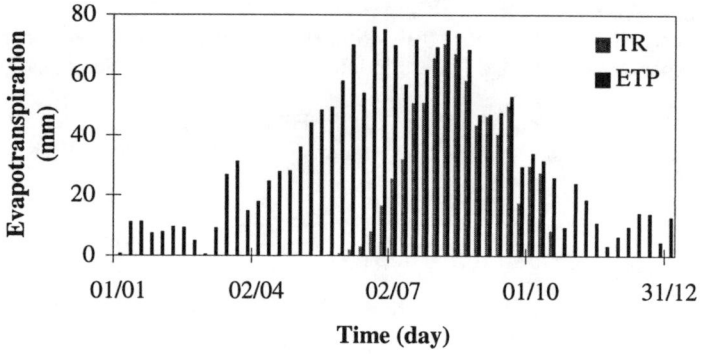

Figure 4 - Potential evapotranspiration and actual transpiration evolution

CONCLUSIONS

We have presented a modeling of the water flow in the soil and the transfer in soil-plant-atmosphere system. The model has been applied on an experimental drained plot where maize is cultivated. The evolution in time and in space of the hydrodynamic variables and of hydrological cycle by such a kind of model is quite interesting in comparison with conceptual modeling. Moreover the growing interest focused on qualitative modeling, mainly nutrients and pesticides, which is inherent in quantitative one, encourage us to carry on the realization a such kind of physically-based modeling. It seems very pertinent to follow up our investigations and to develop this model in this way, and to apply it at the watershed scale.

REFERENCES

BEVEN K., 1989, Changing ideas in hydrology: the case of physically-based models. Journal of Hydrology. 105,157-172

CAMPBELL G., 1974, A simple method for determining unsaturated conductivity from moisture retention data. Soil-Science 117,311-314

GUIRESSE A. M., 1996, Personal communication.

HUTSON J.L. AND CASS A., 1987, A retentivity function for use in soil-water simulation model. Journal of soil science, 38,105-113

HUTSON J.L. AND WAGENET R.J., 1989, LEACHM: Leaching Estimation And Chemistry Model. A process-based model of water and solute movement, transformations, plant uptake and chemical reactions in the saturated zone, Department of Soil, Crop and Atmospheric Sciences, Cornell University, Ithaca, New York

KORETA R., 1996, Sur le devenir des herbicides dans le sol: cas de l'atrazine et de la simazine en sol de boulbènes, de la colonne de sol à la parcelle drainée. Ph D from l'Institut National Polytechnique de Toulouse.

MORIN G, FORTIN J.P., LARDEAU J.P., SOCHANSKA W. AND PAQUETTE S., 1981, Modèle CEQUEAU, Manuel d'utilisation, Rapport Scientifique 93, Université du Québec, Québec

NIMAH M.N. AND HANKS R.J., 1973, Model for estimation of soil water, plant and atmospheric interrelations: I. Description and sensitivity, Soil Sci.Amer.Proc., 37,522-527.

TILLOTSON W.R., ROBBINS C.W., WAGENET R.J. AND HANKS R.J., 1980, Soil water, solute, and plant growth simulation, Bulletin 502, Utah State Agr.Exp.Stn.,Utah, 53p.

WATERSHED-CHANGE INDUCED UNCERTAINTY ON RUNOFF FREQUENCY FOR WATER RESOURCES MANAGEMENT

STEFANO PAGLIARA[1] and BEN CHIE YEN[2]
[1] Ist. di Idraulica, Univ. di Pisa, Italy; presently Visiting Fulbright Scholar, Dept. of Civil Eng. Univ. of Illinois at Urbana-Champaign, Urbana, IL 61801 USA.
[2] Prof. of Civil Eng. Univ. of Illinois at Urbana-Champaign, Urbana, IL 61801 USA

ABSTRACT

The effect of changes in a watershed on the occurrence frequency of runoff is considered. Measured runoff data are actually for different non-stationary watershed conditions; frequency analysis of this unadjusted, biased data set yields a discharge-return period relationship for neither early nor later watershed conditions. The probability distribution function that fits best to the inconsistent biased data is different from that for the adjusted unbiased data series.
A hypothetical example assuming a linear increase in watershed urbanization depicts the errors involved if the inconsistency in the data is not detected and adjusted.

INTRODUCTION

Many watersheds, during the years, suffer changes due to natural or man-made causes. Examples include the changes due to construction of reservoirs, levees, irrigation diversions, ponds, modification of vegetation cover, afforestation or deforestation, fires, volcanic explosions and alteration in land use, particularly for urbanization. There have been several studies on the effect of urbanization on watershed runoff. For example using a conjunctive surface-subsurface hydrodynamic flow model Akan and Yen (1983) demonstrated the changes of surface and subsurface runoffs for changing impervious level of an idealized watershed. Urbanization is often a major problem for the consistency of runoff data series for small watersheds. It involves substantial changes in the land uses and in the morphology of the original channel network, increasing the peak discharge and volume of the flow.

The fact that the period of major urbanization (this century) coincides with the period of intensive hydrologic data acquisition has the consequence that many runoff data recorded over the world are affected by this kind of nonstationary hydrologic situation. Such measured data are physically inconsistent. Subsequently, the calculated runoff frequency distribution is affected and the estimation of the design flood is in error if the data are not properly adjusted. For a correct determination of the design flood the frequency analysis should be performed with the measured data adjusted to the watershed corresponding to the design condition such that inconsistency of data series is eliminated, and the analysis is performed using the adjusted unbiased data series. Likewise, in environmental impact assessment often the flood frequency relationship for a given past or future watershed condition is desired. Some methods are available to detect and quantify the hydrologic impact of changes that have occurred in a watershed, e.g. urbanization changes can be detected by using the double mass curve technique (Riggins and Yen, 1995). Effects of non-homogeneity in hydrologic time series have been studied by Yevjevich and Jeng (1969) using normal and log-normal probability density functions. The objective of this paper is to show that use of unadjusted measured runoff data yields a flood frequency that is not representative of the watershed condition at any given time, whereas adjustment should be made on the measured data to yield an unbiased data series corresponding to a specified watershed condition in order to yield the frequency analysis for the given watershed condition.

DATA AND METHODOLOGY

A search has found no data sufficiently accurate and detailed enough on the urbanization process that could be used for this study. Therefore, the annual maximum series data given in Fig. 8.I.3(b) by Chow (1964, p. 8.19) in his well-known Handbook of Applied Hydrology is adopted here as "measured" data, reproduced as black dots in Fig. 1, as a hypothetical example of runoff from a watershed subject to urbanization. For simplicity let us assume the effect of watershed urbanization is linear with respect to time, producing a 2% increment in discharge every year. Obviously, this is a considerable idealistic simplification of a complex real situation that is generally not easy to quantify.

From the measured data series shown in Fig. 1 we can create the data series corresponding to the pre-urbanized situation of 1970 (data series if no change in the watershed had occurred) and the post-urbanized situation of 1990 (data series if all the change in the basin had occurred prior to the beginning of the data series). As conventionally done in frequency analysis, the data series of measured and pre- and post-urbanized conditions are each fit to a probability distribution function (the Gumbel distribution is reported as an example) as shown in Fig. 2. In this figure the computed curve for the measured data are in the middle while the frequency curve for the pre- and post-change conditions are lower and higher, respectively.

To illustrate the extreme cases, suppose the measured time series of data occurred not randomly as shown in Fig. 1 but with the magnitudes monotonically decreasing (Case 2) or increasing (Case 3), shown as black dots in Fig. 3a and 3b, respectively. Again, the effect of urbanization is assumed linear with a 2% annual increment in discharge as before. Frequency analysis is applied to each of the measured, pre- and post-urbanized data sets. The resultant frequency curves are shown in Fig. 4 as Gumbel plots.

The hypothetical "measured" data adopted from Chow (1964) and shown in Fig. 1 are utilized here merely as a simple example. The corresponding analysis of actual data, such as 1937-1975 data of Saline Branch at Urbana, Illinois, yield similar results as shown in Fig. 5. In this figure the data for the pre- and post-urbanization conditions were obtained hypothetically by assuming a linear increment equal to 1% for each year of observation.

DISCUSSION

Figure 2 clearly illustrates that the inconsistent, biased measured data does not yield a flood frequency relationship for any given time of a changing watershed. As expected, for a given return period the flood for the pre-change condition is smaller than the post-change condition. For return periods within the data range the flood magnitude given by the unadjusted

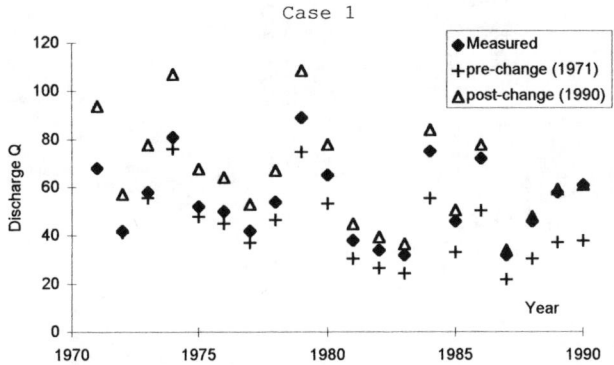

Fig.1 Measured and adjusted data for watershed having pre-change and post-change conditions

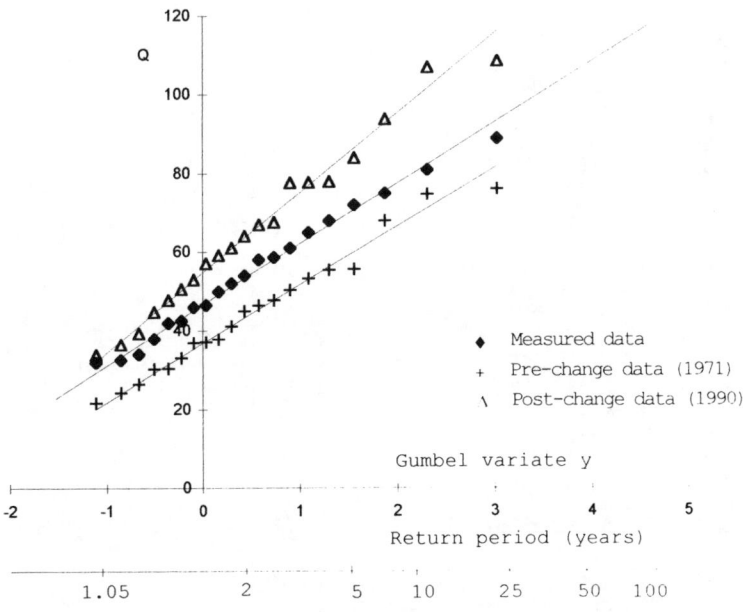

Fig.2 Gumbel frequency plot for data in Fig.1

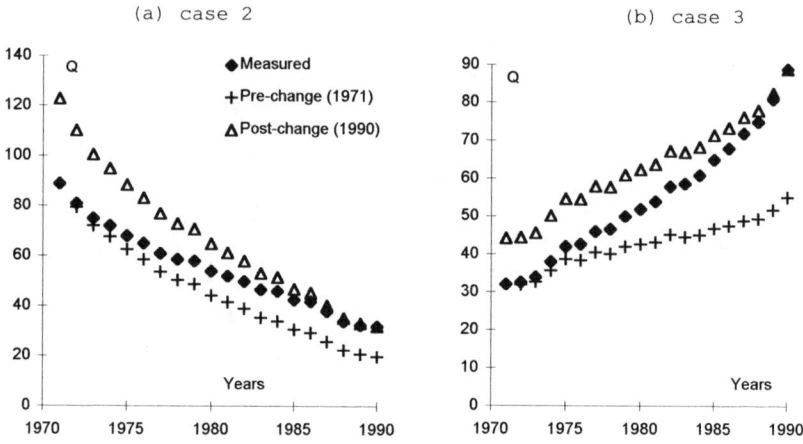

Fig. 3 Hypothetical monotonically decreasing and increasing data time series

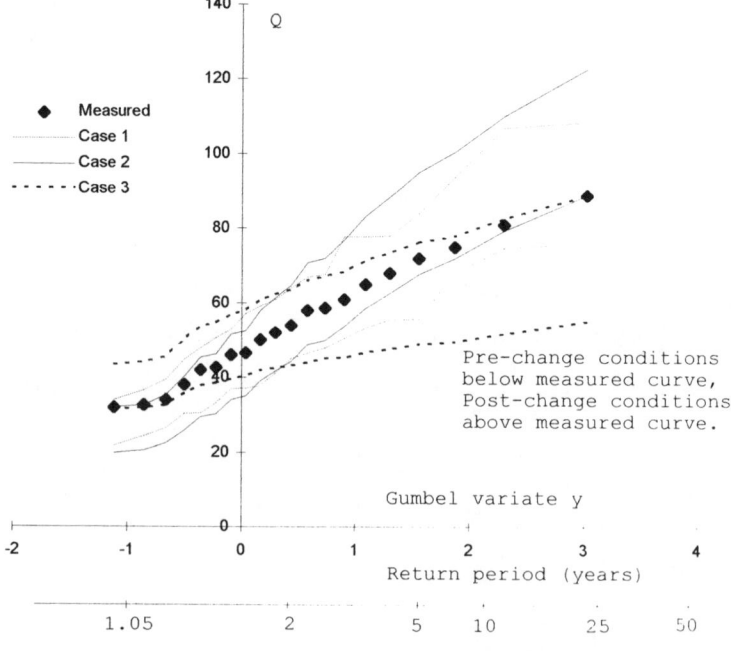

Fig.4 Gumbel frequency plot for data in Fig.3

measured data lies between those of pre- and post-change conditions. However, extrapolating beyond the data range, whether the frequency relationship given by the unadjusted data will over- or under-estimated the flood magnitude depends on the random large floods occurring mostly late, near the post- change condition, or early close to the pre-change condition. This is demonstrated by the extreme cases 2 and 3 shown in Fig. 4. For data of increasing time series (Case 3), floods of large return periods beyond the data range are consistently over-predicted for post-change watershed condition and much more so for pre-change condition. Conversely, for data of decreasing time series (Case 2), large return-period floods beyond the data range are consistently under-estimated. This deficiency is also illustrated in Fig. 5.

Moreover, identifying the best-fit probability distribution function for runoff data is an important issue in hydrology. It is interesting to note that the type of distribution function that would fit the unadjusted measured data is always different from the best-fit function for the adjust, consistent, unbiased data of pre-change or post-change conditions. This is demonstrated in Fig. 6. The data series used for this figure are essentially the same as those shown in Fig. 3 except the measured data are slightly modified such that the new "measured" data follow exactly the Gumbel (EVI) distribution. On the Gumbel plot of Fig. 6 the frequency curve for the measured data follows exactly a straight line, whereas the frequency curves for the adjusted, consistent and unbiased data sets are all nonlinear (in this example non-Gumbel) indicating they do not follow the distribution function of the measured data (Gumbel distribution). In other words, great attempts trying to find improved data fitting without proper adjustment of the data may be a futile effort.

CONCLUSIONS

Changes in a watershed, such as urbanization, alter the hydraulic response of the watershed in producing runoff from rainfall. Consequently, the flood frequency relationship of the watershed is modified. Measured flood data of a changing watershed reflects the hydraulic response of the changing conditions, and hence frequency analysis of the unadjusted data does not yield the flood frequency relationship for any specified watershed condition at a given time. Simple examples presented in this paper demonstrate that measured data should be adjusted to specified watershed condition in order to determine the flood frequency relationship for that condition. It has also been shown that the probability distribution function that fits the unadjusted measured data is different from the probability distribution function for the adjusted data for specified watershed condition at a given time.

REFERENCES

Akan, A.O., and Yen, B.C. " Hydrologic changes associated with urbanization, " Proc. IAHR 20th Congress, Moscow, Russia, Vol.6, pp.20-26, 1983.

Chow, V.T. "Section 8-I: Statistical and probability analysis of hydrologic data, Part I: Frequency analysis," Handbook of Applied Hydrology, ed. by V.T. Chow, McGraw- Hill Book Co., New York 1964.

Riggins, R.E., and Yen B.C. " Detection of effect of urbanization on storm runoff, " Watershed Management, (Proc. Symp. at San Antonio, TX) ed. by T.J. Ward, pp. 408-418, ASCE, Aug. 1995.

Yevjevich, V. and Jeng, R.I." Properties of non-homogeneous hydrologic time series, " Hydrology Paper no.32, Colorado State University, April 1969.

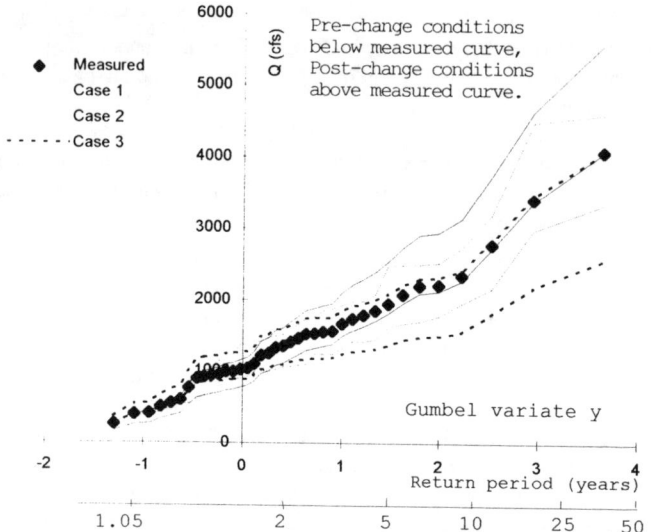

Fig.5 Gumbel frequency plot of Saline Branch at Urbana for measured runoff data and modified monotonically decreasing and increasing data series for pre-change and post-change conditions

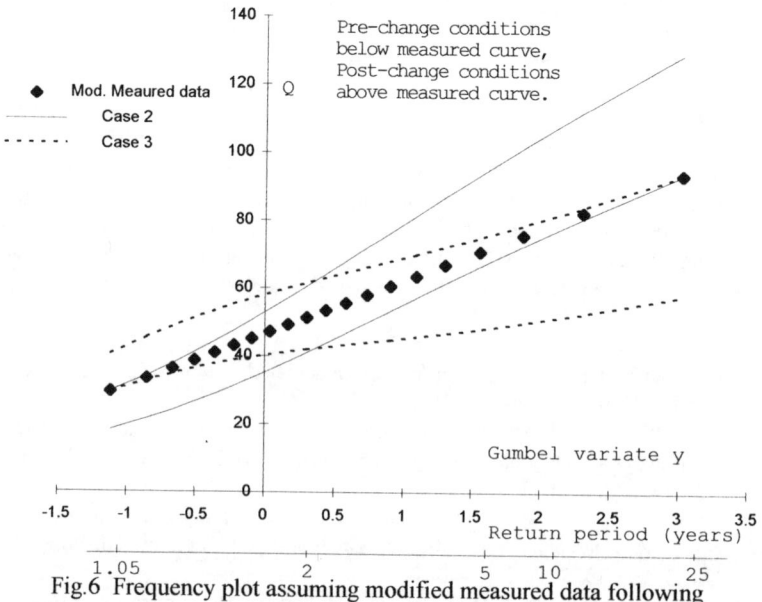

Fig.6 Frequency plot assuming modified measured data following Gumbel distribution

A Physically Based Model for Large River Basins.

W.T.SLOAN, J.EWEN, C.G.KILSBY, C.S.FALLOWS AND P.E.O'CONNELL
University of Newcastle Upon Tyne, UK.

INTRODUCTION

Hydrologists are increasingly called upon to develop water balance models at large catchment to continental scales for water resource management and design purposes. This has been prompted by two main concerns. Firstly, the consensus in the scientific community of an impending change in the earth's climate (Houghton et al, 1996) has resulted in concerns about the impact this might have on water resources. Secondly, there is a need to assess the impact of more immediate anthropogenic induced change, such as changing landuse, or the growing demands on water resources imposed by increasing populations or industrial activity. Physically based distributed models have a structure which is consistent with predicting the hydrological regimes which would prevail under alternative land use and climatic regimes (O'Connell, 1994). Their distributed structure can accommodate any prescribed pattern of landuse or physical structure, while their physical basis allows changes to be made to their parameters which reflect changes in these patterns.

Distributed physically based models have been used successfully in a number of impact assessment studies at the catchment scale (e.g. Adams, 1995; Dunn, 1995). The application of these models to very large river basins has historically been limited by the amount of computer processing required. The capabilities of computers are continually being improved, and the size of catchments to which traditional distributed hydrological models, such as SHETRAN (Ewen, 1995), can be applied is increasing accordingly. However, it still remains impractical to apply them to very large basins or at the continental scale. This does not deter the majority of researchers from attempting to retain a physical basis to models designed for application at this scale (eg. Vorosmarty et al, 1989) but it is generally accepted that they require prudent simplification (Dooge, 1986). The nature and extent of these simplifications is a topic which is vigorously debated in the hydrological literature under the headings of upscaling or the scale problem (

a substantial review of this is given in Bloschl (1996). Approaches include, *inter alia*: applying descriptions of processes known to occur at the point scale at a very much larger scale (Vorosmarty, 1990); and the development of new models which can characterise the physics at large scales based on parameters which can easily be identified at that scale (Beven, 1996). The former of these two approaches is subject to criticism on two fronts (Beven, 1996); firstly, there is little conclusive evidence to suggest that process descriptions applicable at the point scale can adequately describe the physics at the larger scale, and secondly, this approach relies on the use of effective parameters, the identification of which is fraught with practical and conceptual problems and often reduces the model to little more than an empirical one. The latter approach, whilst being a laudable ultimate goal and worthy of further research, has had little practical success. In the context of impact assessment it also suffers from the constraint that nothing can be implied about the impacts of environmental change at the small scale which is often a requirement of water resource managers and policy makers.

THE UP MODEL

The UP (Upscaling with a Physical basis) model described here has been developed for application to a variety of scales ranging from an individual catchment to the continental scale, and for a range of purposes including environmental impact assessment and interaction with a GCM. The intention is to bypass some of the scaling problems encountered by other approaches by employing a systematic procedure for scaling from the physical process descriptions at the point scale, up through the catchment scale to the meso- and macroscales. The approach differs from others in that it attempts to keep to a minimum any assumptions made about the physics at anything but the scale at which the process descriptions are known to be valid.

FROM THE POINT TO THE CATCHMENT SCALE.

This part of the scaling procedure is best explained by, initially, considering one component of the hydrological cycle in isolation.

A requisite of any comprehensive hydrological model, independent of scale, is the ability to estimate groundwater discharge. Under most naturally occurring groundwater conditions the flux of fluid at the point scale is known to be adequately described by Darcy's law. Combining this with the principle that mass is conserved yields equations describing the fluid flow in aquifers, for example the Boussinesq equation for unconfined aquifers. Due to the heterogeneous nature of most aquifers, a general law, which is equivalent to Darcy's law but is applicable at the larger scales, has not been found and is never likely to be. As a result, it is standard practice to solve the groundwater equations for the complete aquifer, normally using a numerical scheme with as detailed a description of the aquifer properties as is possible. Rather than seek to invent a scaled up version of Darcy's law, or search for some new law applicable at larger scales, the UP model employs this standard practice to simulate ground water discharges for a range of scenarios which are deemed to be characteristic of the range of possible recharge rates. The aim is to predict discharges without recourse to the detailed model which is achieved by summarising the results for the detailed scenarios. This summary yields a simplified model in which discharge is given as a function of the total storage in the aquifer. This function allows the aquifer at the catchment scale to be modelled as a lumped system: for known recharge rates it is possible to predict the discharge by keeping track of the total storage.

A similar procedure is used to produce simplified models for each of the other processes involved in producing runoff from a catchment. These simple lumped models are combined to produce a lumped catchment scale model which is computationally efficient whilst retaining a physical basis. Figure 1 shows a schematic diagram of this lumped catchment model described as an UP element. An UP element represents a drainage basin by a number of conceptual water storage compartments, with the associated fluxes calculated from simple lumped models, derived by summarising more detailed distributed models. Typically an UP element would be applied to basins whose surface area is of the order of 100km².

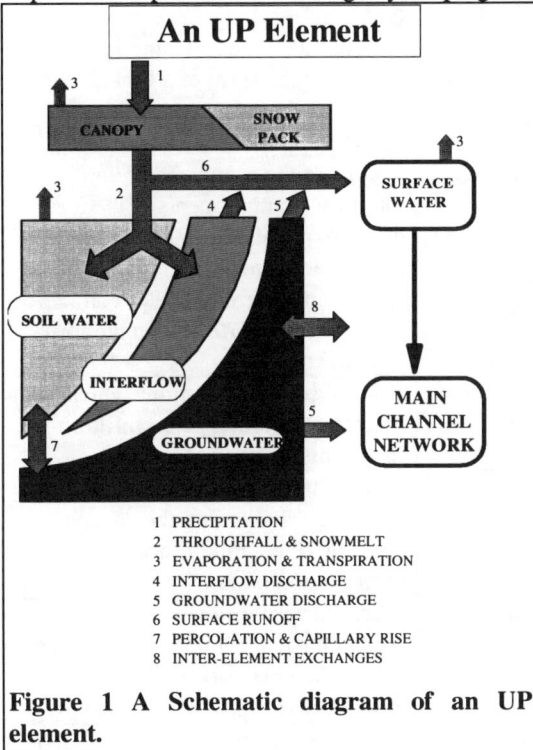

Figure 1 A Schematic diagram of an UP element.

1 PRECIPITATION
2 THROUGHFALL & SNOWMELT
3 EVAPORATION & TRANSPIRATION
4 INTERFLOW DISCHARGE
5 GROUNDWATER DISCHARGE
6 SURFACE RUNOFF
7 PERCOLATION & CAPILLARY RISE
8 INTER-ELEMENT EXCHANGES

It is often difficult to produce models from the detailed summaries which can be cast in a simple continuous functional form, so the UP system relies heavily on discrete transfer functions and lookup tables. The use of transfer functions to describe hydrological processes in a lumped manner is well documented (Chow et al 1988). Lookup tables summarise a continuous function by storing values of the function calculated at a number of discrete values of the independent variables. The function is then evaluated at any arbitrary value of the independent variables by interpolating between these discrete values. Table 1 lists the various components of an UP element, the major processes modelled, the detailed distributed models used and the nature of the lumped models which summarise their results.

Compartment	Detailed Model	Lumped Model
Groundwater	Numerical solution of 2-D Boussinesq equation.	1. Discharge as a function of storage in the form of a lookup table. 2. Saturated area as a function of storage in the form of a lookup table.
Interflow	Numerical solution of a 2-D model in which flows are topographically driven.	1. Discharge as a function of storage in the groundwater and interflow compartments in the form of a lookup table. 2. Saturated area as a function of storage in the groundwater and interflow compartments in the form of a lookup table.
Unsaturated Zone	Numerical solution of Richards equation.	A set of transfer functions which predict percolation given infiltration for a range of storages.
Surface water	Particle tracking of surface water over the saturated area.	A set of transfer functions which predict discharge into the channel given effective rainfall for a range of saturation patterns.
Channel	Analytical solution to the St Venant equations	A set of transfer functions which predict the discharge at the outlet of the channel network given the inputs for a range of flow regimes.
Canopy	Penman Monteith model.	Functions which predict aggregated throughfall and evapotranpiration given canopy storage and soil moisture content.

In scaling from the point to the catchment scale, the procedure of constructing an UP element uses only physically based models applied at the scale at which they are valid

FROM THE CATCHMENT TO THE LARGE BASIN SCALE.

The next part of the scaling procedure requires that the large basin which the UP model is being applied to is delineated into smaller areas, the boundaries of which can either concur with the boundaries of natural watersheds or with a regular grid. Each of these areas is modelled using an UP element.

The procedure described in the previous section for scaling from the point to the catchment scale yielded a lumped, computationally efficient, physically based model which is capable of simulating the major processes involved in producing runoff from an individual small catchment. The element is developed on an individual basin and is based on a detailed analysis of distributed physically based model predictions. It is impractical to carry out this procedure on all but a few of the sub-basins which constitute a large basin. The UP model therefore makes the assumption that the simple constituent lumped models which characterise an UP element can be transposed to UP elements with similar physical properties. As a result, the detailed analysis need only be conducted on a few catchments which are representative of the range of physical characteristics and hydrological regimes encountered in the large basin. Prediction at the large basin scale is achieved essentially through an aggregation process reather than scaling.

APPLICATION OF UP

The validation of the UP model is a two stage process, both of which involve comparing observed and predicted discharges. The first is to validate the approach, outlined in this paper, for scaling from the point to the catchment scale, and the second involves applying the model to a very large river basin. The first stage was conducted by applying the UP model to the Tyne river basin in north east England (Kilsby et al, 1995). This catchment has been the subject of several physically based modelling studies in the past (Dunn et al, 1992) and the University of Newcastle Upon Tyne hold a comprehensive set of data which describe its physical characteristics and hydrological regimes. The second stage of the validation process is underway and involves an application of the model to the Arkansas-Red river basin in the USA which covers approximately 570000 km^2. This is being conducted within GCIP as part of the UK NERC Terrestrial Initiative in Global Environmental Research.

CONCLUSIONS

UP is a physically based model which has been designed to model impacts of environmental change at a variety of scales ranging from an individual catchment to the continental scale. It is unique in its treatment of scaling hydrological processes in that it the uses only physically based models applied at the scale at which they are valid. The model has been validated on the Tyne river basin and is in the process of being applied to the Arkansas-Red river basin The UP model provides an important insight into large river basin hydrology and it will prove a valuable aid to water resource management, particularly under conditions of changing climate and land use change.

REFERENCES

Adams, R., Dunn, S.M., Lunn, R. Mackay, R. and O'Callaghan, J.R., 1995. Validating the Nelup hydrological model for river basin planning. J. Env. Plann. and Manag., 38(1).

Beven K., 1995 Linking Parameters across scales: subgrid parameterisations and scale dependent hydrological models. Hydrological Processes. Vol. 9. 507-525.

Bloschl, G. and Silvapalan, M., 1995. Scale Isuues in Hydrological modelling: A review. Hydrological Processes. Vol. 9. 251-290 (1995)

Chow, V.T., Maidment, R.D. Mays L.W., 1988. Applied Hydrology. McGraw-Hill International Editions.

Dunn, S., Savage, D., Mackay, R., 1992. Hydrological simulation of the Rede catchment using the Systeme Hydrologique European (SHE). In Land use change the causes and consequences,(Ed M.C. Whitby) ITE Symposium no 27, HMSO London, pp 137-146.

Ewen, J., (1995). Contaminant transport component of the catchment modelling system SHETRAN Chapter in Solute Modelling in Catchment Systems (Ed. S Trudgill)

Houghton, J.T., Meira Filho, Callander, B.A., Harris N., Kattenberg, A. and Maskell K., 1996. Climate Change 1995. The Science of Climate Change. Cambridge University Press.

Kilsby, C.G., Ewen, J., Sloan, W.T., Burton, A., O'Connell, P.E., 1995. Large scale hydrological modelling; a physically based approach. in Annales Geophysicae (Supplement to Vol 13) Book of Abstracts for XX General Assembly, European Geophysical Society, Hamburg, Germany.

O'Connell P.E., 1995. Capabilities and limitations of regional hydrological models. In Scenario Studies for the Rural Environment. (Eds J.F.Th. Schoute et al.) Kluwer Academic Publishers.

Vorosmarty, C.J., Moore, B.,Grace, A.L., Gildea, P.M., Melillo, J.M., Peterson, B.J., Rastatter,E.B. and Steudler P.A., 1989. Continental scale models of water balance and fluvial transport: An application to South America. Global Biogeo. Cycles. 3(3) 241-264.

WATER AND ENERGY BALANCES AND THEIR SPATIAL DISTRIBUTION IN A CATCHMENT OF COMPLEX LAND USE

Y. JIA and N. TAMAI
Department of Civil Engineering, University of Tokyo
Hongo 7-3-1, Bunkyo-ku, Tokyo 113, Japan

ABSTRACT

An integrated and distributed modeling of water and energy transfer is conducted in this study. Spatial heterogeneity of physical variables and scale problems have drawn attentions of both hydrologists and meteorologists in recent years. How to consider them in hydrology models and meteorology models is far from mature. Though statistic method is commonly studied at present (e.g., Avissar, 1992), nesting or explicit subgrid method is also observed (e.g., Seth et al.,1994). A similar nesting method is applied in this study to a sub-catchment of the Tama river at the middle reach (Tokyo, Japan) which is partly urbanized. The study results show that there are large spatial variations of water and energy budgets which had to be considered in the water management of the catchment. Close relations are also shown between water balance and energy balance. The computed discharge is favorably matched by the observation at the outlet of the catchment.

MODEL STRUCTURE

For medium-sized or large-sized river basins of complex land use like the studied one (see Fig.3), the statistic method is difficult to be applied because the properties are totally different in areas of different land uses, e.g., urban cover and grassland. Because of the limitations of computation time and capacity, the grid size can't be fine enough to represent the heterogeneity of land use, whereas it may lead to big error to represent the whole grid by using the dominant land use. Therefore, the nesting method will be a reasonable choice which consider distribution of different land uses inside a grid.

The diagram of model structure inside a grid is shown in Fig.1. The land use is summarized into 3 groups---water group, soil-vegetation group and urban group. The soil-vegetation group consists of bare soil, tall vegetation (forest or urban trees) and short vegetation (grass or crops). The urban group consists of urban cover and urban canopy. Therefore 6 types of land use are considered at the same time.

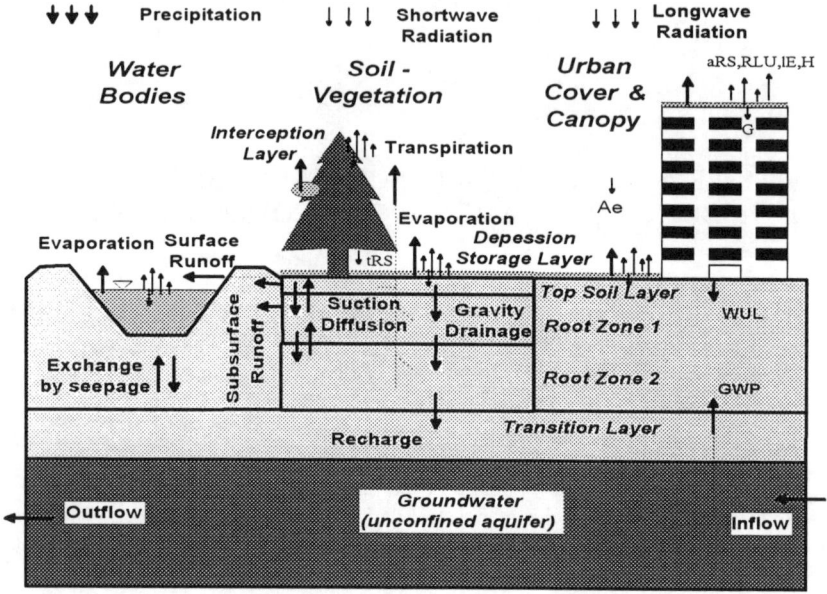

Fig. 1 Diagram of Model Structure inside a Grid

WATER BALANCE

$$P + WUL = GWP + E + R1 + R2 + RG + \Delta S \qquad (1)$$

in which P is precipitation, WUL is leakage of water use, GWP is pumped groundwater, E is evapotranspiration, R1 is surface runoff, R2 is subsurface runoff, RG is groundwater outflow and ΔS is storage change in soil layers and in aquifer. For soil-vegetation group, 6 vertical layers---interception layer, depression storage layer, top soil layer, first root zone, second root zone and transition soil layer & unconfined groundwater layer are included in the model structure. Moisture diffusion caused by suction difference is also considered among the 3 unsaturated soil layers in addition to gravity drainage. Water balances in different land uses are connected together by assuming there is same groundwater storage.

Evapotranspiration is constituted by interception of vegetation canopies; evaporation from surfaces of waters, soil, urban cover and urban canopy, and from top soil layer; transpiration from the dry fraction of leaves with the source from the 3 soil layers. The Rutter model (1971) is followed to compute interception. The following Penman equation is adopted to compute potential evaporation which is based to compute actual evaporation from soil by using a wetting function.

$$Ep = \left[(Rn - G)\Delta + \rho_a C_p \delta e / r_a \right] / \lambda(\Delta + \gamma) \qquad (2)$$

The following Penman-Monteith equation is used to calculate transpiration with the canopy resistance referred to Dickinson et al (1991):

$$Etr = \left[(Rn - G)\Delta + \rho_a C_p \delta e / r_a\right] / \lambda\left[\Delta + \gamma(1 + r_c / r_a)\right] \qquad (3)$$

In eq.(2) and eq.(3), R_n is net radiation, G is heat flux into soil, Δ is the gradient of saturated vapor pressure and temperature, ρ_a is the air density, C_p is the specific heat of air, δe is vapor pressure deficit, r_a is the aerodynamic resistance, r_c is the canopy resistance, λ is the latent heat of water, γ is the psytrometric constant.

Surface runoff for water group is assumed to be the net rainfall (subtracted by evaporation). For urban group, surface runoff will occur after the depression storage reaches maximum. For soil-vegetation group, it consists of two parts: one is rainfall excess (during heavy rain periods) which is computed according to a generalized Green-Ampt model for multi-layered soil with unsteady rainfall (Jia and Tamai, 1996); the other is saturation excess (in other periods) calculated according to water balances in soil layers and unconfined groundwater layers.

Subsurface runoff is computed according to land slope and soil hydraulic conductivity.
Groundwater flow is estimated by storage function. Outflow coefficient is based on topography and geology (Ando,1981).

ENERGY BALANCE

$$RSN + RLN + Ae = lE + H + G \qquad (4)$$

where RSN is net short-wave radiation, RLN is net long-wave radiation, Ae is artificial energy consumption, lE is latent heat flux, H is sensible heat flux, G is heat flux into soil.

Net short-wave radiation equals to incident short-wave radiation RS subtracted by short-wave reflection αRS. Short-wave reflection coefficient α is considered to be variable and is related to solar zenith angle, soil moisture content or canopy heights.

Net long-wave radiation equals to downward long-wave radiation RL subtracted by upward long-wave radiation RLU, and is related to air temperature and surface temperature.

Energy Consumption: statistic energy consumption indices on 7 types of urban land use are used to consider impacts of human lives on energy balance in urban area. A half of energy consumption is assumed to be emitted to land surfaces and the other half to air.

Latent Heat Flux

$$lE = \lambda \cdot E \qquad (5)$$

where E is evapotranspiration and λ is the latent heat of water

Sensible Heat Flux

$$H = \rho_a Cp(Ts - T) / r_a \qquad (6)$$

where ρ_a is the density of air, C_p is the specific heat of air, T_s is land surface temperature, T is air temperature and r_a is the aerodynamic resistance.

Heat Flux into Soil is calculated by the heat balance equation (4).

Surface Temperature Ts: the Force-Restore Method (FRM) is used to solve surface temperature with the calculation of coefficient α according to the Hu and Islam (1995) method which not only ensures minimum distortion of FRM to sinusoidal diurnal forcing but also makes distortion to higher harmonics negligible.

APPLICATION AND VERIFICATION

STUDY REGION
The model mentioned above is applied to a sub-catchment of the Tama river in Tokyo, Japan (see Fig.2) for water and energy balances in 1992, using a grid resolution of 1km × 1km and a time step of 1 hour. The basin area is 579km^2.

Land use data and topography data are based on the Fine Digital Information System (FDIS) of Japan Geography Institute which is renewed once five years. The land use distribution in 1989 is shown in Fig.3, using a grid resolution of 100m × 100m. It is considered to be able to represent the land use in 1992 because urbanization speed is quite low in recent years in the region. The land use is quite versatile with as many as 15 types (summarized into 6 main types in Fig.3) and several types of totally different characteristic may coexist inside a area of 1km^2, which had to be considered in the modeling of water and energy balances.

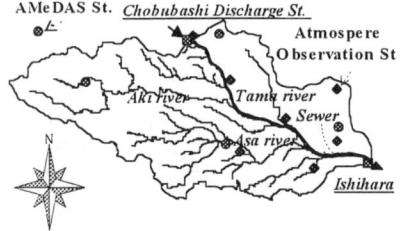

Fig2 Middle-reach Catchment of Tama River   Fig.3 Land Use Distribution in 1992

RESULTS AND DISCUSSION
Annual water balance and energy balance in 1992 are shown in Table 1 & Table 2 respectively. Distributions of evapotranspiration, runoff, latent heat flux and sensible heat flux are respectively shown in Fig.4 and Fig.5 as examples. Because medium and long-term water and energy balances are aimed at present, hourly runoff is directly integrated to obtain monthly and annual values without routing of river flow. From these tables and figures, we can see that: (1) In this basin the general runoff coefficient is estimated as 0.46 and the groundwater outflow is quite low because of groundwater pumping. (2) Both water budgets and energy budgets have big spatial variations because of complex land use. For example, the annual evapotranspiration varies from minimum the 300mm in eastern urban areas to the maximum 900mm in western forest areas, whereas the annual runoff has a converse variation. The annual latent heat flux varies from the minimum 800MJ/m2.year in urban areas to the maximum 2200MJ/m2.year in forest areas, whereas annual sensible heat flux has a converse variation. (3) Distributions of water budgets and energy budgets have similar patterns. For example, spatial distribution of latent heat flux is similar to that of evapotranspiration and sensible heat flux similar to runoff. This means a close relation between water balance and energy balance.

Table 1 Annual water balance in 1992   (mm/year)

P	WUL	GWP	E	R1+R2	Q	ΔS
1554.5	46.5	187.1	727.7	432.0	277.8	-23.2

Table 2 Annual energy balance in 1992   (MJ/m².year)

RSN	RLN	Ae	lE	H	G
4138.8	-1672.5	148.7	1788.8	822.6	3.6

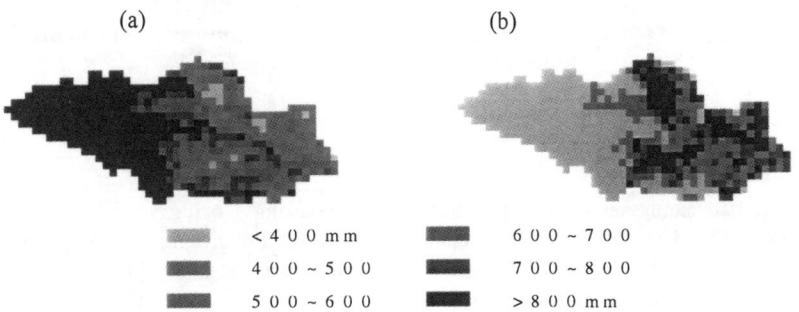

```
        < 4 0 0 mm         6 0 0 ~ 7 0 0
        4 0 0 ~ 5 0 0      7 0 0 ~ 8 0 0
        5 0 0 ~ 6 0 0      > 8 0 0 mm
```

Fig. 4 Distribution of Water Budgets in 1992. (a) evapotranspiration (b) runoff

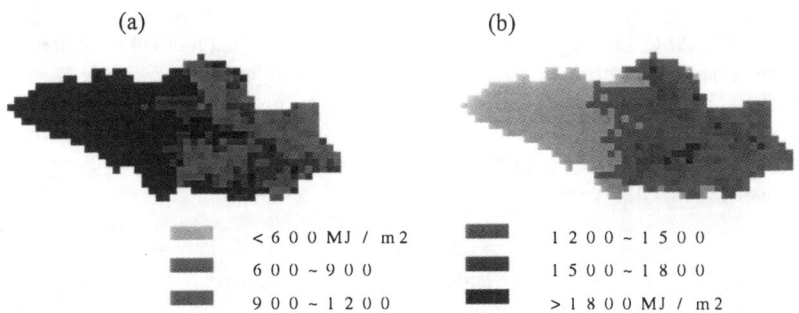

```
        < 6 0 0 MJ / m2    1 2 0 0 ~ 1 5 0 0
        6 0 0 ~ 9 0 0      1 5 0 0 ~ 1 8 0 0
        9 0 0 ~ 1 2 0 0    > 1 8 0 0 MJ / m2
```

Fig. 5 Distribution of Heat Fluxes in 1992. (a) latent heat flux (b) sensible heat flux

VERIFICATION

The runoff and groundwater outflow, added by upstream inflow at Chobubashi station and pumped groundwater, and subtracted by diverted water to outside and water use leakage, the computed discharge at Ishihara station is obtained. The computed annual average discharge is 26.10m³/s with a relative error of 4.1% from the observed 27.21m³/s. The comparison of computed monthly discharge with the observed is shown in Fig.6.

CONTINUOUS, DISTRIBUTED-PARAMETER HYDROLOGIC MODELING WITH CASC2D

FRED L. OGDEN AND SHARIKA U. S. SENARATH
Department of Civil and Environmental Engineering, University of Connecticut
Storrs, Connecticut 06269, U. S. A.

ABSTRACT

Distributed-parameter hydrologic models are increasingly being used to assess the impact of land-use and/or climatic change on watershed hydrography. These studies require the modeling of the soil moisture storage as well as the water-fluxes to and from the soil layer over an extended period of time. This requires re-casting the traditional role of a rainfall-runoff model from event-based to continuous form. The two-dimensional, physically-based, distributed-parameter hydrologic model CASC2D (Julien *et al.* 1995) used in this study has adequate parameterization to simulate infiltration into a deep soil layer (Ogden and Saghafian, 1997). Hence, in this study, evapotranspiration is added to CASC2D using two different approaches. The first method considers bare-ground evaporation from the land-surface using a formulation suggested in Deardorff (1978). The second method estimates the evapotranspiration from a vegetated land-surface using the Penman-Monteith equation (Monteith, 1965, 1981). Continuous simulations are found to improve calibration uniqueness for distributed parameter, physically-based hydrologic models.

FORMULATION

Variants of the Penman-Monteith and Deardorff representation are widely used in land-surface schemes of climate models and distributed hydrologic models (e.g., Dickinson *et al.*, 1986; Beven, 1979). The following sections describe in some detail the formulations employed for evapotranspiration modeling. Wherever possible, parsimony is used as the main selection criteria as dictated by readily available data sources. The primary U.S. meteorological data sources include the NOAA/NCDC Hourly Surface Airways Observations, and the NOAA/NCDC SAMSON CD-ROM data set.

NET INCOMING SHORTWAVE RADIATION
When available, global shortwave radiation data from nearby sites may be used as input. However, the number of such sites in the U.S. is limited. If shortwave radiation data are not available, the net incoming shortwave radiation is calculated from celestial geometry.

meteorological observations and the land-surface shortwave albedo. The direct incoming solar irradiance, $R_{s,direct}$ is calculated as follows:

$$R_{s,direct} = R_{s,horiz.,direct,ground} \frac{\cos i}{\sin \lambda} \quad (1)$$

where $R_{s,horiz.,direct,ground}$ is the direct solar irradiance falling on ground. The cosine of the angle between the direct solar radiation and the normal to the slope ($cos\ i$), and the sine of the solar altitude ($sin\ \lambda$) in (1) are defined in Pielke (1984). The diffuse shortwave radiation is obtained by using the relationship proposed by Kondrat'yev (1965). Both the direct and diffuse shortwave formulations take into account the local land-surface slope calculated from the digital terrain model used in CASC2D.

NET INCOMING LONGWAVE RADIATION
The net incoming longwave radiation is calculated from the infrared surface albedo, ground and air temperatures, a cloudiness correction factor, and ground and atmospheric emissivities. Parameterizations for the cloudiness-correction and atmospheric and surface emissivities are given in Bras (1990).

SOIL HEAT CONDUCTION
In this analysis, following Kasahara and Washington (1971), the soil heat flux at the surface (positive when directed into the soil) is represented as a function of sensible heat flux:

$$G = \frac{1}{3} \rho_a c_p c_H u_a (T_g - T_a) \quad (2)$$

where, T_g and T_a are the ground and air temperatures, respectively, c_p is the specific heat at constant pressure and is equal to 1.013 kJ/(kg°C), u_a is the wind speed at the reference level z and c_H is the dimensionless heat or moisture transfer coefficient applicable to bare soil. Following Deardorff (1978) a value of 0.0025 is selected for c_H. The formulation for air density, ρ_a is given in Shuttleworth (1993).

GROUND TEMPERATURE
The ground temperature appears explicitly in outgoing longwave energy flux-term, and implicitly in the Deardorff (1978) and Penman-Monteith evaporation formulae, and hence is extremely important for radiation and evapotranspiration calculations. The ground temperature fluctuates considerably during the diurnal cycle, and is not generally measured or provided by meteorological or weather stations. An approximate estimate of the ground temperature is obtained by solving the surface energy balance equation using an iterative Newton-Raphson technique. For bare-ground conditions, the land surface energy balance is solved assuming that the energy lost due to temporary storage,

advection and biochemical usage are negligible. The evaporation rate at the ground surface is represented mathematically as follows:

$$E = \rho_a c_H u_a \alpha^* [q_{sat}(T_g) - q_a] \qquad (3)$$

where $q_{sat}(T_g)$ is the saturation specific humidity and is given in Bras (1990). The wetness factor, α^* is a function of soil moisture content. Following Budyko (1948) and Manabe (1969) the wetness factor is estimated by using the following two relationships:

$$\alpha^* = \begin{cases} 1.0 & W \geq W_c \\ \dfrac{W}{W_c} & W_{wilt} < W < W_c \end{cases} \qquad (4)$$

where W is the moisture available in a 40 cm upper soil layer at the beginning of each time step, and W_c represents the fraction of soil field capacity at which potential evaporation ceases. The latter is represented mathematically as follows:

$$W_c = 0.75 \; W_{fc} = 0.75 \; n_e \; d_{soil} \qquad (5)$$

where W_{fc} is the soil field capacity, n_e is the effective porosity and d_{soil} is the depth of the upper soil layer. The 0.75 is taken from Williamson et al. (1987) and denotes the fraction of the field capacity at which potential evaporation ceases. Evaporation is assumed to cease when the soil water content reaches wilting-point soil moisture content, W_{wilt}.

EVAPOTRANSPIRATION PARAMETERIZATION

The Penman-Monteith equation is one of the most advanced resistance-based models available for the prediction of evapotranspiration from a vegetated land-surface (Shuttleworth, 1993). Evapotranspiration estimates are obtained by using the canopy resistance and aerodynamic features of the vegetated surface and several other meteorological and aerodynamic characteristics at the site of interest. As pointed out by Monteith (1965), several simplifying assumptions have been utilized to derive the Penman-Monteith evaporation formula. Despite those simplifying assumptions, Lemeur and Zhang (1990) found that the Penman-Monteith formulation is more suitable than the CRAE (Morton, 1983) and the Advection-Aridity (Brutsaert and Stricker, 1979) models for evapotranspiration modeling in arid watersheds.

Currently the model does not have the capability to calculate the canopy's leaf temperature. Therefore, the ground temperature is substituted for leaf temperature where required. The rate of water diffusion from the ground surface due to turbulence is controlled by aerodynamic resistance, which is a function of wind speed and height of the vegetation cover, and is mathematically represented as:

$$r_a = \frac{\ln\left[\frac{(z_u - d)}{z_{om}}\right] \ln\left[\frac{(z_e - d)}{z_{ov}}\right]}{(0.40)^2 U_z} \quad (6)$$

where z_u and z_e denote the heights of the wind speed and humidity measurements, respectively; U_z is the wind speed in m/s. Following Brutsaert (1975) it is assumed that: z_{om} is equal to $0.123\ h_c$; and z_{ov} is equal to $0.0123\ h_c$. In addition, following Monteith (1981), it is assumed that: $d = 0.67\ h_c$; where, h_c is the mean height of the crop. The Penman-Monteith equation is extremely sensitive to the value of canopy resistance. As pointed out by Lemeur and Zhang (1990), a 10% error in canopy resistance will give rise to a 10% error in the estimated transpiration.

RESULTS

Rainfall, runoff, and watershed data from the Goodwin Creek Experimental Watershed (Blackmarr, 1995) in Panola County, Mississippi, are used to calibrate and verify the continuous simulation features of CASC2D. The predominant runoff production mechanism in Goodwin Creek watershed is Hortonian, which makes the data set ideal for calibration and verification of Hortonian runoff models such as CASC2D. Meteorological observations from Memphis, Tennessee, are used as required. Memphis is the nearest NOAA climate station contained on NCDC Samson CD-ROM.

A test period of April 1, 1982 through August 31, 1982 is selected because this period contains 14 major runoff producing events. The Penman-Monteith ET method is selected, and GIS maps of vegetation height, canopy resistance, surface shortwave albedo, and wilting-point soil water content are created based on available data (Blackmarr, 1995). Figure 1 illustrates the percentage basin-averaged soil moisture content. The values fluctuate considerably due to the diurnal variability of evapo-transpiration (ET). At present, the model does not account for the seasonal variability of canopy resistance. However, the results suggest that the seasonality effects of canopy resistance is worth considering (i.e., due to the high

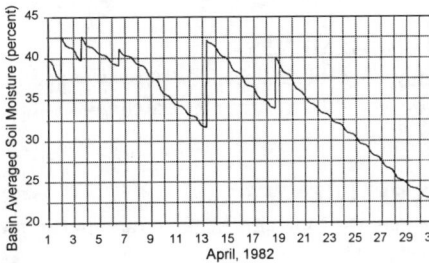
Temporal Variation of Soil Moisture (Goodwin Creek Watershed, MS)

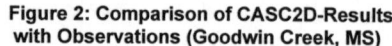

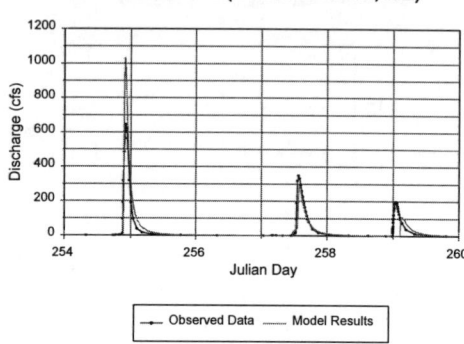

Figure 2: Comparison of CASC2D-Results with Observations (Goodwin Creek, MS)

dependence of ET on canopy resistance). Three hydrographs simulated by CASC2D are compared with observed hydrographs in Figure 2. The simulations begin on Julian day 90. CASC2D over-estimates the peak-discharge on the first event on Figure 2 by approximately forty percent. This represents an under-estimation of ET during the previous four-week long rainfall hiatus period. The second runoff event is simulated quite accurately, while the third event is slightly under-estimated. These estimates are expected to improve with the availability of data sets which account for the seasonal and spatial effects of soil and vegetation properties. In addition, analysis of results reveal that the continuous simulations make non-unique calibrations unlikely for simulations with high-resolution rainfall data.

CONCLUSIONS

A new evapotranspiration parameterization is added to CASC2D with the intent of enhancing its soil moisture storage accounting capabilities for continuous simulations. The new parameterization is capable of representing the moisture flux from a vegetated or a non-vegetated surface using a physics-based approach. The modeling work also required the conversion of CASC2D from an event-based ephemeral model into a perennial model. However, the accuracy of the new parameterization is highly dependent on the availability of solar and meteorological data, and GIS data-sets of several spatially varied surface and vegetation properties. One of the prime concerns regarding the use of distributed-parameter physically-based watershed modeling approaches is that the large number of spatially-varied parameters create a high potential for a non-unique model calibration. In using a continuous simulation methodology, the uniqueness of calibration is promoted, with the detraction of increased data requirements and their associated uncertainties.

ACKNOWLEDGEMENTS

This study was supported by the U. S. Army Corps of Engineers, Waterways Experiment Station through project DACA39-96-K-0012 and U. S. Army Research Office through contract DAAH04-96-1-0026.

APPENDIX REFERENCES

Beven, K., "A Sensitivity Analysis of the Penman-Monteith Actual Evapo-transpiration Estimates," J. Hydrol., 44, 169-190, 1979.

Blackmarr, W.A., ed. "Documentation of Hydrologic, Geomorphic, and Sediment Transport Measurements on the Goodwin Creek Experimental Watershed, Northern Mississippi, for the Period 1982-1993- Preliminary Release", USDA-ARS, National Sedimentation Laboratory, Oxford, MS, 38655, 1995.

Bras, R.L., "Hydrology: An Introduction to Hydrologic Sciences," Addison-Wesley, Reading, Massachusetts, 1990.

Brutsaert, W., "Comments on Surface Roughness Parameters and the Height of Dense Vegetation," J. Meteorol. Soc. Japan, 53, 96-97, 1975.

Brutsaert, W. and Stricker, H., "An Advection-Aridity Approach to Estimate Actual Evapotranspiration," Water Resour. Res., 15, 443-450, 1979.

Budyko, M. J., "Evaporation under Natural Conditions," Gimiz, Leningrad, (JPST, Jerusalem, 1969), 1948.

Deardorff, J.W., "Efficient Prediction of Ground Surface Temperature and Moisture, With Inclusion of a Layer of Vegetation," J. Geophys. Res., 83, 1889-1903, 1978.

Dickinson, R.E., A. Henderson-Sellers, P.J. Kennedy and M.F. Wilson, "Biosphere-Atmosphere Transfer Scheme (BATS) for the NCAR Community Climate Model," National Center for Atmospheric Research, NCAR/TN-275+STR, 1986.

Julien, P.Y, B. Saghafian, and F.L. Ogden, "Raster-Based Hydrologic Modeling of Spatially-Varied Surface Runoff", Water Resources Bull., 31, 523-536, 1995.

Kasahara, A., and W.M. Washington, "General Circulation Experiments with a Six-Layer NCAR Model, Including Orography, Cloudiness and Surface Temperature Calculation," J. Atmos. Sci., 28, 657-701, 1971.

Kondrat'yev, K. Ya., "Radiative Heat Exchange in the Atmosphere," Trans. from Russian by O. Tedder, Pergamon Press, Oxford, 1965.

Lemeur, R., and L. Zhang, "Evaluation of Three Evapotranspiration Models in Terms of Their Applicability for an Arid Region," J. Hydrol., 114, 395-411, 1990.

Manabe, S., "Climate and the Ocean Circulation: 1. The Atmospheric Circulation and the Hydrology of the Earth's Surface," Mon. Wea. Rev., 97, 739-774, 1969.

Monteith, J.L., "Evaporation and Environment," Symp. Soc. Exp. Biol., XIX, 205-234, 1965.

Idem., "Evaporation and Surface Temperature," Q. J. R. Meteorol. Soc., 107, 1-27, 1981.

Morton, F.I., "Operational Estimates of Areal Evapotranspiration and Their Significance to the Science and Practice of Hydrology," J. Hydrol., 66, 1-76, 1983.

Ogden, F.L., and B. Saghafian, "Green & Ampt Infiltration with Redistribution", Submitted for review, ASCE J. of Irrigation and Drainage Engineering, Sept. 1996.

Pielke, R.A., "Mesoscale Meteorological Modeling," Academic Press, New York, 1984.

Shuttleworth, W.J., "Evaporation," In: Handbook of Hydrology, D.R. Maidment, Editor-in-Chief, 4.1-4.53, 1993.

Williamson D.L., J.T. Kiehl, V. Ramanathan, R.E. Dickinson and J.J. Hack, "Description of NCAR Community Climate Model (CCM1)," National Center for Atmospheric Research, NCAR/TN-285+STR, 1987.

OPTIMIZATION OF PIPED SYSTEMS

DR. B.B. SHARP
Burnell Research Laboratory, Vic., Australia

INTRODUCTION

One may have to be an optimist to develop an optimization of a system and then proceed to implement it, depending as it does on the economic fluctuations in the future for as long as 50 years.   50 years because this might be regarded as the the minimum life of a pipeline in service.

The optimum diameter of a pumping (force) main has long been regarded as the most fundamental example of an economic decision of a system accounting for the present costs of components, interest rates on borrowings and future power costs.

It is an exercise in faith.   Faith that the long term fluctuations of these factors move in concert so that the decision NOW proves true as well, many years hence. Although examples can be produced to prove that this is correct, many hesitate to accept it in the design process. Today there are many additonal factors being considered and including topics under the heading of Risk Engineering.

Even with these new factors, there are still significant matters that are omitted from the design process mainly because much of the hydraulic design focusses on steady state conditions.   The importance of the ever present dynamic effects of water hammer is another factor worthy of consideration.

A REVIEW OF SOME BASIC STEADY FLOW CONCEPTS

A paper by Sharp (1985) referred to the economics of pumping with the importance of utilisation and variability of demand providing some new principles and these are summarised below.

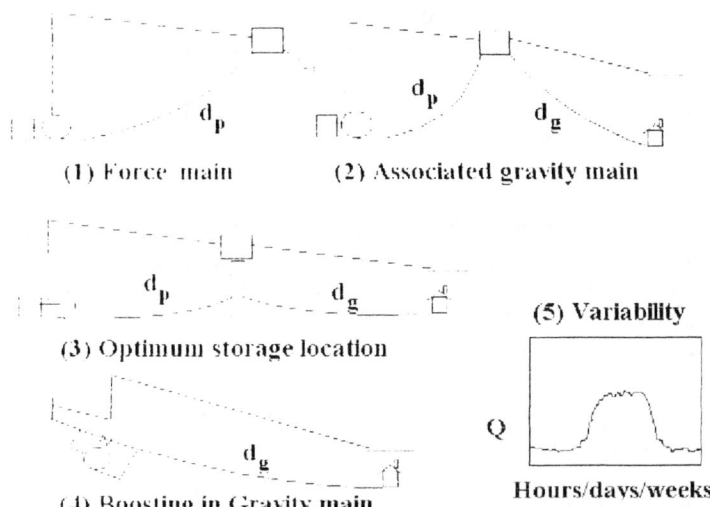

In (1) the economic diameter is found by formula, while in (2) the conservation of flow in and out of the storage enables a simple ratio between the force main and gravity main diameters to be found. The location of the storage in (3) can be determined on economic principles and if there is significant variability as in the figure (5), it can be proved that it is economic in (4) to boost for peak flows and use a gravity main diameter for lower mean flows.

There are other examples such as the use of two pumps, a small and a large so that the force main in (1) is always utilised.

The above relates to apparently simple systems but Sharp (1992), showed that these have relevance to complex systems as well. The state of the art of complex networks optimisation was fully reviewed by Walski (1985) and significant contributions have been made in modelling by Gessler, Jeppson, and Morgan and Coulter in the same proceedings. However unsteady flow effects receive little mention (slowly varying demands are quasi-static).

ADDITIONAL CONSIDERATIONS

1. Operational factors.

Whereas the type and hence strength of pipelines have been taken into consideration in determining the economic sizes of pipelines it seems that in general practice it relates only to steady state flows. However when the size is determined, the flow then provides a design velocity. Immediately it should be clear that there are water hammer implications. Sadly these are often only studied when troubles arise. It is inevitable that an estimation of the water hammer will have a direct bearing on the strength and perhaps material type of pipeline and hence there should be an iterative process to determine the final economic diameter.

It is felt that pumping will be ultimately introduced into most gravity systems and so the skilful designer should be alert to the consequences of inadequate materials or strength in the long term use of the system. It is also true that the operational requirements of valves in the system are related to the system's tolerance to dynamic events.

The approach of determining an economic diameter of a pipeline is therefore intimately related to the hydraulics of the system.

2. Demand requirements

A classic case recently outlined by Sharp (1996) of the concept and reality of a system in operation over 50 years, showed how the needs of the community and the quality of the supply led to large changes in the system. The changes meant bigger pipelines and extensive dis-infection of the water. Whereas the chlorinaton in use was modified and seemingly to meet the quality requirements, it was also true that the slime control, which had profound effects on the hydraulic efficiency of the pipelines, was as an important a factor and relating to the economic processes which impacted on the design of the pipelines.

These uncertainties may divert the designer from serious economic evaluation but decisions have to be made and costing the alternatives should include consideration of all the system variants.

3. Optimization of components.

In addition to the pipeline, the choice of components can be just as important. Optimization is as much a matter of the costs involved as well as the minimization of damaging pressures.

Reference has been made to the water hammer implications and it should be clear that components such as valves depend on dynamic events and not only steady state flows. It has taken a long time for valve manufacturers to realise that a linear motion in closing (or opening) leads to excessive water hammer. The optimum closure rate is exponential (see Sharp, 1981, 1996) which may be approximated by two stage stroking, while some types of valves have better features in the way the liquid is displaced during operation.

If these ideas are not recognised then the possibility of an economic decision re pipe size might be totally negated.

4. Recognition of Authority

Piped systems are generally designed by one group and then operated by another. There may be intervention by other groups such as fire fighting authorities or construction bodies requiring water.

Many cases of inadequate systems derive from the involvement of authorities not familiar with the limitations of the system as designed. Perhaps the worst example is the change of a basic gravity system into a pumping system by the attachment of a pump. It is probably time for the designer to anticipate such factors in the initial economic design.

5. Risk Engineering

There would be about five categories of risk depending on the Good Direct Knowledge and Statistical Evidence of components associated with the model. Historically the economic analysis of systems can only proceed effectively with a sound data base of the variability of demand over monthly and yearly periods and aware of the demand factors for all types of use.

There are cases where the intimate relationship between the water use and electric power availability have determined the degree of dependence and risk so that some loss is expected to be the more economically sensible option than providing absolutely positive supply. This can apply either way, where the certainty of water availability depends on alternative electric supplies, or the certainty of maintaining electric power generation depends on alternative water sources. Ultimately, the economics of piped (water) systems is now a much more complex issue when more than the basic steady state hydraulics is considered in developing a solution.

REFERENCES

Sharp, B.B. (1985), Economics of Pumping and the Utilisation Factor, Proc. A.S.C.E., J. Hyd. Eng., Vol. 111, No. 11, pp 1386-1396

Sharp, B.B. (1992), Economic Pipe Sizing - Complex Networks, Pipeline Systems, Ed., by Coulbeck, B. and Evans, E., Kluwer Academic Publishers, pp 37-41.

Sharp, B.B. (1981), *Water Hammer - Problems and Solutions*, Edward Arnold, London.

Sharp, B.B. and Sharp, D.B. (1996), *Water Hammer - Practical Solutions*, Arnold, Hodder Headline, London.

Sharp, B.B. (1996), The Morgan-Whyalla Pipeline: Concept and Reality, Proc., "Engineering Tomorrow Today-The Darwin Summit", National Conf., Darwin, Northern Territory, Australia, 21-24 April.

Walski, T.M., (1985), Computer Applications in Water Resources, Ed. by H.C. Torno, Proc. of Specialty Conf. of A.S.C.E., Buffalo, N.Y., State of the art Pipe Network Optimization, pp 559-568.

Gessler, J. (1985), ibid, Pipe Network Optimization by Enumeration, pp 572-581.

Morgan, D.R. and Coulter, I.C., (1985), ibid, Water Distribution Design with Multiple Demands, pp 582-590.

MATHEMATICAL MODELLING OF WATER QUALITY IN DISTRIBUTION SYSTEMS

DOMENICO PIANESE, FRANCESCO PIROZZI, LUCIO TAGLIALATELA
Department of Hydraulic and Environmental Engineering "Girolamo Ippolito".
University of Naples "Federico II". Via Claudio, 21 - 80125, Napoli (Italy).

INTRODUCTION
In the last few years many mathematical models have been developed for the evaluation of space and time variabilities of water quality parameters in water distribution systems. The differences between these models stem mainly from the flow conditions and phenomena considered and also the nature of the substances examined. Most of these models refer to a steady state condition and the transportation of substances in the pipes is usually considered to be due only to advection while molecular and eddy diffusion tend to be ignored. Furthermore, the substances contained in the water are divided into two classes: conservative and non-conservative. For the former, variations in concentration are due exclusively to the mixing, in every node, with the water flows deriving from the pipes flowing into the node itself; whereas the latter are subject to reactions causing their transformation, to decay phenomena and to volatilisation processes in the tanks located along the system.
The present paper describes a model previously proposed and used by the authors (*Pianese et al.*, 1995). This model differs from already existing ones because it is more complete, has shorter processing times and makes it possible to adopt physically-based reasoning criteria in order to follow the development of each single process within the network. In order to verify the results of the model, its application to an experimental system is presented and the results of numerical processing are compared with the experimental measurements of the concentrations detected in the system nodes.

LITERATURE REVIEW
One of the first models proposed for the study of quality characteristics was presented by *Shah and Sinai* (1988). This model is used to evaluate the chloride concentration at each node in a water system fed by three different sources with a different saline content. The authors propose a nodal model to search for the water discharges circulating on every pipe of the system and the piezometric heads in the nodes, which can be solved using an iterative method with an algorithm derived from Newton-Raphson's. At each iteration, the values of chloride concentration are calculated in all the system nodes, neglecting the diffusion and dispersion phenomena occurring within the system. This repeated determination of the saline concentration makes the model solution unnecessarily elaborate as the substances contained in the water can normally be said to behave as ideal tracers and, thus, do not influence the motion field. In these conditions the procedure can be divided into two distinct phases (or modules): the first phase includes the preliminary search for the distance-covered directions and the

values of the water discharges flowing through each pipe of the distribution system; the second phase includes the determination of the concentration value of the substances carried by the water flow in each section of the system.

Wood and Ormsbee's model (1989) is particularly useful in systems fed by different sources for evaluating the contribution of each source to the water quality in each node. The authors propose an eulerian approach for the quality module on condition that: i) the substances contained in the water are conservative; ii) their transportation is made up only of the advection action; iii) the confluent waters are completely mixed in each system node. The application of mass transfer for a generic conservative substance makes it possible to evaluate, in each node j, the contribution coefficient $F_c(i,j)$, which represents the percentage with which the feeding source i concurs in forming the flow arriving in node j. By the $F_c(i,j)$ coefficients, it is possible to evaluate the concentration of every conservative substance in each section of the water system.

In the study conducted by *Maione et al.* (1991) the same hypotheses as in the previous model are assumed, apart from the one regarding the nature of the substances carried in the pipes, for which the concentration variability along the sides of the water system through an exponential decay expression is adopted. The derived system of linear equations is solved with the Gauss-Sidel method, using a computing time interval equal to the shortest travel-time inside the pipes. This assumption represents a serious limitation of the model as the computing intervals cannot usually be consistent with the travel times inside all the water system pipes (*Pianese et al.*, 1995).

The paper by *Elmaalouf and Kim* (1995) resumes part of the study conducted by *Elmaalouf* (1992) and proposes a computing model that can be used to search for the contribution coefficients and then determine the concentrations of substances subject to decay phenomena in all the system nodes. The authors claim that the comparison between the model results and some experimental measurements shows a good approximation, especially in the sections whose water flow velocity is lower, whereas the model used to simulate the decay of the carried substances is less suitable in pipes characterised by high velocity.

One of the most complete models quoted in literature is undeniably the one proposed by *Rossman et al.* (1994) with particular reference to the variability of chlorine concentration in the water distribution systems. The model uses a lagrangian approach and is based on the subdivision of the water volume contained in each pipe into a given number of segments (elementary cells), which have the same velocity inside the pipe as that of the flow and behave like completely stirred tanks. The model consists of a set of conservation equation of chlorine mass, written: i) in each elementary cell; ii) in the nodes located upstream of each pipe; iii) in the water system tanks. The modelling of the water quality variations phenomena is particularly accurate whereas only advection transportation is considered and molecular and eddy diffusion phenomena are neglected. For each pipe, the length of the cells is obtained by multiplying the water flow velocity in the pipe by the shortest travel-time inside the pipes. In turn, the number of cells is equal to the ratio between the length of the pipe and that of the cells. It is possible to demonstrate (*Pianese et al.*, 1997) that this option may cause very substantial errors in the lengths of the pipes actually considered in the calculations and may also imply that in every single node the chlorine inlet deriving from the confluent pipes is not well modelled. The set of governing equations is solved by means of an explicit procedure that calculates, at a generic time t, the chlorine concentration in any side of the water system, and considers the boundary condition as the chlorine concentration in the side's initial node at a time $t-\Delta t$. However, this procedure may give rise to parasite numerical variations that can substantially perturb the solutions.

MODEL DESCRIPTION

The proposed model is essentially based on two modules. The first module evaluates the macroscopic characteristics of the water flows circulating in each pipe of the system and, in particular, it calculates: the water discharges outflowing from the nodes; the water discharges flowing along the pipe and the mean flow velocities; the absolute piezometric heads in the nodes and in all points of interest located in the network; the free-surface elevations in the tanks whose free-surface is variable and any water discharges flowing out from the spillways located in them. The above-mentioned variables can be evaluated both with reference to *steady-state conditions* and to conditions of timely slow variations (*pseudo-steady conditions*, described through sequences of steady state conditions), and also for *unsteady compressible or incompressible flow conditions*. The calculation model used for these evaluations is based on a node-oriented algorithm (*Pianese & Masini*, 1994), which can be used for complex networks, meshed and/or branched, characterised by all those devices usually found in a drinking-water distribution system to enable its correct operation and management. In *steady-state conditions*, the solution is achieved through the iterative solution of a set of equations describing the balance of water masses entering and leaving the nodes as a function of the piezometric heads in the nodes.

The second module aims to identify the space-time variations of the quality parameters related to the water flowing in the system and is based on a set of equations describing the mass balances related to each of the examined substances. In particular, the module evaluates the mass variations due to: a) water mixing; b) advection; c) dispersion (due to turbulence); d) molecular diffusion; e) reactions occurring in the mixing with other substances dissolved in the water or present on the inner pipe walls (incrustations, biofilms, etc.); f) volatilisation developing inside the tanks; g) self-decay.

The balance equations that can be used to model the phenomena are:

a) the balance equation of mass already present and introduced into the tank (written under the hypothesis of neglecting the effects induced in the Δt of the calculation by the variations of the water volume cantained in the tanks and the advection, dispersion and diffusion phenomena that occur in the tanks)

$$C_t|_v = \frac{V_v C_{t-\Delta t}|_v + \Delta t \sum_{n=1}^{N_{j1}} Q(n,j)C|_{s=L} + \Delta t M_v - \Delta t(K_{1v} + K_{2v})C_{t-\Delta t}|_v - \Delta t K_{2v}C_{ev}}{V_v + \Delta t \sum_{z=1}^{N_{j2}} Q(j,z) + \Delta t Q_{str}} \quad (1)$$

b) the mass balance equation of substances flowing in the nodes:

$$Cj = \frac{\sum_{n=1}^{N_{j1}} Q(n,j).Cn + Mj}{\sum_{z=1}^{N_{j2}} Q(j,z) + Q_o(j)} \quad (2)$$

c) the mass balance equation of substances present in an elementary cell

$$\frac{\partial C}{\partial t} = -U\frac{\partial C}{\partial s} + D_T \frac{\partial^2 C}{\partial s^2} + D_M \frac{\partial^2 C}{\partial s^2} - K_1 C - K_2(C - C_e) \qquad (3)$$

where: C=concentration at time t and progressive s; U = mean flow velocity; D_T and D_M = eddy diffusion and molecular diffusion coefficients; K_1 and K_{1v} = decay constants in the pipe and in the tank; K_2 and K_{2v} = reaction coefficient with substances on the internal walls of pipes and tanks; C_e and C_{ev} = concentration in the pipe and tank internal walls; N_{j1} and N_{j2} = number of pipes with terminal in node j with flow in input and output from j, respectively; $Q(n,j)$ = flow discharges from node n to node j ($n=1,2,...,N_{j1}$); C_n = concentration at the terminal section of the pipe connecting nodes n and j; M_j = mass directly flowing into j; $Q(j,z)$ = water discharge flowing from node j towards node z adjacent to it and hydraulically located downstream ($z=1,2,..., N_{j1}$); $Q_o(j)$ = water discharge flowing from node j and directly distributed to the users; Δt = time interval used in the calculations; $C_t|_v$ and $C_{t-\Delta t}|_v$ = concentration in the tank at times t and t-Δt, respectively; V_v= mean water volume in the tank V during the interval Δt; Q_{str} = flow discharged from the tank spillway; $C|_{s=L}$ =concentration in the terminal section of the pipe with flow entering the tank; M_v=mass introduced into the tank v.

MODEL APPLICATION

In order to verify the proposed model, the authors referred to the experimental values reported in the work of *Elmaalouf and Kim* (1995) regarding the experimental distribution system shown in Fig. 1 and described in detail in the personal communication sent to the authors by *Elmaalouf* (1996).
The experimental system comprises seven nodes and nine pipes whose geometrical characteristics are also reported in Figure 1. The system is supplied via two pumping stations situated in nodes 1 and 7 which produce flows of 0.424 l/s and 0.426 l/s, respectively. The case examined is the one which in the work of Elmaalouf and Kim is defined as *"scenario 1"*, in which concentrated flows of 0.2836 l/s and 0.5664 l/s are produced in nodes 2 and 5, respectively.
Using the hydraulic module described in the previous paragraph and adopting the formula of Hazen-Williams with a roughness coefficient of 150, it was possible to calculate the values of the discharges in the various pipes in the network, their relative velocities (Table I) and the piezometric heads in the nodes (Table II).
The tables also indicate the values obtained by *Elmaalouf* (1996) using the model described in paragraph 2 above. It should be borne in mind that these values turn out to be different from the ones indicated in *Elmaalouf and Kim*'s work (1995) which, although they refer to the same tests, are affected by an approximation error committed in converting the discharges into units of the international measurement system. Examination of the tables shows an almost perfect agreement of the numerical data.

Table I - *Values of the discharges and mean flow velocities in the pipes*

Section	1	2	3	4	5	6	7	8	9
Q (l/s)	0.424	0.187	0.047	0.254	0.067	0.149	0.263	0.163	0.426
V (m/s)	1.486	0.657	0.372	2.004	0.526	1.177	0.922	1.291	0.841
Q* (l/s)	0.424	0.188	0.047	0.254	0.066	0.149	0.262	0.163	0.426
V* (m/s)	1.487	0.658	0.369	2.002	0.524	1.176	0.920	1.289	0.841

N.B. *The values marked with* * *are the ones reported by Elmaalouf (1996)*

Table II - *Piezometric heads and concentrations evaluated from the authors (H,C) and Elmaalouf (H*,C*)*

Node	1	2	3	4	5	6	7
H (m)	46.655+	45.425	45.245	45.552	44.102	45.721	46.130+
H* (m)	-	45.436	45.256	45.561	44.111	45.727	-
C (mg/l)	10.00	9.31	7.67	2.98	5.08	3.00	3.00
C* (mg/l)	10.00	9.30	7.65	3.00	5.09	3.00	3.00

N.B. The values marked with + refer to the section immediately downstream of the pump

The quality module was verified by considering that, in steady state conditions, the flow from the two pumping stations contains a benzene concentration of 10 mg/l and 3 mg/l, respectively. Assuming that the benzene behaves like a conservative substance, we can ignore the last two terms in expression (3), regarding the reactions that may take place both in the bulk liquid and with the substances adhering to the internal walls of the pipes, and exclude Eq. (1) from the set of equations. The calculations were carried out assuming the following values for the parameters contained in (1'): $D_T = 4 \cdot 10^{-3}$ m$^2 \cdot$s^{-1}; $D_M = 5 \cdot 10^{-10}$ m$^2 \cdot$s^{-1}. The concentrations calculated in nodes 2 - 6 are reported in Table II and turn out to be very close to the ones obtained by *Elmaalouf* (1996) using the model he proposed for analysis of steady state conditions. The agreement with the experimental measurements taken in the same nodes is satisfactory almost throughout, apart from node 5 where we find a percentage deviation of about 36.5 % (Fig.2).

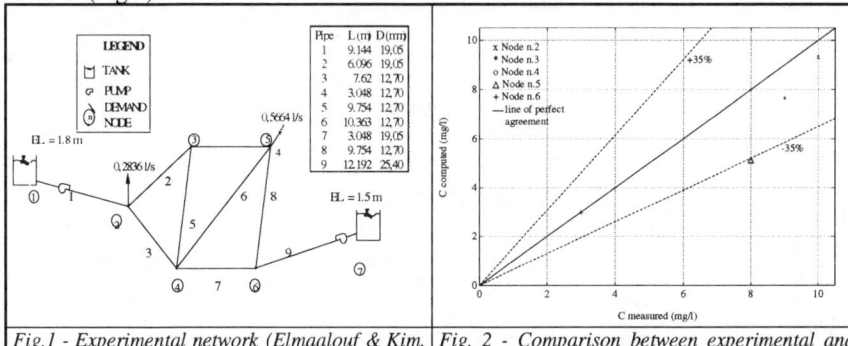

Fig.1 - Experimental network (Elmaalouf & Kim, 1995)	Fig. 2 - Comparison between experimental and numerical results.

In conclusion, the proposed model provides results that are quite satisfactory in the case of non-conservative substances and steady-state conditions. Obviously, the models full potential must be further verified by examining cases in which the water system is characterised by reactions between the substances contained in the bulk liquid or present on the pipe walls have a considerable effect on water quality (*Pianese et al.*, 1997). Furthermore, the model has to be verified in transient conditions, for example investigating the trend over time and space of the chlorine concentration in various sections of the water system where it flows into one or more tanks, following a pre-determined law of temporal variation.

CONCLUSION

In recent years, growing importance has been attached to the opinion that, in the planning and management of distribution networks, it is not sufficient to satisfy water demand under different operating conditions, but it is also necessary to take into account those aspects related to the quality of the water distributed. Therefore, the planning phase should make allowance for the siting of stations for verifying water quality parameters in the most important points along the system. Then, the management phase could envisage a number of on-line corrective interventions on the quality of water circulating within the pipes. In both cases, it is vitally important to make provision for an instrument that can evaluate water composition variability over both time and space. The model should not only represent every single phenomenon occurring within the hydraulic system but should also be sufficiently reliable and have short processing times so as to make it possible to adapt the control and regulation devices to the changeable conditions which can arise in the system itself.

The present paper has introduced a water quality simulation model in water distribution systems that, compared to already existing ones, supplies a more complete and accurate description of the phenomena occurring and, at the same time, assures shorter processing times. The model is based on the application of the mass balance equations to the different items of a water distribution system and has been solved using a lagrangian procedure, thanks to which it is possible to easily follow the development, over time and space, of every single phenomenon. The model has been applied in steady state conditions to an experimental distribution system taken from the literature material and has shown a good agreement with the experimental results, thus confirming its excellent potential.

ACKNOWLEDGEMENTS

This study was partly supported by the *Consiglio Nazionale delle Ricerche - Gruppo per la Difesa dalle Catastrofi Idrogeologiche - Linea 1* (italian National Research Council - Group for Prevention from Hydrogeological Disasters - Line 1) and partly funded (40%) from M.U.R.S.T. - Gruppo "Affidabilità" (Water Systems Reliability research group, working under the supervision of *Ministero dell'Università e della Ricerca Scientifica e Tecnologica*).

The writers would like to tank Dr. Sami G. Elmaalouf for the detailed informations given on the geometrical characteristic of his experimental apparatus.

REFERENCES

Elmaalouf, S.G. (1992). *"A Comprehensive Steady State Quality Model for the Assessment of Contaminants Behaviour in Water Distribution Systems."* Thesis presented in partial fulfilment of the requirements for the degree of Master of Science. California State Univers.

Elmaalouf, S.G., & Y.C. Kim (1995). *"Field Examination of a Distribution System Water Quality Model."* Proceedings of XXVI IAHR Congress. London. Vol. 1, p. 342-347.

Elmaalouf, S.G. (1996). *Personal Communication* about the results shown in the paper "Field Examination of a Distribution System Water Quality Model." (see above).

Maione U., Mignosa P., & M.G. Tanda (1991). *"La Determinazione delle Caratteristiche Chimico-Fisiche delle Acque nelle Reti di Distribuzione Idrica".* «Scritti in onore di Girolamo Ippolito», Lacco Ameno (NA), Italy, Vol. II, p. 5-17 (in italian).

Pianese, D., & P. Masini (1994). *"Sui Transitori che si Sviluppano nelle Reti Idrauliche in Pressione Munite di Luci di Efflusso Regolabili e di Dispositivi di Attenuazione dei Fenomeni di Moto Vario".* Proceedings of the Seminar «Moto Vario nei Sistemi Acquedottistici». Bari (Italy), 30-31 Maggio, p. 159-183. (in italian).

Pianese, D., Pirozzi, F., & L. Taglialatela (1995). *"Variabilità Spazio-Temporali delle Caratteristiche di Qualità delle Acque Convogliate nei Sistemi Acquedottistici."* Proceedings of the Conference «Sistemi Idropotabili Integrati», Bologna (Italy), 21-22 November, p. 115-142 (in italian).

Pianese, D., Pirozzi, F., & L. Taglialatela (1997). *"Influence of Distinct Processes on the Quality of Water in Distribution Systems."* Proceedings of XXVII IAHR Congress. San Francisco, 10-15 August.

Rossman, L.A., Clark, R.M., & W.M. Grayam (1994). *"Modeling Chlorine Residuals in Drinking-Water Distribution Systems".* Journal of Enviromnetal Engineering., ASCE, Vol.120, No.4, p. 803-820.

Shah, M., & G. Sinai (1988). *"Steady State Model for Diluition in Water Netwoks."*. Journal of Hydraulic Engineering, ASCE, Vol. 114, No.2, p. 192-206.

Wood, D. J., & L.E. Ormsbee (1989). *"Supply Identification for Water Distribution Systems".* Journal American Water Works Association, July, p. 74-80.

INFLUENCE OF DISTINCT PROCESSES ON THE QUALITY OF WATER IN DISTRIBUTION SYSTEMS

DOMENICO PIANESE, FRANCESCO PIROZZI, LUCIO TAGLIALATELA
Department of Hydraulic and Environmental Engineering "Girolamo Ippolito" - University of Naples "Federico II" - Via Claudio, 21 - Naples, Italy

INTRODUCTION

A new mathematical model has been introduced *(Pianese et al., 1995, 1997)* to evaluate the variations, in time and space, of the characteristics of the water flowing through a complex water distribution system composed of free water surface tanks, pumping stations, and pipes connecting the tanks to one another and bringing water to the delivery nodes. The model is able to give a highly detailed description of all the phenomena that may arise in the water system, such as: mixing; transportation of the inlet substances (through advection, eddy diffusion, molecular diffusion); volatilisation in tanks; reactions with substances contained in the water and/or present on the pipes' internal walls.
The capabilities of the proposed model in order to evaluate the concentrations, in steady-state conditions, of non-conservative substances in a water system have been already illustrated *(Pianese et al., 1997)*. The present paper reports the results of a sensitivity analysis conducted with reference to a case-study already described in literature, in order to explain the influence, even on steady-state concentration values, of a series of phenomena frequently neglected by other models.

MATHEMATICAL MODELLING OF PROCESSES

MAIN CHARACTERISTICS
This mathematical model consists *(see Pianese et al., 1995)* of the classical equations that can be obtained from the execution of mass balances performed, for different elements of the distribution system, on the substances present and transiting within finite or infinitesimal water volumes. These are as follows:
a) for water volumes flowing within the connecting pipes:

$$\frac{\partial C}{\partial t} = -U\frac{\partial C}{\partial s} + D_T\frac{\partial^2 C}{\partial s^2} + D_M\frac{\partial^2 C}{\partial s^2} - K_1 C - K_2(C - C_e) \tag{1}$$

b) for the network internal nodes:

$$C_j = \frac{\sum_{n=1}^{Nj_1} Q(n,j) \cdot C_n + M_j}{\sum_{z=1}^{Nj_2} Q(j,z) + Q_o(j)} \tag{2}$$

c) for the tanks serving the system:

$$C_t\Big|_v = \frac{V_v \cdot C_{t-\Delta t}\Big|_v + \Delta t \cdot \sum_{n=1}^{N_{j1}} Q(n,j) \cdot C\Big|_{s=L} + \Delta t \cdot M_v - \Delta t \cdot (K_{1v} + K_{2v}) \cdot C_{t-\Delta t}\Big|_v - \Delta t \cdot K_{2v} C_{ev}}{V_v + \Delta t \cdot \sum_{z=1}^{N_{j2}} Q(j,z) + \Delta t \cdot Q_{str}} \quad (3)$$

where: C=concentration; U=flow velocity; D_T and D_M=eddy and molecular diffusion coefficients; K_1 and K_{1v}=decay constants in pipes and tanks; K_2 and K_{2v}=reaction coefficient with substances on the internal walls of pipes and tanks; C_e and C_{ev}=concentration in the pipe and tank internal walls; N_{j1} and N_{j2}=number of pipes with flow in input and output from j; $Q(n,j)$=flow from n to j; C_n=concentration at the terminal section of the pipe connecting n and j; M_j and M_v=mass directly flowing into j and into the tank v; $Q(j,z)$= water discharge flowing from j towards z, hydraulically located downstream; $Q_o(j)$=flow distributed to the users at the node j; $C_t|_v$ and $C_{t-\Delta t}|_v$=concentration in the tank at t and t-Δt; V_v=mean water volume in the tank during Δt; Q_{str}=flow from the tank spillway; $C|_{s=L}$=concentration in the terminal node of the pipe with flow entering the tank; M_v=mass introduced into the tank V.

LAGRANGIAN APPROACH

The solution process used in order to properly take advection phenomena into account is based on a lagrangian approach in which the water volumes flowing through the different network pipes are first subdivided into cells. This process is shown in Fig. 1.

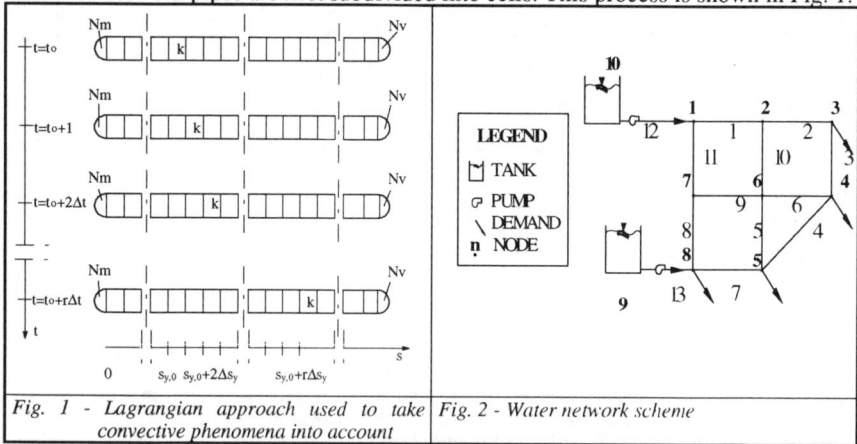

Fig. 1 - Lagrangian approach used to take convective phenomena into account

Fig. 2 - Water network scheme

In practice, this approach considers, for each flow condition, the subdivision of the water volume present in each R system pipe into a number of segments (elementary cells) P_y, (where P_y varies, in general, from one pipe to another and, for the same pipe, time after time if the flow is not strictly in steady-state conditions), following each cell from the moment it passes into the pipe until the moment it leaks, as the cell moves within the pipes. Consequently, with U_y as the mean flow velocity, the number of subdivisions P_y will have to respect the following condition:

$$P_y = \frac{L_y}{\Delta S_y} = \frac{L_y}{\Delta t \cdot U_y} \quad , \text{ with } y = 1,2,\dots,R \quad (4)$$

Nevertheless, as it is generally impossible to obtain the equality expressed by (4) for all the network pipes at the same time, it will be sufficient to obtain only an approximate equality. This approximation is given by the expression:

$$\forall y, \quad P_y \cong \frac{L_y}{\Delta t \cdot U_y} \quad \Leftrightarrow \quad \forall y, \quad \left| \frac{L_y}{\Delta t \cdot U_y \cdot P_y} - 1 \right| \leq \varepsilon \tag{5}$$

with a sufficiently small value of ε (for example: $\varepsilon = 0.01 \div 0.05$).
Ultimately, this type of lagrangian approach will make it possible to solve, for each cell, the following equation, instead of the equation of type (1):

$$\frac{\partial C}{\partial t} = D_T \frac{\partial^2 C}{\partial s^2} + D_M \frac{\partial^2 C}{\partial s^2} - K_1 C - K_2 (C - C_e) \tag{1'}$$

MODEL APPLICATION
DESCRIPTION OF THE CASE-STUDY

The model described in the previous paragraph was applied for testing purposes to the water network examined by *Elmaalouf (1992)*. This network, schematically shown in Fig. 2, is made up of R=13 pipes and N=10 nodes, two of which are external to the network. The system is fed by two pumping stations located in nodes 9 and 10. There are four outflowing nodes with flows variable from 113 l/s (4 cfs) to 284 l/s (10 cfs). The geometric features of the pipes are reported in the following Table I:

Table I - Geometric features of the network sections

Pipe	1	2	3	4	5	6	7	8	9	10	11	12	13
D (mm)	406.4	355.6	101.6	304.8	355.6	355.6	304.8	406.4	406.4	304.8	304.8	609.6	355.6
L (m)	762.5	610.0	457.5	1525	762.5	610.0	915.0	915.0	976.0	457.5	549.0	1220	915.0

The roughness formula taken for reference by the author is that of Hazen-Williams with the coefficient $C_{H-W}=120$. The pumping station located next to node 9 rises the necessary flows from a constant level tank with a height (referred to a datum) of 6.10 m, and is provided with a pump whose total power output is 223.5 Kw (300 hp). The pumping station located next to node 10 rises water from a constant level tank, with a height of 3.05 m in relation to the datum, and is provided with a pump whose total power output is 521.5 Kw (700 hp).
The use of the hydraulic module *(Pianese et al., 1997)* starting from the above-mentioned values leads to the data reported in the second line of the following Table II, which are in total agreement with the values found by Elmaalouf, except for node 3 where the differences can be attributed, in the writers' opinion, only to a press error (for this aspect, see also the results reported in Table 4 of the original paper).

Table II - Comparison of the grade values obtained by the writers (H) and by Elmaalouf (H*)

Node	1	2	3	4	5	6	7	8	9	10
H (m)	91.732	76.127	62.787	64.573	70.458	72.753	79.131	79.181	95.218+	99.070+
H*(m)	91.735	76.226	59.900	64.447	70.431	72.739	79.117	79.166	-	-

N.B. The values marked with + are those related to the section downstream of the pump

The mean velocities with which the water flows along the various pipes are reported, together with the discharges, in Table III below:

Table III - Mean flow velocities and discharges flowing through the pipes

Pipe	1	2	3	4	5	6	7	8	9	10	11	12	13
Q (l/s)	383.3	279.6	4.1	73.1	95.9	214.7	119.1	15.6	206.9	103.7	191.4	574.7	248.1
Q (cfs)	13.51	9.85	0.15	2.58	3.38	7.57	4.20	0.55	7.29	3.65	6.74	20.25	8.75
Q* (cfs)	13.49	9.76	0.24	2.60	3.40	7.64	4.21	0.55	7.31	3.73	6.76	20.25	8.75
V (m/s)	2.955	2.816	0.505	1.001	0.965	2.162	1.632	0.120	1.595	1.421	2.623	1.969	2.498

N.B. The values obtained by Elmaalouf (1992) are marked with *

As regards the evaluation of the quality characteristics of the waters flowing into the different pipes of the system, Elmaalouf refers to his own calculation model which can only take into account stationary conditions. In particular, the author studies the effect of normal concentrated inlets of substance into nodes 9 and 10. The inlets considered by the author are such as to give way to inlet concentrations of 160 mg/l (node 9) and 100 mg/l (node 10). The substance taken for reference in the application is conservative and, therefore, is not subject to decay phenomena or to reactions with other substances perhaps dissolved in the water or present on the pipes' internal walls, or to volatilisation processes. Moreover, as the model used by the author does not take into account the effects deriving from dispersion and diffusion phenomena, these aspects are completely neglected in his evaluations.

SENSITIVITY ANALYSIS

The calculation model described in the previous paragraph has been applied many times to the network reported in Fig. 2, in order to: i) verify its reliability in the light of the results previously obtained by *Elmaalouf (1992)* for steady-state conditions; ii) evaluate the differences possibly arising because the diffusion and dispersion phenomena, neglected by *Elmaalouf*, are here considered; iii) evaluate, under the hypothesis that the inlet substance is non-conservative (for example, it is made up of chlorine), the differences as they arise, since they also take into account the other processes (reactions with other substances dissolved in the bulk of water; reactions with other substances perhaps present on the pipes' internal walls; volatilisation processes in the tanks). The calculations were carried out considering a maximum error of ε lower than 5% (in particular, it was considered that $\varepsilon = 0.0499$). Consequently, taking a calculation time interval lower than one minute for reference ($\Delta t = 54.32$ s), the number of cells obtained for the different network sections turned out to vary from a minimum of 4 to a maximum of 140.

In view of the previously illustrated objectives, six different runs were carried out. The results of these are reported in the lines from the third to the eighth of Table IV.

Run No.1 In Run No.1 the model reliability was evaluated, in relation to the achievable results, for conservative substances and with no dispersion or diffusion phenomena, using the previously described lagrangian modelling. The results obtained on the grounds of the model proposed by the writers, reported in the third line of Table IV, fully coincide with those obtained by *Elmaalouf (1992)*, reported in the second line of Table IV. This once again demonstrates the capabilities of the proposed model to take into due account both the mixing and the advection phenomena.

Table IV - Comparison of the concentration values obtained by Elmaalouf (C^*) and by the writers in relation to the different phenomena as they were taken into account (C^I, C^{II}, ...)

Node	1	2	3	4	5	6	7	8	9	10
C^* (mg/l)	100.00	100.00	100.26	111.01	134.60	102.98	104.50	160.00	160.00	100.00
C^I (mg/l)	100.00	100.00	100.16	111.02	134.58	103.01	104.51	160.00	160.00	100.00
C^{II} (mg/l)	100.00	100.00	100.16	111.01	134.58	102.99	104.49	160.00	160.00	100.00
C^{III} (mg/l)	99.62	99.45	99.45	109.32	133.28	101.78	103.44	159.61	160.00	100.00
C^{IV} (mg/l)	90.74	82.10	74.62	67.98	93.68	74.32	84.39	138.76	160.00	100.00
C^V (mg/l)	90.74	82.10	74.61	67.97	93.68	74.31	84.39	138.76	160.00	100.00
C^{VI} (mg/l)	90.39	81.65	74.09	69.97	92.81	73.45	83.56	138.42	160.00	100.00

Run No.2 Run No.2 aimed to evaluate the possible effects deriving from the presence of dispersion and diffusion phenomena. In particular, the evaluations are effected with reference to a value of $D_T = 4 \times 10^{-3}$ m^{2}s^{-1} and to a value of $D_M = 0.5 \times 10^{-9}$ m^{2}s^{-1}. As can be observed from the results reported in the fourth line of Table IV (values C^{II}), the effects of dispersion and diffusion turn out to be almost completely negligible, at least for the present case. It is therefore proven that, at least in steady-state conditions, it is definitely possible to refer to quite simple calculation models while completely neglecting both dispersion and diffusion phenomena.

Run No.3 Run No.3 aimed to evaluate, for the present case, the influence exerted only by the reactions which may occur between the substances flowing into the system and those already existing in the water. In particular, notwithstanding the high concentration values considered, reference has been made, purely as an example, to the case where the inlet substance consisted of chlorine. In the example, according to laboratory results obtained by water samples drawn up on site *(Rossman et al., 1994)*, a decay constant has been assumed $K_1 = 0.55$ day^{-1}.
As can be observed from the results reported in the fifth line of Table IV (values of C^{III}) the differences, deriving from the fact that the reactions between the inlet substances and those already contained in the water were taken into account, turned out to be small, amounting at most to a few percent of the real value.
Run No.4 The main objective of Run n.4 was to evaluate the influence on the normal concentration values of the reactions developing between the inlet substances and those present on the pipes' internal walls. The evaluation is achieved, purely as an example, under the hypothesis that the substance let into in the tanks located next to nodes 9 and 10 is chlorine, and that a biofilm develops on the pipes' internal walls and can prime a chlorine consumption due to the materials deposited there. More particularly, following the procedure indicated by *Rossman et al. (1994)* made it possible to consider that from equation (1') is obtained:

$$\frac{\partial C}{\partial t} = D_T \frac{\partial^2 C}{\partial x^2} + D_M \frac{\partial^2 C}{\partial x^2} - K_1 C - \frac{K_W K_f}{r(K_W + K_f)} C \qquad (1'')$$

in which K_W is the chlorine decay constant on the pipes' internal wall; K_f is evaluated in turn on the grounds of the water kinematic viscosity v, and the molecular diffusivity constant D_M, with the same expression reported by *Rossman et al. (1994)*. Considering $v=1.14 \times 10^{-6}$ m^{2}s^{-1} and $K_W = 0.15$ m day^{-1}, the results reported in the sixth line of Table IV are obtained (values of C^{IV}). Comparing the obtained results with those reported in the preceding lines clearly points out the considerable difference exerted, even on steady-state concentration values, by the presence of biofilms on the walls.
Run No.5 Run No.5 aimed only to evaluate the influence of the parameter K_W on the values assumed by the chlorine concentrations. In particular, a value of $K_W = 0.45$ m day^{-1} was considered, which is equal to the maximum value of the range of variation as hypothesised by *Rossman et al. (1994)* for this parameter. In order to demonstrate the very small influence of this parameter, the results obtained for $K_W = 0.45$ m day^{-1}, reported in the seventh line of Table IV (values of C^V), almost fully coincide with those obtained for $K_W = 0.15$ m day^{-1} (sixth line of Table IV).
Run No.6 Run No.6 shows the global effect, on the concentrations obtained in stationary conditions, of the different phenomena taken into account by the mathematical model. For this, the following parameter values have been assumed: $K_W = 0.45$ m day^{-1}; $v=1.14 \times 10^{-6}$ m^{2}s^{-1}; $K_1 = 0.55$ day^{-1}; $D_T = 4 \times 10^{-3}$ m^{2}s^{-1}; $D_M = 0.5 \times 10^{-9}$ m^{2}s^{-1}. As can be observed from the values reported in the eighth line of Table IV (values of C^{VI}), even in normal conditions the results obtained by taking into account all the phenomena turn out to be very different from those obtained under the hypothesis of just mixing the substances coming from different feeding sources.

MODELLING OF TRANSIENTS
In order to demonstrate the analysis capability of the proposed model even in unsteady conditions, we now consider the case in which the water distribution system shown in Fig. 2, in the same operating conditions enumerated above, is subjected to a sudden inlet of chlorine into the two tanks. In the relative calculations, the two tanks are assigned a capacity equal to half of the overall water volume distributed during one day. It is assumed that the chlorine is introduced into the two tanks almost simultaneously. In particular, the following inlet law is considered for tank 9 (*t* in second and M_C in Kg s^{-1}): $t=0 \rightarrow M_C=5 \times 10^{-5}$; $t=3600 \rightarrow M_C=5 \times 10^{-5}$; $t=5400 \rightarrow M_C=2.5 \times 10^{-4}$; $t=7200 \rightarrow M_C=5 \times 10^{-5}$; $t=86400 \rightarrow M_C=5 \times 10^{-5}$. Whereas the chlorine is introduced into the tank in node 10 according to the following law: $t=0 \rightarrow$

$M_C=0.10 \times 10^{-3}$; $t=1800 \rightarrow M_C=1 \times 10^{-4}$; $t=3600 \rightarrow M_C=3.5 \times 10^{-4}$; $t=5400 \rightarrow M_C=1.0 \times 10^{-4}$; $t=86400 \rightarrow M_C=1 \times 10^{-4}$. The results of the model, obtained using the values of D_M, D_I, K_I, K_W and K_f already used for reference in Run No.6, are reported for two distinct network paths in the following diagrams in Figs. 3a and 3b.

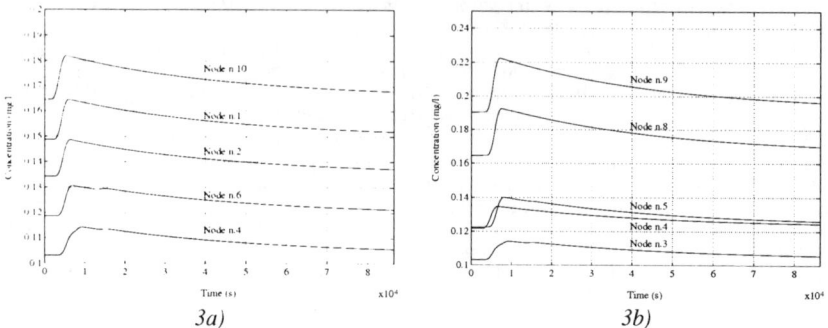

3a)   3b)

Fig.3 - Chlorine time variations in different nodes of the water system illustrated in the Fig.2

Examination of these diagrams clearly shows how, in the absence of mixing phenomena, the residual chlorine concentrations gradually decrease, mainly because of the effect of the reactions developing with the other substances present in the water and on the pipes' internal walls.

CONCLUSIONS

In modelling the variations in the water quality characteristics as the water flows through the various elements making up a water distribution system, frequently only a few processes are taken into account (typically, mixing phenomena). The other phenomena (such as, dispersion, diffusion, decay and reaction with the substances contained in the water or present on the pipes' internal walls) are frequently held to be of secondary importance and are thus neglected by many of the existing models. The present paper has used a case study already employed by other authors and reported in the literature to show how, in the case of non-conservative substances, the phenomena that are normally neglected can considerably modify the concentration values of the substances dissolved in the waters flowing through the system. In particular, the sensitivity analysis conducted has definitively clarified, on the one hand, the small influence of molecular diffusion and turbulent dispersion phenomena and, on the other, the considerable influence of the reactions occurring between the substances introduced into the system, those already previously dissolved in water and any other substances present on the pipes' internal walls. The paper has also shown the significant effect of these reactions, also during the transient following rapid introduction of such substances.

REFERENCES

Elmaalouf, S.G. (1992). *"A Comprehensive Steady State Quality Model for the Assessment of Contaminants Behaviour in Water Distribution Systems."* Thesis presented in partial fullfillment of the requirements for the degree of Master of Science.California State Univers.

Pianese, D., Pirozzi, F., and L. Taglialatela (1995). *"Variabilità spazio-temporali delle caratteristiche di qualità delle acque convogliate nei sistemi acquedottistici."* Proceedings of the Seminar "Sistemi idropotabili integrati". Bologna (Italy), 21-22 November, p. 115-142 (in italian).

Pianese, D., Pirozzi, F., and L. Taglialatela (1997). *"Mathematical Modeling of Water Quality in Distribution Systems."* Proceedings of XXVII IAHR Congress. San Francisco, 10-15 August 1997.

Rossman, L. A., Clark, R. M., and W. M. Grayam (1994). *"Modeling Chlorine Residuals in Drinking-Water Distribution Systems."* Journal of Environmental Engineering, ASCE, Vol. 120, 4, p. 803-820.

RELIABILITY OF ALGORITHMS FOR WATER QUALITY ANALYSIS IN HYDRAULIC NETWORKS

SAMI ELMAALOUF
The Levantine Engineers Society, P.O. Box 13181, Los Angeles, CA 90013, USA
YOUNG C. KIM
Professor and Chairperson, California State University, Los Angeles, CA 90032, USA

ABSTRACT
Numerous algorithms for analyzing water quality parameters in hydraulic networks are now available. These algorithms are based upon steady-state or dynamic hydraulic simulations. Three analytical methods in this paper are analytically evaluated for the same pipe network. Furthermore, two of these analytical methods are compared to an experimental model. The reliability of these algorithms is investigated and evaluated by analytical and experimental comparisons.

INTRODUCTION
A reliable model is essential for predicting and monitoring water quality variations in hydraulic networks. Various mathematically oriented models have appeared in recent years for predicting water quality variations in distribution systems. These models are based on steady-state and dynamic approaches. The steady-state approach uses mass balance formulations around each mode to solve for quality operating parameters. Dynamic analyses, on the other hand, take the variations of system supplies and demands into consideration. Temporal and spatial variations take place, as a consequence, in water quality analysis.

In this paper the algorithms of three methods are evaluated for applications to water quality analysis in pipe systems. The reliability of these algorithms is examined by modelling one distribution network (Figure 1) with each algorithm and comparing the results of each solution to one another. The reliability of algorithms is further evaluated by modelling another distribution network (Figure 2) using two of the three referenced algorithms and comparing the predicted quality parameters to experimental results.

ALGORITHMS FOR WATER QUALITY ANALYSIS
The evaluated algorithms for water quality analysis in hydraulic networks are briefly described hereon. Detailed presentations for each method can be found elsewhere.[1,2,3,4]

Algorithm 1 - This method is based upon nodal mass balance formulations and was described in two previous publications.[1,2] The method is summarized as follows.

The contribution of supply sources to loads of discharge at various nodes in water distribution systems, can be determined when a nodal mass continuity is reached. The source contribution problem can be formulated as

$$[Q_{sc}] \cdot f_c = [Q_{fo}] \tag{1}$$

where $[Q_{sc}]$ is the source flow contribution matrix; f_c is the supply source contribution factor and $[Q_{fo}]$ is the source outflow forcing matrix. If the total inflow from a source node is equal to the demand at a particular node, then Q_{sc} is set to unity. Otherwise

$$Q_{sc} = \Sigma Q_i + q \tag{2}$$

where Q_i is the inflow into a junction node and q is the external demand or supply at a junction node.

After a successful determination of the supply source contribution factor and the correct assembly of the source flow contribution matrix, $[Q_{sc}]$, we have to formulate the constituent concentration problem.

Given the source flow contribution matrix and knowing the concentration of a certain chemical at the supply source (simply by actual field quality measurement at the supply source), the constituent concentration problem formulation is

$$[Q_{sc}] \cdot C_c = [C_s] \tag{3}$$

where C_c is the constituent concentration at the junction nodes and C_s is the constituent concentration injected from a supply source(s).

Algorithm 2 - This method was described in many previous publications.[3] It is described as

$$\frac{\partial C_{ij}}{\partial t} = \frac{q_{ij}}{A_{ij}} \frac{\partial C_{ij}}{\partial x_{ij}} + \theta(C_{ij}) \tag{4}$$

where C_{ij} is the concentration of substance in link i,j as a function of distance and time (i.e., $C_{ij} = C_{ij}(x_{ij}, t)$); x_{ij} is the distance along link i,j; q_{ij} is the flow rate in link i,j at time t; A_{ij} is the cross-sectional area of link i,j; $\theta(C_{ij})$ is the rate of reaction of constituent within link i,j.

Algorithm 3 - This method was described in a previous publication.[4] The nodal concentration can be expressed as

$$C_{cj} = \frac{\sum_{n=1}^{N_{j1}} Q(n,j).C_{cn} + M_{cj}}{\sum_{z=1}^{N_{j2}} Q(j,z) + Q_u(j)} \qquad (5)$$

where C_{cj} is the constituent concentration at junction node j; C_{cn} is the constituent concentration at the end section of conduit connecting node j to node n; and M_{cj} is the injected constituent mass at node j. The algorithm involves temporal and spatial variations in modelling of water quality in distribution systems.

COMPARISON OF SOLUTIONS

The three algorithms above were compared by solving for a chlorine concentration problem in Network 1 (Figure 1).[4] Three loading cases were considered. Case No. 1 assumed the initial concentrations at nodes 1, 6 and 15 as 0.19 mg/l, 0.1 mg/l and 0.03 mg/l, respectively. Case No. 2 assumed the initial concentrations at nodes 1, 6 and 15 as 0.28 mg/l, 0.17 mg/l and 0.03 mg/l, respectively. Case No. 3 assumes the initial concentrations at nodes 1, 6 and 15 as 0.21 mg/l, 0.14 mg/l and 0.03 mg/l, respectively. The three algorithms were used distinctly to determine the concentrations at all nodes, under each case. A summary of results is as follows.

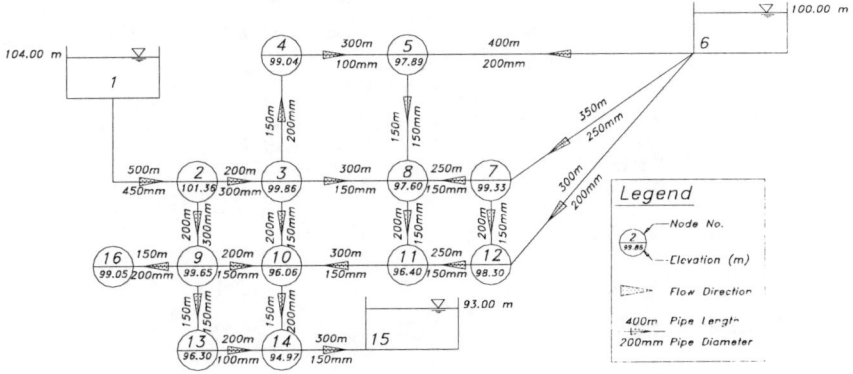

Figure 1 Network 1

Table 1 Comparisons of Concentrations for Network 1

Node No.	Case No. 1 (mg/l)			Case No. 2 (mg/l)			Case No. 3 (mg/l)		
	Alg.1	Alg.2	Alg.3	Alg.1	Alg.2	Alg.3	Alg.1	Alg.2	Alg.3
1	0.19	0.19	0.19	0.28	0.28	0.28	0.21	0.21	0.21
2	0.19	0.18	0.14	0.28	0.25	0.14	0.21	0.19	0.16
3	0.19	0.17	0.12	0.28	0.22	0.12	0.21	0.17	0.13
4	0.18	0.16	0.09	0.26	0.18	0.09	0.20	0.14	0.11
5	0.10	0.10	0.05	0.17	0.13	0.05	0.14	0.11	0.07
6	0.10	0.10	0.10	0.17	0.17	0.17	0.14	0.14	0.14
7	0.10	0.09	0.06	0.17	0.15	0.06	0.14	0.13	0.09
8	0.11	0.11	0.04	0.18	0.14	0.05	0.15	0.11	0.06
9	0.19	0.17	0.12	0.28	0.22	0.12	0.21	0.17	0.13
10	0.18	0.14	0.07	0.26	0.15	0.07	0.20	0.12	0.08
11	0.10	0.09	0.03	0.17	0.09	0.03	0.15	0.08	0.04
12	0.10	0.09	0.05	0.17	0.13	0.05	0.14	0.11	0.08
13	0.19	0.15	0.09	0.28	0.16	0.09	0.21	0.12	0.10
14	0.18	0.13	0.06	0.27	0.07	0.05	0.20	0.05	0.06
15	0.03	0.03	0.03	0.03	0.03	0.03	0.03	0.03	0.03
16	0.19	0.16	0.09	0.28	0.18	0.09	0.21	0.14	0.10

EXPERIMENTAL PROGRAM

An experimental model was prepared for the network shown in Figure 2.

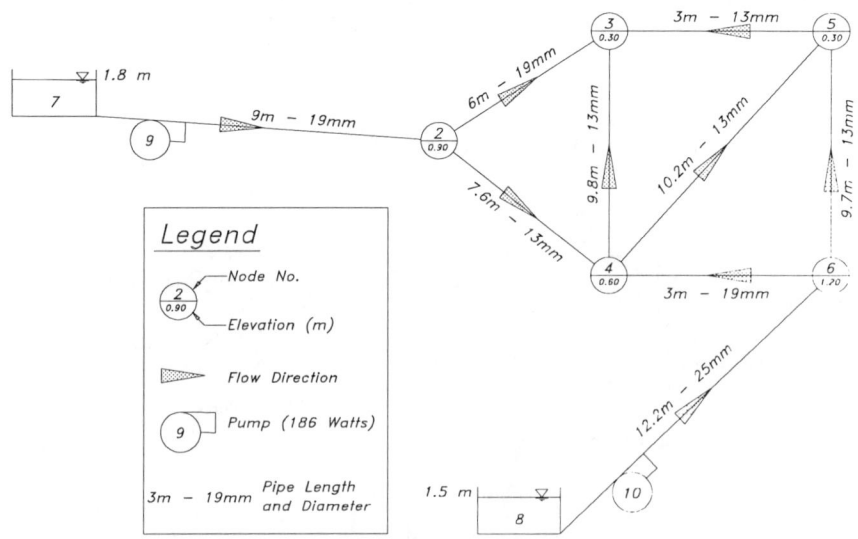

Figure 2 Network 2

The experimental system comprised nine pipes, five connecting junction nodes, two constant-head supply sources, five ball valves at each node to control the outflow at node points, and two 186 Watt (1/4 hp) centrifugal pumps in lines 1 and 9. For this model, the ball valves at nodes 3 and 4 were calibrated to discharge 0.57 l/s and 0.28 l/s, respectively. All other valves were fully closed. A steady (constant) head was maintained at both feeding sources by adding distilled water constantly. The system was operating continuously (an attempt to simulate a steady-state flow in the network). The flow rates were checked at discharging nodes by means of a magnetic flow meter. Hydraulic data was gathered, simultanuously.

Network 2 was stopped and drained. A chemical (Benzene-C_6H_6) was introduced to the supply sources at dissimilar proportions, with concentrations of 10 mg/l in supply source No.1 and 3 mg/l in supply source No.2. The system was started again and water samples were collected. Chemical analyses followed.

The first two algorithms were used distinctly to determine the concentrations at all nodes, and were then compared to the experimental data. A summary of results is as follows.

Table 2 Comparisons of Concentrations for Network 2

Node No.	Experimental Data (mg/l)	Algorithm 1 (mg/l)	Algorithm 2 (mg/l)
2	10.00	10.00	10.00
3	9.00	7.72	7.72
4	7.00	4.01	4.00
5	3.00	3.35	3.35
6	3.00	3.00	3.00
7	10.00	10.00	10.00
8	3.00	3.00	3.00
9	10.00	10.00	10.00
10	3.00	3.00	3.00

CONCLUSION

This paper presented a comparison of various results from analytical water quality algorithms. The comparisons of concentrations for the first network ranged from poor to good and did not provide a conclusive evidence that any of these algorithms is reliable. The agreement between the three algorithms decreased somewhat when the network is larger. Algorithm 1 and 2 yielded very similar results when they were utilized in Network 2. Both analytical results, however, were not very close to actual experimental data at some junction nodes.

The authors evaluated analytical hydraulic results from Network 2 and compared them to experimental results. The comparisons were excellent and therefore provided an evidence that the distribution system used in this study was modelled adequately.

Modelling water quality in distribution systems is a complex undertaking. Although all mass balance relationships are satisfied in the three presented algorithms--yielding correct mathematical solutions for the quality operating parameters--confidence in the reliability of the model and the utilization of reasonable data are essential. More models and experimental studies with more complex networks than those presented here are recommended in order to further verify and validate the actual applicability of the presented methods.

ACKNOWLEDGMENTS

The authors are grateful to Bannaoun Engineers-Constructors for permission to publish this paper. Appreciation is due to Ing. Francesco Pirozzi for his continuous support. The authors also thank Ms. Loretta Rosas and Mr. Joe Kao for their assistance in preparing this paper.

REFERENCES

1. Elmaalouf, S. G.. *A Comprehensive Steady State Quality Model for the Assessment of Contaminants Behaviour in Water Distribution Systems.* Master of Science Thesis, Civil Engineering Dept, California State University, Los Angeles, 1992.
2. Elmaalouf, S. G., Kim, Y. C.. *Field Examination of a Distribution System Water Quality Model.* ASCE Hydraulic Conference, San Francisco, July 1993.
3. EPANET V1.1e. A hydraulic and Water Quality Analyzer. USEPA, Risk Reduction Engineering Laboratory, Cincinnati, Ohio, 1996.
4. Pianese, D., Pirozzi, F., Taglialatela, L.. *Variabilita Spazio-Temporali Delle Caratteristiche Di Qualita Delle Acque Convogliate Nei Sistemi Acquedottistici.* Technical Report No. 783, Universita Degli Studi Di Napoli Federico II, Naples, Italy, Nov. 1995.

ENHANCEMENT OF IRRIGATION SYSTEMS IN DEVELOPING COUNTRIES. A "HOLISTIC" APPROACH

SAMI ELMAALOUF
The Levantine Engineers Society, P.O. Box 13181, Los Angeles, CA 90013, USA
YOUNG C. KIM
Professor and Chairperson, California State University, Los Angeles, CA 90032, USA

ABSTRACT
A "holistic" design approach has been developed to improve an irrigation system. Understanding the demographic, agricultural and hydraulic parameters are key elements in the project. The proposed branching network services an approximate area of 415 Hectares with irrigation water from a single supply source. Prior to designing the network, only 10 percent of the total serviced area was irrigated effectively. This design approach may be used as a template for future similar projects. Enhancement of irrigation water management is a principal benefit of the developed approach.

INTRODUCTION
Stringent measures were imposed by the Lebanese government in the late 1980's to deplete and ultimately stop the cultivation and trafficking of illicit drugs mainly in two counties (Baalbeck and Hermel Counties) in the northern region of the Province of Bekaa (Lebanon). These measures were prompted by international efforts headed by leading industrial nations, represented by the United Nations (U.N. Drug Control Programme).

In the early 1990's, the United Nations (U.N. Development Programme) conceived a master plan that would improve the infrastructure of this area by primarily enhancing the agricultural infrastructure, through adequate water management practices, agricultural loans, training, education, and rehabilitation of unused lands that have a potential to become fertile.

This paper presents and discusses the development of an approach that was used for designing an irrigation water distribution system that will service an approximate area of 415 Hectares upon implementation. This approach will further serve as a template or criterion for other similar projects in the region.

DEMOGRAPHIC, AGRICULTURAL AND HYDRAULIC EVALUATIONS
The proposed irrigation water distribution project is in the Village of Chaat, Baalbeck

County. This county, as well as other counties within this region in the Province of Bekaa, survives mainly on agricultural revenues. With the absence of effective government during the war years (1975-1990), the area became gradually infested with the cultivation of two illicit plants: *Cannabis sativa* and *Papaver somniferum*, commonly known as Hashish and Opium Poppy, respectively. Such crop proved to bring large revenues to the area, although expensive and obsolete irrigation water distribution and management were utilized. No attempts were made by local farmers and landlords to improve such inefficient irrigation practices.

In the aftermath of the harsh measures that were imposed by the local government and the United Nations to halt planting of illicit drugs, it was anticipated that the area will become poorer. It was therefore determined that efficient water systems and adequate agricultural practices are vital to revive the general area and improve the quality of life of people in the region.

In an attempt to improve irrigation water practices in the Village of Chaat, the Ahla Reservoir Irrigation Water Distribution System (Figure 1) was proposed. One supply source feeds this network. The proposed irrigation area in question was subdivided to 14 zones. Each zone was provided with one or more discharge node to adequately supply irrigation water to different parcels. The proposed zones took the parcels limits into consideration, so no conflict in irrigation per owner will be created.

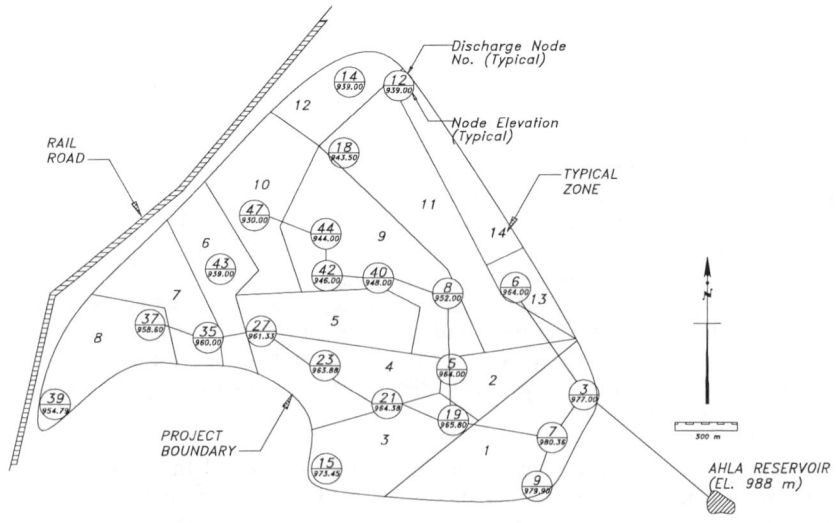

Figure 1 Ahla Reservoir Water Distribution System

CLIMATE
The climate within the region has the following properties.

Table 1 Climatic Properties

Range	Relative Humidity (%)	Wind Speed (Km/day)	Air Temperature (°C)	Rainfall (mm)
Minimum	39.0	156.0	2.7	0.0
Average	51.0	230.0	10.2	33.6
Maximum	63.0	311.0	17.7	96.0

LAND USE AND WATER REQUIREMENTS
The land use within the areas in question is primarily agricultural. The total water capacity of the single source should be capable of meeting maximum daily water demands in the 14 zones. The maximum estimated daily water demand for each discharge scenario is about 140 liters per second.

Table 2 Water Demands Based on Land Use (Zones)

Zone No.	Area of Each Zone (Hectares)	Percent of Total Area (%)	Maximum Daily Water Demands (l/s)
1	50	12.04	10
2	16	3.85	10
3	32	7.71	10
4	34	8.19	10
5	27	6.52	10
6	23	5.54	10
7	24	5.78	10
8	28	6.75	10
9	56	13.49	10
10	32	7.71	10
11	50	12.04	10
12	18	4.34	10
13	7	1.70	10
14	18	4.34	10
Total:	415	100.00	140

Table 3 Water Supply and Demand Summary

Supply	Demand: Crop*
Ahla Reservoir	Vegetables (40%) Legumes (10% - 20%) Wheat (10% - 20%) Fodder Crops (10%) Fruit Trees (10%) Curcubitacea (0% - 30%)
Flowrate = 3.5 to 5.0 Mm^3/year	Demand = 2.2 to 4.4 Mm^3/year**

* Crop Percentages vary with each season.
** Water demands vary per season and crop.

AHLA RESERVOIR - WATER SUPPLY

Field tests were conducted to verify the applicability of data dating back to 1976. Tests during February 1995 agreed with existing data. Water supply data from Ahla Reservoir is summarized in Table 4.

Table 4 Average Annual Water Volume in Ahla Reservoir[1]

Month	Average Volume ($m^3 \times 10^4$)	Drought Volume ($m^3 \times 10^4$)
January	30.9	20.2
February	29.8	19.3
March	35.6	25.3
April	37.4	27.4
May	40.7	31.8
June	38.9	28.8
July	38.7	24.1
August	36.6	23.2
September	33.7	30.1
October	32.9	28.0
November	30.1	24.1
December	30.5	23.5
Total:	415.8	305.8

WATER DISTRIBUTION SYSTEM

The proposed branching network (Figure 1) was designed to operate without artificial energy (pumping) in order to minimize operation cost. An operation schedule was prepared. The schedule outlines which nodes must be opened at a specific time in the irrigation process. The network requires only one operator to alternate the opening and closure of flow control valves located at each major junction (discharge) node. Secondary nodes (irrigation plugs) are coordinated, operated and maintained by land owners and the network operator. The operation manual was put together after evaluating hydraulic (pressure, flowrate, surge) and agricultural (crop, water demands, soil properties, evapotranspiration) considerations.

In order to evaluate, design, and calibrate the water distribution system, a computer model was developed. The computer model is a skeletonization of the network.

Topographic data was obtained from the Lebanese Army maps and in-house survey of the area to determine junction node elevations. All proposed conveying pipes are uPVC pipes (Class 2 - DIN 8062 Standard). Water demands were determined from crop requirements taking into consideration soil-specific parameters and crop species. The design criteria of the Ahla Reservoir Irrigation Water Distribution System is summarized in Table 5.

Table 5 Water Distribution System Design Criteria

Supply	Demand: Crop*	Transmission and Distribution	System Pressures
Ahla Reservoir	Vegetables (40%) Legumes (10% - 20%) Wheat (10% - 20%) Fodder Crops (10%) Fruit Trees (10%) Curcubitacea (0% - 30%)	Design Vel. ≤ 2 m/s Min. Dia. = 75 mm Max. Dia. = 500 mm	Min. = 0.01 MPa Max. = 0.60 MPa

* Percentages of different crops vary with each season.

The model simulated the system under many arbitrary flow discharge patterns. The simulations, with given parameters, can predict velocities, pressures, and headlosses in the system for each distinct loading scenario. These values and operating parameters were evaluated to globally optimize the original network design. System optimization allowed a well calibrated and economical system.

WATER SYSTEM SIMULATIONS
Computer solutions were used to evaluate and optimize the proposed network and to determine the levels of operation and calibration.[2] The highest static pressure is reached at the lowest elevation in the network and is nearly 0.4 MPa (4 Bars). The lowest static pressure is about 0.05 MPa (0.5 Bars). The hydraulic network was calibrated to provide a stable static pressure at each node during all discharge schemes and operation alternatives. In areas where the pressure is equal to or greater than 0.1 MPa (1 Bar), it was possible to utilize a sprinkler system. For areas with lower pressures, it was decided to use flooding techniques for irrigation. A summary of the computer solutions is as follows.

Table 6 Summary of Hydraulic Results at Open Nodes

Discharge Pattern No. 1				Discharge Pattern No. 2			
Node*	Elev. (m)	Demand (L/s)	Pressure (MPa)	Node*	Elev. (m)	Demand (L/s)	Pressure (MPa)
3	977.00	10.00	0.10	6	964.00	10.00	0.21
5	964.00	10.00	0.10	8	952.00	10.00	0.31
7	980.36	10.00	0.05	9	979.90	10.00	0.05
8	952.00	10.00	0.20	12	939.00	10.00	0.33
12	939.00	10.00	0.33	14	939.00	10.00	0.30
15	975.62	10.00	0.09	15	973.45	10.00	0.09
19	965.80	10.00	0.09	19	965.80	10.00	0.20
21	964.38	10.00	0.10	23	963.80	10.00	0.19
23	963.80	10.00	0.09	27	961.33	10.00	0.20
27	961.33	10.00	0.10	39	954.79	10.00	0.24
35	960.00	10.00	0.10	42	946.00	10.00	0.36
39	954.79	10.00	0.14	43	939.00	10.00	0.30
42	946.00	10.00	0.26	44	944.00	10.00	0.37
44	944.00	10.00	0.28	47	930.00	10.00	0.41

* Open Junction Node per discharge pattern.

CONCLUSION

Significant efforts have been proposed by the international community to assist in stopping the cultivation and trafficking of illegal drugs within two counties in the northern outskirts of the Province of Bekaa, Lebanon. This paper discussed the development of an approach that was utilized to enhance irrigation practices in these counties. This approach will assist local farmers to economically irrigate their crop.

The proposed template approach involved the design of a distribution system that aims to enhance current obsolete irrigation practices in the village of Chaat (Baalbeck County, Bekaa, Lebanon). The network will assist in distributing adequate water to different farmers within the project's area. This improvement in the irrigation infrastructure provides sufficient water to 14 zones that used to be traditionally planted with illicit crop. It is hoped that such improvements will revitalize and sustain the agricultural sector within the village and help the overall mission that inhibits farmers from going back to cultivating more profitable, yet destructive and illicit crop.

ACKNOWLEDGMENTS

The authors are grateful to the United Nations Development Programme (Baalbeck, Lebanon) for permission to publish this paper. Appreciation is due to Bannaoun Engineers-Constructors for its moral and financial support. The authors also thank Mr. Joe Kao for his assistance.

REFERENCES

1. Van Eleeuwen, N.H.. *Developpement Hydo-Agricole De La Bekaa Centrale, Liban*. Food and Agriculture Organization (FAO) United Nations, Rome, 1976.
2. EPANET V1.1. A hydraulic and Water Quality Analyzer. USEPA, Risk Reduction Engineering Laboratory, Cincinnati, Ohio, 1993.

A STUDY ON WATER MANAGEMENT OF AN IRRIGATION SCHEME IN SRI LANKA

Dr. Shahane De Costa
Dept. of Civil Engineering, Open University of Sri Lanka

ABSTRACT

A study of the irrigation and farming practices of an irrigation scheme in Sri Lanka's Angunukolapallessa area has been made and it has been pointed out that even though the broader aspect of water management of coping with scarcity and abundance has been achieved by collection of excess water and the regulated discharge of same through implementation of irrigation schemes inclusive of tanks, reservoirs, channels etc., to effectively and efficiently deal with scarcity & abundance of water the detail aspects of water management must also be looked into. It is also shown that to reach the higher levels of effective and efficient water management the involvement, education and training of all concerned is a necessity.

1.INTRODUCTION

Sri Lanka located in the tropics has a warm climate through out the year enabling agriculture to take place continuously if the water requirement constraint is overcome. It is well known that two thirds of Sri Lanka receives rainwater in abundance during some periods of the year while water is scarce in the same area during other periods of the year. In certain area it goes to the extremes of, flood during one period and drought during another.

In order to over come this situation water management has been taking place for the past 2000 years, namely the collection of excess rainwater and the controlled and regulated use of same during the year. Even though proof of this broad aspect of water management is visible from the irrigation tanks and reservoirs constructed dating from far back as ancient kingdoms due to the increasing demand for the scarce resource water, the detail aspect of water management must also be looked into.

Most irrigation scheme in Sri Lanka have been constructed to facilitate paddy cultivation. Paddy being a crop that is very sensitive to water increases the need for effective water management.

One noticeable present day problem of water management in irrigation schemes is that the tail farmers not receiving sufficient water, while the head farmers receive excessive water irrespective of the quality of water being discharged at the sluices.

Generally an irrigation scheme is implemented taking into consideration that in return say certain quantity of paddy would be harvested and supplied to the countries food market. However in the situation where the tail farmers not receiving sufficient water they are compelled to grow other crops such on chili or onions where the water requirement is lesser, thereby negating one aspect of the expected return of the irrigation scheme, namely the paddy requirement of the food market. This behavior causes other disturbances such as reducing the expected price of chili in the market for chili farmers due to the higher than expected production. Therefore chili farmers in other areas earn lesser.

The absence of detail water management leads not only to scarcity of water to some farmers but also leads to problems as mentioned above. The following study is presented to illustrate this.

2. THE CASE STUDY

2.1 Location

The Walawe irrigation project located in the South of Sri Lanka in the Hambantota district consists of a 40.8 km long right bank main canal endeavoring to irrigate 8000 ha which is divided into 19 tracks. The main canal branches into two canals namely the Gajamangama and the Bata ata of which Gajamangama branch canal is 11.2 km long and services tracks 15 to 19 which fall into the Angunakolapellessa division. A schematic location map of same is indicated in Fig-1.

2.2 Farming Practices

The average cultivable area of a track is 425 ha and this is divided into 425 farm lots. It is expected that all farmers cultivate paddy throughout the year. Due to the lack of water for tail farmers specially in the Yala season namely April to September some of the farmers cultivate crops such as chilies, where the water requirement is low compared to paddy, while some others cultivate paddy only in part of their land. The head end farmers presuming that in the days to come the water levels in the canals will be low flood their land no sooner water is available. This results in lower paddy yield due to excessive water. From an interview survey it has been clarified that there have been instances during the dry season where according to the perception of farmers at the head end of the canal insufficient water have been supplied triggering the excess water intake at the start.

Therefore, this not only reduces the total average annual income of all the farmers in the area due either to the lower yield or not cultivating their whole land but also reduces incomes of chili farmers in other areas due to the unexpected increase of chili in the market. Even though there is no formal water management organization for fields in existence, farmer organizations are in existence and well attended meetings take place in which information on water management activities are given.

2.3 Related Problems

The main observation made is that the broad aspect of water management takes place very correctly, that is the collection of all water and the controlled discharge of same at the main sluices to serve the total expected irrigable cultivatable lands. However, even though the quantity of water flowing in the branch canals are sufficient, due to improper water management at field canals all farmers don't receive the anticipated quantity of water.

This problem as mentioned above is due to the head farmers opening their gates and taking excessive water even to their detriment of lesser yield, there by reducing the flow of the field canals resulting in the tail farmers not getting sufficient water even though at the head of the canal sufficient water has been discharged.

The unsatisfactory detail management of the available water for cultivation purposes not only creates disturbances to the socio-economic environment but also to the socio-hygienic environment by deteriorating the biological environment through the formation of stagnant pools of water which cause air pollution and provide breeding grounds for mosquitoes that spread disease, specially in the environs of plots where excessive water has been obtained.

Another major related problem is due to the lack of flowing water at the tail the tail part of the field canals posses higher siltation. This in turns reduces the flowing capacity of water and works in a vicious circle. The low flow of the tail end of canals results in frequent wetting and drying of the side walls in that

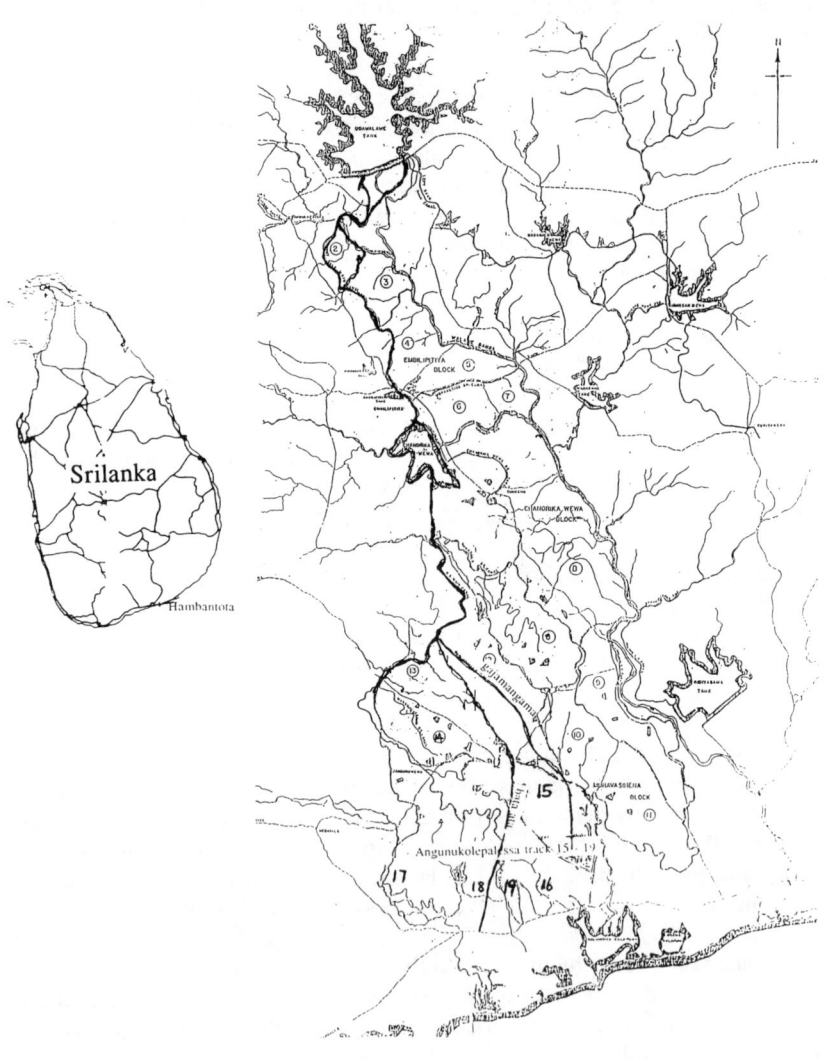

FIG - 1 Schematic Location Map

area which gives rise to erosion of the side walls as well.

The low flow of water in the tail part of the field canals also gives rise to vegetal growth which again reduces the flow of water. All this implies that detail water management is a must and any short comings, if exist gives rise to further problems.

3.ORGANIZATION AND MANAGEMENT OF WATER

From the above case study it could be seen that eventhough the first step of coping with water scarcity and abundance has been achieved in the Angunukolapallessa area to reap proper benefits detail water management practices are necessary. Here the correct amount of water reaches the head of the field canals which means that institutional water management takes place to a sufficient degree. However the final distribution process throws up water management problems.

Many solutions could be proposed to solve this problem. One is the implementation of rules and regulations which dissuades the excessive intake of water even though difficult to implement. Another is the management of water extending to the field canals and individual plots, such as the implementation of a rotational distribution scheme where water is received in turns. For example a set of farmers either at the head or the tail initially close their intakes while another set of farmers keep their intakes open and receive water for some time. There after the farmers who received water close their intakes for the set that did not receive water to open intakes and receive water. This rotational distribution of water could be performed to a well planed time schedule either decided upon by the institutions or the farmer organizations. However for the success of rotational distribution or any water management activity the corporation of all farmers is essential. Therefore at the person to person level a proper educational and training system has to be set up to high light the need for effective water management by the individual farmers.

The rotational water management or any effective water management technique if executed correctly once, to service all farmers head or tail will result in, the optimum use of water, the possible enhancement of the cropping intensity over the farmland and the increasing of the yield which in turn will give rise to higher annual incomes to the farmers. The results of which would be directly visible to the farmers themselves that, thereafter naturally effective and efficient water management would take place.

4. SUMMARY AND CONCLUSION

It could be concluded that basically there exist many levels of water management. One is the broad aspect of coping with scarcity and abundance which is the collection of excess water and the controlled discharge of same while the other is the detailed water management process in which the active corporation of all individuals are necessary. Both of these aspects are essential for the effective and efficient use of this scarce resource.

A clear example of which is in the case study presented above where the first level of coping with scarcity and abundance had been achieved but not the next the detailed water management, as water was collected controlled and discharged enabling sufficient water to flow at the head of field canals but however the tail end farmers did not receive sufficient water. Therefore it could be said that the individualization of water management or the message of effective and efficient water management must reach every individual in order that every individual benefits. It also could be concluded that a good awareness & training of farmers in water management activities is essential for the effective & efficient use of water.

The use of aquifers in Saudi Arabia to reclaim and store wastewaters

ACHI M. ISHAQ AND AMIR ALI KHAN
Department of Civil Engineering
King Fahd University of Petroleum & Minerals
Dhahran, Saudi Arabia

ABSTRACT

In recent years there has been an alarming depletion of groundwater resources in the aquifers of the Kingdom of Saudi Arabia due to increasing agricultural and urban water requirements. Efforts are being made to increase the use of reclaimed wastewater. The sixth development plan of the Kingdom envisages the use of 310 MCM of reclaimed treated wastewater by the year 1998-1999. Wastewater recharge is one of the most promising techniques available for reclaiming treated wastewater. Expected water quality improvements and economic considerations strongly favor the undertaking of large scale recharge projects to replenish the depleting water tables in the Kingdom. Alluvial aquifers and the outcrops of large aquifers would be suitable sites for recharge. The use of aquifers as multi-year reservoirs of the reclaimed wastewater could significantly reduce evaporation losses and construction costs associated with surface storage. The recharged water can be withdrawn and can be used for irrigation and possibly for potable use.

INTRODUCTION

In the past 2 decades, the Kingdom of Saudi Arabia has undergone massive urbanization and agricultural development, both of which require large volumes of water. The domestic demand for water, according to the Ministry of Agriculture and Water (Water Atlas of Saudi Arabia, 1984), is expected to increase from the present 12,000 million cubic meters (MCM) to 17,000 MCM by the year 2000. A net annual deficit (difference between supply and demand) of 16,000 MCM is projected for the year 2010. This shall be met by exploiting nonrenewable groundwater resources (Al-Ibrahim, 1990). Currently, the major part of this demand, around 80%, is met by mining non renewable fossil groundwater resources.

According to the Ministry of Agriculture and Water (Water Atlas of Saudi Arabia, 1984), the age of groundwater in the Kingdom ranges from the very recent to 40,000 yrs old. The water in the principal aquifers is relatively old, indicating that the addition of new water through recharge is very small.

The mining of fossil water has led to the water tables falling drastically in many aquifers. The continued mining of these aquifers can lead to disastrous consequences in the future. The renewable aquifers are recharged by the meager rains and through the use of a few recharge dams.

While there is an acute water shortage, treated wastewater is being wasted. Currently, other than 150 MCM of treated wastewater all of the reclaimed wastewater is discharged into the oceans or wasted. It is estimated that the present 1,000 MCM cubic meters of wastewater generated in the Kingdom will increase to a volume of 1,500 MCM by the year 2000. In an arid country like the Kingdom this could be put to better use. There is a need for effective water resources management. This has been realized by the national planners and efforts are being made to increase the use of reclaimed wastewater. The sixth development plan (1994-1999) envisages that, of the water demand of 17,500 MCM, 310 MCM will be met by reclaimed treated wastewater in the year 1998-1999. Taken against the 150 MCM of treated wastewater used during the year 1993-1994 this means a 15.6 % average annual growth in the use of treated wastewater (Sixth Development Plan 1415-1420, 1994). The greater portion of this reclaimed water shall be used for landscape irrigation and agriculture. It could possibly also be used for potable purposes in the near future. A promising technique for reclaiming treated wastewater is wastewater recharge.

Wastewater recharge offers several advantages. It is more economical to recharge and use recharged water than to use water from other sources such as desalination or other traditional tertiary treatment techniques. The use of reclaimed wastewater for agriculture may lead to a reduced need for applied commercial fertilizers due to the presence of nitrogen and phosphorous compounds in the reclaimed water (Moore et al., 1985). It is a supply source that is secure even during times of drought (Guymon and Hromadka, 1985).

AQUIFERS AS UNDERGROUND RESERVOIRS

One of the greatest advantages of using wastewater recharge as the technique for reclaiming wastewaters is that it will simultaneously provide a solution to the problem of storing the reclaimed water.

The dominant climatic features in the central part of the Kingdom are, high mean temperature, low humidity, and low advective winds. This combination of features results in high evaporation rates. An annual evaporation rate of 3126 millimeters has been reported in the Kharj area. A moderate increase in evaporation can be expected southward into the Rub-al Khali area. Evaporation rates are low along the Arabian Gulf Coast (Water Atlas of Saudi Arabia, 1984). Thus, the use of aquifers for underground storage would significantly reduce the evaporation losses and construction costs associated with surface storage.

The aquifers could also be used as an underground distribution system and thus save on costs associated with a surface distribution network.

When reclaimed wastewater is to be used for potable use, the inclusion of aquifers into the recycling chain introduces "nature" into the recycling chain making the reclaimed water more acceptable to people.

RECHARGE METHODS AND SITES

Treated wastewater can be recharged either through injection wells or through spreading basins. Desert soils in the kingdom have high infiltration and percolation rates, making them quite suitable for spreading basins. Surface spreading is most effective where there are no impeding layers between the land surface and the aquifer, thus alluvial aquifers and the outcrops of principal aquifers would be ideal sites for a spreading operation.

Fig. 1 (Ministry of Agriculture and Water) shows the alluvial aquifers in the Kingdom. The alluvial deposits fill many drainage areas on the western coastal plains of the Kingdom. The aquifers are generally unconfined but may be semi-confined or confined at some places. The transmissivity generally varies from 10^2 and 10^4 square meter per day (Water Atlas of Saudi Arabia, 1984).

Fig. 2 (Ministry of Agriculture and Water) shows the outcrops of the principal aquifers. The Kingdom's 9 principal aquifers provide a dependable supply of water for most parts of central and eastern Saudi Arabia. They range in geologic age from Cambrian to Tertiary (Water Atlas of Saudi Arabia, 1984).

WATER QUALITY TRANSFORMATIONS

The potential for contaminant removal by wastewater recharge operations have been aptly summarized by Culp et al. (1979). Most reclaimed water contaminants are substantially removed during vertical percolation through soil and during horizontal movement in the aquifers. Notable exceptions are total dissolved solids, hardness, nitrates, and a few heavy metals. Though equally good removals can be expected in the Kingdom the exact removals for various sites in the Kingdom would have to be determined through laboratory and field studies.

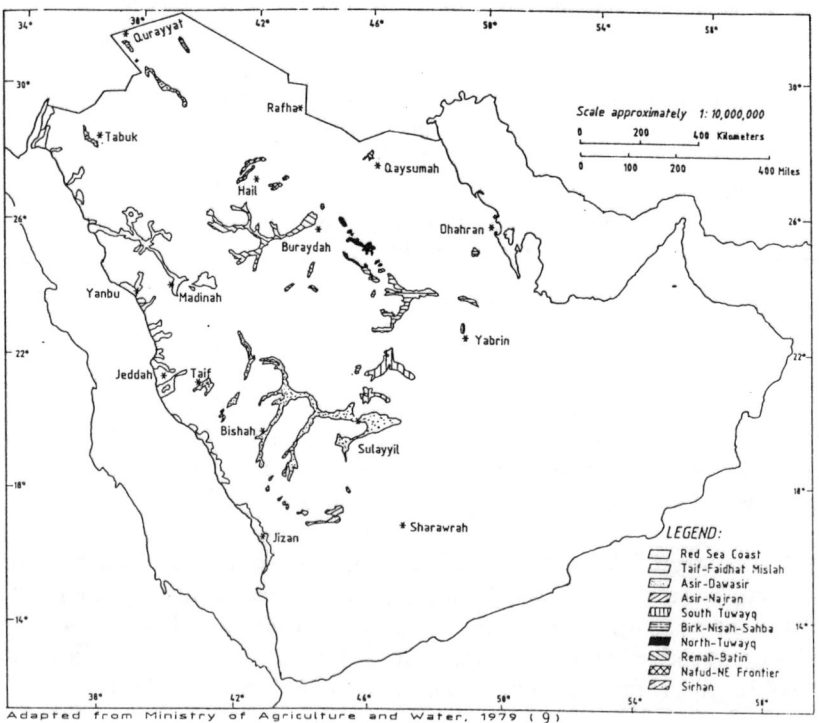

Fig. 1. Location of areas where alluvial deposits are water bearing

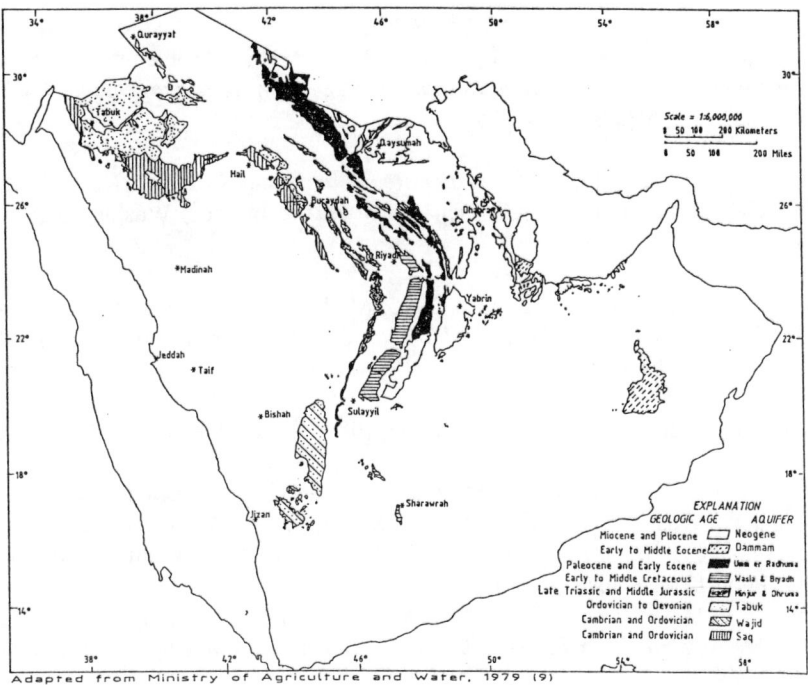

Fig. 2. Outcrop areas of principal aquifers

CONCLUSIONS

It can be concluded without doubt that artificial recharge especially wastewater recharge is among the most promising water reuse/reclamation process currently available to the Kingdom. The use of aquifers as multi-year reservoirs of the reclaimed wastewater could significantly reduce evaporation losses and construction costs associated with surface storage. The benefits associated with wastewater recharge for the variety of reasons for which it can be used are many and have been validated by the many pilot and full scale projects that have been implemented to date.

Expected water quality improvements and economic considerations strongly favor the undertaking of large scale recharge projects to replenish the depleting water tables in the Kingdom. Alluvial aquifers and the outcrops of the principal aquifers would be suitable sites for recharge. The recharged water can be withdrawn to be used for irrigation and possibly for potable use.

REFERENCES

Al-Ibrahim, A.A., 1990. Water Use in Saudi Arabia: Problems and Policy Implications. *Journal of Water Resources and Planning Management*, ASCE, Vol. 116, No. 3, May/June, p. 375.

Culp, Wesner and Culp, 1979. *Water Reuse and Recycling*. Office of Water Research and Technology, U.S. Department of the Interior, Washington, DC, July.

Guymon, G.L. and Hromadka, T.V., 1985, Modeling of Groundwater Response to Artificial Recharge, *Artificial Recharge of Groundwater*, Takashi Asano, ed., Butterworth Publishers.

Ministry of Agriculture and Water (MAW), 1984. *Water Atlas of Saudi Arabia*.

Moore, C.V., Olson, K.D. and Marino, M.A., 1985. On-farm Economics of Reclaimed Wastewater Irrigation, *Irrigation with Reclaimed Municipal Wastewater - A Guidance Manual*, Stuart Pettygrove and Takashi Asano, eds., Lewis Publishers, Chelsea, Mich.

Sixth Development Plan 1415-1420, 1994. Ministry of Planning Press, Riyadh, Saudi Arabia.

The Salt Water Intrusion in the Maryout Aquifer

PROF. DR. RAWYA M. KANSOH
Faculty of Engineering, Alexandria University
Alexandria, Egypt

ABSTRACT
The north side of the lake Maryout reservoir is bound by the Mediterranean sea, the east and south by the Nile Delta reservoir and the west side by limestone stratum.
The lake consists of mainly three basins. The main one is at the north of an area about 6000.00 feddans, the north western basin is about 3000.00 feddans, while the southern basin is 7000.00 feddans (most of the southern is reclaimed) and the Northern basin are about 9000 feddans.
The main sources of the surface water for the lake is the Omoum Drain and its tributaries and the Nubariya canal in addition to the four pump stations for sewer water. The main outlet is El Mex pump station. The water level in the lakes - 2.3 (97.7).

1. INVESTIGATION OF THE SALINITY IN THE MARYOUT AQUIFER AND SALT WATER INTRUSION PHENOMENA
The iso-salinity contour map, figure (1), shows that the groundwater salt content ranges between 70,000 mg/l to 35 000 mg/l. High salinity, particularly at the west side of the aquifer is due to the high rate of extracted water and evaporation. The salinity concentration increases northwards. This map was developed by the Ministry of Land Reclamation according to chemical analysis. The map does not show the depths at which the samples were collected. It must be noticed that the salinity of 45000 mg/l is more than the salinity of the sea water itself (35000 mg/l).

2- CONCEPTUAL MODEL WITH DIFFERENT BOUNDARY CONDITION:
A single aquifer with shallow drawdown may be effectively simulated using a two dimensional areal model and cross section model . It is proposed that modeling will be carried out using finite element computer program SUTRA produced by US geological survey for confined aquifer and modified for leaky aquifer for this purpose. This is a computer program which simulates groundwater flow and the transport of dissolved substances in two dimensions. The model employs a hybrid finite element and integrated finite difference method to approximate the governing equations that describe the interdependent processes. Output from the model will be shown as contours chloride concentrations using the computer program SURFER.

3. GRID LAYOUT

The plan area covered by the model is shown in figure (4). The area is 15000m by 18000m. The modeled area was selected to cover the area of influence of the lake and to include the various relatively remote monitoring points, The finite element mesh used by the numerical program is shown in the figure.

3.2 CROSS SECTIONS

The domain of interest which is simulated extends from the shoreline 18000 m inland from the shoreline with a uniform depth of 50 m. The finite element grid used in the simulation of this aquifer consists of 2000 nodes. (2 slices) and 1000 elements. Every slice represents a vertical cross-section through the aquifer. The aquifer is assumed to be homogeneous with steady flow.

3.3 APPLICATION OF THE MODIFIED TWO DIMENSIONAL MODEL FOR STUDYING THE HYDROLOGY OF THE AQUIFER (BOUNDARY CONDITIONS)

A single aquifer with variable potential may be effectively simulated using a two dimensional plan model, for steady state conditions, i.e., the conditions pertaining after a long period . Output from the model may be shown as contours of chloride concentrations. The new version of SUTRA which has been modified to account for leakage from the lake, with automatic generation of data file, is used to study the current problem.

3.3.1 Boundary condition for the Aquifer
Figure (2) describes the cross section of conceptual model with physical parameters, while figure (3) shows the potential and salinity B.Cs. for cross section of the conceptional model.

3.3.1.1 Boundary condition for the cross section

The aquifer is bounded by the Mediterranean sea at the north. For the groundwater flow equation, the specified potential head boundary (Dirichlet boundary condition) is used with considering the density effects. The equivalent potential head, ϕ, is a function of the depth of the aquifer and the difference between the density of salt and fresh waters. For the salt transport equation, the normal concentration gradient over the surface is set to zero at the window (top portion) to allow for convective mass transport out of the aquifer. The salt concentration in the region contained between the window and the impervious layer is treated as a specified value of salt concentration (c = 35,000) The inland boundary is treated as specified boundary with 70000 pp as specified salt concentration The top boundary is divided into parts one is head depended boundary underneath the lake and other no flow boundary due the big thickness of the clay cap and there is no shallow head in this clay .

4. APPLICATION OF THE TWO DIMENSIONAL MODEL FOR STUDYING CIRCULATION OF THE SEA WATER (SWICHA)

Trials are made to determine the direction of the different flows underneath the lake. Finite element model is designed to compute the salt concentration in the aquifer under the lake. The boundary conditions is shown in the figure (4).

4.1 SALINITY

The salinity near the lake is 5000 mg/l, while the sea water is 35000 mg/. In the areal model, it is clear that the transition zone varies from 35,000 to 70,000 ppm, this is due to many factors such as:
1. Geological formulation of the Maryout aquifer, Stanley and Warne 1993[5],
2. The phenomenon of the upward flow of salt water from the main aquifer, because of the circulation of sea water,
3. The high rate of evaporation from the lake.
4. The subsequent chemical reactions with adjacent entrapping limestone reservoir rocks,
5. The separation of the water boring strata by relatively impervious clay layers.

Figure (4) shows the distribution of the salinity in the areal model.

4.2 CROSS SECTION

From the unit cross section, it is clear from Figure (5) describing the transition zone and flow direction.

4.3 THE RELATION BETWEEN THE SEA AND THE LAKE

Prior to this study, it seamed to be logical that the sea will recharge to the lake, but after this study it become apparent that the contribution from the sea is minimal due to the salt water circulation.

5- CONCLUSIONS

1. Study has proven that there is no significant effect of the sea on the lake due to two factors namely.
   a) The permeability of the clay layer is reasonably low
   b) The salt water circulation.
2. The salinity is relatively high due to chain of reaction in the salt water since centuries ago, where "historically " the lake was filled over its content of salt water. Also due to the existence of limestone layer on the west side of the lake.

6- REFERENCES

1. Amer, A.; M. **1982** " *Analysis of Groundwater Systems*" For Workshop on Analysis of Water Resources Systems Cairo University- Massachusetts Institute Of Technology-Technological Planning Program.
2. FRU-CON Construction Corporation. *Evaluation of Pump Test executed by Mclean Grove East & West Treatment Plant December 1988*

3. Hefny, K. et al **1978"** *Study Of The Safety Factor For The Nile Delta Aquifer And The Nile Valley Aquifer Project"* No 2 Research Institute For Ground Water, Water Research Center, Ministry Of Public Works And Water Resources (in Arabic).
4. Lester, B.; **1991"***A Three-Dimensional Finite-element Code For Analyzing Seawater Intrusion In Coastal Aquifers"* Institute For Ground -Water Research And Education; Colorado School Of Mines ; U.S.A; 178p.
5. Stanley , D. J. and Warne, A. G.; **1993** " *Nile Delta : Recent Geological Evolution And Human Impact* " Science .Vol. 260.: 628-634
6. Technical Memorandun No3 ; **1996"***Lake Maryout geotechnical data report " Alexandria waste water project- phase*" USAID project No 263-0100.

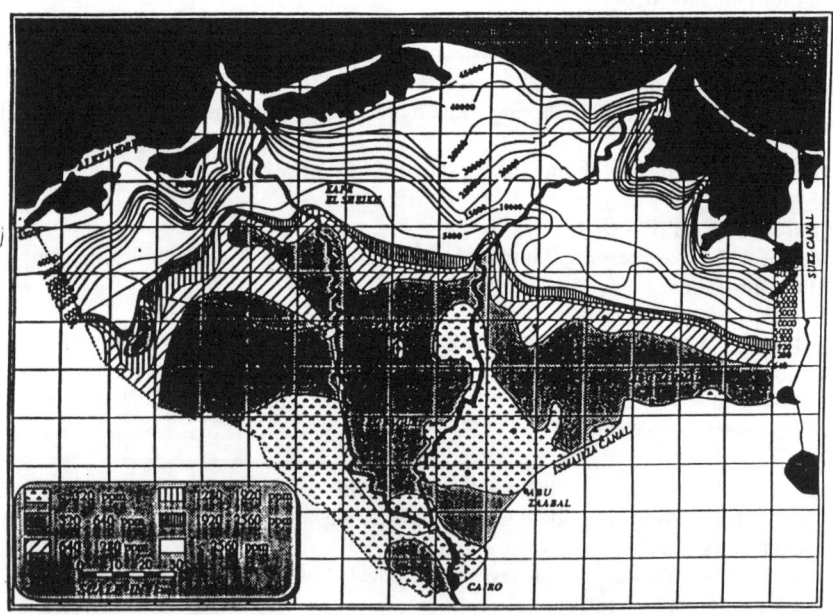

Fig. (1) Contour Map of Groundwater Salinity (1976)

SALT WATER INTRUSION IN MARYOUT AQUIFER

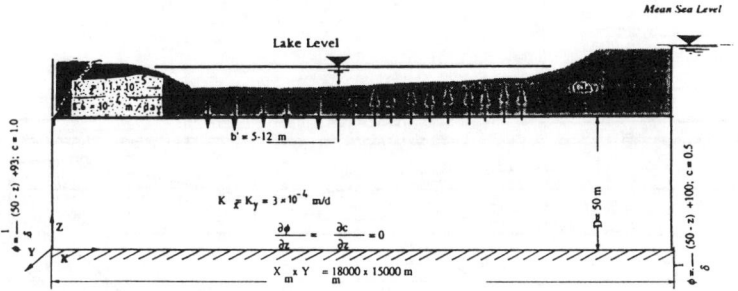

Fig. (2) Physical Model Parameters Used In The Analysis

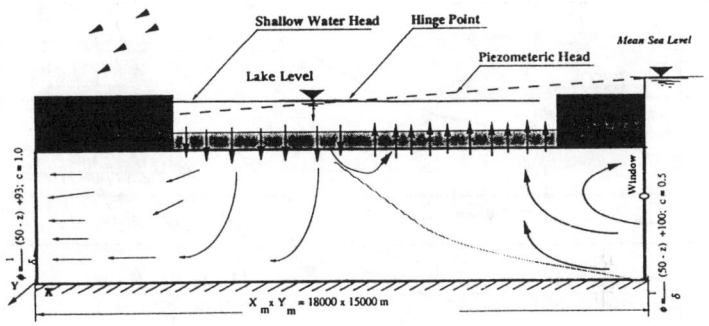

Fig. (3) Physical Model Parameters Used In The Analysis

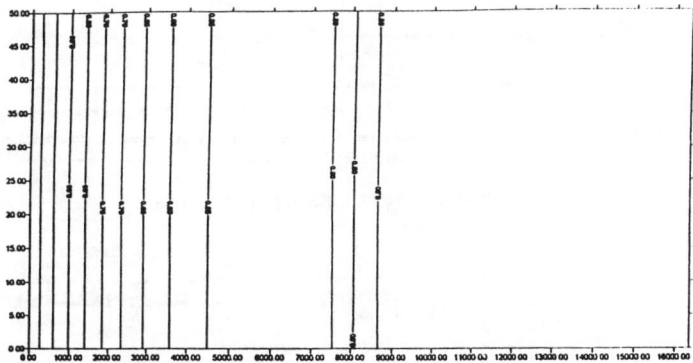

Fig. (5) Relative Salinity in the cross- sec

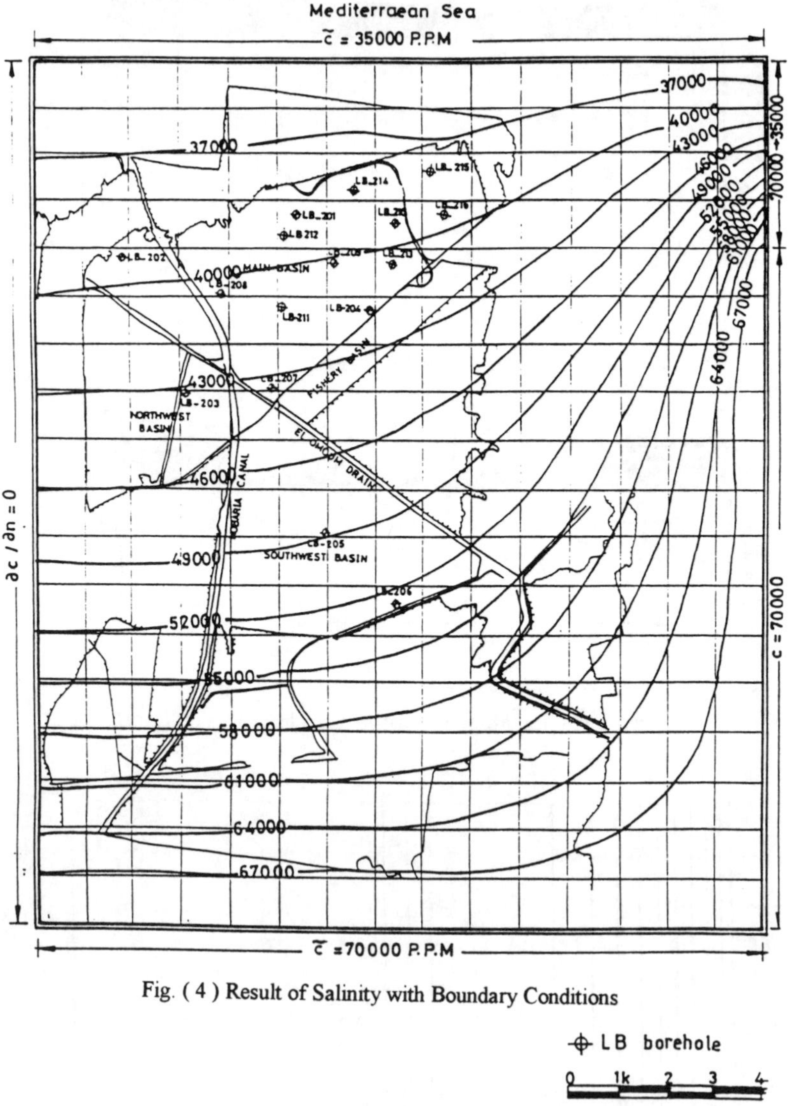

Fig. (4) Result of Salinity with Boundary Conditions

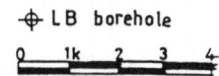

International River Basins: Forging a Consensus

BARBARA A. MILLER and VAHID ALAVIAN
RANKIN International, Inc., Knoxville, US
GEOFFERY MATTHEWS
World Bank, Washington DC, US
LAURA L. COLE
Tennessee Valley Authority, Knoxville, US

ABSTRACT

The paper provides an overview of the current legal framework and system of customs which influence the use and development of international waters, discusses the critical issues in a few specific international river basins, and presents a framework for forging a consensus for the management of shared water resources.

INTRODUCTION

Increasingly, fresh water is being recognized as a scarce and limited resource. Supplying water at an affordable cost and acceptable quality is becoming a challenge. Water demand is also increasing due to growth in population, urbanization, industrialization, agriculture, and income. Competition among water uses has therefore intensified. This situation has the potential to become explosive when water resources are shared among nations. International waters include rivers and lakes which border two or more countries, rivers which flow from one country to another, and shared groundwater resources. It is estimated that more than 40% of the world's population live in approximately 200 shared river basins. While most basins are shared by two countries, there are 13 major rivers with 5 or more riparian countries (see Table 1) (Barrett, 1994).

The potential for conflict over shared water resources is most acute where water is scarce, water demands are rapidly growing, and political relations are tense for other reasons. Water disputes, therefore, are most serious in the Middle East (Nile, Jordan, and Tiger-Euphrates Rivers) and south Asia (Indus and Ganges-Brahmaputra). Other countries, however, recognize the benefits of cooperatively developing shared water resources. None-the-less, water interdependence is a fact of life for many countries. As shown in Table 2, there are 20 countries which import more than 50 % of their water from outside their borders. It is clear, therefore, that the issue of 'shared waters' is one which will increase in significance as the demand for water begins to outstrip the available supply.

Table 1. River Basins Shared by Five or More Countries *(after Barrett, 1994)*

Region	No. of Countries	River Basin
Africa	6	Chad, Volta
	7	Zambezi
	9	Niger, Nile, Congo
Americas	5	La Plata
	7	Amazon
Asia	6	Ganges-Brahmaputra, Mekong
Europe	5	Elbe
	7	Rhine

INTERNATIONAL WATER LAW AND CUSTOM

A considerable body of international water law has developed over the years. A unique feature of this law is that it cannot be enforced by a third party and must rely upon self-enforcement and the opinion of the world community (McCaffrey, 1993). Observance of international law, however, is generally in a states' self-interest and most obligations are adhered to. International water law derives from two main sources: treaties and custom.

TREATIES

The Food and Agricultural Organization of the United Nations currently lists 3,707 agreements relating to international water resources. Most are bilateral in nature and relate to a specific river or lake. Many are also "incomplete," as the signatory parties do not represent all of the countries affected. Although the number of multilateral water agreements is growing, most still relate to specific watercourses (McCaffrey, 1993).

Treaties are generally based on four theories regarding water rights. The first is the theory of *absolute sovereignty*, meaning a country is free to use water within its boundaries to meet its needs and other riparians do not have the right to restrain such use. This theory favors upland states and is not widely accepted today. In the second theory, an international river belongs to all riparians and lower riparians have the right to demand continuation of natural flow from upstream. This benefits powerful lower riparians and is no longer considered valid. A third theory supports the development of an integrated, watershed-based program for the optimum use of international waters for the benefit of all. This theory is not always practical. The fourth recognizes a country's sovereignty over the international waters within its own boundaries, but restricts this sovereignty to ensure the rights of other riparians to a reasonable share of water (Kirmani, 1990).

INTERNATIONAL CUSTOM

While most treaties are site-specific in nature, rules of general applicability are expressed by international organizations, such as the: Institut de Droit International; International

Law Association (ILA); and International Law Commission (ILC) of the United Nations. The most widely known articles, the Helsinki Rules on the Uses of the Waters of International Rivers, were prepared by ILA in 1966. The fundamental principal of these rules is that of *equitable utilization* or that "Each basin State is entitled, within its territory, to a reasonable and equitable share in the beneficial uses of the waters of an international drainage basin." In 1991, the ILC proposed Draft Articles of the Law of the Non-Navigational Uses of International Watercourses, which sets forth the fundamental principals of: *equitable utilization*; *appreciable harm* or the obligation not to cause harm to other riparian states; and the obligation to exchange hydrologic and other relevant information. "Reasonable and equitable" is not clearly defined and is negotiated on a case-by-case basis.

The relative dominance of these principals can be debated depending on one's objectives, relative strength, and position in the watershed. Historically, the downstream country first develops its water resources, as the lower reaches of a river are typically more amenable to agricultural than the upland reaches. As the upstream countries begin to utilize their water resources, downstream countries generally object. Accusations relating to *equitable utilization* vs. *appreciable harm* are then argued, for what one country sees as its acquired right, the other riparian sees as a harmful limit on future development.

Table 2. Dependence on Imported Surface Water *(source: Gleick, 1993)*

Country	Percent of Total Flow Originating Outside of Border	Country	Percent of Total Flow Originating Outside of Border
Egypt	97	Syria	79
Hungary	95	Congo	77
Mauritania	95	Sudan	77
Botswana	94	Paraguay	70
Bulgaria	91	Czechoslovakia	69
Netherlands	89	Niger	68
Gambia	86	Iraq	66
Cambodia	82	Albania	53
Romania	82	Uruguay	52
Luxembourg	80	Germany	51

INTERNATIONAL WATERS: SELECTED CASES

While international law provides a framework for addressing water disputes, in reality law is only one of many factors influencing the outcome of major water controversies. The examples of on-going international water conflicts provided below illustrate the complexity of the issues and the context in which current international law is applied.

THE NILE RIVER

The Nile River runs through nine riparian countries in northern and eastern Africa: Burundi, Egypt, Ethiopia, Kenya, Rwanda, Sudan, Tanzania, Uganda, and Zaire. Egypt, the downstream riparian country, is the primary beneficiary of the Nile and is totally dependent upon its waters. It is expected that due to Egypt's high population growth, increased irrigation activity, and worsening water quality, the country will not have sufficient water to meet its demand by the turn of the century. The primary agreement which dictates allocations on the Nile is the 1959 Nile Waters Agreement between Egypt and Sudan. This agreement basically allocates 66 % of the flow to Egypt, 22 % to Sudan, 10% to losses, and none to upstream riparian countries (Smith and Al-Rawahy, 1990).

As Egypt's water situation becomes more critical and upland countries begin to demand their share of the Nile's waters, the potential for conflict - and the need for cooperation - will intensify. To date, upstream countries have been too poor, distracted by civil war, or dominated by Egypt to focus on water resources development. However, several upstream countries now have plans for water projects which will alter the flow to Egypt. Furthermore, proposed schemes to increase the flow to Egypt involve storage of Nile waters in upland countries with lower evaporation rates. Although the 1959 Niles Waters Agreement includes procedures to settle claims of upstream riparian states through equal flow reductions to Sudan and Egypt, no formal claims have yet been made and these procedures have not yet been tested (Kirmani and Rangeley, 1994).

THE TIGRIS AND EUPHRATES RIVERS

Water conflicts also occur in the Tigris-Euphrates River Basin, which is shared by Turkey, Syria, and Iraq. The Tigris and Euphrates rise in the hills of eastern Turkey. The Tigris receives about half of its water from Iraq, while 90 % of the Euphrates is supplied within Turkey. Current attention is focused on the Euphrates River and the Southeastern Anatolia Project (GAP) being developed by Turkey. Turkey views the project, which ultimately will include 25 irrigation systems, 22 dams, and 19 hydro plants, as central to its long-term economic development. Recent completion of the Ataturk Dam, the backbone of GAP, has triggered international debate. Syria and Iraq, the downstream riparians, fear loss of as much as 40 % and 60% of their share of the Euphrates and are concerned about the quality of the remaining water (Economist, 1996).

The disagreement centers around different interpretations of international law. Syria and Iraq consider the Tigris-Euphrates as international waters and demand they be shared accordingly. Conversely, Turkey alludes to the 'transboundary' nature of the water (i.e., crossing common borders) and plans what it considers to be 'rational and optimal utilization' of the water. Turkey regards the Tigris-Euphrates as the foundation for cooperation and prosperity for all three countries and has proposed: unified data collection and management; rehabilitation of existing projects; agricultural joint ventures;

a Peace Water Pipeline to provide water to Syria, Iraq, Jordan, and Saudi Arabia; sale of electricity to Syria and Iraq; environmental impact studies; and construction of canals to supplement flows in the Euphrates with water from the Tigris. However, as Turkey has self-financed and controls the GAP project, Syria and Iraq fear that their downstream rights are not adequately protected. Intense negotiations continue today (Tekeli, 1990).

THE ARAL SEA

The Aral Sea, the fourth largest inland lake in the world until 1960, is fed by the Amu and Syr Rivers of Central Asia. The primary riparian states are: Kazakhstan, Kirghistan, Tajikistan, Turkmenistan, and Uzbekistan. Due to huge diversions of water from the Amu and Syr for irrigation, inflows to the sea have been reduced more than 90 %, from 45 million acre feet (MAF) in 1960 to 4 MAF in 1989. Consequently, the Aral's surface area has shrunk by nearly one half, its volume by 75%, and the sea level has fallen by 16 meters (48 ft). Salinity in the Sea has increased nearly threefold. The region is beset by associated social, ecological, and health problems: the fishing industry has been destroyed; cotton harvests have declined due dust storms, salinity, and waterlogging; incidences of disease and infant mortality are high due to air and water pollution; and, the ecology of the region is devastated (Kirmani and Rangley, 1994).

Since their independence, the riparian countries have signed agreements and established international institutions for the coordinated management of their common water resources. These efforts are supported by international agencies such as the World Bank, UNEP, and the UNDP. However, serious problems remain. Uzbekistan, Kazakhstan, and Turkmenistan use most of the water. Alhough they acknowledge the "Aral tragedy," they cannot afford to reduce irrigation and undermine their own economies. The republics agreed in 1993 to donate 1 % of their GNP to an Aral Sea Fund, but due to weak economies, the money has not been forthcoming. Rehabilitation, or even stabilization, of the Aral Sea remains a challenge for international cooperation. Although the interest and mechanisms for cooperation are in place, the problems are overwhelming and the solutions are not easy.

A FRAMEWORK FOR FORGING CONSENSUS

As illustrated above, resolving international water conflicts can be extremely difficult, even when cooperation is desired. Self-interest generally dictates water negotiations. Water conflicts are typically resolved when the states enjoy good relations with each other, one state is more powerful than the other, or it is mutually beneficial to do so (McCaffrey, 1993). Otherwise, disputes are likely to remain unresolved. Clearly, the current framework for settling international water disputes is inadequate, particularly given the increasing pressures on water. Examination of current agreements and on-going conflicts, however, points to several factors which are key to forging a consensus to integrated development of shared waters:

- Development of a common set of objectives by all riparian parties affected. This means riparian states must move beyond self-interest to mutual interest in the development of a region or watershed.
- Formulation of projects that meet the real development needs of a country and are aimed at improving prosperity at the grass-roots level. This may result in smaller-scale, locally focused projects as opposed to larger-scale structural approaches. Water delivery should also be viewed as a means (i.e., to improved agriculture or health), not as an end unto itself.
- Watershed-based planning and development. Water is best managed in an integrated fashion within its natural hydrographic units, not by arbitrary political boundaries.
- Creation of a joint mechanism, such as a river basin commission, with experts from all basin states, to guide the planning, development, and long-term management of water agreements and projects.
- Involvement of a third party facilitator, such as the World Bank or the UN, which commands international respect, possesses recognized expertise in the development field, and has the ability to harness financial resources.
- Creation of a mechanism for the on-going collection, analysis, monitoring, and communication of technical data. Timely and accurate data on the current condition and projected trends is essential to sound decision-making.
- Financial resources for planning, development, and long-term maintenance. Where necessary, international financial resources should be brought to bear to assist with capital expenditures, but economic incentives and cost-recovery should be incorporated into project plans to ensure long-term self-sufficiency and adequate project maintenance.

REFERENCES

Barrett, S. 1994. *Conflict and Cooperation in Managing International Water Resources.* Policy Research Working Paper No. 1303. The World Bank, Washington, D.C.

Gleick, P. H. (ed.) 1993. *Water in Crisis: A Guide to the World's Fresh Water Resources.* Pacific Institute and Stockholm Environmental Institute. Oxford University Press, N.Y.

The Economist. 1996. *Water in the Middle East: As Thick as Blood.* Jan. 5th, 1996.

Kirmani, S. S. 1990. *Water, Peace, and Conflict Management: The Experience of the Indus and Mekong River Basins.* Water International, 15: 200-205.

Kirmani, S. and R. Rangeley. 1994. International Inland Waters: Concepts for a More Active World Bank Role. Technical Paper No. 239. The World Bank, Washington, D.C.

McCaffrey, S. C. 1993. *Water, Politics, and International Law* in *Water in Crisis: A Guide to the World's Fresh Water Resources.* P. H. Gleick (ed). Pacific Institute and Stockholm Environmental Institute. Oxford University Press, New York.

S. E. Smith and H. M. Al-Rawahy. *The Blue Nile: Potential for Conflict and Alternatives for Meeting Future Demands.* Water International, 15.

Tekeli, S. 1990. *Turkey Seeks Reconciliation for the Water Issue Induced by the Southeastern Anatolia Project (GAP).* Water International, 15: 206-216.

2r:iahr.paper6.doc

SUBJECT INDEX

Page number refers to the first page of paper

Abutments, 141
Acceleration, 734
Accuracy, 178, 332
Advection, 40
Aerial surveys, 262
Africa, 64, 381, 607
Aggradation, 147
Agriculture, 250, 683, 893
Air entrainment, 69
Algorithms, 326, 482, 583, 887
Alluvial fans, 141
Amplitude, 16
Application methods, 695
Aqueducts, 607
Aquifers, 905, 911
Arid lands, 296, 482
Arizona, 302, 711
ASCE Committees, 308
Asia, 633
Assessments, 393
Automation, 302, 308, 326, 673, 700, 711

Bayesian analysis, 554
Beaches, 16, 46
Bedforms, 10, 28, 734
Border irrigation, 274
Bridge abutments, 190, 196
Bridge failure, 110, 117, 124, 196
Bridge foundations, 110
Bridges, 130, 141, 154, 166, 178, 208, 214, 220, 226, 453, 781
Bridges, highway, 458
Bridges, piers, 135, 184
Budgets, 1
Buildings, 214
Buoyancy, 172

Calibration, 338, 864
California, 279, 332, 651, 683, 799
Canada, 387
Canals, 302, 308, 326, 689, 695, 706
Capacity, 706
Case reports, 274, 464, 706
Caspian Sea, 417, 663
Catchments, 98, 859
Channel beds, 729
Channel bends, 793
Channel design, 69
Channel flow, 10, 196, 746, 752, 763, 793, 829, 835
Channel improvements, 805
Channel stabilization, 147
Channels, waterways, 34
Chaos, 40
Chlorine, 875, 881
Climatic changes, 572
Climatology, 470
Clogging, 92
Coastal management, 417
Coastal morphology, 46
Coastal processes, 46, 621
Cohesive soils, 124
Collective bargaining, 399
Colombia, 470
Competition, 256, 536
Computation, 226, 633, 657, 817, 835
Computer models, 46, 154, 405
Computer networks, 399
Computer programming, 411
Computer programs, 633
Computer software, 338
Computerized simulation, 338
Configuration, 734

Conflict, 349
Constraints, 1
Contaminants, 887
Contraction, 117, 787
Control, 130, 166
Control systems, 308, 320, 326
Coupled systems, 639
Critical flow, 787
Crops, 256, 560
Crossings, 124, 190, 458
Culverts, 458
Currents, 758
Cycles, 52
Cylinders, 238

Dam breaches, 729
Dam failure, 357
Dam safety, 357
Dams, 369, 500, 669, 805
Dams, earth, 464
Data analysis, 494
Databases, 274
Decision support systems, 274
Degradation, 147
Deltas, 663, 911
Design, 92, 98, 453, 458, 494, 554, 615, 657, 683, 893
Design criteria, 178, 689, 695
Detention basins, 58
Developing countries, 1
Dikes, 554, 805
Discharge, 447, 453, 530, 669, 847
Dissipation, 34
Distributed processing, 864
Diversion, 615
Drainage, 250, 279, 285, 899
Drainage systems, 69
Droughts, 1, 387, 482, 589, 595, 645
Dynamics, 10, 22, 28, 34, 46

Ecology, 381, 651
Economics, 256, 673

Ecosystems, 381
Edge effect, 16
Embedment, 184
Energy budget, 859
Energy dissipation, 746
Environmental impacts, 399, 853
Erosion, 160, 464, 805
Estimating, 58
Estimation, 262, 268, 482, 494, 524, 601, 752
Europe, 363, 375, 572, 621
Evaluation, 135
Evaporation, 104, 285
Evapotranspiration, 262, 560, 566, 841, 864
Excitation, 16
Experimentation, 232
Expert systems, 274

Failures, 184, 429, 506, 811
Farms, 695, 700, 706
Filters, digital, 512
Flexibility, 673, 677, 689, 695, 700, 706, 711, 717
Flood control, 58, 314, 500, 651, 657, 799
Flood damage, 117, 214, 464
Flood forecasting, 411, 476, 494, 621, 627
Flood frequency, 458, 494
Flood Management, 639
Flood plains, 190, 196, 214, 639, 746, 775, 793
Flooding, 411, 417, 811
Floods, 110, 357, 387, 572, 663, 669, 805, 829, 847
Florida, 262, 560
Flow characteristics, 740
Flow distribution, 734
Flow measurement, 332
Flow patterns, 208
Flow rates, 326, 683, 870

WATER QUALITY TRANSFORMATIONS

The potential for contaminant removal by wastewater recharge operations have been aptly summarized by Culp et al. (1979). Most reclaimed water contaminants are substantially removed during vertical percolation through soil and during horizontal movement in the aquifers. Notable exceptions are total dissolved solids, hardness, nitrates, and a few heavy metals. Though equally good removals can be expected in the Kingdom the exact removals for various sites in the Kingdom would have to be determined through laboratory and field studies.

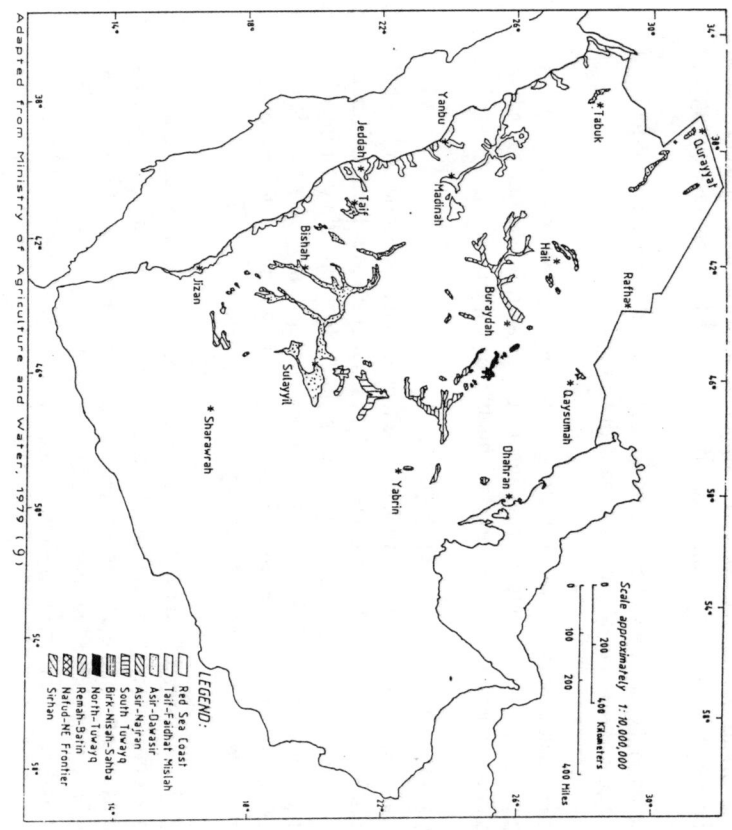

Fig. 1. Location of areas where alluvial deposits are water bearing

USE OF AQUIFERS IN SAUDI ARABIA

AQUIFERS AS UNDERGROUND RESERVOIRS

One of the greatest advantages of using wastewater recharge as the technique for reclaiming wastewaters is that it will simultaneously provide a solution to the problem of storing the reclaimed water.

The dominant climatic features in the central part of the Kingdom are, high mean temperature, low humidity, and low advective winds. This combination of features results in high evaporation rates. An annual evaporation rate of 3126 millimeters has been reported in the Kharj area. A moderate increase in evaporation can be expected southward into the Rub-al Khali area. Evaporation rates are low along the Arabian Gulf Coast (Water Atlas of Saudi Arabia, 1984). Thus, the use of aquifers for underground storage would significantly reduce the evaporation losses and construction costs associated with surface storage.

The aquifers could also be used as an underground distribution system and thus save on costs associated with a surface distribution network.

When reclaimed wastewater is to be used for potable use, the inclusion of aquifers into the recycling chain introduces "nature" into the recycling chain making the reclaimed water more acceptable to people.

RECHARGE METHODS AND SITES

Treated wastewater can be recharged either through injection wells or through spreading basins. Desert soils in the kingdom have high infiltration and percolation rates, making them quite suitable for spreading basins. Surface spreading is most effective where there are no impeding layers between the land surface and the aquifer, thus alluvial aquifers and the outcrops of principal aquifers would be ideal sites for a spreading operation.

Fig. 1 (Ministry of Agriculture and Water) shows the alluvial aquifers in the Kingdom. The alluvial deposits fill many drainage areas on the western coastal plains of the Kingdom. The aquifers are generally unconfined but may be semi-confined or confined at some places. The transmissivity generally varies from 10^2 and 10^4 square meter per day (Water Atlas of Saudi Arabia, 1984).

Fig. 2 (Ministry of Agriculture and Water) shows the outcrops of the principal aquifers. The Kingdom's 9 principal aquifers provide a dependable supply of water for most parts of central and eastern Saudi Arabia. They range in geologic age from Cambrian to Tertiary (Water Atlas of Saudi Arabia, 1984).

SUBJECT INDEX 925

Flow resistance, 752, 769
Flow visualization, 763
Fluid dynamics, 817
Flumes, 723
Food supply, 256
Footings, 196
Forecasting, 417, 530, 645
Fouling, 875, 881
Fractal analysis, 34
Frequency analysis, 847
Fresh water, 393
Froude number, 214
Furrow irrigation, 717

Gates, 22, 75, 314, 320, 326, 669
Geographic information systems, 621
Geomorphology, 141, 154
Glacial till, 633
Gravel, 769
Gravity, 172
Groins, structures, 244, 829
Ground water, 250, 607, 899
Ground-water management, 566
Ground-water recharge, 58, 905
Guidelines, 375

Heat transfer, 104
Heavy metals, 86
Hydraulic conductivity, 518
Hydraulic design, 500
Hydraulic models, 435, 781
Hydraulic performance, 226
Hydraulic roughness, 775
Hydraulic structures, 494, 506
Hydraulics, 154, 178, 196, 238, 320, 399, 823, 893
Hydrodynamics, 135, 308, 417, 441, 657, 787, 811
Hydroelectric power generation, 536
Hydrographs, 500, 651, 657
Hydrologic models, 627, 864

Hydrology, 470, 476, 512, 518, 524, 560, 578, 595, 601, 841, 853, 859

Ice, 633
Ice cover, 752, 758
Inertia, 40
Infiltration, 81, 86, 92, 98, 268
Inflow, 524, 548, 899
Infrastructure, 399
Inlets, waterways, 28, 220
Inspection, 110
International commissions, 387
International development, 393
International factors, 363
International waters, 917
Iron compounds, 86
Irrigation, 250, 256, 262, 268, 296, 302, 320, 332, 338, 369, 695, 700, 711, 717, 899
Irrigation districts, 677
Irrigation systems, 893
Irrigation water, 279, 290, 673, 689
Israel, 349
Italy, 22

Jets, 232
Jordan, 349

Kinetics, 244
Knowledge-based systems, 274
Korea, 357

Laboratory tests, 166, 190, 202, 723
Lakes, 911
Land usage, 859
Laws, 917
Legal factors, 917
Levees, 811

Management, 274, 369, 447
Mathematical models, 875, 881
Maximum flood, 476

Meandering streams, 734, 746, 763, 835
Measurement, 723, 763
Meteorology, 262
Methodology, 338
Mixing, 40, 763
Modeling, 429, 441, 530, 566
Models, 34, 52, 117, 417, 488, 512, 583, 769, 775, 811, 841, 853, 887
Moments, 172, 476
Momentum transfer, 740, 758
Monte Carlo method, 441, 447
Morphology, 10, 805
Multiple objective analysis, 548
Municipal water, 488

National Weather Service, 601
Navier-Stokes equations, 835
Navigation, 369
Negotiations, 349
Networks, 893
Nile River, 530, 805, 911
Nitrogen, 290
Noise, 512
Nonlinear analysis, 22, 46
Nonlinear systems, 10
Numerical analysis, 104, 823
Numerical models, 124, 226, 787, 793, 799

Object-oriented languages, 405, 411
Ocean waves, 16
Offshore platforms, 208
Open channel flow, 226, 238, 308, 320, 338, 411, 723, 758, 823
Open channels, 740, 817
Operation, 578
Optimal control methods, 314
Optimal design, 75, 423
Optimal use, 343
Optimization, 314, 542, 548, 554, 589, 870

Oregon, 220
Outflows, 314
Overflow, 75, 811
Overtopping, 357

Parameters, 435, 560
Particle interactions, 40
Particles, 172
Peak floods, 657
Percolation, 673
Permeability, 104
Permits, 130
Piers, 117, 160, 166, 178, 202, 208, 226
Piles, 52, 220
Pipe networks, 423, 429, 435
Pipelines, 711
Piping systems, 870
Planning, 1, 110, 595
Plants, 841
Poland, 81
Policies, 393
Ponds, 296, 657
Porous pavements, 86, 104
Potable water, 58, 875, 881
Precipitation, atmospheric, 518, 601
Predictions, 46, 232, 572
Privatization, 536
Probabilistic methods, 464, 500
Probability, 476, 506, 578
Probability distribution, 554
Probability distribution functions, 847
Projects, 711
Public benefits, 130
Pumping, 870

Queueing, 488

Radar, 601
Radial flow, 268
Rain gages, 601

SUBJECT INDEX

Rain water, 64, 92
Rainfall, 98, 482
Rainfall duration, 518
Rainfall intensity, 518
Rainfall-runoff relationships, 566
Random processes, 500
Random waves, 16
Recirculation, 817
Recycling, 279, 296
Rehabilitation, 464, 677
Reliability, 423, 470, 506, 578, 677, 887
Reliability analysis, 429, 447
Research, 135, 160, 494
Research needs, 124, 141
Reservoir design, 470
Reservoir management, 589
Reservoir operation, 314, 524, 542, 548, 589, 615
Reservoir storage, 669
Reservoir system regulation, 711
Reservoir systems, 700
Reservoirs, 524, 706, 911
Residences, 488
Restoration, 453
Retention basins, 75
Return flow, 290, 332
Reynolds stress, 244
Riprap, 166, 172, 178, 184
Risk, 500, 506, 663
Risk allocation, 458
Risk analysis, 441, 482, 621, 645
River basin development, 917
River basins, 343, 349, 363, 369, 381, 405, 566, 583, 651, 853
River beds, 769
River flow, 572, 583, 627, 829
River regulation, 805
River systems, 52, 578, 627
Rivers, 10, 34, 141, 202, 357, 441, 447, 639, 657, 663, 775, 781, 793, 823

Roads, 458
Rock properties, 160
Roughness, 268, 746, 769
Roughness coefficient, 752
Runoff, 58, 518, 578, 673, 847, 864
Runoff coefficient, 64
Runoff forecasting, 633

Safety, 130, 453
Salinity, 250
Salt water, 279, 285
Salt water intrusion, 911
Sand, 52
Saudi Arabia, 905
Scale models, 124, 220, 799
Scheduling, 689, 695, 717
Scour, 110, 117, 124, 130, 135, 141, 147, 154, 160, 166, 178, 184, 190, 196, 202, 208, 214, 220, 226, 232, 781
Sea level, 22, 417, 621, 663
Sediment transport, 10, 46, 52, 135, 166, 202, 226, 453
Selenium, 285
Sensitivity analysis, 560
Sewage effluents, 81
Sewers, 75, 320
Shafts, 69
Shallow water, 268
Shear, 823
Shear stress, 208, 244, 752, 763
Simulation, 244, 274, 308, 411, 435, 536, 548, 615, 627, 639, 817
Simulation models, 28, 268, 423, 560
Sliding, 326
Slope stability, 464
Snowmelt, 572, 633
Soil permeability, 98
Soil saturation zones, 899
Soil water, 864
Solar ponds, 285
South America, 470

Spillway capacity, 669
Spillways, 464
Stability analysis, 147, 172
Stabilization, 296
Statistics, 476, 482, 500, 506, 512, 554, 633, 769
Steady flow, 787
Stochastic models, 595, 645
Stochastic processes, 470, 488, 548
Storm surges, 154, 220, 417
Storms, 22
Stormwater, 98
Stormwater management, 81, 86
Strategic planning, 375
Stream channels, 453
Streamflow, 566, 595, 645
Streams, 147, 160
Submerged flow, 829
Subsidence, 117, 621
Subsurface drainage, 717
Supercritical flow, 799
Surface drainage, 717
Surface waters, 911
Surge, 729
Suspended sediments, 28

Taiwan, 441
Tanks, 75
Testing, 302
Thailand, 627
Three-dimensional flow, 244
Three-dimensional models, 835
Tidal waters, 28, 154
Tides, 22
Time series analysis, 512, 530, 645
Topography, 639
Trenches, 98
Trends, 250
Turbulence, 244, 775, 817
Turbulent flow, 232, 723, 729, 746, 752, 758, 763, 835
Two-dimensional analysis, 811

Two-dimensional models, 220, 639, 781, 817, 829

Uncertainty analysis, 453, 506, 536, 542, 554, 560, 847
United States, 135, 387, 677
Unsaturated flow, 841
Unsteady flow, 729, 740
Urban areas, 64
Urban runoff, 69, 81, 86, 92, 104
Urbanization, 859
Utilities, 536

Validation, 639
Valves, 711
Variability, 518, 524, 870, 881
Vegetation, 775, 864
Velocity, 75, 208, 232, 734, 763
Velocity distribution, 202, 723, 793
Velocity profile, 238, 758
Vortices, 202, 238, 823

Washington, 781
Wastewater treatment, 905
Wastewater use, 296, 905
Water allocation policy, 405, 583
Water balance, 262, 542, 853, 859
Water conservation, 332, 589
Water costs, 589
Water demand, 64, 256, 393, 405, 488, 542
Water discharge, 314
Water distribution, 423, 429, 435, 607, 683, 700, 706, 717, 875, 881, 887
Water flow, 40, 69, 615, 734, 841
Water hammer, 870
Water law, 387
Water levels, 308
Water management, 1, 285, 343, 349, 363, 375, 572, 615, 859, 893, 899
Water policy, 1, 375

SUBJECT INDEX 929

Water pollution, 875, 881, 887
Water quality, 81, 92, 250, 441, 615, 677, 875, 881, 887, 905
Water quality control, 290, 447
Water reclamation, 905
Water resources, 64, 530, 583
Water resources development, 262, 343, 393
Water resources management, 363, 369, 381, 387, 393, 399, 536, 542, 578, 595, 607, 847, 853, 917
Water reuse, 279, 285, 290, 296
Water rights, 343, 349
Water shortage, 542, 583, 607
Water storage, 64, 92, 905

Water supply, 1, 58, 64, 302, 583, 607, 645, 673, 677, 683, 689, 695
Water supply systems, 429
Water surface profiles, 740, 799
Water use, 343, 363, 405, 488, 700
Water waves, 729
Watersheds, 369, 566, 841, 847
Wave propagation, 729
Wave spectra, 16
Weirs, 615
Wetlands, 290, 381
Wind velocity, 752

Yield, 256, 524, 589

AUTHOR INDEX
Page number refers to the first page of paper

Abbott, M. B., 399
Ahmed, A. F., 805
Aibara, Toshifumi, 202
Aigner, Detlef, 75
Alavian, Vahid, 917
Alghariani, Saad A., 64, 607
Ali, Kamil H., 232
Ali, Kamil H. M., 208
Annandale, G. W., 160
Asaeda, Takashi, 104
Attia, K. M., 805
Aya, Shirou, 829

Babarutsi, S., 817
Banovec, Primož, 375
Barneveld, H. J., 775
Bautista, E., 302
Bayazit, Mehmetcik, 343
Bezerra Campos, José Nilson, 524
Bhattacharya, D., 512
Biggs, B. J. F., 769
Blaszczyk, Pawel, 81
Blodgett, James C., 117
Blondeaux, P., 16
Bodla, Muhammad Abid, 274
Bourget, Lisa, 387
Bradley, J. B., 781
Bravo, Hector R., 723
Brown, Thomas C., 405
Buchanan, Timothy L., 58
Buchberger, Steven G., 488
Burt, C. M., 677

Carvajal, Luis, 470
Caussade, Bernard, 841
Cervinka, V., 285
Chen, Chang-Shian, 615

Chevray, Rene, 40
Chiew, Yee-Meng, 184
Chrisochoides, A., 196
Chu, V. H., 817
Chung, Wen-Shiang, 441
Clamen, Murray, 387
Clements, L. Lynn, 464
Clemmens, A. J., 302
Clemmens, Albert, 332
Cole, Laura L., 917
Coleman, S. E., 166
Cross, Paul R., P.E., 673

Davis, Stanley R., 130
de Araújo, José Carlos, 524
de Assis de Souza Filho, Francisco, 524
De Costa, Shahane, 899
de Swart, H. E., 28
De Vriend, H. J., 10
Debol'skaya, E. I., 758
Decouttere, C., 621
Deodhar, M. J., 669
Diaz, Gustavo E., 405
Diener, J., 285
Dong, Zengnan, 752
Dunne, Paul, 645
Durrans, S. Rocky, 494
Dyner, Isaac, 536

Elmaalouf, Sami, 887, 893
El-Sayed, Ahmed H., 530
English, Marshall, 256
Errih, Mohamed, 530
Ewen, J., 853

Fallows, C. S., 853
Fernández, Bonifacio, 482
Feyen, Jan, 308
Fidell, M., 677
Finch, C., 285
Finnerty, Bryce, 601
Franco, Carlos, 536
Frederiksen, Harald D., 1
French, Richard H., 58
Froehlich, David C., 172
Fujita, Ichiro, 829
Fukuda, Tetsuro, 290

Galant, M., 417
Gambolati, G., 621
Garcia-Navarro, P., 268
Gauzer, B., 572
Gomez, Shawn M., 40
Gonella, M., 621
Goring, D. G., 769
Gottlieb, O., 22
Goulter, Ian C., 423, 429
Graf, W. H., 238
Grattan, S. R., 279
Green, Thaddeus, 566
Grindeland, T. R., 781

Hager, Willi H., 214, 320, 729
Hagerty, D. J., 124
Han, Kun-Yeun, 447, 811
Harada, Hirokazu, 202
Hasfurther, Victor, 458
Haugh, Joseph S., 464
Heo, Jun H., 506
Herath, S., 627
Herrera, Rafael, 326
Hinckley, John A., Jr., 411
Horlacher, Hans-B., 75
Hosoda, T., 823
Hunt, J. H., 154

Imamoto, H., 763
Ishaq, Achi M., 905
Ishigaki, T., 763
Ismail, N., 296

Jayaratne, B. L., 740
Jha, R., 627
Jia, Y., 859
Johnson, Dennis, 601
Johnson, P. A., 160
Johnson, Peggy A., 453
Jones, J. S., 160
Jones, J. Sterling, 117, 135

Kan, K., 740
Kansoh, Rawya M., 911
Karim, Othman A., 208, 232
Katopodes, Nikolaos D., 314
Kawahara, Y., 244, 740
Kelbaugh, Linda, 130
Kent, Edward J., P.E., 226
Khalsa, Ram Dhan, 338
Khan, Amir Ali, 905
Kilsby, C. G., 853
Kim, Sang-Ho, 447
Kim, Young C., 887, 893
Kimura, I., 823
Kirshen, Paul H., 393
Kliot, Nurit, 349
Klopstra, D., 775
Kobayashi, Tomonao, 202
Kohli, Alexander, 214
Konovalov, V. G., 633
Kotb, 805
Kouchakzadeh, S., 190
Kudou, Tadashi, 69
Kuo, Albert Y., 441
Kuo, Jan-Tai, 441
Kuroda, Masaharu, 290
Kute, Sunil, 669
Kuznetsov, D. S., 758

Lagasse, P. F., 147
LaPotin, Perry J., 411
Lauber, Guido, 729
Lauchlan, C. S., 166
Law, Adrian W. K., 787
Lee, Jong-Tae, 811
Legret, Michel, 92
Leonard, Edward F., 40
Lim, Foo-Hoat, 184
Lin, S. Samuel, 464
Lin, Youn-Jan, 615
Lindstrom, Shane, 711
Liu, Fubo, 308
Lyssenko, P., 417

Maison, Philippe, 841
Manson, J. Russell, 835
Mao, Zeyu, 752
Martin, M., 285
Matthews, Geoffery, 917
Mei, C. C., 22
Meinecke, John W., 723
Melville, B. W., 124, 166
Menezes, F., 285
Merriam, John L., 689, 695, 700, 706
Mesa, Oscar, 470
Mikkelsen, Peter Steen, 98
Miller, Barbara A., 917
Miller, Daniel P., 381
Miyawaki, Nobuyuki, 829
Mizell, Steve A., 58
Montoya, Santiago, 536
Morris, Johnny L., P.E., 110
Mu, Gaofeng, 752
Mueller, D. S., 124
Mueller, David S., 135
Mulikov, R. R., 663
Mulvihill, Michael E., 799
Munoz, R., 285
Muramoto, Y., 823
Musiake, K., 627

Muškatirović, Jelisaveta, 363
Muto, Y., 763

Nandagiri, Lakshman, 262
Nassiri, M., 817
Nikora, V. I., 769
Nix, Stephan J., 494
Nowakowska-Blaszczyk, Alina, 81

Obeysekera, Jayantha, 560
O'Connell, P. E., 853
O'Connor, Brian A., 208, 232
Ogden, Fred L., 864
Ogihara, Kunihiro, 69
Omurtag, Ahmet C., 40
Ostroumova, L. P., 663
Ottemoeller, Stephen H., 683
Ozawa, Kouichi, 734

Pagan-Ortiz, Jorge E., 110, 130
Pagliara, Stefano, 847
Paola, Chris, 52
Paquier, A., 639
Park, Jae-Hong, 811
Parker, G., 124
Parker, Gary, 141
Parola, A. C., 124
Parrish, John B., 338
Pedroza, Edmundo, 326
Pender, Gareth, 835
Peng, J., 244
Peters, D., 285
Pianese, Domenico, 875, 881
Pinheiro, Adilson, 841
Pirotton, M. J. J., 657
Pirozzi, Francesco, 875, 881
Playan, E., 268
Pohl, Reinhard, 500
Polonsky, V. F., 663
Ports, Michael, 220
Poveda, Germán, 470

Raemy, Felix, 320
Raimbault, Georges, 92
Rajasekaram, Veerakcuddy, 542
Rameshwaran, P., 746
Rao, A. R., 512
Reifsnider, E., 677
Replogle, John, 332
Richardson, Everett V., 117
Richardson, John E., P.E., 226
Rinaldi, Massimo, 453
Rinaldo, Andrea, 34
Rodriguez-Iturbe, Ignacio, 34
Ruiz C., Victor M., 326

Salas, Jose D., 506
Salazar, José, 470
Sammarco, P., 22
Sanders, Brett F., 314
Sansalone, John J., 86
Schade, Trent G., 488
Schall, J. D., 154
Schumm, S. A., 147
Schuttelaars, H. M., 28
Senarath, Sharika U. S., 864
Sharp, B. B., 870
Shea, Conor, 220
Shen, Hsieh Wen, 595
Shiau, Jenq Tzong, 595
Shim, S. B., 583
Shipton, S., 399
Shmueli, Deborah, 349
Sigrist, B., 639
Slijkhuis, K. A. H., 554
Sloan, W. T., 853
Smith, Ricardo, 536
Smith, S. P., 160
Soliman, M. M., 805
Southgate, Howard N., 46
Starosolszky, Ö., 572
Steinman, Franci, 375
Stephenson, David, 589
Stickel, Victor G., Jr., 40

Stonestreet, Scott E., 799
Strand, R. J., 302
Strzepek, Kenneth M., 393
Sturm, T. W., 196
Styles, S. W., 677
Styles, Stuart, 689, 695, 700, 706, 717

Taglialatela, Lucio, 875, 881
Takaki, S., 793
Takamatu, H., 793
Takeo, N., 763
Talaat, A. M., 805
Tamai, N., 244, 740, 859
Tamai, Nobuyuki, 734
Tasker, Gary, 645
Teatini, P., 621
Tickle, Kevin S., 435
Tingsanchali, Tawatchai, 542
Tomasi, L., 621
Tomić, Saša, 494
Toro-Escobar, Carlos M., 141
Townsend, R. D., 190
Tran, H. H., 22
Trout, Thomas J., 250
Tu, H., 793
Tung, Yeou-Koung, 458

Umbrell, E. R., 160
Ungate, Christopher D., 369
Usher, J. S., 124

van Gelder, P. H. A. J. M., 554
van Mullem, Joseph A., 518
van Noortwijk, J. M., 775
van Schilfgaarde, Jan, 250
van Velzen, E. H., 775
van Zee, Randy, 560
Varas, Eduardo, 476
Velikanov, A. L., 578
Vick, J. C., 651
Vikulov, Y. G., 663

AUTHOR INDEX

Vittori, G., 16
Vogel, Richard M., 566
Voigt, Richard L., Jr.,, 141
Vrijling, J. K., 554
Vu Thanh, C. A., 104

Wacker, A. Mainard, 458
Wahlin, Brian, 332
Walton, R., 781
Wang, Jing, 752
Wang, Y. C., 548
Wang, Yongtian, 752
Wasantha Lal, A. M., 560
Willetts, B. B., 746
Williams, P. B., 651
Woo, Hyoseop, 357

Xu, Chengchao, 423, 429, 435

Yasunaga, R., 823
Yen, Ben Chie, 847
Yeon, G. B., 583
Yoon, K. H., 583
Yoon, Sung Bum, 178
Yoon, Tae-Hoon, 178
Young, Edward S. C., 615
Yu, C. S., 621
Yu, Shaw L., 411
Yulistiyanto, B., 238

Zafirakou-Koulouris, Antigoni, 566
Zevenbergen, L. W., 147, 154
Zvegintsev, A., 417